Student Solutions Manual

to accompany

Organic Chemistry

Tenth Edition

Neil T. Allison
University of Arkansas

Francis A. Carey
University of Virginia

Robert M. Giuliano
Villanova University

Robert C. Atkins
James Madison University

STUDENT SOLUTIONS MANUAL TO ACCOMPANY ORGANIC CHEMISTRY, TENTH EDITION

Published by McGraw-Hill Education, 2 Penn Plaza, New York, NY 10121. Copyright © 2017 by McGraw-Hill Education. All rights reserved. Printed in the United States of America. No part of this publication may be reproduced or distributed in any form or by any means, or stored in a database or retrieval system, without the prior written consent of McGraw-Hill Education, including, but not limited to, in any network or other electronic storage or transmission, or broadcast for distance learning.

Some ancillaries, including electronic and print components, may not be available to customers outside the United States.

This book is printed on acid-free paper.

1 2 3 4 5 6 7 8 9 0 RHR/RHR 19 18 17 16

ISBN 978-1-259-63638-7
MHID 1-259-63638-0

mheducation.com/highered

CONTENTS

PREFACE

For this tenth edition of the *Student Solutions Manual,* originally created by Robert Atkins and Francis Carey, it was my hope to maintain their original focus—that the study of organic chemistry is aided by use of this manual. It is my intention that the new and updated solutions offered in this manual are as meaningful and useful as the original solutions and that they seamlessly blend into the existing problems.

I have had many excellent organic chemistry teachers in the past, including my graduate advisor, William M. Jones, and my undergraduate mentor, Ed Waali. I count them as superior teachers and true friends. In starting each semester's organic chemistry class, I draw on their examples. When I teach undergraduate classes, the last one with 510 students, I discuss mastering organic chemistry from the first day of class. Of course students have many learning styles, one size does not fit all, but I believe in all cases learning organic chemistry takes figuring out how to solve problems, not just memorizing steps. In my 35 years at the University of Arkansas, I have taught many thousands of students, most of whom have gone on to successful careers in numerous different professions. Lately, most go into health professions and many of these become physicians. I like to emphasize that what "sticks" from my class is the process of how a student ultimately proceeds, with data in hand, to solve problems and that this may carry over to these many professions. For instance, a student who wishes to become a physician may consider that she will use many pieces of data (symptoms and a myriad of lab tests) in determining the proper treatment for a patient. This is not unlike a student solving a problem in organic chemistry who must understand the many concepts and specific details of each reaction step to successfully complete a problem.

In addition to the solutions to the text questions, the Self-Test at the conclusion of each chapter is designed to test the student's mastery of the material. The answers to the self-test questions are found in the Appendix at the back of this book.

After accomplishing this update, which included many new solutions to problems, I continue to have a deep respect for Bob Atkins and Frank Carey in what they have accomplished for over 20 years together. I would also like to acknowledge the talented folks at McGraw-Hill for both this and the last two editions. This includes Mary Hurley and Lora Neyens with whom I communicated with often, on critical updates of the text. My now adult children, Betsy, Joseph, and Alyse, also had to endure time taken away from them with this endeavor. I would like to thank the most, however, my wife, the love of my life, Amelia Allison. She has a strong and steady soul and provided unwavering support and understanding for the many long hours invested in this work.

Neil T. Allison, *University of Arkansas*

TO THE STUDENT

Before beginning the study of organic chemistry, a few words about "how to do it" are in order. You've probably heard that organic chemistry is difficult; there's no denying that it will take a lot of time to master and for many it will become actually enjoyable. For most, it need not be overwhelming, though, when approached with the right frame of mind and with sustained effort. In other words "stay on top of it."

Organic chemistry tends to "build" on itself. That is, once you have mastered a reaction or concept, you will find many being used again and again later on. In this way it is quite different from general chemistry, which tends to be much more compartmentalized. You will continually find previously learned material cropping up and being used to explain and to help you understand new topics. Often, for example, you will see the preparation of one class of compounds using reactions of other classes of compounds studied earlier in the year.

How to keep track of everything? It might be possible to memorize every bit of information presented to you, but you would still lack a fundamental understanding of the subject. It is far better to *generalize* as much as possible.

You will find that the early chapters of the text will emphasize concepts of *reaction theory*. These will be used, as the various classes of organic molecules are presented, to describe *mechanisms* of organic reactions. A relatively few fundamental mechanisms suffice to describe almost every reaction you will encounter. Once learned and understood, these mechanisms provide a valuable means of categorizing the reactions of organic molecules.

You will need to learn numerous facts in the course of the year, however. For example, chemical reagents necessary to carry out specific reactions must be learned. You might find a study aid known as *flash cards* helpful. Just filling out flash cards is an excellent learning opportunity! These cards take many forms, but one idea is to use 3 x 5 index cards. As an example of how the cards might be used, consider the reduction of alkenes (compounds with carbon–carbon double bonds) to alkanes (compounds containing only carbon–carbon single bonds). The front of the card might look like this:

$$H_2C=CH_2 \xrightarrow{\text{reagent?}} H_3C-CH_3$$
$$\text{Alkenes} \hspace{4cm} \text{Alkanes}$$

The reverse of the card would show the reagents necessary for this reaction:

$$\text{Reactant?} \xrightarrow[\text{Pt or Pd catalyst}]{H_2} \text{Product?}$$

The card can actually be studied in two ways. You may ask yourself: What reagents will convert alkenes into alkanes? Or, using the back of the card: What chemical reaction is carried out with hydrogen and a platinum or palladium catalyst? This is by no means the only way to use the cards—be creative! Again, the process of making up the cards will help you to remember.

Although study aids such as flash cards will prove helpful, there is only one way to truly master the subject matter in organic chemistry—*do the problems*! The more you work, the more you will learn. Almost certainly the grade you receive will be a reflection of your ability to solve problems. Don't just think over the problems, either; write them out as if you were handing them in to be graded. Also, be careful of how you use the *Solutions Manual*. The solutions contained in this book are intended to provide explanations to help you understand the problem. Be sure to write out *your* solution to the problem first and only then look it up to see if you have done it correctly. If you miss or struggle with problems, it is important to understand *why* your solution is not correct. Reworking problems at a later date with this understanding will help.

Students frequently feel that they understand the material but don't do as well as expected on tests. One way to overcome this is to "test" yourself. Each chapter in the *Solutions Manual* has a self-test at the end. Work the problems in these tests *without* looking up how to solve them in the text. You'll find it is much harder this way, but it is also a closer approximation to what will be expected of you when taking a test in class.

Success in organic chemistry depends on skills in analytical reasoning. Many of the problems you will be asked to solve require you to proceed through a series of logical steps to the correct answer. Most of the individual concepts of organic chemistry are fairly simple; stringing them together in a coherent fashion is where the challenge lies. By doing exercises conscientiously you should see a significant increase in your overall reasoning ability. Enhancement of their analytical powers is just one fringe benefit enjoyed by those students who are fully involved in the course rather than simply attending it.

Gaining a mastery of organic chemistry is hard work. We hope that the hints and suggestions outlined here will be helpful to you and that you will find your efforts rewarded with a knowledge and understanding of an important area of science.

Neil T. Allison
Francis A. Carey
Robert Giuliano
Robert C. Atkins

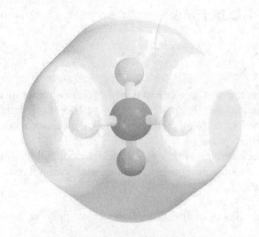

CHAPTER 1
Structure Determines Properties

Table of Contents

SOLUTIONS TO TEXT PROBLEMS

In Chapter Problems

1.1 The number of electrons in an element is equal to the number of protons in its nucleus and is given by its atomic number Z. For carbon, $Z = 6$. Therefore, carbon has six electrons. Of these, only the two $2s$ and two $2p$ electrons are valence electrons. The two $1s$ electrons are not valence electrons.

Silicon, like carbon, is in group 4 of the periodic table. The number of valence electrons of main-group elements is the same as the group number. Therefore, silicon has four valence electrons.

1.2 Electron configurations of elements are derived by applying the following principles:

(a) The number of electrons in a neutral atom is equal to its atomic number Z.

(b) The maximum number of electrons in any orbital is two.

(c) Electrons are added to orbitals in order of increasing energy, filling the $1s$ orbital before any electrons occupy the $2s$ level. The $2s$ orbital is filled before any of the $2p$ orbitals, and the $3s$ orbital is filled before any of the $3p$ orbitals.

(d) All the $2p$ orbitals $(2p_x, 2p_y, 2p_z)$ are of equal energy, and each is singly occupied before any is doubly occupied. The same holds for the $3p$ orbitals.

With this as background, the electron configuration of the third-row elements is derived as follows $[2p^6 = 2p_x^2\, 2p_y^2\, 2p_z^2]$:

Na	$(Z = 11)$	$1s^2 2s^2 2p^6 3s^1$
Mg	$(Z = 12)$	$1s^2 2s^2 2p^6 3s^2$
Al	$(Z = 13)$	$1s^2 2s^2 2p^6 3s^2\, 3p_x^1$
Si	$(Z = 14)$	$1s^2 2s^2 2p^6 3s^2\, 3p_x^1\, 3p_y^1$
P	$(Z = 15)$	$1s^2 2s^2 2p^6 3s^2\, 3p_x^1\, 3p_y^1\, 3p_z^1$
S	$(Z = 16)$	$1s^2 2s^2 2p^6 3s^2\, 3p_x^2\, 3p_y^1\, 3p_z^1$
Cl	$(Z = 17)$	$1s^2 2s^2 2p^6 3s^2 3p_x^2\, 3p_y^2\, 3p_z^1$
Ar	$(Z = 18)$	$1s^2 2s^2 2p^6 3s^2\, 3p_x^2\, 3p_y^2\, 3p_z^2$

1.3 A sodium atom (Na) has 11 electrons. A sodium ion (Na^+) has one less, or 10 electrons. The $+2$ ion that contains 10 electrons must have 12 protons in its nucleus. The element with $Z = 12$ is magnesium. Mg^{2+} is isoelectronic with Na^+.

A -2 ion with 10 electrons must have 8 protons in its nucleus. The element with $Z = 8$ is oxygen.

O^{2-} is isoelectronic with Na^+.

1.4 The electron configurations of the designated ions are:

	Ion	Z	Number of Electrons in Ion	Electron Configuration of Ion
(b)	He^+	2	1	$1s^1$

(c)	H⁻	1	2	$1s^2$
(d)	O⁻	8	9	$1s^2 2s^2 2p^5$
(e)	F⁻	9	10	$1s^2 2s^2 2p^6$
(f)	Ca^{2+}	20	18	$1s^2 2s^2 2p^6 3s^2 3p^6$

Those with a noble gas configuration are H⁻, F⁻, and Ca^{2+}.

1.5 A positively charged ion is formed when an electron is removed from a neutral atom. The following equation represents the ionization of carbon and the electron configurations of the neutral atom and the ion:

$$C \longrightarrow C^+ + e^-$$

$$1s^2 2s^2 2p_x^1 2p_y^1 \qquad 1s^2 2s^2 2p_x^1$$

A negatively charged carbon is formed when an electron is added to a carbon atom. The additional electron enters the $2p_z$ orbital.

$$C + e^- \longrightarrow C^-$$

$$1s^2 2s^2 2p_x^1 2p_y^1 \qquad 1s^2 2s^2 2p_x^1 p_y^1 p_z^1$$

Neither C^+ nor C^- has a noble gas electron configuration.

1.6 (b) In order to have four bonds to carbon, CH_4O must have a carbon–oxygen bond.

Combine carbon, oxygen, and four hydrogens   to write a Lewis formula for methanol

(c) Three of the four bonds involving carbon are to hydrogen atoms; the fourth is to fluorine. Carbon contributes four electrons, fluorine seven, and the three hydrogens contribute one each. The total number of valence electrons in CH_3F is 14.

Combine carbon, fluorine, and three hydrogens   to write a Lewis formula for methyl fluoride

The 14 valence electrons of CH_3F are distributed among four covalent bonds and three unshared pairs of fluorine.

(d) The only possible structure for C_2H_5F in which each carbon has four bonds must contain a carbon–carbon bond.

$$H \cdot \cdot \overset{\overset{\displaystyle H \quad H}{|}}{\underset{\underset{\displaystyle H \quad H}{}}{\ddot{C} \cdot \cdot \ddot{C} \cdot \cdot \ddot{F}:}}$$

Combine two carbons, one fluorine, and five hydrogens

$$H:\overset{\overset{\displaystyle H}{}}{\underset{\underset{\displaystyle H}{}}{\ddot{C}}}:\overset{\overset{\displaystyle H}{}}{\underset{\underset{\displaystyle H}{}}{\ddot{C}}}:\ddot{F}: \quad \text{or} \quad H-\overset{\overset{\displaystyle H}{|}}{\underset{\underset{\displaystyle H}{|}}{C}}-\overset{\overset{\displaystyle H}{|}}{\underset{\underset{\displaystyle H}{|}}{C}}-\ddot{F}:$$

to write a Lewis formula for ethyl fluoride

The number of valence electrons in C_2H_5F is 20; eight contributed by two carbons, seven by fluorine, and a total of five by the hydrogens. These 20 valence electrons are distributed among 7 covalent bonds plus three unshared electron pairs of fluorine in C_2H_5F.

1.7 (b) Hydrogen cyanide (HCN) has 10 valence electrons; one from hydrogen, four from carbon, and five from nitrogen.

$$\text{Combine} \quad H \cdot \quad \cdot \ddot{C} \cdot \quad \cdot \ddot{N}: \quad \text{to give} \quad H:\overset{\cdot}{\underset{\cdot}{C}}:\ddot{N}:$$

In order to satisfy the octet rule, pair the remaining unpaired electrons of carbon with those of nitrogen so as to give a triple bond in:

$$H:C:::N: \quad \text{or} \quad H-C\equiv N:$$

1.8 The partial positive charge of hydrogen is greatest when the atom it is attached to is most electronegative. Among the atoms to which hydrogen is bonded in the compounds CH_4, NH_3, H_2O, SiH_4, and H_2S, oxygen is the most electronegative. Therefore, $\delta+$ is largest for hydrogen in H_2O.

Hydrogen will bear a partial negative charge when the atom to which it is bonded is less electronegative than hydrogen. The electronegativity of hydrogen is 2.1 versus 1.8 for silicon (text Table 1.3). Therefore, hydrogen bears a partial negative charge in SiH_4.

1.9 The Pauling electronegativity scale (text Table 1.3) may be used to solve this problem. For instance, for the bond $H-O$, hydrogen is 2.1 and oxygen has a higher value of 3.5. This can be qualitatively represented two different ways as shown. The $+$ part of the arrow indicates the more electropositive atom.

$$\overset{+\longrightarrow}{H-O} \quad \overset{+\longrightarrow}{H-N} \quad \overset{+\longrightarrow}{C-O} \quad \overset{+\longrightarrow}{C=O} \quad \overset{+\longrightarrow}{C-N} \quad \overset{+\longrightarrow}{C=N} \quad \overset{+\longrightarrow}{C\equiv N}$$

Alternatively, the δ^+ symbol indicates the electropositive atom whereas the δ^- indicates the electronegative atom.

$$\overset{\delta^+ \quad \delta^-}{H-O} \quad \overset{\delta^+ \quad \delta^-}{H-N} \quad \overset{\delta^+ \quad \delta^-}{C-O} \quad \overset{\delta^+ \quad \delta^-}{C=O} \quad \overset{\delta^+ \quad \delta^-}{C-N} \quad \overset{\delta^+ \quad \delta^-}{C=N} \quad \overset{\delta^+ \quad \delta^-}{C\equiv N}$$

The periodic table may also be helpful in assigning electronegativity. Electronegativity increases from left to right across a row. It decreases down a column or group. For instance, for C—O, carbon is to the left of oxygen on the periodic table; thus, carbon is less electronegative as indicated by the arrow or δ^+ symbol.

1.10 As drawn, the formula has five bonds to nitrogen. These five bonds require 10 electrons in the valence shell of nitrogen, which violates the octet rule.

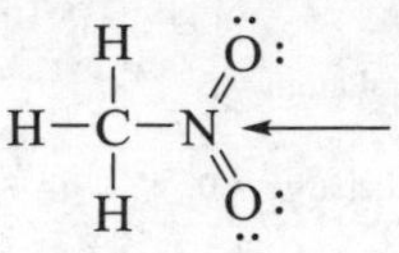

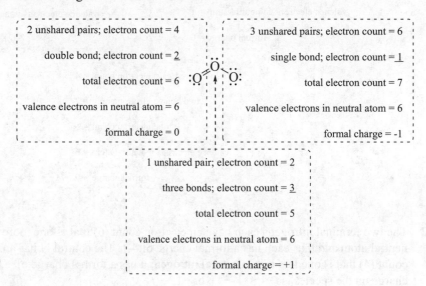

Five bonds to nitrogen is incorrect. 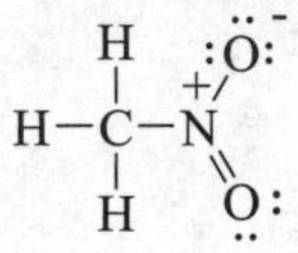   The correct structure for nitromethane
                                             has four bonds to nitrogen with a
                                             formal positive charge.

1.11 (*b*) The formal charges in ozone are calculated as follows:

2 unshared pairs; electron count = 4

double bond; electron count = 2

total electron count = 6

valence electrons in neutral atom = 6

formal charge = 0

3 unshared pairs; electron count = 6

single bond; electron count = 1

total electron count = 7

valence electrons in neutral atom = 6

formal charge = -1

1 unshared pair; electron count = 2

three bonds; electron count = 3

total electron count = 5

valence electrons in neutral atom = 6

formal charge = +1

(*c*) The formal charges in nitrous acid are calculated as follows:

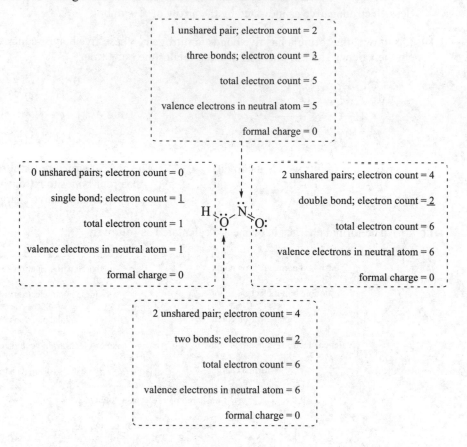

1.12 The two terminal nitrogens each have an electron count (6) that is one more than a neutral atom and thus each has a formal charge of −1. The central N has an electron count (4) that is one less than a neutral nitrogen; it has a formal charge of +1. The net charge on the species is (−1 + 1 − 1), or −1.

$$:\overset{-}{\ddot{N}}=\overset{+}{N}=\overset{-}{\ddot{N}}:$$

1.13 (*b*) First count the valence electrons in C_3H_7Cl.

Carbon: 3 atoms × 4 valence electrons/atom = 12 valence electrons

Hydrogen: 7 atoms × 1 valence electron/atom = 7 valence electrons

Chlorine: 1 atom × 7 valence electrons/atom = 7 valence electrons

Total = 26 valence electrons

Because chlorine can be bonded only to one other atom, only two connectivities are possible.

$$-\overset{|}{\underset{|}{C}}-\overset{|}{\underset{|}{C}}-\overset{|}{\underset{|}{C}}-Cl \quad \text{and} \quad -\overset{|}{\underset{|}{C}}-\overset{|}{\underset{|}{\underset{Cl}{C}}}-\overset{|}{\underset{|}{C}}-$$

Place one hydrogen on each of the bonds available to carbon.

$$
\begin{array}{ccccc}
& H & H & H & \\
& | & | & | & \\
H- & C- & C- & C- & Cl \quad \text{and} \\
& | & | & | & \\
& H & H & H &
\end{array}
\qquad
\begin{array}{ccccc}
& H & H & H & \\
& | & | & | & \\
H- & C- & C- & C- & H \\
& | & | & | & \\
& H & Cl & H &
\end{array}
$$

Each structural formula has 10 bonds, which account for 20 electrons. Place the remaining six electrons as unshared pairs on chlorine to complete its octet.

$$
\begin{array}{ccccc}
& H & H & H & \\
& | & | & | & \\
H- & C- & C- & C- & \ddot{\underset{\cdot\cdot}{C}l}: \quad \text{and} \\
& | & | & | & \\
& H & H & H &
\end{array}
\qquad
\begin{array}{ccccc}
& H & H & H & \\
& | & | & | & \\
H- & C- & C- & C- & H \\
& | & | & | & \\
& H & :\ddot{\underset{\cdot\cdot}{C}l}: & H &
\end{array}
$$

There are two constitutional isomers having the formula C_3H_7Cl.

(c) Proceed as in parts (a) and (b) of this problem.

The number of valence electrons contained in a molecular formula of C_3H_8O is 26 (12 contributed by three carbons, 8 contributed by eight hydrogens, and 6 contributed by one oxygen).

Three connectivities are possible in which carbon has four bonds. Two of these connectivities have a continuous chain of three carbons. The third has the atoms connected in the order CCOC.

$$
\begin{array}{ccc}
| & | & | \\
-C- & C- & C-O-H \\
| & | & |
\end{array}
\qquad
\begin{array}{ccc}
| & | & | \\
-C- & C- & C- \\
| & | & | \\
  & O & \\
  & | & \\
  & H &
\end{array}
\qquad
\begin{array}{cccc}
| & | & & | \\
-C- & C- & O- & C- \\
| & | & & |
\end{array}
$$

Complete the number of available C—H bonds; there are seven in the first two structures, eight in the third.

$$
\begin{array}{ccc}
H & H & H \\
| & | & | \\
H-C- & C- & C-O-H \\
| & | & | \\
H & H & H
\end{array}
\qquad
\begin{array}{ccc}
H & H & H \\
| & | & | \\
H-C- & C- & C-H \\
| & | & | \\
H & O & H \\
 & | & \\
 & H &
\end{array}
\qquad
\begin{array}{cccc}
H & H & & H \\
| & | & & | \\
H-C- & C- & O- & C-H \\
| & | & & | \\
H & H & & H
\end{array}
$$

Each structural formula has 11 bonds, accounting for 22 electrons. The remaining four electrons are assigned to oxygen as two unshared pairs. The octet rule is satisfied.

$$
\begin{array}{ccc}
H & H & H \\
| & | & | \\
H-C- & C- & C-\ddot{\underset{\cdot\cdot}{O}}-H \\
| & | & | \\
H & H & H
\end{array}
\qquad
\begin{array}{ccc}
H & H & H \\
| & | & | \\
H-C- & C- & C-H \\
| & | & | \\
H & :\ddot{\underset{\cdot\cdot}{O}}: & H \\
 & | & \\
 & H &
\end{array}
\qquad
\begin{array}{cccc}
H & H & & H \\
| & | & & | \\
H-C- & C- & \ddot{\underset{\cdot\cdot}{O}}- & C-H \\
| & | & & | \\
H & H & & H
\end{array}
$$

There are three constitutional isomers having the formula C_3H_8O.

1.14 (b) Formaldoxime has the connectivity:

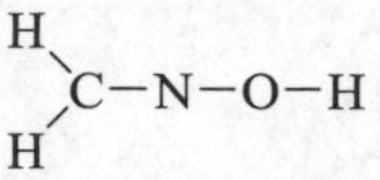

Like nitrosomethane in part (a), formaldoxime has a total of 18 valence electrons. The five bonds identified by the connectivity information account for 10 electrons. Adding the remaining eight electrons as pairs to O, then N, then C (decreasing order of electronegativity) leads to complete octets for O and N, but leaves C with only six electrons. Therefore, share one of the electron pairs of N with C to give a C=N double bond to give the Lewis formula of formaldoxime.

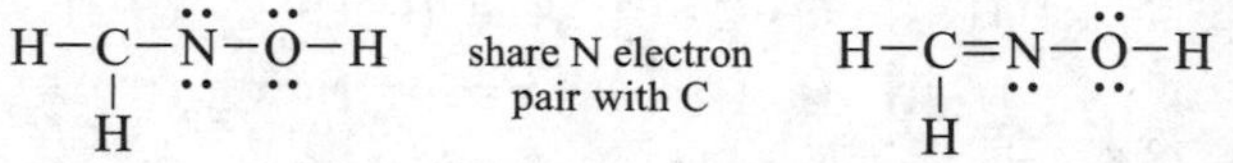

Formaldoxime

1.15 Add unshared pairs to cysteine and serotonin so as to give N, O, and S completed octets.

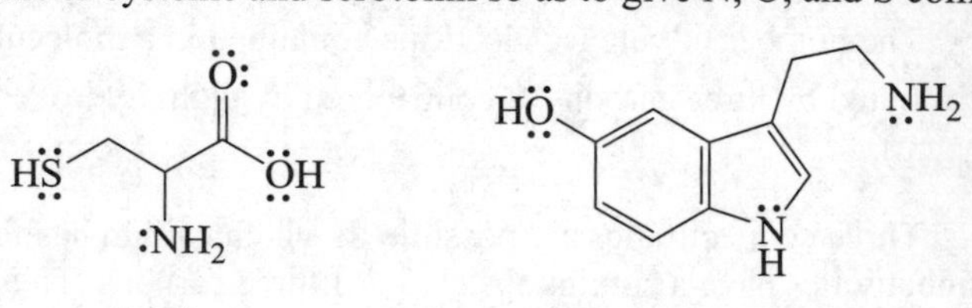

Cysteine   Serotonin

Each bond in a chain corresponds to a carbon with sufficient hydrogens to give it four bonds. Thus, cysteine is $C_3H_7NO_2S$ and serotonin is $C_{10}H_{12}N_2O$. Alternatively, one can write in the atoms and count them.

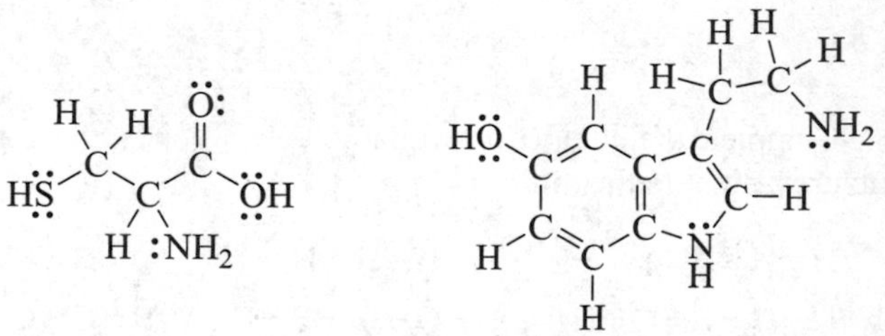

1.16 The resonance structures for carbonate are shown here. If one considers the vertical carbon–oxygen bond in each of the resonance structures, the left structure has a double bond whereas the other two structures have single bonds. If one considers the remaining two carbon–oxygen bonds in each structure, each has the same count: one double and two single bonds. Thus, all three carbon–oxygen bonds would be predicted to have equivalent lengths.

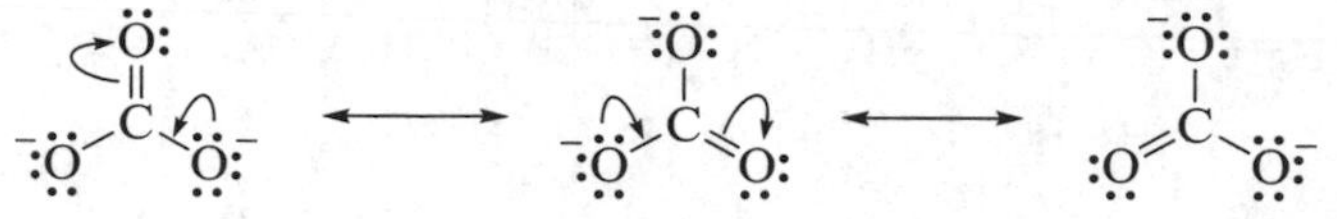

1.17 (b) Carbon has only six valence electrons and is positively charged in the original structural formula. Move the unshared electron pair from nitrogen to a shared position between nitrogen and carbon. The double bond becomes a triple bond.

$$H-\overset{+}{\underset{}{C}}=\overset{\cdot\cdot}{N}-H \longleftrightarrow H-C\equiv\overset{+}{N}-H$$

The new formula has one more bond than the original formula so is more stable and makes a greater contribution to the resonance hybrid. The octet rule is satisfied for both carbon and nitrogen in the new Lewis formula.

(c) Moving an unshared pair from oxygen to between carbon and oxygen converts C—O to C=O. Moving a pair of electrons to nitrogen converts C=N to C—N.

$$H-\overset{:\overset{\cdot\cdot}{O}:^-}{\underset{N-H}{C}} \longleftrightarrow H-\overset{:\overset{\cdot\cdot}{O}}{\underset{:N-H}{C}}$$

The number of bonds is the same in both the original and final structural formulas. The original formula, which has its negative charge on oxygen, is more stable than the second one, which has its negative charge on nitrogen. Oxygen is more electronegative than nitrogen and can bear a negative charge better.

(d) Reorganizing the electrons as indicated by the curved arrows changes the electron counts and formal charges of the two oxygens.

$$:\overset{\cdot\cdot}{O}\overset{\overset{\cdot\cdot}{N}}{}\overset{\cdot\cdot}{O}: \longleftrightarrow :O\overset{\overset{\cdot\cdot}{N}}{}\overset{\cdot\cdot}{O}:^-$$

The two Lewis structures, however, are equally stable and contribute equally to the resonance hybrid.

1.18 Adenosine triphosphate (ATP) has a molecular formula of $C_{10}H_{16}N_5O_{13}P_3$. Each oxygen has two unshared electron pairs and each nitrogen has one unshared electron pair. There are a total of 31 unshared electron pairs in this molecule.

Adenosine triphosphate
$C_{10}H_{16}N_5O_{13}P_3$

1.19 Nitrogen, a second-row element, in the formula $(CH_3)_3N=CH_2$ has 5 bonds accounting for 10 electrons. This violates the octet rule; the formula is incorrect. Phosphorus, a third row element, can have more bonds because *d* orbitals are available for bonding.

1.20 There are four B—H bonds in BH_4^-. The four electron pairs surround boron in a tetrahedral orientation. The H—B—H angles are 109.5°.

1.21 (*b*) Nitrogen in ammonium ion is surrounded by eight electrons in four covalent bonds. These four bonds are directed toward the corners of a tetrahedron.

$$H \blacktriangleleft \overset{\overset{\displaystyle H}{|}}{\underset{\displaystyle H}{N}}{}^+ - H$$

Each HNH angle is 109.5°.

(*c*) Double bonds are treated as a single unit when deducing the shape of a molecule using the VSEPR model. Thus azide ion is linear.

$$:\!\overset{..}{N}\!=\!\overset{+}{N}\!=\!\overset{..}{N}\!:^-$$

The NNN angle is 180°.

(*d*) Because the double bond in carbonate ion is treated as if it were a single unit, the three sets of electrons are arranged in a trigonal planar arrangement around carbon.

$$\overset{\displaystyle \overset{..}{O}:}{\underset{:\overset{..}{O}^- \qquad \overset{..}{O}:^-}{\overset{\|}{C}}}$$

The OCO angle is 120°.

1.22 (*b*) Water is a bent molecule, and so the individual O—H bond dipole moments do not cancel. Water has a dipole moment.

Individual OH bond
moments in water

Direction of net
dipole moment

(*c*) Methane, CH_4, is perfectly tetrahedral, and so the individual (small) C—H bond dipole moments cancel. Methane has no dipole moment.

(*d*) Methyl chloride has a dipole moment.

Individual bond dipole
moments in CH_3Cl

Direction of net
dipole moment

(*e*) Oxygen is more electronegative than carbon and attracts electrons from it. Formaldehyde has a dipole moment.

Individual bond dipole
moments in formaldehyde

Direction of molecular
dipole moment

(*f*) Nitrogen is more electronegative than carbon. Hydrogen cyanide has a dipole moment.

Individual bond dipole
moments in HCN

Direction of molecular
dipole moment

1.23 (*b*) As in part (*a*), breaking the C—O bond in the direction indicated leaves a positive charge on carbon. The electron pair in the original bond become an unshared pair on oxygen.

1.24 (*b*) In the reverse process, an unshared pair of electrons on oxygen becomes the pair of electrons in a C—O bond.

1.25 Following the electrons creates an O—H bond, breaks a C—H bond, changes a C—C single bond to a double bond, and changes a C=O double bond to a single bond.

1.26 (*b*) In an acid–base reaction, a proton is transferred from the Brønsted acid, in this case HCl, to the Brønsted base, in this case $(CH_3)_3N$. Remember to use curved arrows to track the movement of electrons, not atoms.

| Trimethylamine (base) | Hydrogen chloride (acid) | Trimethylamonium ion (conjugate acid) | Chloride ion (conjugate base) |

1.27 Hydride ion is a strong base and will remove a proton from water. The conjugate acid of hydride ion is hydrogen (H_2).

| Hydride ion (base) | Water (acid) | Hydrogen (conjugate acid) | Hydroxide ion (conjugate base) |

1.28 The strength of an acid can be expressed by its pK_a, given by the expression:

$$pK_a = -\log K_a$$

The K_a of salicylic acid is 1.06×10^{-3}; its pK_a is 2.97.

1.29 Because the pK_a of HCN is given as 9.1, its $K_a = 10^{-9.1}$. In more conventional notation, $K_a = 8 \times 10^{-10}$.

1.30 (*b*, *c*) The relative strength of two bases can be determine by comparing the pK_a's of their respective conjugate acids. Remember that the stronger base is derived from the weaker conjugate acid.

Recall that the smaller pK_a is associated with the stronger acid. Ammonia ($pK_a = 36$) is a weaker acid than acetylene ($HC≡CH$, $pK_a = 26$); therefore, amide ion ($H_2\ddot{N}:^-$) is a stronger base than acetylide ion ($HC≡C:^-$).

Similar reasoning leads to the conclusion that because acetylene is a weaker acid than ethanol (CH_3CH_2OH, $pK_a = 16$), acetylide ion is a stronger base than ethoxide ion ($CH_3CH_2\ddot{O}:^-$). Ordering the conjugate acid increasing acidity left to right gives the order of strongest base to weakest.

	Strongest base	⟶	Weakest base
Base	$:\bar{\ddot{N}}H_2$	$H-C≡C:^-$	$CH_3CH_2\ddot{O}:^-$
Conjugate acid	$:NH_3$	$H-C≡C-H$	$CH_3CH_2\ddot{O}-H$
pKa of conjugate acid	36	26	16
	Weakest acid	⟶	Strongest acid

1.31 Bond strength weakens going down a group in the periodic table. Because sulfur lies below oxygen, the H—S bond is weaker than the H—O bond. We would expect H_2S to be a stronger acid than H_2O, and this prediction is borne out by their respective pK_a's:

	H_2S	H_2O
pK_a	9	15.7

Because the stronger acid forms the weaker conjugate base, HS^- is a weaker base than HO^-.

1.32 (*b*) In part (*a*) of the problem, you determined that $(CH_3)_2\overset{+}{\underset{..}{O}}H$ is a stronger acid than $(CH_3)_3\overset{+}{N}H$. Recalling that the weaker acid forms the stronger conjugate base, $(CH_3)_3N:$ is a stronger base than $(CH_3)_2\underset{..}{O}:$.

1.33 To compare the strength of bases, look up the pK_a's of their conjugate acids. Ethanol has a pK_a of 16; *tert*-butyl alcohol has a pK_a of 18. Ethanol is a stronger acid than *tert*-butyl alcohol. Therefore *tert*-butoxide ion is a stronger base than ethoxide.

1.34 The Brønsted acid–base equilibria for the ionization of hypochlorous and hypobromous acid are

| Hypochlorous acid | Water | Hypochlorite ion | Hydronium ion |

Hypobromous acid $\qquad$ Water $\qquad$ Hypobromite ion $\qquad$ Hydronium ion

The positive character (δ^+) of the proton in HOCl is greater than the proton in HOBr because Cl is more electronegative than Br. The greater electronegativity of Cl compared to Br also stabilizes ClO^- compared to BrO^- and makes the equilibrium for the ionization of HOCl more favorable. Therefore, hypochlorous acid is a stronger acid than hypobromous acid. (This prediction is borne out by the measured pK_a's: HOCl = 7.5; HOBr = 8.7. The trend continues with HOI, which has a pK_a of 10.6.)

1.35 The three oxygens in nitrate ion share a total charge of -2. The average formal charge on each oxygen atom is $-\frac{2}{3} = -0.67$.

1.36 Writing the two resonance forms of each conjugate base reveals that they are equivalent.

Oxygen is more electronegative than sulfur and will bear a greater share of the negative charge.

1.37 (b) Begin by writing the equation for the acid–base reaction between acetic acid and fluoride ion. Acetic acid will donate a proton to fluoride ion, converting acetic acid to its conjugate base (acetate ion) and fluoride ion to its conjugate acid, (hydrogen fluoride).

Acetic acid $\qquad$ Fluoride ion $\qquad$ Acetate ion $\qquad$ Hydrogen fluoride

pK_a = 4.7
(weaker acid) $\qquad\qquad\qquad\qquad\qquad$ pK_a = 3.1
(stronger acid)

From the respective pK_a's of the acids, you can see that the weaker acid (acetic acid) is on the left and the stronger acid (hydrogen fluoride) is on the right. Therefore, the equilibrium lies to the left. The equilibrium constant for the process is

$$K_{eq} = \frac{10^{-pK_a} \text{ of acetic acid (reactant)}}{10^{-pK_a} \text{ of hydrogen fluoride (product)}} = \frac{10^{-4.7}}{10^{-3.5}} = 10^{-1.6}$$

(c) Hydrobromic acid will donate a proton to ethanol, converting hydrobromic acid to its conjugate base (bromide ion) and ethanol to its conjugate acid, ethyloxonium ion.

$$:\overset{..}{\underset{..}{Br}}-H \quad + \quad H\overset{..}{\underset{..}{O}}\diagup \quad \rightleftharpoons \quad :\overset{..}{\underset{..}{Br}}:^- \quad + \quad H_2\overset{+}{\underset{..}{O}}\diagup$$

Hydrobromic acid   Ethanol   Bromide ion   Ethyloxonium ion

$pK_a = -5.8$ $pK_a = -2.4$
(stronger acid)   (weaker acid)

From the respective pK_a's of the acids, you can see that the stronger acid (hydrobromic acid) is on the left and the weaker acid (ethyloxonium ion) is on the right. Therefore, the equilibrium lies to the right. The equilibrium constant for the process is

$$K_{eq} = \frac{10^{-pK_a} \text{ of HBr (reactant)}}{10^{-pK_a} \text{ of ethyloxonium ion (product)}} = \frac{10^{5.8}}{10^{2.4}} = 10^{3.4}$$

1.38 Sulfuric acid is a strong acid; the pK_a of the first ionization of H_2SO_4 is -4.8. In an aqueous solution of a strong acid, the predominant acid species is H_3O^+. Hydrogen sulfate ion is a weak acid ($pK_a = 2.0$) and will only undergo ionization to a small extent. The relative amounts of the species present will be

$$[H_3O^+] > [HSO_4^-] > [SO_4^{2-}] > [H_2SO_4]$$

1.39 Comparing the pK_a's of phenol and water reveals that, indeed, the stronger acid (phenol) is on the left in the acid–base equation and the equilibrium constant is greater than 1.

$$\bigcirc\!\!-\overset{..}{\underset{..}{O}}H \quad + \quad H\overset{..}{\underset{..}{O}}:^- \quad \rightleftharpoons \quad \bigcirc\!\!-\overset{..}{\underset{..}{O}}:^- \quad + \quad H_2\overset{..}{\underset{..}{O}}:$$

Phenol   Hydroxide   Phenoxide   Water

$pK_a = 10$ $pK_a = 15.1$
(stronger acid)   (weaker acid)

The same comparison for the reaction of phenol and hydrogen carbonate ion reveals that the stronger acid (carbonic acid) is on the right and the equilibrium constant is less than one.

Phenol   +   Hydrogen $\rightleftharpoons$ Phenoxide   +   Carbonic acid
                                    carbonate

$pK_a = 10$ $pK_a = 6.4$
(weaker acid)   (stronger acid)

1.40 (*b*) Boron trifluoride is a Lewis acid and will accept a pair of electrons from dimethyl sulfide to form a Lewis acid–Lewis base complex.

$$:\overset{..}{\underset{..}{F}}-B\overset{:\overset{..}{\underset{..}{F}}:}{\underset{:\overset{..}{\underset{..}{F}}:}{}} \quad + \quad :\overset{..}{S}\diagdown \quad \rightleftharpoons \quad :\overset{..}{\underset{..}{F}}-\overset{:\overset{..}{\underset{..}{F}}:}{\underset{:\overset{..}{\underset{..}{F}}:}{B}}\overset{-}{\underset{}{}}\overset{+}{S}:\diagdown$$

(c) Trimethylamine reacts with boron trifluoride is a similar fashion to form a Lewis acid–Lewis base complex.

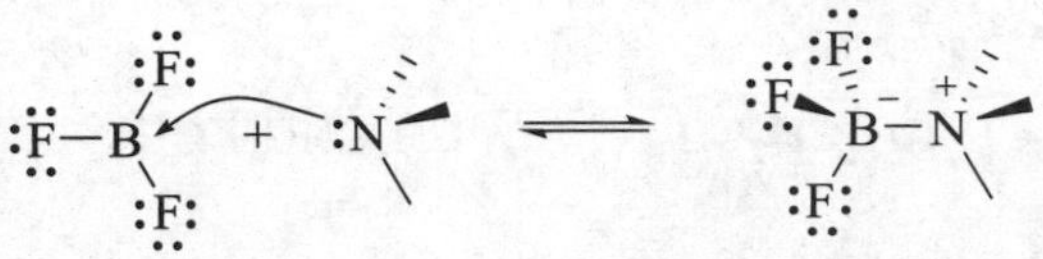

End of Chapter Problems

Structural Formulas

1.41 (a) Each carbon has four valence electrons, each hydrogen has one, and chlorine has seven. Hydrogen and chlorine each can form only one bond, and so the only stable structure must have a carbon–carbon bond. There are 18 valence electrons in C_2H_3Cl, and the framework of five single bonds accounts for only ten electrons. Six of the remaining eight electrons are used to complete the octet of chlorine as three unshared pairs, and the last two are used to form a carbon–carbon double bond.

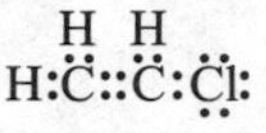

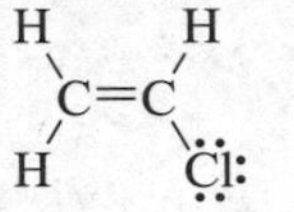

(b) All of the atoms except carbon (H, Br, Cl, and F) are monovalent; therefore, they can only be bonded to carbon. The problem states that all three fluorines are bonded to the same carbon, and so one of the carbons is present as a CF_3 group. The other carbon must be present as a CHBrCl group. Connect these groups together to give the structure of halothane.

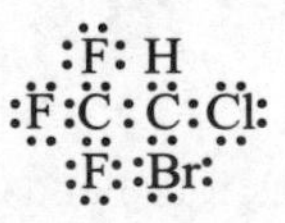

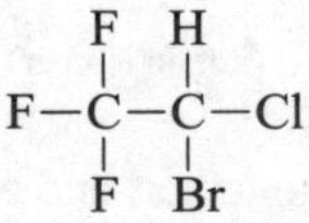

(Unshared electrons pairs omitted for clarity)

(c) As in part (b), all of the atoms except carbon are monovalent. Because each carbon bears one chlorine, two $ClCF_2$ groups must be bonded together.

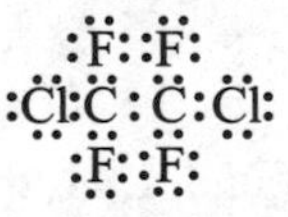

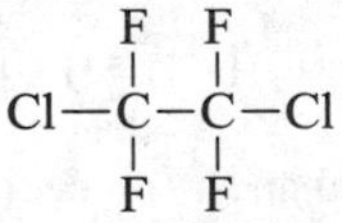

(Unshared electrons pairs omitted for clarity)

1.42 (a) To generate constitutionally isomeric structures having the molecular formula C_4H_{10}, you need to consider the various ways in which four carbon atoms can be bonded together. These are

C—C—C—C   and   C—C—C
                         |
                         C

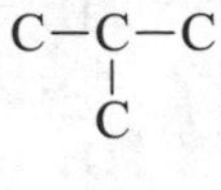

Filling in the appropriate hydrogens gives the correct structures:

$CH_3CH_2CH_2CH_3$ and CH_3CHCH_3
                                    |
                                    CH_3

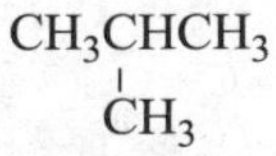

Continue with the remaining parts of the problem using the general approach outlined for part (a).

(*b*) C_5H_{12}

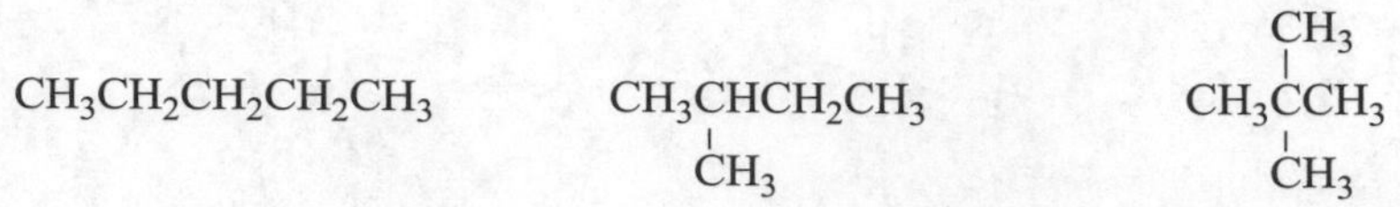

$CH_3CH_2CH_2CH_2CH_3$ $CH_3CHCH_2CH_3$ CH_3CCH_3
                                                     $|$ CH_3
                                                   CH_3 $|$
                                                                                  CH_3CCH_3
                                                                                         $|$
                                                                                       CH_3

(*c*) $C_2H_4Cl_2$

                              CH_3CHCl_2 $ClCH_2CH_2Cl$

(*d*) C_4H_9Br

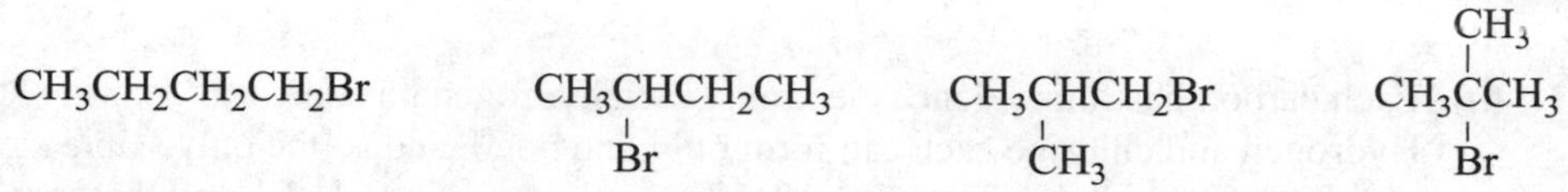

$CH_3CH_2CH_2CH_2Br$ $CH_3CHCH_2CH_3$ CH_3CHCH_2Br CH_3CCH_3
                                                   $|$ $|$ $|$
                                                  Br CH_3 CH_3 ... Br

(*e*) C_3H_9N

$CH_3CH_2CH_2NH_2$ $CH_3CH_2NHCH_3$ $H_3C{-}N$ CH_3CHCH_3
                                                                           CH_3 $|$
                                                                           CH_3 NH_2

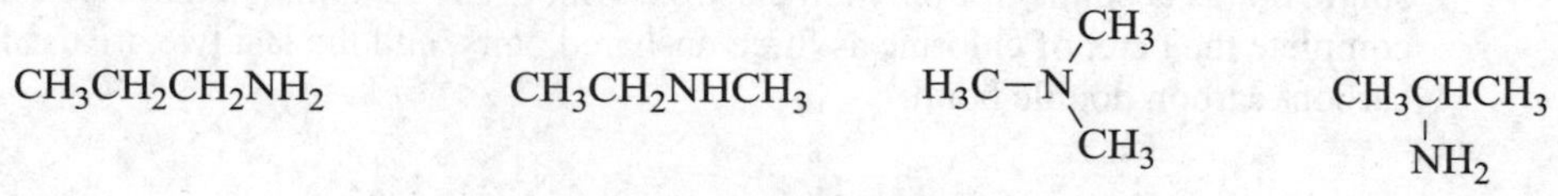

Note that when the three carbons and the nitrogen are arranged in a ring, the molecular formula based on such a structure is C_3H_7N, not C_3H_9N as required.

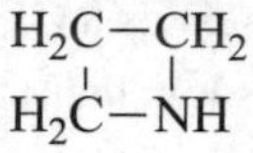

$$H_2C{-}CH_2$$
$$H_2C{-}NH$$

(not an isomer)

1.43 (*a*) All three carbons must be bonded together, and each one has four bonds; therefore, the molecular formula C_3H_8 uniquely corresponds to

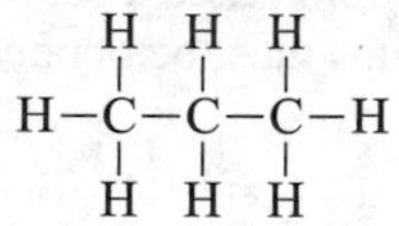

$$
\begin{array}{ccc}
H & H & H \\
| & | & | \\
H{-}C{-}C{-}C{-}H \\
| & | & | \\
H & H & H
\end{array}
$$

(*b*) With two fewer hydrogen atoms than the preceding compound, C_3H_6 must either contain a carbon–carbon double bond or its carbons must be arranged in a ring; thus, the following structures are constitutional isomers:

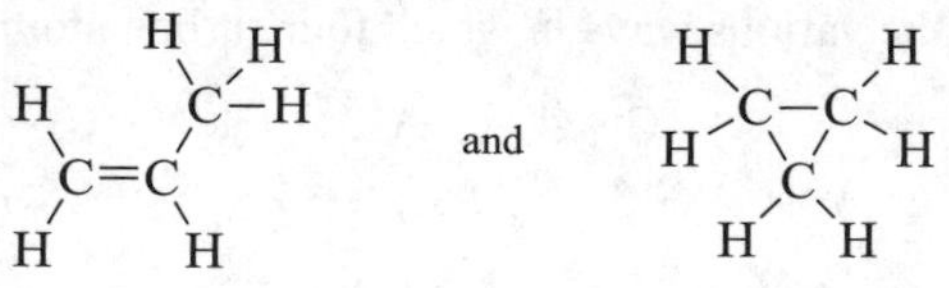

and

(*c*) The molecular formula C_3H_4 is satisfied by the structures

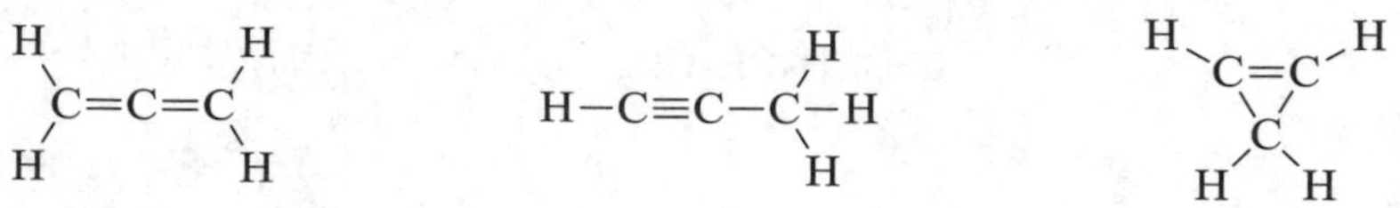

$$H_2C{=}C{=}CH_2 \qquad H{-}C{\equiv}C{-}CH_3$$

1.44 (*a*) The connectivities of C_3H_6O that contain only single bonds must have a ring as part of their structure.

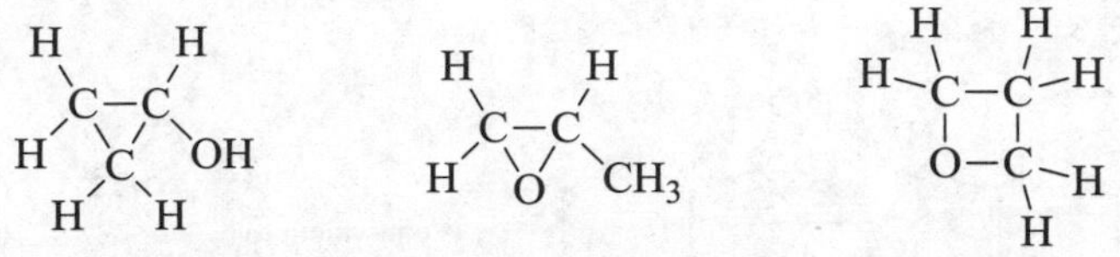

(b) Structures corresponding to C_3H_6O are possible in noncyclic compounds if they contain a carbon–carbon or carbon–oxygen double bond.

$$CH_3CH_2\overset{\overset{O}{\|}}{C}H \qquad CH_3\overset{\overset{O}{\|}}{C}CH_3 \qquad CH_3CH{=}CHOH \qquad CH_3OCH{=}CH_2$$

$$CH_3\underset{\underset{OH}{|}}{C}{=}CH_2 \qquad H_2C{=}CHCH_2OH$$

1.45 The structures, written in a form that indicates hydrogens and unshared electrons, are as shown.

Remember: A neutral carbon has four bonds, a neutral nitrogen has three bonds plus one unshared electron pair, and a neutral oxygen has two bonds plus two unshared electron pairs. Halogen substituents have one bond and three unshared electron pairs. The molecular formulas reveal that (a) and (b) are constitutional isomers.

(a) $C_{10}H_{16}$ is equivalent to

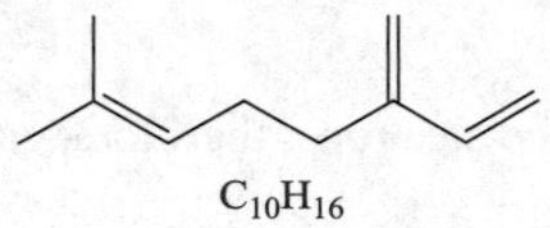

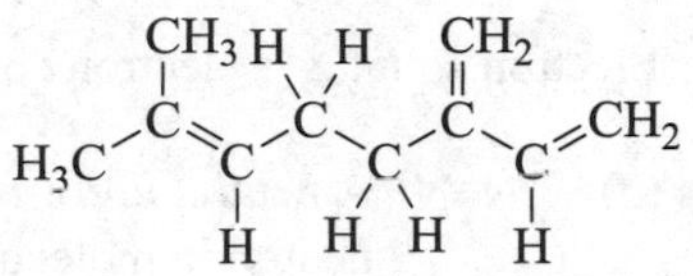

(b) 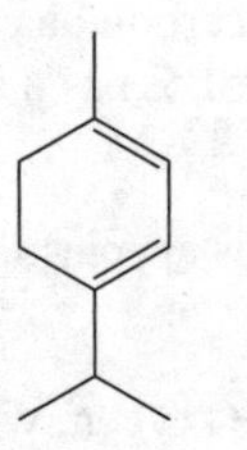 $C_{10}H_{16}$ is equivalent to

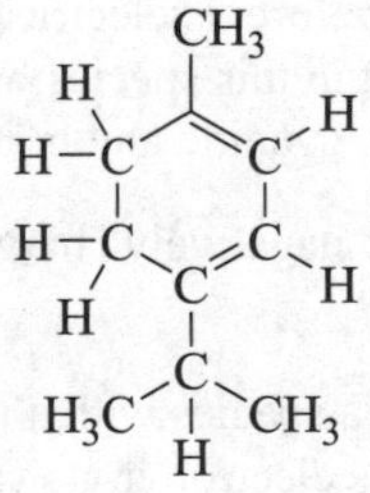

(c) OH $C_7H_{16}O$ is equivalent to

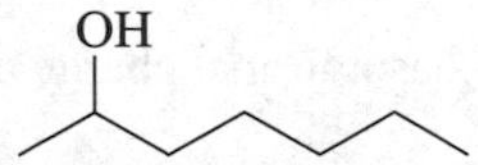

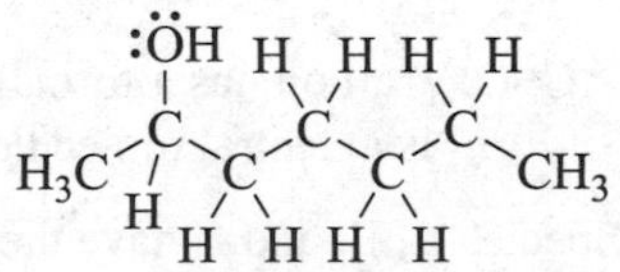

(d) O $C_7H_{14}O$ is equivalent to

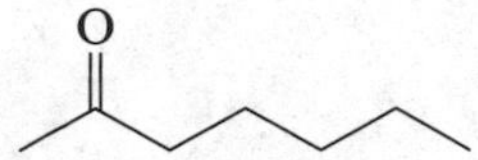

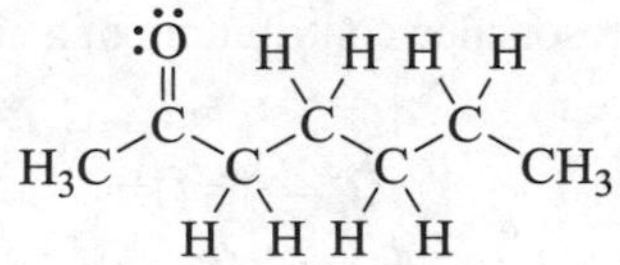

(e) 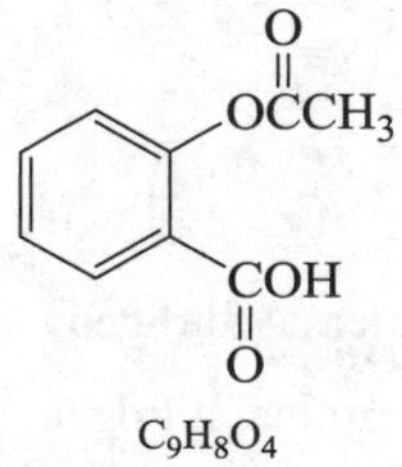 $C_9H_8O_4$ is equivalent to

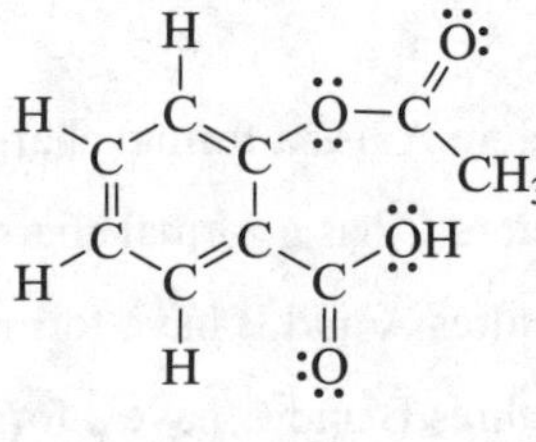

(f)

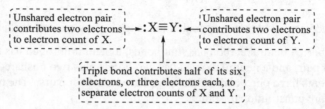

$C_{16}H_8Br_2N_2O_2$

is equivalent to

Formal Charge and Resonance

1.46 All these species are characterized by the formula $:X{\equiv}Y:$ and each atom has an electron count of 5.

| Unshared electron pair contributes two electrons to electron count of X. | → $:X{\equiv}Y:$ ← | Unshared electron pair contributes two electrons to electron count of Y. |

Triple bond contributes half of its six electrons, or three electrons each, to separate electron counts of X and Y.

Electron count X = electron count Y = 2 + 3 = 5

(a) $:N{\equiv}N:$ A neutral nitrogen atom has five valence electrons: therefore, each atom is electrically neutral in molecular nitrogen.

(b) $:C{\equiv}N:$ Nitrogen, as before, is electrically neutral. A neutral carbon has four valence electrons, and so carbon in this species, with an electron count of 5, has a unit negative charge. The species is cyanide anion; its net charge is −1.

(c) $:C{\equiv}C:$ There are two negatively charged carbon atoms in this species. It is a dianion; its net charge is −2.

(d) $:N{\equiv}O:$ Here again is a species with a neutral nitrogen atom. Oxygen, with an electron count of 5, has one less electron in its valence shell than a neutral oxygen atom. Oxygen has a formal charge of +1; the net charge is +1.

(e) $:C{\equiv}O:$ Carbon has a formal charge of −1; oxygen has a formal charge of +1. Carbon monoxide is a neutral molecule.

1.47 (a) Species A, B, and C have the same molecular formula, the same atomic positions, and the same number of electrons. They differ only in the arrangement of their electrons. They are therefore resonance contributors of a single compound.

$$\underset{\text{A}}{\overset{H}{\underset{H}{:}C-\overset{+}{N}{\equiv}N:}} \qquad \underset{\text{B}}{\overset{H}{\underset{H}{}}C=\overset{+}{N}={\overset{..}{\underset{..}{N}}:^{-}}} \qquad \underset{\text{C}}{\overset{H}{\underset{H}{}}\overset{+}{C}-\overset{..}{N}=\overset{..}{N}:^{-}}$$

(b) Structure A has a formal charge of −1 on carbon.

(c) Structure C has a formal charge of +1 on carbon.

(d) Structures A and B have formal charges of +1 on the internal nitrogen.

(e) Structures B and C have a formal charge of −1 on the terminal nitrogen.

(f) All resonance contributors of a particular species must have the same net charge. In this case, the net charge on A, B, and C is 0.

(g) Both A and B have the same number of covalent bonds, but the negative charge is on a more electronegative atom in B (nitrogen) than it is in A (carbon). Structure B is more stable.

(h) Structure B is more stable than structure C. Structure B has one more covalent bond, all of its atoms have octets of electrons, and it has a lesser degree of charge separation than C. The carbon in structure C does not have an octet of electrons.

(i) The CNN unit is linear in A and B but bent in C according to VSEPR. This is an example of how VSEPR can fail when comparing resonance structures.

1.48 (a) These two structures are resonance contributors because they have the same atomic positions and the same number of electrons.

$$:\overset{2-}{\ddot{N}} - \overset{+}{N} \equiv N: \longleftrightarrow :\ddot{N} = \overset{+}{N} = \ddot{N}:^-$$

16 valence electrons   16 valence electrons
(net charge = -1)   (net charge = -1)

(b) The two structures have different numbers of electrons and, therefore, cannot be resonance contributors of the same species.

$$:\overset{2-}{\ddot{N}} - \overset{+}{N} \equiv N: \qquad :N - \overset{2+}{N} = \ddot{N}:^-$$

16 valence electrons   14 valence electrons
(net charge = -1)   (net charge = +1)

(c) These two structures have different numbers of electrons; they are not resonance contributors.

$$:\overset{2-}{\ddot{N}} - \overset{+}{N} \equiv N: \qquad :\overset{2-}{\ddot{N}} - \overset{-}{\ddot{N}} - \ddot{N}:^{2-}$$

16 valence electrons   20 valence electrons
(net charge = -1)   (net charge = -5)

1.49 (a) Structure C has five covalent bonds to nitrogen (ten electrons). The octet rule limits nitrogen to a maximum of eight electrons in its valence shell.

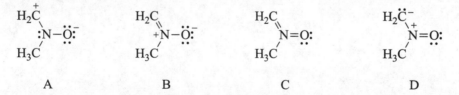

          A   B   C   D

(b) Of the three remaining structures, B and D have four bonds and each is more stable than A, which has only three. B is more stable than D because its negative charge is on a more electronegative element (oxygen versus carbon).

(c) Use curved arrows to show movement of electrons among the resonance contributors.

$$\underset{\text{A}}{\overset{H_2\overset{+}{C}}{\underset{H_3C}{>}}\ddot{N} - \ddot{O}:^-} \longleftrightarrow \underset{\text{B}}{\overset{H_2C}{\underset{H_3C}{>}}\overset{+}{N} - \ddot{O}:^-} \longleftrightarrow \underset{\text{D}}{\overset{H_2\ddot{C}^-}{\underset{H_3C}{>}}\overset{+}{N} = \ddot{O}:}$$

1.50 In structure A, the only contributing structure that can be written that satisfies the octet rule is A:

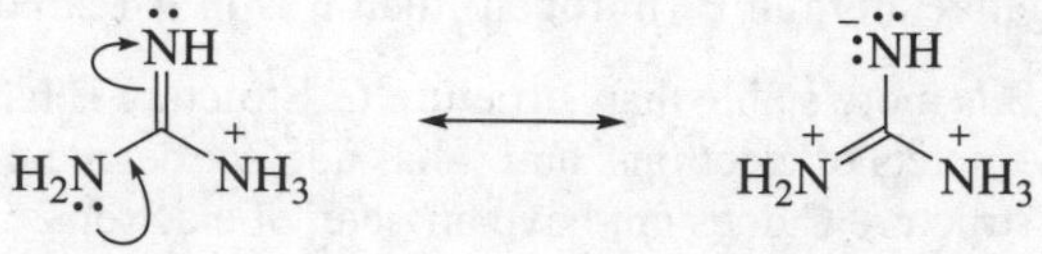

Contributing structure A′ is less stable than A because it has more charge separation than A.

In structure B, electron delocalization disperses the charge without increasing charge separation. All three resonance forms are equally stable.

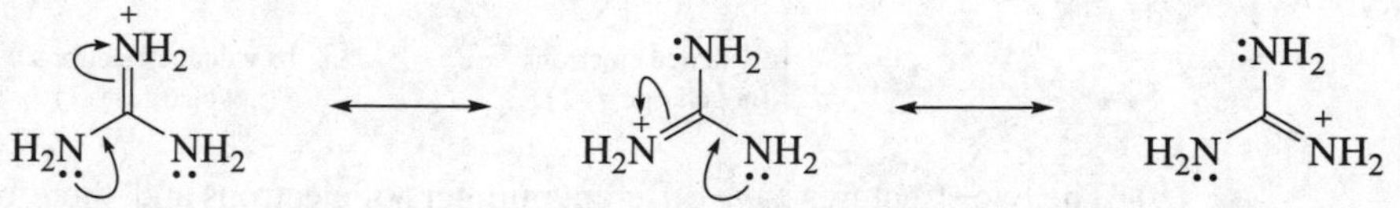

Structure B is stabilized by electron delocalization more than A.

1.51 Notice that in all of the cases in which the more stable Lewis formula contains more bonds than the original one, the additional bond leads to a structure in which the octet rule is satisfied.

(a)

$$H_3C-\overset{..}{N}=\overset{+}{N}: \longleftrightarrow H_3C-\overset{+}{N}\equiv N:$$

One more bond than
original Lewis formula

(b)

$$H-C\overset{:\overset{..}{O}:^-}{\underset{\overset{+}{O}-H}{}} \longleftrightarrow H-C\overset{:\overset{..}{O}}{\underset{:\overset{..}{O}-H}{}}$$

No separation
of opposite charges

(c)

$$H_2\overset{+}{C}-\overset{-}{C}H_2 \longleftrightarrow H_2C=CH_2$$

One more bond and
no charge separation

(d)

$$H_2\overset{+}{C}-CH=CH-\overset{-}{C}H_2 \longleftrightarrow H_2C=CH-CH=CH_2$$

One more bond and
no charge separation

(e)

$$H_2\overset{+}{C}-CH=CH-\overset{..}{O}:^- \longleftrightarrow H_2C=CH-CH=\overset{..}{O}:$$

One more bond and
no charge separation

(f)

$$H_2\overset{-}{C}-C\overset{\overset{..}{O}:}{\underset{H}{}} \longleftrightarrow H_2C=C\overset{:\overset{..}{O}:^-}{\underset{H}{}}$$

Negative charge on more
electronegative element

(g)

$$H-\overset{+}{C}=\overset{..}{O}: \longleftrightarrow H-C\equiv\overset{+}{O}:$$

One more bond

(h)

$$H_2\overset{+}{C}-\overset{..}{O}H \longleftrightarrow H_2C=\overset{+}{O}H$$

One more bond

(i)

$$H_2\overset{-}{C}-\overset{\overset{+}{N}H_2}{N}: \longleftrightarrow H_2C=N\overset{:NH_2}{\underset{..}{}}$$

No separation
of opposite charges

Dipole Moment

1.52 (a) The molecular formula OCS tells us that there are 16 valence electrons in carbon oxysulfide. Carbon contributes four, oxygen six, and sulfur six. The connectivity O—C—S accounts for four electrons. Apportioning the remaining 12 equally between oxygen and sulfur completes the octet of both but leaves carbon with only four electrons.

$$:\ddot{O}-C-\ddot{S}:$$

Use one of oxygen's unshared pairs to form a double bond to carbon. Do the same with one of sulfur's unshared pairs to give a Lewis formula in which all three atoms have complete octets.

$$:\ddot{O}=C=\ddot{S}:$$

(b) Double bonds are treated the same way as single bonds when applying VSEPR. The electrons in the O=C bond are farthest removed from the electrons in the C–S bond when the angle between them is 180°. Carbon oxysulfide is *linear*.

(c) According to Table 1.3 in the text, oxygen, with an electronegativity of 3.5, is the most electronegative atom of the three while the electronegativities of carbon and sulfur are both 2.5. Thus, electron density is drawn toward oxygen and away from the C=S group.

$$O=C=S$$

1.53 The direction of a bond dipole is governed by the electronegativity of the atoms it connects.

In each of the parts to this problem, the more electronegative atom is partially negative and the less electronegative atom is partially positive. Electronegativities of the elements are given in Table 1.3 of the text.

(a) Chlorine is more electronegative than hydrogen.

$$H-Cl$$

(b) Iodine is more electronegative than hydrogen.

$$H-I$$

(c) Oxygen is more electronegative than hydrogen.

$$H \quad O \quad H$$

(d) Oxygen is more electronegative than either hydrogen or chlorine.

$$H \quad O \quad Cl$$

1.54 The direction of a bond dipole is governed by the electronegativity of the atoms involved. Among the halogens, the order of electronegativity is F > Cl > Br > I. Fluorine therefore attracts electrons away from chlorine in FCl, and chlorine attracts electrons away from iodine in ICl.

$$F-Cl \qquad\qquad I-Cl$$
$$\mu = 0.9 \text{ D} \qquad \mu = 0.7 \text{ D}$$

Chlorine is the positive end of the dipole in F—Cl and the negative end in I—Cl.

1.55 (*a*) Fluorine is more electronegative than chlorine, and so its bond to hydrogen is more polar, as the measured dipole moments indicate.

$$H\text{—}F \quad \text{is more polar than} \quad H\text{—}Cl$$
$$\mu = 1.7 \text{ D} \qquad\qquad\qquad \mu = 1.1 \text{ D}$$

(*b*) Boron trifluoride is planar. Its individual B—F bond dipoles cancel. It has no dipole moment.

$$H\text{—}F \quad \text{is more polar than} \quad \underset{F}{\overset{F}{B}}\text{—}F$$
$$\mu = 1.7 \text{ D} \qquad\qquad\qquad \mu = 0 \text{ D}$$

(*c*) A carbon–chlorine bond is strongly polar; carbon–hydrogen and carbon–carbon bonds are only weakly polar.

$$\underset{H_3C}{\overset{H_3C}{\underset{}{H_3C}}}C\text{—}\ddot{\underset{..}{C}l}\text{:} \quad \text{is more polar than} \quad \underset{H_3C}{\overset{H_3C}{\underset{}{H_3C}}}C\text{—}H$$
$$\mu = 2.1 \text{ D} \qquad\qquad\qquad\qquad \mu = 0.1 \text{ D}$$

(*d*) A carbon–fluorine bond in CCl_3F opposes the polarizing effect of the chlorines. The carbon–hydrogen bond in $CHCl_3$ reinforces it. $CHCl_3$ therefore has a larger dipole moment.

$$\underset{Cl}{\overset{Cl}{\underset{}{Cl}}}C\text{—}H \quad \text{is more polar than} \quad \underset{Cl}{\overset{Cl}{\underset{}{Cl}}}C\text{—}F$$
$$\mu = 1.0 \text{ D} \qquad\qquad\qquad\qquad \mu = 0.5 \text{ D}$$

(*e*) Oxygen is more electronegative than nitrogen; its bonds to carbon and hydrogen are more polar than the corresponding bonds formed by nitrogen.

$$H_3C\overset{\ddot{O}}{\underset{..}{}}H \quad \text{is more polar than} \quad H\overset{\ddot{N}}{\underset{H}{}}H$$
$$\mu = 1.7 \text{ D} \qquad\qquad\qquad\qquad \mu = 1.3 \text{ D}$$

(*f*) The Lewis structure for CH_3NO_2 has a formal charge of +1 on nitrogen, making it more electron-attracting than the uncharged nitrogen of CH_3NH_2.

$$H_3C\text{—}\underset{O^-}{\overset{O}{N}}+ \quad \text{is more polar than} \quad H_3C\text{—}\underset{H}{\overset{H}{\ddot{N}}}\text{:}$$
$$\mu = 3.1 \text{ D} \qquad\qquad\qquad\qquad \mu = 1.3 \text{ D}$$

Acids and Bases

1.56 Hydrogen peroxide is more acidic than water because its conjugate base is a weaker base than hydroxide. In the conjugate base of hydrogen peroxide, the electronegative neutral oxygen withdraws electron density from the oxygen bearing the formal charge. Thus the negative charge on the terminal oxygen is not as available compared to that of hydroxide.

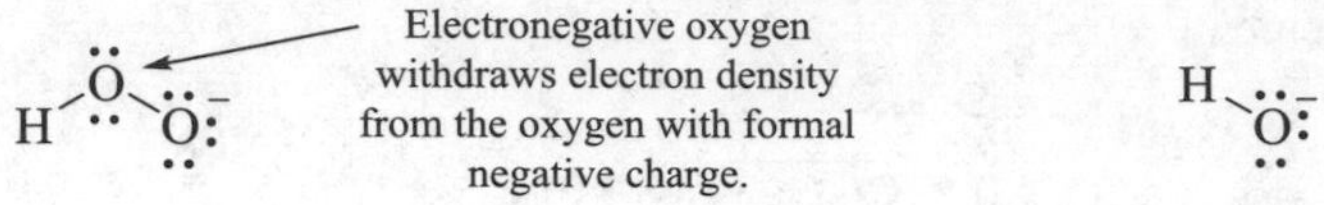

Electronegative oxygen withdraws electron density from the oxygen with formal negative charge.

Hydroperoxide
(conjugate base of peroxide)

Hydroxide
(conjugate base of water)

1.57 (*a*) From Table 1.8, the pK_a of the nonzwitterionic form of montelukast can be approximated by comparing the carboxylic acid group to acetic acid (pK_a = 4.7). The zwitterionic form can be approximated by comparing it to the pyridinium ion (pK_a = 5.2). Because the carboxylic acid group is a stronger acid than the pyridinium ion, the zwitterionic structure is favored.

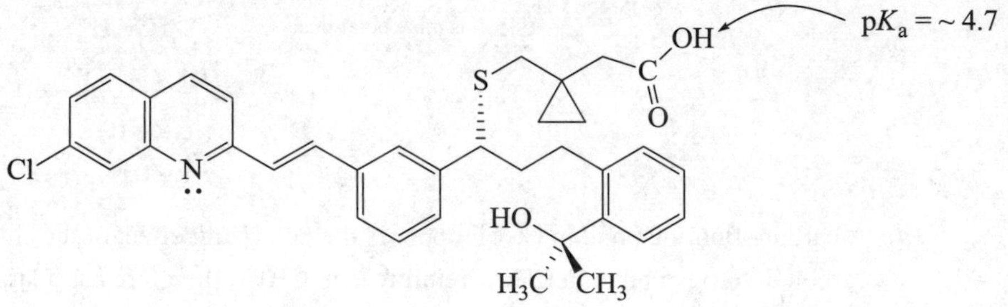

pK_a = ~ 4.7

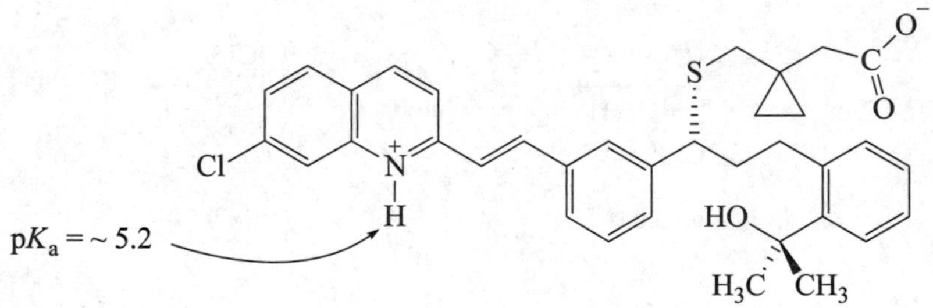

pK_a = ~ 5.2

(*b*) One equivalent of NaOH will deprotonate the N-H proton of montelukast.

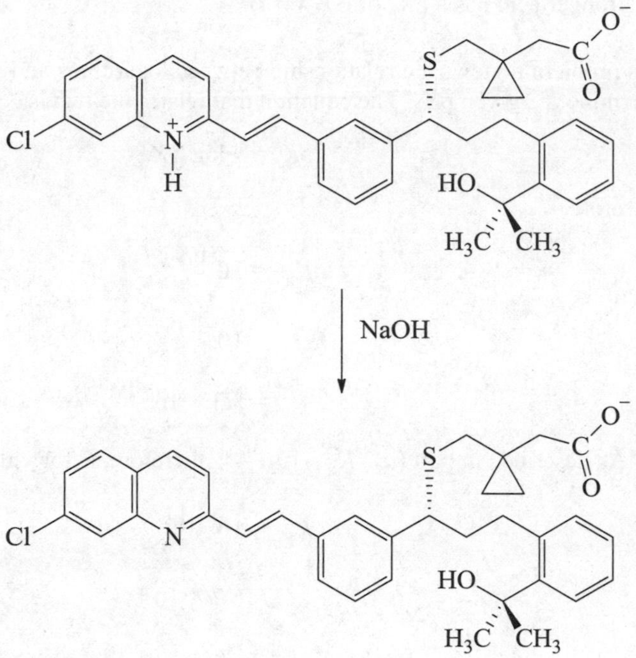

NaOH

(*c*) One equivalent of HCl with protonate the carboxylate (−COO⁻) group to give the carboxylic acid (−COOH).

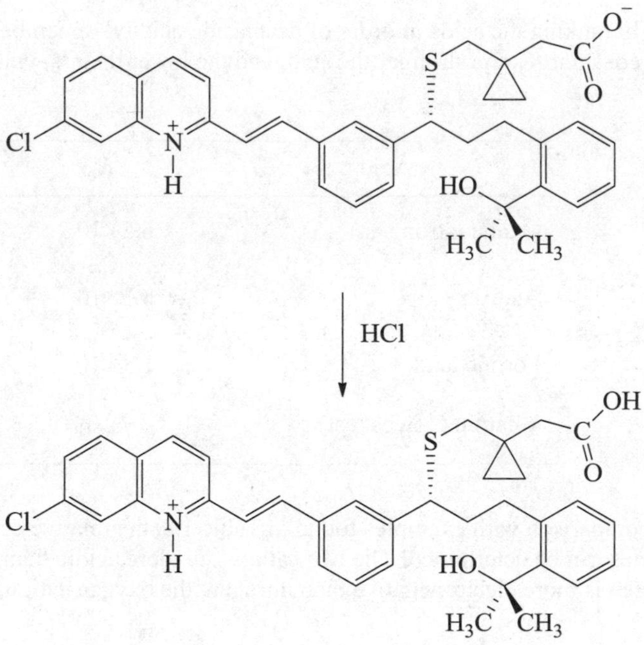

HCl

1.58 (*a*) Because 10^{-pK_a} equals K_a, - the ratio is $\dfrac{10^{-2}}{10^{-8}} = 10^6$. K_a for the acid with a pK_a of 2 is 1 million times larger than for the acid with a pK_a of 8.

(*b*) The difference of the K_a's between the two acids is 10,000. Since $pK_a = -\log K_a$ this corresponds to a difference in pK_a's $= -\log(10,000) = -4$. The weaker acid has a pK_a of 5 so the stronger acid has a pK_a of $5 - 4 = 1$.

1.59 (*a*) This problem reviews the relationship between logarithms and exponential numbers. We need to determine K_a, given pK_a. The equation that relates the two is

$$pK_a = -\log_{10} K_a$$

Therefore

$$K_a = 10^{-pK_a}$$

$$= 10^{-3.48}$$

$$= 3.3 \times 10^{-4}$$

(*b*) As described in part (*a*), $K_a = 10^{-pK_a}$; therefore, K for vitamin C is given by the expression

$$K_a = 10^{-4.17}$$

$$= 6.7 \times 10^{-5}$$

(*c*) Similarly, $K_a = 1.8 \times 10^{-4}$ for formic acid (pK_a 3.75).

(*d*) $K_a = 6.5 \times 10^{-2}$ for oxalic acid (pK_a 1.19).

In ranking the acids in order of decreasing acidity, remember that the larger the equilibrium constant K_a, the stronger the acid; and the lower the pK_a value, the stronger the acid.

Acid	K_a	pK_a
Oxalic (strongest)	6.5×10^{-2}	1.19
Aspirin	3.3×10^{-4}	3.48
Formic acid	1.8×10^{-4}	3.75
Vitamin C (weakest)	6.7×10^{-5}	4.17

1.60 By comparison with examples found in Table 1.8, the relative acidities of these four compounds and species can be determined. The two cations are more acidic than the neutral compounds, and because oxygen is more electronegative than nitrogen, the oxygen cation is more acidic than the nitrogen cation. Likewise, the neutral compound containing oxygen is more acidic than the one containing nitrogen.

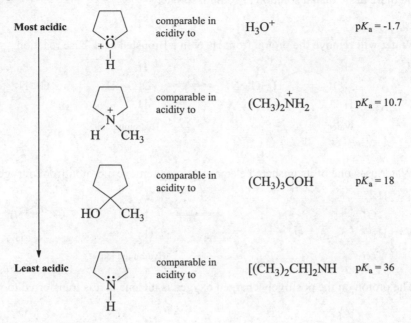

1.61 As in the previous problem, the species given in the problem can be compared with examples in Table 1.8. The best way to determine the basicity of an anion is to consider the acidity of its conjugate acid, remembering that the weaker the acid the stronger the conjugate base.

Most basic ———— is the conjugate base of ———— pK_a = 26

is the conjugate base of O^- ———— pK_a = 16

is the conjugate base of S^- ———— pK_a = 10.7

Least basic is the conjugate base of ———— pK_a = 4.7

1.62 In each case, the more basic solution will be the one that contains the stronger base. Only the anions need be considered; sodium ion is a spectator ion and can be ignored. Remember that the stronger base is the one with the weaker conjugate acid.

(*a*) HCN (pK$_a$ = 9.1) is a weaker acid than HF (pK$_a$ = 3.1), and a solution of CN$^-$ (from NaCN) will be more basic than a solution of F$^-$ (from NaF).

(*b*) Carbonic acid (pK$_a$ = 6.4) is a weaker acid than acetic acid (pK$_a$ = 4.7). A solution of sodium carbonate will be more basic than a solution of sodium acetate.

(*c*) Methanethiol (CH$_3$SH, pK$_a$ = 10.7) is the conjugate acid of methanethiolate ion. Methanethiol is a weaker acid than sulfuric acid $(pK_a = -4.8)$, and a solution of sodium methanethiolate will be more basic than a solution of sodium sulfate.

1.63 (*a*) Water will remove the proton from HCN in a Brønsted acid–base reaction.

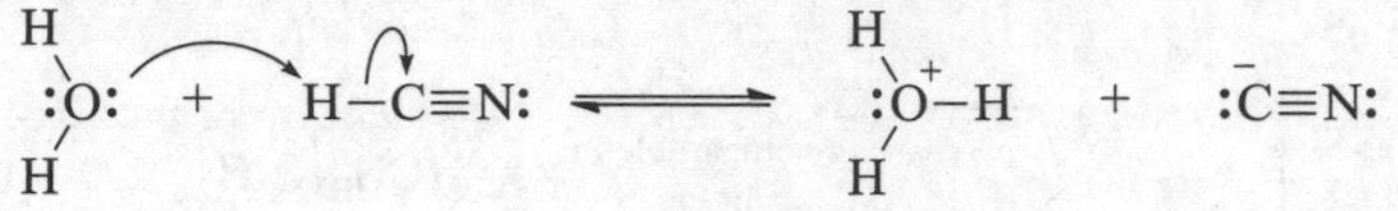

Water   Hydronium ion

(*b*) Water uses one of its unshared electron pairs to remove a proton from nitrogen.

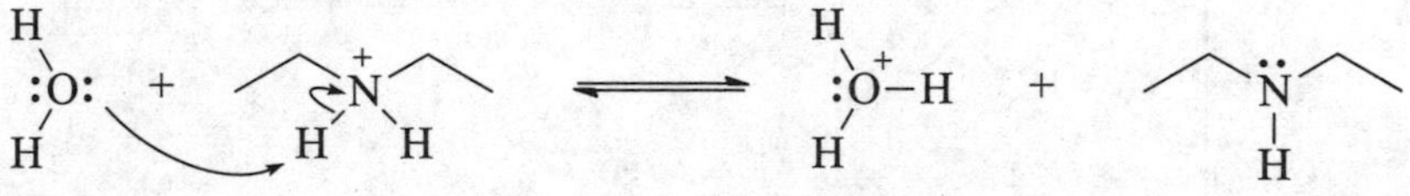

Water   Hydronium ion

(*c*) The proton on the positively charged oxygen is the one that is transferred to water.

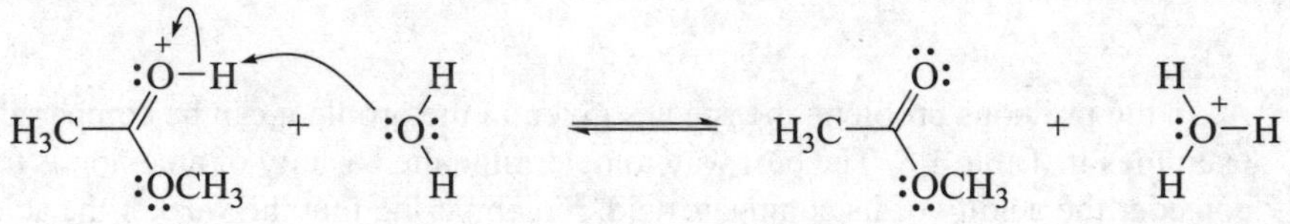

Water   Hydronium ion

1.64 Water can act as both a base, as in the previous problem, or as an acid as in this problem. Water donates one of its protons and forms hydroxide ion in each case. Remember to use curved arrows to track electron movement, not the movement of atoms.

(*a*)

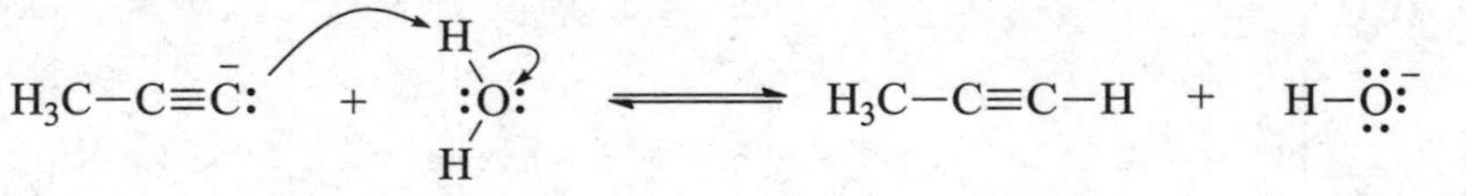

Water   Hydroxide ion

(*b*)

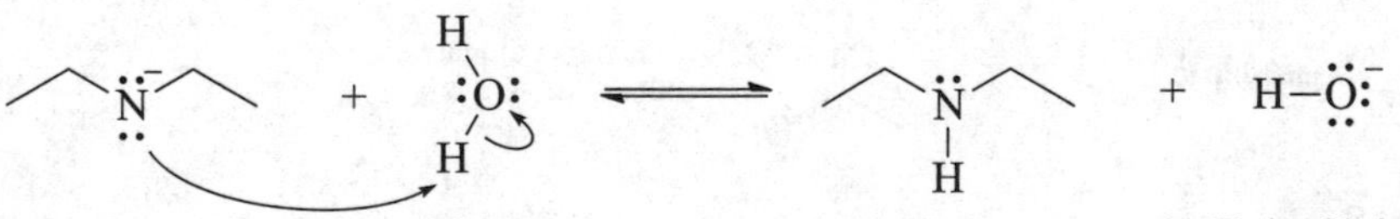

Water   Hydroxide ion

(*c*)

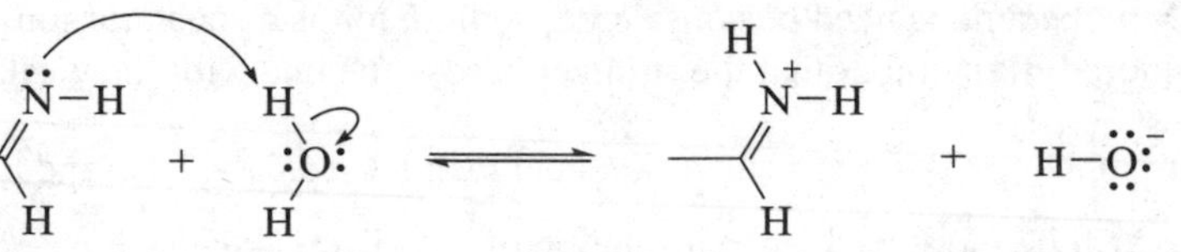

Water   Hydroxide ion

1.65 (a)

Base	Acid	Conjugate acid	Conjugate base
	$pK_a = 10.7$	$pK_a = 18$	

The equilibrium constant can be calculated from the ratio of the two K_a's of the acids in the preceding equation. The K_a's can be obtained from the pK_a's in Table 1.8 by rearranging the equation $pK_a = -\log K_a$ to $K_a = 10^{-pK_a}$. Taking the ratio of the acid and conjugate acid K_a's gives the equilibrium constant K_{eq}:

$$K_{eq} = \frac{K_a \text{ (acid)}}{K_a \text{ (conjugate acid)}} = \frac{10^{-10.7}}{10^{-18}} = 10^{7.3} = 2 \times 10^7$$

(b)

Base	Acid	Conjugate acid	Conjugate base
	$pK_a = 4.7$	$pK_a = 10.2$	

$$K_{eq} = \frac{K_a \text{ (acid)}}{K_a \text{ (conjugate acid)}} = \frac{10^{-4.7}}{10^{-10.2}} = 10^{5.5} = 3.2 \times 10^5$$

(c)

Base	Acid	Conjugate acid	Conjugate base
	$pK_a = 17$	$pK_a = 36$	

$$K_{eq} = \frac{K_a \text{ (acid)}}{K_a \text{ (conjugate acid)}} = \frac{10^{-17}}{10^{-36}} = 1 \times 10^{19}$$

(d)

Base	Acid	Conjugate acid	Conjugate base
	$pK_a = 26$	$pK_a = 36$	

$$K_{eq} = \frac{K_a \text{ (acid)}}{K_a \text{ (conjugate acid)}} = \frac{10^{-26}}{10^{-36}} = 1 \times 10^{10}$$

1.66 (a) The acid on the left is acetylene. Amide ion is the base, which abstracts a proton from acetylene.

Their pK_a's are shown in the equation.

$pK_a = 26$ $pK_a = 36$

Acetylene is a stronger acid than ammonia. The equilibrium lies to the right. The equilibrium constant is of reactant acid of product acid

$$K_{eq} = \frac{10^{-pK_a} \text{ of reactant acid}}{10^{-pK_a} \text{ of product acid}} = \frac{10^{-26}}{10^{-36}} = 10^{10}$$

(b) Contrast this part of the problem with part (a). Methoxide ion is a much weaker base than the amide ion of part (a).

$pK_a = 26$ $pK_a = 15.2$

Here, the weaker acid is on the left. The equilibrium lies to the left. The equilibrium constant is more than 10^{20} times smaller than that of part (a):

$$K_{eq} = \frac{10^{-26}}{10^{-15.2}} = 10^{-10.8} = 1.6 \times 10^{-11}$$

(c) Table 1.8 tells us that the most acidic protons in the reactant (2,4-pentanedione) are those in the CH_2 group.

$pK_a = 9$ $pK_a = 15.2$

2,4-Pentanedione is a stronger acid than methanol. The equilibrium lies to the right and the equilibrium constant is:

$$K_{eq} = \frac{10^{-9}}{10^{-15.2}} = 10^{6.2} = 1.6 \times 10^6$$

(d) The acidic proton in part (c) was part of a CH_2 flanked by two C=O groups. In ethyl acetate (given in Table 1.8), the CH_2 is flanked by only one C=O group and is far less acidic.

$pK_a = 25.6$ $pK_a = 15.2$

The stronger acid is on the right. Therefore, the equilibrium lies to the left and the equilibrium constant is

$$K_{eq} = \frac{10^{-25.6}}{10^{-15.2}} = 10^{-10.4} = 4.0 \times 10^{-11}$$

(e) As in part (d), ethyl acetate is the reactant acid. The base in this case is far stronger than the base in part (d), however.

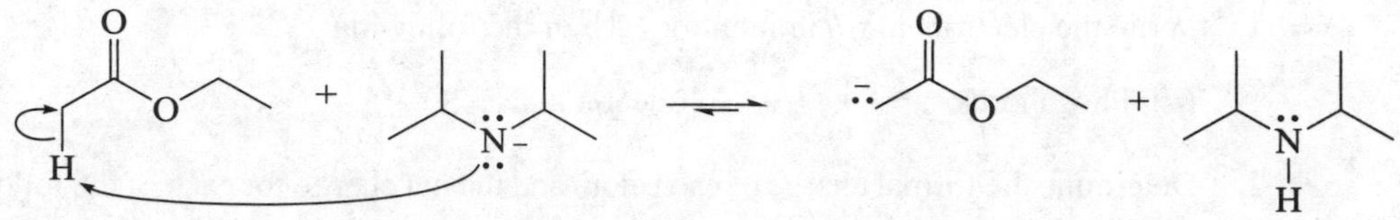

pK_a = 25.6

pK_a = 36

The stronger acid is on the left. The equilibrium lies far to the right.

$$K_{eq} = \frac{10^{-25.6}}{10^{-36}} = 10^{10.4} = 2.5 \times 10^{10}$$

1.67 The conjugate base of squaric acid has the structure

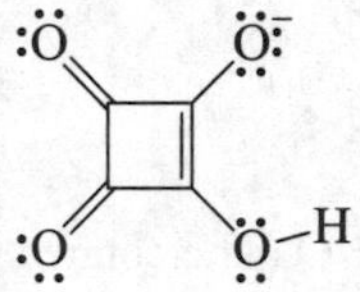

Move electron pairs are indicated, beginning at the negatively charged oxygen.

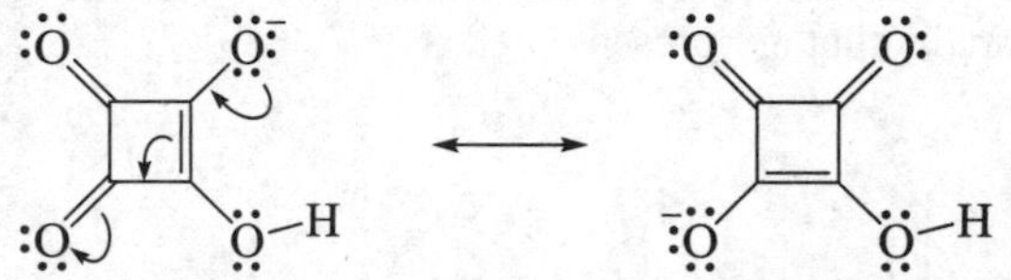

1.68 By following the electron flow, we see that the C—Br bond breaks with both electrons in that bond becoming an unshared pair of bromide ion. A new bond is formed between sulfur and the carbon from which the bromine departs. The electron pair responsible for this new bond is derived from the double bond between carbon and sufur in the compound on the left (thiourea). The electron pair lost from the carbon in C=S is compensated for by C=N double-bond formation involving a lone pair of nitrogen.

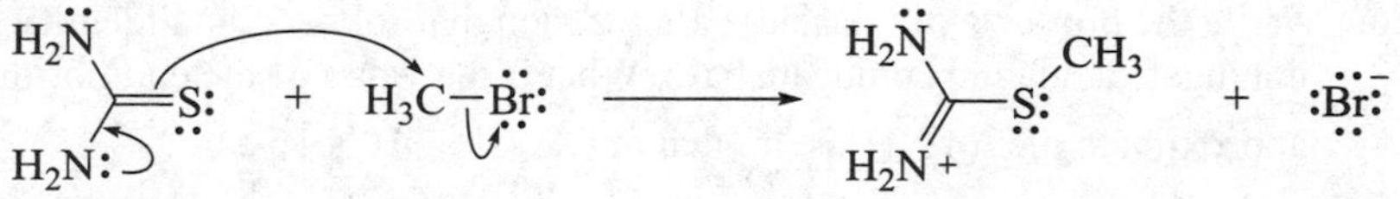

Descriptive Passage and Interpretive Problems 1

ANSWERS TO INTERPRETIVE PROBLEMS

1.69 B; **1.70** D; **1.71** A; **1.72** B; **1.73** A; **1.74** C

SELF-TEST

1. Write the electron configuration for each of the following:

   (a) Phosphorus   (b) Sulfide ion in Na_2S

2. Determine the formal charge of each atom and the net charge for each of the following species:

   (a) $:\ddot{N}=C=\ddot{S}:$ (b) $:O\equiv N-\ddot{O}:$ (c)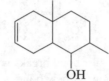

3. Write a second Lewis structure that satisfies the octet rule for each of the species in Problem 2, and determine the formal charge of each atom. Which of the Lewis structures for each species in this and Problem 2 is more stable?

4. Write a correct Lewis structure for each of the following. Be sure to show explicitly any unshared pairs of electrons.

   (a) Methylamine, CH_3NH_2

   (b) Acetaldehyde, C_2H_4O (The connectivity is CCO; all the hydrogens are connected to carbon.)

5. What is the molecular formula of each of the structures shown? Clearly draw any unshared electron pairs that are present.

   (a)   (c)

                                         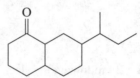

   (b)   (d)

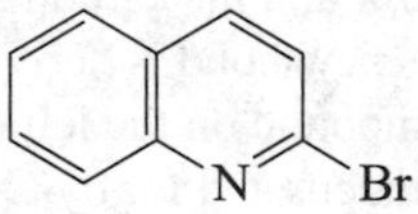

6. Write the products of the acid–base reaction that follows; and identify the acid, base, conjugate acid, and conjugate base. What is the value of the equilibrium constant, K_{eq}? The approximate pK_a of NH_3 is 36; that of CH_3CH_2OH is 16·

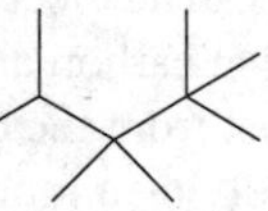

   $$CH_3CH_2O^- \quad + \quad NH_3 \quad \rightleftharpoons$$

   Show all unshared electron pairs and formal charges, and use curved arrows to track electron movement.

7. Account for the fact that all three sulfur–oxygen bonds in SO_3 are the same by drawing the appropriate Lewis structure(s).

8. The cyanate ion contains 16 valence electrons, and its three atoms are connected in the order OCN. Write the most stable Lewis structure for this species, and assign a formal charge to each atom. What is the net charge of the ion?

9. Using the VSEPR method,

(a) Describe the geometry at each carbon atom and the oxygen atom in the following molecule:

$$CH_3OCH=CHCH_3$$

(b) Deduce the shape of NCl_3, and draw a three-dimensional representation of the molecule. Is NCl_3 polar?

10. Assign the shape of each of the following as either linear or bent.

(a) CO_2 (b) NO_2^+ (c) NO_2^-

11. Consider structures A, B, C, and D:

A   B   C   D

(a) Which structure (or structures) contains a positively charged carbon?

(b) Which structure (or structures) contains a positively charged nitrogen?

(c) Which structure (or structures) contains a positively charged oxygen?

(d) Which structure (or structures) contains a negatively charged carbon?

(e) Which structure (or structures) contains a negatively charged nitrogen?

(f) Which structure (or structures) contains a negatively charged oxygen?

(g) Which structure is the most stable?

(h) Which structure is the least stable?

12. Given the following information, write a Lewis structure for urea, CH_4N_2O. The oxygen atom and both nitrogen atoms are bonded to carbon, there is a carbon–oxygen double bond, and none of the atoms bears a formal charge. Be sure to include all unshared electron pairs.

13. Draw a second resonance contributor for each of the species shown. Be sure to show all unshared electron pairs, and indicate the formal charge (if any) on each atom. Is the first or the second structure more stable?

(a) $^-:\!\ddot{O}\!-\!C\!\equiv\!N\!:$ (b) $:\!\ddot{O}\!=\!\ddot{N}\!-\!\ddot{C}H_2$

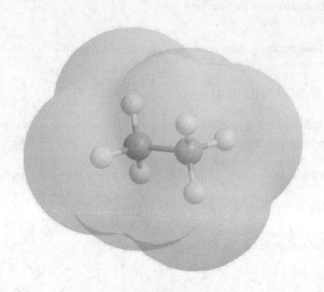

CHAPTER 2

Alkanes and Cycloalkanes: Introduction to Hydrocarbons

Table of Contents

SOLUTIONS TO TEXT PROBLEMS

In Chapter Problems

2.1 A $1s$ orbital from one helium atom is combined with that of a second helium atom to give two new molecular orbitals—one bonding and one antibonding—as shown in the diagram. Because each helium $1s$ atomic orbital has two electrons, two electrons are placed in each of the new molecular orbitals as shown. Although not part of the requested answer, the orbital shapes are also included in the diagram.

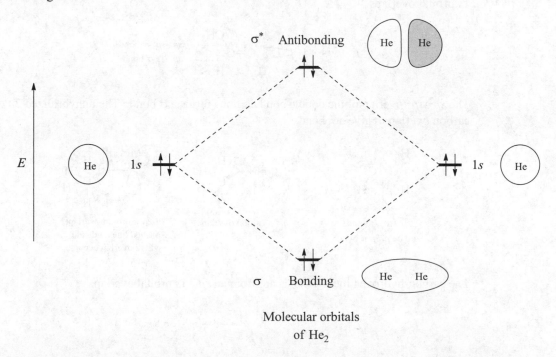

Molecular orbitals
of He_2

As shown in the preceding diagram, diatomic helium would have electrons in antibonding orbitals, which is destabilizing. The energy required to place the two electrons in the antibonding orbital offsets the gain in bonding energy from the two electrons in the bonding orbital.

2.2 A carbon atom is sp^3-hybridized when it is directly bonded to four atoms. Each of the three carbons of propane is bonded to four atoms, either carbon or hydrogen:

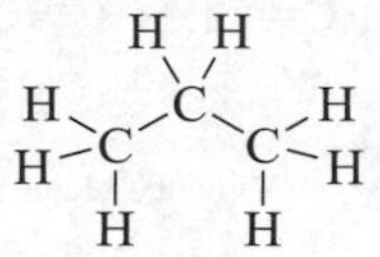

$$\begin{array}{ccccc}
 & H & H & & \\
 & | & | & & \\
H & C & C & & H \\
H-\!\!&C&-\!\!C\!\!&-&H \\
 & | & | & & \\
 & H & H & &
\end{array}$$

Each carbon–carbon σ bond arises from overlap of two half-filled sp^3 hybrid orbitals, one from each carbon atom. Each of the eight carbon–hydrogen σ bonds are from overlap of a half filled carbon sp^3 hybrid orbital with a hydrogen $1s$ orbital having one electron.

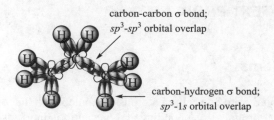

2.3   The methyl carbon is sp^3-hybridized. The carbons of the double bond are both sp^2-hybridized. The carbon-methyl bond is due to an sp^2-sp^3 orbital overlap. The carbon-hydrogen bonds are from sp^2-$1s$ orbital overlaps.

The sp^2-sp^2 overlap of the double bond carbons forms a σ bond. The unhybridized $2p$ orbitals on carbon overlap to give a π bond.

The carbon-hydrogen methyl bonds are from an sp^3–$1s$ orbital overlap.

2.4   The carbons of the double bond are both sp^2-hybridized. The carbons of the triple bond are sp-hybridized. The σ bond indicated with the arrow is formed by overlap of an sp^2 and an sp hybrid orbital.

$$H_2C{=}CH{-\!\!-}C{\equiv}CH$$

sp^2-sp σ bond

Vinylacetylene

There is a total of seven σ bonds in vinylacetylene: three sp^2–$1s$ C-H bonds, one sp–$1s$ C-H bond, one sp^2–sp^2 C-C bond (the σ component of the double bond), one sp–sp C-C bond (the σ component of the triple bond), and the sp^2–sp C-C bond described earlier. In addition, there are three π bonds: one in the double bond and two in the triple bond.

2.5   An unbranched alkane (*n*-alkane) of 28 carbons has 26 methylene (CH_2) groups flanked by a methyl (CH_3) group at each end. The condensed formula is $CH_3(CH_2)_{26}CH_3$.

2.6 The alkane represented by the bond-line formula has 11 carbons. The general formula for an alkane is C_nH_{2n+2} and thus there are 24 hydrogens. The molecular formula is $C_{11}H_{24}$; the condensed formula is $CH_3(CH_2)_9CH_3$.

2.7 In addition to $CH_3(CH_2)_4CH_3$ and $(CH_3)_2CHCH_2CH_2CH_3$, there are three more isomers. One has a five-carbon chain with a one-carbon (methyl) branch:

$$CH_3CH_2\overset{\overset{\displaystyle CH_3}{|}}{C}HCH_2CH_3 \quad \text{or}$$

The remaining two isomers have two methyl branches on a four-carbon chain.

$$CH_3\overset{\overset{\displaystyle CH_3}{|}}{C}H\underset{\underset{\displaystyle CH_3}{|}}{C}HCH_3 \quad \text{or} \qquad\qquad CH_3CH_2\overset{\overset{\displaystyle CH_3}{|}}{\underset{\underset{\displaystyle CH_3}{|}}{C}}CH_3 \quad \text{or}$$

2.8 (*b*) Octacosane is not listed in Table 2.2, but its structure can be deduced from its systematic name. The suffix *-cosane* pertains to alkanes that contain 20–29 carbons in their longest continuous chain. The prefix *octa-* means "eight." *Octacosane* is therefore the unbranched alkane having 28 carbon atoms. Its condensed formula is $CH_3(CH_2)_{26}CH_3$.

(*c*) The alkane in Problem 2.6 has an unbranched chain of 11 carbon atoms and is named *undecane*.

2.9 The ending *-hexadecane* reveals that the longest continuous carbon chain has 16 carbon atoms.

There are four methyl groups (represented by *tetramethyl-*), and they are located at carbons 2, 6, 10, and 14.

2,6,10,14-Tetramethylhexadecane
(phytane)

2.10 (*b*) The systematic name of the unbranched C_5H_{12} isomer is *pentane* (Table 2.2).

$$CH_3CH_2CH_2CH_2CH_3$$

IUPAC name: pentane

A second isomer, $(CH_3)_2CHCH_2CH_3$, has four carbons in the longest continuous chain and so is named as a derivative of butane. Because it has a methyl group at C-2, it is *2-methylbutane*.

$$CH_3\underset{\underset{\displaystyle CH_3}{|}}{C}HCH_2CH_3$$

IUPAC name: 2-methylbutane

The remaining isomer, $(CH_3)_4C$, has three carbons in its longest continuous chain and so is named as a derivative of propane. There are two methyl groups at C-2, and so it is a 2,2-dimethyl derivative of propane.

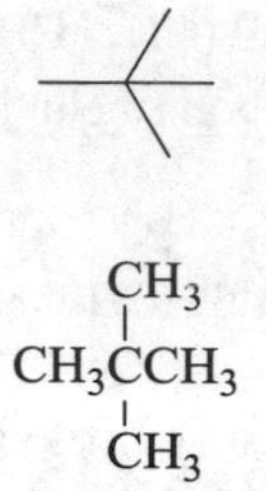

$$CH_3CCH_3 \text{ with } CH_3 \text{ above and } CH_3 \text{ below}$$

IUPAC name: 2,2-dimethylpropane

(c) First write out the structure in more detail, and identify the longest continuous carbon chain.

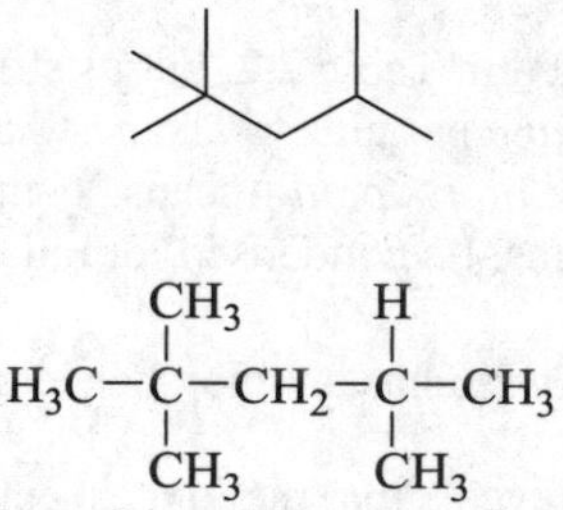

$$H_3C-C-CH_2-C-CH_3$$

There are five carbon atoms in the longest chain, and so the compound is named as a derivative of pentane. This five-carbon chain has three methyl substituents attached to it, making it a trimethyl derivative of pentane. Number the chain in the direction that gives the lowest numbers to the substituents at the first point of difference.

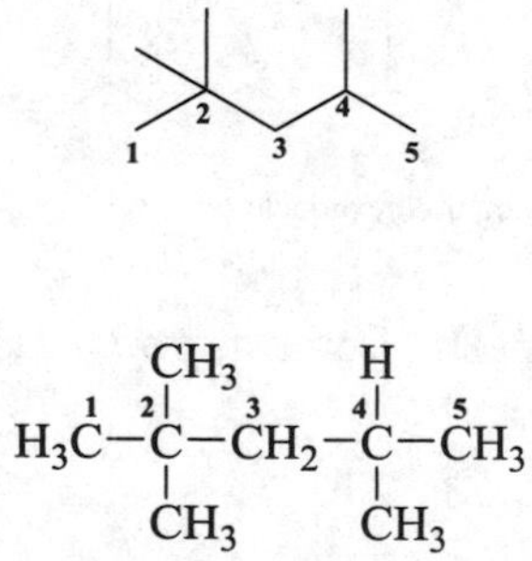

not

$$H_3\overset{1}{C}-\overset{2}{C}-\overset{3}{C}H_2-\overset{4}{C}-\overset{5}{C}H_3$$

2,2,4-Trimethylpentane (correct)

$$H_3\overset{5}{C}-\overset{4}{C}-\overset{3}{C}H_2-\overset{2}{C}-\overset{1}{C}H_3$$

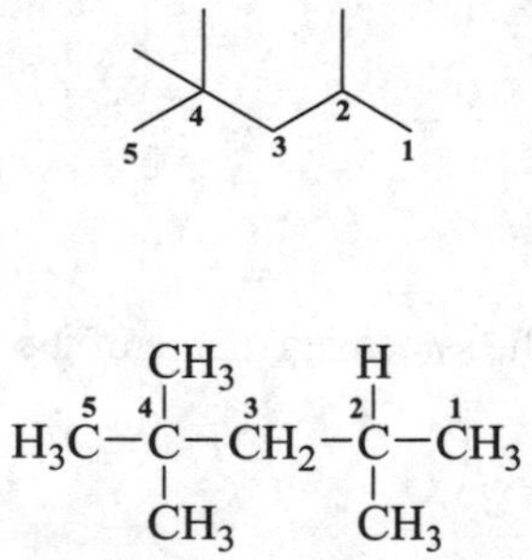

2,4,4-Trimethylpentane (incorrect)

(d) The longest continuous chain in $(CH_3)_3CC(CH_3)_3$ contains four carbon atoms.

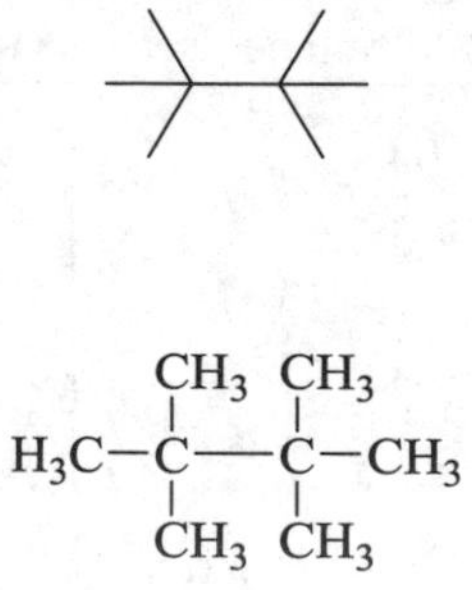

$$H_3C-C-C-CH_3$$

The compound is named as a tetramethyl derivative of butane; it is *2,2,3,3-tetramethylbutane.*

2.11 There are three C_5H_{11} alkyl groups with unbranched carbon chains. One is primary, and two are secondary. The IUPAC name of each group is given beneath the structure. Remember to number the alkyl groups from the potential point of attachment.

$$CH_3CH_2CH_2CH_2CH_2-$$

Pentyl group (primary)

$$\overset{4}{C}H_3\overset{3}{C}H_2\overset{2}{C}H_2\overset{1}{C}HCH_3$$

1-Methylbutyl group (secondary)

$$\overset{3}{C}H_3\overset{2}{C}H_2\overset{1}{C}HCH_2CH_3$$

1-Ethylpropyl group (secondary)

Four alkyl groups are derived from $(CH_3)_2CHCH_2CH_3$. Two are primary, one is secondary, and one is tertiary.

$$\overset{4}{C}H_3\overset{3}{\underset{|}{C}H}\overset{2}{C}H_2\overset{1}{C}H_2-$$
$$CH_3$$

3-Methylbutyl group (primary)

$$-\overset{1}{C}H_2\overset{2}{\underset{|}{C}H}\overset{3}{C}H_2\overset{4}{C}H_3$$
$$CH_3$$

2-Methylbutyl group (primary)

$$\overset{1}{C}H_3\overset{2}{\underset{|}{C}}\overset{3}{C}H_2CH_3$$
$$CH_3$$

1,1-Dimethylpropyl group (tertiary)

$$\overset{3}{C}H_3\overset{2}{\underset{|}{C}H}\overset{1}{C}HCH_3$$
$$CH_3$$

1,2-Dimethylpropyl group (secondary)

2.12 (*b*) Begin by finding the longest continuous chain. The compound is named as a derivative of hexane, because this chain has six carbons.

The chain is numbered so as to give the lowest number to the substituent that appears closest to the end of the chain. In this case, it is numbered right to left so that the substituents are located at C-2 and C-4 rather than at C-3 and C-5. In alphabetical order, the groups are ethyl and methyl; they are listed in alphabetical order in the name. The compound is *4-ethyl-2-methylhexane.*

(*c*) The longest continuous chain is shown in the structure; it contains ten carbon atoms. The structure also shows the numbering scheme that gives the lowest number to the substituent nearest the end of the chain.

In alphabetical order, the substituents are ethyl (at C-8), isopropyl (at C-4), and two methyl groups (at C-2 and C-6). The alkane is *8-ethyl-4-isopropyl-2,6-dimethyldecane.* The systematic name for the isopropyl group (1-methylethyl) may also be used, and the name becomes *8-ethyl-2,6-dimethyl-4-(1-methylethyl)decane.* Notice in both names that the "di" prefix is ignored when alphabetizing.

2.13 (*b*) There are ten carbon atoms in the ring in this cycloalkane; thus, it is named as a derivative of cyclodecane.

Cyclodecane

The numbering pattern of the ring is chosen so as to give the lowest number to the substituent at the first point of difference between them. Thus, the carbon bearing two methyl groups is C-1, and the ring is numbered counterclockwise, placing the isopropyl group on C-4 (numbering clockwise would place the isopropyl on C-8). Listing the substituent groups in alphabetical order, the correct name is *4-isopropyl-1,1-dimethylcyclodecane.* Alternatively, the systematic name for isopropyl (1-methylethyl) could be used, and the name would become *1,1-dimethyl-4-(1-methylethyl)cyclodecane.*

(*c*) When two cycloalkyl groups are attached by a single bond, the compound is named as a cycloalkyl-substituted cycloalkane. This compound is *cyclohexylcyclohexane.*

2.14 Drawing out each carbon and hydrogen can help determine that the molecular formula of hopane is $C_{30}H_{52}$.

Hopane

2.15 The alkane that has the most carbons (nonane) has the highest boiling point (151°C). Among the others, all of which have eight carbons, the unbranched isomer (octane) has the highest boiling point (126°C) and the most branched one (2,2,3,3-tetramethylbutane) the lowest (106°C). The remaining alkane, 2-methylheptane, boils at 116°C.

2.16 All hydrocarbons burn in air to give carbon dioxide and water. To balance the equation for the combustion of cyclohexane (C_6H_{12}), first balance the carbons and the hydrogens on the right side. Then balance the oxygens on the left side.

$$+ \quad 9O_2 \quad \longrightarrow \quad 6CO_2 \quad + \quad 6H_2O$$

Cyclohexane   Oxygen   Carbon dioxide   Water

2.17 (b) Icosane (Table 2.2) is $C_{20}H_{42}$. It has four more methylene (CH_2) groups than hexadecane, the last unbranched alkane in Table 2.3. Its calculated heat of combustion is therefore (4×653 kJ/mol) higher.

Heat of combustion of icosane = heat of combustion of hexadecane + (4×653 kJ/mol)

$$= 10{,}701 \text{ kJ/mol} + 2612 \text{ kJ/mol}$$

$$= 13{,}313 \text{ kJ/mol}$$

The measured value for the heat of combustion is 13,316 kJ/mol)

2.18 Two factors that influence the heats of combustion of alkanes are, in order of decreasing importance, (1) the number of carbon atoms and (2) the extent of chain branching. Pentane, 2-methylbutane, and 2,2-dimethylpropane are all C_5H_{12}; hexane is C_6H_{14}. Hexane has the largest heat of combustion. Branching leads to a lower heat of combustion; 2,2-dimethylpropane is the most branched and has the lowest heat of combustion.

Hexane	$CH_3(CH_2)_4CH_3$	Heat of combustion 4163 kJ/mol (995.0 kcal/mol)
Pentane	$CH_3CH_2CH_2CH_2CH_3$	Heat of combustion 3536 kJ/mol (845.3 kcal/mol)
2-Methylbutane	$(CH_3)_2CHCH_2CH_3$	Heat of combustion 3529 kJ/mol (843.4 kcal/mol)
2,2-Dimethylpropane	$(CH_3)_4C$	Heat of combustion 3514 kJ/mol (839.9 kcal/mol)

2.19 To determine the enthalpy change for the reaction, subtract the standard enthalpy of formation (ΔH_f°) of 2 mol of cyclopropane from ΔH_f° of cyclohexane.

2 mol (39.30 kJ/mol)   1 mol (-124.6 kJ/mol)

-124.6 kJ - 78.6 kJ = -202.2 kJ

2.20 In the first reaction, a carbon–oxygen bond is replaced by a carbon–chlorine bond.

Both oxygen and chlorine are more electronegative than carbon, so the oxidation number of carbon does not change in this reaction. In the second reaction, a carbon–hydrogen bond is replaced by a carbon–bromine bond.

Bromine is more electronegative than carbon; hydrogen is less electronegative. The oxidation number of carbon has increased, and carbon has been oxidized. Reduction of bromine has also occurred; the oxidation number of bromine in Br_2 is 0; in HBr it is -1.

2.21 (*b*)   The CH_3 carbon is unchanged in this reaction; however, the carbon of CH_2Br is bonded to the electronegative Br making the carbon partially positive whereas the carbon of CH_2Li is bonded to an electropositive Li making the carbon partially negative. The change from a partially positive carbon to a partially negative carbon is considered a reduction. Thus, a reducing reagent is required for this reaction.

$$\begin{array}{ccc}
& H & & & H \\
& | \; \delta^+ \; \delta^- & & & | \; \delta^- \; \delta^+ \\
CH_3 & C-Br & \longrightarrow & CH_3 & C-Li \\
& | & & & | \\
& H & & & H
\end{array}$$

(*c*)   This reaction can be considered as an addition of a proton (H^+) to the left carbon and an addition of a hydroxide (HO^-) to the right carbon. The left carbon is thus reduced and the right carbon is oxidized, resulting in no net oxidation.

Alternatively, hydrogen is slightly more electropositive than carbon and oxygen is more electronegative than carbon. Using this method, one carbon is more electronegative and one more electropositive, relative to the starting material. Neither an oxidation reagent nor reducing reagent is required.

$$\begin{array}{ccc}
& & & H \;\; OH \\
& & & | \;\;\; | \\
H_2C=CH_2 & \longrightarrow & & H_2C-CH_2
\end{array}$$

(*d*)   An oxygen has been added to the molecule such that each carbon contains a new bond to oxygen. Each carbon has been oxidized, so an oxidizing reagent is required for this reaction.

End of Chapter Problems

Structure and Bonding

2.22   Each ring decreases the number of hydrogens in the general molecular formula of the hydrocarbon by two.

(*a*)   Cycloalkanes have the general formula C_nH_{2n}. Compare hexane (C_6H_{14}) with cyclohexane (C_6H_{12}), for example:

$$CH_3CH_2CH_2CH_2CH_2CH_3$$

Hexane
(C_6H_{14})

Cyclohexane
(C_6H_{12})

(*b*)   The general formula of an alkene has two fewer hydrogens than that of an alkane and is C_nH_{2n}. A typical example is 1-hexene (C_6H_{12}).

$$H_2C=CHCH_2CH_2CH_2CH_3$$

1-Hexene
(C_6H_{12})

(*c*) A triple bond decreases the general molecular formula by four hydrogens; the general molecular formula of an alkyne is C_nH_{2n-2}. An example is 1-hexyne (C_6H_{10}).

$$HC\equiv CCH_2CH_2CH_2CH_3$$

1-Hexyne
(C_6H_{10})

(*d*) A compound that has both a ring and a double bond would have the general molecular formula C_nH_{2n-2}. Cyclohexene is an example:

Cyclohexene
(C_6H_{10})

2.23 A hydrocarbon with a molecular formula C_5H_8 fits the general formula C_nH_{2n-2}. Therefore, the compound must have two rings, two double bonds, one double bond and one ring, or one triple bond.

(*a*) The hydrocarbon cannot be a cycloalkane because a cycloalkane with five carbon atoms would have the formula C_5H_{10}.

(*b–d*) Each of the remaining possibilities (one ring and one double bond, two double bonds, and one triple bond) are plausible for a formula C_5H_8.

2.24 (*a*) Isomers are different compounds that have the same molecular formula. Butane and 2-methylpropane both have the molecular formula C_4H_{10} and are isomers. The molecular formula of cyclobutane is C_4H_8 and that of 2-methylbutane is C_5H_{12}.

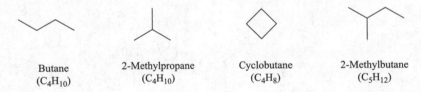

Butane	2-Methylpropane	Cyclobutane	2-Methylbutane
(C_4H_{10})	(C_4H_{10})	(C_4H_8)	(C_5H_{12})

(*b*) The general molecular formula for a compound with either a ring or a double bond is C_nH_{2n}. Cyclopentane and 1-pentene are isomers; the molecular formula of both compounds is C_5H_{10}.

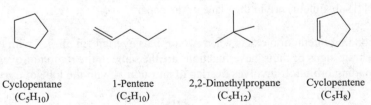

Cyclopentane	1-Pentene	2,2-Dimethylpropane	Cyclopentene
(C_5H_{10})	(C_5H_{10})	(C_5H_{12})	(C_5H_8)

(*c*) Any combination of four rings or double bonds will give a compound with the general molecular formula C_nH_{2n-6}. A compound with two triple bonds will also have the same molecular formula. All four compounds shown are isomers because they all have the molecular formula C_7H_8.

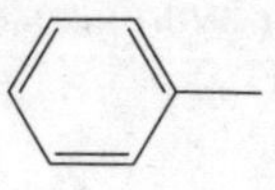

One ring; three
double bonds

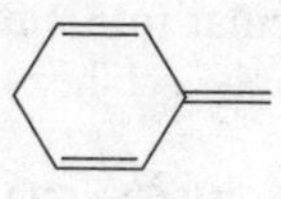

One ring; three
double bonds

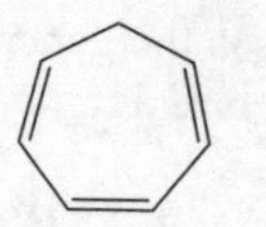

One ring; three
double bonds

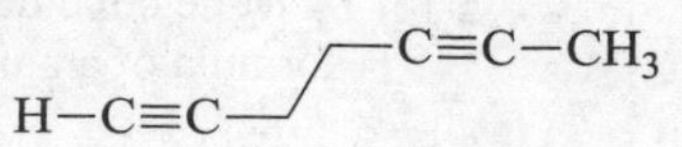

Two triple bonds

2.25 (*a*) Methylene groups are —CH_2—. $ClCH_2CH_2CH_2CH_2Cl$ is therefore the $C_4H_8Cl_2$ isomer in which all the carbons belong to methylene groups.

(*b*) The $C_4H_8Cl_2$ isomers that lack methylene groups are

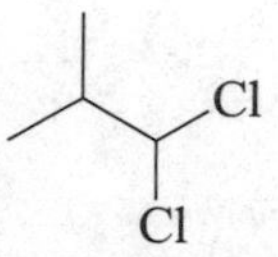

and

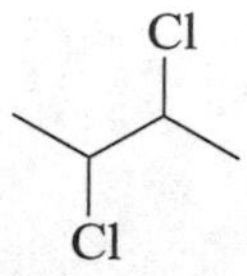

2.26 Carbon-1 is bonded to two atoms: one hydrogen and one carbon (C-2). It also has two π bonds to carbon-2. Carbon-2 is similarly bonded to two atoms: C-1 and C-3. It has two π bonds to carbon-1. Carbon-3 is bonded to three atoms: two carbons (C-2 and C-4) and one hydrogen. It has one π bond to carbon-4. Carbon-4 is similarly bonded to three atoms: two carbons (C-3 and C-5) and one hydrogen. It has one π bond to carbon-3. Carbon-5 is bonded to four atoms: three hydrogens and one carbon (C-4). It does not have any π bonds.

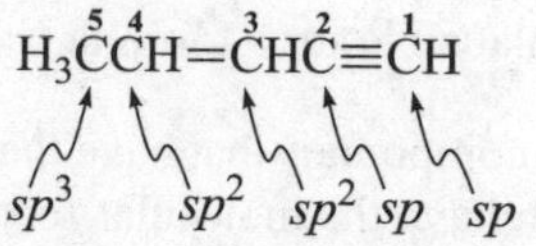

$$H_3\overset{5}{C}\overset{4}{C}H=\overset{3}{C}H\overset{2}{C}\equiv\overset{1}{C}H$$

sp^3 sp^2 sp^2 sp sp

Drawing each hybrid carbon showing all of the C–C–C bonds gives the following structure.

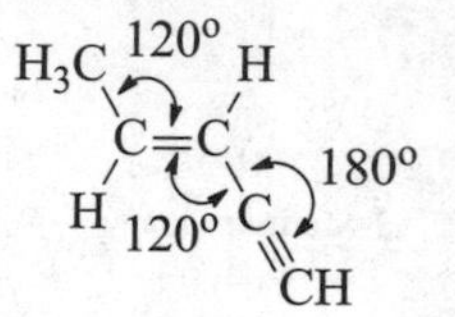

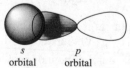

The sp^2-hybridized carbons have a trigonal planar geometry; sp-hybridized carbons have a linear geometry. Note that all of these atoms (with the exception of some hydrogen atoms on the methyl (CH_3) substituent) are in the plane of the paper.

2.27 A bonding interaction exists when two orbitals overlap "in phase" with each other, that is, when the algebraic signs of their wave functions are the same in the region of overlap. The following orbital is a bonding orbital. It involves overlap of an *s* orbital with the lobe of a *p* orbital of the same sign.

s
orbital

p
orbital

On the other hand, the overlap of an *s* orbital with the lobe of a *p* orbital of opposite sign is antibonding.

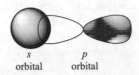

s
orbital

p
orbital

Overlap in the manner shown next is nonbonding. Both the positive lobe and the negative lobe of the *p* orbital overlap with the spherically symmetrical *s* orbital. The bonding overlap between the *s* orbital and one lobe of the *p* orbital is exactly canceled by an antibonding interaction between the *s* orbital and the lobe of opposite sign.

s
orbital

p
orbital

2.28 The end-to-end overlap of two *p* orbitals corresponds to a σ bond as the overlap is symmetrical along the internuclear axis. The side-to-side overlap of two parallel *p* orbitals gives rise to a π bond.

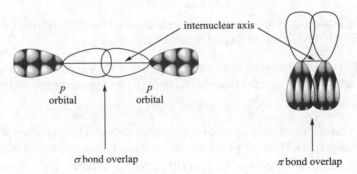

internuclear axis

p
orbital

p
orbital

σ bond overlap

π bond overlap

2.29 Because it is an alkane, the sex attractant of the tiger moth has a molecular formula of C_nH_{2n+2}. The number of carbons and hydrogens may be calculated from its molecular weight.

$$(\text{Atomic weight of carbon})n + (\text{atomic weight of hydrogen})(2n+2) = 254$$
$$12n + 1(2n + 2) = 254$$
$$14n = 252$$
$$n = 18$$

The molecular formula of the alkane is $C_{18}H_{38}$. In the problem, the sex attractant is stated as a 2-methyl-branched alkane. It is therefore 2-methylheptadecane, $(CH_3)_2CH(CH_2)_{14}CH_3$.

2.30 The molecular formula for this pheromone is $C_{15}H_{24}$. Hybridization related to the number of hydrogens and carbons bonded to each carbon. Four means the carbon is sp^3, three sp^2, and two groups gives a sp hybridized carbon. There are no sp hybridized carbons in this molecule.

Nomenclature

2.31 (a) The alkane contains 13 carbons. Because all alkanes have the molecular formula C_nH_{2n+2}, the molecular formula must be $C_{13}H_{28}$.

    (b) The longest continuous chain is indicated and numbered as shown.

    In alphabetical order, the substituents are ethyl (at C-5), methyl (at C-2), and methyl (at C-6). The IUPAC name is *5-ethyl-2,6-dimethylnonane*.

    (c) Fill in the hydrogens in the alkane to identify the various kinds of groups present. There are five *methyl* (CH_3) groups, five *methylene* (CH_2) groups, and three *methine* (CH) groups in the molecule.

    (d) A primary carbon is attached to one other carbon. There are five primary carbons (the carbons of the five CH_3 groups). A secondary carbon is attached to two other carbons, and there are five of these (the carbons of the five CH_2 groups). A tertiary carbon is attached to three other carbons, and there are three of these (the carbons of the three methine groups). A quaternary carbon is attached to four other carbons. None of the carbons is a quaternary carbon.

2.32 The IUPAC name for pristane reveals that the longest chain contains 15 carbon atoms (as indicated by *-pentadecane*). The chain is substituted with four methyl groups at the positions indicated in the name.

Pristane
(2,6,10,14-Tetramethylpentadecane)

2.33 (a) The longest continuous chain contains nine carbon atoms. Begin the problem by writing and numbering the carbon skeleton of nonane.

    Now add two methyl groups (one to C-2 and the other to C-3) and an isopropyl group (to C-6) to give a structural formula for 6-isopropyl-2,3-dimethylnonane.

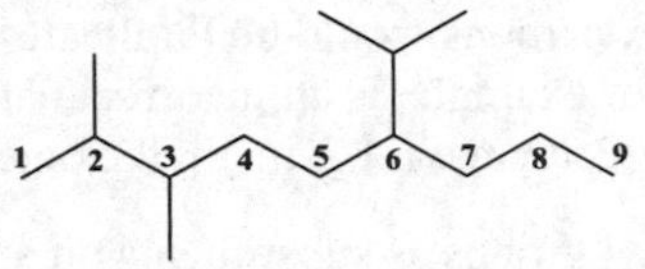

or

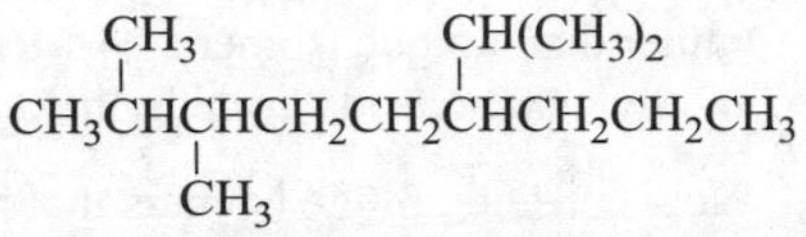

$$CH_3CHCHCH_2CH_2CHCH_2CH_2CH_3$$

with CH_3 and $CH(CH_3)_2$ substituents and a CH_3 below.

(b) To the carbon skeleton of heptane (seven carbons) add a *tert*-butyl group to C-4 and a methyl group to C-3 to give 4-*tert*-butyl-3-methylheptane.

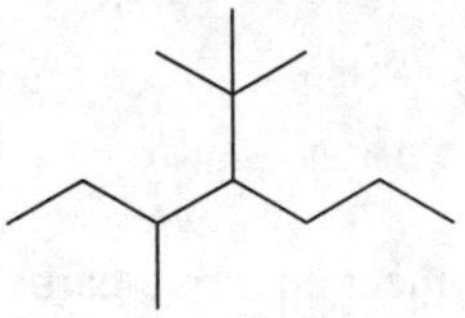

or

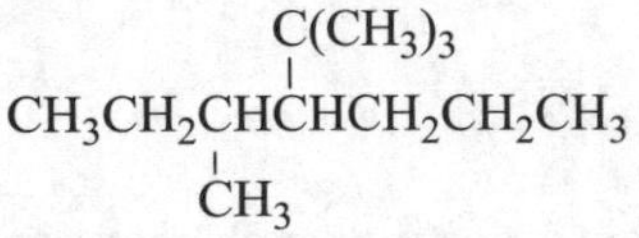

$$CH_3CH_2CHCHCH_2CH_2CH_3$$

with $C(CH_3)_3$ and CH_3 substituents.

(c) An isobutyl group is —$CH_2CH(CH_3)_2$. The structure of 4-isobutyl-1,1-dimethylcyclohexane is as shown.

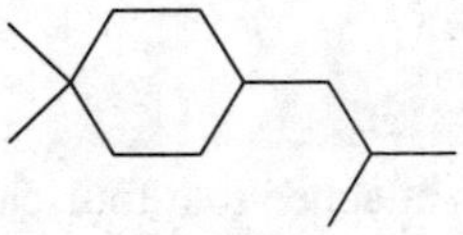

or

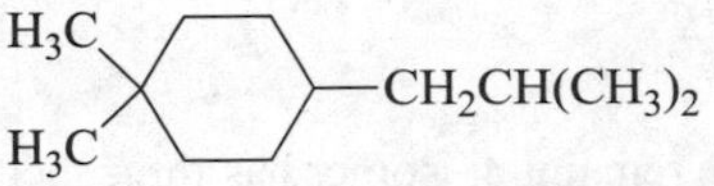

(d) A *sec*-butyl group is $CH_3CHCH_2CH_3$. *sec*-Butylcycloheptane has a *sec*-butyl group on a

seven-membered ring.

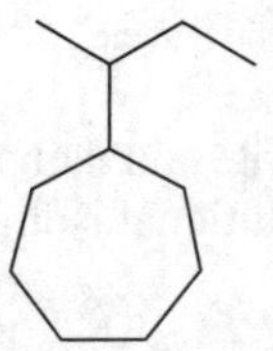

or

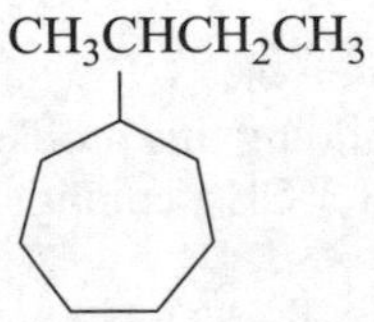

(e) A cyclobutyl group is a substituent on a five-membered ring in cyclobutylcyclopentane.

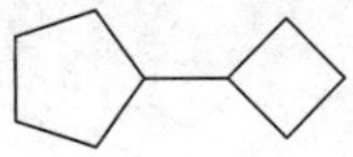

2.34 It is best to approach problems of this type systematically. Because the problem requires all the isomers of C_7H_{16} to be written, begin with the unbranched isomer heptane.

Heptane

Two isomers have six carbons in their longest continuous chain. One bears a methyl substituent at C-2, the other a methyl substituent at C-3.

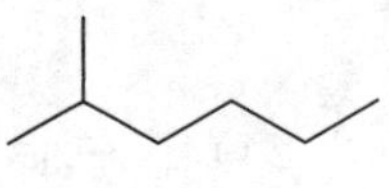

2-Methylhexane 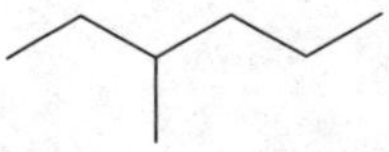   3-Methylhexane

Other structures bearing a continuous chain of six carbons would be duplicates of these isomers rather than unique isomers. "4-Methylhexane," for example, is an incorrect name for 3-methylhexane, and "5-methylhexane" is an incorrect name for 2-methylhexane.

Now consider all the isomers that have two methyl groups as substituents on a five-carbon continuous chain.

2,2-Dimethylpentane   3,3-Dimethylpentane   2,3-Dimethylpentane   2,4-Dimethylpentane

There is one isomer characterized by an ethyl substituent on a five-carbon chain:

3-Ethylpentane

The remaining isomer has three methyl substituents attached to a four-carbon chain.

2,2,3-Trimethylbutane

2.35 In the course of doing this problem, you will write and name the 17 alkanes that, in addition to octane, $CH_3(CH_2)_6CH_3$, comprise the 18 constitutional isomers of C_8H_{18}.

(*a*) The easiest way to attack this part of the exercise is to draw a bond-line depiction of heptane and add a methyl branch to the various positions.

2-Methylheptane   3-Methylheptane   4-Methylheptane

Other structures bearing a continuous chain of seven carbons would be duplicates of these isomers rather than unique isomers. "5-Methylheptane," for example, is an incorrect name for 3-methylheptane, and "6-methylheptane" is an incorrect name for 2-methylheptane.

(*b*) Six of the isomers named as derivatives of hexane contain two methyl branches on a continuous chain of six carbons.

2,2-Dimethylhexane   2,3-Dimethylhexane   2,4-Dimethylhexane   2,4-Dimethylhexane

3,3-Dimethylhexane   2,3-Dimethylhexane

One isomer bears an ethyl substituent:

3-Ethylhexane

(c) Four isomers are trimethyl-substituted derivatives of pentane:

2,2,3-Trimethylpentane   2,3,3-Trimethylpentane   2,2,4-Trimethylpentane   2,3,4-Trimethylpentane

Two bear an ethyl group and a methyl group on a continuous chain of five carbons:

3-Ethyl-2-methylpentane   3-Ethyl-3-methylpentane

(d) Only one isomer is named as a derivative of butane:

2,2,3,3-Tetramethylbutane

2.36 (a) This compound is an unbranched alkane with 27 carbons. Table 2.2 in the text indicates that alkanes with 20–29 carbons have names ending in -*cosane*. Thus, we add the prefix *hepta*- ("seven") to *cosane* to name the alkane $CH_3(CH_2)_{25}CH_3$ as *heptacosane*.

(b) The alkane $(CH_3)_2CHCH_2(CH_2)_{14}CH_3$ has 18 carbons in its longest continuous chain. It is named as a derivative of *octadecane*. There is a single substituent, a methyl group at C-2. The compound is *2-methyloctadecane*.

(c) Write the structure out in more detail to reveal that it is *3,3,4-triethylhexane*.

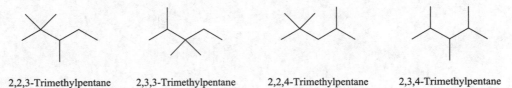

$(CH_3CH_2)_3CCH(CH_2CH_3)_2$ is rewritten as

(d) Each line of a bond-line formula represents a bond between two carbon atoms. Hydrogens are added so that the number of bonds to each carbon atom totals four.

is the same as $CH_3CH_2CHCH_2C(CH_3)_3$
                                                    |
                                                  CH_2CH_3

The IUPAC name is *4-ethyl-2,2-dimethylhexane.*

(*e*)   is the same as $CH_3CH_2CHCH_2CHCH_2CH_3$
                                                        | |
                                                      CH_3 CH_3

The IUPAC name is *3,5-dimethylheptane.*

(*f*) Number the chain in the direction shown to give *3-ethyl-4,5,6 trimethyloctane.* When numbered in the opposite direction, the locants are also 3, 4, 5, and 6. In the case of ties, however, choose the direction that gives the lower number to the substituent that appears first in the name. "Ethyl" precedes "methyl" alphabetically.

2.37 (*a*) The group $CH_3(CH_2)_{10}CH_2$— is an unbranched alkyl group with 12 carbons. It is a *dodecyl group.* The carbon at the point of attachment is directly attached to only one other carbon. It is a primary alkyl group.

(*b*) The longest continuous chain from the point of attachment is six carbons; it is a hexyl group bearing an ethyl substituent at C-3. The group is a *3-ethylhexyl group.* It is a primary alkyl group.

$$—\overset{1}{C}H_2\overset{2}{C}H_2\overset{3}{C}H\overset{4}{C}H_2\overset{5}{C}H_2\overset{6}{C}H_3$$
$$\qquad\qquad\;\; |$$
$$\qquad\qquad CH_2CH_3$$

(*c*) By writing the structural formula of this alkyl group in more detail, we see that the longest continuous chain from the point of attachment contains three carbons. It is a *1,1-diethylpropyl* group. Because the carbon at the point of attachment is directly bonded to three other carbons, it is a tertiary alkyl group.

$$—C(CH_2CH_3)_3 \qquad \text{is rewritten as} \qquad \begin{matrix} & CH_2CH_3 \\ & \overset{1}{|}\overset{2}{}\;\overset{3}{} \\ —\!\!&CCH_2CH_3 \\ & | \\ & CH_2CH_3 \end{matrix}$$

(*d*) This group contains four carbons in its longest continuous chain. It is named as a butyl group with a cyclopropyl substituent at C-1. It is a *1-cyclopropylbutyl* group and is a secondary alkyl group.

$$—\overset{1}{C}H\overset{2}{C}H_2\overset{3}{C}H_2\overset{4}{C}H_3$$

(*e, f*) A two-carbon group that bears a cyclohexyl substituent is a *cyclohexylethyl* group. Number from the potential point of attachment when assigning a locant to the cyclohexyl group.

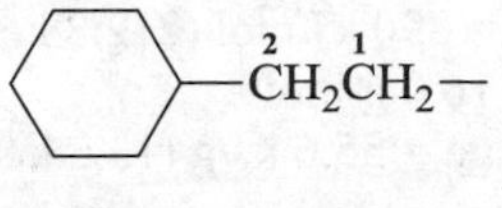

2-Cyclohexylethyl
(primary)

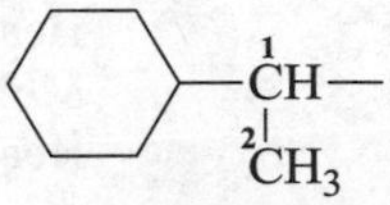

1-Cyclohexylethyl
(secondary)

2.38 The C_4H_9 alkyl groups are named according to the 2004 IUPAC recommendations as follows:

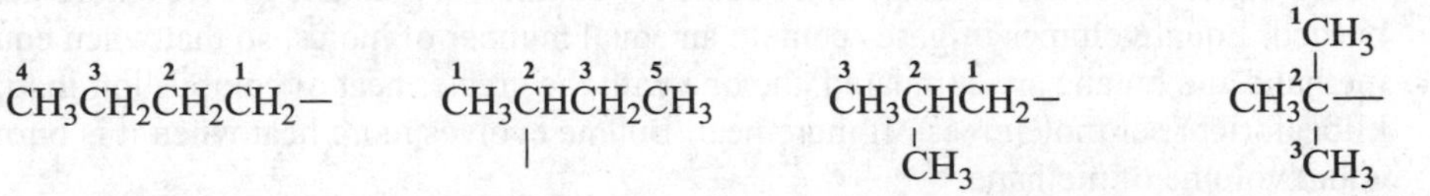

Butan-1-yl

Butan-2-yl

2-Methylpropan-1-yl

2-Methylpropan-2-yl

Reactions

2.39 When any hydrocarbon is burned in air, the products of combustion are carbon dioxide and water.

(*a*) $CH_3(CH_2)_8CH_3 \;+\; \frac{31}{2} O_2 \;\longrightarrow\; 10CO_2 \;+\; 11H_2O$

Decane
($C_{10}H_{22}$)

Oxygen

Carbon
dioxide

Water

(*b*) $+\;\; 15O_2 \;\longrightarrow\; 10CO_2 \;+\; 10H_2O$

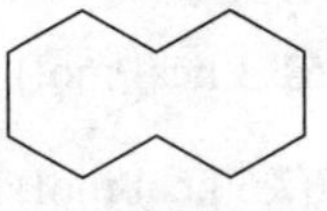

Cyclodecane
($C_{10}H_{20}$)

Oxygen

Carbon
dioxide

Water

(*c*) $+\;\; 15O_2 \;\longrightarrow\; 10CO_2 \;+\; 10H_2O$

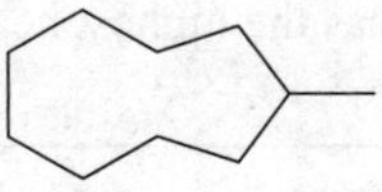

Methylcyclononane
($C_{10}H_{20}$)

Oxygen

Carbon
dioxide

Water

(*d*) $+\; \frac{29}{2} O_2 \;\longrightarrow\; 10CO_2 \;+\; 9H_2O$

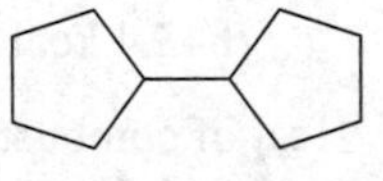

Cyclopentylcyclopentane
($C_{10}H_{18}$)

Oxygen

Carbon
dioxide

Water

2.40 To determine the quantity of heat evolved per unit mass of material, divide the heat of combustion by the molecular weight.

Methane   Heat of combustion = 890 kJ/mol (212.8 kcal/mol)
                 Molecular weight = 16.0 g/mol
                 Heat evolved per gram = 55.6 kJ/g (13.3 kcal/g)

Butane   Heat of combustion = 2876 kJ/mol (687.4 kcal/mol)
                 Molecular weight = 58.0 g/mol
                 Heat evolved per gram = 49.6 kJ/g (11.8 kcal/g)

When equal masses of methane and butane are compared, methane evolves more heat when it is burned. Equal volumes of gases contain an equal number of moles, so that when equal volumes of methane and butane are compared, the one with the greater heat of combustion in kilojoules (or kilocalories) per mole gives off more heat. Butane evolves more heat when it is burned than does an equal volume of methane.

2.41 When comparing heats of combustion of alkanes, two factors are important:

1. The heats of combustion of alkanes increase as the number of carbon atoms increases.

2. An unbranched alkane has a greater heat of combustion than a branched isomer.

(*a*) In the group hexane, heptane, and octane, three unbranched alkanes are being compared. Octane (C_8H_{18}) has the most carbons and has the greatest heat of combustion. Hexane (C_6H_{14}) has the fewest carbons and the lowest heat of combustion. The measured values in this group are as follows:

Hexane   Heat of combustion 4163 kJ/mol (995.0 kcal/mol)

Heptane   Heat of combustion 4817 kJ/mol (1151.3 kcal/mol)

Octane   Heat of combustion 5471 kJ/mol (1307.5 kcal/mol)

(*b*) 2-Methylpropane has fewer carbons than either pentane or 2-methylbutane and so is the member of the group with the lowest heat of combustion. 2-Methylbutane is a 2-methyl-branched isomer of pentane and so has a lower heat of combustion. Pentane has the highest heat of combustion among these compounds.

2-Methylpropane	$(CH_3)_3CH$	Heat of combustion 2868 kJ/mol (685.4 kcal/mol)
2-Methylbutane	$(CH_3)_2CHCH_2CH_3$	Heat of combustion 3529 kJ/mol (843.4 kcal/mol)
Pentane	$CH_3CH_2CH_2CH_2CH_3$	Heat of combustion 3536 kJ/mol (845.3 kcal/mol)

(*c*) 2-Methylbutane and 2,2-dimethylpropane each have fewer carbons than 2-methylpentane, which therefore has the greatest heat of combustion. 2,2-Dimethylpropane is more highly branched than 2-methylbutane; 2,2-dimethylpropane has the lowest heat of combustion.

2,2-Dimethylpropane	$(CH_3)_4C$	Heat of combustion 3514 kJ/mol (839.9 kcal/mol)
2-Methylbutane	$(CH_3)_2CHCH_2CH_3$	Heat of combustion 3529 kJ/mol (843.4 kcal/mol)
2-Methylpentane	$(CH_3)_2CH_2CH_2CH_2CH_3$	Heat of combustion 4157 kJ/mol (993.6 kcal/mol)

(d) Chain branching has a small effect on heat of combustion; the number of carbons has a much larger effect. The alkane with the most carbons in this group is 3,3-dimethylpentane; it has the greatest heat of combustion. Pentane has the fewest carbons in this group and has the smallest heat of combustion.

Pentane	$CH_3CH_2CH_2CH_2CH_3$	Heat of combustion 3527 kJ/mol (845.3 kcal/mol)
3-Methylpentane	$(CH_3CH_2)_2CHCH_3$	Heat of combustion 4159 kJ/mol (994.1 kcal/mol)
3,3-Dimethylpentane	$(CH_3CH_2)_2C(CH_3)_2$	Heat of combustion 4804 kJ/mol (1148.3 kcal/mol)

(e) In this series, the heat of combustion increases with increasing number of carbons. Ethylcyclopentane has the lowest heat of combustion; ethylcycloheptane has the greatest.

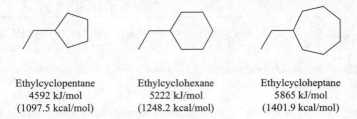

Ethylcyclopentane	Ethylcyclohexane	Ethylcycloheptane
4592 kJ/mol	5222 kJ/mol	5865 kJ/mol
(1097.5 kcal/mol)	(1248.2 kcal/mol)	(1401.9 kcal/mol)

2.42 (a) The equation for the hydrogenation of ethylene is given by the sum of the following three reactions:

$$(1) \quad H_2(g) + \tfrac{1}{2} O_2(g) \longrightarrow H_2O(l) \qquad \Delta H° = -286 \text{ kJ}$$

$$(2) \quad H_2C{=}CH_2(g) + 3O_2(g) \longrightarrow 2CO_2(g) + 2H_2O(l) \qquad \Delta H° = -1410 \text{ kJ}$$

$$(3) \quad 3H_2O(l) + 2CO_2(g) \longrightarrow CH_3CH_3(g) + \tfrac{7}{2}O_2(g) \qquad \Delta H° = +1560 \text{ kJ}$$

$$\text{Sum:} \quad H_2C{=}CH_2(g) + H_2(g) \longrightarrow CH_3CH_3(g) \qquad \Delta H° = -136 \text{ kJ}$$

Equations (1) and (2) are the combustion of hydrogen and ethylene, respectively, and $\Delta H°$ values for these reactions are given in the statement of the problem. Equation (3) is the reverse of the combustion of ethane, and its value of $\Delta H°$ is the negative of the heat of combustion of ethane.

(b) Again, we need to collect equations of reactions for which the $\Delta H°$ values are known.

$$(1) \quad H_2(g) + \tfrac{1}{2} O_2(g) \longrightarrow H_2O(l) \qquad \Delta H° = -286 \text{ kJ}$$

$$(2) \quad HC{\equiv}CH(g) + \tfrac{5}{2} O_2(g) \longrightarrow 2CO_2(g) + H_2O(l) \qquad \Delta H° = -1300 \text{ kJ}$$

$$(3) \quad 2H_2O(l) + 2CO_2(g) \longrightarrow H_2C{=}CH_2(g) + 3O_2(g) \qquad \Delta H° = +1410 \text{ kJ}$$

$$\text{Sum:} \quad HC{\equiv}CH(g) + H_2(g) \longrightarrow H_2C{=}CH_2(g) \qquad \Delta H° = -176 \text{ kJ}$$

Equations (1) and (2) are the combustion of hydrogen and acetylene, respectively. Equation (3) is the reverse of the combustion of ethylene, and its value of $\Delta H°$ is the negative of the heat of combustion of ethylene.

The value of $\Delta H°$ for the hydrogenation of acetylene to ethane is equal to the sum of the two reactions just calculated:

$$HC\equiv CH(g) + H_2(g) \longrightarrow H_2C=CH_2(g) \qquad \Delta H° = -176 \text{ kJ}$$

$$H_2C=CH_2(g) + H_2(g) \longrightarrow CH_3CH_3(g) \qquad \Delta H° = -136 \text{ kJ}$$

$$\text{Sum:} \quad HC\equiv CH(g) + 2H_2(g) \longrightarrow CH_3CH_3(g) \qquad \Delta H° = -312 \text{ kJ}$$

(*c*) We use the equations for the combustion of ethane, ethylene, and acetylene as shown.

(1) $\quad 2H_2C=CH_2(g) + 6O_2(g) \longrightarrow 4CO_2(g) + 4H_2O(l) \qquad \Delta H° = -2820 \text{ kJ}$

(2) $\quad 2CO_2(g) + H_2O(l) \longrightarrow HC\equiv CH(g) + \frac{5}{2}O_2(g) \qquad \Delta H° = +1300 \text{ kJ}$

(3) $\quad 3H_2O(l) + 2CO_2(g) \longrightarrow CH_3CH_3(g) + \frac{7}{2}O_2(g) \qquad \Delta H° = +1560 \text{ kJ}$

$$\text{Sum:} \quad 2H_2C=CH_2(g) \longrightarrow CH_3CH_3(g) + HC\equiv CH(g) \quad \Delta H° = +40 \text{ kJ}$$

The value of $\Delta H°$ for reaction (1) is twice that for the combustion of ethylene because 2 moles of ethylene are involved.

2.43 Greater stability is *not* the reason branched isomers have lower boiling points. Branching affects boiling point because of the effect on a molecule's shape. A branched isomer is more spherical and has a smaller surface than an unbranched isomer. The smaller surface area results in fewer induced-dipole/induced-dipole intermolecular attractions and, thus, a lower boiling point.

Branched isomers are more stable because of stronger intramolecular van der Waals forces. Although the trends in boiling point and stability are parallel, there is not a cause and effect relationship between them. One trend is not the reason for the other.

2.44 Branched alkanes actually give off less heat, not more, on combustion. Branched isomers of an alkane are more stable than unbranched ones. A compound that is more stable has a lower heat of combustion and will give off less heat than a less stable isomer.

2.45 (*a*) In the reaction

$$2CH_3Cl + Si \longrightarrow (CH_3)_2SiCl_2$$

bonds between carbon and an atom more electronegative than itself (chlorine) are replaced by bonds between carbon and an atom less electronegative than itself (silicon). Carbon is reduced; silicon is oxidized.

(*b*) Silicon has a tetrahedral shape in $(CH_3)SiCl_2$. This shape is best described by the sp^3-hybridization bonding model. Silicon is in the third row of the periodic table so the principal quantum number, *n*, equals 3.

2.46 (*a*) The reaction between pentane and fluorine can be described by the equation

$$CH_3CH_2CH_2CH_2CH_3 + 16F_2 \longrightarrow 5CF_4 + 12HF$$

| Pentane | Fluorine | Carbon tetrafluoride | Hydrogen fluoride |

(*b*) Each carbon of pentane becomes bonded to four fluorines in the reaction shown. Fluorine is more electronegative than carbon, and each carbon atom has become oxidized.

2.47 Two atoms appear in their elemental state: Na on the left and H_2 on the right. The oxidation state of an atom in its elemental state is 0. Assign an oxidation state of $+1$ to the hydrogen in the OH group of CH_3CH_2OH because H is less electronegative than O. H changes from $+1$ on the left to 0 on the right; it is reduced. Na changes from 0 on the left to $+1$ on the right; it is oxidized.

$$2CH_3CH_2\overset{+1}{O}H \quad + \quad 2Na^0 \quad \longrightarrow \quad 2CH_3CH_2O\overset{+1}{Na} \quad + \quad H_2^{\ 0}$$

2.48 (*a*, *b*) An oxidizing agent is required when the reaction to be carried out is an oxidation. Carbon is oxidized when the oxygen content of the molecule has increased, as in the top reaction. A reducing agent brings about a reduction. Reduction occurs when the hydrogen content of the molecule increases or the oxygen content (or bonds to oxygen) decreases, or both. The second and third reactions are reductions.

2.49 (*a*) The hydrogen content increases in going from $CH_3C{\equiv}CH$ to $CH_3CH{=}CH_2$. The organic compound $CH_3C{\equiv}CH$ is *reduced*.

(*b*) *Oxidation* occurs because a C—O bond has replaced a C—H bond in going from starting material to product.

(*c*) There are two carbon–oxygen bonds in the starting material and four carbon–oxygen bonds in the products. *Oxidation* occurs.

$$HO{-}CH_2CH_2{-}OH \quad \longrightarrow \quad 2\,H_2C{=}O$$

Two C-O bonds   Four C-O bonds

(*d*) Although the oxidation state of carbon is unchanged in the process

overall, *reduction* of the organic compound has occurred. Its hydrogen content has increased and its oxygen content has decreased. Nitrogen is reduced.

(*e*) In the reactant, chlorine is more electronegative than carbon which means the carbon is electropositive but in the reactant, lithium is more electropositive than carbon so this means that carbon is electronegative. This is a reduction.

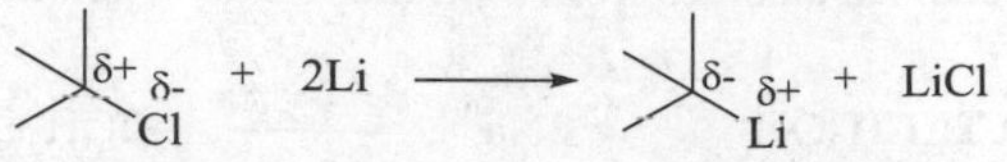

$$
\underset{\text{Cl}}{\overset{\delta+\ \delta-}{\diagup\!\!\!\diagdown}} \quad + \quad 2\text{Li} \quad \longrightarrow \quad \underset{\text{Li}}{\overset{\delta-\ \delta+}{\diagup\!\!\!\diagdown}} \quad + \quad \text{LiCl}
$$

(*f*) Replacing a hydrogen with chlorine on the indicated carbon results in an oxidation.

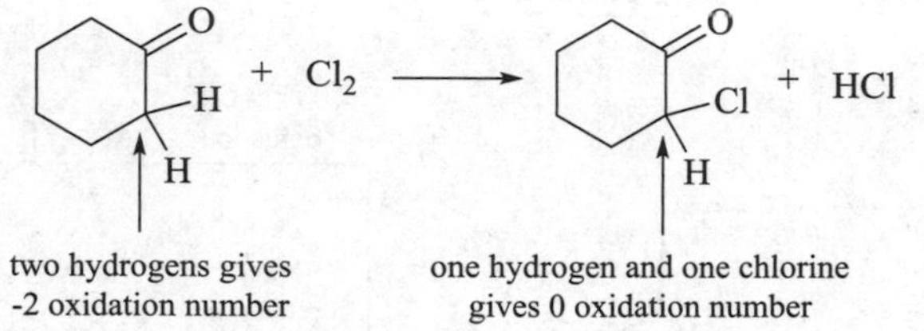

$$
\text{(cyclohexanone with two H's)} \quad + \quad \text{Cl}_2 \quad \longrightarrow \quad \text{(cyclohexanone with Cl and H)} \quad + \quad \text{HCl}
$$

two hydrogens gives   one hydrogen and one chlorine
-2 oxidation number   gives 0 oxidation number

Answers to Interpretive Problems 2

    2.50 C;   **2.51** B;   **2.52** C;   **2.53** C

SELF-TEST

1.   Write the structure of each of the four-carbon alkyl groups. Give the common name and the systematic name for each.

2.   How many σ bonds are present in each of the following?

    (*a*) Nonane   (*b*) Cyclononane

3.   Classify each of the following reactions according to whether the organic substrate is oxidized, reduced, or neither.

    (*a*) CH_3CH_3 + Br_2 $\xrightarrow{\text{light}}$ CH_3CH_2Br + HBr

    (*b*) CH_3CH_2Br + HO^- $\longrightarrow$ CH_3CH_2OH + Br^-

(c) $CH_3CH_2OH \xrightarrow[\text{heat}]{H_2SO_4} H_2C{=}CH_2 + H_2O$

(d) $H_2C{=}CH_2 + H_2 \xrightarrow{Pt} CH_3CH_3$

4. (a) Write a structural formula for 3-isopropyl-2,4-dimethylpentane.

(b) How many methyl groups are there in this compound? How many isopropyl groups?

5. Give the IUPAC name for each of the following substances:

(a)   (b)

6. The compounds in each part of the previous question contain _____ primary carbon(s), _____ secondary carbon(s), and _____ tertiary carbon(s).

7. Using the method outlined in text Section 2.16, give an IUPAC name for each of the following alkyl groups, and classify each one as primary, secondary, or tertiary.

(a) $(CH_3)_2CHCH_2\overset{\displaystyle |}{C}HCH_3$ (b) $(CH_3CH_2)_3C{-}$ (c) $(CH_3CH_2)_3CCH_2{-}$

8. Write a balanced chemical equation for the complete combustion of 2,3-dimethylpentane.

9. Write structural formulas and give the names of all the constitutional isomers of C_5H_{10} that contain a ring.

10. Each of the following names is incorrect. Give the correct name for each compound.

(a) 2,3-Diethylhexane   (c) 2,3-Dimethyl-3-propylpentane

(b) (2-Ethylpropyl)cyclohexane

11. Which C_8H_{18} isomer

(a) Has the highest boiling point?

(b) Has the lowest boiling point?

(c) Has the greatest number of tertiary carbons?

(d) Has only primary and quaternary carbons?

12. Draw the constitutional isomers of C_7H_{16} that have five carbons in their longest chain, and give an IUPAC name for each of them.

13. Given the following heats of combustion (in kilojoules per mole) for the homologous series of unbranched alkanes: hexane (4163), heptane (4817), octane (5471), nonane (6125), estimate the heat of combustion (in kilojoules per mole) for pentadecane.

14.   How many σ and π bonds are present in each of the following?

(*a*) $CH_3CH=CHCH_3$ (*c*)

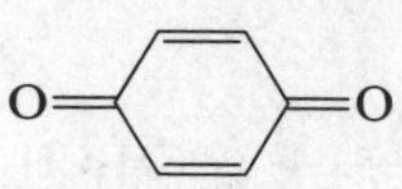

(*b*) $HC\equiv CCH_2CH_3$ (*d*)

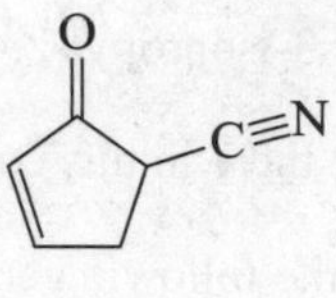

15.   Give the hybridization of each carbon atom in the preceding problem.

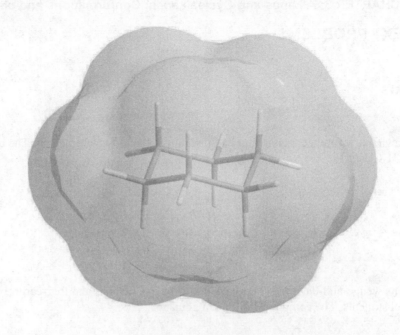

CHAPTER 3

Alkanes and Cycloalkanes: Conformations and cis-trans Stereoisomers

Table of Contents

SOLUTIONS TO TEXT PROBLEMS

In Chapter Problems

3.1 (*b*) The sawhorse formula contains four carbon atoms in an unbranched chain. The compound is butane, $CH_3CH_2CH_2CH_3$.

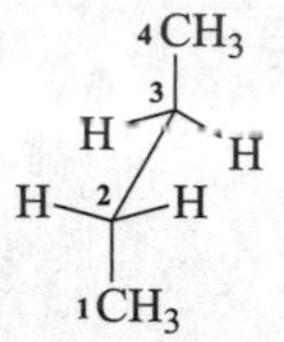

(*c*), (*d*) The wedge-and-dash drawings have a total of five carbons and four-carbon chains with a methyl substituents. The compound is 2-methylbutane.

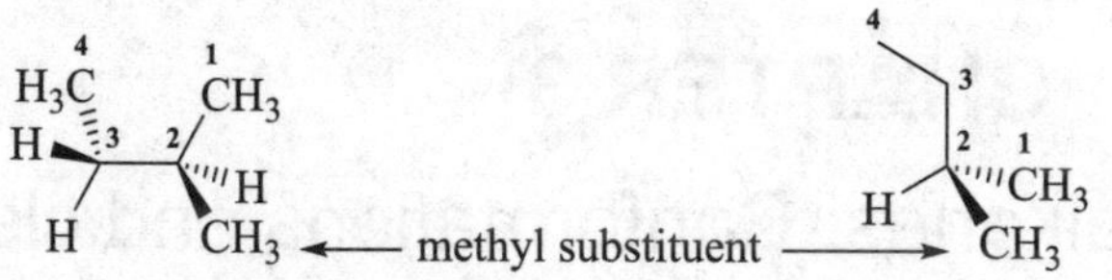

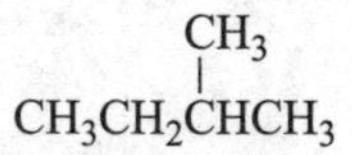

(*e*) The bond-line formula contains four carbon atoms in the longest chain. There are two methyl substituents on C-2 and one methyl substituent on C-3. The compound is 2,2,3-trimethylbutane.

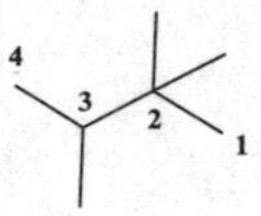

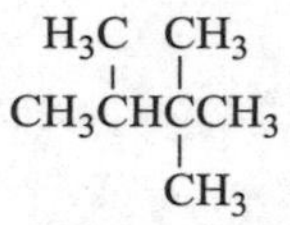

2,2,3-Trimethylbutane

3.2 Red spheres gauche: 60° and 300°. Red spheres anti: 180°. Gauche and anti relationships occur only in staggered conformations; therefore, ignore the eclipsed conformations (0°, 120°, 240°, 360°).

3.3 All the staggered conformations of propane are equivalent to one another, and all its eclipsed conformations are equivalent to one another. The energy diagram resembles that of ethane in that it is a symmetrical one.

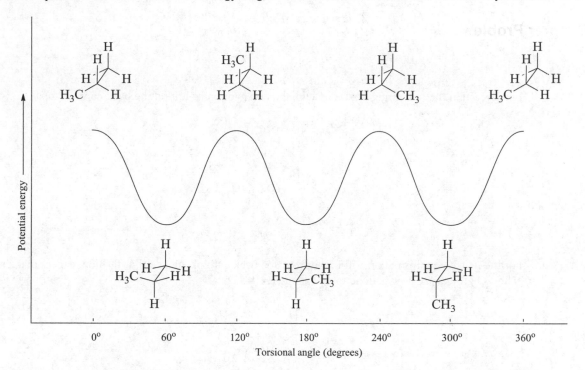

The activation energy for bond rotation in propane is expected to be somewhat higher than that in ethane because of van der Waals strain between the methyl group and a hydrogen in the eclipsed conformation. This strain is, however, less than the van der Waals strain between the methyl groups of butane, which makes the activation energy for bond rotation less for propane than for butane.

3.4 To work this problem, consider the carbons marked 1 and 2 in acetylcholine. The front carbon (1) in the Newman projection is bonded to two hydrogen atoms and a nitrogen atom. These can be placed on the Newman projection in any of the three positions on the front carbon. To construct an anti conformation example, the nitrogen atom is placed at the position at the top of the front carbon.

The two hydrogens can then be filled in as shown.

Placement of the oxygen atom in the back carbon is critical. Because the nitrogen atom in the front is already in place, the oxygen should be positioned anti to the nitrogen as shown.

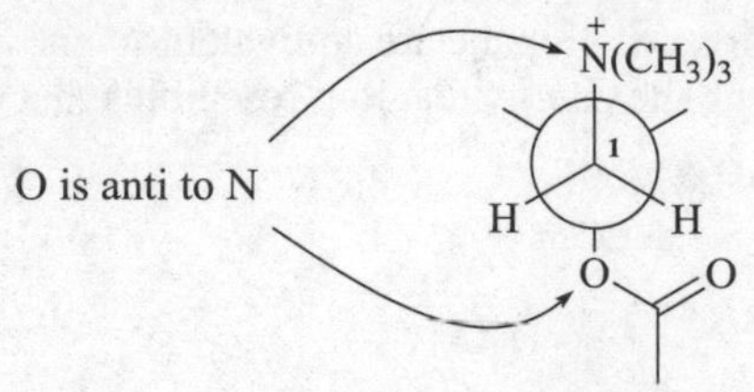

O is anti to N

Finally, fill in the remaining positions with the two hydrogens left on the back carbon (2)

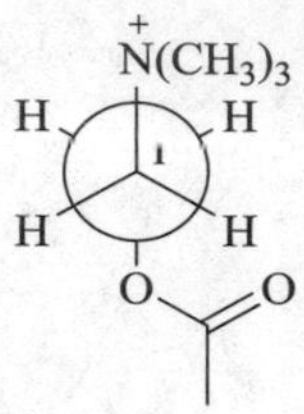

Anti

For the gauche conformation, fill in the front and back carbons in a similar fashion, except make sure that the N and O groups are now gauche to one another.

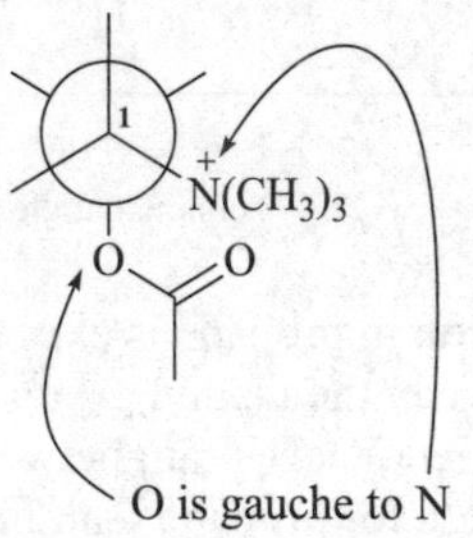

O is gauche to N

Filling in the hydrogen atoms gives the final answer, gauche acetylcholine.

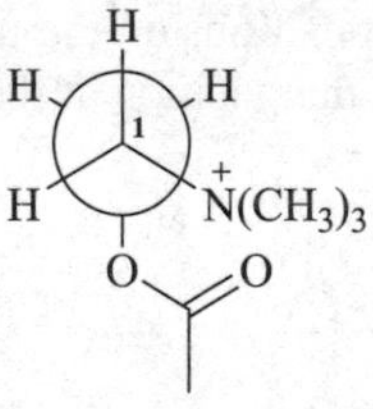

Gauche

3.5   A value of -21.2 kJ/mol was cited in the preceding paragraph as the E_{strain} value for the lowest energy conformation (anti) of butane. The activation energy E_a for rotation about the C(2)—C(3) bond is given by:

$E_a = E_{strain}(\text{eclipsed}) - E_{strain}(\text{anti})$

$E_a = (+9.3 \text{ kJ/mol}) - (-21.2 \text{ kJ/mol})$

$E_a = 30.5 \text{ kJ/mol}$

This energy difference is in reasonable agreement with the 25 kJ/mol value cited in the text.

3.6 Ethylcyclopropane and methylcyclobutane are isomers (both are C_5H_{10}). The less stable isomer has the higher heat of combustion. Ethylcyclopropane has more angle strain and is less stable (has higher potential energy) than methylcyclobutane.

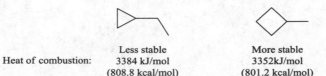

	Less stable	More stable
Heat of combustion:	3384 kJ/mol	3352kJ/mol
	(808.8 kcal/mol)	(801.2 kcal/mol)

3.7 (*b*) To be gauche, substituents X and A must be related by a 60° torsion angle. If A is axial as specified in the problem, X must therefore be equatorial. The bold lines in the drawing show the A—C—C—X torsion angle to be 60°.

X and A are gauche

(*c*) For substituent X at C-1 to be anti to C-3, it must be equatorial. The bold lines in the drawing show the X—C—C—C3 torsion angle to be 180°.

(*d*) When X is axial at C-1, it is gauche to C-3.

3.8 (*b*) According to the numbering scheme given in the problem, a methyl group is axial when it is "up" at C-1 but is equatorial when it is up at C-4. Because substituents are more stable when they occupy equatorial rather than axial sites, a methyl group that is up at C-1 is less stable than one that is up at C-4.

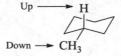

(*c*) An alkyl substituent is more stable in the equatorial position. An equatorial substituent at C-3 is "down."

Up ⟶ H

Down ⟶ CH₃

3.9 A *tert*-butyl group is much larger than a methyl group and has a greater preference for the equatorial position. The most stable conformation of 1-*tert*-butyl-1-methylcyclohexane has an axial methyl group and an equatorial *tert*-butyl group.

1-*tert*-Butyl-1-methylcyclohexane

3.10 The four constitutional isomers of *cis*- and *trans*-1,2-dimethylcyclopropane that do not contain double bonds are

| 1,1-Dimethylcyclopropane | Ethylcyclopropane | Methylcyclobutane | Cyclopentane |

3.11 The *cis* and *trans* stereoisomers of chrysanthemic acid may be conveniently drawn using the wedge-and-dash convention.

cis   *trans*

A perspective drawing may also be used.

cis   *trans*

3.12 When comparing two stereoisomeric cyclohexane derivatives, the more stable stereoisomer is the one with the greater number of its substituents in equatorial orientations. Rewrite the structures as chair conformations with the greatest number of equatorial methyl groups to see how many are axial and and equatorial.

cis-1,3,5-Trimethylcyclohexane

In the more stable chair conformation of *cis*-1,3,5-trimethylcyclohexane, all of the methyl groups are equatorial. It is more stable than *trans*-1,3,5-trimethylcyclohexane (shown in the following), which has one axial methyl group in its more stable chair conformation.

trans-1,3,5-Trimethylcyclohexane

3.13 In each of these problems, a *tert*-butyl group is the larger substituent and will be equatorial in the most stable conformation. Draw a chair conformation of cyclohexane, add an equatorial *tert*-butyl group, and then add the remaining substituent so as to give the required cis or trans relationship to the *tert*-butyl group.

(*b*)   Begin by drawing a chair cyclohexane with an equatorial *tert*-butyl group. In *cis*-1-*tert*-butyl-3-methylcyclohexane, the C-3 methyl group is equatorial.

(*c*)   In *trans*-1-*tert*-butyl-4-methylcyclohexane, both the *tert*-butyl and the C-4 methyl group are equatorial.

(*d*)   Again, the *tert*-butyl group is equatorial; however, in *cis*-1-*tert*-butyl-4-methylcyclohexane, the methyl group on C-4 is axial.

3.14 Five bond cleavages are required to convert cubane to a noncyclic skeleton; cubane is pentacyclic.

3.15 Hydrocarbon A is a spiro compound, encompassing ten carbons. It is, therefore, a *spirodecane*. Four carbons are unique to the cyclopentane ring, five to the cyclohexane ring, making it a substituted *spiro[4.5]decane*. Numbering proceeds through the smaller ring to the larger and substituents are listed in alphabetical order.

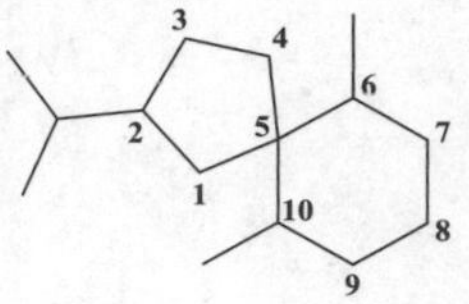

2-Isopropyl-6,10-dimethylspiro[4,5]decane

3.16 The two bond cleavages shown convert camphene to a noncyclic species; therefore, camphene is bicyclic. (Other pairs of bond cleavages are possible and lead to the same conclusion.)

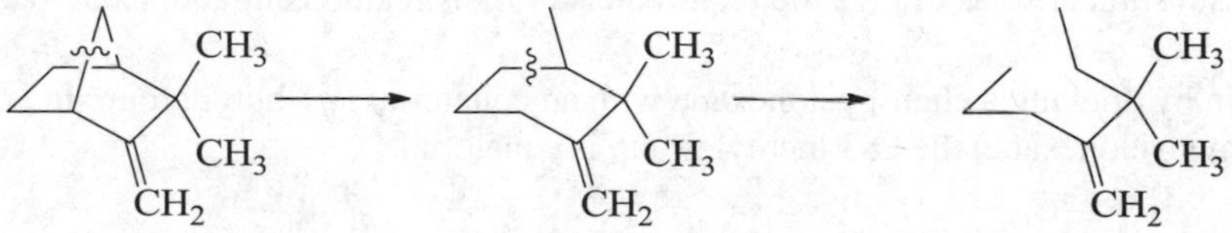

3.17 (*b*)   The structure of bicyclo[2.2.1]heptane was given in part (*a*) of this problem in the text. Numbering begins at a bridgehead position and continues in the direction of the largest bridge. The smallest bridge is numbered last. 1,7,7-Trimethylbicyclo[2.2.1]heptane has the structure shown at the right.

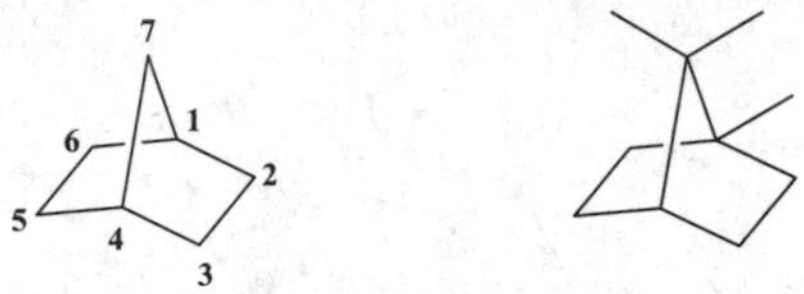

(*c*)   The three bridges in bicyclo[3.1.1]heptane contain three carbons, one carbon, and one carbon. The structure can be written in a form that shows the actual shape of the molecule or one that simply emphasizes its constitution.

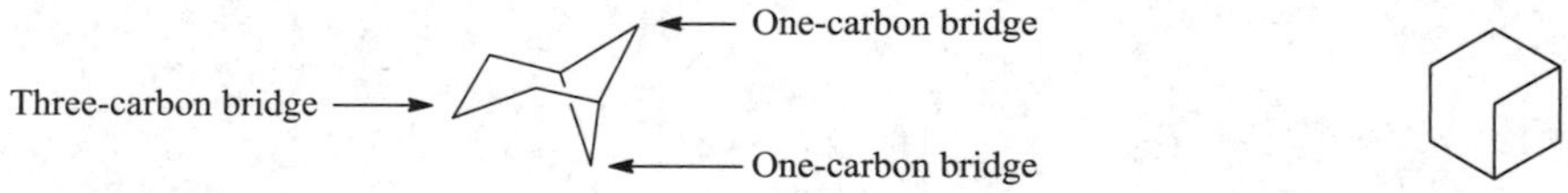

3.18 Geosmin has two cyclohexane rings that are connected by one common bond (this is considered a fused-ring system). These two rings as drawn do not give any information regarding stereochemistry, so one must rely on the relative positions specified by the substituents. The wedge-and-dash notation indicates that one methyl group is *trans* to the hydroxyl group and the other methyl group is *cis* to the hydroxyl group.

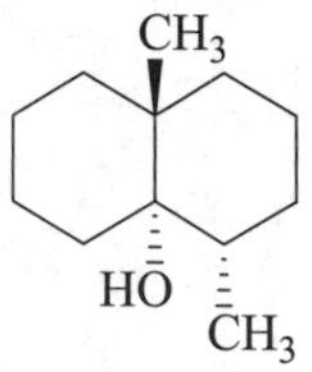

Geosmin

There is also a hydrogen atom that was not specified in the original structure. It is understood to be there, but for clarity it is drawn in with a wedge in the structure on the right. This hydrogen is *cis* to the top methyl group as shown.

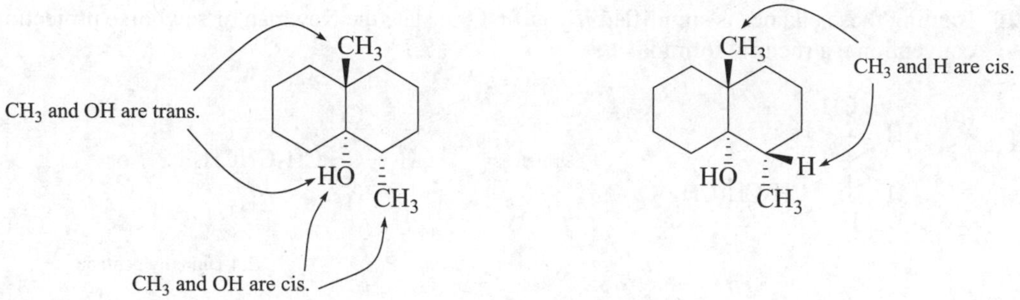

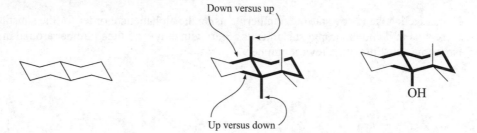

Consider the decalin chair conformation skeleton that follows on the left. The bond that is common to both rings has substituent positions that are already *trans*. This can be understood by looking at the middle structure where the right six-membered ring of decalin has a cyclohexane ring that is bold. The relative positions of the substituents on each of the two carbon atoms that are shared by both rings are designated as down and up. An up position for a methyl substituent versus a down position for the hydroxyl position would give a *trans* orientation as shown.

To draw in the remaining methyl group, the relative positions of the two indicated groups are considered. Because the second methyl group has a *cis* relationship to the hydroxyl group, this substituent must be placed in the down position as indicated. The hydrogen, if specified, is then filled into the remaining axial position as shown. Note that the hydrogen is *cis* to the axial methyl group.

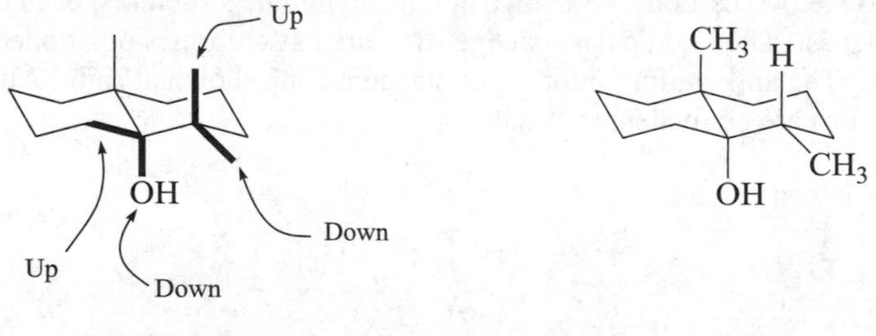

3.19 Because the two conformations are of approximately equal stability when R = H, it is reasonable to expect that the most stable conformation when R = CH₃ will have the CH₃ group equatorial.

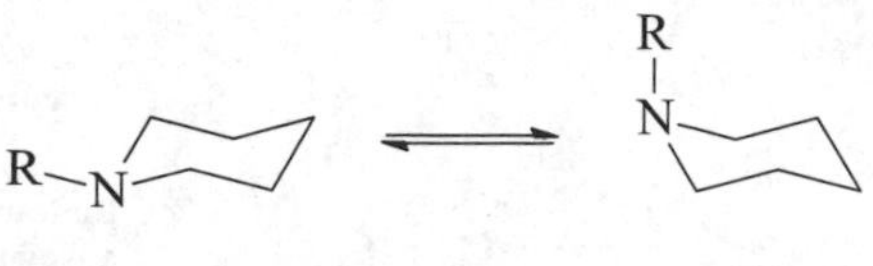

R = H: Both conformations are similar in energy.
R = CH₃: Most stable conformation has the CH₃ substituent equatorial.

End of Chapter Problems

Nomenclature and Terminology

3.20 Naming these alkanes is simplified if you first translate the Newman or sawhorse projection into a more conventional structural formula.

(a)

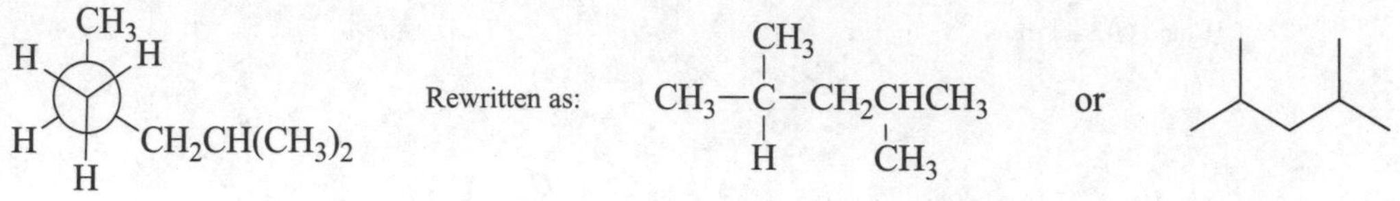

Rewritten as: $CH_3-\overset{\overset{\displaystyle CH_3}{|}}{\underset{\underset{\displaystyle H}{|}}{C}}-CH_2CHCH_3$ or

2,4-Dimethylpentane

(b)

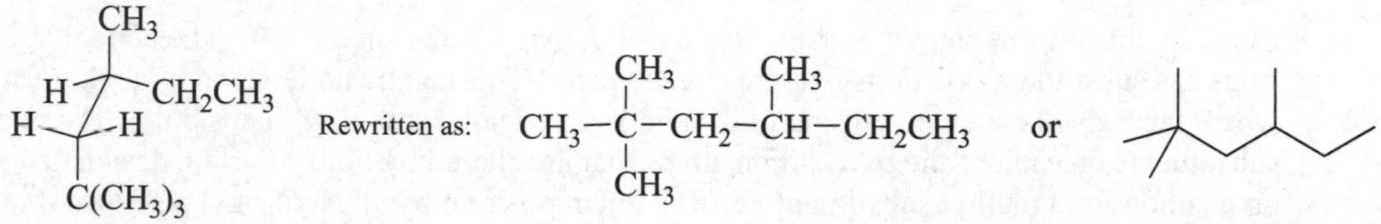

Rewritten as: $CH_3-\overset{\overset{\displaystyle CH_3}{|}}{\underset{\underset{\displaystyle CH_3}{|}}{C}}-CH_2-\overset{\overset{\displaystyle CH_3}{|}}{CH}-CH_2CH_3$ or

2,2,4-Trimethylhexane

(c)   Name cyclic alkanes by starting numbering using the alphabetical order for the substituents. In this case there are ethyl and isopropyl substituents. Start with ethyl and then number around the ring to give the isopropyl substituent the lowest number.

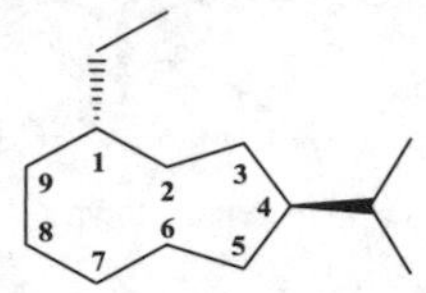

trans-1-Ethyl-4-isopropylcyclononane

3.21   First write out the structural formula of 1,2-dichloroethane in order to identify the substituent groups attached to the two carbons. As shown at left and middle structures, each carbon bears one chlorine and two hydrogens. The middle dash-wedge structure has chlorines positioned on opposite sides, anti, to each other. The anti conformation is the staggered one shown at right. All other staggered conformations are equivalent to this one.

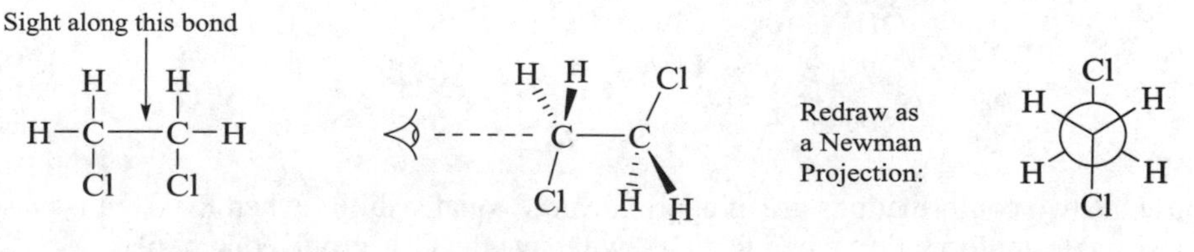

For the gauche conformation, repeat this process.

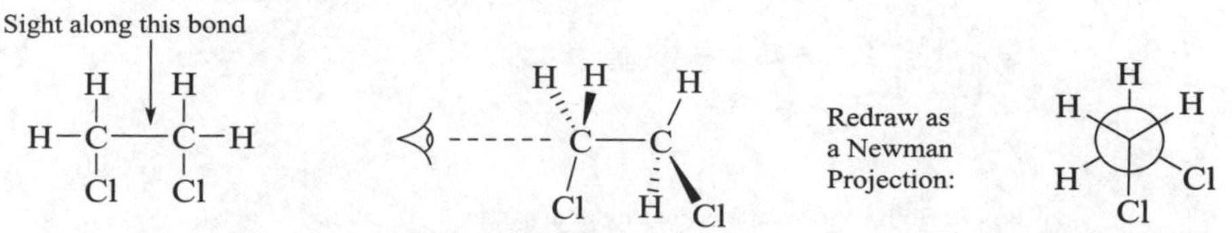

The torsion angle between chlorine substituents is 60° in the gauche conformation and 180° in the anti conformation of $ClCH_2CH_2Cl$.

3.22 Bromine is attached to C-1. Sight down the C-1—C-2 bond and convert the structural drawing to a Newman projection.

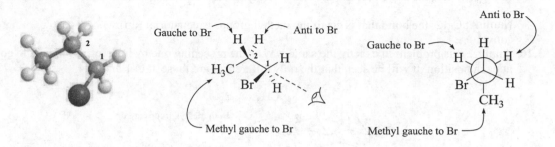

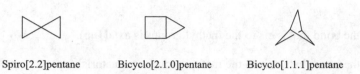

3.23 (*a*) Three isomers of C_5H_8 contain two rings and have no alkyl substituents:

Spiro[2.2]pentane	Bicyclo[2.1.0]pentane	Bicyclo[1.1.1]pentane

(*b*) Five isomers of C_6H_{10} contain two rings and have no alkyl substituents:

Spiro[2.3]hexane Bicyclo[2.2.0]hexane Bicyclo[3.1.0]hexane Bicyclo[2.1.1]hexane Cyclopropylcyclopropane

3.24 (*a*) The reactant is a spiro hydrocarbon. Its IUPAC name is spiro[2.5]octane.

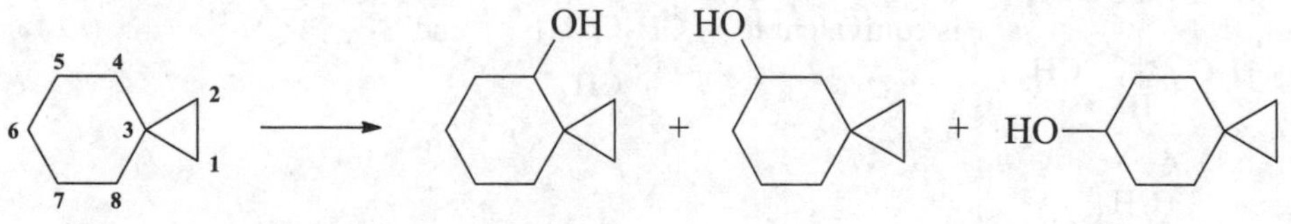

Alcohol A Alcohol B Alcohol C

(*b*) Carbons 4, 5, and 6 are oxidized in the formation of alcohols A, B, and C, respectively.

(*c*) The alcohols are constitutional isomers.

3.25 This problem is primarily an exercise in correctly locating equatorial and axial positions in cyclohexane rings that are joined together into a steroid skeleton. Parts (*a*) through (*e*) are concerned with positions 1, 4, 7, 11, and 12 in that order. The following diagram shows the orientation of axial and equatorial bonds at each of those positions.

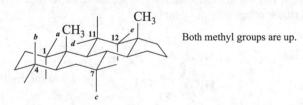

Both methyl groups are up.

(*a*) At C-1, the bond that is cis to the methyl groups is equatorial (up).

(*b*) At C-4, the bond that is cis to the methyl groups is axial (up).

(c)   At C-7, the bond that is trans to the methyl groups is axial (down).

(d)   At C-11, the bond that is trans to the methyl groups is equatorial (down).

(e)   At C-12, the bond that is cis to the methyl groups is equatorial (up).

3.26 Analyze this problem in exactly the same way as the preceding one by locating the axial and equatorial bonds at each position. It will be seen that the only differences are those at C-1 and C-4.

Both methyl groups are up.

(a)   At C-1, the bond that is cis to the methyl groups is axial (up).

(b)   At C-4, the bond that is cis to the methyl groups is equatorial (up).

(c)   At C-7, the bond that is trans to the methyl groups is axial (down).

(d)   At C-11, the bond that is trans to the methyl groups is equatorial (down).

(e)   At C-12, the bond that is cis to the methyl groups is equatorial (up).

3.27 (a)   By rewriting the structures in a form that shows the order of their atomic connections, it is apparent that the two structures are constitutional isomers.

is equivalent to CH_3CCH_3 and   (2,2-Dimethylpropane)

is equivalent to $CH_3CHCH_2CH_3$ and   (2-Methylbutane)

(b)   Both models represent alkanes of molecular formula C_6H_{14}. In each one, the carbon chain is unbranched. The two models are different conformations of the same compound, $CH_3CH_2CH_2CH_2CH_2CH_3$ (hexane).

(c)   The two compounds have the same constitution; both are $(CH_3)_2CHCH(CH_3)_2$. The Newman projections represent different staggered conformations of the same molecule: in one, the hydrogens are anti to each other; whereas in the other they are gauche.

and

are different conformations of 2,3-dimethylbutane.

Hydrogens at C-2   Hydrogens at C-2
and C-3 are anti.   and C-3 are gauche.

(*d*)   The compounds differ in their connectivity. They are constitutional isomers. Although the compounds have different stereochemistry (one is cis, the other trans), they are not stereoisomers. Stereoisomers must have the same constitution.

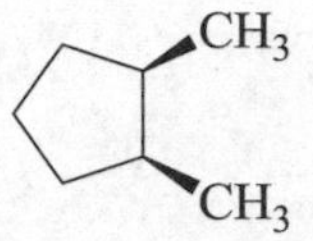

cis-1,2-Dimethylcyclopentane 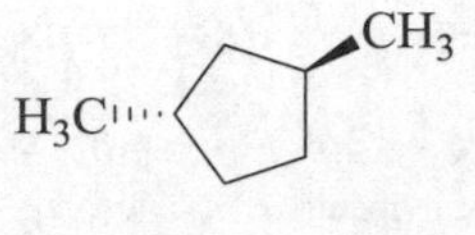   *trans*-1,3-Dimethylcyclopentane

(*e*)   Both structures are *cis*-1-ethyl-4-methylcyclohexane (the methyl and ethyl groups are both "up"). In the structure on the left, the methyl is axial and the ethyl equatorial. The orientations are opposite to these in the structure on the right. The two structures are ring-flipped forms of each other—different conformations of the same compound.

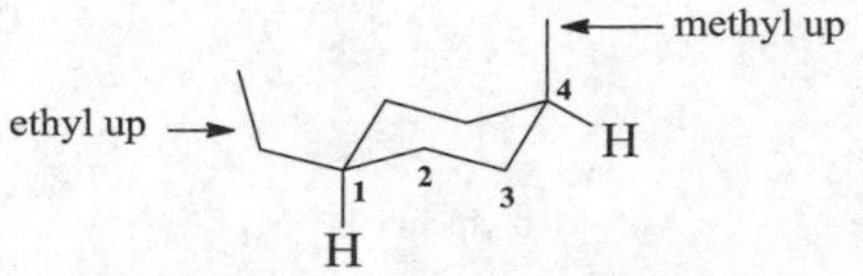

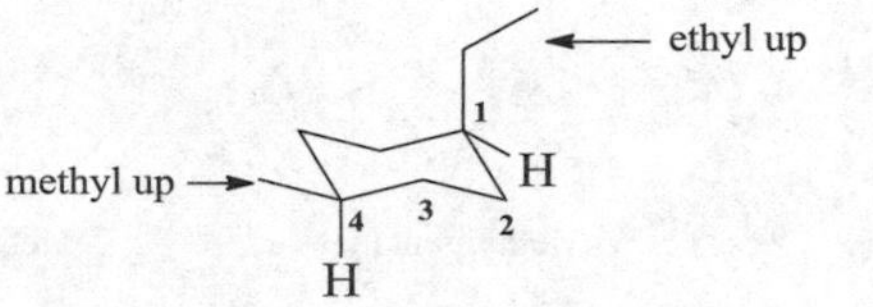

(*f*)   The methyl and the ethyl groups are cis in the first structure but trans in the second. The two compounds are stereoisomers; they have the same constitution but differ in the arrangement of their atoms in space.

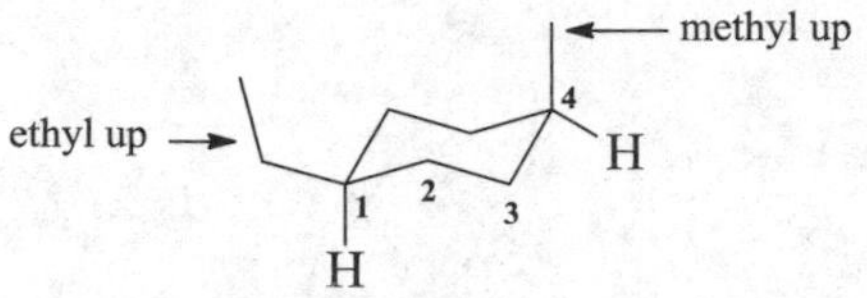

cis-1-Ethyl-4-methylcyclohexane 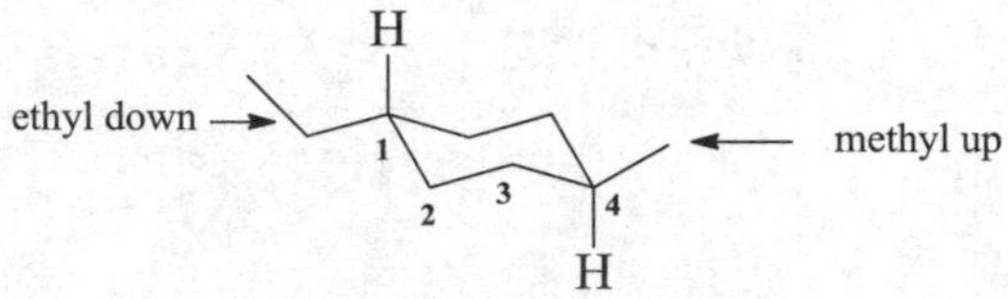   *trans*-1-Ethyl-4-methylcyclohexane

Do not be deceived because the six-membered rings look like ring-flipped forms. Remember, chair–chair interconversion converts all the equatorial bonds to axial and vice versa. Here the ethyl group is equatorial in both structures.

(*g*)   The two structures have the same constitution but differ in the arrangement of their atoms in space; they are stereoisomers. They are not different conformations of the same compound, because they are not related by rotation about C—C bonds. Consider the cyclohexane ring in the structure designated by bold lines. In the first structure the methyl group is trans to the indicated methylene groups, whereas in the second it is cis to these groups.

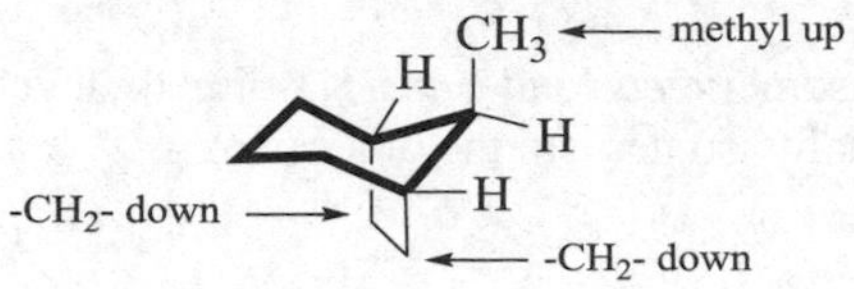

Methyl is trans to the indicated methylene groups 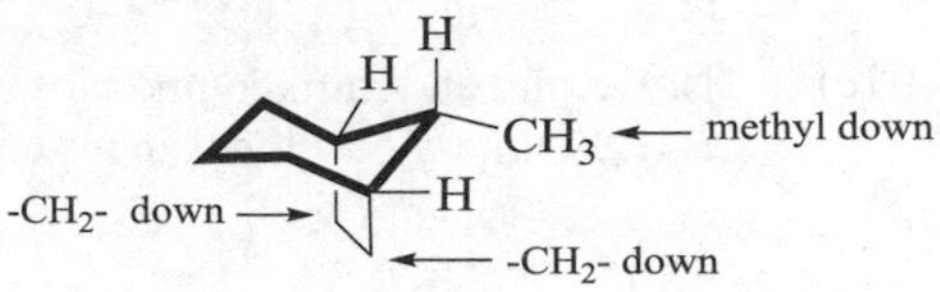   Methyl is cis to indicated methylene groups

3.28 (*a*)   Isomers are different compounds that have the same molecular formula. 2,2-Dimethylhexane has a molecular formula of C_8H_{18} and is not an isomer of 2,2-dimethylpentane or 2,3-dimethylpentane, both of which have the molecular formula C_7H_{16}.

$$(CH_3)_3CCH_2CH_2CH_3 \qquad (CH_3)_3CCH_2CH_2CH_2CH_3 \qquad (CH_3)_2CHCHCH_2CH_3$$
$$\qquad\qquad\qquad\qquad\qquad\qquad\qquad\qquad\qquad\qquad\qquad\qquad\qquad | $$
$$\qquad\qquad\qquad\qquad\qquad\qquad\qquad\qquad\qquad\qquad\qquad\qquad\qquad CH_3$$

2,2-Dimethylpentane   2,2-Dimethylhexane   2,3-Dimethylpentane
(C_7H_{16})   (C_8H_{18})   (C_7H_{16})

The two C_7H_{16} alkanes differ in the order in which their atoms are connected. They are constitutional isomers.

(*b*)   The alkane in the middle is the same as the one at the left, simply written in the opposite direction. Both are 2-methylpentane. 3-Methylpentane is a constitutional isomer of 2-methylpentane.

2-Methylpentane   2-Methylpentane   3-Methylpentane

(*c*)   Pentane has a molecular formula of C_5H_{12}; it is not an isomer of cyclopentane or 1,1-dimethylcyclopropane, which have molecular formulas of C_5H_{10} and are constitutional isomers of each other.

                                                   H_3C CH_3

Pentane   Cyclopentane   1,1-Dimethylcyclopropane

(*d*)   *cis*- and *trans*-1,2-Dimethylcyclobutane are stereoisomers. They have the same constitution but differ in the arrangement of atoms in space. Both are constitutional isomers of *trans*-1,3-dimethylcyclobutane.

H_3C CH_3 CH_3 H_3C

                        H_3C H_3C

cis-1,2-Dimethylcyclobutane   *trans*-1,3-Dimethylcyclobutane   *trans*-1,2-Dimethylcyclobutane

(*e*)   The second and third compounds are stereoisomeric *cis*- and *trans*-1,3-dimethylcyclohexanes. *cis*-1,2-Dimethylcyclohexane is constitutionally isomeric to the other two.

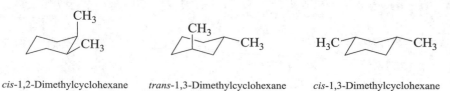

cis-1,2-Dimethylcyclohexane   *trans*-1,3-Dimethylcyclohexane   *cis*-1,3-Dimethylcyclohexane

3.29   According to VSEPR, a double bond is treated as one unit. Therefore, the "electron –pair geometry" in the sulfoxide is tetrahedral, and a pyramidal geometry of the groups bonded to sulfur is predicted. The two isomers are stereoisomers: one has an axial oxygen, the other an equatorial oxygen. The *tert*-butyl group is equatorial in both stereoisomers.

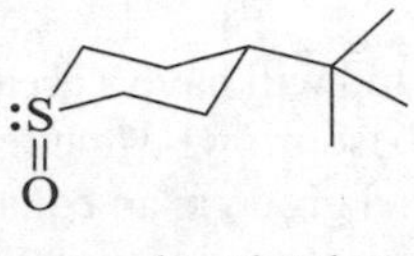

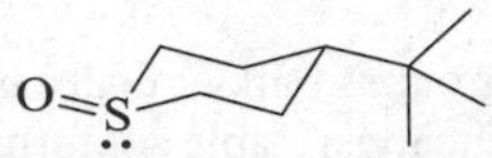

Oxygen and *tert*-butyl are cis   Oxygen and *tert*-butyl are trans

3.30 Conformational representations of the two different forms of glucose are drawn in the usual way. An oxygen atom is present in the six-membered ring, and we are told in the problem that the ring exists in a chair conformation.

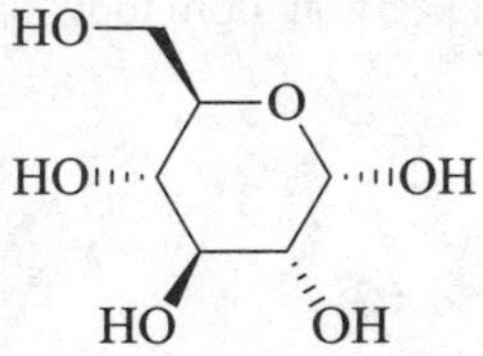

written in its most stable conformation as

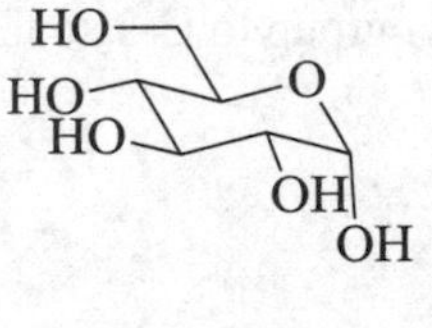

One axial substituent

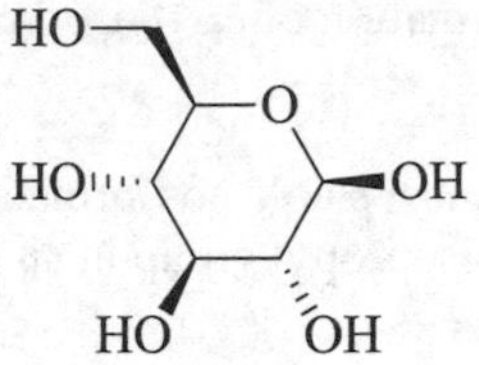

written in its most stable conformation as

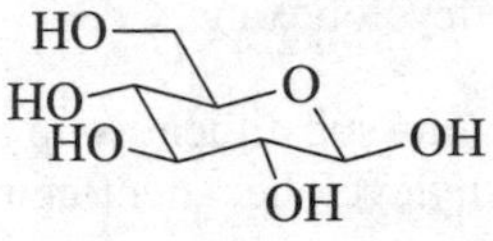

All substituents equatorial

The two structures are not interconvertible by ring flipping; therefore, they are not different conformations of the same molecule. Remember, ring flipping transforms all axial substituents to equatorial ones and vice versa. The two structures differ with respect to only one substituent; they are *stereoisomers* of each other.

Conformations: Relative Stability

3.31 The structural formula of 2,2,5,5-tetramethylhexane is $(CH_3)_3CCH_2CH_2C(CH_3)_3$. First write out the structural formula of 2,2,5,5-tetramethylhexane in order to identify the substituent groups attached to C-3 and C-4. As shown at left, both C-3 and C-4 have two hydrogens and a *tert*-butyl group attached. The most stable conformation has the large *tert*-butyl groups anti to each other.

Sight along this bond

Redraw as a Newman Projection with C-3 as the front carbon:

Anti conformation of 2,2,5,5-tetramethylhexane

tert-butyl substituent   tert-butyl substituent

The bond-line drawing of 2,2,5,5-tetramethylhexane is an alternative way of expressing the structural formula but does not imply the conformation. Bond line drawing of 2,2,5,5-tetramethylhexane is:

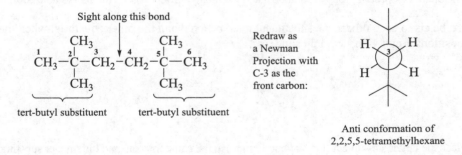

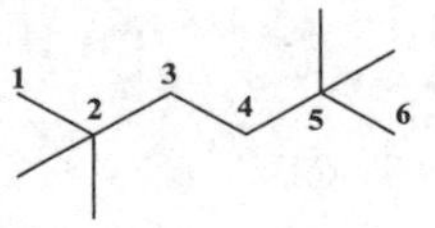

3.32   (a)   An isopropyl group is bulkier than a methyl group and will have a greater preference for an equatorial orientation in the most stable conformation of *cis*-1-isopropyl-3-methylcyclohexane. Draw a chair conformation of cyclohexane, and place an isopropyl group in an equatorial position.

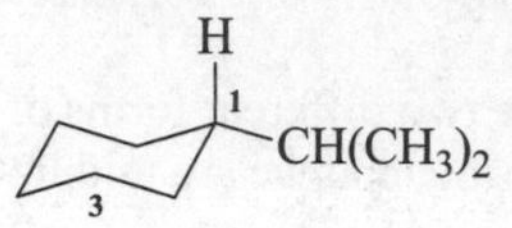

Notice that the equatorial isopropyl group is down on the carbon atom to which it is attached. Add a methyl group to C-3 so that it is also down.

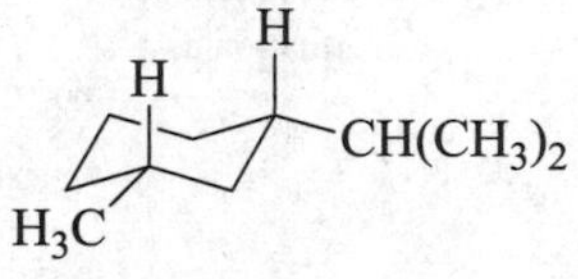

Both substituents are equatorial in the most stable conformation of *cis*-1-isopropyl-3-methylcyclohexane.

(*b*)   One substituent is up and the other is down in the most stable conformation of *trans*-1-isopropyl-3-methylcyclohexane. Begin as in part (*c*) by placing an isopropyl group in an equatorial orientation on a chair conformation of cyclohexane.

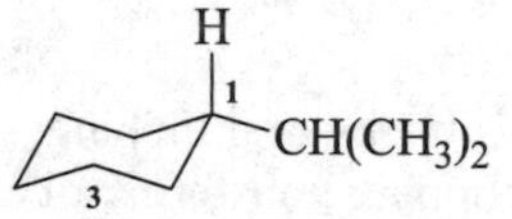

To be trans to the C-1 isopropyl group, the C-3 methyl group must be up.

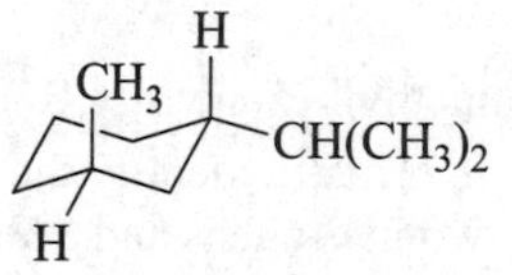

The bulkier isopropyl group is equatorial and the methyl group axial in the most stable conformation.

(*c*)   To be cis to each other, one substituent must be axial and the other equatorial when they are located at positions 1 and 4 on a cyclohexane ring.

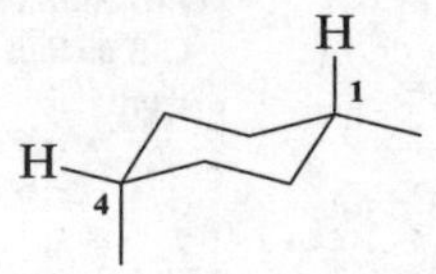

Place the larger substituent (the *tert*-butyl group) at the equatorial site and the smaller substituent (the ethyl group) at the axial one.

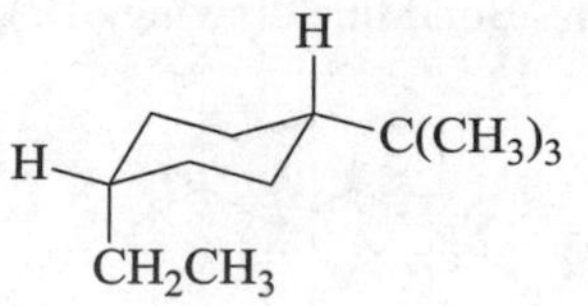

(*d*) First write a chair conformation of cyclohexane, then add two methyl groups at C-1, and draw in the axial and equatorial bonds at C-3 and C-4. Next, add methyl groups to C-3 and C-4 so that they are cis to each other. This can be accomplished two different ways: either the C-3 and C-4 methyl groups are both up or they are both down.

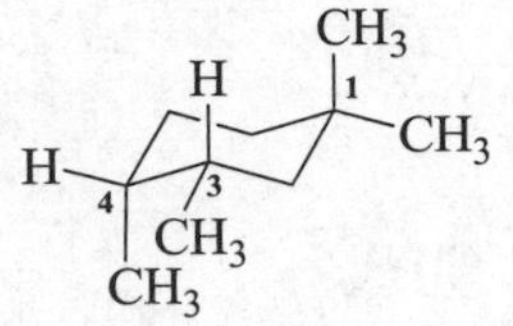

More stable chair conformation: C-3 methyl group is equatorial; no van der Waals strain between axial C-1 and C-3 methyl groups

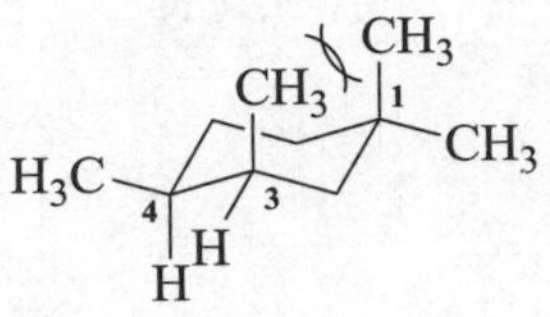

Less stable chair conformation: C-3 methyl group is axial; strong van der Waals strain between axial C-1 and C-3 methyl groups

3.33 (*a*) First write out the structural formula of 2,2-dimethylbutane in order to identify the substituent groups attached to C-2 and C-3. As shown at left, C-2 bears three methyl groups, and C-3 bears two hydrogens and a methyl group. The most stable conformation is the staggered one shown at right. All other staggered conformations are equivalent to this one.

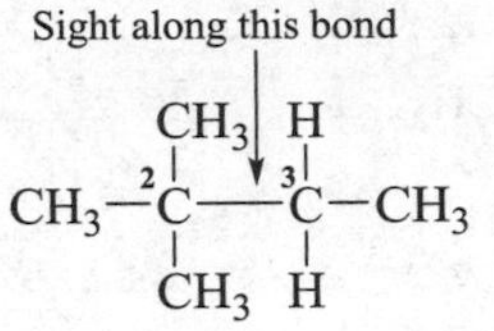

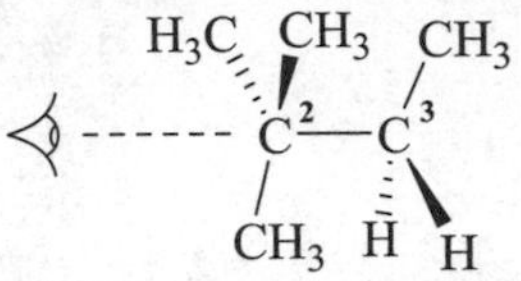

Redraw as a Newman projection:

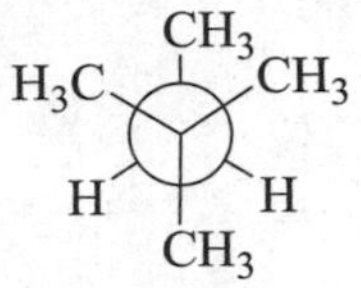

(*b*) The constitution of 2-methylbutane and its two most stable conformations are shown.

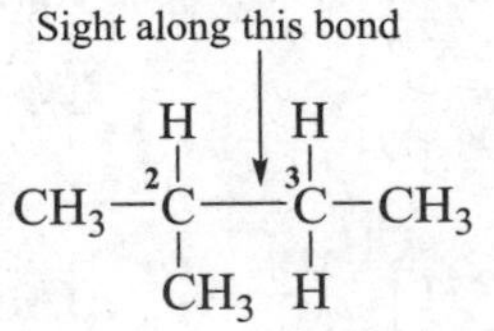

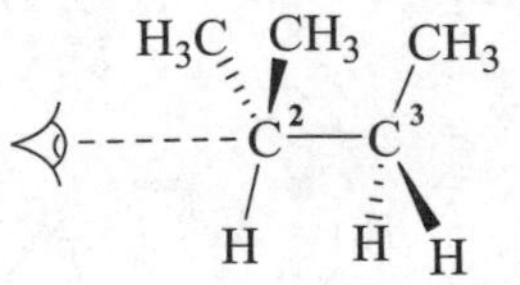

and

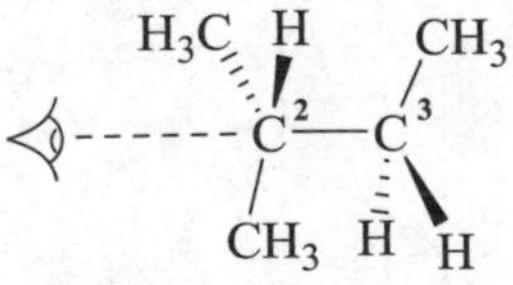

Redraw as Newman projections:

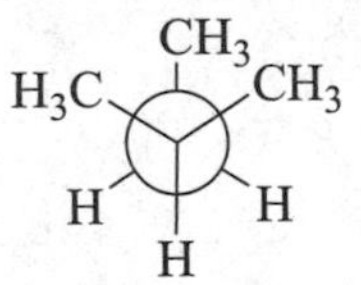

and

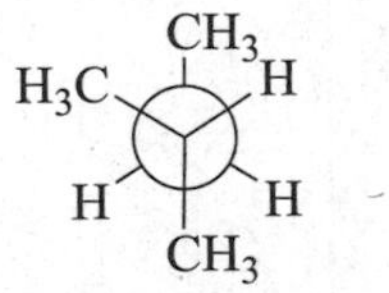

Both conformations are staggered. In one Newman projection (left), the methyl group at the back is gauche to the two methyl groups in the front. In the other (right), it is gauche to one and anti to the other.

(*c*) The constitution of 2,3-dimethylbutane and its two most stable conformations are shown.

The hydrogens at C-2 and C-3 may be gauche to one another (left), or they may be anti (right).

3.34 The 2-methylbutane conformation with one gauche $CH_3 \cdots CH_3$ and one anti $CH_3 \cdots CH_3$ relationship is more stable than the one with two gauche $CH_3 \cdots CH_3$ relationships. The more stable conformation has less van der Waals strain.

3.35 All the staggered conformations about the C-2—C-3 bond of 2,2-dimethylpropane are equivalent to one another and of equal energy; they represent potential energy minima. All the eclipsed conformations are equivalent and represent potential energy maxima.

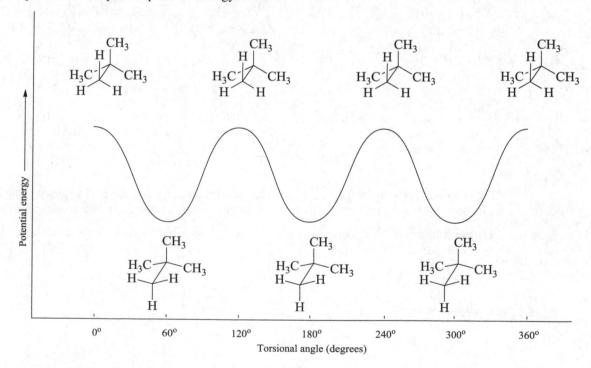

The shape of the potential energy profile for internal rotation in 2,2-dimethylpropane more closely resembles that of ethane than that of butane.

3.36 The potential energy diagram of 2-methylbutane more closely resembles that of butane than that of ethane in that the three staggered forms are not all of the same energy. Similarly, not all of the eclipsed forms are of equal energy.

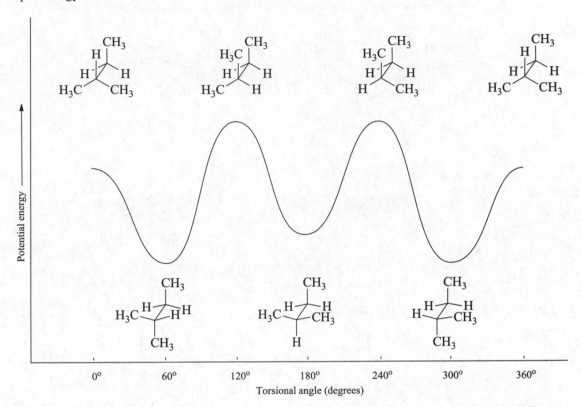

3.37 The structure shown in the text is not the most stable conformation because the bonds of the methyl group are eclipsed with those of the ring carbon to which it is attached. The most stable conformation has the bonds of the methyl group and its attached carbon in a staggered relationship.

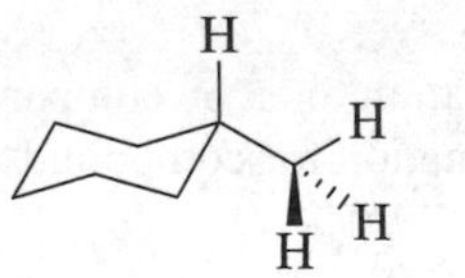

Bonds of methyl group eclipsed
with those of attached carbon

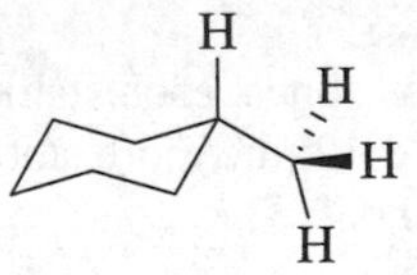

Bonds of methyl group staggered
with those of attached carbon

3.38 Conformation B is more stable than A. The methyl groups are rather close together in A, resulting in van der Waals strain between them. In B, the methyl groups are farther apart.

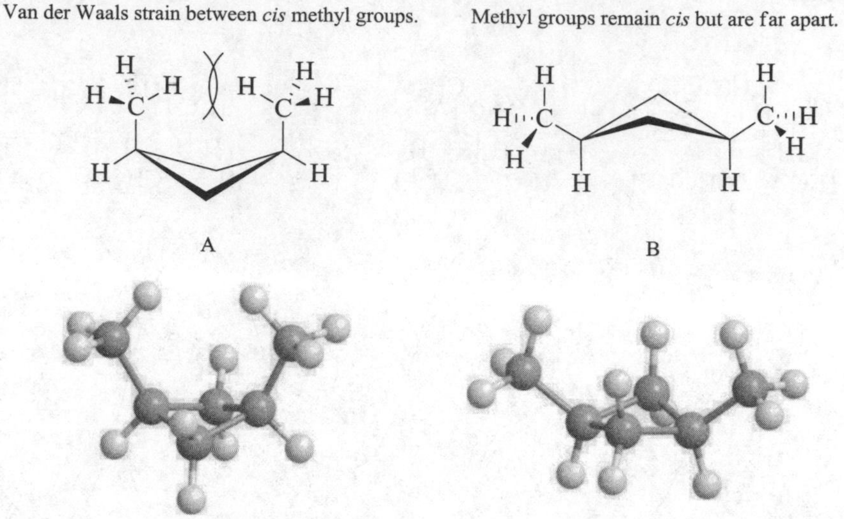

Van der Waals strain between *cis* methyl groups.   Methyl groups remain *cis* but are far apart.

A

B

3.39 The order of decreasing stability is:

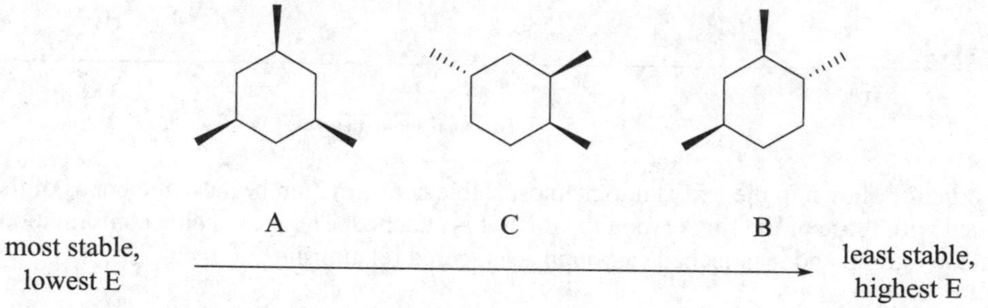

A   C   B

most stable,   least stable,
lowest E   highest E

To rank these cyclohexanes, each stable chair conformation must be compared between compounds. In the solution to problem 3.12, the more stable chair conformation for compound A was represented with methyl groups all equatorial.

A, *cis*-1,3,5-Trimethylcyclohexane

For compound B, draw the projection formula as a chair conformation.

Check to see if this is the most stable conformation by writing its ring-flipped form.

Less stable conformation: two axial methyl groups   More stable conformation: one axial methyl group

The ring-flipped form, with two equatorial methyl groups and one axial methyl group, is more stable than the originally drawn conformation, with two axial methyl groups and one equatorial methyl group.

Cyclohexane B is less stable than A, because of the diaxial repulsions raise the potential energy by +7.3 kJ/mole (1.7 kcal/mol). In addition, the gauche equatorial methyl groups +3.3 kJ/mol (0.8 kcal/mol) also increase the potential energy, similar to gauche vs. anti butane (section 3.2)

1,3-diaxial repulsions
7.3 kJ/mol (1.7 kcal/mol)

gauche methyl groups
3,3 kJ/mol (0.8 kcal/mol)

B

For compound C, the more stable conformation has three equatorial methyl groups as in compound A. However, two of these methyl groups are gauche to each other which raises the potential energy by approximately 3.3 kJ/mol (0.8 kcal/mol). Thus, although C is more stable than B, it is less stable than A.

redrawn:

gauche methyl groups
3,3 kJ/mol (0.8 kcal/mol)

C

3.40 Begin by writing each of the compounds in its most stable conformation. Compare them by examining their conformations for sources of strain, particularly van der Waals strain arising from groups located too close together in space.

(a) Its axial methyl group makes the cis stereoisomer of 1-isopropyl-2-methylcyclohexane less stable than the trans.

cis-1-Isopropyl-2-methylcyclohexane
(less stable stereoisomer)

trans-1-Isopropyl-2-methylcyclohexane
(more stable stereoisomer)

The axial methyl group in the cis stereoisomer is involved in 1,3-diaxial repulsions with the C-4 and C-6 axial hydrogens indicated in the drawing.

(b)   Both groups are equatorial in the cis stereoisomer of 1-isopropyl-3-methylcyclohexane; cis is more stable than trans in 1,3-disubstituted cyclohexanes.

cis-1-Isopropyl-3-methylcyclohexane
(more stable stereoisomer; both
groups are equatorial)

trans-1-Isopropyl-3-methylcyclohexane
(less stable stereoisomer; methyl group
is axial and involved in repulsions
with axial hydrogens at C-1 and C-5)

(c)   The more stable stereoisomer of 1,4-disubstituted cyclohexanes is the trans; both alkyl groups are equatorial in *trans*-1-isopropyl-4-methylcyclohexane.

cis-1-Isopropyl-4-methylcyclohexane
(less stable stereoisomer; methyl
group is axial and involved in
repulsions with axial
hydrogens at C-2 and C-6)

trans-1-Isopropyl-4-methylcyclohexane
(more stable stereoisomer; both
groups are equatorial)

(d)   The first stereoisomer of 1,2,4-trimethylcyclohexane is the more stable one. All its methyl groups are equatorial in its most stable conformation. The most stable conformation of the second stereoisomer has one axial and two equatorial methyl groups.

More stable stereoisomer

All methyl groups equatorial in
most stable conformation

Less stable stereoisomer

One axial methyl group in most
stable conformation

(*e*)   The first stereoisomer of 1,2,4-trimethylcyclohexane is the more stable one here, as it was in part (*d*). All its methyl groups are equatorial, but one of the methyl groups is axial in the most stable conformation of the second stereoisomer.

More stable stereoisomer

All methyl groups equatorial in most stable conformation

Less stable stereoisomer

One axial methyl group in most stable conformation

(*f*)   Each stereoisomer of 2,3-dimethylbicyclo[3.2.1]octane has one axial and one equatorial methyl group. The first one, however, has a close contact between its axial methyl group and both methylene groups of the two-carbon bridge. The second stereoisomer has repulsions with only one axial methylene group; it is more stable.

Less stable stereoisomer
(more van der Waals strain)

More stable stereoisomer
(less van der Waals strain)

3.41   First write structural formulas showing the relative stereochemistries and the preferred conformations of the two stereoisomers of 1,1,3,5-tetramethylcyclohexane.

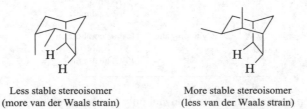

written in its most stable conformation as

cis-1,1,3,5-Tetramethylcyclohexane

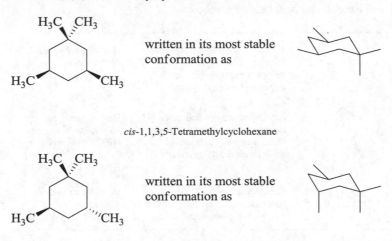

written in its most stable conformation as

trans-1,1,3,5-Tetramethylcyclohexane

The cis stereoisomer is more stable than the trans. It exists in a conformation with only one axial methyl group, while the trans stereoisomer has two axial methyl groups in close contact with each other. The trans stereoisomer is destabilized by van der Waals strain.

3.42 Both structures have approximately the same degree of angle strain and of torsional strain. Structure B has more van der Waals strain than A because two pairs of hydrogens (shown here) approach each other at distances that are rather close.

A:
More stable stereoisomer

Van der Waals strain
destabilizes B.

Properties

3.43 (*a*) The heat of combustion is highest for the hydrocarbon with the greatest number of carbons. Thus, cyclopropane, even though it is more strained than cyclobutane or cyclopentane, has the lowest heat of combustion.

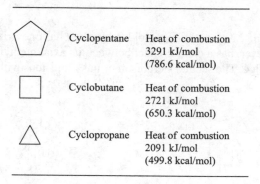

	Cyclopentane	Heat of combustion 3291 kJ/mol (786.6 kcal/mol)
	Cyclobutane	Heat of combustion 2721 kJ/mol (650.3 kcal/mol)
	Cyclopropane	Heat of combustion 2091 kJ/mol (499.8 kcal/mol)

A comparison of heats of combustion can only be used to assess relative stability when the compounds are isomers.

(*b*)   All these compounds have the molecular formula C_7H_{14}. They are isomers, and so the one with the most strain will have the highest heat of combustion.

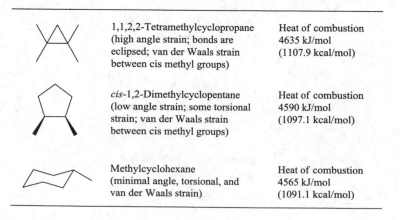

	1,1,2,2-Tetramethylcyclopropane (high angle strain; bonds are eclipsed; van der Waals strain between cis methyl groups)	Heat of combustion 4635 kJ/mol (1107.9 kcal/mol)
	cis-1,2-Dimethylcyclopentane (low angle strain; some torsional strain; van der Waals strain between cis methyl groups)	Heat of combustion 4590 kJ/mol (1097.1 kcal/mol)
	Methylcyclohexane (minimal angle, torsional, and van der Waals strain)	Heat of combustion 4565 kJ/mol (1091.1 kcal/mol)

(*c*) These hydrocarbons all have different molecular formulas. Their heats of combustion decrease with decreasing number of carbons, and comparisons of relative stability cannot be made.

	Cyclopropylcyclopropane (C_6H_{10})	Heat of combustion 3886 kJ/mol (928.8 kcal/mol)
	Spiro[2.2]pentane (C_5H_8)	Heat of combustion 3296 kJ/mol (787.8 kcal/mol)
	Bicyclo[1.1.0]butane (C_4H_6)	Heat of combustion 2648 kJ/mol (633.0 kcal/mol)

(*d*) Bicyclo[3.3.0]octane and bicyclo[5.1.0]octane are isomers, and their heats of combustion can be compared on the basis of their relative stabilities. The three-membered ring in bicyclo[5.1.0]-octane imparts a significant amount of angle strain to this isomer, making it less stable than bicyclo[3.3.0]octane. The third hydrocarbon, bicyclo[4.3.0]nonane, has a greater number of carbons than either of the others and has the largest heat of combustion.

	Bicyclo[4.3.0]nonane (C_9H_{16})	Heat of combustion 5562 kJ/mol (1350.9 kcal/mol)
	Bicyclo[5.1.0]octane (C_8H_{14})	Heat of combustion 5089 kJ/mol (1216.3 kcal/mol)
	Bicyclo[3.3.0]octane (C_8H_{14})	Heat of combustion 5016 kJ/mol (1198.9 kcal/mol)

3.44 When comparing isomers, one that is less stable has a higher heat of combustion than one that is more stable. Both methyl groups can adopt an equatorial conformation in *trans*-1,2-dimethylcyclohexane, and thus trans is the more stable stereoisomer. Notice, however, that the two methyl groups are gauche to one another in the 1,2 isomer. Van der Waals strain between the gauche methyl groups is the reason for the higher heat of combustion compared with the more stable stereoisomers of the 1,3 and 1,4 isomers.

These methyl groups are gauche to each other

trans-1,2-Dimethylcyclohexane
(more stable stereoisomer)

cis-1,3-Dimethylcyclohexane
(more stable stereoisomer)

trans-1,4-Dimethylcyclohexane
(more stable stereoisomer)

The same reasoning applies to the less stable stereoisomer, *cis*-1,2-dimethylcyclohexane. One methyl group is axial while the other is equatorial in the cis stereoisomer, but once again the methyl groups are gauche to each other and there is van der Waals strain between them.

These methyl groups are gauche to each other

cis-1,2-Dimethylcyclohexane
(less stable stereoisomer)

3.45 All the individual bond dipole moments cancel in the anti conformation of $ClCH_2CH_2Cl$, and this conformation has no dipole moment. Because $ClCH_2CH_2Cl$ has a dipole moment of 1.12 D, it can exist entirely in the gauche conformation or it can be a mixture of anti and gauche conformations, but it cannot exist entirely in the anti conformation. Statement 1 is false.

Gauche
(can have a dipole moment)

Anti
(cannot have a dipole moment)

3.46 When all of the chlorines are bonded to one carbon the individual C-Cl dipole moments give the greatest molecular dipole moment.

$$CH_3-C \stackrel{Cl}{\underset{Cl}{\leftarrow}} Cl$$

3.47 When the two equatorial individual C-Cl dipole moments oppose each other on carbons 1 and 4 (opposite ends of cyclohexane) gives the smallest molecular dipole moment.

Answers to Interpretive Problems 3

3.48 B;   **3.49** D;   **3.50** A;   **3.51** B;   **3.52** B;   **3.53** C

SELF-TEST

1. Draw Newman projections for both the gauche and the anti conformations of 1-chloropropane, $CH_3CH_2CH_2Cl$. Sight along the C-1, C-2 bond (the chlorine is attached to C-1).

2. Write Newman projection formulas for

   (*a*)   The least stable conformation of butane.

   (*b*)   Two different staggered conformations of $Cl_2CHCHCl_2$.

3. Give the correct IUPAC name for the compound represented by the following Newman projection.

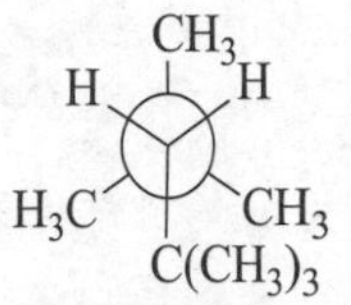

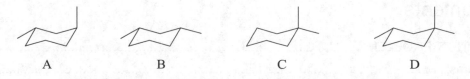

4. Write the structure of the most stable conformation of the *less* stable stereoisomer of 1-*tert*-butyl-3-methylcyclohexane.

5. Draw the most stable conformation of the following substance:

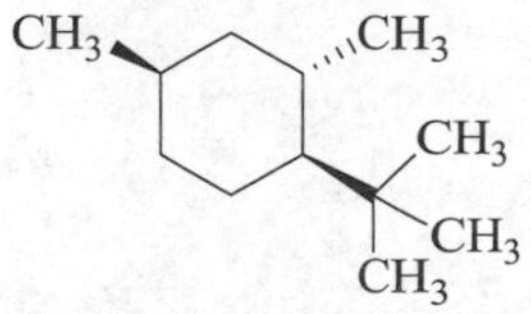

Which substituents are axial and which equatorial?

6. A wedge-and-dash representation of a form of ribose (called β-D-ribopyranose) is shown here. Draw the most stable chair conformation of this substance.

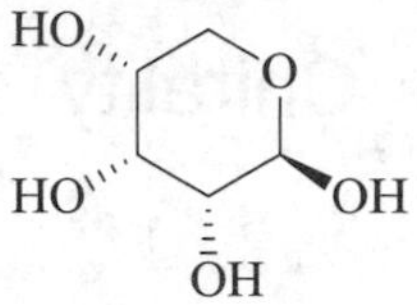

7. Consider compounds A, B, C, and D.

    A   B   C   D

(*a*) Which one is a constitutional isomer of two others?

(*b*) Which two are stereoisomers of one another?

(*c*) Which one has the highest heat of combustion?

(*d*) Which one has the stereochemical descriptor trans in its name?

8. Draw clear depictions of two nonequivalent chair conformations of *cis*-1-isopropyl-4-methylcyclohexane, and indicate which is more stable.

9. Which has the lower heat of combustion, *cis*-1-ethyl-3-methylcyclohexane or *cis*-1-ethyl-4-methylcyclohexane?

10. The hydrocarbon shown is called twistane. Classify twistane as monocyclic, bicyclic, etc. What is the molecular formula of twistane?

11. Sketch an approximate potential energy diagram similar to those shown in the text (Figures 3.4 and 3.7) for rotation about a carbon–carbon bond in 2-methylpropane. Does the form of the potential energy curve more closely resemble that of ethane or that of butane?

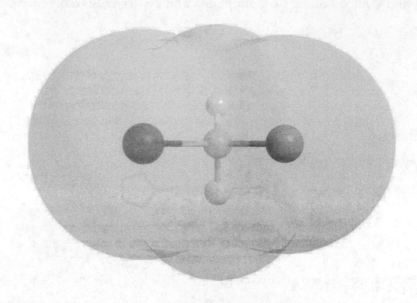

CHAPTER 4
Chirality

Table of Contents

SOLUTIONS TO TEXT PROBLEMS

In Chapter Problems

4.1 (*c*) Carbon-2 is a chirality center in 1-bromo-2-methylbutane. The four groups attached to it, H, CH_3, CH_3CH_2, and $BrCH_2$, are different from each other.

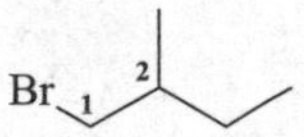

(*d*) There are no chirality centers in 2-bromo-2-methylbutane.

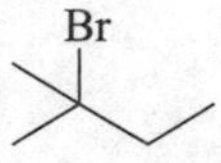

4.2 (*b*) Carbon-2 is a chirality center in 1,1,2-trimethylcyclobutane.

A chirality center; the four groups to which it is directly bonded [H, CH_3, CH_2, and $C(CH_3)_2$] are all different from one another.

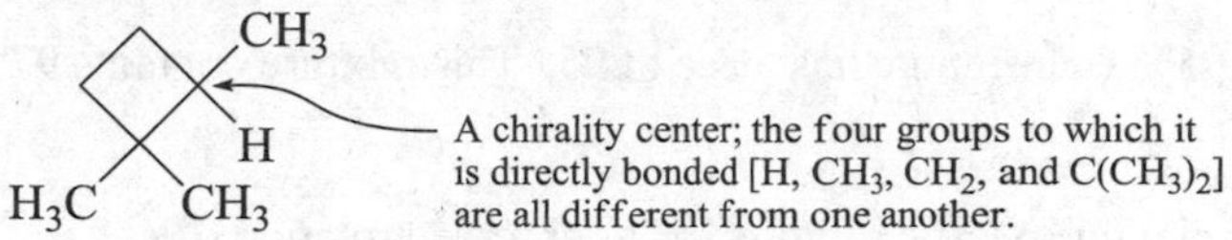

1,1,3-Trimethylcyclobutane, however, has no chirality centers.

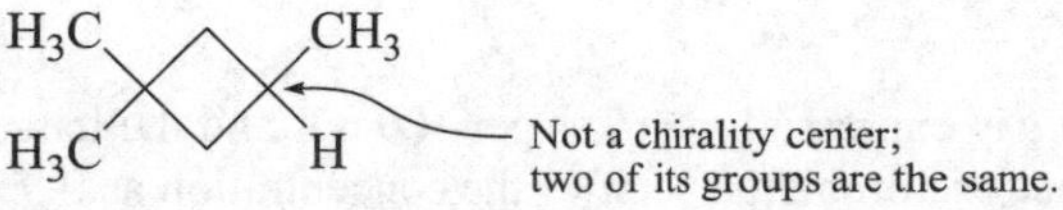

Not a chirality center; two of its groups are the same.

4.3 (*b*) A plane of symmetry passes through the methyl group, C-7, and C-4. The compound is achiral.

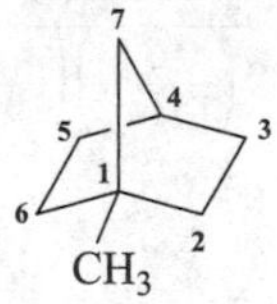

(*c*) (*d*) There are no planes of symmetry in these molecules.

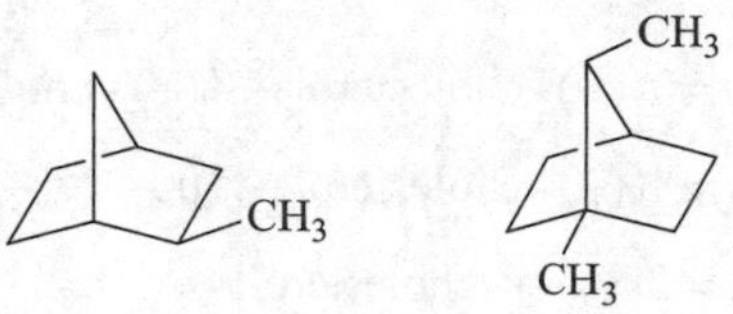

4.4 (*a*) *trans*-1,3-Cyclobutanediol, when the carbon atoms are in the same plane, has a plane of symmetry that passes through the two carbons containing the hydroxy substituents. Rotating the molecule where the four carbons are in the plane of the paper, the plane of symmetry is then perpendicular to the paper as shown.

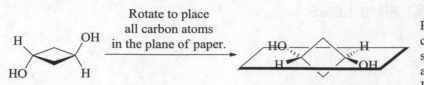

Rotate to place all carbon atoms in the plane of paper.

Plane of symmetry contains the two substituted carbons and their hydroxy groups. It is perpendicular to the paper.

(*b*) *cis*-1,3-Cyclobutanediol does not contain a center of symmetry. It does have two planes of symmetry and is thus achiral.

4.5 ee = (% of major enantiomer) – (% of the minor enantiomer)

Limonene that is 95% enantiopure has an ee of 95. This mixture contains 97.5% of one enantiomer and 2.5% of other.

4.6 (a) The equation relating specific rotation [α] to observed rotation α is

$$[\alpha] = \frac{100\alpha}{cl}$$

The concentration *c* is expressed in grams per 100 mL and the length *l* of the polarimeter tube in decimeters. Because the problem specifies the concentration as 0.3 g/15 mL and the path length as 10 cm, the specific rotation [α] is

$$[\alpha] = \frac{100(-0.78°)}{100\left(\dfrac{0.3\,\text{g}}{15\,\text{mL}}\right)\left(\dfrac{10\,\text{cm}}{10\,\text{cm/dm}}\right)}$$

$$= -39°$$

(b) From the previous problem, the specific rotation of natural cholesterol is [α] = –39°. The mixture of natural (–)-cholesterol and synthetic (+)-cholesterol specified in this problem has a specific rotation [α] of –13°. The optical purity is 33.3% [(–13°/–39°) (100)].

$$\text{Optical purity} = \%(-)\text{-cholesterol} - \%(+)\text{-cholesterol}$$

$$33.3\% = \%(-)\text{-cholesterol} - [100 - \%(-)\text{-cholesterol}]$$

$$133.3\% = 2[\%(-)\text{-cholesterol}]$$

$$66.7\% = \%(-)\text{-cholesterol}$$

The mixture is two-thirds natural (–)-cholesterol and one-third synthetic (+)-cholesterol.

4.7 Draw the molecular model so that it is in the same format as the drawings of (+) and (–)-2-butanol in the text.

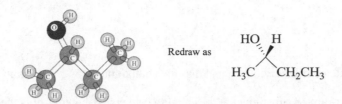

Reorient the molecule so that it can be compared with the drawings of (+) and (−)-2-butanol.

The molecular model when redrawn matches the text's drawing of (+)-2-butanol.

4.8 (*b*) The solution to this problem is analogous to the sample solution given in the text to part (*a*).

(+)-1-Fluoro-2-methylbutane

Order of precedence: $CH_2F > CH_3CH_2 > CH_3 > H$

The lowest-ranked atom (H) at the chirality center points away from us in the drawing. The three higher-ranked substituents trace a clockwise path from CH_2F to CH_2CH_3 to CH_3.

$$H_3C \qquad CH_2F$$
$$CH_2CH_3$$

The absolute configuration is *R*; the compound is (*R*)-(+)-1-fluoro-2-methylbutane.

(*c*) The highest-ranked substituent at the chirality center of 1-bromo-2-methylbutane is CH_2Br, and the lowest-ranked atom is H. Of the remaining two, ethyl outranks methyl.

(+)-1-Bromo-2-methylbutane

Order of precedence: $CH_2Br > CH_2CH_3 > CH_3 > H$

The lowest-ranked atom (H) is directed toward you in the drawing, and therefore the molecule needs to be reoriented so that H points in the opposite direction.

Turn 180°

The three highest-ranking substituents trace a counterclockwise path when the lowest-ranked atom (H) is held away from you.

The absolute configuration is S, and thus the compound is (S)-$(+)$-1-bromo-2-methylbutane.

(d) The highest-ranked substituent at the chirality center of 3-buten-2-ol is the hydroxyl group, and the lowest-ranked atom is H. Of the remaining two, vinyl outranks methyl.

(+)-3-Buten-2-ol

Order of precedence: $\quad HO > H_2C{=}CH > CH_3 > H$

The lowest-ranked atom (H) is directed away from you in the drawing. We see that the order of decreasing precedence appears in a counterclockwise manner.

The absolute configuration is S, the compound's name is (S)-$(+)$-3-buten-2-ol.

4.9 (b) The chirality center is the carbon that bears the methyl group. Its substituents are

$$-CF_2CH_2 \quad > \quad -CH_2CF_2 \quad > \quad -CH_3 \quad > \quad -H$$

Highest rank		Lowest rank

When the lowest-ranked substituent points away from you, the remaining three must appear in descending order of precedence in a counterclockwise fashion in the S enantiomer. (S)-1,1-difluoro-2-methylcyclopropane is therefore

4.10 (a) Begin by converting the Fischer projection to a perspective (three-dimensional) representation, remembering that the vertical bonds at the chirality center are directed away from you and the horizontal bonds are directed toward you.

Now rotate the three-dimensional representation so that the lowest-ranked atom (H) points away from you. You can then see that the order of decreasing precedence $OH > CH_2OH > CH_2CH_3 > H$ traces a clockwise path. The absolute configuration is R.

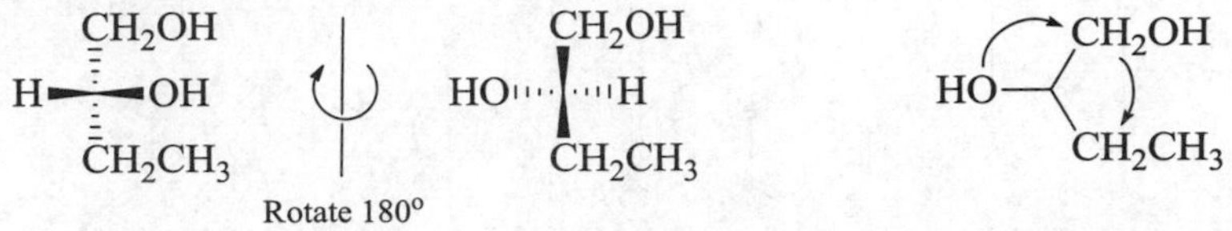

(b) Proceed in the same manner as part (a) of this problem was solved.

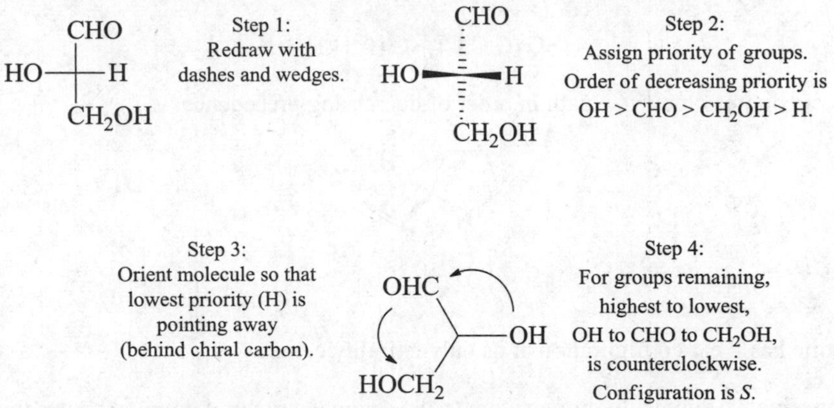

4.11 (b) As in part (a), exchanging any two groups in a Fischer projection converts it to the mirror image. Thus, exchanging the positions of CH_3 and CH_2CH_3 in the Fischer projection of (R)-2-butanol converts the configuration of the chirality center to S.

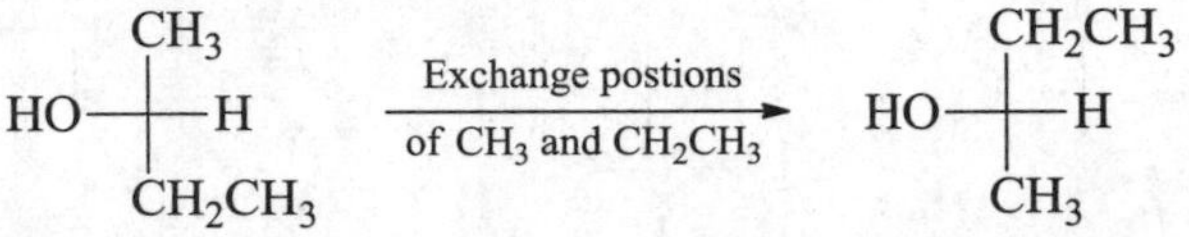

(R)-2-Butanol   (S)-2-Butanol

(c) Switching the positions of three groups gives a Fischer projection that looks different but leaves the configuration of the chirality center unchanged.

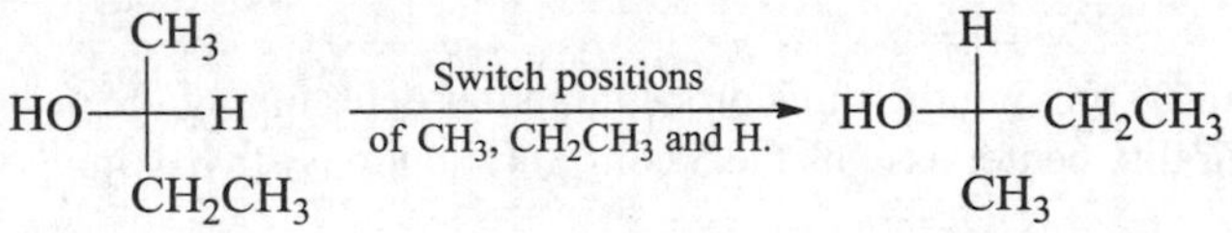

(R)-2-Butanol   (R)-2-Butanol

(d, e) Switching H with OH and CH_3 and CH_2CH_3 has the same effect as rotating the Fischer projection 180°. The resulting projection has the same configuration at the chirality center as the original.

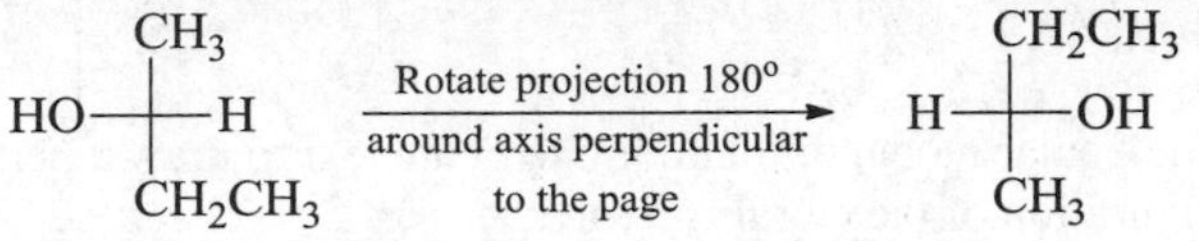

(R)-2-Butanol   (R)-2-Butanol

4.12 The structural formulas for the assigned compounds given in the text are supplemented here by adding hydrogen atoms to the appropriate chirality centers.

(-)-Nicotine: Because the hydrogen atom at the chirality center points forward in the original structural drawing, turn the drawing so that the bond to hydrogen points away from you.

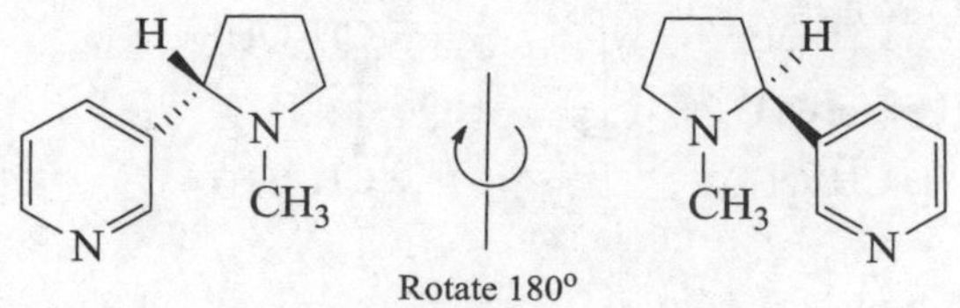

Rotate 180°

In order of decreasing precedence the atoms attached to the chirality center are:

$$N > C\ (C,C,C) > C\ (C,H,H) > H$$

and trace a counter-clockwise path in order of decreasing precedence.

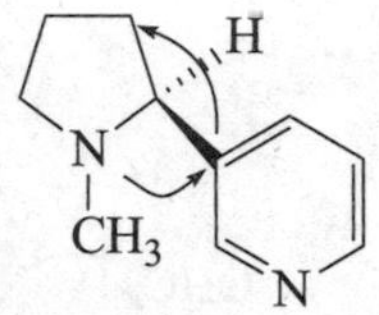

(-)-Nicotine has the *S* configuration at its only chirality center.

(-)-Adrenaline: Adding a hydrogen atom to the original structural drawing shows that the molecule is already properly oriented (H pointing away). The order of decreasing precedence

$$O > C\ (N,H,H) > C(C,C,C) > H$$

traces a clockwise path. The configuration at the chirality center is *R*.

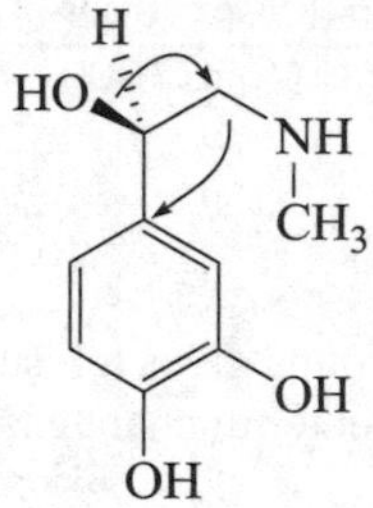

(-)-Thyroxine: Applying the same procedure after determining the order of precedence of atoms connected to the chirality center assigns the *S* configuration to (-)-thyroxine.

$$N > C(O,O,O) > C(C,H,H) > H$$

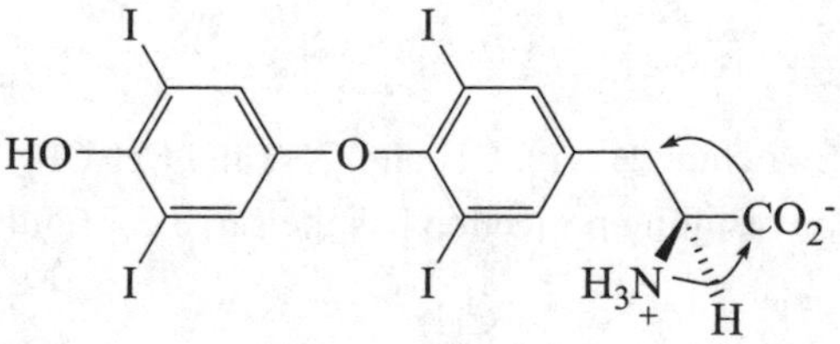

4.13   Begin by using the molecular model of thalidomide in the text to draw a perspective view, paying particular attention to the configuration of the chirality center.

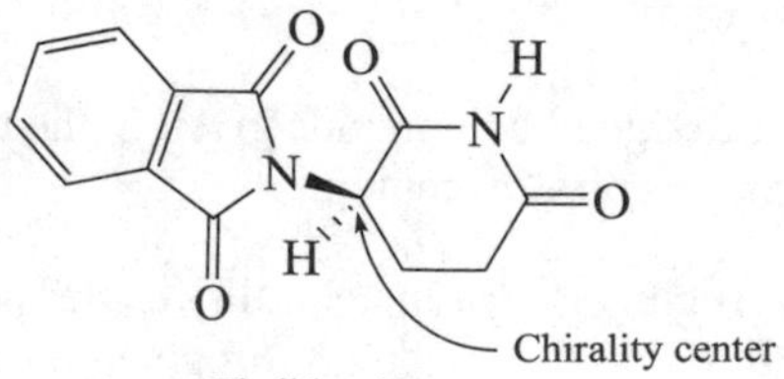

Chirality center

Thalidomide

The four groups attached to the chirality center, in order of decreasing precedence, are

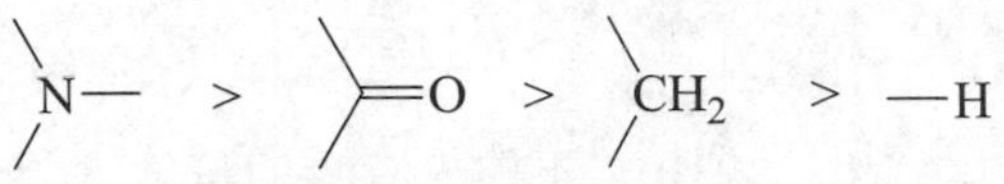

$$\text{N}- \quad > \quad \text{=O} \quad > \quad \text{CH}_2 \quad > \quad -\text{H}$$

These groups trace a clockwise path with the lowest-ranked group (H) pointing away from us. The molecular model of thalidomide is the *R* enantiomer.

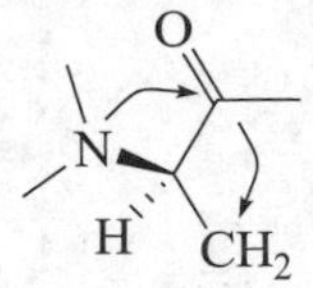

4.14 In this case, the enantiomers are more easily interconverted by rotation about the single bond that connects the two benzene rings. When substituents occupy the 2,6 and 2′,6′ positions, the steric hindrance is high and the energy barrier for rotation is high. With only hydrogens at these positions, the energy barrier for rotation is low.

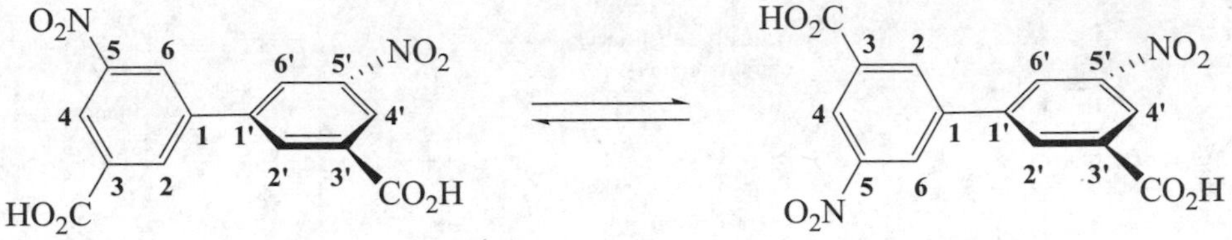

4.15

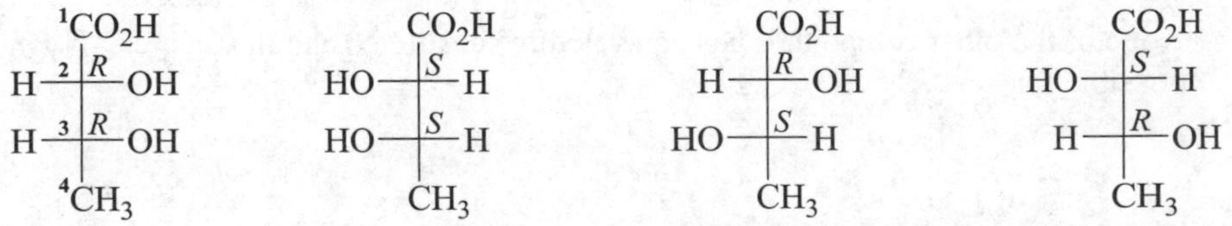

Erythro   Erythro   Threo   Threo

4.16 The erythro stereoisomers are characterized by Fischer projections in which analogous substituents, in this case OH and NH$_2$, are on the same side when the carbon chain is vertical. There are two erythro stereoisomers that are enantiomers of each other:

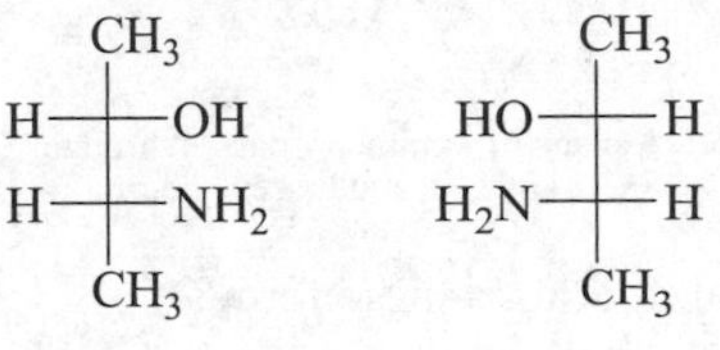

Erythro   Erythro

Analogous substituents are on opposite sides in the threo isomer:

$$
\begin{array}{cccc}
& CH_3 & & CH_3 \\
H & \!-\!\!\!-\!\!\!-\! OH & HO & \!-\!\!\!-\!\!\!-\! H \\
H_2N & \!-\!\!\!-\!\!\!-\! H & H & \!-\!\!\!-\!\!\!-\! NH_2 \\
& CH_3 & & CH_3 \\
& \text{Threo} & & \text{Threo}
\end{array}
$$

4.17   3-Amino-2-butanol has four stereoisomeric forms

($2R,3R$) and its enantiomer ($2S,3S$)

($2R,3S$) and its enantiomer ($2S,3R$)

In the text, we are told that the ($2R,3R$) stereoisomer is a liquid. Its enantiomer ($2S,3S$) has the same physical properties and so must also be a liquid. The text notes that the ($2R,3S$) stereoisomer is a solid (mp 49°C). Its enantiomer ($2S,3R$) must therefore be the other stereoisomer that is a crystalline solid.

4.18   Examine the structural formula of each compound for equivalently substituted chirality centers. The only one capable of existing in a meso form is 2,4-dibromopentane.

Equivalently substituted
chirality centers

$$
\begin{array}{ccl}
& CH_3 & \\
H & \!-\!\!\!-\!\!\!-\! Br & \\
\text{-----} \; H & \!-\!\!\!-\!\!\!-\! H \; \text{-----} & \leftarrow \text{Plane of} \\
& & \text{symmetry} \\
H & \!-\!\!\!-\!\!\!-\! Br & \\
& CH_3 &
\end{array}
$$

2,4-Dibromopentane   Fischer projection of
                                            meso-2,4-dibromopentane

None of the other compounds has equivalently substituted chirality centers. No meso forms are possible for

2,3-Dibromopentane   3-Bromo-2-pentanol   4-Bromo-2-pentanol

4.19   The cis stereoisomer of 1,3-dimethylcyclohexane has a plane of symmetry, and so it is an achiral substance— it is a meso form.

A plane of symmetry passes through
C-2 and C-5 and bisects the ring.

The trans stereoisomer is chiral. It is not a meso form.

4.20   A molecule with three chirality centers has eight (2^3) stereoisomers. The eight combinations of R and S chirality centers are

	Chirality center		*Chirality center*
	1 2 3		**1 2 3**
Isomer 1	$R\,R\,R$	Isomer 5	$S\,S\,S$
Isomer 2	$R\,R\,S$	Isomer 6	$S\,S\,R$
Isomer 3	$R\,S\,R$	Isomer 7	$S\,R\,S$

Isomer 4	$S\,R\,R$	Isomer 8	$R\,S\,S$

4.21 2-Ketohexoses have three chirality centers. They are marked with asterisks in the structural formula.

No meso forms are possible, and so there is a total of eight (2^3) stereoisomeric 2-ketohexoses.

4.22 The tartaric acids incorporate two equivalently substituted chirality centers. (+)-Tartaric acid, as noted in the text, is the 2R,3R stereoisomer. There will be two additional stereoisomers, the enantiomeric (−)-tartaric acid (2S,3S) and an optically inactive meso form.

(2S,3S)-Tartaric acid
(optically active)
mp 170°C, [a]_D -12°

meso-Tartaric acid
(optically inactive)
mp 140°C

4.23 No. Pasteur separated an optically inactive racemic mixture into two optically active enantiomers. A meso form is achiral, is identical to its mirror image, and is incapable of being separated into optically active forms.

4.24 The more soluble salt must have the opposite configuration at the chirality center of 1-phenylethylamine, that is, the S configuration. The malic acid used in the resolution is a single enantiomer, S. In this particular case the more soluble salt is therefore (S)-1-phenylethylammonium (S)-malate.

4.25 The electron pair is the lowest priority so this points away from us. The benzene carbon counts as [—C(C,C,C)] and outranks the carbon with two hydrogens [—C(C,H,H)]. The methyl group ranks third [—C(H,H,H)]. The curved arrows from highest to lowest priority proceed counterclockwise so the phosphorus atom has an S configuration.

Rotate molecule placing electron pair behind P.

End of Chapter Problems

Molecular Chirality

4.26 There are 15 isomeric $C_6H_{14}O$ alcohol structures. Eight of these have chirality centers. The chiral isomers are characterized by carbons (*) that bear four different groups. The remaining molecules are *achiral*.

The primary alcohol 1-hexanol is achiral.

1-Hexanol

Two of the primary alcohols with one branch are chiral:

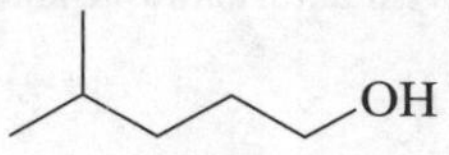

4-Methyl-1-pentanol 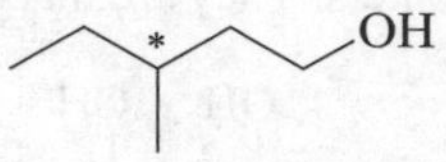   3-Methyl-1-pentanol 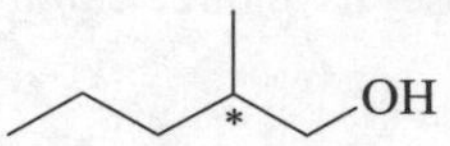   2-Methyl-1-pentanol

One of the primary alcohols with two branches is chiral:

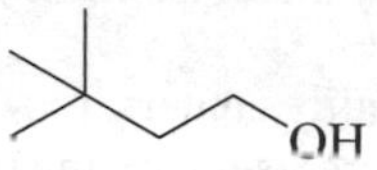

3,3-Dimethyl-1-butanol 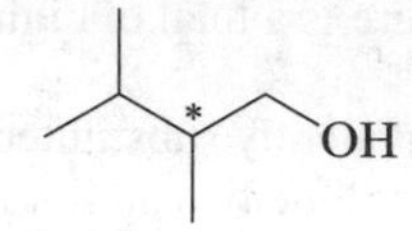   2,3-Dimethyl-1-butanol 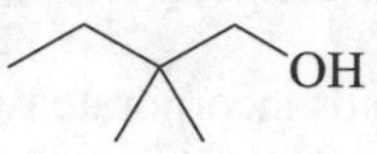   2,2-Dimethyl-1-butanol

There are two secondary alcohols without branches. Both are chiral:

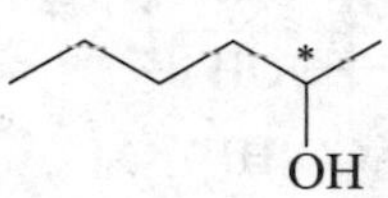

2-Hexanol 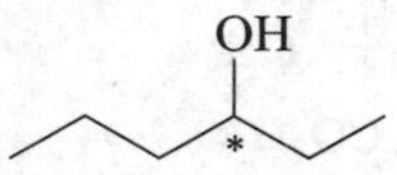   3-Hexanol

All of the secondary alcohols with one branch are chiral and one of these has two chirality centers:

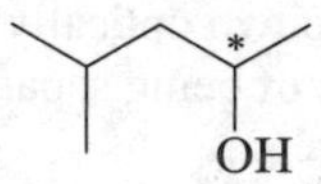

4-Methyl-2-pentanol 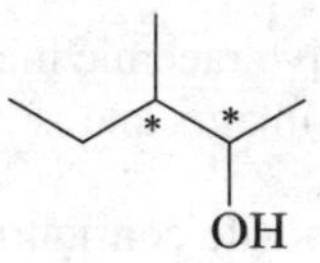   3-Methyl-2-pentanol 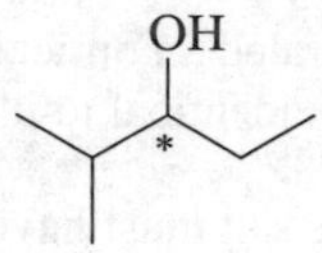   2-Methyl-3-pentanol

The remaining isomers are a secondary alcohol and two tertiary alcohols with two branches. None have chirality centers.

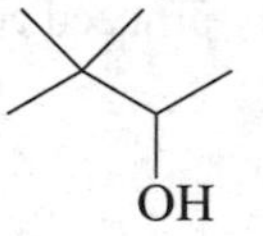

3,3-Dimethyl-2-butanol 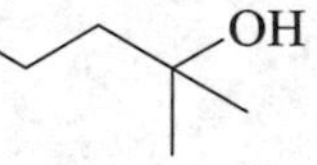   2-Methyl-2-pentanol 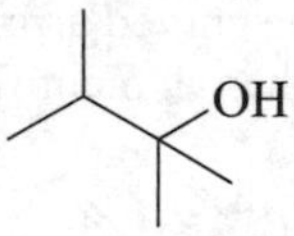   2,3-Dimethyl-2-butanol

4.27 (*a*) There are seven stereoisomers for trichlorocyclobutane.

There are two pairs of enantiomers:

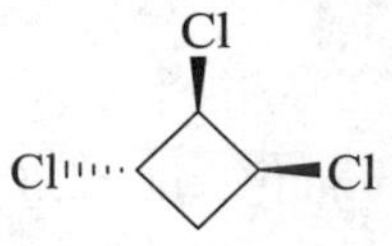

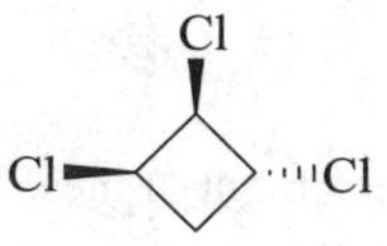

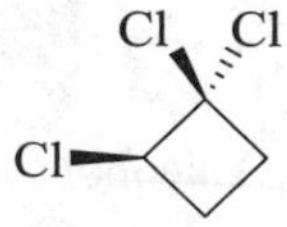

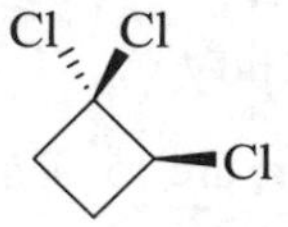

Three trichlorocyclobutanes are meso:

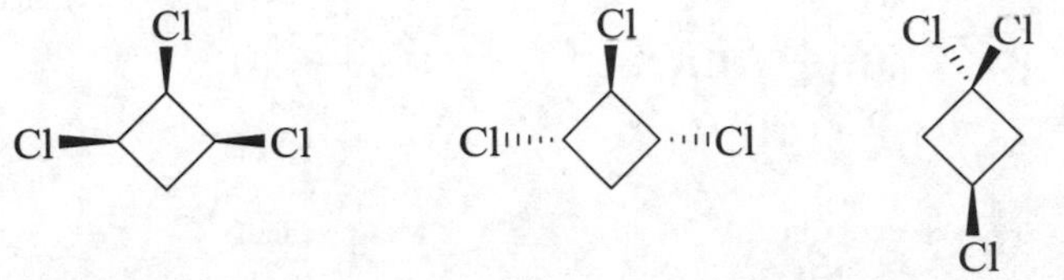

(*b*) All total, there are twelve stereoisomers for trichlorocyclopentane.

There are four pairs of enantiomers:

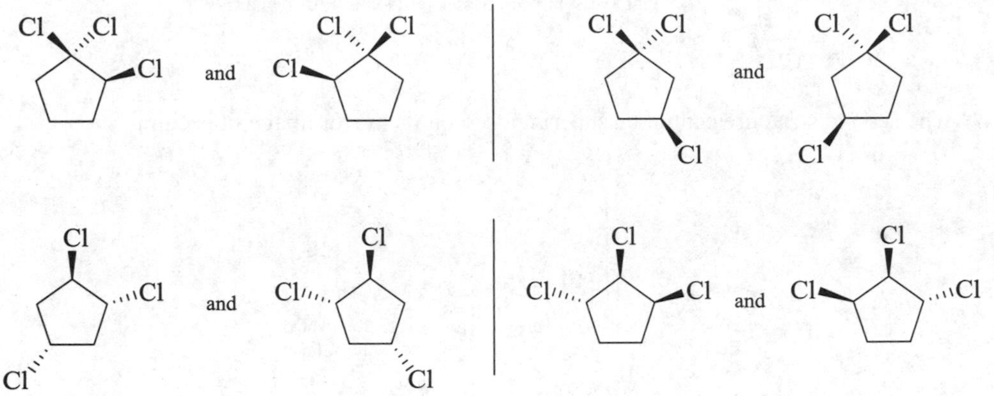

Four trichlorocyclopentanes are meso:

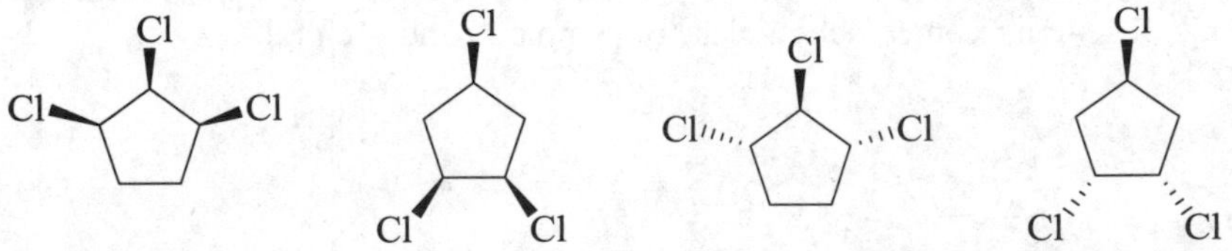

4.28 (*a*) Carbon-2 is a chirality center in 3-chloro-1,2-propanediol. Carbon-2 has two equivalent substituents in 2-chloro-1,3-propanediol and is not a chirality center.

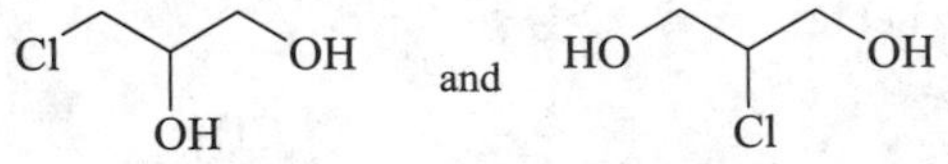

3-Chloro-1,2-propanediol   2-Chloro-1,3-propanediol
        Chiral   Achiral

(*b*) The primary bromide is achiral; the secondary bromide contains a chirality center and is chiral.

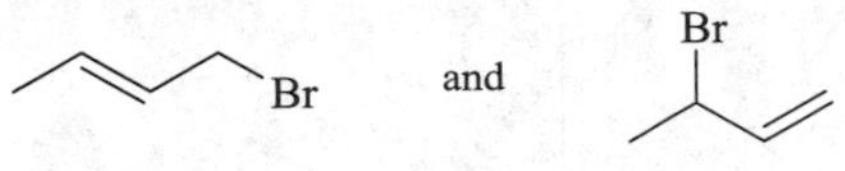

        Achiral   Chiral

(*c*) Both stereoisomers have two equivalently substituted chirality centers, and so we must be alert for the possibility of a meso stereoisomer. The structure at the left is chiral. The one at the right has a plane of symmetry and is the achiral meso stereoisomer.

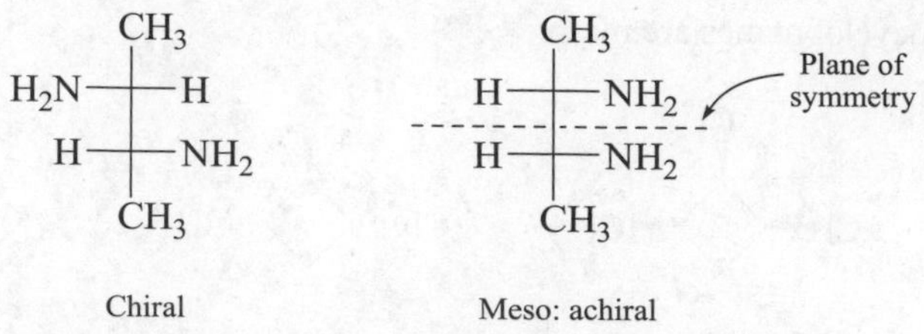

Chiral   Meso: achiral

(*d*) The first structure is achiral; it has a plane of symmetry.

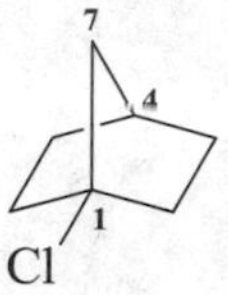

Plane of symmetry passes through C 1, C-4, and C-7.

The second structure cannot be superimposed on its mirror image; it is chiral.

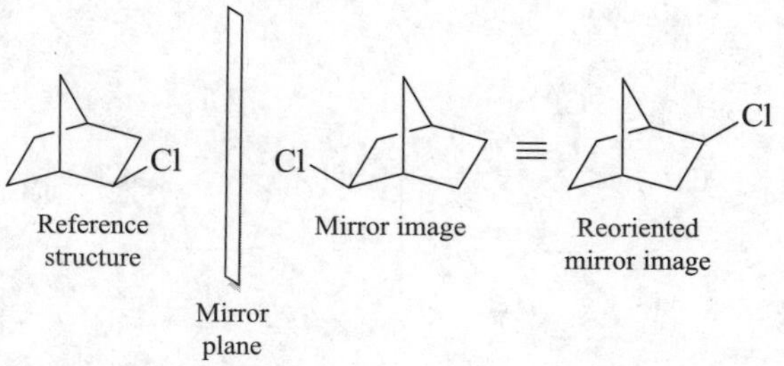

Reference   Mirror image   Reoriented
structure   mirror image

Mirror
plane

(*e*) *cis*-1,3-Cyclohexandiol has a plane of symmetry that passes through the molecule at C-2 and C-5. It is achiral. The trans isomer lacks a plane of symmetry and is chiral.

(*f*) *cis*-3-Methylcyclohexanol lacks a plane of symmetry and is chiral. *trans*-4-Methylcyclohexanol has a plane of symmetry that passes through the molecule at C-1 and C-4. It is achiral. Note also that it does not possess a chirality center.

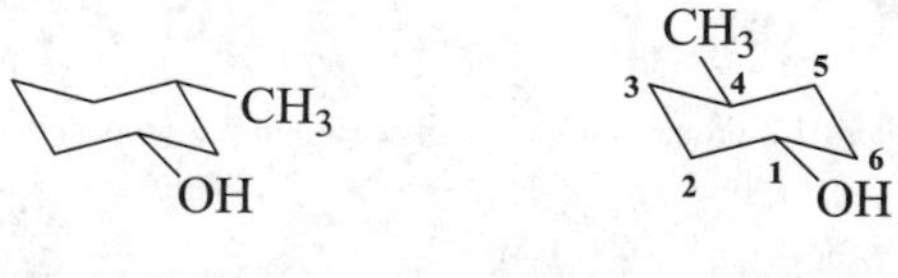

(*g*) The first structure is chiral, it cannot be superimposed on its mirror image.

(g)

Reference   Mirror image   Reoriented
structure   mirror image

Mirror
plane

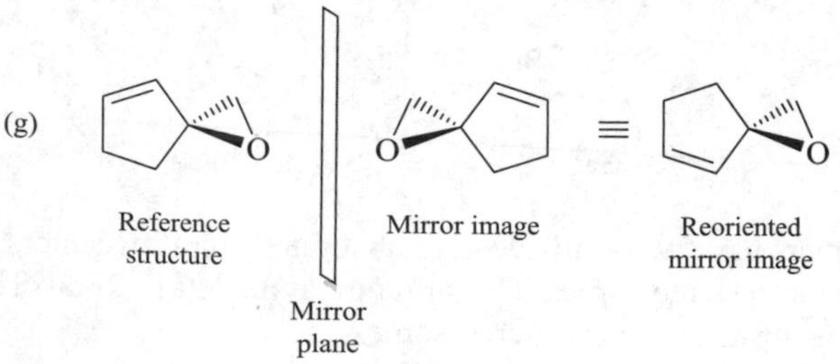

The second structure is achiral; it has a plane of symmetry.

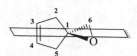

Plane of symmetry passes through C-1, C-3 to C-4 bond; C-6, and O.

4.29 There are four stereoisomers of 2,3-pentanediol, represented by the Fischer projections shown. All are chiral.

CH₃ .... CH₃ .... CH₃ .... CH₃

H—OH ... HO—H ... H—OH ... HO—H

H—OH ... HO—H ... HO—H ... H—OH

CH₂CH₃ .. CH₂CH₃ .. CH₂CH₃ .. CH₂CH₃

Enantiomeric erythro isomers .... Enantiomeric threo isomers

2,4-Pentanediol has three stereoisomers. The meso form is achiral; both threo forms are chiral.

CH₃ .... CH₃ .... CH₃

H—OH ... H—OH ... HO—H

H—H .... H—H .... H—H

H—OH ... HO—H ... H—OH

CH₃ .... CH₃ .... CH₃

meso-2,4-Pentanediol ..... Enantiomeric threo isomers

4.30 Isomers A and E possess a plane of symmetry. Isomer A has plane of symmetry that passes through the methyl group, C-1, C-4, and C-7. Isomer E has a plane of symmetry that bisects bond C2—C3, C5—C7, and atom C-7. Both A and E are achiral. A and E are constitutional isomers of each other and the remaining isomers B, C, and D.

A .... E

Isomers C and D are nonsuperimposable mirror images of each other. They are enantiomers.

C .... D

Mirror plane

Isomer B has the same connectivity as C and D but is not identical or a mirror image. It is a diastereomer of C and D.

B

4.31 The chirality centers are indicated by the arrows.

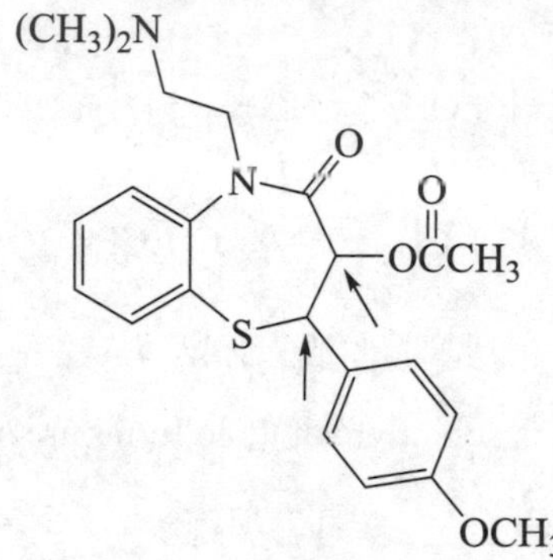

Diltiazem

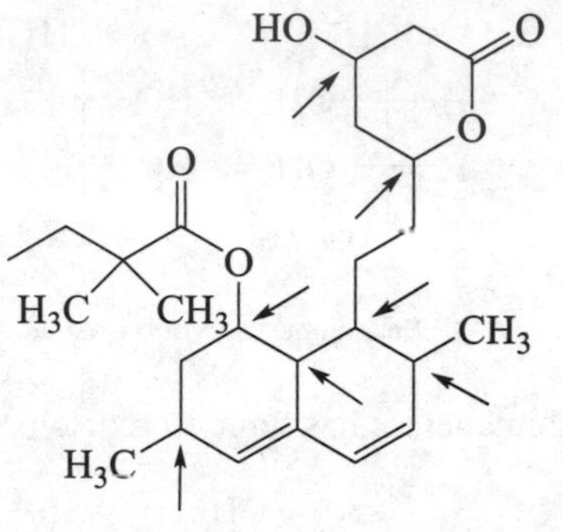

Simvastatin (Zocor)

4.32 Substituted biphenyls of this type possess a chirality axis when $A \neq B$ *and* $X \neq Y$. These conditions are met by only the third example in the table, answer (*c*), where $A = (CH_3)_3C-$, $B = H$ and $X = (CH_3)_3C-$, $Y = H$.

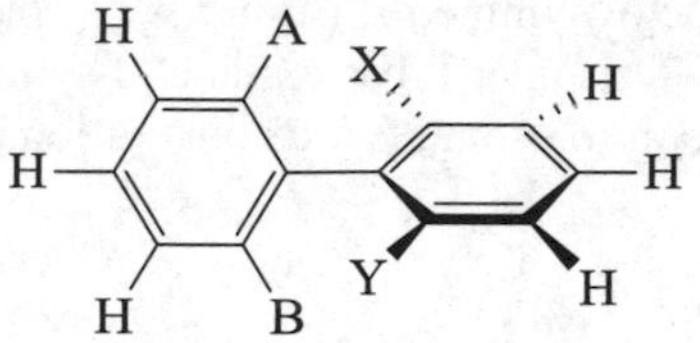

R,S-Configurational Notation

4.33 Among the atoms attached to the chiralty center, the order of decreasing precedence is $Br > Cl > F > H$. When the molecule is viewed with the hydrogen pointing away from us, the order $Br \rightarrow Cl \rightarrow F$ appears clockwise in the *R* enantiomer, counterclockwise in the *S* enantiomer.

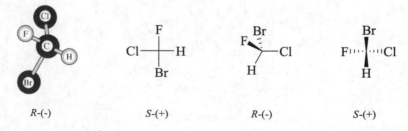

R-(−) S-(+) R-(−) S-(+)

4.34 (*a*) Among the isotopes of hydrogen, T has the highest mass number (3), D next (2), and H lowest (1). Thus, the order of rank at the chirality center in the reactant is $CH_3 > T > D > H$. The order of rank in the product is $HO > CH_3 > T > D$.

Orient with lowest-ranked substituent away from you.

The order of decreasing precedence in the reactant is counterclockwise; the configuration is S. The order of decreasing precedence in the product is clockwise; the configuration is R.

(b) Retention of configuration means that the three-dimensional arrangement of bonds at the chirality center is the same in the reactant and the product. The R and S descriptors change because the order of precedence changes in going from reactant to product; for example, CH_3 is the highest-ranked substituent in the reactants but becomes the second-highest-ranked in the product.

4.35 (a) The order of substituent precedence at the chirality center is

$$HO > CH_2CH_2 > CH_3 > H$$

The molecule is oriented so that the lowest-ranking substituent is directed away from you and the order of decreasing precedence is clockwise. $(-)$-2-Octanol has the R configuration.

(b) In order of decreasing sequence rule precedence, the four substituents at the chirality center of monosodium L-glutamate are

$$\overset{+}{N}H_3 > CO_2^- > CH_2 > H$$

When the molecule is oriented so that the lowest-ranking atom (hydrogen) is directed away from you, the other three substituents are arranged as shown.

The order of decreasing rank is counterclockwise; the absolute configuration is S.

4.36 Rotate the cyclohexane until the lowest priority hydrogens are behind the carbons that contain the hydroxyl and bromine substituents. Consider the C—OH carbon first and number from that carbon in both directions around the ring. The hydroxide oxygen is the highest priority. Both C-2 and C-2' are equivalent, both having –C[C,H,H], so proceed to the next carbon. C-3 is a higher priority than C-3' since Br has a higher atomic number than C. The highest to lowest priority group proceed clockwise. That carbon has an R configuration.

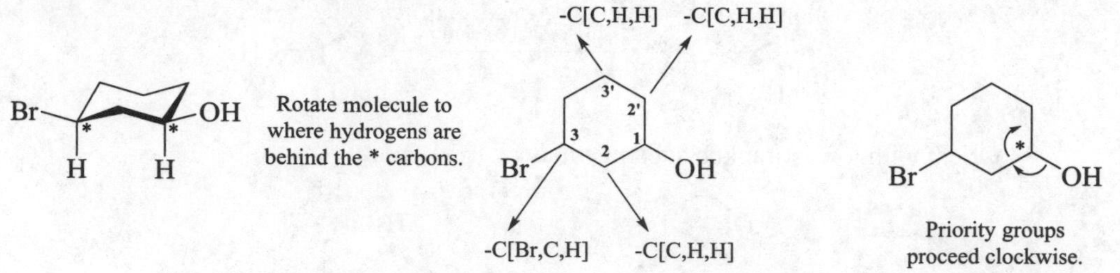

For the other chirality center, repeat the process to find an *S* configuration.

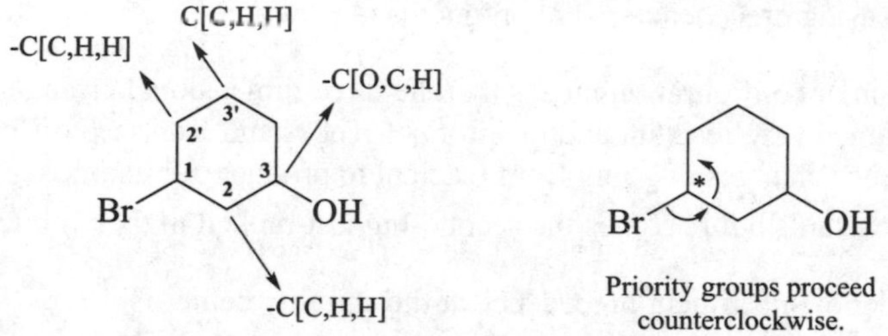

The name of this molecule is (1*R*,3*S*)-3-bromocyclohexanol.

4.37 The highest ranked atom attached to C-2 is nitrogen and the lowest ranked is hydrogen. Of the other two atoms attached to C-2, C-1 [C(O,O,O)] outranks C-3 [C(O,C,H)]. C-2 has the *S* configuration when the order of decreasing precedence N > C-1 > C-3 appears counterclockwise when the H is back and N is forward. C-3 has the *R* configuration when the order of decreasing precedence O > C-2 > C-3 appears clockwise when H is back and O is forward.

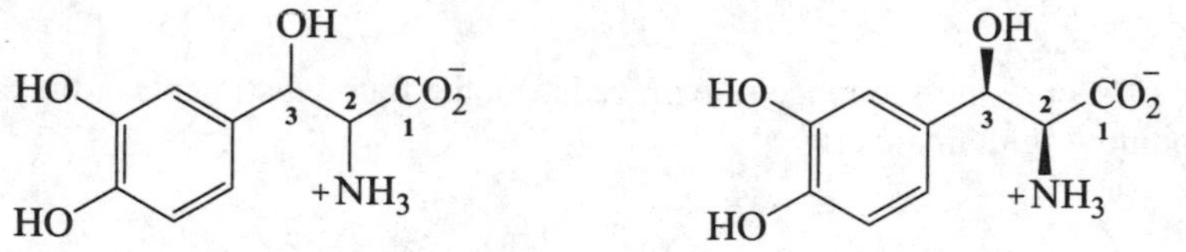

Structural Relationships

4.38 Two compounds can be stereoisomers only if they have the *same* constitution. Thus, you should compare first the constitution of the two structures and then their stereochemistry. One way to compare constitutions is to assign a systematic (IUPAC) name to each molecule. Also remember that enantiomers are nonsuperimposable mirror images, and diastereomers are stereoisomers that are not enantiomers.

(*a*) The two structures have the same constitution. Test them for superimposability. To do this, we need to place them in comparable orientations.

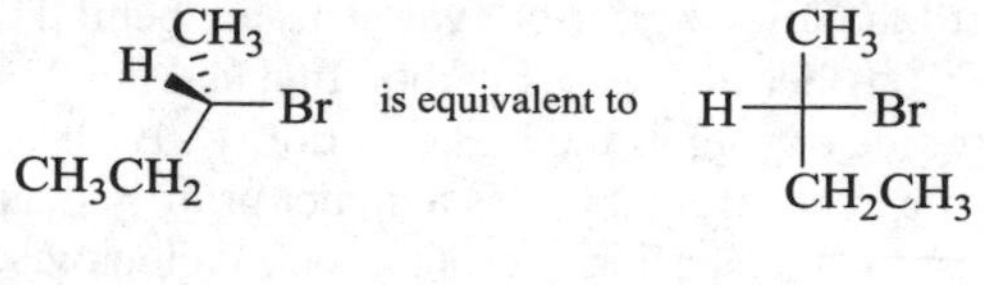

and

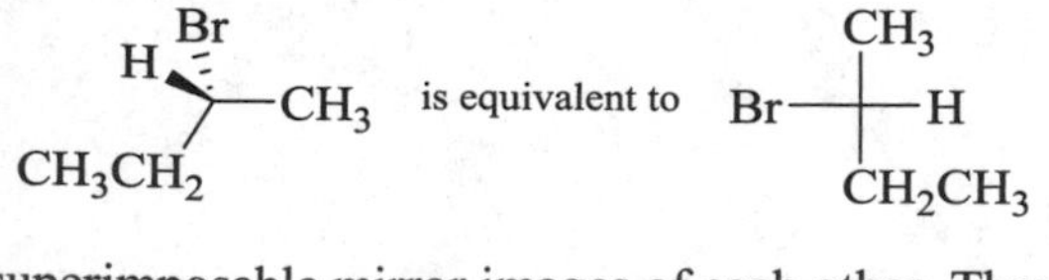

The two are nonsuperimposable mirror images of each other. They are enantiomers.

To check this conclusion, work out the absolute configuration of each using the Cahn–Ingold–Prelog system.

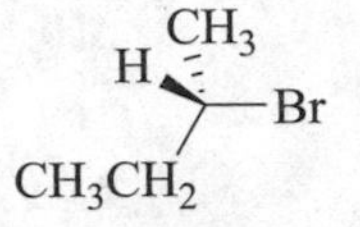

(S)-2-Bromobutane 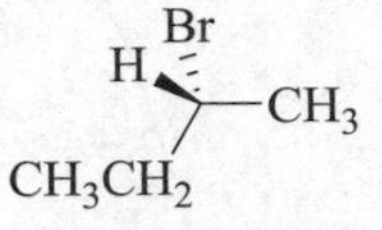   (R)-2-Bromobutane

(b) As drawn, the two structures are mirror images of each other; however, they represent an achiral molecule. The two structures are superimposable mirror images and are not stereoisomers but identical.

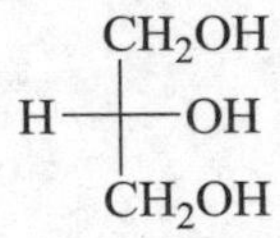

(c) The two structures are enantiomers because they are nonsuperimposable mirror images. Checking their absolute configurations reveals one to be R, the other S. Both have the E configuration at the double bond.

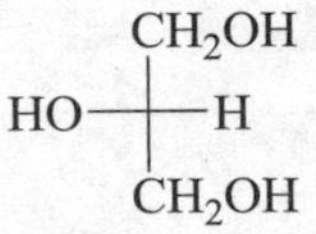

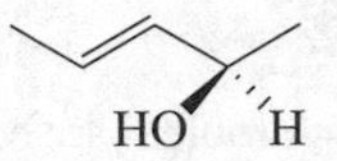

(2R,3E)-3-Penten-2-ol   (2S,3E)-3-Penten-2-ol

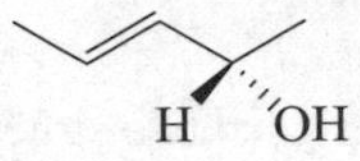

(d) These two compounds differ in the order in which their atoms are joined together; they are constitutional isomers.

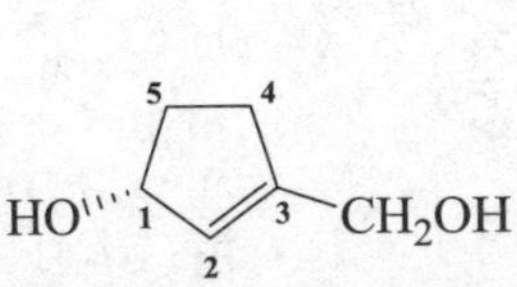

3-Hydroxymethyl-2-cyclopenten-1-ol 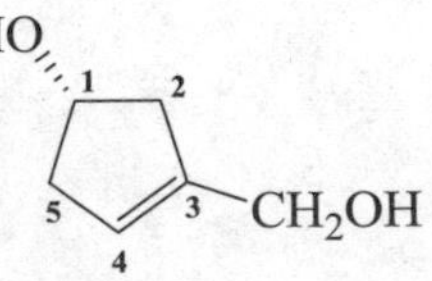   3-Hydroxymethyl-3-cyclopenten-1-ol

(e) Because cis-1,3-dimethylcyclopentane has a plane of symmetry, it is achiral and cannot have an enantiomer. The two structures given in the problem are identical.

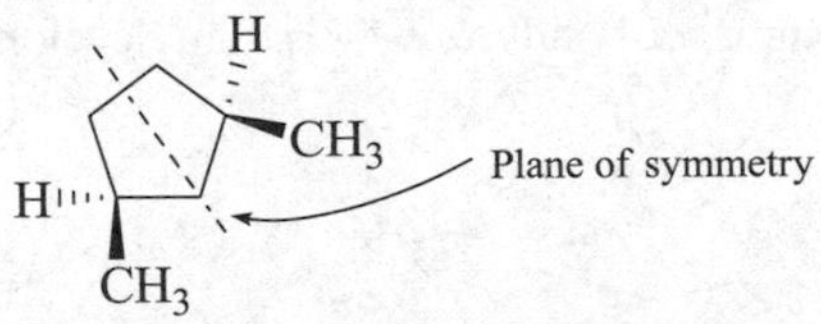

(f) To compare these compounds, reorient the first structure so that it may be drawn as a Fischer projection. The first step in the reorientation consists of a 180° rotation about an axis passing through the midpoint of the C-2–C-3 bond.

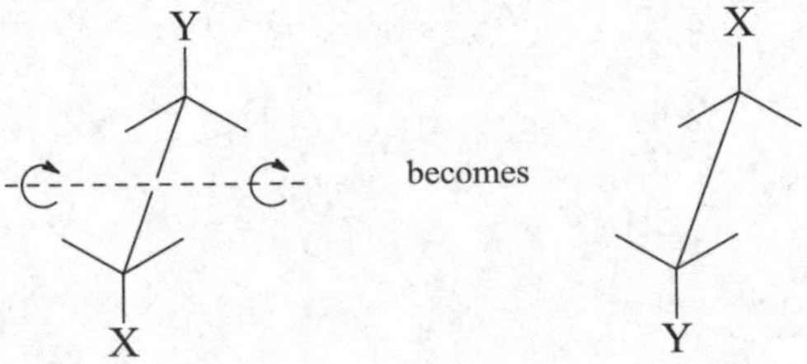

becomes

Thus

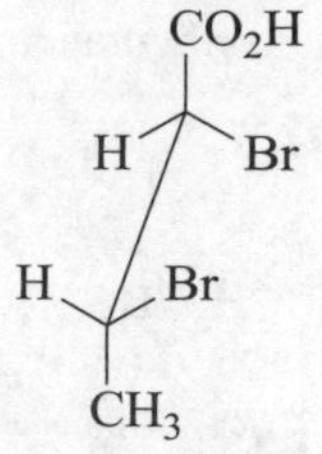

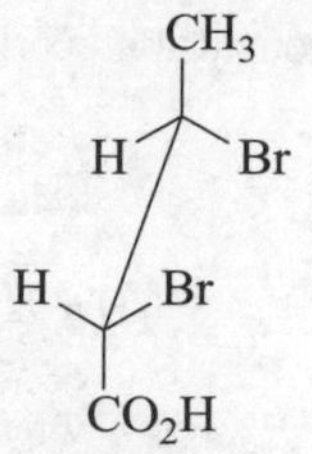

Reference structure

Now rotate the "back" carbon of the reoriented structure to give the necessary alignment for a Fischer projection.

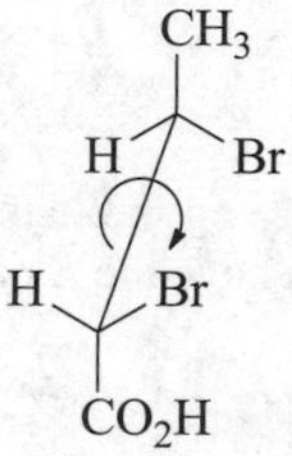

becomes 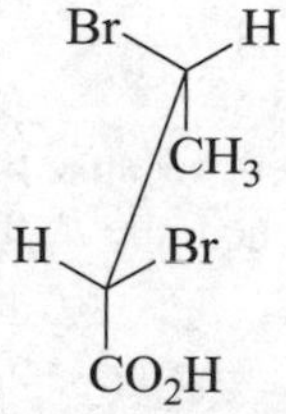   which is the same as

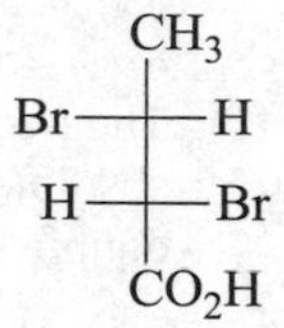

This reveals that the original two structures in the problem are equivalent.

(g) These two structures are nonsuperimposable mirror images of a molecule with two nonequivalent chirality centers; they are enantiomers.

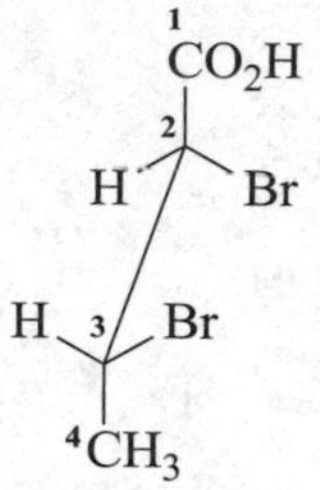

2R,3R   2S,3S

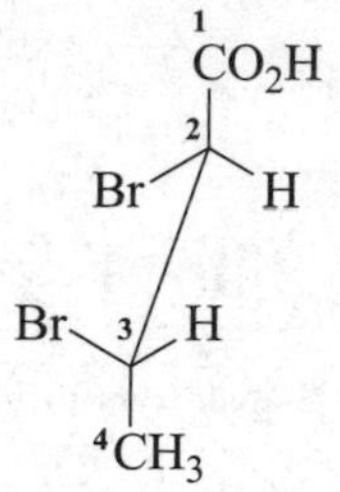

(h) These two structures, *cis*- and *trans*-4-*tert*-butylcyclohexyl iodide, are diastereomers.

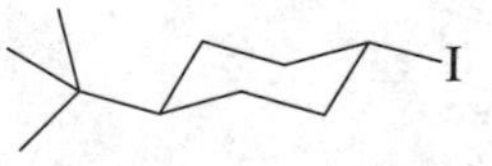

Trans 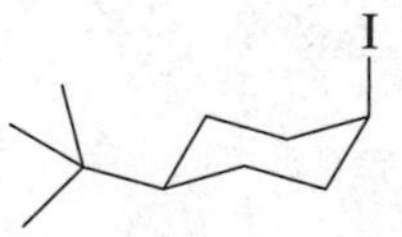   Cis

(i) The two structures are identical.

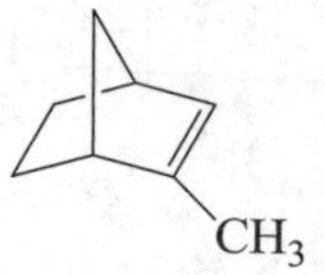

Reference structure 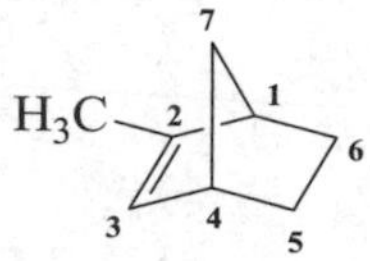   is equivalent to 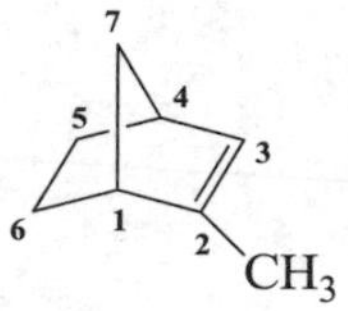   Identical to reference structure

(j) As represented, the two structures are mirror images of each other, but because the molecule is achiral (it has a plane of symmetry), the two must be superimposable. They represent the same compound.

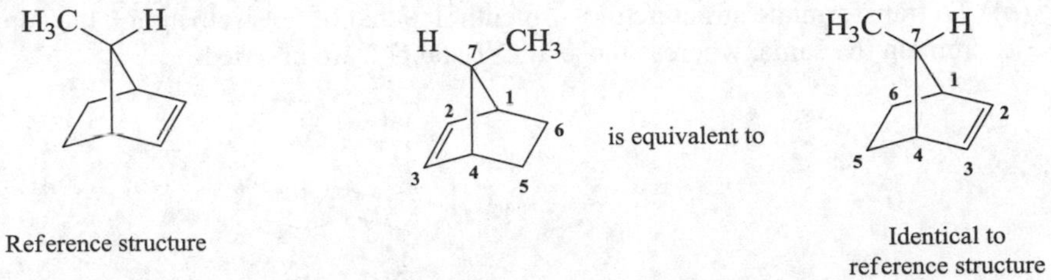

Reference structure   is equivalent to   Identical to
reference structure

The plane of symmetry passes through C-7 and bisects the C-2–C-3 bond and the C-5–C-6 bond.

4.39 (a) Muscarine has three chirality centers, and so eight (2^3) stereoisomers have this constitution.

(b) The three substituents on the ring (at C-2, C-3, and C-5) can be thought of as being either up (U) or down (D) in a perspective drawing. Thus, the eight possibilities are

UUU, UUD, UDU, DUU, UDD, DUD, DDU, DDD

Of these, six have one substituent trans to the other two.

(c) Muscarine is

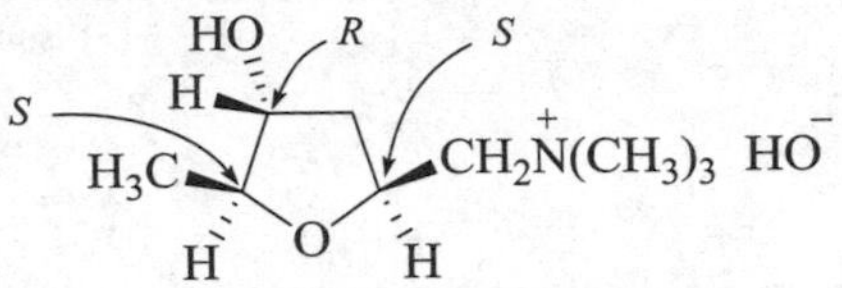

4.40 (a) The first step is to write the constitution of menthol, which we are told is 2-isopropyl-5-methylcyclohexanol.

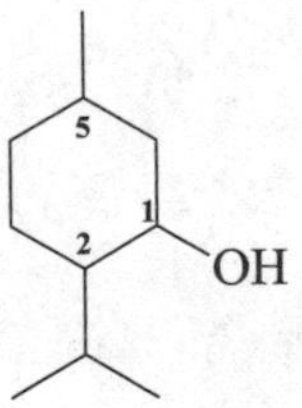

2-Isopropyl-5-methylcyclohexanol

Because the configuration at C-1 is R in (–)-menthol, the hydroxyl group must be "up" in our drawing.

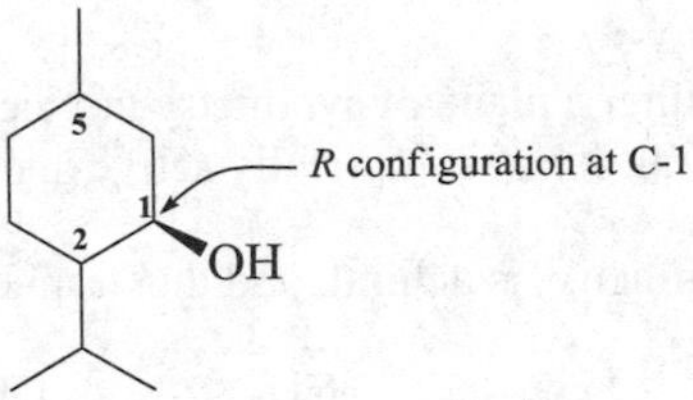

Because menthol is the most stable stereoisomer of this constitution, all three of its substituents must be equatorial. We therefore draw the chair form of the preceding structure, which has the hydroxyl group equatorial and up, placing isopropyl and methyl groups so as to preserve the R configuration at C-1.

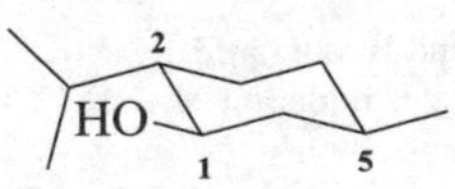

(-)-Menthol

(b) To transform the structure of (−)-menthol to that of (+)-isomenthol, the configuration at C-5 must remain the same, whereas those at C-1 and C-2 are inverted.

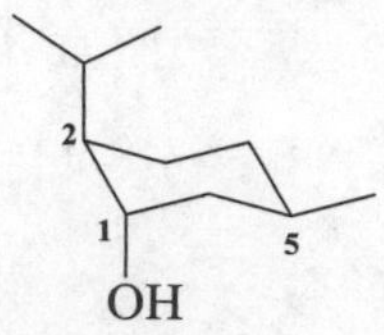

(+)-Isomenthol

(+)-Isomenthol is represented here in its correct configuration, but the conformation with two axial substituents is not the most stable one. The ring-flipped form is the preferred conformation of (+)-isomenthol:

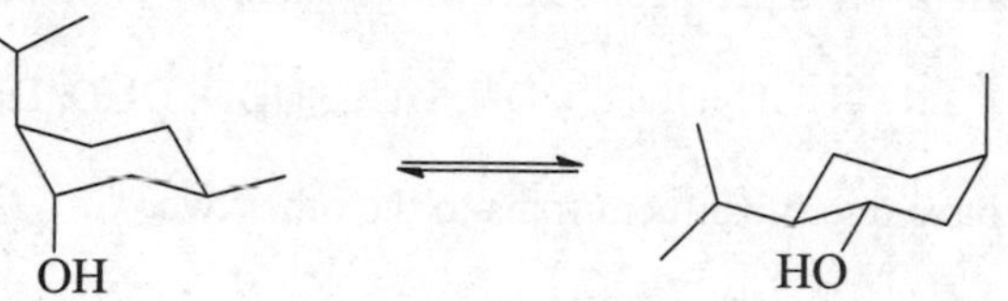

Most stable conformation of
(+)-isomenthol

Optical Activity

4.41 Because the only information available about the compound is its optical activity, examine the two structures for chirality, recalling that only chiral substances can be optically active.

The structure with the six-membered ring has a plane of symmetry passing through C-1 and C-4. It is achiral and cannot be optically active.

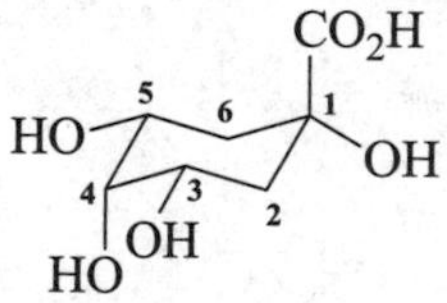

Achiral; $[\alpha]_D$ 0°

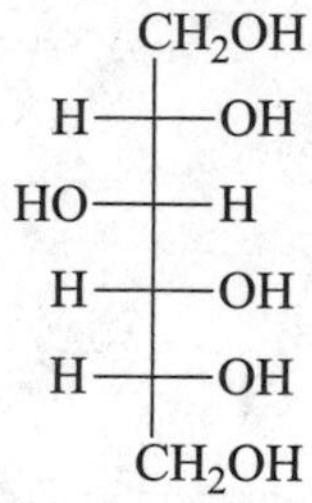

Chiral; can be optically active

The open-chain structure has neither a plane of symmetry nor a center of symmetry; it is not superimposable on its mirror image and so is chiral. It can be optically active and is more likely to be the correct choice.

4.42 Compound B has a center of symmetry, is achiral, and thus cannot be optically active.

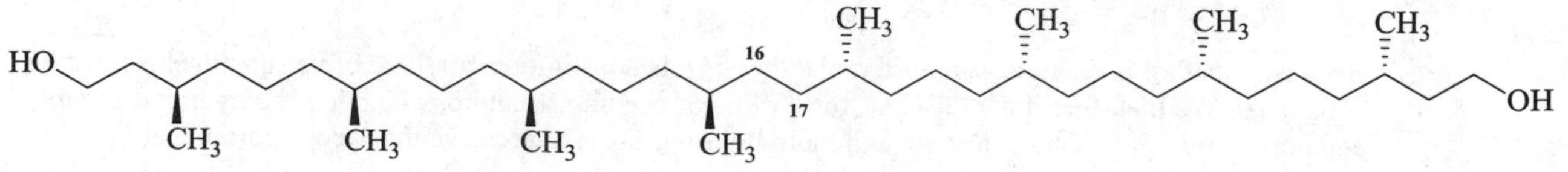

Compound B: not optically active
(center of symmetry is midpoint of C-16 - C-17 bond)

The diol in the problem is optically active, and so it must be chiral. Compound A is the naturally occurring diol.

4.43 (*a*) The equation that relates specific rotation $[\alpha]_D$ to observed rotation α is

$$[\alpha]_D = \frac{100\alpha}{cl}$$

where c is concentration in grams per 100 mL and l is path length in decimeters.

$$[\alpha]_D = \frac{100(-5.20°)}{(2.0\text{g}/100 \text{ mL})(2 \text{ dm})}$$

$$= -130°$$

(*b*) The optical purity of the resulting solution is 10/15, or 66.7%, because 10 g of optically pure fructose has been mixed with 5 g of racemic fructose. The specific rotation will therefore be two thirds (10/15) of the specific rotation of optically pure fructose:

$$[\alpha]_D = \tfrac{2}{3}(-130°) = -87°$$

Answers to Interpretive Problems 4

4.44 C; **4.45** B; **4.46** B; **4.47** B; **4.48** B; **4.49** B; **4.50** D

SELF-TEST

1. For each of the following pairs of drawings, identify the molecules as chiral or achiral and tell whether each pair represents molecules that are enantiomers, diastereomers, or identical.

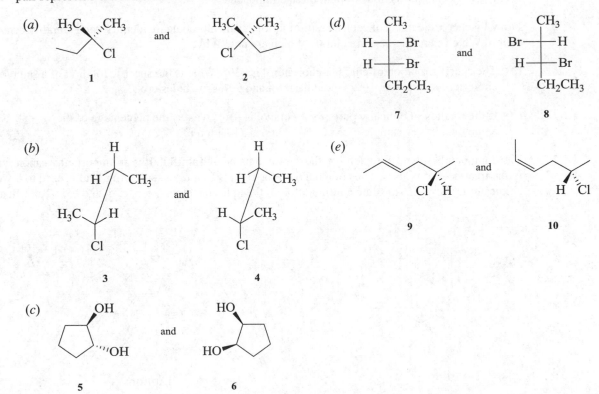

2. Specify the configuration of each chirality center in the preceding problem, using the Cahn–Ingold–Prelog *R–S* system.

3. Predict the number of stereoisomers possible for each of the following constitutions. For which of these will meso forms be possible?

(*a*)

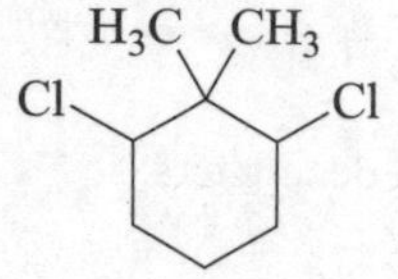

(*b*)

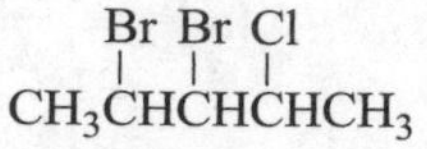

$$\underset{\text{CH}_3\text{CHCHCHCH}_3}{\overset{\text{Br Br Cl}}{| \quad | \quad |}}$$

(*c*)

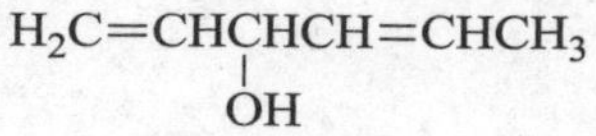

$$\text{H}_2\text{C}=\text{CHCHCH}=\text{CHCH}_3$$
$$\underset{\text{OH}}{|}$$

4. Using the skeletons provided as a guide,

(*a*) Draw a perspective view of (2*R*,3*R*)-3-chloro-2-butanol.

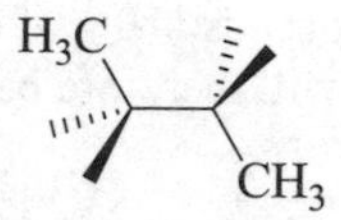

(*b*) Draw a sawhorse diagram of (*R*)-2-bromobutane.

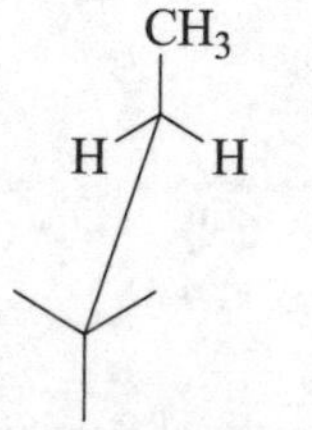

(*c*) Draw Fischer projections of both these compounds.

5. Draw Fischer projections of each stereoisomer of 2,3-dichlorobutane. Identify each chirality center as *R* or *S*. Which stereoisomers are chiral? Which are not? Why?

6. (*a*) The specific rotation of pure (–)-cholesterol is –39°. What is the specific rotation of a sample of cholesterol containing 10% (+)-cholesterol and 90% (–)-cholesterol.?

(*b*) If the rotation of optically pure (*R*)-2-octanol is –10°, what is the percentage of the *S* enantiomer in a sample of 2-octanol that has a rotation of –4°?

7. The compound shown below left has the common name of Tamiflu. It is an important weapon for physicians in the treatment of viruses. The compound on the right is Lipitor, a statin drug, and is used to lower blood chlolesterol. Give the absolute configuration of the indicated atoms, a, b, c, d, and e, in Tamiflu and Lipitor.

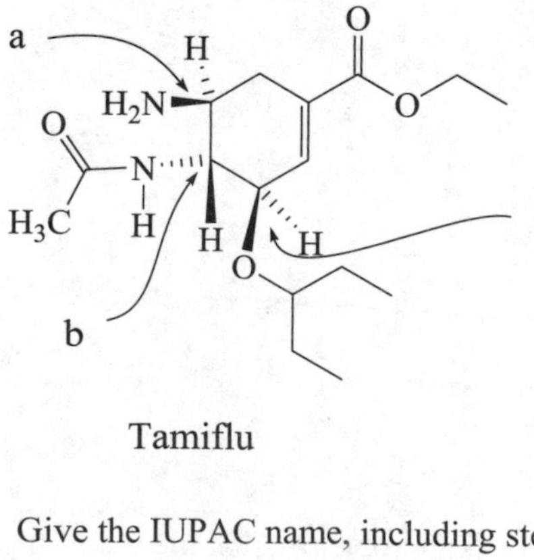

Tamiflu

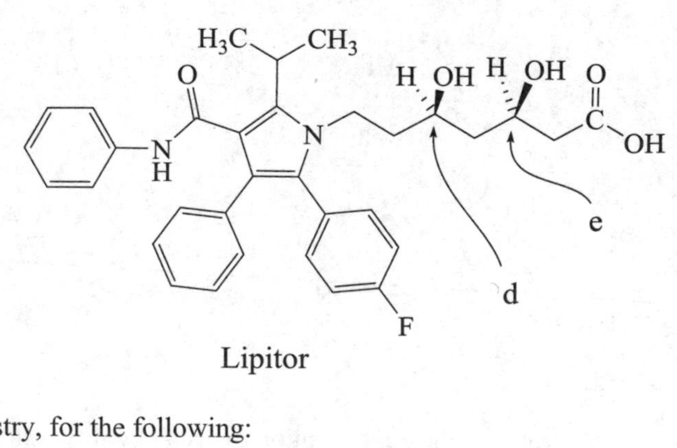

Lipitor

8. Give the IUPAC name, including stereochemistry, for the following:

(a)

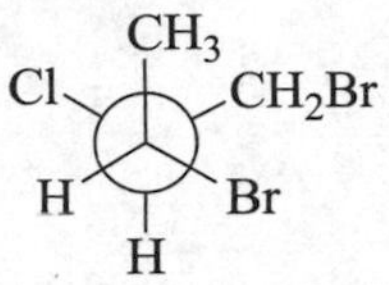

(b)

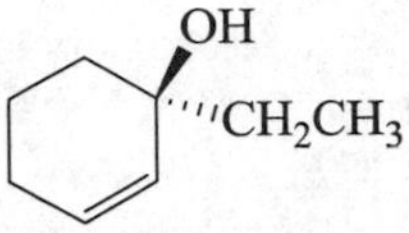

9.   The compound shown below is morphine, a highly potent, addictive analgesic drug that works directly on the central nervous system to relieve pain. Give the number of chirality centers present in morphine and circle those chirality centers that you have identified.

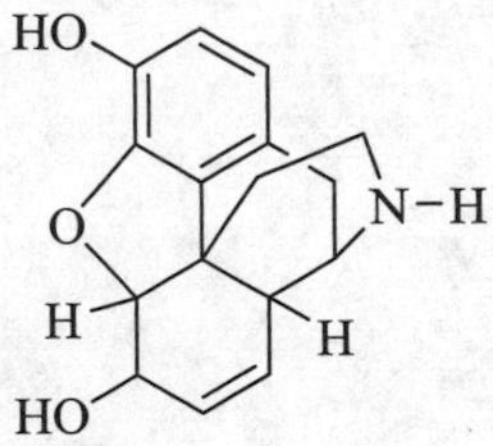

10.   How many stereoisomers are possible for 2-methyl-3-pentanol? What is their relationship (enantiomers, diastereomers)?

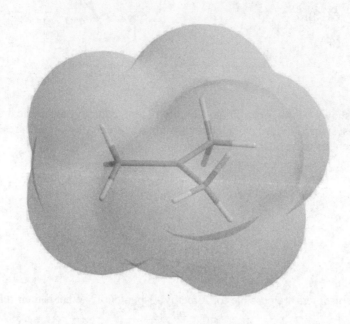

CHAPTER 5

Alcohols and Alkyl Halides: Introduction to Reaction Mechanisms

Table of Contents

SOLUTIONS TO TEXT PROBLEMS

In Chapter Problems

5.1 (b)   Thiols have the general formula RSH. The two thiols having the molecular formula C_3H_8S are

$$CH_3CH_2CH_2SH \quad \text{and} \quad \underset{\underset{SH}{|}}{CH_3CHCH_3}$$

5.2   There are four primary amines with the formula C_4H_{11}.

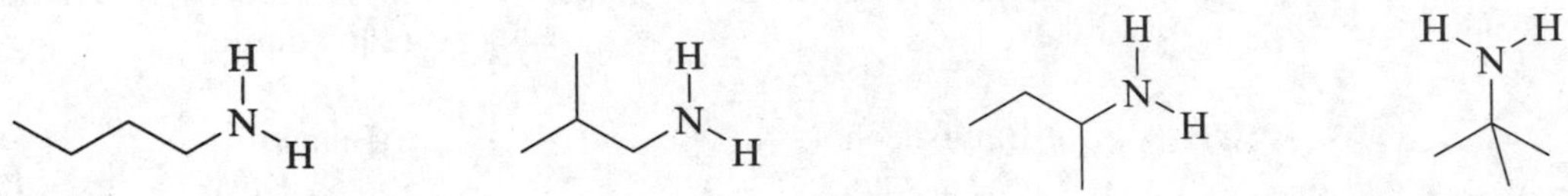

Three are secondary amines:

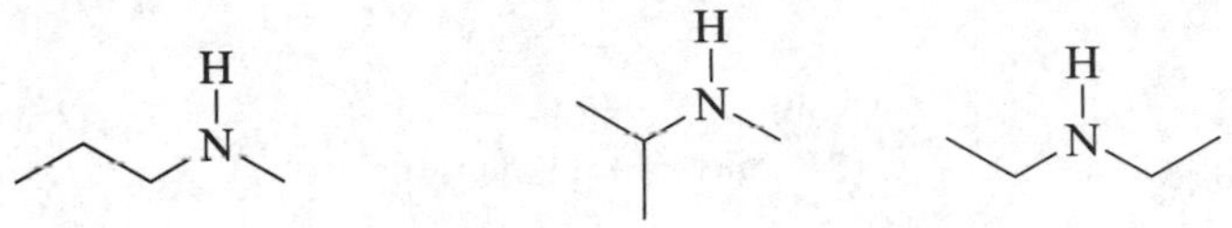

One amine is tertiary:

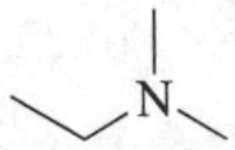

5.3   The functional groups in elenolic acid that contain a carbonyl group are an aldehyde, carboxylic acid, and an ester.

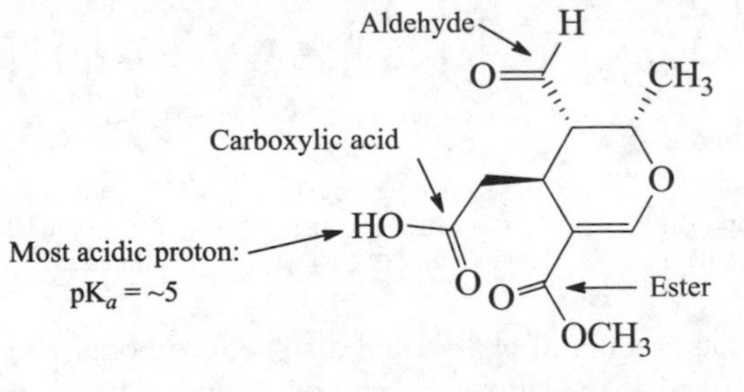

Elenolic acid

5.4   There are four C_4H_9 alkyl groups, and so there are four C_4H_9Cl alkyl chlorides. Each may be named by both the functional class and substitutive methods. The functional class name uses the name of the alkyl group followed by the halide as a second word. The common names for the C_4H_9 alkyl groups are acceptable in the IUPAC system, and functional class names using them are given first in the following list. The substitutive name modifies the name of the corresponding alkane to show the location of the halogen atom.

	Functional class IUPAC name	**Substitutive IUPAC name**
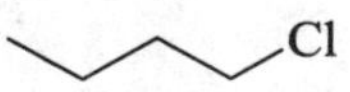	*n*-Butyl chloride (butyl chloride)	1-Chlorobutane

sec-Butyl chloride (1-methylpropyl chloride)	2-Chlorobutane
Isobutyl chloride (2-methylpropyl chloride)	1-Chloro-2-methylpropane
tert-Butyl chloride (1,1-dimethylethyl chloride)	2-Chloro-2-methylpropane

5.5 The isomeric $C_4H_{10}O$ alcohols may also be named using both the functional class and substitutive methods, as in the previous problem.

	Functional class IUPAC name	**Substitutive IUPAC name (2004 name)**
	n-Butyl alcohol (butyl alcohol)	1-Butanol (butan-1-ol)
	sec-Butyl alcohol (1-methylpropyl alcohol)	2-Butanol (butan-2-ol)
	Isobutyl alcohol (2-methylpropyl alcohol)	2-Methyl-1-propanol (2-methylpropan-1-ol)
	tert-Butyl alcohol (1,1-dimethylethyl alcohol)	2-Methyl-2-propanol (2-methylpropan-2-ol)

5.6 The isomeric $C_4H_{10}O$ alcohols are classified as primary, secondary, or tertiary according to the degree of substitution of the carbon that bears the hydroxyl group.

$$CH_3CH_2CH_2-\overset{\overset{\displaystyle H}{|}}{\underset{\underset{\displaystyle H}{|}}{C}}-OH \qquad H_3C-\overset{\overset{\displaystyle H}{|}}{\underset{\underset{\displaystyle OH}{|}}{C}}-CH_2CH_3 \qquad (CH_3)_2CH-\overset{\overset{\displaystyle H}{|}}{\underset{\underset{\displaystyle H}{|}}{C}}-OH \qquad H_3C-\overset{\overset{\displaystyle CH_3}{|}}{\underset{\underset{\displaystyle CH_3}{|}}{C}}-OH$$

Primary alcohol | Secondary alcohol | Primary alcohol | Tertiary alcohol
(one alkyl group | (two alkyl groups | (one alkyl group | (three alkyl groups
bonded to —CH₂OH) | bonded to >CHOH) | bonded to —CH₂OH) | bonded to ≥COH)

5.7 Dipole moment is the product of charge and distance. Although the electron distribution in the carbon–chlorine bond is more polarized than that in the carbon–bromine bond, this effect is counterbalanced by the longer carbon–bromine bond distance.

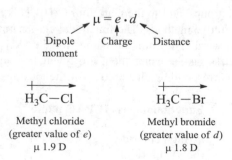

$$\mu = e \cdot d$$

Dipole moment   Charge   Distance

H_3C-Cl H_3C-Br

Methyl chloride   Methyl bromide
(greater value of *e*)   (greater value of *d*)
μ 1.9 D   μ 1.8 D

5.8   There are two alcohols and one ether compound that have the formula C_3H_8O. Ethyl methyl ether is a gas at room temperature because it cannot form intermolecular hydrogen bonds.

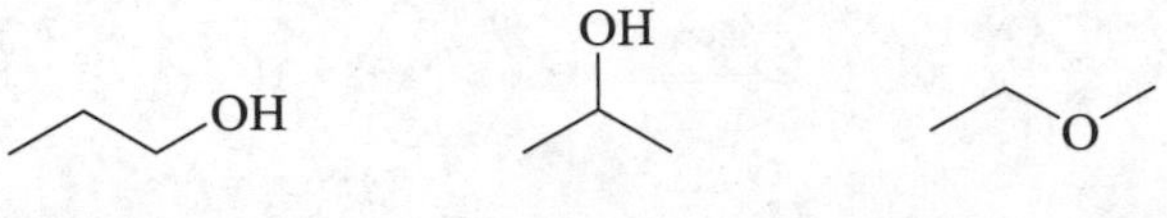

Propan-1-ol	Propan-2-ol	Methoxyethane
1-Propanol	2-Propanol	Ethyl methyl ether
	Isopropanol	

5.9   (*b*)   Hydrogen chloride converts tertiary alcohols to tertiary alkyl chlorides.

3-Ethyl-3-pentanol	Hydrogen chloride	3-Chloro-3-ethylpentane	Water

(*c*)   1-Tetradecanol is a primary alcohol having an unbranched 14-carbon chain. Hydrogen bromide reacts with primary alcohols to give the corresponding primary alkyl bromide.

1-Tetradecanol	Hydrogen bromide	1-Bromotetradecane	Water

5.10   The overall reaction for the reaction of aqueous hydrogen chloride with *tert*-butyl chloride is

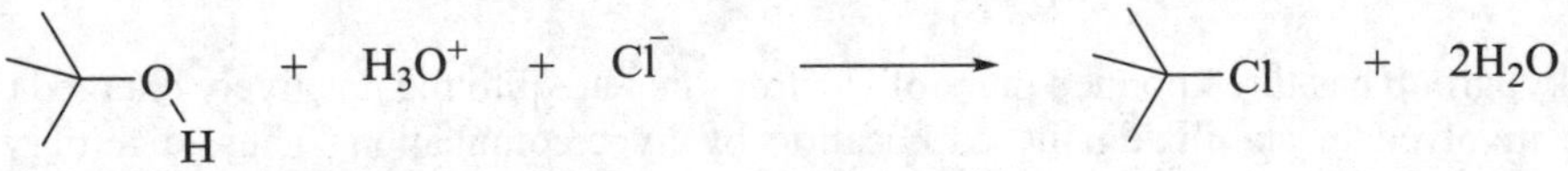

has the mechanistic steps

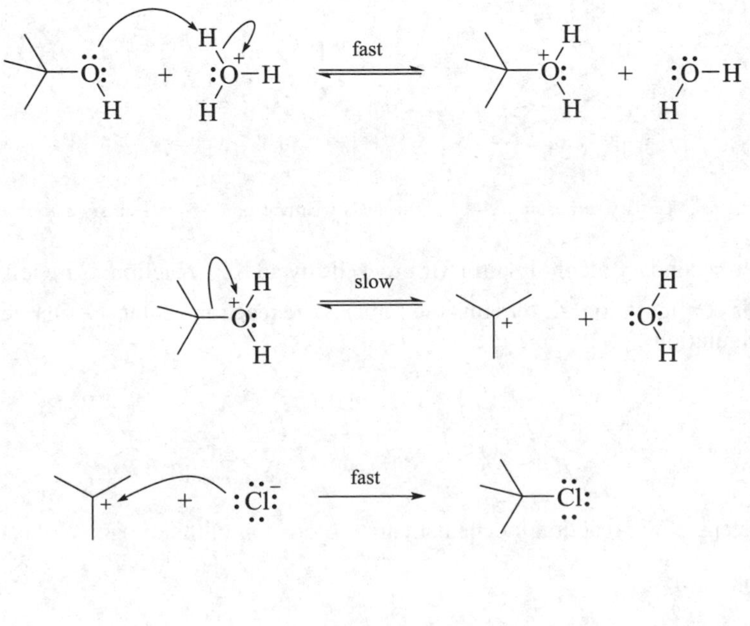

5.11 The rate determining, or slow, step is the formation of a carbocation by dissociation of the alkyloxonium ion.

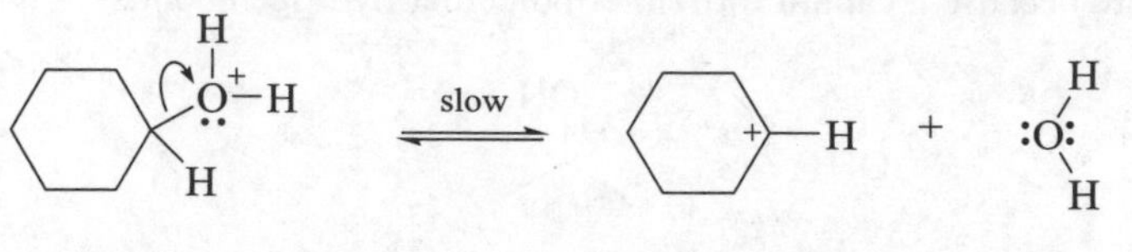

           Cyclohexyloxonium ion   Cyclohexyl cation   Water

5.12 There are four different $C_6H_{17}^+$ carbocations that have the given skeleton.

Tertiary carbocations are more stable than secondary, and secondary more stable than primary. A tertiary carbon is connected to three other carbons. Therefore, the most stable carbocation with this structure is:

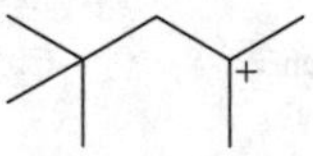

5.13 Carbocations are stabilized by substituents that release electrons to the positively charged carbon. The presence of three strongly electronegative fluorines makes the CF_3 group strongly electron attracting, not electron releasing.

Thus, the carbocation $(CF_3)_3C^+$ is destabilized. $(CH_3)_3C^+$ is a more stable carbocation.

5.14 Each alkyl group has three bonded pairs of electrons that are β to the positively charged carbon of a carbocation and are involved in stabilizing the carbocation by hyperconjugation. Thus, a tertiary carbocation has nine bonded pairs of electrons β to the charged carbon, a secondary carbocation has six, and a primary carbocation has three.

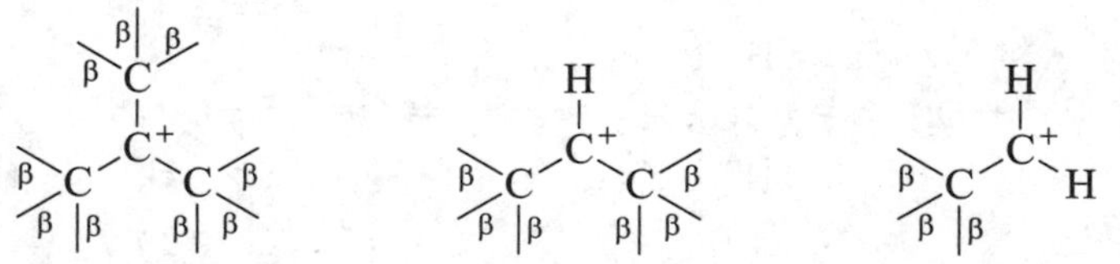

      Tertiary carbocation   Secondary carbocation   Primary carbocation

5.15 Reaction of the secondary alcohol with HBr proceeds by an S_N1 reaction. The left chirality center is not affected so this configuration, *R*, remains the same. The reaction center, C–OH, in the starting material also has an *R* configuration.

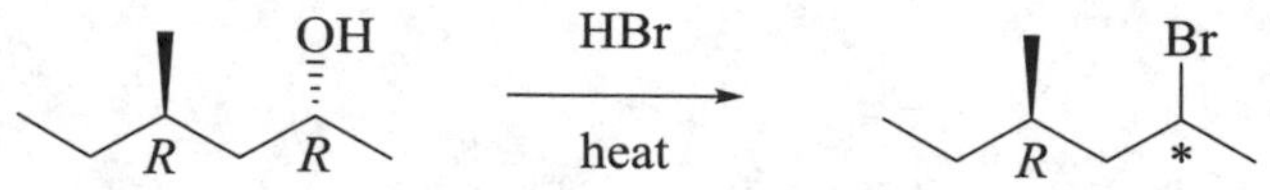

The first two steps of the reaction mechanism are protonation of the oxygen and carbocation formation.

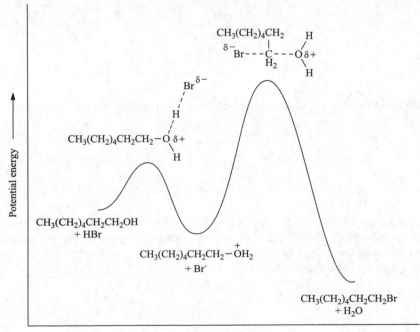

The carbocation formed reacts with bromide. More inversion product is predicted because the leaving group, in this case water, partially blocks one face of the carbocation. The configuration of each chirality center is shown.

5.16 Primary alcohols react with hydrogen halides in two steps, not three. The first step is a rapid protonation of the alcohol to form an intermediate alkyloxonium ion. In the second step, the halide ion, bromide ion in this case, pushes off a water molecule to give the alkyl bromide product.

5.17 For the case of 1-butanol as shown in the equation, chloride ion bonds to carbon, the C—O bond breaks, and the pair of electrons in that bond become part of a double bond to sulfur, displacing a neutral pyridine molecule.

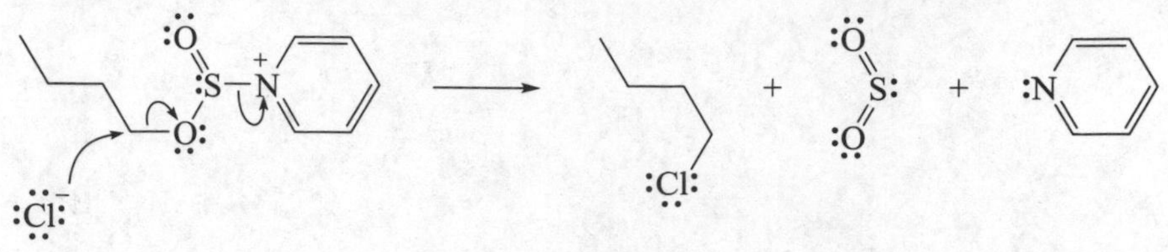

5.18 Methanesulfonates are prepared from their respective alcohols using methanesulfonyl chloride as a reactant. The configuration of the carbon bearing the oxygen does not change since the C-O bond is not broken.

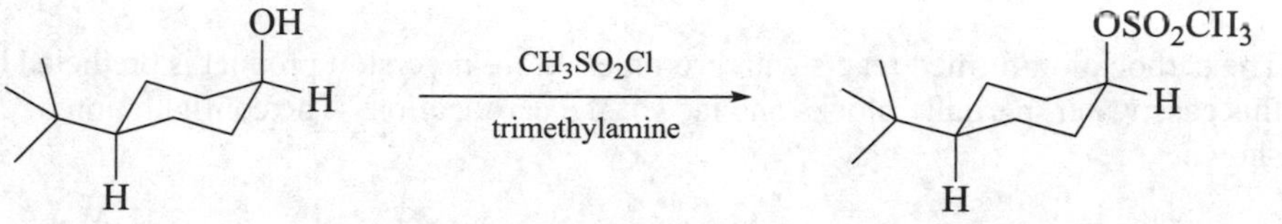

cis-4-*tert*-butylcyclohexanol

cis-4-*tert*-butylcyclohexyl methanesulfonate

End of Chapter Problems

Structure and Nomenclature

5.19 (*a*)   Cyclobutanol has a hydroxyl group attached to a four-membered ring.

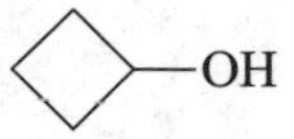

Cyclobutanol

(*b*)   *sec*-Butyl alcohol is the functional class name for 2-butanol.

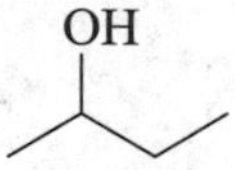

sec-Butyl alcohol

(*c*)   The hydroxyl group is at C-3 of an unbranched seven-carbon chain in 3-heptanol.

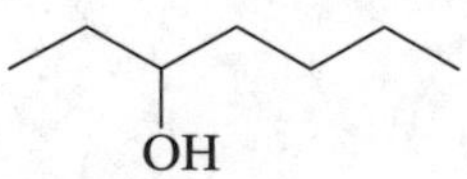

3-Heptanol

(*d*)   A chlorine at C-2 is on the opposite side of the ring from the C-1 hydroxyl group in *trans*-2-chlorocyclopentanol. Note that it is not necessary to assign a number to the carbon that bears the hydroxyl group; naming the compound as a derivative of cyclopentanol automatically requires the hydroxyl group to be located at C-1.

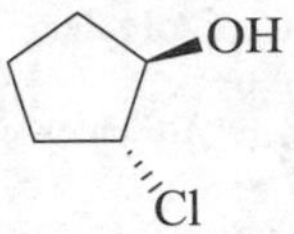

trans-2-Chlorocyclopentanol

(e)   This compound is an alcohol in which the longest continuous chain that incorporates the hydroxyl function has eight carbons. It bears chlorine substituents at C-2 and C-6 and methyl and hydroxyl groups at C-4.

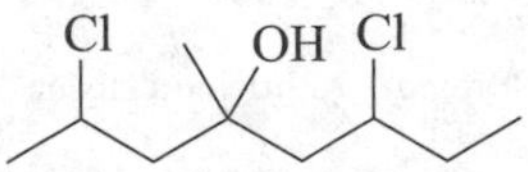

2,6-Dichloro-4-methyl-4-octanol

(f)   The hydroxyl group is at C-1 in *trans*-4-*tert*-butylcyclohexanol; the *tert*-butyl group is at C-4. The structures of the compound can be represented as shown at the left; the structure at the right depicts it in its most stable conformation.

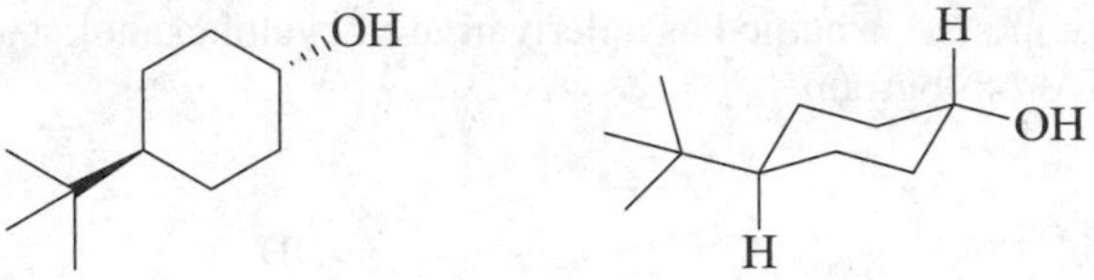

trans-4-*tert*-Butylcyclohexanol

(g and h)   The cyclopropyl group is on the same carbon as the hydroxyl group in 1-cyclopropylethanol (on the left) on adjacent carbons in 2-cyclopropylethanol (on the right).

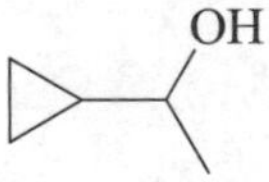

1-Cyclopropylethanol 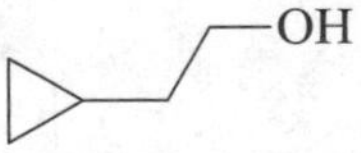   2-Cyclopropylethanol

5.20   (a)   This compound has a five-carbon chain that bears a methyl substituent and a bromine. The numbering scheme that gives the lower number to the substituent closest to the end of the chain is chosen. Bromine is therefore at C-1, and methyl is a substituent at C-4.

$$CH_3CHCH_2CH_2CH_2Br$$
$$\overset{|}{C}H_3$$

1-Bromo-4-methylpentane

(b)   This compound has the same carbon skeleton as the compound in part (a) but bears a hydroxyl group in place of the bromine and so is named as a derivative of 1-pentanol.

$$CH_3CHCH_2CH_2CH_2OH$$
$$\overset{|}{C}H_3$$

4-Methyl-1-pentanol

(c)   This molecule is a derivative of ethane and bears three chlorines and one bromine. The name 2-bromo-1,1,1-trichloroethane gives a lower number at the first point of difference than 1-bromo-2,2,2-trichloroethane.

$$Cl_3CCH_2Br$$

2-Bromo-1,1,1-trichloroethane

(*d*)   This compound is a constitutional isomer of the preceding one. Regardless of which carbon the numbering begins at, the substitution pattern is 1,1,2,2. Alphabetical ranking of the halogens therefore dictates the direction of numbering. Begin with the carbon that bears bromine.

$$\underset{\underset{\displaystyle Cl}{\displaystyle |}}{Cl_2CHCHBr}$$

1-Bromo-1,2,2-trichloroethane

(*e*)   This is a trifluoro derivative of ethanol. The direction of numbering is dictated by the hydroxyl group, which is at C-1 in ethanol.

$$CF_3CH_2OH$$

2,2,2-Trifluoroethanol

(*f*)   Here the compound is named as a derivative of cyclohexanol, and so numbering begins at the carbon that bears the hydroxyl group.

cis-3-*tert*-Butylcyclohexanol

(*g*)   This alcohol has its hydroxyl group attached to C-2 of a three-carbon continuous chain; it is named as a derivative of 2-propanol.

2-Cyclopentyl-2-propanol

(*h*)   The six carbons that form the longest continuous chain have substituents at C-2, C-3, and C-5 when numbering proceeds in the direction that gives the lowest numbers to substituents at the first point of difference. The substituents are cited in alphabetical order.

5-Bromo-2,3-dimethylhexane

Had numbering begun in the opposite direction, the numbers would be 2,4,5 rather than 2,3,5.

(*i*)   Hydroxyl controls the numbering because the compound is named as an alcohol.

4,5-Dimethyl-2-hexanol

5.21 (a)

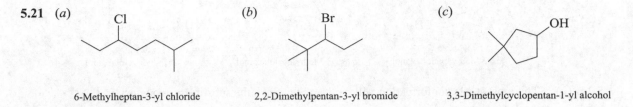

6-Methylheptan-3-yl chloride   2,2-Dimethylpentan-3-yl bromide   3,3-Dimethylcyclopentan-1-yl alcohol

5.22 Primary alcohols are alcohols in which the hydroxyl group is attached to a carbon atom that has one alkyl substituent and two hydrogens. Four primary alcohols have the molecular formula $C_5H_{12}O$. The functional class name for each compound is given in parentheses.

1-Pentanol
(pentyl alcohol)

2-Methyl-1-butanol
(2-methylbutyl alcohol)

3-Methyl-1-butanol
(3-methylbutyl alcohol)

2,2-Dimethyl-1-propanol
(2,2-dimethylpropyl alcohol)

Secondary alcohols are alcohols in which the hydroxyl group is attached to a carbon atom that has two alkyl substituents and one hydrogen. There are three secondary alcohols of molecular formula $C_5H_{12}O$:

2-Pentanol
(1-methylbutyl alcohol)

3-Pentanol
(1-ethylpropyl alcohol)

3-Methyl-2-butanol
(1,2-dimethylpropyl alcohol)

Only 2-methyl-2-butanol is a tertiary alcohol (three alkyl substituents on the hydroxyl-bearing carbon):

2-Methyl-2-butanol
(1,1-dimethylpropyl alcohol)

5.23 The first methylcyclohexanol to be considered is 1-methylcyclohexanol. The preferred chair conformation will have the larger methyl group in an equatorial orientation, whereas the smaller hydroxyl group will be axial.

Most stable conformation of 1-methylcyclohexanol

In the other isomers, methyl and hydroxyl will be in a 1,2, 1,3, or 1,4 relationship and can be cis or trans in each. We can write the preferred conformation by recognizing that the methyl group will always be equatorial and the hydroxyl either equatorial or axial.

trans-2-Methylcyclohexanol   *cis*-3-Methylcyclohexanol   *trans*-4-Methylcyclohexanol

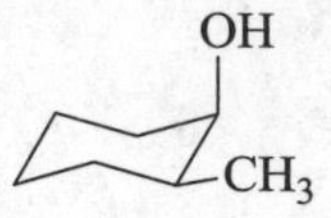

cis-2-Methylcyclohexanol

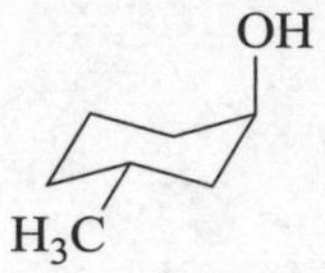

trans-3-Methylcyclohexanol

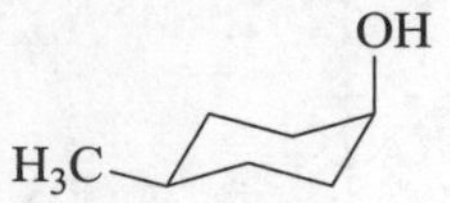

cis-4-Methylcyclohexanol

5.24 (*a*) The most stable conformation will be the one with all the substituents equatorial.

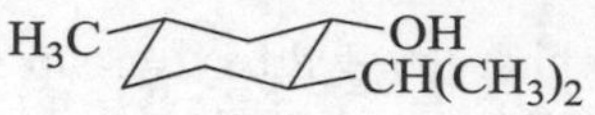

The hydroxyl group is trans to the isopropyl group and cis to the methyl group.

(*b*) All three substituents need not always be equatorial; instead, one or two of them may be axial. Because neomenthol is the second most stable stereoisomer, we choose the structure with *one* axial substituent. Furthermore, we choose the structure with the smallest substituent (the hydroxyl group) as the axial one. Neomenthol is shown as follows:

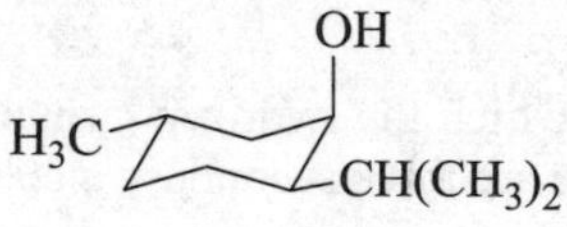

Functional Groups

5.25 The oxygen and two of the carbons of C_3H_5ClO are part of the structural unit that characterizes epoxides. The problem specifies that a methyl group (CH_3) is *not* present; therefore, add the remaining carbon and the chlorine as a —CH_2Cl unit, and fill in the remaining bonds with hydrogens.

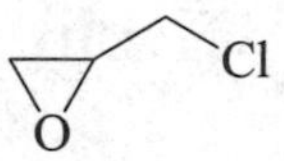

Epichlorohydrin

5.26 (*a*)

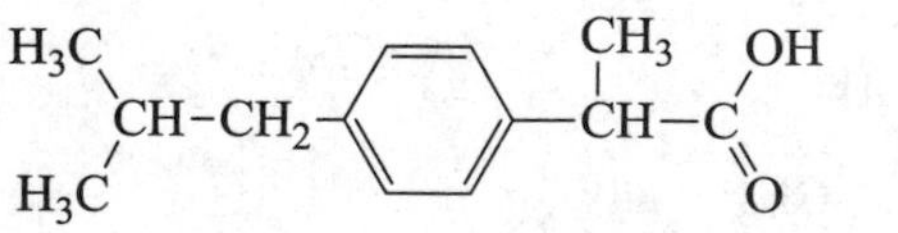

Ibuprofen

(*b*)

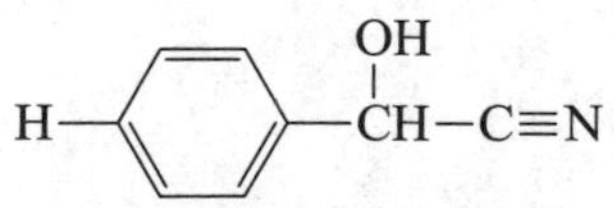

Mandelonitrile

5.27 Isoamyl acetate is

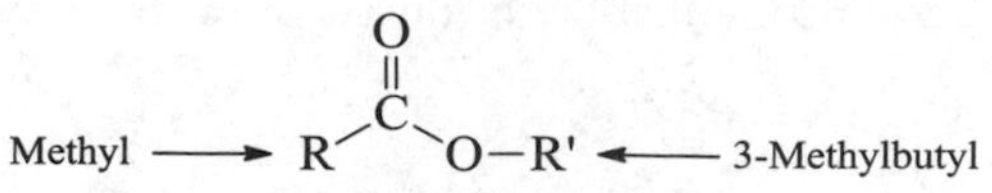

an ester functional group

which is:

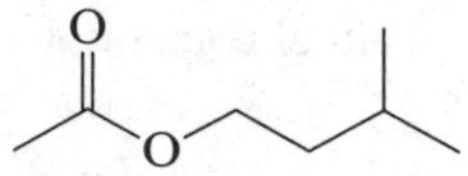

5.28 Thiols are characterized by the —SH group. *n*-Butyl mercaptan is $CH_3CH_2CH_2CH_2SH$.

5.29 α-Amino acids have the general formula

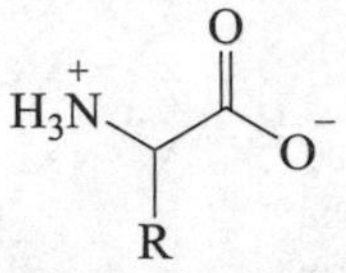

$$H_3\overset{+}{N}\underset{R}{\overset{\displaystyle O}{|}}O^-$$

The individual amino acids in the problem have the structures shown:

(a)

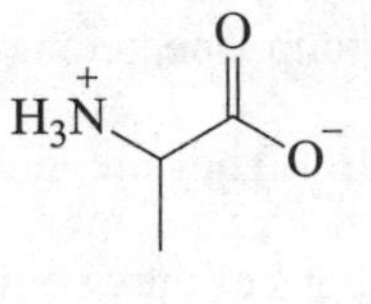

Alanine

(b)

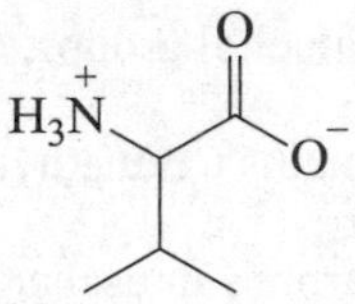

Valine

(c, d) An isobutyl group is $(CH_3)_2CHCH_2$—, and a sec-butyl group is $CH_3CH_2CHCH_3$

The structures of leucine and isoleucine are

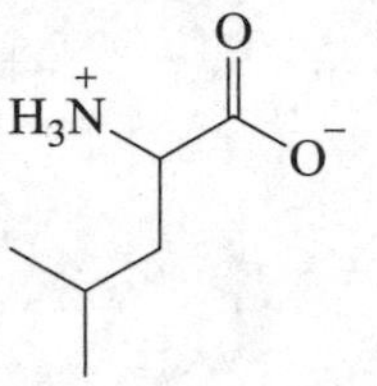

Leucine

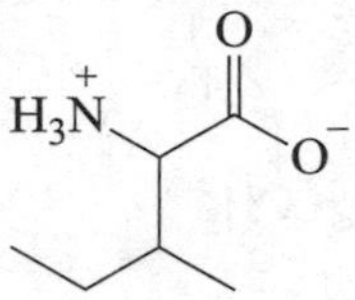

Isoleucine

(e–g) The functional groups that characterize alcohols, thiols, and carboxylic acids are —OH, —SH, and —CO₂H, respectively. The structures of serine, cysteine, and aspartic acid are

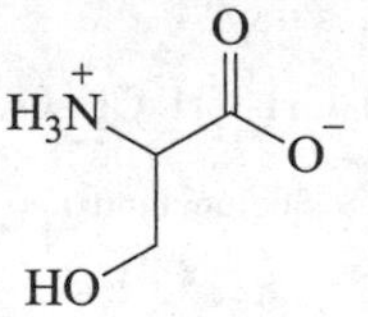

Serine

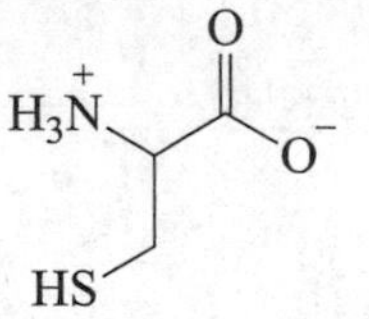

Cysteine

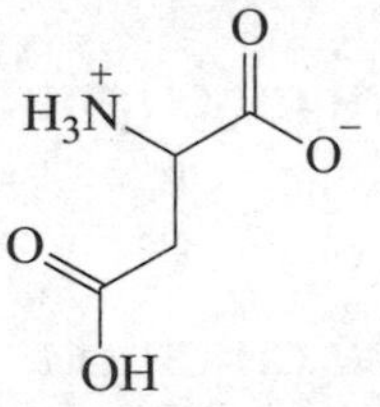

Aspartic acid

5.30   The functional groups to which the four oxygens belong are classified as

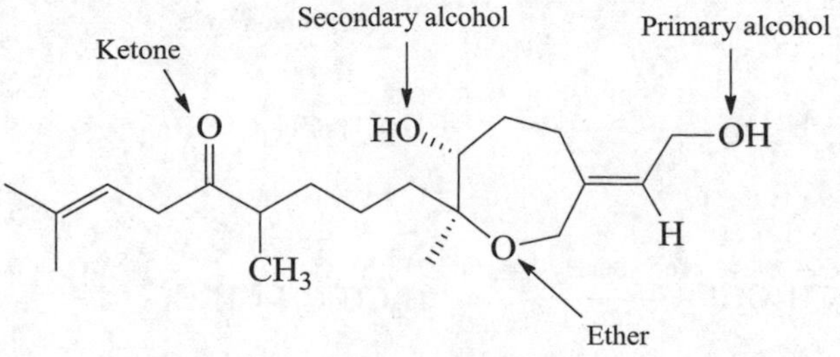

Ketone

Secondary alcohol

Primary alcohol

Ether

CH_3

5.31   One nitrogen belongs to a tertiary amine function; the other is an amide nitrogen.

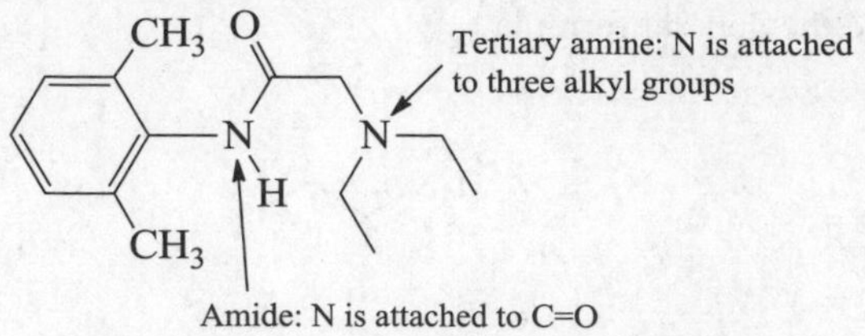

Tertiary amine: N is attached
to three alkyl groups

Amide: N is attached to C=O

5.32 *Uscharidin* has the structure shown.

(a) There are two alcohol groups, one aldehyde group, one ketone group, and one ester functionality.

(b) Uscharidin contains ten methylene groups (CH_2). They are indicated in the structure by *.

(c) The primary carbons in uscharidin are the carbons of the two methyl groups.

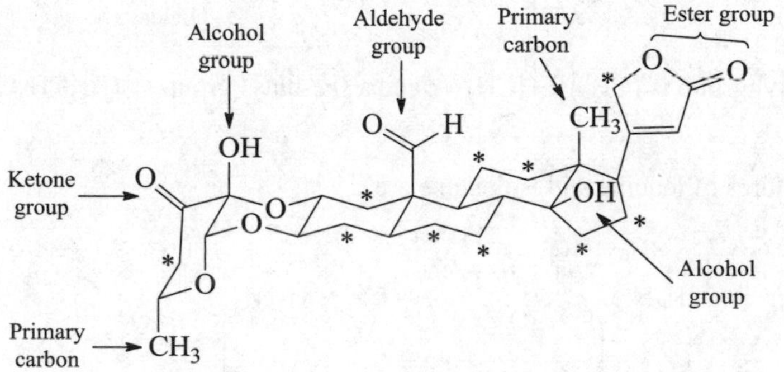

Alcohol group

Aldehyde group

Primary carbon

Ester group

Ketone group

Alcohol group

Primary carbon

Reactions and Mechanisms

5.33 This problem illustrates the reactions of a primary alcohol with the reagents described in the chapter.

(a) $CH_3CH_2CH_2CH_2OH$ + $NaNH_2$ $\longrightarrow$ $CH_3CH_2CH_2CH_2O^- Na^+$ + NH_3

Sodium butoxide

(b) $CH_3CH_2CH_2CH_2OH$ $\xrightarrow[\text{heat}]{\text{HBr}}$ $CH_3CH_2CH_2CH_2Br$

1-Bromobutane

(c) $CH_3CH_2CH_2CH_2OH$ $\xrightarrow[\text{heat}]{\text{NaBr, }H_2SO_4}$ $CH_3CH_2CH_2CH_2Br$

1-Bromobutane

(d) $CH_3CH_2CH_2CH_2OH$ $\xrightarrow{\text{PBr}_3}$ $CH_3CH_2CH_2CH_2Br$

1-Bromobutane

(e) $CH_3CH_2CH_2CH_2OH$ $\xrightarrow{\text{SOCl}_2}$ $CH_3CH_2CH_2CH_2Cl$

1-Chlorobutane

(*f*) $CH_3CH_2CH_2CH_2OH$ $\xrightarrow[\text{pyridine}]{CH_3SO_2Cl}$ $CH_3CH_2CH_2CH_2OSO_2CH_3$

Butyl methanesulfonate

5.34 (*a*) This reaction was used to convert the primary alcohol to the corresponding bromide in 60% yield.

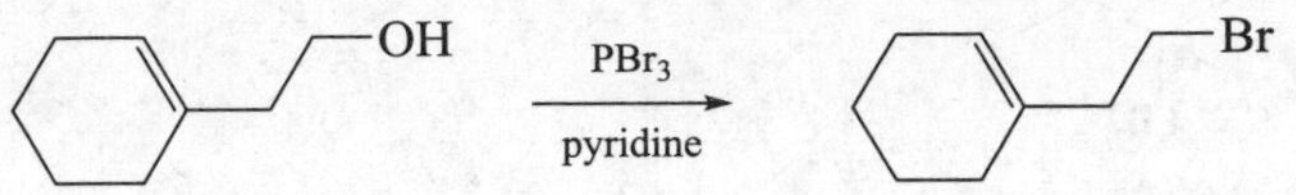

(*b*) Thionyl chloride treatment of this secondary alcohol gave the chloro derivative in 59% yield.

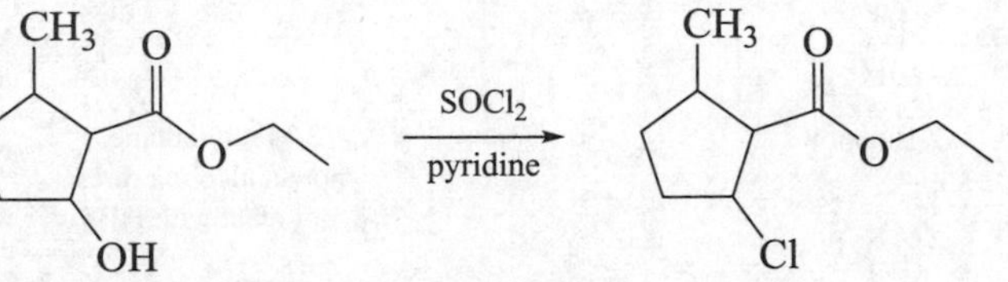

(*c*) The starting material is a tertiary alcohol and reacted readily with hydrogen chloride to form the corresponding chloride in 67% yield.

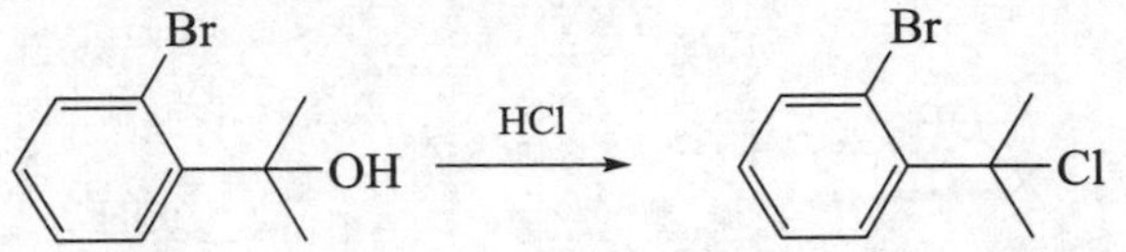

(*d*) Both primary alcohol functional groups were converted to primary bromides; the yield was 88%.

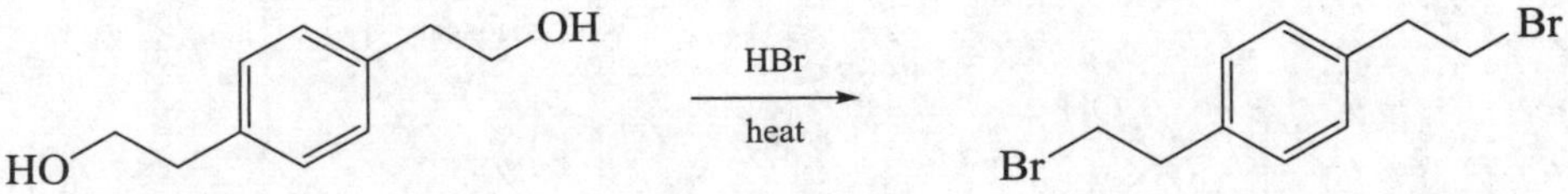

(*e*) *p*-Toluenesulfonyl chloride (TsCl) reacts with alcohols to give *p*-toluenesulfonates.

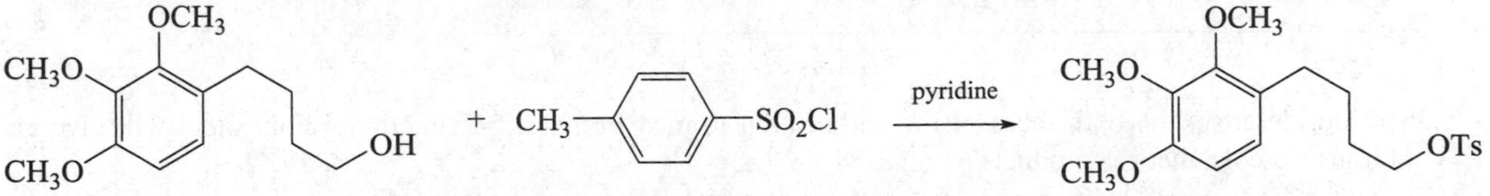

p-Toluenesulfonyl
chloride
(TsCl)

5.35 The order of reactivity of alcohols with hydrogen halides is tertiary > secondary > primary.

$$ROH + HBr \longrightarrow RBr + H_2O$$

Reactivity of alcohols with hydrogen bromide:

Part	More reactive	Less reactive
(*a*)	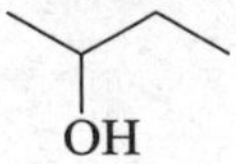 2-Butanol: secondary	 1-Butanol: primary

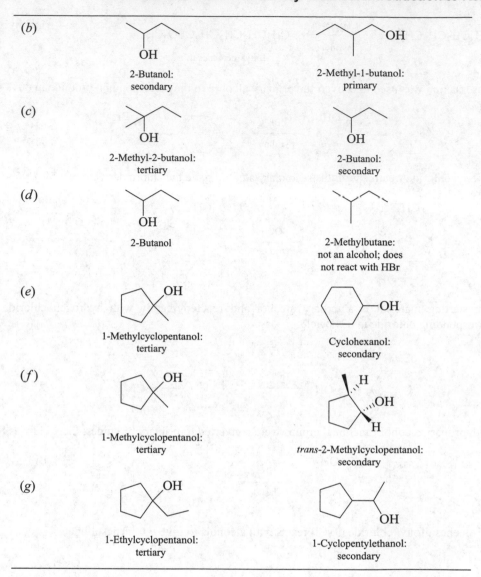

(b)

2-Butanol:
secondary

2-Methyl-1-butanol:
primary

(c)

2-Methyl-2-butanol:
tertiary

2-Butanol:
secondary

(d)

2-Butanol

2-Methylbutane:
not an alcohol; does
not react with HBr

(e)

1-Methylcyclopentanol:
tertiary

Cyclohexanol:
secondary

(f)

1-Methylcyclopentanol:
tertiary

trans-2-Methylcyclopentanol:
secondary

(g)

1-Ethylcyclopentanol:
tertiary

1-Cyclopentylethanol:
secondary

5.36 Chloride attack on both faces of the carbocation formed from 1,3,5-Trimethylcyclohexanol will give cis and trans stereoisomers as products.

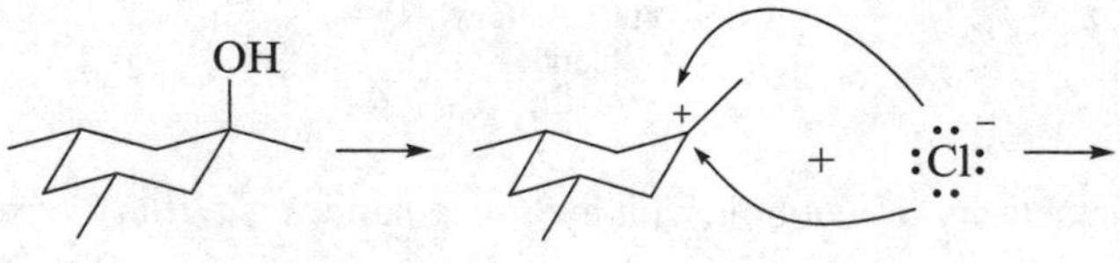

1,3,5-Trimethylcyclohexanol

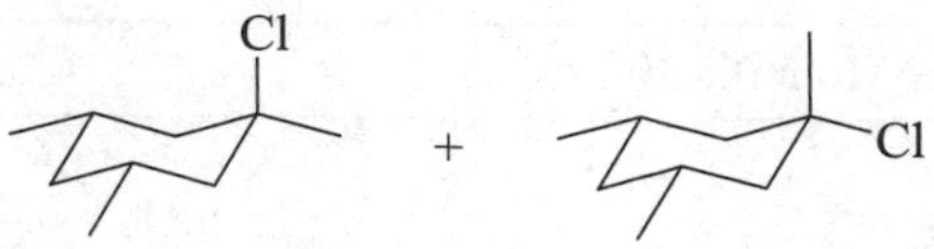

Chloride reacting with the top and bottom faces of the carbocation formed from 1,4,4-trimethylcyclohexanol gives only one isomer, 1-chloro-1,4,4-trimethylcyclohexane.

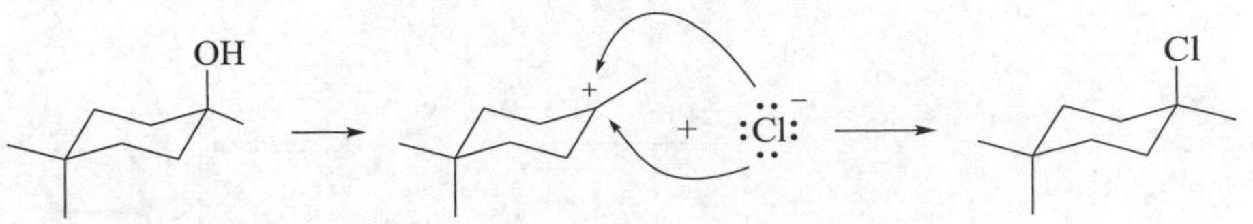

1,4,4-Trimethyl-
cyclohexanol

5.37 Hydrogen chloride undergoes S_N2 reactions with primary alcohols slower than S_N1 reactions with tertiary and secondary alcohols so the primarly alcohol, cyclohexylmethanol, will react slowest.

Cyclohexylmethanol

The tertiary alcohol, 1-methylcyclohexanol, will form its carbocation faster than a secondary alcohols in the S_N1 reaction so this compound will react the fastest.

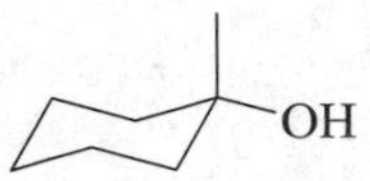

1-Methylcyclohexanol

The remaining two secondary alcohols are enantiomers. Ring flipping one of the cyclohexanes and reflection gives the other. These enantiomers have identical reactivity.

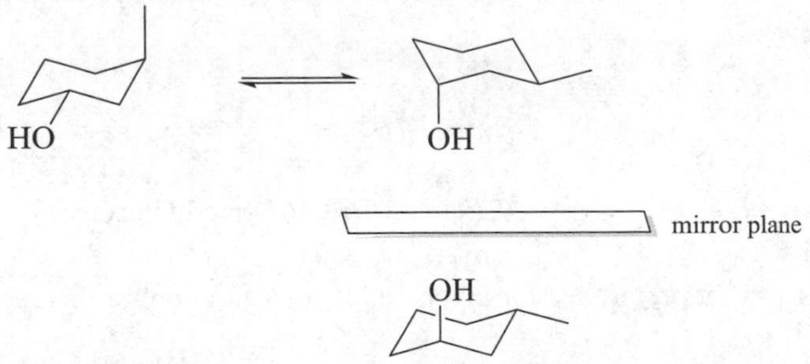

mirror plane

5.38 Reaction with methanesulfonyl chloride in the presence of a mild base, trimethylamine, will give the product.

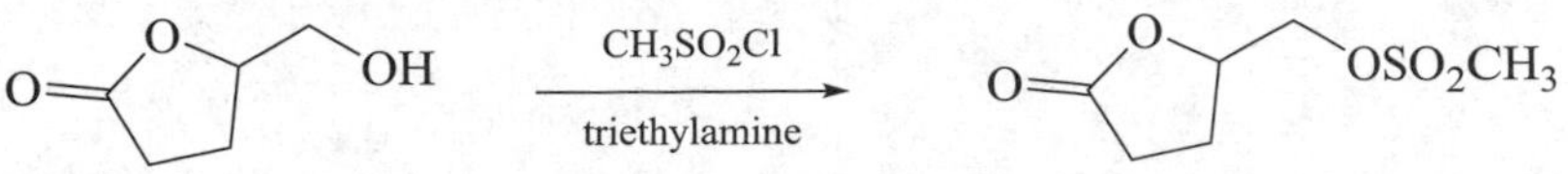

5.39 Loss of water from bicyclo[2.2.1]heptan-1-ol gives a tertiary carbocation. In contrast to other tertiary carbocations we have studied, this one results in angle strain since it cannot achieve a trigonal planar geometry Angle strain in the transition state leading to the non- trigonal planar carbocation raises the E_a.

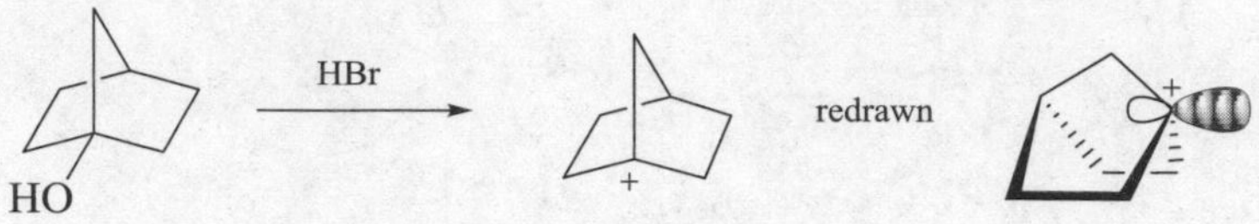

Bicyclo[2.2.1]heptan-1-ol   Bicyclo[2.2.1]heptyl cation

5.40 (*a*)   There is a slight preference for inversion of configuration for this reaction.

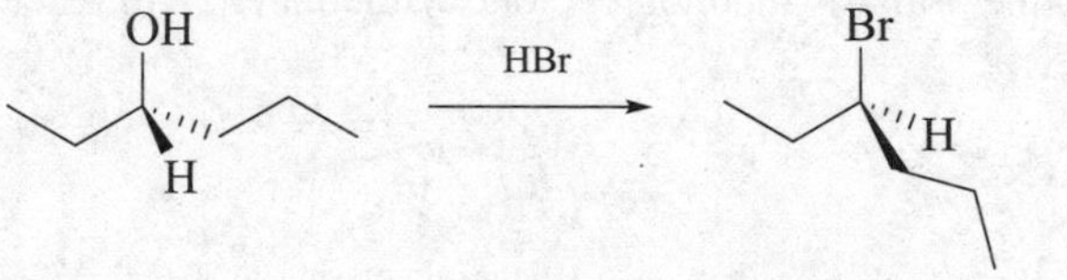

(*R*)-3-bromohexane

(*b*)   This reaction that proceeds with inversion of configuration

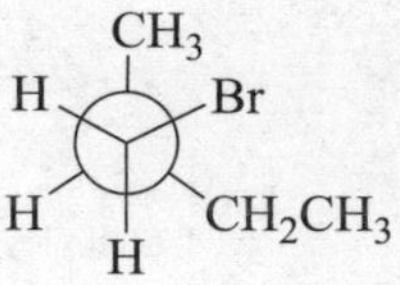

(2*S*)-1-Bromo-2-methylbutane

(*c*)   There is a slight preference for inversion of configuration for this reaction.

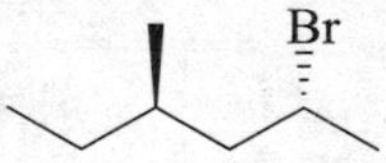

(2*R*,4*R*)-2-Bromo-4-methylhexane

(*d*)   There is a slight preference for inversion of configuration for this reaction.

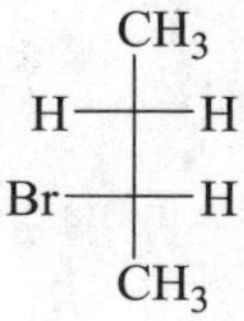

(*S*)-2-Bromobutane

5.41 The observed product suggests a carbocation rearrangement. Since primary carbocations are too high in energy to be generated, a rearrangement occurs concomitant with loss of water.

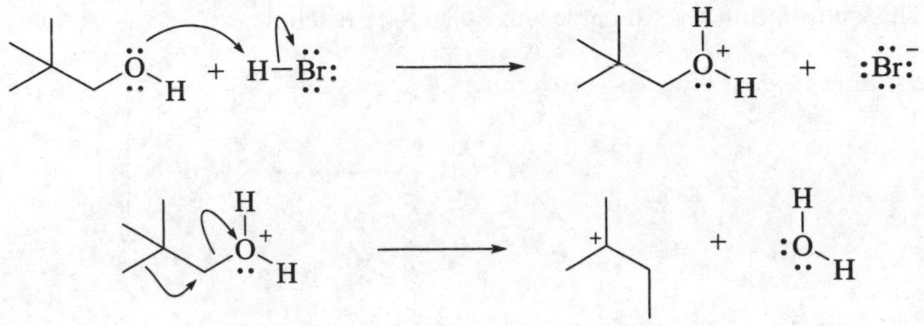

Reaction of the tertiary carbocation with bromide gives the product.

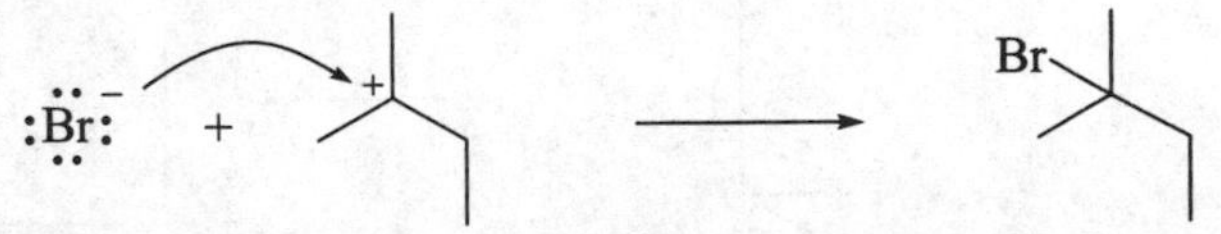

5.42 (*a*) Protonation by HBr of *trans*-4-methylcyclohexanol precedes the rate determining step. Formation of a secondary carbocation depends only on the protonated alcohol so this is the unimolecular slow step of the reaction.

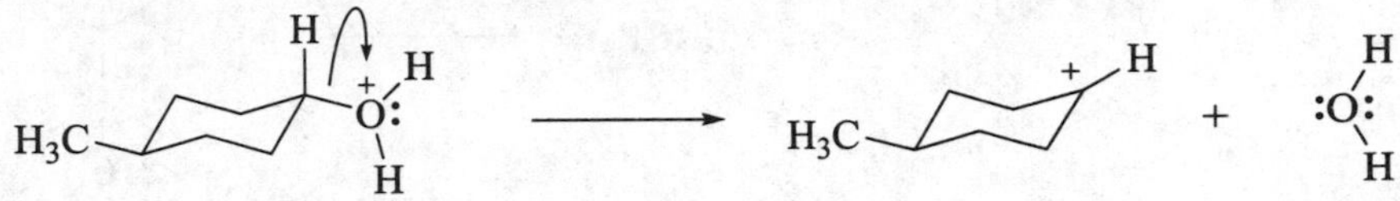

(*b*) Reaction on opposite faces of the carbocation gives a mixture of cis- and trans-1-bromo-4-methylcyclohexane.

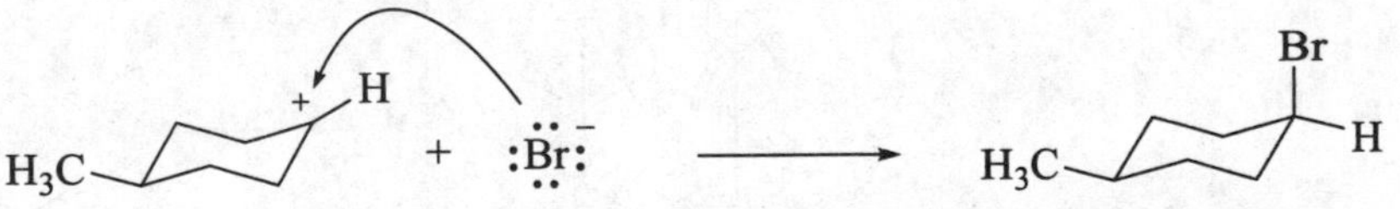

cis-1-Bromo-4-methylcyclohexane

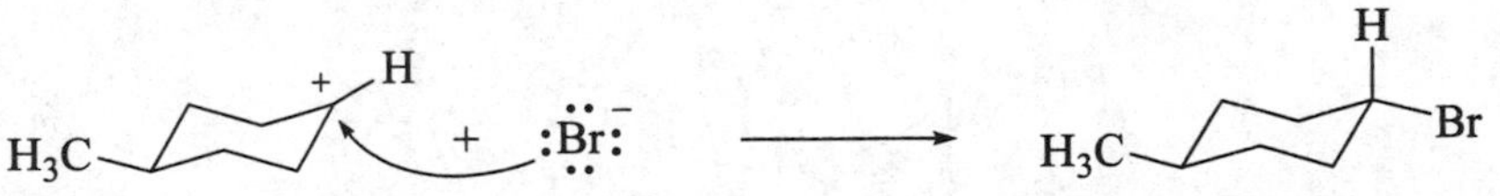

trans-1-Bromo-4-methylcyclohexane

5.43 This substitution reaction proceeds by an S_N1 reaction.

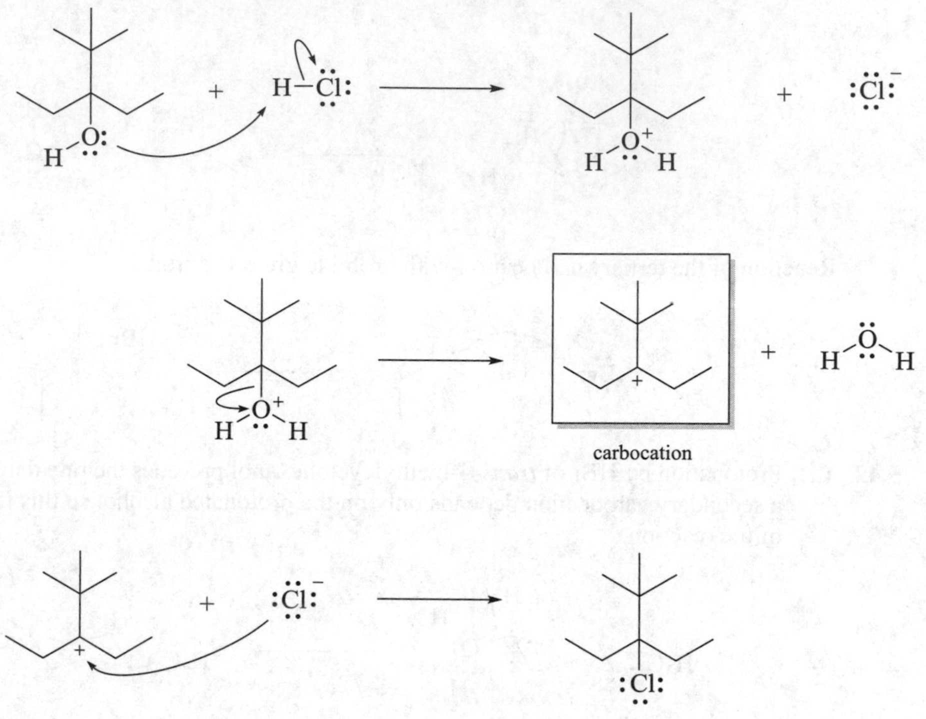

carbocation

The structural change in the second product suggests a carbocation rearrangement.

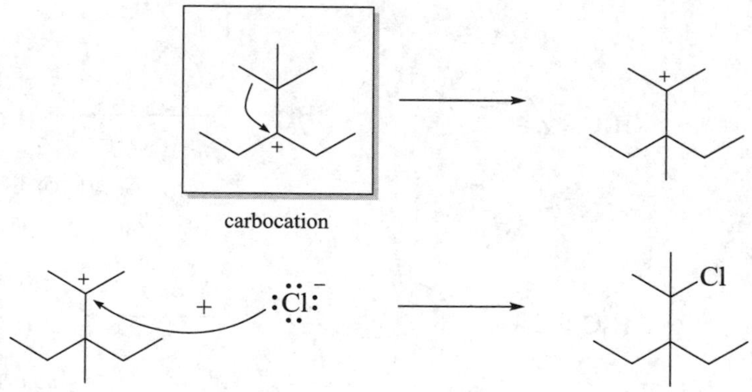

carbocation

5.44 The first step in the reaction of an alcohol with HBr is protonation of the OH oxygen. The problem states that the reaction is in aqueous medium, so the proton donor is H_3O^+.

The alcohol is primary, so the reaction cannot involve a carbocation, and bond formation to Br and loss of H_2O from the oxonium ion occur in the same bimolecular step.

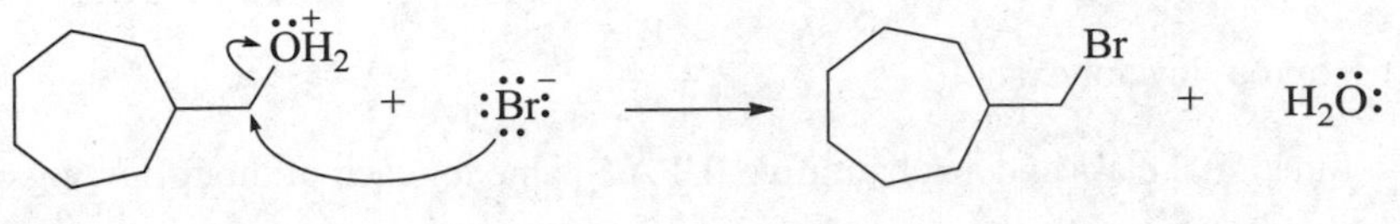

The mechanism is S_N2.

5.45 Methanol cannot react by the S_N1 mechanism because CH_3^+ is far too unstable to be an intermediate in a chemical reaction. The mechanism is S_N2. The first step is proton transfer from HBr to methanol. In the second step, bromide acts as a nucleophile to displace water from methyloxonium ion.

Step 1:

| Methanol | Hydrogen bromide | | Methyloxonium ion | Bromide ion |

Step 2:

| Bromide ion | Methyloxonium ion | | Methyl bromide | Water |

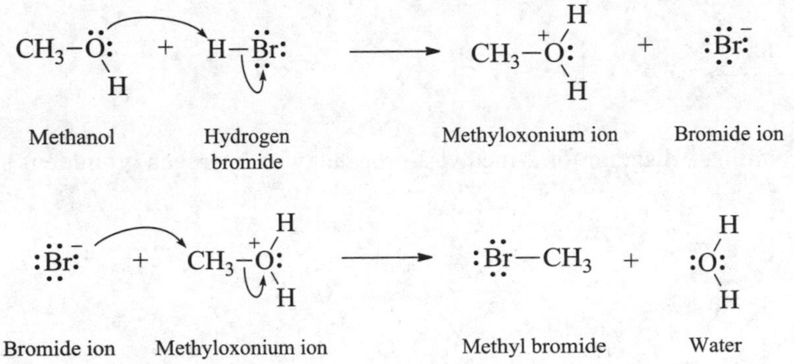

Answers to Interpretive Problems 5

5.46 D; **5.47** A; **5.48** A; **5.49** C; **5.50** A; **5.51** A; **5.52** B

SELF-TEST

1. The compound shown is an example of the broad class of organic compounds known as **steroids**. What functional groups does the molecule contain?

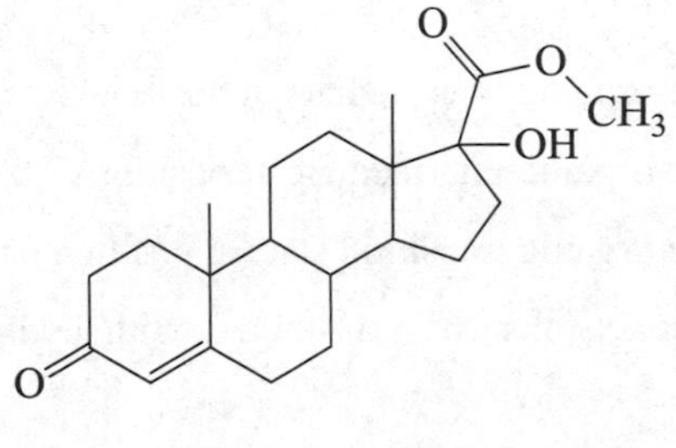

2. Give the correct substitutive IUPAC name for each of the following compounds:

(a)

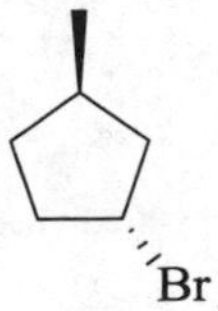

(b)

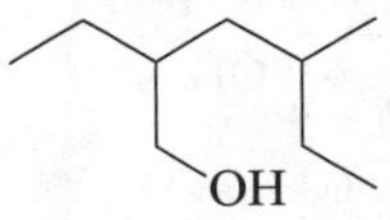

3.   Draw the structures of the following substances:

(a)   2-Chloro-1-iodo-2-methylheptane

(b)   *cis*-3-Isopropylcyclohexanol

4.   Give both a functional class and a substitutive IUPAC name for each of the following compounds:

(a)   (b)

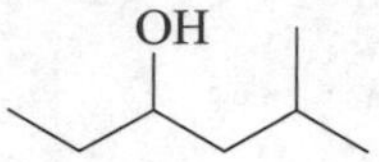

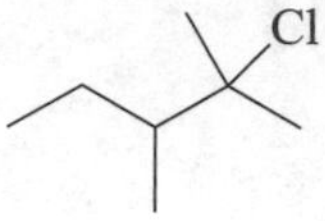

5.   Supply the missing component for each of the following reactions:

(a) $CH_3CH_2CH_2OH$ $\xrightarrow{SOCl_2}$?

(b)   ? $\xrightarrow{HBr}$ (structure with Br)

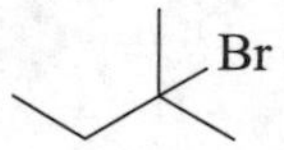

6.   The reaction coordinate diagram for 2-methyl-2-propanol with hydrogen bromide is shown. Use this to answer (a)-(c).

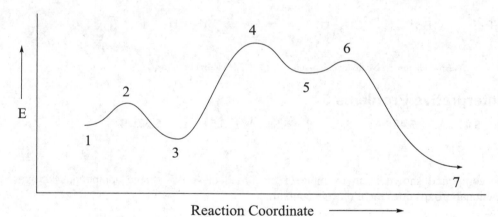

(a)   Give the structure of the organic intermediate at position 3.

(b)   Give the structure of the organic intermediate at position 5

(c)   Give the structure of the organic transition state at position 6.

7.   Choose the compound that reacts fastest and slowest with hydrogen chloride. Give the reasons for your choices.

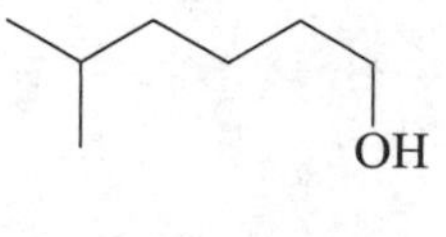

5-Methylhexan-1-ol 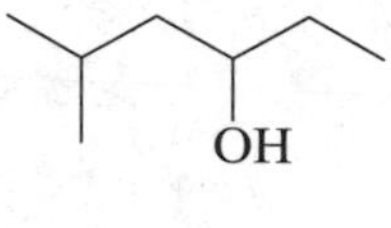   5-Methylhexan-3-ol 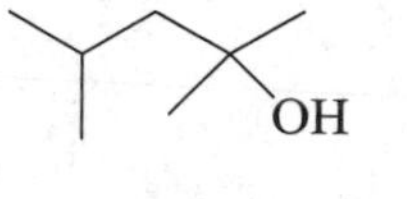   2,4-Dimethylpentan-2-ol 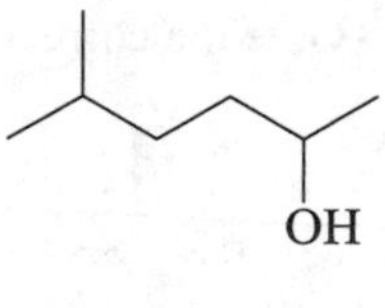   5-Methylhexan-2-ol

8.   (a)   Write out each of the elementary steps in the reaction of *tert*-butyl alcohol with hydrogen bromide. Use curved arrows to show electron movement in each step.

(b)   Draw the structure of the transition state representing the unimolecular dissociation of the alkyloxonium ion in the preceding reaction.

(*c*)   How does the mechanism of the reaction between 1-butanol and hydrogen bromide differ from the reaction in part (*a*)?

9.   Which compound reacts faster with sodium bromide and sulfuric acid? Why?

2-methyl-3-pentanol   or   3-methyl-3-pentanol

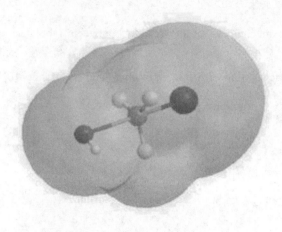

CHAPTER 6
Nucleophilic Substitution

Table of Contents

SOLUTIONS TO TEXT PROBLEMS

In Chapter Problems

6.1 Identify the nucleophilic anion in each reactant. The nucleophilic anion replaces bromine as a substituent on carbon.

(b) Potassium ethoxide is a source of the nucleophilic anion $CH_3CH_2O^-$.

$$CH_3CH_2\overset{..}{\underset{..}{O}}:^- \;+\; CH_3\overset{..}{\underset{..}{Br}}: \longrightarrow CH_3CH_2\overset{..}{\underset{..}{O}}CH_3 \;+\; :\overset{..}{\underset{..}{Br}}:^-$$

| Ethoxide ion (nucleophile) | Methyl bromide | Ethyl methyl ether (product) | Bromide ion |

(c)

| Benzoate ion (nucleophile) | Methyl bromide | Methyl benzoate (product) | Bromide ion |

(d) Lithium azide is a source of azide ion.

$$:\overset{..}{\underset{..}{N}}=\overset{+}{N}=\overset{..}{\underset{..}{N}}:^-$$

It reacts with methyl bromide to give methyl azide.

$$^-:\overset{..}{\underset{..}{N}}=\overset{+}{N}=\overset{..}{\underset{..}{N}}:^- \;+\; CH_3\overset{..}{\underset{..}{Br}}: \longrightarrow \;^-:\overset{..}{\underset{..}{N}}=\overset{+}{N}=\overset{..}{N}-CH_3 \;+\; :\overset{..}{\underset{..}{Br}}:^-$$

| Azide ion (nucleophile) | Methyl bromide | Methyl azide (product) | Bromide ion |

(e) The nucleophilic anion in KCN is cyanide ($:N\equiv C:^-$). The carbon atom is negatively charged and is normally the site of nucleophilic reactivity.

$$:N\equiv\overset{..}{C}:^- \;+\; CH_3\overset{..}{\underset{..}{Br}}: \longrightarrow :N\equiv C-CH_3 \;+\; :\overset{..}{\underset{..}{Br}}:^-$$

| Cyanide ion (nucleophile) | Methyl bromide | Methyl cyanide (product) | Bromide ion |

(f) The anion in sodium hydrogen sulfide (NaSH) is $H\overset{..}{\underset{..}{S}}:^-$.

$$H\overset{..}{\underset{..}{S}}:^- \;+\; CH_3\overset{..}{\underset{..}{Br}}: \longrightarrow H\overset{..}{\underset{..}{S}}-CH_3 \;+\; :\overset{..}{\underset{..}{Br}}:^-$$

| Hydrogen sulfide ion (nucleophile) | Methyl bromide | Methanethiol (product) | Bromide ion |

(g) Sodium iodide is a source of the nucleophilic anion iodide ion, $:\overset{..}{\underset{..}{I}}:^-$. The reaction of sodium iodide with alkyl bromides is usually carried out in acetone to precipitate the sodium bromide formed and prevent the reverse reaction.

$$:\overset{..}{\underset{..}{I}}:^{-} + CH_3\overset{..}{\underset{..}{Br}}: \xrightarrow{\text{acetone}} :\overset{..}{\underset{..}{I}}{-}CH_3 + :\overset{..}{\underset{..}{Br}}:^{-}$$

Iodide ion   Methyl bromide   Methyl iodide   Bromide ion
(nucleophile)   (product)

6.2   Write out the structure of the starting material. Notice that it contains a primary bromide and a primary chloride. Bromide is a better leaving group than chloride and is the one that is displaced faster by the nucleophilic cyanide ion.

$$ClCH_2CH_2CH_2Br \xrightarrow[\text{ethanol-water}]{\text{NaCN}} ClCH_2CH_2CH_2C{\equiv}N$$

1-Bromo-3-chloropropane   4-Chlorobutanenitrile

6.3   No, the two-step sequence is not consistent with the observed behavior for the hydrolysis of methyl bromide. The rate-determining step in the two-step sequence shown is the first step, ionization of methyl bromide to give methyl cation.

1. $CH_3Br \xrightarrow{\text{slow}} \overset{+}{C}H_3 + Br^{-}$

2. $\overset{+}{C}H_3 + HO^{-} \xrightarrow{\text{fast}} CH_3OH$

In such a sequence, the nucleophile would not participate in the reaction until after the rate-determining step is past, and the reaction rate would depend only on the concentration of methyl bromide and be independent of the concentration of hydroxide ion.

$$\text{Rate} = k[CH_3Br]$$

The predicted kinetic behavior is first order. Second-order kinetic behavior is actually observed for methyl bromide hydrolysis, so the proposed two-step mechanism cannot be correct.

6.4   Inversion of configuration occurs at the chirality center. When shown in a Fischer projection, this corresponds to replacing the leaving group on one side by the nucleophile on the opposite side.

$$
\begin{array}{c}
CH_3 \\
H{-}\!\!\!-\!\!\!-Br \\
CH_2(CH_2)_4CH_3
\end{array}
\xrightarrow[S_N2]{\text{NaOH}}
\begin{array}{c}
CH_3 \\
HO{-}\!\!\!-\!\!\!-H \\
CH_2(CH_2)_4CH_3
\end{array}
$$

(S)-(+)-2-Bromooctane   (R)-(-)-2-Octanol

6.5   The example given in the text illustrates inversion of configuration in the S_N2 hydrolysis of (S)-(+)-2-bromooctane, which yields (R)-(-)-2-octanol. The hydrolysis of (R)-(-)-2-bromooctane exactly mirrors that of its enantiomer and yields (S)-(+)-2-octanol.

Hydrolysis of racemic 2-bromooctane gives racemic 2-octanol. Remember, optically inactive reactants must yield optically inactive products.

6.6   C-2 of isopropyl bromide is partially bonded to both Br and I in the S_N2 transition state. Iodide attacks from the side opposite the C–Br bond.

Iodide ion   Isopropyl bromide S_N2 Transition state   Isopropyl iodide   Bromide ion

6.7   Sodium iodide in acetone is a reagent that converts alkyl chlorides and bromides into alkyl iodides by an S_N2 mechanism. Pick the alkyl halide in each pair that is more reactive toward S_N2 displacement.

(*b*)   The less crowded alkyl halide reacts faster in an S_N2 reaction. 1-Bromopentane is a primary alkyl halide and so is more reactive than 3-bromopentane, which is secondary.

1-Bromopentane
(primary; more reactive in S_N2)

3-Bromopentane
(secondary; less reactive in S_N2)

(*c*)   Both halides are secondary, but fluoride is a poor leaving group in nucleophilic substitutions. Alkyl chlorides are more reactive than alkyl fluorides.

2-Chloropentane
(more reactive)

2-Fluoropentane
(less reactive)

(*d*)   A secondary alkyl bromide reacts faster under S_N2 conditions than a tertiary one.

2-Bromo-5-methylhexane
(secondary; more reactive in S_N2)

2-Bromo-2-methylhexane
(tertiary; less reactive in S_N2)

(*e*)   The number of carbons does not matter as much as the degree of substitution at the reaction site. The primary alkyl bromide is more reactive than the secondary.

1-Bromodecane
(primary; more reactive in S_N2)

2-Bromopropane
(secondary; less reactive in S_N2)

6.8   Dibromopentane has two reactive C—Br functions; one is primary, the other secondary. Primary alkyl halides react faster than secondary halides in S_N2 reactions with anionic nucleophiles. Therefore, the anionic nitrogen bonds to the primary carbon (C-1), not the secondary carbon (C-4) of 1,4-dibromopentane.

6.9   The reactivity of an alkyl halide in an S_N1 reaction is dictated by the ease with which it ionizes to form a carbocation. A more stable carbocation is formed faster than a less stable one. Tertiary alkyl halides are the most reactive, methyl halides the least reactive.

(*b*) Cyclopentyl iodide ionizes to form a secondary carbocation, while the carbocation from 1-methylcyclopentyl iodide is tertiary. The tertiary halide is more reactive.

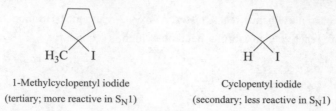

1-Methylcyclopentyl iodide
(tertiary; more reactive in S_N1)

Cyclopentyl iodide
(secondary; less reactive in S_N1)

(*c*) Cyclopentyl bromide ionizes to a secondary carbocation. 1-Bromo-2,2-dimethylpropane is a primary alkyl halide and is therefore less reactive.

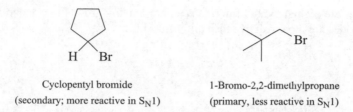

Cyclopentyl bromide
(secondary; more reactive in S_N1)

1-Bromo-2,2-dimethylpropane
(primary, less reactive in S_N1)

(*d*) Iodide is a better leaving group than chloride in both S_N1 and S_N2 reactions.

tert-Butyl iodide
(more reactive)

tert-Butyl chloride
(less reactive)

6.10 The problem specifies that the relative rate data are for substitutions that take place by the S_N1 mechanism; therefore, compare *tert*-butyl cation with the carbocations derived from A and B.

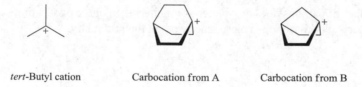

tert-Butyl cation   Carbocation from A   Carbocation from B

The most stable coplanar geometry for the three bonds attached to a positively charged carbon is easily accommodated in *tert*-butyl cation, but not in the carbocations from A and B because of their bicyclic frameworks. Angle strain in the transition states of A and B raises E_a for ionization and decreases their solvolysis rates. This strain is much larger in B than A because its bridge in A contains one carbon less than A.

6.11 The alkyl halide is tertiary and so undergoes hydrolysis by the S_N1 mechanism. The carbocation intermediate can be captured by water at either face. A mixture of the axial and the equatorial alcohols is formed. The same two substitution products are formed from *trans*-1,4-dimethylcyclohexyl bromide because it undergoes hydrolysis via the same carbocation intermediate.

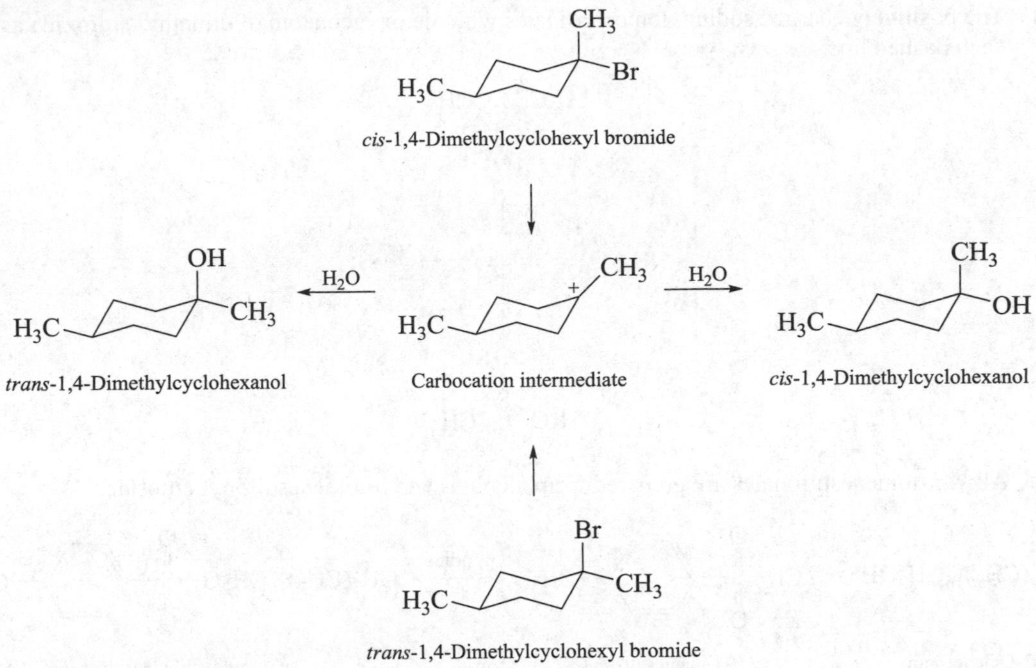

CH₃
H₃C—⟨ ⟩—Br

cis-1,4-Dimethylcyclohexyl bromide

↓

OH
H₃C—⟨ ⟩—CH₃   ←(H₂O)—   CH₃ (+) (Carbocation intermediate) H₃C—⟨ ⟩   —(H₂O)→   CH₃ H₃C—⟨ ⟩—OH

trans-1,4-Dimethylcyclohexanol   Carbocation intermediate   *cis*-1,4-Dimethylcyclohexanol

↑

Br
H₃C—⟨ ⟩—CH₃

trans-1,4-Dimethylcyclohexyl bromide

6.12 Write chemical equations illustrating each rearrangement process.

Hydride shift:

Secondary carbocation → Tertiary carbocation

Methyl shift:

Secondary carbocation → Secondary carbocation

Rearrangement by a hydride shift is observed because it converts a secondary carbocation to a more stable tertiary one. A methyl shift gives a secondary carbocation—in this case the same carbocation as the one that existed prior to rearrangement.

6.13 Diethyl ether is an aprotic solvent because it is not a hydrogen-bond donor—it does not have an –OH or –NH functional group.   A dielectric constant of 4 is low, so diethyl ether would be a nonpolar aprotic solvent.

6.14   The positively charged sodium ion coordinates with the oxygen atom of dimethyl sulfoxide as indicated by each dashed line.

6.15   Alkyl *p*-toluenesulfonates are prepared from alcohols and *p*-toluenesulfonyl chloride.

1-Octadecanol   *p*-Toluenesulfonyl   Octadecyl *p*-toluenesulfonate
                          chloride

6.16   As in part (*a*), identify the nucleophilic anion in each part. The nucleophile replaces the *p*-toluenesulfonate (tosylate) leaving group by an S_N2 process. The tosylate group is abbreviated as OTs.

(*b*) $:\!\ddot{\mathrm{I}}\!:^-$ + $CH_3(CH_2)_{16}CH_2\ddot{O}Ts$ $\longrightarrow$ $CH_3(CH_2)_{16}CH_2\ddot{I}\!:$ + $Ts\ddot{O}\!:^-$

Iodide ion   Octadecyl *p*-toluenesulfonate   Octadecyl iodide   *p*-Toluenesulfonate
                                                                                    anion

(*c*) $:\!N\!\equiv\!C\!:^-$ + $CH_3(CH_2)_{16}CH_2\ddot{O}Ts$ $\longrightarrow$ $CH_3(CH_2)_{16}CH_2C\!\equiv\!N\!:$ + $Ts\ddot{O}\!:^-$

Cyanide   Octadecyl *p*-toluenesulfonate   Octadecyl cyanide   *p*-Toluenesulfonate
ion   anion

(*d*) $H\!-\!\ddot{S}\!:^-$ + $CH_3(CH_2)_{16}CH_2\ddot{O}Ts$ $\longrightarrow$ $CH_3(CH_2)_{16}CH_2\ddot{S}\!-\!H$ + $Ts\ddot{O}\!:^-$

Hydrogen   Octadecyl *p*-toluenesulfonate   1-Octadecanethiol   *p*-Toluenesulfonate
sulfide ion   anion

(*e*)

$\ddot{S}\!:^-$ + $CH(CH_2)_{16}CH_2\ddot{O}Ts$ $\longrightarrow$ $CH_3(CH_2)_{16}CH_2\ddot{S}$ + $Ts\ddot{O}\!:^-$

Butanethiolate ion   Octadecyl *p*-toluenesulfonate   Butyl octadecyl sulfide   *p*-Toluenesulfonate
                                                                                                  anion

6.17 The hydrolysis of (*S*)-(+)-1-methylheptyl *p*-toluenesulfonate proceeds with inversion of configuration, giving the *R* enantiomer of 2-octanol.

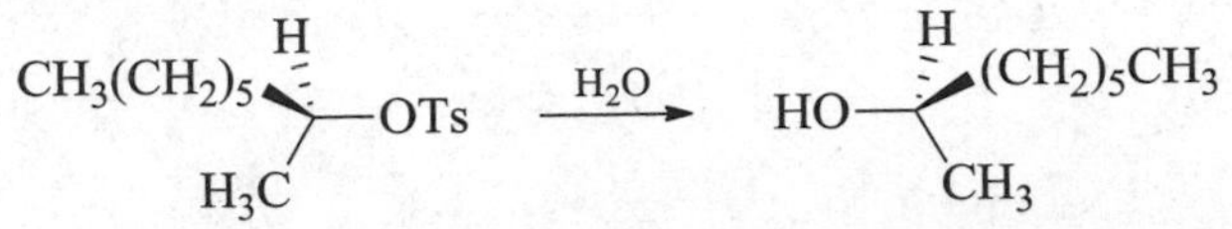

(*S*)-(+)-1-Methylheptyl
p-toluenesulfonate

(*R*)-(-)-2-Octanol

In Section 6.10 of the text, we are told that optically pure (*S*)-(+)-1-methylheptyl *p*-toluenesulfonate is prepared from optically pure (*S*)-(+)-2-octanol having a specific rotation $[\alpha]_D^{25}$ +9.9°. The conversion of an alcohol to a *p*-toluenesulfonate proceeds with complete *retention* of configuration. Hydrolysis of this *p*-toluenesulfonate with *inversion* of configuration therefore yields optically pure (*R*)-(−)-2-octanol of $[\alpha]_D^{25}$ -9.9°.

6.18 A retrosynthetic analysis for the preparation of 2-bromo-3-methylbutane includes converting the alcohol to an alkyl sulfonate. The good sulfonate leaving group allows an S_N2 reaction to the desired alkyl bromide, 3-bromo-3-methylbutane.

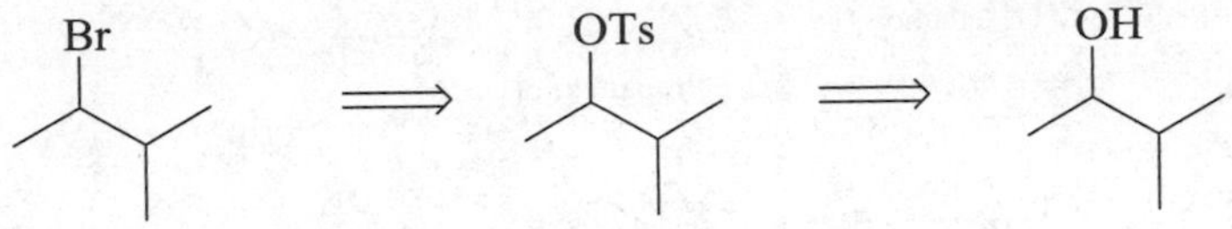

2-Bromo-3-methylbutane

3-Methyl-2-butanol

A reaction sequence for this retrosynthetic analysis is:

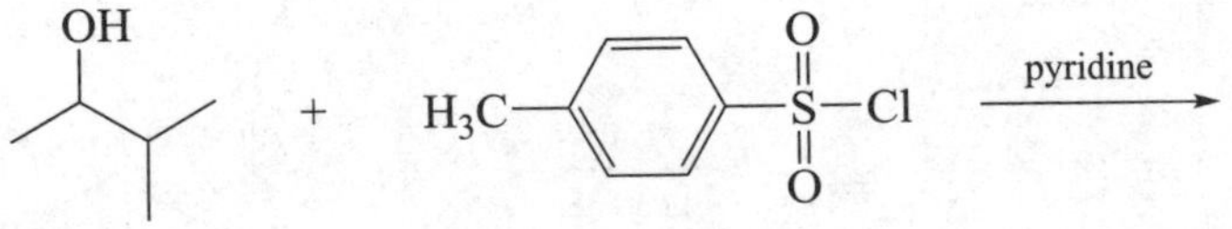

p-Toluenesulfonyl chloride
(TsCl)

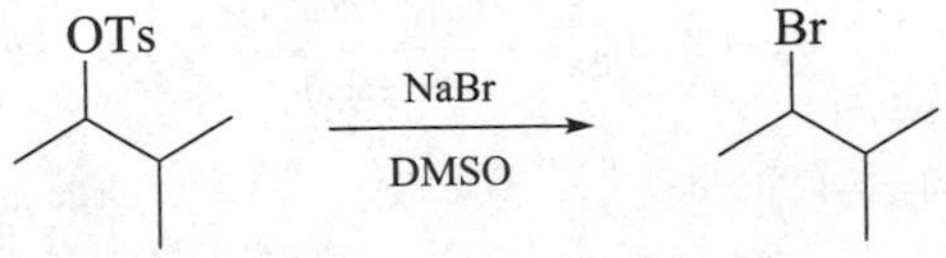

End of Chapter Problems

Predict the Products

6.19 1-Bromopropane is a primary alkyl halide, and so it will undergo predominantly S_N2 displacement regardless of the basicity of the nucleophile.

(*a*)

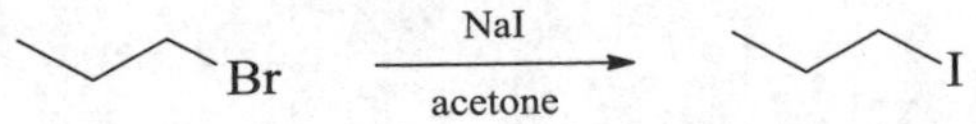

1-Bromopropane

1-Iodopropane

(b)

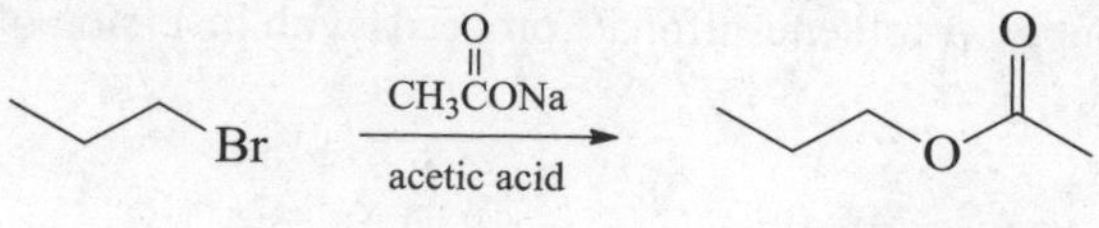

Propyl acetate

(c)

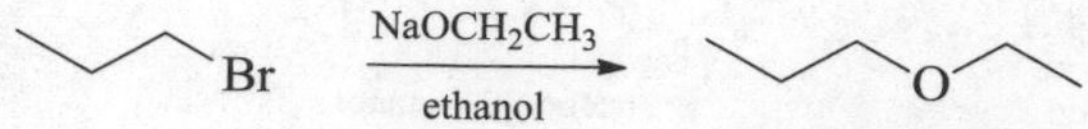

Ethyl propyl ether

(d)

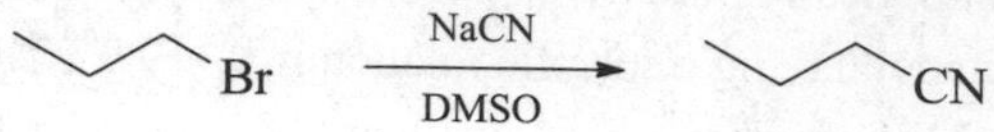

Butanenitrile

(e)

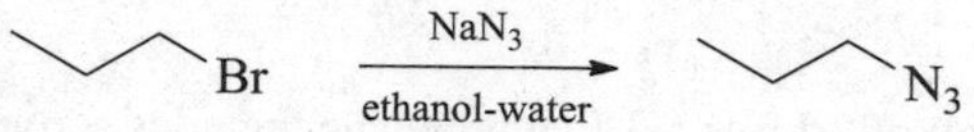

1-Azidopropane

(f)

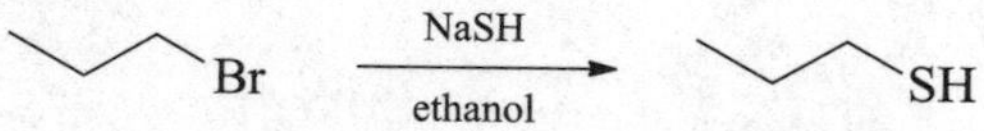

1-Propanethiol

(g)

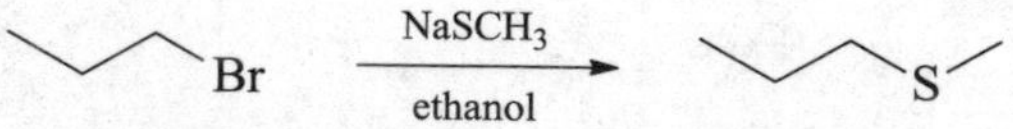

Methyl propyl sulfide

6.20 (a) The substrate is a primary alkyl bromide and reacts with sodium iodide in acetone to give the corresponding iodide.

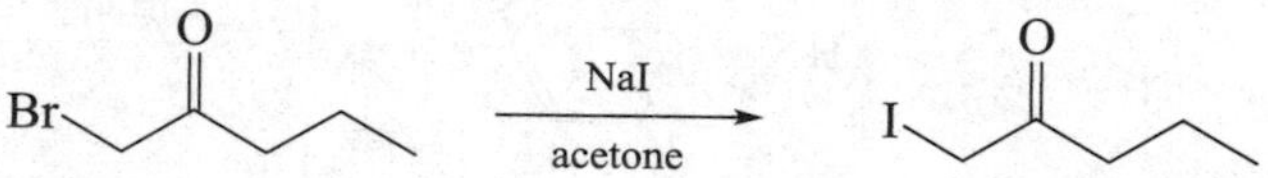

2-Bromo-2-pentanone   2-Iodo-2-pentanone

(b) Primary alkyl chlorides react with sodium acetate to yield the corresponding acetate esters.

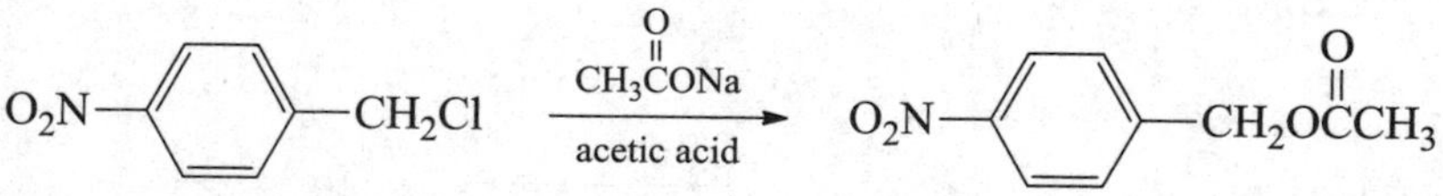

p-Nitrobenzyl chloride   p-Nitrobenzyl acetate (78%-82%)

(c) The only leaving group in the substrate is bromide. Neither of the carbon–oxygen bonds is susceptible to cleavage by nucleophilic attack.

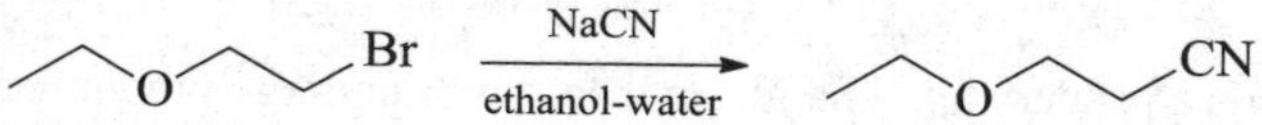

2-Bromoethyl ethyl ether   2-Cyanoethyl ethyl ether
                                                         (52%-58%)

(d) Hydrolysis of the primary chloride yields the corresponding alcohol.

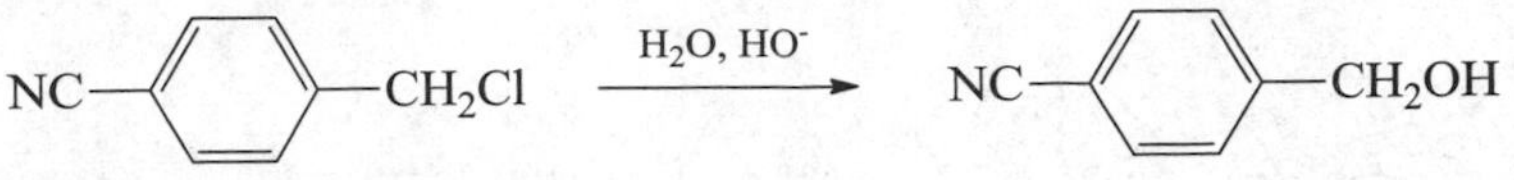

p-Cyanobenzyl chloride   *p*-Cyanobenzyl alcohol (85%)

(*e*) The substrate is a primary chloride.

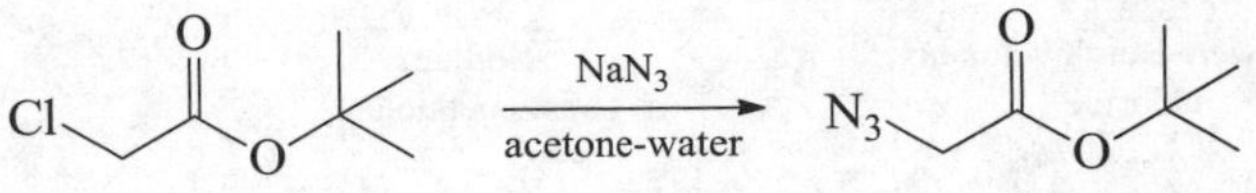

tert-Butyl chloroacetate   *tert*-Butyl azidoacetate (92%)

(*f*) Primary alkyl tosylates yield iodides on treatment with sodium iodide in acetone.

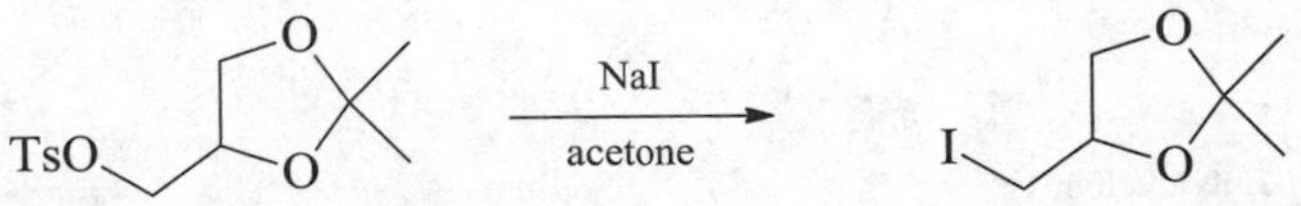

(2,2-Dimethyl-1,3-dioxolan-4-yl)-   (2,2-Dimethyl-5-(iodomethyl)-
methyl *p*-toluenesulfonate   1,3-dioxolane (60%)

(*g*) Sulfur displaces bromide from ethyl bromide.

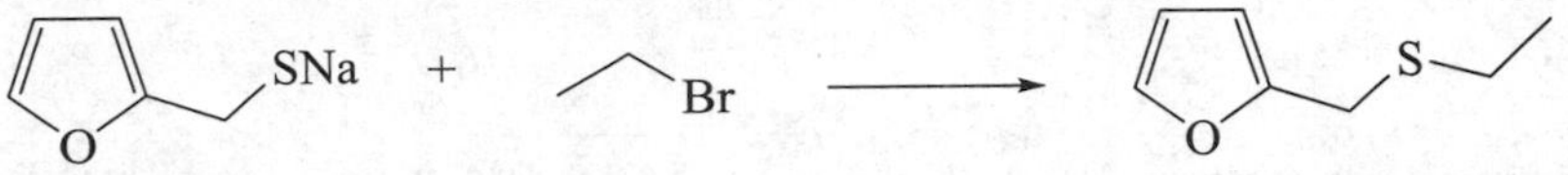

Sodium (2-furyl)-   Ethyl bromide   Ethyl (2-furyl)methyl sulfide (80%)
methanethiolate

(*h*) The first reaction is one in which a substituted alcohol is converted to a *p*-toluenesulfonate. This is followed by an S_N2 displacement with lithium iodide.

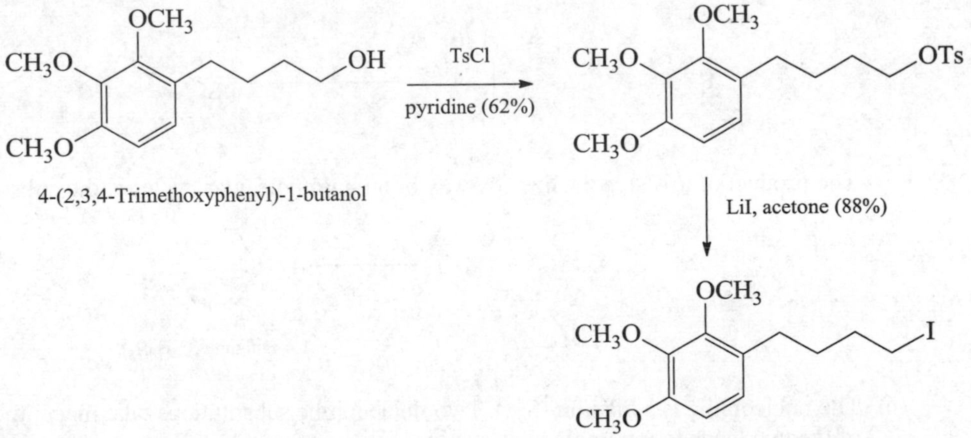

4-(2,3,4-Trimethoxyphenyl)-1-butanol

4-(2,3,4-Trimethoxyphenyl)butyl iodide

6.21 The two products are diastereomers. They are formed by bimolecular nucleophilic substitution (S_N2). In each case, a good nucleophile ($C_6H_5S^-$) displaces chloride from a secondary carbon with inversion of configuration.

(*a*) The trans chloride yields a cis substitution product.

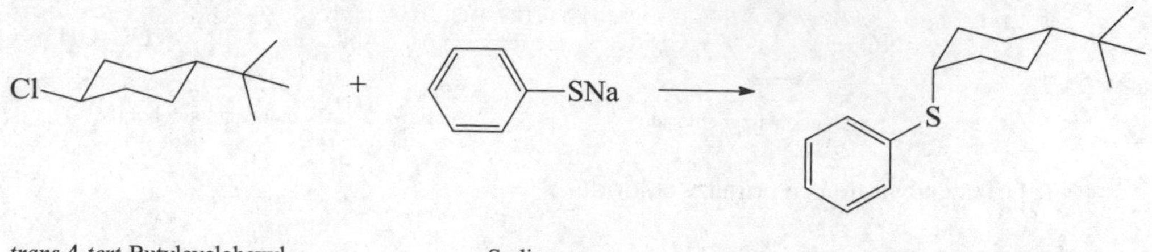

trans-4-*tert*-Butylcyclohexyl chloride   Sodium benzenethiolate   *cis*-4-*tert*-Butylcyclohexyl phenyl sulfide

(*b*) The cis chloride yields a trans substitution product.

cis-4-*tert*-Butylcyclohexyl chloride   Sodium benzenethiolate   *trans*-4-*tert*-Butylcyclohexyl phenyl sulfide

6.22 (*a*) The starting material incorporates both a primary chloride and a secondary chloride. The nucleophile (iodide) attacks the less hindered primary position.

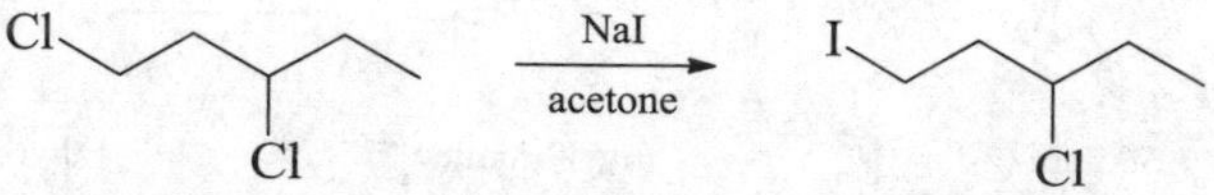

1,3-Dichloropentane   3-Chloro-1-iodopentane ($C_5H_{10}Cl$)

(*b*) Nucleophilic substitution of the first bromide by sulfur occurs in the usual way.

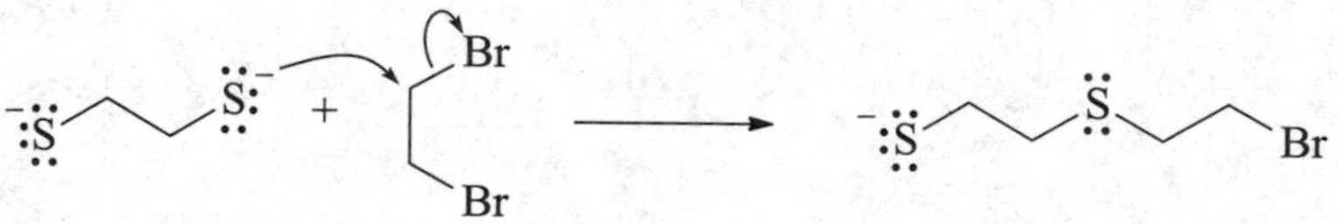

The product of this step cyclizes by way of an intramolecular nucleophilic substitution.

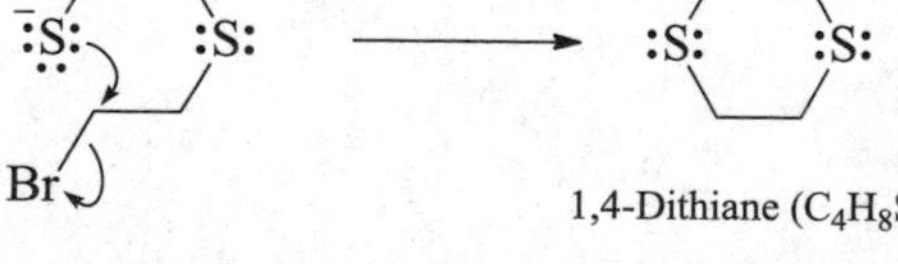

1,4-Dithiane ($C_4H_8S_2$)

(*c*) The nucleophile is a dianion (S^{2-}). Two nucleophilic substitutions take place; the second of the two leads to intramolecular cyclization.

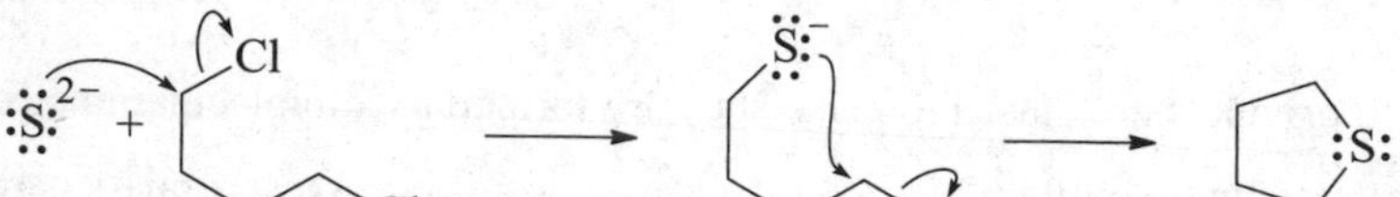

Thiolane (C_4H_8S)

6.23 (*a*) The two most stable Lewis structures (resonance contributors) of thiocyanate are

$$:\ddot{S}-C\equiv N: \longleftrightarrow :\ddot{S}=C=\ddot{N}:^{-}$$

(b) The two Lewis structures indicate that the negative charge is shared by two atoms: S and N. Thus, thiocyanate ion has two potentially nucleophilic sites, and the two possible products are

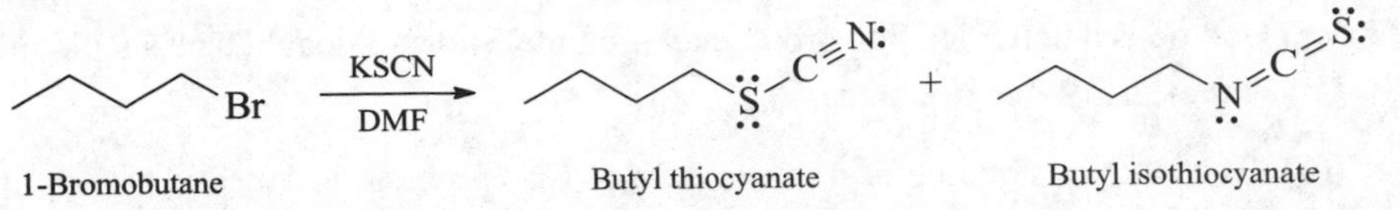

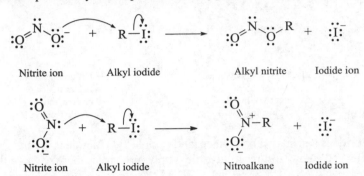

| 1-Bromobutane | | Butyl thiocyanate | Butyl isothiocyanate |

6.24 Nitrite ion has two potentially nucleophilic sites, oxygen and nitrogen.

| :O=N—O:⁻ + R—I: ⟶ :O=N—O—R + :I:⁻ |

Nitrite ion   Alkyl iodide   Alkyl nitrite   Iodide ion

Nitrite ion   Alkyl iodide   Nitroalkane   Iodide ion

Thus, an alkyl iodide can yield either an alkyl nitrite or a nitroalkane depending on whether the oxygen or the nitrogen of nitrite ion attacks carbon. Both do, and the product from 2-iodooctane is a mixture of

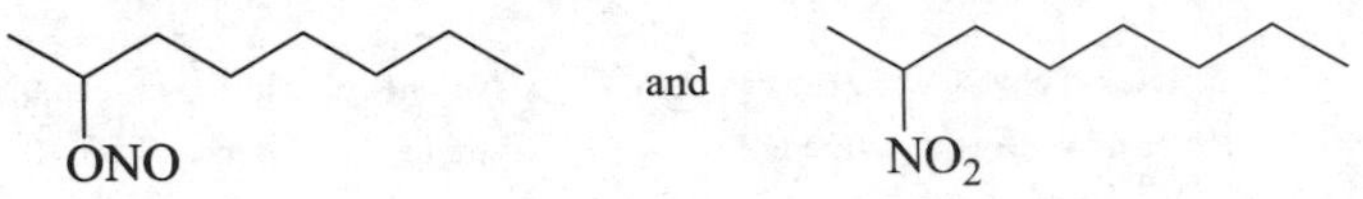

and

6.25 Using the unshared electron pair on its nitrogen, triethylamine acts as a nucleophile in an S_N2 reaction toward ethyl iodide.

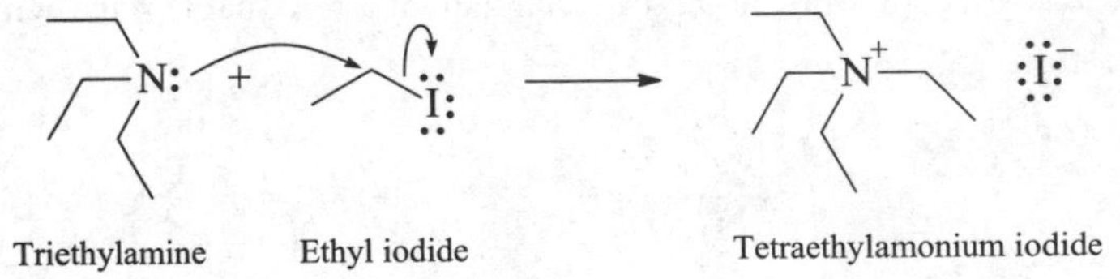

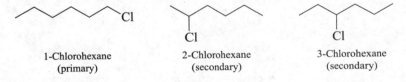

Triethylamine   Ethyl iodide   Tetraethylamonium iodide

The product of the reaction is a salt and has the structure shown. The properties given in the problem (soluble in polar solvents, high melting point) are typical of those of an ionic compound.

Rate and Mechanism

6.26 1-Chlorohexane is a primary alkyl halide; 2-chlorohexane and 3-chlorohexane are secondary.

1-Chlorohexane   2-Chlorohexane   3-Chlorohexane
(primary)   (secondary)   (secondary)

Primary and secondary alkyl halides react with potassium iodide in acetone by an S_N2 mechanism, and the rate depends on steric hindrance to attack on the alkyl halide by the nucleophile.

(a) Primary alkyl halides are more reactive than secondary alkyl halides in S_N2 reactions. 1-Chlorohexane is the most reactive isomer.

(b) Substituents at the carbon adjacent to the one that bears the leaving group slow down the rate of nucleophilic displacement. In 2-chlorohexane, the group adjacent to the point of attack is CH_3.

In 3-chlorohexane, the group adjacent to the point of attack is CH_2CH_3. 2-Chlorohexane has been observed to be more reactive than 3-chlorohexane by a factor of 2.

6.27 (*a*) Iodide is a better leaving group than bromide, and so 1-iodobutane should undergo S_N2 attack by cyanide faster than 1-bromobutane.

     (*b*) The reaction conditions are typical for an S_N2 process. The methyl branch in 1-chloro-2-methylbutane sterically hinders attack at C-1. The unbranched isomer, 1-chloropentane, reacts faster.

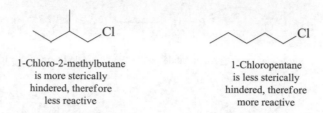

1-Chloro-2-methylbutane   1-Chloropentane
is more sterically   is less sterically
hindered, therefore   hindered, therefore
less reactive   more reactive

     (*c*) Hexyl chloride is a primary alkyl halide, and cyclohexyl chloride is secondary. Azide ion is a good nucleophile, and so the S_N2 reactivity rules apply; primary is more reactive than secondary.

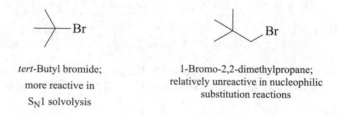

Hexyl chloride is primary,   Cyclohexyl chloride is secondary
therefore more reactive in S_N2 therefore less reactive in S_N2

     (*d*) 1-Bromo-2,2-dimethylpropane is too hindered to react with the weakly nucleophilic ethanol by an S_N2 reaction; and because it is a primary alkyl halide, it is less reactive in S_N1 reactions. *tert*-Butyl bromide will react with ethanol by an S_N1 mechanism at a reasonable rate owing to formation of a tertiary carbocation.

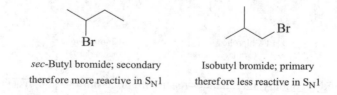

tert-Butyl bromide;   1-Bromo-2,2-dimethylpropane;
more reactive in   relatively unreactive in nucleophilic
S_N1 solvolysis   substitution reactions

     (*e*) Solvolysis of alkyl halides in aqueous formic acid is faster for those that form carbocations readily. The S_N1 reactivity order applies here: secondary > primary.

sec-Butyl bromide; secondary   Isobutyl bromide; primary
therefore more reactive in S_N1 therefore less reactive in S_N1

     (*f*) 1-Chlorobutane is a primary alkyl halide and so should react by an S_N2 mechanism. Sodium methoxide is more basic than sodium acetate and is a better nucleophile. Reaction will occur faster with sodium methoxide than with sodium acetate.

     (*g*) Azide ion is a very good nucleophile, whereas *p*-toluenesulfonate is a very good leaving group but a very poor nucleophile. In an S_N2 reaction with 1-chlorobutane, sodium azide will react faster than sodium *p*-toluenesulfonate.

6.28 Both aspects of this reaction—its slow rate and the formation of a rearranged product—have their origin in the positive character developed at a primary carbon. The alcohol is protonated and the carbon–oxygen bond of the resulting alkyloxonium ion begins to break:

2,2-Dimethyl-1-propanol   Alkyloxonium ion

As positive character develops at the primary carbon, a methyl group migrates. Rearrangement gives a tertiary carbocation, which is captured by bromide to give the product.

2-Bromo-2-methylbutane

6.29 Alkyl chlorides arise by the reaction sequence

Primary alcohol p-Toluenesulfonyl chloride   Pyridine   Primary alkyl p-toluenesulfonate   Pyridinium chloride

Primary alkyl p-toluenesulfonate   Primary alkyl chloride

The reaction proceeds to form the alkyl p-toluenesulfonate as expected, but the chloride anion formed in this step subsequently acts as a nucleophile and displaces p-toluenesulfonate from RCH_2OTs.

6.30 Iodide ion is both a better nucleophile than cyanide and a better leaving group than bromide. The two reactions shown are therefore faster than the reaction of cyclopentyl bromide with sodium cyanide alone.

Cyclopentyl bromide   Cyclopentyl iodide   Cyclopentyl cyanide

6.31 (a) If each act of exchange (substitution) occurred with retention of configuration, there would be no observable racemization; $k_{rac} = 0$.

(R)-(-)-2-Iodooctane
(*indicates radioactive label)

[(R)-(-)-2-Iodooctane]*

Therefore,

$$\frac{k_{rac}}{k_{exch}} = 0.$$

(*b*) If each act of exchange proceeds with inversion of configuration, (*R*)-(−)-2-iodooctane will be transformed to radioactively labeled (*S*)-(+)-2-iodooctane.

(*R*)-(-)-2-Iodooctane   [(*S*)-(+)-2-Iodooctane]*

Starting with 100 molecules of (*R*)-(−)-2-iodooctane, the compound will be completely racemized when 50 molecules have become radioactive. Therefore,

$$\frac{k_{rac}}{k_{exch}} = 2.$$

(*c*) If radioactivity is incorporated in a stereorandom fashion, then 2-iodooctane will be 50% racemized when 50% of it has reacted. Therefore,

$$\frac{k_{rac}}{k_{exch}} = 1.$$

In fact, Hughes found that the rate of racemization was twice the rate of incorporation of radioactive iodide. This experiment provided strong evidence for the belief that bimolecular nucleophilic substitution proceeds stereospecifically with inversion of configuration.

6.32 (*a*) Methyl halides are unhindered and react rapidly by the S$_N$2 mechanism.

(*b*) Sodium ethoxide is a good nucleophile and will react with unhindered primary alkyl halides by the S$_N$2 mechanism.

(*c*) The tertiary halide *tert*-butyl bromide will undergo solvolysis by the S$_N$1 mechanism.

(*d*) In a stereospecific reaction, stereoisomeric reactants yield products that are stereoisomers of each other. Reactions that occur by the S$_N$2 mechanism are stereospecific.

(*e*) The unimolecular mechanism S$_N$1 involves the formation of carbocation intermediates.

(*f*) Rearrangements are possible when carbocations are intermediates in a reaction. Thus, reactions occurring by the S$_N$1 mechanism are most likely to have a rearranged carbon skeleton.

(*g*) Iodide is a better leaving group than bromide, and alkyl iodides react faster than alkyl bromides by any of the two mechanisms S$_N$1, S$_N$2.

Stereochemistry

6.33 This reaction has been reported in the chemical literature and proceeds as shown (91% yield):

(*S*)-1-Bromo-2-methylbutane   (*S*)-1-Iodo-2-methylbutane

Notice that the configuration of the product is the *same* as the configuration of the reactant. This is because the chirality center is not involved in the reaction. When we say that S$_N$2 reactions proceed with inversion of configuration, we refer only to the carbon at which substitution takes place, not a chirality center somewhere else in the molecule.

6.34 The compound that reacts with *trans*-4-*tert*-butylcyclohexanol is a sulfonyl chloride and converts the alcohol to the corresponding sulfonate.

Compound A

Reaction of compound A with lithium bromide in acetone effects displacement of the sulfonate leaving group by bromide with inversion of configuration (S_N2 substitution).

Compound A   *cis*-1-Bromo-4-*tert*-butylcyclohexane
                                                          Compound B

6.35 Iodide ion reacts with (*R*)-2-chlorobutane with inversion of configuration by an S_N2 process to give (*S*)-2-iodobutane.

Iodide ion   (*R*)-2-Chlorobutane   (*S*)-2-Iodooctane   Chloride ion

When some of the (*R*)-2-chlorobutane has reacted, the solution contains both (*R*)-2-chlorobutane and (*S*)-2-iodobutane. Iodide is a better leaving group than chloride, so the (*S*)-2-iodobutane that is formed reacts with iodide faster than the starting (*R*)-2-chlorobutane does.

(*S*)-2-Iodooctane   Iodide ion   Iodide ion   (*R*)-2-Iodobutane

Thus, the (*S*)-2-iodobutane undergoes racemization faster than the (*R*)-2-chlorobutane reacts, and the 2-iodobutane that is isolated after all the (*R*)-2-chlorobutane has reacted is racemic.

6.36 (*a*) The reaction of an alcohol with a sulfonyl chloride gives a sulfonate. The oxygen of the alcohol remains in place and is the atom to which the sulfonyl group becomes attached.

(*S*)-(+)-2-Butanol   (*S*)-*sec*-Butyl methanesulfonate

(b) Sulfonate is similar to iodide in its leaving-group behavior. The product in part (a) is attacked by NaSCH$_2$CH$_3$ in an S$_N$2 reaction to give a product with optical rotation $\alpha_D = -25°$. Inversion of configuration occurs at the chirality center.

(S)-sec-Butyl methanesulfonate   (R)-(-)-sec-Butyl ethyl sulfide

(c) In this part of the problem, we deduce the stereochemical outcome of the reaction of 2-butanol with PBr$_3$. We know the absolute configuration of (+)-2-butanol (S) from the statement of the problem and the configuration of (–)-sec-butyl ethyl sulfide (R) from part (b). We are told that the sulfide formed from (+)-2-butanol via the bromide has a positive rotation of α_D of +23. It must therefore have the opposite configuration of the product in part (b).

(S)-(+)-2-Butanol   2-Bromobutane   (S)-(+)-sec-Butyl ethyl sulfide

Because the reaction of the bromide with NaSCH$_2$CH$_3$ proceeds with inversion of configuration at the chirality center and because the final product has the same configuration as the starting alcohol, the conversion of the alcohol to the bromide must proceed with inversion of configuration.

(S)-(+)-2-Butanol   (R)-(-)-2-Bromobutane

(d) The conversion of (S)-(+)-2-butanol to sec-butyl methanesulfonate does not involve any of the bonds to the chirality center, and so it must proceed with 100% retention of configuration. Assuming that the reaction of the methanesulfonate with NaSCH$_2$CH$_3$ proceeds with 100% inversion of configuration, we conclude that the maximum rotation of sec-butyl ethyl sulfide is the value given in the statement of part (b), that is, -25°. Because the sulfide produced in part (c) has a rotation of +23°, it is 92% optically pure. It is reasonable to assume that the loss of optical purity occurred in the conversion of the alcohol to the bromide, rather than in the reaction of the bromide with NaSCH$_2$CH$_3$. If the bromide is 92% optically pure and has a rotation of –38°, optically pure 2-bromobutane therefore has a rotation of $\frac{38}{0.92}$ or ±41°.

Synthesis

6.37 (a) In this synthesis, a primary alkyl chloride must be converted to a primary alkyl iodide. This is precisely the kind of transformation for which sodium iodide in acetone is used.

Isobutyl chloride   Sodium   Isobutyl iodide   Sodium
                            iodide   chloride

(b) First introduce a leaving group into the molecule by converting isopropyl alcohol to an isopropyl halide. Then convert the resulting isopropyl halide to isopropyl azide by nucleophilic substitution.

| Isopropyl alcohol | Isopropyl bromide | Isopropyl azide |

(c) First write out the structure of the starting material and of the product so as to determine their relationship in three dimensions.

(R)-sec-Butyl alcohol   (S)-sec-Butyl azide

The hydroxyl group must be replaced by azide with inversion of configuration. First, however, a leaving group must be introduced, and it must be introduced in such a way that the configuration at the chirality center is not altered. The best way to do this is to convert (R)-sec-butyl alcohol to its corresponding p-toluenesulfonate.

(R)-sec-Butyl alcohol   (R)-sec-Butyl p-toluenesulfonate

Next, convert the p-toluenesulfonate to the desired azide by an S$_N$2 reaction.

(R)-sec-Butyl p-toluenesulfonate   (S)-sec-Butyl azide

(d) This problem is carried out in exactly the same way as the preceding one, except that the nucleophile in the second step is HS⁻.

(R)-sec-Butyl alcohol   (R)-sec-Butyl p-toluenesulfonate   (S)-2-Butanethiol

6.38 In each case the synthesis requires formation of a C—O bond, so there are two possible C—O—C disconnections to evaluate.

(a) Disconnection 1 generates methyl bromide and potassium *tert*-butoxide. An S$_N$2 reaction between the two should work well to give the desired compound.

$$H_3C \text{-}\!\!\vdots\text{-} O\text{—}C(CH_3)_3 \overset{1}{\Longrightarrow} CH_3Br \ + \ KOC(CH_3)_3$$

Disconnection 2 generates potassium methoxide and *tert*-butyl bromide. Tertiary alkyl halides react with alkoxide bases by elimination. Attempts to prepare the ether this way will fail.

$$H_3C-O\!\mid\!C(CH_3)_3 \xrightarrow{2} CH_3OK + BrC(CH_3)_3$$

(b) The reasoning here is similar to part (a). The reaction of sodium methoxide with cyclopentyl bromide (disconnection 1) will result in the formation of cyclopentene by elimination. Reaction of potassium cyclopentoxide with methyl bromide (disconnection 2) will yield the desired ether.

(c) The desired ether linkage is formed by an S_N2 reaction. The combination represented in retrosynthesis 1 is a poor choice because of steric hindrance in the alkyl bromide (neopentyl bromide). Retrosynthesis 2 is the correct choice; ethyl bromide is extremely reactive in S_N2 reactions.

$$(CH_3)_3CCH_2\!\mid\!O-CH_2CH_3 \xrightarrow{1} (CH_3)_3CCH_2-Br + KO-CH_2CH_3$$

$$(CH_3)_3CCH_2-O\!\mid\!CH_2CH_3 \xrightarrow{2} (CH_3)_3CCH_2-OK + Br-CH_2CH_3$$

6.39 (a) In order to introduce the cyano group, we must first replace the hydroxyl of citronellol by a suitable leaving group such as bromide or p-toluenesulfonate. Retrosynthetically:

(S)-(-)-Citronellol

(b) The reagents needed for each synthesis are indicated in the following. In each case the cyano group is introduced by an SN2 reaction.

6.40 (a) To convert *trans*-2-methylcyclopentanol to *cis*-2-methylcyclopentyl acetate, the hydroxyl group must be replaced by acetate with inversion of configuration. Hydroxide is a poor leaving group and so must first be converted to a good leaving group. The best choice is *p*-toluenesulfonate, because this can be prepared by a reaction that alters none of the bonds to the chirality center.

trans-2-Methylcyclopentanol

trans-2-Methylcyclopentyl
p-toluenesulfonate

Treatment of the *p*-toluenesulfonate with potassium acetate in acetic acid will proceed with inversion of configuration to give the desired product.

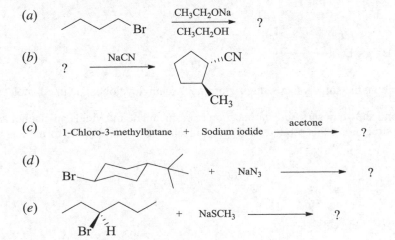

(b) To decide on the best sequence of reactions, we must begin by writing structural formulas to determine what kinds of transformations are required.

We already know from part (*a*) how to convert *trans*-2-methylcyclopentanol to *cis*-2-methylcyclopentyl acetate. So it is necessary is to design a synthesis of *trans*-2-methylcyclopentanol. Therefore,

Hydroboration–oxidation converts 1-methylcyclopentene to the desired alcohol by anti-Markovnikov syn hydration of the double bond. The resulting alcohol is then converted to its *p*-toluenesulfonate and treated with acetate ion as in part (*a*) to give *cis*-2-methylcyclopentyl acetate.

Answers to Interpretive Problems 6

    6.41 D;   **6.42** A;   **6.43** D;   **6.44** B;   **6.45** B;   **6.46** C;   **6.47** C

SELF-TEST

1.   Write the correct structure of the reactant or product omitted from each of the following. Clearly indicate stereochemistry where it is important.

(*a*)

$$\xrightarrow[\text{CH}_3\text{CH}_2\text{OH}]{\text{CH}_3\text{CH}_2\text{ONa}} \quad ?$$

(*b*)

$$? \xrightarrow{\text{NaCN}}$$

(*c*)   1-Chloro-3-methylbutane + Sodium iodide $\xrightarrow{\text{acetone}}$?

(*d*)

    + NaN$_3$ $\longrightarrow$?

(*e*)

    + NaSCH$_3$ $\longrightarrow$?

(*f*)

$$
\begin{array}{c}
CH_3 \\
H\!-\!\!-\!Br \\
H\!-\!\!-\!F \\
CH_2CH_3
\end{array}
\quad \xrightarrow{\text{NaSH}} \quad ?
$$

2.   Choose the best pair of reactants to form the following ether by an S_N2 reaction:

3.   Outline the chemical steps necessary to convert

(*a*)

$$
\begin{array}{c}
CH_3 \\
H\!-\!\!-\!OTs \\
CH_2CH_3
\end{array}
\quad \text{to} \quad
\begin{array}{c}
CH_3 \\
H\!-\!\!-\!CN \\
CH_2CH_3
\end{array}
$$

(*b*)
$$
\underset{\text{(S)-2-Pentanol}}{}\quad \text{to} \quad \overset{\displaystyle SCH_3}{\underset{\text{(R)-}CH_3\overset{|}{C}HCH_2CH_2CH_3}{}}
$$

4.   Hydrolysis of 3-chloro-2,2-dimethylbutane yields 2,3-dimethyl-2-butanol as the major product. Explain this observation, using structural formulas to outline the mechanism of the reaction.

5.   Identify the class of reaction (e.g., S_N1, S_N2), and write the kinetic and chemical equations for

(*a*) The solvolysis of *tert*-butyl bromide in methanol

(*b*) The reaction of chlorocyclohexane with sodium azide (NaN_3)

6.   (*a*) Provide a brief explanation why the halogen exchange reaction shown is an acceptable synthetic method:

$$
\text{Br-} \quad + \quad \text{NaI} \quad \xrightarrow{\text{acetone}} \quad \text{I-} \quad + \quad \text{NaBr}
$$

(*b*) Briefly explain why the reaction of 1-bromobutane with sodium azide occurs faster in dimethyl sulfoxide $[(CH_3)_2S{=}O]$ than in water.

7.   Write chemical structures for compounds A through D in the following sequence of reactions. Compounds A and C are alcohols.

$$
A \xrightarrow{\text{NaNH}_2} B
$$

$$
C \xrightarrow{\text{HBr, heat}} D
$$

$$
B + D \longrightarrow
$$

8.   Write a mechanism describing the solvolysis (S_N1) of 1-bromo-1-methylcyclohexane in ethanol.

9.   Solvolysis of the compound shown occurs with carbocation rearrangement and yields an alcohol as the major product. Write the structure of this product, and give a mechanism to explain its formation.

$$
\xrightarrow{\text{H}_2\text{O}} \quad ?
$$

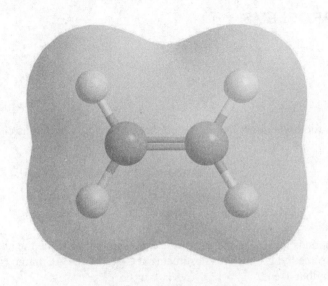

CHAPTER 7

Structure and Preparation of Alkenes:
Elimination Reactions

Table of Contents

SOLUTIONS TO TEXT PROBLEMS

In Chapter Problems

7.1   (*b*)   Writing the structure in more detail, we see that the longest continuous chain contains four carbon atoms.

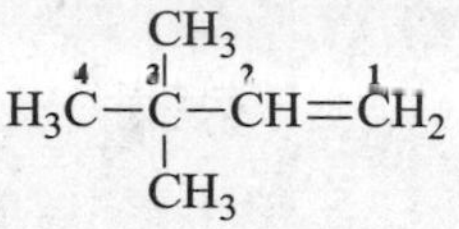

$$H_3\overset{4}{C}-\underset{\underset{CH_3}{|}}{\overset{\overset{CH_3}{|}}{\underset{3}{C}}}-\overset{2}{C}H=\overset{1}{C}H_2$$

The double bond is located at the end of the chain, and so we name the alkene as a derivative of 1-butene. Two methyl groups are substituents at C-3. The IUPAC name is 3,3-dimethyl-1-butene or 3,3-dimethylbut-1-ene.

(*c*)   Expanding the structural formula reveals the molecule to be a methyl-substituted derivative of hexene.

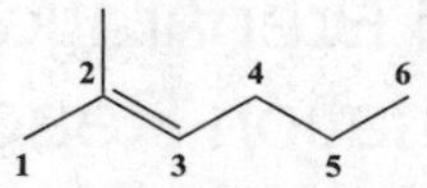

2-Methyl-2-hexene
or
2-Methylhex-2-ene

(*d*)   In compounds containing a double bond and a halogen, the double bond takes precedence in numbering the longest carbon chain.

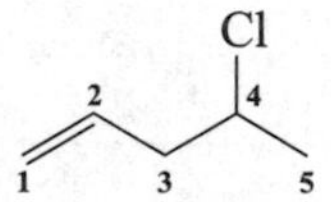

4-Chloro-1-pentene
or
4-Chloropent-1-ene

(*e*)   When a hydroxyl group is present in a compound containing a double bond, the hydroxyl takes precedence over the double bond in numbering the longest carbon chain.

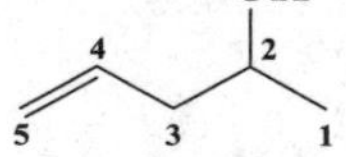

4-Penten-2-ol
or
Pent-4-en-2-ol

7.2   There are three sets of nonequivalent positions on a cyclopentene ring, identified as *a*, *b*, and *c* on the cyclopentene structure shown:

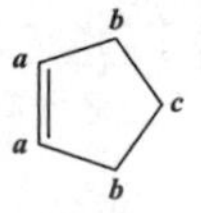

Thus, there are three different monochloro-substituted derivatives of cyclopentene. The carbons that bear the double bond are numbered C-1 and C-2 in each isomer, and the other positions are numbered in sequence in the direction that gives the chlorine-bearing carbon its lower number. 1-Chlorocyclopentene is a vinylic chloride. 3-Chlorocyclopentene is an allylic chloride.

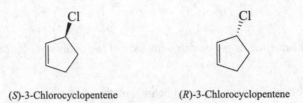

1-Chlorocyclopentene	3-Chlorocyclopentene	4-Chlorocyclopentene

There are two stereoisomers for the allylic chloride, 3-chlorocyclopentene. They are enantiomers.

(*S*)-3-Chlorocyclopentene   (*R*)-3-Chlorocyclopentene

7.3   There are only *two sp²*-hybridized carbons, the two connected by the double bond. The other six carbons are *sp³*-hybridized. There are three *sp²–sp³* σ bonds and three *sp³–sp³* σ bonds.

7.4   Consider first the C_5H_{10} alkenes that have an unbranched carbon chain:

1-Pentene	*cis*-2-Pentene	*trans*-2-Pentene

There are three additional isomers. These have a four-carbon chain with a methyl substituent.

2-Methyl-1-butene	2-Methyl-2-butene	3-Methyl-1-butene

7.5   No. They are constitutional isomers. The double bond connects C-2 and C-3 in *cis*-2-hexene and C-3 and C-4 in *trans*-3-hexene. Stereoisomers have the same constitution but differ in the arrangement of atoms in space.

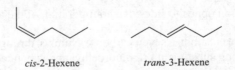

cis-2-Hexene	*trans*-3-Hexene

7.6   First, identify the constitution of 9-tricosene. Referring back to Table 2.2 in Section 2.14 of the text, we see that tricosane is the unbranched alkane containing 23 carbon atoms. 9-Tricosene, therefore, contains an unbranched chain of 23 carbons with a double bond between C-9 and C-10. Because the problem specifies that the pheromone has the cis configuration, the first 8 carbons and the last 13 must be on the same side of the C-9–C-10 double bond.

$$CH_3(CH_2)_7 \quad (CH_2)_{12}CH_3$$

cis-9-Tricosene

7.7 (b) One of the carbons of the double bond bears a methyl group and a hydrogen; methyl is of higher rank than hydrogen. The other doubly bonded carbon bears the groups —CH_2CH_2F and —$CH_2CH_2CH_2CH_3$. At the first point of difference between these two, fluorine is of higher atomic number than carbon, and so —CH_2CH_2F is of higher precedence.

Higher ranked substituents are on the same side of the double bond; the alkene has the *Z* configuration.

(c) One of the carbons of the double bond bears a methyl group and a hydrogen; as we have seen, methyl is of higher rank. The other doubly bonded carbon bears —CH_2CH_2OH and —$C(CH_3)_3$.

Let's analyze these two groups to determine their order of precedence.

—C(C,H,H)   —C(C,C,C)

Lower priority   Higher priority

We examine the atoms one by one at the point of attachment before proceeding down the chain. Therefore, —$C(CH_3)_3$ outranks —CH_2CH_2OH.

Higher ranked groups are on opposite sides; the configuration of the alkene is *E*.

(d) The cyclopropyl ring is attached to the double bond by a carbon that bears the atoms (C,C,H) and is therefore of higher precedence than an ethyl group —C(C,H,H).

Higher ranked groups are on opposite sides; the configuration of the alkene is *E*.

7.8 (b) The longest continuous chain has seven carbons. Number through the double bond to give it the lowest number. The remaining substituent is 2-fluoroethyl and is bonded to the longest chain at C-3.

(*Z*)-3-(2-Fluoroethyl)-2-heptene

(c) The longest continuous chain has five carbons. The alcohol takes precedence over ("outrank") the double bond in numbering the chain. The tert-butyl substituent is bonded to C-3 of the longest chain.

(E)-3-*tert*-Butyl-3-penten-1-ol

(d) The longest continuous chain has five carbons. The chain is numbered through the double bond to give it the lowest number. The cyclopropyl group is bonded to C-3 of the longest chain.

(E)-3-Cyclopropyl-2-pentene

7.9 The dipole moments of (E)-2,3-dichloro-2-butene cancel; the molecule is nonpolar. The dipole moments of *trans*-1-chloropropene are additive, as are the dipole moments of 1,1-dichloro-2-methylpropene. The latter compound has twice as many dipoles and is the most polar.

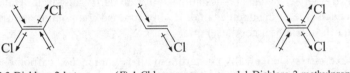

(E)-2,3-Dichloro-2-butene;   (E)-1-Chloropropene   1,1-Dichloro-2-methylpropene;
     least polar   most polar

7.10 A trisubstituted alkene has three carbons directly attached to the doubly bonded carbons. There are three trisubstituted C_6H_{12} isomers, two of which are stereoisomers.

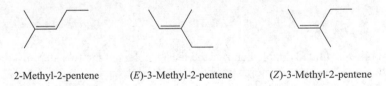

2-Methyl-2-pentene   (E)-3-Methyl-2-pentene   (Z)-3-Methyl-2-pentene

7.11 The most stable C_6H_{12} alkene is the one with the most highly substituted double bond; the least stable one has the least-substituted double bond.

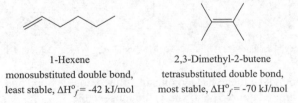

1-Hexene   2,3-Dimethyl-2-butene
monosubstituted double bond,   tetrasubstituted double bond,
least stable, $\Delta H°_f$ = -42 kJ/mol   most stable, $\Delta H°_f$ = -70 kJ/mol

7.12   Apply the two general rules for alkene stability to rank these compounds. First, more highly substituted double bonds are more stable than less substituted ones. Second, when two double bonds are similarly constituted, the trans stereoisomer is more stable than the cis. The predicted order of decreasing stability is therefore

2-Methyl-2-butene
(trisubstituted)
most stable

(E)-2-Pentene
(disubstituted)

(Z)-2-Pentene
(disubstituted)

1-Pentene
(monosubstituted)
least stable

7.13   Begin by writing the structural formula corresponding to the IUPAC name given in the problem. A bond-line depiction is useful here.

3,4-Di-*tert*-butyl-2,2,5,5-tetramethyl-3-hexene

The alkene is extremely crowded and destabilized by van der Waals strain. Bulky *tert*-butyl groups are cis to one another on each side of the double bond. Highly strained compounds are often quite difficult to synthesize, and this alkene is a good example.

7.14   Approach this exercise systematically by starting with an unbranched carbon chain and placing double bonds between appropriate carbons in the chain. Examine each constitution for the possibility of cis/trans stereoisomers. Of the three isomers with an unbranched chain:

1-Pentene   *cis*-2-Pentene   *trans*-2-Pentene

1-pentene (monosubstituted double bond) is the least stable. Of the two stereoisomers with a disubstituted double bond, *trans*-2-pentene is more stable than *cis*-2-pentene.

The remaining three isomers have a four-carbon chain bearing a methyl branch. No stereoisomers are possible here.

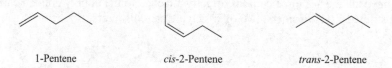

2-Methyl-1-butene   2-Methyl-2-butene   3-Methyl-1-butene

Of these three, 3-methyl-1-butene has a monosubstituted double bond and is the least stable; 2-methyl-2-butene has a trisubstituted double bond and is the most stable.

Within each substitution pattern, rank the alkenes according to whether the chain is branched (more stable) or unbranched (less stable) to give the order of increasing stability. Also shown are the experimentally determined standard enthalpies of formation.

1-Pentene	3-Methyl-1-butene	cis-2-Pentene	trans-2-Pentene	2-Methyl-1-butene	2-Methyl-2-butene
monosubstituted	monosubstituted	disubstituted	disubstituted	disubstituted	trisubstituted
unbranched	branched	unbranched	unbranched	branched	branched

ΔH°_f (kJ/mol):

-49	-52	-53	-58	-61	-68

ΔH°_f (kcal/mol):

-11.7	-12.4	-12.7	-13.1	-14.6	-16.2

As can be seen, the predicted order of stability matches the experimentally determined order. The difference between 3-methyl-1-butene and cis-2-pentene, however, encourages us to be alert for overlaps in specific cases.

7.15 Use the zigzag arrangement of bonds in the parent skeleton figure to place E and Z bonds as appropriate for each part of the problem. From the sample solution to parts (a) and (b), the ring carbons have the higher priorities. Thus, an E double bond will have ring carbons arranged ⌐_ and a Z double bond _/ .

(c)

(Z)-3-Methylcyclodecene

(e)

(Z)-5-Methylcyclodecene

(d)

(E)-3-Methylcyclodecene

(f)

(E)-5-Methylcyclodecene

7.16 Write out the structure of the alcohol, recognizing that the alkene is formed by loss of a hydrogen and a hydroxyl group from adjacent carbons.

(b, c) Both 1-propanol and 2-propanol give propene on acid-catalyzed dehydration.

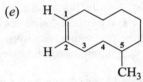

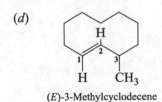   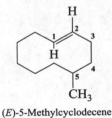

1-Propanol   Propene   2-Propanol (*d*) Carbon-3 has no hydrogens in 2,3,3-trimethyl-2-butanol. Elimination can involve only the hydroxyl group at C-2 and a hydrogen at C-1.

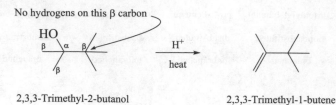

2,3,3-Trimethyl-2-butanol   2,3,3-Trimethyl-1-butene

7.17 (*b*) Elimination can involve loss of a hydrogen from the methyl group or from C-2 of the ring in 1-methylcyclohexanol.

1-Methylcyclohexanol $\xrightarrow{-H_2O}$ Methylenecyclohexane (a disubstituted alkene; minor product)   +   1-Methylcyclohexene (a trisubstituted alkene; major product)

According to the Zaitsev rule, the major alkene is the one corresponding to loss of a hydrogen from the alkyl group that has the smaller number of hydrogens. Thus, hydrogen is removed from the methylene group in the ring rather than from the methyl group, and 1-methylcyclohexene is formed in greater amounts than methylenecyclohexane.

(*c*) The two alkenes formed are as shown in the equation.

   $\xrightarrow{-H_2O}$ Compound has a trisubstituted double bond; less stable.   +   Compound has a tetrasubstituted double bond; more stable.

The more highly substituted alkene is formed in greater amounts, as predicted by Zaitsev's rule.

7.18 2-Pentanol can undergo dehydration in two different directions, giving either 1-pentene or 2-pentene. 2-Pentene is formed as a mixture of the cis and trans stereoisomers.

2-Pentanol $\xrightarrow[\text{heat}]{H^+}$ 1-Pentene   +   *cis*-2-Pentene   +   *trans*-2-Pentene

7.19   (*b*) The site of positive charge in the carbocation is the carbon atom that bears the hydroxyl group in the starting alcohol.

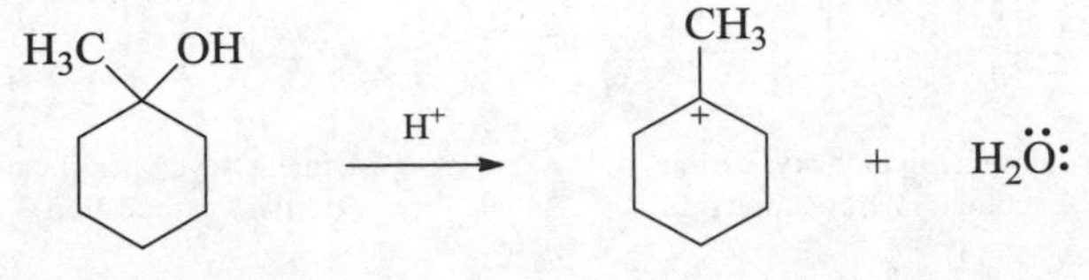

1-Methylcyclohexanol

Water may remove a proton from the methyl group, as shown in the following equation:

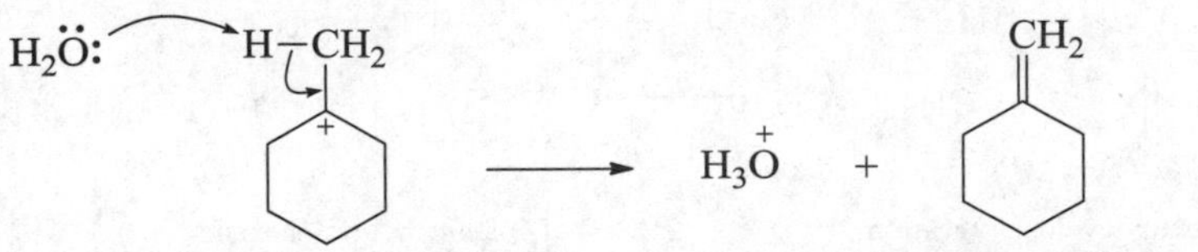

Methylenecyclohexane

Loss of a proton from the ring gives the major product 1-methylcyclohexene.

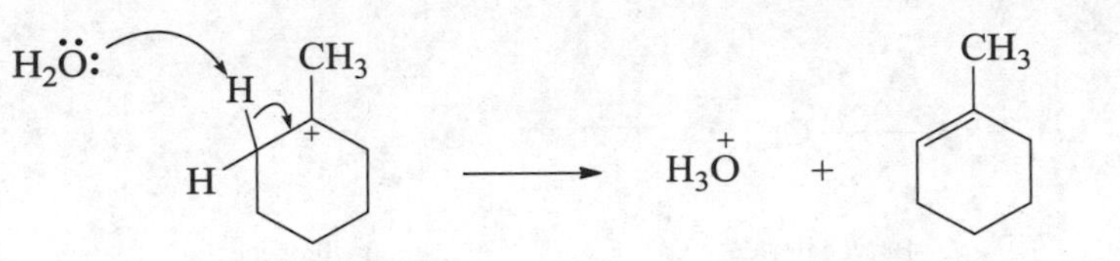

1-Methylcyclohexane

(*c*) Loss of the hydroxyl group under conditions of acid catalysis yields a tertiary carbocation.

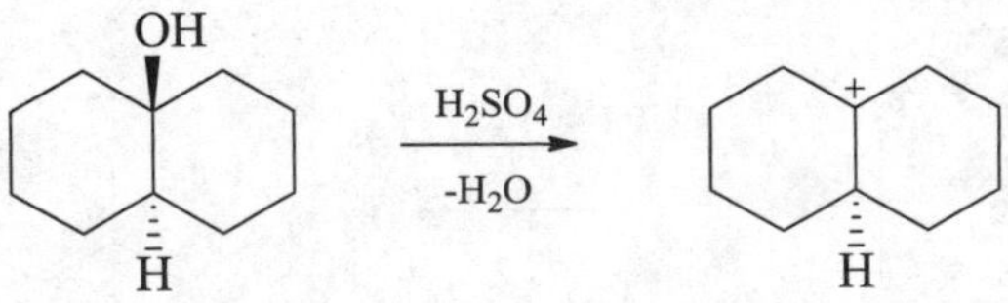

Water may remove a proton from an adjacent methylene group to give a trisubstituted alkene.

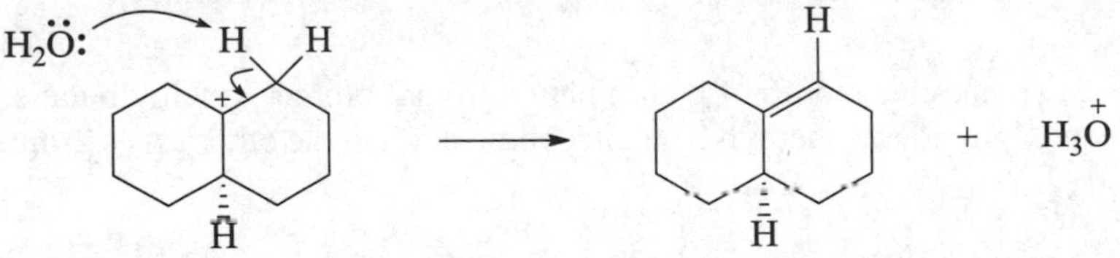

Removal of the methine proton gives a tetrasubstituted alkene.

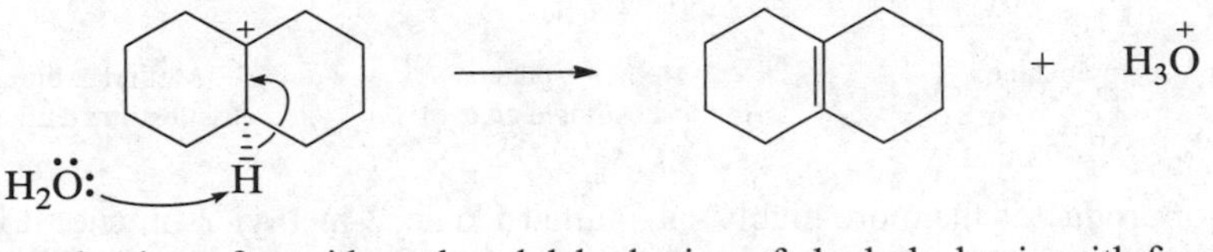

7.20   In writing mechanisms for acid-catalyzed dehydration of alcohols, begin with formation of the carbocation intermediate:

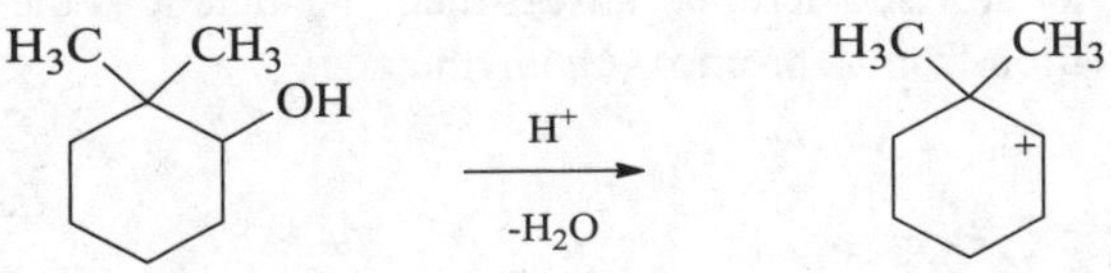

2,2-Dimethylcyclohexanol   2,2-Dimethylcyclohexyl cation

This secondary carbocation can rearrange to a more stable tertiary carbocation by a methyl group shift.

2,2-Dimethylcyclohexyl cation   1,2-Dimethylcyclohexyl cation
(secondary carbocation)   (tertiary carbocation)

Loss of a proton from the 1,2-dimethylcyclohexyl cation intermediate yields 1,2-dimethylcyclohexene.

1,2-Dimethylcyclohexyl cation   1,2-Dimethylcyclohexene

7.21 (*b*) All the β hydrogens of *tert*-butyl chloride are equivalent. Loss of any of these hydrogens along with the chlorine yields 2-methylpropene as the only alkene.

tert-Butyl chloride   2-Methylpropene

(*c*) All the β hydrogens of 3-bromo-3-ethylpentane are equivalent. Therefore, β elimination can give only 3-ethyl-2-pentene.

3-Bromo-3-ethylpentane   3-Ethyl-2-pentene

(*d*) There are two possible modes of β elimination from 2-bromo-3-methylbutane. Elimination in one direction provides 3-methyl-1-butene; elimination in the other gives 2-methyl-2-butene.

2-Bromo-3-methylbutane   3-Methyl-1-butene   2-Methyl-2-butene
                                (monosubstituted alkene)   (trisubstituted alkene)

The major product is the more highly substituted alkene, 2-methyl-2-butene. It is the more stable alkene and corresponds to removal of a hydrogen from the carbon that has the fewer hydrogens.

(*e*) Regioselectivity is not an issue here, because 3-methyl-1-butene is the only alkene that can be formed by β elimination from 1-bromo-3-methylbutane.

1-Bromo-3-methylbutane   3-Methyl-1-butene

(*f*) Two alkenes may be formed here. The more highly substituted one is 1-methylcyclohexene, and this is predicted to be the major product in accordance with Zaitsev's rule.

1-Iodo-1-methylcyclohexane   Methylenecyclohexane
(disubstituted)   1-Methylcyclohexene
(trisubstituted;
major product)

7.22 Elimination in 2-bromobutane can take place between C-1 and C-2 or between C-2 and C-3. Three alkenes are capable of being formed: 1-butene and the stereoisomers *cis*-2-butene and *trans*-2-butene.

2-Bromobutane   1-Butene   *cis*-2-Butene   *trans*-2-Butene

As predicted by Zaitsev's rule, the most stable alkene predominates. The major product is *trans*-2-butene.

7.23 There is only one hydrogen that can be removed during dehydrohalogenation of $ClF_2CCHClF$. The leaving group on the other carbon can be either F or Cl. Cl is the better leaving group of the two; therefore, the reaction is:

1,2-Dichloro-1,1,2-trifluoroethane   1-Chloro-1,2,2-trifluoroethane

7.24 The flow of electrons begins with an unshared pair of hydroxide and ends as an unshared pair of chloride.

1,2-Dichloro-1,1,2-trifluoroethane   1-Chloro-1,2,2-trifluoroethane

7.25 On treatment with strong bases such as sodium ethoxide, alkyl halides undergo dehydrohalogenation by an E2 mechanism. Three alkenes are possible.

Major product

According to the Zaitsev rule, the major product is the isomer with the more substituted double bond. 1,2-Dimethylcyclohexene has a tetrasubstituted double bond; it is the major product.

7.26   First, write the structural formula for each compound in its most stable conformation.

β hydrogens anti coplanar to Br

A

all β hydrogens gauche to Br

B

In its most stable conformation, compound A has two β hydrogens anti coplanar to Br, so each has the proper geometry for E2 elimination. Compound B has no β hydrogens anti coplanar to Br. Compound A undergoes dehydrohalogenation by the E2 mechanism faster than B.

7.27   These problems relate to isotope effects on the rate of elimination by the E2 mechanism.

(b)   As in part (a), the β C–H bond of $(CH_3)_2CHCD_2Br$ breaks faster than the β C–D bond of $(CH_3)_2CDCH_2Br$.

(c)   As in parts (a) and (b), the alkyl halide that has H at the β carbon reacts faster than the one with D at its β carbon.

$$CD_3CD_2\overset{\overset{\beta|}{CD_2}}{\underset{\underset{H}{|}}{C}}CH_2Br \quad \text{faster rate of E2 than} \quad CH_3CH_2\overset{\overset{\beta|}{CH_3}}{\underset{\underset{D}{|}}{C}}CH_2Br$$

H at β carbon   D at β carbon

7.28   The rate-determining step in the E1 elimination of 2-bromo-2-methylbutane is ionization to form a carbocation. A proton is lost from the β carbon in the next step, which is fast.

In the E1 mechanism shown, hydrogen is lost after the rate-determining step, an isotope effect of 3-8 is not expected. Only a slight effect may be observed, $k_H/k_D = 1\text{-}2$.

7.29   (b)   An S_N2 reaction of the unhindered primary bromide and a good nucleophile, that is also a strong base, gives an ether as the major product.

Potassium cyclohexanolate   Ethyl bromide   Cyclohexyl ethyl ether   Potassium bromide

(c) Reaction of *sec*-butyl bromide under methanol solvolysis conditions gives predominately an S_N1 substitution product.

sec-Butyl bromide

sec-Butyl methyl ether
2-Methoxybutane

(d) Elimination reaction of *sec*-butyl bromide under methanol solvolysis conditions with the added strong anionic base, sodium methoxide, gives alkenes as the major products.

sec-Butyl bromide

trans-2-butene   *cis*-2-butene   1-butene

(Alkene mixture is major product)

7.30 An S_N2 reaction between an alkoxide and an alkyl halide to give an ether in good yields depends on having either a primary or methyl halide as one of the reactants.

(a) The two possibilities for C-O bond formation are shown. Either indicated C—O bonds must be formed.

The two reactions of alkyl halides with alkoxides that could form these bonds will give alkenes as major products by an E2 reaction. The starting alkyl halides are secondary so are too hindered to give an S_N2 reaction with these strong alkoxide bases.

Propene

(Alkene mixture is major product)

(b) The two possibilities for reaction between an alkoxide and an alkyl halide are shown in the product. Either C-O bonds must be formed.

One possible reaction will give propene as the major product by an E2 reaction using a secondary alkyl halide. The other reaction gives the desired S_N2 reaction because the primary alkyl halide is less hindered.

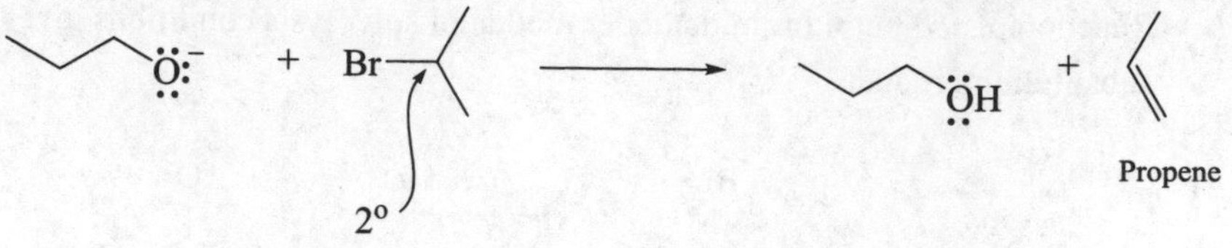

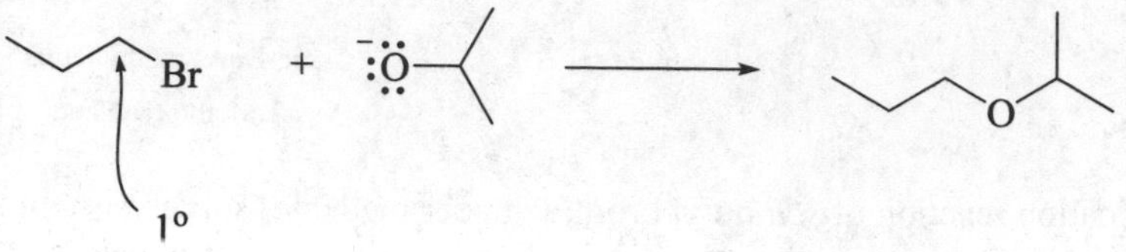

Propene

End of Chapter Problems

Structure and Nomenclature

7.31   (*a*)   1-Heptene is

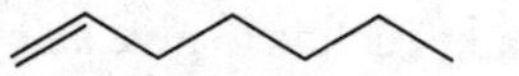

(*b*)   3-Ethyl-2-pentene is

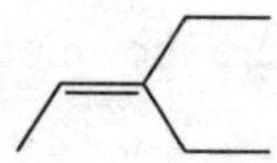

(*c*)   *cis*-3-Octene is

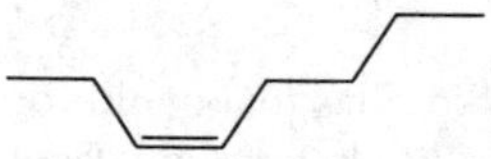

(*d*)   *trans*-1,4-Dichloro-2-butene is

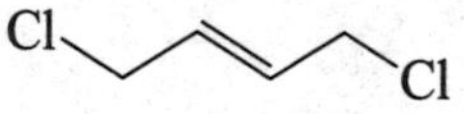

(*e*)   (*Z*)-3-Methyl-2-hexene is

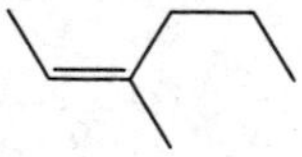

(*f*)   (*E*)-3-Chloro-2-hexene is

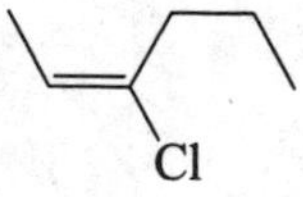

(*g*)   1-Bromo-3-methylcyclohexene is

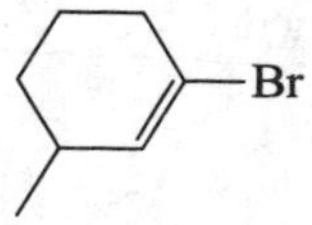

(*h*)   1-Bromo-6-methylcyclohexene is

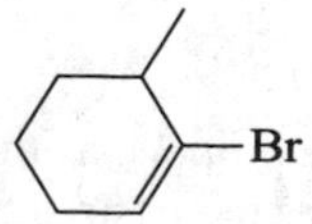

(i)   4-Methyl-4-penten-2-ol is

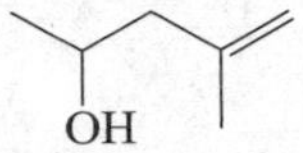

(j)   A vinyl group is —CH=CH$_2$.
       Vinylcycloheptane is

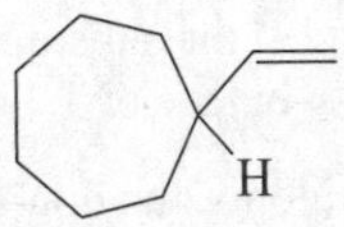

(k)   An allyl group is —CH$_2$CH=CH$_2$.
       1,1-Diallylcyclopropane is

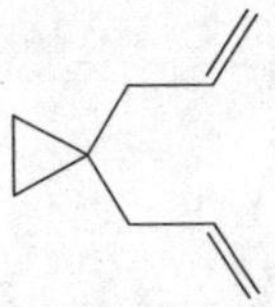

(l)   An isopropenyl substituent is —C̈=CH$_2$.
       with CH$_3$ on the C
       trans-1-Isopropenyl-3-methylcyclohexane is

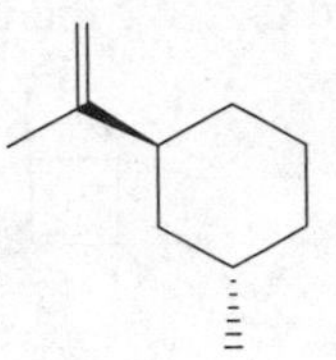

       or

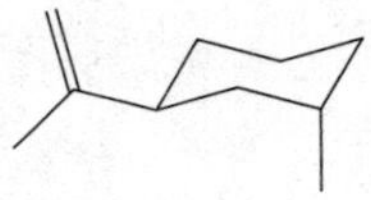

7.32   Alkenes with tetrasubstituted double bonds have four alkyl groups attached to the doubly bonded carbons. There is only one alkene of molecular formula C$_7$H$_{14}$ that has a tetrasubstituted double bond. Acceptable IUPAC names are 2,3-dimethyl-2-pentene or 2,3-dimethylpent-2-ene.

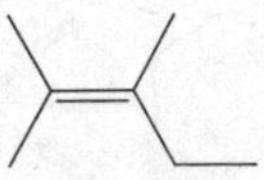

2,3-Dimethyl-2-pentene
or
2,3-Dimethylpent-2-ene

7.33   (a)   The longest chain that includes the double bond in (CH$_3$CH$_2$)$_2$C=CHCH$_3$ contains five carbon atoms, and so the parent alkene is a pentene. The numbering scheme that gives the double bond the lowest number is

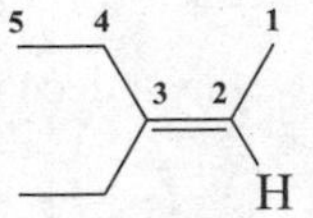

The compound is named 3-ethyl-2-pentene or 3-ethylpent-2-ene.

(b)   Write out the structure in detail, and identify the longest continuous chain that includes the double bond.

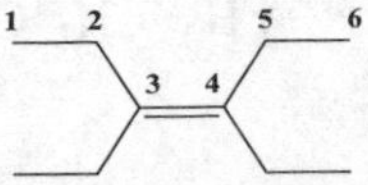

The longest chain contains six carbon atoms, and the double bond is between C-3 and C-4. The compound is named as a derivative of 3-hexene. There are ethyl substituents at C-3 and C-4. Acceptable IUPAC names are 3,4-diethyl-3-hexene or 3,4-diethylhex-3-ene.

(c)   Write out the structure completely.

The longest carbon chain contains four carbons. Number the chain so as to give the lowest numbers to the doubly bonded carbons, and list the substituents in alphabetical order. This compound may be named 1,1-dichloro-3,3-dimethyl-1-butene or 1,1-dichloro-3,3-dimethylbut-1-ene.

(*d*) The longest chain has five carbon atoms, the double bond is at C-1, and there are two methyl substituents. Acceptable IUPAC names are 4,4-dimethyl-1-pentene or 4,4-dimethylpent-1-ene.

(*e*) We number this trimethylcyclobutene derivative so as to provide the lowest number for the substituent at the first point of difference, remembering also that the double bond is between C-1 and C-2. We therefore number

rather than

Acceptable IUPAC names are 1,4,4-trimethylcyclobutene or 1,4,4-trimethylcyclobut-1-ene.

(*f*) The cyclohexane ring has a 1,2-cis arrangement of vinyl substituents. The compound is *cis*-1,2-divinylcyclohexane.

(*g*) Name this compound as a derivative of cyclohexene. It may be named 1,2-divinylcyclohexene or 1,2-divinylcyclohex-1-ene.

7.34 (a) The structure of 2,6,10,14-tetramethyl-2-pentadecene is:

(b) The allyl group is —$CH_2CH=CH_2$. Allyl isothiocyanate is:

$$S=C=N-CH_2-CH=CH_2$$

(c) The isopropenyl group is 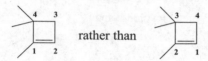. Replacing R on grandisol gives:

HOCH$_2$CH$_2$— ▯ —C(CH$_3$)=CH$_2$
H$_3$C   H

7.35 (*a*) Multifidene has two chirality centers and three double bonds. Neither the ring double bond nor the double bond of the C-4 vinyl (—CH=CH$_2$) substituent can give rise to stereoisomers, but the butenyl side chain can be either *E* or *Z*. Eight (2^3) stereoisomers are therefore possible. We can rationalize them as

—CH=CHCH$_2$CH$_3$
CH=CH$_2$

Stereoisomer	C-3	C-4	Butenyl double bond	
1	*R*	*R*	*E*	enantiomers
2	*S*	*S*	*E*	
3	*R*	*R*	*Z*	enantiomers
4	*S*	*S*	*Z*	
5	*R*	*S*	*E*	enantiomers
6	*S*	*R*	*E*	
7	*R*	*S*	*Z*	enantiomers
8	*S*	*R*	*Z*	

(*b*) Given the information that the alkenyl substituents are cis to each other, the number of stereoisomers is reduced by half. Four stereoisomers are therefore possible.

(*c*) Knowing that the butenyl group has a *Z* double bond reduces the number of possibilities by half. Two stereoisomers are possible.

(*d*) The two stereoisomers are

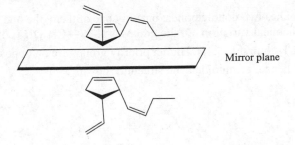

and

(*e*) These two stereoisomers are enantiomers. They are nonsuperimposable mirror images.

Mirror plane

7.36 Natural sphingosine has the S and R configurations at its two stereogenic centers, as indicated, and the E geometry of the double bond. There are seven other stereoisomers with this same constitution, three more with the E geometry of the alkene, and a set of four that have the Z double bond.

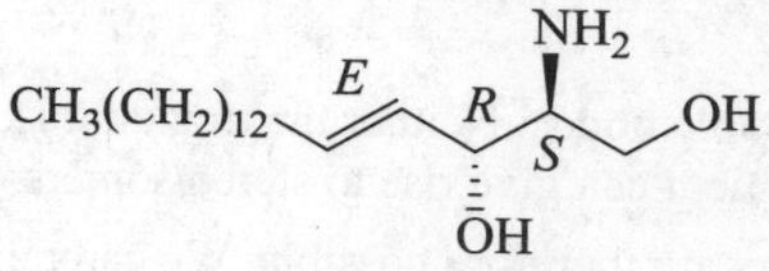

Sphingosine

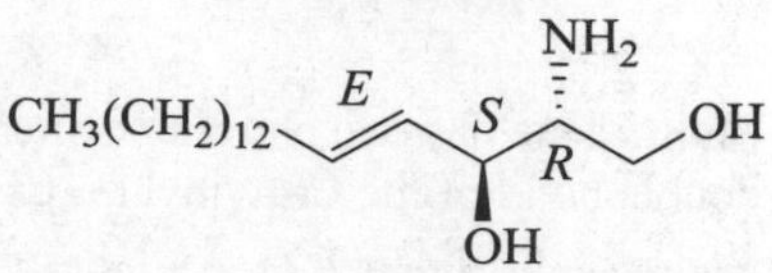

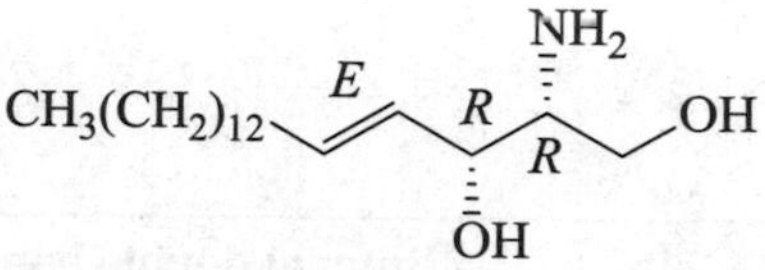

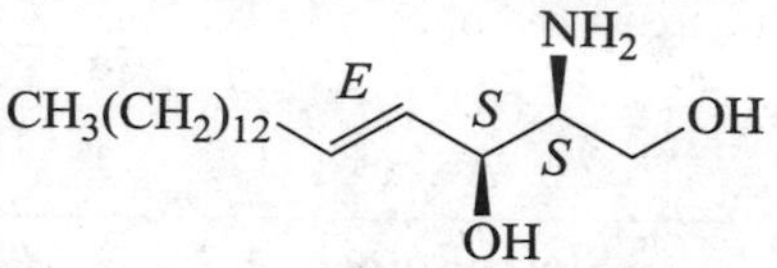

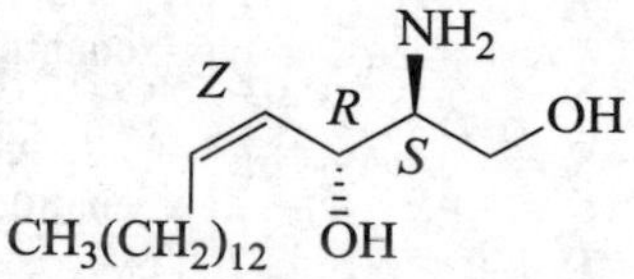

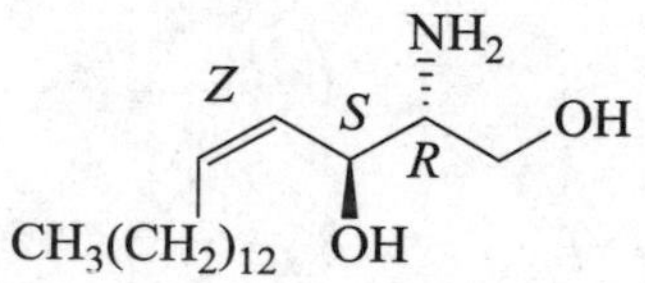

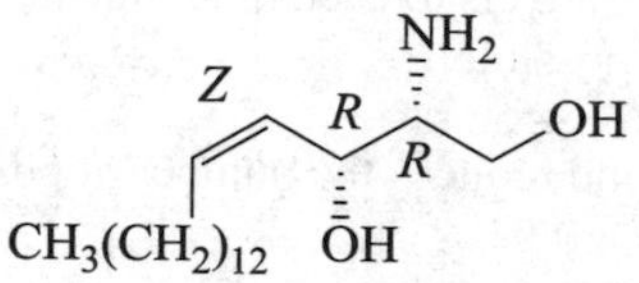

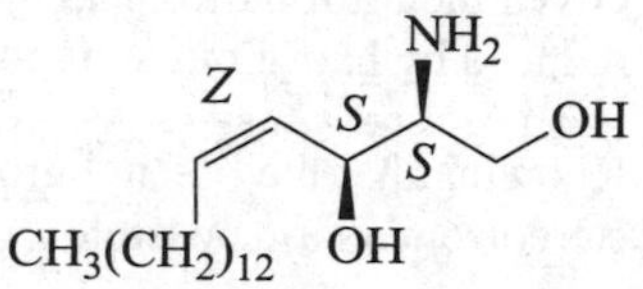

7.37 (*a*) The E configuration means that the higher priority groups are on opposite sides of the double bond.

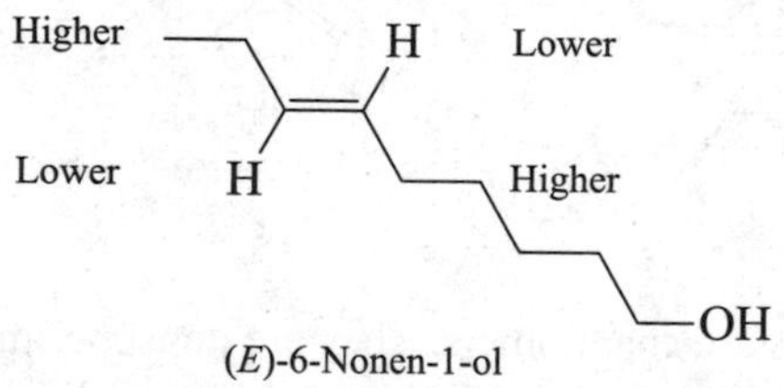

(*E*)-6-Nonen-1-ol

(*b*) Geraniol has two double bonds, but only one of them, the one between C-2 and C-3, is capable of stereochemical variation. Of the groups at C-2, —CH$_2$OH is of higher priority than —H. At C-3, —CH$_2$CH$_2$ outranks CH$_3$. Higher priority groups are on opposite sides of the double bond in the E isomer; hence, geraniol has the structure shown.

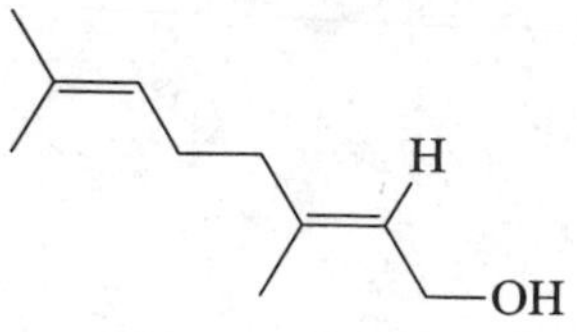

Geraniol

(c) Because nerol is a stereoisomer of geraniol, it has the same constitution and differs from geraniol only in having the *Z* configuration of the double bond.

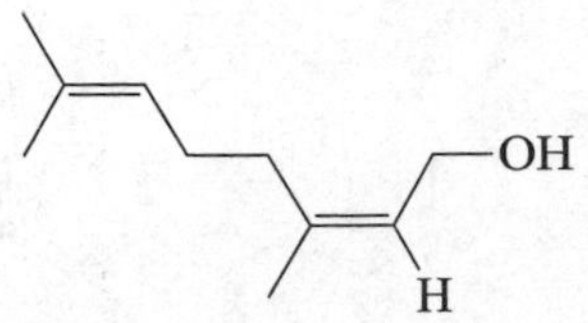

Nerol

(d) Beginning at the C-6, C-7 double bond, we see that the propyl group is of higher priority than the methyl group at C-7. Because the C-6, C-7 double bond is *E*, the propyl group must be on the opposite side of the higher priority group at C-6, where the CH_2 fragment has a higher priority than hydrogen. We therefore write the stereochemistry of the C-6, C-7 double bond as

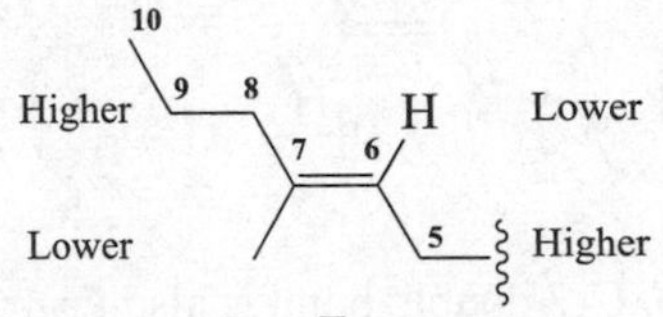

E

At C-2, CH_2OH is of higher priority than H; and at C-3, CH_2CH_2C- is of higher priority than CH_2CH_3. The double-bond configuration at C-2 is *Z*. Therefore

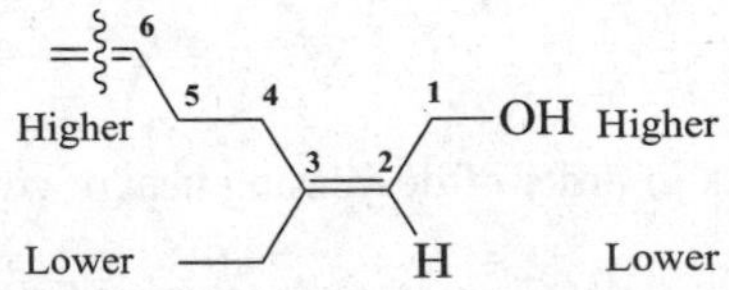

Z

Combining the two partial structures, we obtain for the full structure of the codling moth's sex pheromone

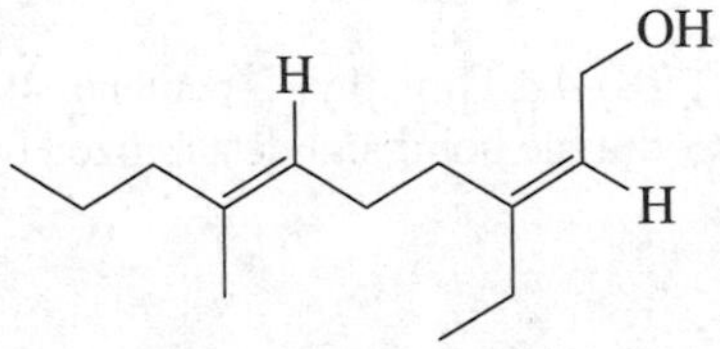

The compound is (2*Z*,6*E*)-3-ethyl-7-methyl-2,6-decadien-1-ol.

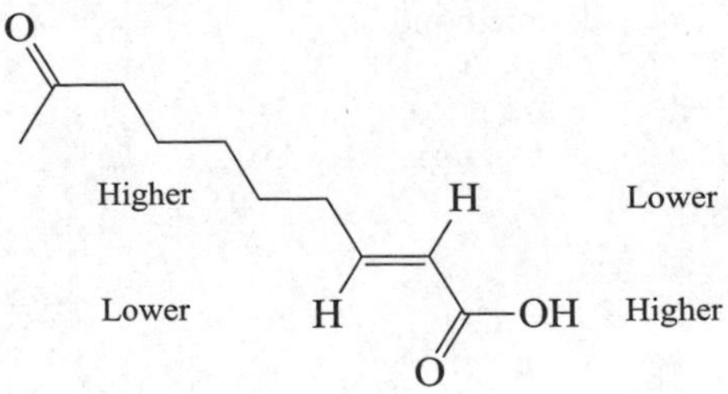

(e) The sex pheromone of the honeybee is (*E*)-9-oxo-2-decenoic acid, with the structure (*f*) Looking first at the C-2, C-3 double bond of the cecropia moth's growth hormone.

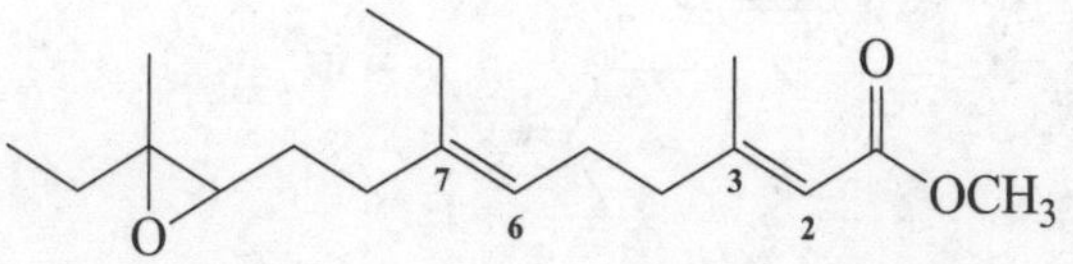

we find that its configuration is *E*, because the higher priority groups are on opposite sides of the double bond.

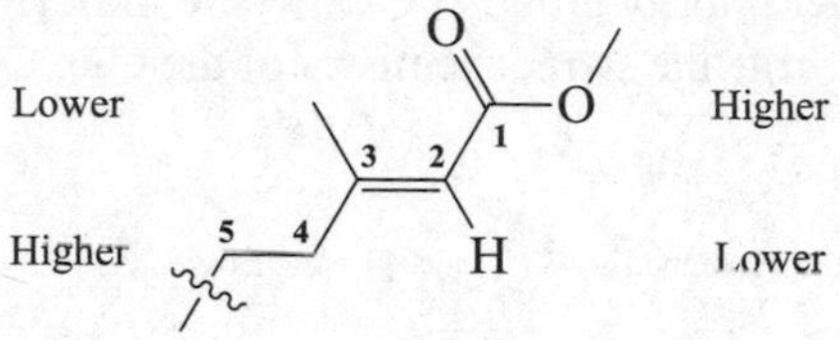

The configuration of the C-6, C-7 double bond is also *E*.

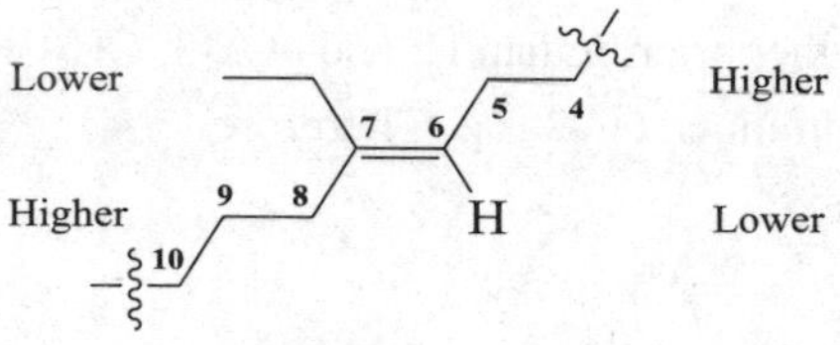

7.38 The alkenes are listed as follows in order of decreasing heat of combustion:

(*e*)    2,4,4-Trimethyl-2-pentene; 5293 kJ/mol (1264.9 kcal/mol). Highest heat of combustion because it is C_8H_{16}; all others are C_7H_{14}.

(*a*)    1-Heptene; 4658 kJ/mol (1113.4 kcal/mol). Monosubstituted double bond; therefore least stable C_7H_{14} isomer.

(*d*)   (*Z*)-4,4-Dimethyl-2-pentene; 4650 kJ/mol (1111.4 kcal/mol). Disubstituted double bond, but destabilized by van der Waals strain.

(*b*) 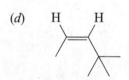   2,4-Dimethyl-1-pentene; 4638 kJ/mol (1108.6 kcal/mol). Disubstituted double bond.

(*c*) 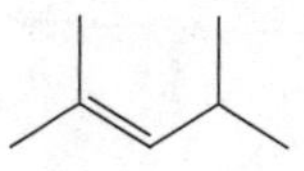 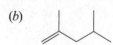   2,4-Dimethyl-2-pentene; 4632 kJ/mol (1107.1 kcal/mol). Trisubstituted double bond.

7.39 (*a*) 1-Methylcyclohexene is more stable; it contains a *trisubstituted* double bond, whereas 3-methylcyclohexene has only a disubstituted double bond.

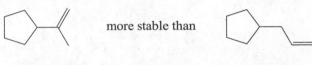

1-Methylcyclohexene   3-Methylcyclohexene

(*b*) Both isopropenyl and allyl are three-carbon alkenyl groups. The isopropenyl group is $H_2C=CCH_3$ and the allyl group is $—CH_2CH=CH_2$. Isopropenylcyclopentane has a disubstituted double bond and so is predicted to be more stable than allylcyclopentane, in which the double bond is monosubstituted.

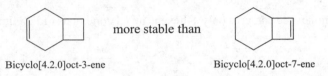

Isopropenylcyclopentane   Allylcyclopentane

(*c*) A double bond in a six-membered ring is less strained than a double bond in a four-membered ring; therefore, bicyclo[4.2.0]oct-3-ene is more stable.

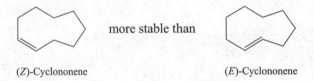

Bicyclo[4.2.0]oct-3-ene   Bicyclo[4.2.0]oct-7-ene

(*d*) Cis double bonds are more stable than trans double bonds when the ring is smaller than 11-membered. (*Z*)-Cyclononene has a cis double bond in a nine-membered ring, and is therefore more stable than (*E*)-cyclononene.

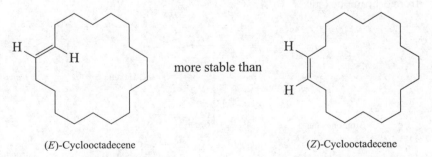

(*Z*)-Cyclononene   (*E*)-Cyclononene

(*e*) Trans double bonds are more stable than cis when the ring is large. Here the rings are 18-membered. Therefore, (*E*)-cyclooctadecene is more stable than (*Z*)-cyclooctadecene.

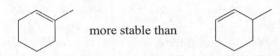

(*E*)-Cyclooctadecene   (*Z*)-Cyclooctadecene

7.40 (*a*) Carbon atoms that are involved in double bonds are sp^2-hybridized, with ideal bond angles of 120°. Incorporating an sp^2- hybridized carbon into a three-membered ring leads to more angle strain than incorporation of an sp^3-hybridized carbon with an ideal bond angle of 109°. 1-Methylcyclopropene has two sp^2-hybridized carbons in a three-membered ring and so has substantially more angle strain than methylenecyclopropane.

1-Methylcyclopropene   Methylenecyclopropane

The higher degree of substitution at the double bond in 1-methylcyclopropene is not sufficient to offset the increased angle strain, and so 1-methylcyclopropene is less stable than methylenecyclopropane.

(b) 3-Methylcyclopropene has a disubstituted double bond and two sp^2-hybridized carbons in its three-membered ring. It is the least stable of the isomers.

3-Methylcyclopropene

Reactions

7.41 In all parts of this exercise, write the structure of the alkyl halide in sufficient detail to identify the carbon that bears the halogen and the β-carbon atoms that bear at least one hydrogen. These are the carbons that become doubly bonded in the alkene product.

(a) 1-Bromohexane can give only 1-hexene under conditions of E2 elimination.

$$\text{Br} \xrightarrow[\text{E2}]{\text{base}}$$

1-Bromohexane   1-Hexene (only alkene)

(b) 2-Bromohexane can give both 1-hexene and 2-hexene on dehydrobromination. The 2-hexene fraction is a mixture of cis and trans stereoisomers.

$$\xrightarrow[\text{E2}]{\text{base}}$$

2-Bromohexane   1-Hexene   +   cis-2-Hexene   +   trans-2-Hexene

(c) Both a cis–trans pair of 2-hexenes and a cis–trans pair of 3-hexenes are capable of being formed from 3-bromohexane.

$$\xrightarrow[\text{E2}]{\text{base}}$$

3-Bromohexane

cis-2-Hexene   +   trans-2-Hexene

cis-3-Hexene   +   trans-3-Hexene

(d) Dehydrobromination of 2-bromo-2-methylpentane can involve one of the hydrogens of either a methyl group (C-1) or a methylene group (C-3). Elimination of a β hydrogen from either methyl group gives the same product.

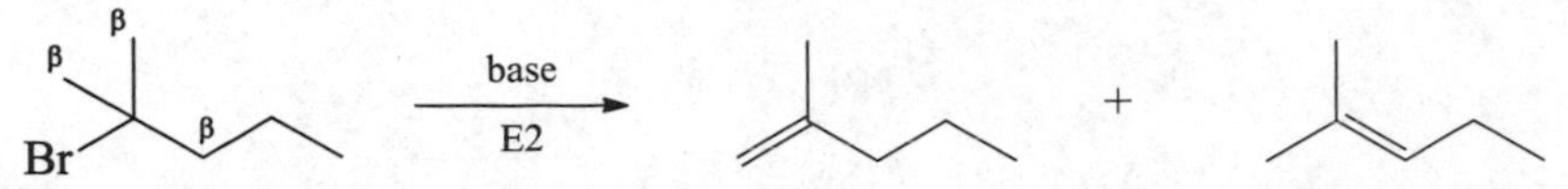

2-Bromo-2-methylpentane   2-Methyl-1-pentene   2-Methyl-2-pentene

Neither alkene is capable of existing in stereoisomeric forms, so these two are the only products of E2 elimination.

(*e*) 2-Bromo-3-methylpentane can undergo dehydrohalogenation by loss of a proton from either C-1 or C-3. Loss of a proton from C-1 gives 3-methyl-1-pentene.

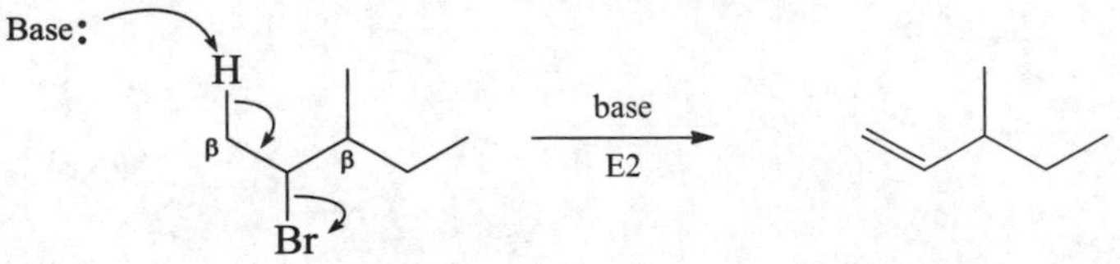

2-Bromo-3-methylpentane   3-Methyl-1-pentene

Loss of a proton from C-3 gives a mixture of (*E*)- and (*Z*)-3-methyl-2-pentene.

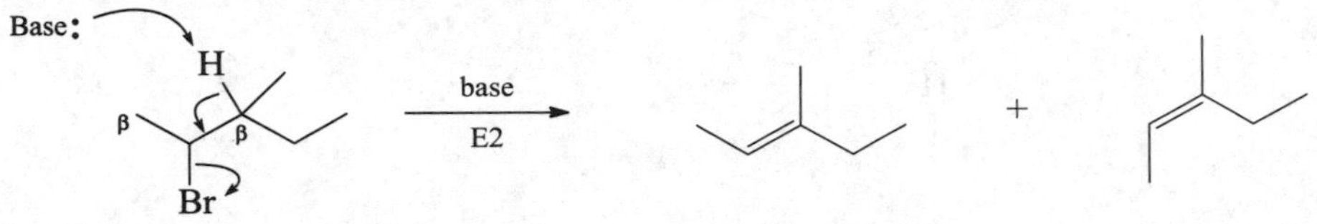

2-Bromo-3-methylpentane   (*E*)-3-Methyl-2-pentene   (*Z*)-3-Methyl-2-pentene

(*f*) Three alkenes are possible from 3-bromo-2-methylpentane. Loss of the C-2 proton gives 2-methyl-2-pentene.

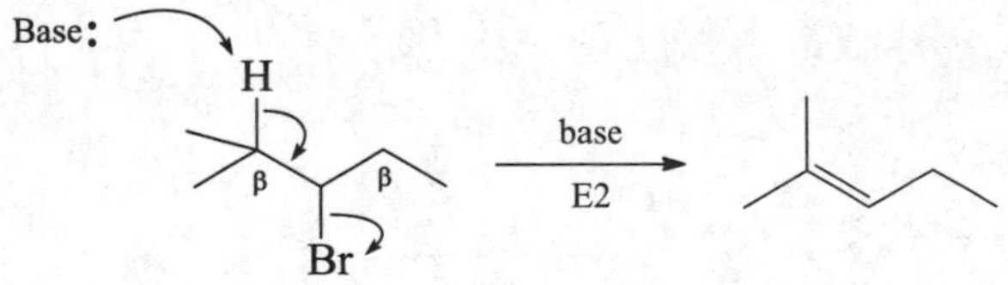

3-Bromo-2-methylpentane   2-Methyl-2-pentene

Abstraction of a proton from C-4 can yield either (*E*)- or (*Z*)-4-methyl-2-pentene.

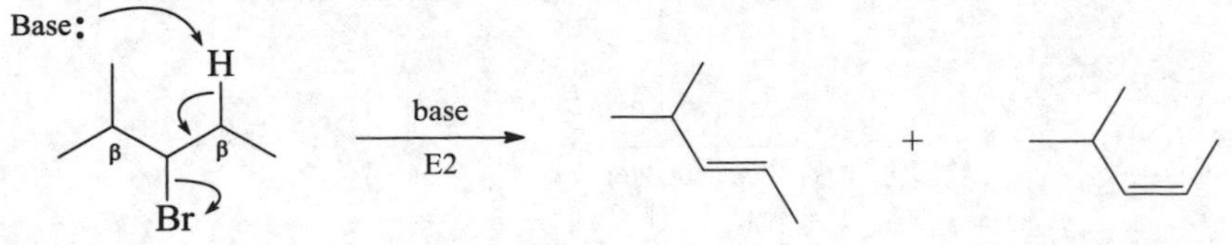

3-Bromo-2-methylpentane   (*E*)-4-Methyl-2-pentene   (*Z*)-4-Methyl-2-pentene

(*g*) Proton abstraction from the C-3 methyl group of 3-bromo-3-methylpentane yields 2-ethyl-1-butene.

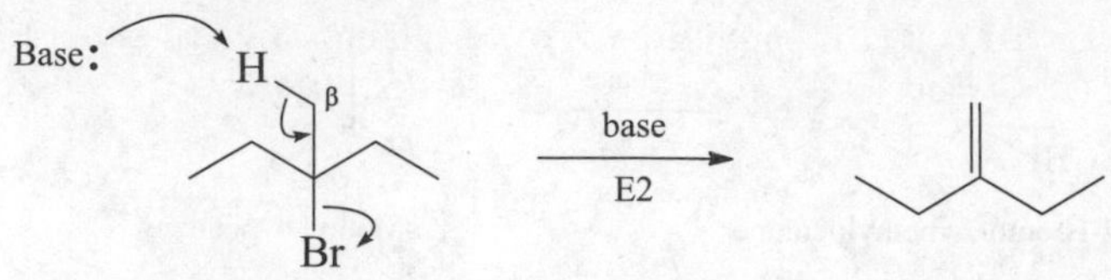

3-Bromo-3-methylpentane   2-Ethyl-1-butene

Stereoisomeric 3-methyl-2-pentenes are formed by proton abstraction from C-2. Loss of a proton from either β carbon gives the same two stereoisomeric products.

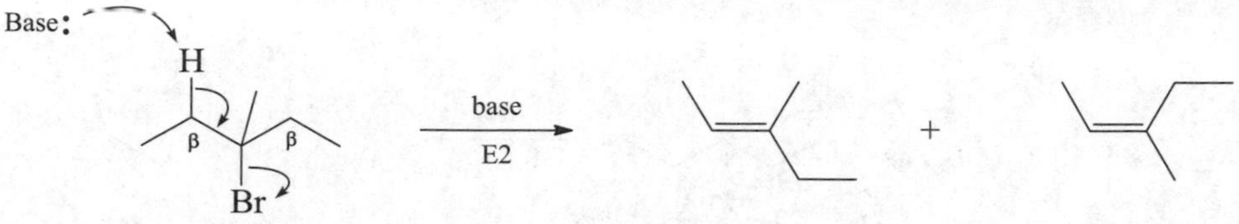

3-Bromo-3-methylpentane   (*E*)-3-Methyl-2-pentene   (*Z*)-3-Methyl-2-pentene

(*h*) Only 3,3-dimethyl-1-butene may be formed under conditions of E2 elimination from 3-bromo-2,2-dimethylbutane.

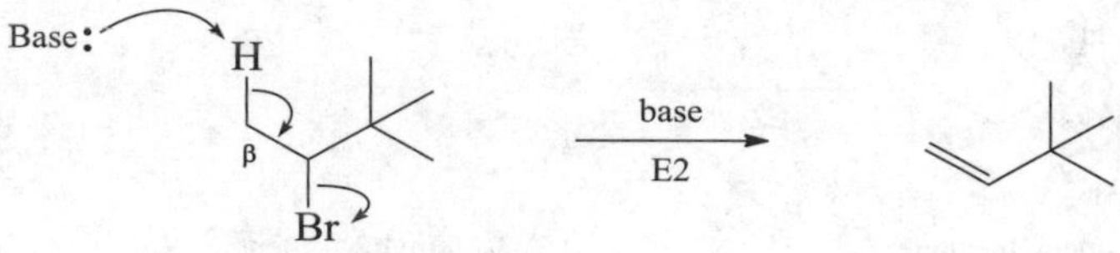

3-Bromo-2,2-dimethylbutane   3,3-Dimethyl-1-butene

7.42 (*a*) The reaction that takes place with 1-bromo-3,3-dimethylbutane is an E2 elimination involving loss of the bromine at C-1 and abstraction of the proton at C-2 by the strong base potassium *tert*-butoxide, yielding a single alkene.

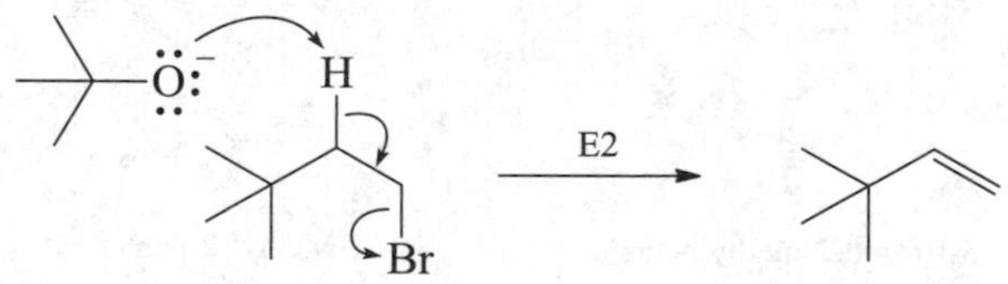

1-Bromo-3,3-dimethylbutane   3,3-dimethyl-1-butene

(*b*) Two alkenes are capable of being formed in this β elimination reaction.

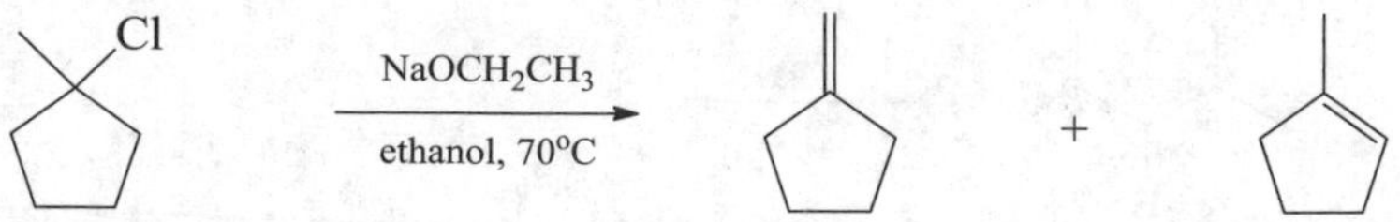

1-Methylcyclopentyl choride   Methylenecyclopentane   1-Methylcyclopentene

The more highly substituted alkene is 1-methylcyclopentene; it is the major product of this reaction. According to Zaitsev's rule, the major alkene is formed by proton removal from the β carbon that has the fewest hydrogens.

(c) Acid-catalyzed dehydration of 3-methyl-3-pentanol can lead either to 2-ethyl-1-butene or to a mixture of (E)- and (Z)-3-methyl-2-pentene.

3-Methyl-3-pentanol   2-Ethyl-1-butene   (E)-3-Methyl-2-pentene   (Z)-3-Methyl-2-pentene

The major product is a mixture of the trisubstituted alkenes, (E)- and (Z)-3-methyl-2-pentene. Of these two stereoisomers, the E isomer is slightly more stable and is expected to predominate.

(d) Acid-catalyzed dehydration of 2,3-dimethyl-2-butanol can proceed in either of two directions.

2,3-Dimethyl-2-butanol   2,3-Dimethyl-1-butene   2,3-Dimethyl-2-butene
                                      (disubstituted)   (tetrasubstituted)

The major alkene is the one with the more highly substituted double bond, 2,3-dimethyl-2-butene. Its formation corresponds to Zaitsev's rule in that a proton is lost from the β carbon that has the fewest hydrogens.

(e) Only a single alkene is capable of being formed on E2 elimination from this alkyl iodide. Stereoisomeric alkenes are not possible, and because both β hydrogens are equivalent, regioisomers cannot be formed either.

3-Iodo-2,4-dimethylpentane   2,4-Dimethyl-2-pentene

(f) Despite the structural similarity of this alcohol to the alkyl halide in the preceding part of this problem, its dehydration is more complicated. The initially formed carbocation is secondary and can rearrange to a more stable tertiary carbocation by a hydride shift.

2,4-Dimethyl-3-pentanol   Secondary carbocation   Tertiary carbocation
                                   (less stable)   (more stable)

The tertiary carbocation, once formed, can give either 2,4-dimethyl-1-pentene or 2,4-dimethyl-2-pentene by loss of a proton.

Tertiary carbocation   2,4-Dimethyl-1-pentene   2,4-Dimethyl-2-pentene
                                (disubstituted)   (trisubstituted)

The proton is lost from the methylene group in preference to the methyl group. The major alkene is the more highly substituted one, 2,4-dimethyl-2-pentene.

7.43   In all parts of this problem you need to reason backward from an alkene to an alkyl bromide of molecular formula $C_7H_{13}Br$ that gives *only* the desired alkene under E2 elimination conditions. Recall that the carbon–carbon double bond is formed by loss of a proton from one of the carbons that becomes doubly bonded and a bromine from the other.

(*a*)   Cycloheptene is the only alkene formed by an E2 elimination reaction of cycloheptyl bromide.

Cycloheptyl bromide   Cycloheptene

(*b*)   (Bromomethyl)cyclohexane is the correct answer. It gives methylenecyclohexane as the *only* alkene under E2 conditions.

(Bromomethyl)cyclohexane   Methylenecyclohexane

1-Bromo-1-methylcyclohexane is not correct. It gives a mixture of 1-methylcyclohexene and methylenecyclohexane on elimination.

1-Bromo-1-methylcyclohexane   Methylenecyclohexane   1-Methylcyclohexene

(*c*)   In order for 4-methylcyclohexene to be the only alkene, the starting alkyl bromide must be 1-bromo-4-methylcyclohexane. Either the cis or the trans isomer may be used, although the cis will react more readily because its more stable conformation (equatorial methyl) has an axial bromine.

cis-1-Bromo-4-methylcyclohexane   *trans*-1-Bromo-4-methylcyclohexane   4-Methylcyclohexene
(more stable conformation of cis isomer;   (less stable conformation of trans isomer
reacts faster)   is required for E2;
                                    reacts slower)

1-Bromo-3-methylcyclohexane is incorrect; its dehydrobromination yields a mixture of 3-methylcyclohexene and 4-methylcyclohexene.

cis- or trans-1-Bromo-3-methylcyclohexane   3-Methylcyclohexene   4-Methylcyclohexene

(d) The alkyl bromide must be primary in order for the desired alkene to be the only product of E2 elimination.

2-Cyclopentylethyl bromide   Vinylcyclopentane

If 1-cyclopentylethyl bromide were used, a mixture of regioisomeric alkenes would be formed, with the desired vinylcyclopentane being the minor component of the mixture.

1-Cyclopentylethyl bromide   Ethylidenecyclopentane   Vinylcyclopentane
                                   (major product)   (minor product)

(e) Either cis- or trans-1-bromo-3-isopropylcyclobutane would be appropriate here.

cis- or trans-1-Bromo-3-isopropylcyclobutane   3-Isopropylcyclobutene

(f) The desired alkene is the exclusive product formed on E2 elimination from 1-bromo-1-tert-butylcyclopropane.

1-Bromo-1-tert-butylcyclopropane   1-tert-Butylcyclopropene

7.44 (a) Both 1-bromopropane and 2-bromopropane yield propene as the exclusive product of E2 elimination.

$$CH_3CH_2CH_2Br \quad \text{or} \quad CH_3CHCH_3 \xrightarrow[\text{E2}]{\text{base}} CH_3CH=CH_2$$
$$\phantom{CH_3CH_2CH_2Br \quad \text{or} \quad CH_3CH}|$$
$$\phantom{CH_3CH_2CH_2Br \quad \text{or} \quad CH_3CH}Br$$

1-Bromopropane   2-Bromopropane   Propene

(b) 2-Methylpropene is formed on dehydrobromination of either tert-butyl bromide or isobutyl bromide.

$$(CH_3)_3CBr \quad \text{or} \quad (CH_3)_2CHCH_2Br \xrightarrow[\text{E2}]{\text{base}} (CH_3)_2CH=CH_2$$

tert-Butyl bromide   Isobutyl bromide   2-Methylpropene

(c) A tetrabromoalkane is required as the starting material to form a tribromoalkene under E2 elimination conditions. Either 1,1,2,2-tetrabromoethane or 1,1,1,2-tetrabromoethane is satisfactory.

$$\text{Br}_2\text{CHCHBr}_2 \quad \text{or} \quad \text{BrCH}_2\text{CBr}_3 \xrightarrow[\text{E2}]{\text{base}} \text{BrCH}=\text{CBr}_2$$

1,1,2,2-Tetrabromoethane   1,1,1,2-Tetrabromoethane   1,1,2-Tribromoethene

(*d*) The bromine substituent may be at either C-2 or C-3.

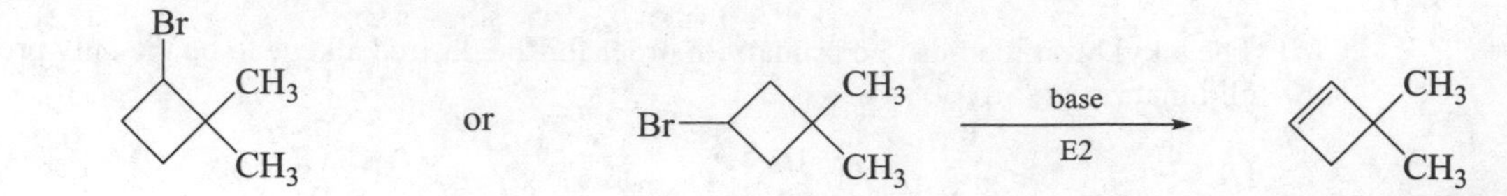

2-Bromo-1,1-dimethylcyclobutane   3-Bromo-1,1-dimethylcyclobutane   3,3-dimethylcyclobutene

7.45 (*a*) Heating an alcohol in the presence of an acid catalyst ($KHSO_4$) leads to dehydration with formation of an alkene. In this alcohol, elimination can occur in only one direction to give a mixture of cis and trans alkenes.

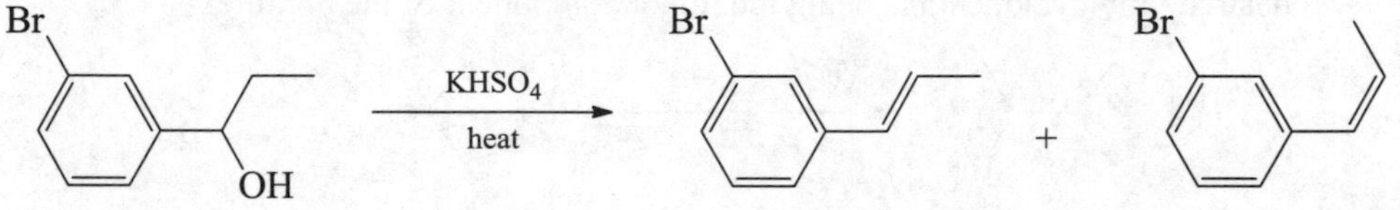

(major)   (Cis-trans mixture)

(*b*) Alkyl halides undergo E2 elimination on being heated with potassium *tert*-butoxide.

$$\text{ICH}_2\text{CH(OCH}_2\text{CH}_3)_2 \xrightarrow[\substack{(\text{CH}_3)_3\text{COH} \\ \text{heat}}]{\text{KOC(CH}_3)_3} \text{H}_2\text{C}=\text{C(OCH}_2\text{CH}_3)_2$$

(*c*) β Elimination can occur only in one direction, to give the alkene shown.

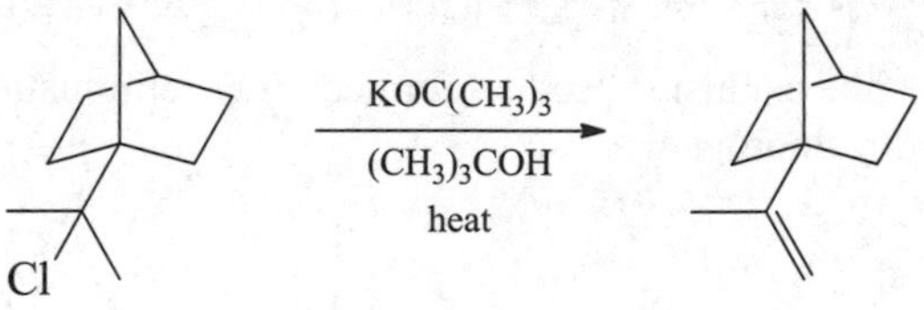

(*d*) The reaction is a conventional one of alcohol dehydration and proceeds as written in 76–78% yield.

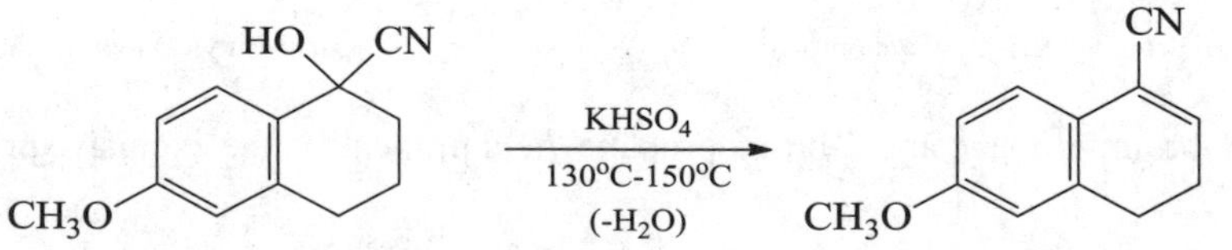

(*e*) Dehydration of citric acid occurs, giving aconitic acid.

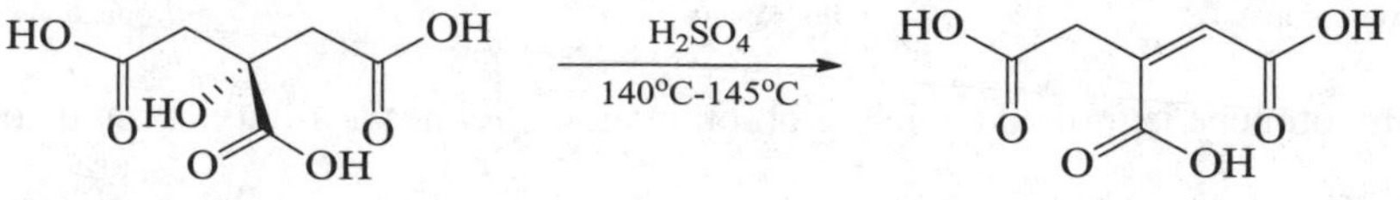

Citric acid   Aconitic acid

(*f*) Sequential double dehydrohalogenation gives the diene.

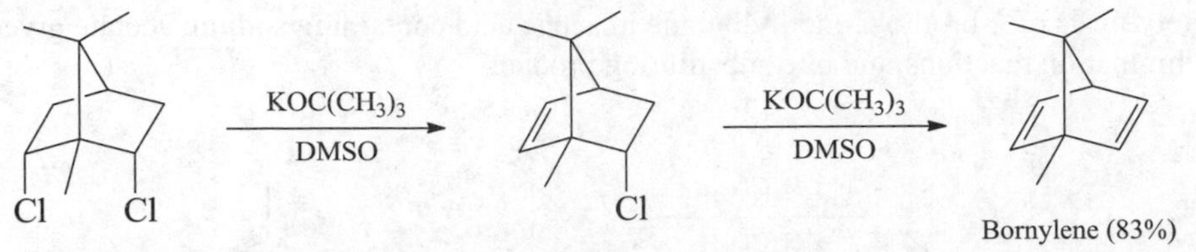

Bornylene (83%)

(g) This example has been reported in the chemical literature, and in spite of the complexity of the starting material, elimination proceeds in the usual way.

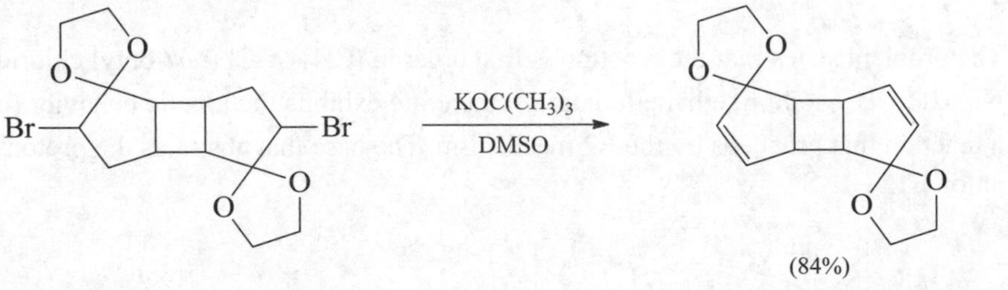

(84%)

(h) Again, we have a fairly complicated substrate, but notice that it is well disposed toward E2 elimination of the axial bromine.

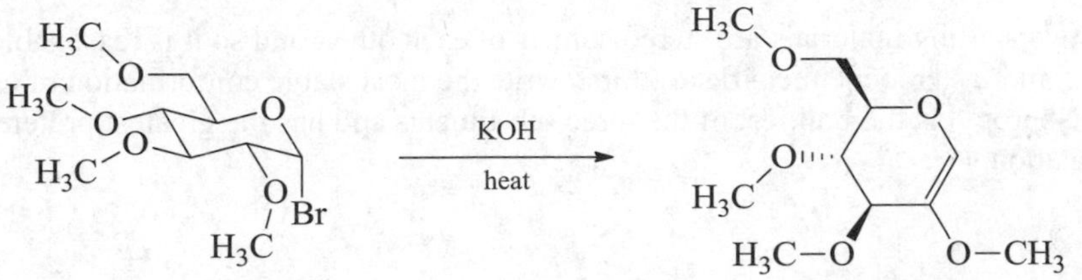

7.46 The first step of the reaction sequence forms a tosylate from the alcohol functional group. Tosylates are good leaving groups so in the second step, an E2 reaction occurs. The regiochemistry is controlled by the β hydrogen that is anti to the leaving group.

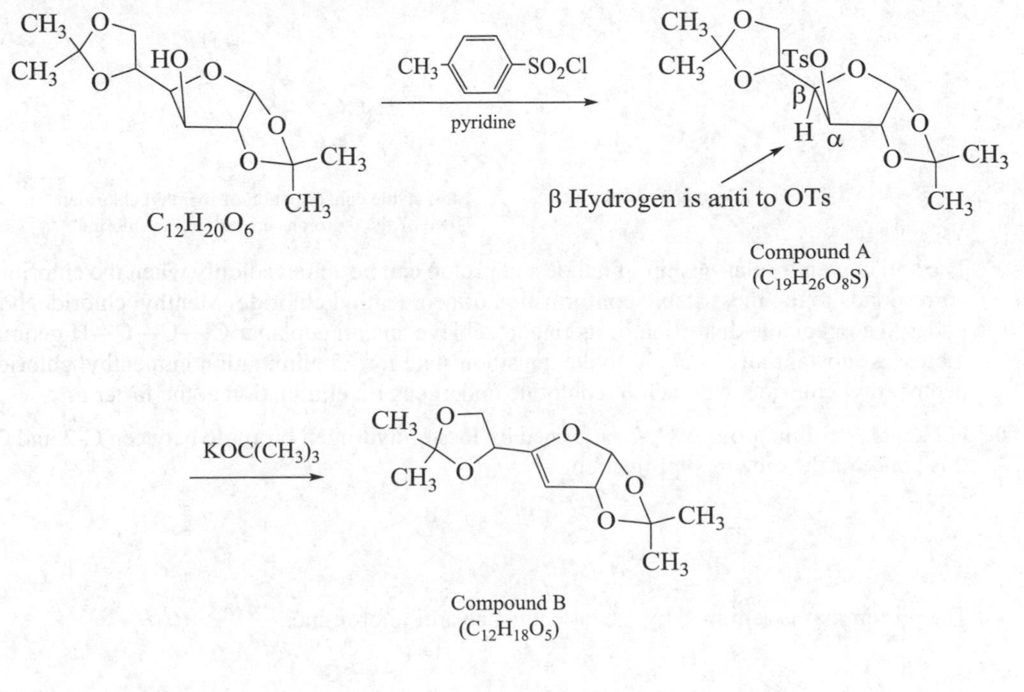

$C_{12}H_{20}O_6$

β Hydrogen is anti to OTs

Compound A
($C_{19}H_{26}O_8S$)

Compound B
($C_{12}H_{18}O_5$)

7.47 Solvolysis of 2-bromo-2-methylbutane in acetic acid containing sodium acetate gives two alkenes by elimination reactions and one substitution product.

2-Bromo-2-methylbutane

Mechanisms

7.48 The problem states that the reaction is first order in $(CH_3)_3CCl$ (*tert*-butyl chloride) and first order in $NaSCH_2CH_3$ (sodium ethanethiolate). It therefore exhibits the kinetic behavior (overall second order) of a reaction that proceeds by the E2 mechanism. The base that abstracts the proton from carbon is the anion $CH_3CH_2S^-$.

7.49 The two starting materials are stereoisomers of each other, and so it is reasonable to begin by examining each one in more stereochemical detail. First, write the most stable conformation of each isomer, keeping in mind that isopropyl is the bulkiest of the three substituents and has the greatest preference for an equatorial orientation.

Menthyl chloride

Most stable conformation of menthyl chloride:
all three β protons are gauche to the chlorine

Neomenthyl chloride

Most stable conformation of menthyl chloride:
two of the β protons are anti to the chlorine

The anti coplanar relationship of halide and proton can be achieved only when the chlorine is axial; this corresponds to the most stable conformation of neomenthyl chloride. Menthyl chloride, however, must undergo appreciable distortion of its ring to achieve an anti coplanar Cl—C—C—H geometry. Strain increases substantially in going to the transition state for E2 elimination in menthyl chloride but not in neomenthyl chloride. Neomenthyl chloride undergoes E2 elimination at the faster rate.

7.50 For Zaitsev elimination, A C=C is formed by loss of hydrogen bromide between C-2 and C-3. Sight down this bond for the Newman projection.

The proton that is removed by the base must be anti to bromine.

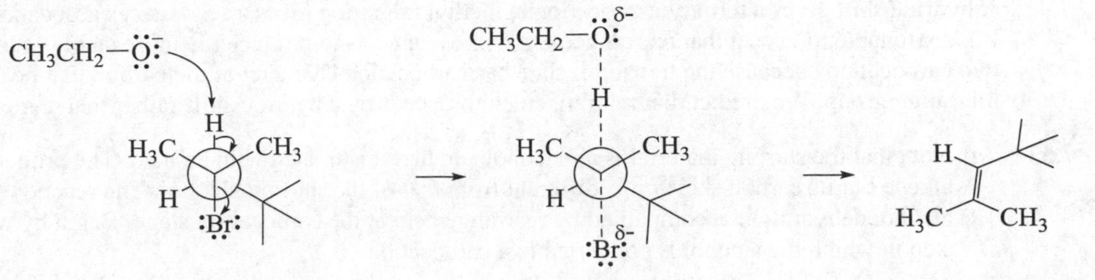

Newman projection for
2-bromo-2,4,4-trimethylpentane

E2 transition state
for Zaitsev elimination

2,4,4-Trimethyl-2-pentene

7.51 Choose B and treat it with a strong base such as $NaOCH_2CH_3$ to ensure that the reaction follows an E2 mechanism. The only possible E2 product is C.

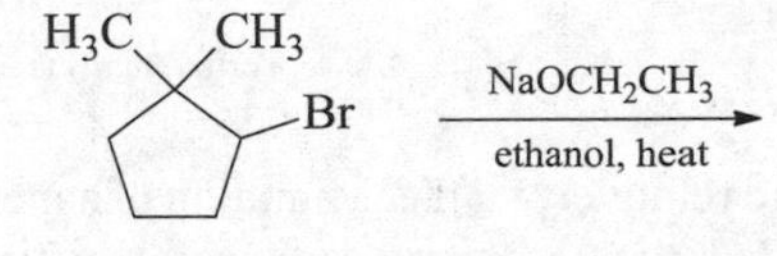

Dehydration of alcohol A would proceed through a carbocation, which could rearrange prior to loss of a proton and yield a mixture of products.

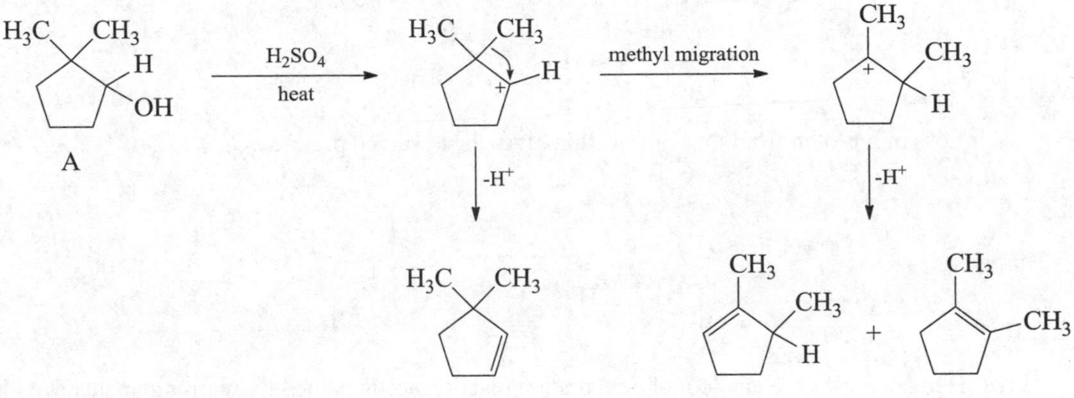

7.52 Begin by writing chemical equations for the processes specified in the problem. First consider rearrangement by way of a hydride shift:

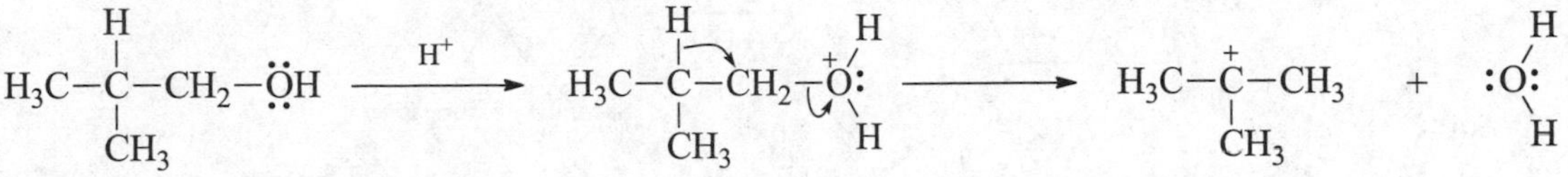

2-Methyl-1-propanol   Isobutyloxonium ion   Tertiary cation   Water

Rearrangement by way of a methyl group shift is as follows:

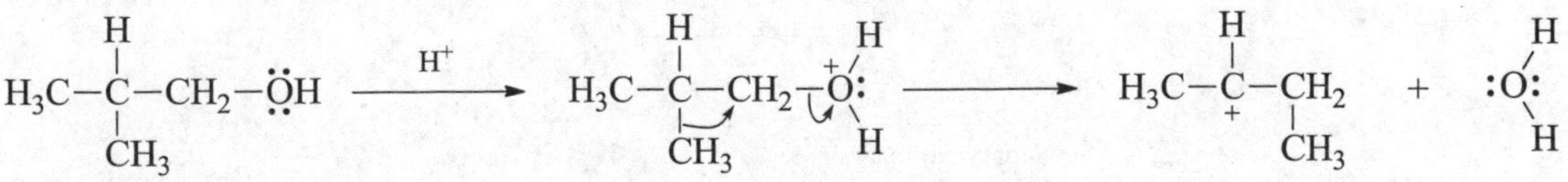

2-Methyl-1-propanol   Isobutyloxonium ion   Secondary cation   Water

A hydride shift gives a tertiary carbocation; a methyl migration gives a secondary carbocation. It is reasonable to expect that rearrangement will occur so as to produce the more stable of these two carbocations because the transition state has carbocation character at the carbon that bears the migrating group. We predict that rearrangement proceeds by a hydride shift rather than a methyl shift.

7.53 (a) Note that the starting material is an alcohol and that it is treated with an acid. The product is an alkene but its carbon skeleton is different from that of the starting alcohol. The reaction is one of alcohol dehydration accompanied by rearrangement at the carbocation stage. Begin by writing the step in which the alcohol is converted to a carbocation.

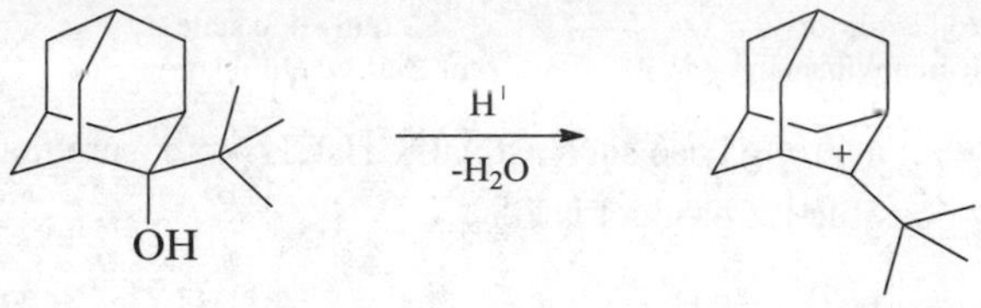

The carbocation is tertiary and relatively stable. Migration of a methyl group from the *tert*-butyl substituent, however, converts it to an isomeric carbocation, which is also tertiary.

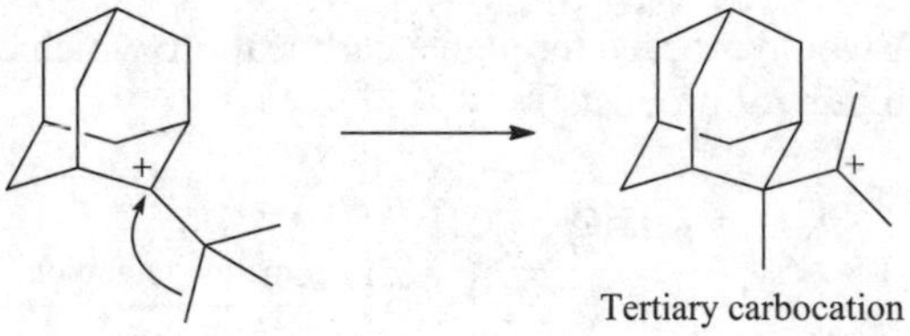

Loss of a proton from this carbocation gives the observed product.

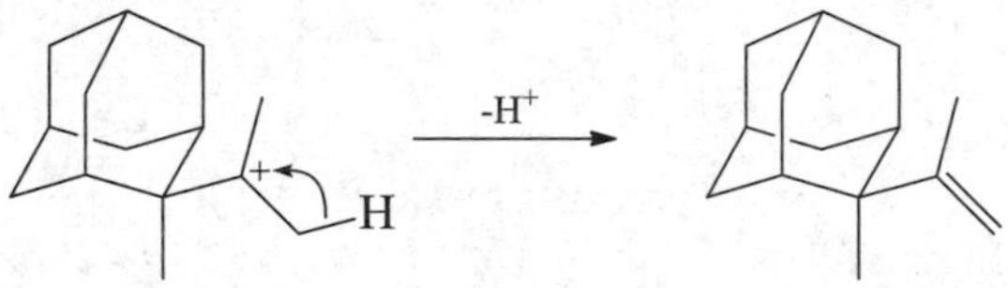

(b) Here also we have an alcohol dehydration reaction accompanied by rearrangement. The initially formed carbocation is secondary.

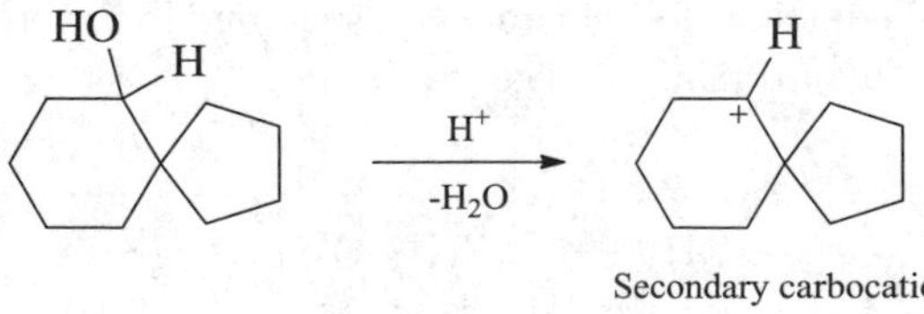

This cation can rearrange to a tertiary carbocation by an alkyl group shift.

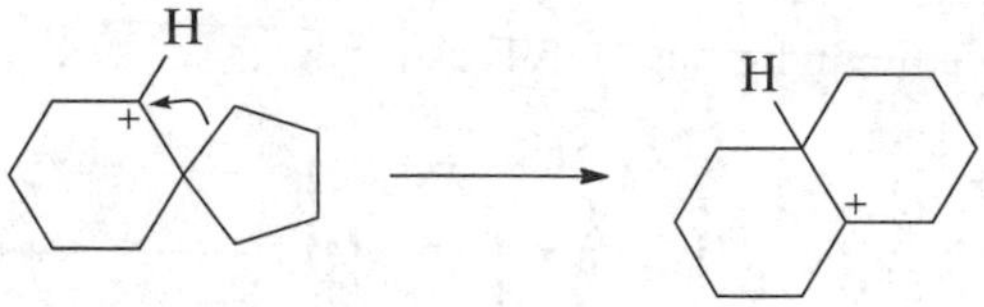

Loss of a proton from the tertiary carbocation gives the observed alkene.

Tertiary carbocation

(c) The reaction begins as a normal alcohol dehydration in which the hydroxyl group is protonated by the acid catalyst and then loses water from the oxonium ion to give a carbocation.

4-Methylcamphenilol   Secondary carbocation

We see that the final product, 1-methylsantene, has a rearranged carbon skeleton corresponding to a methyl shift, and so we consider the rearrangement of the initially formed secondary carbocation to a tertiary ion.

Secondary carbocation   Tertiary carbocation   1-Methylsantene

Deprotonation of the tertiary carbocation yields 1-methylsantene.

7.54 Formation of 1,2-dimethylcyclohexene begins by protonation of the hydroxyl group and loss of water to form a secondary carbocation.

2,2-Dimethylcyclohexanol   Secondary carbocation

The secondary carbocation can, as we have seen, rearrange by a methyl shift. 1,2-Dimethylcyclohexene is formed by loss of a proton.

Secondary carbocation   Tertiary carbocation   1,2-Dimethylcyclohexene

The same secondary carbocation can also rearrange by migration of one of the ring bonds.

Secondary carbocation   Tertiary carbocation

The tertiary carbocation formed by this rearrangement can lose a proton to give the observed byproduct.

Tertiary carbocation   Isopropylidenecyclopentane

7.55 Let's do both part (a) and part (b) together by reasoning mechanistically. The first step in any acid-catalyzed alcohol dehydration is proton transfer to the OH group.

2,2-Dimethyl-1-hexanol

But notice that because this alcohol does not have any hydrogens on its β carbon, it cannot dehydrate directly. Any alkenes that are formed must arise by rearrangement processes. Consider, for example, migration of either of the two equivalent methyl groups at C-2.

Tertiary carbocation

The resulting tertiary carbocation can lose a proton in three different directions.

Tertiary carbocation   3-Methyl-3-heptene   2-Ethyl-1-hexene   3-Methyl-2-heptene
                             (mixture of E and Z isomers)   (mixture of E and Z isomers)

The alkene mixture shown in the preceding equation constitutes part of the answer to part (b). None of the alkenes arising from methyl migration is 2-methyl-2-heptene, the answer to part (a), however.

What other group can migrate? The other group attached to the β carbon is a butyl group. Consider its migration to give a different tertiary carbocation.

butyl group   Tertiary carbocation

Loss of a proton from the carbocation gives the alkene in part (a).

Tertiary carbocation   2-Methyl-2-heptene

A proton can also be lost from one of the methyl groups to give 2-methyl-1-heptene. This is the last alkene constituting the answer to part (b).

Tertiary carbocation   2-Methyl-1-heptene

7.56 (*a*),(*b*) The reaction conditions of tertiary alkyl halides in 80% ethanol/20% water suggests S_N1 and E1 reactions. Both S_N1 and E1 reactions generate carbocations. In addition, formation of the same percentages of products under identical reaction conditions from two different reactants suggests a common carbocation intermediate.

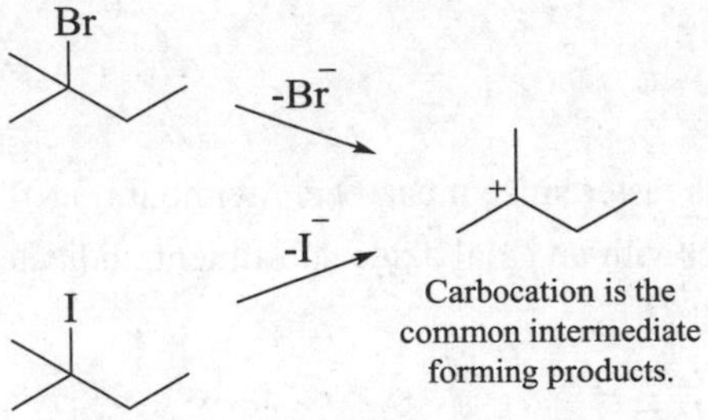

(*c*),(*d*) Formation of the carbocation is the rate determining, or slow step, of the reaction mechanism. Iodide is a better leaving group so 2-iodo-2-methylbutane reacts faster to form both substitution and elimination products..

(*e*) The solvent is made up of ethanol and water. Each of these react with the carbocation intermediate to form the products.

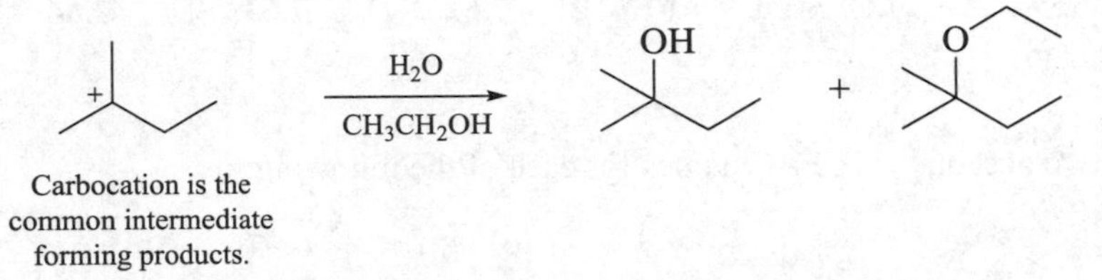

(*f*) Two elimination products are formed from loss of a β hydrogen.

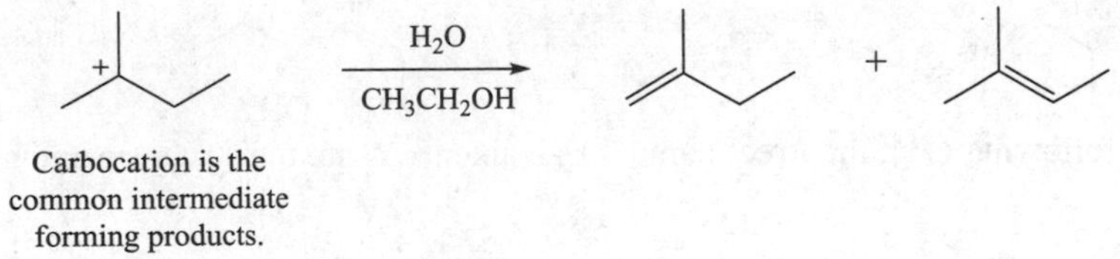

(*g*) The identical common carbocation intermediate <u>and</u> identical reaction conditions will lead to identical ratios of elimination and substitution products.

7.57 The elimination reaction, E2, occurs in the first reaction with loss of the β hydrogen that is anti to the tosylate leaving group in the lower energy cyclohexane chair conformation. A trisubstituted alkene is formed.

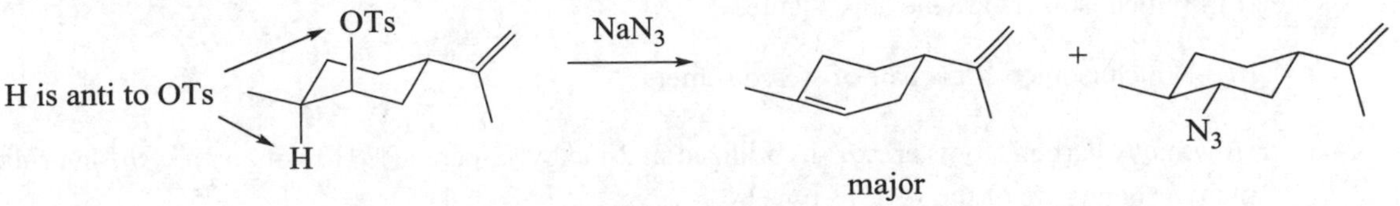

In the reaction with an equatorial tosylate group, the ring flipping to less stable chair conformation must first occur by ring flipping prior to an elimination reaction.

less stable chair;
H is anti to OTs

The S_N2 reaction is faster since it can occur without ring-flipping to a higher energy conformation. The substitution product with an axial azide substituent, indicates inversion of configuration that is expected from an S_N2 reaction.

major

Answers to Interpretive Problems 7

7.58 A; **7.59** A; **7.60** D; **7.61** B; **7.62** D; **7.63** D; **7.64** B; **7.65** D

SELF-TEST

1.   Give two acceptable IUPAC names for each of the following:

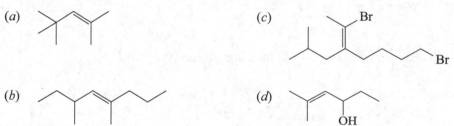

(*a*)   (*c*)

(*b*)   (*d*)

2.   Each of the following is an incorrect name for an alkene. Write the structure and give the correct name for each.

(*a*) 2-Ethyl-3-methyl-2-butene   (*c*) 2,3-Dimethylcyclohexene

(*b*) 2-Chloro-5-methyl-5-hexene   (*d*) 2-Methyl-1-penten-4-ol

3.   (*a*) Write the structures of all the alkenes of molecular formula C_5H_{10}.

(*b*) Which isomer is the most stable?

(*c*) Which isomers are the least stable?

(*d*) Which isomers are a pair of stereoisomers?

4.   How many carbon atoms are sp^2-hybridized in 2-methyl-2-pentene? How many are sp^3-hybridized? How many σ bonds are of the sp^2–sp^3 type?

5.   Write the structure, clearly indicating the stereochemistry, of each of the following:

(*a*) (*Z*)-4-Ethyl-3-methyl-3-heptene

    (b) (E)-1,2-Dichloro-3-methyl-2-hexene

    (c) (E)-3-Methyl-3-penten-1-ol

6. Write structural formulas for two alkenes of molecular formula C_7H_{14} that are stereoisomers of each other and have a trisubstituted double bond. Give systematic names for each.

7. Write structural formulas for the reactant or product(s) omitted from each of the following. If more than one product is formed, indicate the major one.

(a)

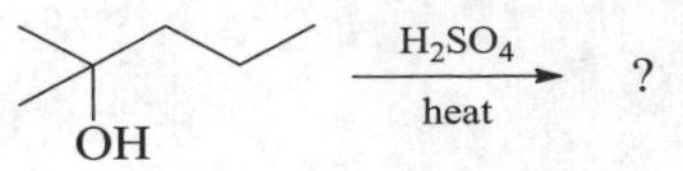

(c)

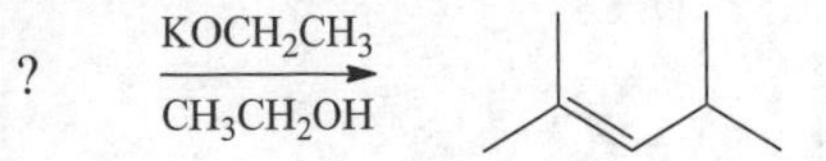

(only alkene formed)

(b)

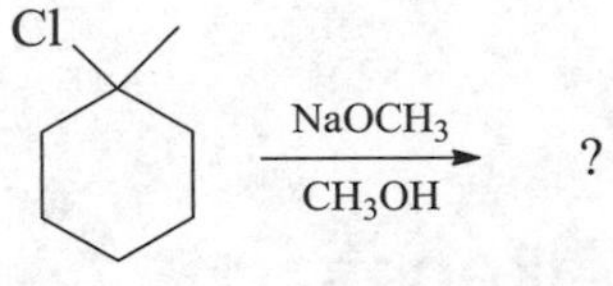

(d)

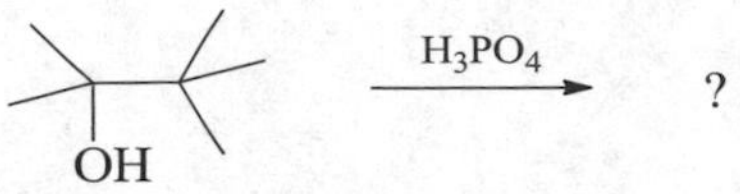

8. Write the structure of the $C_6H_{13}Br$ isomer that is *not* capable of undergoing E2 elimination.

9. Write a stepwise mechanism for the formation of 2-methyl-2-butene from the dehydration of 2-methyl-2-butanol in sulfuric acid.

10. Draw the structures of all the alkenes, including stereoisomers, that can be formed from the E2 elimination of 3-bromo-2,3-dimethylpentane with sodium ethoxide ($NaOCH_2CH_3$) in ethanol. Which of these would you expect to be the major product?

11. Using curved arrows and perspective drawings (of chair cyclohexanes), explain the formation of the indicated product from the following reaction:

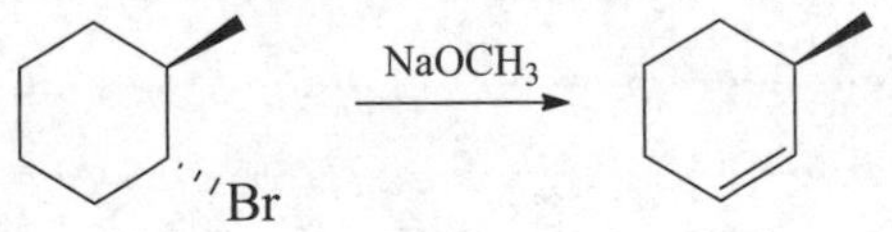

12. Compare the relative rate of reaction of *cis*- and *trans*-1-chloro-3-isopropylcyclohexane with sodium methoxide in methanol by the E2 mechanism.

13. Outline a mechanism for the following reaction:

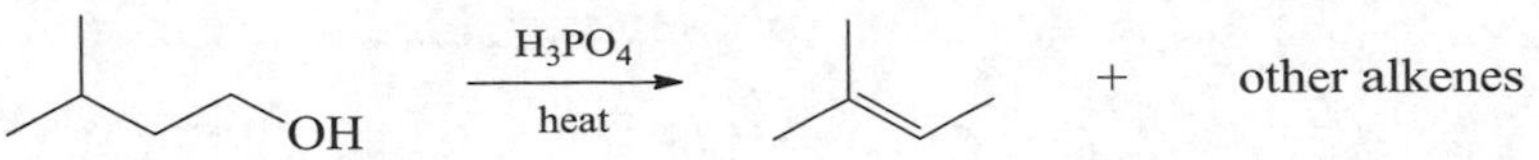

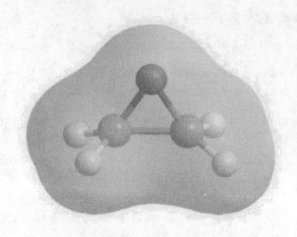

CHAPTER 8

Addition Reactions of Alkenes

Table of Contents

SOLUTIONS TO TEXT PROBLEMS

In Chapter Problems

8.1   Catalytic hydrogenation converts an alkene to an alkane having the same carbon skeleton. Because 2-methylbutane is the product of hydrogenation, all three alkenes must have a four-carbon chain with a one-carbon branch. The three alkenes are therefore:

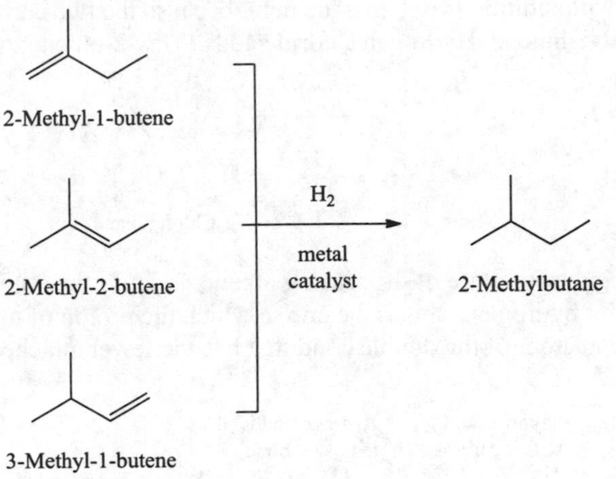

2-Methyl-1-butene

2-Methyl-2-butene $\xrightarrow[\text{catalyst}]{\text{H}_2}$ 2-Methylbutane
                         metal

3-Methyl-1-butene

8.2   No, the *less* stable 1,2-dimethylcyclohexane, with one axial methyl group, is formed as the major product. Hydrogenation is not reversible so products do not equilibrate.

cis-1,2-Dimethyl-
cyclohexane (68%)

(less stable)

trans-1,2-Dimethyl-
cyclohexane (32%)

(more stable)

8.3   The top face of the double bond of α-pinene is shielded by the methyl group. Hydrogen is transferred to the bottom face. The only product is Compound A.

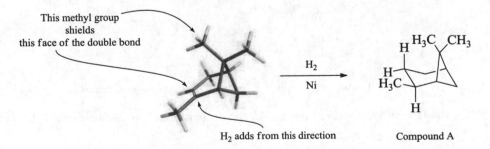

This methyl group
shields
this face of the double bond

$\xrightarrow[\text{Ni}]{\text{H}_2}$

H$_2$ adds from this direction

Compound A

8.4   The most highly substituted double bond is the most stable and has the smallest heat of hydrogenation.

2-Methyl-2-butene:	2-Methyl-1-butene:	3-Methyl-1-butene:
most stable		least stable
(trisubstituted)	(disubstituted)	(monosubstituted)

Heat of hydrogenation:	112 kJ/mol	118 kJ/mol	126 kJ/mol
	(26.7 kcal/mol)	(28.2 kcal/mol)	(30.2 kcal/mol)

8.5   (*b*) Regioselectivity of addition is not an issue here, because the two carbons of the double bond are equivalent in *cis*-2-butene. Hydrogen chloride adds to *cis*-2-butene to give 2-chlorobutane.

HCl →

cis-2-Butene   2-Chlorobutane

(*c*) Begin by writing the structure of the starting alkene. Identify the doubly bonded carbon that has the greater number of hydrogens; this is the one to which the proton of hydrogen chloride adds. Chlorine adds to the carbon atom of the double bond that has the fewer attached hydrogens.

Chlorine adds   Hydrogen adds
to this carbon.   to this carbon.

HCl →

2-Methyl-1-butene   2-Chloro-2-methylbutane

By applying Markovnikov's rule, we see that the major product is 2-chloro-2-methylbutane.

(*d*) One end of the double bond has no hydrogens, but the other end has one. In accordance with Markovnikov's rule, the proton of hydrogen chloride adds to the carbon that already has one hydrogen. The product is 1-chloro-1-ethylcyclohexane.

Chlorine adds   Hydrogen adds
to this carbon.   to this carbon.

HCl →

Ethylidenecyclohexane   1-Chloro-1-ethylcyclohexane

8.6   (*b*) A secondary carbocation is an intermediate in the reaction of *cis*-2-butene with hydrogen chloride.

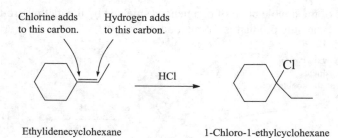

cis-2-Butene   Hydrogen chloride   Secondary carbocation   Chloride ion

Capture of this carbocation by chloride gives 2-chlorobutane.

(*c*) A proton is transferred to C-1 of 2-methyl-1-butene to produce a tertiary carbocation.

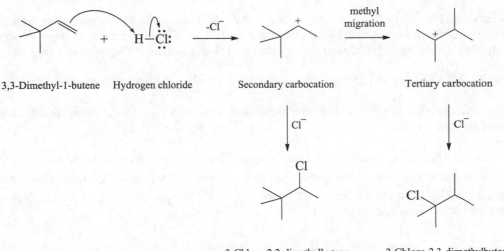

2-Methyl-1-butene   Hydrogen chloride   Tertiary carbocation   Chloride ion

This is the carbocation that leads to the observed product, 2-chloro-2-methylbutane.

(*d*) A tertiary carbocation is formed by protonation of the double bond.

Ethylidenecyclohexane   Hydrogen chloride   Tertiary carbocation   Chloride ion

This carbocation is captured by chloride to give the observed product, 1-chloro-1-ethylcyclohexane.

8.7 The carbocation formed by protonation of the double bond of 3,3-dimethyl-1-butene is secondary. Methyl migration can occur to give a more stable tertiary carbocation. Reaction of chloride with each carbocation intermediate gives the products as shown.

3,3-Dimethyl-1-butene   Hydrogen chloride   Secondary carbocation   Tertiary carbocation

3-Chloro-2,2-dimethylbutane   2-Chloro-2,3-dimethylbutane

The two chloride products are 3-chloro-2,2-dimethylbutane and 2-chloro-2,3-dimethylbutane.

8.8 The presence of hydroxide ion in the second step is incompatible with the acidic medium in which the reaction is carried out. The reaction as shown in step 1

$$(CH_3)_2C{=}CH_2 + H_3O^+ \longrightarrow (CH_3)_3C^+ + H_2O$$

is performed in acidic solution. There are, for all practical purposes, no hydroxide ions in aqueous acid, the strongest base present being water itself. It is quite important to pay attention to the species that are actually present in the reaction medium whenever you formulate a reaction mechanism.

8.9   The more stable the carbocation, the faster it is formed. The more reactive alkene gives a tertiary carbocation in the rate-determining step.

Tertiary carbocation

Secondary carbocation

8.10   The mechanism of electrophilic addition of hydrogen chloride to 2-methylpropene as outlined in text Section 8.4 proceeds through a carbocation intermediate. This mechanism is the reverse of the E1 elimination. The E2 mechanism is concerted, meaning it occurs in a single step.

8.11   In a reversible reaction, the equilibrium constant of the reverse reaction is the reciprocal of the forward reaction. In other words, $K_{reverse} = \dfrac{1}{K_{forward}}$. If $K_{forward} = 9$ for the reaction

then $K_{reverse} = 0.11$.

8.12   A catalyst affects the rate of a reaction, but not its equilibrium constant, by speeding up both the forward and reverse reactions. The usual hydrogenation catalysts (Pd, Pt, Ni) allow the ethane $\rightleftharpoons$ ethylene equilibrium to be established more quickly but do not affect the relative amounts of reactant and product.

8.13   The ethylene product will increase relative to ethyl bromide with increasing temperature.

The solution to this problem lies in using the equation $\Delta G^\circ = \Delta H^\circ - T\Delta S^\circ$. The more negative the ΔG°, the more favored the reaction.

In the first equation, one molecule (ethanol) produces two molecules (ethylene and water). The entropy, ΔS°, of this reaction will thus be substantially positive. This means that the $-T\Delta S^\circ$ term will be influenced by varying the temperature. Increasing the temperature will result in a more negative ΔG°, thus increasing the amounts of ethylene and water.

$$CH_3CH_2{-}OH \ (g) \ \underset{}{\overset{HBr}{\rightleftharpoons}} \ H_2C{=}CH_2 \ (g) \ + \ H_2O \ (g)$$

One molecule   Two molecules

Note that in this first reaction the HBr is only used as a catalyst and is not consumed or produced in the reaction.

In the second reaction, two molecules (ethanol and HBr) produce two molecules (ethyl bromide and water). The entropy of this reaction will be very close to zero. The $-T\Delta S^\circ$ term then is very close to zero and varying the temperature will have a minimal influence on the ΔG° term. Thus, this equilibrium will not change significantly with temperature.

$$CH_3CH_2\text{—}OH \ (g) \ + \ HBr \ (g) \ \rightleftharpoons \ CH_3CH_2\text{—}Br \ (g) \ + \ H_2O \ (g)$$

Two molecules   Two molecules

8.14   (*b*)   The carbon–carbon double bond is symmetrically substituted in *cis*-2-butene. Therefore, the regioselectivity of hydroboration–oxidation is not an issue. Hydration of the double bond gives 2-butanol.

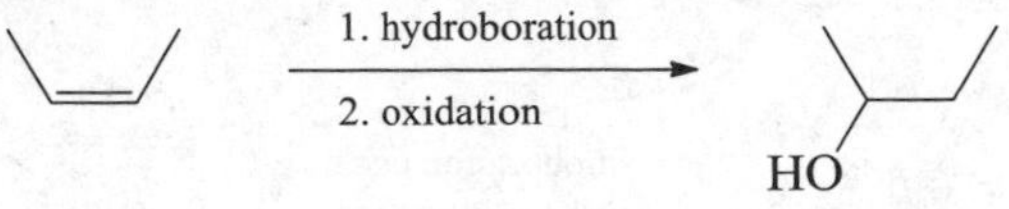

cis-2-Butene   2-Butanol

(*c*)   Hydroboration–oxidation of alkenes leads to hydration of the double bond with a regioselectivity opposite to Markovnikov's rule.

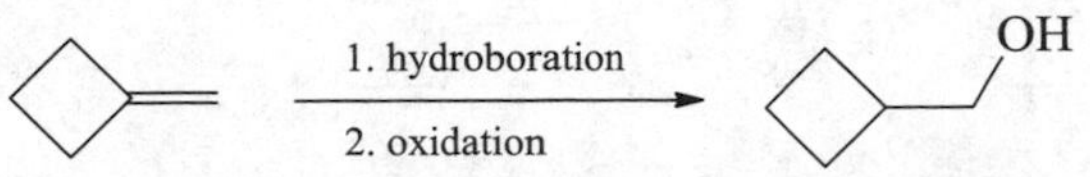

Methylenecyclobutane   Cyclobutylmethanol

(*d*)   Hydroboration–oxidation of cyclopentene gives cyclopentanol.

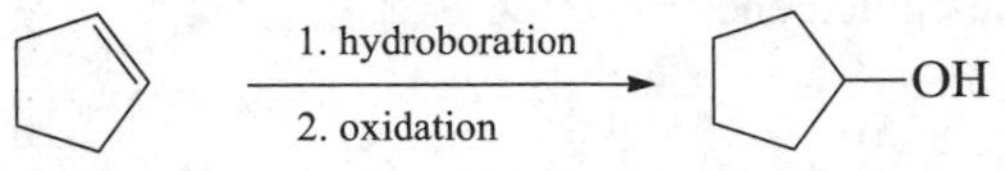

Cyclopentene   Cyclopentanol

(*e*)   When alkenes are converted to alcohols by hydroboration–oxidation, the hydroxyl group is introduced at the less substituted carbon of the double bond.

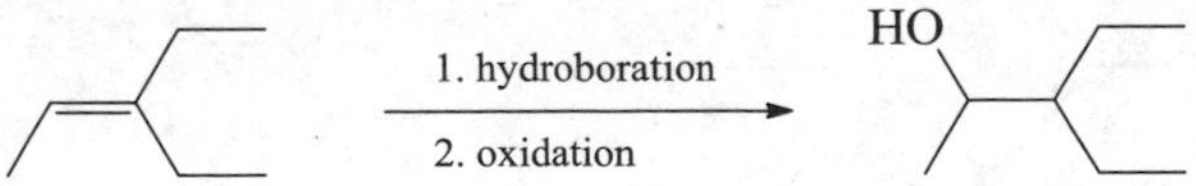

3-Ethyl-2-pentene   3-Ethyl-2-pentanol

(*f*)   The less substituted carbon of the double bond in 3-ethyl-1-pentene is at the end of the chain. It is this carbon that bears the hydroxyl group in the product of hydroboration–oxidation.

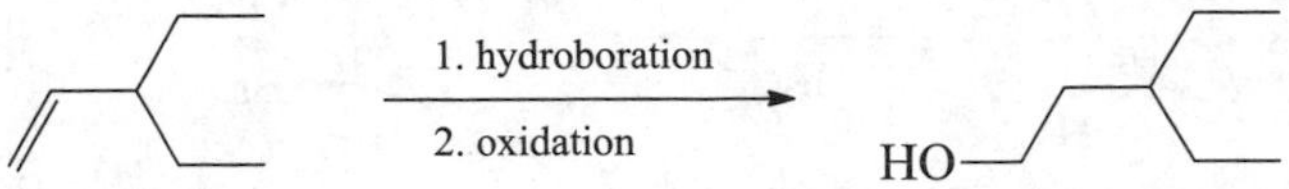

3-Ethyl-1-pentene   3-Ethyl-1-pentanol

8.15 The bottom face of the double bond of α-pinene is less hindered than the top face.

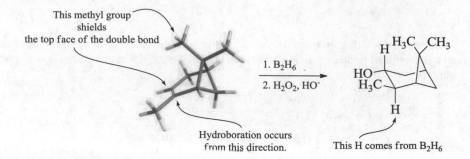

Syn addition of H and OH takes place and with a regioselectivity opposite to that of Markovnikov's rule.

8.16 Alkyl substituents on the double bond increase the reactivity of the alkene toward addition of bromine.

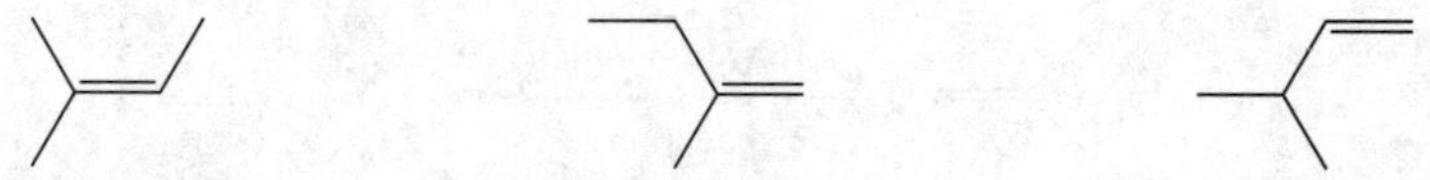

2-Methyl-2-butene
most reactive
(trisubstituted double bond)

2-Methyl-1-butene

(disubstituted double bond)

3-Methyl-1-butene
least reactive
(monosubstituted double bond)

8.17 (*b*) Bromine becomes bonded to the less highly substituted carbon of the double bond, the hydroxyl group to the more highly substituted one.

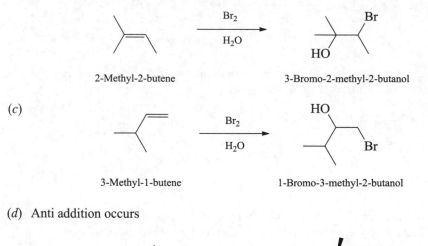

2-Methyl-2-butene

3-Bromo-2-methyl-2-butanol

(*c*)

3-Methyl-1-butene

1-Bromo-3-methyl-2-butanol

(*d*) Anti addition occurs

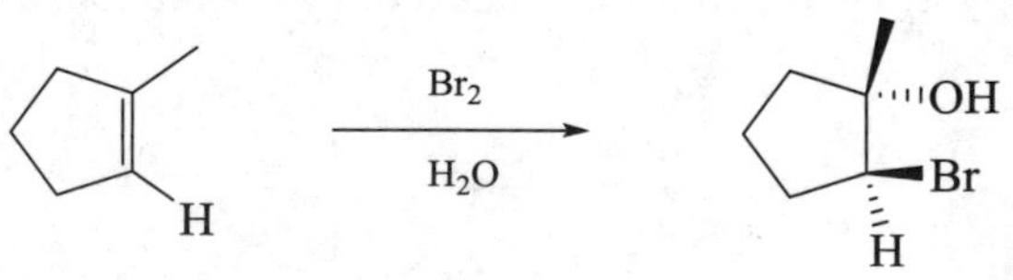

1-Methylcyclopentene

trans-2-Bromo-1-methylcyclopentanol

8.18   The structure of disparlure is as shown.

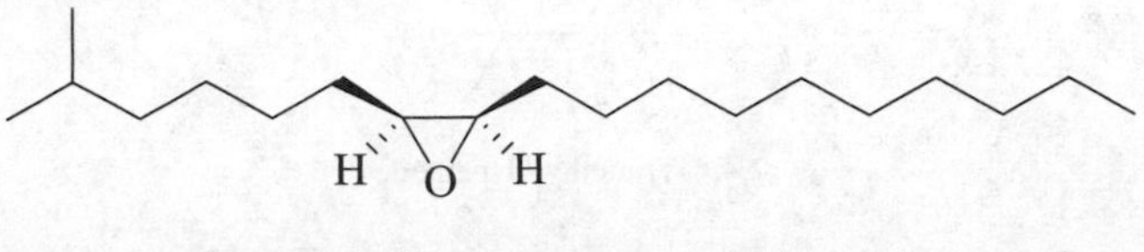

Its longest continuous chain contains 18 carbon atoms, and so it is named as an epoxy derivative of octadecane. Number the chain in the direction that gives the lowest numbers to the carbons that bear oxygen. Thus, disparlure is *cis*-7,8-epoxy-2-methyloctadecane.

8.19   Disparlure can be prepared by epoxidation of the corresponding alkene. Cis alkenes yield cis epoxides upon epoxidation. *cis*-2-Methyl-7-octadecene is therefore the alkene chosen to prepare disparlure by epoxidation.

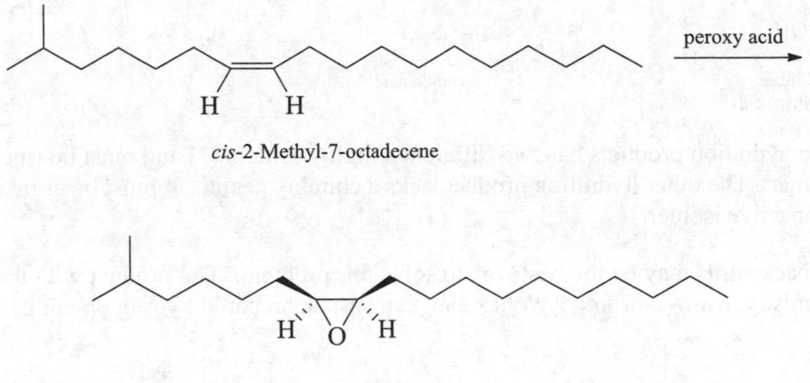

cis-2-Methyl-7-octadecene

Disparlure

8.20   Ozonolysis of 1-methylcyclopentene cleaves the double bond between carbons a and b to give a single compound containing two carbonyl groups in a six-carbon chain.

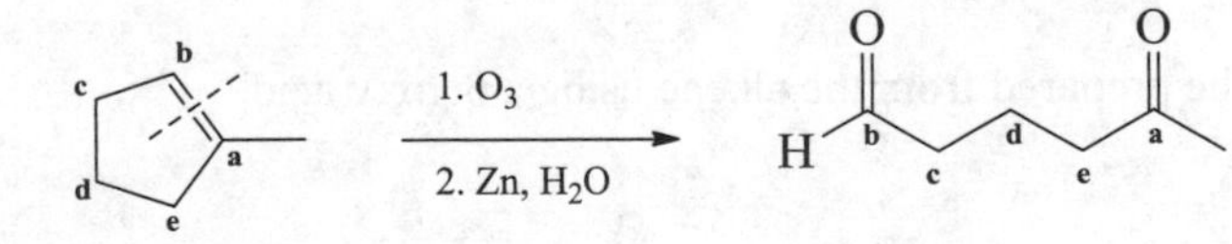

1-Methylcyclopentene   5-Oxohexanal

8.21   The products of ozonolysis are formaldehyde and 4,4-dimethyl-2-pentanone.

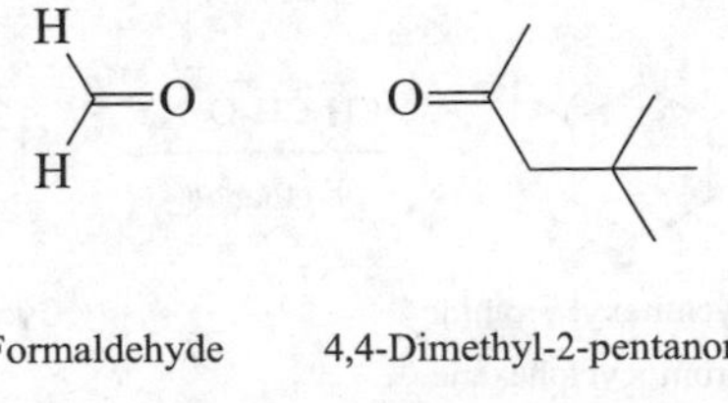

Formaldehyde   4,4-Dimethyl-2-pentanone

The two carbons that were doubly bonded to each other in the alkene become the carbons that are doubly bonded to oxygen in the products of ozonolysis. Therefore, mentally remove the oxygens and connect these two carbons by a double bond to reveal the structure of the starting alkene.

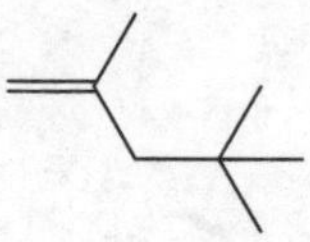

2,4,4-Trimethyl-1-pentene

8.22 Hydration of the double bond of aconitic acid (shown in the center) can occur in two regiochemically distinct ways:

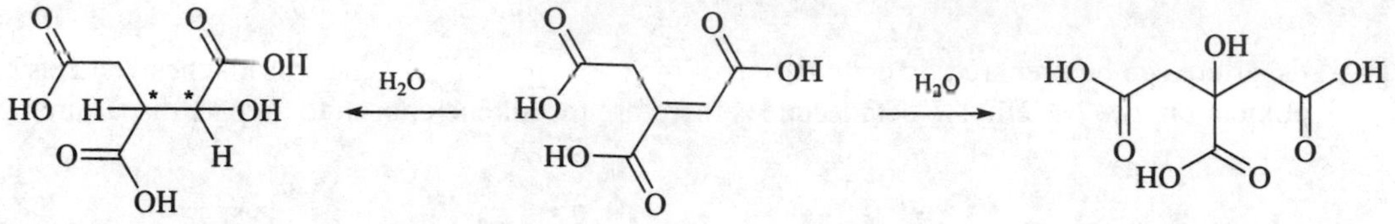

Chiral
(isocitric acid)

Aconitic acid

Achiral
(citric acid)

One of the hydration products has two different chirality centers (*) and must be isocitric acid, the optically active isomer. The other hydration product lacks a chirality center. It must be citric acid, the achiral, optically inactive isomer.

8.23   Working backwards may be the best way to solve this problem. The product with the indicated stereochemistry, *trans*-2-bromocyclohexanol, can best be prepared by ring opening of an epoxide.

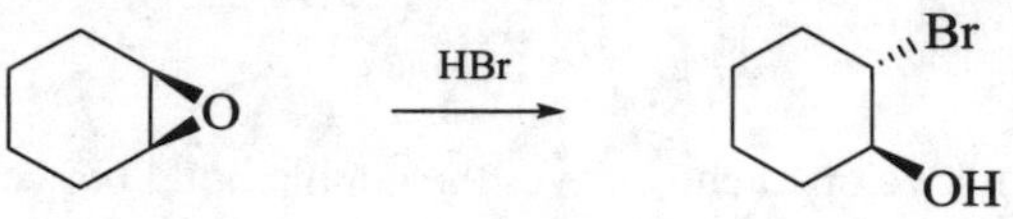

1,2-Epoxycyclohexane   *trans*-2-Bromocyclohexanol

The epoxide can be prepared from the alkene using a peroxy acid.

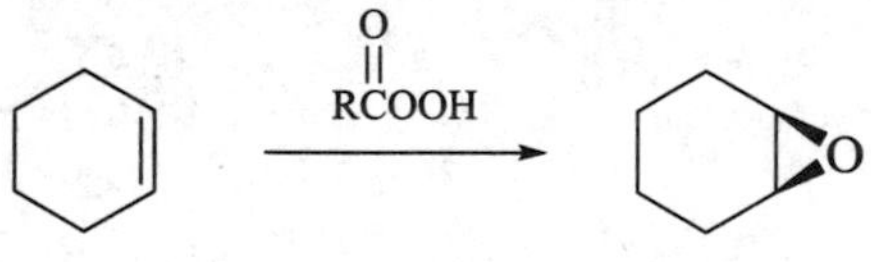

Cyclohexene   1,2-Epoxycyclohexane

Cyclohexene can be prepared from cyclohexylbromide by an E2 reaction.

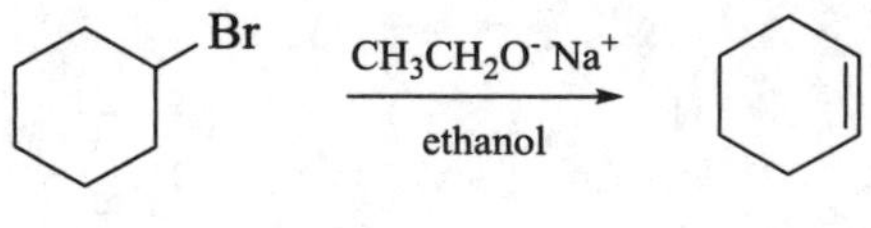

Cyclohexyl bromide
Bromocyclohexane

Cyclohexene

8.24 Working backwards may be the best way to solve this problem.

1,2-Dibromo-3,3-dimethylbutane ⟹ Compound B ⟹ Compound A ⟹ 3,3-Dimethyl-2-butanol

The product 1,2-dibromo-3,3-dimethylbutane can be formed from an alkene, Compound B, in one step.

3,3-Dimethyl-1-butene
Compound B

1,2-Dibromo-3,3-dimethylbutane

3,3-Dimethyl-1-butene can be prepared by an E2 reaction from 3-bromo-2,2-dimethylbutane, Compound A.

3-Bromo-2,2-dimethylbutane
Compound A

3,3-Dimethyl-1-butene

For the first reaction step to convert the alcohol to an alkyl bromide, two possibilities may be considered. One uses PBr$_3$ as the reagent and the other HBr. The latter reagent has the disadvantage of forming a carbocation that would rearrange so PBr$_3$ would be the best reagent for this conversion.

3,3-Dimethyl-2-butanol

3-Bromo-2,2-dimethylbutane
Compound A

An alternative for this last step is to convert the alcohol to an alkyl chloride using thionyl chloride and pyridine. The alkyl chloride product would also undergo E2 reactions to give 3,3-dimethyl-1-butene.

End of Chapter Problems

Reactions of Alkenes

8.25 In all parts of this exercise we deduce the carbon skeleton on the basis of the alkane formed on hydrogenation of an alkene and then determine what carbon atoms may be connected by a double bond in that skeleton. Problems of this type are best done by using carbon skeleton formulas.

(*a*) Product is 2,2,3,4,4-pentamethylpentane. Only one alkene precursor is possible.

2,2,3,4,4-Pentamethylpentane   Alkene precursor

(*b*) Product is 2,3-dimethylbutane. Two alkene precursors are possible.

2,3-dimethylbutane   Alkene precursors

(*c*) Product is methylcyclobutane. This compound may be formed from hydrogenation of three possible alkenes.

Methylcyclobutane   Alkene precursors

8.26 The methyl group in compound B shields one face of the double bond from the catalyst surface; therefore, hydrogen can be transferred only to the bottom face of the double bond. The methyl group in compound A does not interfere with hydrogen transfer to the double bond.

Compound A   Compound B

Thus, hydrogenation of A is faster than that of B because B contains a more sterically hindered double bond.

8.27 Hydrogen can add to the double bond of 1,4-dimethylcyclopentene either from the same side as the C-4 methyl group or from the opposite side. The two possible products are *cis*- and *trans*-1,3-dimethylcyclopentane.

1,4-Dimethylcyclopentene   *cis*-1,3-Dimethylcyclopentane   *trans*-1,3-Dimethylcyclopentane

Hydrogen transfer occurs to the less hindered face of the double bond, that is, trans to the C-4 methyl group. Thus, the major product is *cis*-1,3-dimethylcyclopentane.

8.28 This problem illustrates the reactions of alkenes with various reagents and requires application of Markovnikov's rule to the addition of unsymmetrical electrophiles.

(*a*) Addition of hydrogen chloride to 1-pentene will give 2-chloropentane.

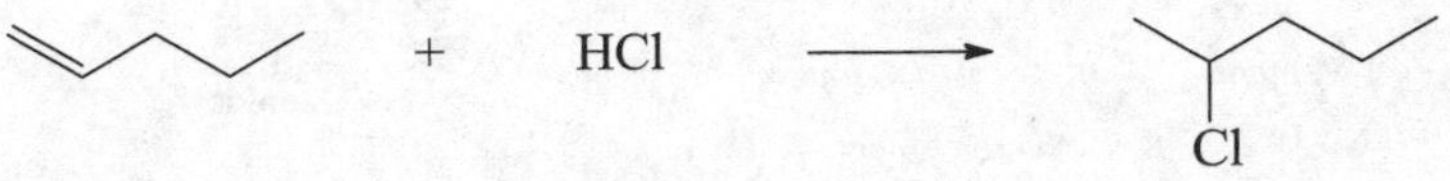

1-Pentene   Hydrogen chloride   2-Chloropentane

(*b*) Dilute sulfuric acid will cause hydration of the double bond with regioselectivity in accord with Markovnikov's rule.

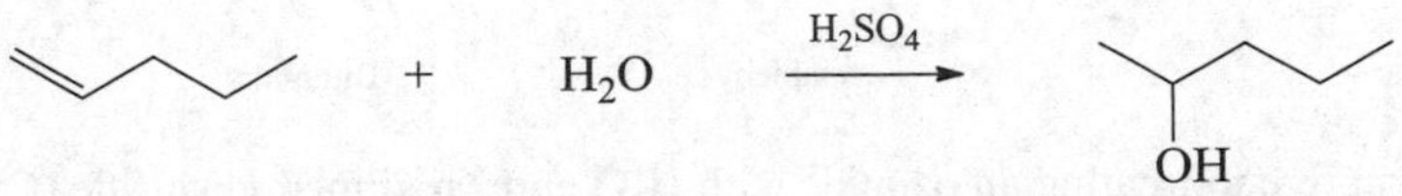

1-Pentene   Water   2-Pentanol

(*c*) Hydroboration–oxidation of an alkene brings about hydration of the double bond opposite to Markovnikov's rule; 1-pentanol will be the product.

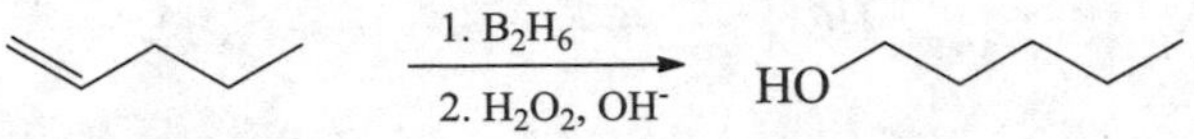

1-Pentene   1-Pentanol

(*d*) Bromine adds across the double bond to give a vicinal dibromide.

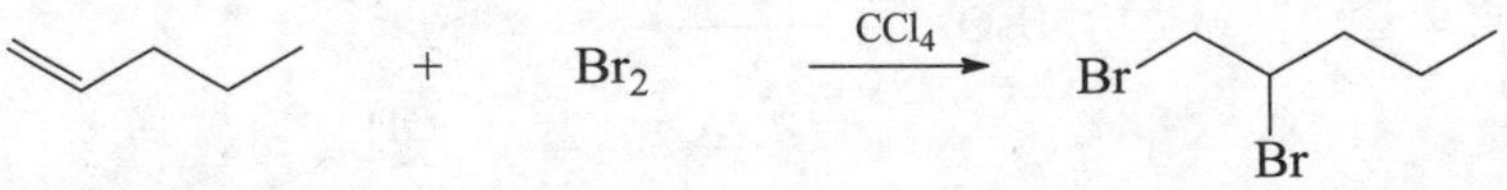

1-Pentene   Bromine   1,2-Dibromopentane

(*e*) Vicinal bromohydrins are formed when bromine in water adds to alkenes. Br adds to the less substituted carbon, OH to the more substituted one.

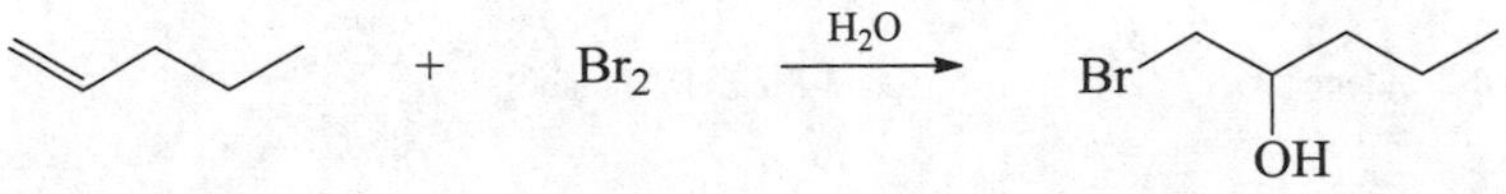

1-Pentene   Bromine   1-Bromo-2-pentanol

(*f*) Epoxidation of the alkene occurs on treatment with peroxy acids.

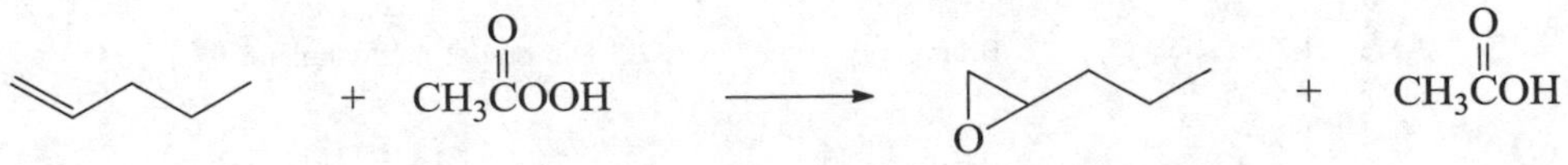

1-Pentene   Peroxyacetic acid   1,2-Epoxypentane   Acetic acid (*g*) Ozone reacts with alkenes to give ozonides.

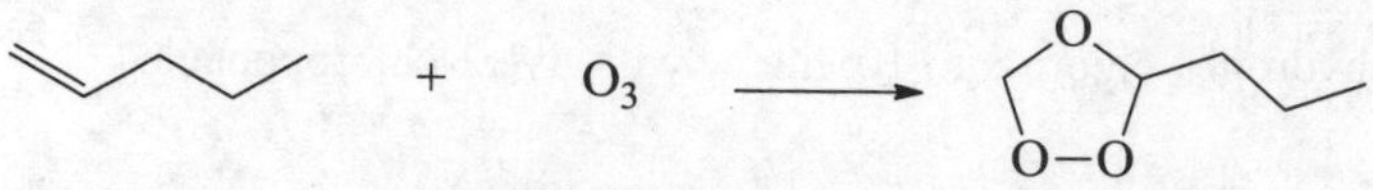

| 1-Pentene | Ozone | Ozonide |

(*h*) When the ozonide in part (*j*) is hydrolyzed in the presence of zinc, formaldehyde and butanal are formed.

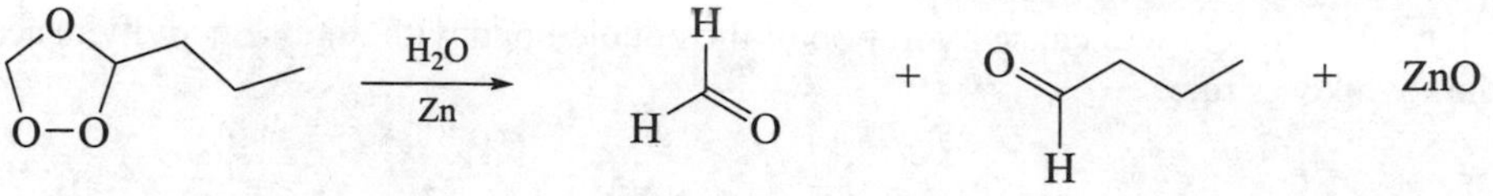

| Ozonide | Formaldehyde | Butanal |

(*i*) As an alternative to treating an ozonide with H_2O and Zn, dimethyl sulfide $(CH_3)_2S$ is often used. The products are the same as in part (*k*) of this problem. $(CH_3)_2SO$ (dimethyl sulfoxide) is also formed.

8.29 When we compare the reactions of 2-methyl-2-butene with the analogous reactions of 1-pentene, we find that the reactions proceed in a similar manner.

(*a*)

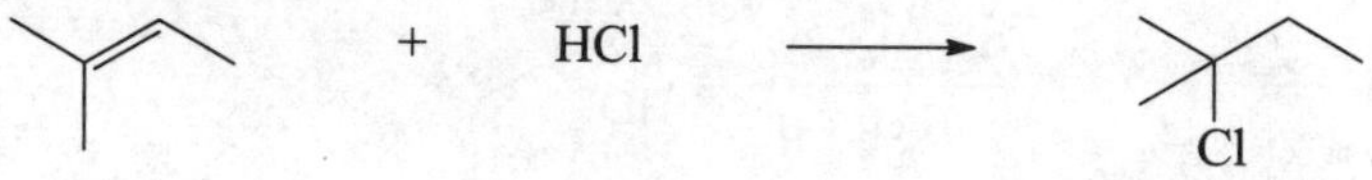

| 2-Methyl-2-butene | Hydrogen chloride | 2-Chloro-2-methylbutane |

(*b*)

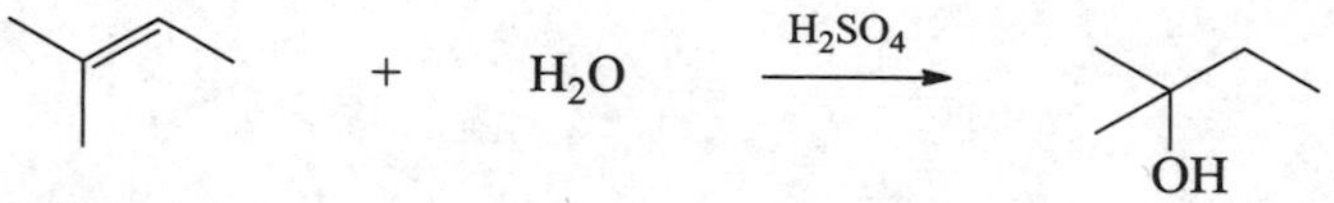

| 2-Methyl-2-butene | Water | 2-Methyl-2-butanol |

(*c*)

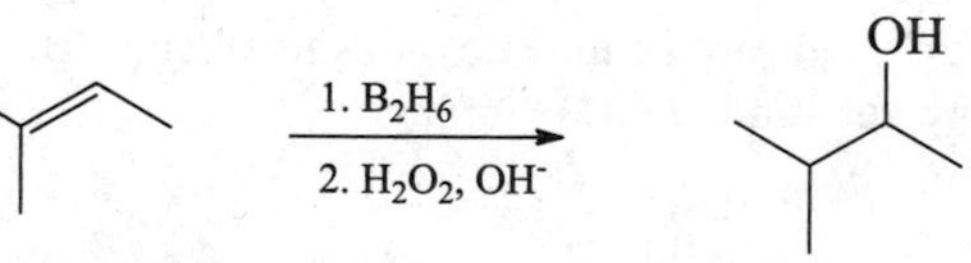

| 2-Methyl-2-butene | 3-Methyl-2-butanol |

(*d*)

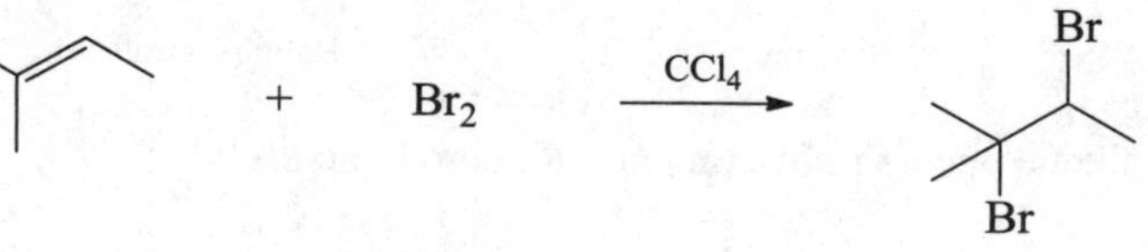

| 2-Methyl-2-butene | Bromine | 2,3-Dibromo-2-methylbutane |

(*e*)

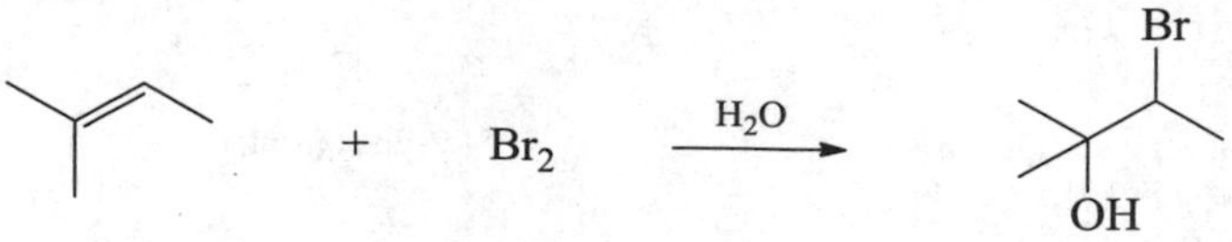

| 2-Methyl-2-butene | Bromine | 3-Bromo-2-methyl-2-butanol |

(*f*)

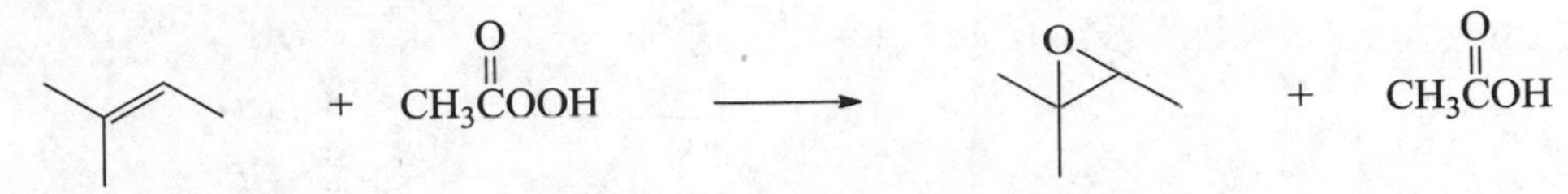

2-Methyl-2-butene   Peroxyacetic acid   2,3-Epoxy-2-methylbutane   Acetic acid

(*g*)

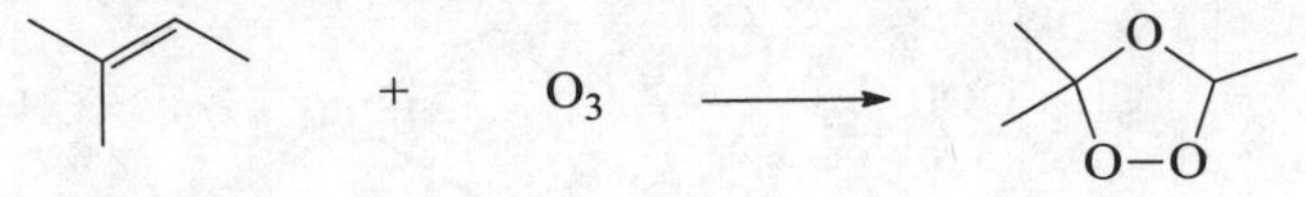

2-Methyl-2-butene   Ozone   Ozonide

(*h*)

Ozonide   Acetone   Acetaldehyde

(*i*)   As an alternative to treating an ozonide with H_2O and Zn, dimethyl sulfide $(CH_3)_2S$ is often used. The products are the same as in part (*h*) of this problem. $(CH_3)_2SO$ (dimethyl sulfoxide) is also formed.

8.30   Cycloalkenes undergo the same kinds of reactions as do noncyclic alkenes.

(*a*)

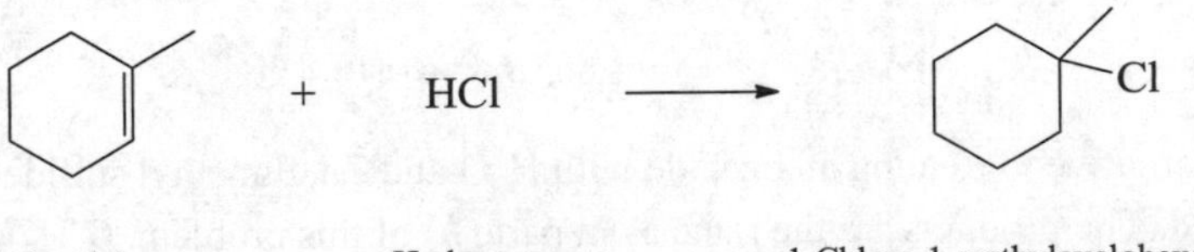

1-Methylcyclohexene   Hydrogen chloride   1-Chloro-1-methylcyclohexane

(*b*)

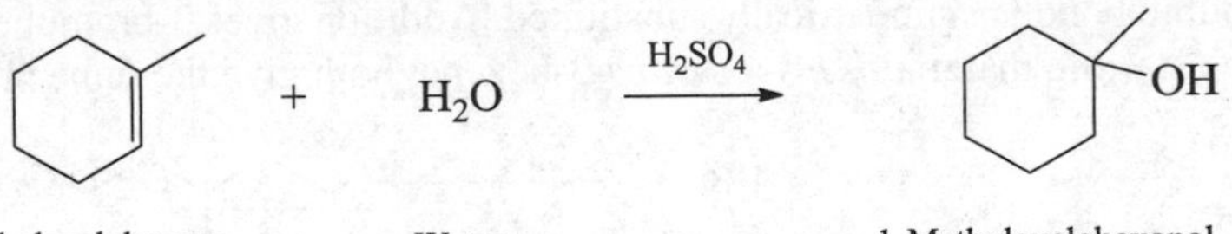

1-Methylcyclohexene   Water   1-Methylcyclohexanol

(*c*)

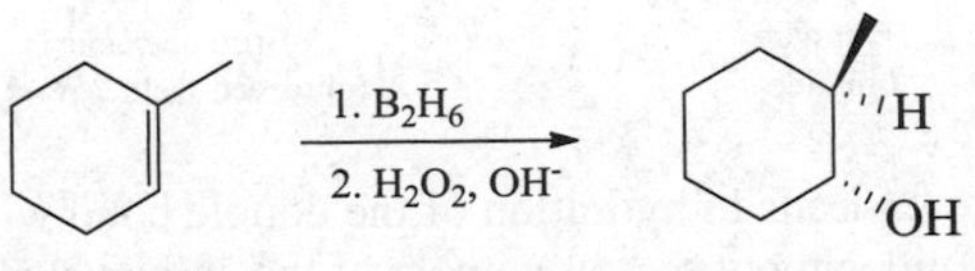

1-Methylcyclohexene   *trans*-2-Methylcyclohexanol

(*d*)

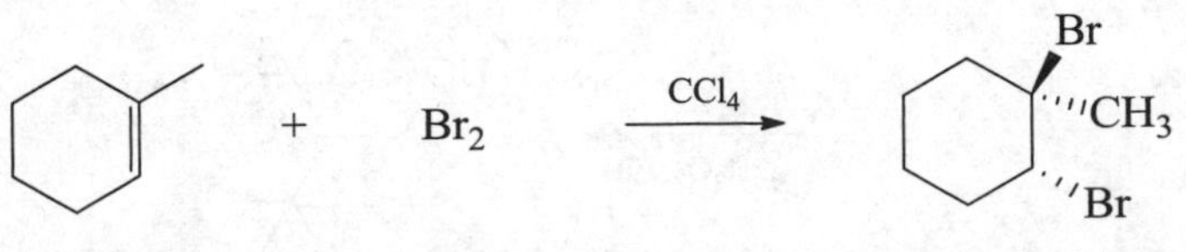

1-Methylcyclohexene   Bromine   *trans*-1,2-Dibromo-1-methylcyclohexane

(e)

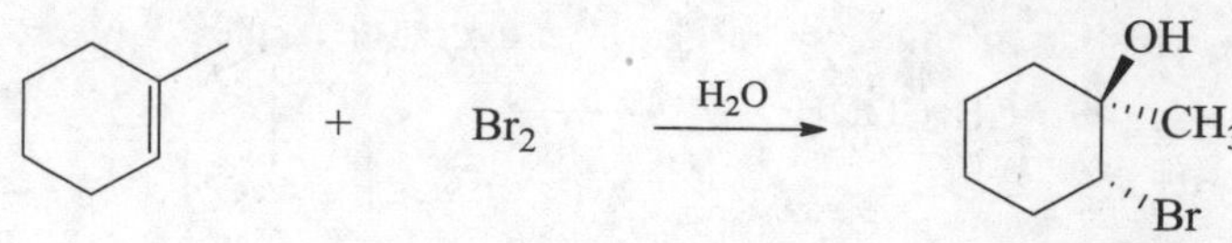

1-Methylcyclohexene   Bromine   *trans*-2-Bromo-1-methylcyclohexanol

(f)

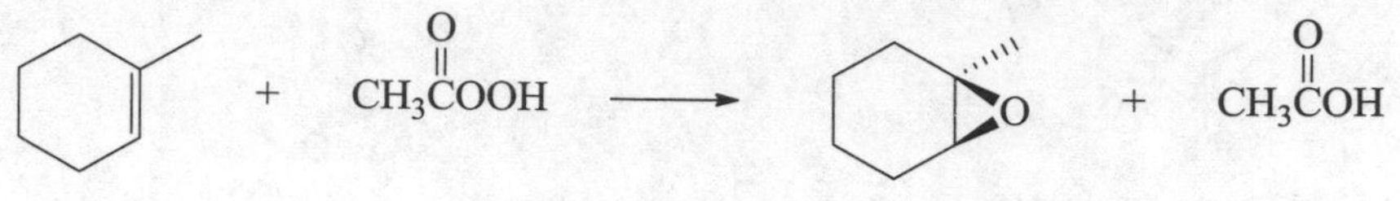

1-Methylcyclohexene   Peroxyacetic acid   1,2-Epoxy-1-methylcyclohexane   Acetic acid

(g)

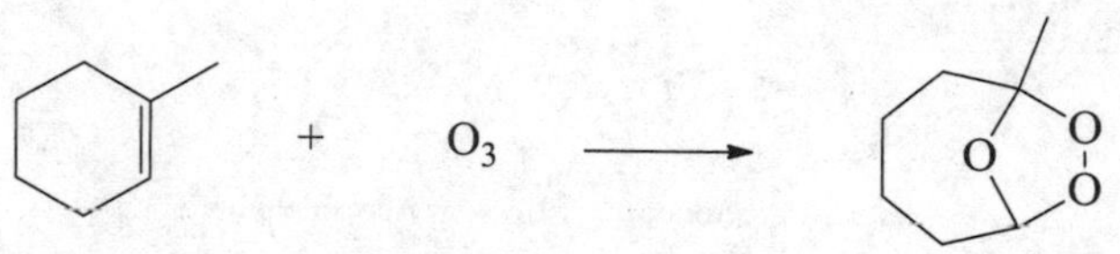

1-Methylcyclohexene   Ozone   Ozonide

(h)

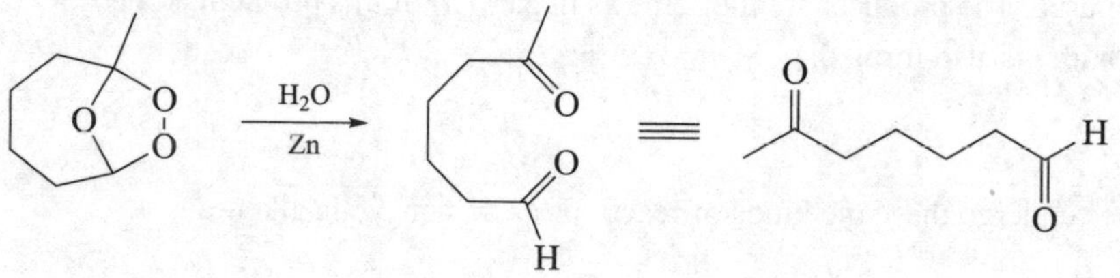

Ozonide   6-Oxoheptanal

(i)   As an alternative to treating an ozonide with H_2O and Zn, dimethyl sulfide $(CH_3)_2S$ is often used. The products are the same as in part (h) of this problem. $(CH_3)_2SO$ (dimethyl sulfoxide) is also formed.

8.31 (a) The double bond is symmetrically substituted. Addition gives 3-bromohexane. (It does not matter whether the starting material is *cis*- or *trans*-3-hexene; both give the same product.)

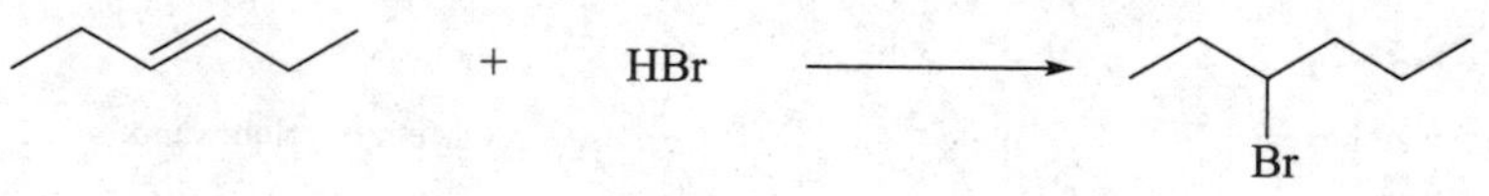

       3-Hexene   Hydrogen   3-Bromohexane
                            bromide   (observed yield 76%)

(b) Hydroboration–oxidation of alkenes leads to hydration of the double bond with a regioselectivity opposite to Markovnikov's rule and without rearrangement of the carbon skeleton.

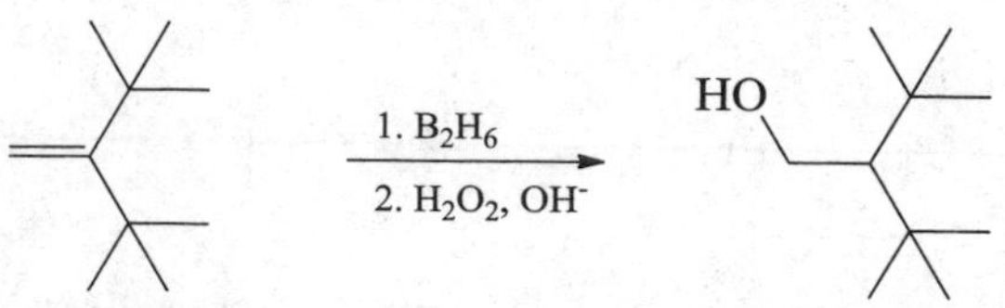

             1. B_2H_6
             2. H_2O_2, OH^-

2-*tert*-Butyl-3,3-dimethyl-1-butene   2-*tert*-Butyl-3,3-dimethyl-1-butanol
                                            (observed yield 65%)

(*c*) Hydroboration–oxidation of an alkene leads to syn hydration of the double bond.

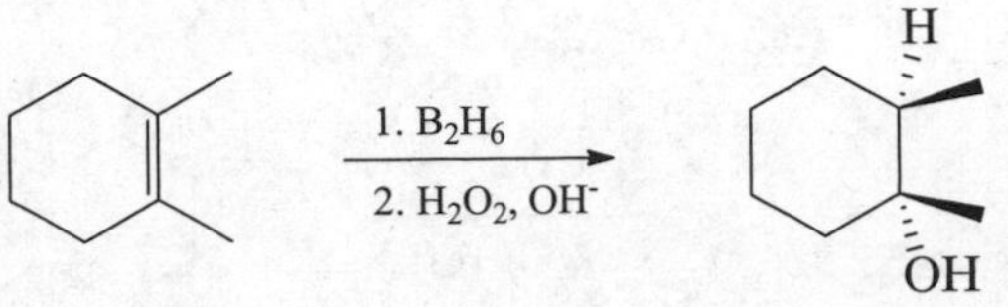

1,2-Dimethylcyclohexane

1. B$_2$H$_6$
2. H$_2$O$_2$, OH$^-$

cis-1,2-Dimethylcyclohexanol
(observed yield 82%)

(*d*) Bromine adds across the double bond of alkenes to give vicinal dibromides.

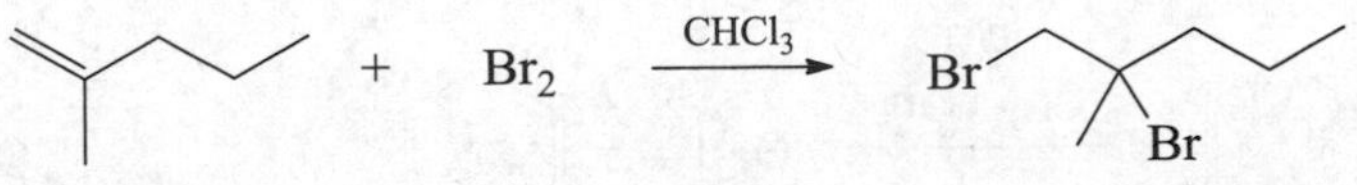

2-Methyl-1-pentene   + Br$_2$ $\xrightarrow{\text{CHCl}_3}$

1,2-Dibromo-2-methylpentane
(observed yield 60%)

(*e*) In aqueous solution, bromine reacts with alkenes to give bromohydrins. Bromine is the electrophile in this reaction and adds to the carbon that has the greater number of attached hydrogens.

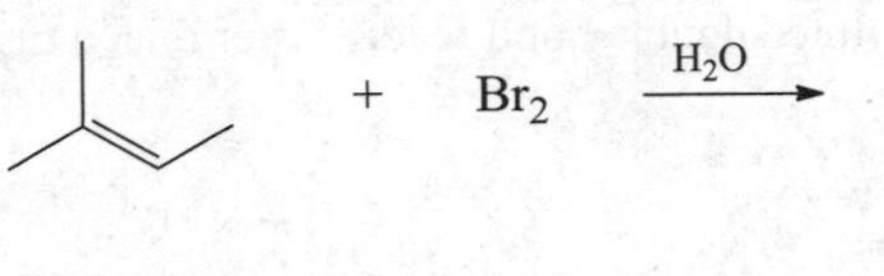

2-Methyl-2-butene   + Br$_2$ $\xrightarrow{\text{H}_2\text{O}}$

3-Bromo-2-methyl-2-butanol
(observed yield 77%)

(*f*) An aqueous solution of chlorine will react with 1-methylcyclopentene by anti addition. Chlorine is the electrophile and adds to the less substituted end of the double bond.

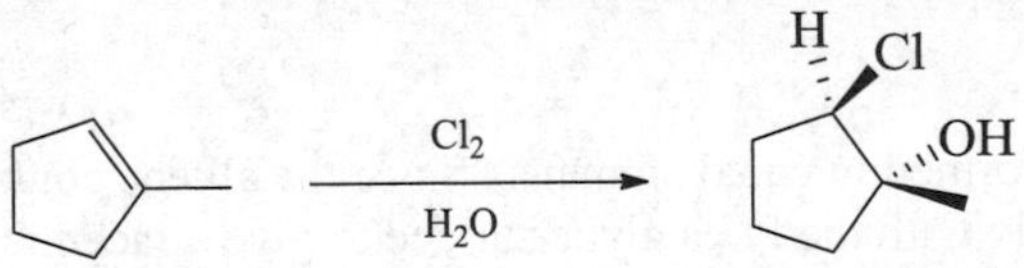

1-Methylcyclopentene $\xrightarrow[\text{H}_2\text{O}]{\text{Cl}_2}$

trans-2-Chloro-1-
methylcyclopentanol

(*g*) Compounds of the type RCOOH are peroxy acids and react with alkenes to give epoxides.

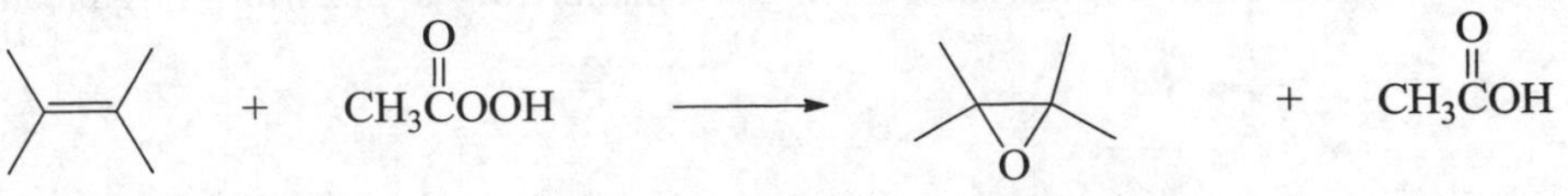

2,3-Dimethyl-2-butene   Peroxyacetic acid

2,3-Epoxy-2,3-dimethylbutane
(observed yield 70–80%)

Acetic acid

(*h*) The double bond is cleaved by ozonolysis. Each of the doubly bonded carbons becomes doubly bonded to oxygen in the product.

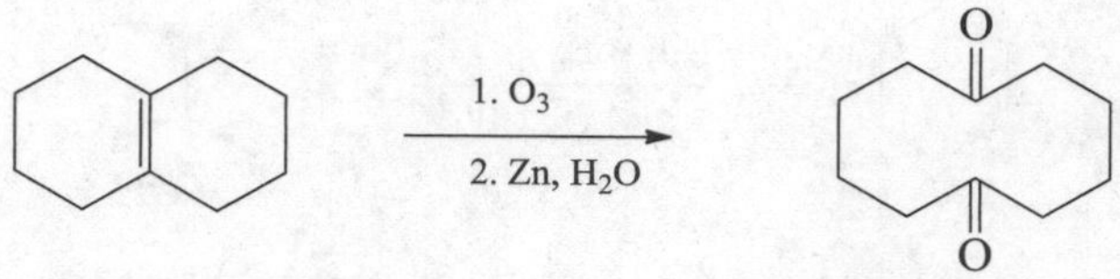

1,2,3,4,5,6,7,8-Octahydronaphthalene

Cyclodecane-1,6-dione
(observed yield 45%)

8.32 The product is epoxide B.

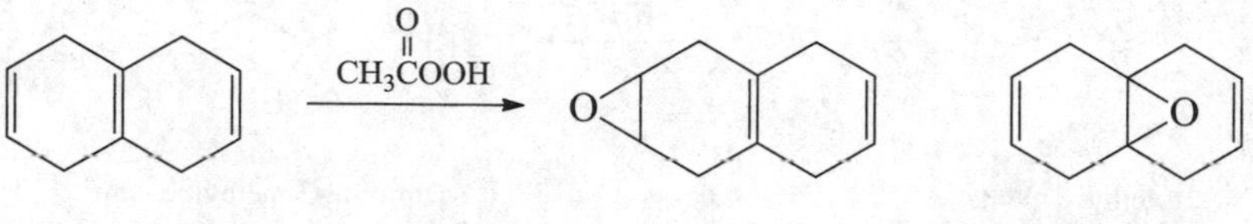

A

B

Major product,
formed faster

Epoxidation is an electrophilic addition; oxygen is transferred to the more electron-rich, more highly substituted double bond. A tetrasubstituted double bond reacts faster than a disubstituted one.

Stereochemistry

8.33 Hydrogenation of (*E*)-3-methyl-2-hexene gives a pair of enantiomers.

(*E*)-3-Methyl-2-hexene

(*S*)-3-Methylhexane

(*R*)-3-Methylhexane

racemic mixture

Both compounds are formed in equal amounts. Since the alkene contains a plane of symmetry, both faces of the alkene react equally with the Pt catalytic surface to give a racemic mixture.

symmetry plane

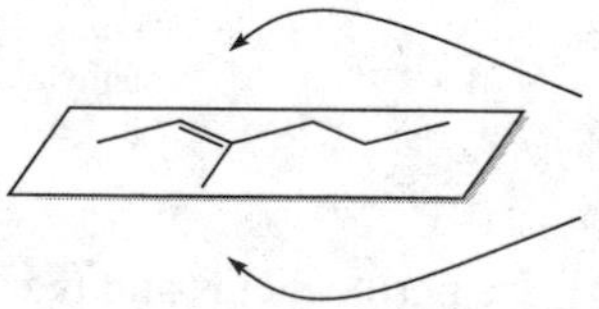

top and bottom faces
equivalently approach
catalytic surface

8.34 Hydrogenation of the alkenes shown will give a mixture of *cis*- and *trans*-1,4,-dimethylcyclohexane.

or

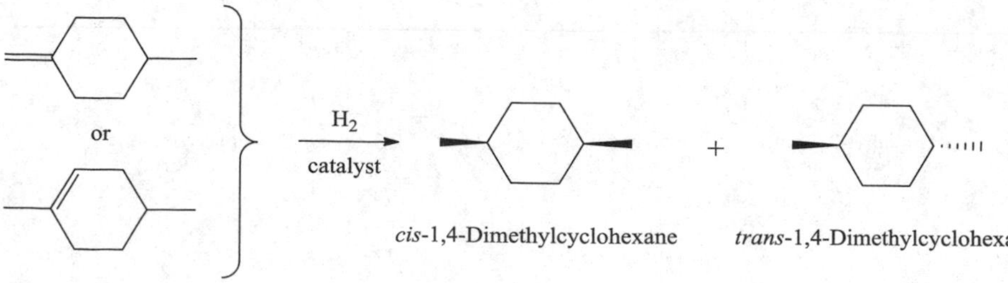

cis-1,4-Dimethylcyclohexane

trans-1,4-Dimethylcyclohexane

Only when the methyl groups are cis in the starting alkene will the cis stereoisomer be the sole product following hydrogenation. Hydrogenation of *cis*-3,6-dimethylcyclohexene will yield exclusively *cis*-1,4-dimethylcyclohexane.

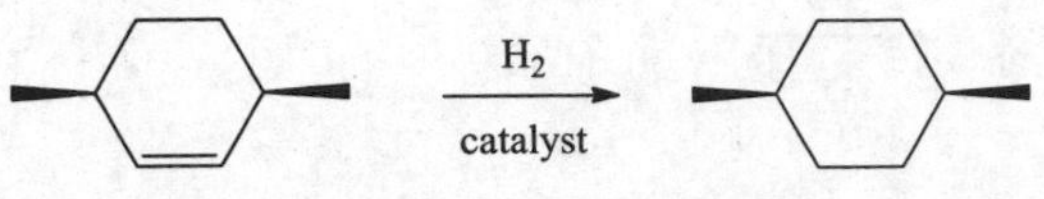

cis-3,6-Dimethylcyclohexene   *cis*-1,4-Dimethylcyclohexane

8.35 The problem presents the following experimental observation:

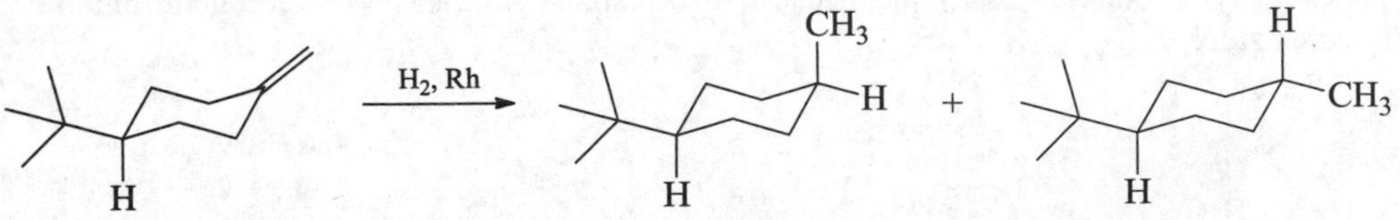

4-*tert*-Butyl(methylene)-
cyclohexane

cis-1-*tert*-Butyl-4-
methylcyclohexane (88%)

trans-1-*tert*-Butyl-4-
methylcyclohexane (12%)

This observation tells us that the predominant mode of hydrogen addition to the double bond is from the equatorial direction. Equatorial addition is the less hindered approach and thus occurs faster.

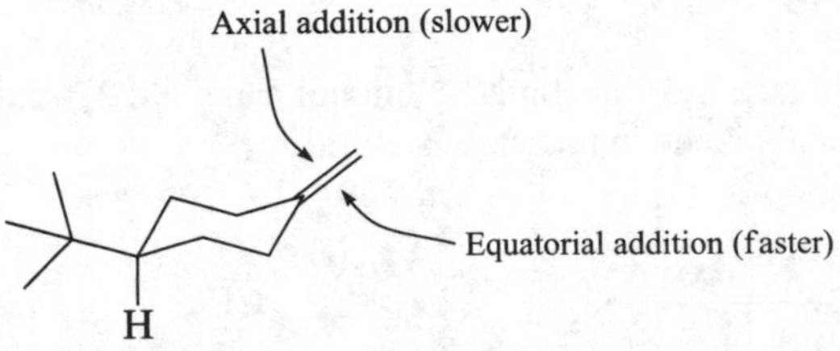

Axial addition (slower)

Equatorial addition (faster)

(*a*) Epoxidation should therefore give the following products:

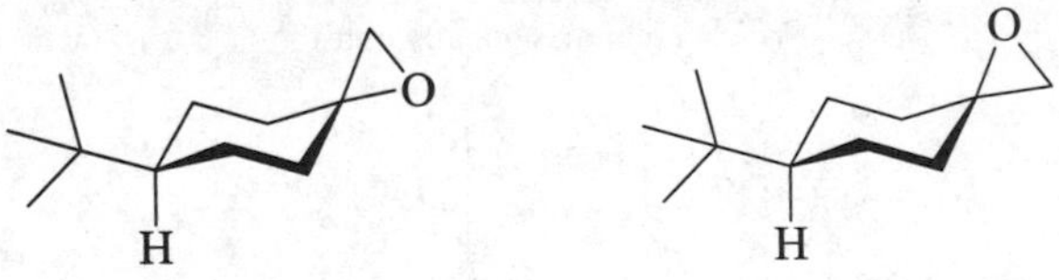

Major product

Minor product

The major product is the stereoisomer that corresponds to transfer of oxygen from the equatorial direction.

(*b*) Hydroboration–oxidation occurs from the equatorial direction.

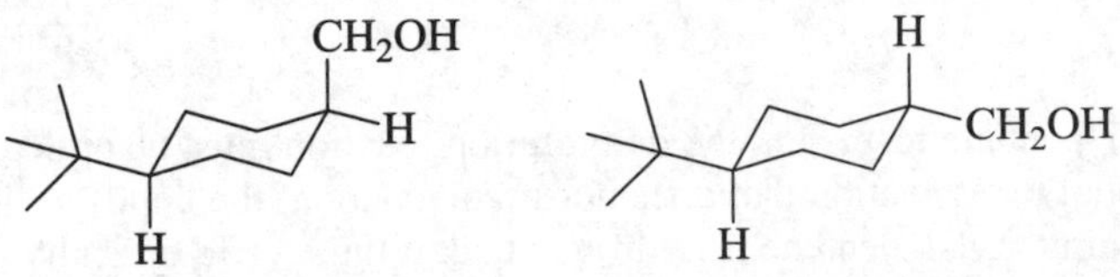

Major product

Minor product

8.36 3-Carene can in theory undergo hydrogenation to give either *cis*-carane or *trans*-carane.

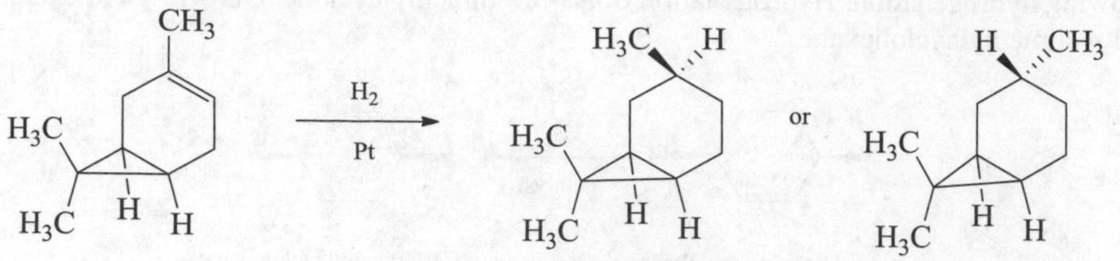

cis-Carane (98%)   *trans*-Carane

The exclusive product is *cis*-carane, because it corresponds to transfer of hydrogen from the less hindered side.

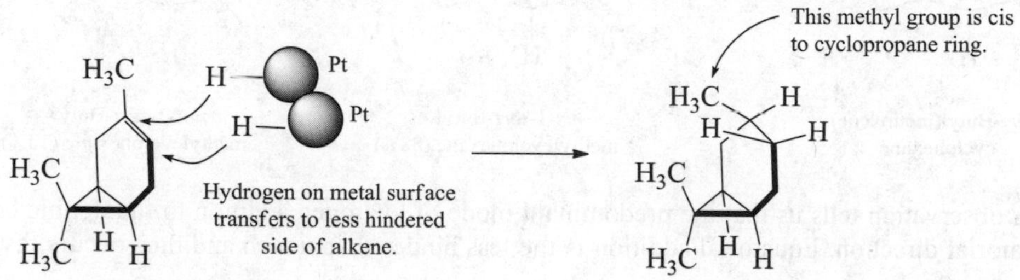

This methyl group is cis to cyclopropane ring.

cis-Carane

8.37   Dehydration of optically pure 2,3-dimethyl-2-pentanol can yield 2,3-dimethyl-1-pentene or 2,3-dimethyl-2-pentene. Only the terminal alkene in this case is chiral.

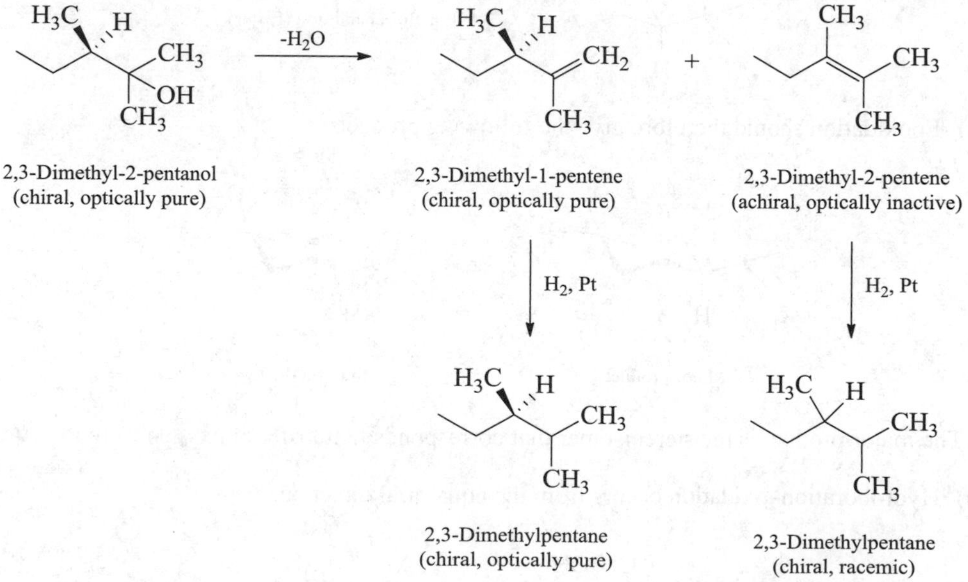

2,3-Dimethyl-2-pentanol   (chiral, optically pure)

2,3-Dimethyl-1-pentene   (chiral, optically pure)

2,3-Dimethyl-2-pentene   (achiral, optically inactive)

H_2, Pt

H_2, Pt

2,3-Dimethylpentane   (chiral, optically pure)

2,3-Dimethylpentane   (chiral, racemic)

The 2,3-dimethyl-1-pentene formed in the dehydration reaction must be optically pure because it arises from optically pure alcohol by a reaction that does not involve any of the bonds to the chirality center. When optically pure 2,3-dimethyl-1-pentene is hydrogenated, it must yield optically pure 2,3-dimethylpentane—again, no bonds to the chirality center are involved in this step.

The 2,3-dimethyl-2-pentene formed in the dehydration reaction is achiral and must yield racemic 2,3-dimethylpentane on hydrogenation.

Because the alkane is 50% optically pure, the alkene fraction must have contained equal amounts of optically pure 2,3-dimethyl-1-pentene and its achiral isomer 2,3-dimethyl-2-pentene.

8.38 (*a*) Oxygen may be transferred to either the front face or the back face of the double bond when (*R*)-3-buten-2-ol reacts with a peroxy acid. The structure of the minor stereoisomer was given in the problem. The major stereoisomer results from addition to the opposite face of the double bond.

(*R*)-3-Buten-2-ol   Minor stereoisomer   Major stereoisomer

(*b*) The two epoxides have the same configuration (*R*) at the secondary alcohol carbon but opposite configurations at the chirality center of the epoxide ring. They are diastereomers.

(*c*) In addition to the two diastereomeric epoxides whose structures are shown in the solution to part (*a*), the enantiomers of each will be formed when racemic 3-buten-2-ol is epoxidized. The relative amounts of the four products will be

Enantiomeric forms of minor stereoisomer, totaling 40%

20%   20%

Enantiomeric forms of major stereoisomer, totaling 60%

30%   30%

8.39 (*a*) Structures A and B are chiral. Structure C has a plane of symmetry and is an achiral meso form.

(*b*) Ozonolysis of the starting material proceeds with the stereochemistry shown. Compound B is the product of the reaction.

Compound B

(*c*) If the methyl groups were cis to each other in the cycloalkene, they would be on the same side of the Fischer projection in the product. Compound C would be formed.

Thermocinistry

Thermochemistry

8.40   2,3-Dimethyl-2-butene has the higher heat of combustion. The balanced equations for combustion of 2,3-dimethyl-2-butene and 1-butene tell us that on a molar basis 2,3-dimethyl-2-butene consumes more oxygen and gives off more carbon dioxide and water than 1-butene does. Therefore, 2,3-dimethyl-2-butene also gives off more heat..

2,3-Dimethyl-2-butene	+ $9O_2$ $\longrightarrow$	$6CO_2$	+	$6H_2O$
	Oxygen	Carbon dioxide		Water

1-Butene	+ $6O_2$ $\longrightarrow$	$4CO_2$	+	$4H_2O$
	Oxygen	Carbon dioxide		Water

Hydrogenation, on the other hand, consumes one mole of H_2 for each alkene and the difference in their heats of hydrogenation reflects only the difference in the potential energy of each alkene.

8.41   We need first to analyze the substitution patterns at the double bonds.

A	1-Pentene		Monosubstituted
B	(E)-4,4-Dimethyl-2-pentene		trans-Disubstituted
C	(Z)-4-Methyl-2-pentene		cis-Disubstituted
D	(Z)-2,2,5,5-Tetramethyl-3-hexene		Two tert-butyl groups cis
E	2,4-Dimethyl-2-pentene		Trisubstituted

Compound (D), having two cis tert-butyl groups, should have the least stable (highest energy) double bond. The remaining alkenes are arranged in order of increasing stability (decreasing heats of hydrogenation) according to the degree of substitution of the double bond: monosubstituted, cis-disubstituted, trans-disubstituted, trisubstituted. The heats of hydrogenation are therefore:

D	A	C	B	E
151 kJ/mol	122 kJ/mol	114 kJ/mol	111 kJ/mol	105 kJ/mol
(36.2 kcal/mol)	(29.3 kcal/mol)	(27.3 kcal/mol)	(26.5 kcal/mol)	(25.1 kcal/mol)
least stable alkene (highest energy)				most stable alkene (lowest energy)

8.42 Both *cis*-2-butene and *trans*-2-butene react with HBr to give the same product (2-bromobutane).

Therefore, the difference in the heats of reaction is due entirely to the difference in the alkenes respective energies. *cis*-2-Butene gives off 5 kJ/mol (1.1 kcal/mol) more heat when it reacts with HBr than *trans*-2-butene does, which makes *cis*-2-butene 5 kJ/mol (1.1 kcal/mol) higher in energy (less stable) than *trans*-2-butene. This energy difference is approximately the same as that obtained from their heats of hydrogenation [4 kJ/mol (1 kcal/mol)] and heats of combustion [3 kJ/mol (0.7 kcal/mol)].

8.43

	Reaction is	Sign of	
		ΔH°	ΔS°
(*a*)	exergonic at all temperatures	−	+
(*b*)	exergonic at low temperature; endergonic at high temperature	−	−
(*c*)	endergonic at all temperatures	+	−
(*d*)	endergonic at low temperature; exergonic at high temperature	+	+

(*a*) A reaction will be exergonic at all temperatures when ΔG° is negative at all temperatures. Therefore, ΔH° should have a negative sign. ΔS° should have a positive sign to ensure that the $-T\Delta S^\circ$ term is also negative. A reaction that is exergonic at all temperatures is one that is exothermic and proceeds with an increase in entropy.

(*b*) This case is one in which an increase in temperature reverses an exergonic reaction to an endergonic one. It must be a reaction in which ΔS° has a negative sign, so that $-T\Delta S^\circ$ becomes positive as T increases and overcomes a negative value for ΔH°.

(*c*) Endergonic at all temperatures is the "worst" case—an endothermic reaction in which entropy decreases.

(*d*) Entropy must increase in order for an endothermic reaction (endergonic at low temperature) to become exergonic at high temperature.

8.44   Each of the three compounds has a different alkene substitution pattern. One is monosubstituted, one is disubstituted, and one trisubstituted. The least substituted alkene gives the highest heat of hydrogenation.

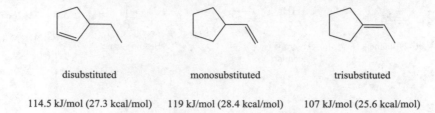

| disubstituted | monosubstituted | trisubstituted |

114.5 kJ/mol (27.3 kcal/mol)   119 kJ/mol (28.4 kcal/mol)   107 kJ/mol (25.6 kcal/mol)

8.45   (*a*) $\Delta G°$ for the iodination of ethylene at 25°

$$H_2C{=}CH_2 \ (g) \quad + \quad I_2 \ (g) \ \rightleftharpoons \ ICH_2CH_2I \ (g)$$

can be calculated from the values given for $\Delta H°$ and $\Delta S°$.

$$\Delta G° = \Delta H° - T\Delta S°$$

$$\Delta G° = -48 \text{ kJ} - (298 \text{ K})(-0.13 \text{ kJ/K})$$

$$\Delta G° = -48 \text{ kJ} + 38.7 \text{ kJ}$$

$$\Delta G° = -9 \text{ kJ}$$

The equilibrium constant K is calculated from $\Delta G°$.

$$\Delta G° = -RT \ \ln K$$

$$-(9 \text{ kJ})(1000 \text{ J/kJ}) = -(8.314 \text{ J/K})(298 \text{ K})\ln K$$

$$-9000 \text{ J} = -(8.314 \text{ J/K})(298 \text{ K})\ln K$$

$$\ln K = \frac{-9000 \text{ J}}{-(8.314 \text{ J/K})(298)} = 3.6$$

$$K = e^{3.6} = 38$$

(*b*) The sign of $\Delta G°$ is negative; the reaction is exergonic.

(*c*) The sign of $\Delta S°$ is negative. Therefore $(-T\Delta S°)$ becomes increasingly positive as the temperature is raised, $\Delta G°$ becomes less negative, and K decreases.

Synthesis

8.46   (*a*) The desired transformation is the conversion of an alkene to a vicinal dibromide.

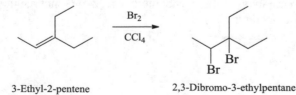

| 3-Ethyl-2-pentene | 2,3-Dibromo-3-ethylpentane |

(*b*) Markovnikov addition of hydrogen chloride is indicated.

3-Ethyl-2-pentene → HCl → 3-Chloro-3-ethylpentane

(*c*) Acid-catalyzed hydration will occur in accordance with Markovnikov's rule to yield the desired tertiary alcohol.

3-Ethyl-2-pentene → H_2O, H_2SO_4 → 3-Ethyl-3-pentanol

(*d*) Hydroboration–oxidation results in hydration of alkenes with a regioselectivity opposite to that of Markovnikov's rule.

3-Ethyl-2-pentene → 1. B_2H_6 2. H_2O_2, OH^- → 3-Ethyl-2-pentanol

(*e*) A peroxy acid will convert an alkene to an epoxide.

3-Ethyl-2-pentene → CH_3COOH → 2,3-Epoxy-3-ethylpentane

(*f*) Hydrogenation of alkenes converts them to alkanes.

3-Ethyl-2-pentene → H_2, Pt → 3-Ethylpentane

8.47 (*a*) Four primary alcohols have the molecular formula $C_5H_{12}O$:

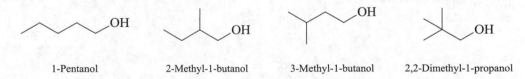

1-Pentanol   2-Methyl-1-butanol   3-Methyl-1-butanol   2,2-Dimethyl-1-propanol

2,2-Dimethyl-1-propanol cannot be prepared by hydroboration–oxidation of an alkene, because no alkene can have this carbon skeleton.

(*b*) Hydroboration–oxidation of alkenes is the method of choice for converting terminal alkenes to primary alcohols.

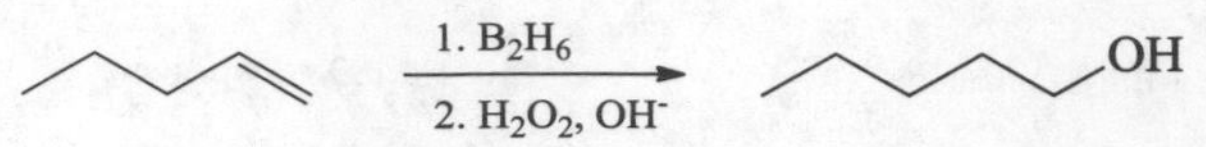

1-Pentene → 1-Pentanol

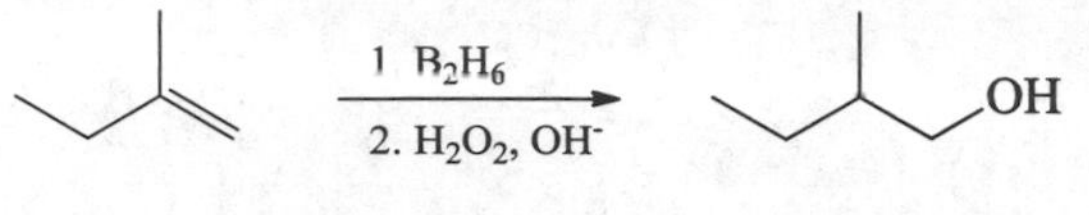

1. B_2H_6
2. H_2O_2, OH^-

2-Methyl-1-butene → 2-Methyl-1-butanol

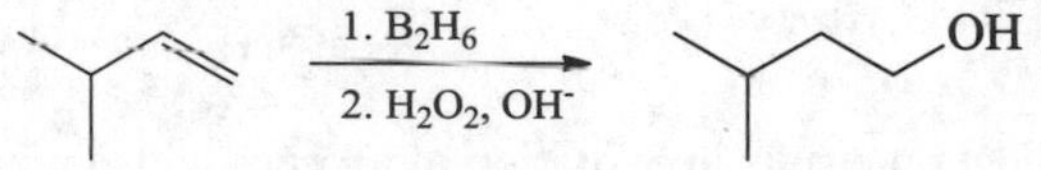

1. B_2H_6
2. H_2O_2, OH^-

3-Methyl-1-butene → 3-Methyl-1-butanol

1. B_2H_6
2. H_2O_2, OH^-

(*c*) The only tertiary alcohol is 2-methyl-2-butanol. It can be made by acid-catalyzed hydration of 2-methyl-1-butene or 2-methyl-2-butene.

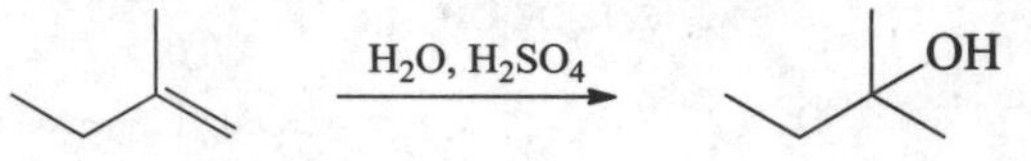

2-Methyl-1-butene → 2-Methyl-2-butanol

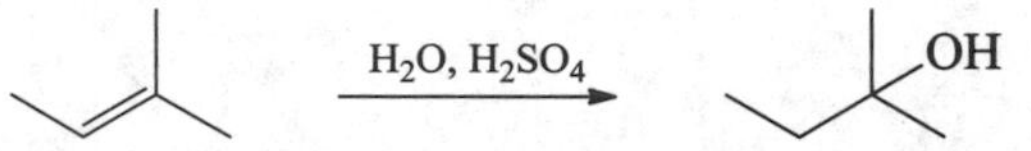

H_2O, H_2SO_4

2-Methyl-2-butene → 2-Methyl-2-butanol

H_2O, H_2SO_4

8.48 Working backwards may be the best way to solve this problem that is expressed in retrosynthesis form.

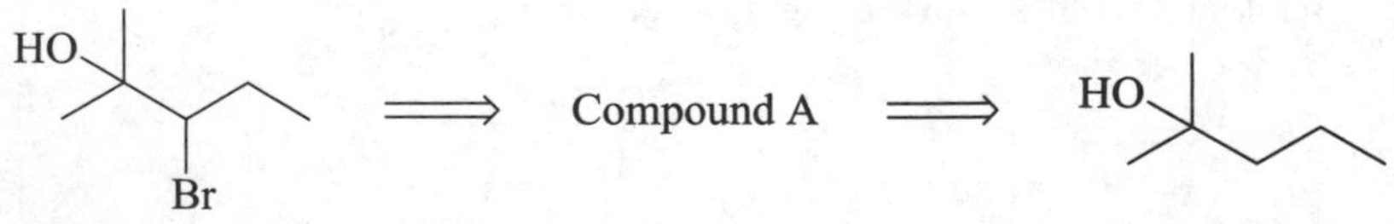

3-Bromo-2-methyl-2-pentanol   Compound A   2-Methyl-2-pentanol

The product with the indicated regiochemistry, having a hydroxyl group bonded to a tertiary carbon and a bromine on an adjacent secondary carbon suggest a vicinal halohydrin synthesis from 2-methyl-2-pentene using bromine and water.

2-Methyl-2-pentene

Compound A

3-Bromo-2-methyl-2-pentanol

The trisubstituted alkene, Compound A, can be prepared in a dehydration step from the starting alcohol.

2-Methyl-2-pentanol

2-Methyl-2-pentene

Compound A

8.49 Inspection of the starting material and product in the retrosynthetic reaction can lead to a strategy for working both forward and backwards for solving this problem.

$C_7H_{16}O_3S$

Compound A

$C_6H_{14}O$

Compound B

Comparison of the starting alkene and the final cyanide product suggests that the terminal carbon of the double bond must be functionalized. Most of the alkene addition reactions that we have studied involve Markovnikov addition with the exception on one, hydroboration-oxidation.

CN must eventually attach to the terminal alkene carbon

Compound B does include an oxygen in its formula so starting with the alkene and working forward from the starting material we now have a functional group on the terminal carbon.

1. BH$_3$ THF

2. H$_2$O$_2$, OH$^-$

$C_6H_{14}O$

Compound B

We can now turn our attention to the last step and consider how to place a cyano group on the molecule.

Earlier we learned that cyanide (CN$^-$) is a good nucleophile in S$_N$2 reactions with primary alkyls. In order to carry out an S$_N$2 reaction we need a compound with a good leaving group. We have two hints with the formulas given. One hint in the formula of Compound A, is that it contains a sulfur atom. We know that sulfonates are good leaving groups and they are prepared from alcohols. A second hint is that Compound

A has one more carbon than B, and two more oxygen atoms. This means that CH_3SO_2Cl must be the reagent used in the conversion of B to A. The reaction completion of the scheme from Compound B is

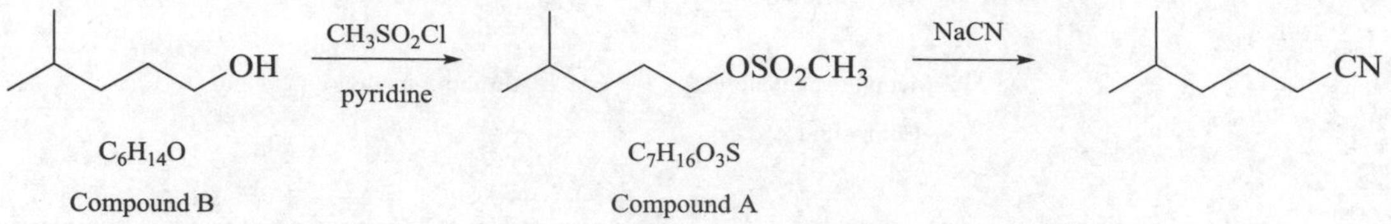

$C_6H_{14}O$

Compound B

$C_7H_{16}O_3S$

Compound A

8.50 (*a*) An alkene is a reasonable retrosynthetic link between a primary alcohol target and a secondary alcohol reactant.

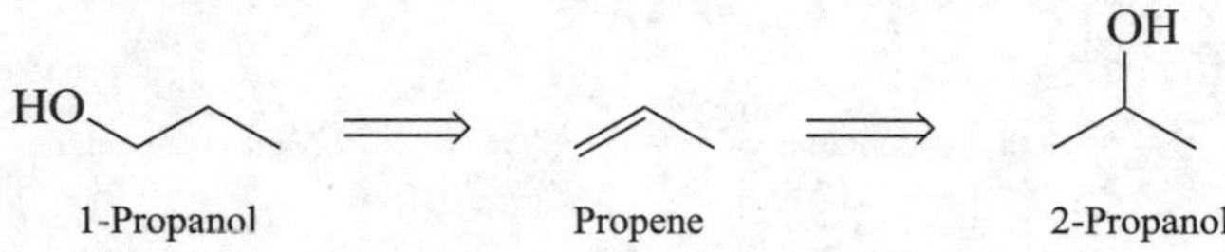

1-Propanol   Propene   2-Propanol

Hydroboration-oxidation converts propene to 1-propanol and can be prepared from 2-propanol by acid-catalyzed dehydration.

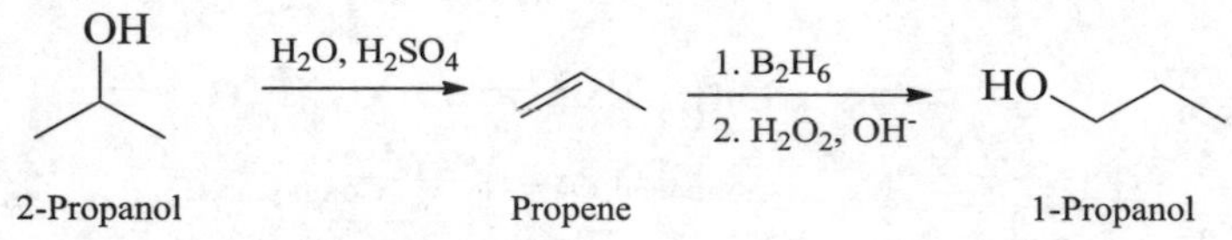

2-Propanol   Propene   1-Propanol

(*b*) Vicinal dibromides are usually prepared by adding elemental bromine to an alkene. Preparation of alkenes from alkyl bromides requires strong base in an E2 reaction.

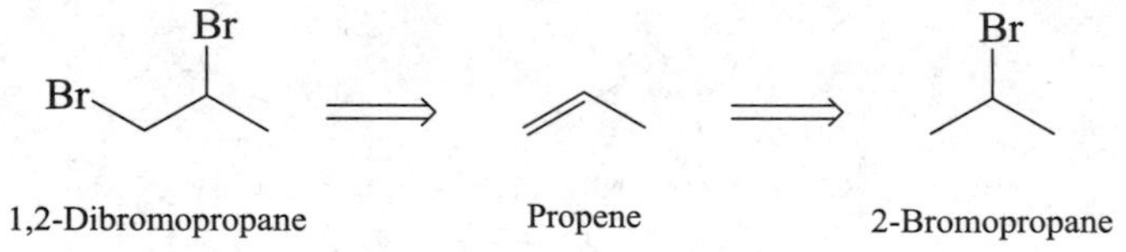

1,2-Dibromopropane   Propene   2-Bromopropane

Propene is prepared from 2-bromopropane and sodium ethoxide by E2 elimination. The correct synthesis is

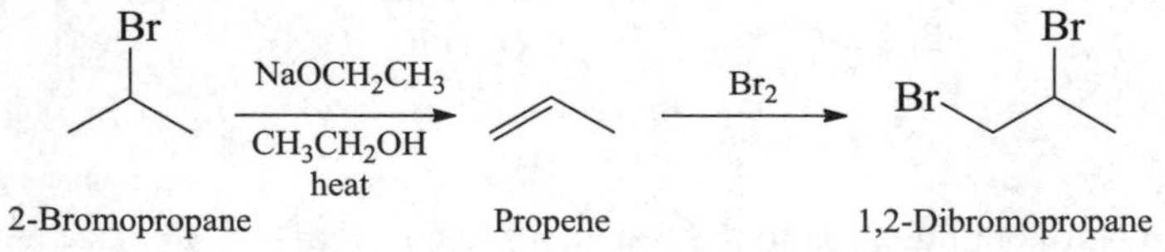

2-Bromopropane   Propene   1,2-Dibromopropane

(*c*) Vicinal bromohydrins are prepared by addition of bromine and water to alkenes. Regiochemistry of the bromohydrin formation dictates that the –OH group is attached to the more substituted carbon. So a reasonable retrosynthetic pathway is

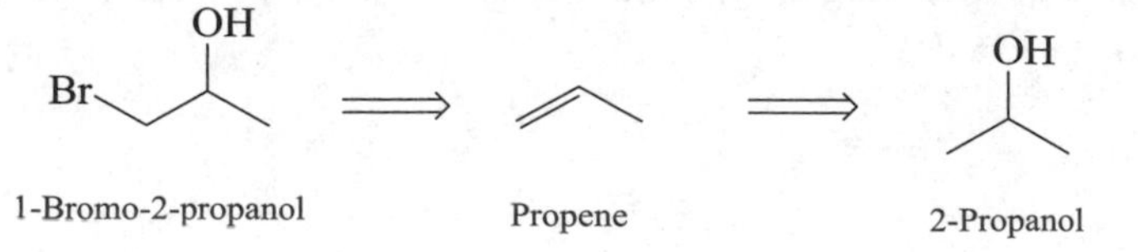

1-Bromo-2-propanol   Propene   2-Propanol

Preparation of alkenes directly from an alcohol takes acid and heat. The correct synthesis is therefore

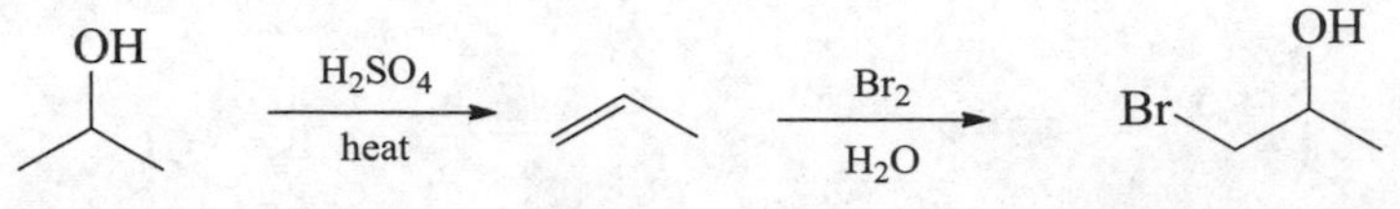

2-Propanol   Propene   1-Bromo-2-propanol

(*d*) As in part (*c*), we see that the product is a vicinal bromohydrin that is prepared from an alkene.

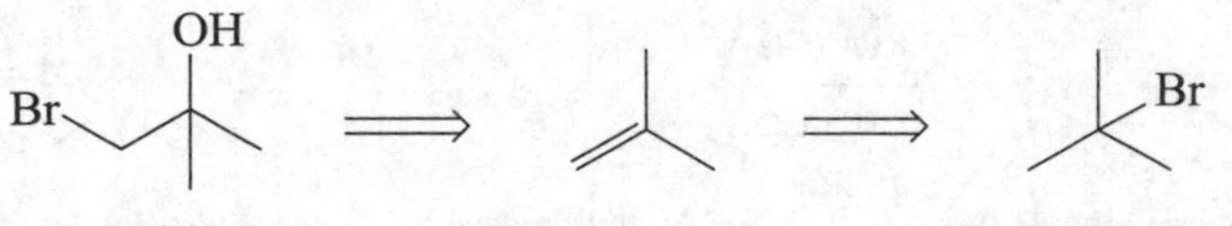

1-Bromo-2-methyl-2-propanol   2-Methylpropene   *tert*-Butyl bromide

The alkene, in turn, can be prepared directly from the starting alkyl bromide with strong base.

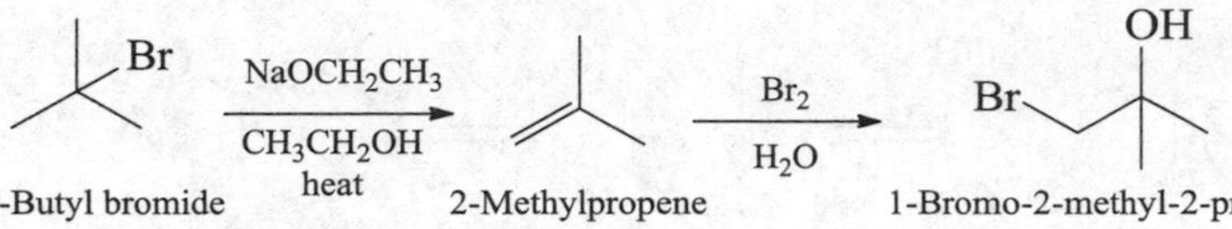

tert-Butyl bromide   2-Methylpropene   1-Bromo-2-methyl-2-propanol

(*e*) Epoxides can be prepared from alkenes using peroxyacetic acid. This leads to the retrosynthetic reasoning

1,2-Epoxypropane   Propene   2-Propanol

Propene can be prepared using acid and heat.

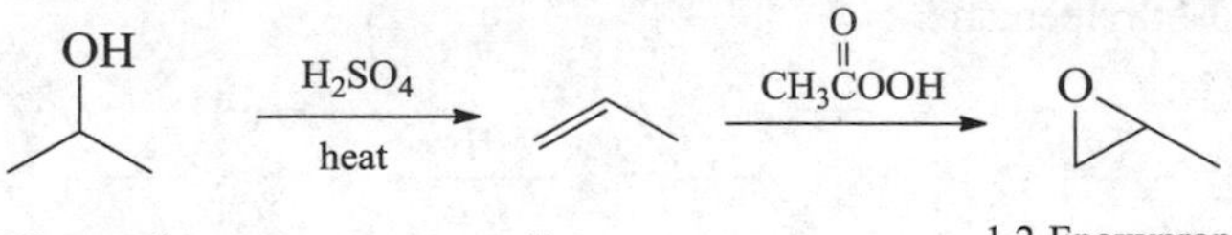

2-Propanol   Propene   1,2-Epoxypropane

(*f*) *tert*-Butyl alcohol and isobutyl alcohol have the same carbon skeleton; all that is required is to move the hydroxyl group from C-1 to C-2. As pointed out in part (*a*) of this problem, we can do it in two efficient steps through a synthesis that involves hydration of an alkene.

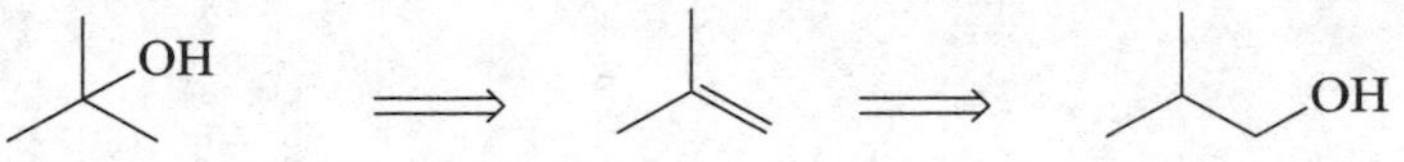

tert-Butyl alcohol   2-Methylpropene   Isobutyl alcohol

Acid-catalyzed hydration of the alkene gives the desired regioselectivity.

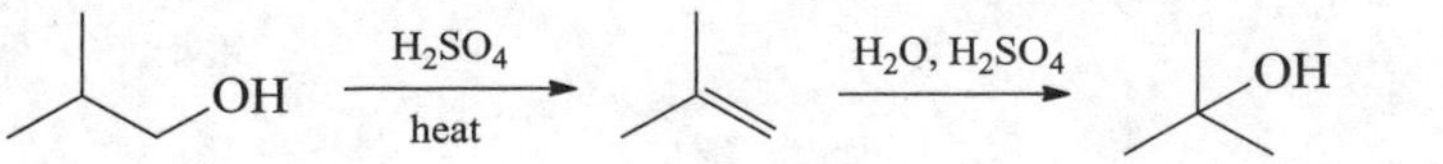

Isobutyl alcohol   2-Methylpropene   *tert*-Butyl alcohol

(*g*) The strategy of this exercise is similar to that of the preceding one. Convert the starting material to an alkene by elimination, followed by electrophilic addition to the double bond.

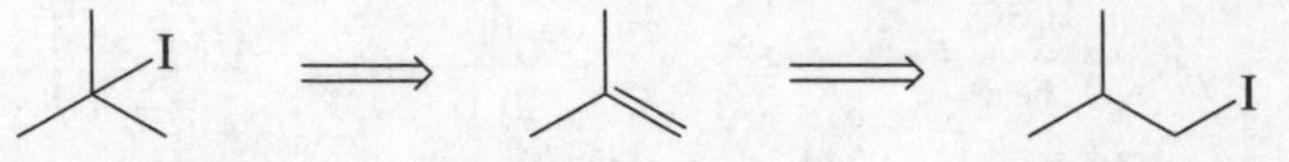

I
I
tert-Butyl iodide
2-Methylpropene
Isobutyl iodide

For an E2 reaction of a primary alkyl halide, a hindered base is typically employed.

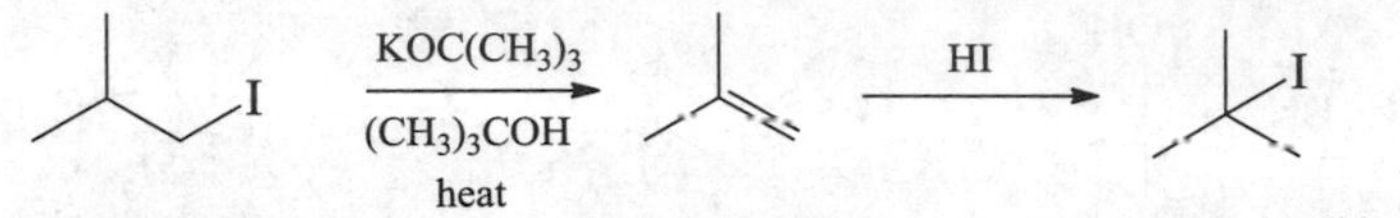

I
KOC(CH3)3
(CH3)3COH
heat
HI
I
Isobutyl iodide
2-Methylpropene
tert-Butyl iodide

(*h*) This problem is similar to the one in part (*d*) in that it requires the preparation of a vicinal halohydrin from an alkyl halide. The strategy is the same.

OH
Cl
Cl

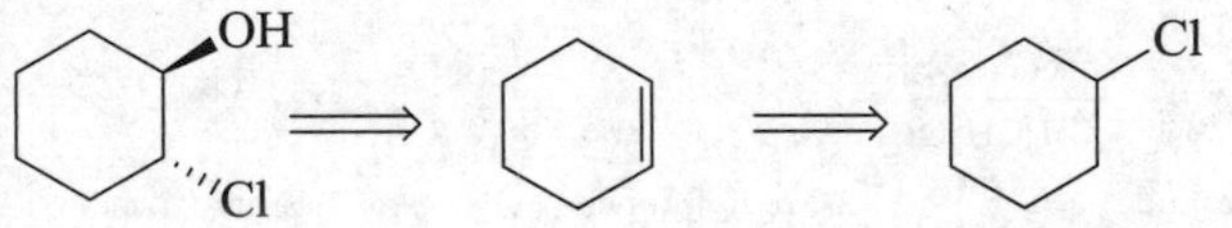

Convert the alkyl halide to an alkene, and then form the chlorohydrin by treatment with elemental chlorine in aqueous solution.

Cl
NaOCH2CH3
CH3CH2OH
heat
Cl2
H2O
OH
Cl

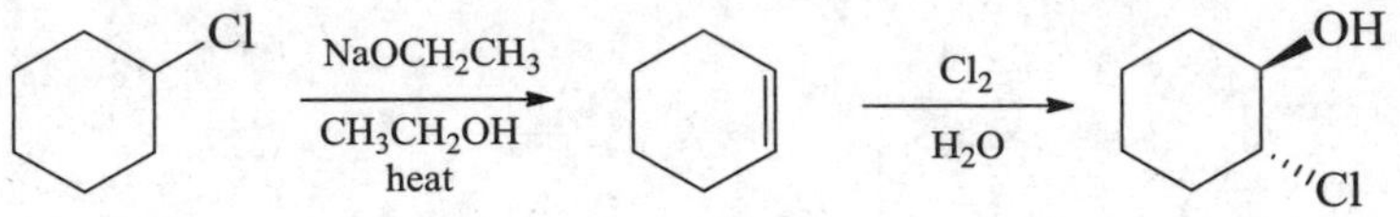

Structure Determination

8.51 The carbon skeleton is revealed by the hydrogenation experiment. Compounds B and C must have the same carbon skeleton as 3-ethylpentane.

or
H2
cat

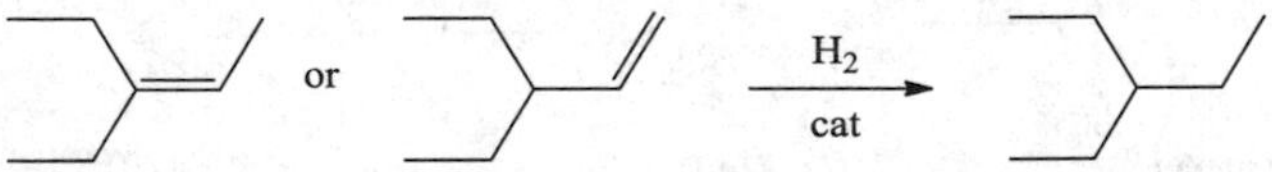

Three alkyl bromides have this carbon skeleton, namely, 1-bromo-3-ethylpentane, 2-bromo-3-ethylpentane, and 3-bromo-3-ethylpentane. Of these three, only 2-bromo-3-ethylpentane will give two alkenes on dehydrobromination.

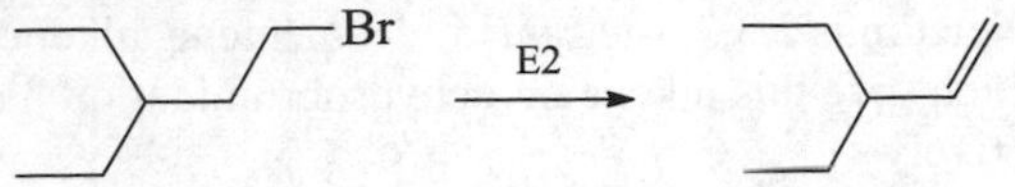

1-Bromo-3-ethylpentane (only product)

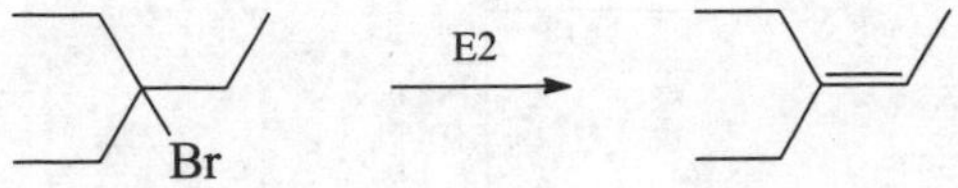

3-Bromo-3-ethylpentane (only product)

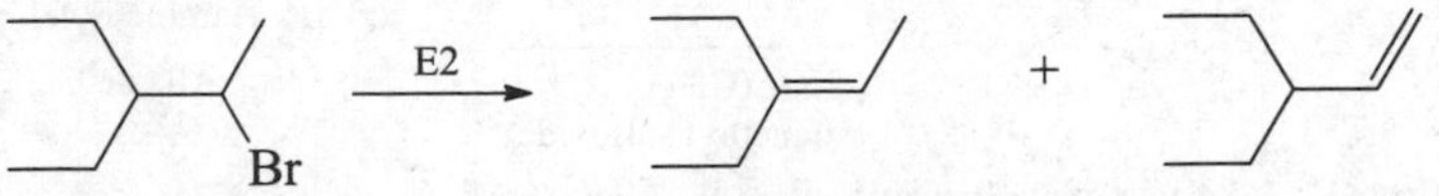

2-Bromo-3-ethylpentane   3-Ethyl-2-pentene   3-Ethyl-1-pentene
(Compound A)   (Compound B)   (Compound C)

Compound A must therefore be 2-bromo-3-ethylpentane. Dehydrobromination of A will follow Zaitsev's rule, so that the major alkene (compound B) is 3-ethyl-2-pentene and the minor alkene (compound C) is 3-ethyl-1-pentene.

8.52 The information that compound B gives 2,4-dimethylpentane on catalytic hydrogenation establishes its carbon skeleton.

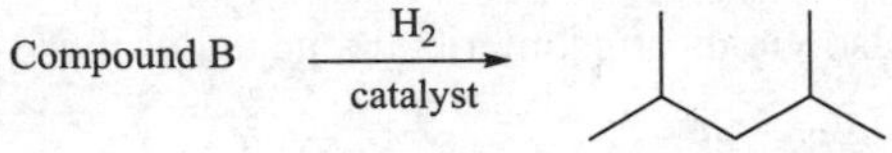

2,4-Dimethylpentane

Compound B is an alkene derived from compound A—an alkyl bromide of molecular formula $C_7H_{15}Br$. We are told that compound A is not a primary alkyl bromide. Compound A can therefore be only

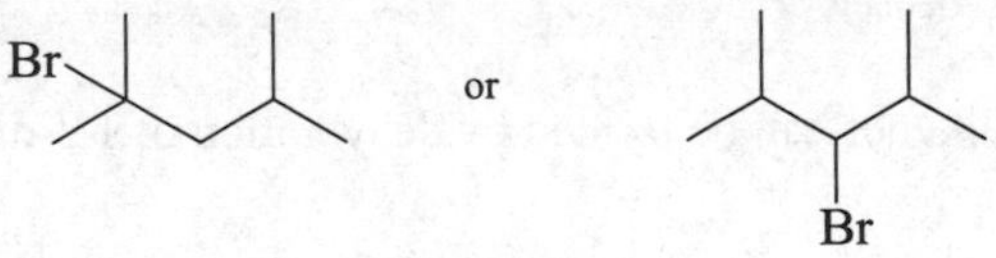

or

Because compound A gives a single alkene on being treated with sodium ethoxide in ethanol, it can only be 3-bromo-2,4-dimethylpentane, and compound B must be 2,4-dimethyl-2-pentene.

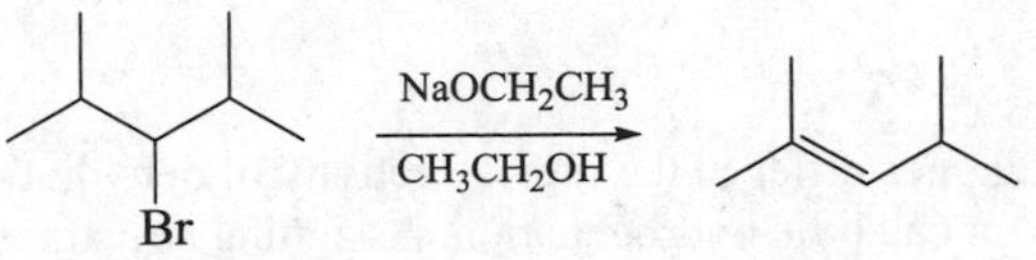

3-Bromo-2,4-dimethylpentane   2,4-Dimethyl-2-pentene
(Compound A)   (Compound B)

8.53 Alkene C must have the same carbon skeleton as its hydrogenation product, 2,3,3,4-tetramethylpentane.

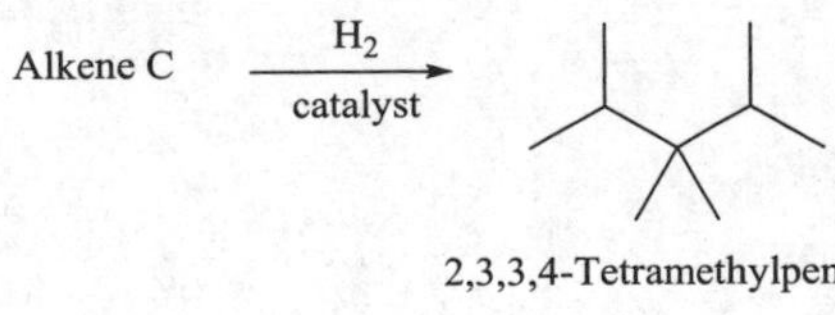

2,3,3,4-Tetramethylpentane

The only alkene with this carbon skeleton is 2,3,3,4-tetramethyl-1-pentene, alkene C. The two alkyl bromides, compounds A and B, which give this alkene on dehydrobromination have their bromine substituents at C-1 and C-2, respectively.

1-Bromo-2,3,3,4-tetramethylpentane

2-Bromo-2,3,3,4-tetramethylpentane

KOC(CH₃)₃
dimethyl sulfoxide

KOC(CH₃)₃
dimethyl sulfoxide

2,3,3,4-Tetramethyl-1-pentene
Alkene C

8.54 The only alcohol (compound A) that can undergo acid-catalyzed dehydration to alkene B without rearrangement is the one shown in the equation.

$\dfrac{KHSO_4}{heat \;-H_2O}$

Alcohol A

Alkene B

Dehydration of alcohol A also yields an isomeric alkene under these conditions.

$\dfrac{KHSO_4}{heat \;-H_2O}$

Alcohol A

Alkene C

8.55 First decide what three alkenes can be formed by dehydration of 1,2-dimethylcyclohexanol.

OH

H_3O^+, heat

A   B   C

Notice that the three alkenes differ in the regioselectivity of dehydration and that each will give 1,2-dimethylcyclohexane on catalytic hydrogenation. Assigning structures to the alkenes on the basis of Zaitsev's rule gives A, B, and C as indicated in the equation. A has the least substituted double bond and is formed in the smallest amount. C has the most substituted double bond and is formed in greatest amount. A is the least stable alkene and is present in the smallest amount at equilibrium. C has the most stable double bond and predominates at equilibrium.

8.56 Electrophilic addition of hydrogen iodide should occur in accordance with Markovnikov's rule.

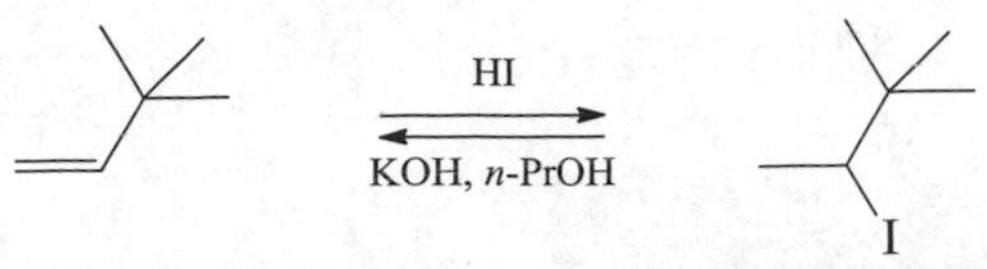

3,3-Dimethyl-1-butene   3-Iodo-2,2-dimethylbutane

Treatment of 3-iodo-2,2-dimethylbutane with alcoholic potassium hydroxide should bring about E2 elimination to regenerate the starting alkene. Hence, compound A is 3-iodo-2,2-dimethylbutane.

The carbocation intermediate formed in the addition of hydrogen iodide to the alkene is one that can rearrange by a methyl group migration.

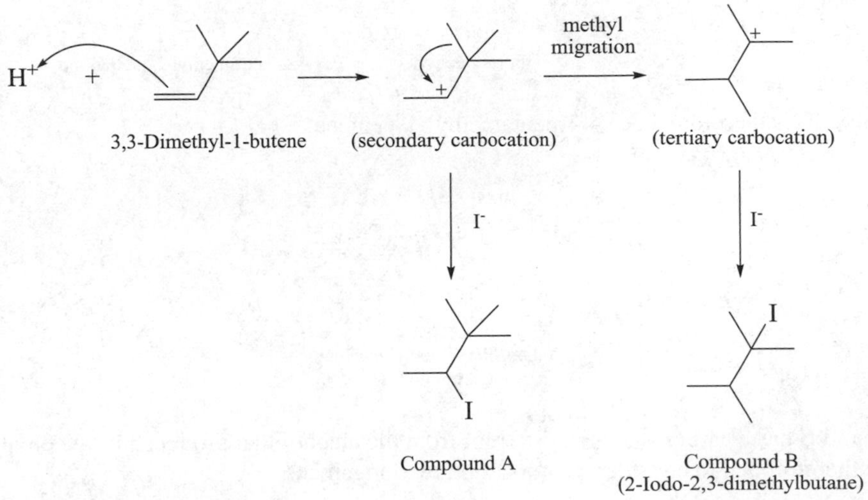

3,3-Dimethyl-1-butene   (secondary carbocation)   (tertiary carbocation)

Compound A   Compound B
(2-Iodo-2,3-dimethylbutane)

A likely candidate for compound B is therefore the one with a rearranged carbon skeleton, 2-iodo-2,3-dimethylbutane. This is confirmed by the fact that compound B undergoes elimination to give 2,3-dimethyl-2-butene.

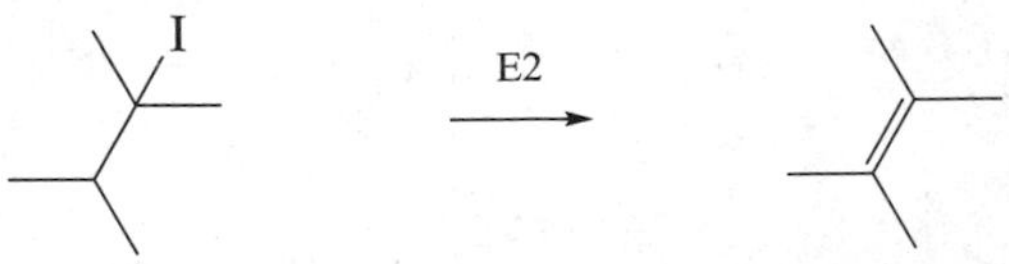

Compound B   2,3-Dimethyl-2-butene
(2-Iodo-2,3-dimethylbutane)

8.57 The ozonolysis data are useful in quickly identifying alkenes A and B.

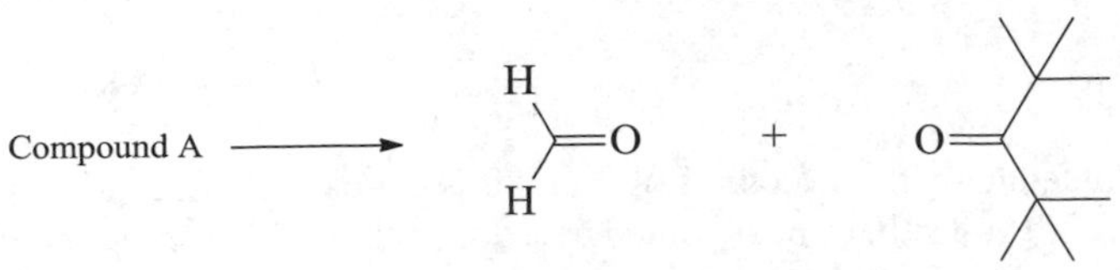

Compound A ⟶

Formaldehyde   2,2,4,4-Tetramethyl-3-pentanone

Compound A is therefore 2-*tert*-butyl-3,3-dimethyl-1-butene.

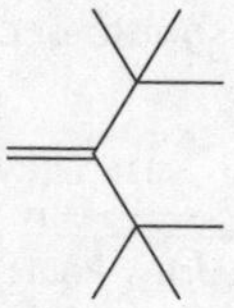

2-*tert*-Butyl-3,3-dimethyl-1-butene
Compound A

Compound B gives the following upon ozonolysis.

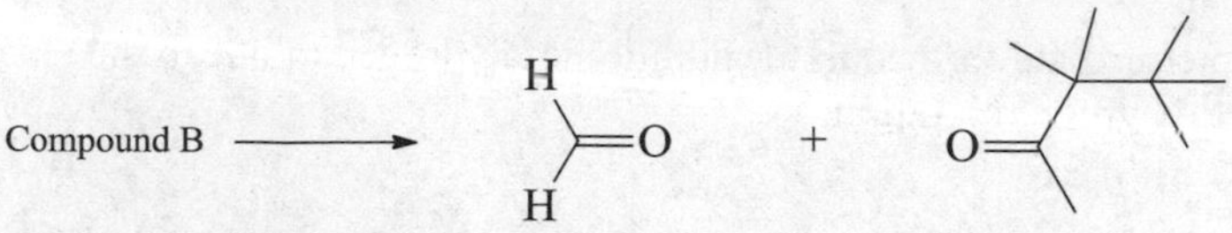

Formaldehyde   3,3,4,4-Tetramethyl-2-pentanone

Compound B is therefore 2,3,3,4,4-pentamethyl-1-pentene.

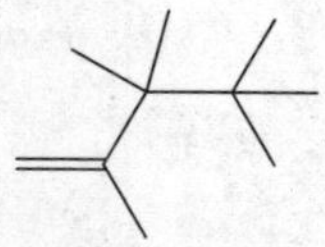

2,3,3,4,4-Pentamethyl-1-pentene
Compound B

Compound B has a carbon skeleton different from the alcohol that produced it by dehydration.
We are therefore led to consider a carbocation rearrangement.

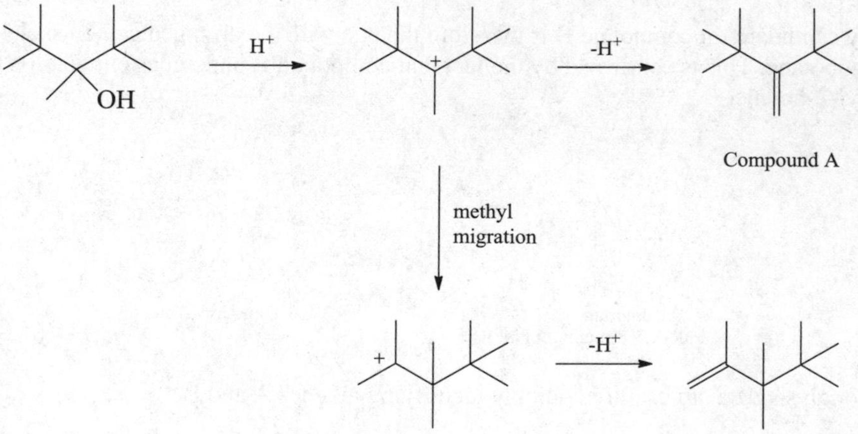

8.58   The important clue to deducing the structures of A and B is the ozonolysis product C. Remembering that the two carbonyl carbons of C must have been joined by a double bond in the precursor B, we write

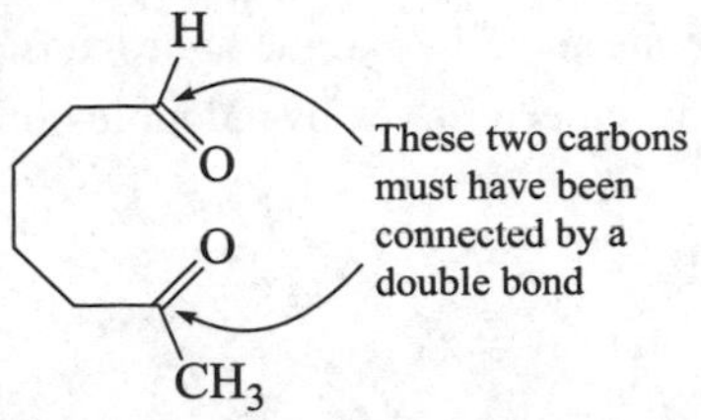

Compound C

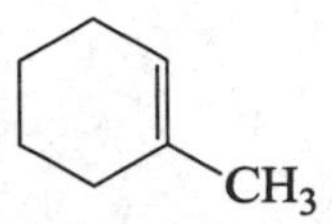

Compound B

The tertiary bromide that gives compound B on dehydrobromination is 1-methylcyclohexyl bromide.

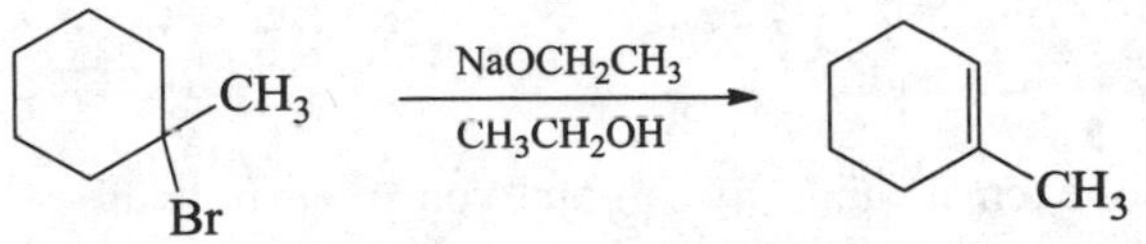

Compound A   Compound B

When tertiary halides are treated with base, they undergo E2 elimination. The regioselectivity of elimination of tertiary halides follows the Zaitsev rule.

8.59 Because santene and 1,3-diacetylcyclopentane (compound A) contain the same number of carbon atoms, the two carbonyl carbons of the diketone must have been connected by a double bond in santene. The structure of santene must therefore be

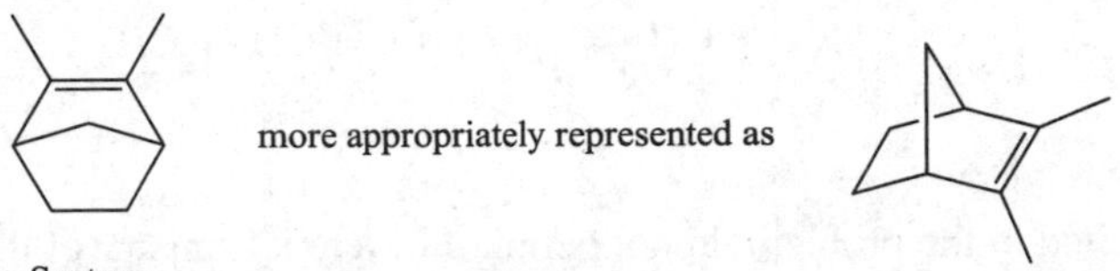

more appropriately represented as

Santene

8.60 (a) Compound A contains nine of the ten carbons and 14 of the 16 hydrogens of sabinene. Ozonolysis has led to the separation of one carbon and two hydrogens from the rest of the molecule. The carbon and the two hydrogens must have been lost as formaldehyde, $H_2C=O$. This H_2C unit was originally doubly bonded to the carbonyl carbon of compound A. Sabinene must therefore have the structure shown in the equation representing its ozonolysis:

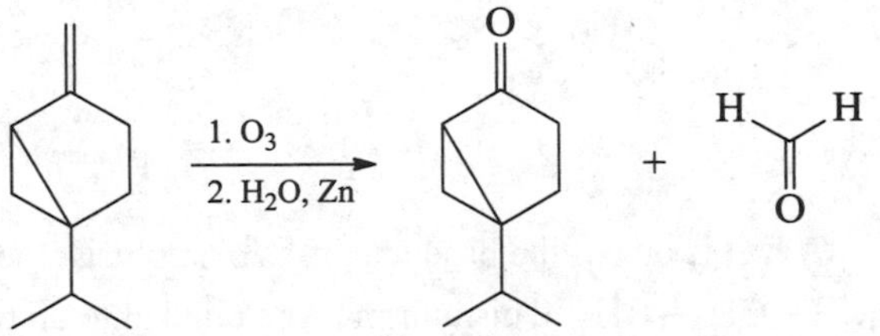

Sabinene   Compound A   Formaldehyde

(*b*) Compound B contains all ten of the carbons and all 16 of the hydrogens of Δ^3-carene. The two carbonyl carbons of compound B must have been linked by a double bond in Δ^3-carene.

Cleaved by ozonolysis

$$\text{1. O}_3$$
$$\text{2. H}_2\text{O, Zn}$$

Δ^3-Carene

Compound B

8.61 The sex attractant of the female housefly consumes one mole of hydrogen on catalytic hydrogenation (the molecular formula changes from $C_{23}H_{46}$ to $C_{23}H_{48}$). Thus, the molecule has one double bond. The position of the double bond is revealed by the ozonolysis data.

$$C_{23}H_{46} \quad \xrightarrow[\text{2. H}_2\text{O, Zn}]{\text{1. O}_3} \quad CH_3(CH_2)_7\overset{\displaystyle O}{\overset{\|}{C}}H \; + \; CH_3(CH_2)_{12}\overset{\displaystyle O}{\overset{\|}{C}}H$$

An unbranched 9-carbon unit and an unbranched 14-carbon unit make up the carbon skeleton, and these two units must be connected by a double bond. The housefly sex attractant therefore has the constitution

$$CH_3(CH_2)_7CH{=}CH(CH_2)_{12}CH_3$$

9-Tricosene

The data cited in the problem do not permit the stereochemistry of this natural product to be determined.

8.62 The hydrogenation data tell us that $C_{19}H_{38}$ contains one double bond and has the same carbon skeleton as 2,6,10,14-tetramethylpentadecane. We locate the double bond at C-2 on the basis of the fact that acetone, $(CH_3)_2C{=}O$, is obtained on ozonolysis. The structures of the natural product and the aldehyde produced on its ozonolysis are as follows:

Ozonolysis cleaves molecule here

Aldehyde obtained on ozonolysis

8.63 Because $O{=}CH{-}CH_2{-}CH{=}O$ is one of the products of its ozonolysis, the sex attractant of the arctiid moth must contain the unit $={}CH{-}CH_2{-}CH={}$ This unit must be bonded to an unbranched 12-carbon unit at one end and an unbranched 6-carbon unit at the other in order to give $CH_3(CH_2)_{10}CH=O$ and $CH_3(CH_2)_4CH=O$ on ozonolysis.

$$CH_3(CH_2)_{10}CH=CHCH_2CH=CH(CH_2)_4CH_3$$

Sex attractant of arctiid moth
(wavy lines show positions of cleavage on ozonolysis)

1. O_3
2. H_2O, Zn

$$CH_3(CH_2)_{10}\overset{O}{\overset{\|}{C}}H \quad + \quad H\overset{O}{\overset{\|}{C}}CH_2\overset{O}{\overset{\|}{C}}H \quad + \quad H\overset{O}{\overset{\|}{C}}(CH_2)_4CH_3$$

The stereochemistry of the double bonds cannot be determined on the basis of the available information.

Mechanism

8.64 (*a*) Phosphoric acid protonates cyclohexene giving cyclohexyl cation. This carbocation reacts with iodide ion to produce iodocyclohexane. The overall reaction corresponds to electrophilic addition of HI.

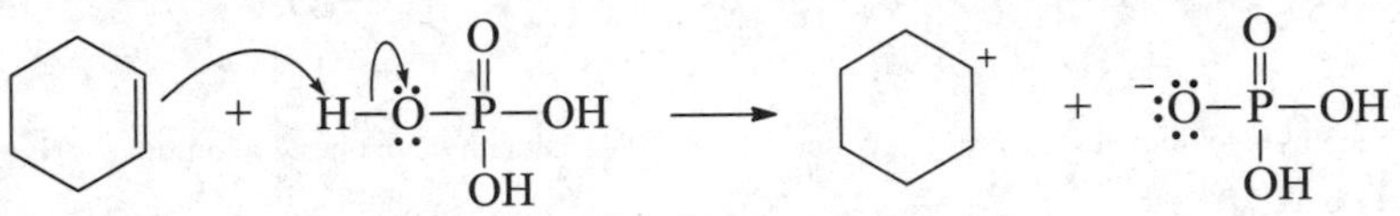

| Cyclohexene | Phosphoric acid | | Cyclohexyl cation | Dihydrogen phosphate ion |

| Cyclohexyl cation | Iodide ion | Iodocyclohexane |

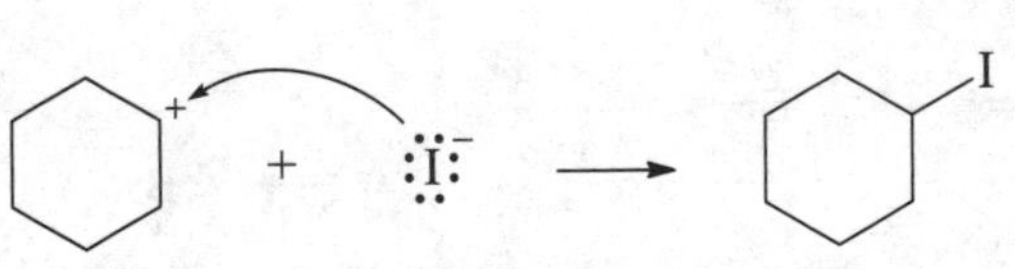

(b) This reaction follows the conventional mechanism for the acid-catalyzed hydration of alkenes. Water can attack the carbocation intermediate from either side giving a mixture of syn and anti addition.

(c) The product, a bromohydrin, is formed when water attacks a bromonium ion.

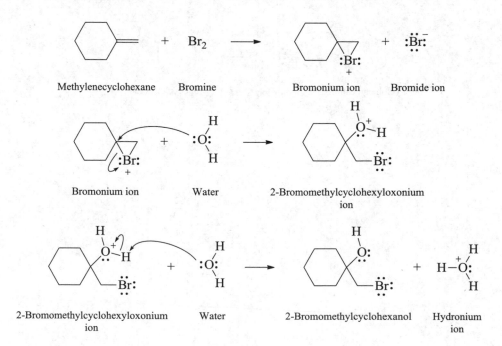

8.65 The first step in the mechanism of acid-catalyzed hydration of alkenes is protonation of the double bond to give a carbocation intermediate.

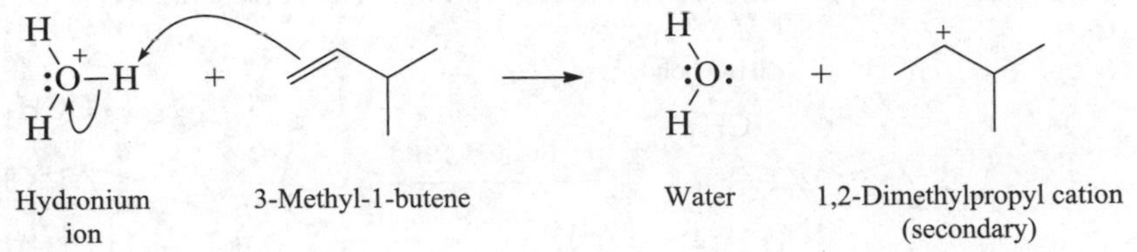

| Hydronium ion | 3-Methyl-1-butene | Water | 1,2-Dimethylpropyl cation (secondary) |

The carbocation formed in this step is secondary and capable of rearranging to a more stable tertiary carbocation by a hydride shift.

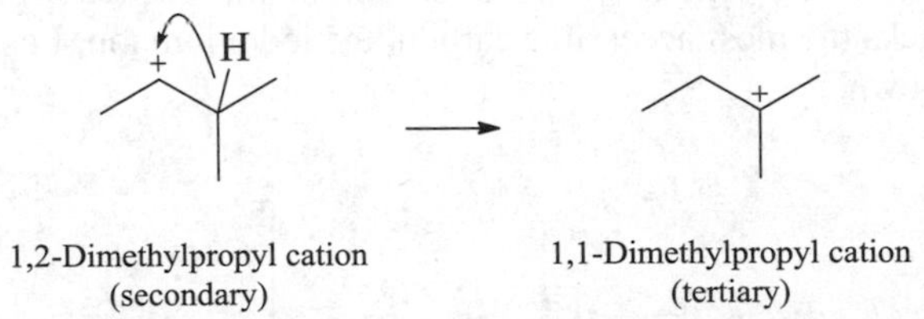

| 1,2-Dimethylpropyl cation (secondary) | 1,1-Dimethylpropyl cation (tertiary) |

The alcohol that is formed when water reacts with the tertiary carbocation is 2-methyl-2-butanol, not 3-methyl-2-butanol.

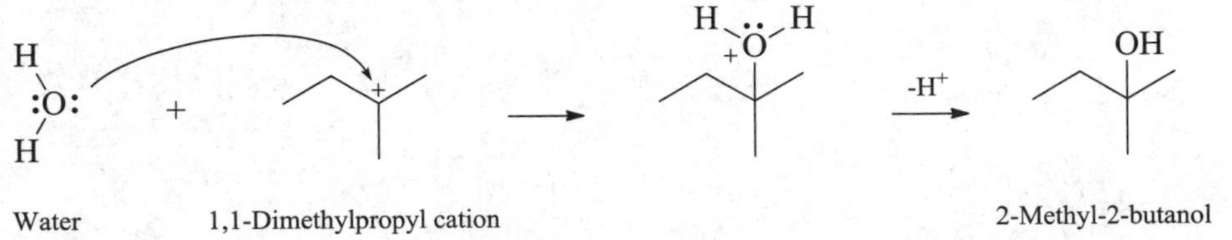

| Water | 1,1-Dimethylpropyl cation | | 2-Methyl-2-butanol |

8.66 In the presence of sulfuric acid, the carbon–carbon double bond of 2-methyl-1-butene is protonated and a carbocation is formed.

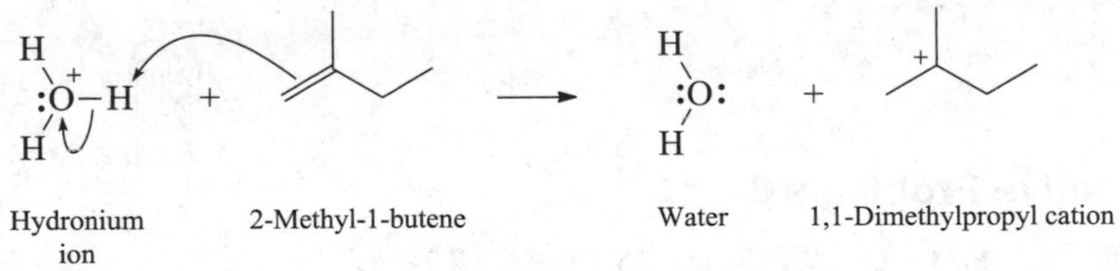

| Hydronium ion | 2-Methyl-1-butene | Water | 1,1-Dimethylpropyl cation |

This carbocation can then lose a proton from its CH_2 group to form 2-methyl-2-butene.

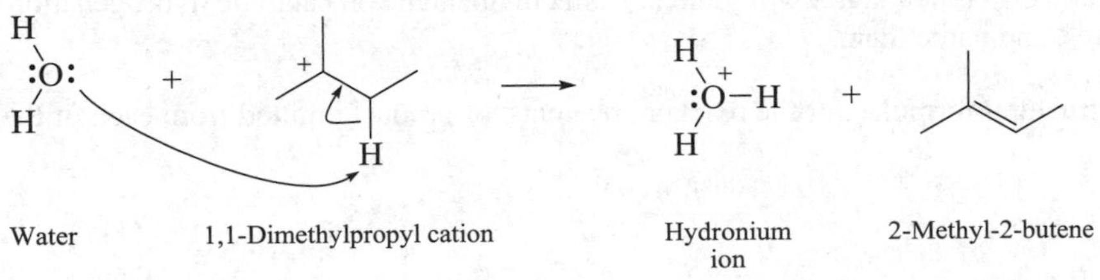

| Water | 1,1-Dimethylpropyl cation | Hydronium ion | 2-Methyl-2-butene |

8.67 The reaction of an alkene with bromine in methanol solution is analogous to vicinal bromohydrin formation in water. The stereochemistry of addition of Br and OCH_3 is anti. Br adds to the less substituted carbon of the double bond, OCH_3 to the more substituted carbon. The correct answer is D

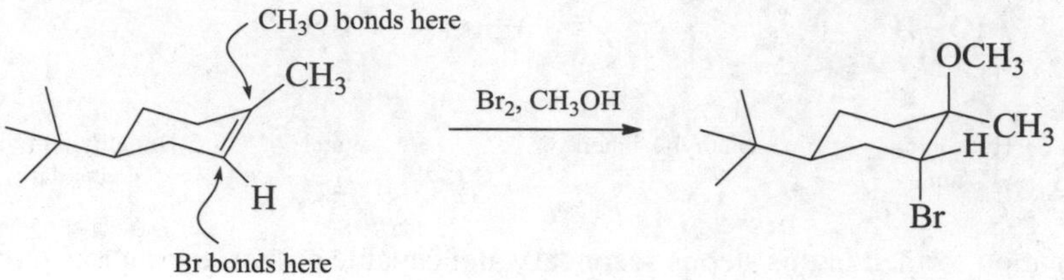

8.68 Reaction of iodine with the alkene forms an iodonium ion intermediate. The carbonyl oxygen acts as a nucleophile and attacks the most accessible carbon, the iodonium ion. Loss of a proton from this intermediate gives the product shown.

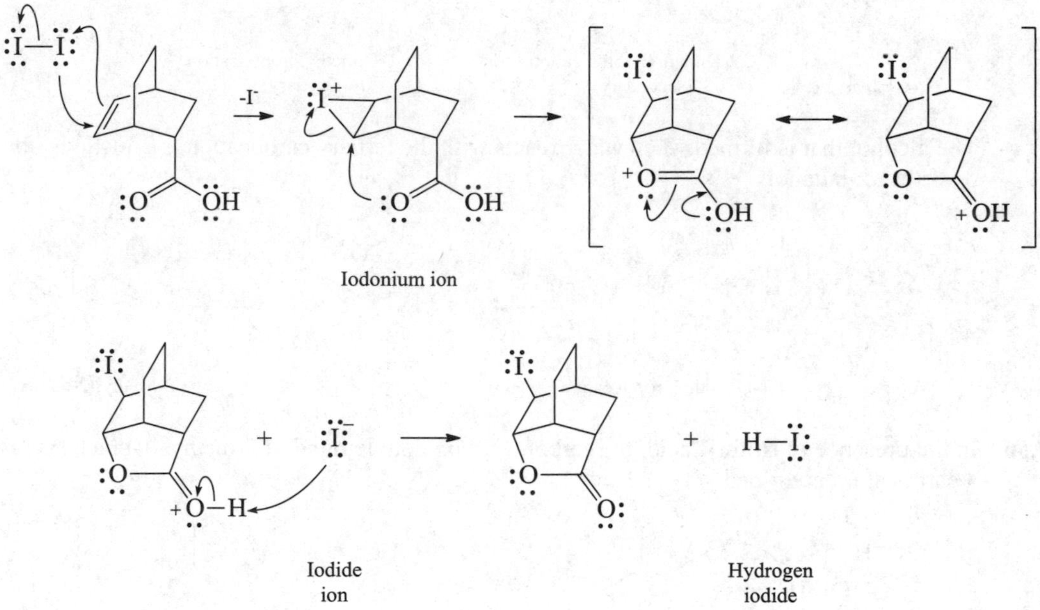

Answers to Interpretive Problems 8

8.69 A; **8.70** B; **8.71** C; **8.72** D; **8.73** A; **8.74** B

SELF-TEST

1. How many different alkenes will yield 2,3-dimethylpentane on catalytic hydrogenation? Draw their structures, and name them.

2. Write structural formulas for the reactant, reagents, or product omitted from each of the following:

   (*a*)

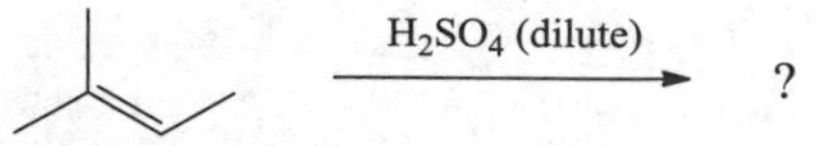

   (*b*)

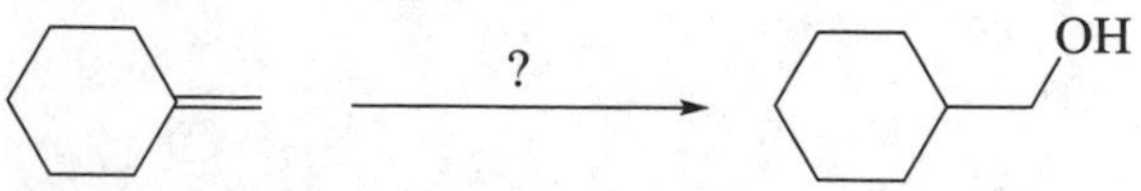

(c)

? ($C_{10}H_{16}$) $\xrightarrow[\text{2. } H_2O, \text{ Zn}]{\text{1. } O_3}$

(d)

+ Br_2 $\xrightarrow{H_2O}$?

3. Provide a sequence of reactions to carry out the following conversions. More than one synthetic step is necessary for each. Write the structure of the product of each synthetic step.

(a)

(b)

4. Chlorine reacts with an alkene to give the 2,3-dichlorobutane isomer whose structure is shown. What are the structure and name of the alkene? Outline a mechanism for the reaction.

5. Write a structural formula, including stereochemistry, for the compound formed from *cis*-3-hexene on treatment with peroxyacetic acid.

6. Give a mechanism describing the elementary steps in the reaction of 2-methyl-1-butene with hydrogen chloride. Use curved arrows to show the flow of electrons.

7. Excluding carbocation rearrangements, what two alkenes give 2-chloro-2-methylbutane on reaction with hydrogen chloride?

8. The reaction of 3-methyl-1-butene with hydrogen chloride gives two alkyl halide products; one is a secondary alkyl chloride and the other is tertiary. Write the structures of the products, and provide a mechanism explaining their formation.

9. A hydrocarbon A (C_6H_{12}) undergoes reaction with HBr to yield compound B ($C_6H_{13}Br$). Treatment of B with sodium ethoxide in ethanol yields C, an isomer of A. Reaction of C with ozone followed by treatment with water and zinc gives acetone, $(CH_3)_2C{=}O$, as the only organic product. Provide structures for A, B, and C, and outline the reaction pathway.

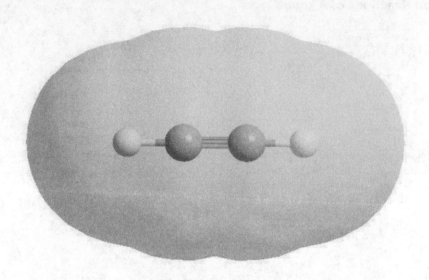

CHAPTER 9
Alkynes

Table of Contents

SOLUTIONS TO TEXT PROBLEMS

In Chapter Problems

9.1 The reaction is an acid–base process; water is the proton donor. Two separate proton-transfer steps are involved.

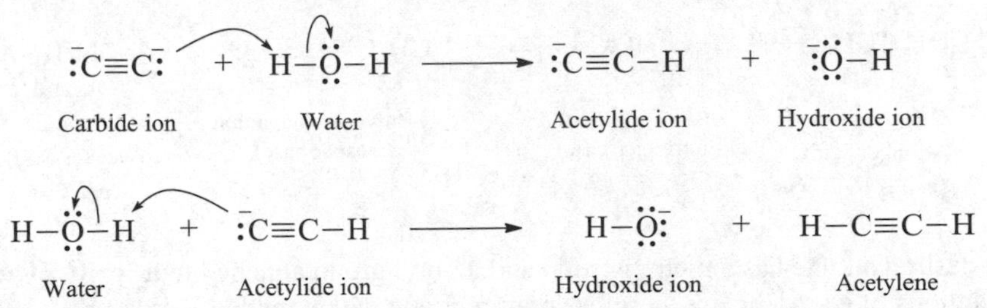

9.2 A triple bond may connect C-1 and C-2 or C-2 and C-3 in an unbranched chain of five carbons.

$$CH_3CH_2CH_2C{\equiv}CH \qquad CH_3CH_2C{\equiv}CCH_3$$

1-Pentyne
or Pent-1-yne

2-Pentyne
or Pent-2-yne

One of the C_5H_8 isomers has a branched carbon chain.

$$CH_3CHC{\equiv}CH$$
$$\quad |$$
$$\quad CH_3$$

3-Methyl-1-butyne
or
3-Methylbut-1-yne

9.3 The bonds become shorter and stronger in the series as the electronegativity increases.

	NH_3	H_2O	HF
Electronegativity:	N (3.0)	O (3.5)	F (4.0)
Bond distance (pm):	N—H (101)	O—H (95)	F—H (92)
Bond dissociation enthalpy (kJ/mol):	N—H (435)	O—H (497)	F—H (568)
Bond dissociation enthalpy (kcal/mol):	N—H (104)	O—H (119)	F—H (136)

9.4 (*b*) A proton is transferred from acetylene to ethyl anion.

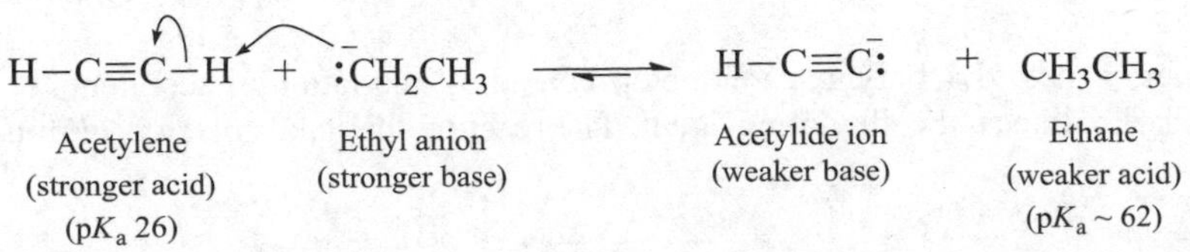

The position of equilibrium lies to the right. Ethyl anion is a very powerful base and deprotonates acetylene quantitatively.

(*c*) Amide ion is not a strong enough base to remove a proton from ethylene. The equilibrium lies to the left.

$$H_2C=C\overset{H}{\underset{H}{\diagup}} + \;:NH_2 \;\rightleftharpoons\; H_2C=C\overset{..}{\underset{H}{\diagup}} + \;:NH_3$$

Ethylene	Amide anion	Vinyl anion	Ammonia
(weaker acid)	(weaker base)	(stronger base)	(stronger acid)
(p$K_a \sim 45$)			(pK_a 36)

(*d*) Alcohols are stronger acids than ammonia; the position of equilibrium lies to the right.

$$CH_3C\equiv CCH_2\overset{..}{\underset{..}{O}}-H \;+\; :NH_2 \;\longrightarrow\; CH_3C\equiv CCH_2\overset{..}{\underset{..}{O}}: \;+\; :NH_3$$

2-Butyn-1-ol	Amide anion	2-Butyn-1-olate anion	Ammonia
(stronger acid)	(stronger base)	(weaker base)	(weaker acid)
(p$K_a \sim 16\text{-}20$)			(pK_a 36)

9.5 (*b*) The desired alkyne has a methyl group and a butyl group attached to a —C≡C— unit. Two alkylations of acetylene are therefore required: one with a methyl halide, the other with a butyl halide.

$$H—\!\!\!\equiv\!\!\!—H \xrightarrow[\text{2. CH}_3\text{Br}]{\text{1. NaNH}_2,\ \text{NH}_3} H_3C—\!\!\!\equiv\!\!\!—H \xrightarrow[\text{2. CH}_3\text{CH}_2\text{CH}_2\text{CH}_2\text{Br}]{\text{1. NaNH}_2,\ \text{NH}_3} H_3C—\!\!\!\equiv$$

Acetylene	Propyne	2-Heptyne

It does not matter whether the methyl group or the butyl group is introduced first; the order of steps shown in this synthetic scheme may be inverted.

(*c*) An ethyl group and a propyl group need to be introduced as substituents on a —C≡C— unit. As in part (*b*), it does not matter which of the two is introduced first.

$$H—\!\!\!\equiv\!\!\!—H \xrightarrow[\text{2. CH}_3\text{CH}_2\text{CH}_2\text{Br}]{\text{1. NaNH}_2,\ \text{NH}_3} \equiv\!\!\!—H \xrightarrow[\text{2. CH}_3\text{CH}_2\text{Br}]{\text{1. NaNH}_2,\ \text{NH}_3} \equiv$$

Acetylene	1-Pentyne	3-Heptyne

9.6 Both 1-pentyne and 2-pentyne can be prepared by alkylating acetylene. All the alkylation steps involve nucleophilic substitution of a methyl or primary alkyl halide.

$$H—\!\!\!\equiv\!\!\!—H \xrightarrow[\text{2. CH}_3\text{CH}_2\text{CH}_2\text{Br}]{\text{1. NaNH}_2,\ \text{NH}_3} \equiv\!\!\!—H$$

Acetylene	1-Pentyne

$$H—\!\!\!\equiv\!\!\!—H \xrightarrow[\text{2. CH}_3\text{CH}_2\text{Br}]{\text{1. NaNH}_2,\ \text{NH}_3} \equiv\!\!\!—H \xrightarrow[\text{2. CH}_3\text{Br}]{\text{1. NaNH}_2,\ \text{NH}_3} \equiv$$

Acetylene	1-Butyne	2-Pentyne

A third isomer, 3-methyl-1-butyne, cannot be prepared by alkylation of acetylene, because it requires a secondary alkyl halide as the alkylating agent. The reaction that takes place is elimination, not substitution.

$$H—\!\!\!\equiv\!\!\!: \;+\; \overset{}{\underset{Br}{\diagdown\!\diagup}} \xrightarrow{\text{E2}} H—\!\!\!\equiv\!\!\!—H \;+\; \diagup\!\!\diagdown$$

Acetylide ion	Isopropyl bromide	Acetylene	Propene

9.7 Each of the dibromides shown yields 3,3-dimethyl-1-butyne when subjected to double dehydrohalogenation with strong base.

2,2-Dibromo-3,3-dimethylbutane   1,1-Dibromo-3,3-dimethylbutane   1,2-Dibromo-3,3-dimethylbutane

9.8 (*b*) The first task is to convert 1-propanol to propene:

1-Propanol   Propene

After propene is available, it is converted to 1,2-dibromopropane and then to propyne as described in the sample solution for part (*a*).

(*c*) Treat isopropyl bromide with a base to effect dehydrohalogenation.

Isopropyl bromide   Propene

Next, convert propene to propyne as in parts (*a*) and (*b*).

(*d*) The starting material contains only two carbon atoms, and so an alkylation step is needed at some point. Propyne arises by alkylation of acetylene, and so the last step in the synthesis is

Acetylene   Propyne

The designated starting material, 1,1-dichloroethane, is a geminal dihalide and can be used to prepare acetylene by a double dehydrohalogenation.

1,1-Dichloroethane   Acetylene

(*e*) The first task is to convert ethyl alcohol to acetylene. Once acetylene is prepared, it can be alkylated with a methyl halide.

Ethyl alcohol   Ethylene   1,2-Dibromoethane   Acetylene   Propyne

9.9 The first task is to assemble a carbon chain containing eight carbons. Acetylene has two carbon atoms and can be alkylated via its sodium salt to 1-octyne. Hydrogenation over platinum converts 1-octyne to octane.

$$HC\equiv CH \xrightarrow[\text{NH}_3]{\text{NaNH}_2} HC\equiv C\colon^- Na^+ \xrightarrow{\text{BrCH}_2(\text{CH}_2)_4\text{CH}_3} HC\equiv C\text{—} \xrightarrow[\text{Pt}]{\text{H}_2 \text{ (2 moles)}}$$

Acetylene   Sodium   1-Octyne   Octane
         acetylide

Alternatively, two successive alkylations of acetylene with $CH_3CH_2CH_2Br$ could be carried out to give 4-octyne ($CH_3CH_2CH_2C\equiv CCH_2CH_2CH_3$), which could then be hydrogenated to octane.

9.10   As the equation preceding the problem in the text shows, *cis*-5-decene can be prepared by reduction of 5-decyne using the Lindlar catalyst.

             5-Decyne   *cis*-5-Decene

The task, then, is to prepare 5-decyne from acetylene. This can be accomplished by two successive alkylations using 1-bromobutane.

Acetylene   1-Hexyne   5-Decyne

9.11   (*b*)   Addition of hydrogen chloride to vinyl chloride gives the geminal dichloride 1,1-dichloroethane.

           Vinyl chloride   1,1-Dichloroethane

(*c*)   Because 1,1-dichloroethane can be prepared by adding 2 moles of hydrogen chloride to acetylene as shown in the sample solution to part (*a*), first convert 1,1-dibromoethane to acetylene by dehydrohalogenation.

     1,1-Dibromoethane   Acetylene   1,1-Dichloroethane

9.12   The enol arises by addition of water to the triple bond.

       2-Butyne   2-Buten-2-ol (enol form)   2-Butanone

The mechanism described in the textbook (Mechanism 9.1) is adapted to the case of 2-butyne hydration as shown:

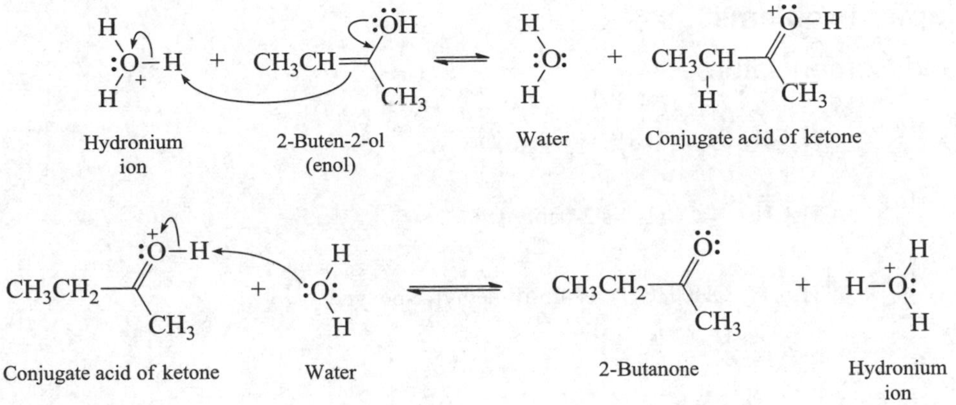

Hydronium ion · 2-Buten-2-ol (enol) · Water · Conjugate acid of ketone

Conjugate acid of ketone · Water · 2-Butanone · Hydronium ion

9.13 Hydration of 1-octyne gives 2-octanone according to the equation that immediately precedes this problem in the text. Prepare 1-octyne as described in the solution to Problem 9.9, and then carry out its hydration in the presence of mercury(II) sulfate and sulfuric acid.

Hydration of 4-octyne gives 4-octanone. Prepare 4-octyne as described in the solution to Problem 9.9.

9.14 Each of the carbons that are part of —CO_2H groups was once part of a —C≡C— unit. The two fragments $CH_3(CH_2)_4CO_2H$ and $HO_2CCH_2CH_2CO_2H$ account for only 10 of the original 16 carbons.

The full complement of carbons can be accommodated by assuming that two molecules of $H_3C(CH_2)_4CO_2H$ are formed, along with one molecule of $HO_2CCH_2CH_2CO_2H$. The starting alkyne is therefore deduced from the ozonolysis data to be as shown:

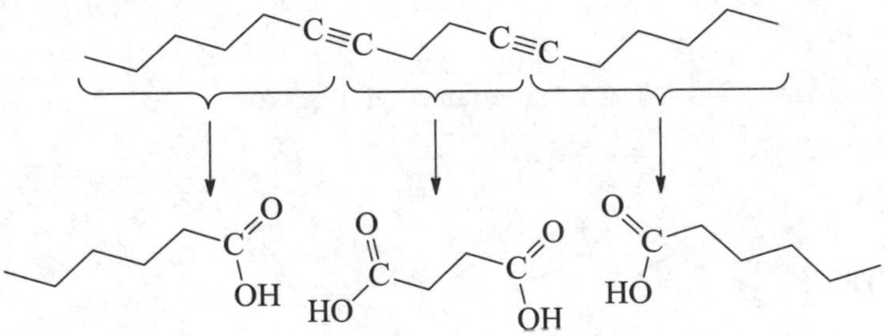

9.15 The structure of the product suggests assembling the carbon chain by alkylation of propyne.

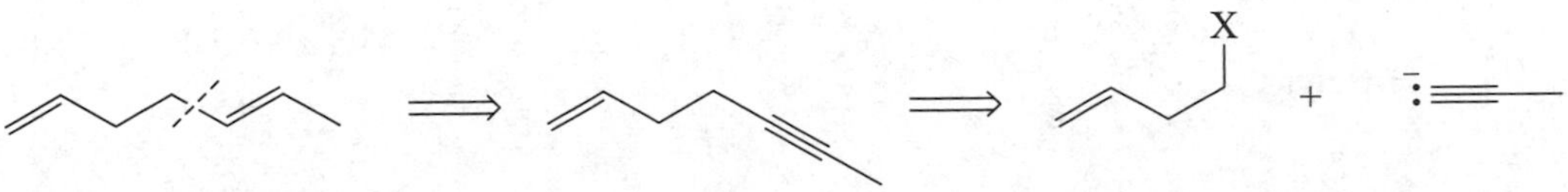

Getting to the final product would therefore require converting a triple bond to a *trans* double bond, leaving the existing double bond intact. Both operations, building the carbon chain and introducing the *trans* double bond are feasible and lead to the synthesis shown.

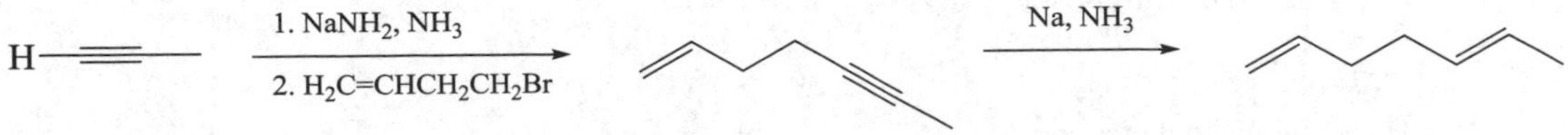

End of Chapter Problems

Structure and Nomenclature

9.16 (a) $\overset{5}{C}H_3\overset{4}{C}H_2\overset{3}{C}H_2\overset{2}{C}\equiv\overset{1}{C}H$ is 1-pentyne

(b) $\overset{5}{C}H_3\overset{4}{C}H_2\overset{3}{C}\equiv\overset{2}{C}\overset{1}{C}H_3$ is 2-pentyne

(c) $\overset{1}{C}H_3\overset{2}{C}\equiv\overset{3}{C}\overset{4}{C}H\overset{5}{C}H\overset{6}{C}H_3$ is 4,5-dimethyl-2-hexyne
$H_3C \quad CH_3$

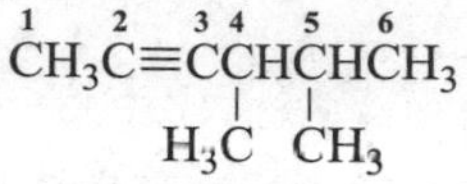

(d) ▷—$\overset{5}{C}H_2\overset{4}{C}H_2\overset{3}{C}H_2\overset{2}{C}\equiv\overset{1}{C}H$ is 5-cyclopropyl-1-pentyne

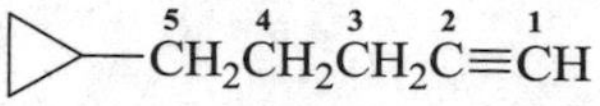

(e) is cyclotridecyne

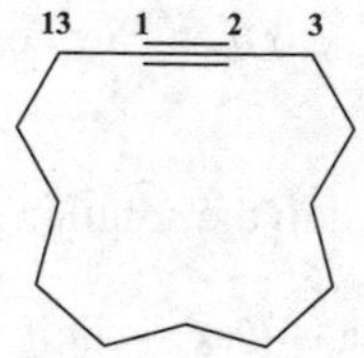

(f) is 4-butyl-2-nonyne
(Parent chain must contain the triple bond)

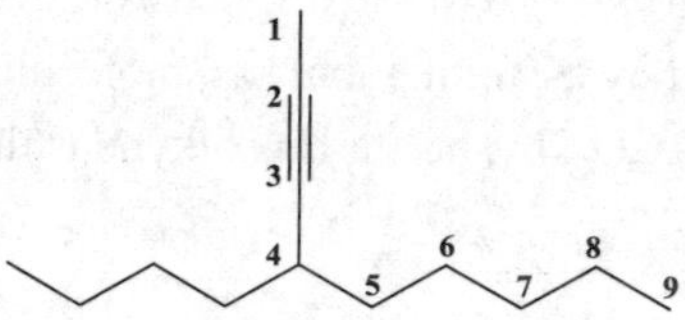

(g) 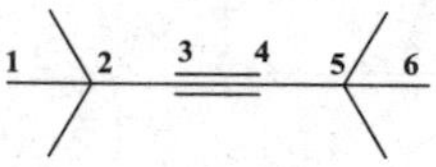 is 2,2,5,5-tetramethyl-3-hexyne

9.17 (a) 1-Octyne is

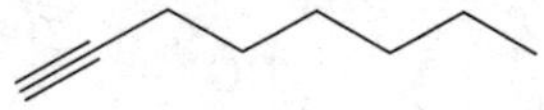

(b) 2-Octyne is

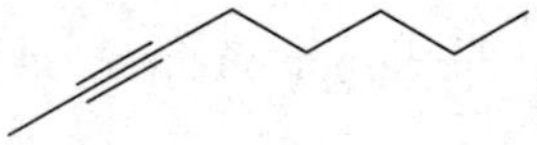

(c) 3-Octyne is

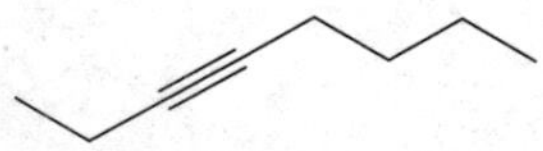

(d) 4-Octyne is

(e) 2,5-Dimethyl-3-hexyne is

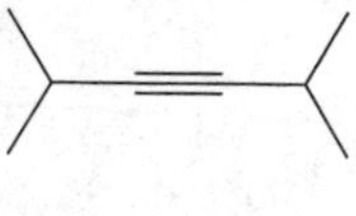

(f) 4-Ethyl-1-hexyne is

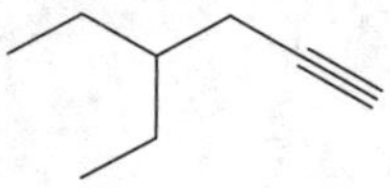

(g) Ethynylcyclohexane is

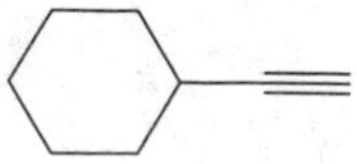

(*h*)

3-Ethyl-3-methyl-1-pentyne is

9.18 Ethynylcyclohexane has the molecular formula C_8H_{12}. All the other compounds are C_8H_{14}.

9.19 The numbering for oropheic acid starts at the carboxylic acid functional group. The structure of this $C_{18}H_{22}O_2$ compound is

17-Octadecene-9,11,13-triynoic acid

Reactions

9.20 Only alkynes with the carbon skeletons shown can give 3-ethylhexane on catalytic hydrogenation.

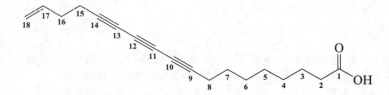

or   or $\xrightarrow[\text{Pt}]{\text{H}_2 \text{ (2 moles)}}$

3-Ethyl-1-hexyne   4-Ethyl-1-hexyne   4-Ethyl-2-hexyne   3-Ethylhexane

9.21 The carbon skeleton of the unknown acetylenic amino acid must be the same as that of homoleucine. The structure of homoleucine is such that there is only one possible location for a carbon–carbon triple bond in an acetylenic precursor.

$\xrightarrow[\text{Pt}]{\text{H}_2 \text{ (2 moles)}}$

$C_7H_{11}NO_2$ Homoleucine

9.22 Hydrogenation over Lindlar palladium converts an alkyne to a cis alkene. Oleic acid therefore has the structure indicated in the following equation:

$$CH_3(CH_2)_7C{\equiv}C(CH_2)_7CO_2H \xrightarrow[\text{Lindlar Pd}]{\text{H}_2}$$

Stearolic acid   Oleic acid

Hydrogenation of alkynes over platinum leads to alkanes.

$$CH_3(CH_2)_7C \equiv C(CH_2)_7CO_2H \xrightarrow[Pt]{H_2} CH_3(CH_2)_{16}CO_2H$$

Stearolic acid   Stearic acid

9.23 The alkane formed by hydrogenation of (*S*)-3-methyl-1-pentyne is achiral; it cannot be optically active.

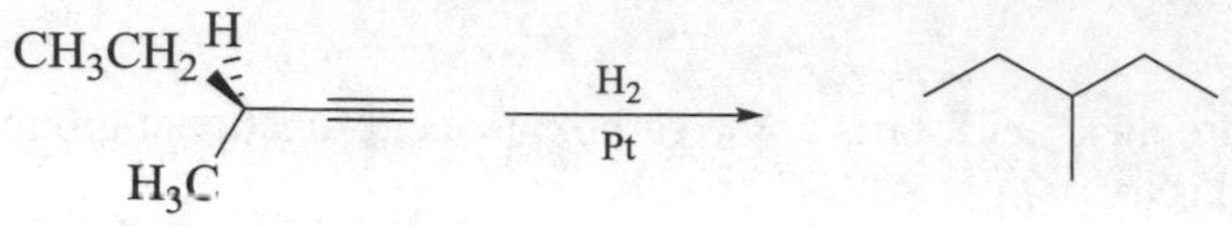

(*S*)-3-Methyl-1-pentyne

3-Methyl-1-pentane
(does not have a chirality
center, optically inactive)

The product of hydrogenation of *(S)*-4-methyl-1-hexyne is optically active because a chirality center is present in the starting material and is carried through to the product.

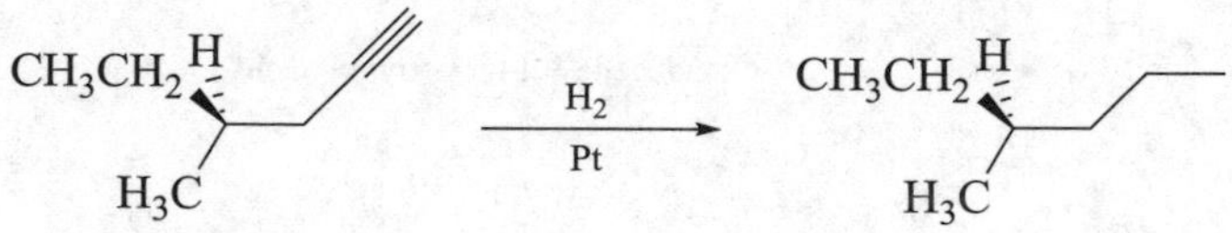

(*S*)-4-Methyl-1-hexyne

(*S*)-3-Methylhexane

Both (*S*)-3-methyl-1-pentyne and (*S*)-4-methyl-1-hexyne yield optically active products when their triple bonds are reduced to double bonds.

9.24 (*a*)

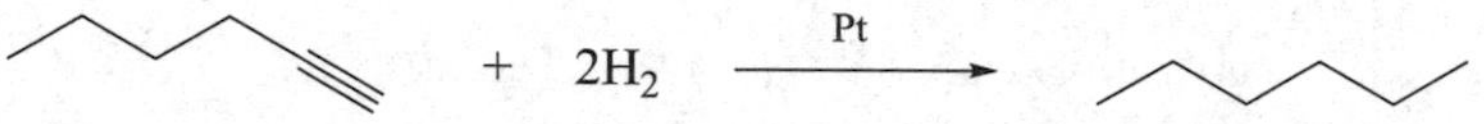

1-Hexyne $+ 2H_2$ $\xrightarrow{Pt}$ Hexane

(*b*)

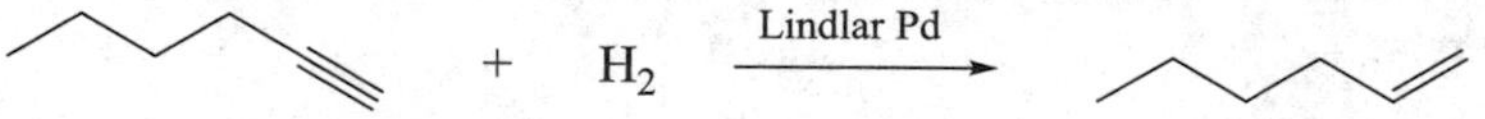

1-Hexyne $+$ H_2 $\xrightarrow{Lindlar\ Pd}$ 1-Hexene

(*c*)

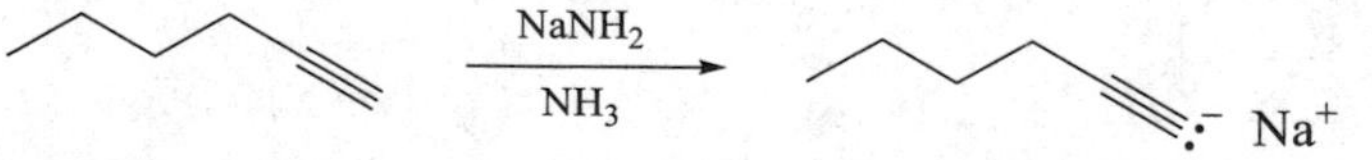

1-Hexyne $\xrightarrow[NH_3]{NaNH_2}$ Sodium 1-hexynide

(*d*)

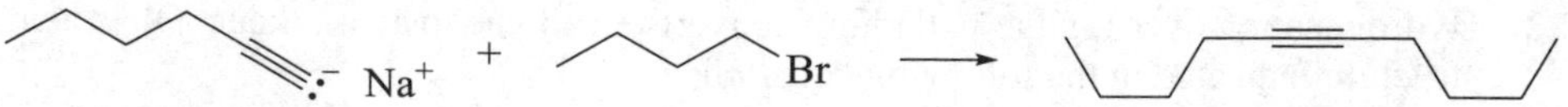

Sodium 1-hexynide $+$ 1-Bromobutane $\longrightarrow$ 5-Decyne

(*e*)

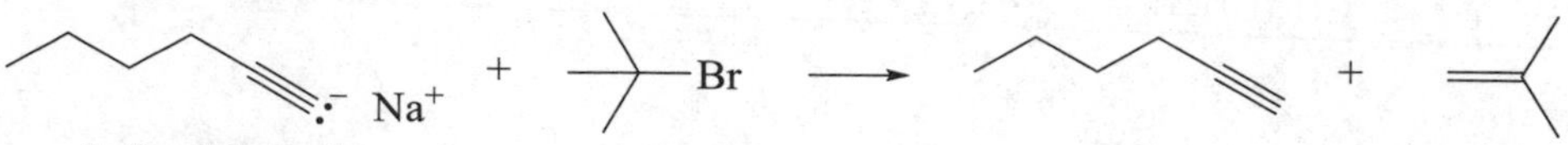

Sodium 1-hexynide $+$ *tert*-Butyl
bromide $\longrightarrow$ 1-Hexyne $+$ 2-Methylpropene

(f) 1-Hexyne $\xrightarrow[\text{(1 mol)}]{\text{HCl}}$ 2-Chloro-1-hexene

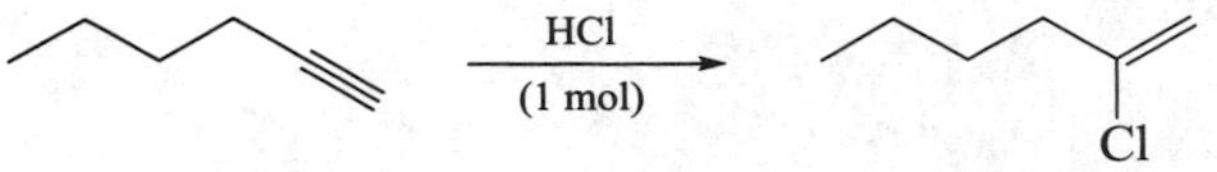

(g) 1-Hexyne $\xrightarrow[\text{(2 mol)}]{\text{HCl}}$ 2,2-Dichlorohexane

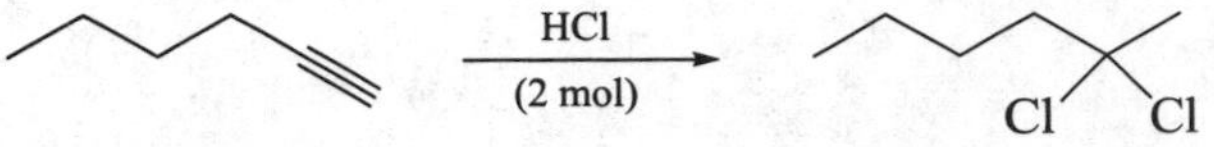

(h) 1-Hexyne $\xrightarrow[\text{(1 mol)}]{\text{Cl}_2}$ (E)-1,2-Dichloro-1-hexene

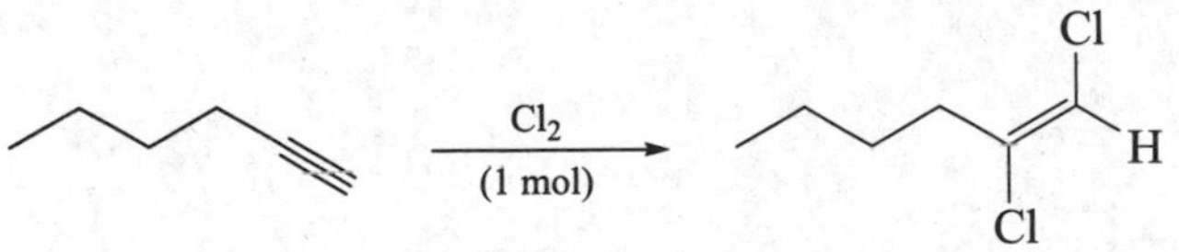

(i) 1-Hexyne $\xrightarrow[\text{(2 mol)}]{\text{Cl}_2}$ 1,1,2,2-Tetrachlorohexane

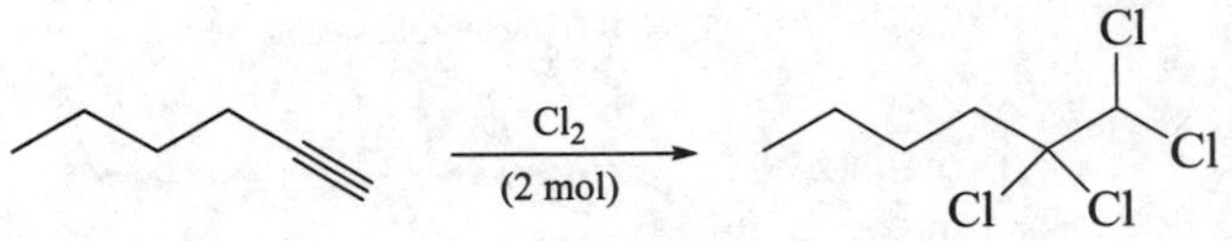

(j) 1-Hexyne $\xrightarrow[\text{HgSO}_4]{\text{H}_2\text{O, H}_2\text{SO}_4}$ 2-Hexanone

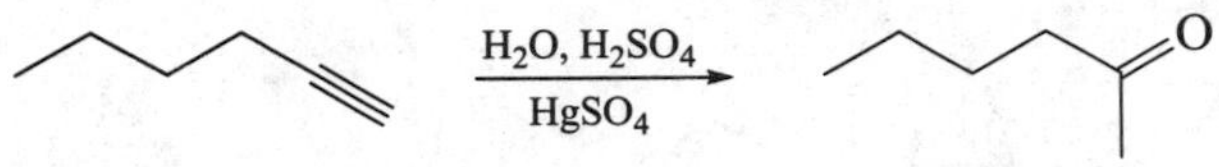

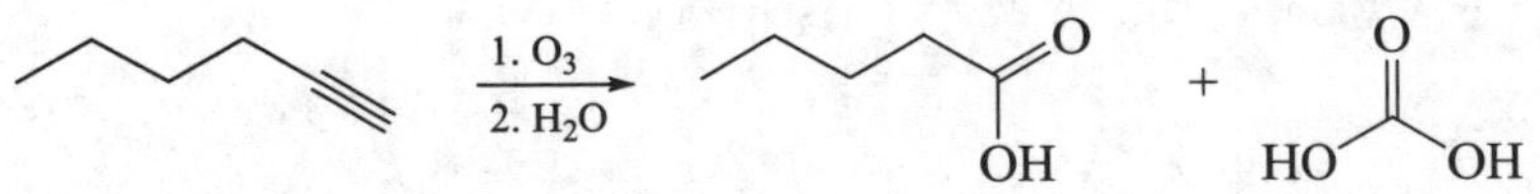

(k) 1-Hexyne $\xrightarrow[\text{2. H}_2\text{O}]{\text{1. O}_3}$ Pentanoic acid + Carbonic acid (not isolated, converts to CO_2 and H_2O)

9.25

(a) 3-Hexyne + $2H_2$ $\xrightarrow{\text{Pt}}$ Hexane

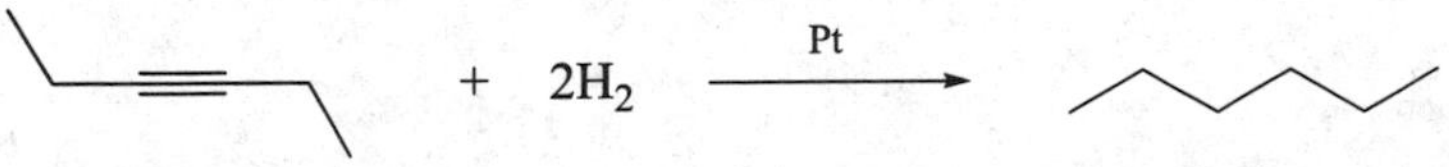

(b) 3-Hexyne + H_2 $\xrightarrow{\text{Lindlar Pd}}$ (Z)-3-Hexene

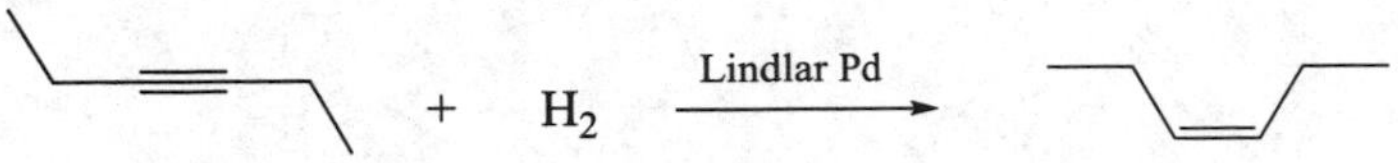

(c) 3-Hexyne $\xrightarrow[\text{(1 mol)}]{\text{HCl}}$ (Z)-3-Chloro-3-hexene

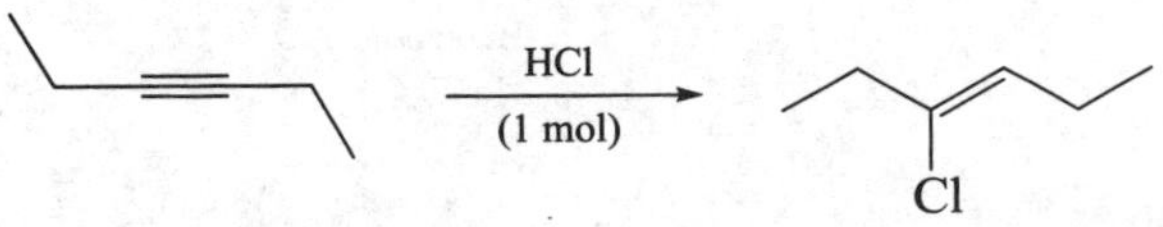

(d)

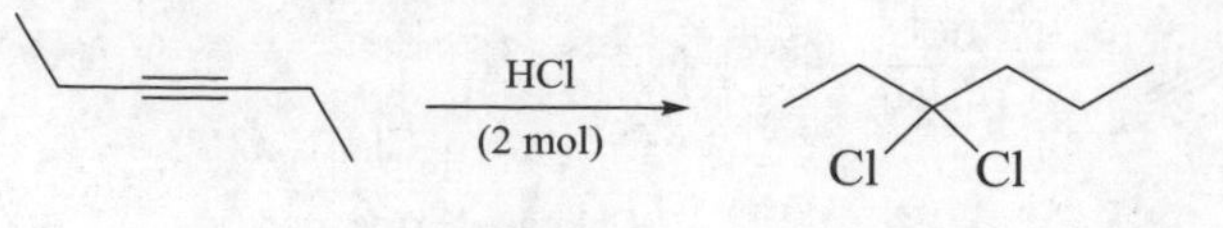

3-Hexyne $\xrightarrow[\text{(2 mol)}]{\text{HCl}}$ 3,3-Dichlorohexane

(e)

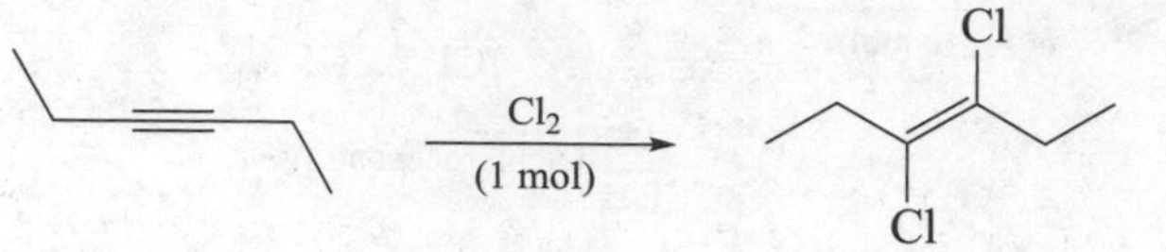

3 Hexyne $\xrightarrow[\text{(1 mol)}]{\text{Cl}_2}$ (E)-3,4-Dichloro-3-hexene

(f)

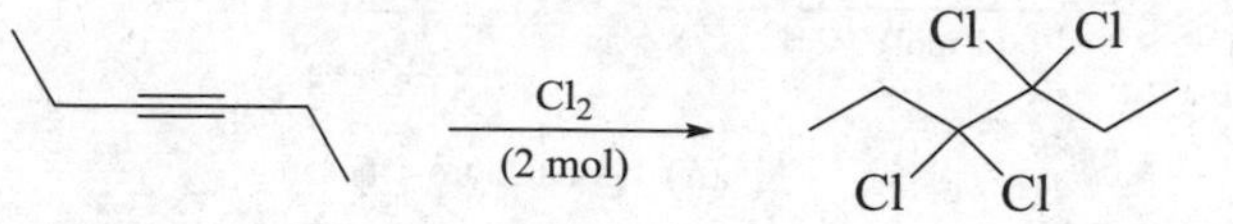

3-Hexyne $\xrightarrow[\text{(2 mol)}]{\text{Cl}_2}$ 3,3,4,4-Tetrachlorohexane

(g)

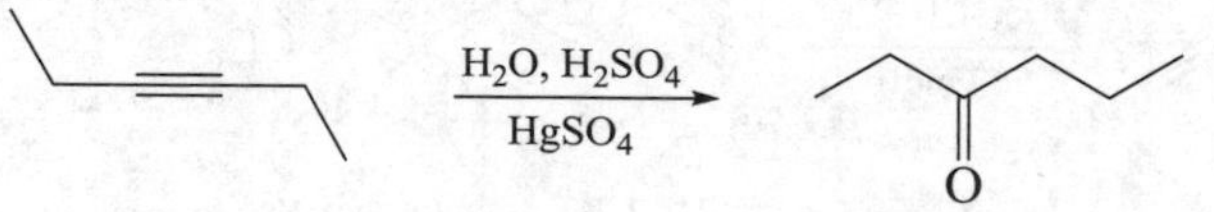

3-Hexyne $\xrightarrow[\text{HgSO}_4]{\text{H}_2\text{O, H}_2\text{SO}_4}$ 3-Hexanone

(h)

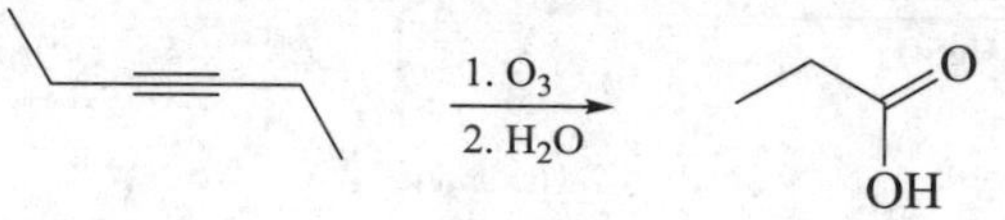

3-Hexyne $\xrightarrow[\text{2. H}_2\text{O}]{\text{1. O}_3}$ Propanoic acid

9.26 The two carbons of the triple bond are similarly but not identically substituted in 2-heptyne, $CH_3C{\equiv}CCH_2CH_2CH_2CH_3$. Two regioisomeric enols are formed, each of which gives a different ketone.

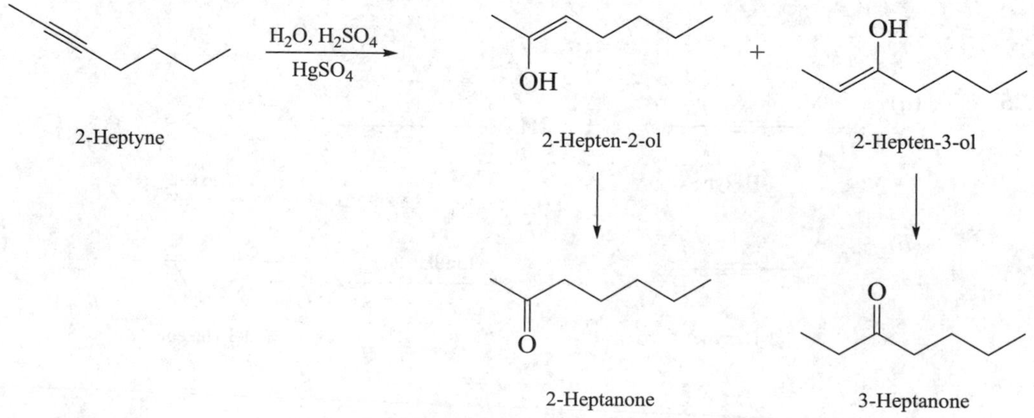

2-Heptyne $\xrightarrow[\text{HgSO}_4]{\text{H}_2\text{O, H}_2\text{SO}_4}$ 2-Hepten-2-ol + 2-Hepten-3-ol

2-Heptanone   3-Heptanone

9.27 (a) The dihaloalkane contains both a primary alkyl chloride and a primary alkyl iodide functional group. Iodide is a better leaving group than chloride and is the one replaced by acetylide.

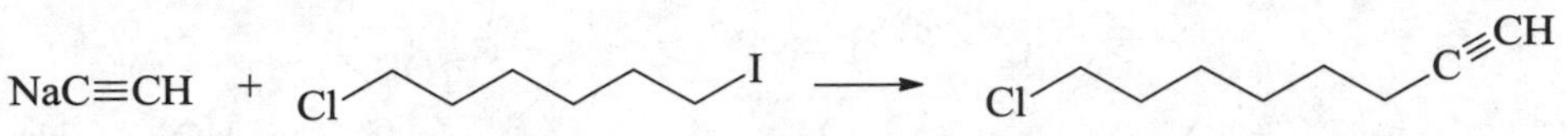

| Sodium acetylide | 1-Chloro-6-iodohexane | 8-Chloro-1-octyne |

(b) Both vicinal dibromide functions are converted to alkyne units on treatment with excess sodium amide.

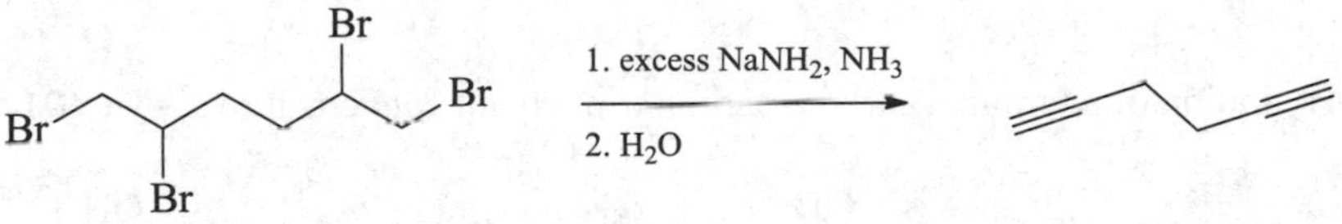

1,2,5,6-Tetrabromohexane 1,5-Hexadiyne

(c) The starting material is a geminal dichloride. Potassium *tert*-butoxide in dimethyl sulfoxide is a sufficiently strong base to convert it to an alkyne.

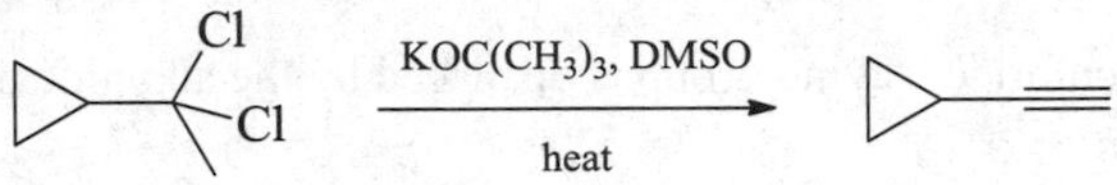

1,1-Dichloro-1-cyclopropylethane Ethynylcyclopropane

(d) Alkyl *p*-toluenesulfonates react similarly to alkyl halides in nucleophilic substitutions. The alkynide nucleophile displaces the *p*-toluenesulfonate leaving group from ethyl *p*-toluenesulfonate.

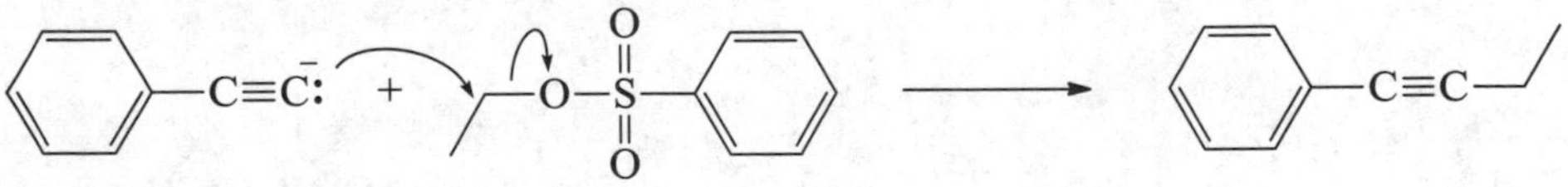

Phenylacetylide ion Ethyl *p*-toluenesulfonate 1-Phenyl-1-butyne

(e) Both carbons of a —C≡C— unit are converted to carboxyl groups (—CO_2H) on ozonolysis.

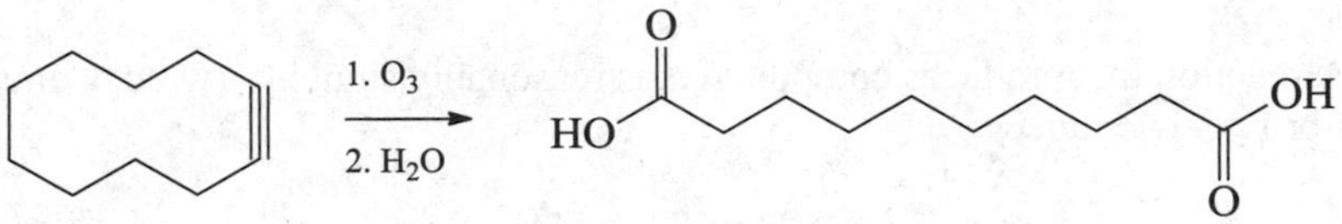

Cyclodecyne Decanedioic acid

(*f*) Ozonolysis cleaves the carbon–carbon triple bond.

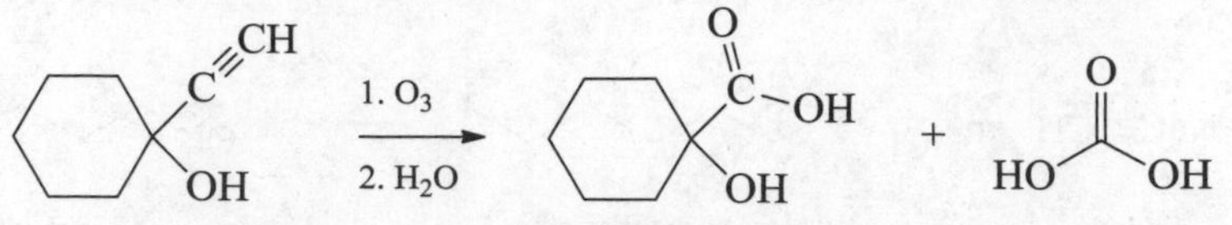

| 1-Ethynylcyclohexanol | 1-Hydroxycyclohexane-carboxylic acid | Carbonic acid |

(*g*) Hydration of a terminal carbon–carbon triple bond converts it to a —$\overset{\text{O}}{\overset{\|}{\text{C}}}CH_3$ group.

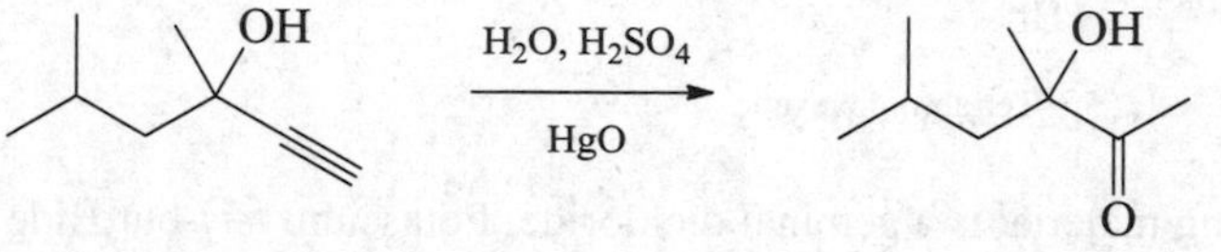

3,5-Dimethyl-1-hexyn-3-ol   3-Hydroxy-3,5-dimethyl-2-hexanone

(*h*) The primary chloride leaving group is displaced by the alkynide nucleophile.

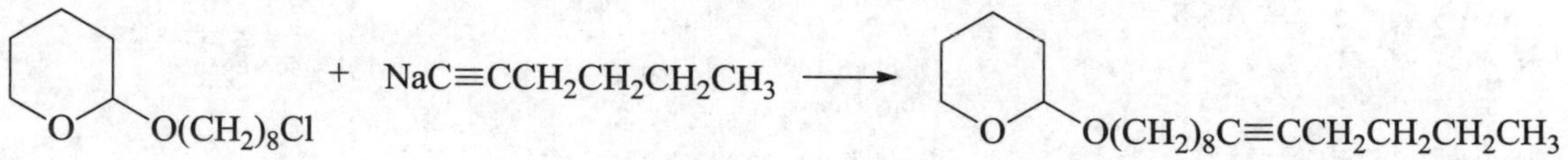

| 8-Chlorooctyl tetrahydropyranyl ether | Sodium 1-hexynide | 9-Tetradecyn-1-yl, tetrahydropyranyl ether |

(*i*) Hydrogenation of the triple bond over the Lindlar catalyst converts the compound to a cis alkene.

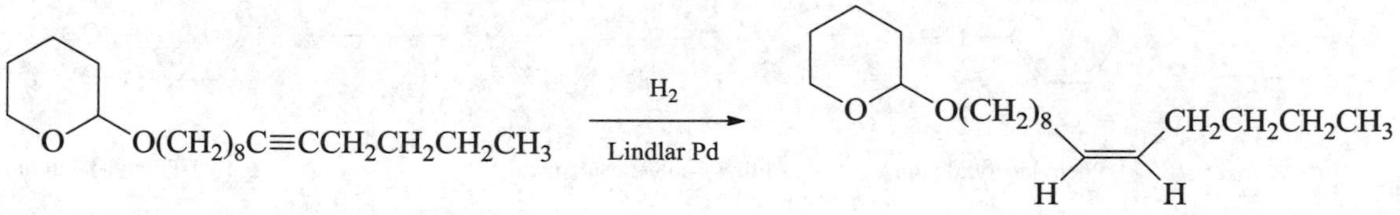

9-Tetradecyn-1-yl, tetrahydropyranyl ether   (*Z*)-9-Tetradecen-1-yl, tetrahydropyranyl ether

9.28 The reaction that produces compound A is reasonably straightforward. Compound A is 14-bromo-1-tetradecyne.

Sodium acetylide   1,12-Dibromododecane   Compound A (C$_{14}$H$_{25}$Br)

Treatment of compound A with sodium amide converts it to compound B. Compound B on ozonolysis gives a diacid that retains all the carbon atoms of B. Compound B must therefore be a cyclic alkyne, formed by an intramolecular alkylation.

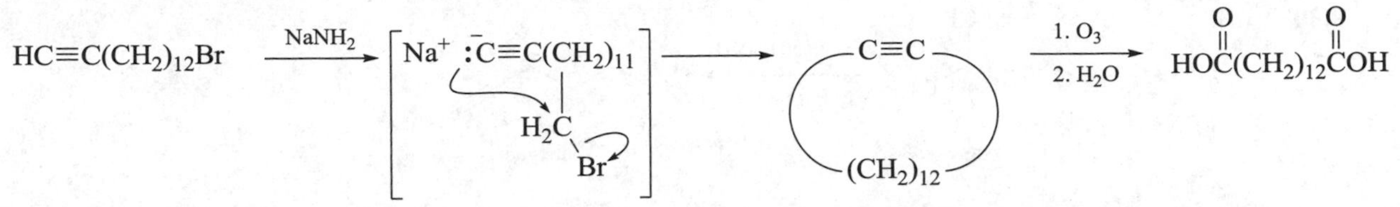

$$HC\equiv C(CH_2)_{12}Br \xrightarrow{NaNH_2} \left[Na^+ \ ^-:C\equiv C(CH_2)_{11} \right]$$

H₂C / Br

Compound A ($C_{14}H_{25}Br$)

Compound B

$\xrightarrow{\text{1. } O_3}_{\text{2. } H_2O}$ $HOC(CH_2)_{12}COH$ (with two C=O)

Compound B is cyclotetradecyne.

Hydrogenation of compound B over Lindlar palladium yields *cis*-cyclotetradecene (compound C).

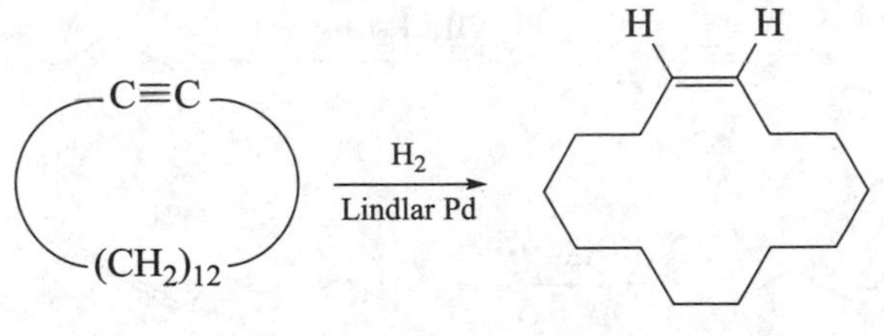

$C\equiv C$ / $(CH_2)_{12}$ $\xrightarrow[\text{Lindlar Pd}]{H_2}$

Compound C ($C_{14}H_{26}$)

Hydrogenation over platinum gives cyclotetradecane (compound D).

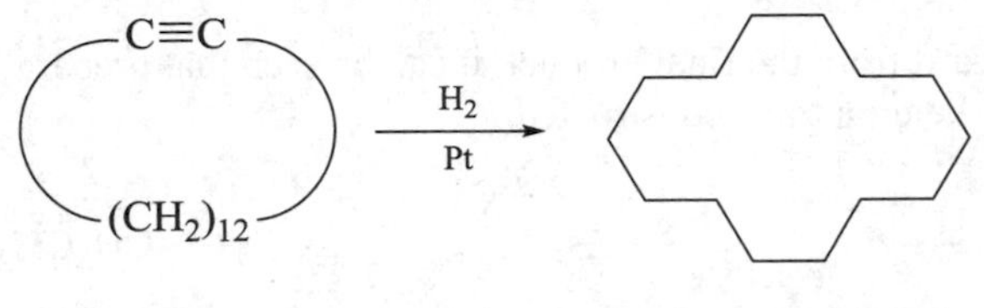

$C\equiv C$ / $(CH_2)_{12}$ $\xrightarrow[\text{Pt}]{H_2}$

Compound D ($C_{14}H_{28}$)

The cis isomer of cyclotetradecene is converted to $O{=}CH(CH_2)_{12}CH{=}O$ on ozonolysis, whereas cyclotetradecane does not react with ozone.

Synthesis

9.29 A single dehydrobromination step occurs in the conversion of 1,2-dibromodecane to $C_{10}H_{19}Br$. Bromine may be lost from C-1 to give 2-bromo-1-decene.

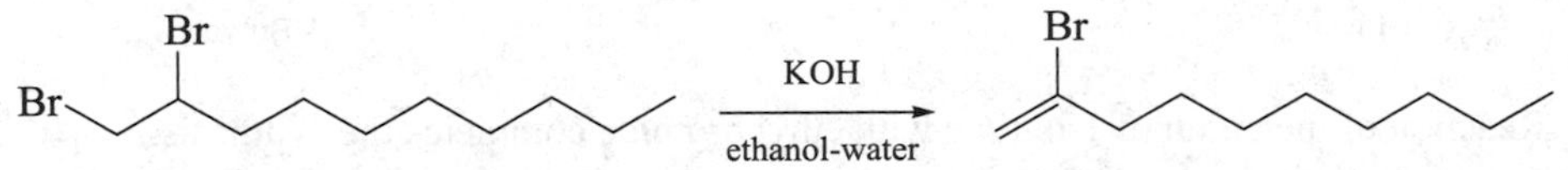

1,2-Dibromodecane $\xrightarrow[\text{ethanol-water}]{KOH}$ 2-Bromo-1-decene

Loss of bromine from C-2 gives (*E*)- and (*Z*)-1-bromo-1-decene.

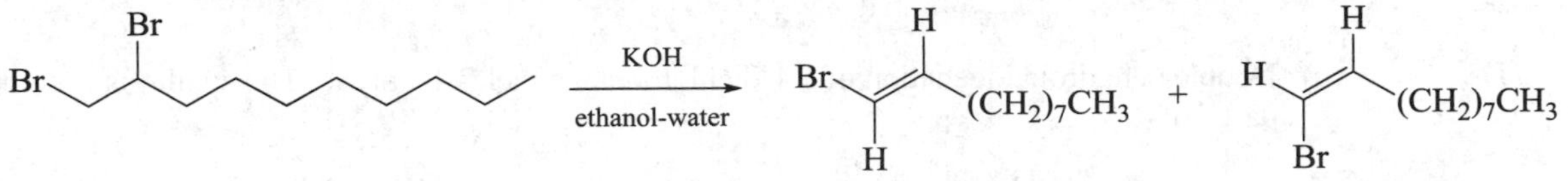

1,2-Dibromodecane $\xrightarrow[\text{ethanol-water}]{KOH}$ (*E*)-1-Bromo-1-decene + (*Z*)-1-Bromo-1-decene

9.30   (*a*)

1,1-Dichlorohexane $\xrightarrow[\text{2. H}_2\text{O}]{\text{1. NaNH}_2, \text{ NH}_3}$ **1-Hexyne**

(*b*)

1-Hexene $\xrightarrow[\text{CCl}_4]{\text{Br}_2}$ **1,2-Dibromohexane** $\xrightarrow[\text{2. H}_2\text{O}]{\text{1. NaNH}_2, \text{ NH}_3}$ **1-Hexyne**

(*c*)

Acetylene $\xrightarrow[\text{NH}_3]{\text{NaNH}_2}$ $\equiv\!:^- \text{Na}^+$ $\xrightarrow{\text{CH}_3\text{CH}_2\text{CH}_2\text{CH}_2\text{Br}}$ **1-Hexyne**

(*d*)

1-Iodohexane $\xrightarrow[\text{DMSO}]{\text{KOC(CH}_3)_3}$ **1-Hexene**

1-Hexene is then converted to 1-hexyne as in part (*b*).

9.31   (*a*)   Working backward from the final product, it can be seen that preparation of 1-butyne will allow the desired carbon skeleton to be constructed.

3-Hexyne $\Longrightarrow$ $+$ CH$_3$CH$_2$Br

The desired intermediate, 1-butyne, is available by halogenation followed by double dehydrohalogenation of 1-butene.

1-Butene $\xrightarrow[\text{CCl}_4]{\text{Br}_2}$ $\xrightarrow[\text{2. H}_2\text{O}]{\text{1. NaNH}_2, \text{ NH}_3}$ **1-Butyne**

Reaction of the anion of 1-butyne with ethyl bromide completes the synthesis.

1-Butyne $\xrightarrow[\text{NH}_3]{\text{NaNH}_2}$ $\equiv\!:^-$ $\xrightarrow{\text{CH}_3\text{CH}_2\text{Br}}$ **3-Hexyne**

(*b*)   Double dehydrohalogenation of 1,1-dichlorobutane yields 1-butyne. The synthesis is completed as in part (*a*).

1,1-Dichlorobutane $\xrightarrow[\text{2. H}_2\text{O}]{\text{1. NaNH}_2, \text{ NH}_3}$ **1-Butyne**

(*c*)   1-Butyne can be prepared from acetylene as follows.

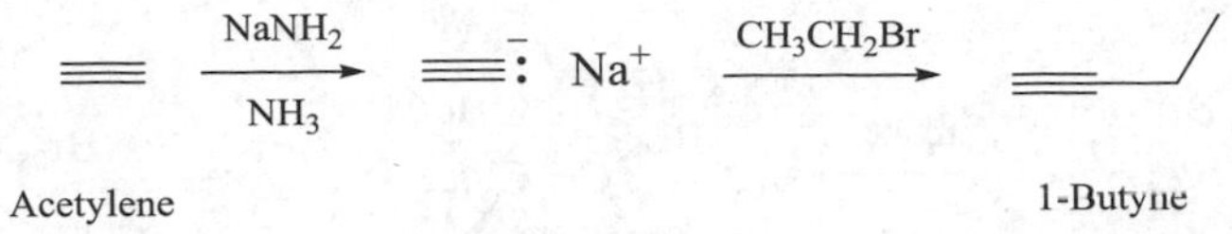

$$\equiv \xrightarrow[\text{NH}_3]{\text{NaNH}_2} \equiv\!\!:^- \text{Na}^+ \xrightarrow{\text{CH}_3\text{CH}_2\text{Br}} \equiv\!\!/$$

Acetylene   1-Butyne

1-Butyne is then converted to 3-hexyne as in part (*a*).

9.32 (*a*) Bromine adds anti to the double bond, so from the *E*-alkene only a meso dibromide is formed.

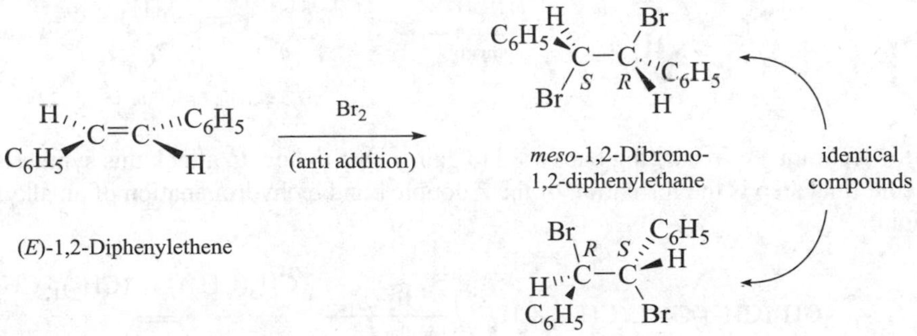

(*E*)-1,2-Diphenylethene

meso-1,2-Dibromo-
1,2-diphenylethane

identical
compounds

The reaction with KOH is an E2 reaction, so the reactive conformation will be one in which the bromine and hydrogen are anti coplanar.

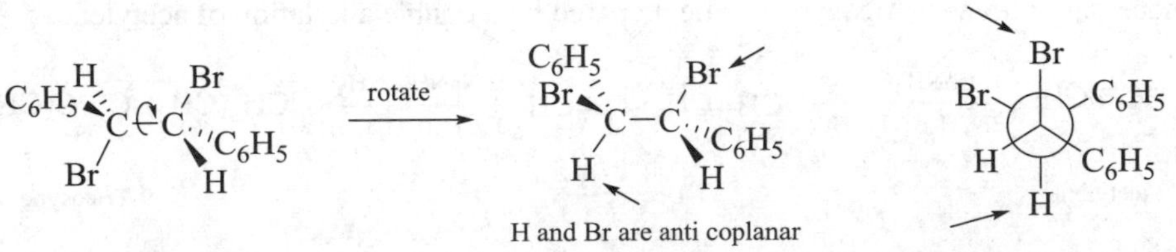

H and Br are anti coplanar

Anti elimination from this conformation will give only the *E*-vinylic bromide.

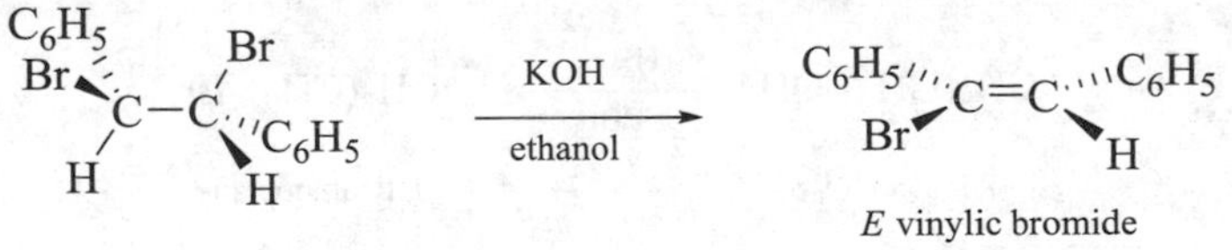

E vinylic bromide

(*b*) If the sequence starts with (*Z*)-1,2-diphenyl-ethene, the dibromides will be *S,S/R,R* (chiral)

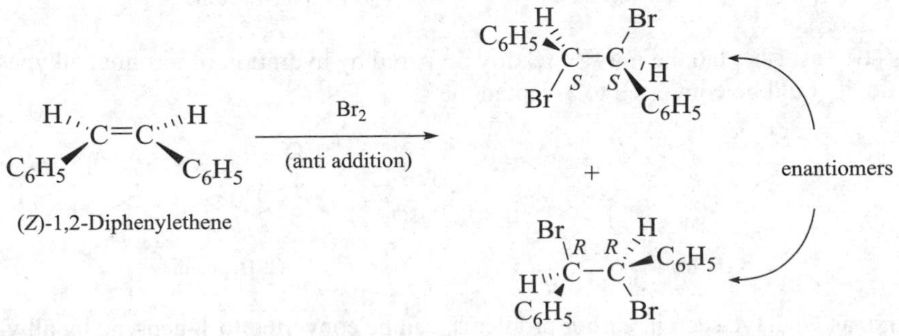

(*Z*)-1,2-Diphenylethene

+   enantiomers

Elimination from the reactive conformation of either the *S,S*- or *R,R*-dibromide gives the *Z*-vinylic bromide.

H and Br are anti coplanar

Z vinylic bromide

9.33 Apply the technique of reasoning backward to gain a clue to how to attack this synthesis problem. A reasonable final step is the formation of the *Z* double bond by hydrogenation of an alkyne over Lindlar palladium.

$$CH_3(CH_2)_7C\equiv C(CH_2)_{12}CH_3 \xrightarrow[\text{Lindlar Pd}]{H_2}$$

9-Tricosyne   (*Z*)-9-Tricosene

The necessary alkyne 9-tricosyne can be prepared by a double alkylation of acetylene.

$$HC\equiv CH \xrightarrow[\text{2. } CH_3(CH_2)_7Br]{\text{1. } NaNH_2,\ NH_3} CH_3(CH_2)_7C\equiv CH \xrightarrow[\text{2. } CH_3(CH_2)_{12}Br]{\text{1. } NaNH_2,\ NH_3} CH_3(CH_2)_7C\equiv C(CH_2)_{12}CH_3$$

Acetylene   1-Decyne   9-Tricosyne

It does not matter which alkyl group is introduced first.

The alkyl halides are prepared from the corresponding alcohols.

$$CH_3(CH_2)_7OH \xrightarrow[\text{or } PBr_3]{HBr} CH_3(CH_2)_7Br$$

1-Octanol   1-Bromooctane

$$CH_3(CH_2)_{12}OH \xrightarrow[\text{or } PBr_3]{HBr} CH_3(CH_2)_{12}Br$$

1-Tridecanol   1-Bromotridecane

9.34 Ketones such as 2-heptanone may be readily prepared by hydration of terminal alkynes. Thus, if we had 1-heptyne, it could be converted to 2-heptanone.

1-Heptyne   2-Heptanone

Acetylene, as we have seen in earlier problems, can be converted to 1-heptyne by alkylation.

$$HC\equiv CH \xrightarrow[\text{NH}_3]{NaNH_2} HC\equiv \overset{-}{C}:\ Na^+$$

Acetylene   Sodium acetylide

$$HC\equiv\overset{..}{C}{:}^{-}\ Na^{+}\quad +\quad Br\diagdown\diagup\diagdown\diagup\diagdown\quad\longrightarrow$$

Sodium acetylide   1-Bromopentane   1-Heptyne

9.35 Attack this problem by first planning a synthesis of 4-methyl-2-pentyne from any starting material in a single step. Two different alkyne alkylations suggest themselves:

$$CH_3C\equiv CCH(CH_3)_2 \begin{cases} (a)\ \text{from}\ CH_3C\equiv\overset{..}{C}{:}^{-}\ \text{and}\ BrCH(CH_3)_2 \\ (b)\ \text{from}\ CH_3I\ \text{and}\ {:}^{-}C\equiv CCH(CH_3)_2 \end{cases}$$

4-Methyl-2-pentyne

Isopropyl bromide is a secondary alkyl halide and cannot be used to alkylate $CH_3C\equiv C{:}^{-}$ according to reaction (a); elimination would occur. A reasonable last step is therefore the alkylation of $(CH_3)_2CHC\equiv CH$ via reaction of its anion with methyl iodide.

The next question that arises from this analysis is the origin of $(CH_3)_2CHC\equiv CH$. One of the available starting materials is 1,1-dichloro-3-methylbutane. It can be converted to $(CH_3)_2CHC\equiv CH$ by a double dehydrohalogenation. The complete synthesis is therefore

$$(CH_3)_2CHCH_2CHCl_2 \xrightarrow[\text{2. H}_2\text{O}]{\text{1. NaNH}_2,\ \text{NH}_3} (CH_3)_2CHC\equiv CH \xrightarrow[\text{2. CH}_3\text{I}]{\text{1. NaNH}_2} (CH_3)_2CHC\equiv CCH_3$$

1,1-Dichloro-3-methylbutane   3-Methyl-1-butyne   4-Methyl-1-pentyne

9.36 (a) 2,2-Dibromopropane is prepared by addition of hydrogen bromide to propyne.

$$\underline{\quad\equiv\quad} + 2HBr \longrightarrow$$

Propyne   Hydrogen bromide   2,2-Dibromopropane

The designated starting material, 1,1-dibromopropane, is converted to propyne by a double dehydrohalogenation.

$$\diagup\diagdown\diagup\overset{\text{Br}}{\underset{\text{Br}}{|}} \xrightarrow[\text{2. H}_2\text{O}]{\text{1. NaNH}_2,\ \text{NH}_3} \underline{\quad\equiv\quad}$$

1,1-Dibromopropane   Propyne

(b) As in part (a), first convert the designated starting material to propyne, and then add 2 moles of hydrogen bromide.

$$\overset{\text{Br}}{\diagup\diagdown\diagup}\diagdown\text{Br} \xrightarrow[\text{2. H}_2\text{O}]{\text{1. NaNH}_2,\ \text{NH}_3} \underline{\quad\equiv\quad} \xrightarrow{2HBr}$$

1,2-Dibromopropane   Propyne   2,2-Dibromopropane

(c) Instead of trying to introduce two additional chlorines into 1,2-dichloropropane by free-radical substitution (a mixture of products would result), convert the vicinal dichloride to propyne, and then add 2 moles of Cl_2.

1,2-Dichloropropane $\xrightarrow[\text{2. H}_2\text{O}]{\text{1. NaNH}_2,\ \text{NH}_3}$ Propyne $\xrightarrow{2\text{Cl}_2}$ 2,2-Dibromopropane

(*d*) The required carbon skeleton can be constructed by alkylating acetylene with ethyl bromide.

Acetylene $\xrightarrow[\text{NH}_3]{\text{NaNH}_2}$ Sodium acetylide $\xrightarrow{\text{CH}_3\text{CH}_2\text{Br}}$ 1-Butyne

Addition of 2 moles of hydrogen iodide to 1-butyne gives 2,2-diiodobutane.

1-Butyne + 2HI $\longrightarrow$ 2,2-Diiodobutane
Hydrogen iodide

(*e*) The six-carbon chain is available by alkylation of acetylene with 1-bromobutane.

$$\text{HC} \equiv \text{CH} \xrightarrow[\text{2. CH}_3\text{CH}_2\text{CH}_2\text{CH}_2\text{Br}]{\text{1. NaNH}_2,\ \text{NH}_3} \text{HC} \equiv \text{CCH}_2\text{CH}_2\text{CH}_2\text{CH}_3$$

Acetylene　　　　　　　　　　　　1-Hexyne

The alkylating agent, 1-bromobutane, is prepared from 1-butene by hydroboration-oxidation, to make the alcohol. 1-Butanol is then treated with hydrogen bromide to make 1-bromobutane.

$$\text{CH}_3\text{CH}_2\text{CH}=\text{CH}_2 \xrightarrow[\text{2. H}_2\text{O}_2,\ \text{OH}^-]{\text{1. B}_2\text{H}_6} \text{CH}_3\text{CH}_2\text{CH}_2\text{CH}_2\text{OH}$$

1-Butene　　　　　　　　　　　　1-Butanol

$$\text{CH}_3\text{CH}_2\text{CH}_2\text{CH}_2\text{OH} \xrightarrow{\text{HBr}} \text{CH}_3\text{CH}_2\text{CH}_2\text{CH}_2\text{Br}$$

1-Butanol　　　　　　　　　　　　1-Bromobutane

Once 1-hexyne is prepared, it can be converted to 1-hexene by hydrogenation over Lindlar palladium or by sodium–ammonia reduction.

$$\text{CH}_3\text{CH}_2\text{CH}_2\text{CH}_2\text{CH}\equiv\text{CH} \xrightarrow[\text{or Na, NH}_3]{\text{H}_2,\ \text{Lindlar Pd}} \text{CH}_3\text{CH}_2\text{CH}_2\text{CH}_2\text{CH}=\text{CH}_2$$

1-Hexyne　　　　　　　　　　　　1-Hexene

(*f*) Prepare 1-hexyne from acetylene as shown in part (*e*). Alkylation of 1-hexyne with 1-bromobutane, gives the necessary ten-carbon chain.

$$\text{HC}\equiv\text{CCH}_2\text{CH}_2\text{CH}_2\text{CH}_3 \xrightarrow[\text{2. CH}_3\text{CH}_2\text{CH}_2\text{CH}_2\text{Br}]{\text{1. NaNH}_2,\ \text{NH}_3} \text{CH}_3\text{CH}_2\text{CH}_2\text{CH}_2\text{C}\equiv\text{CCH}_2\text{CH}_2\text{CH}_2\text{CH}_3$$

1-Hexyne　　　　　　　　　　　　5-Decyne

Hydrogenation of 5-decyne yields decane.

$$CH_3CH_2CH_2CH_2C{\equiv}CCH_2CH_2CH_2CH_3 \xrightarrow[\text{Pt}]{2H_2} CH_3CH_2CH_2CH_2CH_2{-}CH_2CH_2CH_2CH_2CH_3$$

5-Decyne   Decane

(g) A standard method for converting alkenes to alkynes is to add Br$_2$ and then carry out a double dehydrohalogenation.

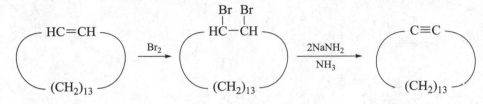

Cyclopentadecene   1,2-Dibromocyclopentadecane   Cyclopentadecyne

(h) Alkylation of the triple bond gives the required carbon skeleton.

1-Ethynylcyclohexene   1-(1-Propynyl)cyclohexene

Hydrogenation over the Lindlar catalyst converts the carbon–carbon triple bond to a cis double bond.

1-(1-Propynyl)cyclohexene   (Z)-1-(1-Propenyl)cyclohexene

Mechanism

9.37 The enol is that of an aldehyde, so there is a hydrogen attached to enol carbon that bears the OH group. The enol form of hexanal is

$$CH_3CH_2CH_2CH_2CH{=}C\overset{\displaystyle H}{\underset{\displaystyle OH}{\big|}}$$

Answers to Interpretive Problems 9

9.38 D;   **9.39** B;   **9.40** A;   **9.41** C;   **9.42** B

SELF-TEST

1. Provide the IUPAC names for the following:

(a)   (b)   (c)

2. Give the structure of the reactant, reagent, or product omitted from each of the following reactions.

(a) $\xrightarrow{\text{HCl (1 mol)}}$?

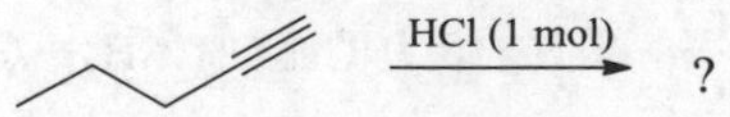

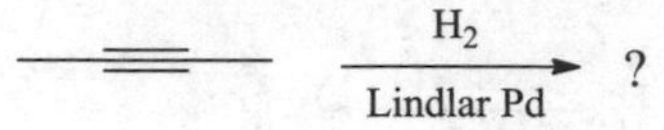

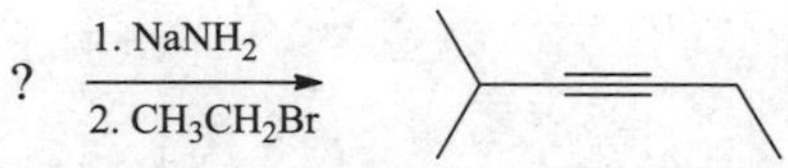

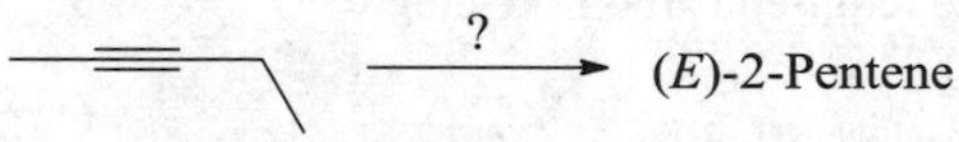

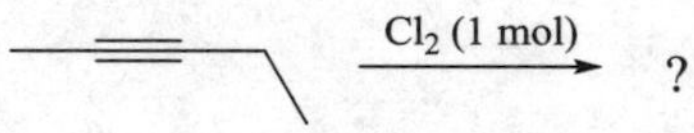

(b) $\xrightarrow{\text{HCl (2 mol)}}$?

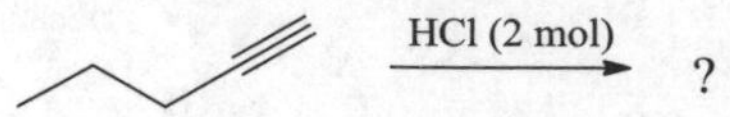

(c) $\xrightarrow{\quad ? \quad}$

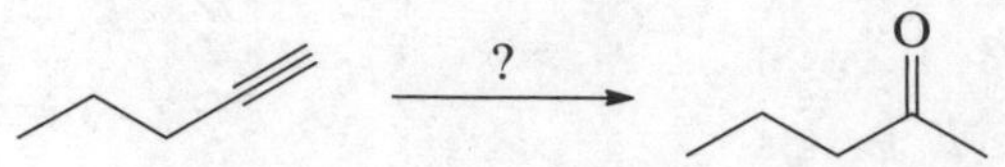

(d) $\xrightarrow[\text{Lindlar Pd}]{\text{H}_2}$?

(e) ? $\xrightarrow[\text{2. CH}_3\text{CH}_2\text{Br}]{\text{1. NaNH}_2}$

(f) $\xrightarrow{\quad ? \quad}$ (E)-2-Pentene

(g) $\xrightarrow{\text{Cl}_2 \text{ (1 mol)}}$?

(h) $\xrightarrow[\text{2. H}_2\text{O}]{\text{1. O}_3}$?

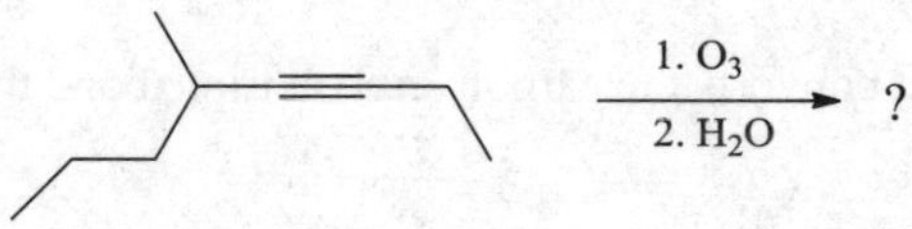

3. Which one of the following two reactions is effective in the synthesis of 4-methyl-2-hexyne? Why is the other not effective?

1.

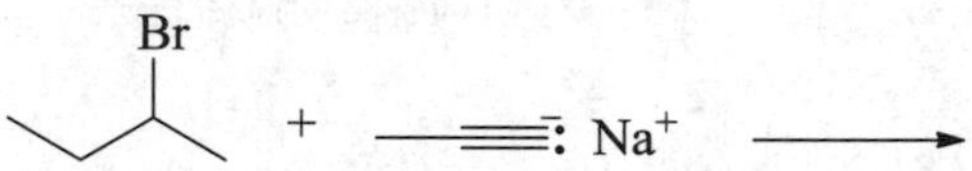

2.

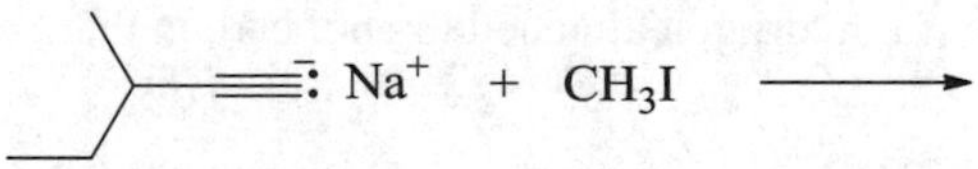

4. Outline a series of steps, using any necessary organic and inorganic reagents, for the preparation of

(a) 1-Butyne from ethyl bromide as the source of all carbon atoms

(b) 3-Hexyne from 1-butyne

(c) 3-Hexyne from 1-butene

(d)   from acetylene

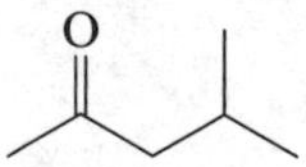

5. Treatment of propyne in successive steps with sodium amide, 1-bromobutane, and hydrogen with Lindlar-Pd yields as the final product _____ .

6. Give the structures of compounds A through D in the following series of equations.

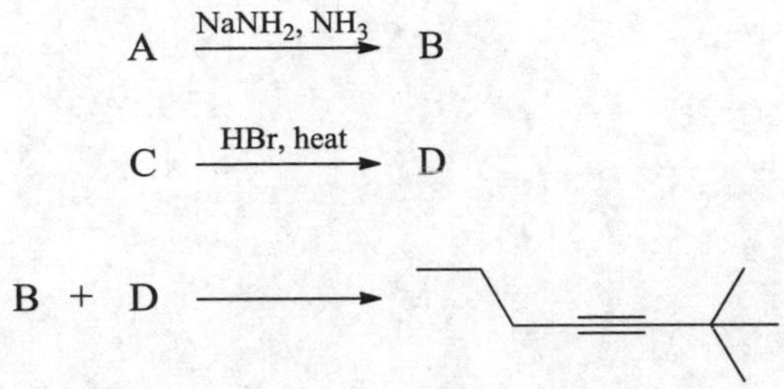

$$A \xrightarrow{\text{NaNH}_2,\ \text{NH}_3} B$$

$$C \xrightarrow{\text{HBr, heat}} D$$

$$B + D \longrightarrow$$

7. What are the structures of compounds E and F in the following sequence of reactions?

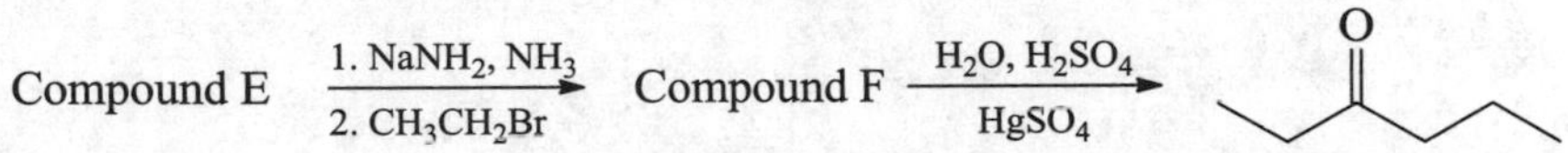

Compound E $\xrightarrow[\text{2. CH}_3\text{CH}_2\text{Br}]{\text{1. NaNH}_2,\ \text{NH}_3}$ Compound F $\xrightarrow[\text{HgSO}_4]{\text{H}_2\text{O, H}_2\text{SO}_4}$

8. Give the reagents that would be suitable for carrying out the following transformation. Two or more reaction steps are necessary.

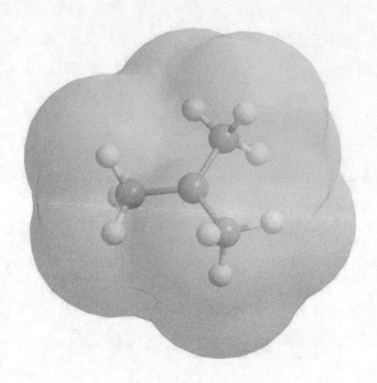

CHAPTER 10

Introduction to Free Radicals

Table of Contents

SOLUTIONS TO TEXT PROBLEMS

In Chapter Problems

10.1 There are three C_5H_{11} alkyl radicals with unbranched carbon chains. One is primary, and two are secondary.

primary radical   secondary radical   secondary radical

Four alkyl radicals are derived from $(CH_3)_2CHCH_2CH_3$. Two are primary, one is secondary, and one is tertiary.

primary radical   primary radical

tertiary radical   secondary radical

The most stable alkyl free radicals are tertiary. Only one tertiary radical of the formula C_5H_{11} is possible.

10.2 Recall that bond dissociation enthalpy is the value of ΔH° for homolysis of a specific covalent bond in a molecule. For the two different kinds of hydrogens in propane, the relevant equations are

$$CH_3CH_2CH_3 \longrightarrow CH_3CH_2\overset{\bullet}{C}H_2 \quad + \quad \cdot H \qquad \Delta H^\circ = +423 \text{ kJ/mol}$$

Propane n-Propyl radical   Hydrogen atom   (101 kcal/mol)

$$CH_3CH_2CH_3 \longrightarrow (CH_3)_2\overset{\bullet}{C}H \quad + \quad \cdot H \qquad \Delta H^\circ = +413 \text{ kJ/mol}$$

Propane   Isopropyl radical   Hydrogen atom   (99 kcal/mol)

The reactant (propane) and one of the products (hydrogen atom) are the same in both equations; therefore, the energy difference between them corresponds to the energy difference between n-propyl and isopropyl radical. Isopropyl radical is 10 kJ/mol (2 kcal/mol) more stable than n-propyl radical.

The situation is different for the case of n-propyl versus isopropyl chloride.

$$CH_3CH_2CH_2\ddot{\underset{\cdot\cdot}{Cl}}: \longrightarrow CH_3CH_2\dot{C}H_2 + \cdot\ddot{\underset{\cdot\cdot}{Cl}}: \qquad \Delta H^\circ = +354 \text{ kJ/mol}$$

n-Propyl chloride   n-Propyl radical   Chlorine atom   (85 kcal/mol)

$$(CH_3)_2CH\ddot{\underset{\cdot\cdot}{Cl}}: \longrightarrow (CH_3)_2\dot{C}H + \cdot\ddot{\underset{\cdot\cdot}{Cl}}: \qquad \Delta H^\circ = +355 \text{ kJ/mol}$$

Isopropyl chloride   Isopropyl radical   Chlorine atom   (81 kcal/mol)

Here, only the chlorine atom is common to both equations. The two radicals are different *and* the two alkyl halides are different. Therefore, it is not possible to assign all of the difference in energy to the two radicals. The alkyl halides may be, and likely are, different in energy as well.

10.3   The reaction for the iodination of methane and the bond dissociation energies (BDE) for each bond broken and made are in given in the equation.

$$H_3C-H \quad + \quad I-I \quad \longrightarrow \quad H_3C-I \quad + \quad H-I$$

BDE				
kJ/mol	439	151	238	298
(kcal/mol)	(105)	(36)	(57)	(71)

The ΔH° of the reaction is equal to the sum of bonds broken minus the sum of bonds formed. The positive value of this reaction indicates that it will not proceed.

$$\Delta H^\circ = \Sigma(\text{BDE of bonds broken}) - \Sigma(\text{BDE of bonds formed})$$

$$\Delta H^\circ = (439 \text{ kJ/mol} + 151 \text{ kJ/mol}) - (238 \text{ kJ/mol} + 298 \text{ kJ/mol}) = +54 \text{ kJ/mol}$$

$$\Delta H^\circ = (105 \text{ kcal/mol} + 36 \text{ kcal/mol}) - (57 \text{ kcal/mol} + 71 \text{ kcal/mol}) = +13 \text{ kcal/mol}$$

10.4   First write the equation for the overall reaction.

Chloromethane   Chlorine   Dichloromethane   Hydrogen
                                                        chloride

The initiation step is dissociation of chlorine to two chlorine atoms.

Chlorine   Two chlorine atoms

A chlorine atom abstracts a hydrogen atom from chloromethane in the first propagation step.

Chloromethane   Chlorine   Chloromethyl   Hydrogen
                 atom   radical   chloride

Chloromethyl radical reacts with Cl_2 in the next propagation step.

Chloromethyl radical   Chlorine   Dichloromethane   Chlorine atom

10.5 Writing the structural formula for ethyl chloride reveals that there are two nonequivalent sets of hydrogen atoms, in either of which a hydrogen is capable of being replaced by chlorine.

$$CH_3CH_2Cl \xrightarrow[\text{light or heat}]{Cl_2} CH_3CHCl_2 + ClCH_2CHCl_2$$

Ethyl chloride   1,1-Dichloroethane   1,2-Dichloroethane

10.6 Propane has six primary hydrogens and two secondary. In the chlorination of propane, the relative proportions of hydrogen atom removal are given by multiplying the number of equivalent hydrogens by the relative rate. Given that a secondary hydrogen is abstracted 3.9 times faster than a primary one, we write the expression for the amount of chlorination at the primary relative to that at the secondary position as

$$\frac{\text{Number of primary hydrogens} \times \text{rate of abstraction of a primary hydrogen}}{\text{Number of secondary hydrogens} \times \text{rate of abstraction of a secondary hydrogen}} = \frac{6 \times 1}{2 \times 3.9} = \frac{0.77}{1.00}$$

Thus, the percentage of propyl chloride formed is 0.77/1.77, or 43%, and that of isopropyl chloride is 57%. (The amounts actually observed are propyl 45%, isopropyl 55%.)

10.7 2-Methylpropane has nine primary hydrogens and one tertiary hydrogen. Yield of 1-chloro-2-methylpropane is given as 63% and that of 2-chloro-2-methylpropane is 37%.

2-Methylpropane   1-Chloro-2-methylpropane (63%)   2-Chloro-2-methylpropane (37%)

Setting up the equation analogous to the text and solving for relative rates of hydrogen abstraction with a chlorine atom gives a 15:1 ratio of reactivity for tertiary vs. primary.

$$\frac{63\% \text{ 1-chloro-2-methylpropane}}{37\% \text{ 2-chloro-2-methylpropane}} = \frac{\text{rate of tertiary H abstraction} \times 1 \text{ tertiary hydrogen}}{\text{rate of primary H abstraction} \times 9 \text{ primary hydrogens}}$$

$$\frac{\text{Rate of tertiary H abstraction}}{\text{Rate of primary H abstraction}} = \frac{63}{37} \times \frac{9}{1} = \frac{567}{37} = \frac{15}{1}$$

10.8   (*b*) Four constitutionally isomeric monochloro derivatives of 3-methylpentane are possible.

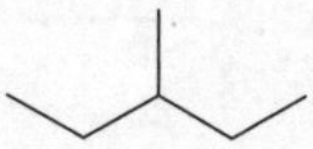

3-Methylpentane

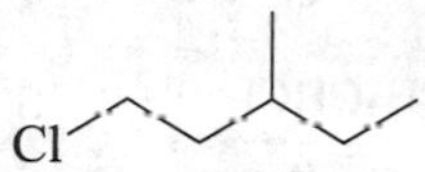

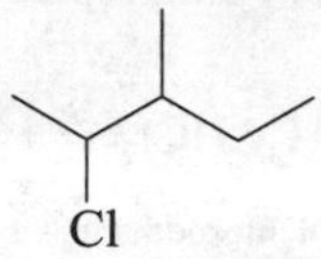

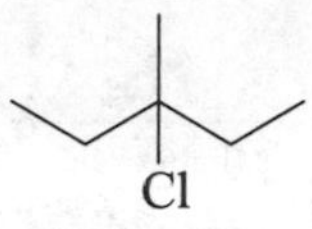

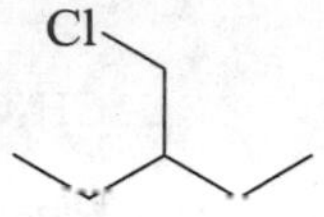

(*c*) Three constitutionally isomeric monochloro derivatives of 2,2-Dimethylbutane are possible.

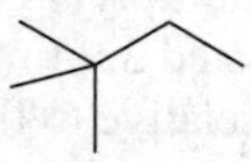

2,2-Dimethylbutane

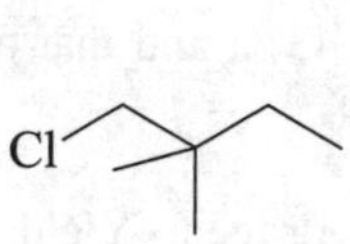

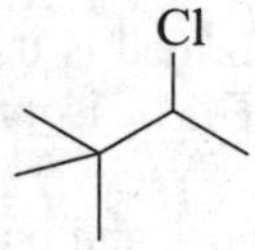

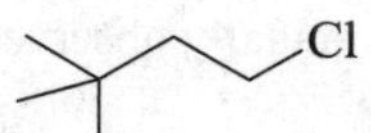

(*d*) Monochlorination of 2,3-dimethylbutane gives three constitutional isomers.

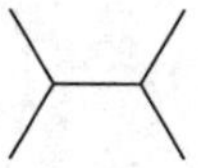

2,3-Dimethylbutane

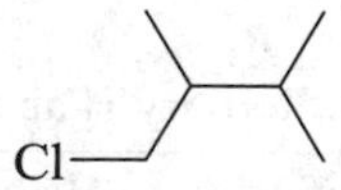

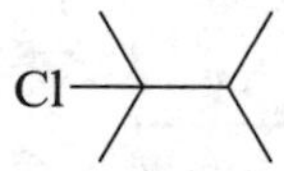

10.9   (*b*) Unlike chlorination, free-radical bromination of alkanes is highly selective. Bromination of 2,2,4-trimethylpentane is highly selective for replacement of the tertiary hydrogen.

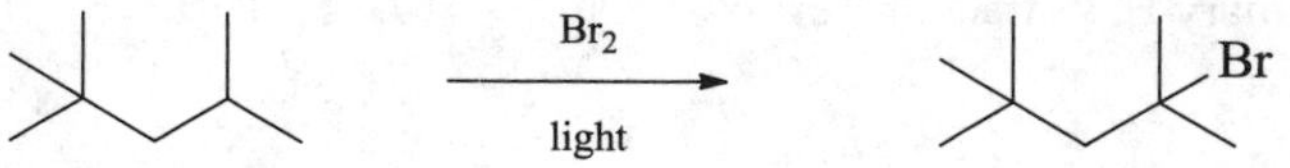

2,2,4-Trimethylpentane   2-Bromo-2,4,4-trimethylpentane (*c*)   As in part (*b*), bromination results in substitution of a tertiary hydrogen.

Tertiary carbon

1-Isopropyl-1-
methylcyclopentane

$\xrightarrow[\text{light}]{\text{Br}_2}$

1-(1-Bromo-1-methylethyl)-
1-methylcyclopentane

10.10 The reaction of allyl bromide (3-bromo-1-propene) with HBr gives two products, depending on the reaction conditions. When peroxides are excluded, 1,2-dibromopropane is formed by Markovnikov addition of HBr.

Allyl bromide

$\xrightarrow{\text{HBr}}$

1,2-Dibromopropane

When peroxides are present the "abnormal" product, according to Kharasch and Mayo, is observed.

Allyl bromide

$\xrightarrow[\text{peroxides}]{\text{HBr}}$

1,3-Dibromopropane

10.11 (*b*) Regioselectivity of addition is not an issue here, because the two carbons of the double bond are equivalent in *cis*-2-butene. Whether peroxides are present, or not, hydrogen bromide adds to *cis*-2-butene to give 2-bromobutane.

cis-2-Butene

$\xrightarrow[\substack{\text{no peroxides} \\ \text{or} \\ \text{peroxides}}]{\text{HBr}}$

2-Bromobutane

(*c*) With Markovnikov addition, bromine adds to the carbon atom of the double bond that has the fewer attached hydrogens.

2-Methyl-1-butene

$\xrightarrow[\text{no peroxides}]{\text{HBr}}$

2-Bromo-2-methylbutane

However, with peroxide present, hydrogen bromide gives the opposite regiochemistry to Markovnikov's addition rule.

2-Methyl-1-butene

$\xrightarrow[\text{peroxides}]{\text{HBr}}$

1-Bromo-2-methylbutane

(*d*) One end of the double bond has no hydrogens, but the other end has one. In accordance with Markovnikov's rule, the proton of hydrogen bromide adds to the carbon that already has one hydrogen. The product is 1-bromo-1-ethylcyclohexane.

Chlorine adds   Hydrogen adds
to this carbon.   to this carbon.

HBr
no peroxides

Br

Ethylidenecyclohexane   1-Bromo-1-ethylcyclohexane

In the presence of peroxide the opposite regiochemistry is observed. The product is 1-bromo-1-ethylcyclohexane.

HBr
peroxides

Br

Ethylidenecyclohexane   (1-Bromoethyl)cyclohexane

10.12 Because $CH_3CH=C(CH_3)_2$ reacts with HBr in the presence of peroxides to give only compound B, we can identify B by recalling that addition of HBr to the double bond of 2-methyl-2-butene in the presence of peroxides occurs opposite to Markovnikov's rule. Therefore, B is 2-bromo-3-methylbutane.

HBr
peroxides

Br

2-Methyl-2-butene   2-Bromo-3-methylbutane
                                         Isomer B

Addition of HBr to 3-methyl-1-butene in the absence of peroxides takes place according to Markovnikov's rule. The product of this reaction is also isomer B.

HBr

Br

3-Methyl-1-butene   2-Bromo-3-methylbutane
                                    Isomer B

We attribute the formation of a second isomer (A) in the electrophilic addition of HBr to 3-methyl-1-butene to a rearrangement of the carbocation formed by protonation of the double bond. The initially formed secondary carbocation can give B (as above) or undergoes a hydride shift to form a more stable tertiary carbocation to give A.

3-Methyl-1-butene   Secondary carbocation   Tertiary carbocation   2-Bromo-2-methylbutane Isomer A

10.13 The desired trans stereochemistry of the product may be achieved by metal–ammonia reduction of 2-heptyne.

2-Heptyne   *trans*-2-Heptene

2-Heptyne is available from propyne by alkylation with 1-bromobutane.

Propyne   2-Heptyne

10.14 Only hydrogen bromide, not hydrogen chloride or hydrogen iodide, adds anti-Markovnikov to double bonds so replacing HBr with HI in the third step of the reaction will not yield 1-iodo-2,3,3-trimethylbutane. However, iodide is an good nucleophile and the bromine in 1-bromo-2,3,3-trimethylbutane is an good leaving group on a primary carbon. An S_N2 reaction in an additional step gives the iodoalkane.

10.15 Look on the vinyl polymer as being formed from a derivative of ethylene in which the two functional groups, a nitrile and an ester, are attached to the same carbon.

End of Chapter Problems

Structure and Bonding

10.16 The radicals formed from homolytic cleavage of C-C bonds correspond to the bond dissociation enthalpies. The more stable radicals that are formed are derived from lower bond dissociation energies (weaker bonds).

(*a*) First write the equations that describe homolytic carbon-carbon bond cleavage in each alkane. We see that two methyl radicals are generated from ethane whereas a methyl radical and a primary ethyl radical is produced by a cleavage of propane.

$$H_3C-CH_3 \longrightarrow \cdot CH_3 + \cdot CH_3 \qquad \text{Higher BDE}$$

Ethane   methyl radical + methyl radical

$$H_3C-CH_2CH_3 \longrightarrow \cdot CH_3 + \cdot CH_2CH_3 \qquad \text{Lower BDE}$$

Propane   methyl radical + ethyl radical

An ethyl radical is more stable than methyl, and so less energy is required to break the carbon-carbon bond in propane than in ethane. The measured carbon-carbon bond dissociation enthalpy in ethane is 375 kJ/mol (90 kcal/mol), and that in propane is 369 kJ/mol (88 kcal/mole).

(b) Writing the equation for carbon–carbon bond cleavage in 2-methylpropane a secondary isopropyl radical and a methyl radical are produced by cleavage of 2-methylpropane. There is a higher bond dissociation energy associated with forming a methyl radical and a primary radical vs. a methyl and more stable secondary isopropyl radical.

$$H_3C-CH_2CH_3 \longrightarrow \cdot CH_3 + \cdot CH_2CH_3 \qquad \text{Higher BDE}$$

Propane   methyl radical + ethyl radical

$$\underset{\displaystyle H_3C-CHCH_3}{\overset{\displaystyle CH_3}{}} \longrightarrow \cdot CH_3 + \cdot CH_2CH_3 \qquad \text{Lower BDE}$$

2-Methylpropane   methyl radical + isopropyl radical

(c) Writing the equation for carbon–carbon bond cleavage in 2,2-dimethylpropane produces a tertiary radical and a methyl radical. There is a higher bond dissociation energy associated with forming a methyl radical and a secondary radical vs. a methyl and more stable tertiary *tert*-butyl radical.

$$\underset{\displaystyle H_3C-CHCH_3}{\overset{\displaystyle CH_3}{}} \longrightarrow \cdot CH_3 + \underset{}{\overset{\displaystyle CH_3}{\cdot CH_2CH_3}} \qquad \text{Higher BDE}$$

2-Methylpropane   methyl radical + isopropyl radical

$$\underset{\displaystyle CH_3}{\overset{\displaystyle CH_3}{H_3C-C-CH_3}} \longrightarrow \cdot CH_3 + \underset{\displaystyle CH_3}{\overset{\displaystyle CH_3}{\cdot C-CH_3}} \qquad \text{Lower BDE}$$

2,2-Dimethylpropane   methyl radical + *tert*-butyl radical

(d) There is a higher bond dissociation energy associated with breaking the cyclopentane C-C bond vs. cyclobutane C-C bond. In both cases, two primary radicals are formed; however, the breaking the cyclobutane C-C bond relieves more angle strain compared to cyclopentane resulting in a lower bond dissociation energy.

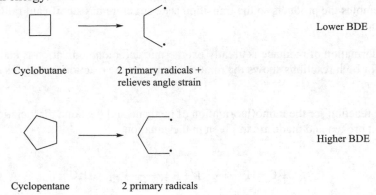

10.17 (a) The reaction between methane and a bromine atom corresponds to the first propagation step for the free radical bromination of methane. The bond dissociation energies (BDE) for each bond broken and made are in given below the equation.

$$H-CH_3 \ + \ \overset{..}{:}\overset{..}{Br}\cdot \ \longrightarrow \ \cdot CH_3 \ + \ H-\overset{..}{Br}\overset{..}{:}$$

BDE			
kJ/mol	439		366
(kcal/mol)	(105)		(87.5)

The ΔH° of the reaction is equal to the sum of bonds broken minus the sum of bonds formed.

$$\Delta H^\circ = \Sigma(\text{BDE of bonds broken}) - \Sigma(\text{BDE of bonds formed})$$

$$\Delta H^\circ = (439 \text{ kJ/mol}) - (366 \text{ kJ/mol}) = +73 \text{ kJ/mol}$$

$$\Delta H^\circ = (105 \text{ kcal/mol}) - (87.5 \text{ kcal/mol}) = +17.5 \text{ kcal/mol}$$

(b) Given the activation energy is 76 kJ/mol (18.3 kJ/mol), a potential energy diagram can be sketched.

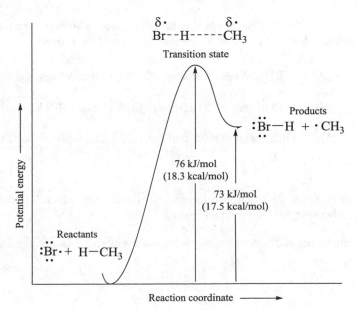

This step in the mechanism is endothermic so this is a late transition state. The transition state lies closer to the products than to the reactants. According to Hammond's postulate, the transition state structure resembles the products so the transition state structure has significant radical character on the methyl carbon.

10.18 Monofluorination of methane is greatly exothermic and monoiodination is endothermic. Inspection of the BDE's for both reactions shows the main influence on the reaction enthalpies resides ins the product bond enthalpies.

(*a*) The reaction for the monofluorination of methane and the bond dissociation energies (BDE) for each bond broken and made are in given in the equation.

$$H_3C-H \quad + \quad F-F \quad \longrightarrow \quad H_3C-F \quad + \quad H-F$$

| BDE kJ/mol | 439 | 159 | 459 | 571 |
| (kcal/mol) | (105) | (38) | (110) | (136) |

The ΔH° of the reaction is equal to the sum of bonds broken minus the sum of bonds formed. The negative value of this reaction indicates that it will proceed.

$\Delta H^\circ = \Sigma$(BDE of bonds broken) $-\Sigma$(BDE of bonds formed)

$\Delta H^\circ = (439 \text{ kJ/mol} + 159 \text{ kJ/mol}) - (459 \text{ kJ/mol} + 571 \text{ kJ/mol}) = -432 \text{ kJ/mol}$

$\Delta H^\circ = (105 \text{ kcal/mol} + 38 \text{ kcal/mol}) - (110 \text{ kcal/mol} + 136 \text{ kcal/mol}) = -103 \text{ kcal/mol}$

(*b*) The reaction for the monoiodination of methane and the bond dissociation energies (BDE) for each bond broken and made are in given in the equation.

$$H_3C-H \quad + \quad I-I \quad \longrightarrow \quad H_3C-I \quad + \quad H-I$$

| BDE kJ/mol | 439 | 151 | 238 | 298 |
| (kcal/mol) | (105) | (36) | (57) | (71) |

The ΔH° of the reaction is equal to the sum of bonds broken minus the sum of bonds formed. The positive value of this reaction indicates that it will not proceed.

$\Delta H^\circ = \Sigma$(BDE of bonds broken) $-\Sigma$(BDE of bonds formed)

$\Delta H^\circ = (439 \text{ kJ/mol} + 151 \text{ kJ/mol}) - (238 \text{ kJ/mol} + 298 \text{ kJ/mol}) = +54 \text{ kJ/mol}$

$\Delta H^\circ = (105 \text{ kcal/mol} + 36 \text{ kcal/mol}) - (57 \text{ kcal/mol} + 71 \text{ kcal/mol}) = +13 \text{ kcal/mol}$

Reactions

10.19 When we compare the reactions of hydrogen bromide with alkenes in the absence and presence of peroxides, the regiochemistry of the products change.

(*a*) Electrophilic addition of hydrogen bromide will give 2-bromopentane

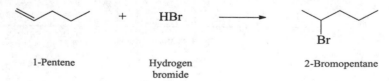

| 1-Pentene | Hydrogen bromide | 2-Bromopentane |

The presence of peroxides will cause free-radical addition of hydrogen bromide, and regioselective addition opposite to Markovnikov's rule will be observed.

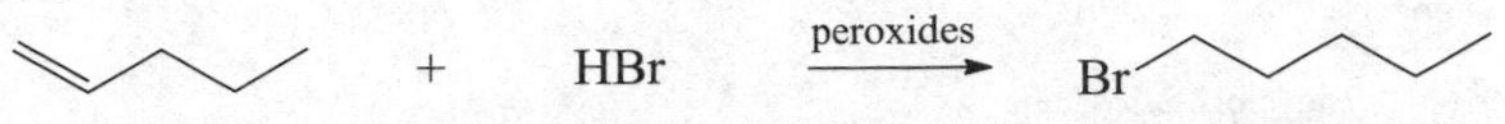

| 1-Pentene | Hydrogen bromide | 1-Bromopentane |

(*b*) Electrophilic addition of hydrogen bromide to 2-methyl-2-butene will give 2-bromo-2-methylbutane.

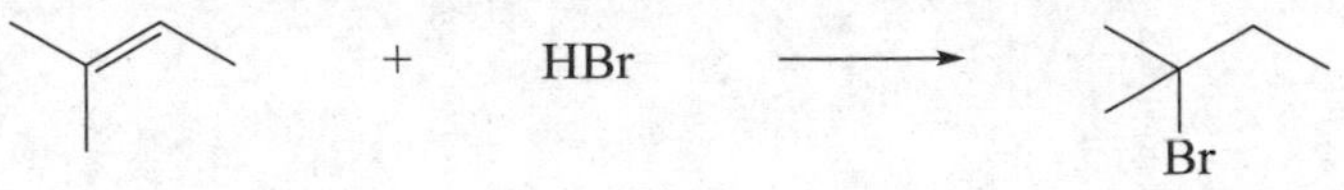

| 2-Methyl-2-butene | Hydrogen bromide | 2-Bromo-2-methylbutane |

The presence of peroxides gives regioselective addition opposite to Markovnikov's rule.

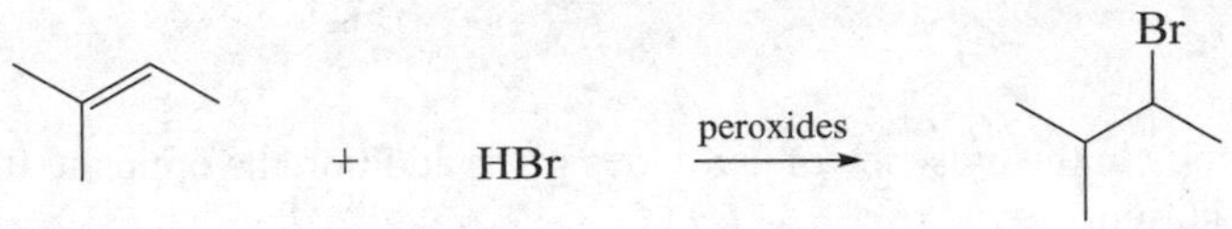

| 2-Methyl-2-butene | Hydrogen bromide | 2-Bromo-3-methylbutane |

(*c*) Electrophilic addition of hydrogen bromide to 1-methylcyclohexene will give 1-bromo-1-methylcyclohexane.

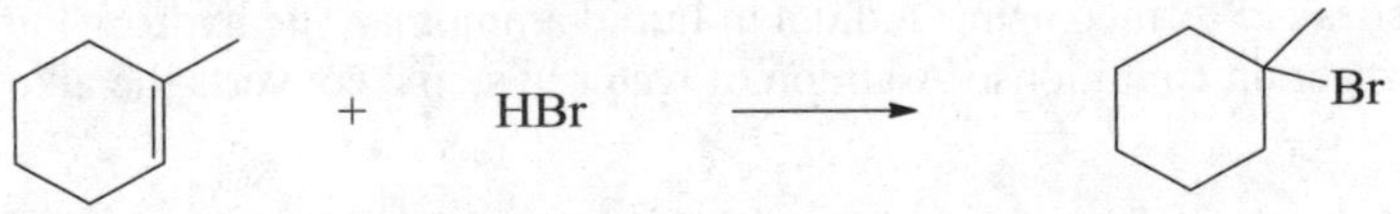

| 1-Methylcyclohexene | Hydrogen bromide | 1-Bromo-1-methylcyclohexane |

The presence of peroxides gives regioselective addition opposite to Markovnikov's rule.

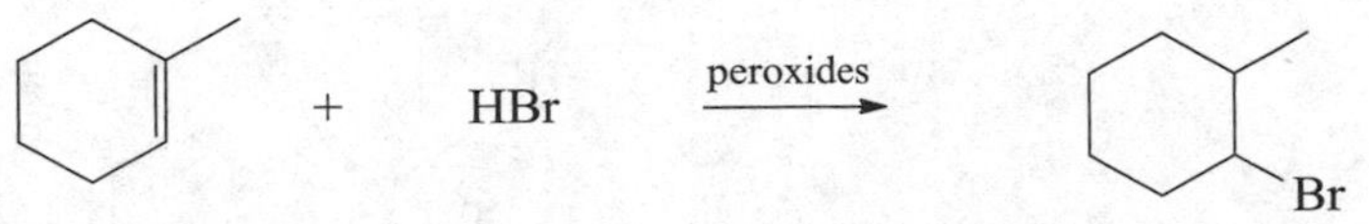

| 1-Methylcyclohexene | Hydrogen bromide | 1-Bromo-2-methylcyclohexane (mixture of cis and trans) |

10.20 (*a*) The monochlorination products of bicyclo[2.2.1]heptane include two that are related as diastereomers.

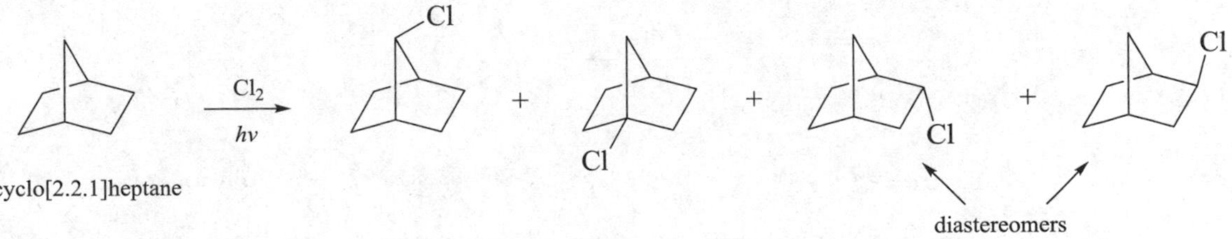

Bicyclo[2.2.1]heptane

diastereomers

Monochlorination of bicyclo[2.2.2]octane only gives two products.

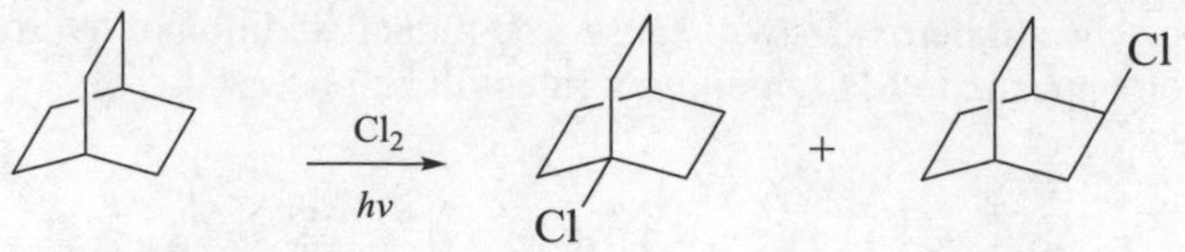

Bicyclo[2.2.2]octane

(*b*) The diastereomer pair each have an enantiomer.

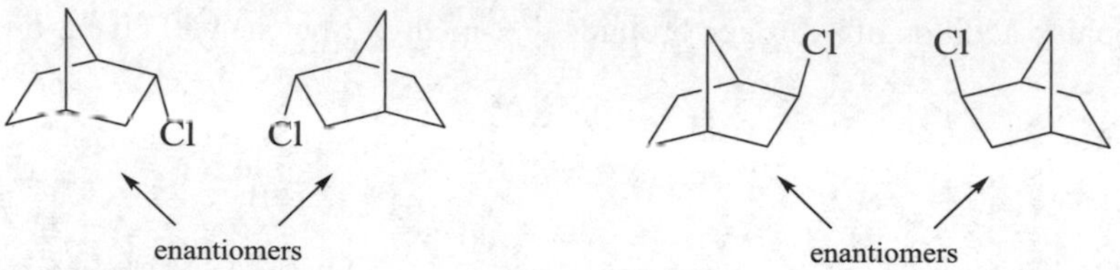

10.21 (*a*) Free radical bromination is selective for tertiary carbons.

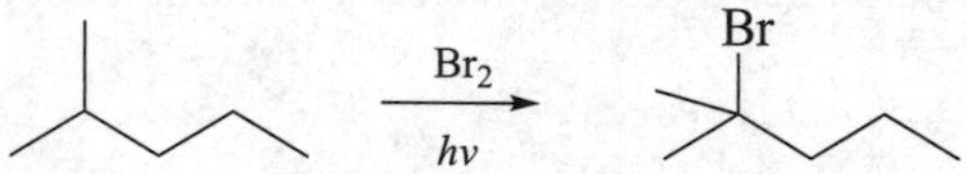

(*b*) Hydrogen bromide in the presence of peroxides gives addition the opposite from that expected with electrophilic addition.

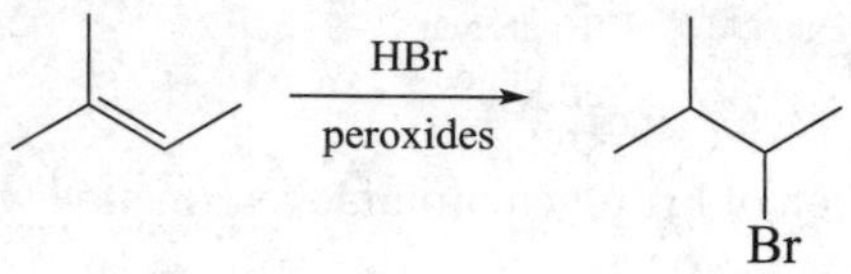

(*c*) Trans alkenes are formed using sodium in liquid ammonia. The hydroxyl group is present as an alkoxide under the reaction conditions. Addition of water in step 2 converts the alkoxide function to hydroxyl.

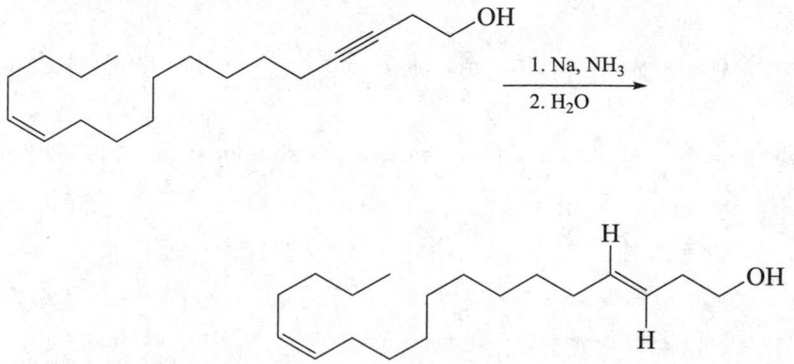

10.22 Free-radical addition to cis-2-pentene gives two constitutional isomers. There are no chirality centers in 3-bromopentane.

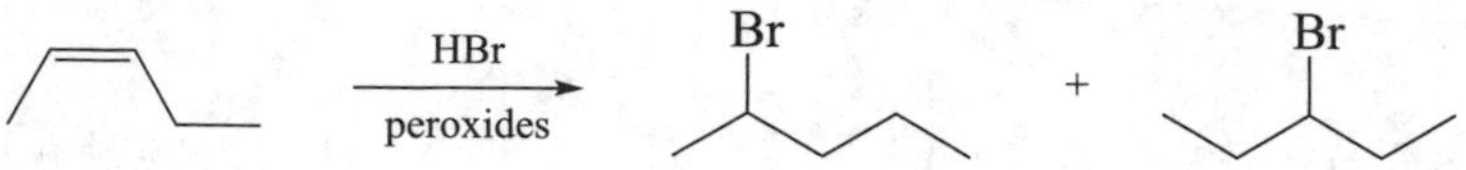

cis-2-Pentene   2-Bromopentane   3-Bromopentane

2-Bromopentane has a chirality center and so an enantiomeric pair is formed.

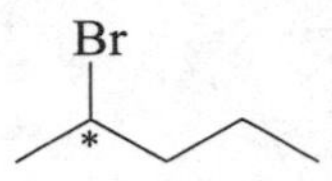

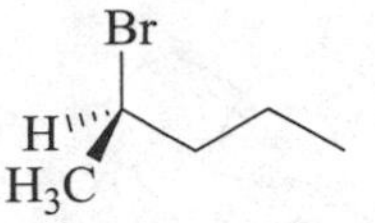

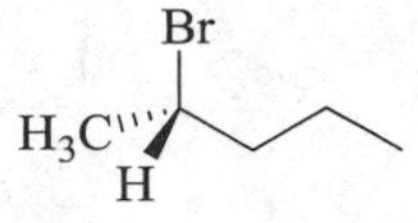

2-Bromopentane   (*R*)-2-Bromopentane   (*S*)-2-Bromopentane

10.23 The four methylene hydrogens of 1,2-dibromoethane are enantiotopic. Replacement of any one of them by some atom or group produces two enantiomers. Light-initated free-radical chlorination will replace one of the hydrogens with a chlorine. The enantiomers are chiral and are formed in equal amounts as a racemic mixture.

$$BrCH_2CH_2Br \;+\; Cl_2 \quad \xrightarrow{\text{light}} \quad BrCH_2\overset{\underset{\displaystyle |}{Cl}}{C}HBr \;+\; HCl$$

The product is a racemic mixture of

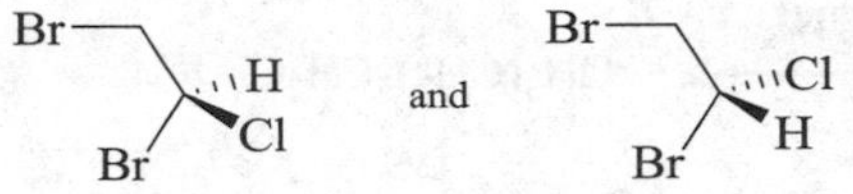

and

10.24 There are only two possible products from free-radical chlorination of the starting alkane:

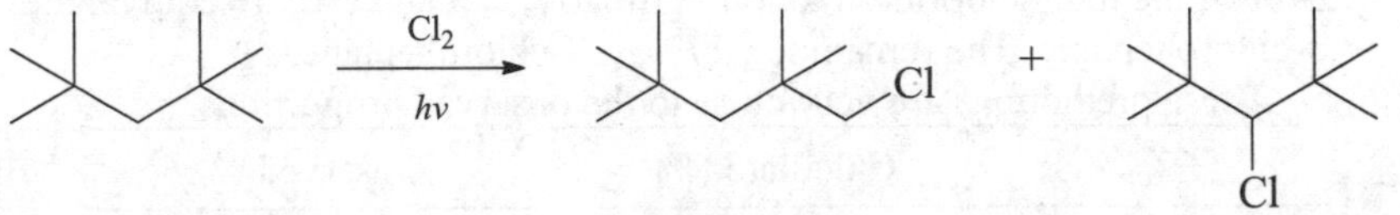

2,2,4,4-Tetramethylpentane   1-Chloro-2,2,4,4-tetramethylpentane (primary)   3-Chloro-2,2,4,4-tetramethylpentane (secondary)

As revealed by their structural formulas, one isomer is a primary alkyl chloride, the other is secondary. The problem states that the major product (compound A) undergoes S_N1 hydrolysis much more slowly than the minor product (compound B). Because secondary halides are much more reactive than primary halides under S_N1 conditions, the major (unreactive) product is the primary alkyl halide 1-chloro-2,2,4,4-tetramethylpentane (compound A) and the minor (reactive) product is the secondary alkyl halide 3-chloro-2,2,4,4-tetramethylpentane (compound B).

10.25 The key to this problem is the fact that one of the alkyl chlorides of molecular formula $C_6H_{13}Cl$ does not undergo E2 elimination. It must therefore have a structure in which the carbon atom that is β to the chlorine bears no hydrogens. This $C_6H_{13}Cl$ isomer is 1-chloro-2,2-dimethylbutane.

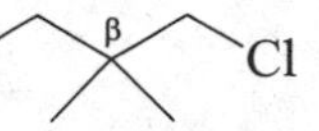

1-Chloro-2,2-dimethylbutane
(cannot form an alkene)

Identifying this monochloride derivative gives us the carbon skeleton. The starting alkane (compound A) must be 2,2-dimethylbutane. Its free-radical halogenation gives three different monochlorides:

2,2-Dimethylbutane   1-Chloro-2,2-dimethylbutane   3-Chloro-2,2-dimethylbutane   1-Chloro-3,3-dimethylbutane
(compound A)

Both 3-chloro-2,2-dimethylbutane and 1-chloro-3,3-dimethylbutane give only 3,3-dimethyl-1-butene on E2 elimination.

3-Chloro-2,2-dimethylbutane   1-Chloro-3,3-dimethylbutane   3,3-Dimethyl-1-butene
(alkene B)

10.26 (*a*)   Heptane has five methylene groups, which on chlorination together contribute 85% of the total monochlorinated product.

$$CH_3(CH_2)_5CH_3 \longrightarrow CH_3(CH_2)_5CH_2Cl \quad + \quad (2\text{-chloro} + 3\text{-chloro} + 4\text{-chloro})$$

Heptane   15%   85%

Because the problem specifies that attack at each methylene group is equally probable, the five methylene groups each give rise to 85/5, or 17%, of the monochloride product.

Because C-2 and C-6 of heptane are equivalent, we calculate that 2-chloroheptane will constitute 34% of the monochloride fraction. Similarly, C-3 and C-5 are equivalent, and so there should be 34% 3-chloroheptane. The remainder, 17%, is 4-chloroheptane.

These predictions are very close to the observed proportions.

	Calculated %	Observed %
2-Chloro	34	35
3-Chloro	34	34
4-Chloro	17	16

(*b*)   There are a total of 20 methylene hydrogens in dodecane, $CH_3(CH_2)_{10}CH_3$. The 19% 2-chloro-dodecane that is formed arises by substitution of any of the four equivalent methylene hydrogens at C-2 and C-11. The total amount of substitution of methylene hydrogens must therefore be

$$\frac{20}{4} \times 19\% = 95\%$$

The remaining 5% corresponds to substitution of methyl hydrogens at C-1 and C-12. The proportion of 1-chlorododecane in the monochloride fraction is 5%.

10.27 (*a*)   Two of the monochlorides derived from chlorination of 2,2,4-trimethylpentane are primary chlorides:

1-Chloro-2,2,4-trimethylpentane   1-Chloro-2,4,4-trimethylpentane

The two remaining isomers are a secondary chloride and a tertiary chloride:

3-Chloro-2,2,4-trimethylpentane   2-Chloro-2,4,4-trimethylpentane

(b)   Substitution of any one of the nine hydrogens designated as x in the structural diagram yields 1-chloro-2,2,4-trimethylpentane. Substitution of any one of the six hydrogens designated as y gives 1-chloro-2,4,4-trimethylpentane.

Assuming equal reactivity of a single x hydrogen and a single y hydrogen, the ratio of the two isomers is then expected to be 9:6. Because together the two primary chlorides total 65% of the monochloride fraction, there will be 39% 1-chloro-2,2,4-trimethylpentane (substitution of x) and 26% 1-chloro-2,4,4-trimethylpentane (substitution of y).

10.28 The three monochlorides are shown in the equation

$$CH_3CH_2CH_2CH_2CH_3 \xrightarrow[\text{light}]{Cl_2} CH_3CH_2CH_2CH_2CH_2Cl + CH_3CHCH_2CH_2CH_3 + CH_3CH_2CHCH_2CH_3$$
$$\qquad\qquad\qquad\qquad\qquad\qquad\qquad\qquad\qquad\qquad\qquad\qquad\qquad\quad | \qquad\qquad\qquad\qquad\quad |$$
$$\qquad\qquad\qquad\qquad\qquad\qquad\qquad\qquad\qquad\qquad\qquad\qquad\qquad\quad Cl \qquad\qquad\qquad\qquad\quad Cl$$

Pentane   1-Chloropentane   2-Chloropentane   3-Chloropentane

Pentane has six primary hydrogens (two CH_3 groups) and six secondary hydrogens (three CH_2 groups). Because a single secondary hydrogen is abstracted three times faster than a single primary hydrogen and there are equal numbers of secondary and primary hydrogens, the product mixture should contain three times as much of the secondary chloride isomers as the primary chloride. The primary chloride 1-chloropentane, therefore, is expected to constitute 25% of the product mixture. The secondary chlorides 2-chloropentane and 3-chloropentane are not formed in equal amounts. Rather, 2-chloropentane may be formed by replacement of a hydrogen at C-2 or at C-4, whereas 3-chloropentane is formed only when a C-3 hydrogen is replaced. The amount of 2-chloropentane is therefore 50%, and that of 3-chloropentane is 25%. We predict the major product to be 2-chloropentane, and the predicted proportion of 50% corresponds closely to the observed 46%.

Synthesis

10.29 The starting material designated in the statement of the problem is isopropyl alcohol. It is readily converted to 1-bromopropane by the two-step sequence shown.

Isopropyl alcohol   Propene   1-Bromopropane
(2-Propanol)   (a)

The 1-bromopropane prepared in (a) is a key starting material for the preparation of the compounds in parts (b-f) of this problem. The nitrile in part (e), for example, is prepared directly from 1-bromopropane by an S_N2 reaction.

1-Bromopropane   Butyronitrile

Likewise, 1-pentyne is the product formed using sodium acetylide as the nucleophile in part (*f*).

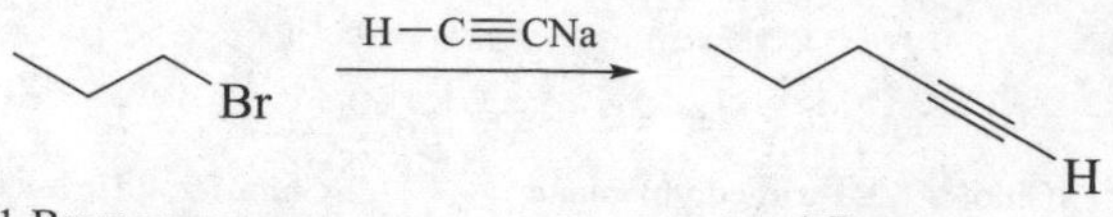

1-Bromopropane   1-Pentyne

Catalytic hydrogenation of 1-pentyne over a Lindlar palladium catalyst gives 1-pentene in (*c*).

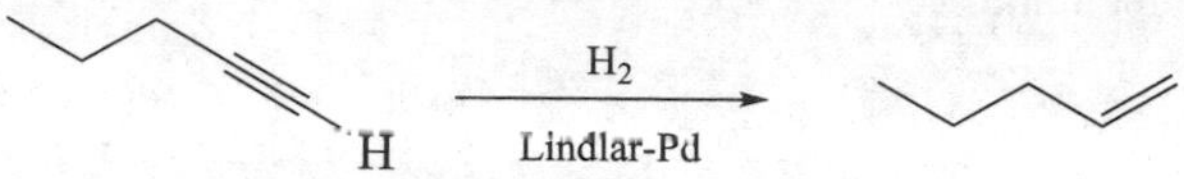

1-Pentyne   1-Pentene

An analogous sequence using sodium propynide gives 2-hexyne, which on catalytic hydrogenation yields (*Z*)-2-hexene (*d*).

1-Bromopropane   1-Pentyne   (*Z*)-2-Hexene

Reduction of 2-hexyne with sodium and ammonia gives (*E*)-2-hexene (*b*).

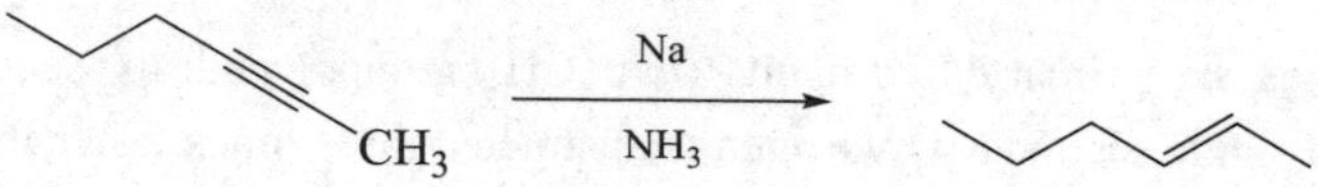

1-Pentyne   (*E*)-2-Hexene

10.30 (*a*)   Unlike chlorination and bromination, direct free-radical iodination of alkanes is not a useful reaction. An indirect method can be represented retrosynthetically as:

$$ RI \implies RBr \text{ or } RCl \implies RH $$

The alkane is subjected to free-radical bromination or chlorination and the resulting alkyl halide converted to the corresponding iodide by nucleophilic substitution (S_N2) with sodium iodide in acetone.

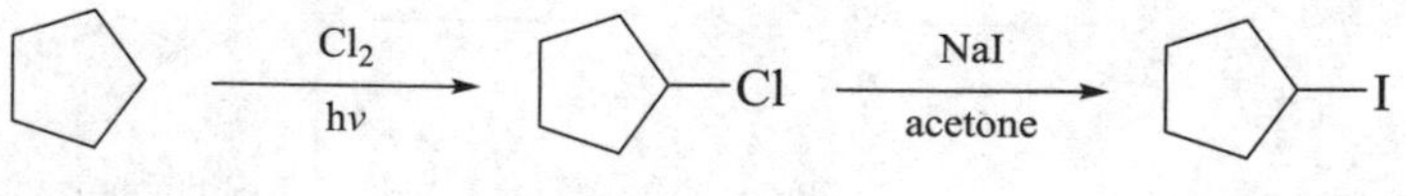

Cyclopentane   Cyclopentyl chloride   Cyclopentyl iodide
                                 (Chlorocyclopentane)   (Iodocyclopentane)

(*b*)   Rather than try to move a bromine from one carbon to an adjacent one, it is better to remove it in one reaction and add a different bromine in a subsequent reaction. For the present case, we can represent this retrosynthetically as:

1-Bromo-2-   2-Methylpropene   2-Bromo-2-
methylpropane   methylpropane

The compound connecting the original bromide and the desired one is the alkene 2-methylpropene. It can be prepared from *tert*-butyl chloride by elimination and converted to 1-bromo-2-methylpropane by free-radical addition of HBr to the double bond.

| 2-Bromo-2-methylpropane | 2-Methylpropene | 1-Bromo-2-methylpropane |

(c) At the outset, we recognize that we need to add two hydrogen atoms and two bromine atoms to the triple bond and also control the stereochemistry of addition. The most reasonable assumption is that hydrogenation will be a *syn* addition and bromination will be *anti*. It is also reasonable to assume that the sequence of reactions will be hydrogenation first, bromine addition second.

meso-2,3-Dibromobutane   *trans*-2-Butene   2-Butyne

This retrosynthetic analysis suggests the following synthesis.

2-Butyne   *trans*-2-Butene   *meso*-2,3-Dibromobutane

(d) We recognize that replacing the bromine atom of 1-bromopentane by an acetylide unit not only provides a way to both extend the chain by two carbons but introduces unsaturation as well

1-Heptene   1-Heptyne   1-Bromopentane

and leads to the following synthesis.

1-Bromopentane   1-Heptyne   1-Heptene

(e) To convert 1,2-dibromopentane to *cis*-2-hexene we need to both extend the carbon chain and introduce a double bond. A suitable retrosynthesis is:

cis-2-Hexene   2-Hexyne   1-Pentyne   1,2-Dibromopentane

The overall strategy is to introduce the triple bond for the purpose of allowing subsequent one-carbon chain extension, followed by reducing the triple bond to a cis double bond.

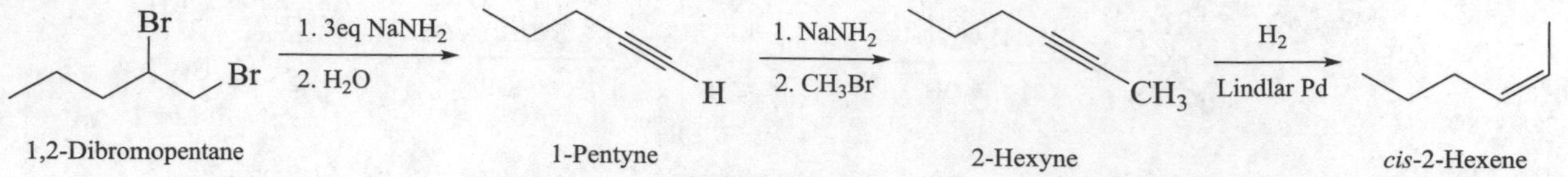

1,2-Dibromopentane   1-Pentyne   2-Hexyne   *cis*-2-Hexene

(*f*) The most obvious final step in the preparation of butyl methyl ether from 1-butene is carbon-oxygen bond formation by nucleophilic substitution.

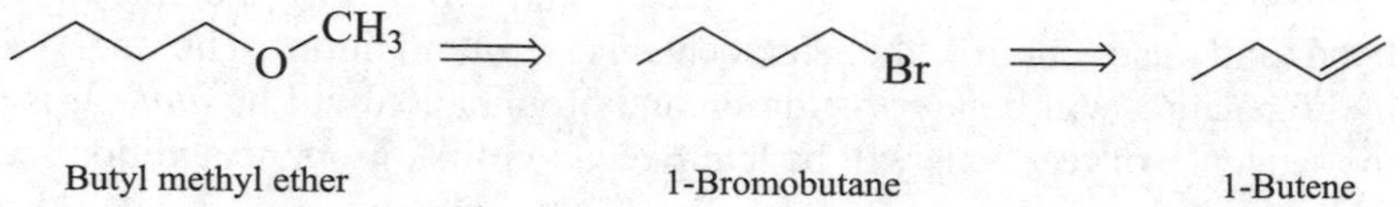

Butyl methyl ether   1-Bromobutane   1-Butene

The synthesis is straightforward.

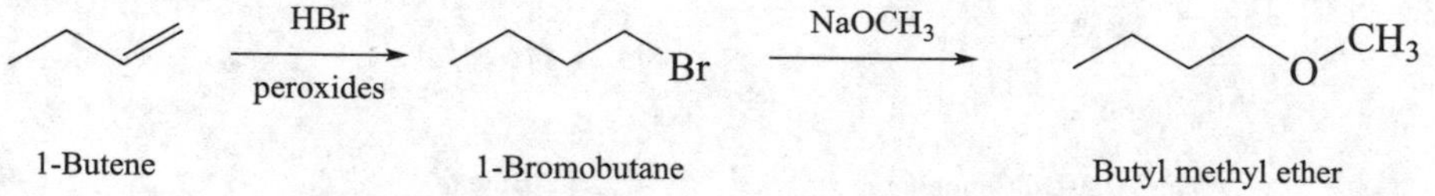

1-Butene   1-Bromobutane   Butyl methyl ether

Alternative, also reasonable, syntheses are possible but would require one or more additional steps.

(*g*) The most efficient retrosynthesis requires two reactions.

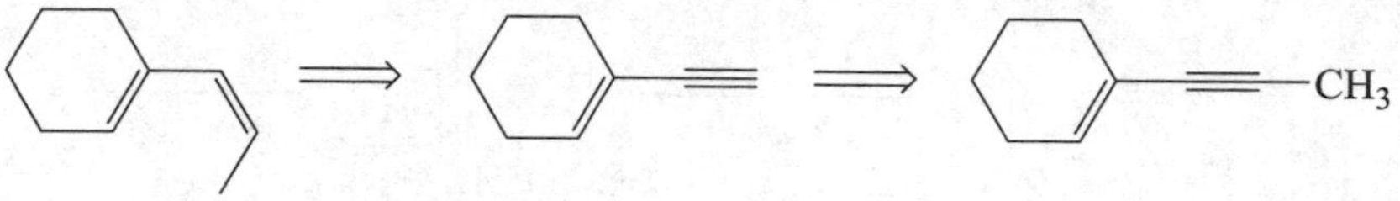

This leads to the synthesis.

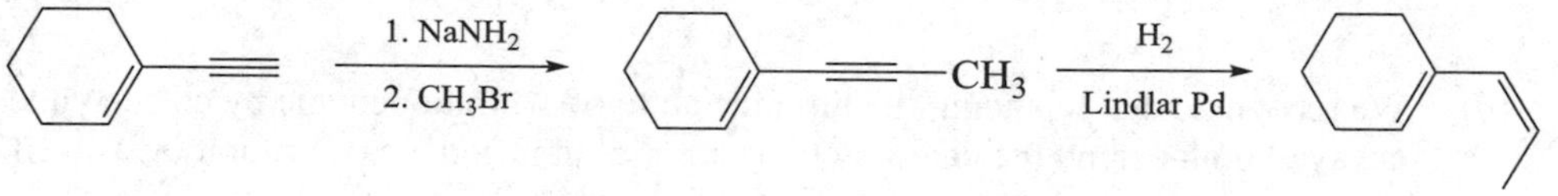

10.31 A retrosynthetic analysis of (*Z*)-9-tricosene from alcohols and acetylene is shown.

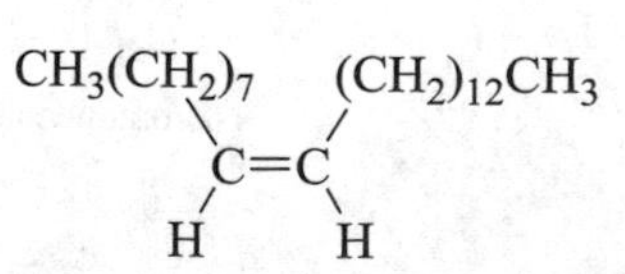

(*Z*)-9-Tricosene

To accomplish this transformation the alcohols are first converted to bromides.

$$CH_3(CH_2)_7-OH \xrightarrow{PBr_3} CH_3(CH_2)_7-Br$$

$$HO-(CH_2)_{12}CH_3 \xrightarrow{PBr_3} Br-(CH_2)_{12}CH_3$$

Acetylene is then converted to acetylide that then reacts with one of the bromides.

$$H-C\equiv C-H \xrightarrow{\text{NaNH}_2} H-C\equiv C:^-$$

$$H-C\equiv C:^- + Br-(CH_2)_{12}CH_3 \longrightarrow H-C\equiv C-(CH_2)_{12}CH_3$$

This is followed by deprotonation of the terminal acetylene and reaction with the other bromide.

$$H-C\equiv C-(CH_2)_{12}CH_3 \xrightarrow{\text{NaNH}_2} {}^-:C\equiv C-(CH_2)_{12}CH_3$$

$$CH_3(CH_2)_7-Br + {}^-:C\equiv C-(CH_2)_{12}CH_3 \longrightarrow CH_3(CH_2)_7-C\equiv C-(CH_2)_{12}CH_3$$

Hydrogenation with Lindlar Pd gives (*Z*)-9-tricosine.

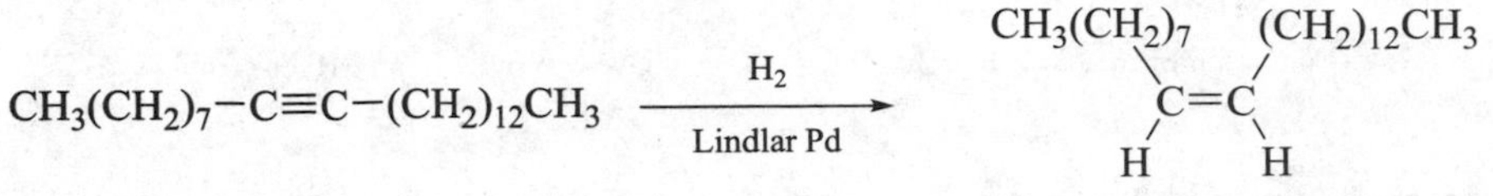

$$CH_3(CH_2)_7-C\equiv C-(CH_2)_{12}CH_3 \xrightarrow[\text{Lindlar Pd}]{H_2}$$

$$\begin{array}{cc} CH_3(CH_2)_7 & (CH_2)_{12}CH_3 \\ \diagdown & \diagup \\ C=C \\ \diagup & \diagdown \\ H & H \end{array}$$

(*Z*)-9-Tricosene

Mechanism

10.32 When peroxides are present, hydrogen bromide adds to alkenes by a free-radical chain mechanism.

Initiation:

Peroxide dissociation is the first part of the initiation stage. The alkoxy radicals so generated then abstract a hydrogen atom from HBr to produce bromine atoms.

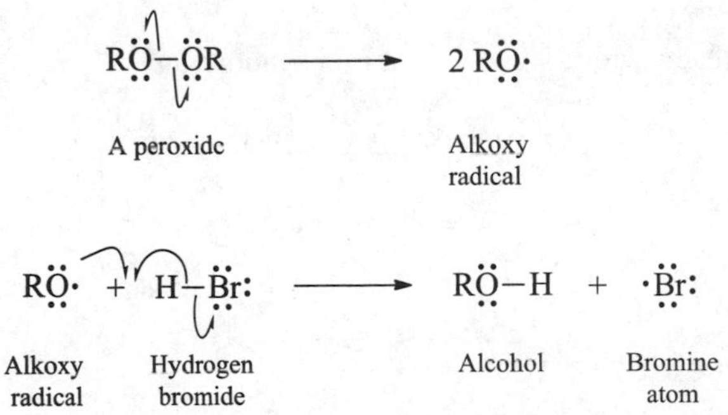

$$R\ddot{O}-\ddot{O}R \longrightarrow 2\ R\ddot{O}\cdot$$

A peroxide   Alkoxy radical

$$R\ddot{O}\cdot + H-\ddot{B}r: \longrightarrow R\ddot{O}-H + \cdot\ddot{B}r:$$

Alkoxy radical   Hydrogen bromide   Alcohol   Bromine atom

Propagation:

The propagation phase begins with a bromine atom adding to the double bond in the direction that produces the more stable alkyl radical. This radical then abstracts a hydrogen atom from HBr to give the alkyl bromide and a bromine atom

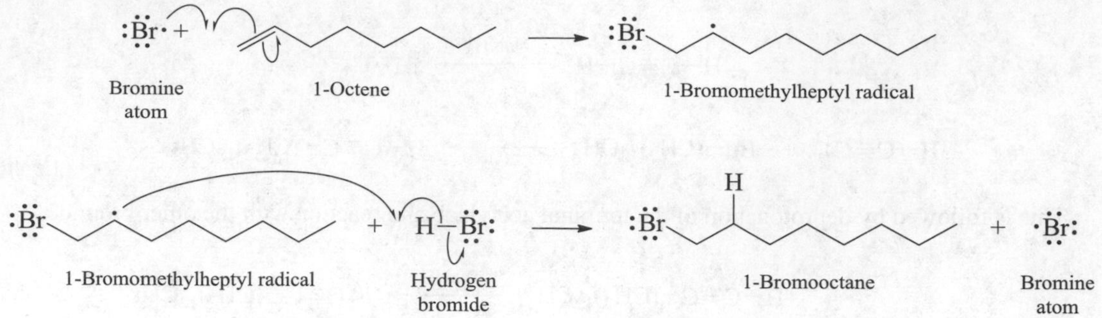

The bromine atom produced in the second propagation reaction adds to a second 1-octene molecule and the two propagation steps repeat over and over.

10.33 The equation for the reaction is

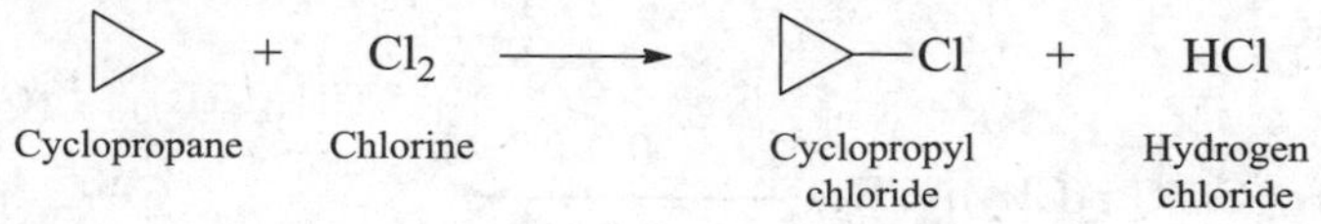

The reaction begins with the initiation step in which a chlorine molecule dissociates to two chlorine atoms.

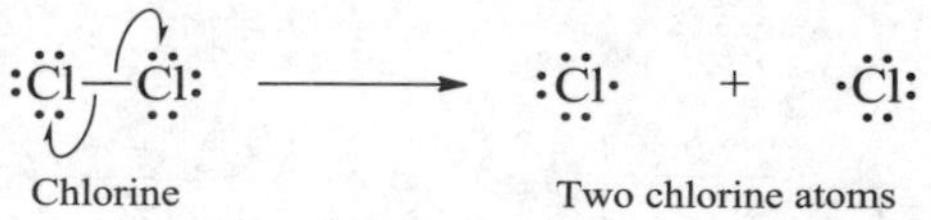

A chlorine atom abstracts a hydrogen atom from cyclopropane in the first propagation step.

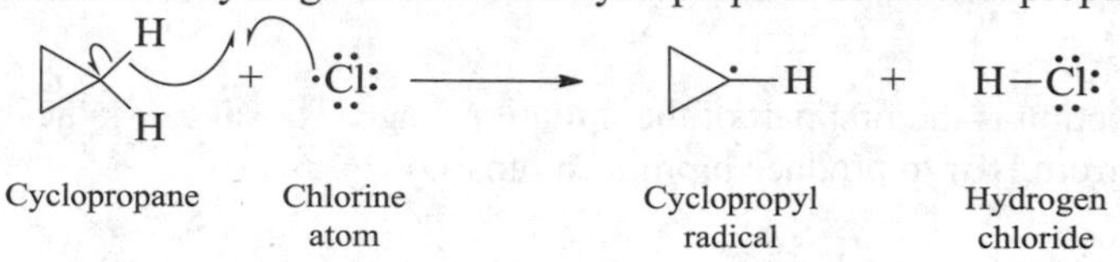

Cyclopropyl radical reacts with Cl_2 in the next propagation step.

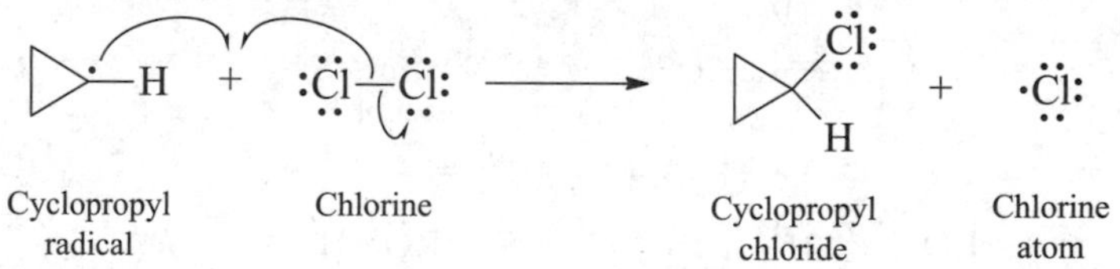

Answers to Interpretive Problems 10

10.34 D;   **10.35** A;   **10.36** A;   **10.37** A;   **10.38** B

SELF-TEST

1.   Write the structural formulas for the reactant, reagent, or product omitted from the following:

     (a)

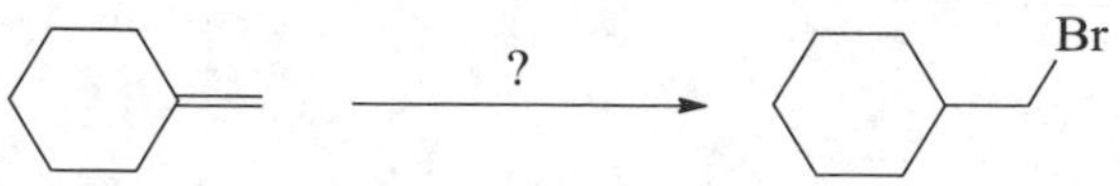

     (b)

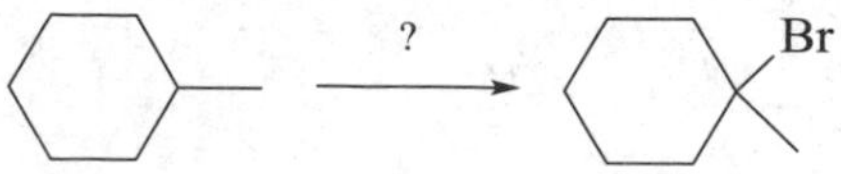

(c)

$$\text{?} \xrightarrow{\text{Na, NH}_3} \diagdown\diagup\diagdown\diagup$$

2. Provide a sequence of reactions to carry out the following conversions. More than one synthetic step is necessary for each. Write the structure of the product of each synthetic step.

    (a)

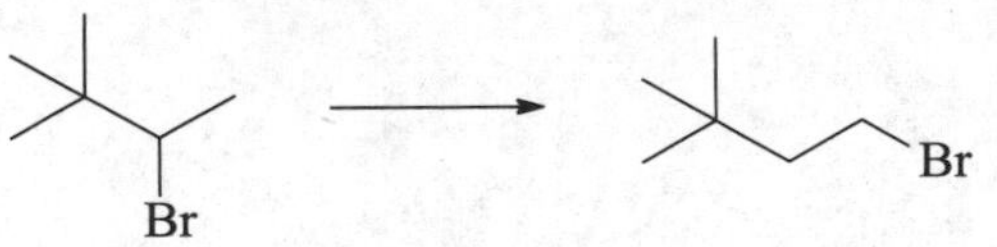

    (b)

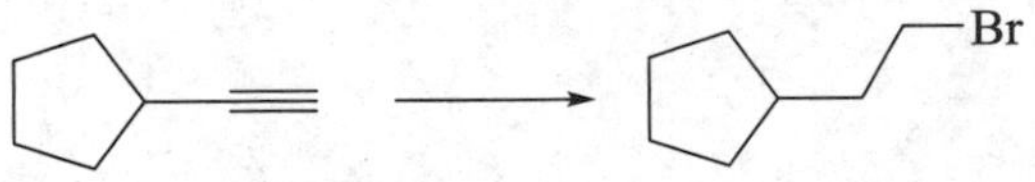

    (c)

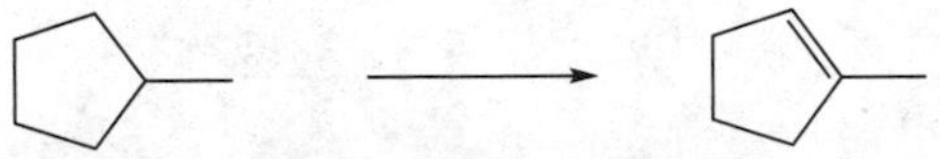

3. Write a balanced chemical equation for the reaction of chlorine with the pentane isomer that gives only one product on monochlorination.

4. Provide a detailed mechanism describing the elementary steps in the reaction of 1-butene with HBr in the presence of peroxides.

5. Give the major organic product formed from the following sequence of reactions.

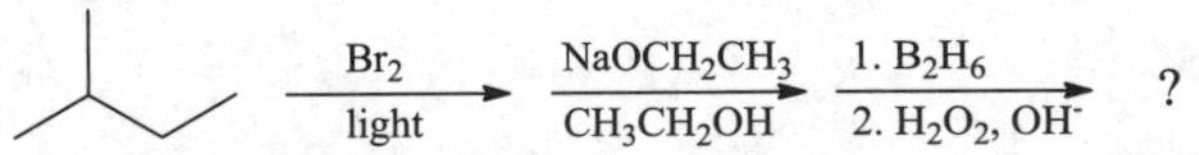

    $$\xrightarrow[\text{light}]{\text{Br}_2} \xrightarrow[\text{CH}_3\text{CH}_2\text{OH}]{\text{NaOCH}_2\text{CH}_3} \xrightarrow[\text{2. H}_2\text{O}_2,\ \text{OH}^-]{\text{1. B}_2\text{H}_6} \text{?}$$

6. Write the propagation steps for the light-initiated reaction of bromine with methylcyclohexane.

7. Using the data in text Table 10.1, calculate the heat of reaction ($\Delta H°$) for the light-initiated reaction of bromine (Br_2) with 2-methylpropane to give 2-bromo-2-methylpropane and hydrogen bromide.

8. (Choose the correct response for each part.) Which species or compound

    (a) Has an odd number of electrons?

    > ethoxide ion   or   ethyl radical

    (b) Undergoes bond cleavage in the initiation step in the reaction by which methane is converted to chloromethane?

    > CH_4 or Cl_2

9. Rank the following radical in terms of increasing potential energy (most stable to least stable).

10. Give the intermediate, including stereochemistry, which immediately precedes the final product in following reaction.

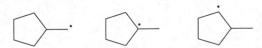

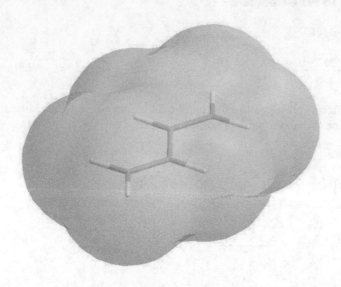

CHAPTER 11
Conjugation in Alkadienes and Allylic Systems

Table of Contents

SOLUTIONS TO TEXT PROBLEMS

In Chapter Problems

11.1   Allylic carbons are those attached to a C=C unit, and allylic hydrogens are directly attached to allylic carbons. The seven hydrogens attached to carbon indicated in the structural drawing of α-terpineol are allylic hydrogens. The OH group is not attached to an allylic carbon, so is not allylic.

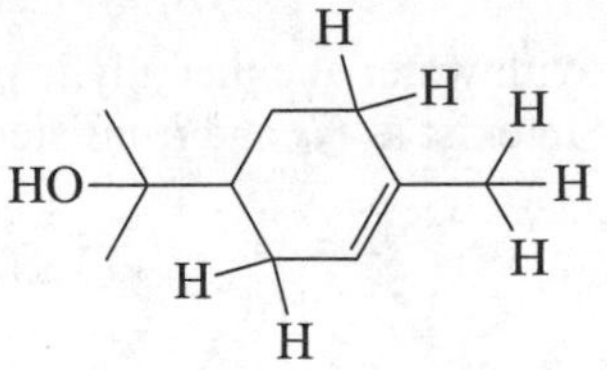

α-Terpineol

11.2 (*b*) Begin with the unshared electron pair and move electrons toward the double bond.

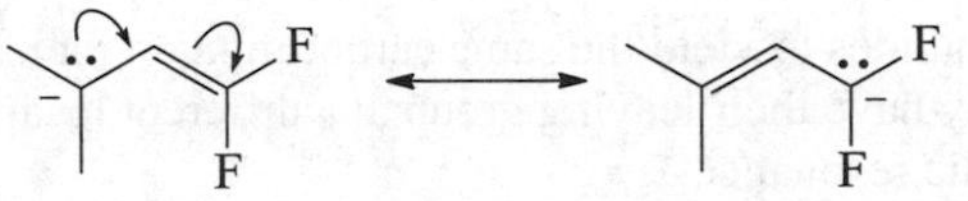

Major contributor

The major contributor is stabilized by the electron-withdrawing power of fluorine. The carbon bearing the two fluorines bears more of the negative charge than the carbon attached to two methyl groups.

(*c*) Resonance contributors of an allylic free radical are generated by moving the two electrons in the double bond separately.

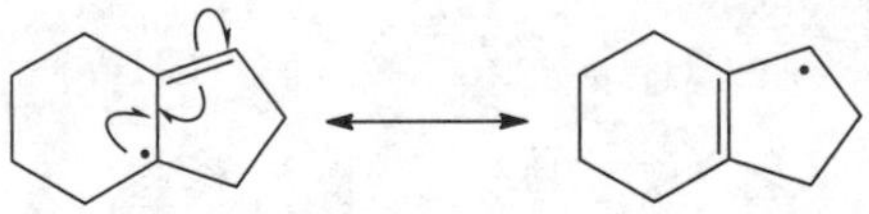

Major contributor

The unpaired electron is on a tertiary carbon in the orginal structure, so it is more stable and bears a greater share of unpaired electron density than the structure on the right where the unpaired electron is on a secondary carbon.

11.3   The problem compares the two compounds in respect to their first-order rate constants for solvolysis. Therefore, assume the mechanism is S_N1 and evaluate the relative stabilities of the respective carbocations.

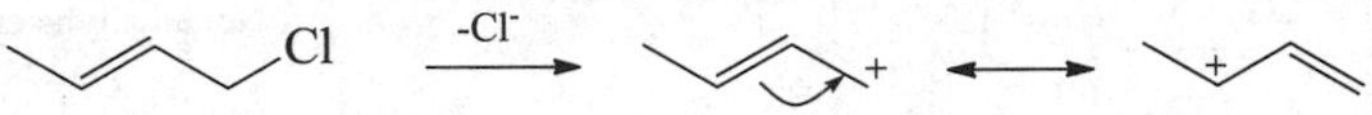

trans-1-Chloro-2-butene

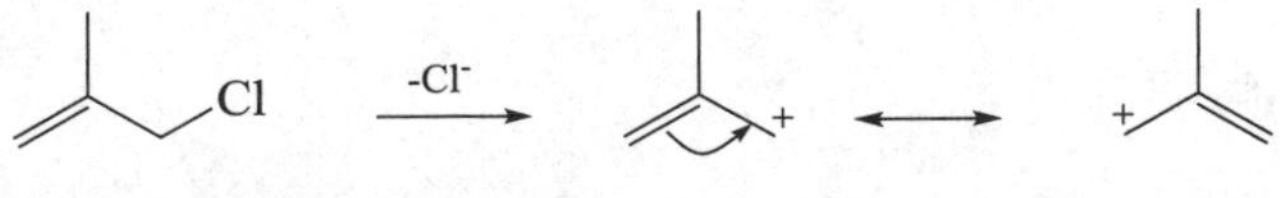

3-Chloro-2-methylpropene

The more reactive allylic halide is the one that forms the more stable carbocation. Of the two contributing structures to the carbocation from ionization of *trans*-1-chloro-2-butene, one is a primary carbocation, the other is secondary. Its secondary carbocation character makes it more stable and formed faster than the corresponding carbocation from 2-chloro-2-methylpropene which is a hybrid of two primary carbocation contributors.

11.4 The carbocation formed in the rate-determining step of an S_N1 reaction of the allylic bromide given can be represented by the two contributing resonance structures:

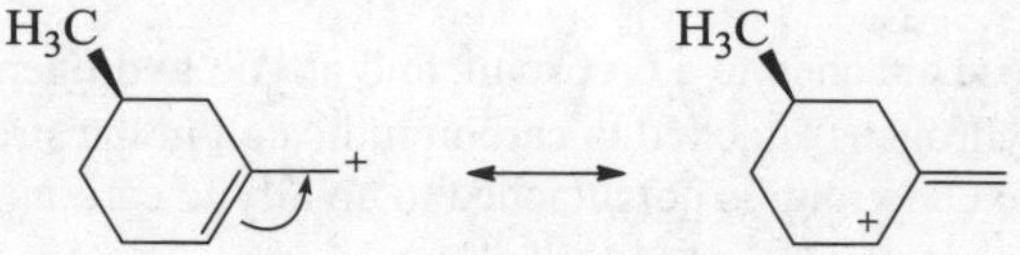

This intermediate can react with water at either allylic position. Two constitutionally isomeric alcohols are possible, one of which can exist as *cis* and *trans* stereoisomers leading to a total of three isomeric products.

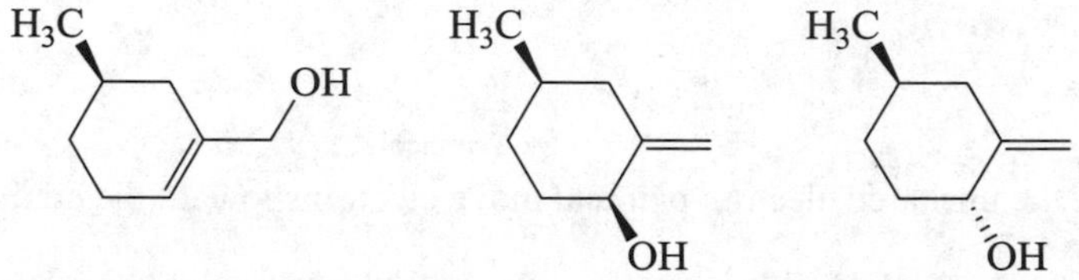

11.5 For two isomeric halides to yield the same carbocation on ionization, they must have the same carbon skeleton. They may have their leaving group at a different location, but the carbocations must become equivalent by allylic resonance.

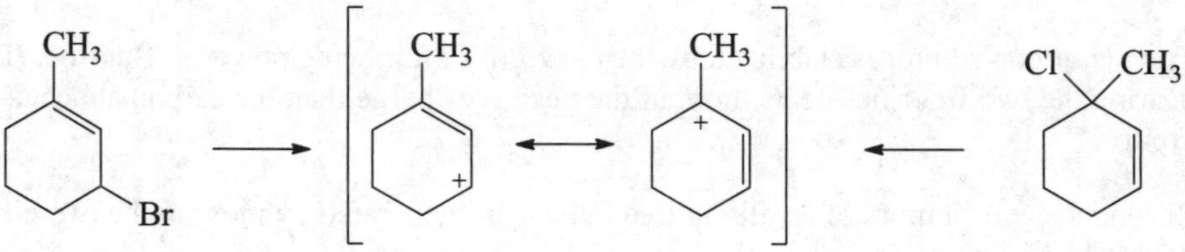

3-Bromo-1-
methylcyclohexene

3-Chloro-3-
methylcyclohexene

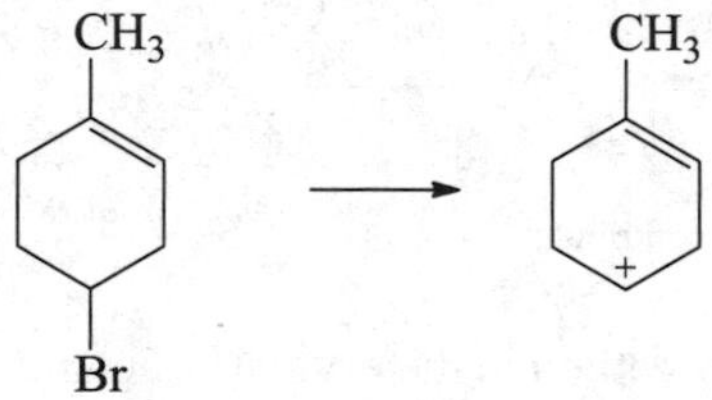

Not an allylic carbocation

4-Bromo-1-
methylcyclohexene

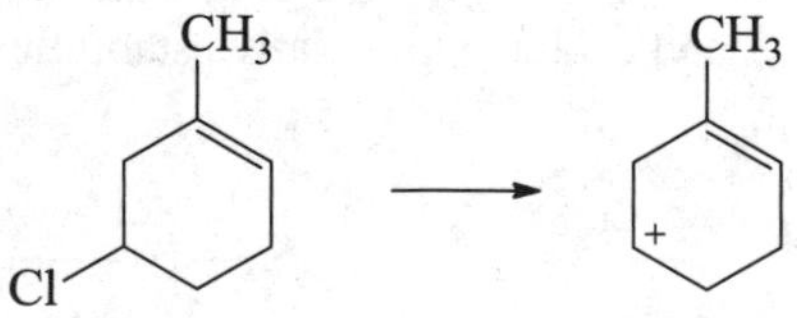

Not an allylic carbocation

5-Chloro-1-
methylcyclohexene

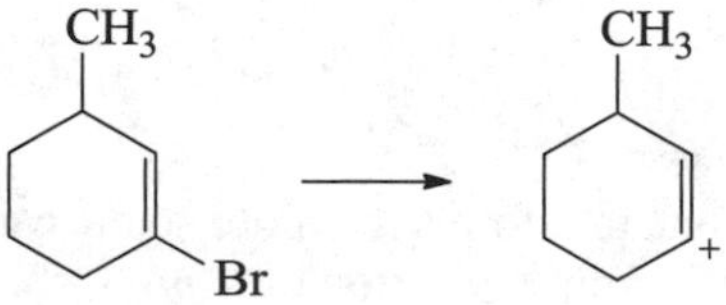

Not an allylic carbocation

1-Bromo-3-
methylcyclohexene

11.6 With strong bases such as sodium ethoxide, elimination by the E2 mechanism competes with S_N2 substitution. Of the two reactants, one is a primary allylic halide and the other secondary.

trans-1-Chloro-2-butene
(primary)

3-Chloro-1-butene
(secondary)

The S_N2/E2 rate ratio is more favorable for the primary halide than the secondary. The lower yield of substitution product for 3-chloro-1-butene results from an increase in the proportion of elimination.

11.7 First, classify the C—H bonds according to type.

Hydrogens of
C-1 and C-6 are primary

Hydrogens of
C-2 and C-5 are secondary and allylic
(weakest C-H bonds)

Hydrogens of
C-3 and C-4 are vinylic
(strongest C-H bonds)

C—H bond strengths increase with increasing *s* character of carbon (Section 9.4). Therefore, the vinylic hydrogens at C-3 and C-4, which are attached to sp^2-hybridized carbon, have higher bond dissociation enthalpies than those attached to sp^3-hybridized carbons (C-1, C-2, C-5, and C-6). Allylic C—H bonds are weaker than non-allylic ones; therefore, the C(2)—H and C(5)—H bonds are the weakest.

11.8 The statement of the problem specifies that in allylic brominations using *N*-bromosuccinimide the active reagent is Br_2. Thus, the equation for the overall reaction is

Cyclohexene   Bromine   3-Bromocyclohexene   Hydrogen
bromine

The propagation steps are analogous to those of other free-radical brominations. An allylic hydrogen is removed by a bromine atom in the first step.

Cyclohexene   Bromine   3-Bromocyclohexene   Hydrogen
bromide

The allylic radical formed in the first step abstracts a bromine atom from Br_2 in the second propagation step.

2-Cyclohexenyl
radical

Bromine

3-Bromocyclohexene

Bromine
atom

11.9   Write both resonance forms of the allylic radicals produced by hydrogen atom abstraction from the alkene.

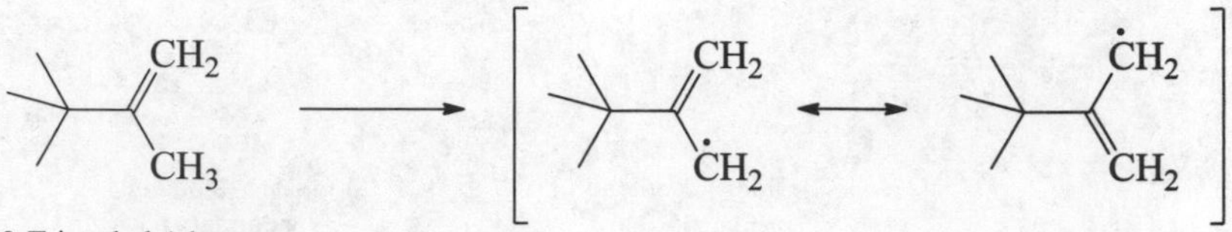

2,3,3-Trimethyl-1-butene

Both resonance forms are equivalent, and so 2,3,3-trimethyl-1-butene gives a single bromide on treatment with *N*-bromosuccinimide (NBS).

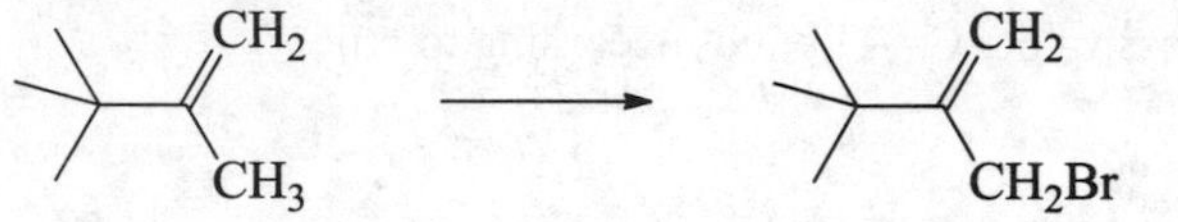

2,3,3-Trimethyl-1-butene   2-(Bromomethyl)-3,3-dimethyl-1-butene

11.10   A reasonable mechanism begins with an acid-base reaction in which ethoxide ion abstracts an allylic proton from the reactant.

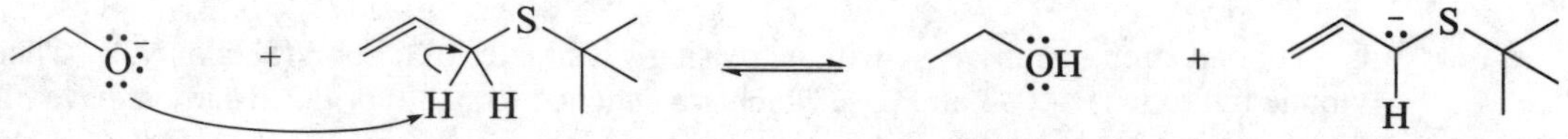

Ethoxide ion   Allyl *tert*-butyl sulfide   Ethanol   Conjugate base of allyl *tert*-butyl sulfide

The negative charge in the conjugate base of allyl *tert*-butyl sulfide is shared by the two allylic carbons.

Conjugate base of allyl *tert*-butyl sulfide resonance structures

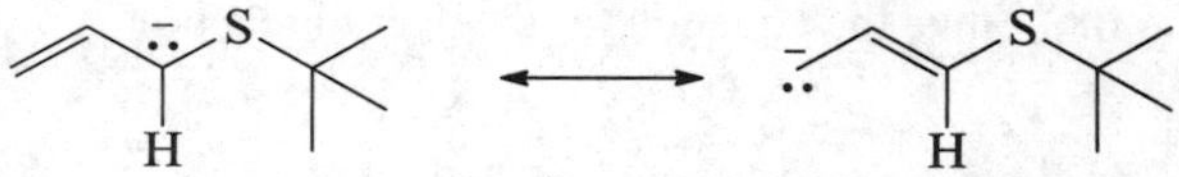

Proton transfer to the primary carbon of the conjugate base gives *tert*-butyl propenyl sulfide.

Ethanol   Conjugate base of allyl *tert*-butyl sulfide   Ethoxide ion   *tert*-Butyl propenyl sulfide

The problem states that *tert*-butyl propenyl sulfide was isolated in 66% yield, so it must be a weaker acid, or have a larger pK_a than allyl *tert*-butyl sulfide.

11.11   (b) The C-1 and C-3 double bonds of cembrene are conjugated with each other.

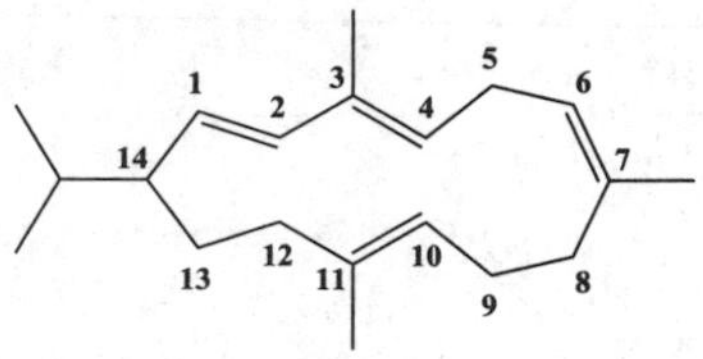

Cembrene

The double bonds at C-6 and C-10 are isolated from each other and from the conjugated diene system.

(*c*) The sex attractant of the dried-bean beetle has a cumulated diene system involving C-4, C-5, and C-6. This allenic system is conjugated with the C-2 double bond.

$$CH_3(CH_2)_6CH_2\overset{6}{C}H=\overset{5}{C}=\overset{4}{C}H \quad H$$

11.12 The more stable the isomer, the lower its heat of combustion. The conjugated diene is the most stable and has the lowest heat of combustion. The cumulated diene is the least stable and has the highest heat of combustion.

(*E*)-1,3-Pentadiene	1,4-Pentadiene	1,2-Pentadiene
Most stable		Least stable
3186 kJ/mol (761 kcal/mol)	3217 kJ/mol (769 kcal/mol)	3251 kJ/mol (776 kcal/mol)

11.13 Compare the mirror-image forms of each compound for superimposability. For 2-methyl-2,3-pentadiene,

and

Reference structure
for 2-methyl-2,3-pentadiene

Mirror image

Rotation of the mirror image 180° around an axis passing through the three carbons of the C=C=C unit demonstrates that the reference structure and its mirror image are superimposable.

Rotate 180°

Mirror image

Reoriented mirror image

2-Methyl-2,3-pentadiene is an achiral allene.

Comparison of the mirror-image forms of 2-chloro-2,3-pentadiene reveals that they are not superimposable. 2-Chloro-2,3-pentadiene is a chiral allene.

Cl ''CH₃ and Cl ''CH₃ | Rotate 180° H₃C Cl
C C ↻ C
C C C
H₃C H H CH₃ H₃C H

Reference structure   Mirror image   Reoriented mirror image
for 2-chloro-2,3-pentadiene

11.14 Both starting materials undergo β elimination to give a conjugated diene system. The dienes with isolated double bonds are not formed.

H₃C X

X = OH   3-Methyl-5-hexen-3-ol
X = Br   4-Bromo-4-methyl-1-hexene

X

CH₃   CH₃   CH₂
+

4-Methyl-1,3-hexadiene   4-Methyl-1,4-hexadiene   2-Ethyl-1,4-pentadiene
(mixture of *E* and *Z* isomers;   (mixture of *E* and *Z* isomers;   (not observed)
major product)   not observed)

11.15 Using HCl as a representative acid, we can write the protonation of C-2 of 1,3-cyclopentadiene as

H   H   H—Cl: H   H
-Cl⁻   H
H   H
H   H   H   H
+
H   H   H   H

1,3-Cyclopentadiene   3-Cyclopentenyl cation

The carbocation that is formed is not allylic; therefore, it cannot be stabilized by allylic resonance.

11.16 The numbers 1,2 and 1,4 refer to the carbons of the conjugated diene unit, not to the entire carbon chain. Therefore,

H   ←—— 1,2-addition
Cl
H
1,4-addition ——→
Cl

As can be seen by the structural formulas, the product of 1,2-addition is 4-chloro-2-hexene whereas the product of 1,4-addition is 2-chloro-3-hexene.

11.17 The two double bonds of 2-methyl-1,3-butadiene are not equivalent, and so two different products of 1,2-addition are possible, along with one 1,4-addition product.

2-Methyl-1,3-butadiene   3,4-Dibromo-3-methyl-1-butene (1,2-addition)   3,4-Dibromo-2-methyl-1-butene (1,2-addition)   1,4-Dibromo-2-methyl-2-butene (1,4-addition)

11.18 (*b*) Cyanobenzoquinone contains two double bonds that could potentially react in a Diels–Alder reaction because they both have electron-attracting carbonyl groups. However, the double bond that is substituted with the cyano group is even more dienophilic because the cyano group is also electron attracting.

11.19 2,3-Di-*tert*-butyl-1,3-butadiene is extremely unreactive because the *s*-cis conformation that is required for the Diels–Alder reaction is too high in energy. The two *tert*-butyl groups of 2,3-di-*tert*-butyl-1,3-butadiene will encounter a high level of steric hindrance in the *s*-cis conformation.

s-Trans (low energy)   rotate C-C single bond   *s*-Cis (high energy)   steric hindrance between *tert*-butyl substituents

11.20 A Diels-Alder reaction occurs to give a bicyclic product. In one the CO_2CH_3 group is *endo*, in the other it is *exo*.

1,3-Cyclopentadiene   Methyl acrylate   Methyl *endo*-bicyclo[2.2.1]hept-2-ene-5-carboxylate   Methyl *exo*-bicyclo[2.2.1]hept-2-ene-5-carboxylate

In keeping with the Alder *endo* rule, the *endo* stereoisomer is the major product.

11.21 First identify the diene and dienophile in this molecule.

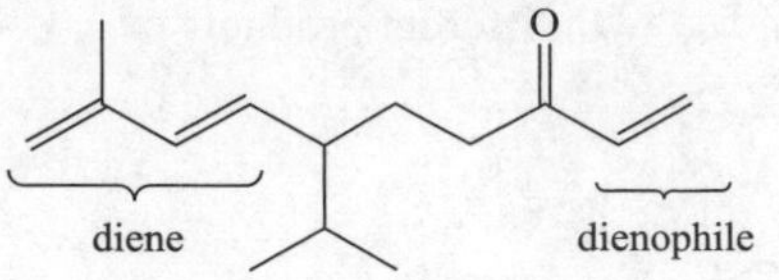

Next rotate the diene to an *s*-cis conformation.

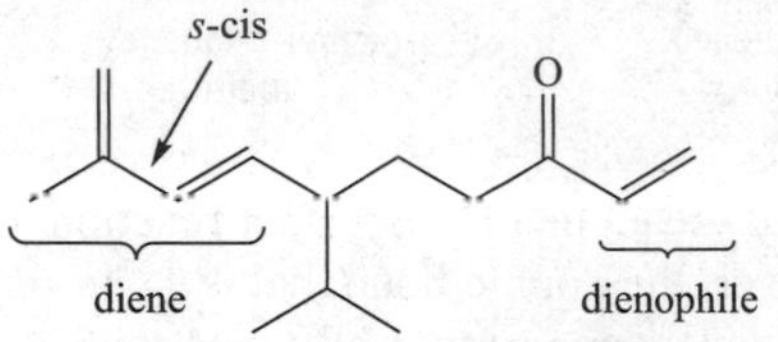

Rearrange the rest of the molecule so that the dienophile is placed in the proper orientation for the Diels-Alder reaction. The curved arrows indicate where bonds are made and broken.

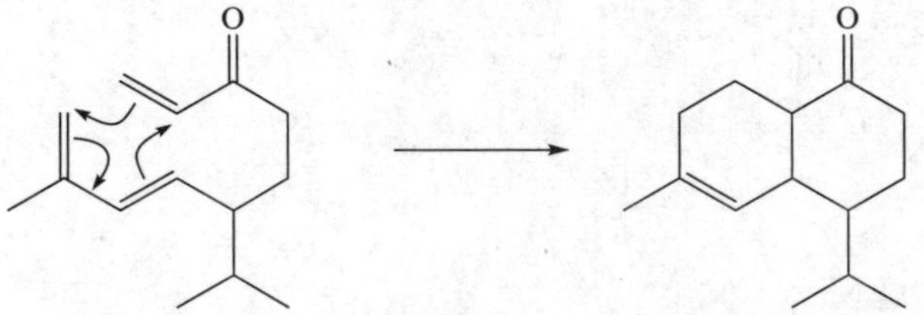

11.22 Use curved arrows to dissociate the Diels-Alder adduct and reveal the diene and dienophile that combine to produce it. Begin at the double bond.

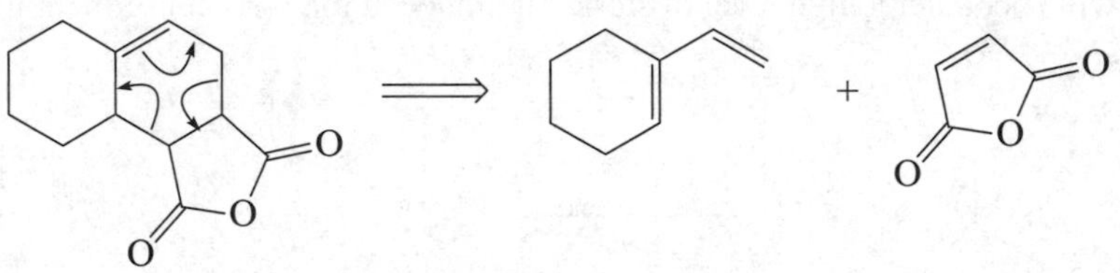

11.23 The retrosynthetic analysis for the preparation of 5,6-dicyanobicyclo[2.2.2]oct-2-ene from cyclohexanol as shown in the text is:

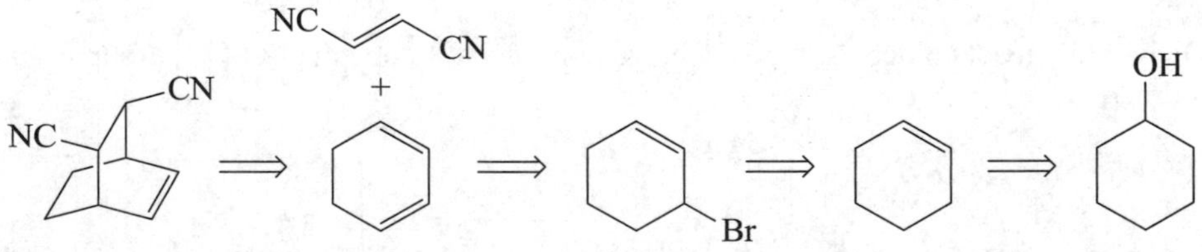

Cyclohexanol is first converted to cyclohexene by dehydration, then to 3-bromocyclohexene by allylic bromination with *N*-bromosuccinimide. This is followed by a dehydrobromination step to give the 1,3-cyclohexadiene which undergoes a Diels-Alder reaction with *trans*-1,2-dicyanoethylene to give the desired product.

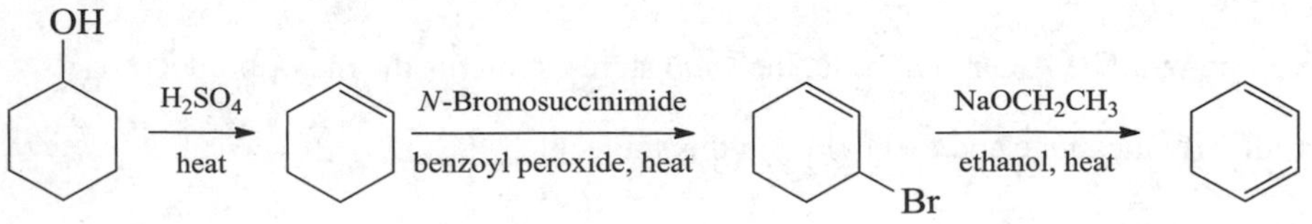

then:

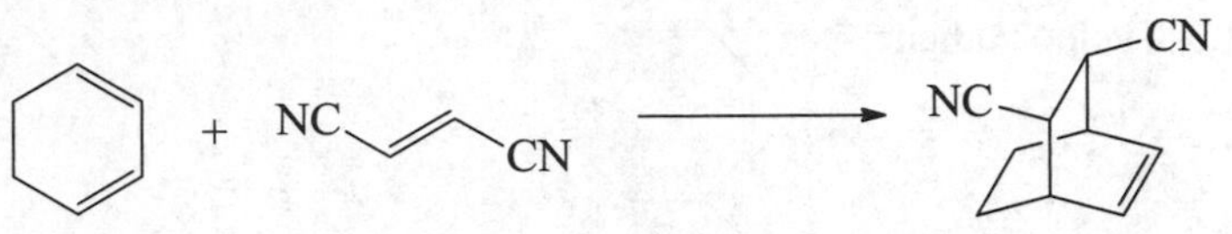

11.24 The problem states that a Cope rearrangement of a divinylcyclopropane was used in the synthesis. *cis*-Divinylcyclopropanes give 1,4-cycloheptadienes.

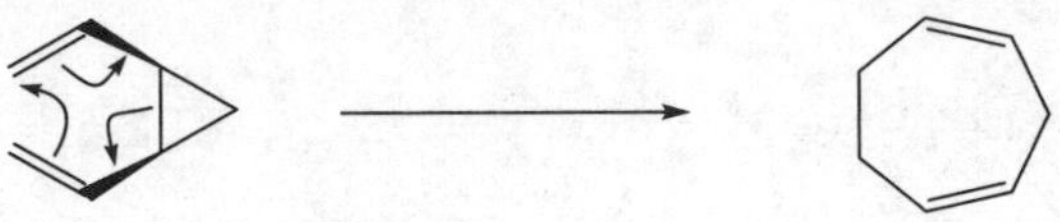

cis-Divinylcyclopropane   1,4-Cycloheptadiene

Using retrosynthetic analysis, the position of the substituent can be placed on one of the vinyl groups to give the product.

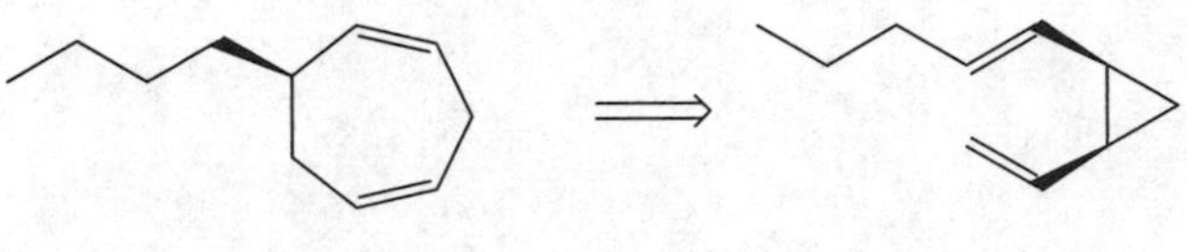

(*R*)-(-)-Dictyopterene C

11.25 Claisen rearrangements can be identified by numbering 1-3, starting from each carbon-carbon double bond. The C3 carbon and oxygen bond identify where the bond is broken and the C1 and C1' carbons identify where a bond will be formed. Double bonds are formed between C2-C3 and C2'-O3'.

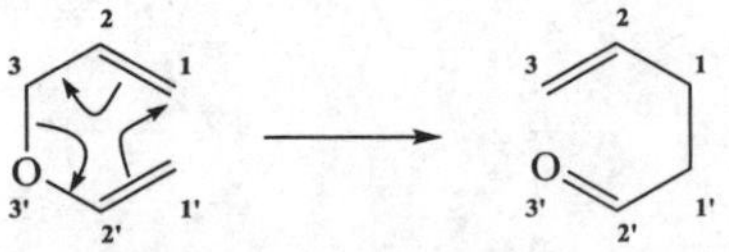

Using this same method the following product is formed:

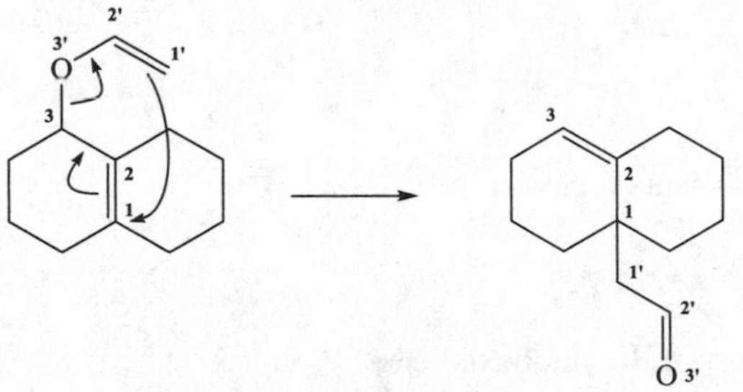

End of Chapter Problems

Structure and Nomenclature

11.26 Dienes and trienes are given IUPAC names by replacing the -ane ending of the alkane with -*adiene* or -*atriene* and locating the positions of the double bonds by number. The stereoisomers are identified as *E* or *Z* according to the rules established in Chapter 7.

(*a*)   3,4-Octadiene:

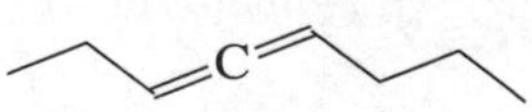

(*b*)   (*E,E*)-3,5-Octadiene:

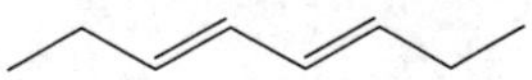

(*c*)   (*Z,Z*)-1,3-Cyclooctadiene:

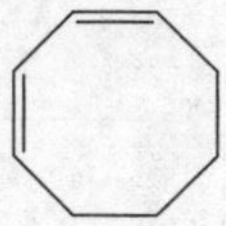

(*d*)   (*Z,Z*)-1,4-Cyclooctadiene:

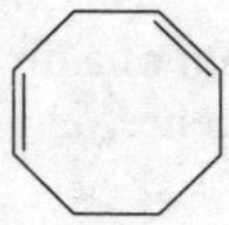

(*e*)   (*E,E*)-1,5-Cyclooctadiene:

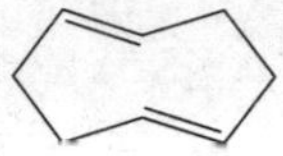

(*f*)   (2*E*,4*Z*,6*E*)-2,4,6-Octatriene: H₃C ... CH₃

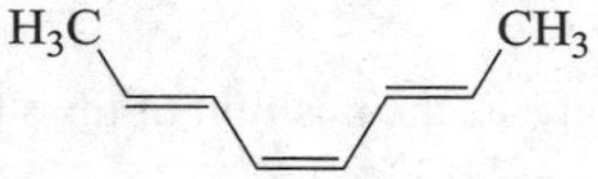

(*g*)   5-Allyl-1,3-cyclopentadiene: 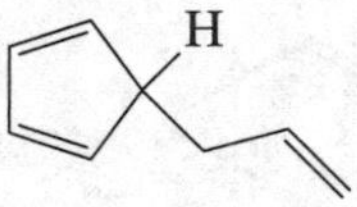   H

(*h*)   *trans*-1,2-Divinylcyclopropane: H ... H

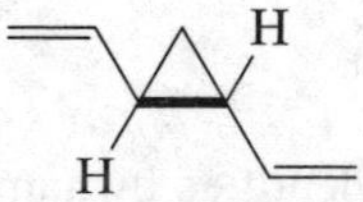

(*i*)   2,4-Dimethyl-1,3-pentadiene:

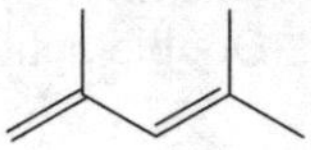

11.27   (*a*)    1,8-Nonadiene

(*b*) 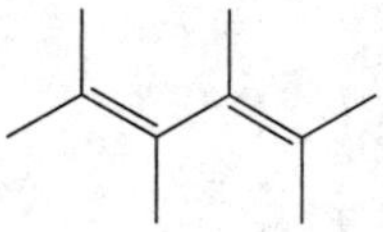   2,3,4,5-Tetramethyl-2,4-hexadiene

(*c*) 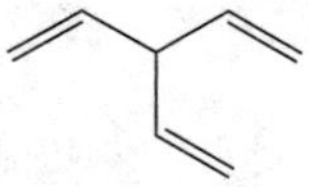   3-Vinyl-1,4-pentadiene

(*d*) 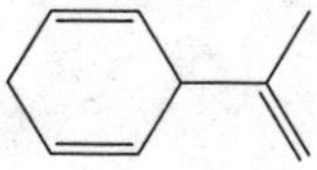   3-Isopropenyl-1,4-cyclohexadiene

(*e*) 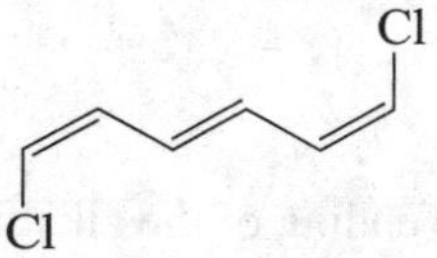   Cl ... Cl   (1*Z*,3*E*,5*Z*)-1,6-Dichloro-1,3,5-hexatriene

(*f*)   H₂C=C=CH−CH=CH−CH₃   1,2,4-Hexatriene (the C-4 double bond could be either *E* or *Z*)

(g)

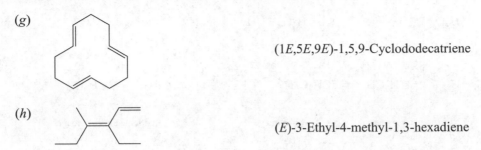

(1*E*,5*E*,9*E*)-1,5,9-Cyclododecatriene

(h)

(*E*)-3-Ethyl-4-methyl-1,3-hexadiene

11.28 The important piece of information that allows us to complete the structure properly is that the ant repellent is an *allenic* substance. The allenic unit cannot be incorporated into the ring because the three carbons must be collinear. The only possible constitution is therefore

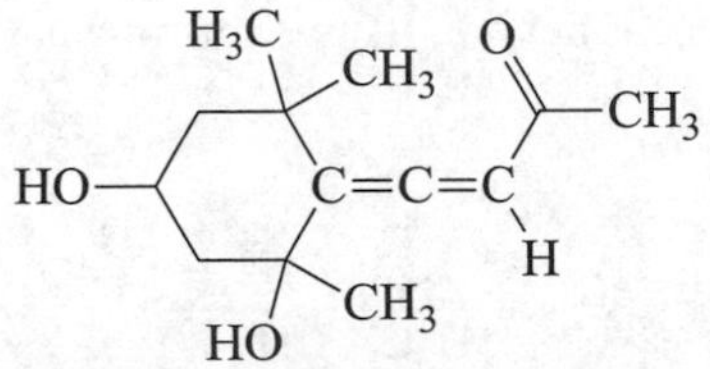

11.29 Compare the mirror-image forms of each compound for superimposability.

(a)

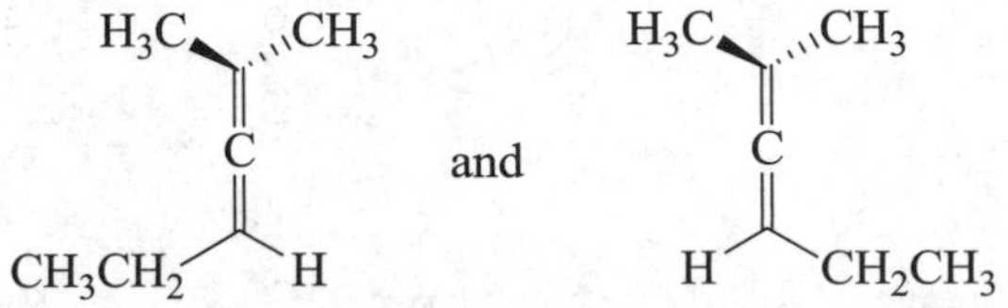

and

Reference structure
for 2-Methyl-2,3-hexadiene

Mirror image

Rotation of the mirror image 180° around an axis passing through the three carbons of the C=C=C unit demonstrates that the reference structure and its mirror image are superimposable.

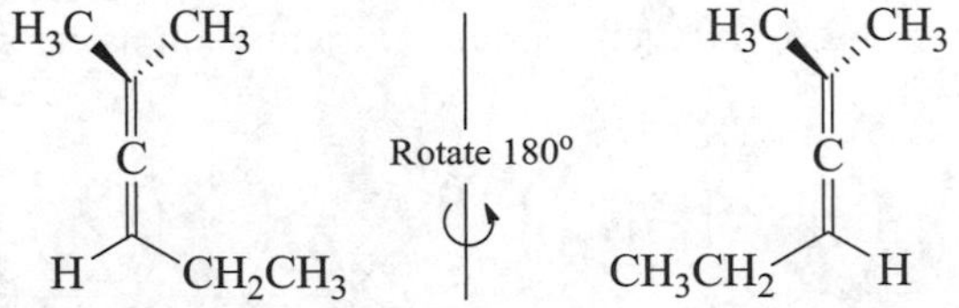

Rotate 180°

Mirror image

Reoriented mirror image

2-Methyl-2,3-hexadiene is an achiral allene.

(b) The two mirror-image forms of 4-methyl-2,3-hexadiene are as shown:

H₃C⸜⸜H
and
H₃C⸜⸜H | Rotate 180°
C   C   ↻
H₃C   CH₂CH₃   CH₃CH₂   CH₃ |

H   CH₃
C
H₃C   CH₂CH₃

Reference structure   Mirror image   Reoriented mirror image
for 4-methyl-2,3-hexadiene

The two structures cannot be superimposed. 4-Methyl-2,3-hexadiene is chiral. Rotation of either representation 180° around an axis that passes through the three carbons of the C=C=C unit leads to superposition of the groups at the "bottom" carbon but not at the "top."

(c) 2,4-Dimethyl-2,3-pentadiene is achiral. Its two mirror-image forms are superimposable.

H₃C   CH₃   H₃C   CH₃   H₃C   CH₃
C   and   C   C
H₃C   CH₃   H₃C   CH₃   H₃C   CH₃

Reference structure   Mirror image   Mirror planes in
for 2,4-dimethyl-2,3-pentadiene   2,4-dimethyl-2,3-pentadiene

The molecule has two planes of symmetry defined by the three carbons of each CH_3CCH_3 unit.

11.30 (a) Carbons 2 and 3 of 1,2,3-butatriene are *sp*-hybridized, and the bonding is an extended version of that seen in allene. Allene is nonplanar; its two CH_2 units must be in perpendicular planes in order to maximize overlap with the two mutually perpendicular *p* orbitals at C-2. With one more *sp*-hybridized carbon, 1,2,3-butatriene has an "extra turn" in its carbon chain, making the molecule planar.

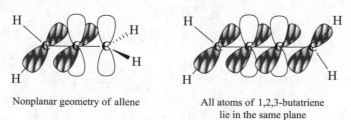

Nonplanar geometry of allene   All atoms of 1,2,3-butatriene
                                       lie in the same plane

(b) The planar geometry of the cumulated triene system leads to the situation in which cis and trans stereoisomers are possible for 2,3,4-hexatriene ($CH_3CH=C=C=CHCH_3$). Cis–trans stereoisomers are diastereomers.

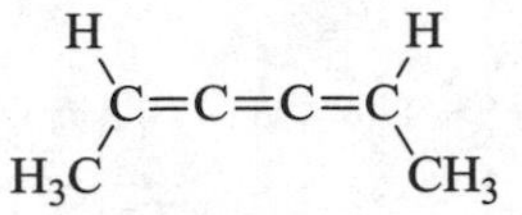

cis-2,3,4-Hexatriene 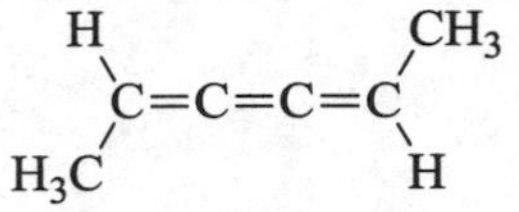   trans-2,3,4-Hexatriene

11.31 The allylic hydrogens in this hydrocarbon are the most acidic. Therefore, remove a proton from the methyl group to form the conjugate base.

Two additional contributors can be generated by allylic resonance of the original structure shown for the conjugate base.

Reactions

11.32 (a) Because the product is 2,3-dimethylbutane, we know that the carbon skeleton of the starting material must be

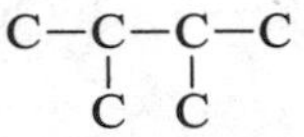

$$C-C-C-C$$
$$||$$
$$CC$$

Because 2,3-dimethylbutane is C_6H_{14} and the starting material is C_6H_{10}, *two* molecules of H_2 must have been taken up and the starting material must have two double bonds. The starting material can only be 2,3-dimethyl-1,3-butadiene.

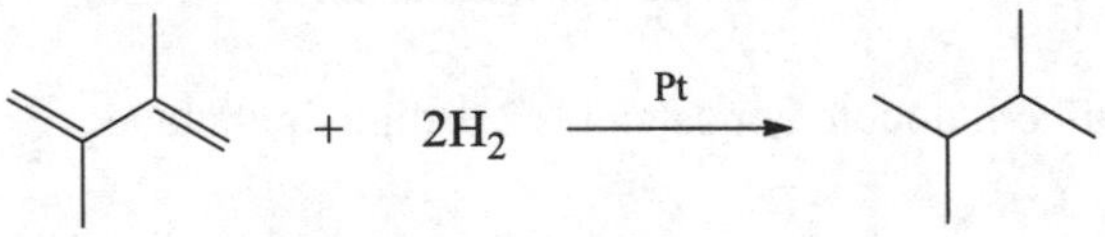

(b) Write the carbon skeleton corresponding to 2,2,6,6-tetramethylheptane.

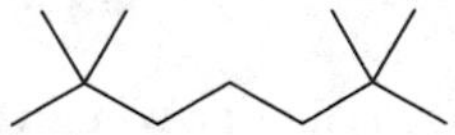

Compounds of molecular formula $C_{11}H_{20}$ have two double bonds or one triple bond. The only compounds with the proper carbon skeleton are the alkyne and the allene shown.

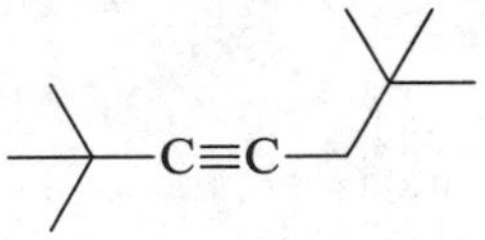

2,2,6,6-Tetramethyl-3-heptyne 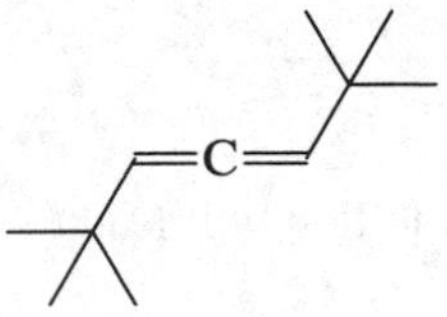   2,2,6,6-Tetramethyl-3,4-heptadiene

11.33 The dienes that give 2,4-dimethylpentane on catalytic hydrogenation must have the same carbon skeleton as that alkane.

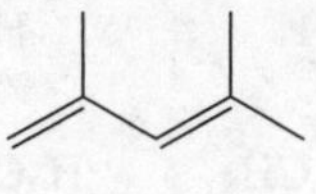

(a)
2,4-Dimethyl-
1,3-pentadiene
(conjugated diene)

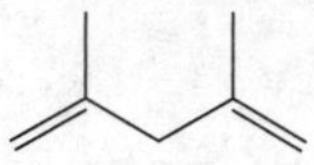

(b)
2,4-Dimethyl-
1,4-pentadiene
(isolated diene)

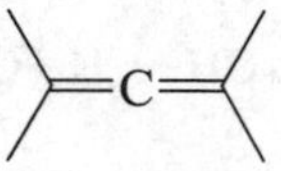

(c)
2,4-Dimethyl-
2,3-pentadiene
(cumulated diene)

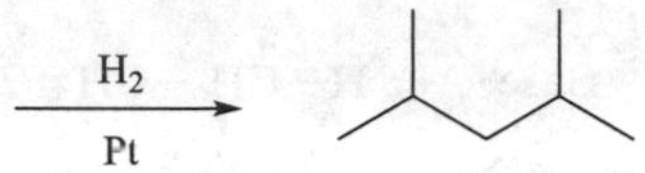

$\xrightarrow[\text{Pt}]{\text{H}_2}$

2,3-Dimethylpentane

11.34 The starting material in all cases is 2,3-dimethyl-1,3-butadiene.

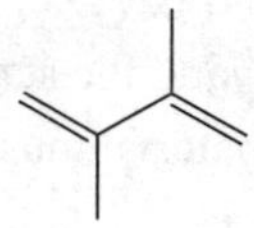

2,3-Dimethyl-1,3-butadiene

(a) Hydrogenation of both double bonds yields 2,3-dimethylbutane.

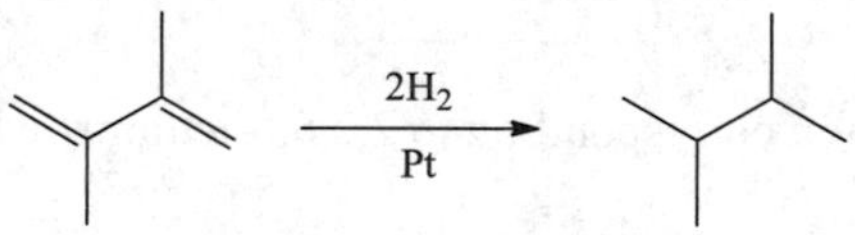

$\xrightarrow[\text{Pt}]{2\text{H}_2}$

(b) 1,2-Addition of 1 mole of hydrogen chloride will give the product of Markovnikov addition to one of the double bonds, 3-chloro-2,3-dimethyl-1-butene.

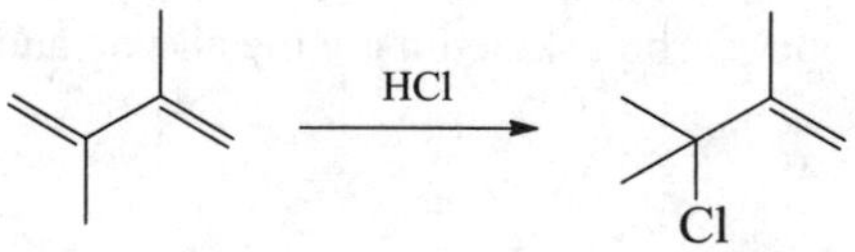

$\xrightarrow{\text{HCl}}$

Cl

(c) 1,4-Addition will lead to double-bond migration and produce 1-chloro-2,3-dimethyl-2-butene.

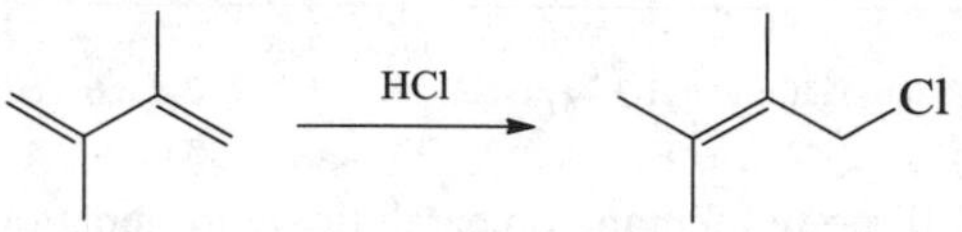

$\xrightarrow{\text{HCl}}$

Cl

(d) The 1,2-addition product is 3,4-dibromo-2,3-dimethyl-1-butene.

(e) The 1,4-addition product will be 1,4-dibromo-2,3-dimethyl-2-butene.

(f) Bromination of both double bonds will lead to 1,2,3,4-tetrabromo-2,3-dimethylbutane irrespective of whether the first addition step occurs by 1,2- or 1,4-addition.

(g) The reaction of a diene with maleic anhydride is a Diels–Alder reaction.

11.35 The starting material in all cases is 1,3-cyclohexadiene.

(a) Cyclohexane will be the product of hydrogenation of 1,3-cyclohexadiene:

(b) 1,2-Addition will occur according to Markovnikov's rule to give 3-chlorocyclohexene.

(c) The product of 1,4-addition is 3-chlorocyclohexene also. 1,2-Addition and 1,4-addition of hydrogen chloride to 1,3-cyclohexadiene give the same product.

(*d*) Bromine can add to one of the double bonds to give 3,4-dibromocyclohexene:

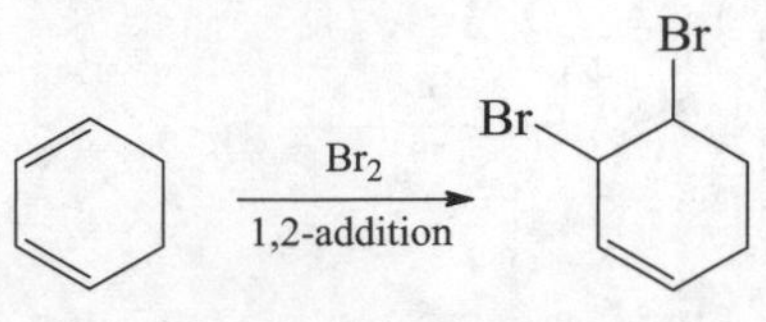

(*e*) 1,4-Addition of bromine will give 3,6-dibromocyclohexene:

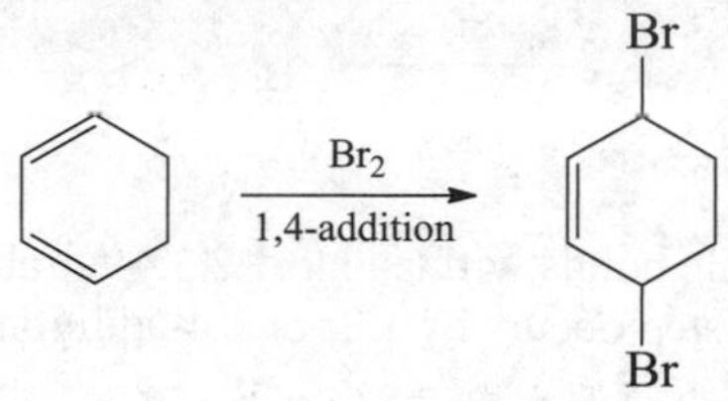

(*f*) Addition of 2 moles of bromine will yield 1,2,3,4-tetrabromocyclohexane.

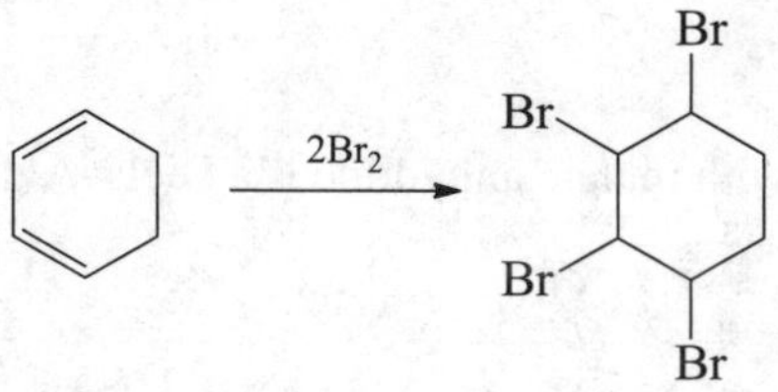

(*g*) The constitution of the Diels–Alder adduct of 1,3-cyclohexadiene and maleic anhydride will have a bicyclo[2.2.2]octyl carbon skeleton.

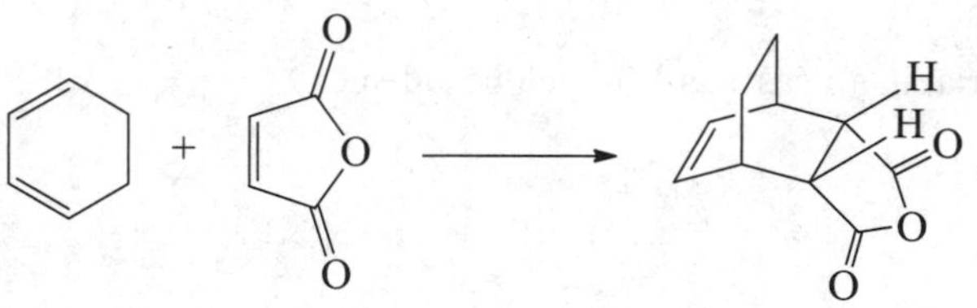

11.36 Think mechanistically. The reaction of alkenes with *N*-bromosuccinimide is a free-radical process proceeding by way of an allylic radical.

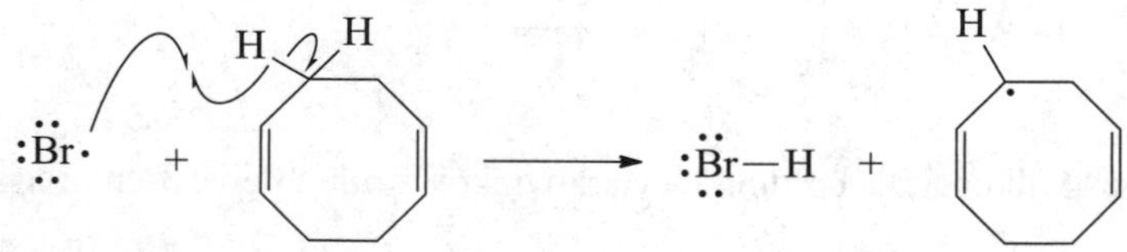

This allylic radical is delocalized with the odd electron shared by two carbons.

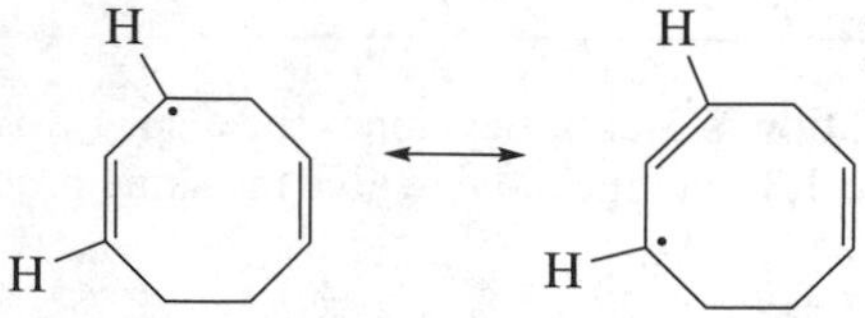

Resonance forms of free-radical intermediate

The two isomeric bromides are formed from the same allylic radical.

3-Bromo-1,5-cyclooctadiene   6-Bromo-1,4-cyclooctadiene

11.37 (*a*) The first alkene has carbonyl groups attached to the carbon–carbon double bond. Carbonyl groups are much stronger electron-attracting groups than the –CH$_2$OH group. Stronger electron-attracting groups make alkenes more reactive as dienophiles in the Diels–Alder reaction.

(*b*) The second dienophile has two carbonyl groups attached to the double bond, while the first has just one. The second dienophile (maleimide) is more reactive.

One carbonyl group
attached to C=C,
less reactive

Two carbonyl groups
attached to C=C,
more reactive

(*c*) The second dienophile is more reactive because the sulfone group (SO$_2$CH$_3$) is a stronger electron-attracting group than the methyl sulfide group (–SCH$_3$) in the first dienophile.

11.38 In the Diels–Alder reaction, the diene must attain the *s*-cis conformation. The first diene cannot adopt the *s*-cis conformation because the single bond between the two double bonds is "locked" by the ring in an *s*-trans conformation. For the second compound, the single bond is outside the ring structure, so it can rotate to the *s*-cis conformation.

s-trans   *s*-cis

11.39 The two Diels–Alder adducts formed in the reaction of 1,3-pentadiene with acrolein arise by the two alignments shown:

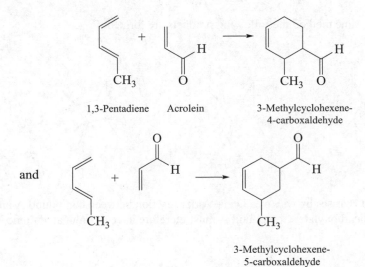

1,3-Pentadiene   Acrolein   3-Methylcyclohexene-
4-carboxaldehyde

and

3-Methylcyclohexene-
5-carboxaldehyde

11.40 Diels-Alder cycloadditions are stereospecific; the stereochemical relationship between substituents in the dienophile is retained in the product.

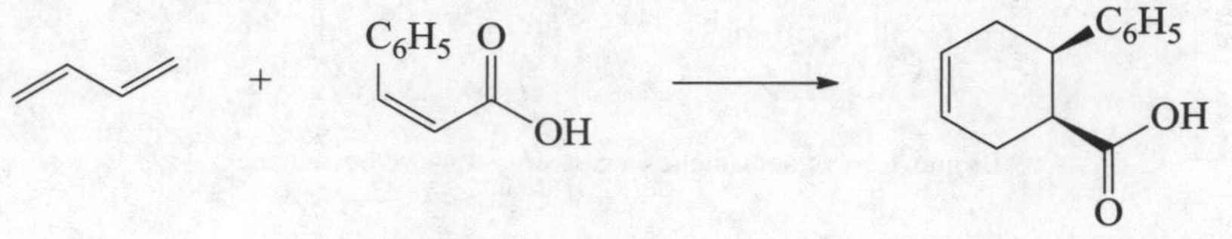

| 1,3-Butadiene | *cis*-Cinnamic acid | *cis*-2-Phenyl-4-cyclohexene-carboxylic acid |

| 1,3-Butadiene | *trans*-Cinnamic acid | *trans*-2-Phenyl-4-cyclohexene-carboxylic acid |

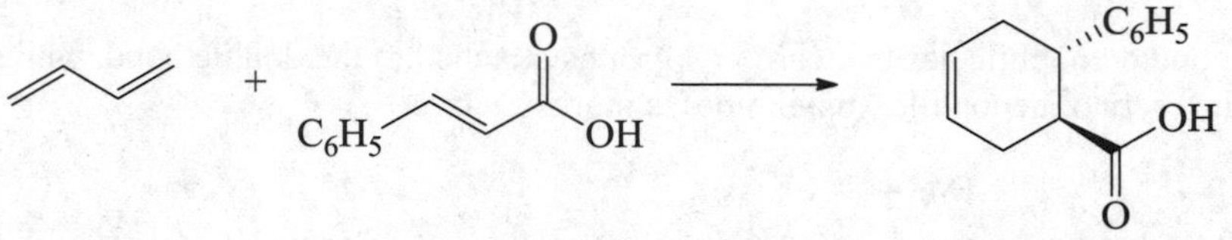

11.41 Arrange 1,3-butadiene in the *s*-cis conformation and use curved arrows to show the cycloaddition.

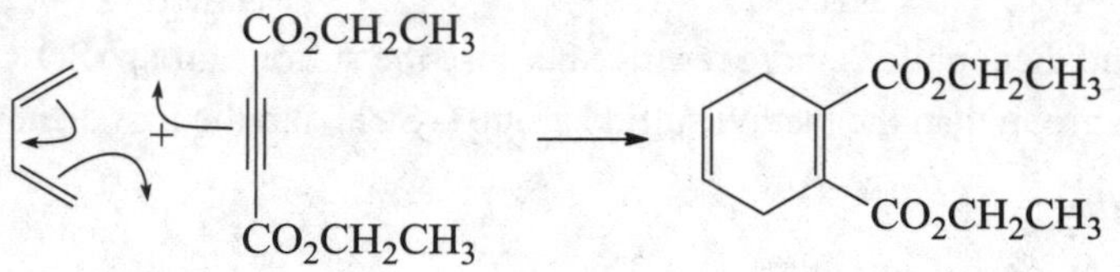

| 1,3-Butadiene | Diethyl acetylenedicarboxylate | Diethyl 1,4-cyclohexadiene-1,2-dicarboxylate |

11.42 Claisen rearrangements can be identified by numbering 1-3, starting from each carbon-carbon double bond. The C3 carbon and oxygen bond identify where the bond is broken and the C1 and C1' carbons identify where a bond will be formed. Double bonds are formed between C2-C3 and C2'-O3'.

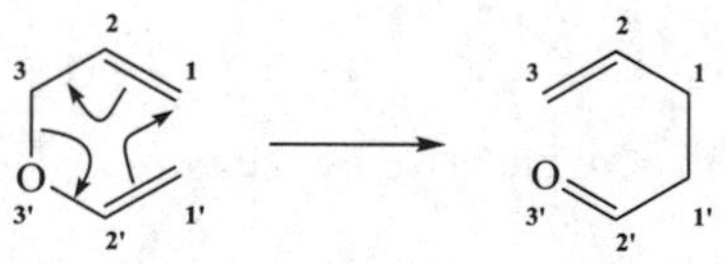

Using this same method the following products are formed:

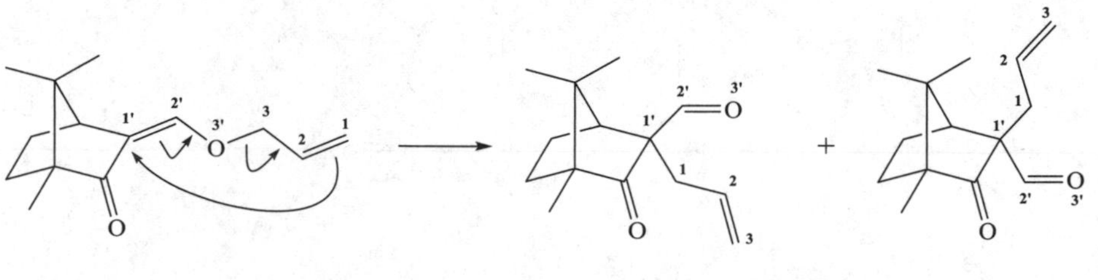

11.43 Compound B arises by way of a Diels–Alder reaction between compound A and dimethyl acetylenedicarboxylate. Compound A must therefore have a conjugated diene system.

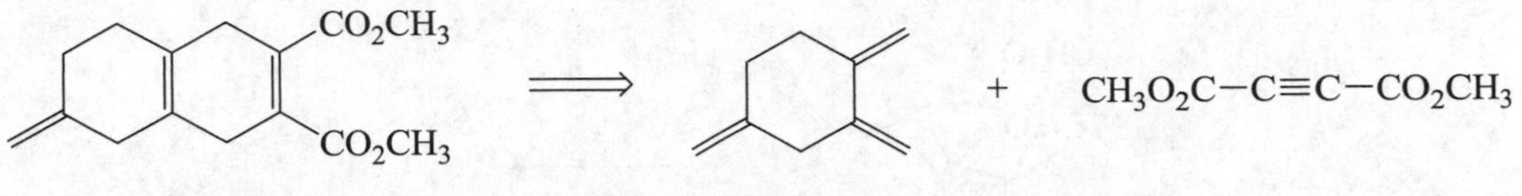

Compound B   Compound A

11.44 To predict the constitution of the Diels–Alder adducts, we can ignore the substituents and simply remember that the fundamental process is

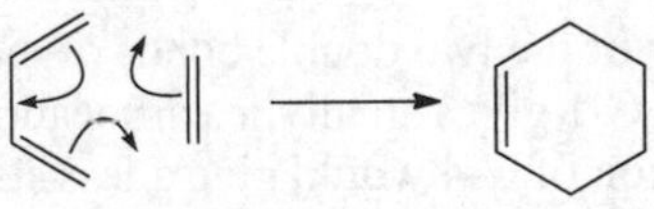

(a)

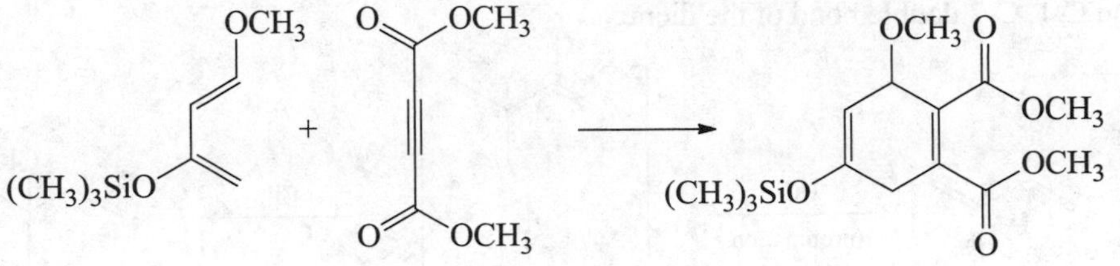

(b)

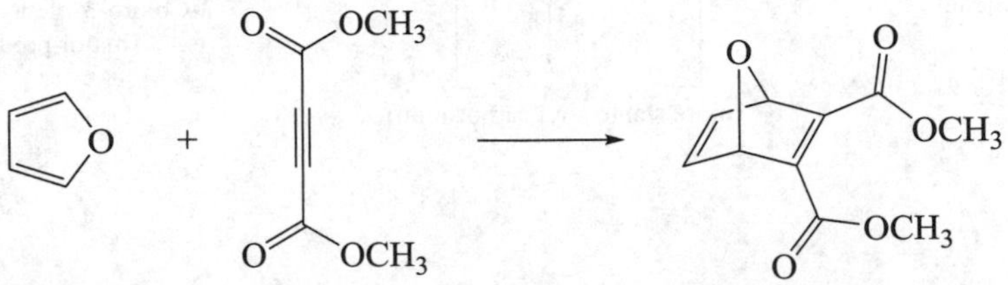

(c)

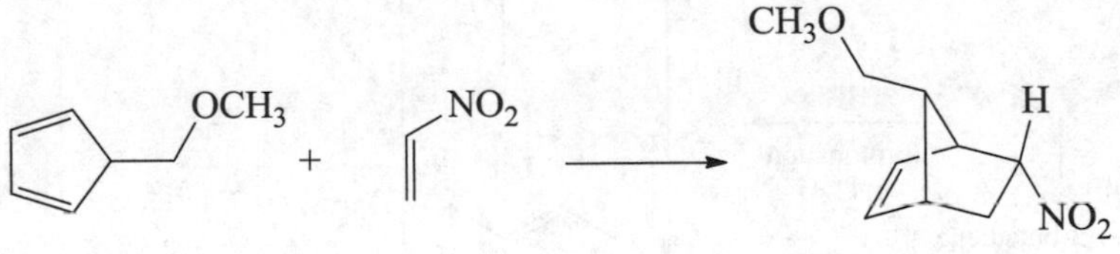

11.45 (a) The intramolecular Diels-Alder reaction yields the following bicyclic ketone.

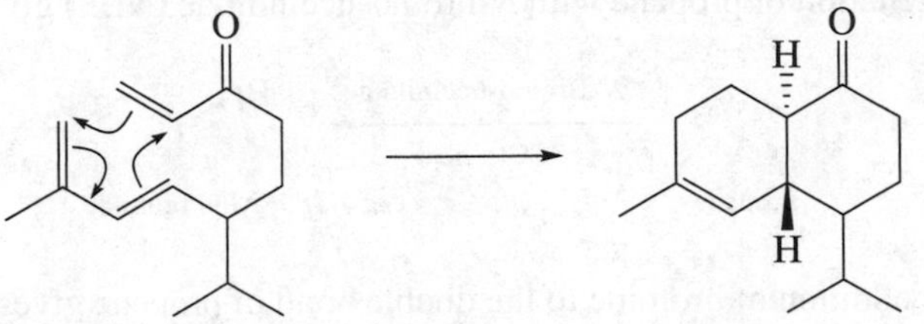

(b) The following bicyclic ester is formed from an intramolecular Diels-Alder reaction. The trans alkene gives the indicated trans configuration about the cyclohexene ring.

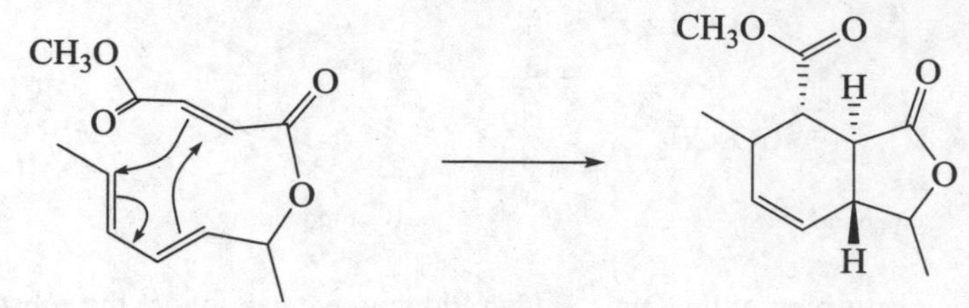

Synthesis

11.46 1,2-Addition can occur at either of the two double bonds of the diene system to give an allyl carbocation as an intermediate. Protonation of C-1 gives an allylic carbocation for which the more stable resonance form is a tertiary carbocation. Protonation of C-4 would give a less stable allylic carbocation for which the more stable resonance form is a secondary carbocation. Thus, the major 1,2-addition product arises by addition to the C-1, C-2 double bond of the diene.

(more stable allyl carbocation)

(less stable allyl carbocation)

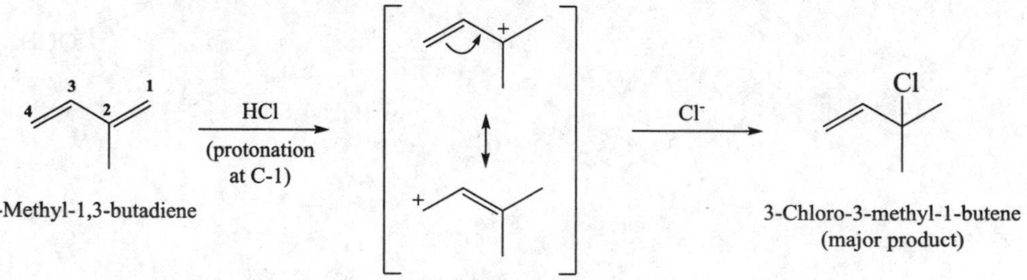

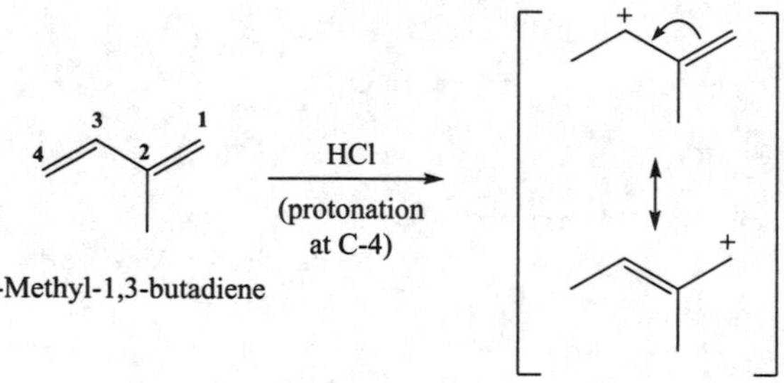

11.47 (*a*) Allylic halogenation of propene with *N*-bromosuccinimide (NBS) gives allyl bromide.

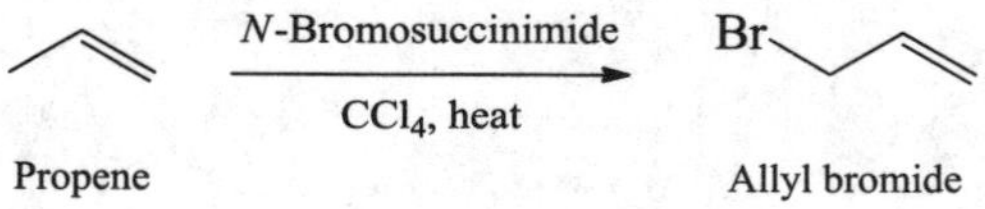

(*b*) Electrophilic addition of bromine to the double bond of propene gives 1,2-dibromopropane.

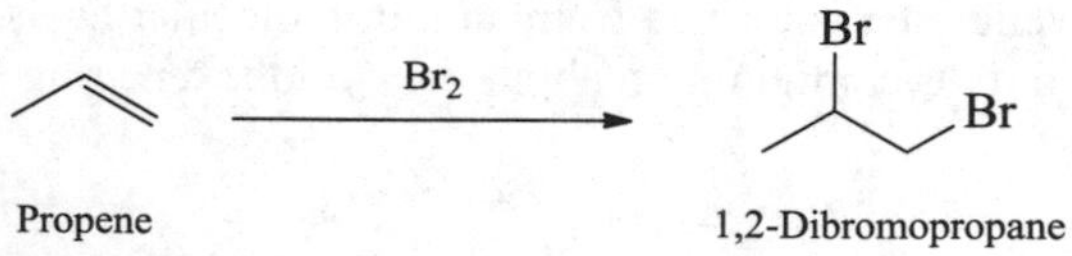

(*c*) 1,3-Dibromopropane is made from allyl bromide from part (*a*) by free-radical addition of hydrogen bromide.

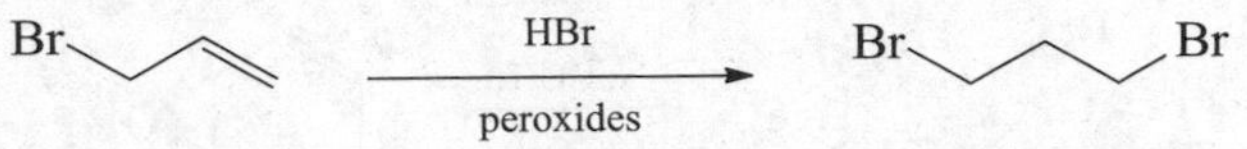

Allyl bromide → 1,3-Dibromopropane

(d) Addition of hydrogen chloride to allyl bromide proceeds in accordance with Markovnikov's rule.

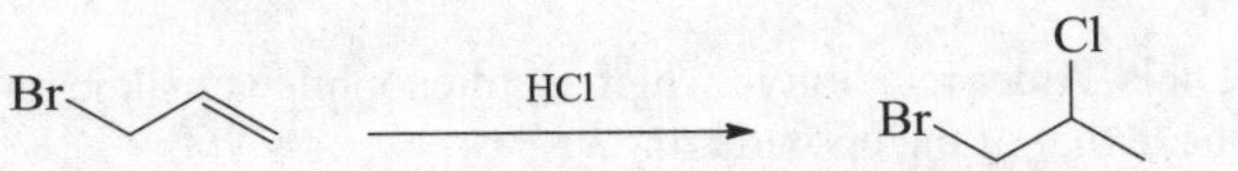

Allyl bromide → 1-Bromo-2-chloropropane

(e) Addition of bromine to allyl bromide gives 1,2,3-tribromopropane.

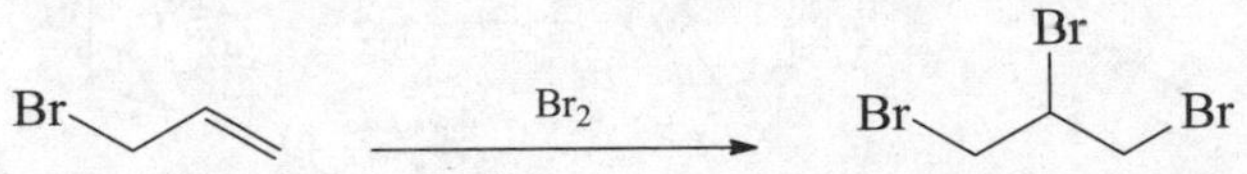

Allyl bromide → 1,2,3-Tribromopropane

(f) Nucleophilic substitution by hydroxide on allyl bromide gives allyl alcohol.

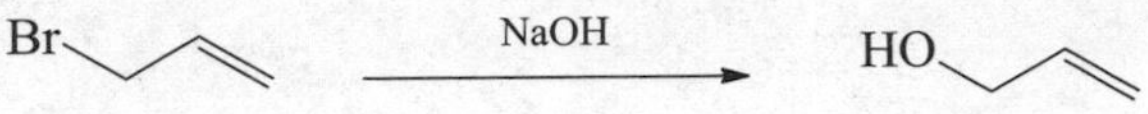

Allyl bromide → Allyl alcohol

(g) Alkylation of sodium acetylide using allyl bromide gives the desired pent-1-en-4-yne.

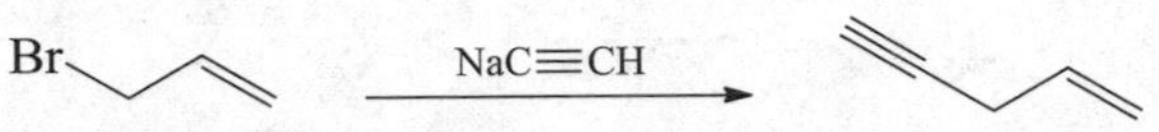

Allyl bromide → Pent-1-en-4-yne

(h) Sodium–ammonia reduction of pent-1-en-4-yne reduces the triple bond but leaves the double bond intact. Hydrogenation over Lindlar palladium could also be used.

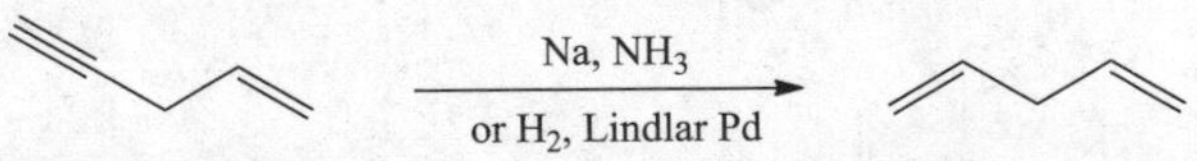

Pent-1-en-4-yne → 1,4-Pentadiene

11.48 Reaction (a) is an electrophilic addition of bromine to an alkene; the appropriate reagent is bromine in carbon tetrachloride.

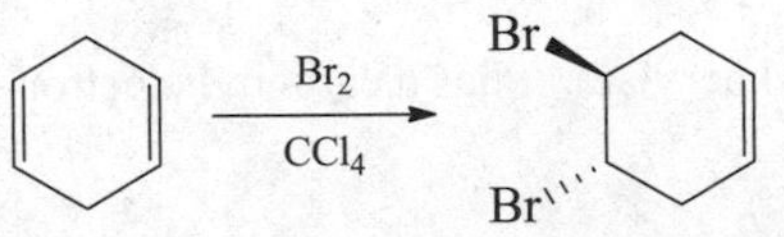

(74%)

Reaction (b) is an epoxidation of an alkene, for which almost any peroxy acid could be used. Peroxybenzoic acid was actually used.

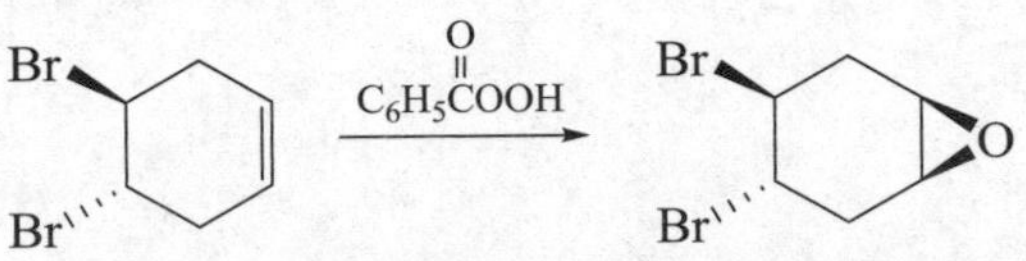

(69%)

Reaction (c) is an elimination reaction of a vicinal dibromide to give a conjugated diene and requires E2 conditions. Sodium methoxide in methanol was used.

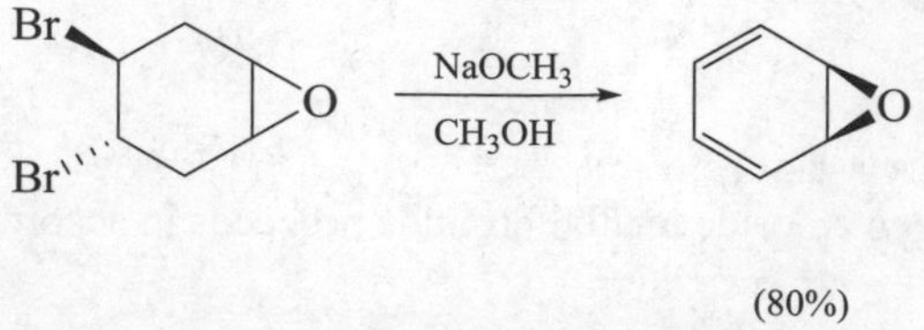

(80%)

Reaction (*d*) is a Diels–Alder reaction in which the dienophile is maleic anhydride. The dienophile adds from the side opposite that of the epoxide ring.

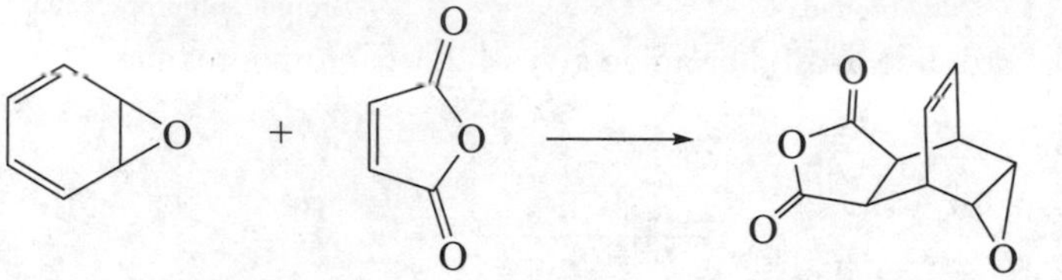

11.49 (*a*) The dienophile is 2-chloroacrylonitrile. It is highly reactive in the Diels–Alder reaction due to the presence of two electron-attracting groups.

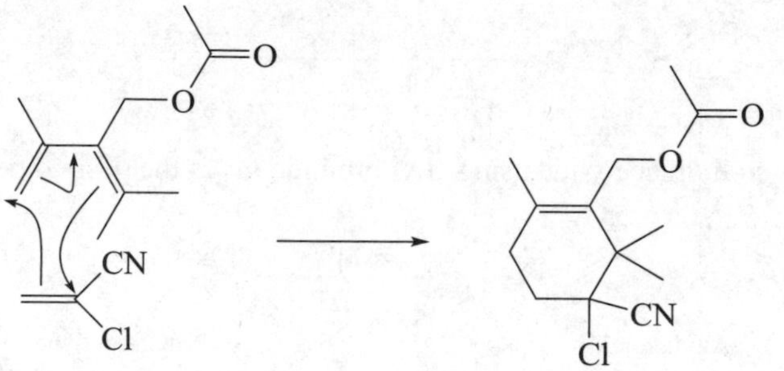

(*b*) The dienophile is 2-methyl-2-cyclopentenone. It has a carbonyl electron-attracting group.

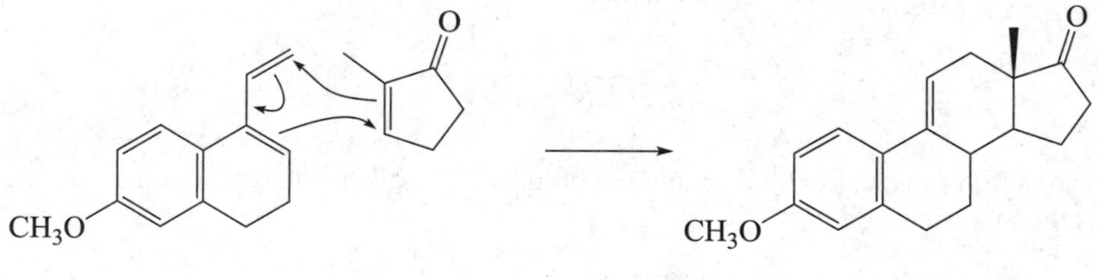

(*c*) The dienophile is methyl acrylate. It has a carbonyl electron-attracting group.

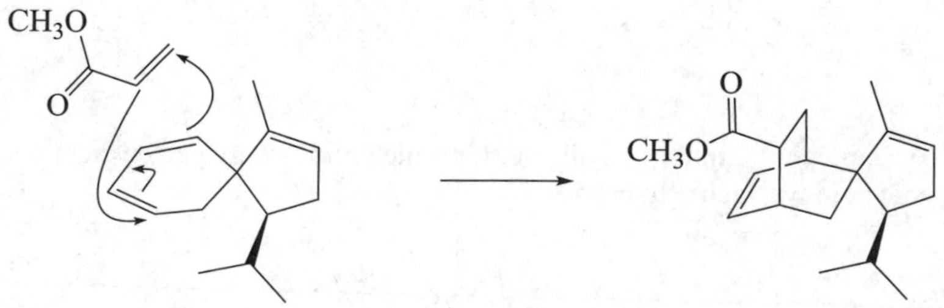

Mechanism

11.50 The reaction is a nucleophilic substitution in which the nucleophile ($C_6H_5S^-$) becomes attached to the carbon (C-1) that bore the chloride leaving group. Allylic rearrangement is not observed since the product from reaction at C-3 of 1-chloro-2-butene is not observed; therefore, it is reasonable to conclude that an allylic carbocation is *not* involved. The mechanism is S_N2.

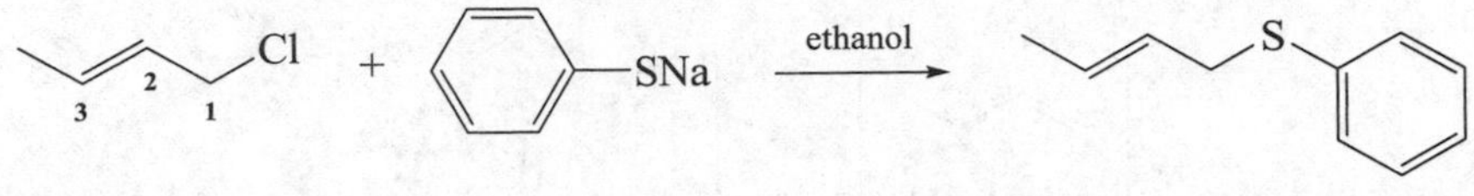

1-Chloro-2-butene   Sodium   2-Butenyl phenyl sulfide
                        benzenethiolate

11.51 (*a*) The first reaction is bromination by way of an allylic radical. This intermediate radical is the source of both isomers.

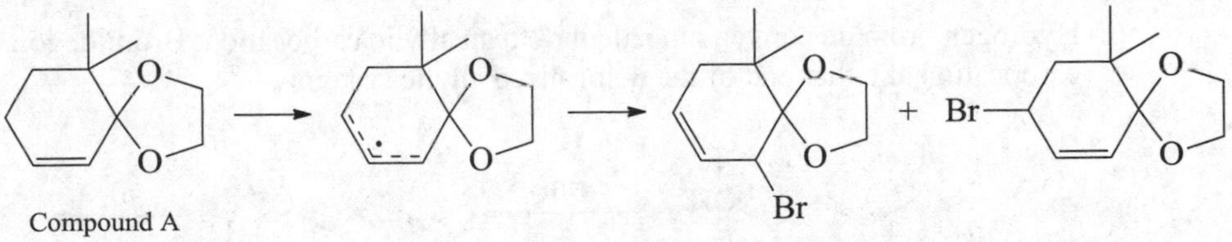

Compound A

(*b*) Silver acetate is a source of Ag⁺ which catalyzes the ionization of the two allylic bromides in the preceding reaction. The same allylic carbocation is formed from each bromide, which reacts with acetate to give the observed product.

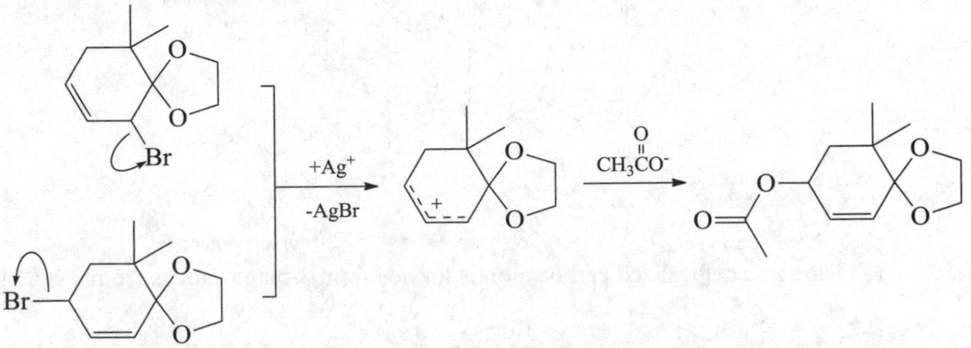

11.52 (*a*) Solvolysis of $(CH_3)_2C=CHCH_2Cl$ in ethanol proceeds by an S_N1 mechanism and involves a carbocation intermediate.

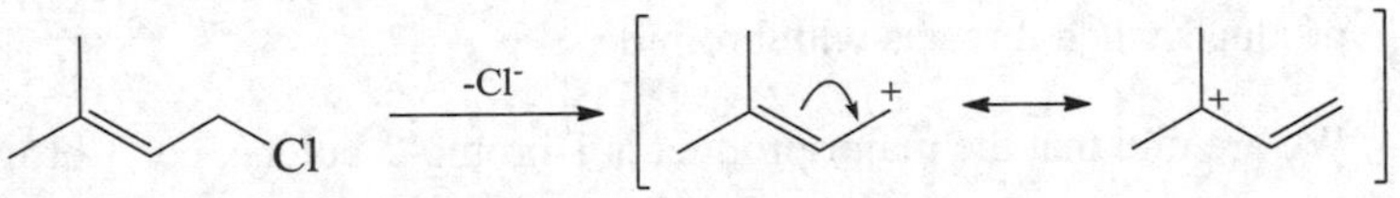

1-Chloro-3-methyl-2-butene

In one resonance form, this carbocation is a tertiary carbocation. It is more stable and is therefore formed faster than allyl cation, $CH_2=CH-\overset{+}{C}H_2$.

(*b*) An allylic carbocation is formed from the alcohol in the presence of an acid catalyst.

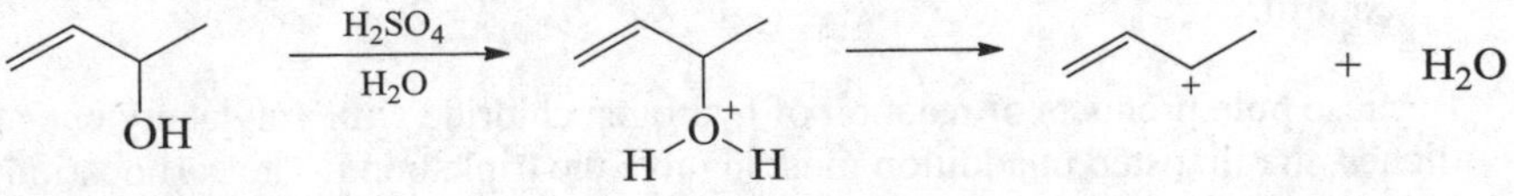

3-Buten-2-ol

This carbocation is a delocalized one and can be captured at either end of the allylic system by water acting as a nucleophile.

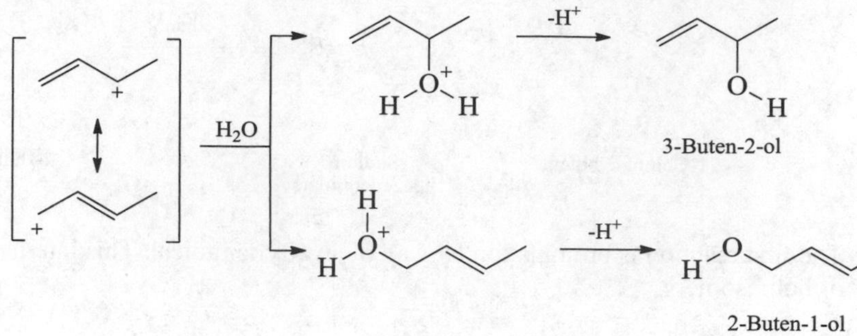

3-Buten-2-ol

2-Buten-1-ol

(*c*) Hydrogen bromide converts the alcohol to an allylic carbocation. Bromide ion captures this carbocation at either end of the delocalized allylic system.

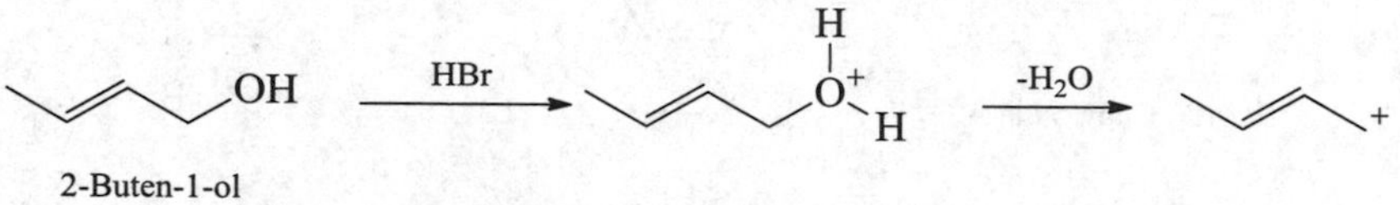

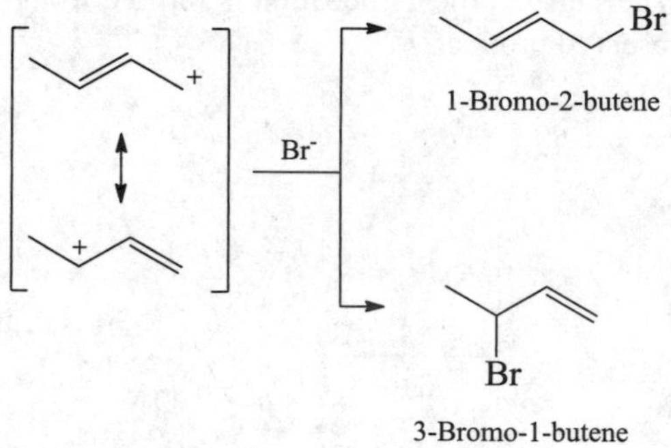

2-Buten-1-ol

1-Bromo-2-butene

3-Bromo-1-butene

(*d*) The same delocalized carbocation is formed from 3-buten-2-ol as from 2-buten-1-ol.

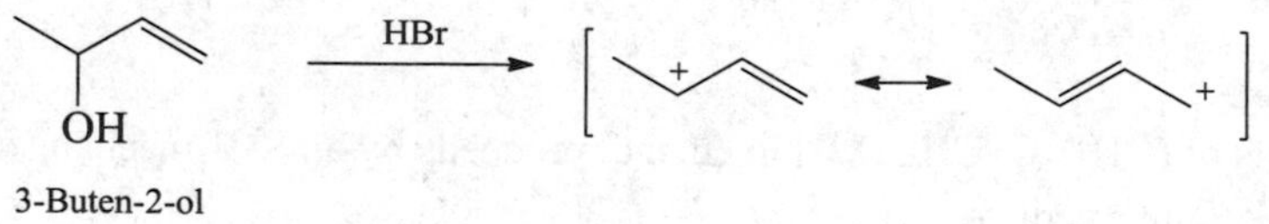

3-Buten-2-ol

Because this carbocation is the same as the one formed in part (*c*), it gives the same mixture of products when it reacts with bromide.

(*e*) We are told that the major product is 1-bromo-2-butene, not 3-bromo-1-butene.

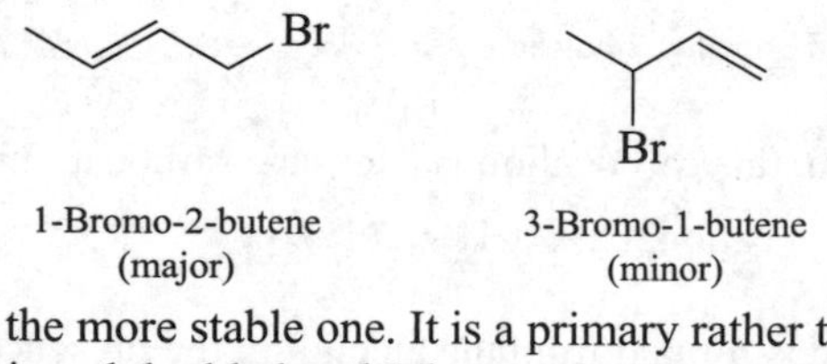

1-Bromo-2-butene
(major)

3-Bromo-1-butene
(minor)

The major product is the more stable one. It is a primary rather than a secondary halide and contains a more substituted double bond. The reaction is therefore governed by thermodynamic control.

11.53 Because both products of reaction of hydrogen chloride with vinylacetylene are chloro-substituted dienes, the first step in addition must involve the triple bond. The carbocation produced is an allylic vinyl cation for which two Lewis structures may be written. Capture of this cation gives the products of 1,2 and 1,4 addition. The 1,2 addition product is more stable because of its conjugated system. The observations of the experiment tell us that the 1,4 addition product is formed faster, although we could not have predicted that.

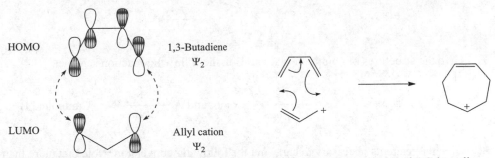

HC≡C–C(H)(CH$_2$) Vinylacetylene →[HCl] [H$_2$C=C$^+$–C(H)(CH$_2$) ↔ H$_2$C=C=C(H)(CH$_2$)$^+$] →[Cl$^-$]

H$_2$C=C(Cl)–C(H)=CH$_2$ +   H$_2$C=C=C(H)(CH$_2$Cl)

2-Chloro-1,3-butadiene   4-Chloro-1,2-butadiene
(1,2 addition)   (1,4 addition)

11.54 (*a*) Electrons flow from ethylene to allyl cation; therefore, focus on the ethylene HOMO and the allyl cation LUMO.

LUMO — Allyl cation Ψ_2

Mismatch → X ← Match

HOMO — Ethylene Ψ_1

There is symmetry mismatch between the two orbitals. Concerted cycloaddition is forbidden.

(*b*) The orbitals to examine are the HOMO of 1,3-butadiene and the LUMO of allyl cation.

HOMO — 1,3-Butadiene Ψ_2

LUMO — Allyl cation Ψ_2

The symmetries of the two orbitals allow bond formation between allyl cation and 1,3-butadiene. The product of cycloaddition of allyl cation to 1,3-butadiene is 4-cycloheptenyl cation.

Answers to Interpretive Problems 11

11.55 A;   **11.56** C;   **11.57** B;   **11.58** A;   **11.59** D

SELF-TEST

1. Give the structures of all the constitutionally isomeric alkadienes of molecular formula C$_5$H$_8$. Indicate which are conjugated and which are allenes.

2. Provide an acceptable IUPAC name for each of the conjugated dienes of the previous problem, *including stereoisomers*.

3. Hydrolysis of 3-bromo-3-methylcyclohexene yields two isomeric alcohols. Draw their structures and the structure of the intermediate that leads to their formation.

4. Give the chemical structure of the reactant, reagent, or product omitted from each of the following:

(a)

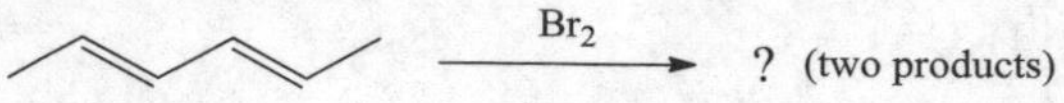

Br_2 → ? (two products)

(b)

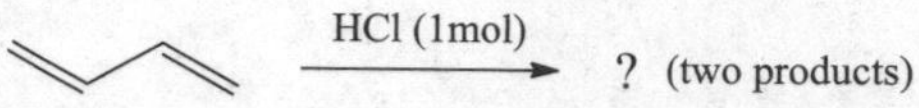

HCl (1mol) → ? (two products)

(c)

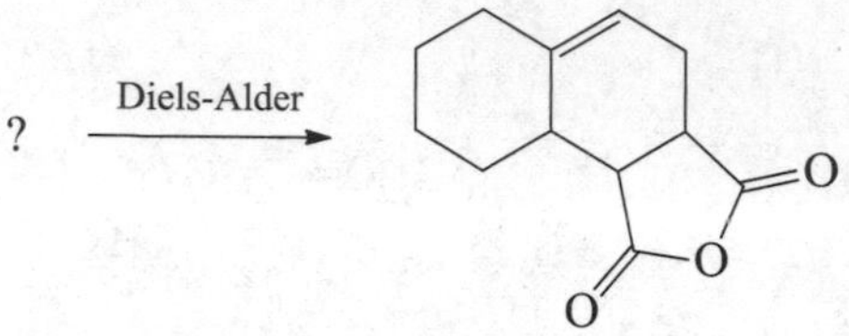

? — Diels-Alder →

(d)

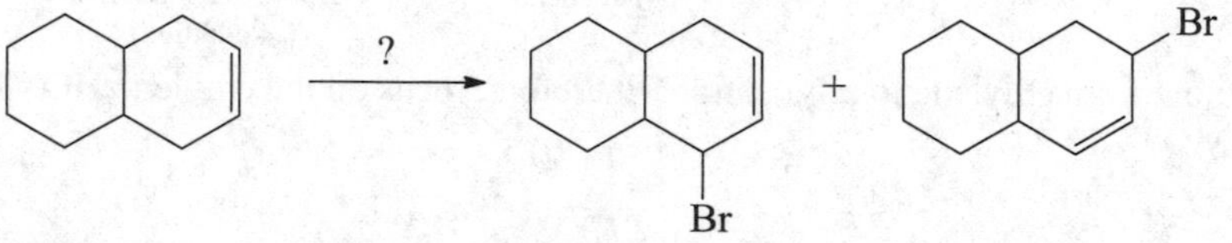

(e)

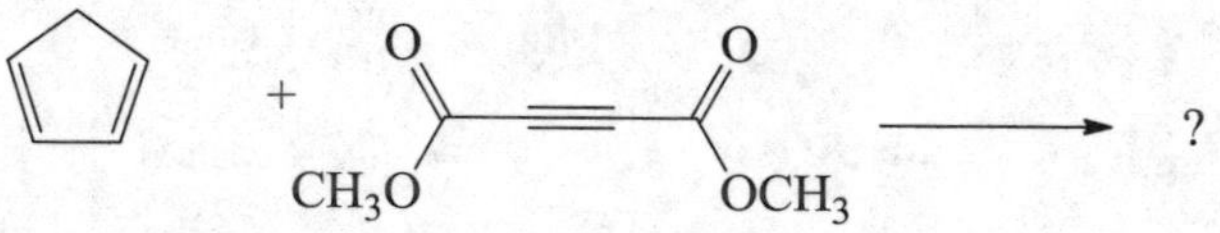

5. One of the isomeric conjugated dienes having the formula C_6H_8 and possessing a 5-membered ring is not able to react with a dienophile in a Diels–Alder reaction. Draw the structure of this compound.

6. Draw the structure of the carbocation formed on ionization of the compound shown. A constitutional isomer of this compound gives the same carbocation; draw its structure.

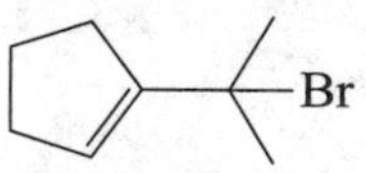

7. Give the structures of compounds A and B in the following reaction scheme.

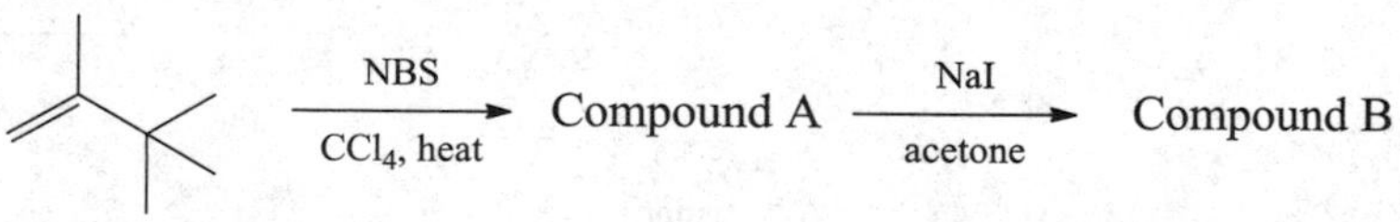

$$\text{NBS} \atop \text{CCl}_4, \text{ heat}$$ → Compound A $$\text{NaI} \atop \text{acetone}$$ → Compound B

8. Give the reagents necessary to carry out the following conversion. Note that more than one reaction step is necessary.

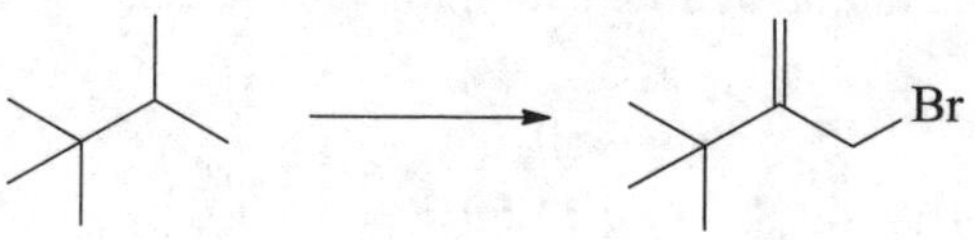

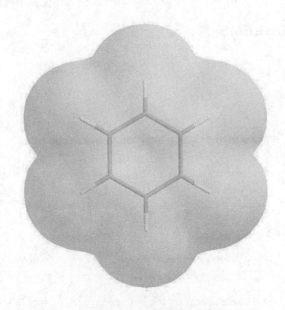

CHAPTER 12

Arenes and Aromaticity

Table of Contents

SOLUTIONS TO TEXT PROBLEMS

In Chapter Problems

12.1 Toluene is $C_6H_5CH_3$; it has a methyl group attached to a benzene ring.

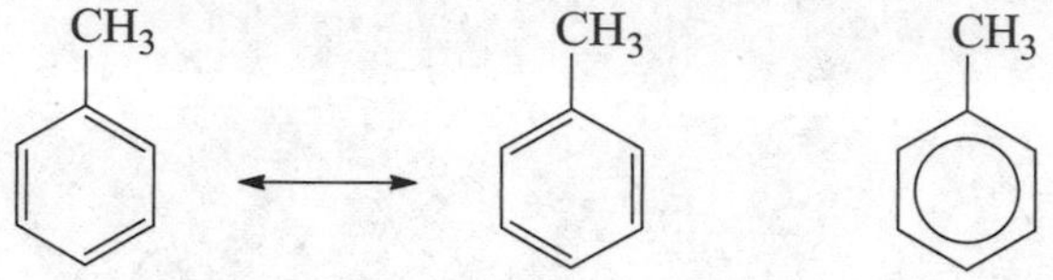

Kekulé forms of toluene

Robinson symbol
for toluene

Benzoic acid has a —CO_2H substituent on the benzene ring.

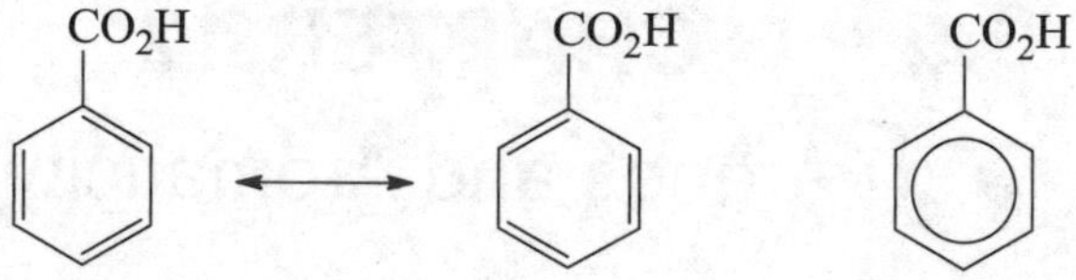

Kekulé forms of benzoic acid

Robinson symbol
for benzoic acid

12.2 Given

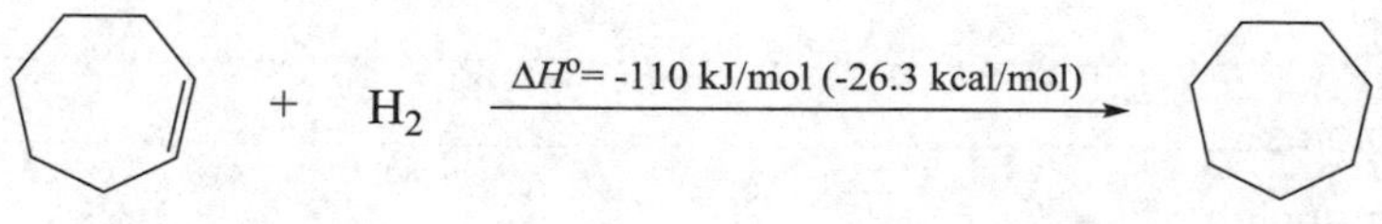

Cycloheptene

ΔH°= -110 kJ/mol (-26.3 kcal/mol)

Cycloheptane

and assuming that there is no resonance stabilization in 1,3,5-cycloheptatriene, we predict that its heat of hydrogenation would be three times that of cycloheptene, or 330 kJ/mol (78.9 kcal/mol). The measured heat of hydrogenation is

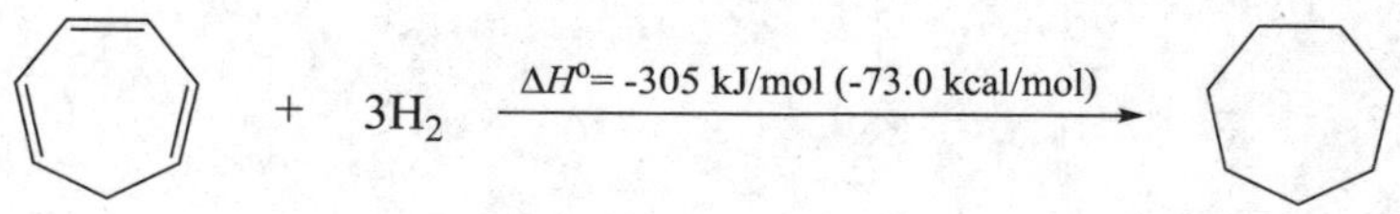

1,3,5-Cycloheptatriene

+ 3H$_2$

ΔH°= -305 kJ/mol (-73.0 kcal/mol)

Cycloheptane

Therefore

Resonance energy= 330 kJ/mol (predicted for no delocalization) – 305 kJ/mol (observed)

= 25 kJ/mol (5.9 kcal/mol)

The value given in the text for the resonance energy of benzene (152 kJ/mol) is six times larger than this. 1,3,5-Cycloheptatriene is *not* aromatic.

12.3 (*b*) The parent compound is styrene, $C_6H_5CH=CH_2$. The desired compound has a chlorine in the meta position.

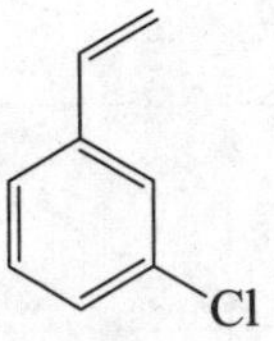

m-Chlorostyrene

(*c*) The parent compound is aniline, $C_6H_5NH_2$. *p*-Nitroaniline is therefore

p-Nitroaniline

12.4 Numbering in biphenyl begins at the atoms that connect the ring and are written with primes in one ring and without primes in the other.

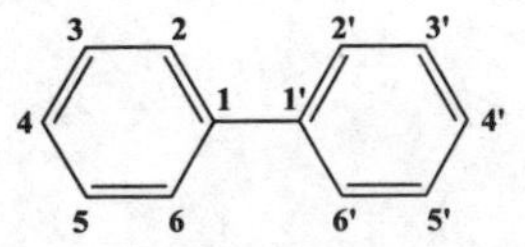

Biphenyl 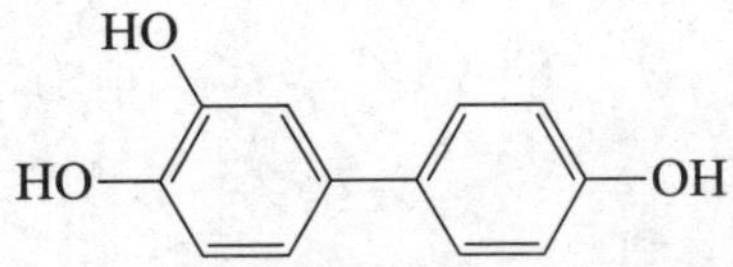   3,4,4'-Trihydroxybiphenyl

12.5 There are only three monochloro derivatives of anthracene. Any other monochloro derivative is equivalent to one of these. Carbons at the ring junction positions already have four bonds so cannot bear a substituent.

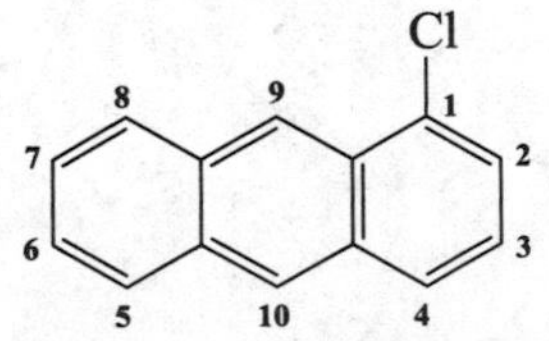

1-Chloroanthracene 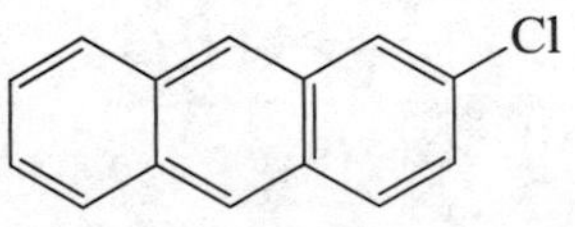   2-Chloroanthracene   9-Chloroanthracene

12.6 The most stable resonance form is the one that has the greatest number of rings that are Kekulé formulations of benzene. A resonance form for chrysene can be drawn in which all four rings correspond to Kekulé structures.

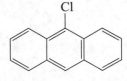

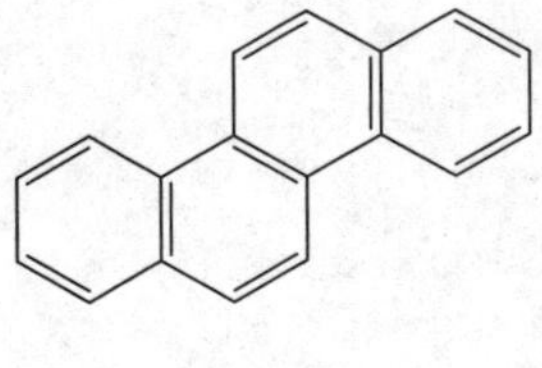

Chrysene

12.7 (*b*) Move electrons in pairs beginning at the negatively charged carbon. The structure at the right is preferred because the negative charge is on oxygen rather than carbon.

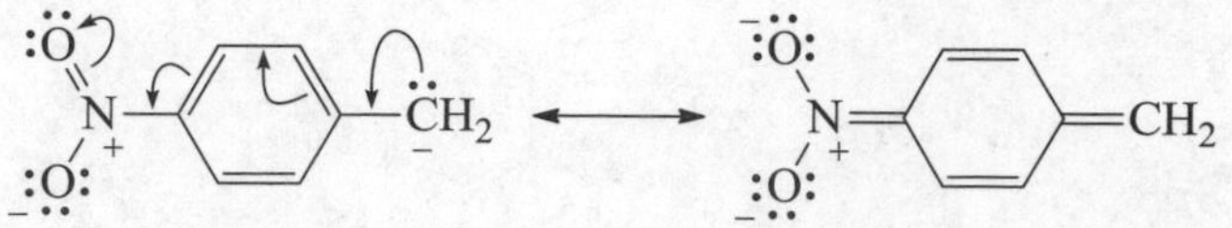

12.8 A first-order reaction suggests an S_N1 mechanism; therefore, compare the relative stabilities of the carbocation intermediates formed during the solvolysis of each compound.

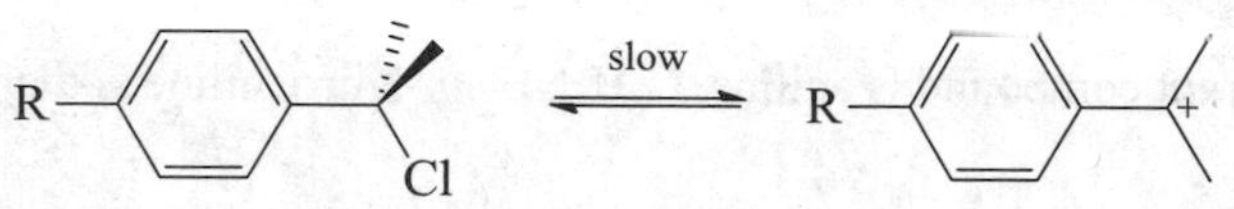

When R = CH$_3$, the carbocation has tertiary carbocation character, and is more stable than the corresponding carbocation where R = H.

Therefore, the carbocation is formed faster (E_a is lower).

12.9 Each of these reactions involves nucleophilic substitution of the S_N2 type at the benzylic position of benzyl bromide.

(*b*)

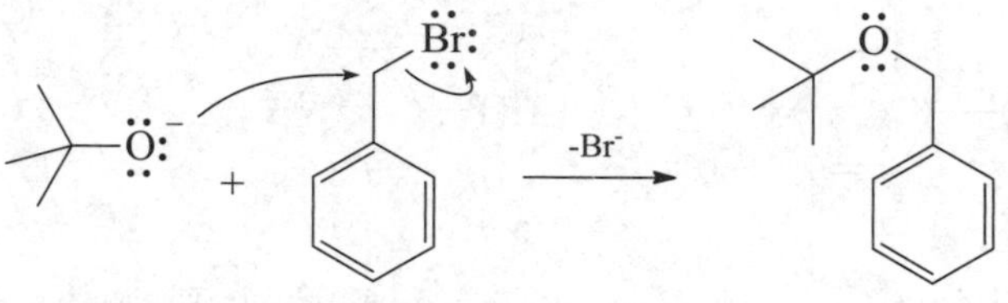

| *tert*-Butoxide ion | Benzyl bromide | Benzyl *tert*-butyl ether |

(*c*)

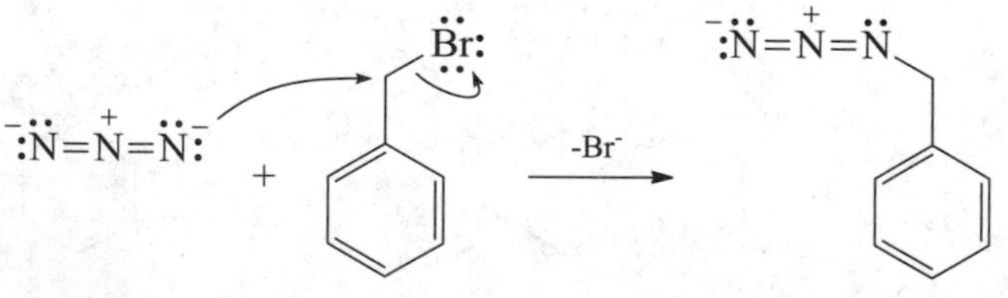

| Azide ion | Benzyl bromide | Benzyl azide |

(*d*)

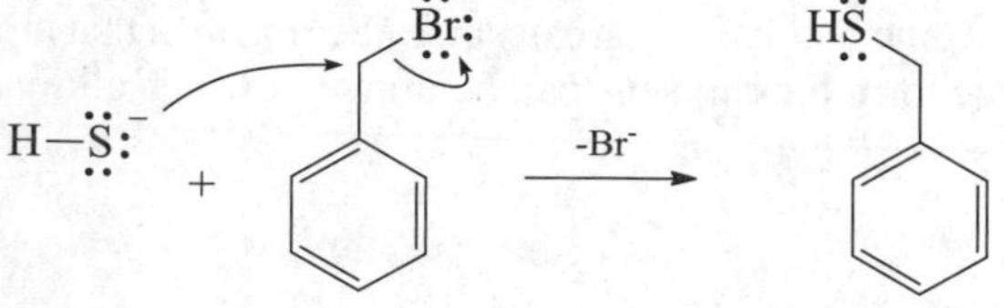

| Hydrogen sulfide ion | Benzyl bromide | Phenylmethanethiol |

(e)

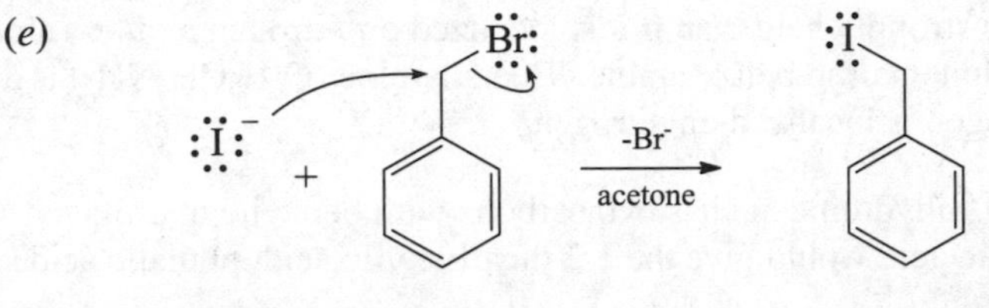

Hydrogen sulfide ion   Benzyl bromide   Benzyl iodide

12.10 A protonation step followed by a deprotonation step successfully aromatizes the remaining six-membered ring.

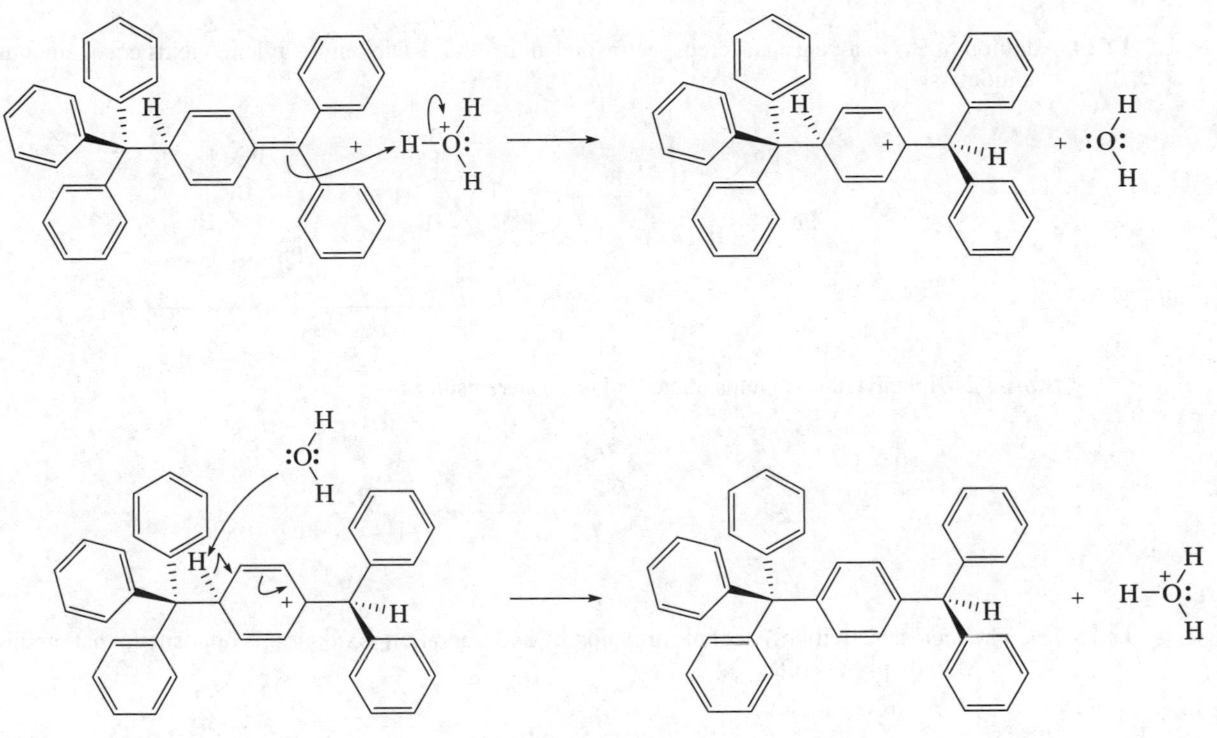

12.11 (b) Only the benzylic hydrogen is replaced by bromine in the reaction of 4-methyl-3-nitroanisole with N-bromosuccinimide (NBS).

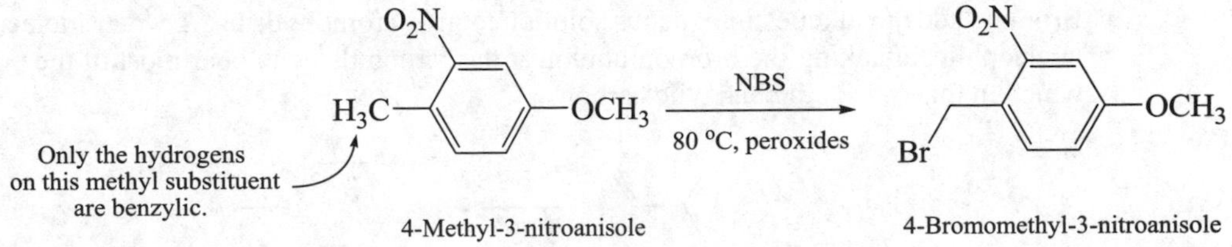

Only the hydrogens on this methyl substituent are benzylic.

4-Methyl-3-nitroanisole

$\xrightarrow[\text{80 °C, peroxides}]{\text{NBS}}$

4-Bromomethyl-3-nitroanisole

12.12 The nitrogen lone pair in N-methylaniline is delocalized into the benzene ring.

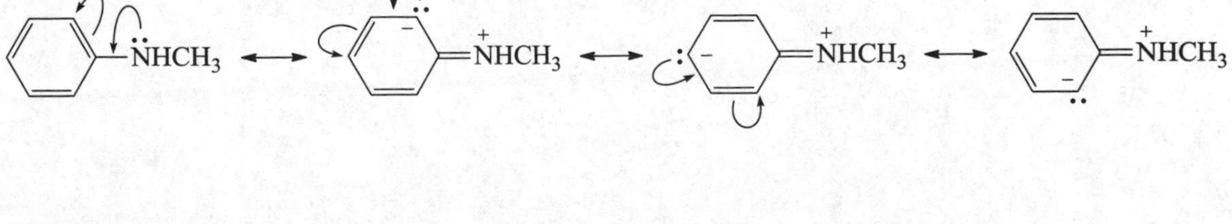

Therefore, the electron pair is more strongly held than if it is localized on nitrogen and less able to be used to bond to a proton than the nitrogen lone pair in benzylamine. Benzylamine ($C_6H_5CH_2NH_2$) is a stronger base. The unshared electron pair on nitrogen is localized on nitrogen.

12.13 (*b*) The five-membered ring in 2,3-dihydroindene has two carbons with benzylic hydrogens. Oxidation of 2,3-dihydroindene with chromic acid would give the 1,2-dicarboxylic acid, phthalic acid.

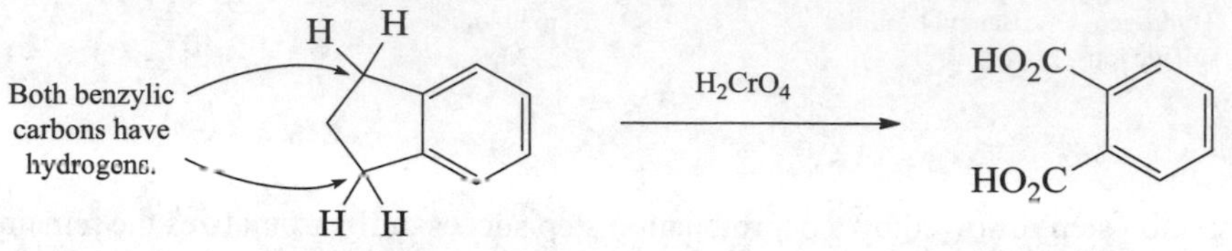

2,3-Dihydroindene   Phthalic acid

12.14 Addition of Br_2 to alkenes is stereospecific and anti. *cis*-1,2-Diphenylethylene yields equal amounts of enantiomers.

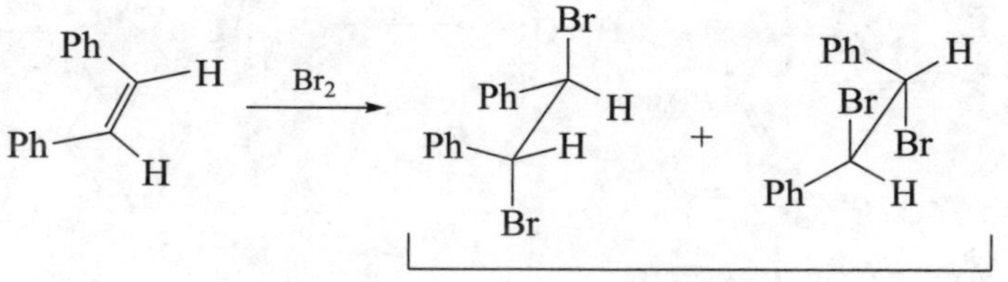

enantiomers

trans-1,2-Diphenylethene yields an achiral *meso* stereoisomer.

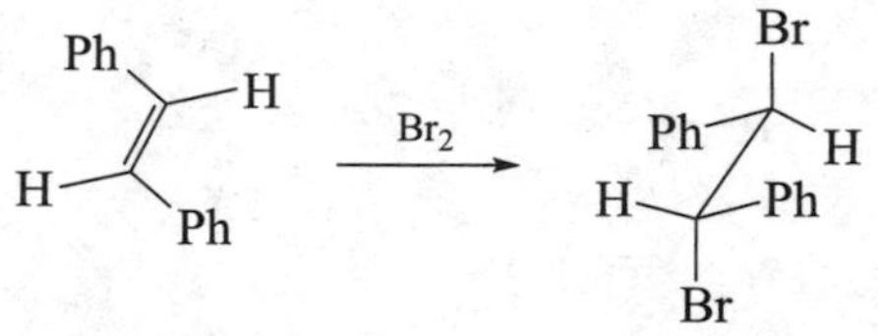

12.15 (*b*) The regioselectivity of alcohol formation by hydroboration–oxidation is opposite to that predicted by Markovnikov's rule.

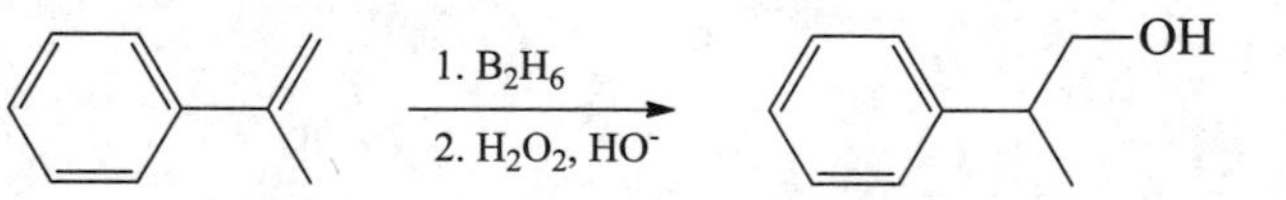

2-Phenylpropene   2-Phenyl-1-propanol (92%)

(*c*) Bromine adds to alkenes in aqueous solution to give bromohydrins. A water molecule acts as a nucleophile, attacking the bromonium ion at the carbon that can bear most of the positive charge, which in this case is the benzylic carbon.

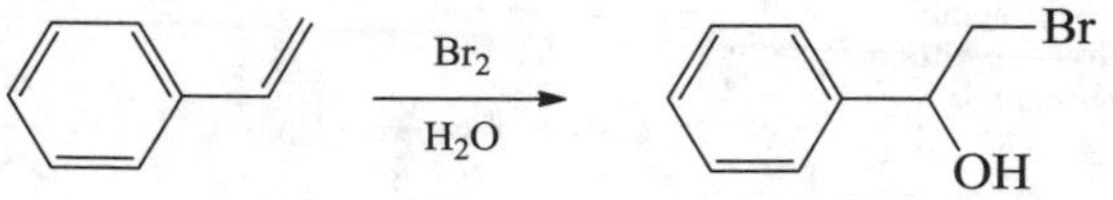

Styrene   2-Bromo-1-phenylethanol (82%)

(*d*) Peroxy acids convert alkenes to epoxides.

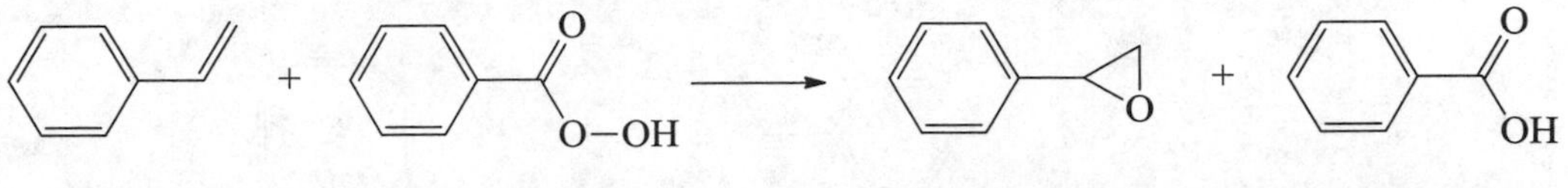

| Styrene | Peroxybenzoic acid | Epoxystyrene (69%-75%) | Benzoic acid |

12.16 Birch reduction will give a 1,4-cyclohexadiene derivative in which both the alkyl and alkoxy groups are substituents on double bonds. The only compound that satisfies this requirement is 1-methoxy-5-methyl-1,4-cyclohexadiene.

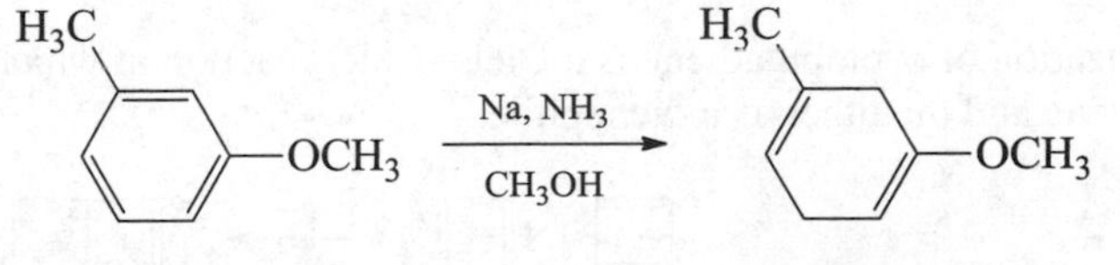

1-Methoxy-5-methyl-
1,4-cyclohexadiene

12.17 The regio- and stereochemistry of *trans*-2-phenylcyclopentanol suggest it can be prepared from 1-phenylcyclopentene by hydroboration-oxidation.

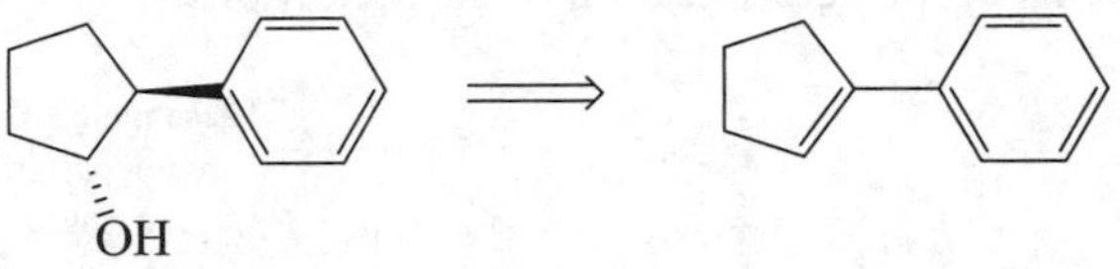

| *trans*-2-Phenylcyclopentanol | | 1-Phenylcyclopentene |

The synthesis of 1-phenylcyclopentene from phenylcyclopentane can be accomplished by benzylic halogenation and elimination.

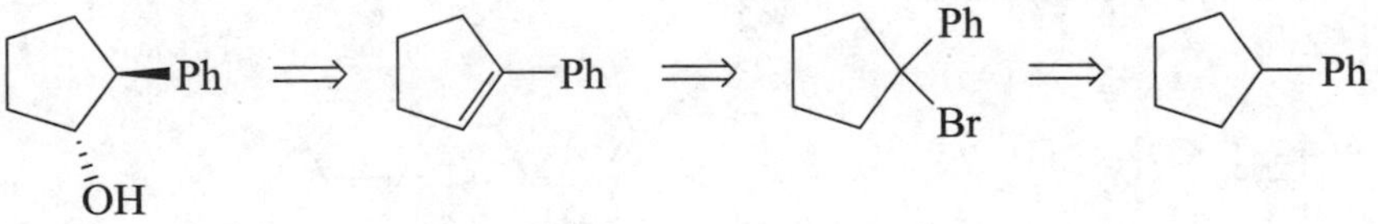

| *trans*-2-Phenyl-cyclopentanol | 1-Phenyl-cyclopentene | 1-Bromo-1-phenylcyclopentane | Phenylcyclopentane |

The synthesis is:

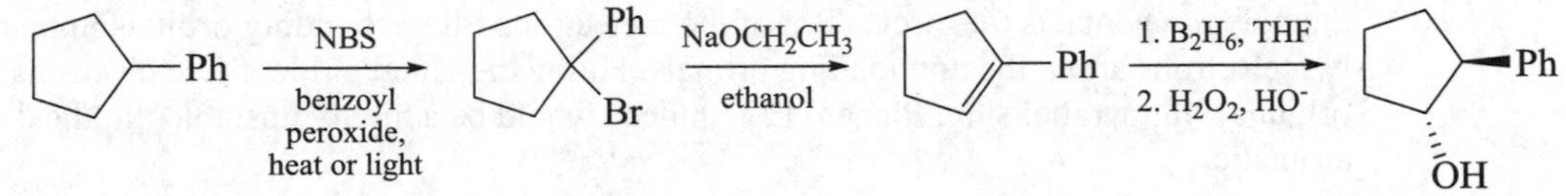

| Phenylcyclopentane | 1-Bromo-1-phenylcyclopentane | 1-Phenyl-cyclopentene | *trans*-2-Phenyl-cyclopentanol |

12.18 Styrene contains a benzene ring and will be appreciably stabilized by resonance, which makes it lower in energy than cyclooctatetraene.

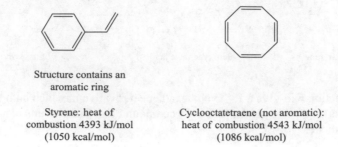

Structure contains an
aromatic ring

Styrene: heat of combustion 4393 kJ/mol (1050 kcal/mol)	Cyclooctatetraene (not aromatic): heat of combustion 4543 kJ/mol (1086 kcal/mol)

12.19 The dimerization of cyclobutadiene is a Diels–Alder reaction in which one molecule of cyclobutadiene acts as a diene and the other as a dienophile.

Diene   Dienophile   Diels-Alder adduct

12.20 (*b*) Compound B is not aromatic for two reasons. Although the compound has six π electrons, they are not all part of a ring. Secondly, the sp^3-hybridized carbon present in the ring acts as an insulator and prevents cyclic delocalization of the π electrons.

These π electrons
are not part of
the ring.

sp^3 carbon atom

(*c*) Compound C is aromatic because it contains two benzene rings. Hückel's rule can only be applied to *monocyclic* compounds.

Compound C
(biphenyl)

12.21 (*b*) [12]Annulene is a 12-membered ring with six conjugated double bonds. As in part (*a*), inscribe a polygon with 12 sides in a circle with its vertex downward. Place an orbital at each point where the polygon contacts the circle. Ten of the π electrons fill the bonding orbitals in pairs. The remaining two electrons are in the nonbonding orbitals. Following Hund's rule, these electrons are in separate orbitals with parallel spin. Planar [12]annulene would be a highly unstable diradical species and is not aromatic.

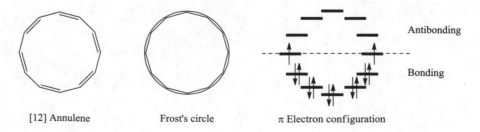

[12] Annulene   Frost's circle   π Electron configuration

12.22 One way to evaluate the relationship between heats of combustion and structure for compounds that are not isomers is to divide the heat of combustion by the number of carbons so that heats of combustion are compared on a "per carbon" basis.

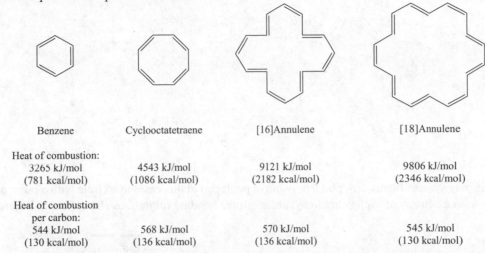

| Benzene | Cyclooctatetraene | [16]Annulene | [18]Annulene |

Heat of combustion:

| 3265 kJ/mol | 4543 kJ/mol | 9121 kJ/mol | 9806 kJ/mol |
| (781 kcal/mol) | (1086 kcal/mol) | (2182 kcal/mol) | (2346 kcal/mol) |

Heat of combustion per carbon:

| 544 kJ/mol | 568 kJ/mol | 570 kJ/mol | 545 kJ/mol |
| (130 kcal/mol) | (136 kcal/mol) | (136 kcal/mol) | (130 kcal/mol) |

As the data indicate (within experimental error), the heats of combustion *per carbon* of the two aromatic hydrocarbons, benzene and [18]annulene, are equal. Similarly, the heats of combustion per carbon of the two nonaromatic hydrocarbons, cyclooctatetraene and [16]annulene, are equal. The two aromatic hydrocarbons have heats of combustion per carbon that are less than those of the nonaromatic hydrocarbons. On a per carbon basis, the aromatic hydrocarbons have lower potential energy (are more stable) than the nonaromatic hydrocarbons.

12.23 As in Problem 12.21, an energy-level diagram can be generated for cycloheptatrienyl cation by inscribing a heptagon in a circle with a point down. The cation has six π electrons that fill the bonding orbitals, yielding an energy-level diagram similar to Figure 12.17 in the text.

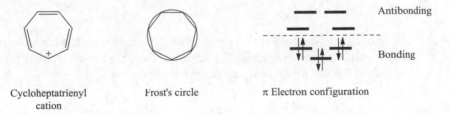

| Cycloheptatrienyl cation | Frost's circle | π Electron configuration |

12.24 Although cycloheptatrienyl radical contains a cyclic, completely conjugated π system, it is not aromatic. There are seven π electrons in the radical, a number that does not satisfy Hückel's rule. The seventh electron occupies an antibonding orbital.

Cycloheptatrienyl radical

To be antiaromatic, a species must contain $4n$ π electrons, where n is a whole number. Cycloheptatrienyl radical has seven π electrons and is neither aromatic nor antiaromatic.

12.25 The seven resonance forms for tropylium cation (cycloheptatrienyl cation) may be generated by moving π electrons in pairs toward the positive charge. The resonance forms are simply a succession of allylic carbocations.

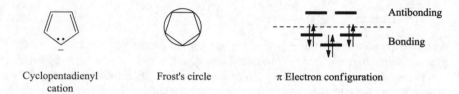

12.26 As in previous problems, inscribe a polygon (a pentagon in this case) in a circle with a point down. The six π electrons of cyclopentadienyl anion fill the bonding orbitals, as shown in text Figure 12.18.

Antibonding

Bonding

| Cyclopentadienyl cation | Frost's circle | π Electron configuration |

12.27 Resonance structures are generated for cyclopentadienyl anion by moving the unshared electron pair from the carbon to which it is attached to a position where it becomes a shared electron pair in a π bond.

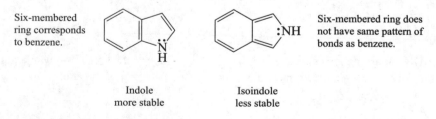

12.28 (*b*) Cyclononatetraenide anion has ten π electrons; it is aromatic. The ten π electrons are most easily seen by writing a Lewis structure for the anion: there are two π electrons for each of four double bonds, and the negatively charged carbon contributes two.

12.29 Indole is more stable than isoindole. Although the bonding patterns in both five-membered rings are the same, the six-membered ring in indole has a pattern of bonds identical to benzene and so is highly stabilized. The six-membered ring in isoindole is not of the benzene type.

Six-membered
ring corresponds
to benzene.

:NH   Six-membered ring does
not have same pattern of
bonds as benzene.

Indole
more stable

Isoindole
less stable

12.30 The prefix *benz-* in benzimidazole (structure given in text) signifies that a benzene ring is fused to an imidazole ring. By analogy, benzoxazole has a benzene ring fused to oxazole. Similarly, benzothiazole has a benzene ring fused to thiazole.

Benzimidazole   Benzoxazole   Benzothiazole

12.31 The problem states that pyrrole is 10^7–10^9 weaker in basicity than pyridine. Therefore, its conjugate acid is 10^7–10^9 times stronger. Subtract 7–9 from the pK_a of the conjugate acid of pyridine.

pK_a of [Pyrrole—H$^+$] = pK_a of [Pyridine—H$^+$] – (7–9)

pK_a of [Pyrrole—H$^+$] = 5.2 – (7–9)

pK_a of [Pyrrole—H$^+$] = –1.8 to –3.8

The conjugate acid of pyrrole is a strong acid. Its pK_a of –1.8 to –3.8 is smaller than that of hydronium ion (–1.7). The conjugate acid of pyridine is a weak acid. Its pK_a is greater than that of hydronium ion.

12.32 Move electrons in pairs from electron-rich sites toward positive charge to convert the original resonance contributor to a resonance form in which the positive charge is on the other nitrogen.

Resonance in imidazolium ion

End of Chapter Problems

Structure and Nomenclature

12.33 Because the problem requires that the benzene ring be monosubstituted, all that needs to be examined are the various isomeric forms of the C_4H_9 substituent.

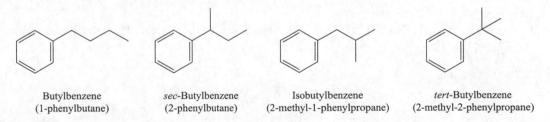

Butylbenzene   *sec*-Butylbenzene   Isobutylbenzene   *tert*-Butylbenzene
(1-phenylbutane)   (2-phenylbutane)   (2-methyl-1-phenylpropane)   (2-methyl-2-phenylpropane)

These are the four constitutional isomers. *sec*-Butylbenzene is chiral and so exists in enantiomeric *R* and *S* forms.

12.34 (*a*) An allyl substituent is —CH$_2$CH=CH$_2$.

Allylbenzene

(*b*) The constitution of 1-phenyl-1-butene is C$_6$H$_5$CH=CHCH$_2$CH$_3$. The *E* stereoisomer is

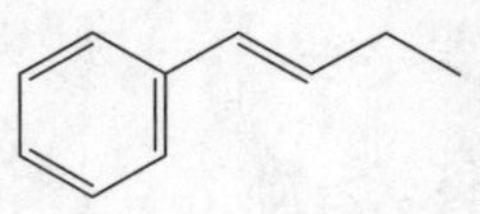

(E)-1-Phenyl-1-butene

The two higher-ranked substituents, phenyl and ethyl, are on opposite sides of the double bond.

(c) The constitution of 2-phenyl-2-butene is $CH_3C{=}CHCH_3$. The *Z* stereoisomer is
$\underset{\displaystyle C_6H_5}{|}$

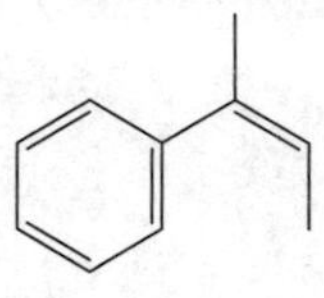

(Z)-2-Phenyl-2-butene

The two higher-ranked substituents, phenyl and methyl, are on the same side of the double bond.

(d) 1-Phenylethanol is chiral and has the constitution $CH_3\underset{\displaystyle OH}{\overset{\displaystyle |}{C}}HC_6H_5$. Among the substituents attached

to the chirality center, the order of decreasing precedence is

$$HO > C_6H_5 > CH_3 > H$$

In the *R* enantiomer, the order of decreasing precedence of the three highest-ranked substituents must appear clockwise when the lowest-ranked substituent is directed away from you.

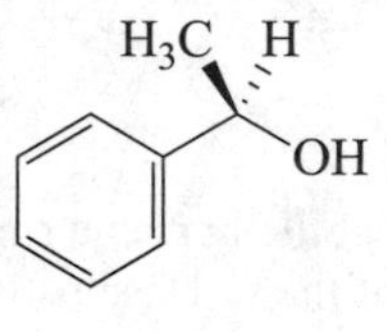

(R)-1-Phenylethanol

(e) A benzyl group is $C_6H_5CH_2{-}$. Benzyl alcohol is, therefore, $C_6H_5CH_2OH$, and *o*-chlorobenzyl alcohol is

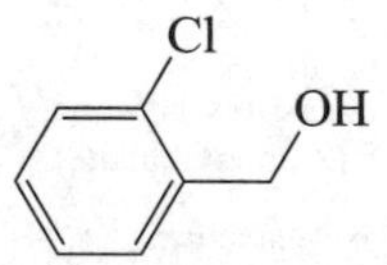

(f) In *p*-chlorophenol, the benzene ring bears chlorine and hydroxyl substituents in a 1,4-substitution pattern.

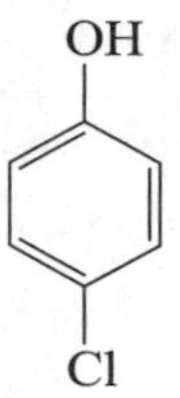

p-Chlorophenol

(g) Benzenecarboxylic acid is an alternative IUPAC name for benzoic acid.

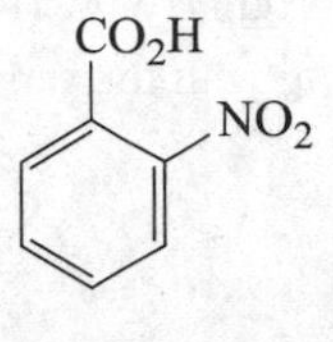

2-Nitrobenzenecarboxylic acid

(h) Two isopropyl groups are in a 1,4 relationship in *p*-diisopropylbenzene.

p-Diisopropylbenzene

(i) Aniline is $C_6H_5NH_2$. Therefore

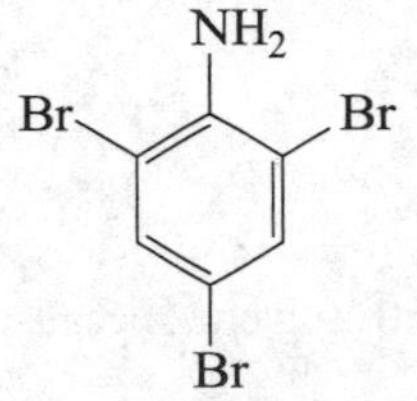

2,4,6-Tribromoaniline

(j) Acetophenone (from text Table 12.1) is $CH_3\overset{O}{\overset{\|}{C}}C_6H_5$. Therefore

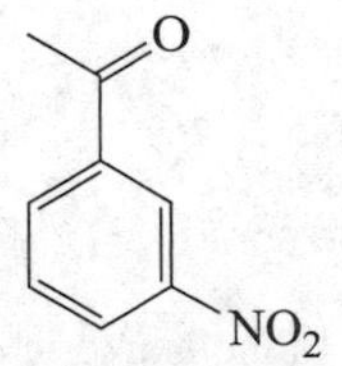

m-Nitroacetophenone

(k) Styrene is $C_6H_5CH{=}CH_2$, and numbering of the ring begins at the carbon that bears the alkenyl side chain.

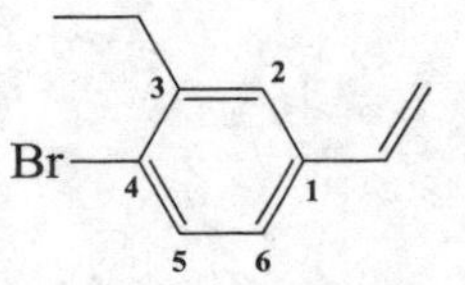

4-Bromo-3-ethylstyrene

12.35 (a) Anisole is the name for $C_6H_5OCH_3$, and allyl is an acceptable name for the group $H_2C=CHCH_2—$. Number the ring beginning with the carbon that bears the methoxy group.

Estragole
4-Allylanisole

(b) Phenol is the name for C_6H_5OH. The ring is numbered beginning at the carbon that bears the hydroxyl group, and the substituents are listed in alphabetical order.

Diosphenol
2,6-Diiodo-4-nitrophenol

(c) Aniline is the name given to $C_6H_5NH_2$. This compound is named as a dimethyl derivative of aniline. Number the ring sequentially beginning with the carbon that bears the amino group.

m-Xylidine
2,6-Dimethylaniline

12.36 (a) There are three isomeric nitrotoluenes because the nitro group can be ortho, meta, or para to the methyl group.

o-Nitrotoluene
(2-nitrotoluene)

m-Nitrotoluene
(3-nitrotoluene)

p-Nitrotoluene
(4-nitrotoluene)

(b) Benzoic acid is $C_6H_5CO_2H$. In the isomeric dichlorobenzoic acids, two of the ring hydrogens of benzoic acid have been replaced by chlorines. The isomeric dichlorobenzoic acids are

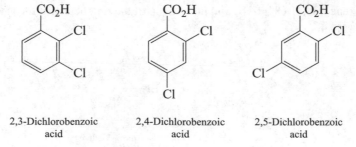

2,3-Dichlorobenzoic
acid

2,4-Dichlorobenzoic
acid

2,5-Dichlorobenzoic
acid

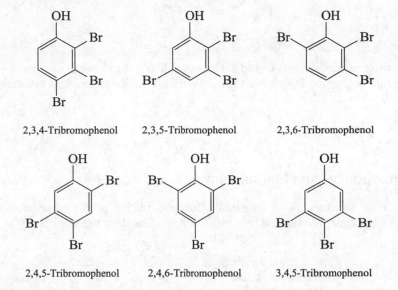

2,6-Dichlorobenzoic acid	3,4-Dichlorobenzoic acid	3,5-Dichlorobenzoic acid

The prefixes *o*-, *m*-, and *p*- may not be used in trisubstituted arenes; numerical prefixes are used. Note also that *benzenecarboxylic* may be used in place of *benzoic*.

(*c*) In the various tribromophenols, we are dealing with tetrasubstitution on a benzene ring. Again, *o*-, *m*-, and *p*- are not valid prefixes. The hydroxyl group is assigned position 1 because the base name is phenol.

2,3,4-Tribromophenol   2,3,5-Tribromophenol   2,3,6-Tribromophenol

2,4,5-Tribromophenol   2,4,6-Tribromophenol   3,4,5-Tribromophenol

(*d*) There are only three tetrafluorobenzenes. The two hydrogens may be ortho, meta, or para to each other.

1,2,3,4-Tetrafluorobenzene   1,2,3,5-Tetrafluorobenzene   1,2,4,5-Tetrafluorobenzene

(*e*) There are only two naphthalenecarboxylic acids.

Naphthalene-1-
carboxylic acid

Naphthalene-2-
carboxylic acid

12.37 The structure and numbering system for quinoline are given in Section 12.21 of the text. *Nitroxoline* has the structural formula

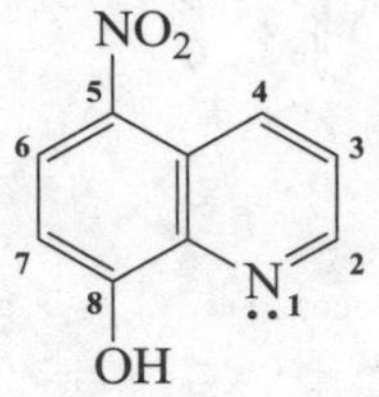

5-Nitro-8-hydroxyquinoline

12.38 The ring system of *acridine* ($C_{13}H_9N$) is analogous to that of anthracene (i.e., tricyclic and linearly fused). Furthermore, the two most stable resonance forms are equivalent to each other. The nitrogen atom must therefore be in the central ring, and the structure of acridine is

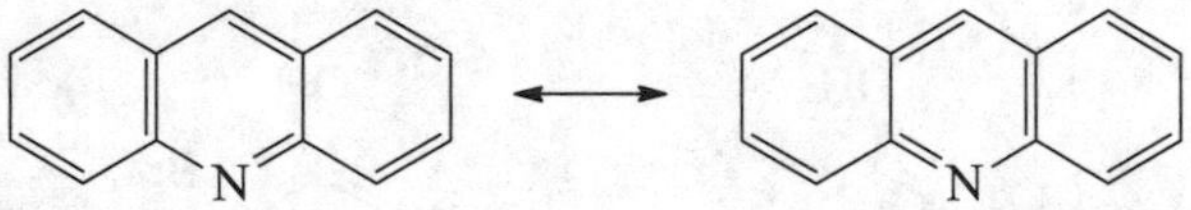

The two resonance forms would not be equivalent if the nitrogen were present in one of the terminal rings. The ring containing the nitrogen has only one resonance form with a benzenoid ring.

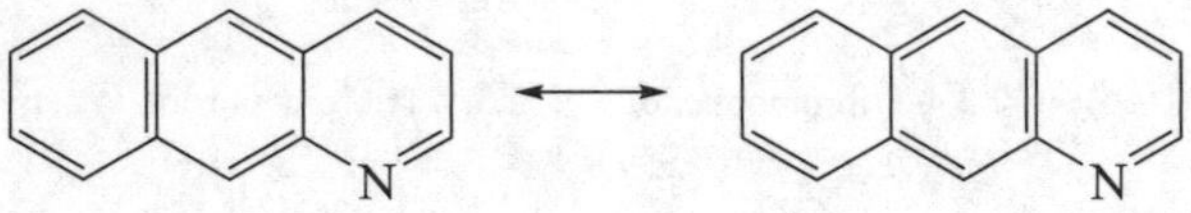

Resonance, Aromaticity, and Mechanism

12.39 (*a*) In the structure shown for naphthalene, one ring but not the other corresponds to a Kekulé form of benzene. We say that one ring is *benzenoid* and the other is not.

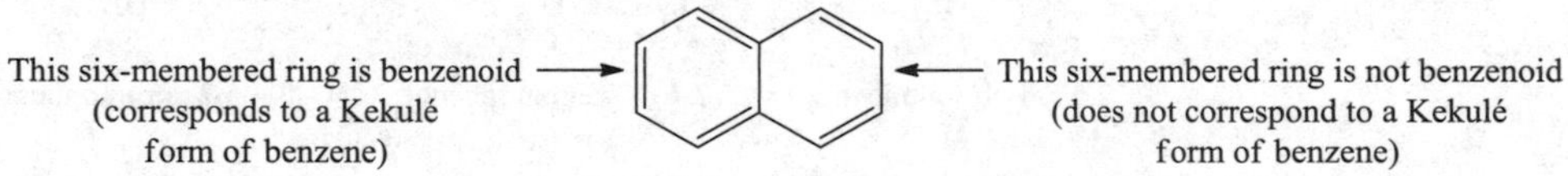

This six-membered ring is benzenoid ⟶   ⟵ This six-membered ring is not benzenoid
(corresponds to a Kekulé   (does not correspond to a Kekulé
form of benzene)   form of benzene)

By rewriting the benzenoid ring in its alternative Kekulé form, *both* rings become benzenoid.

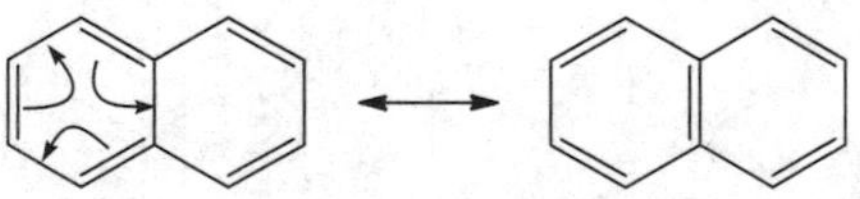

Both rings
are benzenoid

(*b*) Here a cyclobutadiene ring is fused to benzene. By writing the alternative resonance form of cyclobutadiene, the six-membered ring becomes benzenoid.

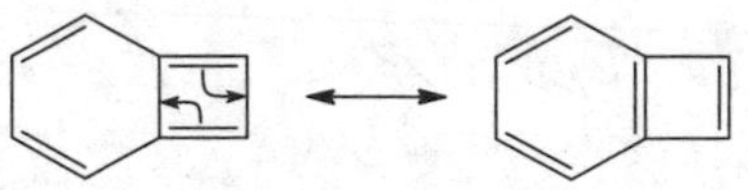

(*c*) The structure portrayed for phenanthrene contains two terminal benzenoid rings and a nonbenzenoid central ring. All three rings may be represented in benzenoid forms by converting one of the terminal six-membered rings to its alternative Kekulé form as shown:

Central ring
not benzenoid

All three rings
benzenoid

12.40 Cyclooctatetraene is not aromatic. 1,2,3,4-Tetramethylcyclooctatetraene and 1,2,3,8-tetramethyl-cyclooctatetraene are constitutional isomers.

1,2,3,4-Tetramethyl-
cyclooctatetratene

1,2,3,8-Tetramethyl-
cyclooctatetratene

Leo A. Paquette at Ohio State University synthesized each of these compounds independently of the other and showed them to be stable enough to be stored separately without interconversion.

12.41 Cyclooctatetraene has eight π electrons and thus does not satisfy the $(4n + 2)$ π electron requirement of the Hückel rule.

Cyclooctatetraene.
Each double bond contributes
two π electrons to give a total of eight.

All of the exercises in this problem involve counting the number of π electrons in the various species derived from cyclooctatetraene and determining whether they satisfy the $(4n + 2)$ π electron rule.

(a) Adding one π electron gives a species $C_8H_8^-$ with nine π electrons. $4n + 2$, where n is a whole number, can never equal nine. This species is therefore *not aromatic*.

(b) Adding two π electrons gives a species $(C_8H_8^{2-})$ with ten π electrons. $4n + 2 = 10$ when $n = 2$. The species $(C_8H_8^{2-})$ *is aromatic*.

(c) Removing one π electron gives a species $(C_8H_8^+)$ with seven π electrons. $4n + 2$ cannot equal 7. The species $(C_8H_8^+)$ *is not aromatic*.

(d) Removing two π electrons gives a species $(C_8H_8^{2+})$ with six π electrons. $4n + 2 = 6$ when $n = 1$. The species $(C_8H_8^{2+})$ *is aromatic*. (It has the same number of π electrons as benzene.)

12.42 (a, b) Cyclononatetraene does not have a continuous conjugated system of π electrons. Conjugation is incomplete because it is interrupted by a CH_2 group. Thus, (a) adding one more π electron or (b) two more π electrons will *not* give an aromatic system.

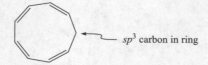

(c) Removing a proton from the CH_2 group permits complete conjugation. The species produced has ten π electrons and is aromatic, because $4n + 2 = 10$ when $n = 2$.

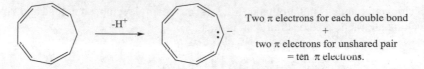

(d) Removing a proton from one of the sp^2-hybridized carbons of the ring does not produce complete conjugation; the CH_2 group remains present to interrupt cyclic conjugation. The anion formed is *not* aromatic.

12.43 (a) The more stable dipolar resonance structure is A because it has an aromatic cyclopentadienide anion bonded to an aromatic cyclopropenyl cation. In structure B, neither ring is aromatic.

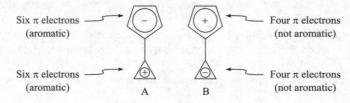

(b) Structure D can be stabilized by resonance involving a dipolar form.

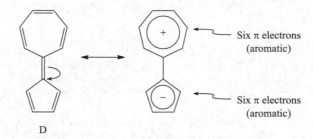

Comparable stabilization is not possible in structure C because neither a cyclopropenyl system nor a cycloheptatrienyl system is aromatic in its anionic form. Both are aromatic as cations.

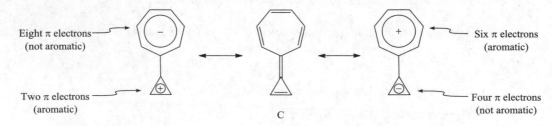

12.44 The major resonance contributor is A, because the electron distribution in each ring satisfies Hückel's rule of being a fully conjugated ring with $(4n + 2)$ π electrons, where $n = 1$.

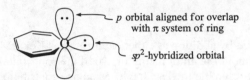

6 π electrons in 7-membered ring — — 6 π electrons in 5-membered ring

Other good resonance structures are possible, although they all have (+) charges in the seven-membered ring and (–) charges in the five-membered ring.

12.45 (*a*) This molecule, called *oxepin*, is *not aromatic*. It is antiaromatic. The three double bonds each contribute two π electrons, and an oxygen atom contributes two π electrons to the conjugated system, giving a total of eight π electrons. Only one of the two unshared pairs on oxygen can contribute to the π system; the other unshared pair is in an sp^2-hybridized orbital and cannot interact with it.

p orbital aligned for overlap with π system of ring

sp^2-hybridized orbital

(*b*) This compound, called *azonine,* has ten electrons in a completely conjugated planar monocyclic π system and therefore satisfies Hückel's rule for $(4n + 2)$ π electrons where $n = 2$. There are eight π electrons from the conjugated tetraene and two electrons contributed by the nitrogen unshared pair.

Two π electrons — Two π electrons

:NH ← Unshared pair on nitrogen is delocalized into the π system of ring

Two π electrons — Two π electrons

(*c*) Borazole, sometimes called *inorganic benzene,* is *aromatic*. Six π electrons are contributed by the unshared pairs of the three nitrogen atoms. Each boron contributes a *p* orbital to maintain the conjugated system but no electrons.

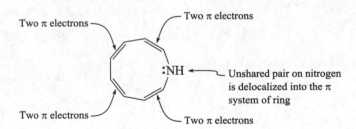

(*d*) This compound has eight π electrons and is *not aromatic*. It is antiaromatic.

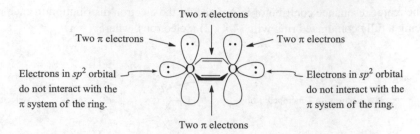

Two π electrons

Two π electrons

Two π electrons

Electrons in sp^2 orbital do not interact with the π system of the ring.

Electrons in sp^2 orbital do not interact with the π system of the ring.

Two π electrons

Reactions and Synthesis

12.46 The product that results from addition to the center ring has two intact benzene rings for which Kekulé structures can be written in all of the resonance forms.

In this compound, two rings correspond to Kekulé benzene. This holds true for all other significant resonance contributors that can be written for this product.

The product formed by addition to either outer ring has Kekulé benzene rings in some of the rings of its resonance forms but not in all of them. In two of the contributors, only one ring corresponds to Kekulé benzene, and they resemble the less stable resonance contributor for naphthalene.

The resonance contributors shown for the product resulting from the addition of bromine to an outer ring of anthracene. Only one ring corresponds to Kekulé benzene for the two contributors on the right.

12.47 As in bromination (Problem 12.46), it is the central ring of anthracene that is most reactive since the product has two intact benzene rings for which Kekulé structures can be written in all of the resonance forms. In the Diels–Alder reaction with the dienophile, maleic anhydride, the double bonds of the central ring of anthracene serve as the diene.

$C_{18}H_{12}O_3$

12.48 The fraction of benzylic substitution is given by the number of benzylic hydrogens (2) multiplied by their relative rate (5.9), divided by the sum of all of the hydrogens multiplied by their relative rates.

2(5.9) 2(4.0)
H   H

$$\frac{2(5.9)}{2(5.9)+2(2.7)+2(4.0)+3(1.0)} = 0.42$$

2(2.7) 3(1.0)

On this basis, the percentage of 1-chloro-1-phenylbutane in the product is calculated to be 42%.

12.49 Birch reduction of benzene and other aromatics yields a derivative of 1,4-cyclohexadiene. When alkyl or alkoxy groups are present on the ring, reduction is regioselective in favoring isomers in which these groups are substituents on the double bond.

2-Methoxynaphthalene

Na, NH₃
ethanol

1,4-Dihydro-
2-methoxynaphthalene

+

1,4-Dihydro-
6-methoxynaphthalene

These are the only structures that retain aromaticity in the non-reduced ring. Neither ring is benzenoid in other dihydro derivatives of 2-methoxynaphthalene containing the 1,4-cyclohexadiene moeity.

12.50 The stability of free radicals is reflected in their ease of formation. Toluene, which forms a benzyl radical, reacts with bromine 64,000 times faster than does ethane, which forms a primary alkyl radical. Ethylbenzene, which forms a secondary benzylic radical, reacts 1 million times faster than ethane.

+ Br· ⟶   + HBr

Ethylbenzene
(most reactive)

Secondary benzylic
radical

+ Br· ⟶   + HBr

Toluene

Primary benzylic
radical

$H_3C{-}CH_3$ + Br· ⟶ $H_3C{-}\overset{\bullet}{C}H_2$ + HBr

Ethane
(least reactive)

Primary
radical

12.51 The dihydronaphthalene in which the double bond is conjugated with the aromatic ring is more stable; thus, 1,2-dihydronaphthalene has a lower heat of hydrogenation than 1,4-dihydronaphthalene.

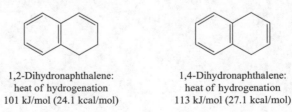

1,2-Dihydronaphthalene:
heat of hydrogenation
101 kJ/mol (24.1 kcal/mol)

1,4-Dihydronaphthalene:
heat of hydrogenation
113 kJ/mol (27.1 kcal/mol)

12.52 The rate-determining step in an S_N1 process is formation of the carbocation. The reactivity order can be determined by examining the stability of the carbocations formed remembering the general principle that the more stable a carbocation is, the faster it is formed. All three carbocations are secondary; however, the one formed from 3-iodocyclopentene is stabilized by conjugation with the double bond and is formed fastest. The carbocation formed from 5-iodo-1,3-cyclopentadiene is also conjugated, but it is a cyclic, fully conjugated π system having four π electrons and thus is *antiaromatic*. It is the least stable, and is formed slowest.

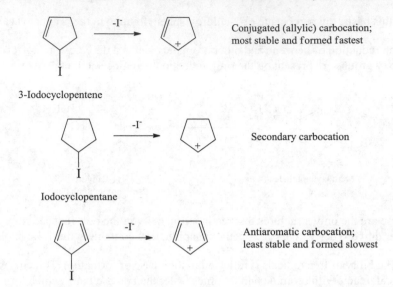

12.53 The process is an acid–base reaction in which cyclopentadiene transfers a proton to amide ion (the base) to give the aromatic cyclopentadienide anion. The sodium ion (Na^+) has been omitted from the equation.

1,3-Cyclopentadiene   Amide ion   Cyclopentadienide   Ammonia
                                                                anion

12.54 The resonance structures for the cyclopentadienide anions formed by loss of a proton from 1-methyl-1,3-cyclopentadiene and 5-methyl-1,3-cyclopentadiene are equivalent.

1-Methyl-1,3-cyclopentadiene   5-Methyl-1,3-cyclopentadiene

$NaNH_2, NH_3$ $NaNH_2, NH_3$

12.55 1,4-Addition to the diene unit of furan gives the unstable product shown.

$$\text{furan} \xrightarrow[\substack{CCl_4 \\ -10\ ^\circ C}]{Br_2} \text{Br-O-Br}$$

$$C_4H_4Br_2O$$

12.56 (*a*) Hydrogenation of isopropylbenzene converts the benzene ring to a cyclohexane unit.

$$\text{Isopropylbenzene} \xrightarrow[\text{Pt}]{H_2\ (3\ mol)} \text{Isopropylcyclohexane}$$

Isopropylbenzene   Isopropylcyclohexane

(*b*) Sodium and ethanol in liquid ammonia is the combination of reagents that brings about Birch reduction of benzene rings. The 1,4-cyclohexadiene that is formed has its isopropyl group as a substituent on one of the double bonds.

$$\text{Isopropylbenzene} \xrightarrow[\text{NH}_3]{Na,\ ethanol} \text{1-Isopropyl-1,4-cyclohexadiene}$$

Isopropylbenzene   1-Isopropyl-1,4-cyclohexadiene

(*c*) Oxidation of the isopropyl side chain occurs. The benzene ring remains intact.

$$\text{Isopropylbenzene} \xrightarrow[\text{H}_2SO_4,\ heat]{Na_2Cr_2O_7} \text{Benzoic acid}$$

Isopropylbenzene   Benzoic acid

(*d*) *N*-Bromosuccinimide (NBS) is a reagent effective for the substitution of a benzylic hydrogen.

$$\text{Isopropylbenzene} \xrightarrow[\substack{\text{benzoyl peroxide} \\ heat}]{NBS} \text{2-Bromo-2-phenylpropane}$$

Isopropylbenzene   2-Bromo-2-phenylpropane

(*e*) The tertiary bromide undergoes E2 elimination to give a carbon–carbon double bond.

$$\text{2-Bromo-2-phenylpropane} \xrightarrow[\text{CH}_3CH_2OH]{NaOCH_2CH_3} \text{2-Phenylpropene}$$

2-Bromo-2-phenylpropane   2-Phenylpropene

12.57 All the specific reactions in this problem have been reported in the chemical literature with results as indicated.

(*a*) Hydroboration–oxidation of alkenes leads to syn anti-Markovnikov hydration of the double bond.

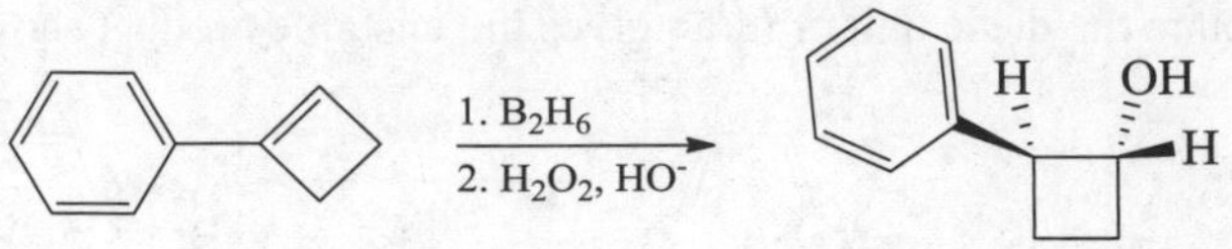

1-Phenylcyclobutene

1. B$_2$H$_6$
2. H$_2$O$_2$, HO$^-$

trans-2-Phenyl-
cyclobutanol (82%)

(b) The compound contains a substituted benzene ring and an alkene-like double bond. When hydrogenation of this compound was carried out, the alkene-like double bond was hydrogenated cleanly.

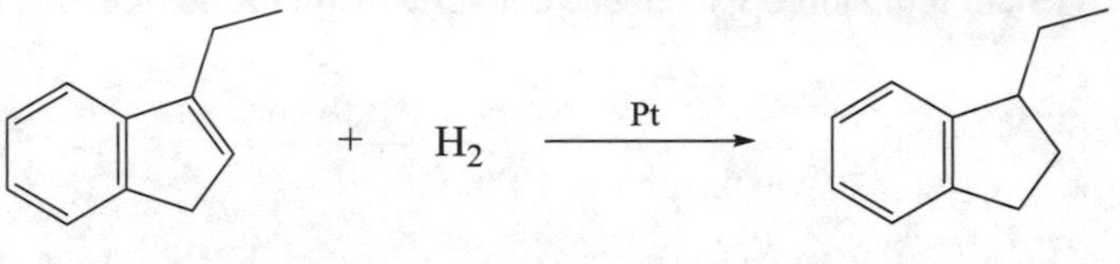

1-Ethylindene   +   H$_2$ $\xrightarrow{\text{Pt}}$ 1-Ethylindane (80%)

(c) Free-radical chlorination will lead to substitution of benzylic hydrogens. The starting material contains four benzylic hydrogens, all of which may eventually be replaced.

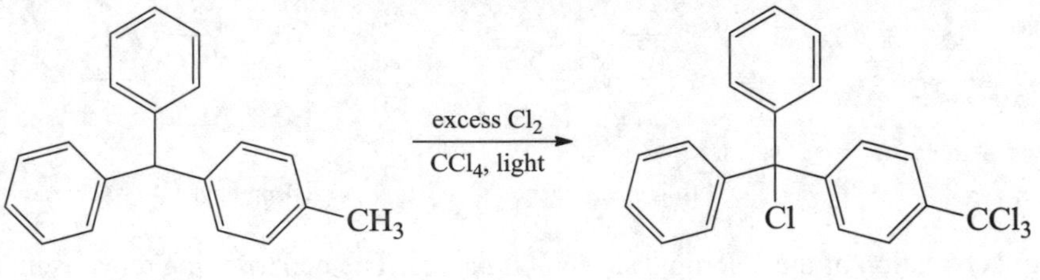

$\xrightarrow[\text{CCl}_4,\text{ light}]{\text{excess Cl}_2}$

CH$_3$ Cl   CCl$_3$

(65%)

(d) Epoxidation of alkenes is stereospecific. Groups that are trans in the alkene remain trans in the epoxide.

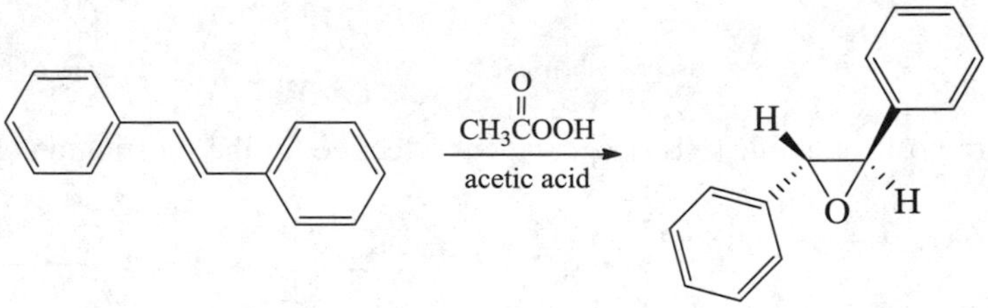

$\xrightarrow[\text{acetic acid}]{\overset{\overset{\text{O}}{\|}}{\text{CH}_3\text{COOH}}}$

(E)-1,2-Diphenylethene   *trans*-1,2-Epoxy-1,2-diphenylethane
(78%-83%)

(e) The reaction is one of acid-catalyzed alcohol dehydration.

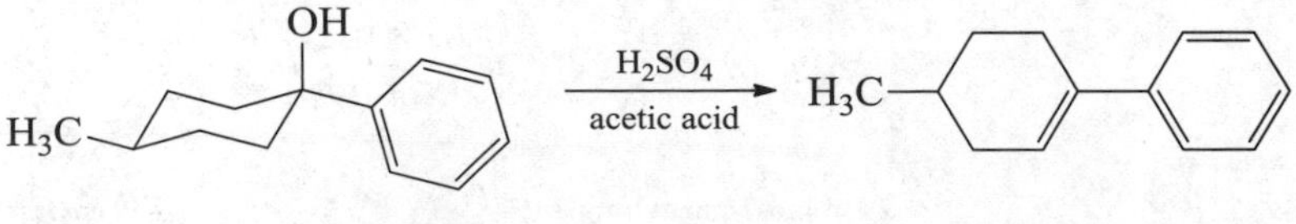

$\xrightarrow[\text{acetic acid}]{\text{H}_2\text{SO}_4}$

cis-4-Methyl-1-phenylcyclohexanol   4-Methyl-1-
phenylcyclohexene (81%)

(f) This reaction illustrates identical reactivity at two equivalent sites in a molecule. Both alcohol functions are tertiary and benzylic and undergo acid-catalyzed dehydration readily.

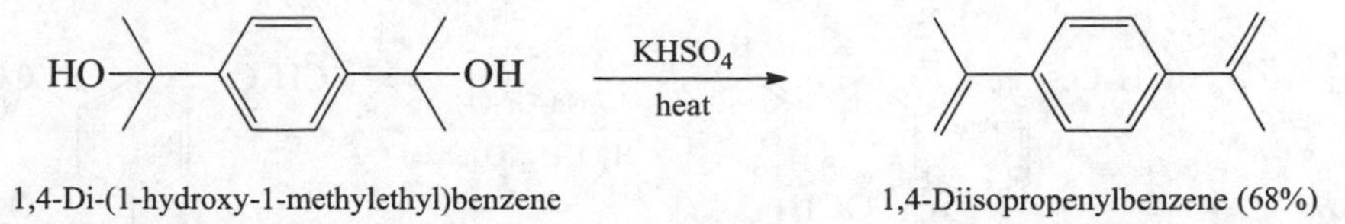

1,4-Di-(1-hydroxy-1-methylethyl)benzene   1,4-Diisopropenylbenzene (68%)

(*g*) The compound shown is DDT (standing for the nonsystematic name *dichlorodiphenyltrichloroethane*). It undergoes β elimination to form an alkene.

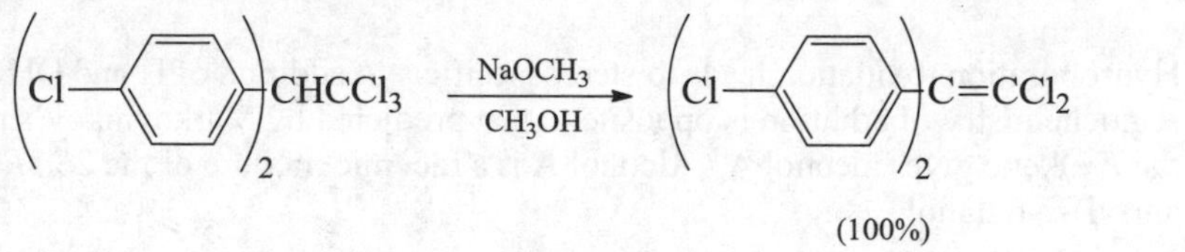

(100%)

(*h*) Alkyl side chains on naphthalene undergo reactions analogous to those of alkyl groups on benzene.

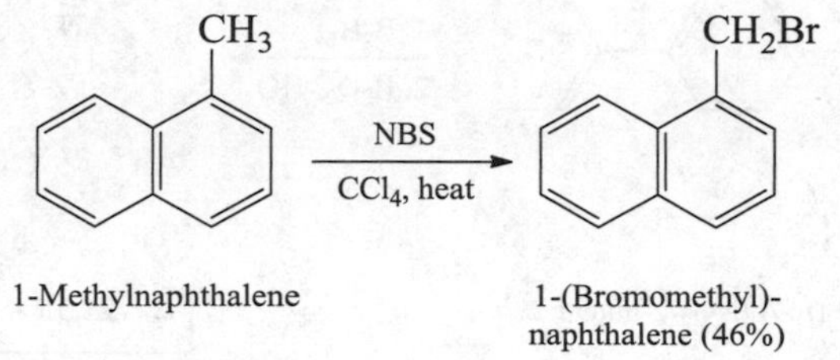

1-Methylnaphthalene   1-(Bromomethyl)-naphthalene (46%)

(*i*) Potassium carbonate is a weak base. Hydrolysis of the primary benzylic halide converts it to an alcohol.

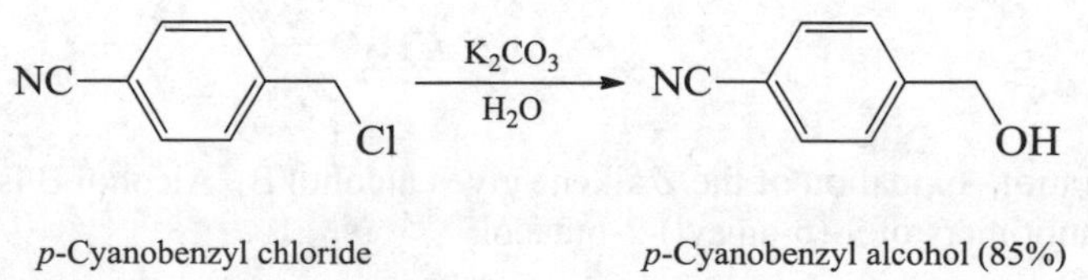

p-Cyanobenzyl chloride   *p*-Cyanobenzyl alcohol (85%)

12.58 Only benzylic (or allylic) hydrogens are replaced by *N*-bromosuccinimide (NBS). Among the four bromines in 3,4,5-tribromobenzyl bromide, three are substituents on the ring and are incapable of being introduced by benzylic bromination. The starting material must therefore have these three bromines already in place.

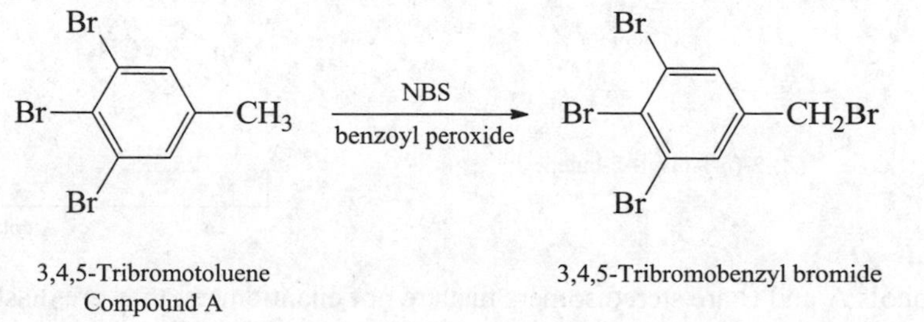

3,4,5-Tribromotoluene
Compound A   3,4,5-Tribromobenzyl bromide

12.59 2,3,5-Trimethoxybenzoic acid has the structure shown. The three methoxy groups occupy the same positions in this oxidation product that they did in the unknown compound. The carboxylic acid function must have arisen by oxidation of the —$CH_2CH=C(CH_3)_2$ side chain. Therefore

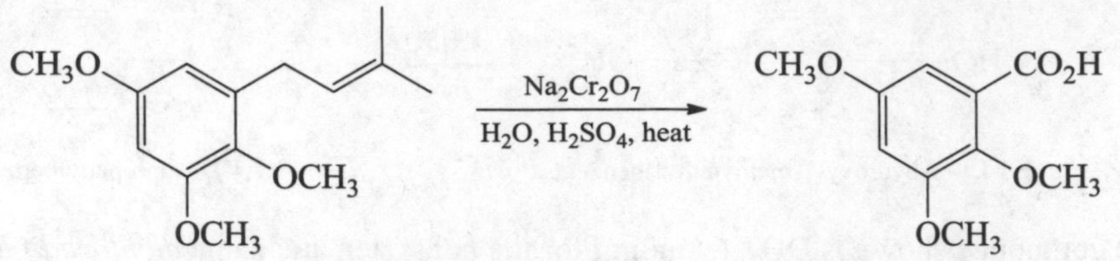

($C_{14}H_{20}O_3$)   2,3,5-Trimethoxybenzoic acid

12.60 Hydroboration–oxidation leads to stereospecific syn addition of H and OH to the double bond. The regiochemistry of addition is opposite to that predicted by Markovnikov's rule. Hydroboration–oxidation of the *E* alkene gives alcohol A. Alcohol A is a racemic mixture of the 2*S*,3*R* and 2*R*,3*S* enantiomers of 3-(*p*-anisyl)-2-butanol.

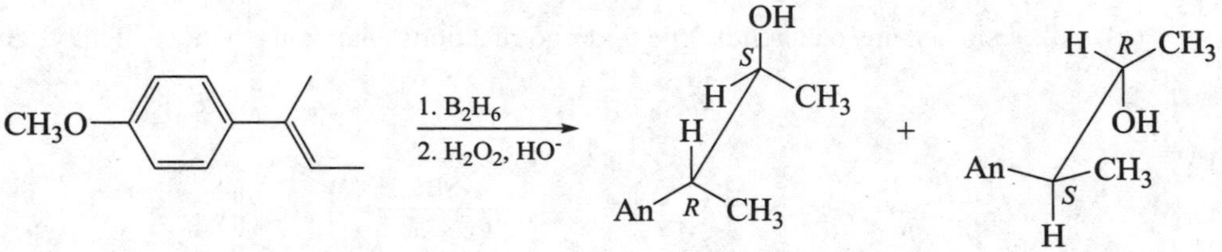

(*E*)-2-(*p*-Anisyl)-2-butene   2*S*,3*R*   2*R*,3*S*

Alcohol A

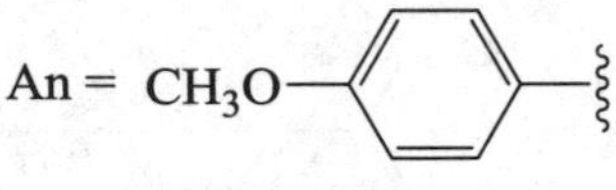

An = CH$_3$O— \ce{<benzene>} —

Hydroboration–oxidation of the *Z* alkene gives alcohol B. Alcohol B is a racemic mixture of the 2*R*,3*R* and 2*S*,3*S* enantiomers of 3-(*p*-anisyl)-2-butanol.

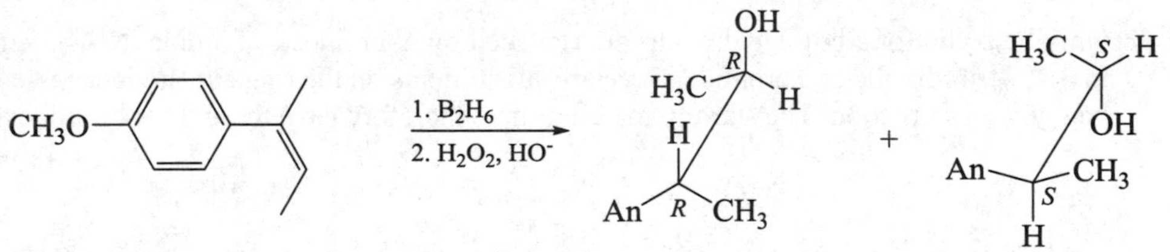

(*Z*)-2-(*p*-Anisyl)-2-butene   2*R*,3*R*   2*S*,3*S*

Alcohol B

Alcohols A and B are stereoisomers that are not enantiomers; they are diastereomers.

12.61 (*a*) The conversion of ethylbenzene to 1-phenylethyl bromide is a benzylic bromination. It can be achieved by using either bromine or *N*-bromosuccinimide (NBS).

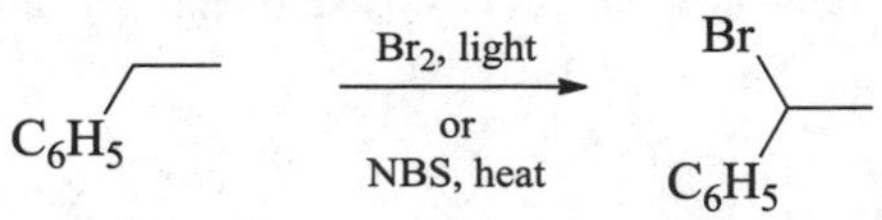

Ethylbenzene   1-Phenylethyl bromide

(*b*) The conversion of 1-phenylethyl bromide to 1,2-dibromo-1-phenylethane

$$\underset{C_6H_5}{\overset{Br}{\diagdown}}\!\!-\!\!\overset{Br}{\diagdown} \quad \Longrightarrow \quad \underset{C_6H_5}{\overset{Br}{\diagdown}}\!\!-$$

cannot be achieved in a single step. We must reason backward from the target molecule, that is, determine how to make 1,2-dibromo-1-phenylethane in one step from any starting material. Vicinal dibromides are customarily prepared by addition of bromine to alkenes. This suggests that 1,2-dibromo-1-phenylethane can be prepared by the reaction

$$C_6H_5\!\!-\!\!CH\!\!=\!\!CH_2 \;+\; Br_2 \;\longrightarrow\; \underset{C_6H_5}{\overset{Br}{\diagdown}}\!\!-\!\!CH_2Br$$

Styrene   1,2-Dibromo-1-phenylethane

The necessary alkene, styrene, is available by dehydrohalogenation of the given starting material, 1-phenylethyl bromide.

$$\underset{C_6H_5}{\overset{Br}{\diagdown}}\!\!- \quad\xrightarrow[\text{CH}_3\text{CH}_2\text{OH}]{\text{NaOCH}_2\text{CH}_3}\quad C_6H_5\!\!-\!\!CH\!\!=\!\!CH_2$$

1-Phenylethyl bromide   Styrene

Thus, by reasoning backward from the target molecule, the synthetic scheme becomes apparent.

$$\underset{C_6H_5}{\overset{Br}{\diagdown}}\!\!- \;\xrightarrow[\text{CH}_3\text{CH}_2\text{OH}]{\text{NaOCH}_2\text{CH}_3}\; C_6H_5\!\!-\!\!CH\!\!=\!\!CH_2 \;\xrightarrow{\text{Br}_2}\; \underset{C_6H_5}{\overset{Br}{\diagdown}}\!\!-\!\!CH_2Br$$

1-Phenylethyl bromide   Styrene   1,2-Dibromo-1-phenylethane

(c) The conversion of styrene to phenylacetylene cannot be carried out in a single step. As was pointed out in Chapter 9, however, a standard sequence for converting terminal alkenes to alkynes consists of bromine addition followed by a double dehydrohalogenation in strong base.

$$C_6H_5\!\!-\!\!CH\!\!=\!\!CH_2 \;\xrightarrow{\text{Br}_2}\; \underset{C_6H_5}{\overset{Br}{\diagdown}}\!\!-\!\!CH_2Br \;\xrightarrow[\text{NH}_3]{\text{NaNH}_2}\; C_6H_5\!\!-\!\!C\!\!\equiv\!\!CH$$

Styrene   1,2-Dibromo-1-phenylethane   Phenylacetylene

(d) The conversion of phenylacetylene to butylbenzene requires both a carbon–carbon bond formation step and a hydrogenation step. The acetylene function is essential for carbon–carbon bond formation by alkylation. The correct sequence is therefore

$$C_6H_5\!\!-\!\!C\!\!\equiv\!\!CH \;\xrightarrow[\text{NH}_3]{\text{NaNH}_2}\; C_6H_5\!\!-\!\!C\!\!\equiv\!\!C\!:^-\,Na^+$$

Phenylacetylene

$$C_6H_5\!\!-\!\!C\!\!\equiv\!\!C\!:^-\,Na^+ \;+\; \diagup\!\!\!\diagdown\!\!\!\diagup\text{Br} \;\longrightarrow\; C_6H_5\!\!-\!\!C\!\!\equiv\!\!C\!\!-\!\!CH_2CH_2CH_3$$

$$C_6H_5\ \text{—}\!\!\equiv\!\!\text{—}\ \xrightarrow[\text{Pt}]{H_2}\ C_6H_5\ \text{—}\!\!\!\!\text{—}$$

Butylbenzene

(e) The transformation corresponds to alkylation of acetylene, and so the alcohol must first be converted to a species with a good leaving group such as its halide derivative.

$$C_6H_5 \text{—}\!\!\!\text{—} OH \xrightarrow{PBr_3} C_6H_5 \text{—}\!\!\!\text{—} Br$$

2-Phenylethanol   2-Phenylethyl bromide

$$C_6H_5 \text{—}\!\!\!\text{—} Br\ +\ HC\!\equiv\!C\!:^-\ Na^+\ \longrightarrow\ C_6H_5 \text{—}\!\!\!\text{—}\!\!\equiv$$

2-Phenylethyl bromide   Sodium   4-Phenyl-1-butyne
                                                acetylide

(f) The target compound is a bromohydrin. Bromohydrins are formed by addition of bromine and water to alkenes.

$$C_6H_5 \text{—}\!\!\!\text{—} Br \xrightarrow[\text{(CH}_3)_3\text{COH}]{KOC(CH_3)_3} C_6H_5 \text{—}\!\!\!\text{=} \xrightarrow[\text{H}_2\text{O}]{Br_2} C_6H_5 \overset{\text{OH}}{\underset{}{\text{—}\!\!\!\text{—}}} Br$$

2-Phenylethyl bromide   Styrene   2-Bromo-1-phenyl-
                                                               ethanol

12.62 The structure and numbering system for pyridine are given in Section 12.21 of the text, where we were told that pyridine is aromatic. Oxidation of 3-methylpyridine is analogous to oxidation of toluene. The methyl side chain is oxidized to a carboxylic acid.

3-Methylpyridine   Niacin

Answers to Interpretive Problems 12

12.63 (a) A negative ρ value indicates a positive charge is present in the rate determining step. This is consistent with an S_N1 reaction.

(b) The fastest rate is predicted for Ar = m-methylphenyl, since it is the only one in the group A-E with a negative value of σ.

(c) The slowest rate is when Ar = m-benzonitrile since it has a σ_{meta} of +0.62. The largest positive value indicates that it would best destabilize a positive charge in the transition state.

(d) To solve for relative rates, use the Hammett equation and solve for k_x in terms of k_0 for the fastest and slowest compounds.

$$\log \frac{k_{m\text{-Me}}}{k_0} = (-0.06)(-5.1) \qquad\qquad \log \frac{k_{m\text{-CN}}}{k_0} = (+0.62)(-5.1)$$

$$\frac{k_{m\text{-Me}}}{k_0} = 10^{(-0.06)(-5.1)} \qquad\qquad \frac{k_{m\text{-CN}}}{k_0} = 10^{(+0.62)(-5.1)}$$

$$\frac{k_{m\text{-Me}}}{k_0} = 10^{0.306} \qquad\qquad \frac{k_{m\text{-CN}}}{k_0} = 10^{-3.16}$$

$$k_{m\text{-Me}} = k_0\,(2.02) \qquad\qquad k_{m\text{-CN}} = k_0\,(6.89 \times 10^{-4})$$

Dividing these equations to solve for relative rates $k_{m\text{-Me}}/k_{m\text{-CN}}$ shows the compound with the *m*-toluene substituent reacting almost 3,000 times faster than the compound having the *m*-benzonitrile substituent:

$$\frac{k_{m\text{-Me}}}{k_{m\text{-CN}}} = \frac{k_0\,(2.02)}{k_0\,(6.89 \times 10^{-4})}$$

$$\frac{k_{m\text{-Me}}}{k_{m\text{-CN}}} = 2.9 \times 10^3$$

12.64 A positive ρ value indicates a negative charge is present in the rate determining step. This is consistent with A (Step 1).

12.65 (*a*) A positive ρ value of +0.8 indicates that a negative charge is present in the rate determining step. The magnitude of this value which is less than one, shows that only a slight negative charge is present at the reaction center. If bond breaking is slower than bond making, a slight negative charge on the carbon in an S_N2 transition state will be formed.

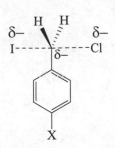

(b) Compound A has the largest σ_{para} value of +0.70, so it will react fastest.

(c) Compound C has the smallest σ_{para} value of -0.14, so it will react slowest.

(d) Solving as in Problem 12.63. We see that compound A, with the *para*-cyanophenyl substituent, reacts 4.7 times faster than Compound C, with the *para*-methylphenyl substituent.

$$\log \frac{k_{p\text{-CN}}}{k_0} = (+0.70)(+0.8) \qquad \log \frac{k_{p\text{-Me}}}{k_0} = (-0.14)(+0.8)$$

$$\frac{k_{p\text{-CN}}}{k_0} = 10^{(+0.70)(+0.8)} \qquad \frac{k_{p\text{-Me}}}{k_0} = 10^{(-0.14)(+0.8)}$$

$$\frac{k_{p\text{-CN}}}{k_0} = 10^{0.56} \qquad \frac{k_{p\text{-Me}}}{k_0} = 10^{-0.11}$$

$$k_{p\text{-CN}} - k_0 \ (3.63) \qquad k_{p\text{-Me}} = k_0 \ (0.77)$$

$$\frac{k_{p\text{-CN}}}{k_{p\text{-Me}}} = \frac{k_0 \ (3.63)}{k_0 \ (0.77)}$$

$$\frac{k_{p\text{-CN}}}{k_{p\text{-Me}}} = 4.70$$

12.66 The greater the magnitude of ρ indicates a greater charge in the transition state. The positive ρ value indicates a negative charge in the transition state of both compounds. The reactant B with the larger ρ value has greater C—H bond breaking in the transition state structure.

SELF-TEST

1.   Give an acceptable IUPAC name for each of the following:

(*a*)

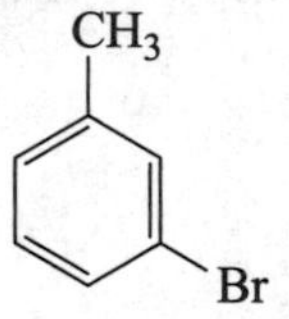

(*c*)

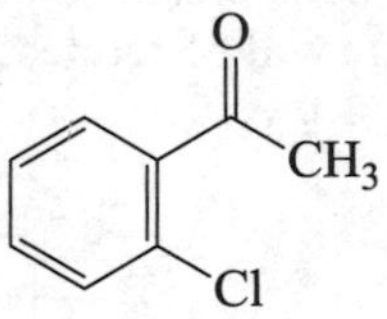

(*b*)

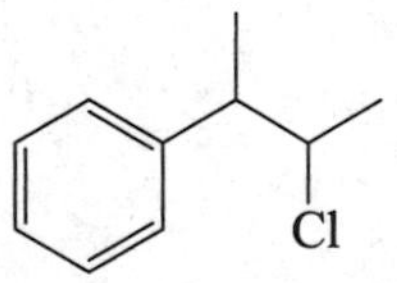

(*d*)

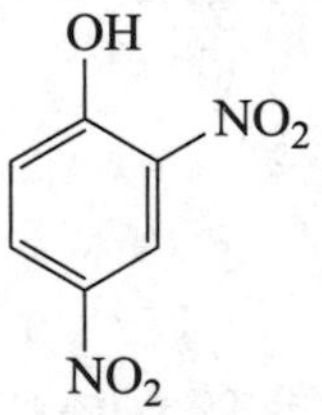

2.   Draw the structure of each of the following:

(*a*) 3,5-Dichlorobenzoic acid   (*c*)   2,4-Dimethylaniline

(*b*) *p*-Nitroanisole   (*d*)   *m*-Bromobenzyl chloride

3.   Write a positive (+) or negative (−) charge at the appropriate position so that each of the following structures contains the proper number of π electrons to permit it to be considered an aromatic ion. For this problem, ignore strain effects that might destabilize the molecule.

(a)

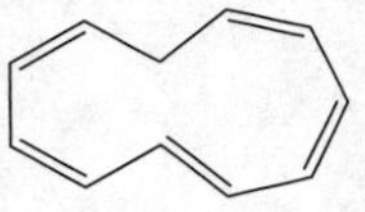

(b)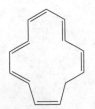

4. For each of the following, determine how many π electrons are counted toward satisfying Hückel's rule. Assuming the molecule can adopt a planar conformation, is it aromatic?

(a)

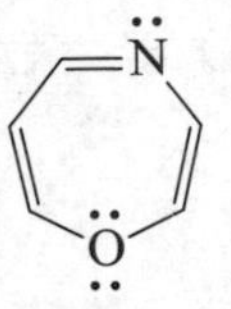

(b)

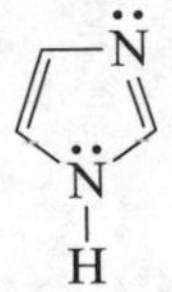

(c)

5. Azulene, shown in the following structure, is highly polar. Draw a dipolar resonance structure to explain this fact.

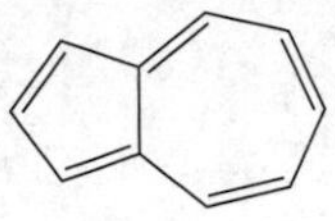

Azulene

6. Give the reactant, reagent, or product omitted from each of the following:

(a)
$$\xrightarrow[\text{peroxides, heat}]{\text{NBS}} ?$$

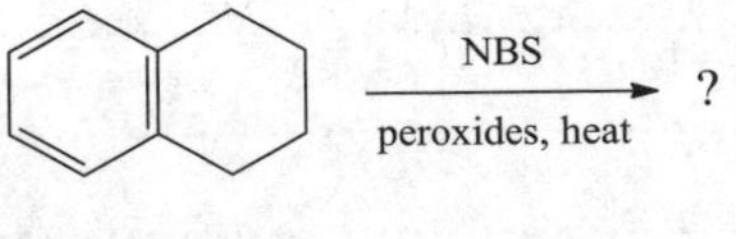

(d)
$$\xrightarrow{?}$$
(product shown: Cl-substituted benzoic acid)

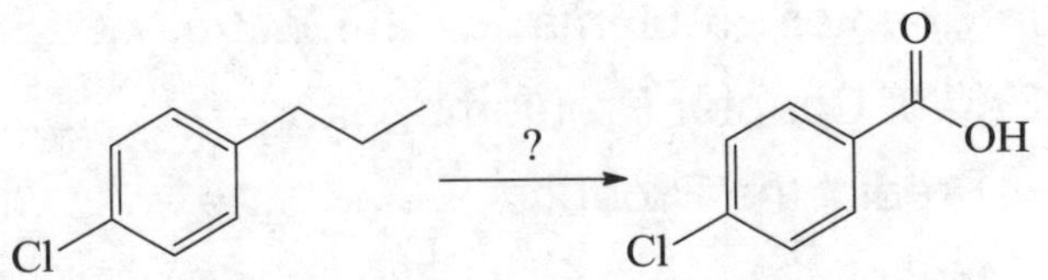

(b)
$$? \xrightarrow[\text{CH}_3\text{OH}]{\text{NaOCH}_3}$$

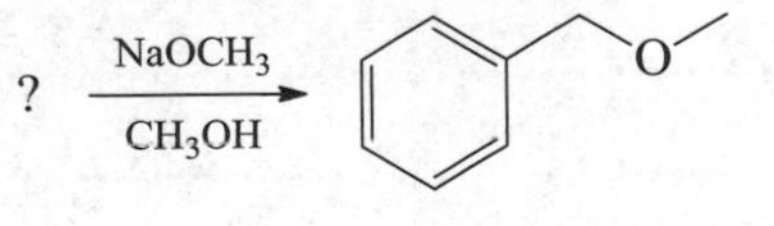

(e)
$$\xrightarrow[]{\overset{\text{O}}{\underset{\|}{\text{CH}_3\text{COOH}}}} ?$$

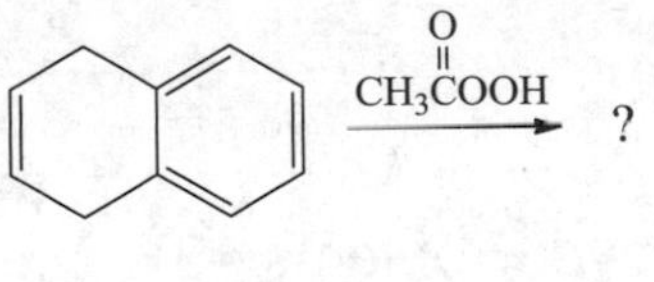

(c)
$$\xrightarrow[\text{H}^+]{\text{H}_2\text{O}} ?$$

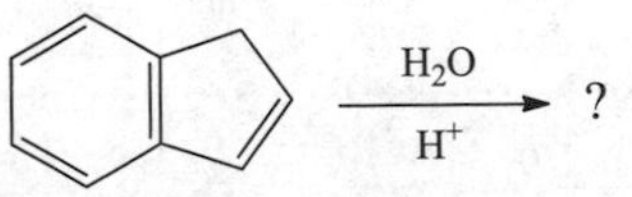

(f)
$$\xrightarrow[\text{H}_2\text{O}]{\text{Br}_2} ?$$

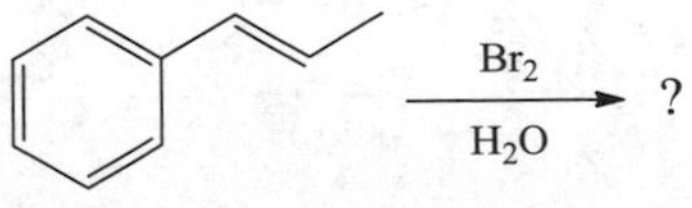

7. Provide two methods for the synthesis of 1-bromo-1-phenylpropane from an aromatic hydrocarbon.

8. Write the structures of the resonance forms that contribute to the stabilization of the intermediate in the reaction of styrene ($C_6H_5CH=CH_2$) with hydrogen bromide in the absence of peroxides.

9. Write one or more resonance structures that represent the delocalization of the following carbocation.

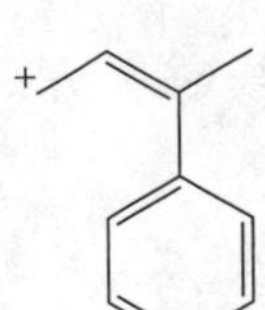

10. An unknown compound, $C_{12}H_{18}$, reacts with sodium dichromate ($Na_2Cr_2O_7$) in warm aqueous sulfuric acid to give *p-tert*-butylbenzoic acid. What is the structure of the unknown?

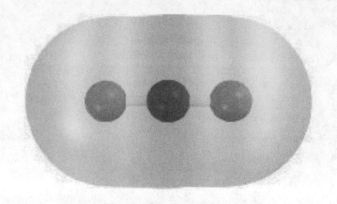

CHAPTER 13

Electrophilic and Nucleophilic Aromatic Substitution

Table of Contents

SOLUTIONS TO TEXT PROBLEMS

In Chapter Problems

13.1   According to the mechanism of electrophilic aromatic substitution, formation of the cyclohexadienyl cation is rate-determining and endothermic. The transition state for an endothermic elementary step in a mechanism more closely resembles the products of that step, in this case the carbocation intermediate, than it resembles the reactants. Therefore, the transition state for electrophilic aromatic substitution more closely resembles the carbocation intermediate.

13.2   Electrophilic aromatic substitution leads to replacement of one of the hydrogens directly attached to the ring. All four of the ring hydrogens of *p*-xylene are equivalent; so it does not matter which one is replaced by the nitro group.

1,4-Dimethylbenzene
(*p*-Xylene)

1,4-Dimethyl-2-
nitrobenzene

13.3   Following the pathway outlined in Mechanism 13.1, the aromatic ring reacts with nitronium ion to give the cyclohexadienyl cation intermediate.

1,4-Dimethyl-   Nitronium
benzene   ion

Cyclohexadienyl cation
intermediate

Loss of a proton from the intermediate gives the product, 1,4-dimethyl-2-nitrobenzene.

Cyclohexadienyl cation   Water
intermediate

1,4-Dimethyl-2-
nitrobenzene

Hydronium ion

13.4   The aromatic ring of 1,2,4,5-tetramethylbenzene has two equivalent hydrogens. Sulfonation of the ring leads to replacement of one of them by −SO₃H.

1,2,4,5-Tetramethylbenzene

2,3,5,6-Tetramethylbenzene-
sulfonic acid

13.5   Acetyl hypoiodite reacts with benzene to give acetate and the cyclohexadienyl cation intermediate.

Benzene   Acetyl hypoiodite   Cyclohexadienyl   Acetate ion
                                            cation intermediate

Acetate then abstracts a proton from the carbocation intermediate to give iodobenzene and acetic acid.

Cyclohexadienyl   Acetate ion   Iodobenzene   Acetic acid
cation intermediate

13.6   The major product is isopropylbenzene.

Benzene   1-Chloropropane   Propylbenzene   Isopropylbenzene
                                          (20% yield)   (40% yield)

Aluminum chloride coordinates with 1-chloropropane to give a Lewis acid–Lewis base complex, which can react with benzene to yield propylbenzene or can rearrange to produce isopropyl cation. Isopropylbenzene arises by reaction of isopropyl cation with benzene.

1-Chloropropane/aluminum chloride   Isopropyl cation
complex

13.7   The species that reacts with the benzene ring is cyclohexyl cation, formed by protonation of cyclohexene.

Cyclohexene   Sulfuric acid   Cyclohexyl cation   Hydrogen sulfate ion

The mechanism for the reaction of cyclohexyl cation with benzene is analogous to the general mechanism for electrophilic aromatic substitution.

Benzene   Cyclohexyl   Cyclohexadienyl cation   Cyclohexylbenzene
                 cation   intermediate

13.8 The reaction of benzene with 2-methylpropene in the presence of an acid catalyst gives *tert*-butylbenzene.

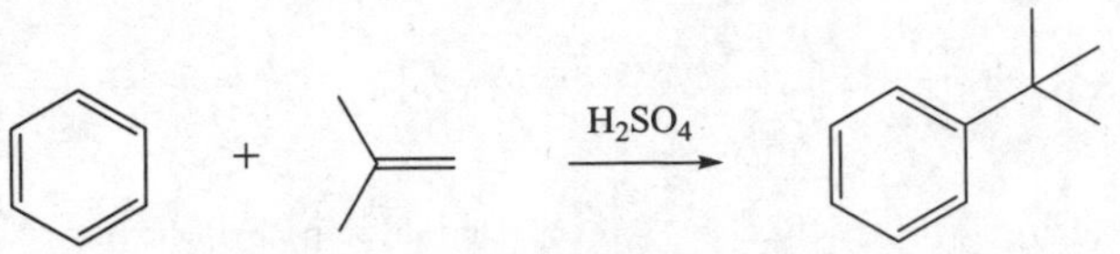

Instead of an alkene, *tert*-butyl alcohol may be used.

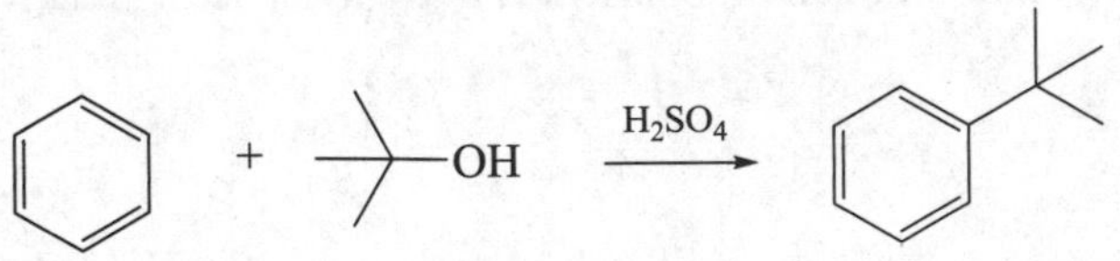

The electrophile that is generated either from the alkene or alcohol for both of these reactions is the same, *tert*-butyl carbocation: $(CH_3)_3C^+$.

13.9 The preparation of cyclohexylbenzene from benzene and cyclohexene was described in text Section 13.6. Cyclohexylbenzene is converted to 1-phenylcyclohexene by benzylic bromination, followed by dehydrohalogenation.

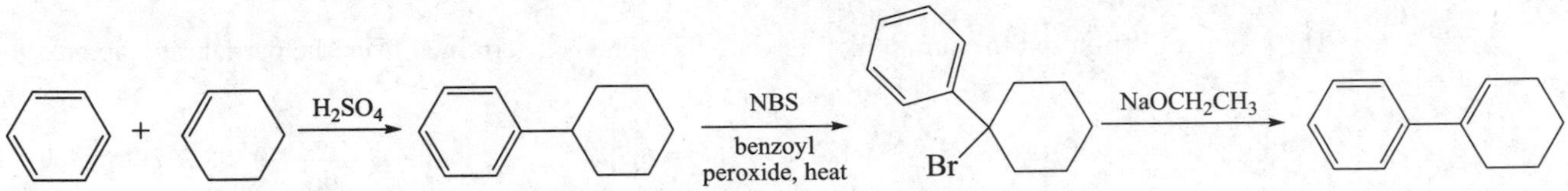

| Benzene | Cyclohexene | | Phenylcyclohexane | | 1-Bromo-1-phenylcyclohexane | | 1-Phenylcyclohexene |

13.10 Treatment of 1,3,5-trimethoxybenzene with an acyl chloride and aluminum chloride brings about Friedel–Crafts acylation at one of the three equivalent positions available on the ring.

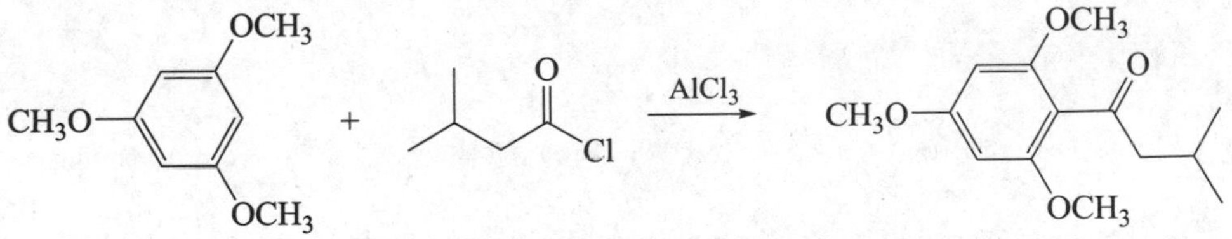

1,3,5-Trimethoxybenzene   3-Methylbutanoyl chloride   Isobutyl 2,4,6-trimethoxyphenyl ketone

13.11 Because the anhydride is cyclic, its structural units are not incorporated into a ketone and a carboxylic acid as two separate product molecules. Rather, they become part of a four-carbon unit attached to benzene by a ketone carbonyl. The acyl substituent terminates in a carboxylic acid functional group.

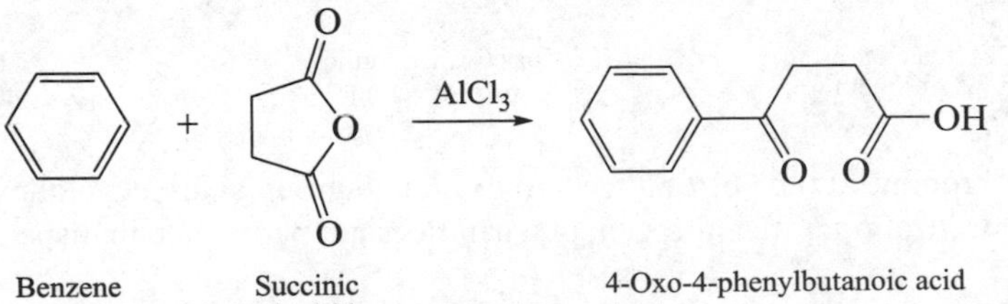

Benzene   Succinic anhydride   4-Oxo-4-phenylbutanoic acid

13.12 (*b*) A Friedel–Crafts alkylation of benzene using 1-chloro-2,2-dimethylpropane would not be a satisfactory method to prepare (2,2-dimethylpropyl)benzene because of the likelihood of a carbocation rearrangement. The best way to prepare this compound is by Friedel–Crafts acylation followed by Clemmensen (as shown below) or Wolff-Kishner reduction.

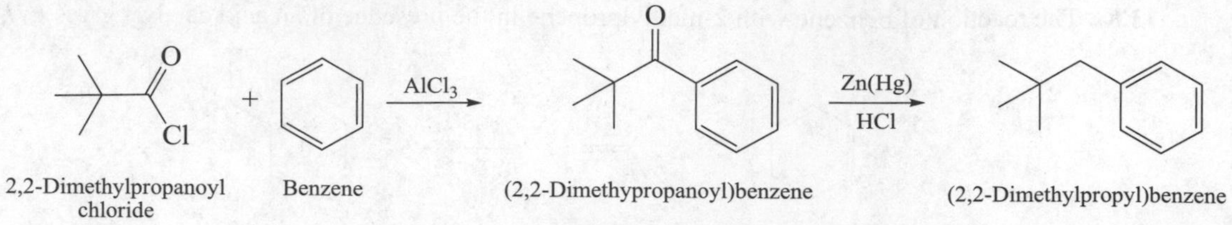

2,2-Dimethylpropanoyl chloride   Benzene   (2,2-Dimethypropanoyl)benzene   (2,2-Dimethylpropyl)benzene

13.13 (*b*) Partial rate factors for nitration of toluene and *tert*-butylbenzene, relative to a single position of benzene, are as shown:

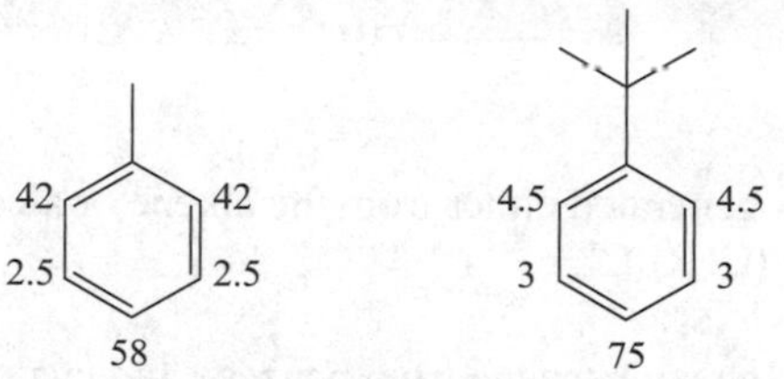

The sum of these partial rate factors is 147 for toluene, 90 for *tert*-butylbenzene. Toluene is 147/90, or 1.7, times more reactive than *tert*-butylbenzene.

(*c*) The product distribution for nitration of *tert*-butylbenzene is determined from the partial rate factors.

$$\text{Ortho: } \frac{2(4.5)}{90} = 10\% \qquad \text{Meta: } \frac{2(3)}{90} = 6.7\% \qquad \text{Para: } \frac{75}{90} = 88.3\%$$

13.14 The compounds shown all undergo electrophilic aromatic substitution more slowly than benzene. Therefore, —CH_2Cl, —$CHCl_2$, and —CCl_3 are *deactivating* substituents.

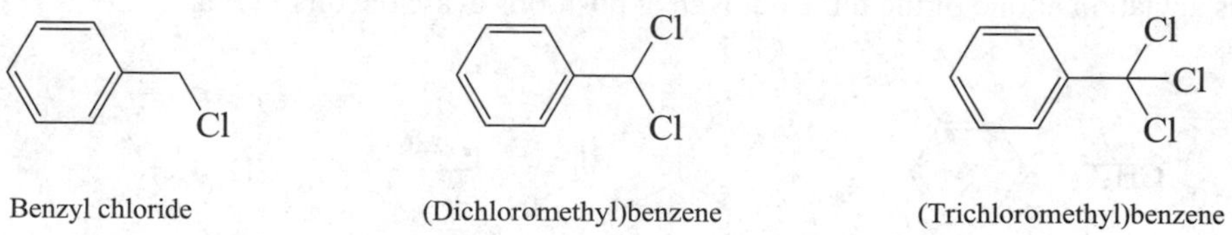

Benzyl chloride   (Dichloromethyl)benzene   (Trichloromethyl)benzene

The electron-withdrawing power of these substituents, and their tendency to direct incoming electrophiles meta to themselves, will increase with the number of chlorines each contains. Thus, the substituent that gives 4% meta nitration (96% ortho + para) contains the fewest chlorine atoms (—CH_2Cl) and the one that gives 64% meta nitration contains the most (—CCl_3).

    —CH_2Cl —$CHCl_2$ —CCl_3

Deactivating, ortho,   Deactivating, ortho,   Deactivating,
para-directing   para-directing   meta-directing

13.15 (*b*) Attack by bromine at the position meta to the amino group gives a cyclohexadienyl cation intermediate in which delocalization of the nitrogen lone pair does not participate in dispersal of the positive charge.

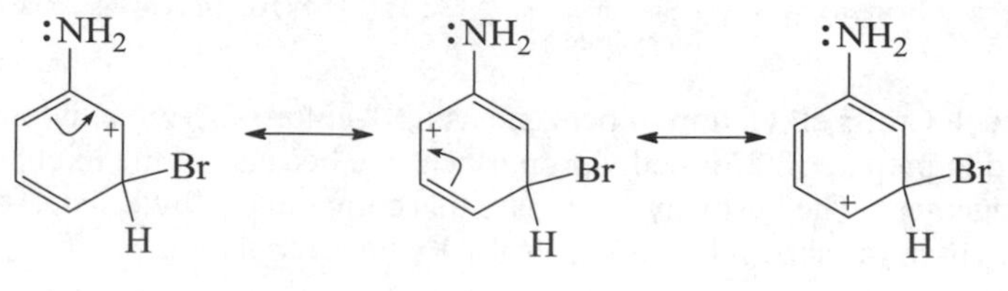

(c) Attack at the position para to the amino group yields a cyclohexadienyl cation intermediate that is stabilized by delocalization of the electron pair of the amino group.

13.16 Electrophilic aromatic substitution in biphenyl is best understood by considering one ring as the functional group and the other as a substituent. An aryl substituent is ortho, para-directing. Nitration of biphenyl gives a mixture of *o*-nitrobiphenyl and *p*-nitrobiphenyl.

Biphenyl   *o*-Nitrobiphenyl   *p*-Nitrobiphenyl
                    (37%)   (63%)

13.17 (b) The carbonyl group attached directly to the ring is a signal that the substituent is a meta-directing group. Nitration of methyl benzoate yields methyl *m*-nitrobenzoate.

Methyl benzoate   Methyl *m*-nitrobenzoate
                        (isolated in 81-85% yield)

(c) The acyl group in 1-phenyl-1-propanone is meta-directing; the carbonyl is attached directly to the ring. The product is 1-(*m*-nitrophenyl)-1-propanone.

1-Phenyl-1-propanone   1-(*m*-Nitrophenyl)-1-propanone
                           (isolated in 60% yield)

13.18 Writing the structures out in more detail reveals that the substituent —$\overset{+}{N}(CH_3)_3$ lacks the unshared electron pair of —$\overset{..}{N}(CH_3)_2$.

This unshared pair is responsible for the powerful activating effect of an —$\overset{..}{N}(CH_3)_2$ group. On the other hand, the nitrogen in —$\overset{+}{N}(CH_3)_3$ is positively charged and in that respect resembles the nitrogen of a nitro group. We expect the substituent —$\overset{+}{N}(CH_3)_3$ to be deactivating and meta-directing.

13.19 The reaction is a Friedel–Crafts alkylation in which 4-chlorobenzyl chloride serves as the carbocation source and chlorobenzene is the aromatic substrate. Alkylation occurs at the positions ortho and para to the chlorine substituent of chlorobenzene.

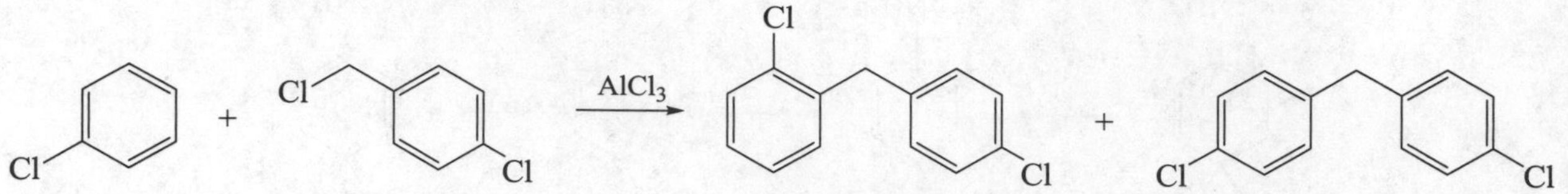

Chlorobenzene   4-Chlorobenzyl chloride   1-Chloro-2-(4'-chlorobenzyl)-benzene   1-Chloro-4-(4'-chlorobenzyl)-benzene

13.20 Bromine reacts with the aromatic ring to form the cyclohexadienyl cation intermediate.

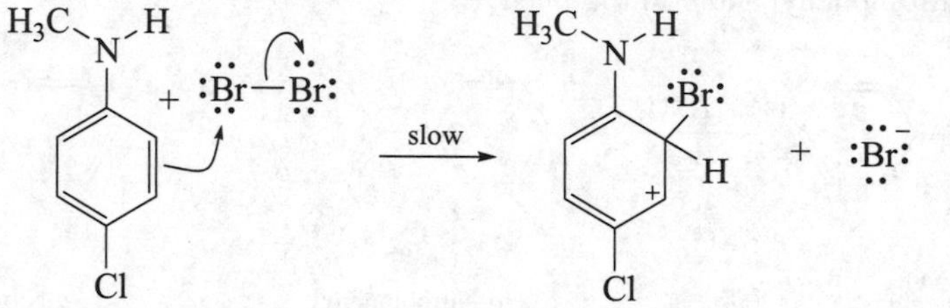

4-Chloro-*N*-methylaniline   Cyclohexadienyl cation intermediate   Bromide ion

Loss of a proton gives the product, 2-bromo-4-chloro-*N*-methylaniline.

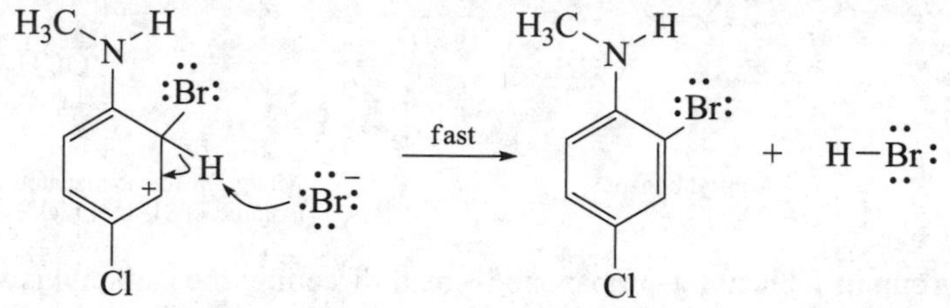

Cyclohexadienyl cation intermediate   2-Bromo-4-chloro-*N*-methylaniline   Hydrogen Bromide

13.21 The general rule is that the regioselectivity of electrophilic aromatic substitution is governed by the most strongly activating substituent present on the ring. In this case that substituent is the hydroxyl group.

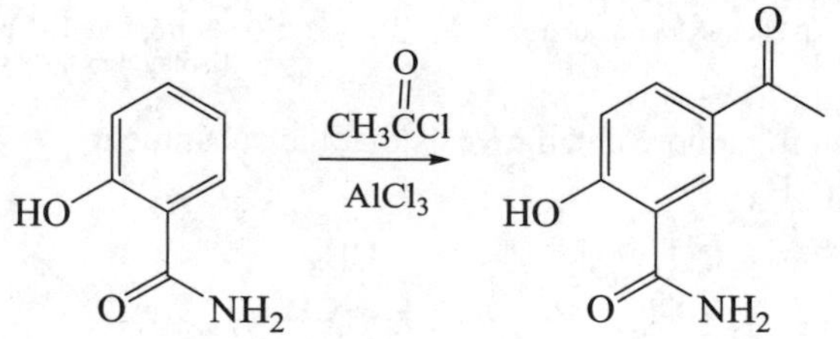

13.22 (*b*) Halogen substituents are ortho, para-directing, and the disposition in *m*-dichlorobenzene is such that their effects reinforce each other. The major product is 2,4-dichloro-1-nitrobenzene. Substitution at the position between the two chlorines is slow because it is more sterically hindered.

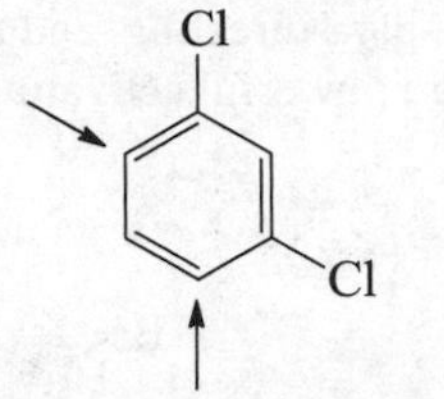

Most reactive positions (arrows)
in electrophilic aromatic
substitution of *m*-dichlorobenzene

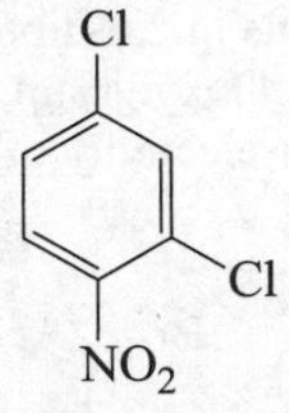

2,4-Dichloro-1-nitrobenzene
(major product of nitration)

(c) Nitro groups are meta-directing. Both nitro groups of *m*-dinitrobenzene direct an incoming electrophile to the same position in electrophilic aromatic substitution. Nitration of *m*-nitrobenzene yields 1,3,5-trinitrobenzene.

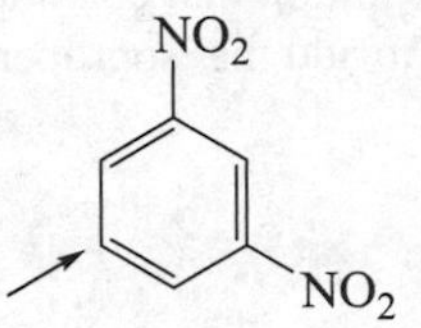

Both nitro groups of
m-dinitrobenzene direct
electrophile to same position (arrow)

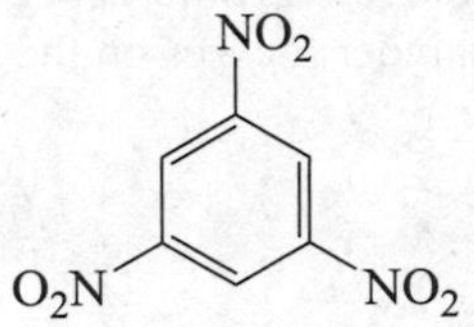

1,3,5-Trinitrobenzene
(principal product of nitration
of *m*-dinitrobenzene)

(d) Methoxyl groups are ortho, para-directing, and carbonyl is meta-directing. The open positions of the ring that are activated by the methoxy group in *p*-methoxyacetophenone are also those that are meta to the carbonyl, so the directing effects of the two substituents reinforce each other. Nitration of *p*-methoxyacetophenone yields 4-methoxy-3-nitroacetophenone.

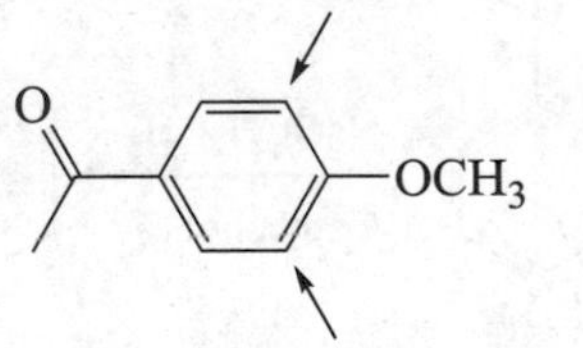

Positions ortho to methoxy
(arrows) are meta to the carbonyl.

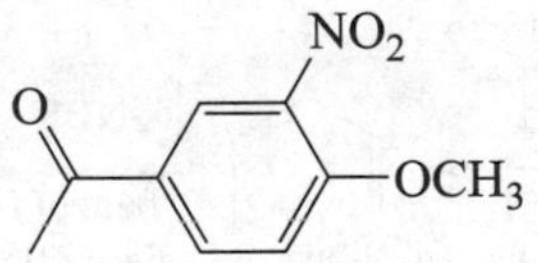

4-Methoxy-3-nitroacetophenone

(e) The methoxy group of *p*-methylanisole activates the positions that are ortho to it; the methyl activates those ortho to itself. Methoxy is a more powerful activating substituent than methyl, so nitration occurs ortho to the methoxy group.

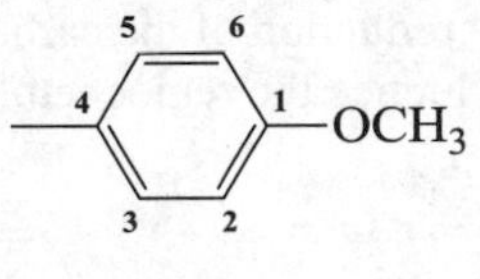

Methyl activates C-3 and C-5;
methoxy activates C-2 and C-6

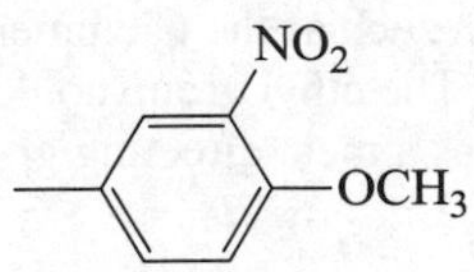

4-Methyl-2-nitroanisole
(major product of nitration)

(*f*) All the substituents in 2,6-dibromoanisole are ortho, para-directing, and their effects are felt at different positions. The methoxy group, however, is a far more powerful activating substituent than bromine, so it controls the regioselectivity of nitration.

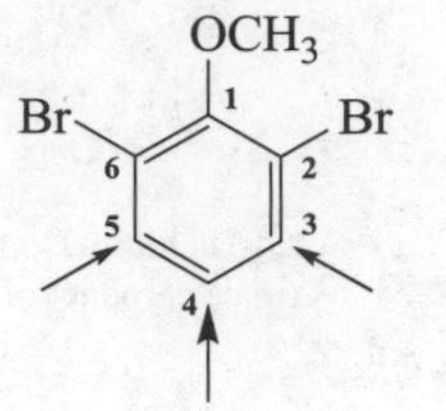

Methoxy directs toward C-4
(heavy arrow); bromines direct
toward C-3 and C-5 (light arrows)

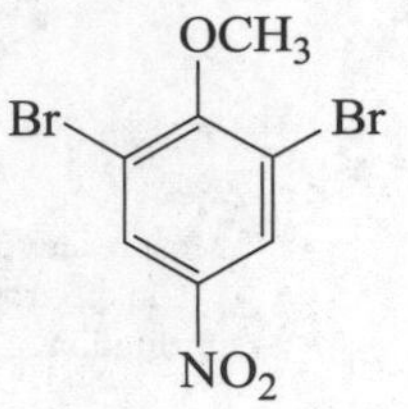

2,6-Dibromo-4-nitroanisole
(principle product of nitration)

13.23 The most reasonable retrosynthesis begins with the recognition that epoxides are customarily prepared from alkenes. The remainder focuses on the steps necessary to add the isopropenyl side chain of compound A to the benzene ring.

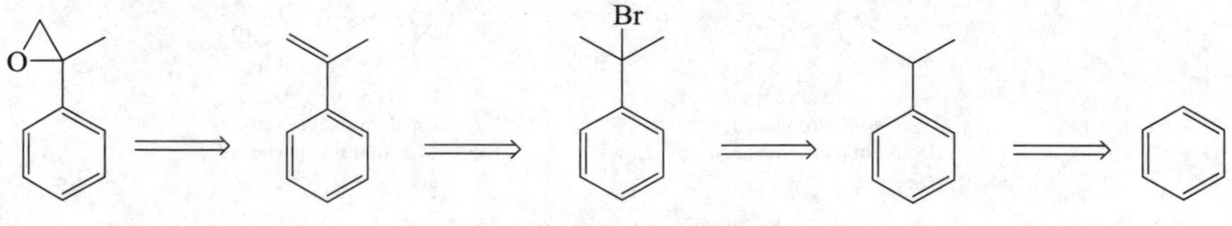

Each reaction in the synthesis is a standard method for the respective transformation.

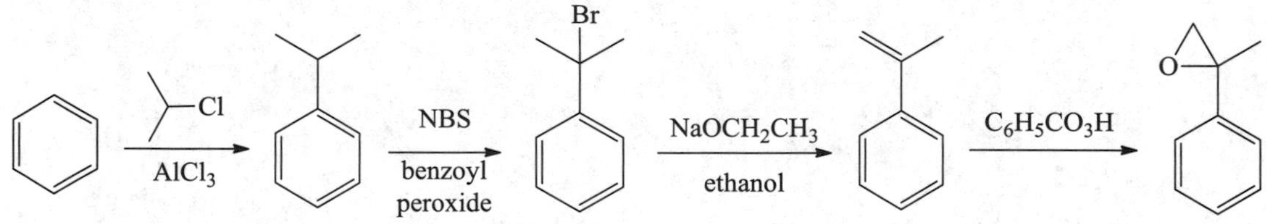

Benzene

Isopropyl-
benzene

(1-Bromo-1-
methylethyl)-
benzene

Isopropenyl-
benzene

(1,2-Epoxy-1-
methylethyl)-
benzene

13.24 Two ortho-para groups are meta to each other in *m*-chloroethane. Therefore, some functional group manipulation such as the Clemmensen or Wolff-Kishner reduction involving the ethyl group should be considered. The ethyl group could arise by reduction of the carbonyl of acetyl in this way, and the acetyl could serve as a meta directing group to influence the regiochemistry. A reasonable retrosynthesis is:

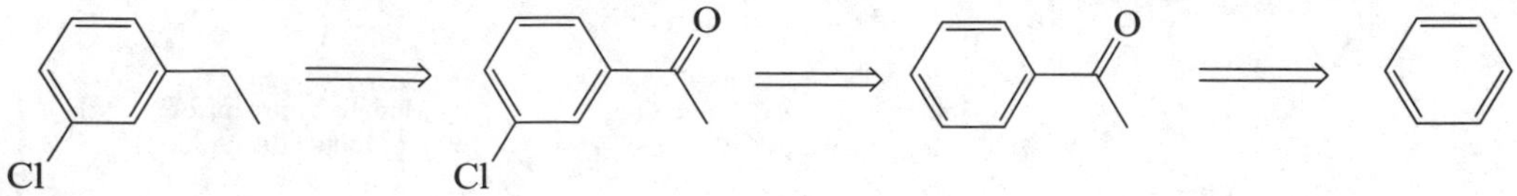

which leads to:

Benzene   Acetophenone   *m*-Chloroacetophenone   *m*-Chloroethylbenzene

13.25 The text points out that C-1 of naphthalene is more reactive than C-2 toward electrophilic aromatic substitution. Thus, of the two possible products of sulfonation, naphthalene-1-sulfonic acid should be formed faster and should be the major product under conditions of kinetic control. Because the problem states that the product under conditions of thermodynamic control is the other isomer, naphthalene-2-sulfonic acid is the major product at elevated temperature.

Naphthalene   Naphthalene-1-sulfonic acid
major product at 0°C;
formed faster   Naphthalene-2-sulfonic acid
major product at 160°C;
more stable

Naphthalene-2-sulfonic acid is the more stable isomer for steric reasons. The hydrogen at C-8 crowds the —SO_3H group in naphthalene-1-sulfonic acid.

13.26 Attack of the electrophile at C-2 leads to a carbocation intermediate, which is a hybrid of three resonance contributors shown, while attack at C-3 gives an ion that has only two resonance contributors. More resonance contributors correlates with a lower-energy carbocation intermediate and a faster rate of reaction.

13.27 (*b*) The negatively charged sulfur in $C_6H_5CH_2SNa$ is a good nucleophile and displaces chloride from 1-chloro-2,4-dinitrobenzene.

1-Chloro-2,4-
dinitrobenzene   Sodium
phenylmethanethiolate   Benzyl
2,4-dinitrophenyl sulfide

(*c*) The nitrogen in methylamine is nucleophilic and displaces chloride from 1-chloro-2,4-dinitrobenzene.

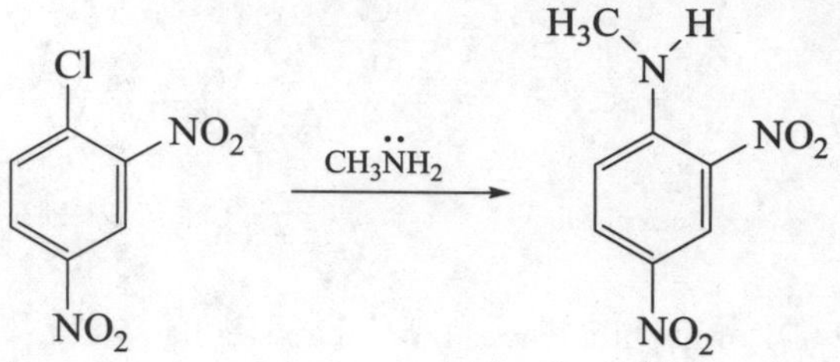

1-Chloro-2,4-
dinitrobenzene

N-Methyl-2,4-
dinitroaniline

13.28 The most stable resonance structure for the cyclohexadienyl anion formed by reaction of methoxide ion with *o*-fluoronitrobenzene involves the nitro group and has the negative charge on oxygen.

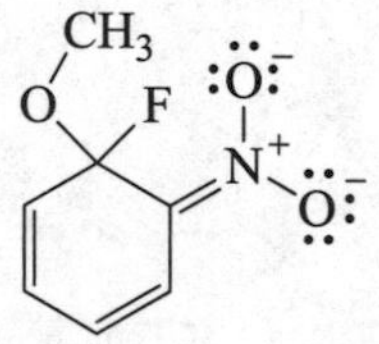

13.29 The positions that are activated toward nucleophilic attack are those that are ortho and para to the nitro group. Among the carbons that bear a bromine leaving group in 1,2,3-tribromo-5-nitrobenzene, only C-2 satisfies this requirement.

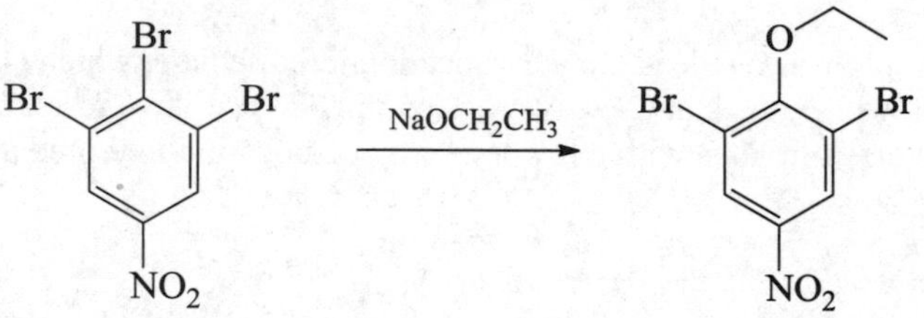

1,2,3-Tribromo-
5-nitrobenzene

1,2,3-Tribromo-2-ethoxy-
5-nitrobenzene

13.30 Nucleophilic addition occurs in the rate-determining step at one of the six equivalent carbons of hexafluorobenzene to give the cyclohexadienyl anion intermediate.

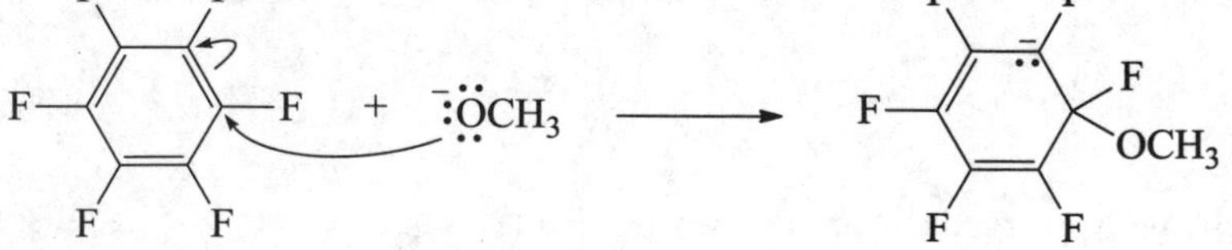

Hexafluorobenzene

Methoxide
ion

Cyclohexadienyl anion
intermediate

Elimination of fluoride ion from the cyclohexadienyl anion intermediate restores the aromaticity of the ring and completes the reaction.

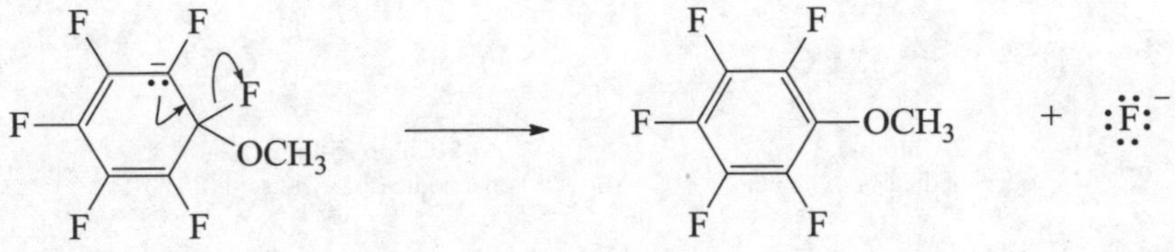

Cyclohexadienyl anion intermediate

2,3,4,5,6-Pentafluoroanisole

Fluoride ion

13.31 4-Chloropyridine is more reactive toward nucleophiles than 3-chloropyridine because the anionic intermediate formed by reaction of 4-chloropyridine can delocalize its charge on nitrogen. Because nitrogen is more electronegative than carbon, the intermediate is more stable.

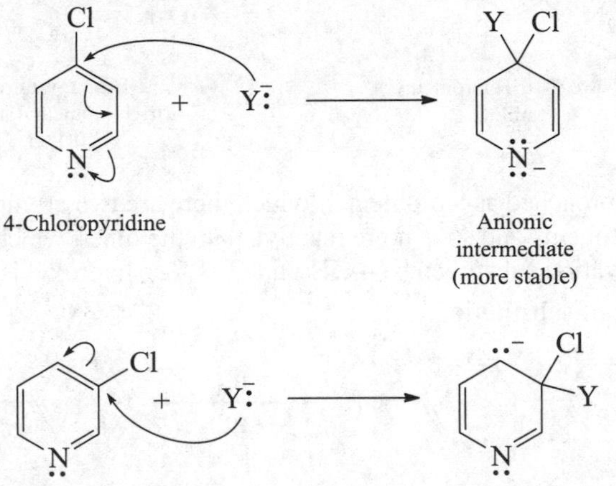

4-Chloropyridine

Anionic intermediate (more stable)

3-Chloropyridine

Anionic intermediate (less stable)

13.32 The negative charge of the intermediate is delocalized among the two nitrogen atoms in the ring, the carbon bearing the cyano group, and the nitrogen of the cyano group.

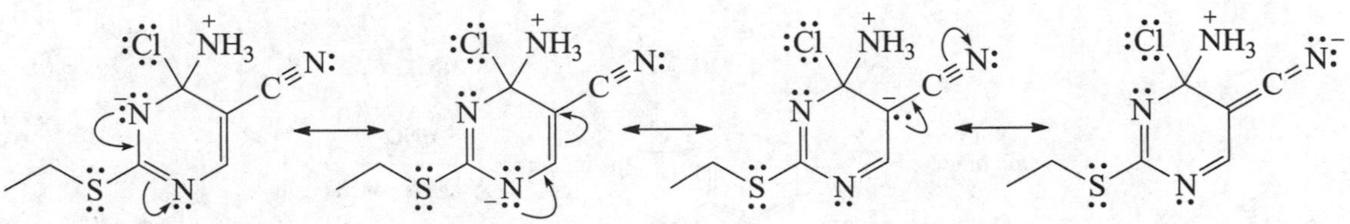

End of Chapter Problems

Predict the Product

13.33 (*a*) Nitration of the ring takes place para to the ortho, para-directing chlorine substituent; this position is also meta to the meta-directing carboxyl groups.

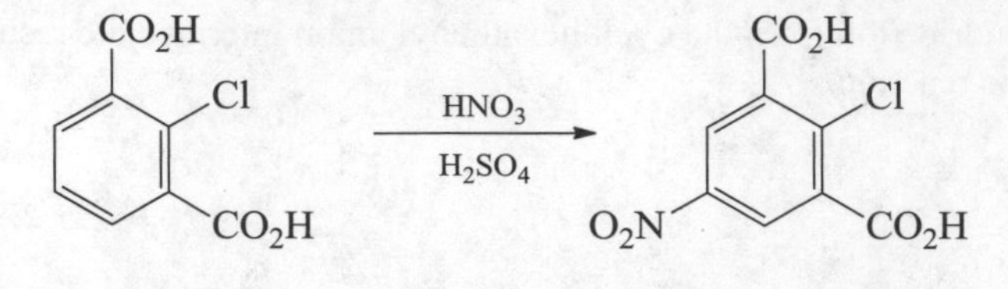

2-Chloro-1,3-
benzenedicarboxylic acid

2-Chloro-5-nitro-1,3-
benzenedicarboxylic acid (86%)

(*b*) Bromination of the ring occurs at the only available position activated by the amino group, a powerful activating substituent and an ortho, para director. This position is meta to the meta-directing trifluoromethyl group and to the meta-directing nitro group.

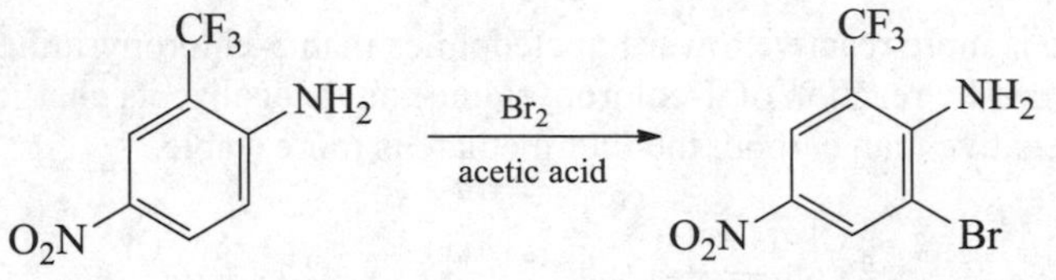

4-Nitro-2-(trifluoromethyl)-
aniline

2-Bromo-4-nitro-6-
(trifluoromethyl)aniline
(81%)

(*c*) This may be approached as a problem in which there are two aromatic rings. One of them bears two activating substituents and so is more reactive than the other, which bears only one activating substituent. Of the two activating substituents (—OH and C_6H_5—), hydroxyl is the more powerful and controls the regioselectivity of substitution.

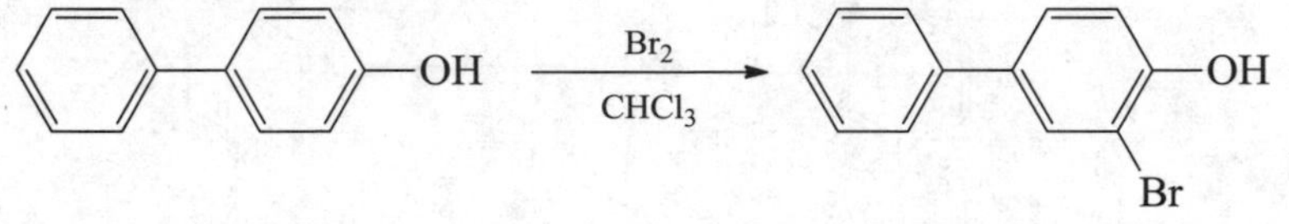

p-Phenylphenol

2-Bromo-4-phenylphenol

(*d*) Both substituents are activating and nitration occurs readily even in the absence of sulfuric acid; both are ortho, para-directing and comparable in activating power. The position at which substitution takes place is governed by the relative sizes of the two alkyl substitutents and is ortho to the smaller isopropyl group, para to the larger tert-butyl group.

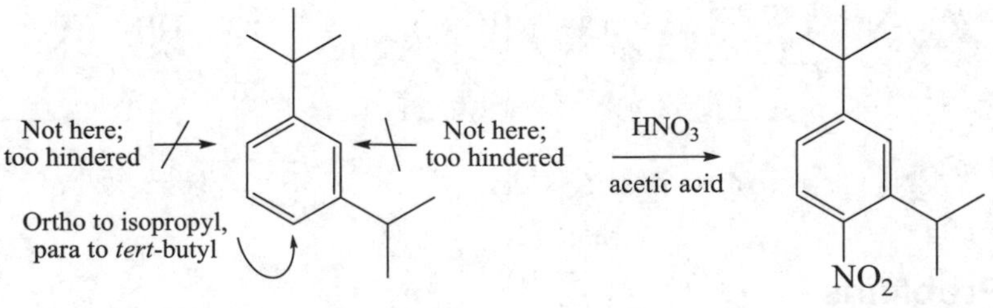

1-*tert*-Butyl-3-
isopropylbenzene

4-*tert*-Butyl-2-isopropyl-
1-nitrobenzene (78%)

(e) Protonation of 1-octene yields a secondary carbocation, which is the electrophile that reacts with benzene.

| Benzene | 1-Octene | 2-Phenyloctane (84%) |

(f) The reaction that occurs with arenes and acid anhydrides in the presence of aluminum chloride is Friedel–Crafts acylation. The methoxy group is the more powerful activating substituent, so acylation occurs para to it.

o-Fluoroanisole   Acetic anhydride   3-Fluoro-4-methoxyacetophenone   Acetic acid
(70–80%)

(g) The isopropyl group is ortho, para-directing, and the nitro group is meta-directing. In this case their orientation effects reinforce each other. Electrophilic aromatic substitution takes place ortho to isopropyl and meta to nitro.

p-Isopropylnitrobenzene   1-Isopropyl-2,4-dinitrobenzene (96%)

(h) In the presence of an acid catalyst (H_2SO_4), 2-methylpropene is converted to *tert*-butyl cation, which then attacks the aromatic ring ortho to the strongly activating methoxy group.

2-Methylpropene   2-*tert*-Butyl-4-methylanisole (98%)

(i) There are two things to consider in this problem: (1) In which ring does bromination occur, and (2) what is the orientation of substitution in that ring? Substitution will take place in the ring that bears the most powerful activating substituent, the hydroxyl group. Both positions ortho to the hydroxyl group are already substituted, so bromination takes place para to it.

3-Benzyl-2,6-dimethylphenol   3-Benzyl-4-bromo-2,6-dimethylphenol (100%)

(*j*) Fluorine is an ortho, para-directing substituent. It undergoes Friedel–Crafts alkylation on being treated with benzyl chloride and aluminum chloride to give a mixture of *o*-fluorodiphenylmethane and *p*-fluorodiphenylmethane.

| Fluorobenzene | Benzyl chloride | | *o*-Fluorodiphenylmethane (15%) | | *p*-Fluorodiphenylmethane (85%) |

(*k*) This reaction is *nucleophilic* aromatic substitution. Of the two bromine atoms, one is ortho and the other meta to the nitro group. Nitro groups activate positions ortho and para to themselves toward nucleophilic aromatic substitution, and so it will be the bromine ortho to the nitro group that is displaced.

| 1,4-Dibromo-2-nitrobenzene | Piperidine | *N*-(4-Bromo-2-nitrophenyl)piperidine |

(*l*) The —$\overset{\overset{\displaystyle O}{\|}}{\underset{\cdot\cdot}{N}HCCH_3}$ substituent is a more powerful activator than the ethyl group. It directs Friedel–Crafts acylation primarily to the position para to itself.

| o-Ethylacetanilide | Acetyl chloride | 4-Acetamido-3-ethylacetophenone (57%) |

(*m*) Clemmensen reduction reduces carbonyl groups of aldehydes and ketones to a CH_2 unit.

| 2,4,6-Trimethylacetophenone | 2-Ethyl-1,3,5-trimethylbenzene (74%) |

(*n*) Bromination of thiophene-3-carboxylic acid occurs at C-5. Reaction does not occur at C-2 because substitution at this position would place a carbocation adjacent to the electron-withdrawing carboxyl group.

| Thiophene-3-carboxylic acid | 5-Bromothiophene-3-carboxylic acid (69%) |

(*o*) This reaction is nucleophilic aromatic substitution by the addition–elimination mechanism. The nucleophile is the anion $C_6H_5CH_2S^-$.

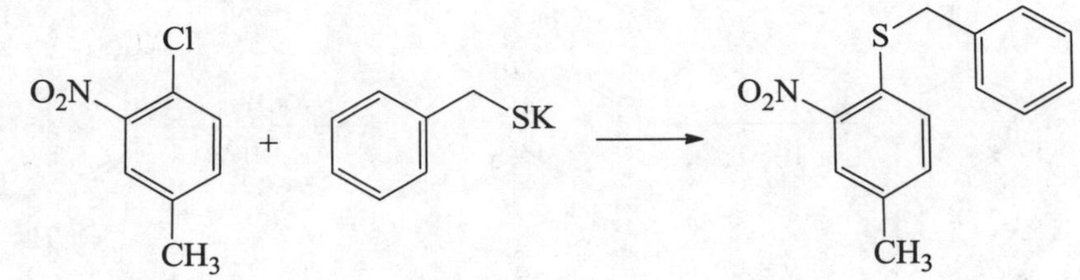

4-Chloro-3-nitrotoluene   Potassium   4-(Benzylthio)-3-nitrotoluene
(phenylmethane)thiolate   (57%)

(*p*) The problem requires you to track the starting material through two transformations. The first of these is nitration of *m*-dichlorobenzene, an electrophilic aromatic substitution.

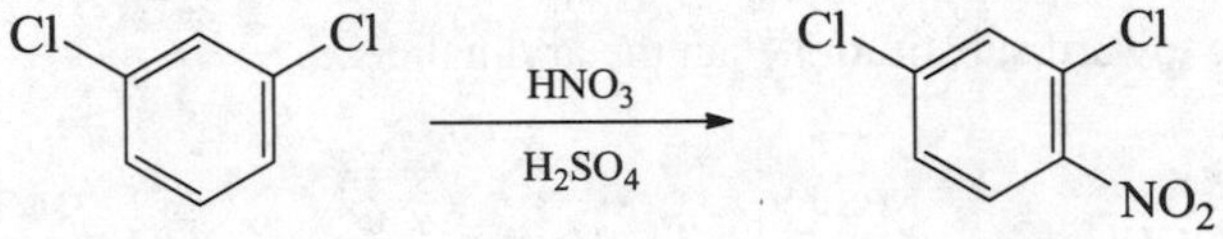

m-Dichlorobenzene   2,4-Dichloro-1-
nitrobenzene

Because the final product of the sequence has four nitrogen atoms ($C_6H_6N_4O_4$), 2,4-dichloro-1-nitrobenzene is an unlikely starting material for the second transformation. Stepwise nucleophilic aromatic substitution of both chlorines is possible but leads to a compound with the wrong molecular formula ($C_6H_7N_3O_2$).

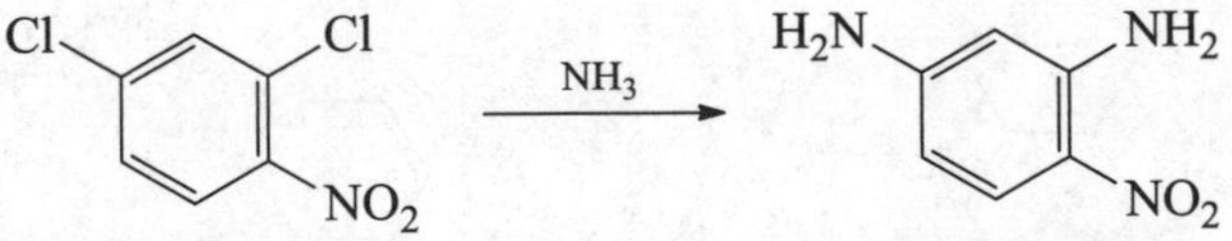

2,4-Dichloro-1-
nitrobenzene   2,4-Diamino-1-
nitrobenzene

To obtain a final product with the correct molecular formula, the original nitration reaction must lead not to a mononitro but to a dinitro derivative.

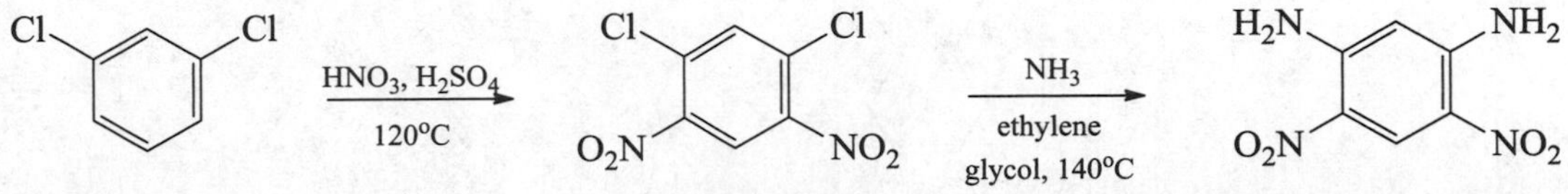

m-Dichlorobenzene   1,4-Dichloro-2,4-
dinitrobenzene   1,5-Diamino-2,4-dinitrobenzene
($C_6H_6N_4O_4$)

This two-step sequence has been carried out with product yields of 70–71% in the first step and 88–95% in the second step.

(*q*) This problem also involves two transformations, nitration and nucleophilic aromatic substitution. Nitration will take place ortho to chlorine (meta to trifluoromethyl).

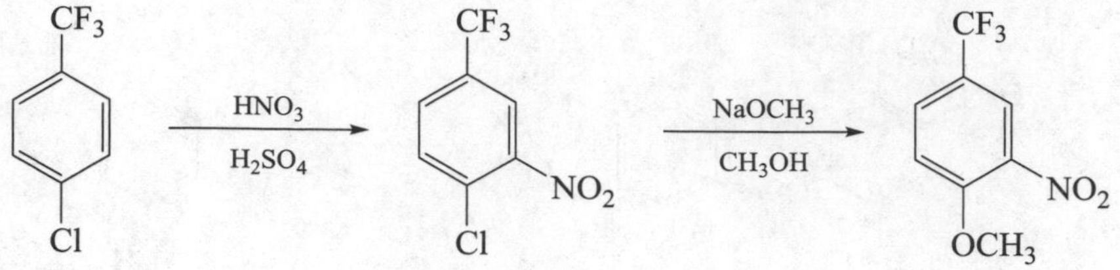

1-Chloro-4-
(trifluoromethyl)-
benzene

1-Chloro-2-nitro-4-
(trifluoromethyl)benzene

2-Nitro-4-
(trifluoromethyl)anisole

(*r*) *N*-Bromosuccinimide (NBS) is a reagent used to substitute benzylic and allylic hydrogens with bromine. The benzylic bromide undergoes S_N2 substitution with the nucleophile, methanethiolate. The alkyl halide is more reactive toward substitution than the aryl halide.

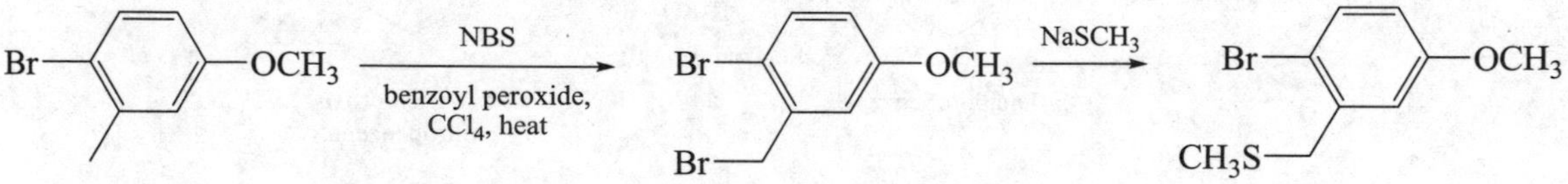

2-Bromo-5-methoxytoluene

2-Bromo-5-methoxybenzyl
bromide

2-Bromo-5-methoxybenzyl
methyl sulfide

13.34 The xylene isomers and their respective acetylation products are:

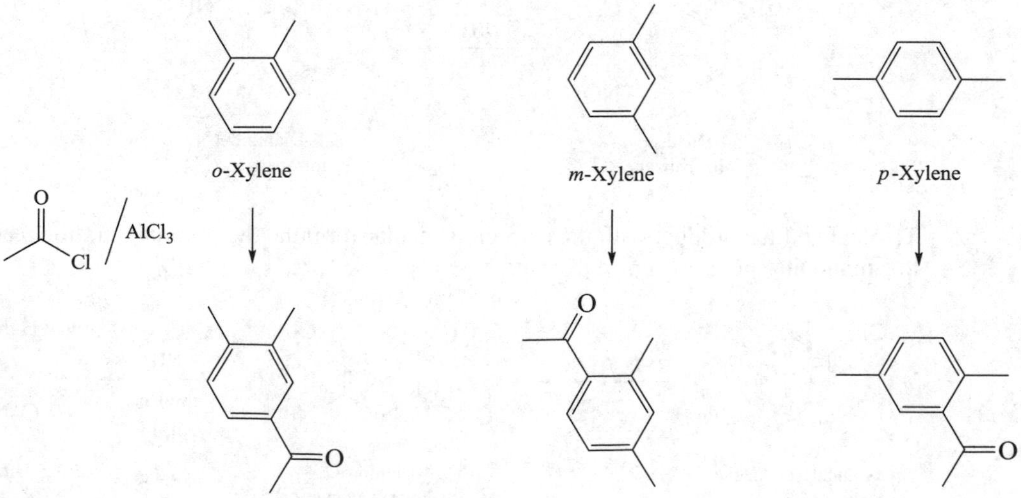

o-Xylene   *m*-Xylene   *p*-Xylene

3,4-Dimethylacetophenone (94 %)   2,4-Dimethylacetophenone (86 %)   2,5-Dimethylacetophenone (99 %)

13.35 The ring that bears the nitrogen in benzanilide is activated toward electrophilic aromatic substitution. The ring that bears the C=O is strongly deactivated.

Benzanilide   N-(o-Chlorophenyl)benzamide   N-(p-Chlorophenyl)benzamide

13.36 Reaction of a nitro-substituted aryl halide with a good nucleophile leads to nucleophilic aromatic substitution. Methoxide will displace fluoride from the ring, preferentially at the positions ortho and para to the nitro group.

1,2,3,4,5-Pentafluoro-   2,3,4,5-Tetrafluoro-   2,3,5,6-Tetrafluoro-
6-nitrobenzene   6-nitroanisole   4-nitroanisole

13.37 (a) Chlorine is ortho, para-directing, carboxyl is meta-directing. The positions that are ortho to the chlorine are meta to the carboxyl, so that both substituents direct an incoming electrophile to the same position. Introduction of the second nitro group at the remaining position that is ortho to the chlorine puts it meta to the carboxyl and meta to the first nitro group.

p-Chlorobenzoic acid   4-Chloro-3,5-dinitrobenzoic acid (90%)

(b) An amino group is one of the strongest activating substituents. The para and both ortho positions are readily substituted in aniline. When aniline is treated with 3 equivalents of bromine, 2,4,6-tribromoaniline is formed in quantitative yield.

Aniline   2,4,6-Tribromoaniline (100%)

(*c*) The positions ortho and para to the amino group in *o*-aminoacetophenone are the ones most activated toward electrophilic aromatic substitution.

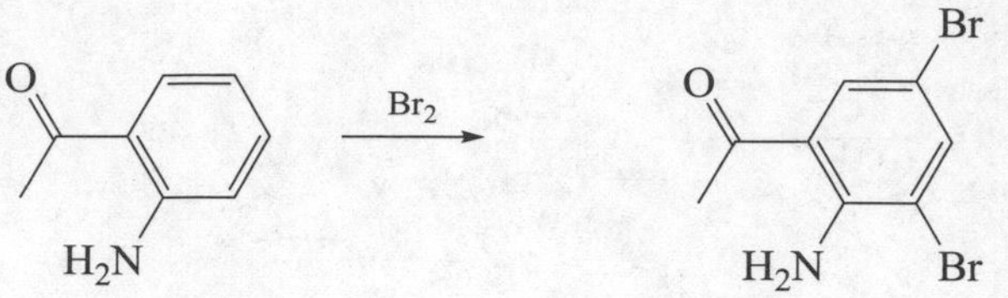

o-Aminoacetophenone   2-Amino-3,5-dibromoacetophenone (65%)

(*d*) Both bromine substituents are introduced ortho to the strongly activating hydroxyl group in *p*-nitrophenol.

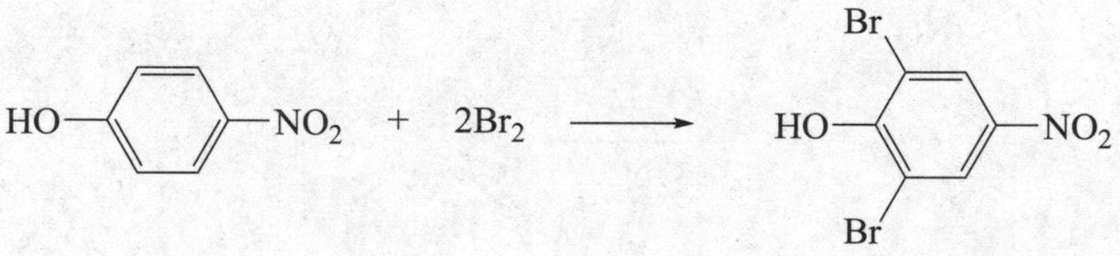

p-Nitrophenol   2,6-Dibromo-4-nitrophenol (96–98%)

(*e*) Friedel–Crafts alkylation occurs when biphenyl is treated with *tert*-butyl chloride and iron(III) chloride (a Lewis acid catalyst); the product of monosubstitution is *p-tert*-butylbiphenyl. The *tert*-butyl substituent is an ortho para director, but the two ortho positions of the ring that bears the *tert*-butyl group are too sterically hindered, so the second alkylation step introduces a *tert*-butyl group at the para position of the second ring.

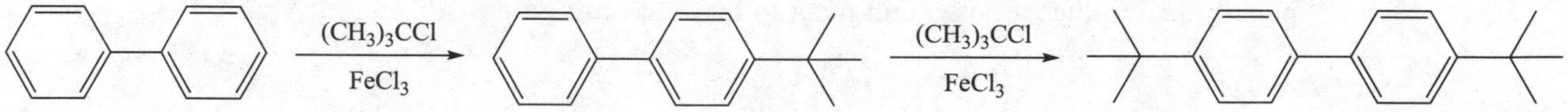

Biphenyl   4,4'-Di-*tert*-butylbiphenyl (70%)

(*f*) Disulfonation of phenol occurs at positions ortho and para to the hydroxyl group. The ortho, para product predominates over the ortho, ortho one for steric reasons.

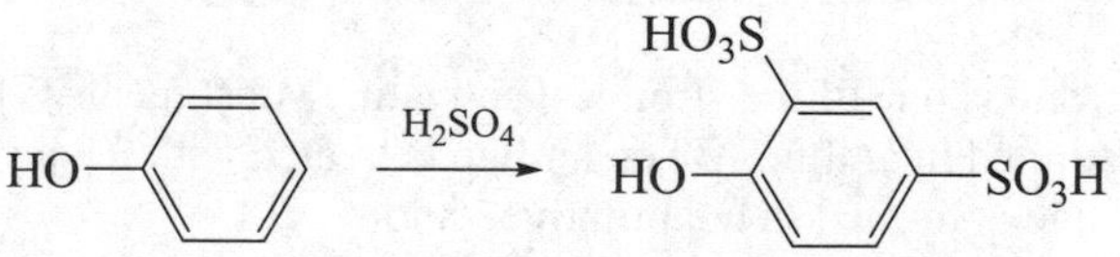

Phenol   2-Hydroxy-1,5-benzenedisulfonic acid

13.38 Dinitration of *p*-chloro(trifluoromethyl)benzene will take place at the ring positions ortho to the chlorine. Compound A is 2-chloro-1,3-dinitro-5-(trifluoromethyl)benzene. Trifluralin is formed by nucleophilic aromatic substitution of chlorine by dipropylamine. Trifluralin is 2,6-dinitro-*N,N*-dipropyl-4-(trifluoromethyl)aniline.

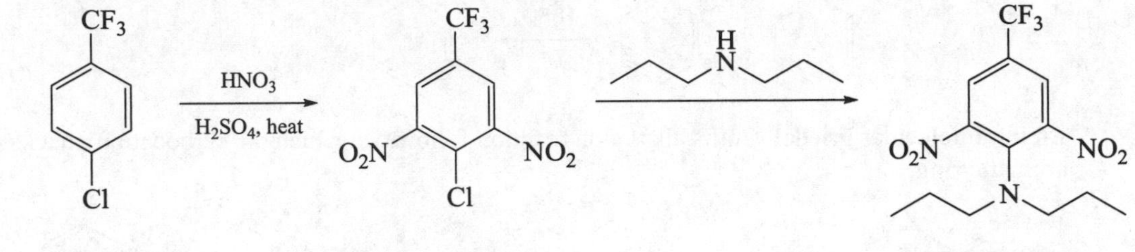

p-Chloro-(trifluoromethyl)benzene

2-Chloro-1,3-dinitro-5-(trifluoromethyl)benzene (compound A)

2,6-Dinitro-*N,N*-dipropyl-4-(trifluoromethyl)aniline (trifluralin)

13.39 Intramolecular Friedel–Crafts acylation reactions that produce five-membered or six-membered rings occur readily. Cyclization must take place at the position ortho to the reacting side chain.

(*a*) A five-membered cyclic ketone, isolated in 46% yield, was formed in the following reaction.

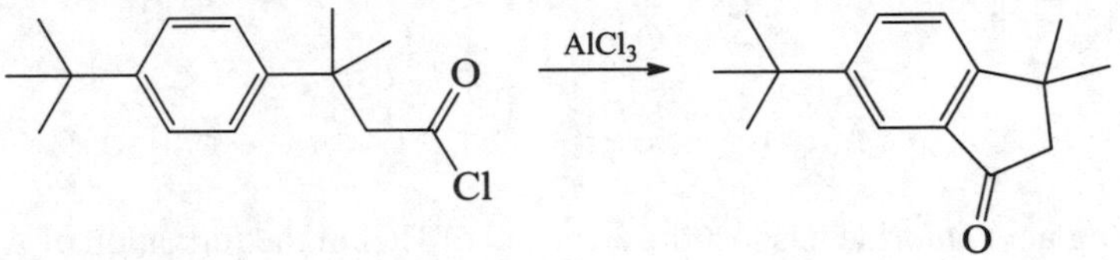

(*b*) This intramolecular Friedel–Crafts acylation takes place to form a six-membered cyclic ketone in 93% yield.

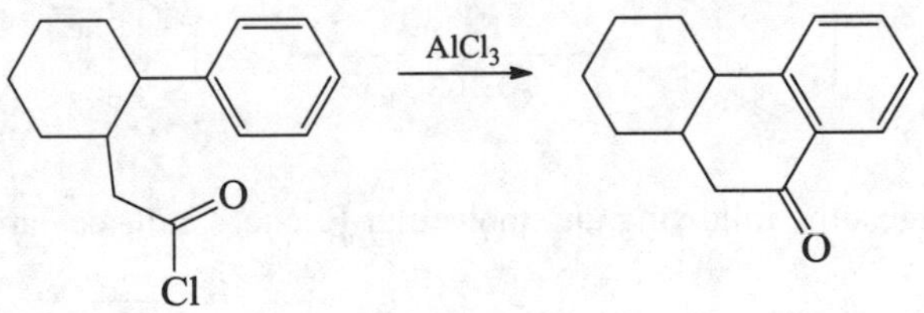

(*c*) In this case two aromatic rings are available for acylation. The more reactive ring is the one that bears the two activating methoxy groups, and cyclization occurs principally at the least hindered position on the dimethoxy benzene ring to give the indicated product in 78% yield.

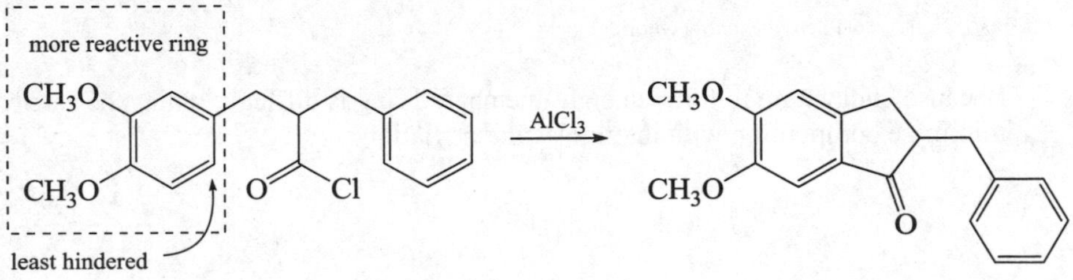

13.40 The alcohol is tertiary and benzylic and gives a carbocation on reaction with sulfuric acid.

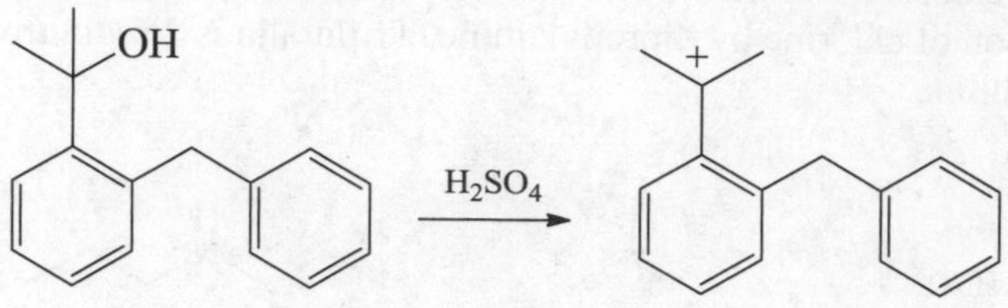

An intramolecular Friedel–Crafts alkylation reaction follows, in which the carbocation attacks the adjacent aromatic ring.

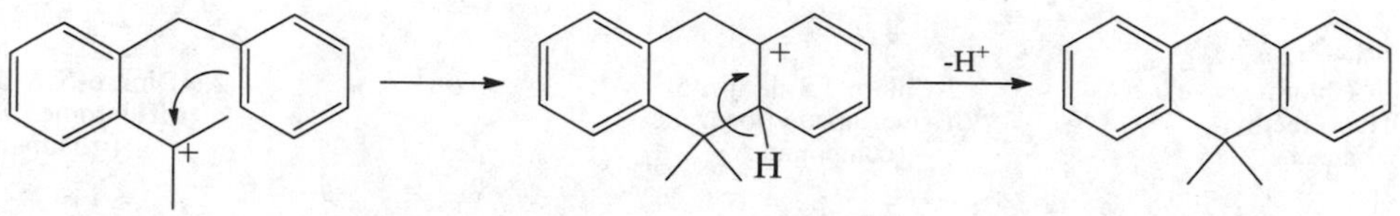

13.41 The relation of compound A to the starting material is

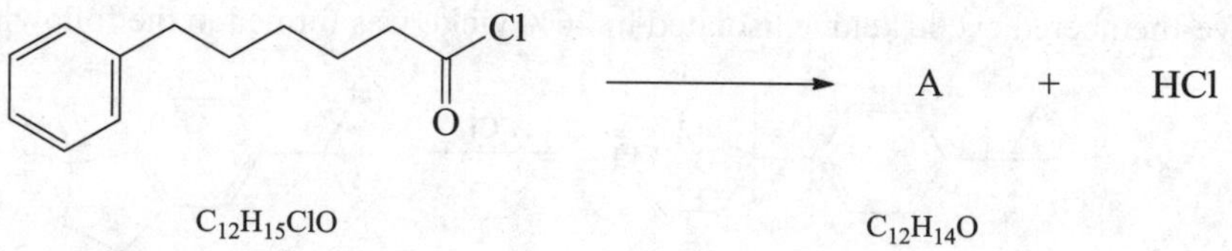

$$C_{12}H_{15}ClO \qquad\qquad C_{12}H_{14}O$$

The starting acyl chloride has lost the elements of HCl in the formation of A. Because A forms benzene-1,2-dicarboxylic acid on oxidation, it must have two carbon substituents ortho to each other on a benzene ring.

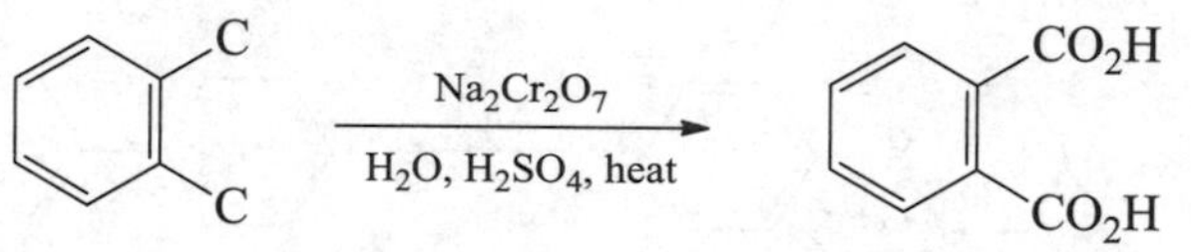

These facts suggest the following intramolecular Friedel-Crafts acylation:

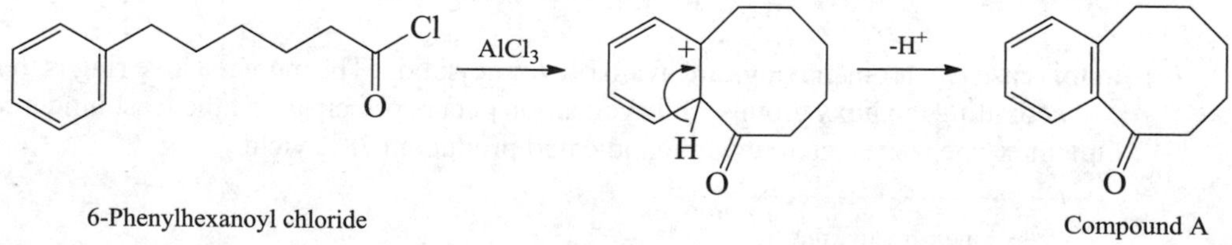

6-Phenylhexanoyl chloride   Compound A

Because cyclization to form an eight-membered ring is difficult, it must be carried out in dilute solution to minimize competition with intermolecular acylation.

Reactivity

13.42 (*a*) Toluene is more reactive than chlorobenzene in electrophilic aromatic substitution reactions because a methyl substituent is activating but a halogen substituent is deactivating. Both are ortho, para-directing, however. Nitration of toluene is faster than nitration of chlorobenzene.

Toluene   *o*-Nitrotoluene   *p*-Nitrotoluene

(*b*) A fluorine substituent is ortho, para-directing and not nearly as deactivating as a trifluoromethyl group. The reaction that takes place is Friedel–Crafts alkylation of fluorobenzene.

Fluorobenzene   Benzyl chloride   *o*-Benzylfluorobenzene (15%)   *p*-Benzylfluorobenzene (85%)

Strongly deactivated aromatic compounds such as (trifluoromethyl)benzene do not undergo Friedel–Crafts reactions.

(*c*) A carbonyl group directly bonded to a benzene ring strongly *deactivates* it toward electrophilic aromatic substitution. An oxygen attached directly to the ring strongly *activates* it toward electrophilic aromatic substitution. Methyl benzoate is much less reactive than benzene, phenyl acetate is much more reactive than benzene.

Phenyl acetate;   Benzene   Methyl benzoate;
most reactive   least reactive

Phenyl acetate is sufficiently reactive to undergo bromination even in the absence of a catalyst.

Phenyl acetate   *o*-Bromophenyl acetate   *p*-Bromophenyl acetate

Bromination of methyl benzoate requires more vigorous conditions; catalysis by iron(III) bromide is required for bromination of deactivated aromatic rings.

(*d*) Acetanilide is strongly activated toward electrophilic aromatic substitution and reacts faster than nitrobenzene, which is strongly deactivated.

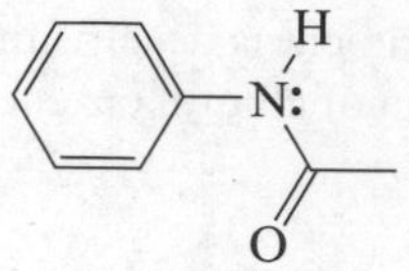

Acetanilide
Nitrogen lone pair stabilizes cyclohexadienyl cation intermediate

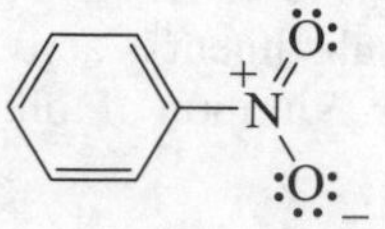

Nitrobenzene
Nitrogen is positively charged and destabilizes cyclohexadienyl cation intermediate

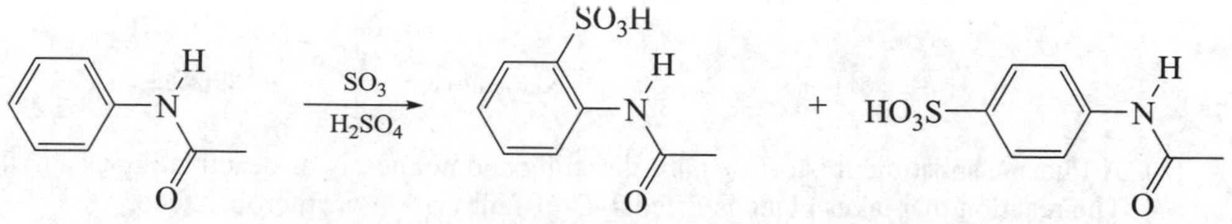

Acetanilide   *o*-Acetamidobenzenesulfonic acid   *p*-Acetamidobenzenesulfonic acid

(*e*) Both reactants are *p*-dialkyl aromatic hydrocarbons and are activated toward Friedel–Crafts acylation. Because electronic effects are comparable, we look to differences in steric factors and conclude that reaction will be faster when the two alkyl groups are methyl rather than *tert*-butyl.

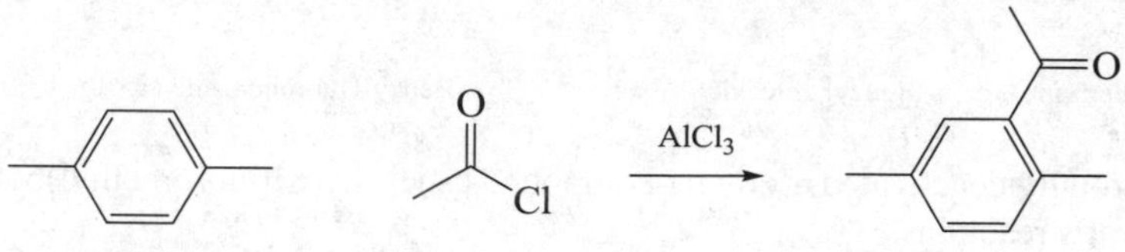

p-Xylene   Acetyl chloride   2,5-Dimethylacetophenone

13.43 (*a*) *o*-Chloronitrobenzene is more reactive than chlorobenzene because the cyclohexadienyl anion intermediate is stabilized by the nitro group.

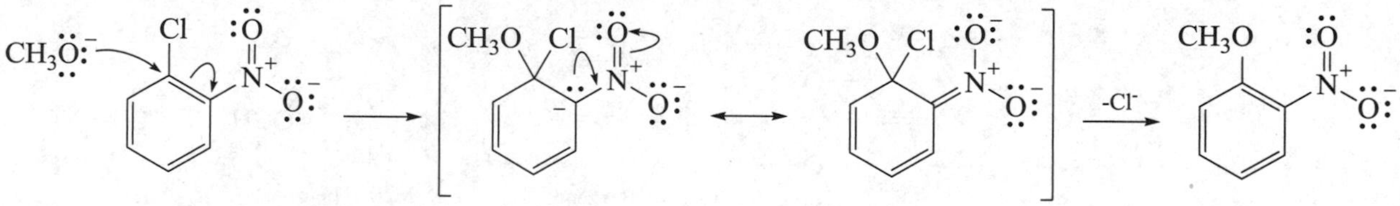

Comparing the rate constants for the two aryl halides in this reaction reveals that *o*-chloronitrobenzene is more than 20 billion times more reactive at 50°C.

(*b*) The cyclohexadienyl anion intermediate is more stable, and is formed faster, when the electron-withdrawing nitro group is ortho to chlorine. *o*-Chloronitrobenzene reacts faster than *m*-chloronitrobenzene. The measured difference is a factor of approximately 40,000 at 50°C.

(c) 4-Chloro-3-nitroacetophenone is more reactive, because the ring bears two powerful electron-withdrawing groups in positions where they can stabilize the cyclohexadienyl anion intermediate.

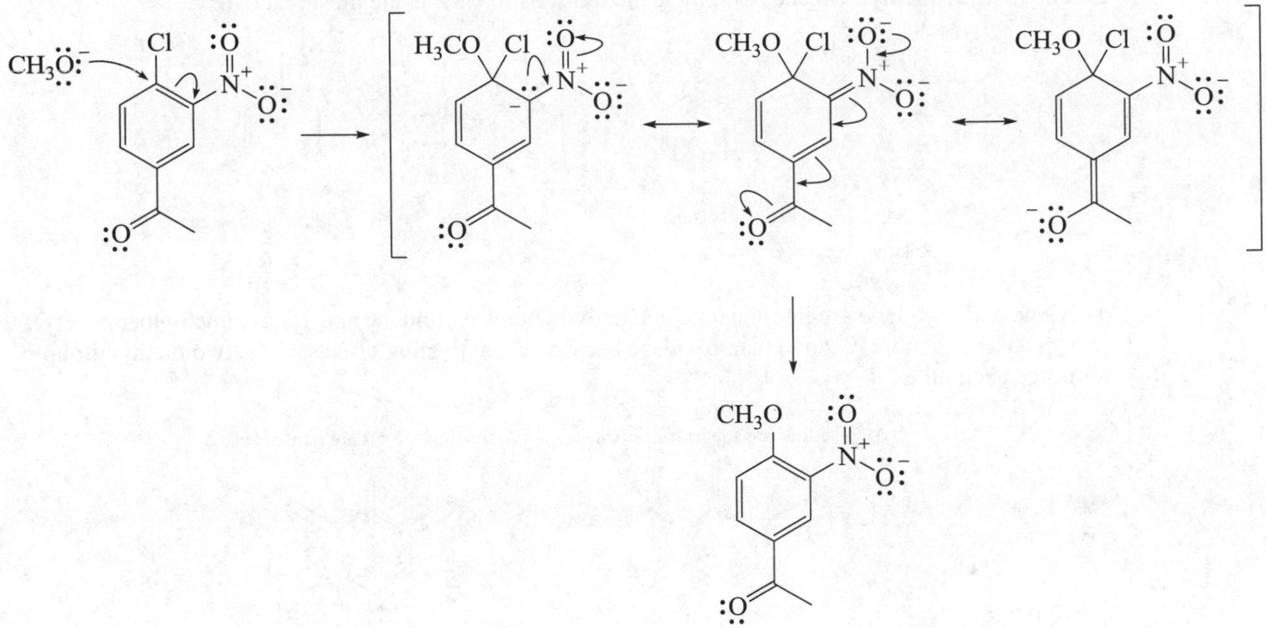

(d) Nitro groups activate aryl halides toward nucleophilic aromatic substitution best when they are ortho or para to the leaving group.

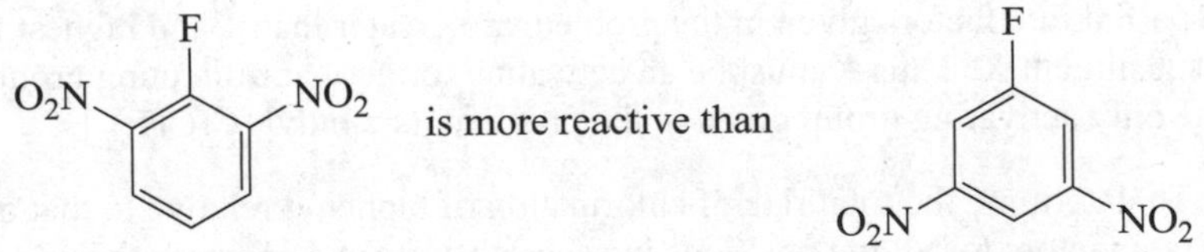

2-Fluoro-1,3-dinitrobenzene is more reactive than 1-Fluoro-3,5-dinitrobenzene

(e) The aryl halide with nitro groups ortho and para to the bromide leaving group is more reactive than the aryl halide with only one nitro group.

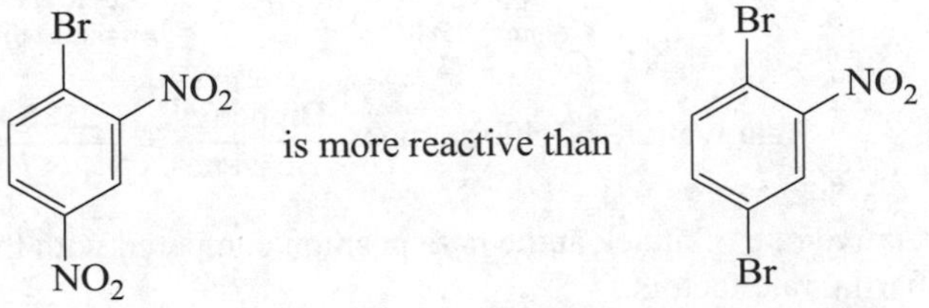

1-Bromo-2,4-dinitrobenzene is more reactive than 1,4-Dibromo-2-nitrobenzene

13.44 Reactivity toward electrophilic aromatic substitution increases with increasing number of electron-releasing substituents. Benzene, with no methyl substituents, is the least reactive, followed by toluene, with one methyl group. 1,3,5-Trimethylbenzene, with three methyl substituents, is the most reactive.

	Benzene	Toluene	1,3,5-Trimethylbenzene
Relative reactivity:	1	60	2×10^7

o-Xylene and *m*-xylene are intermediate in reactivity between toluene and 1,3,5-trimethylbenzene. Of the two, *m*-xylene is more reactive than *o*-xylene because the activating effects of the two methyl groups reinforce each other.

All positions are somewhat activated.   Activating effects are reinforced.

o-Xylene
Relative reactivity: 5×10^2

m-Xylene
Relative reactivity: 5×10^4

13.45 The partial rate factors given in the problem are greater than 1 and highest for the positions ortho and para to the substituent X. Thus X must be an activating, ortho, para-directing group. The most likely candidate for X is the only activating group given in the problem, *tert*-butyl [C(CH$_3$)$_3$].

13.46 (*a*) To determine the total rate of chlorination of biphenyl relative to that of benzene, we add up the partial rate factors for all the positions in each substrate and compare them.

Biphenyl
(sum = 2580)

Benzene
(sum = 6)

Relative rate of chlorination: $\dfrac{\text{Biphenyl}}{\text{Benzene}} = \dfrac{2580}{6} = \dfrac{430}{1}$

(*b*) The relative rate of attack at the para position compared with the ortho positions is given by the ratio of their partial rate factors.

$$\frac{\text{Para}}{\text{Ortho}} = \frac{1580}{1000} = \frac{1.58}{1}$$

Therefore, 15.8 g of *p*-chlorobiphenyl is formed for every 10 g of *o*-chlorobiphenyl.

13.47 The problem stipulates that the reactivity of various positions in *o*-bromotoluene can be estimated by multiplying the partial rate factors for the corresponding positions in toluene and bromobenzene. Therefore, given the partial rate factors:

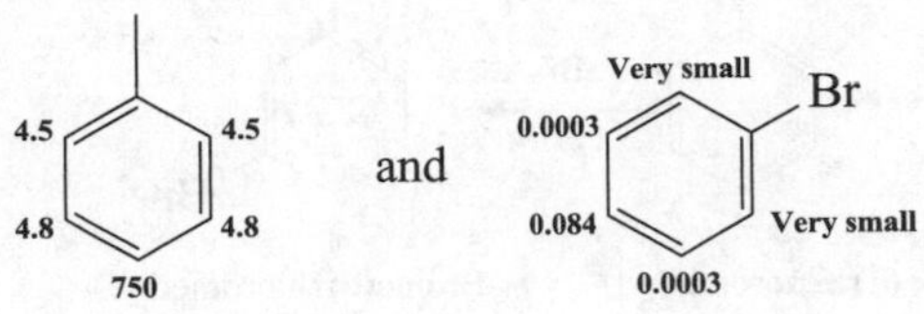

the two are multiplied together to give the combined effects of the two substituents at the various ring positions.

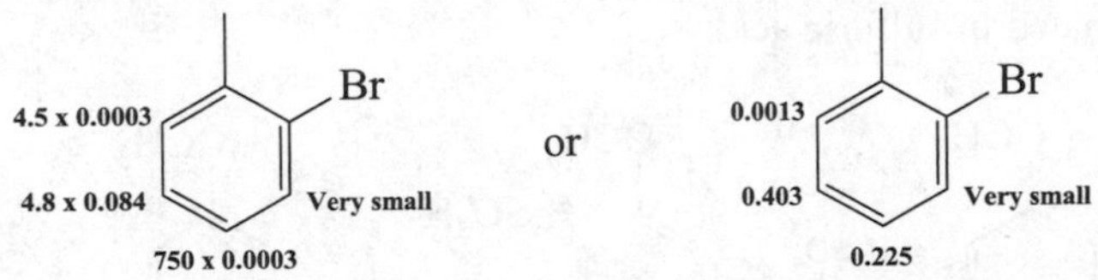

The most reactive position is the one that is para to bromine. The predicted major product is therefore 4-bromo-3-methylacetophenone and is what is observed experimentally.

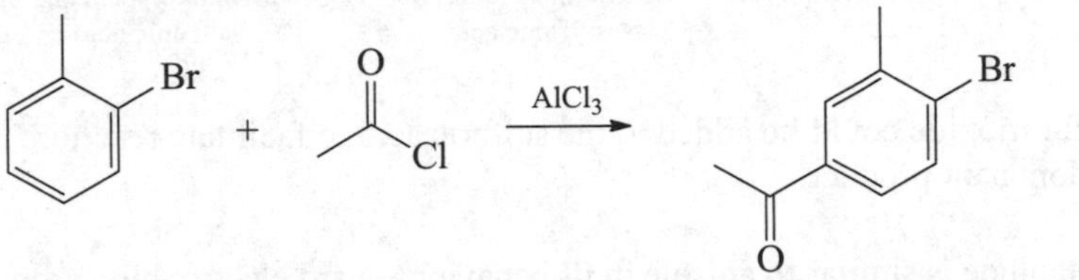

o-Bromotoluene	Acetyl chloride	4-Bromo-3-methylacetophenone

This was first considered to be anomalous behavior on the part of *o*-bromotoluene, but, as can be seen, it is consistent with the individual directing properties of the two substituents.

Synthesis

13.48 (*a*) Nitrobenzene is much less reactive than benzene toward electrophilic aromatic substitution. The nitro group on the ring is a meta director.

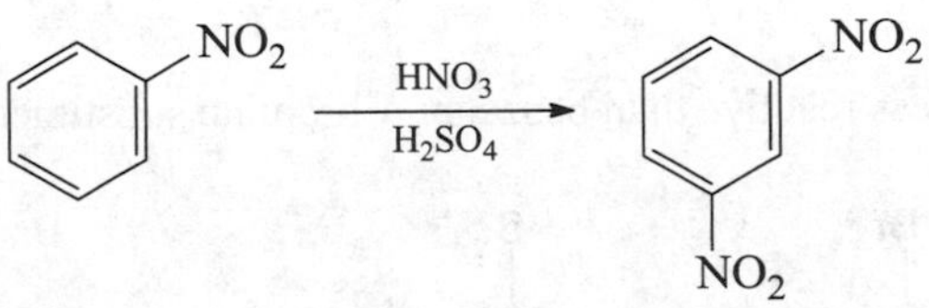

Nitrobenzene	*m*-Dinitrobenzene

(*b*) Toluene is more reactive than benzene in electrophilic aromatic substitution. A methyl substituent is an ortho, para director.

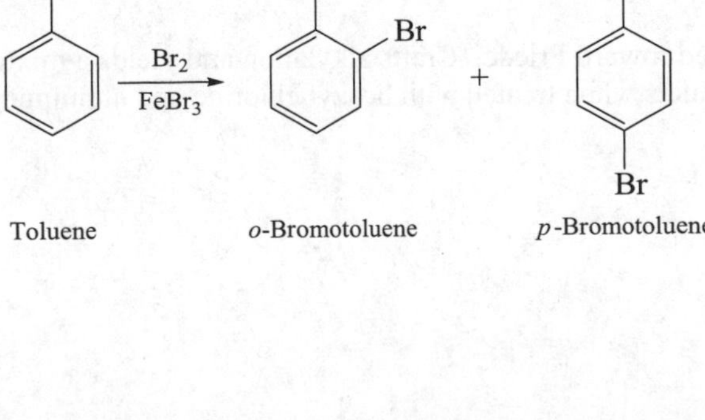

Toluene	*o*-Bromotoluene	*p*-Bromotoluene

(c) Trifluoromethyl is deactivating and meta-directing.

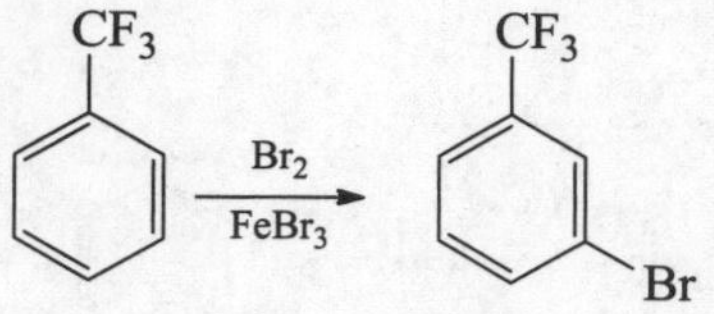

(Trifluoromethyl)-   *m*-Bromo(trifluoromethyl)-
benzene   benzene

(d) Anisole is ortho, para-directing, strongly activated toward electrophilic aromatic substitution, and readily sulfonated in sulfuric acid.

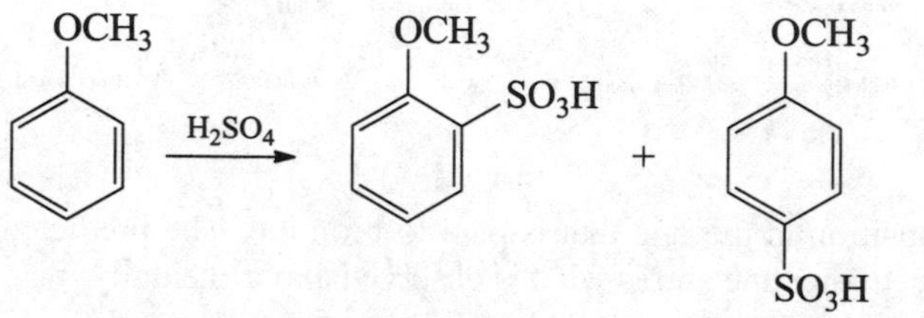

Anisole   *o*-Methoxybenzene-   *p*-Methoxybenzene
                         sulfonic acid   sulfonic acid

Sulfur trioxide could be added to the sulfuric acid to facilitate reaction. The para isomer is the predominant product.

(e) Acetanilide is similar to anisole in its behavior toward electrophilic aromatic substitution, it is an ortho, para director.

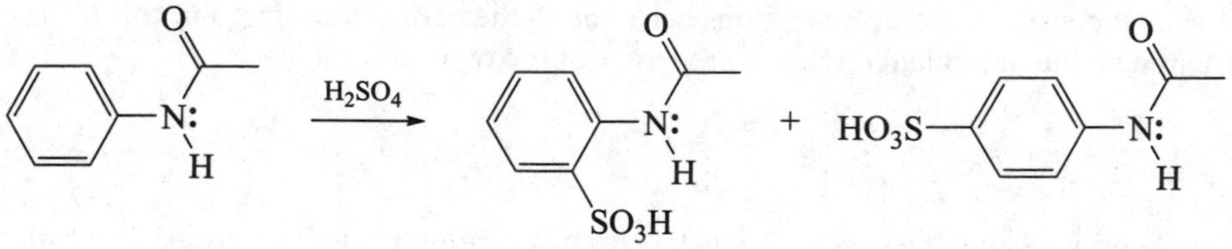

Acetanilide   *o*-Acetamidobenzene-   *p*-Acetamidobenzene-
                            sulfonic acid   sulfonic acid

(f) Bromobenzene is less reactive than benzene. A bromine substituent is ortho, para-directing.

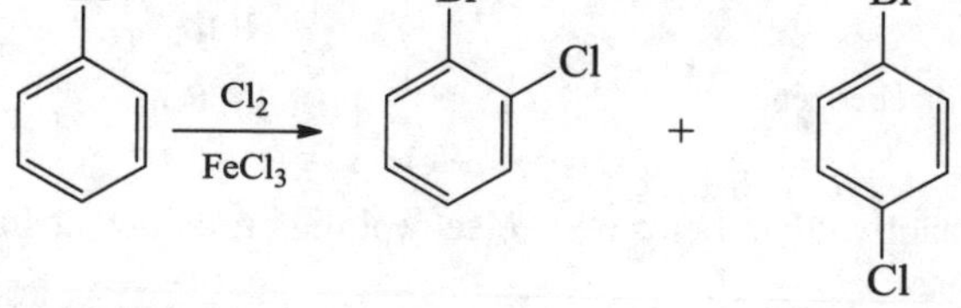

Bromobenzene   *o*-Bromochloro-   *p*-Bromochloro-
                          benzene   benzene

(g) Anisole is activated toward Friedel–Crafts alkylation and yields a mixture of *o*- and *p*-benzylated products when treated with benzyl chloride and aluminum chloride.

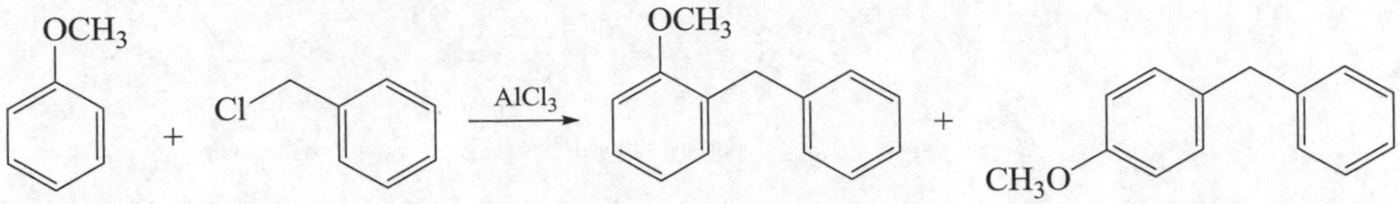

Anisole + Benzyl chloride →(AlCl₃) o-Benzylanisole + p-Benzylanisole

(*h*) This is a Friedel-Crafts acylation reaction. Activation of acyl chlorides with aluminum chloride is necessary.

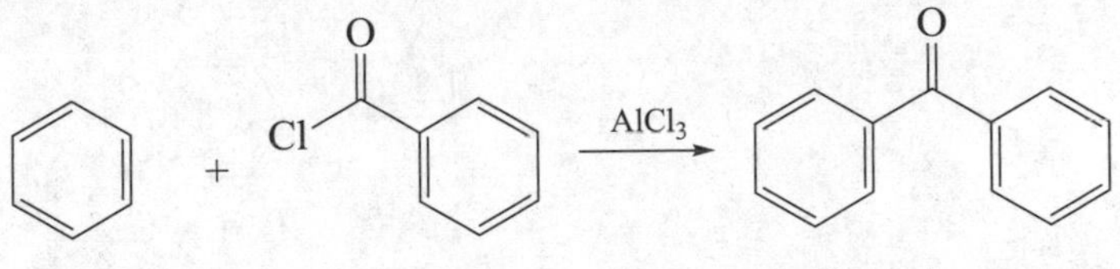

Benzene + Benzoyl chloride →(AlCl₃) Benzophenone

(*i*) A benzoyl substituent is meta-directing and deactivating.

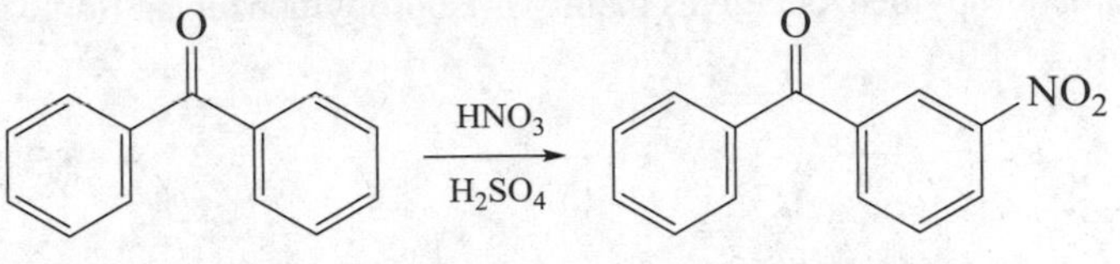

Benzophenone →(HNO₃ / H₂SO₄) *m*-Nitrobenzophenone

(*j*) Clemmensen reduction conditions involve treating a ketone with zinc amalgam and concentrated hydrochloric acid.

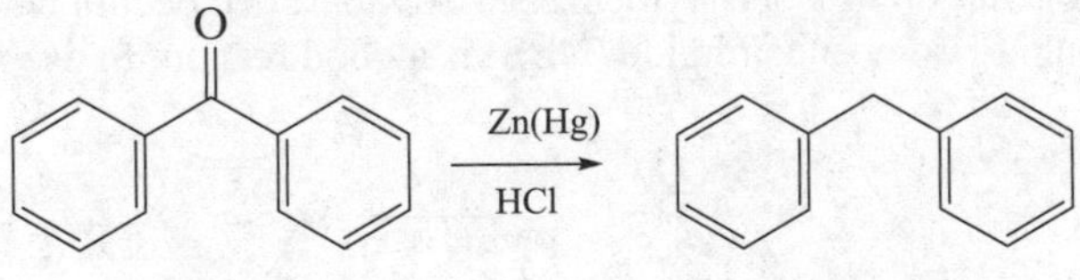

Benzophenone →(Zn(Hg) / HCl) Diphenylmethane

(*k*) Wolff–Kishner reduction uses hydrazine in a basic solution of a high-boiling alcohol solvent to reduce ketone functions to methylene groups.

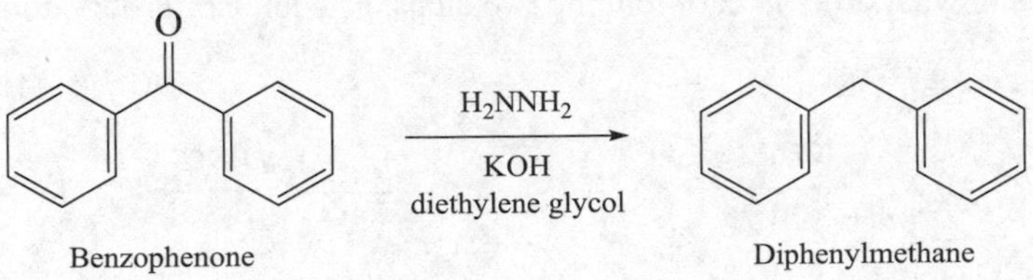

Benzophenone →(H₂NNH₂ / KOH / diethylene glycol) Diphenylmethane

13.49 When carrying out each of the following syntheses, evaluate how the structure of the product differs from that of benzene or toluene; that is, determine which groups have been substituted on the benzene ring or altered in some way. The sequence of reaction steps when multiple substitution is desired is important; recall that some groups direct ortho, para and others meta.

(*a*) Isopropylbenzene may be prepared by a Friedel–Crafts alkylation of benzene with isopropyl chloride (or bromide, or iodide).

Benzene   +   Isopropyl chloride $\xrightarrow{AlCl_3}$ Isopropylbenzene

It would not be appropriate to use propyl chloride and trust that a rearrangement would lead to isopropylbenzene, because a mixture of propylbenzene and isopropylbenzene would be obtained.

Isopropylbenzene may also be prepared by alkylation of benzene with propene in the presence of sulfuric acid.

Benzene   +   Propene $\xrightarrow{H_2SO_4}$ Isopropylbenzene

(b) Because the isopropyl and sulfonic acid groups are para to each other, the first group introduced on the ring must be the ortho, para director, that is, the isopropyl group. We may therefore use the product of part (a), isopropylbenzene, in this synthesis. An isopropyl group is a fairly bulky ortho, para director, so sulfonation of isopropylbenzene gives mainly p-isopropylbenzenesulfonic acid.

Isopropylbenzene $\xrightarrow[H_2SO_4]{SO_3}$ p-Isopropylbenzenesulfonic acid

A sulfonic acid group is meta-directing, so the order of steps must be alkylation followed by sulfonation, rather than the reverse.

(c) Free-radical halogenation of isopropylbenzene, from part (a), occurs with high regioselectivity at the benzylic position. N-Bromosuccinimide (NBS) is a good reagent to use for benzylic bromination.

Isopropylbenzene $\xrightarrow[\text{peroxides}]{NBS}$ 2-Bromo-2-phenylpropane

(d) Toluene is an obvious starting material for the preparation of 4-tert-butyl-2-nitrotoluene. Two possibilities, both involving nitration and alkylation of toluene, present themselves; the problem to be addressed is in what order to carry out the two steps. Friedel–Crafts alkylation must precede nitration.

Toluene $\xrightarrow[AlCl_3]{(CH_3)_3CCl}$ p-tert-Butyltoluene $\xrightarrow[H_2SO_4]{HNO_3}$ 4-tert-Butyl-2-nitrotoluene

Introduction of the nitro group as the first step is unsatisfactory because Friedel–Crafts reactions cannot be carried out on nitro-substituted aromatic compounds because they are too deactivated.

(e) Two electrophilic aromatic substitution reactions need to be performed: chlorination and Friedel–Crafts acylation. The order in which the reactions are carried out is important; chlorine is an ortho, para director, and the acetyl group is a meta director. Because the groups are meta in the desired compound, introduce the acetyl group first.

Benzene → Acetophenone → m-Chloroacetophenone

(f) Reverse the order of steps in part (e) to prepare p-chloroacetophenone.

Benzene → Chlorobenzene → p-Chloroacetophenone

Friedel–Crafts reactions can be carried out on halobenzenes but not on arenes that are more strongly deactivated.

(g) Here again the problem involves two successive electrophilic aromatic substitution reactions, in this case using toluene as the initial substrate. The proper sequence is Friedel–Crafts acylation first, followed by bromination of the ring.

Toluene → p-Methylacetophenone → 3-Bromo-4-methylacetophenone

If the sequence of steps had been reversed, with halogenation preceding acylation, the first intermediate would be p-bromotoluene, Friedel–Crafts acylation of which would give a complex mixture of products because both groups are ortho, para-directing. On the other hand, the orienting effects of the two groups in p-methylacetophenone reinforce each other, so that its bromination is highly regioselective and in the desired direction.

(h) Recalling that alkyl groups attached to the benzene ring by CH_2 may be prepared by reduction of the appropriate ketone, we may reduce 3-bromo-4-methylacetophenone, as prepared in part (g), by the Clemmensen or Wolff–Kishner procedure to give 2-bromo-4-ethyltoluene.

3-Bromo-4-methylacetophenone → 2-Bromo-4-ethyltoluene

Zn(Hg), HCl or H_2NNH_2, KOH, diethylene glycol, heat

(i) Although bromo and nitro substituents are readily introduced by electrophilic aromatic substitution, the only method we have available so far to prepare carboxylic acids is by oxidation of alkyl side chains. Therefore, using toluene as a starting material, convert the methyl group to a carboxyl group by oxidation, then nitrate. Nitro and carboxyl are both meta-directing groups, so the bromination in the last step occurs with the proper regioselectivity.

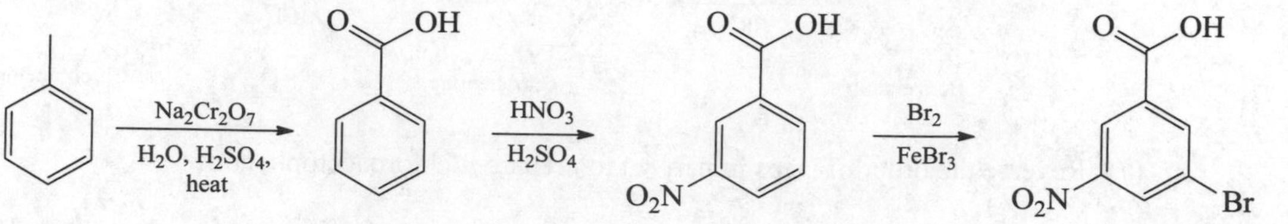

Toluene   Benzoic acid   *m*-Nitrobenzoic acid   3-Bromo-5-nitrobenzoic acid

If bromination were performed before nitration, the bromo substituent would direct an incoming electrophile to positions ortho and para to itself, giving the wrong orientation of substituents in the product.

(*j*) Again toluene is a suitable starting material, with its methyl group serving as the source of the carboxyl substituent. The orientation of the substituents in the final product requires that the methyl group be retained until the final step.

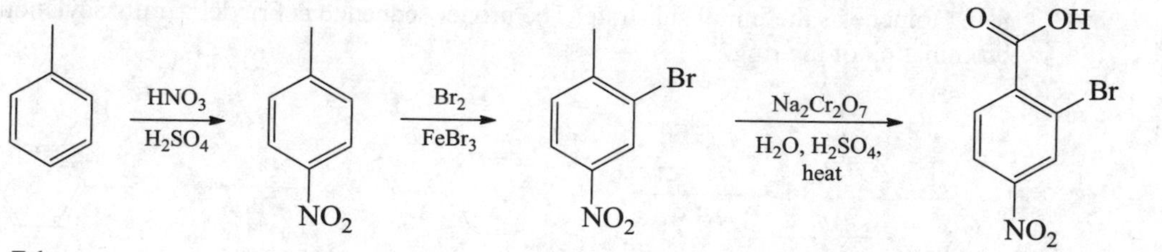

Toluene   *p*-Nitrotoluene   2-Bromo-4-nitrotoluene   2-Bromo-4-nitrobenzoic acid

Nitration must precede bromination, as in the previous part, in order to prevent formation of an undesired mixture of isomers.

(*k*) 1-Phenyloctane cannot be prepared efficiently by direct alkylation of benzene, because of the probability that rearrangement will occur. Indeed, a mixture of 1-phenyloctane and 2-phenyloctane is formed under the usual Friedel–Crafts conditions, along with 3-phenyloctane.

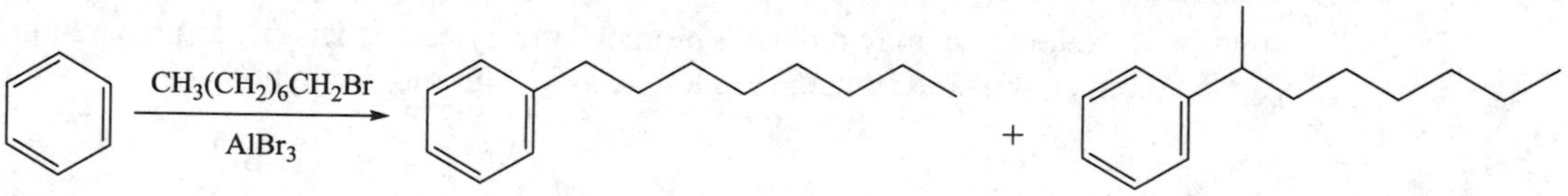

Benzene   1-Phenyloctane (40%)   +   2-Phenyloctane (30%)

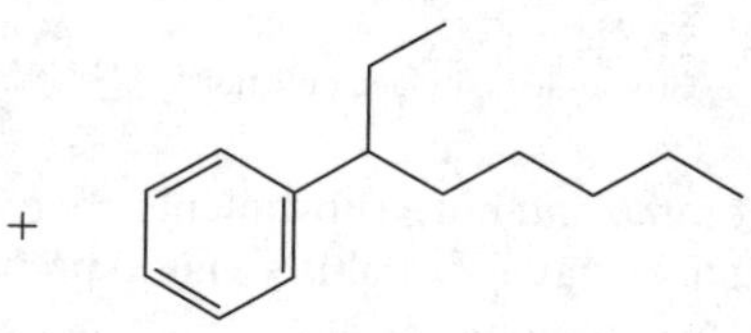

+   3-Phenyloctane (30%)

A method that permits the synthesis of 1-phenyloctane free of isomeric compounds is acylation followed by reduction.

Benzene → 1-Phenyloctanone → 1-Phenyloctane

Alternatively, Wolff–Kishner conditions (hydrazine, potassium hydroxide, diethylene glycol, heat) could be used in the reduction step.

(*l*) Direct alkenylation of benzene under Friedel–Crafts reaction conditions does *not* take place, and so 1-phenyl-1-octene *cannot* be prepared by reaction between benzene and 1-chloro-1-octene.

Having already prepared 1-phenyloctane in part (*k*), however, we can functionalize the benzylic position by bromination and then carry out a dehydrohalogenation to obtain the target compound.

1-Phenyloctane → 1-Bromo-1-phenyloctane → (*E*)-1-Phenyl-1-octene (mixture of *E* + *Z*)

(*m*) 1-Phenyl-1-octyne cannot be prepared in one step from benzene; 1-haloalkynes are unsuitable reactants for a Friedel–Crafts process. In Chapter 9, however, we learned that alkynes may be prepared by double dehydrohalogenation of a vicinal dihalide. Adapting this method to the alkene prepared in part (*l*) gives the desired alkyne.

(*E*)-1-Phenyl-1-octene (mixture of *E* + *Z*) → 1,2-Dibromo-1-phenyloctane → 1-Phenyl-1-octyne

(*n*) Nonconjugated cyclohexadienes are prepared by Birch reduction of arenes. Thus the last step in the synthesis of 1,4-di-*tert*-butyl-1,4-cyclohexadiene is the Birch reduction of 1,4-di-*tert*-butylbenzene.

Benzene → 1,4-Di-*tert*-butylbenzene → 1,4-Di-*tert*-butyl-1,4-cyclohexadiene

13.50 (*a*) Methoxy is an ortho, para-directing substituent. All that is required to prepare *p*-methoxybenzenesulfonic acid is to sulfonate anisole.

Anisole → *p*-Methoxybenzenesulfonic acid

(b) In reactions involving disubstitution of anisole, the better strategy is to introduce the para substituent first. The methoxy group is ortho, para-directing, but para substitution predominates.

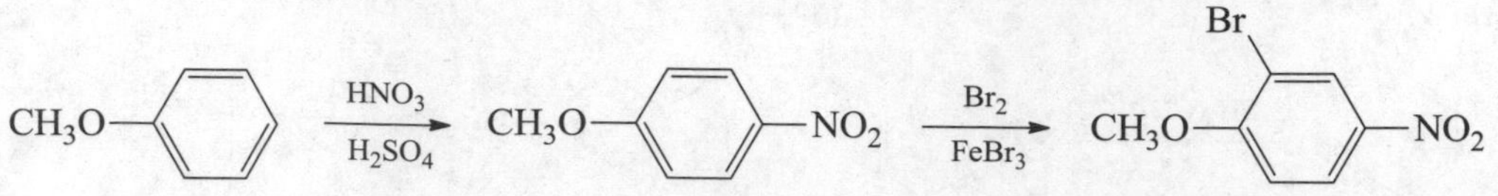

Anisole   p-Nitroanisole   2-Bromo-4-nitroanisole

(c) Reversing the order of the steps used in part (b) yields 4-bromo-2-nitroanisole.

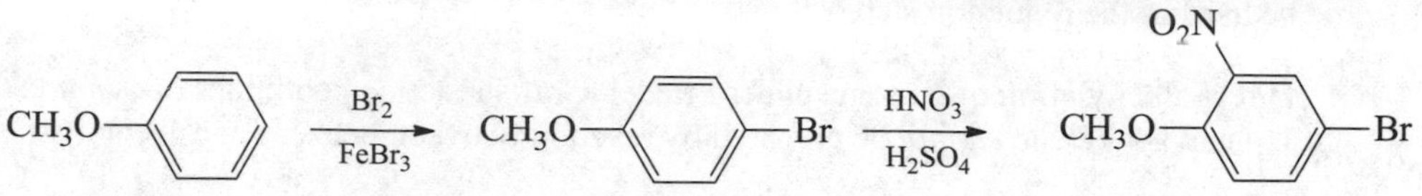

Anisole   p-Bromoanisole   4-Bromo-2-nitroanisole

(d) Direct introduction of a vinyl group onto an aromatic ring is not feasible. p-Methoxystyrene must be prepared indirectly by adding an ethyl side chain and then taking advantage of the reactivity of the benzylic position by bromination (e.g., with N-bromosuccinimide) and dehydrohalogenation.

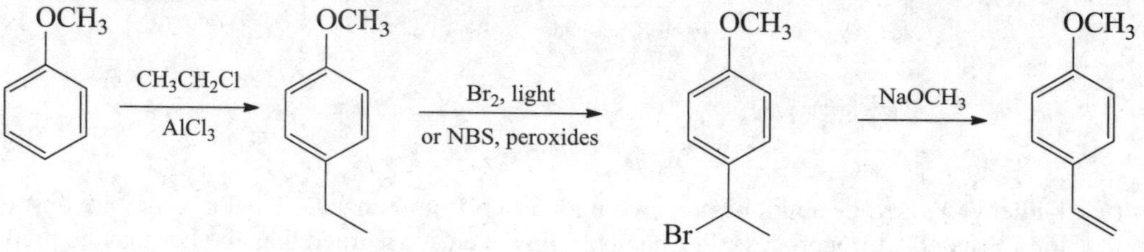

Anisole   p-Ethylanisole   p-(1-Bromoethyl)anisole   p-Methoxystyrene

13.51 Only reaction (b) will give the desired product. 1-Fluoro-2,4-dintrobenzene is activated toward nucleophilic aromatic substitution by the addition-elimination mechanism.

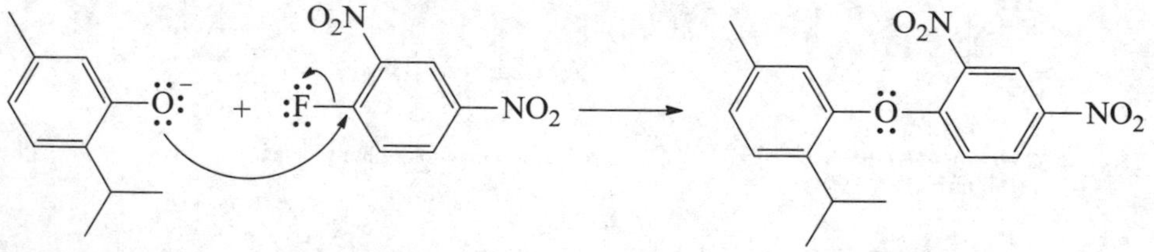

In reaction (a) nitration by electrophilic aromatic substitution will occur, but in the wrong ring. The nitro group already present deactivates its ring toward introduction of a second nitro group.

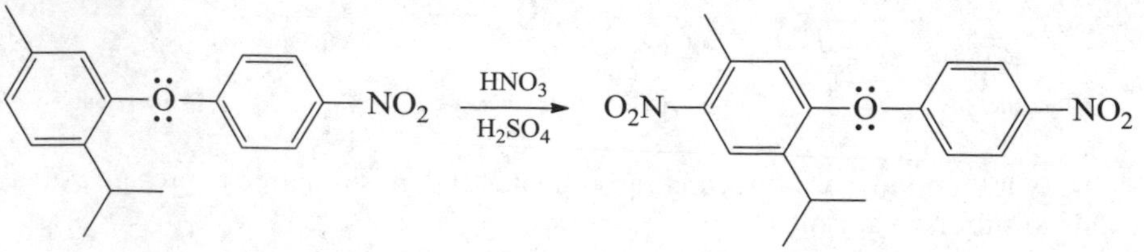

No reaction occurs in (*c*) because the fluoro-substituted ring is not activated toward nucleophilic aromatic substitution. Fluorine and the nitro groups need to be in the same ring in order for substitution to take place.

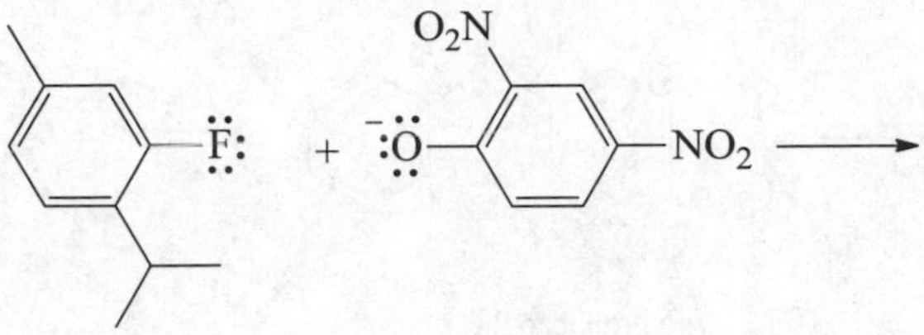

13.52 In a Friedel–Crafts acylation, an acyl chloride or acid anhydride reacts with an arene in the presence of aluminum chloride to yield an aryl ketone. Begin each problem by using the disconnection approach to identify the aromatic reactant.

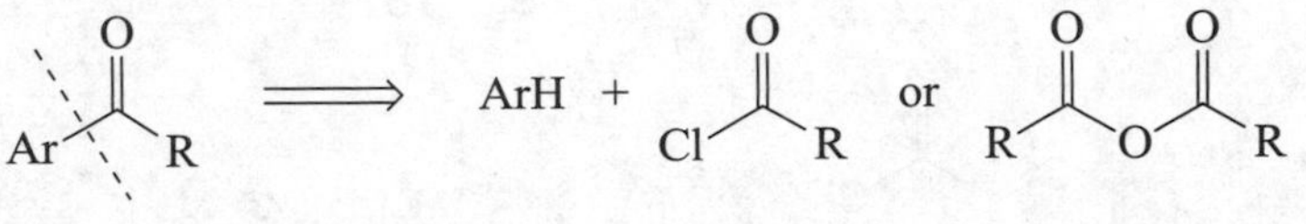

(*a*) A retrosynthesis based on this disconnection strategy suggests that Friedel-Crafts acylation of benzene with phenylacetyl chloride should be suitable for the preparation of benzyl phenyl ketone.

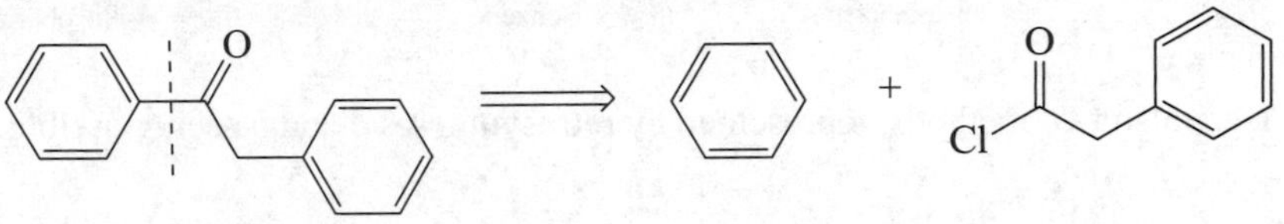

Benzyl phenyl ketone   Benzene   Phenylacetyl chloride

Indeed, in the presence of aluminum chloride this combination gave the desired ketone in 82% yield.

(*b*) Disconnection of the target reveals the arene to be *p*-xylene.

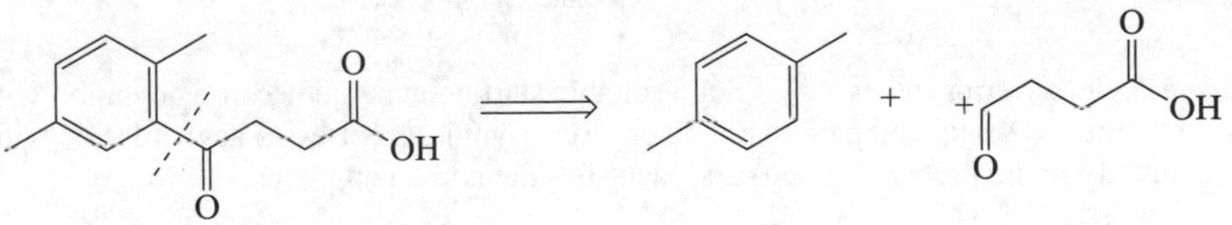

4-Oxo-4-(2,5-dimethylphenyl)butanoic acid   *p*-Xylene   A "succinylium" equivalent

The electrophile is shown as a "succinylium" cation. It, or its synthetic equivalent, is produced when succinic anhydride reacts with the Friedel-Crafts catalyst. The acylation, written in synthetic format is:

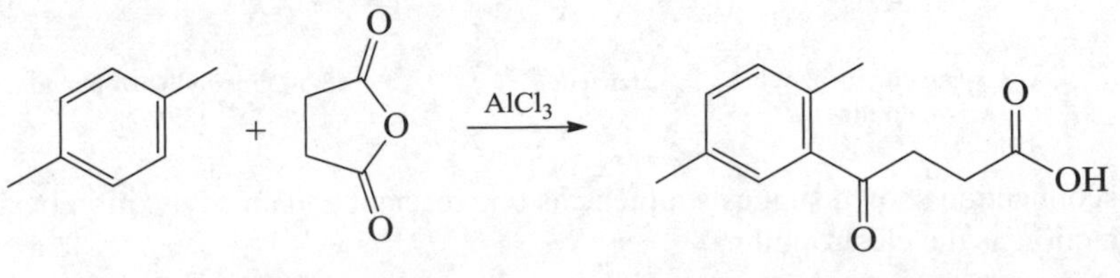

p-Xylene   Succinic anhydride   4-Oxo-4-(2,5-dimethylphenyl)butanoic acid

In the reaction carried out as shown, the yield of the desired product was 55%.

(*c*) Of the two disconnections *a* and *b*, only *b* leads to an effective synthesis.

Nitrobenzene *p*-Nitrobenzophenone Benzene

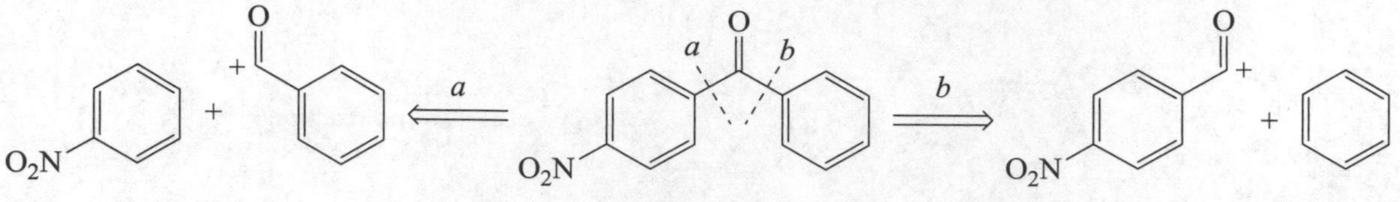

The synthesis involving acylation of nitrobenzene fails because of the inability of strongly deactivated aromatics to undergo Friedel-Crafts reactions. The alternative combination gave the desired product in 87% yield.

Nitrobenzene Benzene *p*-Nitrobenzophenone

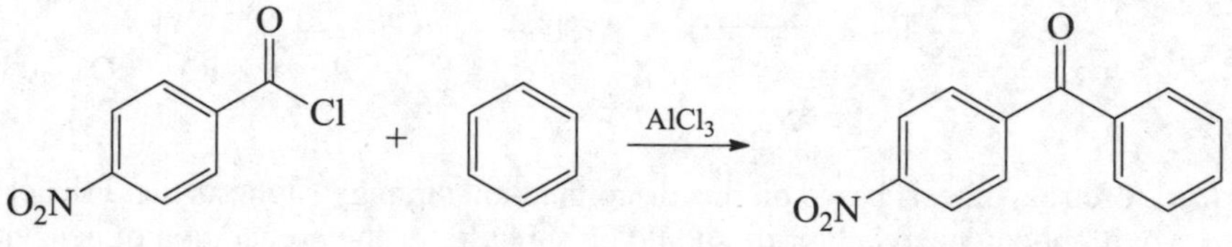

(*d*) Here also two methods, represented by retrosyntheses *a* and *b* seem possible.

m-Xylene 3,5-Dimethylbenzophenone Benzene

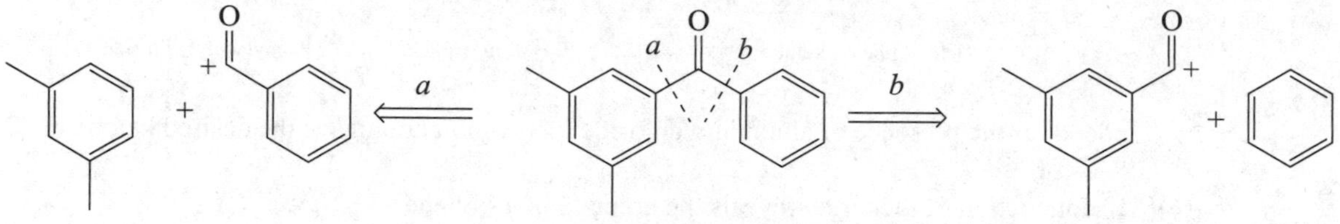

m-Xylene (retrosynthesis *a*) is not a suitable starting material because acylation will take place ortho to one methyl group and para to the other. Retrosynthesis *b* uses an acyl chloride with the three substituents already in the proper locations and identifies the correct approach.

3,5-Dimethylbenzoyl chloride Benzene 3,5-Dimethylbenzophenone (89%)

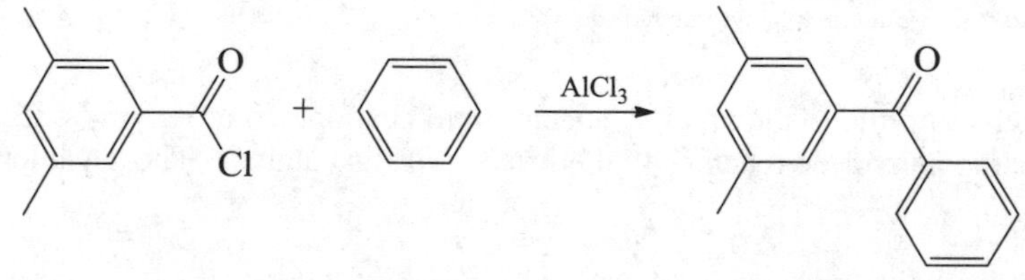

(e) The disconnection shown suggests toluene as one reactant and an acyl cation bearing an ortho carboxylic acid function as the electrophile.

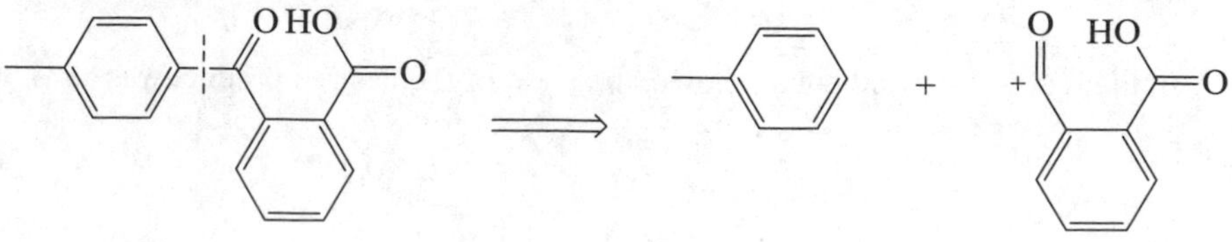

An electrophile of this type is accessible via a cyclic anhydride and leads to the synthesis shown.

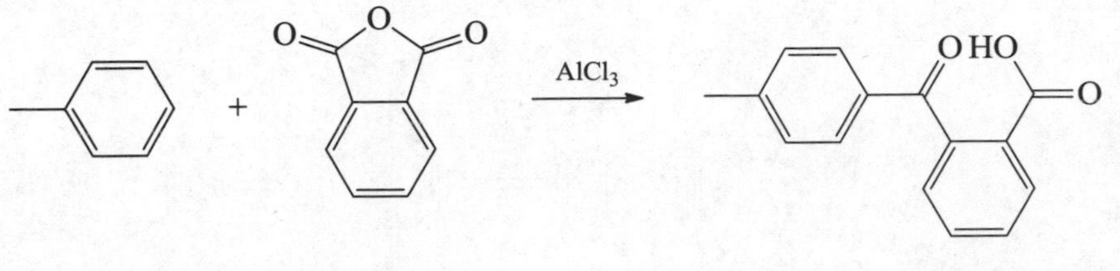

Toluene   Phthalic anhydride   o-(4-Methylbenzoyl)benzoic acid
                                                        (98%)

13.53 The first step is a Friedel–Crafts acylation. Reaction of benzene with a cyclic anhydride introduces both the acyl and carboxyl group into the molecule.

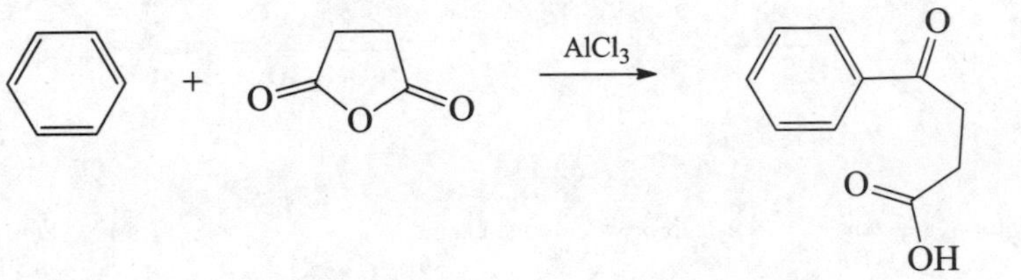

Benzene   Succinic anhydride   4-Oxo-4-phenylbutanoic acid

The second step is a reduction of the ketone carbonyl to a methylene group. A Clemmensen reduction is normally used for this step.

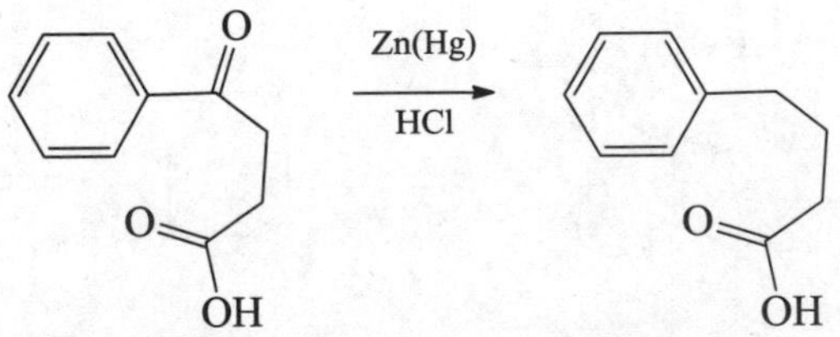

4-Oxo-4-phenylbutanoic acid   Phenylbutanoic acid

The cyclization phase of the process is an intramolecular Friedel–Crafts acylation reaction. It requires conversion of the carboxylic acid to the acyl chloride (thionyl chloride is a suitable reagent) followed by treatment with aluminum chloride.

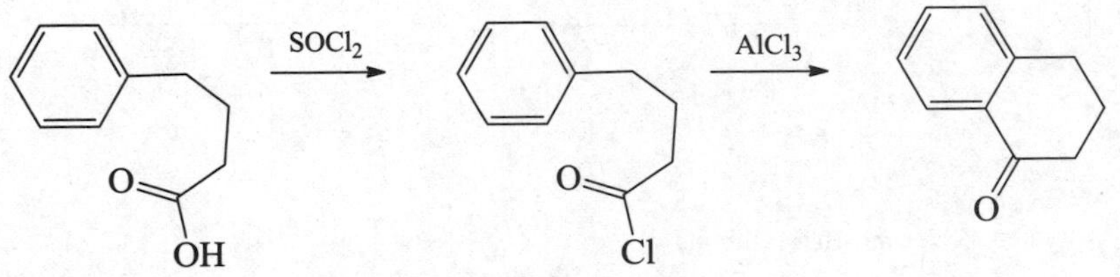

Phenylbutanoic acid   Phenylbutanoyl chloride

13.54 (*a*) The problem to be confronted here is that two meta-directing groups are para to each other in the product. However, by recognizing that the carboxylic acid function can be prepared by oxidation of the isopropyl group, we can construct a retrosynthesis in which the key intermediate has the sulfonic acid group para to the ortho, para-directing isopropyl group.

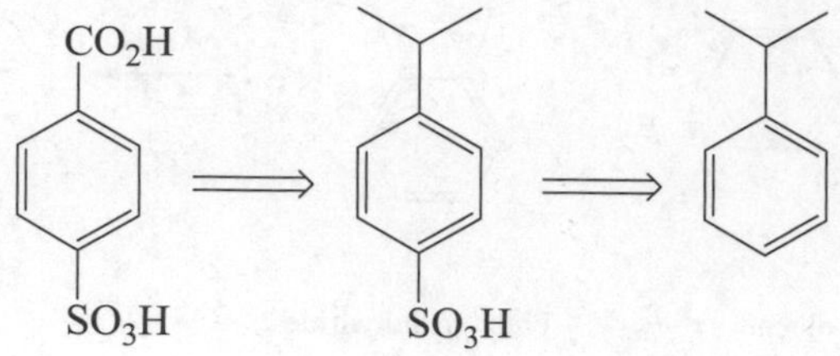

A reasonable synthesis is:

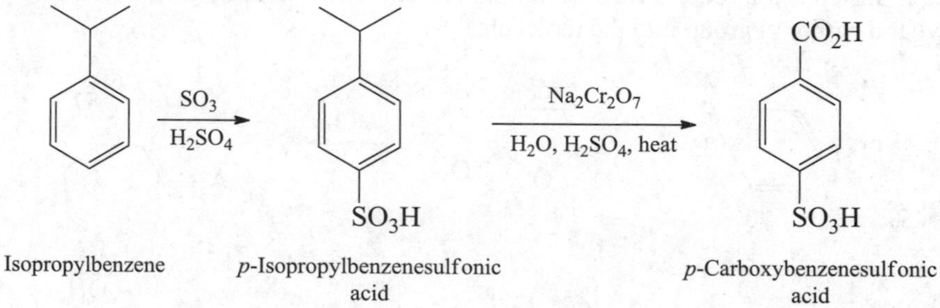

Isopropylbenzene   *p*-Isopropylbenzenesulfonic
                                          acid

p-Carboxybenzenesulfonic
acid

(*b*) In this problem two methyl groups must be oxidized to carboxylic acid functions and a *tert*-butyl group must be introduced, most likely by a Friedel–Crafts reaction. Because Friedel–Crafts alkylations cannot be performed on deactivated aromatic rings, oxidation must *follow*, not precede, alkylation and requires the following retrosynthesis.

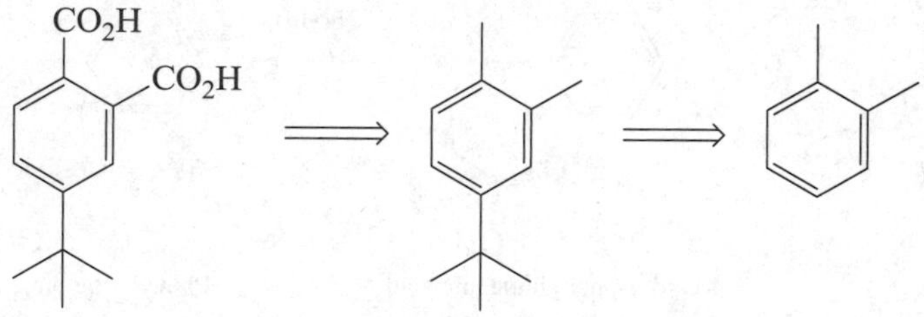

The following reaction sequence is appropriate:

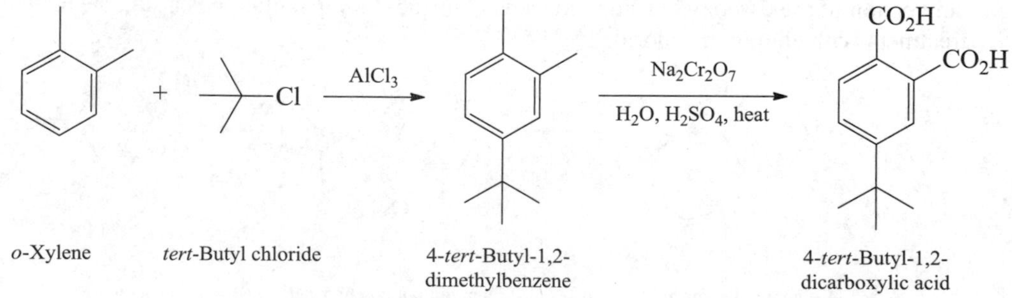

o-Xylene   *tert*-Butyl chloride   4-*tert*-Butyl-1,2-
                                                                    dimethylbenzene

4-*tert*-Butyl-1,2-
dicarboxylic acid

In practice, zinc chloride was used as the Lewis acid to catalyze the Friedel–Crafts reaction (64% yield). Oxidation of the methyl groups occurs preferentially because the *tert*-butyl group has no benzylic hydrogens.

(*c*) The starting material has two equivalent C-H benzenoid positions that can be modified by electrophilic aromatic substitution. Since a ketone functional group is in the product, removal of the carbonyl in the starting material early in the synthesis is necessary. A most reasonable retrosynthetic pathway for the target molecule is the hydrocarbon shown.

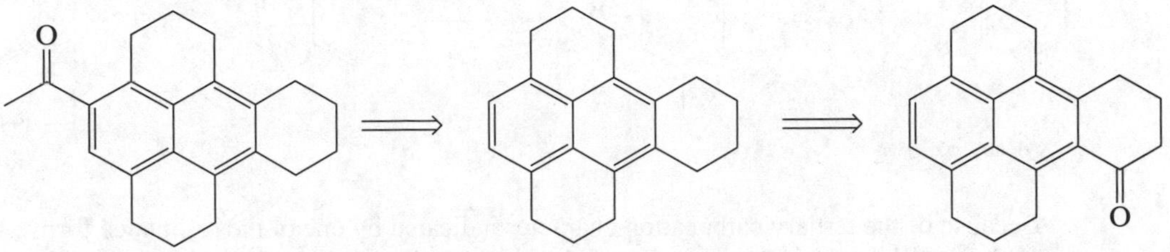

The acetyl group can be attached by Friedel-Crafts acylation to either one of the two available reactive sites of the naphthalene unit. The hydrocarbon itself is readily prepared by a Clemmensen or Wolff-Kishner reduction of the original starting material.

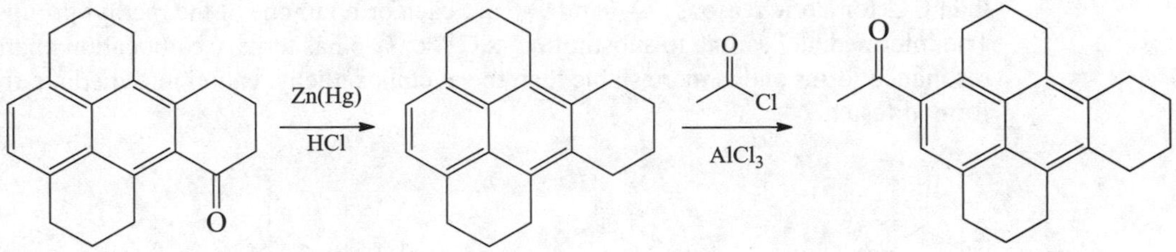

(*d*) The retrosynthesis shown takes into consideration (*a*) the activating ortho, para-directing properties of methoxy groups, and (*b*) the unreactivity of nitro-substituted aromatic rings under Friedel-Crafts alkylation conditions.

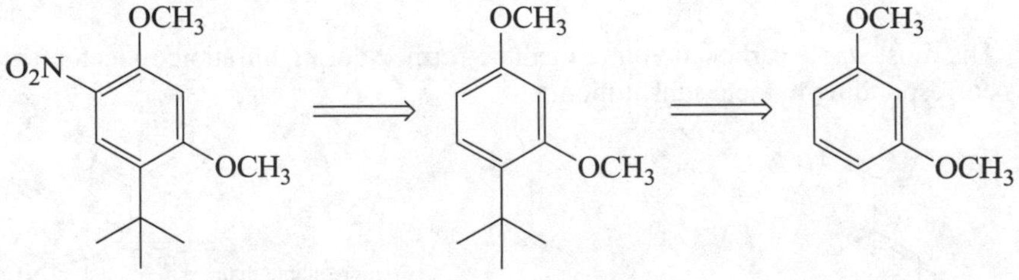

1-*tert*-Butyl-2,4-dimethoxy-5-nitrobenzene   1-*tert*-Butyl-2,4-dimethoxybenzene   1,3-Dimethoxybenzene

Because Friedel–Crafts reactions may not be performed on deactivated aromatic rings, the *tert*-butyl group must be introduced before the nitro group. The correct sequence is therefore:

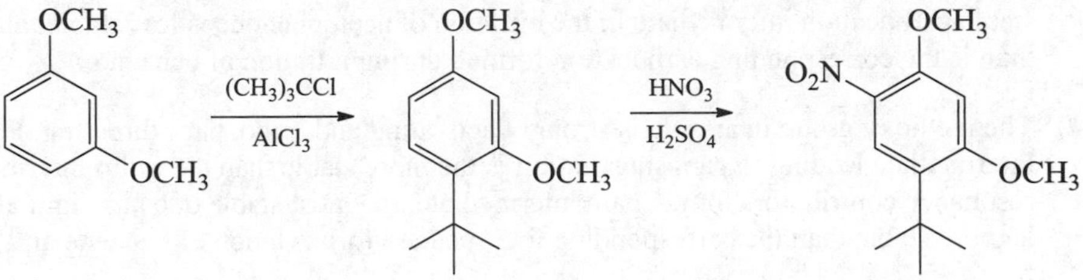

Mechanism

13.55 (*a*) The four ring carbons of *p*-xylene that are available for substitution are equivalent. The three principal resonance contributors to the cyclohexadienyl cation intermediate formed on reaction with bromine are as shown:

p-Xylene

Because of the tertiary carbocation character indicated by one of the resonance forms, this cyclohexadienyl cation intermediate is more stable than the one formed from benzene, and bromination of *p*-xylene is faster than bromination of benzene.

(*b*) Carbons 4 and 6 of *m*-xylene are equivalent and more reactive toward electrophilic aromatic substitution than C-2 for steric reasons. C-4 and C-6 are each ortho to one of the methyl groups and para to the other. The intermediate leading to substitution at C-4 or C-6 has tertiary carbocation character in two of the resonance forms and is more stable than the cyclohexadienyl cation intermediate from benzene and is formed faster.

m-Xylene

(*c*) The most stable carbocation intermediate formed during nitration of acetophenone is the one corresponding to meta substitution.

Acetophenone   more stable than   or

An acyl group is electron-withdrawing and destabilizes a carbocation to which it is attached. The most stable carbocation intermediate in the nitration of acetophenone is less stable, and is formed more slowly, than is the corresponding carbocation formed during nitration of benzene.

(*d*) The methoxy group in anisole is strongly activating and ortho, para-directing. For steric reasons the intermediate leading to para substitution is the more stable than the ortho intermediate. Of the various resonance contributors for the para intermediate, the most stable one has eight electrons around oxygen, is more stable than the corresponding intermediate for acylation of benzene and is formed at a faster rate.

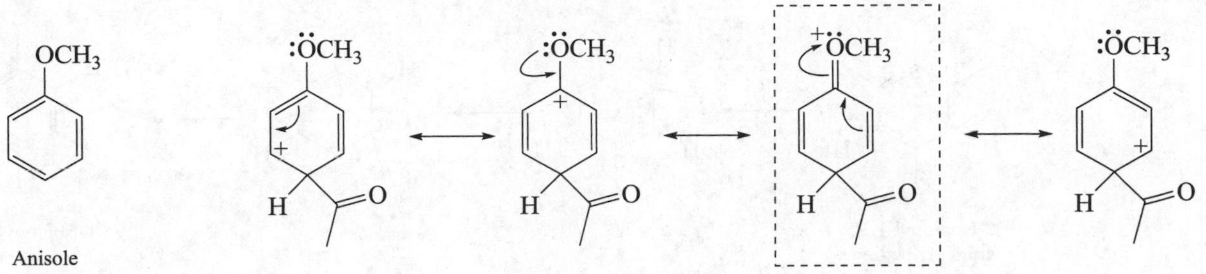

Anisole

(e) An isopropyl group is an activating substituent and is ortho, para-directing. Electrophilic attack at the ortho position is sterically hindered. The most stable intermediate is

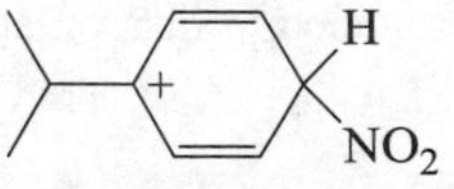

or any of its resonance forms. Because of its tertiary carbocation character, this cation is more stable than the corresponding cyclohexadienyl cation intermediate from benzene.

(f) A nitro substituent is deactivating and meta-directing. The most stable cyclohexadienyl cation formed in the bromination of nitrobenzene is

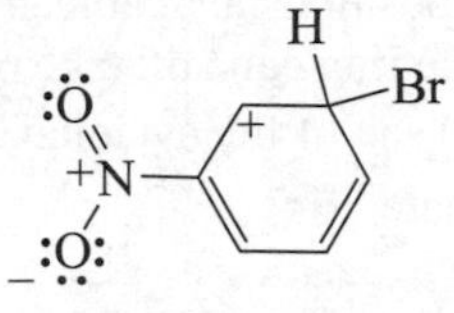

or any of its resonance forms. This ion is less stable than the cyclohexadienyl cation formed during bromination of benzene.

(g) Sulfonation of furan takes place at C-2. The cationic intermediate is more stable than the cyclohexadienyl cation formed from benzene because it is stabilized by electron release from oxygen.

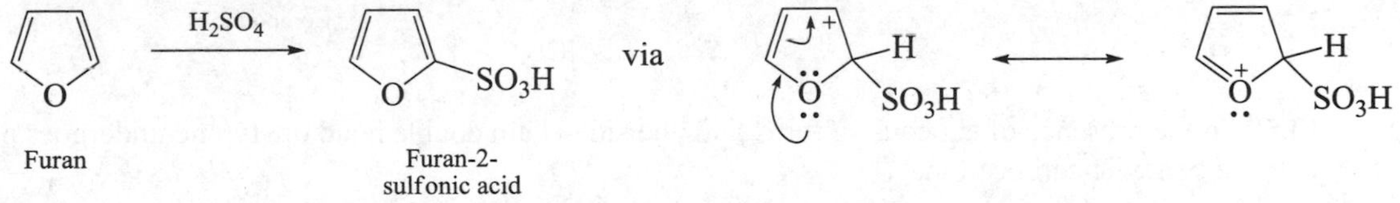

Furan   Furan-2-
                    sulfonic acid

(h) Pyridine reacts with electrophiles at C-3. It is less reactive than benzene, and the carbocation intermediate is less stable than the corresponding intermediate formed from benzene.

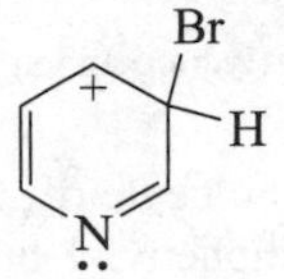

13.56 The isomerization begins with protonation of the aromatic ring, an electrophilic attack by HCl catalyzed by $AlCl_3$. The carbocation produced then rearranges by a methyl shift, and the rearranged cyclohexadienyl cation loses a proton to form the isomeric product.

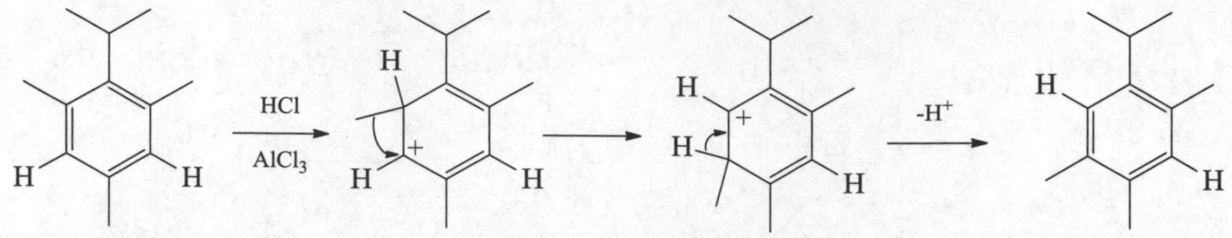

2-Isopropyl-1,3,5-
trimethylbenzene

1-Isopropyl-2,4,5-
trimethylbenzene

The driving force for rearrangement is relief of steric strain between the isopropyl group and one of its adjacent methyl groups.

13.57 On treatment with base, intramolecular nucleophilic aromatic substitution leads to the observed product.

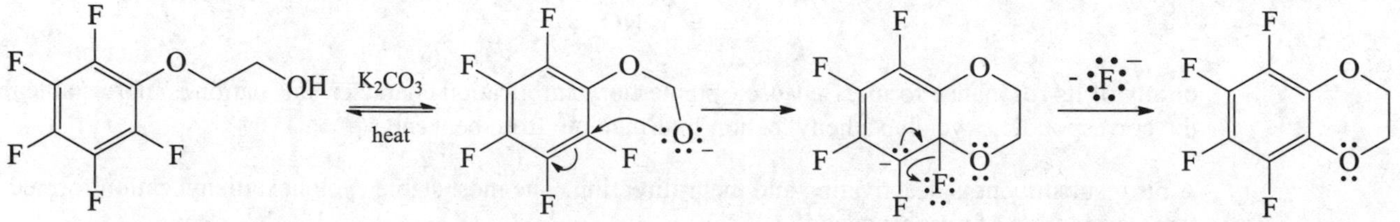

13.58 Although hexamethylbenzene has no positions available at which ordinary electrophilic aromatic *substitution* might occur, electrophilic *attack* on the ring can still take place to form a cyclohexadienyl cation. Compound A is the tetrachloroaluminate (AlCl$_4^-$) salt of the cyclohexadienyl cation; it undergoes deprotonation on being treated with aqueous sodium bicarbonate.

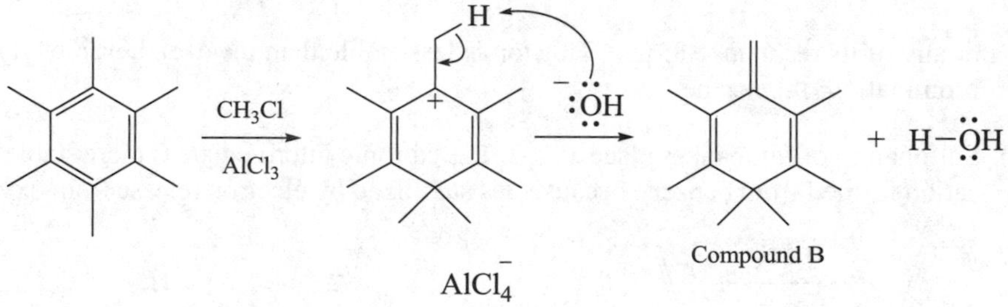

Compound B

13.59 In the presence of aqueous sulfuric acid, the side-chain double bond of styrene undergoes protonation to form a benzylic carbocation.

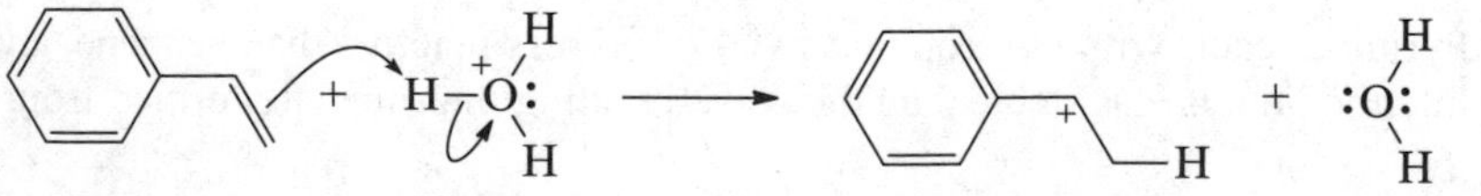

Styrene   Hydronium ion   1-Phenylethyl cation   Water

This carbocation then reacts with a molecule of styrene to give a carbocation, which is then deprotonated to give one of the styrene dimers (1,3-diphenyl-1-butene).

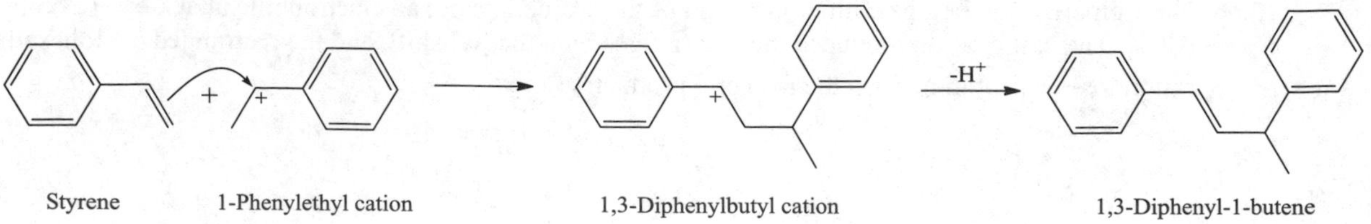

Styrene   1-Phenylethyl cation   1,3-Diphenylbutyl cation   1,3-Diphenyl-1-butene

The same carbocation produced in this step can cyclize in what amounts to an intramolecular Friedel–Crafts alkylation to give the other styrene dimer (1-methyl-1-phenylindane).

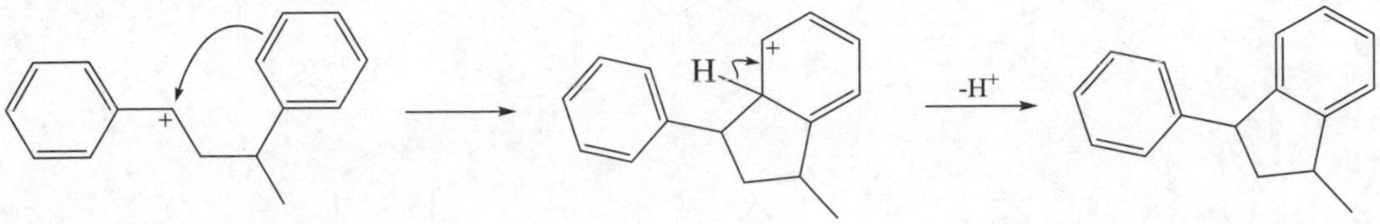

1,3-Diphenylbutyl cation 1-Methyl-3-phenylindane

Answers to Interpretive Problems 13

13.60 C; **13.61** D; **13.62** B; **13.63** A

SELF-TEST

1. Write the three most stable resonance contributors to the cyclohexadienyl cation formed in the ortho bromination of toluene.

2. Give the major product(s) for each of the following reactions. Indicate whether the reaction proceeds faster or slower than the corresponding reaction of benzene.

(*a*) 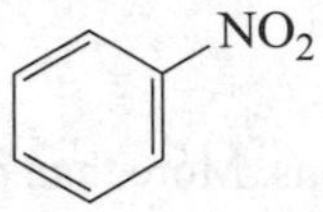 $\xrightarrow[H_2SO_4]{HNO_3}$?

(*c*) 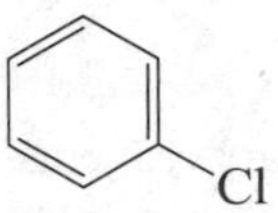 $\xrightarrow[H_2SO_4]{SO_3}$?

(*b*) 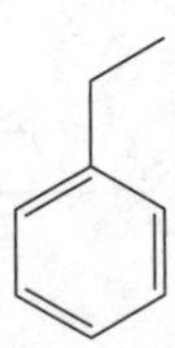 $\xrightarrow[FeBr_3]{Br_2}$?

3. Write the formula of the electrophilic species present in each reaction of the preceding problem.

4. Provide the reactant, reagent, or product omitted from each of the following:

(*a*) ? $\xrightarrow[HCl]{Zn}$ 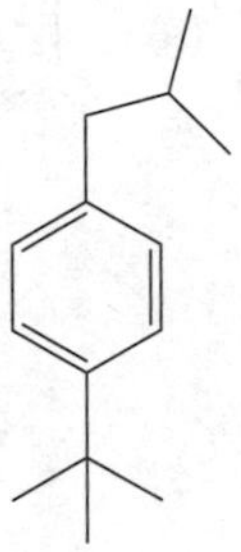

(*d*) 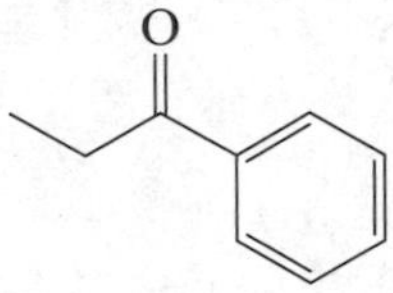 $\xrightarrow[H_2SO_4]{HNO_3}$?

(*b*) 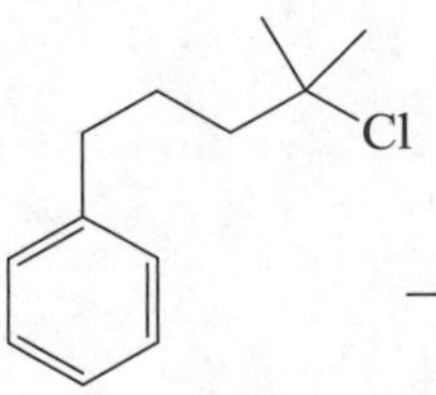 $\xrightarrow{AlCl_3}$?

(*e*) 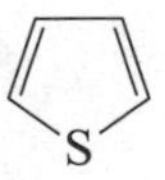 $\xrightarrow[FeCl_3]{Cl_2}$?

(c)

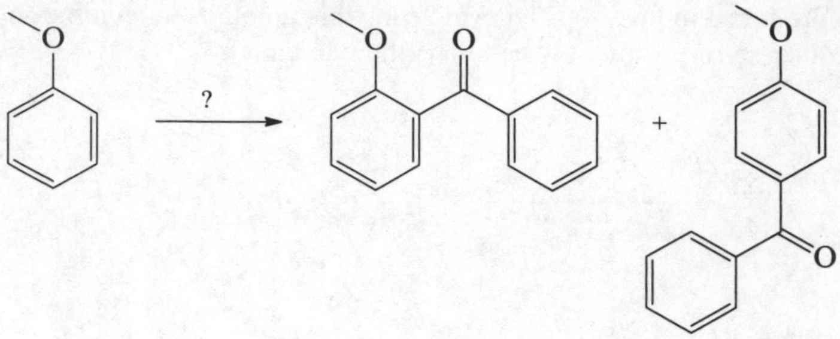

5. Draw the structure(s) of the major product(s) formed by reaction of each of the following compounds with one equivalent of Cl_2 and $FeCl_3$. If two products are formed in significant amounts, draw them both.

(a) 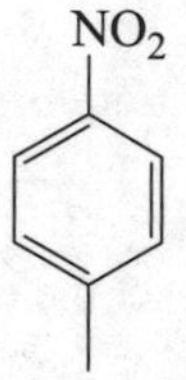   (b) 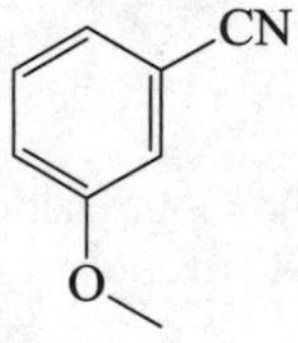   (c) 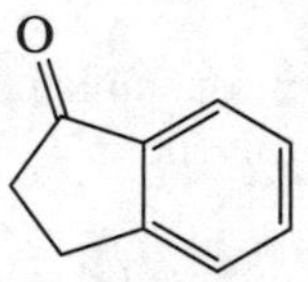   (d)

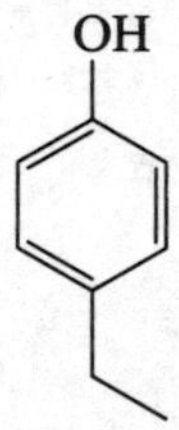

6. Provide the necessary reagents for each of the following transformations. More than one step may be necessary.

(a)

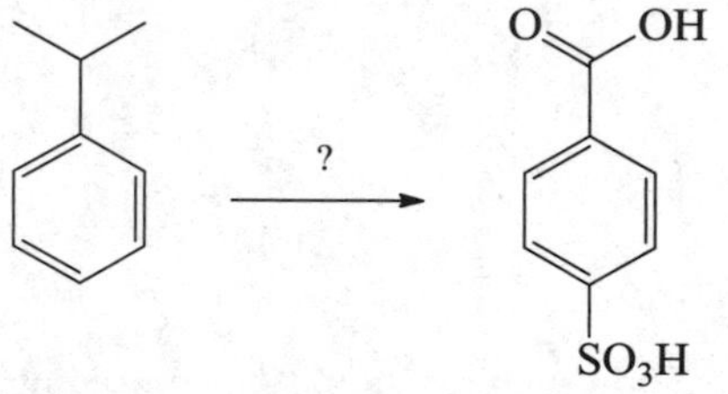

(d)

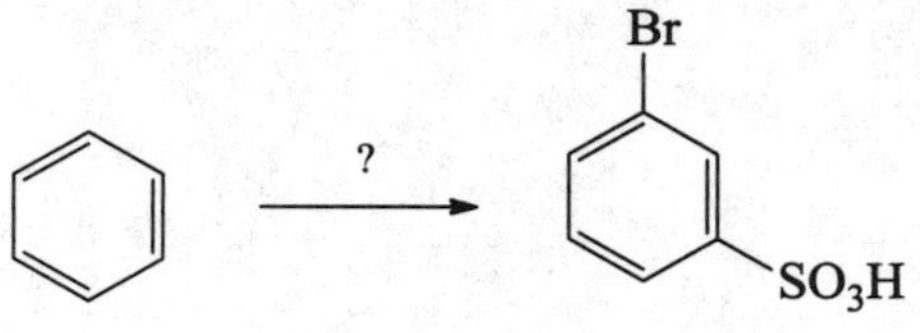

(b)

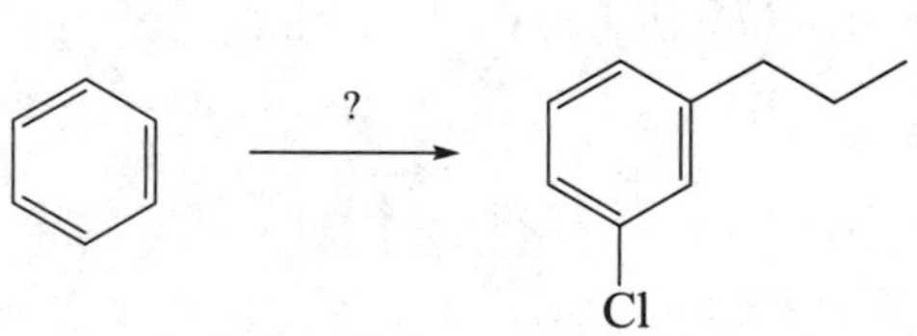

(e)

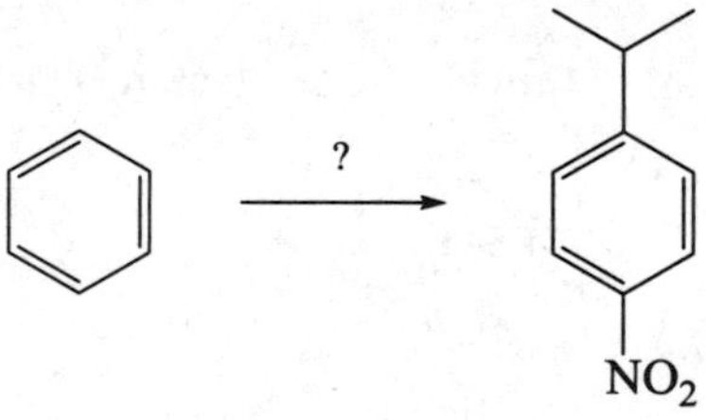

(c)

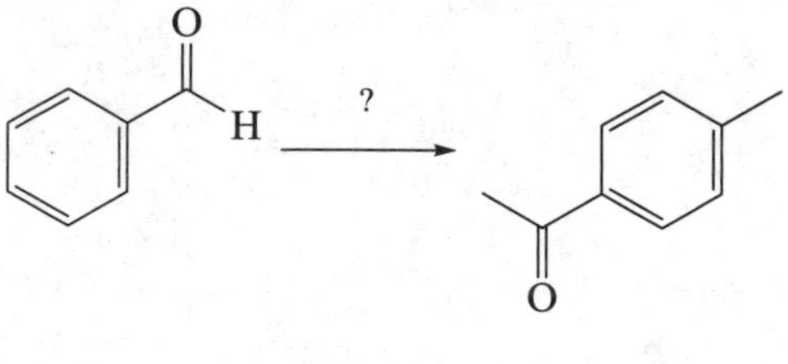

7. Outline a reasonable synthesis of each of the following from either benzene or toluene and any necessary organic or inorganic reagents.

(a) (b) (c)

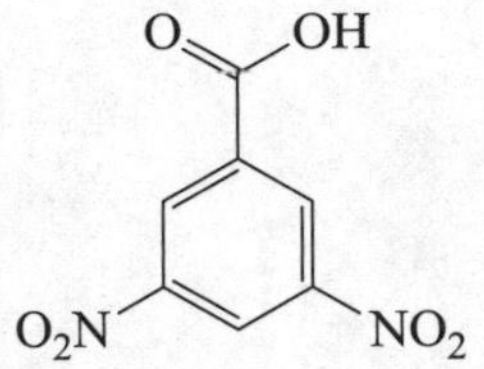

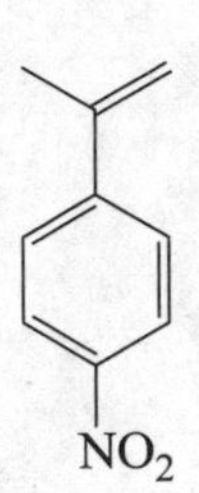

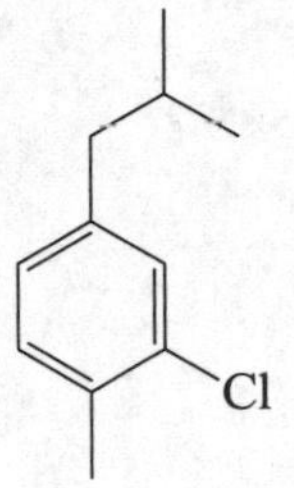

8. Outline a reasonable synthesis of the compound shown using anisole ($C_6H_5OCH_3$) and any necessary inorganic reagents.

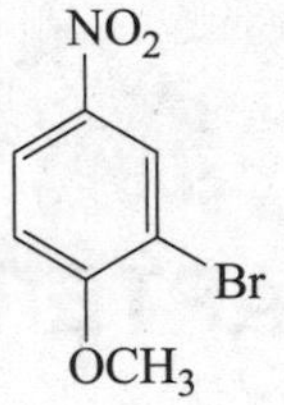

9. Give the product obtained from the following reaction, and draw the structure of the intermediate formed in the reaction.

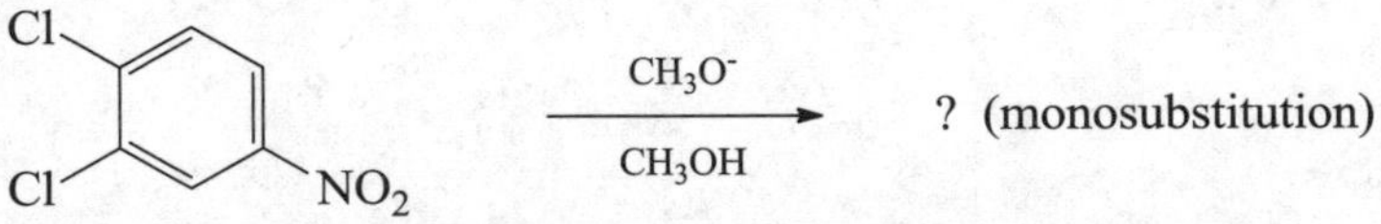

$$\xrightarrow[\text{CH}_3\text{OH}]{\text{CH}_3\text{O}^-} \quad ? \text{ (monosubstitution)}$$

10. Suggest synthetic schemes by which chlorobenzene may be converted into

   (a) 2,4-Dinitroanisole (1-methoxy-2,4-dinitrobenzene)
   (b) p-Isopropylaniline

11. Write a mechanism using resonance structures to show how a nitro group directs ortho, para in nucleophilic aromatic substitution.

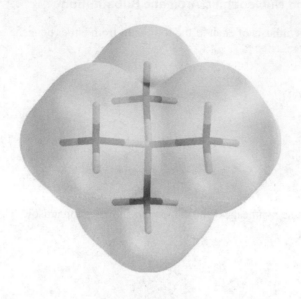

CHAPTER 14

Spectroscopy

Table of Contents

SOLUTIONS TO TEXT PROBLEMS

In Chapter Problems

14.1 The field strength of an NMR spectrometer magnet and the frequency of electromagnetic radiation used to observe an NMR spectrum are directly proportional. Thus, the ratio 4.7 T/200 MHz is the same as 1.41 T/60 MHz. The magnetic field strength of a 60 MHz NMR spectrometer is 1.41 T. The field strength of a 920 MHz instrument is 21.6 T.

14.2 The ratio of ^1H and ^{13}C resonance frequencies remains constant. When the ^1H frequency is 200 MHz, ^{13}C NMR spectra are recorded at 50.4 MHz. Thus, when the ^1H frequency is 100 MHz, ^{13}C NMR spectra will be observed at 25.2 MHz.

14.3 (a) The ^{1}NMR chemical shift (δ of bromoform ($CHBr_3$) recorded at 300 MHz is

$$\delta = \frac{2065 \text{ Hz}}{300 \times 10^6 \text{Hz}} \times 10^6 = 6.88 \text{ ppm}$$

(b) When measured at 400 MHz, the chemical shift would be the same for bromoform, 6.88 ppm.

(c) The frequency of the measured absorption at 400 MHz for bromoform is 2752 Hz. This is usually measured by the spectrometer. However, using the equation in part (a) and replacing the value of 2065 Hz with y, replacing (300×10^6) with (400×10^6) and then solving for y, one may calculate this value.

$$y = \frac{(6.88) \times (400 \times 10^6 \text{Hz})}{(1 \times 10^6)} = 2752 \text{ Hz}$$

14.4 (b) Electronegative atoms such as chlorine exert an electron-withdrawing effect on the protons that are separated from them by the fewest bonds. The effect is cumulative, and two chlorines will exert a greater effect than one. The proton on C-1 will be the least shielded and those on C-3 will be the most shielded.

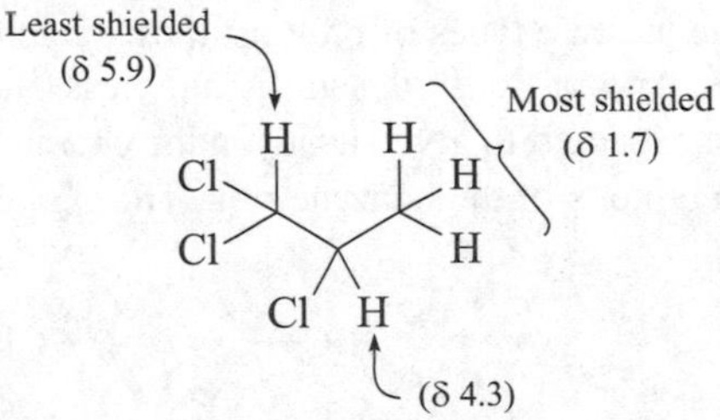

(c) The protons on the carbons adjacent to the oxygen of tetrahydrofuran will be less shielded than those separated from the oxygen by more bonds.

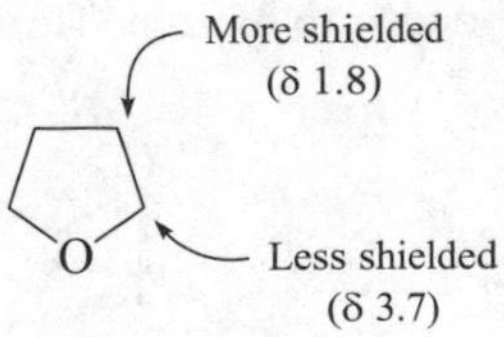

14.5 (b) The reported chemical shifts for benzyl acetate are δ 2.0, 5.1, and 7.2. The chemical shifts from Figure 14.8 are given in the structure on the left. The aryl protons can easily be assigned to δ 7.2. The absorption at δ 2.0 is just in the range for C–H protons adjacent to carbonyl groups. For the remaining benzylic protons, the chemical shift at δ 5.1 does not fall into any range predicted from Figure 14.8 (2.3–2.8 or 3.3–3.7); however, because the benzylic protons are also adjacent to an oxygen, one would predict that this combination would significantly shift the proton signal farther downfield, (i.e., to δ 5.1).

14.6 Protons attached to carbons adjacent to a carbonyl group are less shielded than those separated from the carbonyl by more bonds. Methyl protons are generally more shielded than methylene protons; thus, the chemical shifts of each set of protons in 2-pentanone may be assigned.

14.7 The least shielded proton is the O—H proton of the carboxyl group. Recall (from Chapter 13) that nitro groups are strongly electron-withdrawing and that this effect is felt most strongly ortho and para to the group. Thus of the two sets of protons on the benzene ring, the less shielded of them is the set ortho to the nitro substituent. The most shielded protons in the spectrum are those of the methyl group.

14.8 1,4-Dimethylbenzene has two types of protons: those attached directly to the benzene ring and those of the methyl groups. Aryl protons are significantly less shielded than alkyl protons. As shown in text Figure 14.8, they are expected to give signals in the chemical shift range δ 6.5–8.5. Thus, the signal at δ 7.0 is due to the protons of the benzene ring. The signal at δ 2.2 is due to the methyl protons.

14.9 (*b*) Four nonequivalent sets of protons are bonded to carbon in 1-butanol as well as a fifth distinct type of proton, the one bonded to oxygen. There should be five signals in the ^{1}H NMR spectrum of 1-butanol.

Five different proton environments
in 1-butanol; five signals

(c) Apply the "proton replacement" test to butane.

Butane    1-Chlorobutane 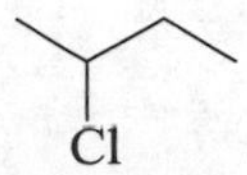   2-Chlorobutane 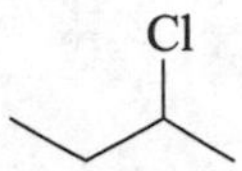   2-Chlorobutane    1-Chlorobutane

Butane has *two* different types of protons; it will exhibit *two* signals in its ^{1}H NMR spectrum.

(d) Like butane, 1,4-dibromobutane has two different types of protons. This can be illustrated by using a chlorine atom as a test group.

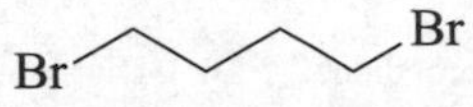

1,4-Dibromobutane 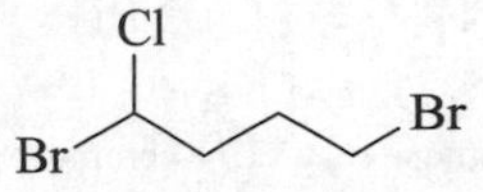   1,4-Dibromo-1-chlorobutane 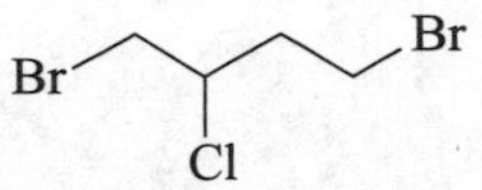   1,4-Dibromo-2-chlorobutane

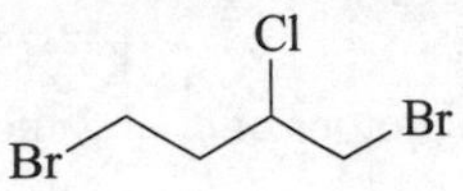

1,4-Dibromo-2-chlorobutane 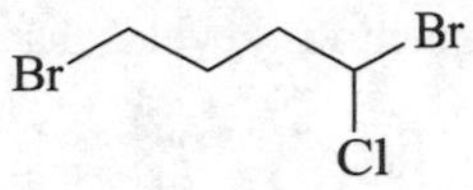   1,4-Dibromo-1-chlorobutane

The ^{1}H NMR spectrum of 1,4-dibromobutane is expected to consist of two signals.

(e) All the carbons in 2,2-dibromobutane are different from one another, and so protons attached to one carbon are not equivalent to the protons attached to any of the other carbons. This compound should have *three* signals in its ^{1}H NMR spectrum.

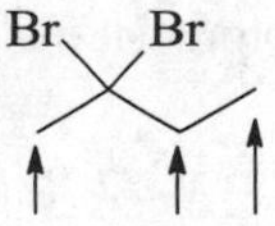

2,2-Dibromobutane has three
nonequivalent sets of protons.

(f) All the protons in 2,2,3,3-tetrabromobutane are equivalent. Its ^{1}H NMR spectrum will consist of one signal.

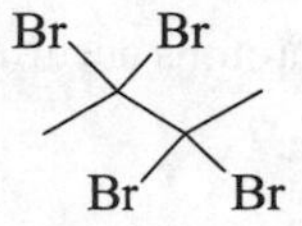

2,2,3,3-Tetrabromobutane

(g) There are *four* nonequivalent sets of protons in 1,1,4-tribromobutane. It will exhibit four signals in its ^{1}H NMR spectrum.

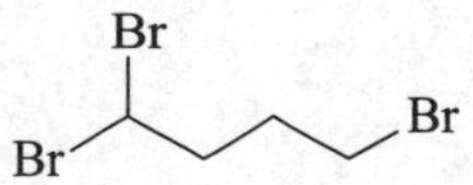

1,1,4-Tribromobutane

(*h*) The seven protons of 1,1,1-tribromobutane belong to three nonequivalent sets, and hence the ^{1}H NMR spectrum will consist of three signals.

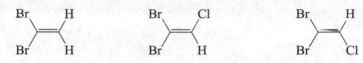

1,1,1-Tribromobutane

14.10 (*b*) Apply the replacement test to each of the protons of 1,1-dibromoethene.

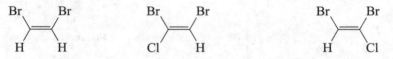

1,1-Dibromoethene   1,1-Dibromo-2-chloroethene   1,1-Dibromo-2-chloroethene

Replacement of one proton by a test group (Cl) gives exactly the same compound as replacement of the other. The two protons of 1,1-dibromoethene are equivalent, and there is only one signal in the ^{1}H NMR spectrum of this compound.

(*c*) The replacement test reveals that both protons of *cis*-1,2-dibromoethene are equivalent.

Br   Br   Br   Br   Br   Br
  \\   /   \\   /   \\   /
   C = C   C = C   C = C
  /   \\   /   \\   /   \\
H   H   Cl   H   H   Cl

cis-1,2-Dibromoethene   (*Z*)-1,2-Dibromo-1-chloroethene   (*Z*)-1,2-Dibromo-1-chloroethene

Because both protons are equivalent, the ^{1}H NMR spectrum of *cis*-1,2-dibromoethene consists of one signal.

(*d*) Both protons of *trans*-1,2-dibromoethene are equivalent; each is cis to a bromine substituent.

Br   H
  \\   /
   C = C
  /   \\
H   Br

trans-1,2-Dibromoethene
(one signal in the ^{1}H NMR spectrum)

(*e*) *Four* nonequivalent sets of protons occur in allyl bromide.

H   H
  \\   /
   C = C
  /   \\
H CH_2−Br

Allyl bromide
(four signals in the ^{1}H NMR spectrum)

(*f*) The protons of a single methyl group are equivalent to one another, but all three methyl groups of 2-methyl-2-butene are nonequivalent. The vinyl proton is unique.

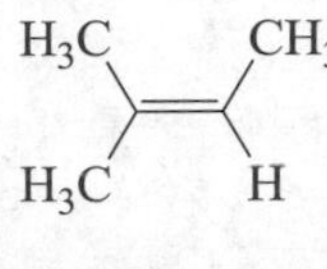

2-Methyl-2-butene
(four signals in the ^{1}H NMR spectrum)

14.11 (*b*) The three methyl protons of 1,1,1-trichloroethane (Cl_3CCH_3) are equivalent. They have the same chemical shift and do not split each other's signals. The ^{1}H NMR spectrum of Cl_3CCH_3 consists of a single sharp peak.

(*c*) Separate signals will be seen for the methylene (CH_2) protons and for the methine (CH) proton of 1,1,2-trichloroethane.

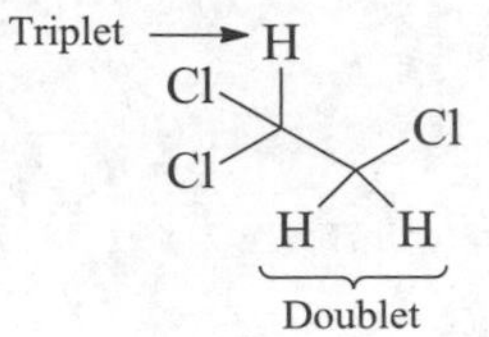

1,1,2-Trichloroethane

The methine proton splits the signal for the methylene protons into a doublet. The two methylene protons split the methine proton's signal into a triplet.

(*d*) Examine the structure of 1,2,2-trichloropropane.

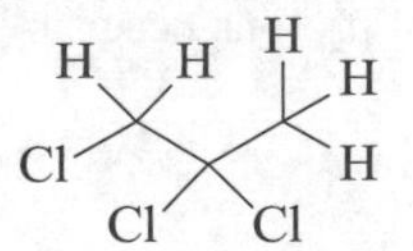

1,2,2-Trichloropropane

The ^{1}H NMR spectrum exhibits a signal for the two equivalent methylene protons and one for the three equivalent methyl protons. Both these signals are sharp singlets. The protons of the methyl group and the methylene group are separated by more than three bonds and do not split each other's signals.

(*e*) The methine proton of 1,1,1,2-tetrachloropropane splits the signal of the methyl protons into a doublet; its signal is split into a quartet by the three methyl protons.

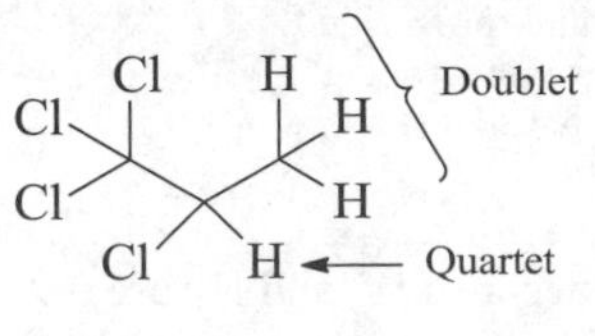

1,1,1,2-Tetrachloropropane

14.12 (*b*)   The ethyl group appears as a triplet–quartet pattern and the methyl group as a singlet.

$$CH_3—CH_2—O—CH_3$$

Triplet   Quartet   Singlet; not vicinal to any
other protons in molecule

(*c*)   The two ethyl groups of diethyl ether are equivalent to each other. The two methyl groups appear as one triplet and the two methylene groups as one quartet.

$$CH_3—CH_2—O—CH_2—CH_3$$

Triplet   Quartet   Quartet   Triplet

(*d*)   The two ethyl groups of *p*-diethylbenzene are equivalent to each other and give rise to a single triplet–quartet pattern.

$$CH_3CH_2 \; \underset{\substack{H \quad\quad H}}{\overset{\substack{H \quad\quad H}}{\bigcirc}} \; CH_2CH_3$$

Three signals:
CH_3 triplet;
CH_2 quartet;
aromatic H singlet

All four protons of the aromatic ring are equivalent, have the same chemical shift, and do not split either each other's signals or any of the signals of the ethyl group.

(*e*)   Four nonequivalent sets of protons occur in this compound:

$$ClCH_2—CH_2—O—CH_2—CH_3$$

Triplet   Triplet   Quartet   Triplet

Vicinal protons in the $ClCH_2CH_2O$ group split one another's signals, as do those in the CH_3CH_2O group.

14.13 (*b*)   The signal of the proton at C-2 is split into a quartet by the methyl protons, and each line of this quartet is split into a doublet by the aldehyde proton. It appears as a doublet of quartets. (*Note:* It does not matter whether the splitting pattern is described as a doublet of quartets or a quartet of doublets. There is no substantive difference in the two descriptions.)

These three protons
split the signal for the
proton at C-2 into a
quartet.

This proton splits the signal
for the proton at C-2 into a
doublet.

14.14   Eight lines are observed because H_a and H_b are not chemical shift equivalent; they split each other into a doublet. Each of these doublets is split again by coupling of each to H_c, giving a total of eight lines. H_a and H_b are nonequivalent because the carbon they are attached to is adjacent to a chirality center. By using the replacement technique discussed in the text, H_a and H_b are diastereotopic.

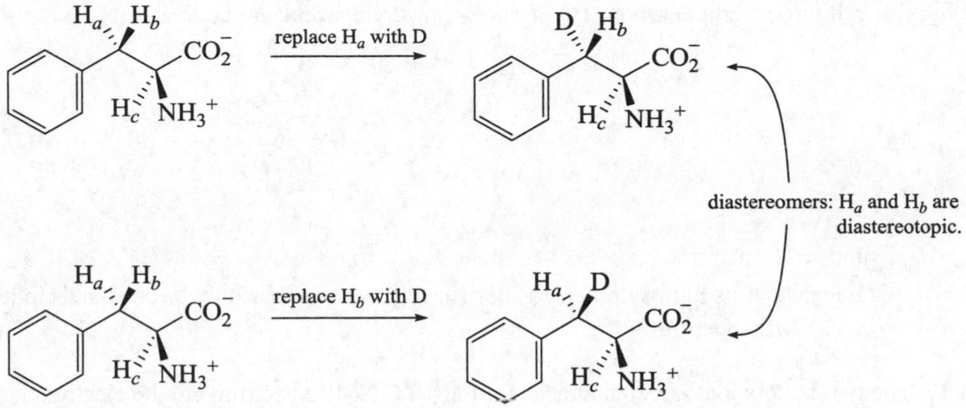

diastereomers: H_a and H_b are diastereotopic.

14.15 The alcohol proton will be split into a triplet in a primary alcohol, a doublet in a secondary alcohol, and will not be split in a tertiary alcohol.

$$RCH_2-O-H \qquad R_2CH-O-H \qquad R_3C-O-H$$

Primary alcohol:	Secondary alcohol:	Tertiary alcohol:
CH$_2$ group splits signal	CH group splits signal	no protons vicinal to O-H
for O-H proton into a triplet.	for O-H proton into a doublet.	proton; therefore, no splitting.

14.16 (*b*) The two methyl carbons (*n*) of the isopropyl group are equivalent.

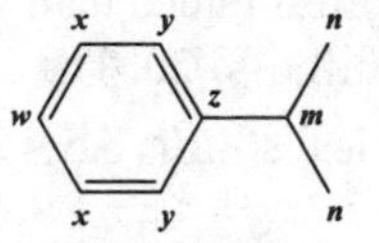

Four different types of carbons occur in the aromatic ring and two different types are present in the isopropyl group. The ^{13}C NMR spectrum of isopropylbenzene contains *six* signals.

(*c*) The methyl substituent (*n*) at C-2 is different from those at C-1 and C-3:

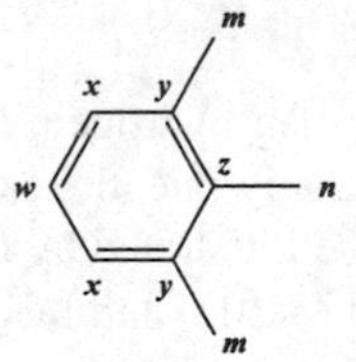

The four nonequivalent ring carbons and the two different types of methyl carbons give rise to a ^{13}C NMR spectrum that contains *six* signals.

(*d*) The three methyl carbons of 1,2,4-trimethylbenzene are different from one another:

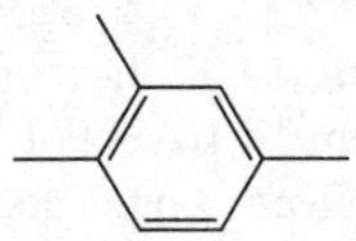

Also, all the ring carbons are different from each other. The nine different carbons give rise to *nine* separate signals.

(*e*) All three methyl carbons (*x*) of 1,3,5-trimethylbenzene are equivalent.

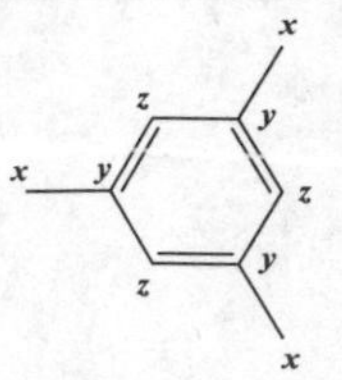

Because of its high symmetry 1,3,5-trimethylbenzene has only *three* signals in its ^{13}C NMR spectrum.

14.17 The two factors that affect chemical shift in a ^{13}C NMR spectrum are the electronegativity of the groups attached to carbon and the hybridization of carbon. The more electronegative chlorine will exert a greater deshielding effect than bromine in 1-bromo-3-chloropropane. The peaks can be assigned as follows:

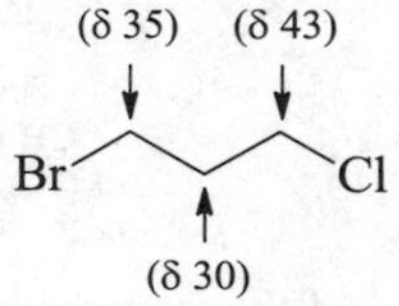

14.18 *sp*^{3}-Hybridized carbons are more shielded than *sp*^{2}-hybridized ones. Carbon *x* is the most shielded, and has a chemical shift of δ 20. The oxygen of the OCH$_3$ group decreased the shielding of carbon *z*; its chemical shift is δ 55. The least shielded is carbon *y* with a chemical shift of δ 157.

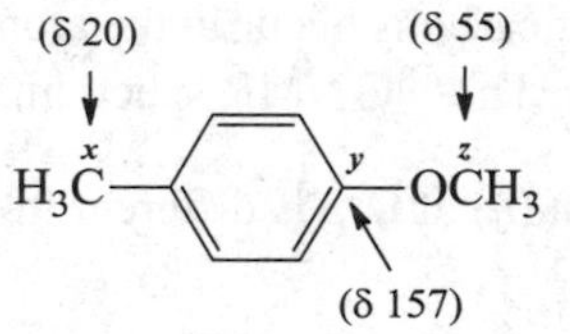

14.19 Carbon is more electronegative than hydrogen. Because the carbonyl of a ketone has two carbon atoms attached and an aldehyde only one, the carbonyl carbon of an aldehyde is expected to be more shielded. This effect can be illustrated by comparing the chemical shifts of the carbonyl carbons of the aldehyde butanal (δ 203) and the ketone 2-butanone (δ 208).

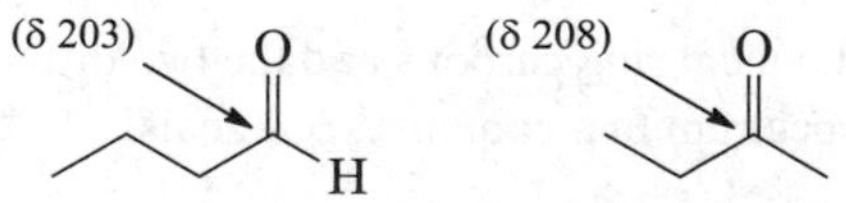

Butanal   2-Butanone

14.20 The ^{13}C NMR spectrum in Figure 14.27 shows nine signals and is the spectrum of 1,2,4-trimethylbenzene from part (*d*) of Problem 14.16. Six of the signals, in the range δ 127–138, are due to the six nonequivalent carbons of the benzene ring. The three signals near δ 20 are due to the three nonequivalent methyl groups.

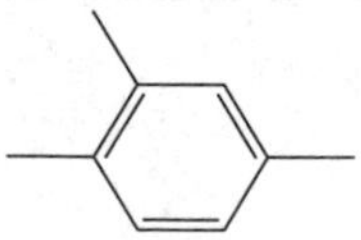

1,2,4-Trimethylbenzene

14.21 The formula $C_6H_{12}O$ represents a molecule with an index of hydrogen deficiency of one, so it has a ring or a double bond. Two carbons have chemical shifts in the alkene carbon region, suggesting that the molecule is an alkene. The two sp^2 carbons have chemical shifts of 115 and 142 ppm, respectively. One of these carbons has two attached protons, so it has a shift of 115. One carbon has a chemical shift of about 70 ppm, which is the region for an sp^3 carbon attached to an oxygen, and it is a CH. The remaining three carbons include two additional CH_2 carbons and one CH_3. On the basis of their chemical shifts, the best answer is 1-hexene-3-ol.

$$\underset{\underset{OH}{|}}{CH_2=CHCHCH_2CH_2CH_3}$$

1-Hexene-3-ol

14.22 Absorption frequencies given in wavenumbers (cm^{-1}) are directly proportional to the energy difference between two adjacent vibrational states. Thus the energy difference is smaller for O–D stretching (2630 cm^{-1}) than for O–H stretching (3600 cm^{-1}).

14.23 In order for an absorption to be observable in an infrared spectrum, a change in molecular dipole moment must occur. The double bond in ethylene, $H_2C=CH_2$, is symmetrically substituted. Thus the double bond stretching vibration does not produce a change in dipole moment and no peak is observed.

14.24 The IR spectrum shows absorption above 3000 cm^{-1} in the sp^2 C–H stretching region indicating a substituted benzene ring. The prominent peak in the 1700–1750 cm^{-1} region suggests the presence of a carbonyl group. The strong absorption at 1200 cm^{-1} due to C—O—C stretching allows the ester to be selected as the compound giving rise to the spectrum in text Figure 14.36.

Benzyl acetate: compound whose IR spectrum is shown in Figure 14.36.

14.25 The energy of electromagnetic radiation is inversely proportional to its wavelength. Because excitation of an electron for the $\pi \rightarrow \pi^*$ transition of ethylene occurs at a shorter wavelength (λ_{max} = 170 nm) than that of *cis,trans*-1,3-cyclooctadiene (λ_{max} = 230 nm), the HOMO–LUMO energy difference in ethylene is *greater*.

14.26 Conjugation shifts λ_{max} to longer wavelengths in alkenes. The conjugated diene 2-methyl-1,3-butadiene has the longest wavelength absorption, λ_{max} = 222 nm. The isolated diene 1,4-pentadiene and the simple alkene cyclopentene both absorb below 200 nm.

2-Methyl-1,3-butadiene
(λ_{max} = 222 nm)

14.27 (*b*) The distribution of molecular-ion peaks in *o*-dichlorobenzene is identical to that in the para isomer. As the sample solution to part (*a*) in the text describes, peaks at *m/z* 146, 148, and 150 are present for the molecular ion.

(*c*) The two isotopes of bromine are ^{79}Br and ^{81}Br. When both bromines of *p*-dibromobenzene are ^{79}Br, the molecular ion appears at *m/z* 234. When one is ^{79}Br and the other is ^{81}Br, *m/z* for the molecular ion is 236. When both bromines are ^{81}Br, *m/z* for the molecular ion is 238.

(*d*) The combinations of ^{35}Cl, ^{37}Cl, ^{79}Br, and ^{81}Br in *p*-bromochlorobenzene and the values of *m/z* for the corresponding molecular ion are as shown.

$$(^{35}\text{Cl}, {}^{79}\text{Br}) \quad m/z = 190$$

$$(^{37}\text{Cl}, {}^{79}\text{Br}) \text{ or } (^{35}\text{Cl}, {}^{81}\text{Br}) \quad m/z = 192$$

$$(^{37}\text{Cl}, {}^{81}\text{Br}) \quad m/z = 194$$

14.28 The base peak in the mass spectrum of alkylbenzenes corresponds to carbon–carbon bond cleavage at the benzylic carbon.

Base peak: $C_9H_{11}^+$ Base peak: $C_8H_9^+$ Base peak: $C_9H_{11}^+$
m/z 119   *m/z* 105   *m/z* 119

14.29 The base peak in each spectrum is the benzylic ion, which rearranges to a tropylium ion. This ion has *m/z* 169/171 in the case of 1-bromo-4-propylbenzene due to the presence of bromine, so the lower spectrum is for this compound. The top spectrum with a base peak of *m/z* 91 for the unsubstituted tropylium ion is for (3-bromopropyl)benzene.

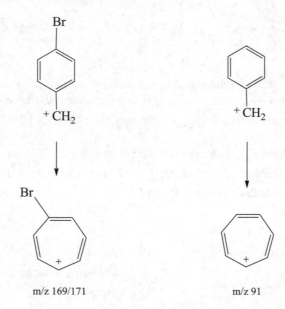

m/z 169/171   m/z 91

14.30 (*b*) The index of hydrogen deficiency is given by the following formula:

$$\text{Index of hydrogen deficiency} = \frac{1}{2}(C_nH_{2n+2} - C_nH_x)$$

The compound given contains eight carbons (C_8H_8); therefore,

$$\text{Index of hydrogen deficiency} = \frac{1}{2}(C_8H_{18} - C_8H_8) = 5$$

The problem specifies that the compound consumes 2 moles of hydrogen, and so it contains two double bonds (or one triple bond). Because the index of hydrogen deficiency is equal to 5, there must be three rings.

(c) Chlorine substituents are equivalent to hydrogens when calculating the index of hydrogen deficiency. Therefore, consider $C_8H_8Cl_2$ as equivalent to C_8H_{10}. Thus, the index of hydrogen deficiency of this compound is 4.

$$\text{Index of hydrogen deficiency} = \frac{1}{2}(C_8H_{18} - C_8H_{10}) = 4$$

Because the compound consumes 2 moles of hydrogen on catalytic hydrogenation, it must therefore contain two rings.

(d) Oxygen atoms are ignored when calculating the index of hydrogen deficiency. Thus, C_8H_8O is treated as if it were C_8H_8.

$$\text{Index of hydrogen deficiency} = \frac{1}{2}(C_8H_{18} - C_8H_8) = 5$$

Because the problem specifies that 2 moles of hydrogen is consumed on catalytic hydrogenation, this compound contains three rings.

(e) Ignoring the oxygen atoms in $C_8H_{10}O_2$, we treat this compound as if it were C_8H_{10}.

$$\text{Index of hydrogen deficiency} = \frac{1}{2}(C_8H_{18} - C_8H_{10}) = 4$$

Because 2 moles of hydrogen is consumed on catalytic hydrogenation, there must be two rings.

(f) Ignore the oxygen, and treat the chlorine as if it were hydrogen. Thus, C_8H_9ClO is treated as if it were C_8H_{10}. Its index of hydrogen deficiency is 4, and it contains two rings.

(g) With nitrogen present, one hydrogen is taken away from the formula. So C_3H_5N is treated as if it were C_3H_4.

$$\text{Index of hydrogen deficiency} = \frac{1}{2}(C_3H_8 - C_3H_4) = 2$$

Because the compound consumes 2 moles of hydrogen on catalytic hydrogenation, it must contain two double bonds (or one triple bond) and cannot therefore contain a ring.

(h) With nitrogen present, one hydrogen is taken away from the formula. So C_4H_5N is treated as if it were C_4H_4.

$$\text{Index of hydrogen deficiency} = \frac{1}{2}(C_4H_{10} - C_4H_4) = 3$$

Because the compound consumes 2 moles of hydrogen on catalytic hydrogenation, it must contain two double bonds (or one triple bond) and one ring.

End of Chapter Problems

^{1}H NMR Spectroscopy

14.31 Because each compound exhibits only a single peak in its ^{1}H NMR spectrum, all the hydrogens are equivalent in each one. Structures are assigned on the basis of their molecular formulas and chemical shifts.

(*a*) This compound has the molecular formula C_8H_{18} and so must be an alkane. The 18 hydrogens (δ 0.9) are contributed by six equivalent methyl groups.

2,2,3,3-Tetramethylbutane

(*b*) A hydrocarbon with the molecular formula C_5H_{10} has an index of hydrogen deficiency of 1 and so is either a cycloalkane or an alkene. Because all ten hydrogens (δ 1.5) are equivalent, this compound must be cyclopentane.

Cyclopentane

(*c*) The chemical shift of the eight equivalent hydrogens in C_8H_8 is δ 5.8, which is consistent with protons attached to a carbon–carbon double bond.

1,3,5,7-Cyclooctatetraene

(*d*) The compound C_4H_9Br has no rings or double bonds. The nine hydrogens (δ 1.8) belong to three equivalent methyl groups.

Br

tert-Butyl bromide

(*e*) The dichloride has no rings or double bonds (index of hydrogen deficiency = 0). The four equivalent hydrogens (δ 3.7) are present as two —CH_2Cl groups. Remember that equivalent protons on adjacent carbons do not split each other's signal.

1,2-Dichloroethane

(*f*) All three hydrogens (δ 2.7) in $C_2H_3Cl_3$ must be part of the same methyl group in order to be equivalent.

$$
\begin{array}{cc}
H & Cl \\
| & | \\
H-C-C-Cl \\
| & | \\
H & Cl
\end{array}
$$

1,1,1-Trichloroethane

(*g*) This compound has no rings or double bonds. To have eight equivalent hydrogens (δ 3.7) it must have four equivalent methylene groups.

$$CH_2Cl$$
$$ClCH_2\overset{|}{\underset{|}{C}}CH_2Cl$$
$$CH_2Cl$$

1,3-Dichloro-2,2-di(chloromethyl)propane

(*h*) A compound with a molecular formula of $C_{12}H_{18}$ has an index of hydrogen deficiency of 4. A likely candidate for a compound with 18 equivalent hydrogens is one with six equivalent CH_3 groups. Thus, 6 of the 12 carbons belong to CH_3 groups, and the other 6 have no hydrogens. The compound is hexamethylbenzene.

A chemical shift of δ 2.2 is consistent with the fact that all of the protons are benzylic hydrogens.

(*i*) The molecular formula of $C_3H_6Br_2$ tells us that the compound has no double bonds and no rings. All six hydrogens (δ 2.6) are equivalent, indicating two equivalent methyl groups. The compound is 2,2-dibromopropane, $(CH_3)_2CBr_2$.

14.32 (*a*) A five-proton signal at δ 7.1 indicates a monosubstituted aromatic ring. With an index of hydrogen deficiency of 4, C_8H_{10} contains this monosubstituted aromatic ring and no other rings or multiple bonds. The triplet–quartet pattern at high field suggests an ethyl group.

Ethylbenzene

(*b*) The index of hydrogen deficiency of 4 and the five-proton multiplet at δ 7.0 to 7.5 are accommodated by a monosubstituted aromatic ring. The remaining four carbons and nine hydrogens are most reasonably a *tert*-butyl group, because all nine hydrogens are equivalent.

tert-Butylbenzene

(*c*) Its molecular formula requires that C_6H_{14} be an alkane. The doublet–septet pattern is consistent with an isopropyl group, and the total number of protons requires that two of these groups be present.

Septet
δ 1.4

H   H
H₃C—⌐   ⌐—CH₃   Doublet
H₃C   CH₃   δ 0.8

2,3-Dimethylbutane

Note that the methine (CH) protons do not split each other, because they are equivalent and have the same chemical shift.

(d) The molecular formula C_6H_{12} requires the presence of one double bond or ring. A peak at δ 5.1 is consistent with —C=CH, and so the compound is a noncyclic alkene. The vinyl proton gives a triplet signal, and so the group C=CHCH₂ is present. The ¹H NMR spectrum shows the presence of the following structural units:

H   ◄——— δ 5.1 (Triplet)
C=C
CH₂   ◄——— δ 2.0 (Allylic)

CH₂CH₃   ◄——— δ 0.9 (Triplet)

CH₃   ◄——— δ 1.6 (Singlet, allylic)
C=C
CH₃   ◄——— δ 1.7 (Singlet, allylic)

Putting all these fragments together yields a unique structure.

Singlet ——► H₃C   H   ◄—— Triplet
                  C=C
Singlet ——► H₃C   CH₂CH₃

                        Pentet Triplet

2-Methyl-2-pentene

(e) The compound $C_4H_6Cl_4$ contains no double bonds or rings. There are no high-field peaks (δ 0.5 to 1.5), and so there are no methyl groups. At least one chlorine substituent must therefore be at each end of the chain. The most likely structure has the four chlorines divided into two groups of two.

Cl
              Cl
Cl
                    Cl
              Cl

Triplet Doublet
δ 4.6   δ 3.9

1,1,4,4-Tetrachlorobutane

(*f*) The molecular formula $C_4H_6Cl_2$ indicates the presence of one double bond or ring. A signal at δ 5.7 is consistent with a proton attached to a doubly bonded carbon. The following structural units are present:

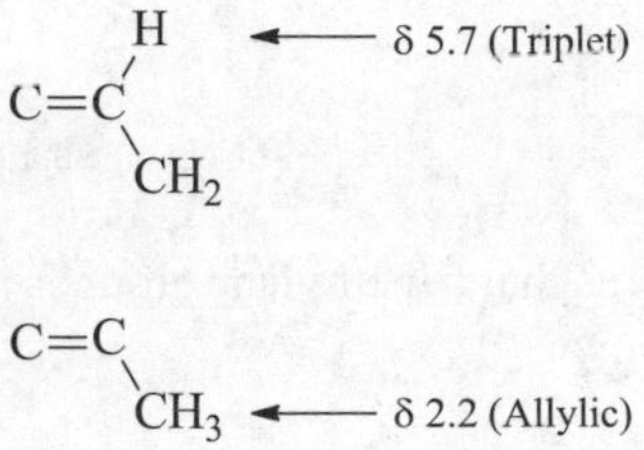

For the methyl group to appear as a singlet and the methylene group to appear as a doublet, the chlorine substituents must be distributed as shown:

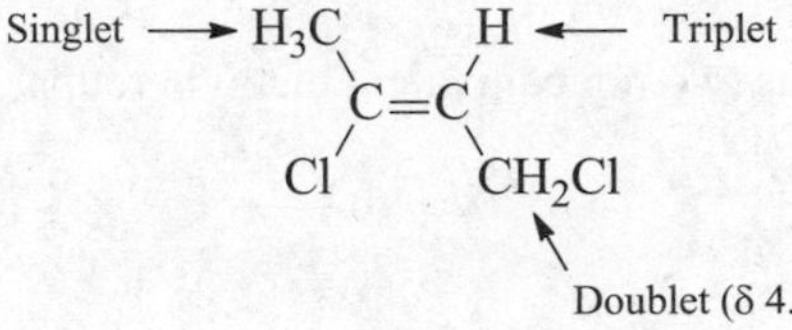

1,3-Dichloro-2-butene

The stereochemistry of the double bond (*E* or *Z*) is not revealed by the ^1H NMR spectrum.

(*g*) A molecular formula of C_3H_7ClO is consistent with the absence of rings and multiple bonds (index of hydrogen deficiency = 0). None of the signals is equivalent to three protons, and so no methyl groups are present. Three methylene groups occur, all of which are different from one another. The compound is therefore

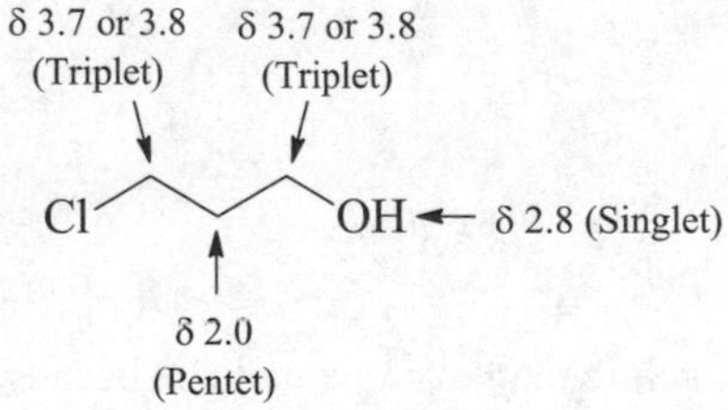

(*h*) The compound has a molecular formula of $C_{14}H_{14}$ and an index of hydrogen deficiency of 8. With a ten-proton signal at δ 7.1, a logical conclusion is that there are two monosubstituted benzene rings. The other four protons belong to two equivalent methylene groups.

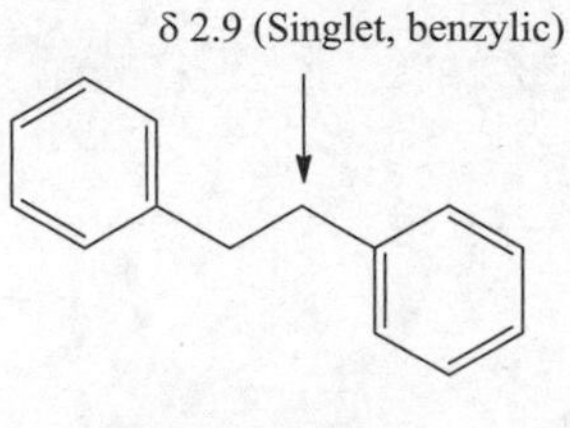

1,2-Diphenylethane

14.33 The compounds of molecular formula C_4H_9Cl are the isomeric chlorides: *n*-butyl, isobutyl, *sec*-butyl, and *tert*-butyl chloride.

(*a*) All nine methyl protons of *tert*-butyl chloride $(CH_3)_3CCl$ are equivalent; its ^1H NMR spectrum has only one peak.

(b) A doublet at δ 3.4 indicates a —CH₂Cl group attached to a carbon that bears a single proton.

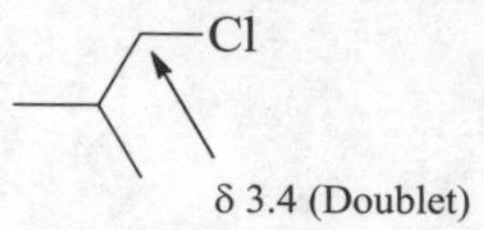

δ 3.4 (Doublet)

Isobutyl chloride

(c) A triplet at δ 3.5 means that a methylene group is attached to the carbon that bears the chlorine.

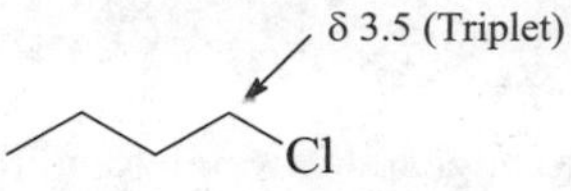

δ 3.5 (Triplet)

n-Butyl chloride

(d) This compound has two nonequivalent methyl groups.

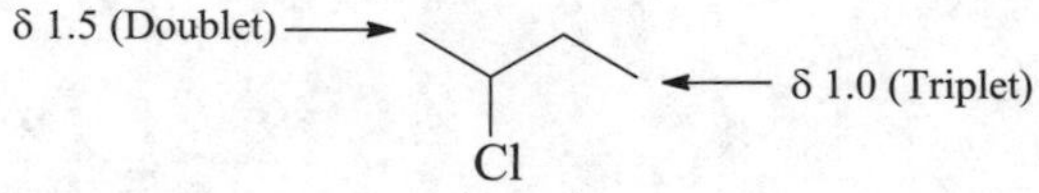

δ 1.5 (Doublet)   δ 1.0 (Triplet)

sec-Butyl chloride

14.34 The hydrogens bonded to the sp^3 carbon in the five-membered ring of fluorene are relatively acidic. The reason for this can be seen by realizing that the central ring of fluorene resembles cyclopentadiene, and recalling from text Chapter 12 (Section 12.20) that cyclopentadiene can lose a proton to form a stable aromatic ion.

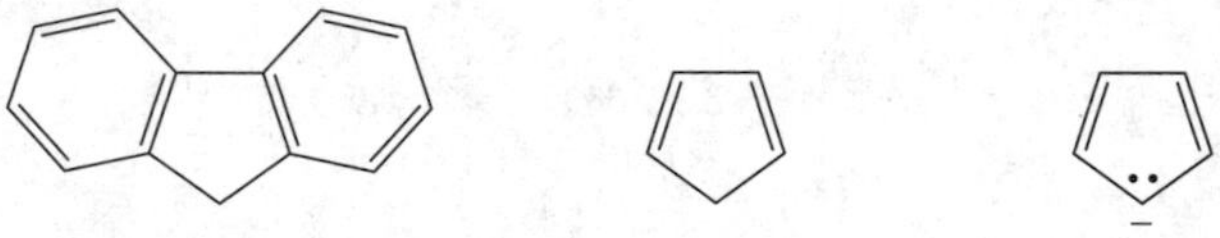

Fluorene   Cyclopentadiene   Cyclopentadienyl anion (aromatic)

Fluorene is even more acidic than cyclopentadiene because its conjugate base is not only an aromatic anion, but it is stabilized by benzylic conjugation with the two benzene rings. A strong base such as $NaOCH_3$ can thus readily remove a proton from fluorene. Reprotonation in the presence of CH_3OD results in exchange of the protons (^1H) with deuterium ($^2H = D$).

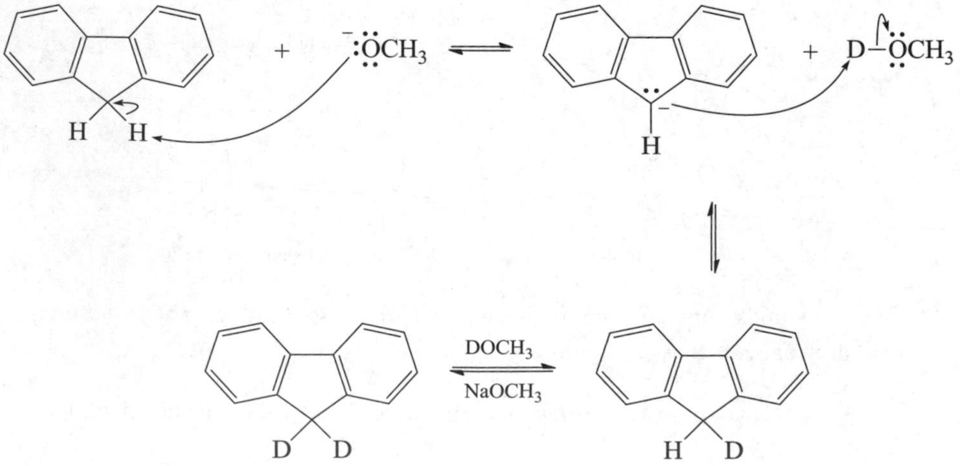

14.35 The most deshielded signal is for H_a, and it shows both the larger trans coupling ($J_{a,b}$) and the smaller cis coupling ($J_{a,c}$). The geminal $J_{b,c}$ coupling is nearly zero in this system, so there are only the two vicinal couplings that are observed. The value of $J_{a,b}$ is 16.4 Hz and the value of $J_{a,c}$ is 9.6 Hz.

14.36 Here are the structures for the four spectra, and some explanation.

(a) 1-Phenyl-1-propanol
(b) 1-Phenyl-2-propanol
(c) 2-Phenyl-1-propanol
(d) 2-Phenyl-2-propanol

The first spectrum corresponds to 1-phenyl-1-propanol. The triplet at about 4.5 ppm corresponds to the proton (labeled A) attached to the benzylic carbon, which also has the hydroxyl group. It is split into a triplet by the adjacent CH_2. The triplet at about 0.9 ppm integrating for three protons is the CH_3 group. The complex pattern integrating for two protons at 1.7 ppm corresponds to the CH_2 group (labeled B). The splitting is due to the presence of the chirality center (benzylic carbon), so the two CH_2 protons are not chemical shift equivalent (they are diastereotopic). They split each other and are further split by protons labeled 1 and 3. The hydroxyl group proton is at 2.4 ppm.

The second spectrum is for 1-phenyl-2-propanol. The upfield doublet at about 1.2 ppm is the methyl group (labeled C) and the singlet at about 1.95 ppm is the hydroxyl group proton (labeled D). The eight-line pattern centered at about 2.7 ppm corresponds to the two benzylic protons (labeled A). Due to the adjacent chirality center, these two protons are not chemical shift equivalent (they are diastereotopic), they split one another, and each signal is further split by the adjacent methine proton (labeled B). The multiplet centered at about 3.7 is the methine proton (B), which is deshielded because the carbon has the hydroxyl group attached to it.

The third spectrum is that of 2-phenyl-1-propanol. Again, the upfield doublet is for the methyl group (labeled C). The six-line pattern at about 2.9 ppm is for the methine proton labeled B. It has five nonequivalent neighbors where the coupling to the CH_2 protons is about the same as the coupling to the CH_3 protons. The doublet at about 3.62 ppm is for the protons labeled A. There is some additional splitting in this doublet that could be due to these being diastereotopic protons; they could also be coupling with the hydroxyl group proton.

The last and simplest spectrum matches 2-phenyl-2-propanol. The six methyl group protons (labeled A) are chemical shift equivalent and not split because the adjacent carbon has no protons. The broad singlet is the hydroxyl group proton.

14.37 [24]Annulene is expected to be antiaromatic because the number of π electrons equals $4n$, not the $4n + 2$ required for aromaticity according to Hückel's rule. As shown in text Figure 14.9*b*, the outer protons are expected to be more shielded in a compound that is antiaromatic.

[24]Annulene

14.38 The *trans* and *cis* isomers of 1-bromo-4-*tert*-butylcyclohexane can be taken as models to estimate the chemical shift of the proton of the CHBr group when it is axial and equatorial, respectively, in the two chair conformations of bromocyclohexane. An axial proton is more shielded ($\delta 3.81$ for *trans*-1-bromo-4-tert-butylcyclohexane) than an equatorial one ($\delta 4.62$ for *cis*-1-bromo-4-tert-butylcyclohexane).

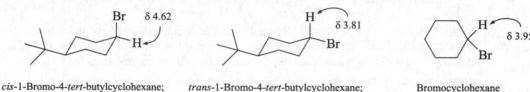

cis-1-Bromo-4-*tert*-butylcyclohexane;   *trans*-1-Bromo-4-*tert*-butylcyclohexane;   Bromocyclohexane
less shielded   more shielded

The difference in chemical shift between these stereoisomers is 0.81 ppm. The corresponding proton in bromocyclohexane is 0.67 ppm more shielded than the equatorial proton in *cis*-1-bromo-4-*tert*-butylcyclohexane. The proportion of bromocyclohexane that has an axial hydrogen is therefore $\frac{0.67}{0.81}$ or 83%. For bromocyclohexane, 83% of the molecules have an equatorial bromine, and 17% have an axial bromine.

^{13}C NMR Spectroscopy

14.39 All these compounds have the molecular formula $C_4H_{10}O$. They have neither multiple bonds nor rings.

(*a*) Two equivalent CH_3 groups occur at $\delta 18.9$. One carbon bears a single hydrogen. The least shielded carbon, presumably the one bonded to oxygen, has two hydrogen substituents. Putting all the information together reveals this compound to be isobutyl alcohol.

$\delta 30.8$

$\delta 18.9$ $\delta 69.4$

OH

Isobutyl alcohol

(b) This compound has four distinct peaks, and so none of the four carbons is equivalent to any of the others. The signal for the least shielded carbon represents CH, and so the oxygen is attached to a secondary carbon. Only one carbon appears at low field; the compound is an alcohol, not an ether. Therefore

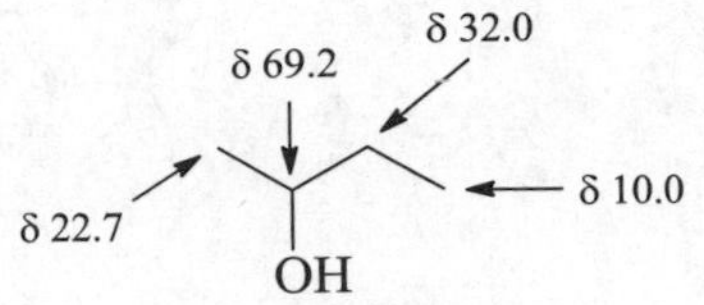

sec-Butyl alcohol

(c) Signals for three equivalent CH_3 carbons indicate that this isomer is *tert*-butyl alcohol. This assignment is reinforced by the observation that the least shielded carbon has no hydrogens attached to it.

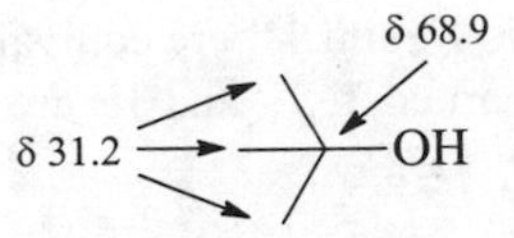

tert-Butyl alcohol

14.40 Each of the carbons in the compound gives its ^{13}C NMR signal at relatively low field; it is likely that each one bears an electron-withdrawing substituent. The compound is

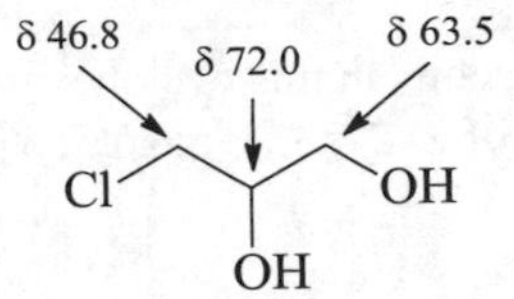

3-Chloro-1,2-propanediol

The isomeric compound 2-chloro-1,3-propanediol

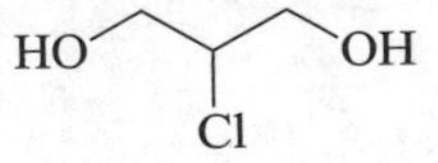

cannot be correct. The C-1 and C-3 positions are equivalent; the ^{13}C NMR spectrum of this compound exhibits only two peaks, not three.

14.41 (a) There are seven nonequivalent carbons. Carbons marked 4 and 4', 5 and 5', 7 and 7' are all equivalent.

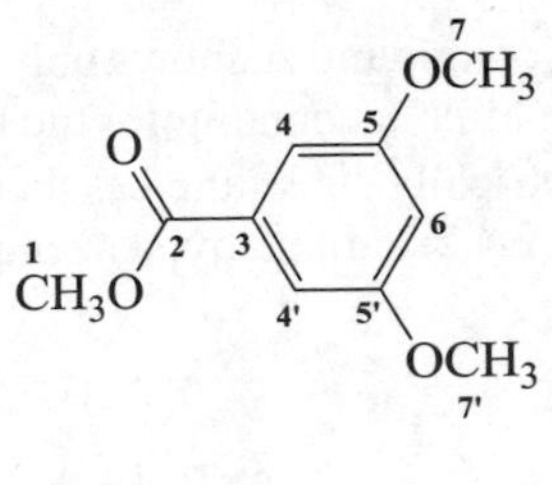

(*b*) All ten carbons in this compound are nonequivalent.

(*c*) There are three nonequivalent carbons in this cyclohexane. Three mirror planes point to the equivalency of the carbons. A mirror plane through carbons 1', 2'', and 3'' reflects carbons 2, 2', 3, 3', and 1, 1''. A mirror plane through carbons 1'', 2, and 3 reflects 1, 1', 2', 2'', and 3', 3''. A mirror plane through carbons 1, 2', and 3' reflects 1', 1'', 2, 2'', and 3, 3''.

Thus, carbons marked 1, 1', and 1'' are equivalent. Those marked 2, 2', and 2'' are equivalent; and those marked 3, 3', and 3'' are equivalent.

(*d*) There are six nonequivalent carbons in this cyclohexane. A mirror plane through carbons 1, 4, and 6 indicate the equivalence of 2, 2''; 3, 3''; and 5, 5''.

14.42 Compounds A and B ($C_{10}H_{14}$) have an index of hydrogen deficiency of 4. Both have peaks in the δ 130–140 range of their ^{13}C NMR spectra, so that the index of hydrogen deficiency can be accommodated by a benzene ring.

The ^{13}C NMR spectrum of compound A shows only a single peak in the upfield region, at δ 20. Thus, the four remaining carbons, after accounting for the benzene ring, are four equivalent methyl groups. The benzene ring is symmetrically substituted as there are only two signals in the aromatic region at δ 132 and 135. Compound A is 1,2,4,5-tetramethylbenzene.

1,2,4,5-Tetramethylbenzene
(compound A)

In compound B the four methyl groups are divided into two pairs. Three different carbons occur in the benzene ring, as noted by the appearance of three signals in the aromatic region (δ 128–135). Compound B is 1,2,3,4-tetramethylbenzene.

1,2,3,4-Tetramethylbenzene
(compound B)

14.43 Here are the structures for the four spectra, and some explanation.

(a)

$BrCH_2CH_2CH_2CH_2CH_3$

1-Bromopentane

(c)

$$CH_3CCH_2CH_3$$

with CH_3 above and Br below the second carbon

2-Bromo-2-methylbutane

(b)

$BrCH_2CH_2CH(CH_3)_2$

1-Bromo-3-methylbutane

(d)

$CH_3CH_2CHCH_2CH_3$
with Br below the central carbon

3-Bromopentane

The first spectrum is for 1-bromopentane. All five carbons are nonequivalent, so five signals are observed.

The second spectrum is for 1-bromo-3-methylbutane. Four signals are observed because the methyl groups are chemical shift equivalent. The carbon attached to bromine is downfield at 40 ppm.

The third spectrum is for 2-bromo-2-methylbutane. It is the only isomer with a carbon that has no protons attached. This carbon, at about 69 ppm, has bromine attached to it.

The fourth spectrum is 3-bromopentane. Due to the symmetry in this molecule, there are only three nonequivalent carbons. The methane carbon at about 62 ppm has bromine attached to it.

^{19}F and ^{31}P NMR Spectroscopy

14.44 (a) The nuclear spin of ^{19}F is $\pm\frac{1}{2}$, that is, the same as that of a proton. The splitting rules for $^{19}F-^{1}H$ couplings are the same as those for $^{1}H-^{1}H$. Thus, the single fluorine atom of CH_3F splits the signal for the protons of the methyl group into a *doublet*.

(b) The set of three equivalent protons of CH_3F splits the signal for fluorine into a *quartet*.

(c) The proton signal in CH_3F is a doublet centered at δ 4.3. The separation between the two halves of this doublet is 45 Hz, which is equivalent to 0.225 ppm at 200 MHz (200 Hz = 1 ppm). Thus, one line of the doublet appears at δ (4.3 + 0.225) and the other at δ (4.3 − 0.225).

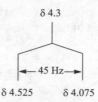

δ 4.3

←—— 45 Hz ——→

δ 4.525   δ 4.075

14.45 Because ^{31}P has a spin of $\pm\frac{1}{2}$, it is capable of splitting the ^1H NMR signal of protons in the same molecule. The problem stipulates that the methyl protons are coupled through three bonds to phosphorus in trimethyl phosphite.

$$CH_3O-P \overset{\displaystyle OCH_3}{\underset{\displaystyle OCH_3}{\big|}}$$

(a) The reciprocity of splitting requires that the protons split the ^{31}P signal of phosphorus. There are nine equivalent protons, and so the ^{31}P signal is split into ten peaks.

(b) Each peak in the ^{31}P multiplet is separated from the next by a value equal to the $^1H-^{31}P$ coupling constant of 12 Hz. There are nine such intervals in a ten-line multiplet, and so the separation is 108 Hz between the highest and lowest field peaks in the multiplet.

Combined Spectra

14.46 Compounds with the molecular formula C_3H_5Br have either one ring or one double bond.

(a) The two peaks at δ 5.4 and 5.6 have chemical shifts consistent with the assumption that each peak is due to a vinyl proton (C=CH). The remaining three protons belong to an allylic methyl group (δ 2.3). The compound cannot be $CH_3CH=CHBr$, because the methyl signal would be split into a doublet. Isomer A can only be

$$H_2C=C\overset{\displaystyle CH_3}{\underset{\displaystyle Br}{<}}$$

2-Bromopropene

(b) Two of the carbons of isomer B have chemical shifts characteristic of sp^2-hybridized carbon. One of these bears two protons (δ 118.8); the other bears one proton (δ 134.2). The remaining carbon is sp^3-hybridized and bears two hydrogens. Isomer B is allyl bromide.

δ 118.8   δ 134.2

$$H_2C=C\overset{\displaystyle H}{\underset{\displaystyle CH_2Br}{<}}$$

δ 32.6

Allyl bromide

(c) All the carbons are sp^3-hybridized in this isomer. Two of the carbons belong to equivalent CH_2 groups, and the other bears only one hydrogen. Isomer C is cyclopropyl bromide.

δ 16.8

δ 12.0 ▷—Br

Cyclopropyl bromide

14.47 A hydrocarbon having a molecular ion, M^1, at m/z 102 has a molecular formula of C_8H_6. This formula corresponds to an index of hydrogen deficiency of 6, suggesting the presence of a phenyl group (C_6H_5) accounting for three double bonds and one ring. Infrared absorptions due to sp^2 C–H stretching can be seen in text Figure 14.50 at 3000 cm^{-1}.

The remaining atoms of the formula (C_2H) suggest an alkyne sidechain. The infrared spectrum supports this observation with absorptions at 2100 cm^{-1} (C≡C stretching) and 3300 cm^{-1} (sp C—H stretching). The compound is phenylacetylene.

Phenylacetylene

14.48 Because the compound has a five-proton signal at δ 7.4 and an index of hydrogen deficiency of 4, we conclude that six of its eight carbons belong to a monosubstituted benzene ring. The IR spectrum exhibits absorption at 3300 cm^{-1}, indicating the presence of a hydroxyl group. The compound is an alcohol. A three-proton doublet at δ 1.5 suggests the presence of a CH_3CH unit. The one-proton doublet of quartets at δ 4.8 along with the δ 2.5 doublet, suggests a CH_3CH unit. The compound is 1-phenylethanol.

δ 4.8
(Doublet
of quartets)

δ 7.3 {

δ 1.5
(Doublet)

δ 2.5
(Doublet)

1-Phenylethanol

14.49 The peak at highest m/z in the mass spectrum of the compound m/z = 134; this is likely to correspond to the molecular ion. Among the possible molecular formulas, $C_{10}H_{14}$ correlates best with the information from the ^{1}H NMR spectrum. What is evident is that there is a signal due to aromatic protons, as well as a triplet–quartet pattern of an ethyl group. A molecular formula of $C_{10}H_{14}$ suggests a benzene ring that bears two ethyl groups. Because the signal for the aryl protons is so sharp, they are probably equivalent. The compound is *p*-diethylbenzene.

δ 7.1
(Singlet)

δ 1.2
(Triplet)

δ 2.6
(Quartet)

p-Diethylbenzene

That the benzene ring is para substituted is supported by the band at 820 cm^{-1} in the infrared spectrum.

14.50 There is a prominent peak in the IR spectrum of the compound near 1700 cm^{-1}, characteristic of C=O stretching vibrations. The ^{1}H NMR spectrum shows only two sets of signals, a triplet at δ 1.1 and a quartet at δ 2.4. The compound contains a CH$_3$CH$_2$ group as its only protons. Its ^{13}C NMR spectrum has three peaks, one of which is at very low field. The signal at δ 211 is in the region characteristic of carbons of C=O groups. If one assumes that the compound contains only carbon, hydrogen, and one oxygen atom and that the peak at highest m/z in its mass spectrum $m/z = 86$ corresponds to the molecular ion, then the compound has the molecular formula C$_5$H$_{10}$O. All the information points to the conclusion that the compound is 3-pentanone.

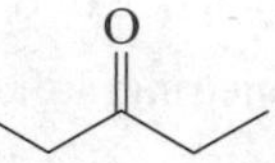

3-Pentanone

14.51 Solving a multi-spectra problem is similar to putting puzzle pieces together. Identify the functional groups and as this is done, start to assemble the structure.

The mass spectrum shows a molecular ion peak at m/z 178/180 indicating the isotope pattern for a single bromine substituent. In the ^{13}C spectrum, there are five absorptions. The vinylic region has absorptions at δ 129 and 137 ppm and each has one proton attached according to the DEPT analysis. In addition, in the alkane region, two absorptions are observed at δ 24 ppm (CH$_2$) and 46 ppm (CH$_3$). The latter absorption at δ 46 ppm indicates an electronegative atom (X), but not a bromine, is attached to the carbon. So far we have the following features:

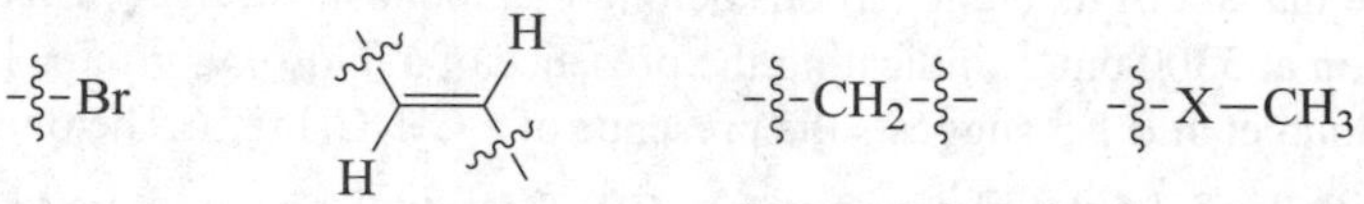

The ^{13}C absorption at 160 ppm does not have any protons and this chemical shift indicates either an ester or a carboxylic acid. Combine this with the IR spectrum data that shows a strong absorption in the carbonyl region at 1730 cm^{-1} indicates an ester functional group. In addition to this, a δ 46 ppm ^{13}C absorption for the CH$_3$ group in addition to the ^{1}H NMR singlet at δ 3.8 ppm (3H integration) would be consistent with a methyl ester functionality. So now we have the following:

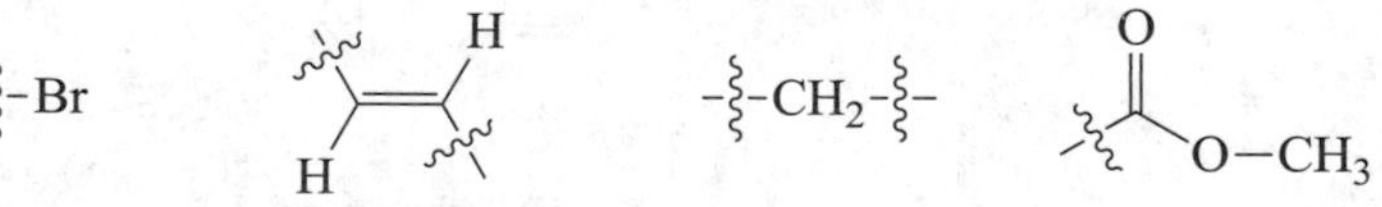

The —CH$_2$— ^{13}C absorption along with the 2H integrated signal in the ^{1}H NMR at 4 ppm is consistent with the bromine atom being attached to the —CH$_2$—. Now we have:

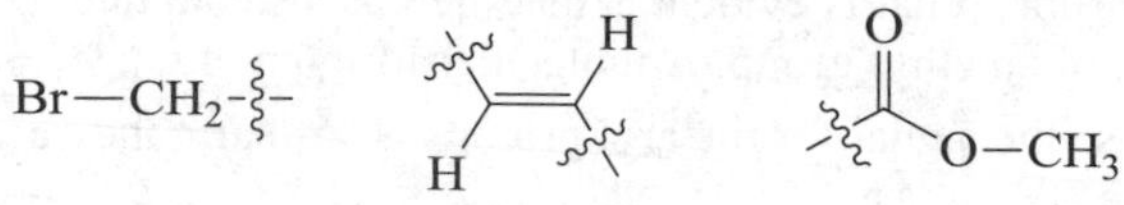

There is only one way to assemble these groups and this would give the following:

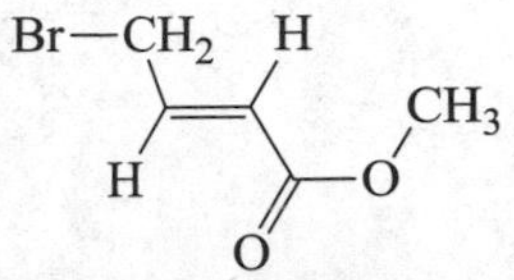

$$Br-CH_2 \quad H$$
$$\diagdown C=C \diagup CH_3$$
$$H \diagup \quad \diagdown C-O$$
$$\quad \quad \quad \| \quad$$
$$\quad \quad O$$

Methyl 4-bromocrotonate

Is this consistent with the coupling in the ^{1}H NMR spectrum? Yes! Although more complex spin-spin splitting is involved, the absorptions can support the above structure. The two absorptions in the ^{1}H NMR spectrum at δ 7.04 ppm and 6.05 are typical for vinyl proton resonances. These show coupling to each other with the same large coupling $J_{a,b}$. For H_a a larger vicinal coupling $J_{a,x}$ indicates two vicinal hydrogens. The smaller triplets as indicated by $J_{b,x}$ indicates allylic coupling. This indicates the structural feature shown below that is consistent with the structure above. In addition, the signal for the —CH$_2$— at δ 4 ppm is a doublet or doublets, indicating one vicinal hydrogen (H_a) and one hydrogen two carbons removed (H_b).

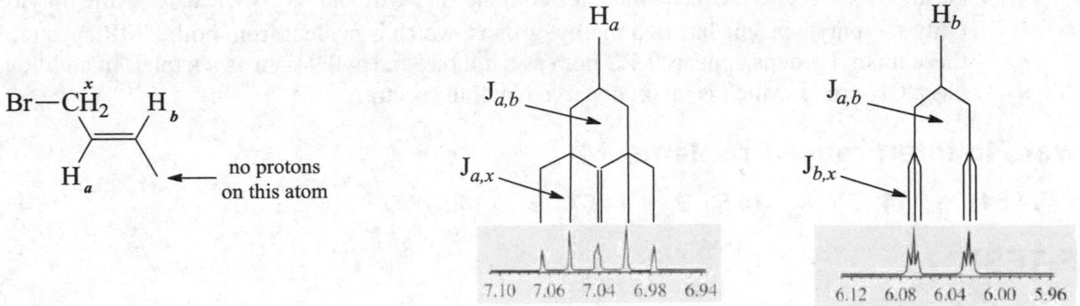

In addition, the mass spectrum shows a base peak at m/z 147/149 corresponds to the acylium ion formed by loss of OCH$_3$ from the molecular ion and still containing bromine.

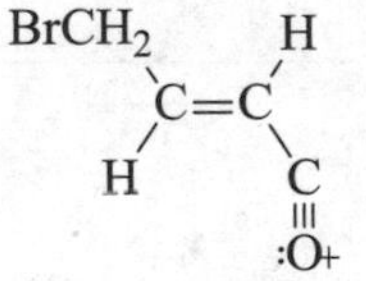

$$BrCH_2 \quad H$$
$$\diagdown C=C \diagup$$
$$H \diagup \quad \diagdown C$$
$$\quad \quad \quad \quad \| \|$$
$$\quad \quad :O+$$
$$\quad \quad \quad \quad m/z \; 147/149$$

14.52 The observed doublet of doublets in the aryl proton region integrating for four protons suggests a para-disubstituted benzene ring

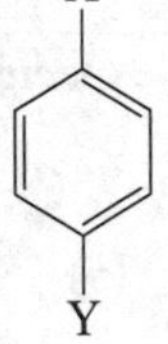

The singlet at 4.8 ppm integrates for two protons and is in the chemical shift region that suggests the carbon is attached to an electronegative element such as oxygen.

The large broad peak at 3600 cm^{-1} in the IR spectrum is for a hydroxyl group, so the molecule likely contains a CH$_2$OH group. The IR spectrum also shows two very strong peaks at about 1350 and 1500 cm^{1}, suggesting a nitro group. The answer is *p*-nitrobenzyl alcohol.

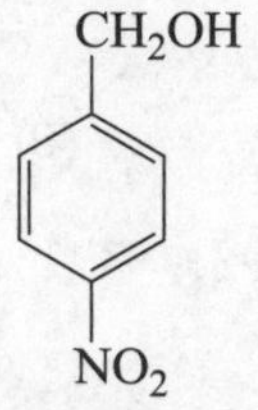

$$CH_2OH$$

[benzene ring structure with CH$_2$OH at top and NO$_2$ at bottom]

$$NO_2$$

14.53 The Friedel–Crafts reaction would be expected to give two possible products based on what we know about this reaction from Section 13.6. The two products are butylbenzene and *sec*-butylbenzene

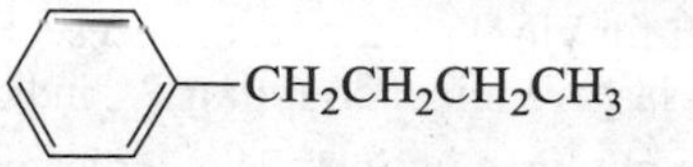

[structure: benzene ring]—$CH_2CH_2CH_2CH_3$ [structure: benzene ring]—$\overset{\displaystyle CH_3}{\underset{\displaystyle CHCH_2CH_3}{|}}$

Butylbenzene 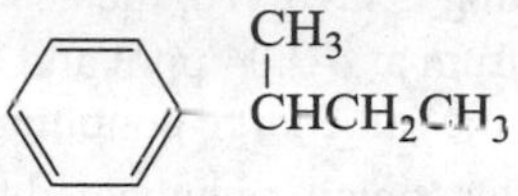   *sec*-Butylbenzene

Of the two, *sec*-butylbenzene matches both the ^{1}H NMR and ^{13}C NMR data, while butylbenzene does not. Only *sec*-butylbenzene has two methyl groups, which is evident from both NMR spectra. In the ^{1}H NMR, these methyl groups appear at 1.2 ppm as a doublet and at 0.9 ppm as a triplet. In addition, *sec*-butylbenzene has a CH group, which is again observed in both spectra.

Answers to Interpretive Problems 14

14.54 D   **14.55** C   **14.56** B   **14.57** B   **14.58** D

SELF-TEST

1.   Complete the following table relating to ^{1}H NMR spectra by supplying the missing data.

	Spectrometer frequency	Chemical shift	
		ppm	**Hz**
(a)	60 MHz	1	336
(b)	300 MHz	4.35	2
(c)	3	3.50	700
(d)	100 MHz	4	of TMS

2.   Indicate the number of signals to be expected and the multiplicity of each in the ^{1}H NMR spectrum of each of the following substances:

(a)   (b)   (c)
    $BrCH_2CH_2CH_2Br$ $CH_3CH_2\overset{\displaystyle Cl}{\underset{\displaystyle Cl}{\overset{|}{\underset{|}{C}}}}CH_2CH_3$ $CH_3OCH_2\overset{\displaystyle O}{\overset{\|}{C}}OCH_3$

3.   Two isomeric compounds having the molecular formula $C_6H_{12}O_2$ both gave ^{1}H NMR spectra consisting of only two singlets. Given the chemical shifts and integrations shown, identify both compounds.

Compound A: δ 1.45 (9H)   Compound B: δ 1.20 (9H)
                     δ 1.95 (3H)   δ 3.70 (3H)

4. Identify each of the following compounds on the basis of the IR and ^{1}H NMR information provided.

   (a) $C_{10}H_{12}O$:   IR: 1710 cm^{-1}

                            NMR: δ 1.0 (triplet, 3H)
                                          δ 2.4 (quartet, 2H)
                                          δ 3.7 (singlet, 2H)
                                          δ 7.2 (singlet, 5H)

   (b) $C_6H_{14}O_2$:   IR: 3400 cm^{-1}

                            NMR: δ 1.2 (singlet, 12H)
                                          δ 2.0 (broad singlet, 2H

   (c) $C_{10}H_{16}O_6$:   IR: 1740 cm^{-1}

                            NMR: δ 1.3 (triplet, 9H)
                                          δ 4.2 (quartet, 6H)
                                          δ 4.4 (singlet, 1H)

   (d) C_4H_7NO: IR: 2240 cm^{-1}
                                        3400 cm^{-1} (broad)

                            NMR: δ 1.65 (singlet, 6H)
                                          δ 3.7 (singlet, 1H)

5. Predict the number of signals and their approximate chemical shifts in the ^{13}C NMR spectrum of the compound shown.

6. How many signals will appear in the ^{13}C NMR spectrum of each of the three C_5H_{12} isomers?

7. The ^{13}C NMR spectrum of an alkane of molecular formula C_6H_{14} exhibits two signals at δ 23 (4C) and δ 37 (2C). What is the structure of this alkane?

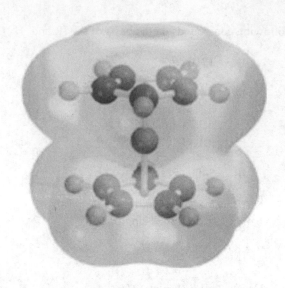

CHAPTER 15
Organometallic Compounds

Table of Contents

SOLUTIONS TO TEXT PROBLEMS

In Chapter Problems

15.1 (*b*) Magnesium bears a cyclohexyl substituent and a chlorine. Chlorine is named as an anion. The compound is cyclohexylmagnesium chloride.

(*c*) Zinc bears a iodomethyl substituent and an iodine. Iodine is named as an anion. The compound is iodomethylzinc iodide.

15.2 Retrosynthetic analysis tells us that the Grignard reagent is prepared from the corresponding bromide, and that the bromide is prepared by addition of HBr to the alkene with a regioselectivity opposite to Markovnikov's rule.

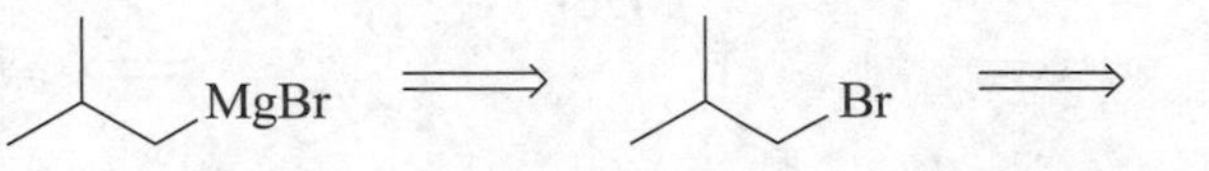

The sequence and necessary reagents are:

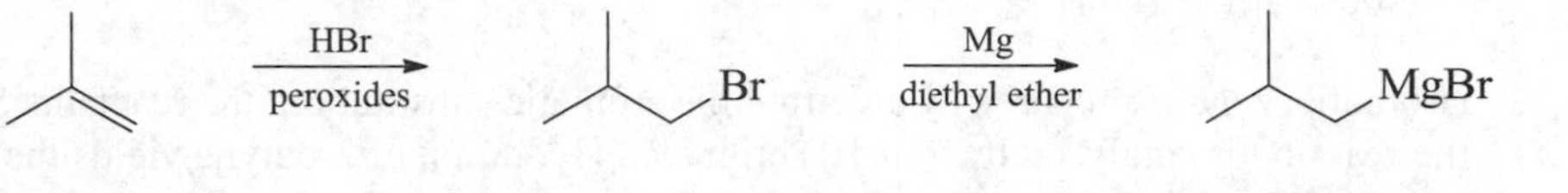

| 2-Methylpropene | 1-Bromo-2-methylpropane | 2-Methylpropylmagnesium bromide |

15.3 Butyllithium can be thought of as equivalent to a butyl anion. It is a strong base and abstracts a proton from the nitrogen of *N,N*-diisopropylamine. The value of *K* is given by the difference in the p*K*$_a$s of the two acids.

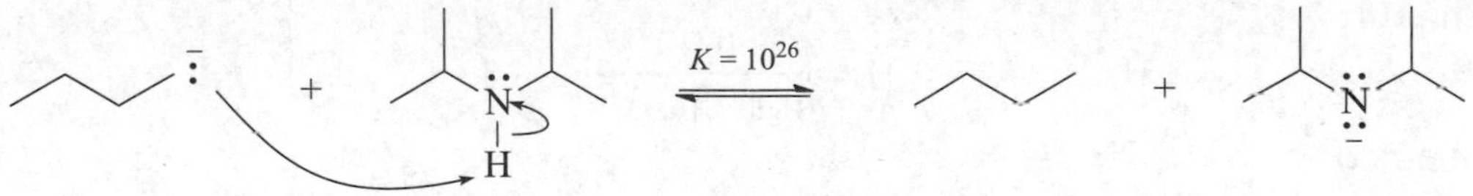

| Butyl anion | *N,N*-Diisopropylamine | Butane | *N,N*-Diisopropylamide ion |
| | p*K*$_a$ = 36 | p*K*$_a$ = 62 | |

15.4 (b) Grignard reagents react with ketones to give tertiary alcohols.

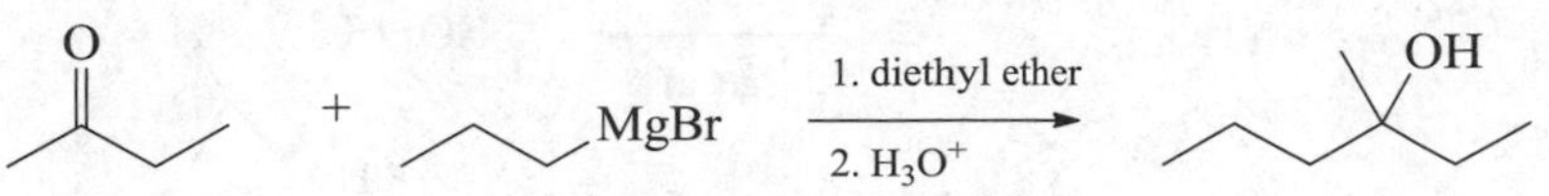

| 2-Butanone | Propylmagnesium bromide | 2-Methyl-3-hexanol |

(c) Grignard reagents react with formaldehyde to give primary alcohols.

| Formaldehyde | *p*-Methoxyphenylmagnesium chloride | *p*-Methoxybenzyl alcohol |

15.5 Disconnecting the phenyl group from the carbon that bears the –OH group suggests combining phenylmagnesium bromide and the four-carbon ketone. Phenylmagnesium bromide is prepared by reaction of bromobenzene with magnesium.

The problem specifies that 2-butyne is one of the reactants, so what remains is to prepare bromobenzene and the four-carbon ketone.

Bromobenzene is prepared by electrophilic aromatic substitution; the reactants are benzene and bromine and the reaction is catalyzed by iron(III) bromide. Hydration of 2-butyne yields the desired ketone.

15.6 The Simmons-Smith reagent (iodomethylzinc iodide) transfers its methyl group to an alkene, forming a cyclopropane. The alkene in this case has a chirality center and this makes the two faces of the double bond diastereotopic. Transfer of CH_2 to one face of the double bond gives a diastereomer of the compound that results from CH_2 transfer to the other face.

15.7 Palladium is a Group 10 transition element, so has 8 4d electrons. Pd^{2+} has two fewer than this, or 6.

The +2 ion in the fourth period with 6 3d electrons is Ni^{2+}. The one in the sixth period is Pt^{2+}.

15.8 Iron has an atomic number of 26 and an electron configuration of $[Ar]4s^2 3d^6$. Thus, it has eight valence electrons and requires ten more to satisfy the 18-electron rule. Five CO ligands, each providing two electrons, are therefore needed. The compound is $Fe(CO)_5$.

15.9 Manganese is a Group 7 transition element so has seven $3d$ electrons. Six electrons from the benzene ligand plus 6 from the three carbon monoxide ligands and the seven from manganese give 19. Subtract 1 for the positive charge for a total of 18. The complex is coordinatively saturated.

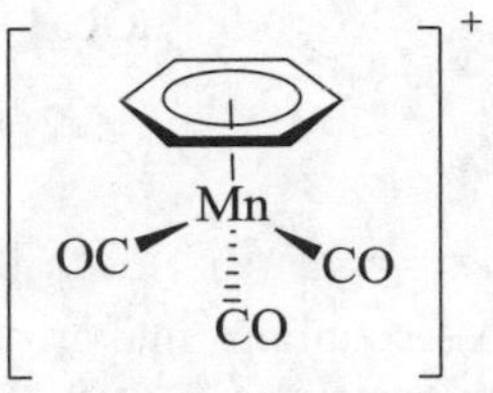

15.10 All four ligands are neutral molecules, so the charge on the complex is the same as the oxidation state of the metal (+1).

15.11 Zirconium is a Group 4 element and has four $4d$ electrons. Add 12 for the two cyclopentadienide ligands and 4 for the two chlorides to give a total of 20. Subtract 4 to balance the charges brought by four ligands to give a total of 16. It is coordinatively unsaturated and zirconium is in the +4 oxidation state.

15.12 In reactions of diorganocuprates, the group attached to copper replaces the halogen substituent of the reactant. The stereochemistry of double bonds is retained.

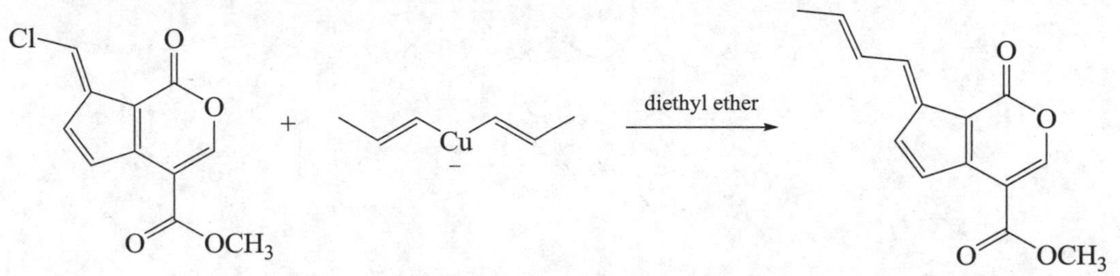

15.13 (*b*) The relation of the target to the starting material suggests, among other possibilities, a disconnection of the type:

which leads to the reactant given in the problem.

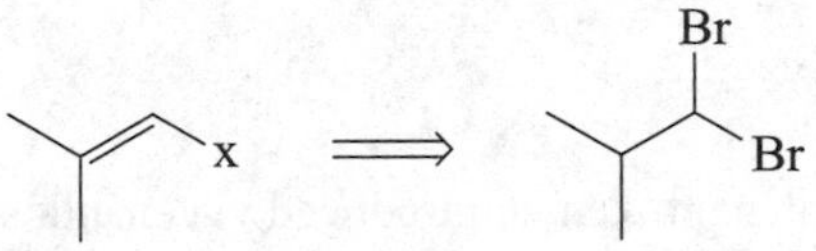

A reasonable synthesis based on this disconnection is:

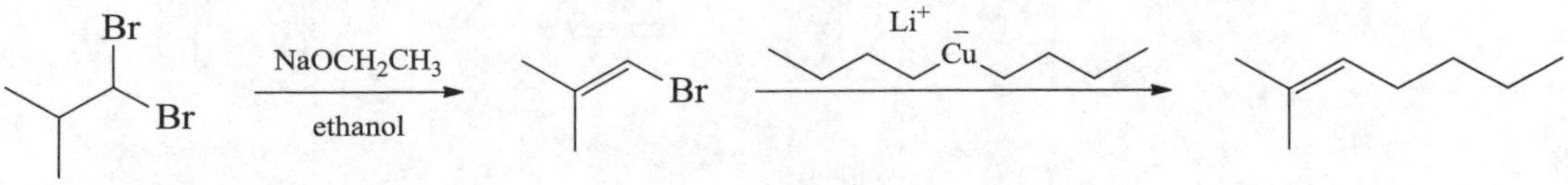

15.14 (*b*) This is an example of a palladium-catalyzed Heck cross-coupling between an aryl halide and an alkene.

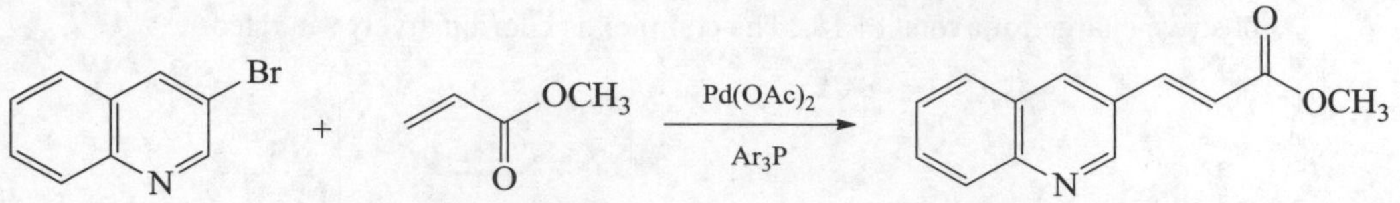

(*c*) Palladium catalyzed cross-couplings find great use in the synthesis of biaryls. The Suzuki method involves an organoboron reagent.

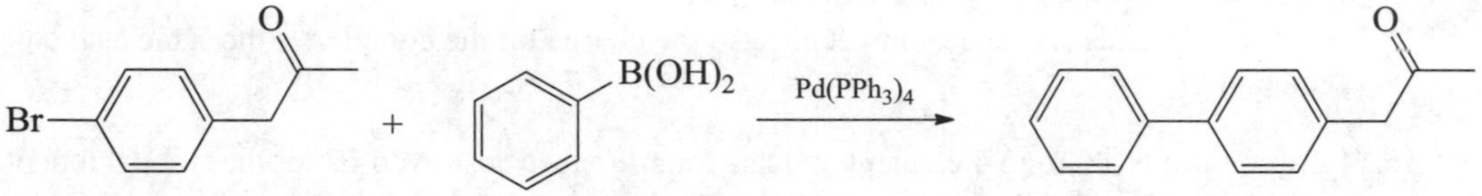

15.15 Completing the disconnection given in the statement of the problem so as to give a allylic bromide that will give humulene by a Suzuki cross-coupling leads to:

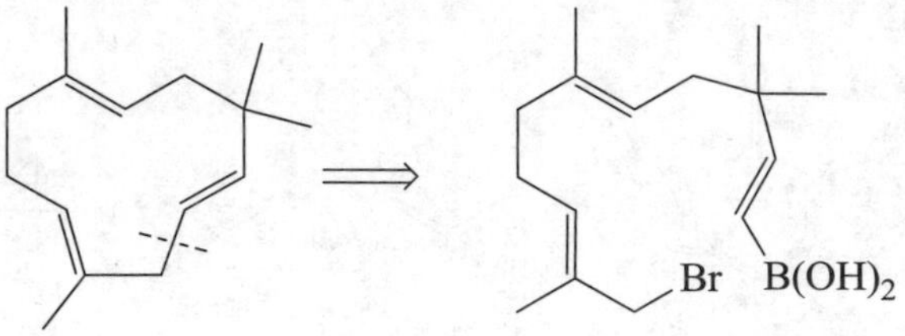

15.16 Hydrogenation can occur from either of two directions. The major product corresponds to hydrogen transfer from the less hindered side.

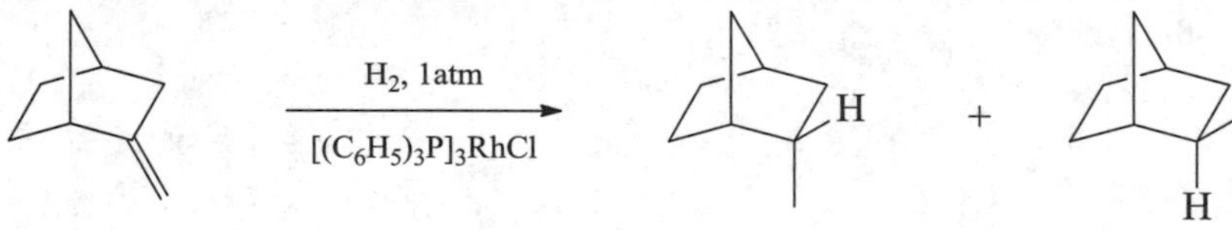

2-Methylene-
bicyclo[2.2.1]heptane

endo-2-Methylbicyclo-
[2.2.1]heptane (73%)

exo-2-Methylbicyclo-
[2.2.1]heptane (27%)

15.17 The chirality center in naproxen is introduced via enantioselective hydrogenation of a double bond. Therefore, retrosynthetically

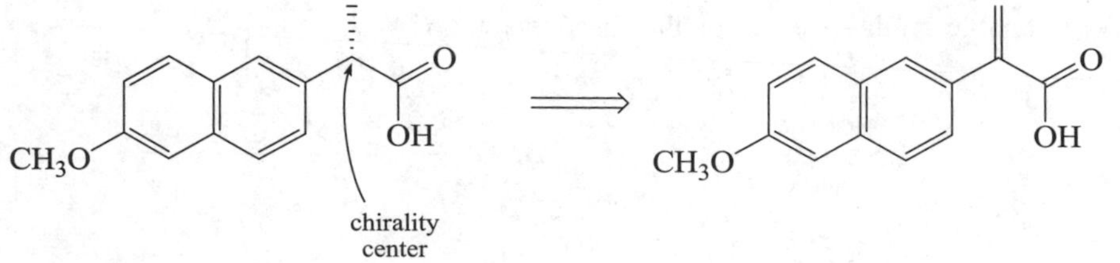

chirality
center

15.18 Cross-metathesis exchanges doubly bonded carbons. The products can be deduced graphically. They are 2-butene and 3-hexene. Both are mixtures of their cis and trans isomers.

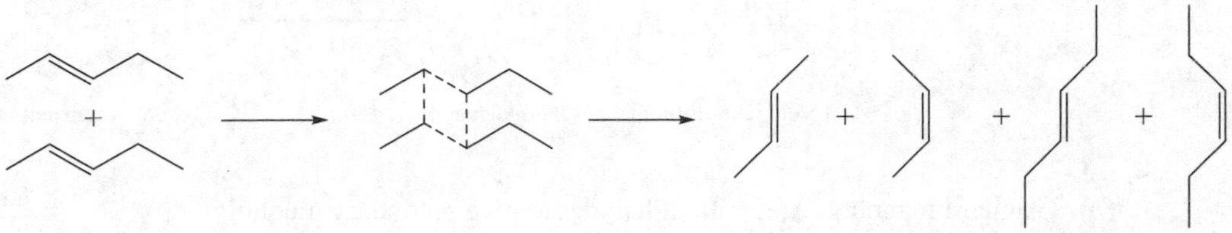

2-Pentene

15.19 The molecular formula of the product ($C_{15}H_{19}NO_2$) differs from the starting material by C_2H_4, which is the same as the molecular formula of ethylene. This suggests that ring-closing metathesis has occurred.

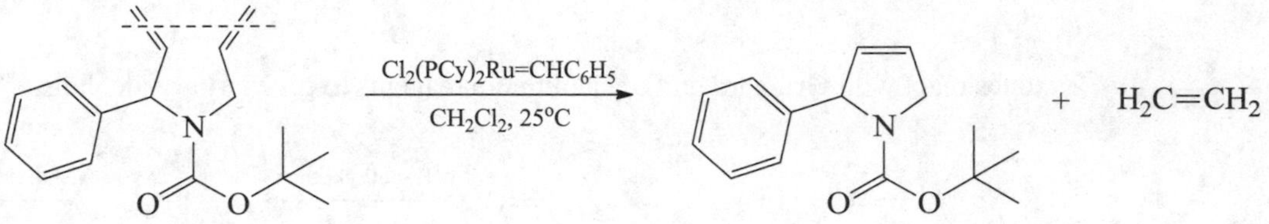

$C_{17}H_{23}NO_2$

$C_{15}H_{19}NO_2$ Ethylene

End of Chapter Problems

Preparation and Reactions of Main-Group Organometallic Compounds

15.20 (*a*) Organolithium reagents are prepared by reaction of the metal with the corresponding alkyl or aryl halide.

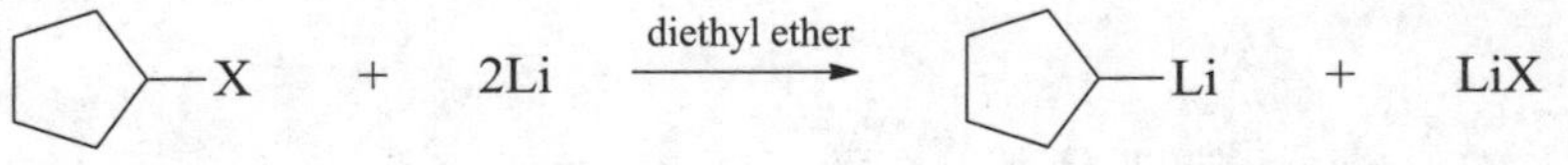

Cyclopentyl halide   Lithium   Cyclopentyllithium   Lithium halide
(X = Cl, Br or I)

(*b*) Grignard reagents are prepared in the same way as alkyl or aryllithiums.

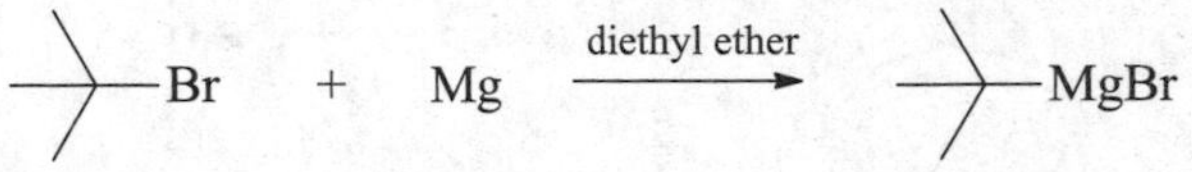

tert-Butyl bromide   Magnesium   *tert*-Butylmagnesium
bromide

(*c*) Acetylenic organolithium reagents are normally prepared by reaction of a terminal alkyne with some readily available alkyllithium. An acid-base reaction takes place in which the terminal alkyne acts as a proton donor toward the alkyllithium acting as a base.

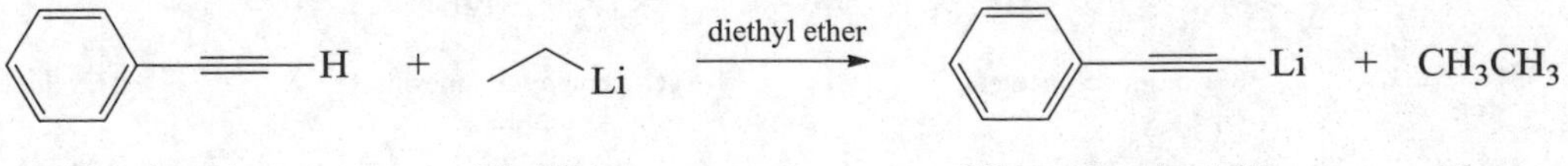

Phenylacetylene   Ethyllithium   Lithium phenylacetylide   Ethane

15.21 (*a*) Organolithium reagents react with formaldehyde to give primary alcohols.

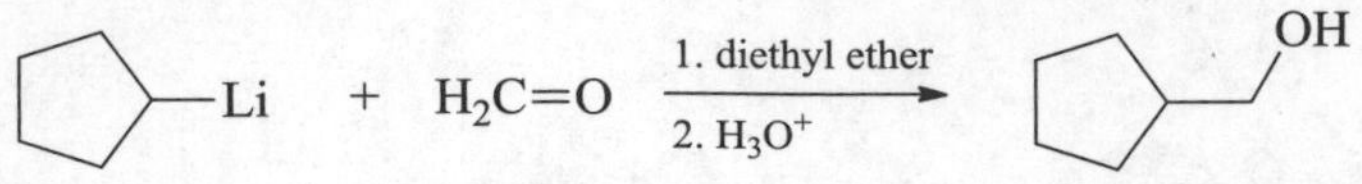

Cyclopentyllithium   Formaldehyde   Cyclopentylmethanol

(*b*) Grignard reagents react with aldehydes to give secondary alcohols.

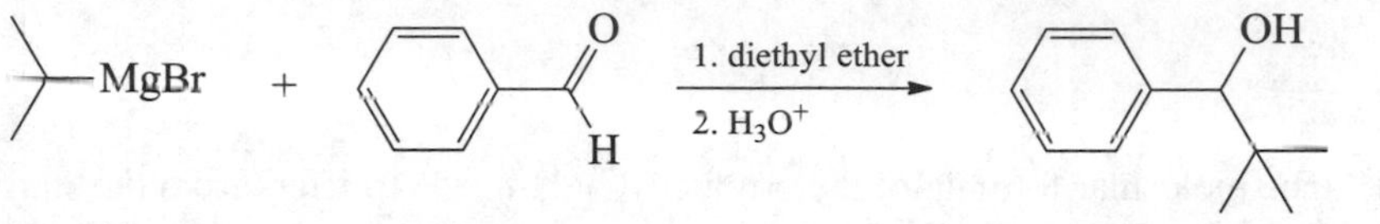

tert-Butylmagnesium   Benzaldehyde   2,2-Dimethyl-1-phenyl-
bromide   1-propanol

(*c*) Ketones react with Grignard and organolithium reagents to give tertiary alcohols.

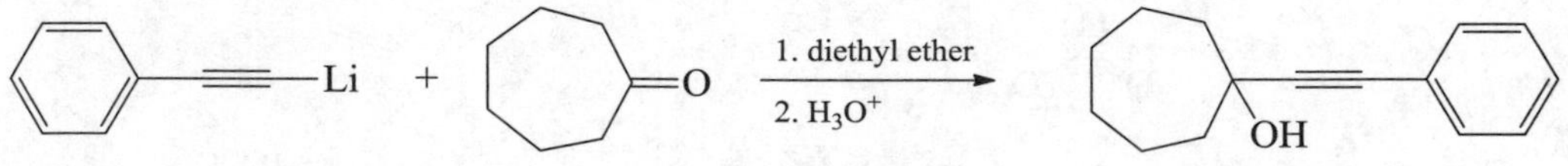

Lithium phenylacetylide   Cycloheptanone   1-(Phenylethynyl)-1-cycloheptanol

15.22 (*a*) Sodium acetylide adds to ketones to give tertiary alcohols.

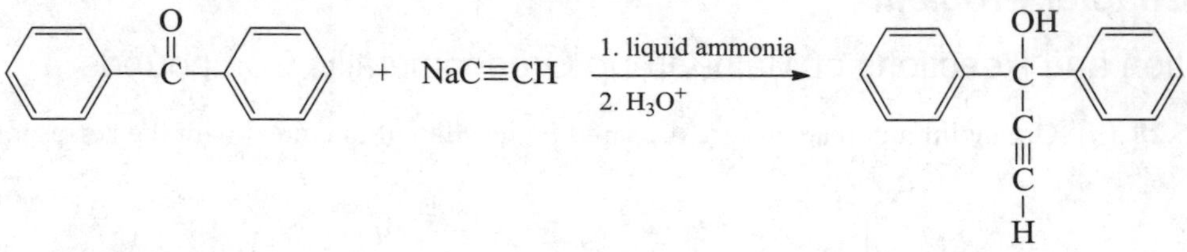

Benzophenone   1,1-Diphenyl-2-propyn-1-ol
                                                 (50%)

(*b*) The substrate is a ketone, which reacts with ethyllithium to yield a tertiary alcohol.

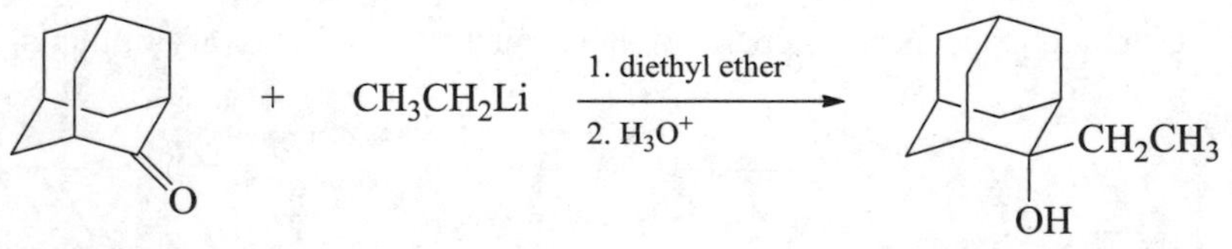

1-Adamantanone   2-Ethyl-2-adamantanol (83%)

(*c*) The first step is conversion of bromocyclopentene to the corresponding Grignard reagent, which then reacts with formaldehyde to give a primary alcohol.

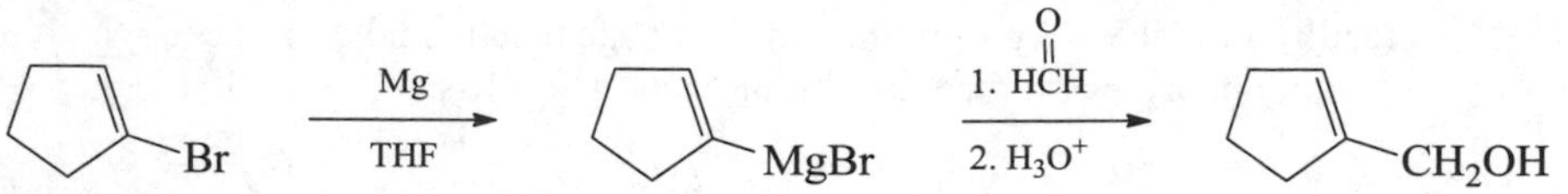

1-Bromocyclopentene   1-Cyclopentenylmagnesium   1-Cyclopentenylmethanol
                               bromide   (53%)

15.23 Phenylmagnesium bromide reacts with 4-*tert*-butylcyclohexanone as shown.

| 4-*tert*-Butylcyclohexanone | | *cis*-4-*tert*-Butyl-1-phenylcyclohexanol | *trans*-4-*tert*-Butyl-1-phenylcyclohexanol |

The phenyl substituent can be introduced either cis or trans to the *tert*-butyl group. The two alcohols are therefore stereoisomers (diastereomers). Dehydration via E1 mechanism of either alcohol yields 4-*tert*-butyl-1- phenylcyclohexene.

cis-4-*tert*-Butyl-1-phenylcyclohexanol   *trans*-4-*tert*-Butyl-1-phenylcyclohexanol   4-*tert*-Butyl-1-phenylcyclohexene

Reactions of Transition-Metal Organometallic Compounds

15.24 (*a*) The reaction is one in which an alkene is converted to a cyclopropane through use of the Simmons–Smith reagent, iodomethylzinc iodide.

Allylbenzene   Benzylcyclopropane (64%)

(*b*) Methylene transfer using the Simmons–Smith reagent is stereospecific. The trans arrangement of substituents in the alkene is carried over to the cyclopropane product.

(*E*)-1-Phenyl-2-butene   *trans*-1-Benzyl-2-methylcyclopropane
                                                          (50%)

(*c*) Lithium dimethylcuprate transfers a methyl group, which substitutes for iodine on the iodoalkene.

2-Iodo-8-methoxybenzonorbornadiene   8-methoxy-2-methylbenzonorbornadiene
                                                              (73%)

(*d*) The starting material is a *p*-toluenesulfonate. *p*-Toluenesulfonates are similar to alkyl halides in their reactivity. Substitution occurs; a butyl group from lithium dibutylcuprate replaces *p*-toluenesulfonate.

(3-Furyl)methyl *p*-toluenesulfonate   Lithium dibutylcuprate   3-Pentylfuran

(*e*) Reaction occurs preferentially with aryl iodides in the Heck reaction.

15.25 If we use the 2-bromobutane given, along with the information that the reaction occurs with inversion of configuration, the stereochemical course of the reaction may be written as

which is equivalent to

The phenyl group becomes bonded to carbon from the opposite side of the leaving group. Applying the Cahn–Ingold–Prelog notational system described in text Section 4.6 to the product, the order of decreasing precedence is

$$C_6H_5 > CH_3CH_2 > CH_3 > H$$

Orienting the molecule so that the lowest-ranked group (H) is away from us, we see that the order of decreasing precedence is clockwise.

The absolute configuration is *R*.

15.26 The substrates are secondary alkyl *p*-toluenesulfonates, and so we expect elimination to compete with substitution. The elimination product is formed in both reactions and has the molecular formula of 4-*tert*-butylcyclohexene. Because the two *p*-toluenesulfonates reactants are diastereomers, it is likely that their substitution products are also diastereomers, especially because they have the same molecular formula.

Assuming that the substitution reactions proceed with inversion of configuration, we conclude that the products are as shown.

trans-4-tert-Butylcyclohexyl
p-toluenesulfonate

cis-1-tert-Butyl-4-methylcyclohexane
($C_{11}H_{22}$)

4-tert-butylcyclohexene
($C_{10}H_{18}$)

cis-4-tert-Butylcyclohexyl
p-toluenesulfonate

trans-1-tert-Butyl-4-methylcyclohexane
($C_{11}H_{22}$)

4-tert-butylcyclohexene
($C_{10}H_{18}$)

The *trans* p-toluenesulfonate gives a higher yield of substitution product than the *cis* isomer. Elimination via an anti orientation of H and OTs is easily accessible in the *cis* isomer, but requires distortion of the cyclohexane ring in the *trans*. Elimination, however, is the major pathway in both isomers.

Synthetic Applications of Organometallic Compounds

15.27 In the solutions to this problem, the Grignard reagent butylmagnesium bromide is used. In each case, the use of butyllithium would be equally satisfactory.

(a) 1-Pentanol is a primary alcohol having one more carbon atom than 1-bromobutane. Retrosynthetic analysis suggests the reaction of a Grignard reagent with formaldehyde.

1-Pentanol

Butylmagnesium halide

Formaldehyde

An appropriate synthetic scheme is

1-Bromobutane

Butylmagnesium
bromide

1-Pentanol

(b) 2-Hexanol is a secondary alcohol having two more carbon atoms than 1-bromobutane. As revealed by retrosynthetic analysis, it may be prepared by reaction of ethanal (acetaldehyde) with butylmagnesium bromide.

2-Hexanol

Butylmagnesium halide

Ethanal
(acetaldehyde)

The correct reaction sequence is

1-Bromobutane   Butylmagnesium bromide   2-Hexanol

(*c*) 1-Phenyl-1-pentanol is a secondary alcohol. Disconnection suggests that it can be prepared from butylmagnesium bromide and an aldehyde; benzaldehyde is the appropriate aldehyde.

1-Phenyl-1-pentanol   Butylmagnesium halide   Benzaldehyde

Butylmagnesium bromide   Benzaldehyde   1-Phenyl-1-pentanol

(*d*) 1-Butylcyclobutanol is a tertiary alcohol. The appropriate ketone is cyclobutanone.

Butylmagnesium bromide   Cyclobutanone   1-Butylcyclobutanol

15.28 In each part of this problem, disconnect the phenyl group of the product to reveal the aldehyde or ketone that reacts with phenyllithium to give the desired carbon skeleton. Recall that reaction of an organolithium reagent with formaldehyde gives a primary alcohol, reaction with higher aldehydes gives a secondary alcohol, and reaction with ketones gives a tertiary alcohol. In addition to constructing the carbon skeleton, some functional group manipulation is also required.

(*a*) The target molecule has a –CH_2Br side chain on the benzene ring. Given that the starting material is phenyllithium, we need to combine (in separate steps) making a C–C bond and incorporating the bromine substituent. A reasonable retrosynthesis is:

Benzyl bromide   Benzyl alcohol   Phenyllithium   Formaldehyde

This retrosynthesis suggests the following synthesis:

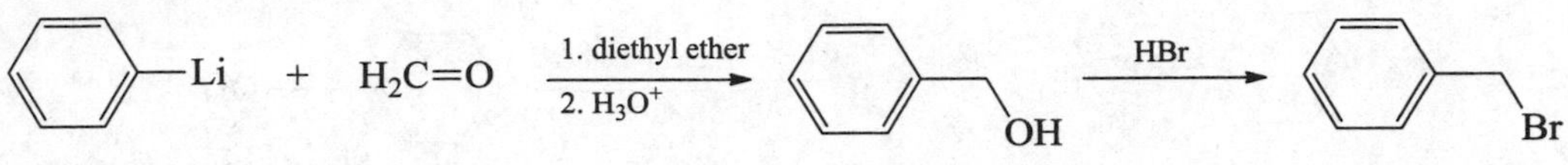

Phenyllithium   Formaldehyde   Benzyl alcohol   Benzyl bromide

(b) This exercise is analogous to the preceding one and involves making a tertiary bromide rather than a primary one.

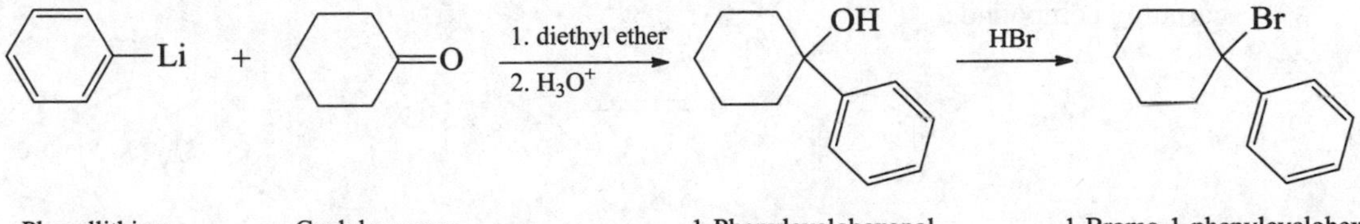

Phenyllithium   Cyclohexanone   1-Phenylcyclohexanol   1-Bromo-1-phenylcyclohexane

(c) Here, both regiochemistry and stereochemistry need to be considered. Working backward from the target, we find that both can be controlled by hydroboration-oxidation of 1-phenylcyclohexene and that 1-phenylcyclohexene is available by dehydration of the tertiary alcohol formed in part (b) of this problem.

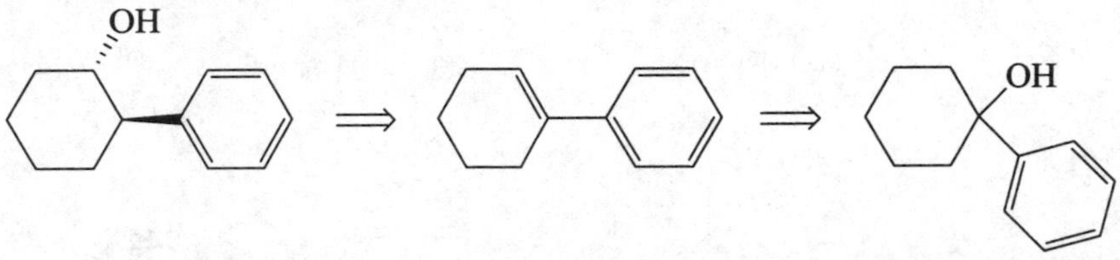

trans-2-Phenylcyclohexanol   1-Phenylcyclohexene   1-Phenylcyclohexanol

Therefore, prepare 1-phenylcyclohexanol as in part (b) and continue as follows.

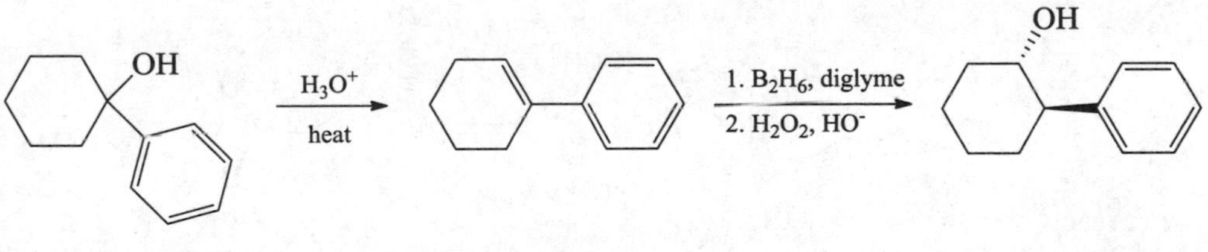

1-Phenylcyclohexanol   1-Phenylcyclohexene   *trans*-2-Phenylcyclohexanol

(d) 2-Phenyl-1,3-butadiene can be prepared by dehydration of a tertiary alcohol, the carbon skeleton of which results from addition of phenyllithium to a ketone.

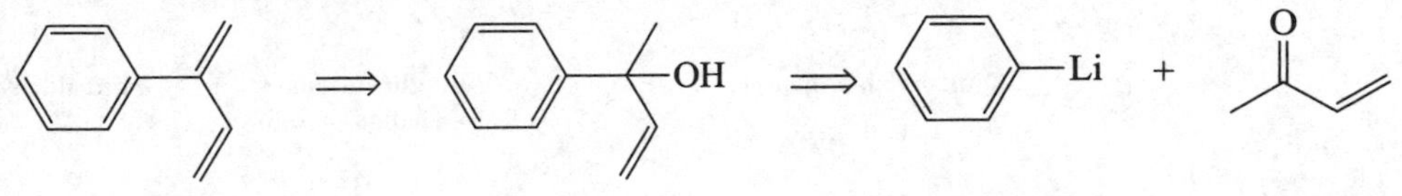

2-Phenyl-1,3-butadiene   2-Phenyl-3-buten-2-ol   Phenyllithium   Methyl vinyl ketone

The synthesis is:

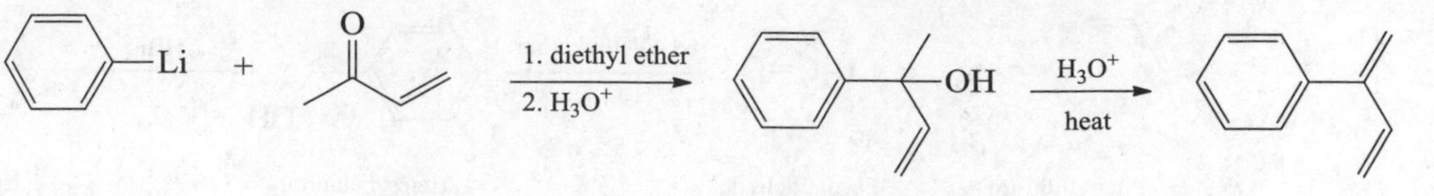

Phenyllithium   Methyl vinyl ketone   2-Phenyl-3-buten-2-ol   2-Phenyl-1,3-butadiene

15.29 In these problems, the principles, of retrosynthetic analysis are applied. The alkyl groups attached to the carbon that bears the hydroxyl group are mentally disconnected to reveal the Grignard reagent and carbonyl compound.

(a)

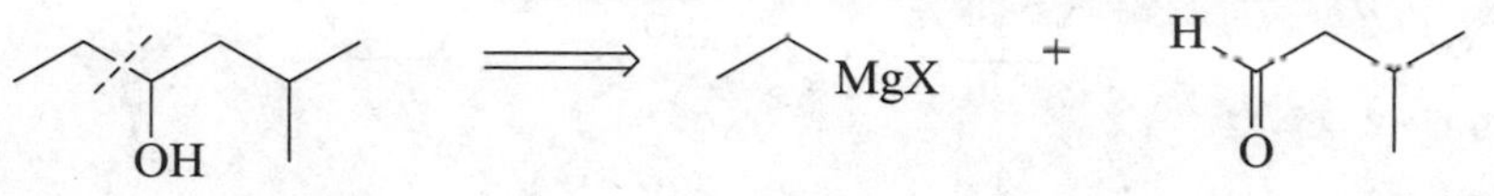

5-Methyl-3-hexanol   Ethylmagnesium halide   3-Methylbutanal

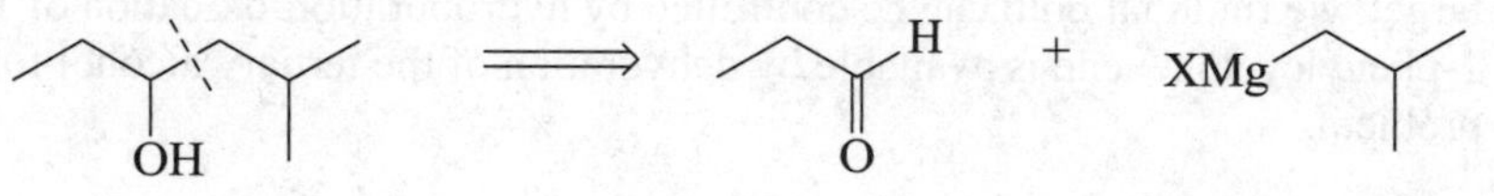

5-Methyl-3-hexanol   Propanal   Isobutylmagnesium halide

(b)

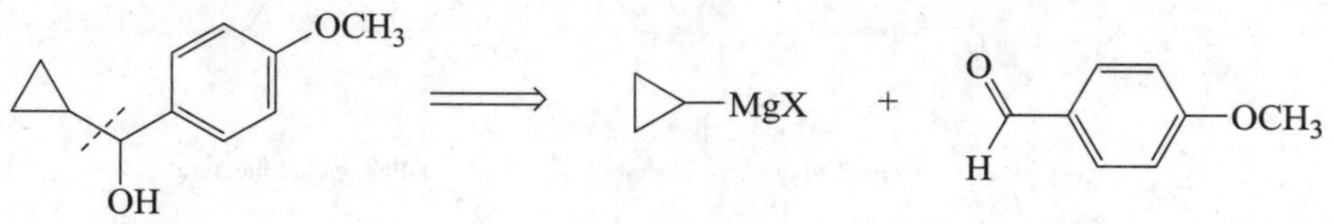

1-Cyclopropyl-1-(p-methoxyphenyl)methanol   Cyclopropyl-magnesium halide   p-Methoxybenzaldehyde

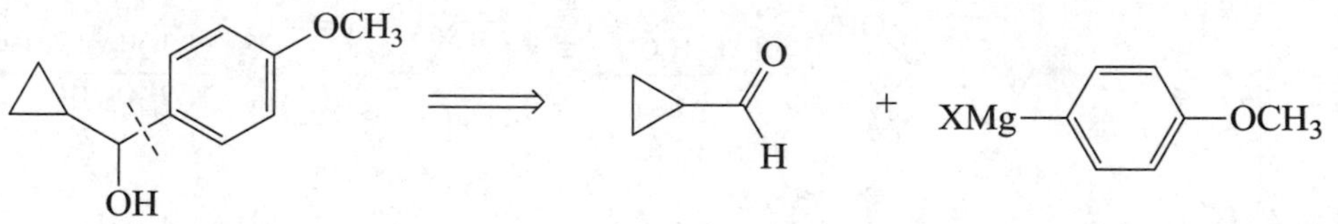

1-Cyclopropyl-1-(p-methoxyphenyl)methanol   Cyclopropane-carbaldehyde   p-Methoxyphenylmagnesium halide

(c)

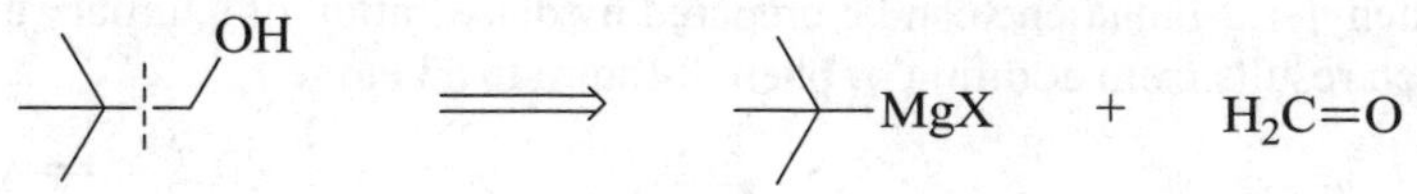

2,2-Dimethyl-1-propanol   tert-Butylmagnesium halide   Formaldehyde

(d)

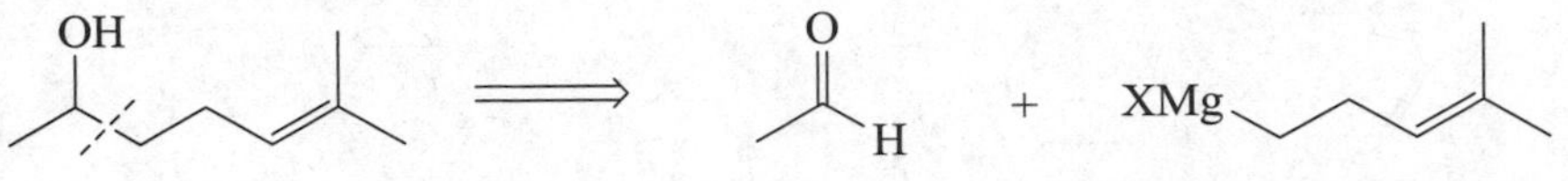

6-Methyl-5-hepten-2-ol ⟹ Ethanal + 4-Methylpent-3-en-1-ylmagnesium halide

6-Methyl-5-hepten-2-ol ⟹ Methylmagnesium halide + 5-Methyl-4-hexenal

(e)

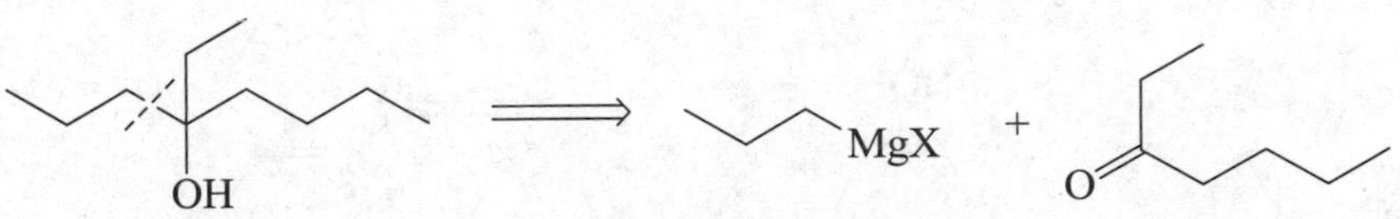

4-Ethyl-4-octanol ⟹ Propylmagnesium halide + 3-Heptanone

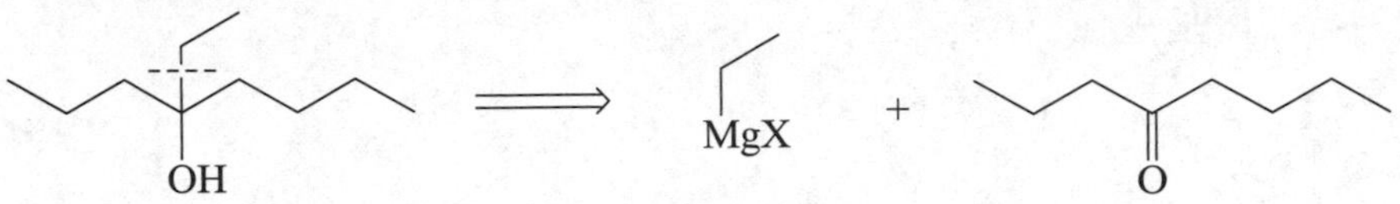

4-Ethyl-4-octanol ⟹ Ethylmagnesium halide + 4-Octanone

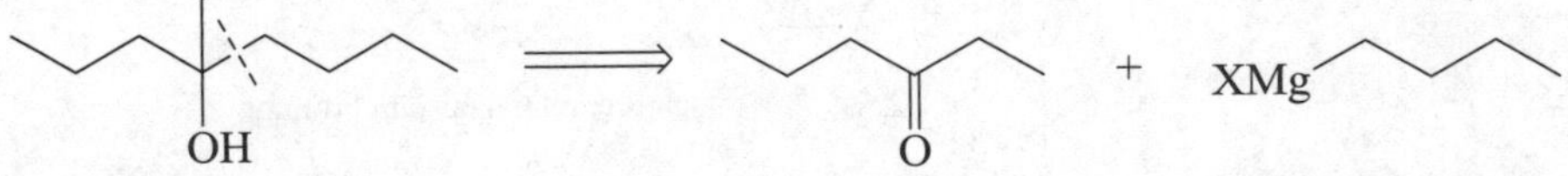

4-Ethyl-4-octanol ⟹ 3-Hexanone + Butylmagnesium halide

15.30 (a) Meparfynol is a tertiary alcohol and so can be prepared by addition of a carbanionic species to a ketone. Use the same reasoning that applies to the synthesis of alcohols from Grignard reagents. On mentally disconnecting one of the bonds to the carbon bearing the hydroxyl group

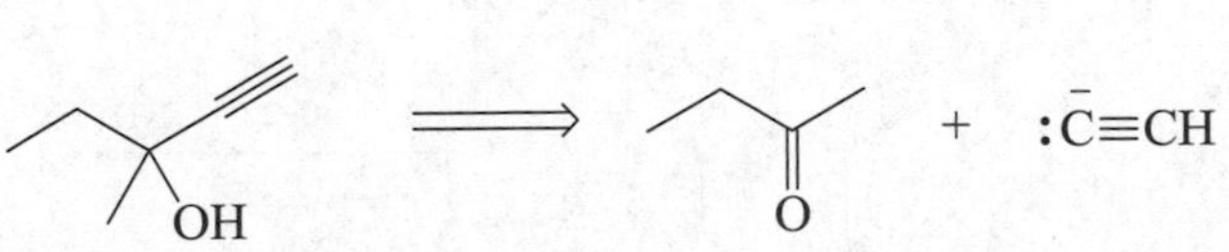

we see that the addition of acetylide ion to 2-butanone will provide the target molecule.

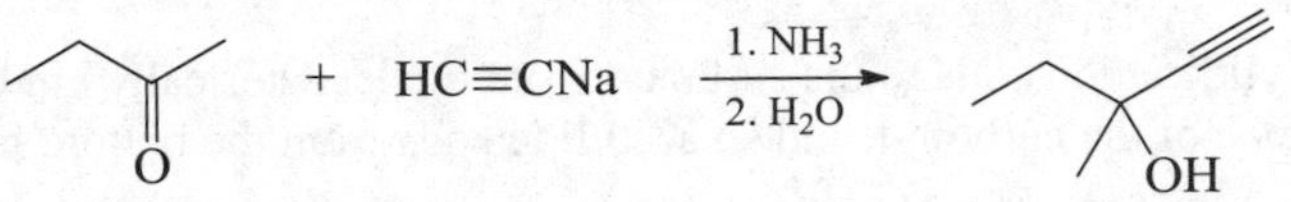

2-Butanone   Sodium acetylide   Meparfynol (94%)

The alternative, reaction of a Grignard reagent with an alkynyl ketone, is not acceptable in this case. The acidic terminal alkyne C—H would transfer a proton to the Grignard reagent before the Grignard reagent could add to the ketone.

(b) Diphepanol is a tertiary alcohol and so may be prepared by reaction of a Grignard or organolithium reagent with a ketone. Retrosynthetically, two possibilities seem reasonable:

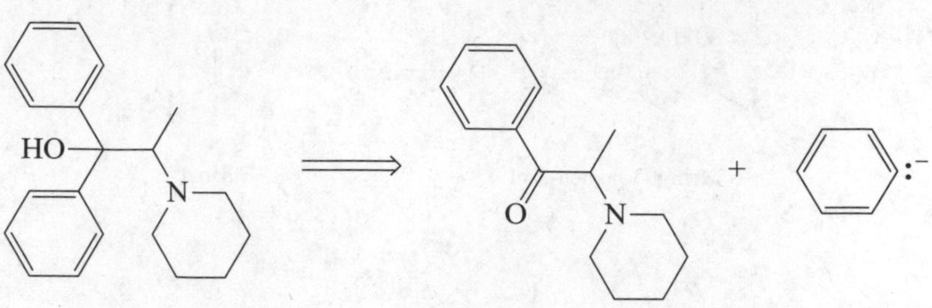

and

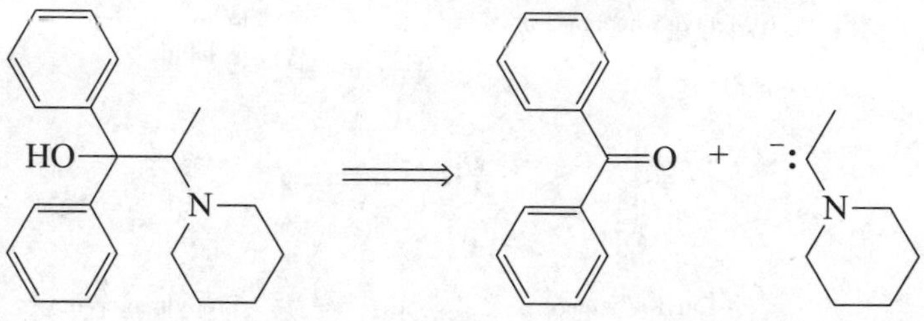

In principle, either strategy is acceptable; in practice, the one involving phenylmagnesium bromide is used.

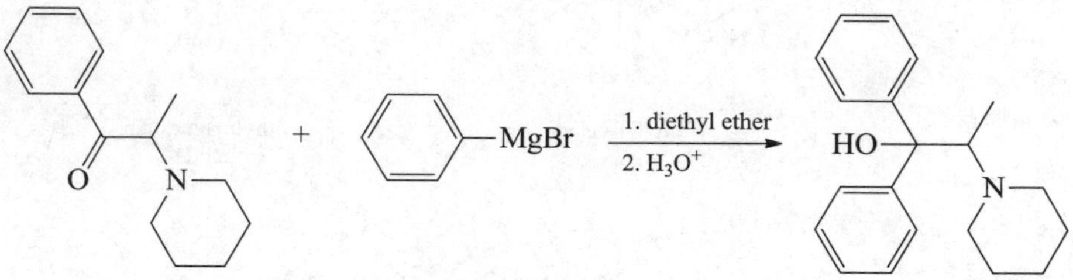

Phenylmagnesium bromide   Diphepanol

(c) A reasonable last step in the synthesis of mestranol is the addition of sodium acetylide to the ketone shown.

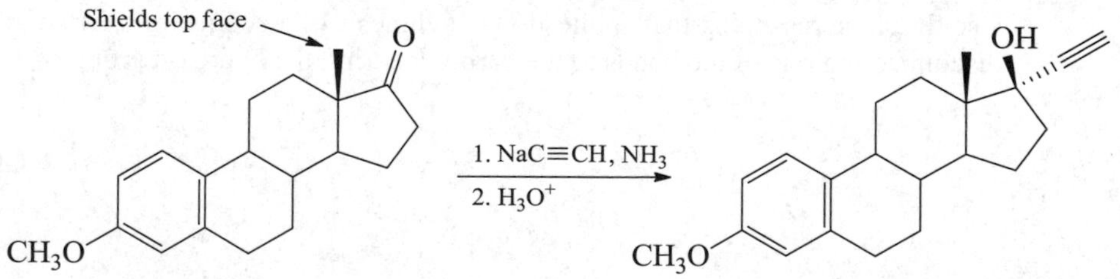

Mestranol

Acetylide anion adds to the carbonyl from the less sterically hindered side. The methyl group shields the top face of the carbonyl, and so acetylide adds from the bottom face to give the diastereomer shown.

15.31 We are told in the statement of the problem that the first step is conversion of the alcohol to the corresponding *p*-toluenesulfonate. This step is carried out as follows:

| 3,8-Epoxy-1-undecanol | *p*-Toluenesulfonyl chloride (TsCl) | 3,8-Epoxyundecyl *p*-toluenesulfonate |

Alkyl *p*-toluenesulfonates react with lithium dialkylcuprates in the same way that alkyl halides do. Treatment of the preceding *p*-toluenesulfonate with lithium dibutylcuprate gives the desired compound.

| 3,8-Epoxyundecyl *p*-toluenesulfonate | | 4,9-Epoxypentadecane |

As actually performed, a 91% yield of the desired product was obtained in the reaction of the *p*-toluenesulfonate with lithium dibutylcuprate.

15.32 (*a*) The Simmons-Smith reaction for cyclopropanation is stereospecific. A cis alkene will yield a cis disubstituted cyclopropane and a trans alkene will give a trans disubstituted cyclopropane. Retrosynthetic analysis for preparing *cis*-1,2-diethylcyclopropane from 3-hexyne is shown. In turn, we learned in Chapter 9 that cis alkenes can be formed by hydrogenation using Lindlar Pd.

| *cis*-1,2-Diethylcyclpropane | (Z)-3-Hexene | 3-Hexyne |

The synthesis is as follows:

| 3-Hexyne | (Z)-3-Hexene | *cis*-1,2-Diethylcyclpropane |

(*b*) The reaction using a trans alkene will yield a trans disubstituted cyclopropane as the retrosynthesis shows. In turn, we learned in Chapter 10 that trans alkenes can be formed by reduction of alkynes using sodium in liquid ammonia.

| *trans*-1,2-Diethylcyclpropane | (E)-3-Hexene | 3-Hexyne |

The synthesis is as follows:

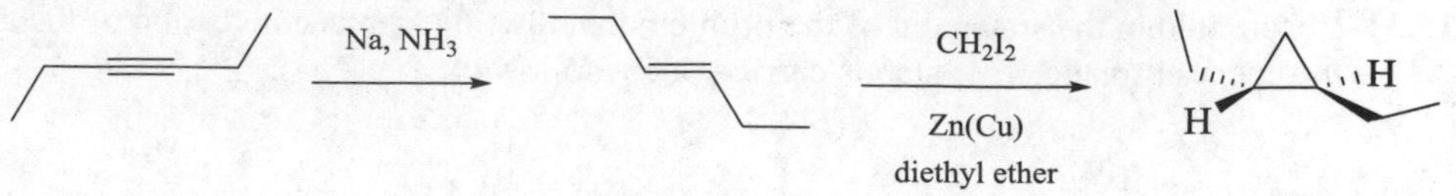
Na, NH3
CH2I2
Zn(Cu)
diethyl ether
H
H

3-Hexyne (*E*)-3-Hexene *trans*-1,2-Diethylcyclpropane

Transition Metal-Catalyzed Reactions

15.33 Hydrogenation of a double bond must generate the chirality center. Therefore, the double bond cannot be part of the four-carbon side chain that becomes the isobutyl group. It must be part of the other substituent.

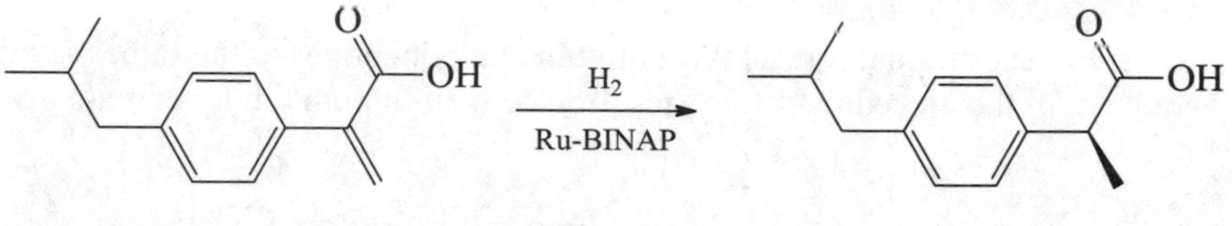
O
OH
H2
Ru-BINAP
O
OH

(*S*)-(+)-Ibuprofen

15.34 The disconnection suggests the possibility of preparing the target molecule by a palladium-catalyzed cross-coupling.

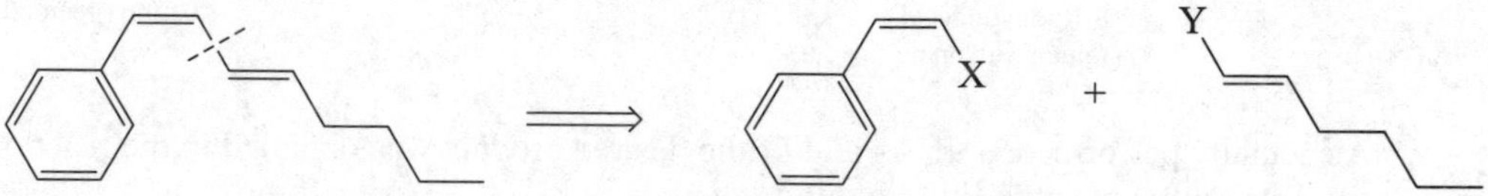
X
+
Y

The problem specifies (*Z*)-$C_6H_5CH{=}CHBr$ as one of the reactants. Hydroboration of catecholborane with 1-hexyne gives the other.

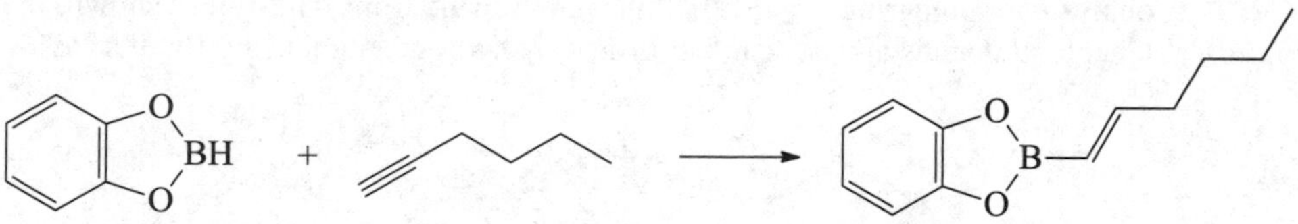
O
BH
O
+
O
B
O

Combining the two in the presence of a catalyst such as $(Ph_3P)_4Pd$ gives the desired product.

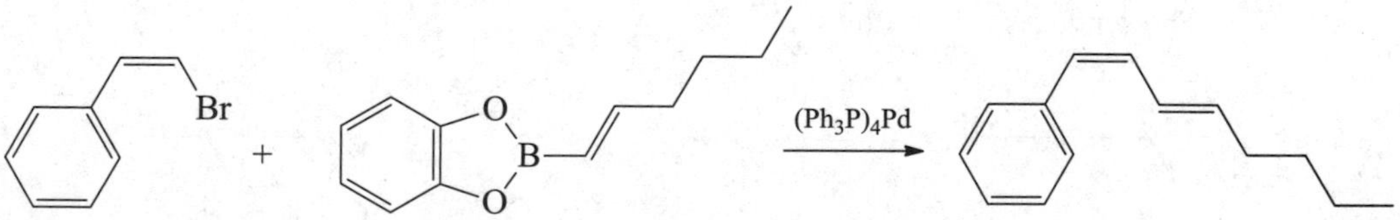
Br
+
O
B
O
(Ph3P)4Pd

15.35 One reactant is a boronate, the other is a vinylic halide, and the reaction is palladium catalyzed. These are typical components of a Suzuki cross-coupling. The reaction proceeds by σ bond formation between the two reactive sites. Double bond configurations are preserved.

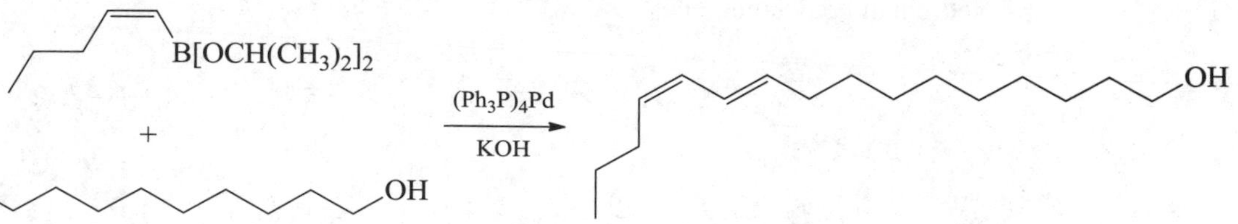
B[OCH(CH3)2]2
+
I
OH
(Ph3P)4Pd
KOH
OH

15.36 Each reactant contains a carbon–carbon double bond and the reaction is carried out in the presence of an olefin-metathesis catalyst. The molecular formula of the product is the sum of the molecular formulas of the reactants, less C_2H_4. These facts suggest cross-metathesis.

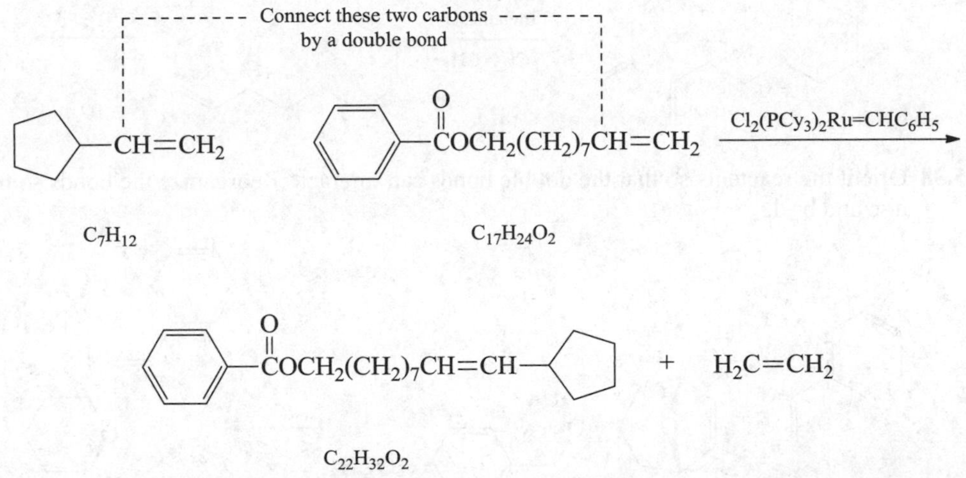

C_7H_{12}

$C_{17}H_{24}O_2$

$C_{22}H_{32}O_2$

15.37 (*a*) Exaltolide is the hydrogenation product of compound A. Compound A is formed by ring-closing metathesis of the reactant. To derive the structure of compound A, rewrite the reactant in a form that approximates the shape of the resulting ring.

$$H_2C=CHCH_2(CH_2)_7\overset{O}{\overset{\|}{C}}OCH_2(CH_2)_3CH=CH_2 \quad \text{is the same as}$$

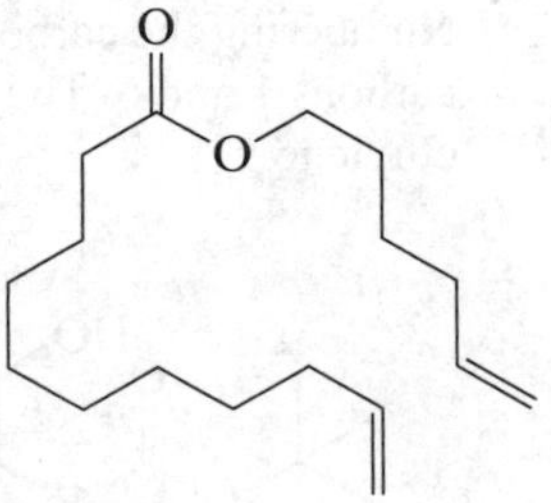

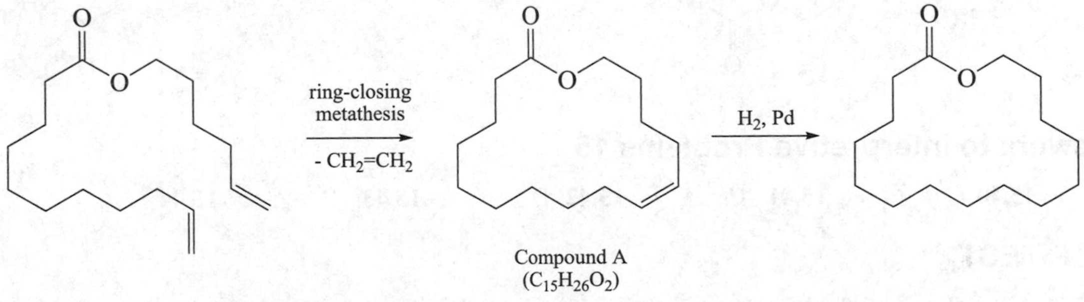

Compound A
($C_{15}H_{26}O_2$)

(*b*) Analyze this ring-closing metathesis in the same way as part (*a*).

$$H_2C=CHCH_2(CH_2)_2\overset{O}{\overset{\|}{C}}OCH_2(CH_2)_8CH=CH_2 \quad \text{is the same as}$$

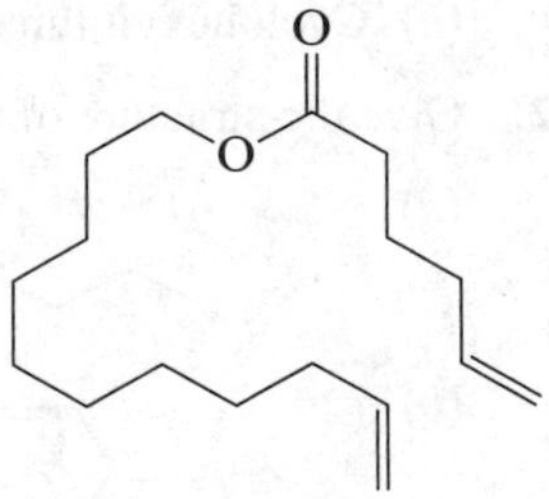

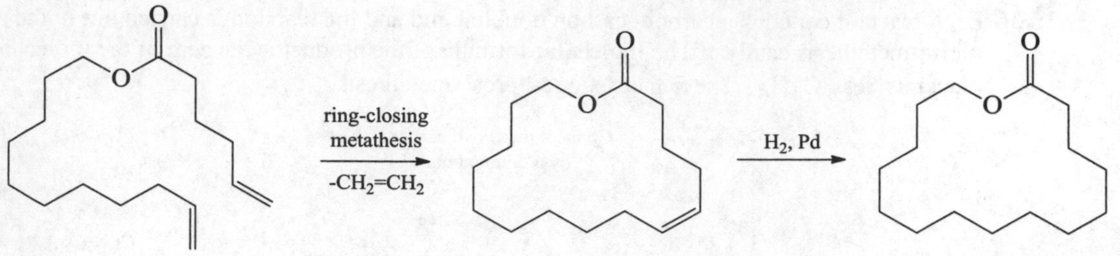

15.38 Orient the reactants so that the double bonds can interact. Reorganize the bonds so that a=b and c=d become a=c and b=d.

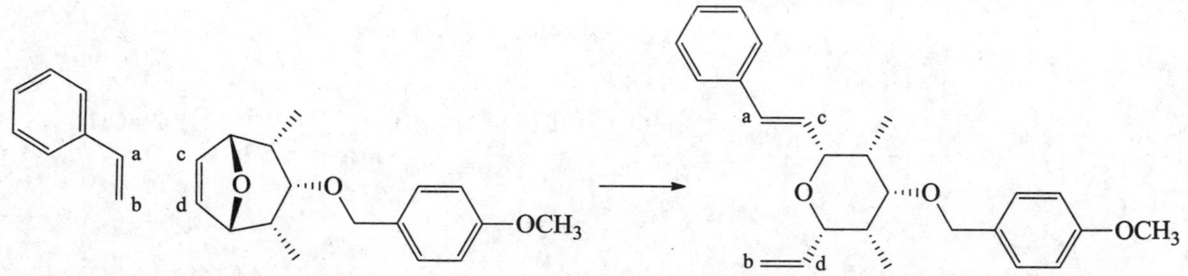

15.39 The precursor is the open-chain compound, which forms the cyclohexene ring by ring-closing metathesis. Numbering the carbons of the ring in the reactant and the product is helpful to see the new bond, between carbons 1 and 6. The two alkene carbons that are not numbered in the reactant are lost as a molecule of ethylene.

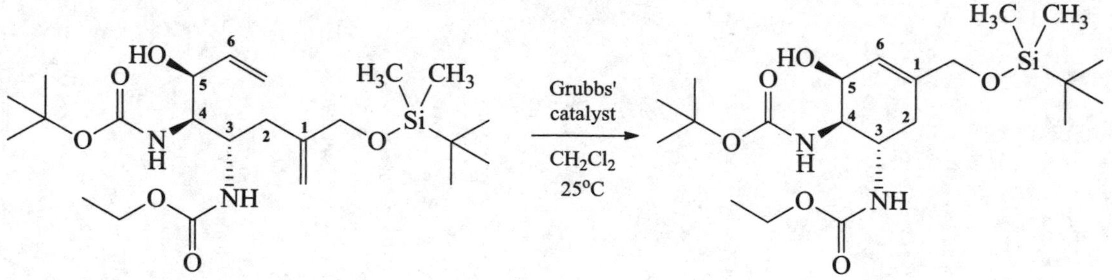

Answers to Interpretive Problems 15

15.40 C   **15.41** C   **15.42** D   **15.43** B   **15.44** A

SELF-TEST

1. Give a method for the preparation of each of the following organometallic compounds, using appropriate starting materials:

   (*a*) Cyclohexyllithium   (*b*) *tert*-Butylmagnesium bromide   (*c*) Lithium dibenzylcuprate

2. Give the structure of the product obtained by each of the following reaction schemes:

   (*a*)

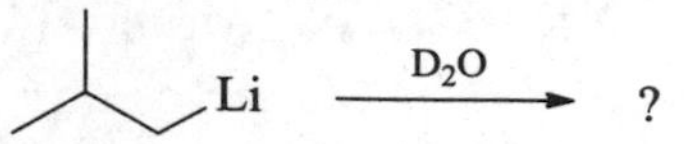

   (*b*)

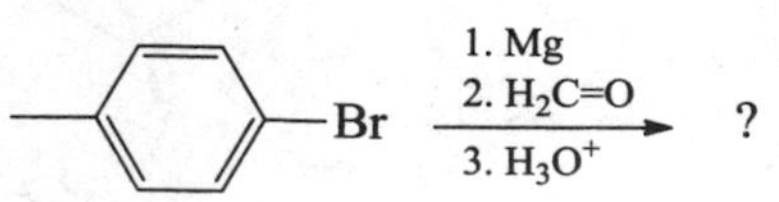

(c)

$$\text{(cyclohexanone)} \xrightarrow[\text{2. } H_3O^+]{\text{1. } CH_3CH_2Li} \quad ?$$

3. Give two combinations of an organometallic reagent and a carbonyl compound that may be used for the preparation of each of the following:

(a) OH

(b) OH

4. Give the structure of the organometallic reagent necessary to carry out each of the following:

(a) CH₃ / I $\xrightarrow{\quad ? \quad}$ CH₃

(b) $\xrightarrow{\quad ? \quad}$

5. Compounds A through F are some common organic solvents. Which ones would be suitable for use in the preparation of a Grignard reagent? For those that are not suitable, give a brief reason why.

$$CH_3CH_2CH_2CH_2OCH_2CH_2CH_2CH_3 \qquad CH_3OCH_2CH_2OCH_3 \qquad HOCH_2CH_2OH$$

A   B   C

$$CH_3\overset{\displaystyle O}{\overset{\displaystyle \|}{C}}OCH_2CH_3 \qquad\qquad \qquad CH_3\overset{\displaystyle O}{\overset{\displaystyle \|}{C}}OH$$

D   E   F

6. Show by a series of chemical equations how to prepare octane from 1-butanol as the source of all its carbon atoms.

7. Synthesis of the following alcohol is possible by two schemes using Grignard reagents. Give the reagents necessary to carry out each of them.

OH

8. Using ethylbenzene and any other necessary organic or inorganic reagents, outline a synthesis of 3-phenyl-2-butanol.

9. Give the structure of the final product of each of the following sequences of reactions.

(a)

$$\text{(benzene)} \xrightarrow[\text{FeBr}_3]{Br_2} \xrightarrow{Mg} \xrightarrow{CH_3\overset{O}{\overset{\|}{C}}CH_2CH_3} \xrightarrow{H_3O^+} \quad ?$$

(b)

$$\text{1-Butene} \xrightarrow{HCl} \xrightarrow{Mg} \xrightarrow{CH_3\overset{O}{\overset{\|}{C}}H} \xrightarrow{H_3O^+} \quad ?$$

(c)

$$H_3C{-}C{\equiv}CH \xrightarrow{NaNH_2} \xrightarrow{\text{(cyclopentanone)}} \xrightarrow{H_3O^+} \quad ?$$

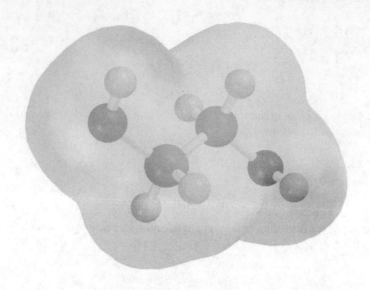

CHAPTER 16
Alcohols, Diols, and Thiols

Table of Contents

SOLUTIONS TO TEXT PROBLEMS

In Chapter Problems

16.1 The two primary alcohols, 1-butanol and 2-methyl-1-propanol, can be prepared by hydrogenation of the corresponding aldehydes.

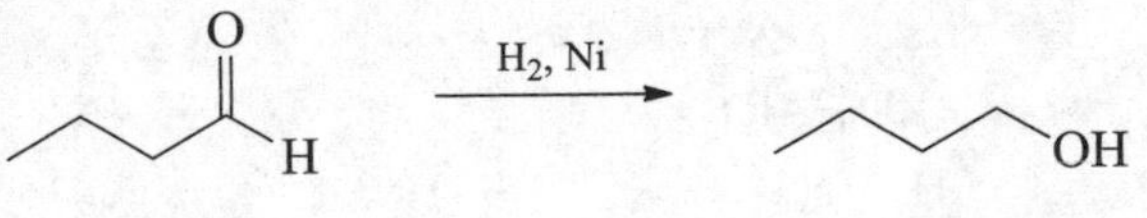

Butanal H_2, Ni   1-Butanol

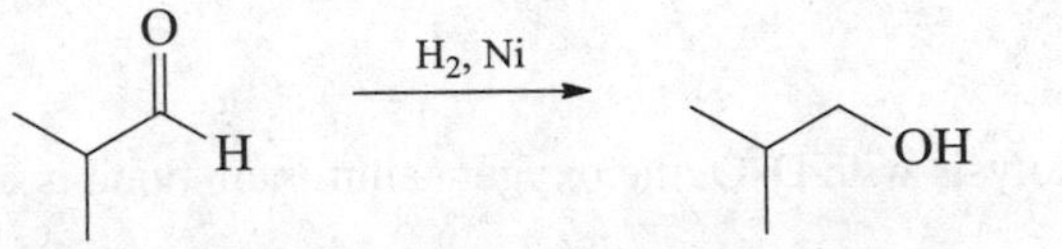

2-Methylpropanal H_2, Ni   2-Methyl-1-propanol

The secondary alcohol 2-butanol arises by hydrogenation of a ketone.

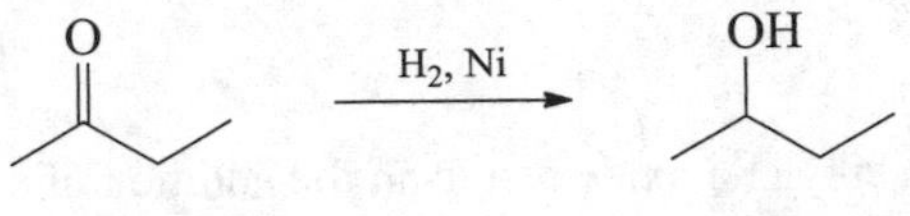

2-Butanone H_2, Ni   2-Butanol

Tertiary alcohols such as 2-methyl-2-propanol, $(CH_3)_3COH$, cannot be prepared by hydrogenation of a carbonyl compound.

16.2 (*b*) A deuterium atom is transferred from $NaBD_4$ to the carbonyl group of acetone.

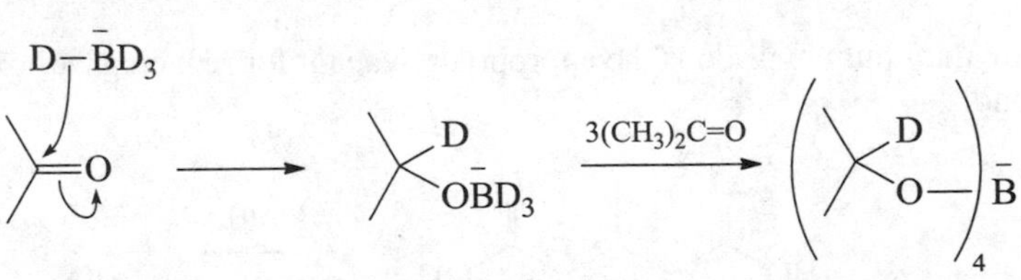

On reaction with CH_3OD, deuterium is transferred from the alcohol to the oxygen of $[(CH_3)_2CDO]_4B^-$.

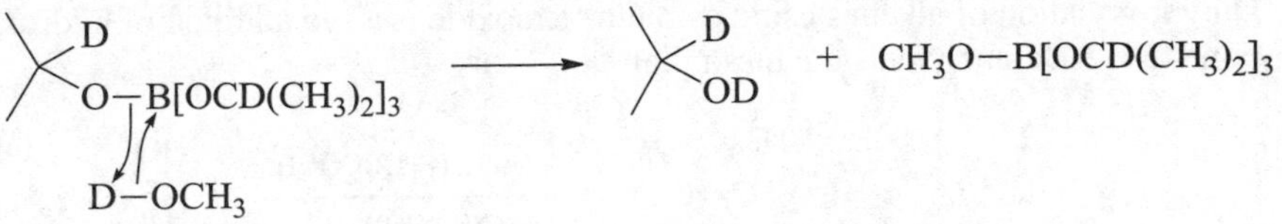

Overall:

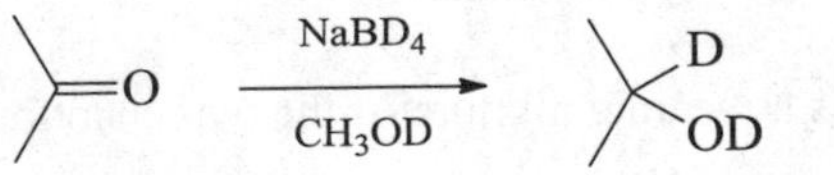

Acetone $NaBD_4$ / CH_3OD 2-Propanol-2-*d*-*O*-*d*

(c) In this case, $NaBD_4$ serves as a deuterium donor to carbon, and CD_3OH is a proton (not deuterium) donor to oxygen.

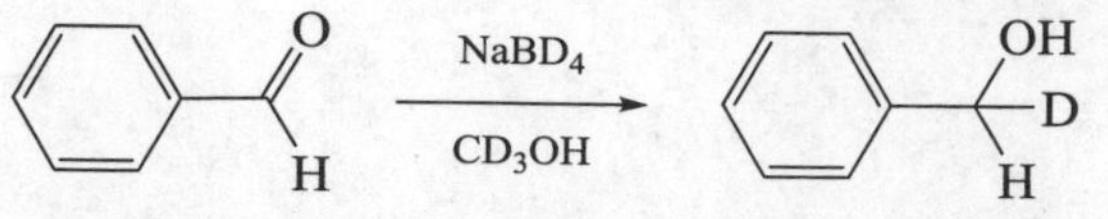

Benzaldehyde   Benzyl alcohol-1-*d*

(d) Lithium aluminum deuteride is a deuterium donor to the carbonyl carbon of formaldehyde.

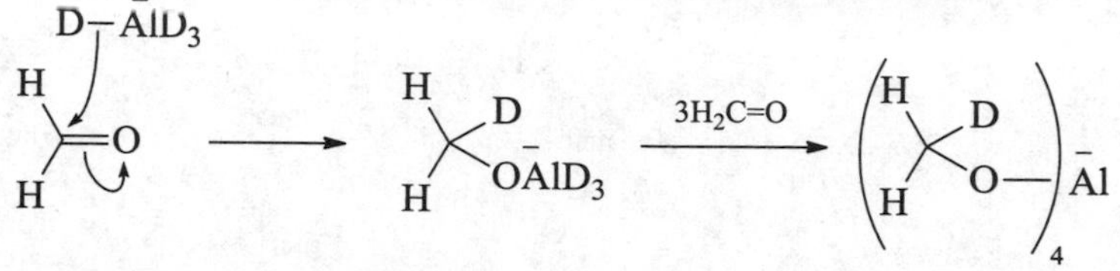

On hydrolysis with D_2O, the oxygen–aluminum bond is cleaved and DCH_2OD is formed.

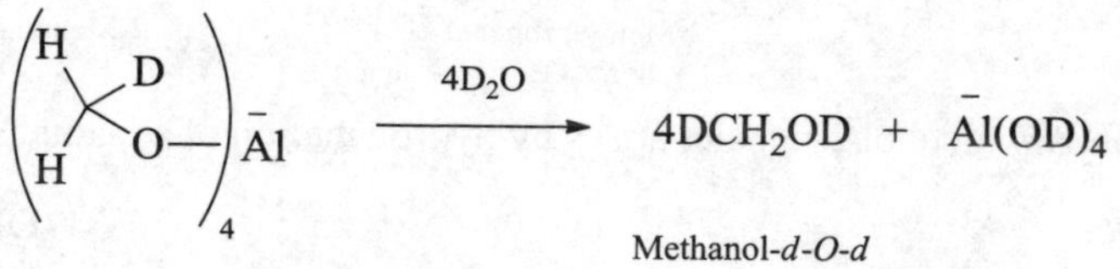

Methanol-*d-O-d*

16.3 (b) Reaction with ethylene oxide results in the addition of a —CH_2CH_2OH unit to the Grignard reagent. Cyclohexylmagnesium bromide (or chloride) is the appropriate reagent.

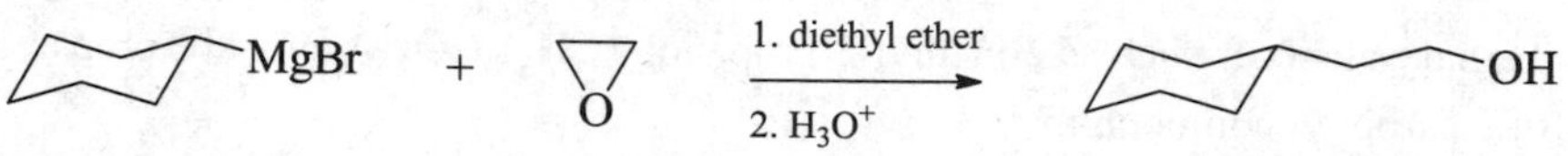

Cyclohexylmagnesium   Ethylene oxide   2-Cyclohexylethanol
bromide

16.4 Lithium aluminum hydride is the appropriate reagent for reducing carboxylic acids or esters to alcohols.

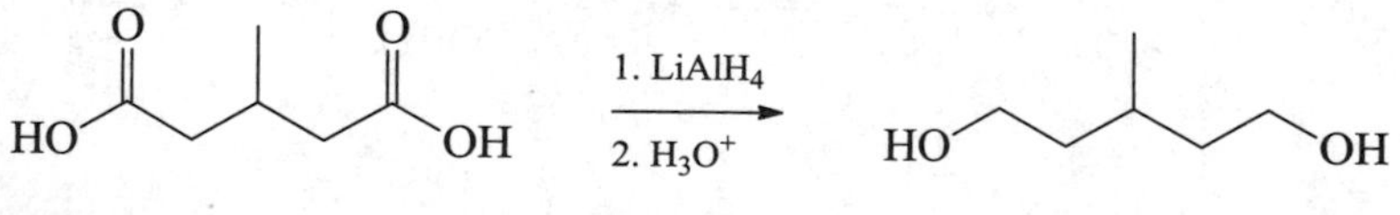

3-Methyl-1,5-pentanedioic acid   3-Methyl-1,5-pentanediol

16.5 Dihydroxylation of alkenes using osmium tetraoxide is a syn addition of hydroxyl groups to the double bond. *cis*-2-Butene yields the meso diol.

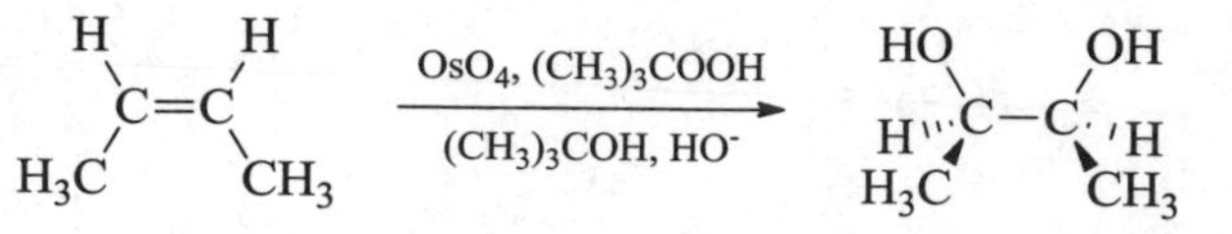

cis-2-Butene   *meso*-2,3-Butanediol

trans-2-Butene yields a racemic mixture of the two enantiomeric forms of the chiral diol.

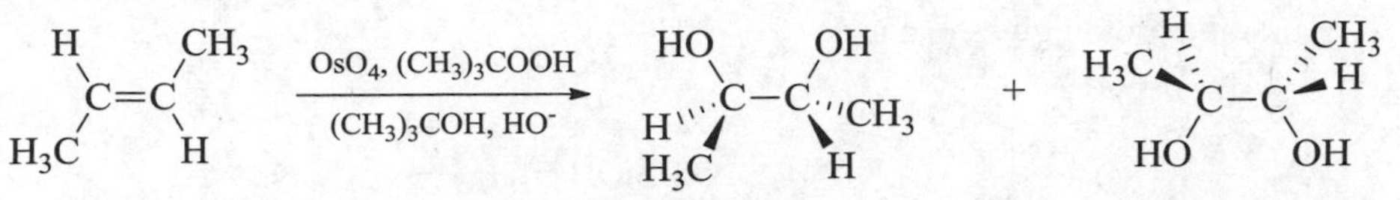

trans-2-Butene   (2R,3R)-2,3-Butanediol   (2S,3S)-2,3-Butanediol

The Fischer projection formulas of the three stereoisomers are

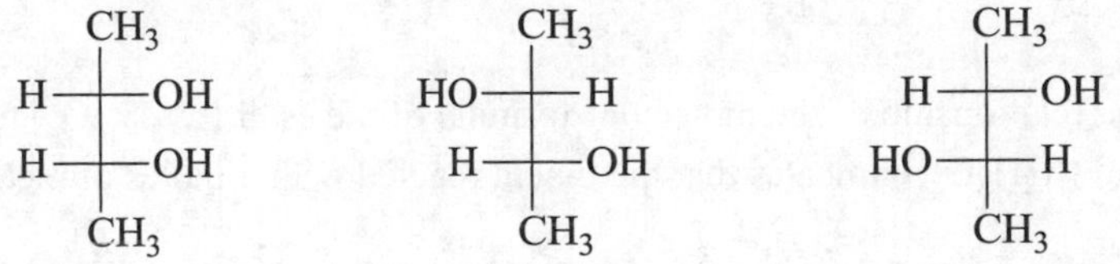

meso-2,3-Butanediol   (2R,3R)-2,3-Butanediol   (2S,3S)-2,3-Butanediol

16.6 Dihydroxylation using the Sharpless method is a syn addition. Therefore, trans-2-butene can give the following two diols.

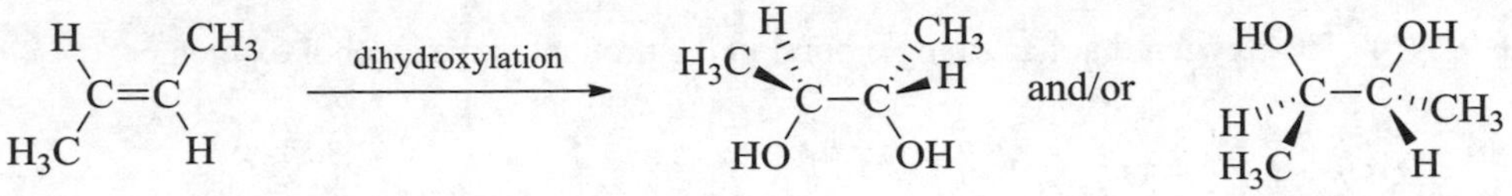

trans-2-Butene   (2S,3S)-2,3-Butanediol   (2R,3R)-2,3-Butanediol

The problem states that the reaction was enantioselective and gave the diol having the (R)-configuration at one carbon. Therefore, the correct structure is the one shown at the right and the configuration at C-3 must also be R.

16.7 The first step is proton transfer to 1,5-pentanediol to form the corresponding alkyloxonium ion.

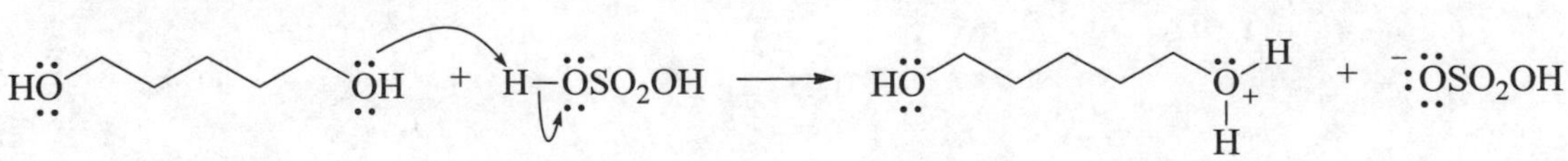

1,5-Pentanediol   Sulfuric acid   Conjugate acid of 1,5-pentanediol   Hydrogen sulfate

Rewriting the alkyloxonium ion gives

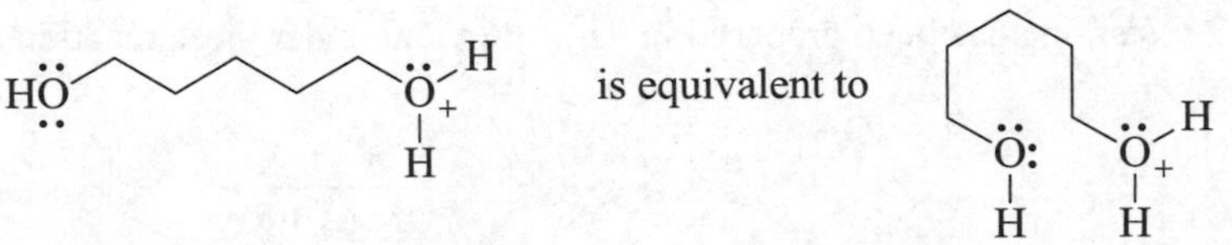

is equivalent to

The alkyloxonium ion undergoes cyclization by intramolecular nucleophilic attack of its alcohol function on the carbon that bears the leaving group.

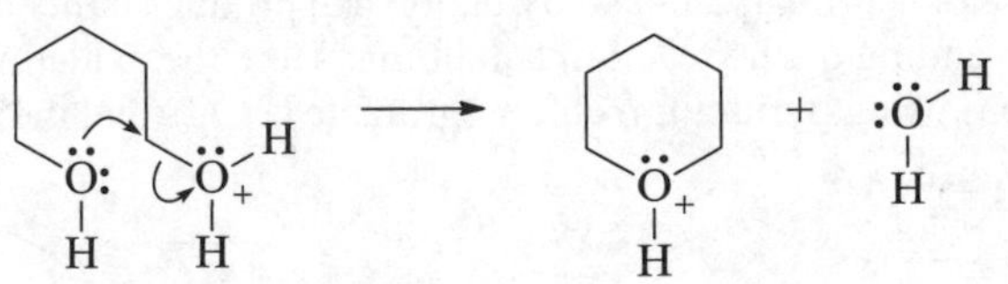

Conjugate acid of   Conjugate acid   Water
1,5-pentanediol   of oxane

Loss of a proton gives oxane.

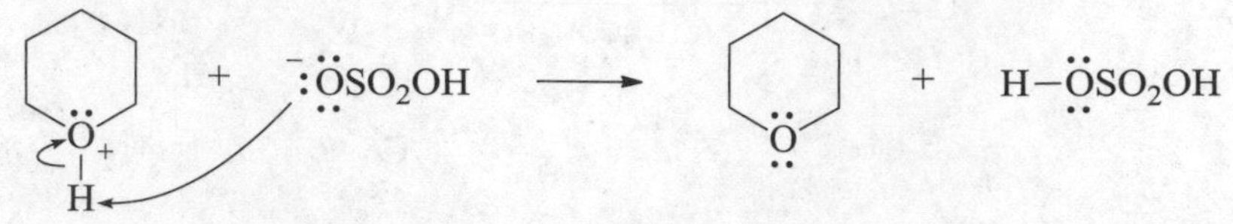

| Conjugate acid of oxane | Hydrogen sulfate | | Oxane | Sulfuric acid |

16.8 (b) The relationship of the molecular formula of the ester ($C_{10}H_{10}O_4$) to that of the starting dicarboxylic acid ($C_8H_6O_4$) indicates that the diacid reacted with 2 moles of methanol to form a diester.

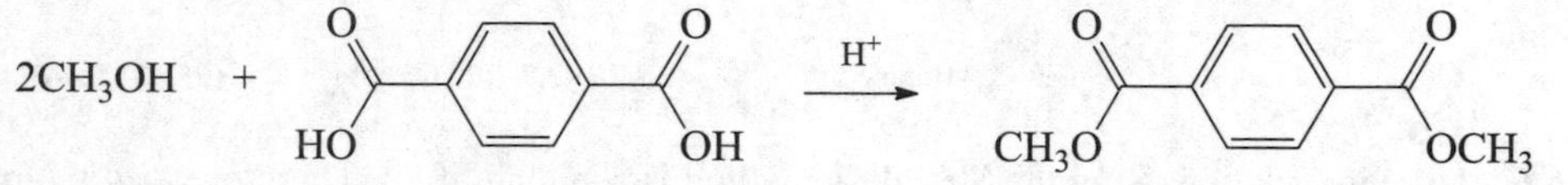

Methanol   1,4-Benzenedicarboxylic acid   Dimethyl 1,4-benzenedicarboxylate

16.9 (b) The oxygen that has a single bond in the product is the alcohol oxygen.

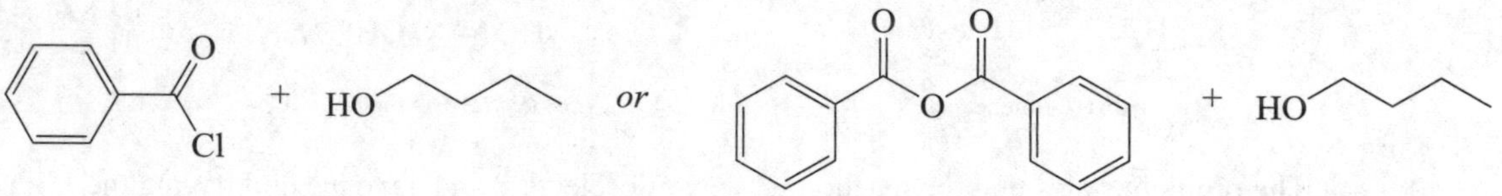

Benzoyl chloride   1-Butanol   Benzoic anhydride   1-Butanol

(c)

Acetyl chloride   *trans*-2-Methylcyclohexanol   Acetic anhydride   *trans*-2-Methylcyclohexanol

16.10 (b) The substrate is a secondary alcohol and so gives a ketone on oxidation with sodium dichromate. 2-Octanone has been prepared in 92–96% yield under these reaction conditions.

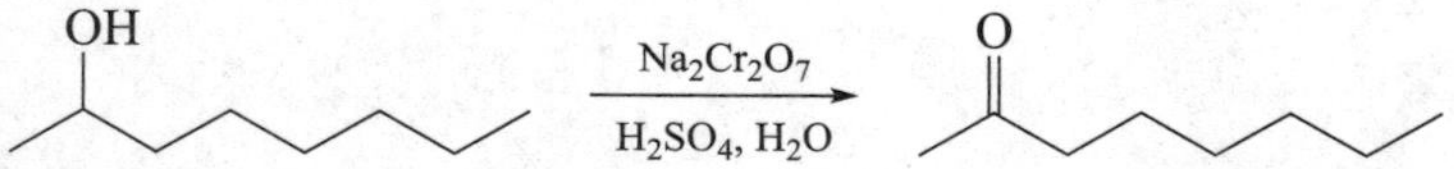

2-Octanol   2-Octanone

(c) The alcohol is primary, and so oxidation can produce either an aldehyde or a carboxylic acid, depending on the reaction conditions. Here the oxidation is carried out under anhydrous conditions using pyridinium chlorochromate (PCC), and the product is the corresponding aldehyde.

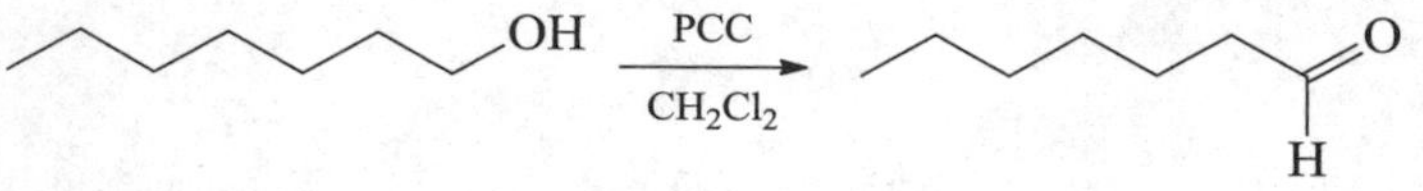

1-Heptanol   Heptanal

16.11 As described in the text and in the statement of the problem, the last step in the reaction is believed to be the unimolecular dissociation of the intermediate to citronellal.

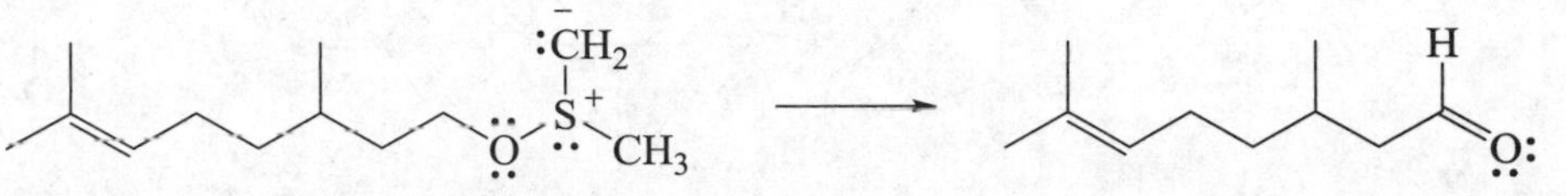

The simplest explanation is that this dissociation takes place in a single step during which the negatively charged carbon abstracts a proton from the carbon attached to oxygen.

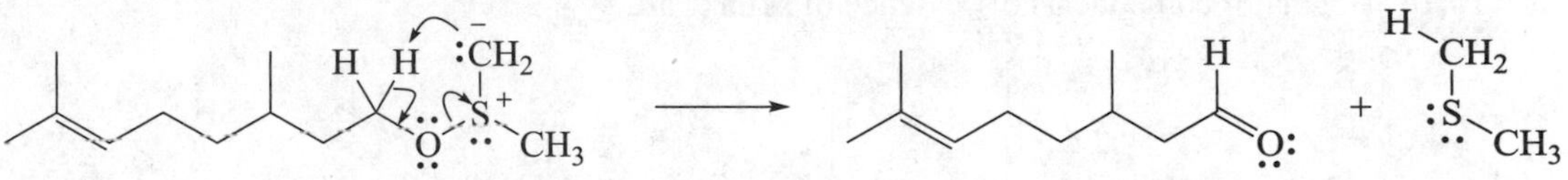

The sulfur-containing product is dimethyl sulfide.

16.12 (*b*) Enzymatic oxidation of CH_3CD_2OH leads to loss of one of the C-1 deuterium atoms to NAD^+. The dihydropyridine ring of the reduced form of the coenzyme will bear a single deuterium.

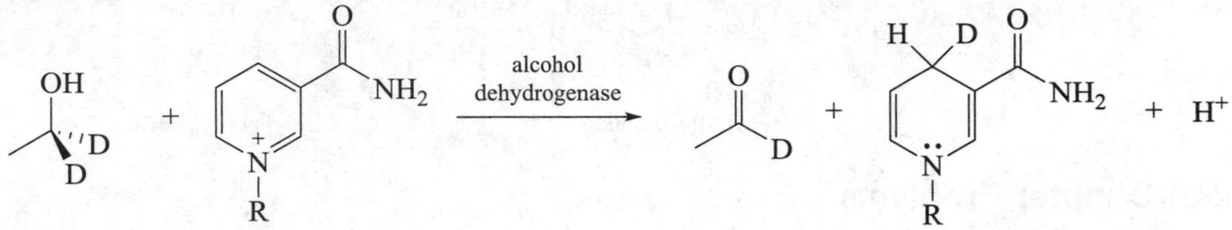

| 1,1-Dideuterio-ethanol | NAD^+ | | 1-Deuterio-ethanal | NADD |

(*c*) The deuterium atom of CH_3CH_2OD is lost as D^+. The reduced form of the coenzyme contains no deuterium.

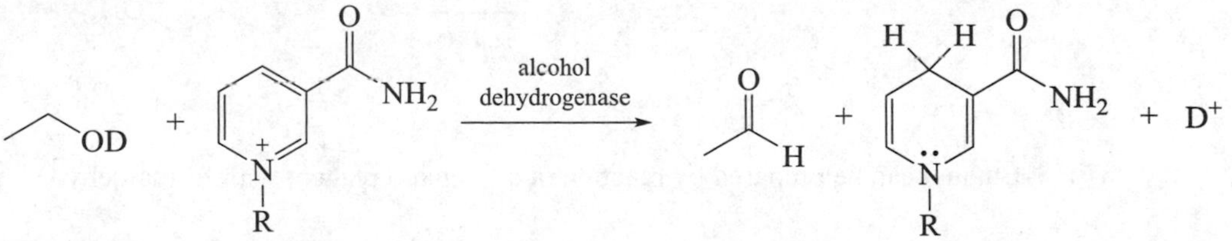

| Ethanol-*O-d* | NAD^+ | | Ethanal | NADH |

16.13 (*b*) Oxidation of the carbon–oxygen bonds to carbonyl groups accompanies their cleavage.

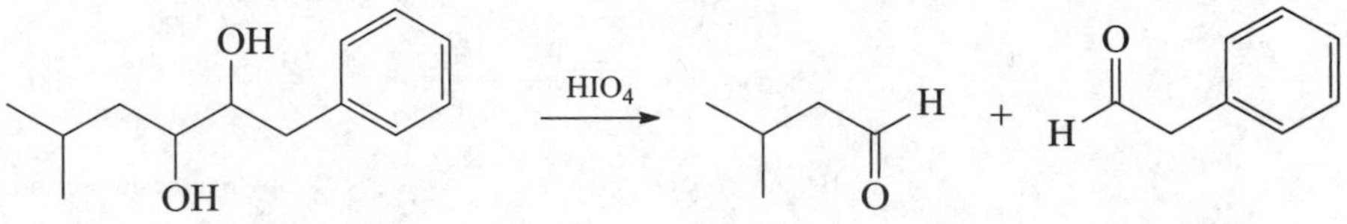

| 5-Methyl-1-phenyl-2,3-hexanediol | | 3-Methylbutanal | 2-Phenylethanal |

(c) The CH$_2$OH group is cleaved from the ring as formaldehyde to leave cyclopentanone.

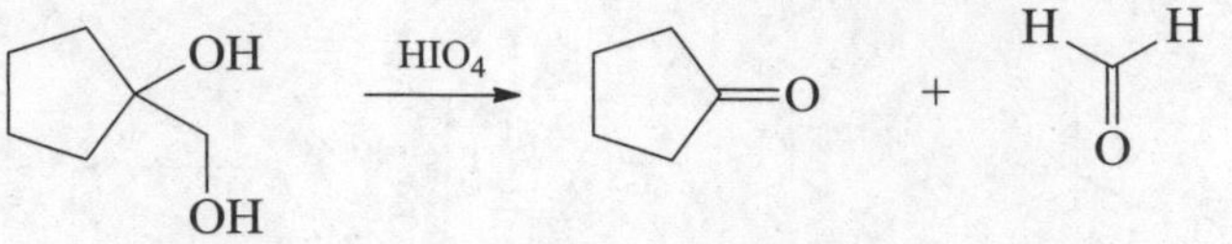

1-(Hydroxymethyl)-   Cyclopentanone   Formaldehyde
cyclopentanol

16.14 The major components of "essence of skunk" are

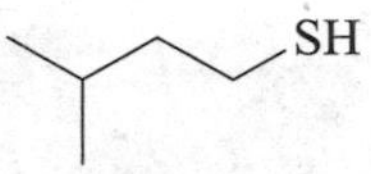

3-Methyl-1-butanethiol 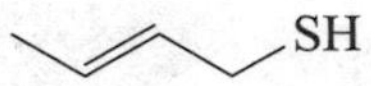   *trans*-2-Butene-1-thiol

16.15 (b) Sulfides may be prepared from the corresponding alkyl halide by nucleophilic substitution using the conjugate base of RSH.

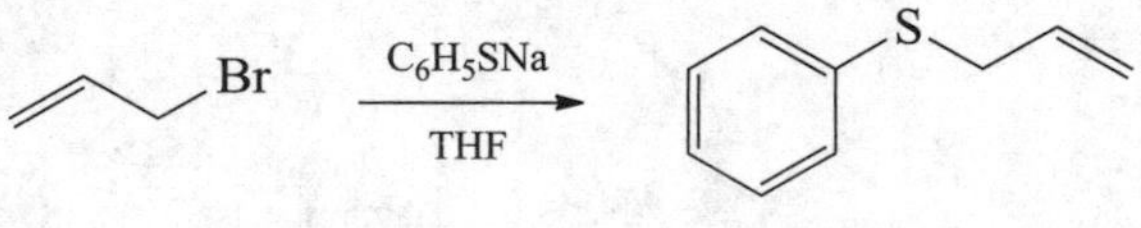

Allyl bromide   Allyl phenyl sulfide

End of Chapter Problems

Preparation of Alcohols, Diols, and Thiols

16.16 (a) The appropriate alkene for the preparation of 1-butanol by a hydroboration–oxidation sequence is 1-butene. Remember, hydroboration–oxidation leads to hydration of alkenes with a regioselectivity opposite to that seen in acid-catalyzed hydration.

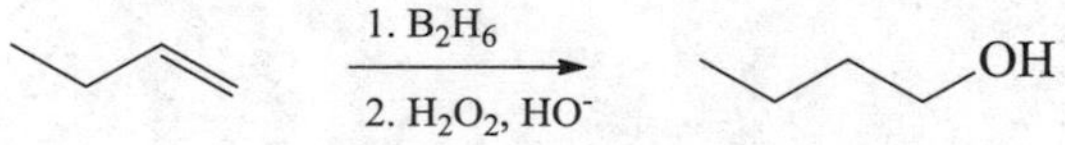

1-Butene   1-Butanol

(b) 1-Butanol can be prepared by reaction of a Grignard reagent with formaldehyde.

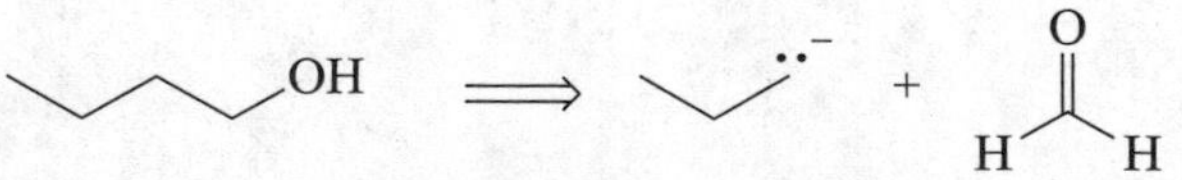

An appropriate Grignard reagent is propylmagnesium bromide.

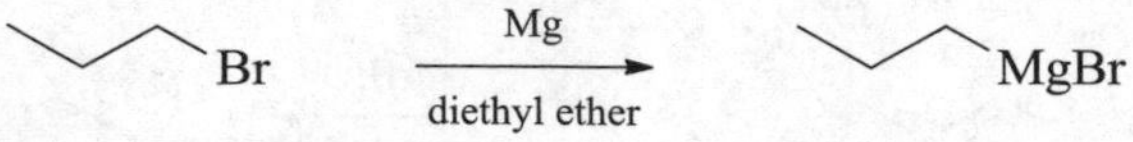

1-Bromopropane   Propylmagnesium bromide

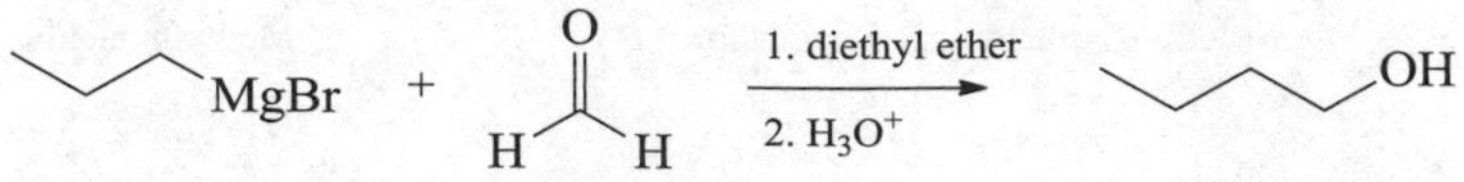

1-Butanol

(c) Alternatively, 1-butanol may be prepared by the reaction of a Grignard reagent with ethylene oxide.

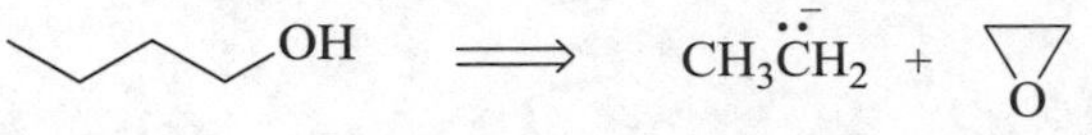

In this case, ethylmagnesium bromide would be used.

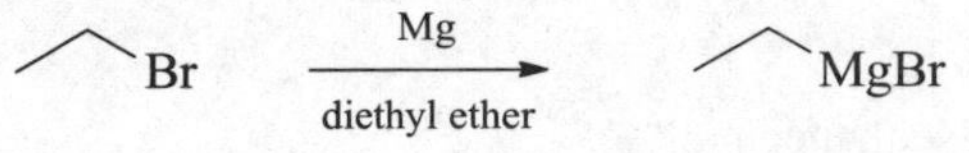

Ethyl bromide   Ethylmagnesium bromide

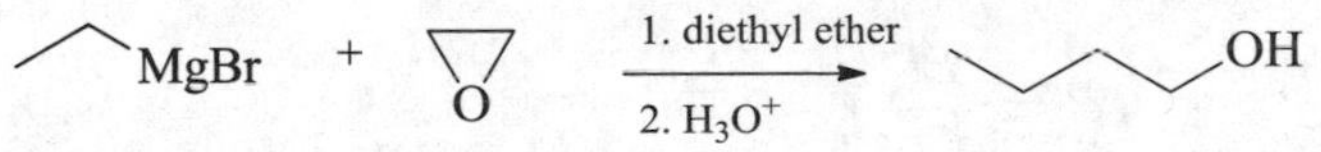

Ethylene oxide   1-Butanol

(d) Primary alcohols may be prepared by reduction of the carboxylic acid having the same number of carbons. Among the reagents we have discussed, the only one that is effective in the reduction of carboxylic acids is lithium aluminum hydride. The four-carbon carboxylic acid butanoic acid is the proper substrate.

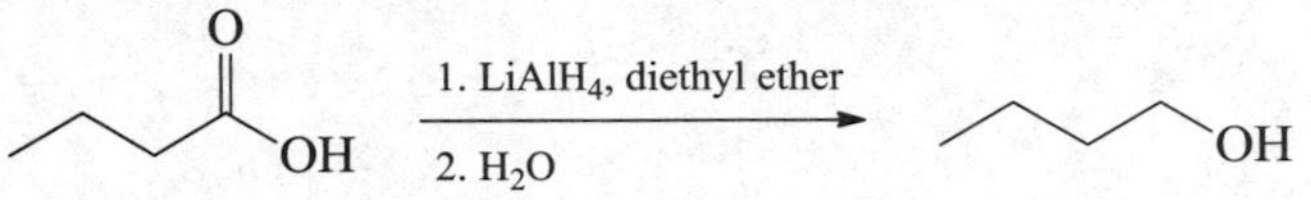

Butanoic acid   1-Butanol

(e) Because 1-butanol is a primary alcohol having four carbons, butanal must be the aldehyde that is hydrogenated. Suitable catalysts are nickel, palladium, platinum, and ruthenium.

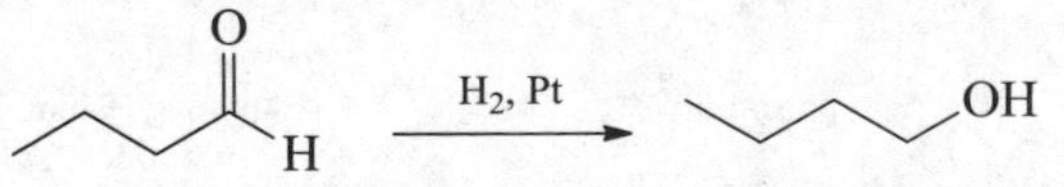

Butanal   1-Butanol

(f) Sodium borohydride reduces aldehydes and ketones efficiently. It does not reduce carboxylic acids, and its reaction with esters is too slow to be of synthetic value.

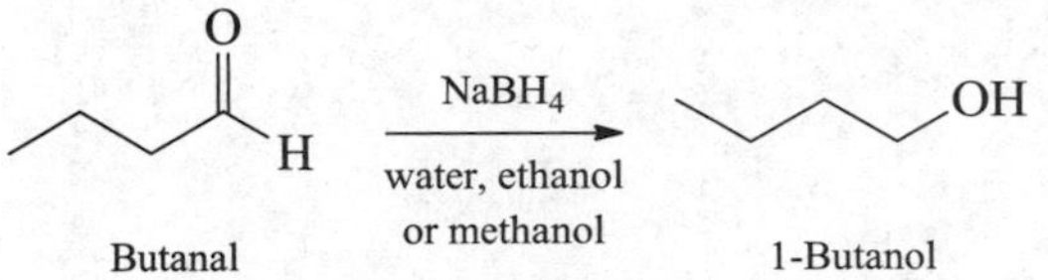

Butanal   1-Butanol

16.17 (a) Both (Z)- and (E)-2-butene yield 2-butanol on hydroboration–oxidation.

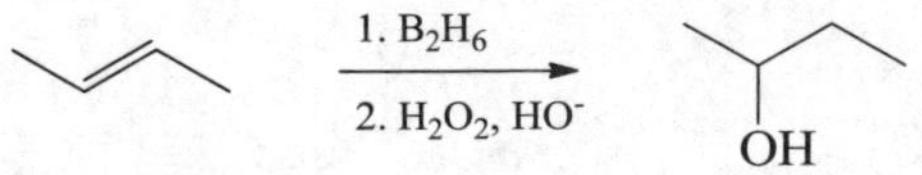

(E)-2-Butene
(or (Z)-butene)   2-Butanol

(b) Disconnection of one of the bonds to the carbon that bears the hydroxyl group reveals a feasible route using a Grignard reagent and propanal.

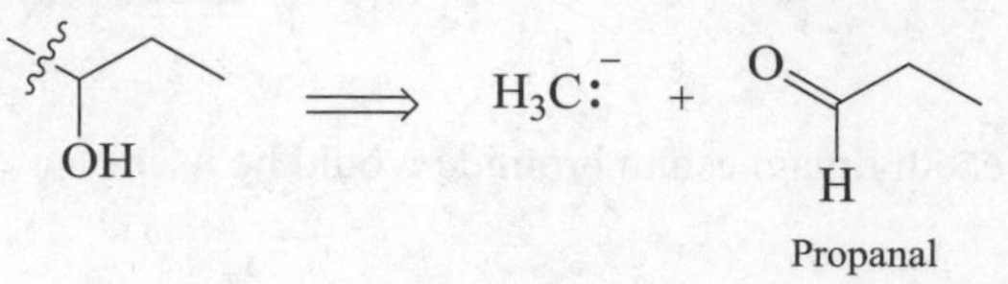

Propanal

The synthetic sequence is

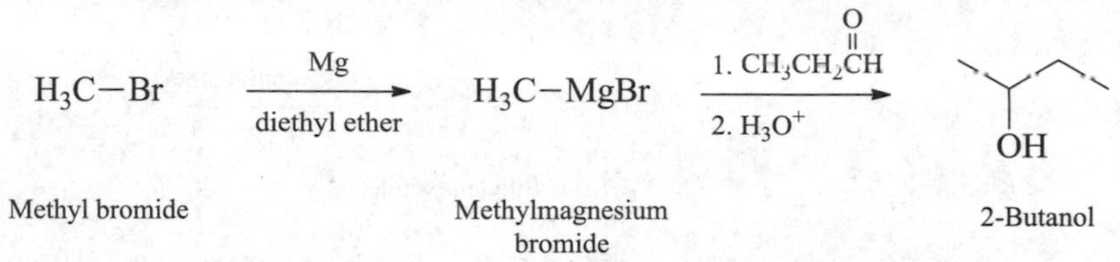

Methyl bromide   Methylmagnesium   2-Butanol
                                      bromide

(c) Another disconnection is related to a synthetic route using a Grignard reagent and acetaldehyde.

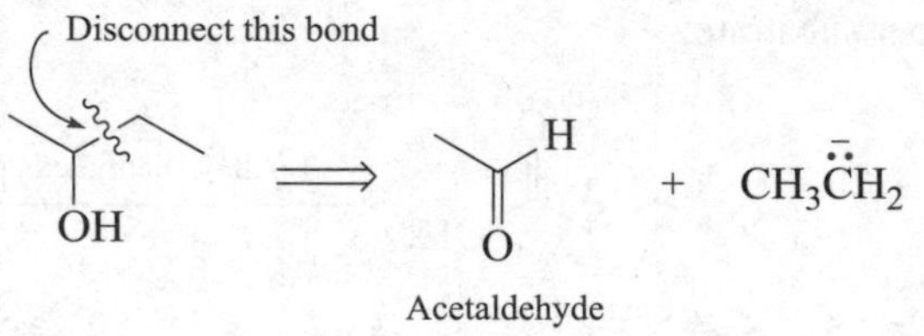

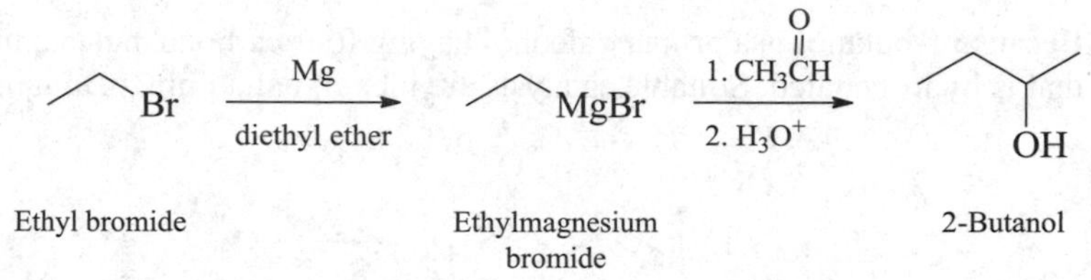

Disconnect this bond

Acetaldehyde

Ethyl bromide   Ethylmagnesium   2-Butanol
                                      bromide

(d–f) Because 2-butanol is a secondary alcohol, it can be prepared by reduction of a ketone having the same carbon skeleton, in this case 2-butanone. All three reducing agents indicated in the equations are satisfactory.

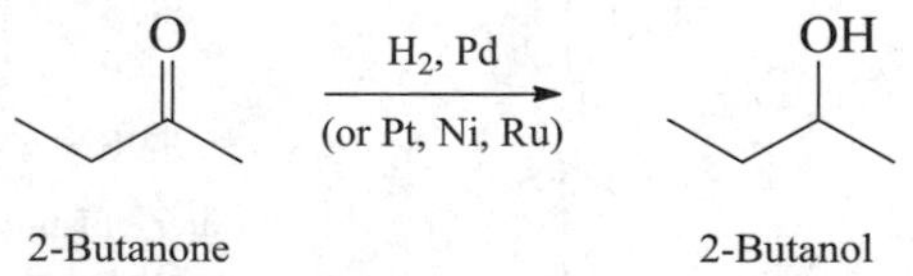

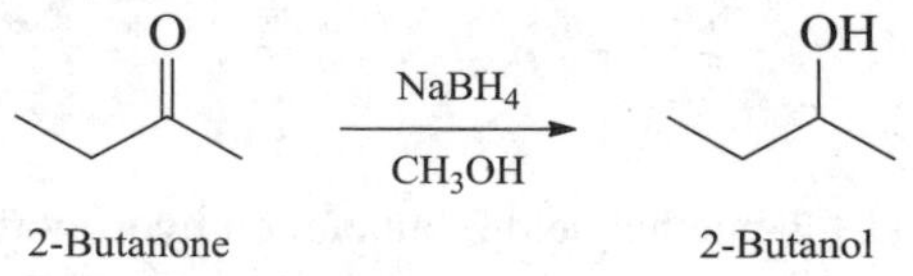

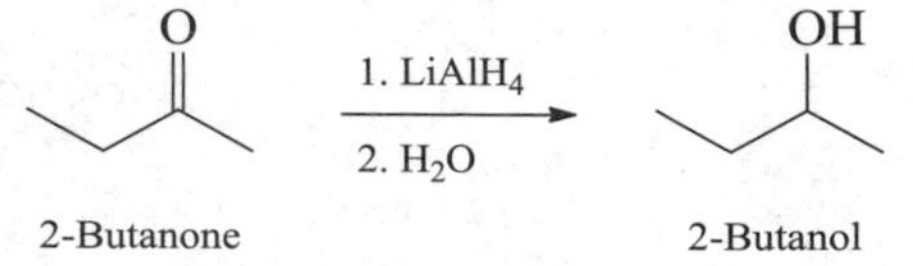

2-Butanone   2-Butanol

2-Butanone   2-Butanol

2-Butanone   2-Butanol

16.18 (*a*) All of the primary alcohols having the molecular formula $C_5H_{12}O$ may be prepared by reduction of aldehydes. The appropriate equations are

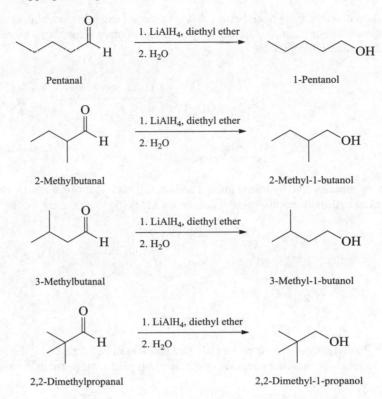

Pentanal   1. LiAlH₄, diethyl ether   2. H₂O   1-Pentanol

2-Methylbutanal   1. LiAlH₄, diethyl ether   2. H₂O   2-Methyl-1-butanol

3-Methylbutanal   1. LiAlH₄, diethyl ether   2. H₂O   3-Methyl-1-butanol

2,2-Dimethylpropanal   1. LiAlH₄, diethyl ether   2. H₂O   2,2-Dimethyl-1-propanol

(*b*) The secondary alcohols having the molecular formula $C_5H_{12}O$ may be prepared by reduction of ketones.

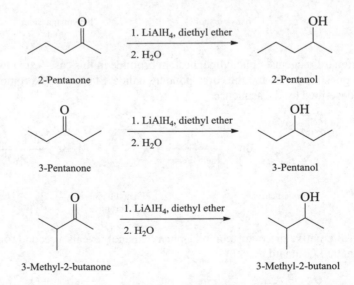

2-Pentanone   1. LiAlH₄, diethyl ether   2. H₂O   2-Pentanol

3-Pentanone   1. LiAlH₄, diethyl ether   2. H₂O   3-Pentanol

3-Methyl-2-butanone   1. LiAlH₄, diethyl ether   2. H₂O   3-Methyl-2-butanol

(*c*) As with the reduction of aldehydes in part (*a*), reduction of carboxylic acids yields primary alcohols. For example, 1-pentanol may be prepared by reduction of pentanoic acid.

Pentanoic acid   1. LiAlH₄, diethyl ether   2. H₂O   1-Pentanol

The remaining primary alcohols, 2-methyl-1-butanol, 3-methyl-1-butanol, and 2,2-dimethyl-1-propanol, may be prepared in the same way.

(*d*) As with carboxylic acids, esters may be reduced using lithium aluminum hydride to give primary alcohols. For example, 2,2-dimethyl-1-propanol may be prepared by reduction of methyl 2,2-dimethylpropanoate.

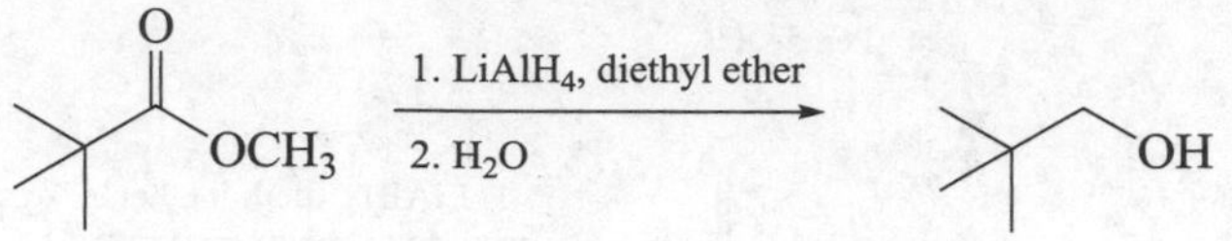

Methyl
2,2-dimethylpropanoate

2,2-Dimethyl-1-propanol

16.19 Glucose contains five hydroxyl groups and an aldehyde functional group. Its hydrogenation will not affect the hydroxyl groups but will reduce the aldehyde to a primary alcohol.

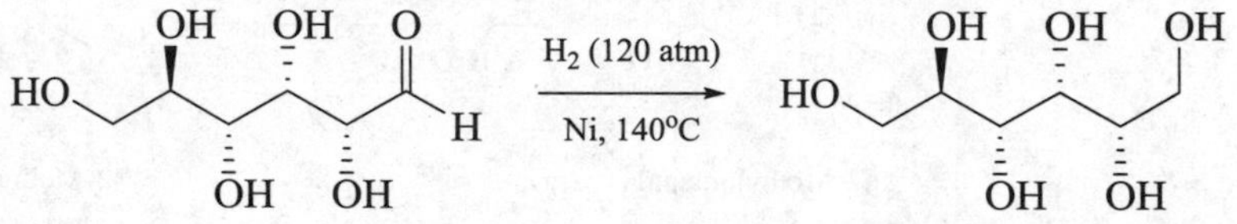

Glucose

Sorbitol

16.20 (*a*) 1-Phenylethanol is a secondary alcohol and so can be prepared by the reaction of a Grignard reagent with an aldehyde. One combination is phenylmagnesium bromide and ethanal (acetaldehyde).

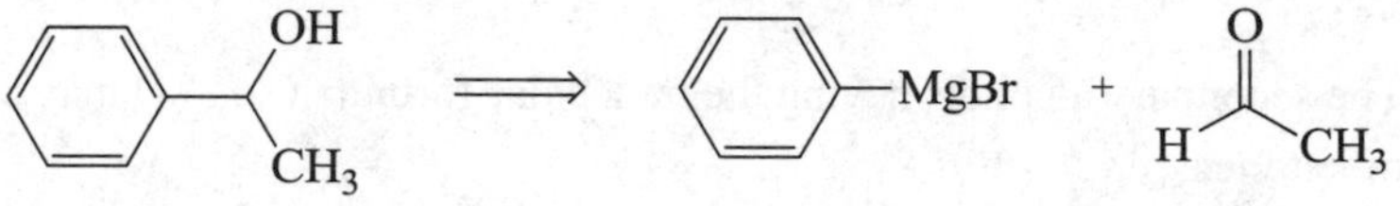

1-Phenylethanol

Phenylmagnesium
bromide

Ethanal
(acetaldehyde)

Grignard reagents—phenylmagnesium bromide in this case—are always prepared by reaction of magnesium metal and the corresponding halide. Starting with bromobenzene, a suitable synthesis is described by the sequence

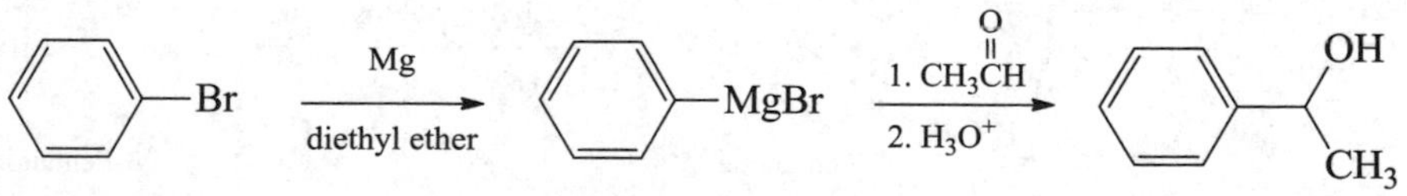

Bromobenzene

Phenylmagnesium
bromide

1-Phenylethanol

(*b*) An alternative disconnection of 1-phenylethanol reveals a second route using benzaldehyde and a methyl Grignard reagent.

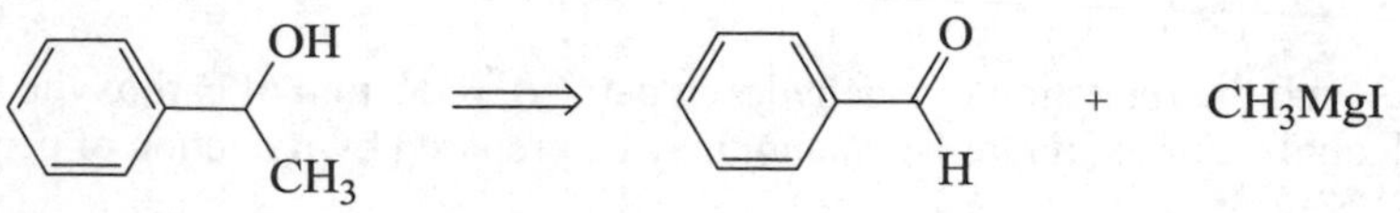

1-Phenylethanol

Benzaldehyde

Methylmagnesium
iodide

Equations representing this approach are

$$CH_3I \xrightarrow[\text{diethyl ether}]{Mg} CH_3MgI \xrightarrow[\text{2. } H_3O^+]{\text{1. } C_6H_5\overset{O}{\overset{\|}{C}}H}$$

Iodomethane   Methylmagnesium   1-Phenylethanol
                         iodide

(*c*) Aldehydes are, in general, obtainable by oxidation of the corresponding primary alcohol. By recognizing that benzaldehyde can be obtained by oxidation of benzyl alcohol with PCC, we write

$$\xrightarrow[\text{CH}_2\text{Cl}_2]{\text{PCC}} \qquad \xrightarrow[\text{2. } H_3O^+]{\text{1. } CH_3MgI, \text{ diethyl ether}}$$

Benzyl alcohol   Benzaldehyde   1-Phenylethanol

(*d*) As listed in Table 12.1, acetophenone is a frequently encountered common name for methyl phenyl ketone. The conversion of acetophenone to 1-phenylethanol is a reduction.

$$\xrightarrow{\text{reducing agent}}$$

Acetophenone   1-Phenylethanol

Any of a number of reducing agents could be used. These include
  1. $NaBH_4$, CH_3OH
  2. $LiAlH_4$ in diethyl ether, then H_2O
  3. H_2 and a Pt, Pd, Ni, or Ru catalyst

(*e*) Benzene can be employed as the starting material in a synthesis of 1-phenylethanol. Friedel–Crafts acylation of benzene gives acetophenone, which can then be reduced as in part (*d*).

$$+ \quad \underset{\text{Cl}}{\overset{O}{\overset{\|}{C}}}\text{CH}_3 \xrightarrow{\text{AlCl}_3}$$

Benzene   Acetyl chloride   Acetophenone

Acetic anhydride ($CH_3\overset{O}{\overset{\|}{C}}O\overset{O}{\overset{\|}{C}}CH_3$) can be used in place of acetyl chloride.

16.21 2-Phenylethanol is an ingredient in many perfumes, to which it imparts a rose-like fragrance. Numerous methods have been employed for its synthesis.

(*a*) As a primary alcohol having two more carbon atoms than bromobenzene, it can be formed by reaction of a Grignard reagent, phenylmagnesium bromide, with ethylene oxide.

$$\Longrightarrow \qquad -\text{MgBr} \quad + \quad \underset{O}{\triangle}$$

2-Phenylethanol   Phenylmagnesium   Ethylene oxide
                                         bromide

The desired reaction sequence is therefore

Bromobenzene   Phenylmagnesium   2-Phenylethanol
                                  bromide

(b) Hydration of styrene with a regioselectivity contrary to that of Markovnikov's rule is required. This is accomplished readily by hydroboration–oxidation.

Styrene   2-Phenylethanol

(c) Reduction of aldehydes yields primary alcohols.

2-Phenylethanal   2-Phenylethanol

Among the reducing agents that could be (and have been) used are
  1. $NaBH_4$, CH_3OH
  2. $LiAlH_4$ in diethyl ether, then H_2O
  3. H_2 and a Pt, Pd, Ni, or Ru catalyst

(d) The only reagent that is suitable for the direct reduction of carboxylic acids to primary alcohols is lithium aluminum hydride.

2-Phenylethanoic acid   2-Phenylethanol

16.22 (a) The alcohol needs to be converted to a substrate with a leaving group for a nucleophilic substitution reaction with KSH. An allyl bromide formed from free radical halogenation is suitable for this reaction.

Propene   2-Propen-1-thiol

(b) The alcohol, 1-hexanol, can be synthesized by the reaction of ethylene oxide with the Grignard reagent prepared from 1-bromobutane.

1-Bromobutane   1-Hexanol

(c) 2-Hexanol can be synthesized by the reaction of the Grignard reagent used in part (b) with acetaldehyde.

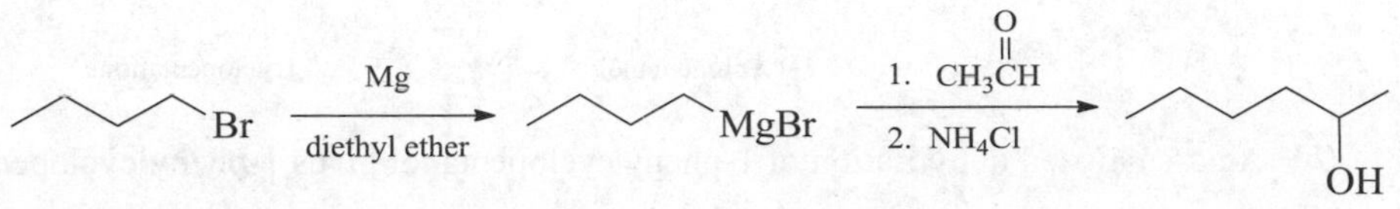

1-Bromobutane   2-Hexanol

(d) The product is a vicinal diol. Vicinal diols can be prepared from alkenes by dihydroxylation with osmium tetroxide. The procedure used here includes a co-oxidant, *tert*-butyl hydroperoxide, and catalytic osmium tetroxide. The alkene that is required, 2-methylpropene, is prepared by dehydration of *tert*-butyl alcohol.

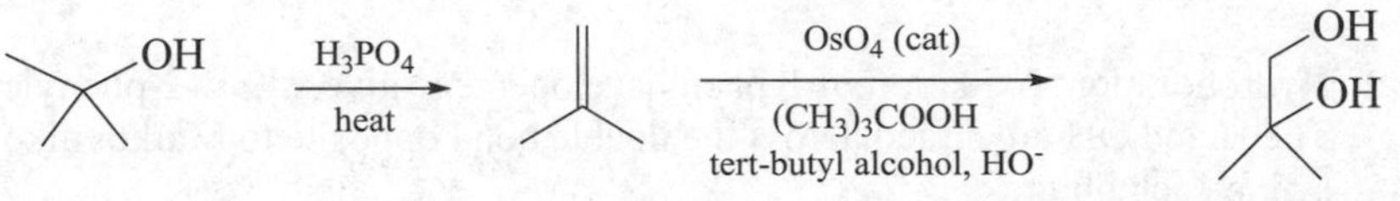

tert-Butyl alcohol   2-Methyl-1,2-propanediol

(e) The primary alkyl chloride can be prepared from the alcohol, which in turn can be made from the addition of the Grignard reagent derived from bromobenzene to ethylene oxide.

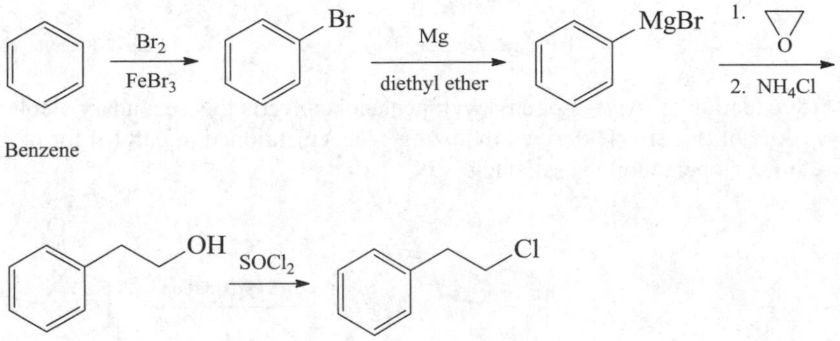

Benzene

1-Chloro-2-phenylethane

16.23 (a) Because 1-phenylcyclopentanol is a tertiary alcohol, a likely synthesis would involve reaction of a ketone and a Grignard reagent. Thus, a reasonable last step is treatment of cyclopentanone with phenylmagnesium bromide.

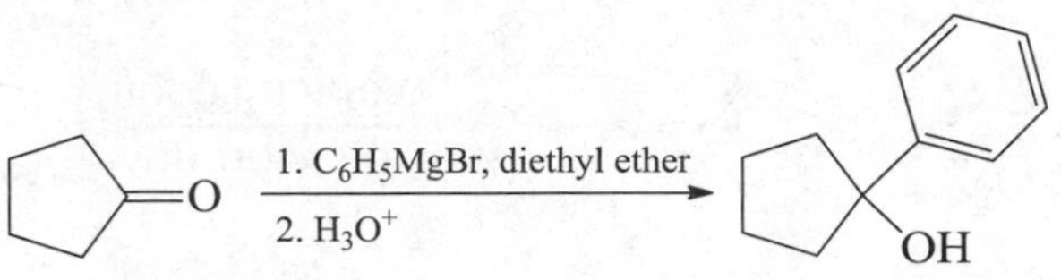

Cyclopentanone   1-Phenylcyclopentanol

Cyclopentanone is prepared by oxidation of cyclopentanol. Any one of a number of oxidizing agents would be suitable. These include PDC or PCC in CH_2Cl_2 or chromic acid (H_2CrO_4) generated from $Na_2Cr_2O_7$ in aqueous sulfuric acid.

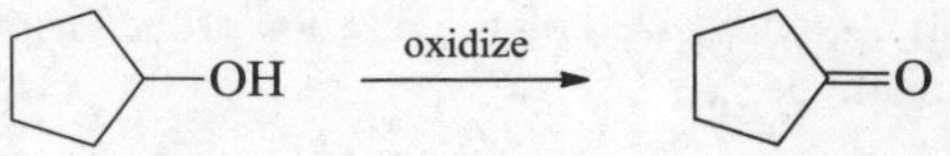

Cyclopentanol   Cyclopentanone

(b) Acid-catalyzed dehydration of 1-phenylcyclopentanol gives 1-phenylcyclopentene.

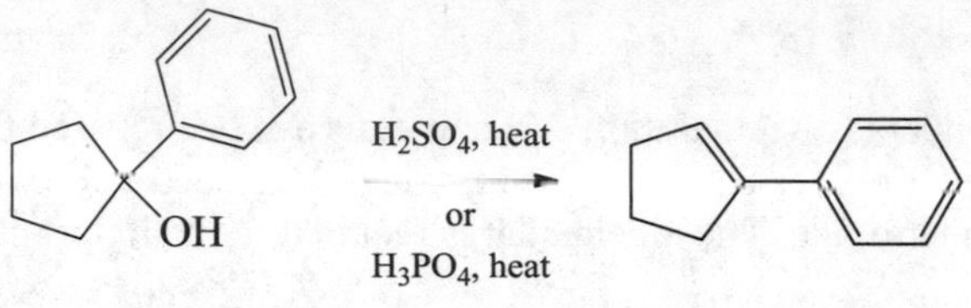

1-Phenylcyclopentanol   1-Phenylcyclopentene

(c) Hydroboration–oxidation of 1-phenylcyclopentene gives *trans*-2-phenylcyclopentanol. The H and OH are added across the double bond opposite to Markovnikov's rule and syn to each other.

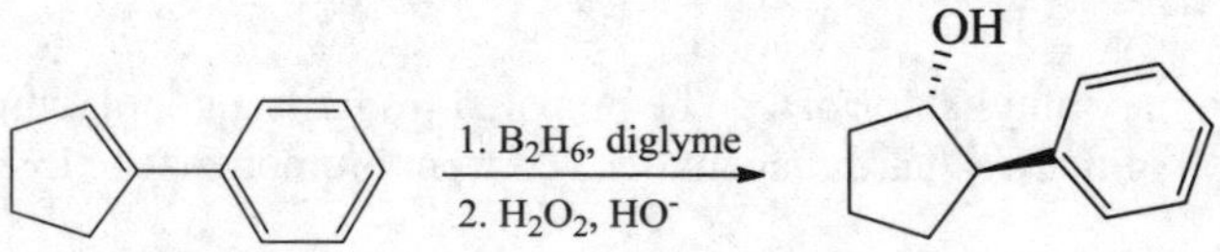

1-Phenylcyclopentene   *trans*-2-Phenylcyclopentanol

(d) Oxidation of *trans*-2-phenylcyclopentanol converts this secondary alcohol to the desired ketone. Any of the Cr(VI)-derived oxidizing agents mentioned in part (a) for oxidation of cyclopentanol to cyclopentanone is satisfactory.

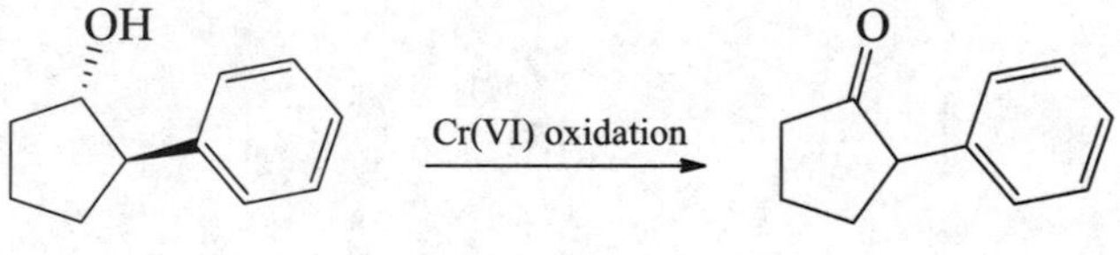

trans-2-Phenylcyclopentanol   2-Phenylcyclopentanone

(e) The standard procedure for preparing *cis*-1,2-diols is by dihydroxylation of alkenes with osmium tetraoxide.

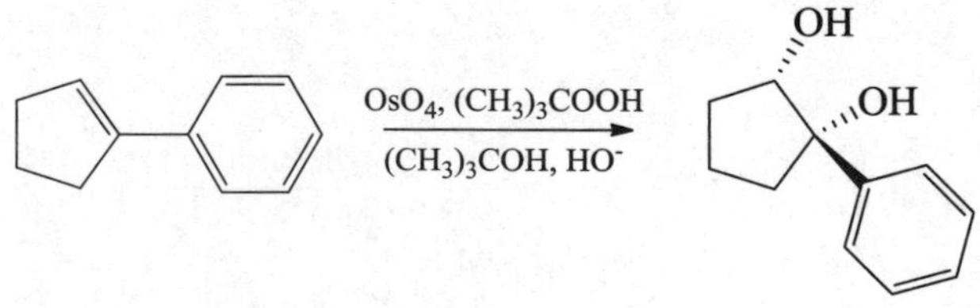

1-Phenylcyclopentene   1-Phenyl-*cis*-1,2-cyclopentanediol

(f) The first reaction step for the desired compound is accomplished either by ozonolysis of 1-phenylcyclopentene from part (b):

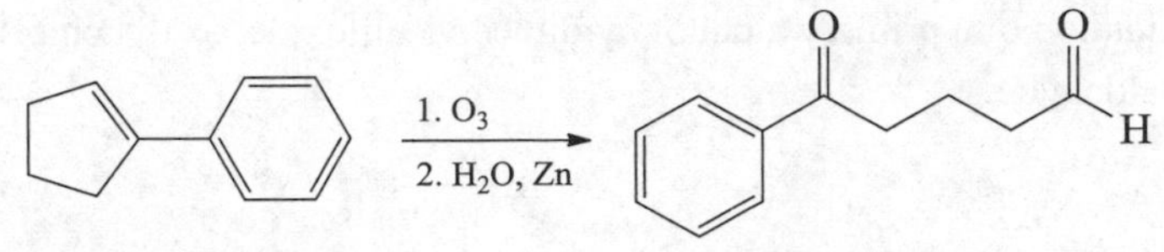

1-Phenylcyclopentene → 5-Oxo-1-phenyl-1-pentanone

or by periodic acid cleavage of the diol in part (*e*):

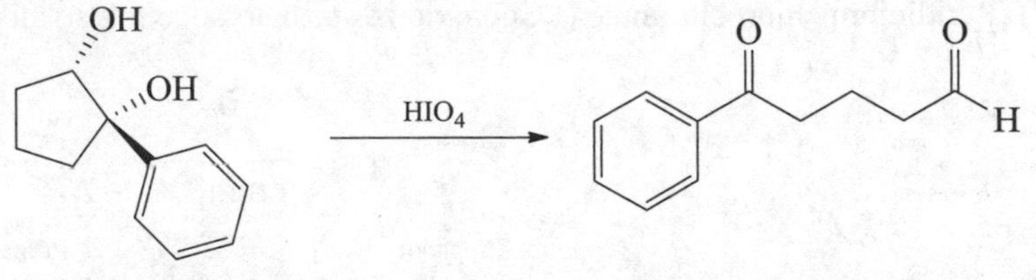

1-Phenyl-*cis*-1,2-cyclopentanediol → 5-Oxo-1-phenyl-1-pentanone

Reduction of both carbonyls in 5-oxo-1-phenyl-1-pentanone gives the desired diol.

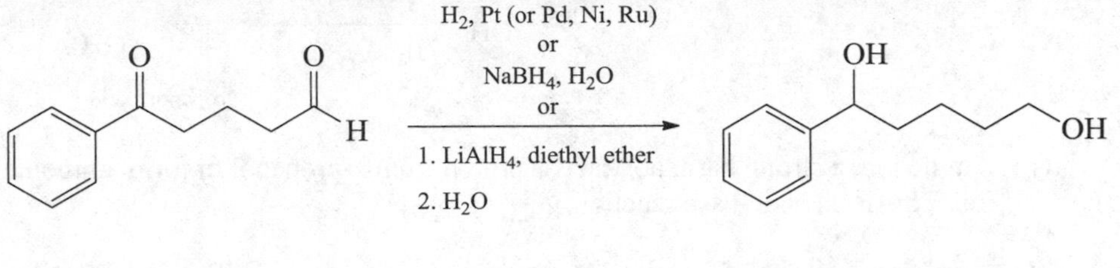

5-Oxo-1-phenyl-1-pentanone → 1-Phenyl-1,5-pentanediol

Reactions

16.24 (*a*) Chromic acid (H_2CrO_4) is used because we do not want the oxidation to stop at the aldehyde stage. Chromic acid can be generated using $Na_2Cr_2O_7$ in sulfuric acid.

(*b*) Secondary alcohols can be oxidized to ketones with chromic acid, pyridinium chlorochromate (PCC), pyridinium dichromate (PDC), or by dimethyl sulfoxide, oxalyl chloride ($COCl)_2$, then triethylamine.

(*c*) Oxidation of a primary alcohol to an aldehyde can be carried out with PDC or PCC, or by dimethyl sulfoxide, oxalyl chloride ($COCl)_2$, then triethylamine, but not with chromic acid.

(*d*) This reaction is oxidative cleavage of a vicinal diol, and is carried out with periodic acid, HIO_4.

16.25 (*a, b*) Primary alcohols react in two different ways on being heated with acid catalysts: they can condense to form dialkyl ethers or undergo dehydration to yield alkenes. Ether formation is favored at lower temperature, and alkene formation is favored at higher temperature.

1-Propanol → Dipropyl ether + Water

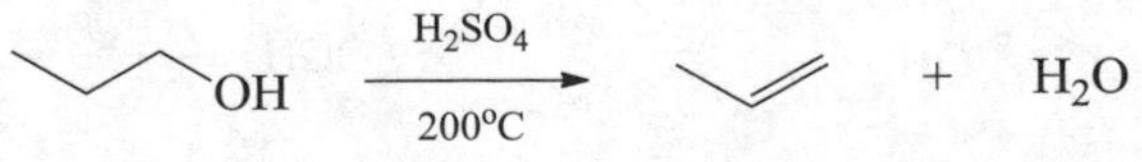

1-Propanol → Propene + Water

(*c*) Oxidation of a primary alcohol by dimethyl sulfoxide, oxalyl chloride (COCl)$_2$, then triethylamine gives an aldehyde.

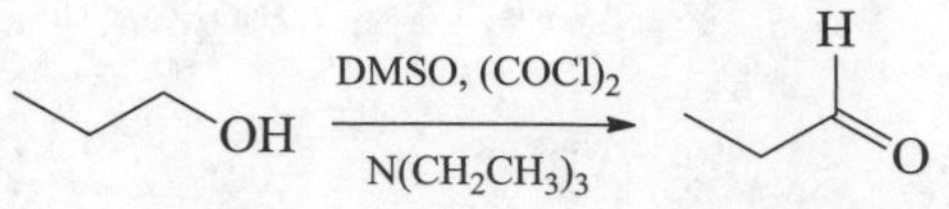

1-Propanol   Propanal

(*d*) Pyridinium chlorochromate (PCC) oxidizes primary alcohols to aldehydes.

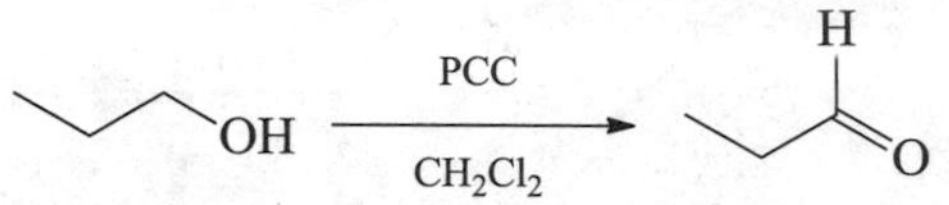

1-Propanol   Propanal

(*e*) Potassium dichromate in aqueous sulfuric acid oxidizes primary alcohols to carboxylic acids.

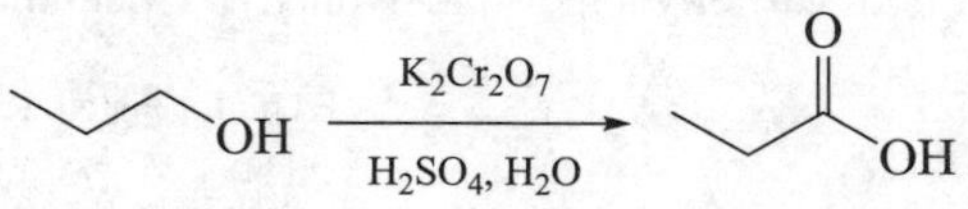

1-Propanol   Propanoic acid

(*f*) Amide ion, a strong base, abstracts a proton from 1-propanol to form ammonia and 1-propanolate ion. This is an acid–base reaction.

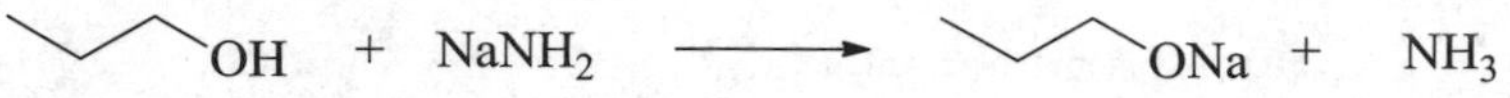

1-Propanol   Sodium amide   Sodium 1-propanolate   Ammonia

(*g*) With acetic acid and in the presence of an acid catalyst, 1-propanol is converted to its acetate ester.

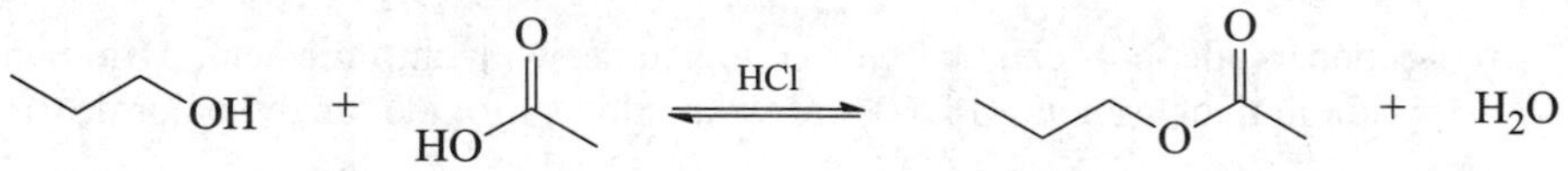

1-Propanol   Acetic acid   Propyl acetate   Water

This is an equilibrium process that slightly favors products.

(*h*) Alcohols react with *p*-toluenesulfonyl chloride to give *p*-toluenesulfonates.

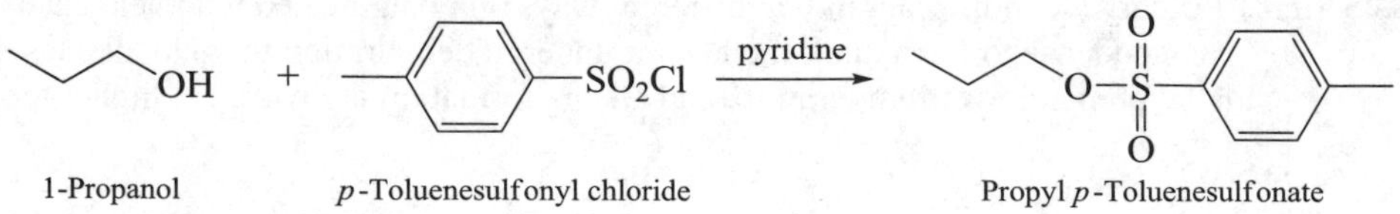

1-Propanol   *p*-Toluenesulfonyl chloride   Propyl *p*-Toluenesulfonate

(*i*) Acyl chlorides convert alcohols to esters.

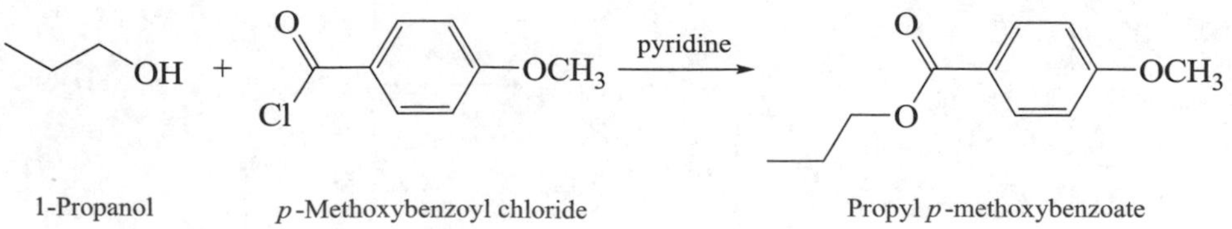

1-Propanol   *p*-Methoxybenzoyl chloride   Propyl *p*-methoxybenzoate (*j*) The reagent is benzoic anhydride. Acid anhydrides react with alcohols to give esters.

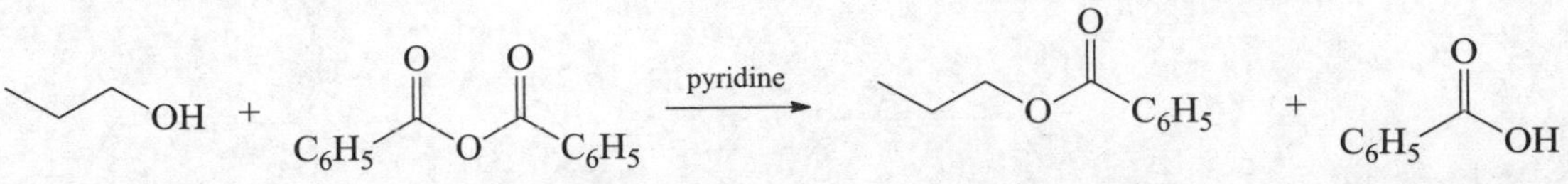

| 1-Propanol | Benzoic anhydride | | Propyl benzoate | | Benzoic acid |

(*k*) The reagent is succinic anhydride, a cyclic anhydride. Esterification occurs, but in this case the resulting ester and carboxylic acid functions remain part of the same molecule.

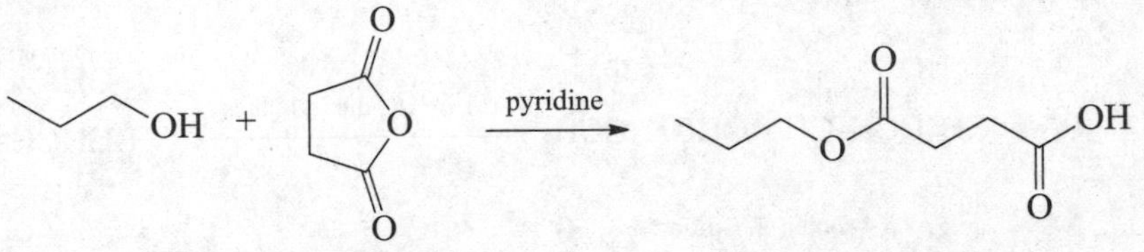

1-Propanol   Succinic anhydride   Hydrogen propyl succinate

16.26 (*a*) On being heated in the presence of sulfuric acid, tertiary alcohols undergo elimination.

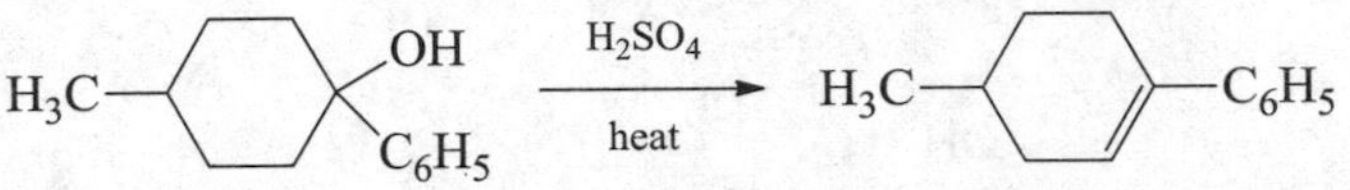

4-Methyl-1-
phenylcyclohexanol

4-Methyl-1-
phenylcyclohexene (81%)

(*b*) The combination of reagents specified converts alkenes to vicinal diols.

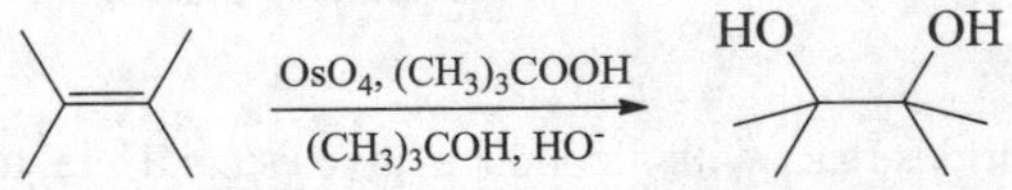

2,3-Dimethyl-2-butene

2,3-Dimethyl-2-butanediol
(72%)

(*c*) Hydroboration–oxidation of the double bond takes place with a regioselectivity that is opposite to Markovnikov's rule. The elements of water are added in a stereospecific syn fashion.

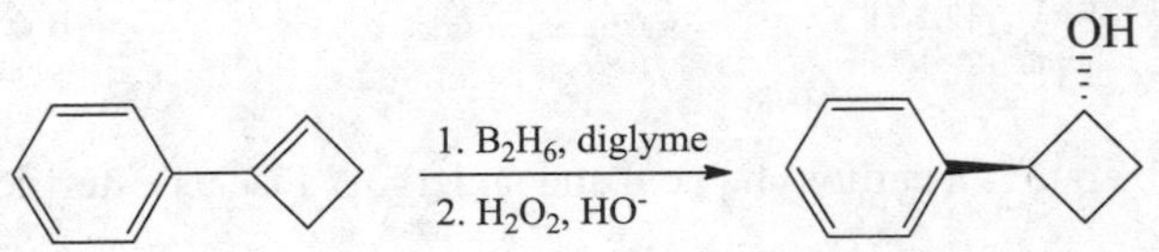

1-Phenylcyclobutene

trans-2-Phenylcyclobutanol
(82%)

(*d*) Lithium aluminum hydride reduces carboxylic acids to primary alcohols but does not reduce carbon–carbon double bonds.

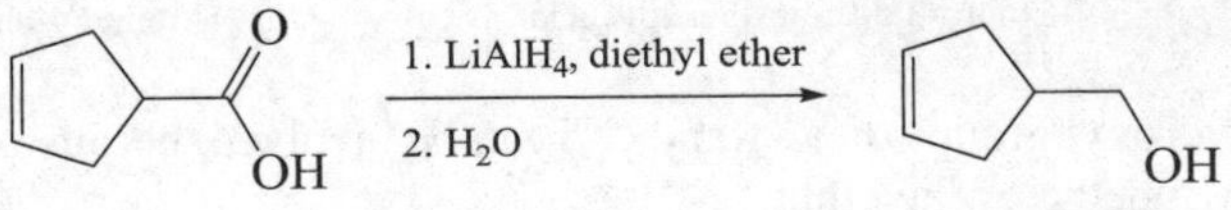

Cyclopentene-4-
carboxylic acid

(3-Cyclopentenyl)-
methanol

(e) Chromic acid oxidizes the secondary alcohol to the corresponding ketone but does not affect the triple bond.

3-Octyn-2-ol   3-Octyn-2-one (80%)

(f) Lithium aluminum hydride reduces carbonyl groups efficiently but does not normally react with double bonds.

$$\underset{\text{4-Octen-2,7-dione}}{CH_3\overset{O}{\overset{\|}{C}}CH_2CH=CHCH_2\overset{O}{\overset{\|}{C}}CH_3} \xrightarrow[\text{2. } H_2O]{\text{1. LiAlH}_4\text{, diethyl ether}} \underset{\text{4-Octen-2,7-diol (75%)}}{CH_3\overset{OH}{\overset{|}{C}H}CH_2CH=CHCH_2\overset{OH}{\overset{|}{C}H}CH_3}$$

(g) Alcohols react with acyl chlorides to yield esters. The O—H bond is broken in this reaction; the C—O bond of the alcohol remains intact on ester formation.

trans-3-Methylcyclohexanol   3,5-Dinitrobenzoyl chloride   trans-3-Methylcyclohexyl
                                                                 3,5-dinitrobenzoate (74%)

(h) Acid anhydrides react with alcohols to give esters. Here, too, the spatial orientation of the C—O bond remains intact.

exo-Bicyclo[2.2.1]-   Acetic anhydride   exo-Bicyclo[2.2.1]hept-   Acetic acid
heptan-2-ol   2-yl acetate (90%)

(i) The substrate is a carboxylic acid and undergoes Fischer esterification with methanol.

4-Chloro-3,5-dinitrobenzoic acid   Methyl 4-chloro-3,5-dinitrobenzoate (96%)

16.27 Only the hydroxyl groups on C-1 and C-4 can be involved, because only these two can lead to a five-membered cyclic ether.

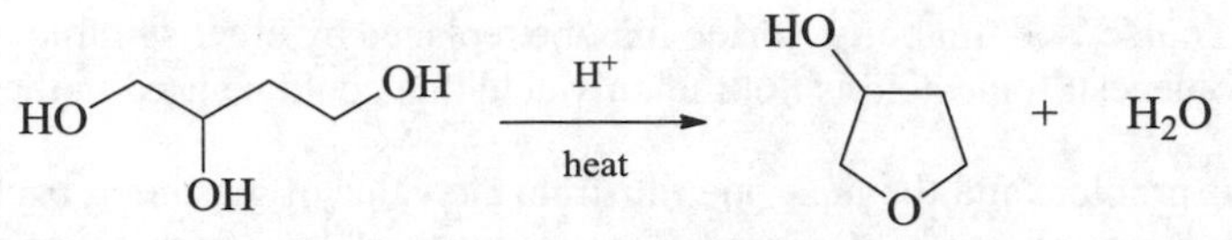

1,2,4-Butanetriol   3-Hydroxyoxolane ($C_4H_8O_2$)

Any other combination of hydroxyl groups would lead to a strained three-membered or four-membered ring and is unfavorable under conditions of acid catalysis.

16.28 (*a*) Cysteine contains an —SH group and is a thiol. Oxidation of thiols gives rise to disulfides.

$$2RSH \xrightarrow{\text{oxidize}} RSSR$$

Thiol   Disulfide

Biological oxidation of cysteine gives the disulfide cystine.

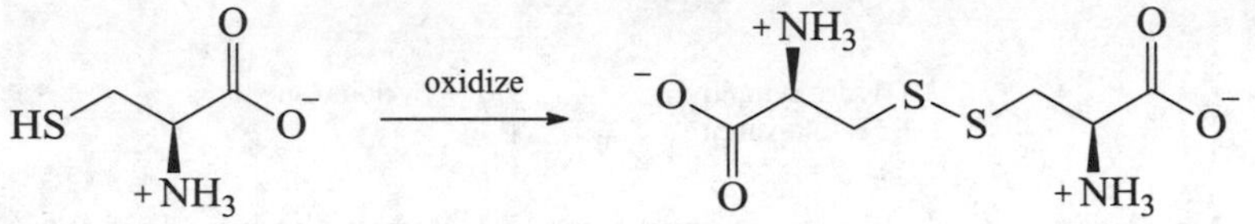

Cysteine   Cystine

(*b*) Oxidation of a thiol yields a series of acids, including a sulfinic acid and a sulfonic acid.

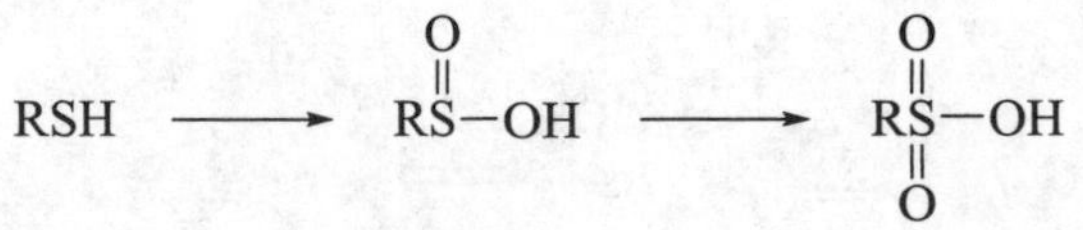

Thiol   Sulfinic acid   Sulfonic acid

Biological oxidation of cysteine can yield, in addition to the disulfide cystine, cysteine sulfinic acid and the sulfonic acid cysteic acid.

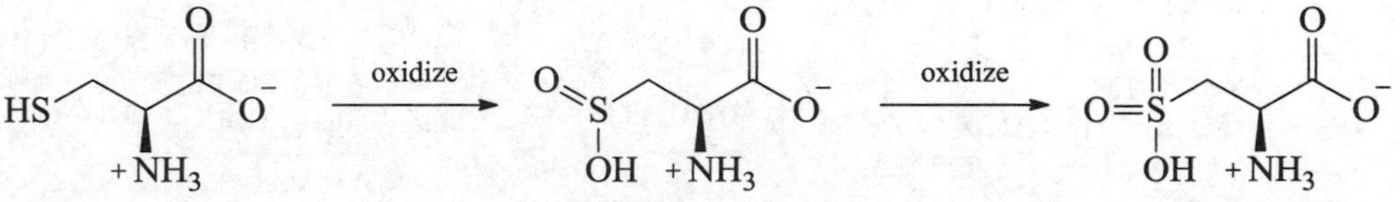

Cysteine   Cysteine sulfinic acid ($C_3H_7NO_4S$)   Cysteic acid ($C_3H_7NO_5S$)

Synthesis

16.29 (*a*) The task of converting a ketone to an alkene requires first the reduction of the ketone to an alcohol and then dehydration. In practice, the two-step transformation has been carried out in 54% yield by treating the ketone with sodium borohydride and then heating the resulting alcohol with *p*-toluenesulfonic acid.

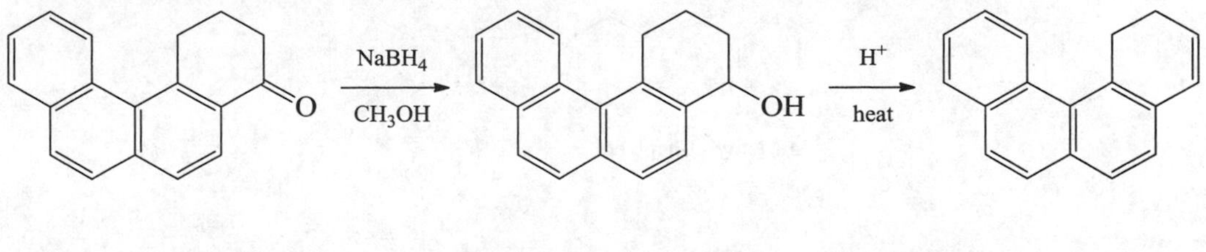

Of course, sodium borohydride may be replaced by other suitable reducing agents, and *p*-toluenesulfonic acid is not the only acid that could be used in the dehydration step.

(*b*) This problem and the next one illustrate the value of reasoning backward. The desired product, cyclohexanol, can be prepared cleanly from cyclohexanone. Once cyclohexanone is recognized to be a key intermediate, the synthetic pathway becomes apparent—what is needed is a method to convert the indicated starting material to cyclohexanone.

The reagent ideally suited to this task is periodic acid. The synthetic sequence to be followed is therefore

1-(Hydroxymethyl)-   Cyclohexanone   Cyclohexanol
cyclohexanol

(*c*) No direct method allows a second hydroxyl group to be introduced at C-2 of 1-phenylcyclohexanol in a single step. We recognize the product as a vicinal diol and recall that such compounds are available by dihydroxylation of alkenes. This tells us that we must first dehydrate the tertiary alcohol, then dihydroxylate the resulting alkene.

The syn stereoselectivity of the dihydroxylation step ensures that the product will have its hydroxyl groups cis, as the problem requires.

1-Phenylcyclohexanol   1-Phenylcyclohexene   1-Phenyl-*cis*-1,2-
cyclohexanediol

16.30 Because the target molecule is an eight-carbon secondary alcohol and the problem restricts our choices of starting materials to alcohols of five carbons or fewer, we are led to consider building up the carbon chain by a Grignard reaction.

4-Methyl-3-heptanol

The disconnection shown leads to a three-carbon aldehyde and a five-carbon Grignard reagent. Starting with the corresponding alcohols, the following synthetic scheme seems reasonable.

First, propanal is prepared.

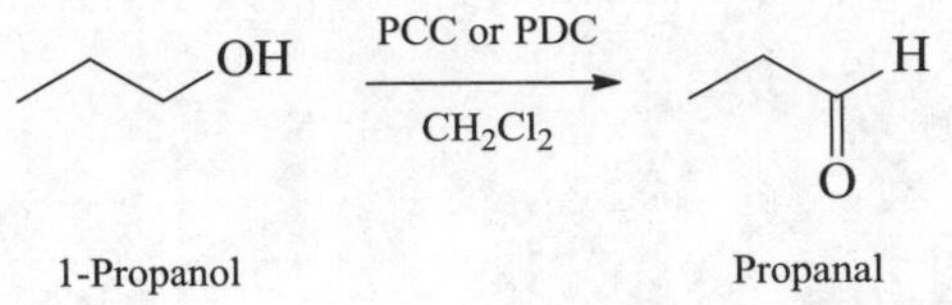

1-Propanol   Propanal

After converting 2-pentanol to its bromo derivative, a solution of the Grignard reagent is prepared.

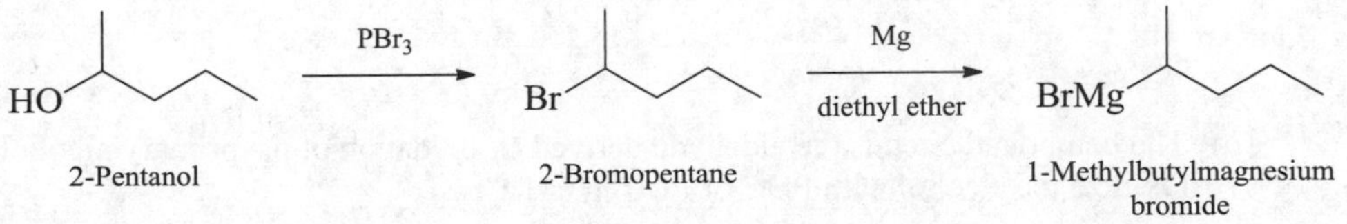

2-Pentanol   2-Bromopentane   1-Methylbutylmagnesium
                                                              bromide

Reaction of the Grignard reagent with the aldehyde yields the desired 4-methyl-3-heptanol.

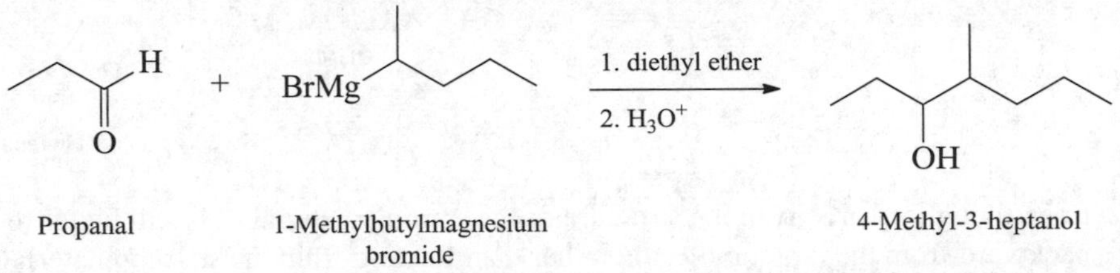

Propanal   1-Methylbutylmagnesium   4-Methyl-3-heptanol
                        bromide

16.31 (*a*) Retrosynthetically, we can see that the cis carbon–carbon double bond is available by hydrogenation of the corresponding alkyne over the Lindlar catalyst.

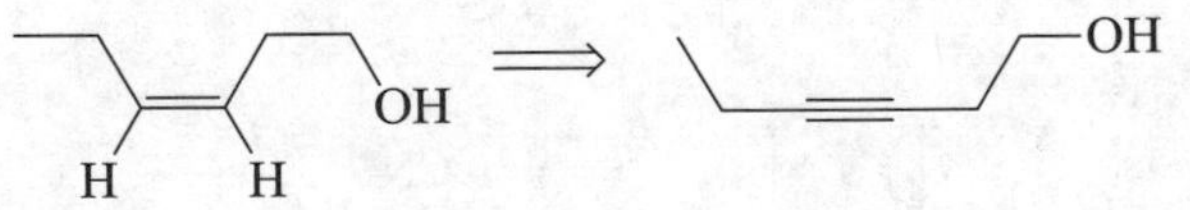

The —CH_2CH_2OH unit can be appended to an alkynide anion by reaction with ethylene oxide.

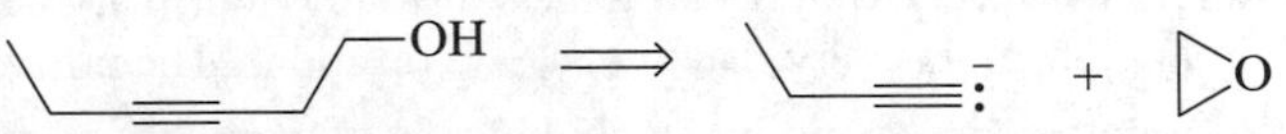

The alkynide anion is derived from 1-butyne by alkylation of acetylene. This analysis suggests the following synthetic sequence:

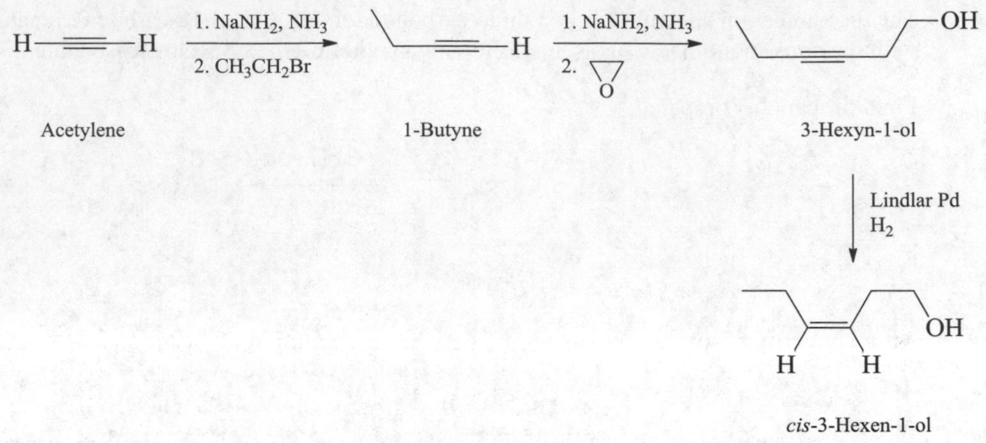

(b) The compound cited is the aldehyde derived by oxidation of the primary alcohol in part (a). Oxidize the alcohol with PDC or PCC in CH_2Cl_2.

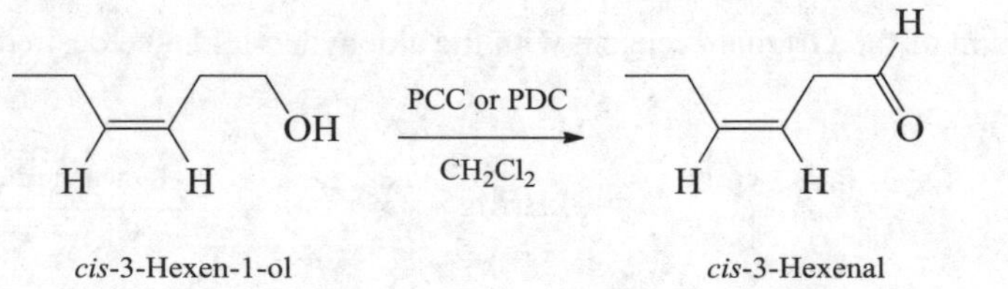

16.32 Even though we are given the structure of the starting material, it is still better to reason backward from the target molecule rather than forward from the starting material.

The desired product contains a cyano (—CN) group. The only method we have seen so far for introducing such a function into a molecule is by nucleophilic substitution. The last step in the synthesis must therefore be

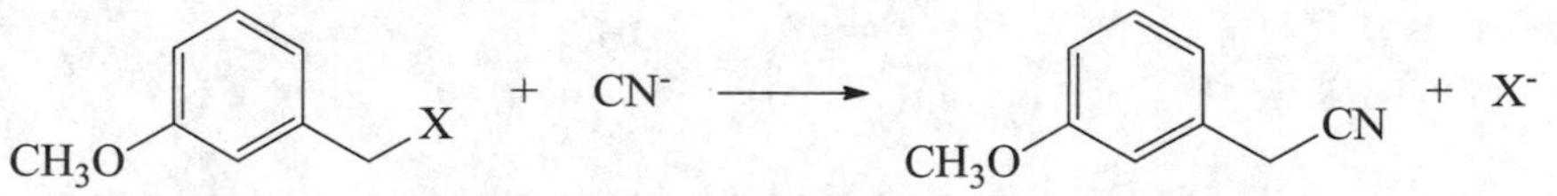

This step should work very well because the substrate is a primary benzylic halide, cannot undergo elimination, and is very reactive in S_N2 reactions. The primary benzylic halide can be prepared from the corresponding alcohol by any of a number of methods.

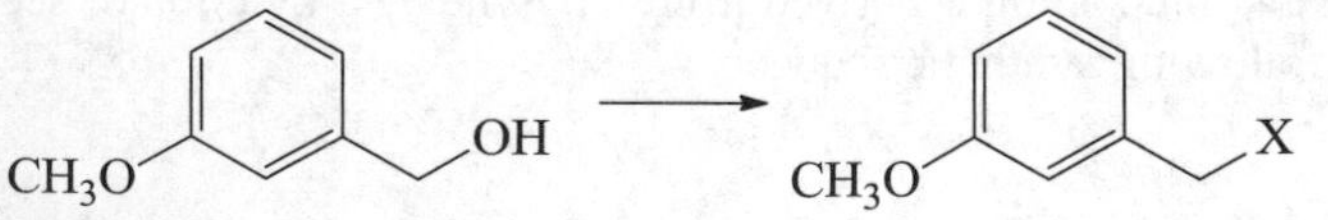

Suitable reagents include HBr, PBr₃, or SOCl₂.

Now we only need to prepare the primary alcohol from the given starting aldehyde, which is accomplished by reduction.

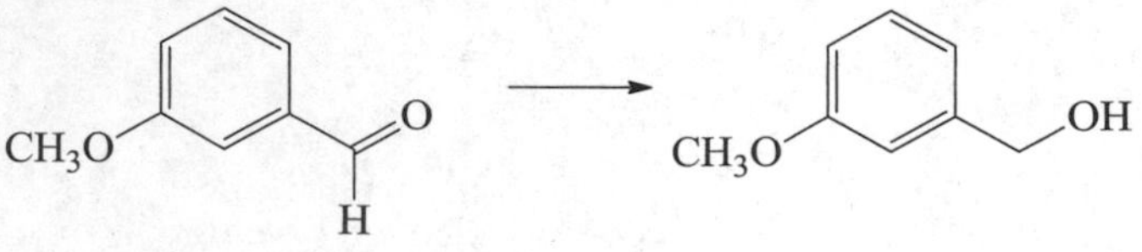

Reduction can be achieved by catalytic hydrogenation, with lithium aluminum hydride, or with sodium borohydride. The actual sequence of reactions as carried out is as shown.

m-Methoxybenzaldehyde — *m*-Methoxybenzyl alcohol — *m*-Methoxybenzyl bromide — *m*-Methoxybenzyl cyanide

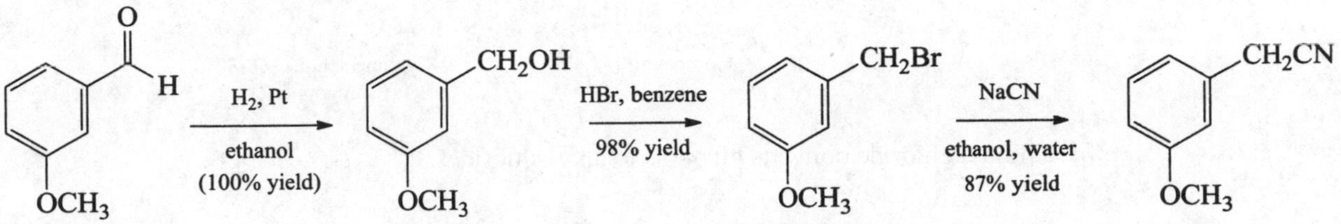

Another three-step synthesis, which is reasonable but does not involve an alcohol as an intermediate, is

m-Methoxybenzaldehyde — *m*-Methoxytoluene — *m*-Methoxybenzyl bromide — *m*-Methoxybenzyl cyanide

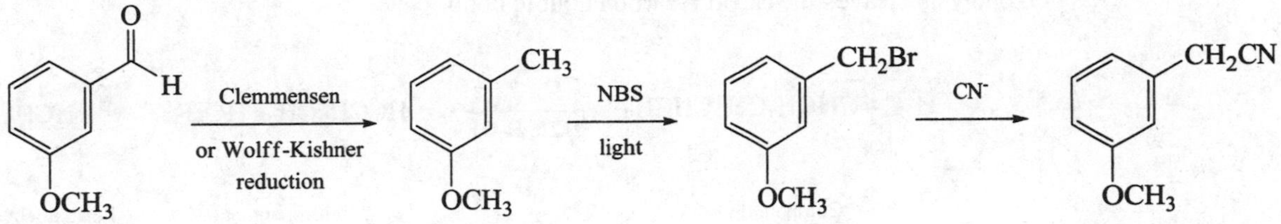

16.33 (*a*) Addition of hydrogen chloride to cyclopentadiene takes place by way of the most stable carbocation. In this case, it is an allylic carbocation.

3-Chlorocyclopentene (80-90%)
(Compound A)

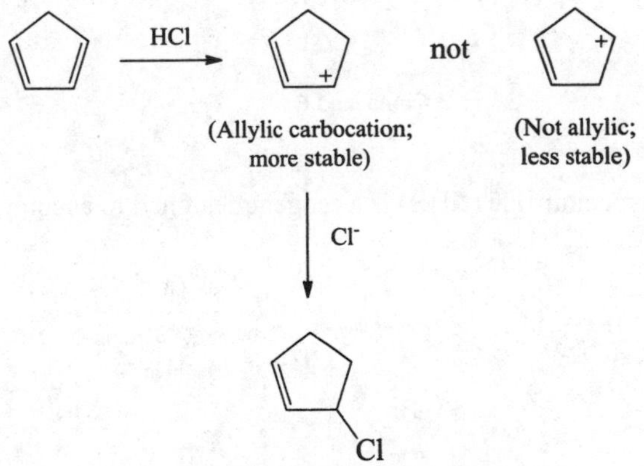

Hydrolysis of 3-chlorocyclopentene gives the corresponding alcohol. Sodium bicarbonate in water is a weakly basic solvolysis medium.

Compound A — 2-Cyclopenten-1-ol (88%) (Compound B)

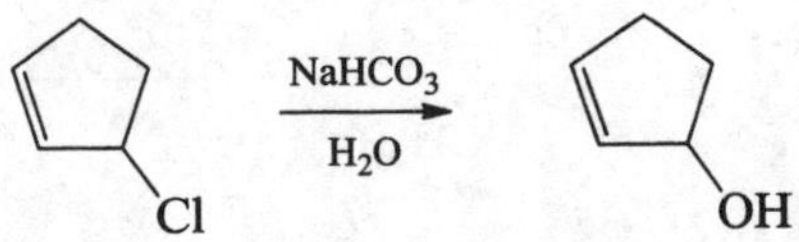

Oxidation of compound B (a secondary alcohol) gives the ketone 2-cyclopentenone.

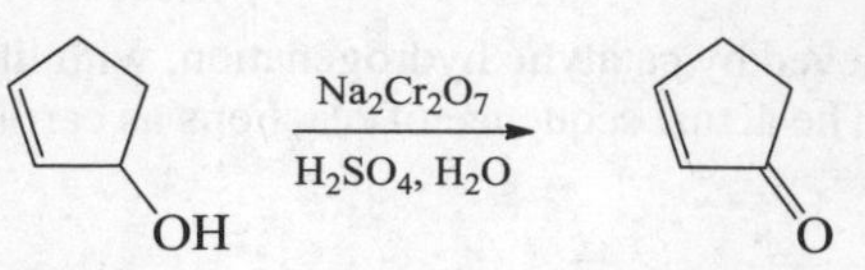

Compound B

2-Cyclopentenone (60-80%)
(Compound C)

(b) Thionyl chloride converts alcohols to alkyl chlorides.

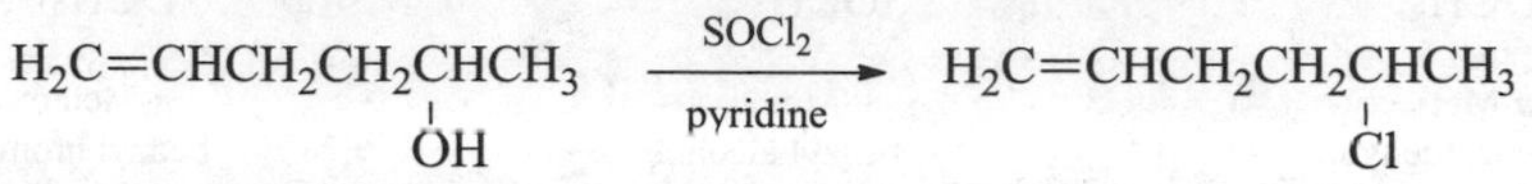

5-Hexen-2-ol

5-Chloro-1-hexene
(Compound D)

Ozonolysis cleaves the carbon–carbon double bond.

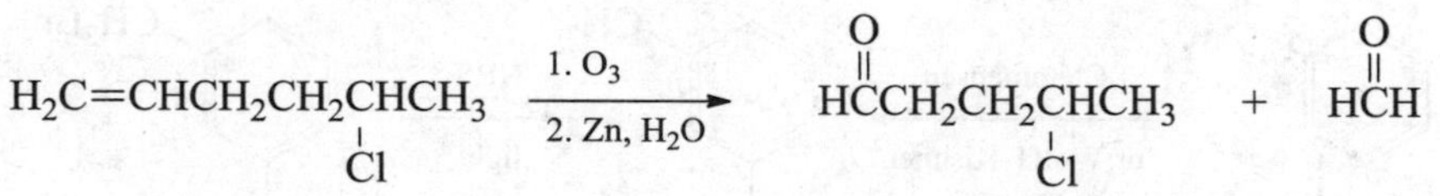

Compound D

4-Chloropentanal
(Compound E)

Formaldehyde

Reduction of compound E yields the corresponding alcohol.

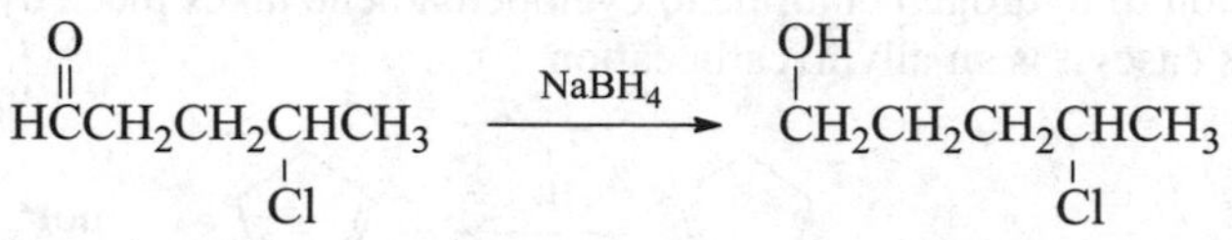

Compound E

4-Chloro-1-pentanol
(Compound F)

(c) *N*-Bromosuccinimide (NBS) is a reagent designed to accomplish benzylic bromination.

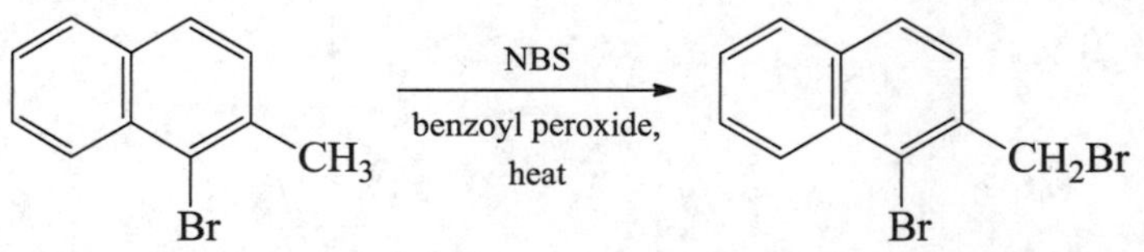

1-Bromo-2-methylnaphthalene

1-Bromo-2-(bromomethyl)naphthalene
(Compound G)

Hydrolysis of the benzylic bromide gives the corresponding benzylic alcohol. The bromine that is directly attached to the naphthalene ring does not react under these conditions.

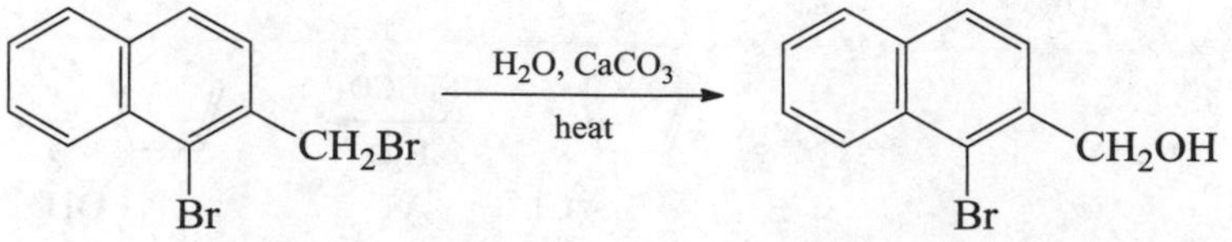

Compound G

(1-Bromo-2-naphthyl)methanol
(Compound H)

Oxidation of the primary alcohol with PCC gives the aldehyde.

Compound H   1-Bromonaphthalene-2-carboxaldehyde
                                                        (Compound I)

16.34 The formation of an alkanethiol by reaction of an alkyl halide or alkyl *p*-toluenesulfonate with
the conjugate base of H_2S is a substitution that occurs by the S_N2 pathway and will proceed with
inversion of configuration. The *p*-toluenesulfonate is formed from the corresponding alcohol by
a reaction that does not involve any of the bonds to the chirality center. Therefore, begin with
(*S*)-2-butanol.

(*S*)-2-Butanol   (*S*)-*sec*-Butyl   (*R*)-2-Butanethiol
                                   p-toluenesulfonate

Structure Determination

16.35 The difference between the two ethers is that 1-*O*-benzylglycerol contains a vicinal diol function,
but 2-*O*-benzylglycerol does not. Periodic acid will react with 1-*O*-benzylglycerol but not with
2-*O*-benzylglycerol.

1-*O*-Benzylglycerol   2-Benzyloxyethanal   Formaldehyde

2-*O*-Benzylglycerol

16.36 The compound is 2,5-dimethyl-2,5-hexanediol.

The ratio of carbon to hydrogen in the molecular formula is C_nH_{2n+2} ($C_8H_{18}O_2$), so the compound has no
double bonds or rings. It cannot be a vicinal diol, because it does not react with periodic acid. All peaks in
the NMR spectrum are singlets. The 12-proton singlet at δ 1.2 corresponds to the four equivalent methyl
groups and the four-proton singlet at δ 1.6 to the two equivalent methylene groups. No nonequivalent
protons can be vicinal because no splitting is observed. The two-proton singlet at δ 2.3 is due to the two
equivalent hydroxyl protons.

16.37 The molecular formula of the compound $C_8H_{10}O$ corresponds to an index of hydrogen deficiency
of 4. The 4 hydrogen signal at δ 7.2 in the ^1H NMR spectrum suggests these unsaturations are

due to a disubstituted benzene ring. That the ring is para-substituted is supported by the symmetry of the signal; it is a pair of doublets, not a quartet.

The broad band in the IR spectrum at 3300 cm^{-1} is the O—H stretching vibration of an alcohol. The presence of an alcohol is confirmed by the disappearance of the broad triplet at δ 2.5 following addition of D_2O as the hydroxyl proton undergoes rapid exchange with deuterium. The doublet at δ 4.6 with area of 2H indicates a –CH_2 group bonded to –OH and the benzene ring. The singlet at δ 2.4 with an area of 3H is typical for a methyl group bonded to the benzene ring. The two doublets centered above and below δ 7.3 are indicative of a para substituted benzene ring. These data all point to a structure of 4-methylbenzyl alcohol.

16.38 (*a*) This compound has only two different types of carbons. One type of carbon comes at low field and is most likely a carbon bonded to oxygen and three other equivalent carbons. The spectrum leads to the conclusion that this compound is *tert*-butyl alcohol.

(*b*) Four different types of carbons occur in this compound. The $C_4H_{10}O$ isomers that have four nonequivalent carbons are $CH_3CH_2CH_2CH_2OH$, $CH_3CHCH_2CH_3$ and $CH_3OCH_2CH_2CH_3$.
$\qquad\qquad\qquad\qquad\qquad\qquad\qquad\qquad\quad |$
$\qquad\qquad\qquad\qquad\qquad\qquad\qquad\qquad\;\; OH$

The lowest-field signal, the one at δ 69.2 from the carbon that bears the oxygen substituent, is a methine (CH). The compound is therefore 2-butanol.

$$CH_3CHCH_2CH_3$$
$$|$$
$$OH$$

(*c*) This compound has two equivalent CH_3 groups, as indicated by the signal at δ 18.9. Its lowest-field carbon is a CH_2, and so the group CH_2O must be present. The compound is 2-methyl-1-propanol.

16.39 The compound has only three carbons, none of which is a CH_3 group. Two of the carbon signals arise from CH_2 groups; the other corresponds to a CH group. The only structure consistent with the observed data is that of 3-chloro-1,2-propanediol.

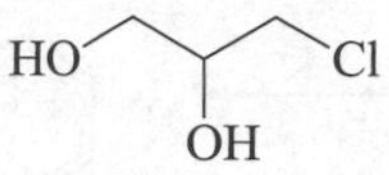

HO⌒⌒Cl
OH

The structure $HOCH_2CHCH_2OH$ cannot be correct. It would exhibit only two peaks in its ^{13}C NMR
  |
  Cl

spectrum, because the two terminal carbons are equivalent to each other.

16.40 The observation of a peak at m/z 31 in the mass spectrum of the compound suggests the presence
of a primary alcohol. This fragment is most likely $H_2C=\overset{+}{O}H$. On the basis of this fact and the
appearance of four different carbons in the ^{13}C NMR spectrum, the compound is 2-ethyl-1-butanol.

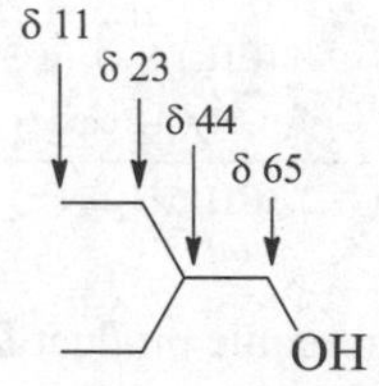

δ 11
  δ 23
    δ 44
      δ 65

OH

Answers to Interpretive Problems 16

16.41 A; **16.42** A; **16.43** B; **16.44** C; **16.45** B; **16.46** B; **16.47** D

SELF-TEST

1. For each of the following reactions, give the structure of the missing reactant or reagent.

(a)

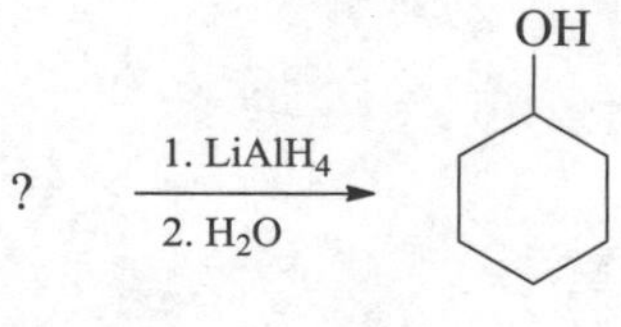

OH

? $\xrightarrow[\text{2. H}_2\text{O}]{\text{1. LiAlH}_4}$

(b)

? →

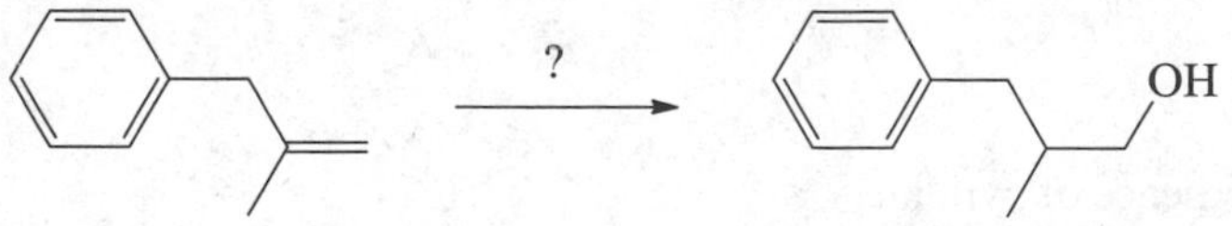

OH

(c)

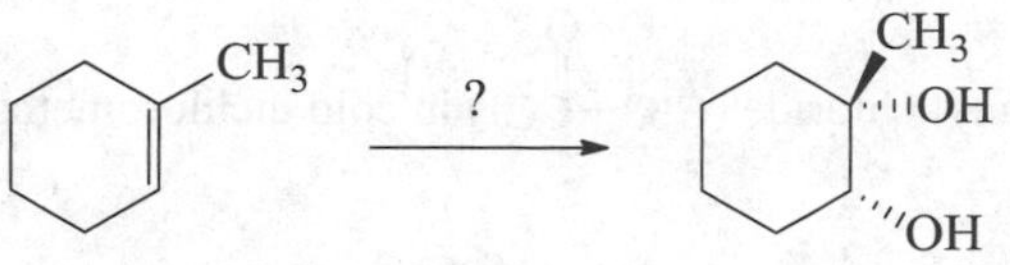

CH₃

? →

CH₃
''''OH
''OH

(d)

Br

? →

SH

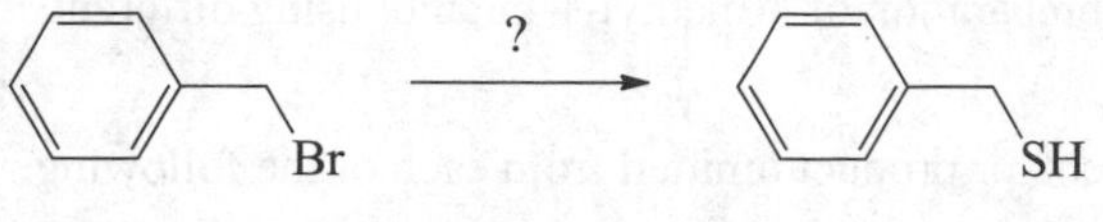

2. For the following reactions of 2-phenylethanol, $C_6H_5CH_2CH_2OH$, give the correct reagent
or product(s) omitted from the equation.

(a)

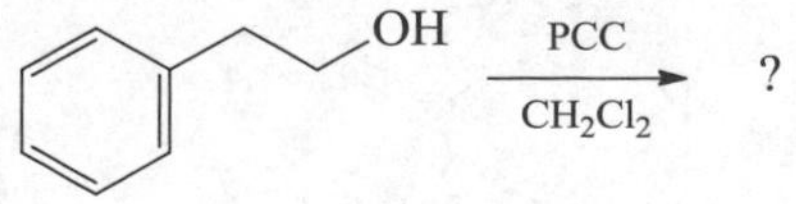

OH $\xrightarrow[\text{CH}_2\text{Cl}_2]{\text{PCC}}$?

(b)

?

(c)

2

$\xrightarrow[\text{heat}]{\text{H}^+}$ H_2O + ?

(d)

?

(e)

$\begin{array}{c} 1.\ (CH_3)_2\overset{+}{S}\text{-}\overset{-}{O},\ (COCl)_2 \\ CH_2Cl_2,\ cold \\ \hline 2.\ (CH_3CH_2)_3N \end{array}$?

3. Write the structure of the major organic product formed in the reaction of 2-propanol with each of the following reagents:

(a) Sodium amide (NaNH$_2$)

(b) Potassium dichromate (K$_2$Cr$_2$O$_7$) in aqueous sulfuric acid, heat

(c) PDC in dichloromethane

(d) Acetic acid ($CH_3\overset{O}{\overset{\|}{C}}OH$) in the presence of dissolved hydrogen chloride

(e) $H_3C\text{—}\overset{}{\bigcirc}\text{—}SO_2Cl$ in the presence of pyridine

(f) $CH_3CH_2\text{—}\overset{}{\bigcirc}\text{—}\overset{O}{\overset{\|}{C}}Cl$ in the presence of pyridine

(g) $CH_3\overset{O}{\overset{\|}{C}}O\overset{O}{\overset{\|}{C}}CH_3$ in the presence of pyridine

(h) Dimethyl sulfoxide ((CH$_3$)$_2\overset{+}{S}\text{—}\overset{-}{O}$), oxalyl chloride (Cl$\overset{O}{\overset{\|}{C}}$—$\overset{O}{\overset{\|}{C}}$Cl) in cold dichloromethane,

followed by triethylamine.

4. Outline two synthetic schemes for the preparation of 3-methyl-1-butanol using different Grignard reagents.

5. Give the structure of the reactant, reagent, or product omitted from each of the following. Show stereochemistry where important.

(a)

$\xrightarrow{\text{HIO}_4}$?

(b)

? (a diol) $\xrightarrow[\text{heat}]{\text{H}^+}$

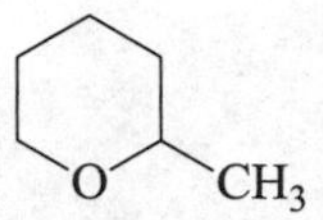

(c)

$$? \xrightarrow[\substack{\text{OsO}_4,\ \text{HO}^- \\ (\text{CH}_3)_3\text{COOH} \\ (\text{CH}_3)_3\text{COH}}]{} \text{2,3-Butanediol (chiral diastereomer)}$$

6. Give the reagents necessary to carry out each of the following transformations:

(a) Conversion of benzyl alcohol ($C_6H_5CH_2OH$) to benzaldehyde ($C_6H_5CH=O$).

(b) Conversion of benzyl alcohol to benzoic acid ($C_6H_5CO_2H$).

(c) Conversion of $H_2C=CHCH_2CH_2CO_2H$ to $H_2C=CHCH_2CH_2CH_2OH$.

(d) Conversion of cyclohexene to *cis*-1,2-cyclohexanediol.

7. Provide structures for compounds A to C in the following reaction scheme:

$$A\ (C_5H_{12}O_2) \xrightarrow[\text{H}^+,\ \text{H}_2\text{O}]{\text{K}_2\text{Cr}_2\text{O}_7} B\ (C_5H_8O_3) \xrightarrow[]{\text{CH}_3\text{OH},\ \text{H}^+} C\ (C_6H_{10}O_3)$$

A: $\downarrow$ H$^+$, heat

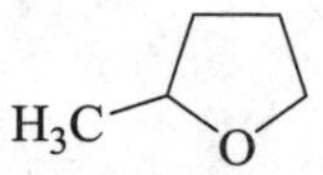

C: 1. LiAlH$_4$ / 2. H$_2$O $\rightarrow$

A + CH$_3$OH

8. Using any necessary organic or inorganic reagents, outline a scheme for each of the following conversions.

(a)

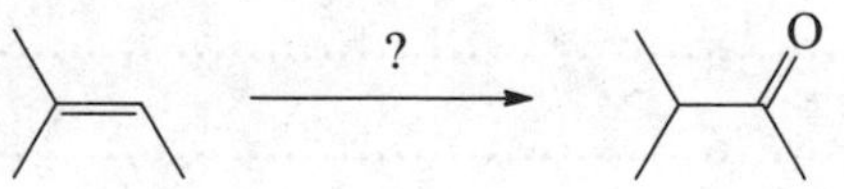

(b)

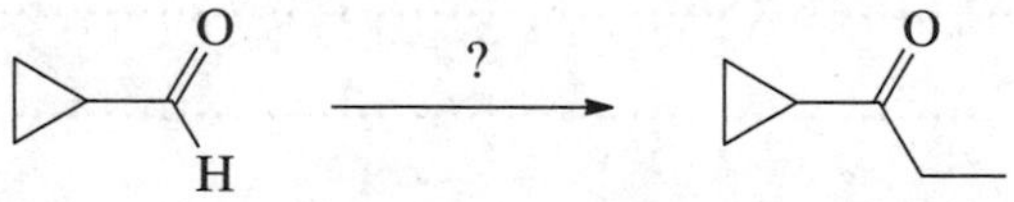

(c)

$$C_6H_5CH_3 \xrightarrow{?} C_6H_5CH_2CH_2CO_2CH_2CH_3$$

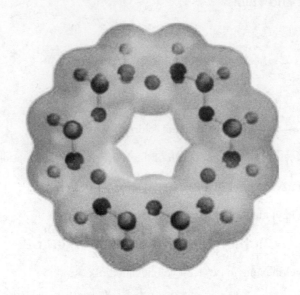

CHAPTER 17

Ethers, Epoxides, and Sulfides

Table of Contents

SOLUTIONS TO TEXT PROBLEMS

In Chapter Problems

17.1 (b) Oxirane is the IUPAC name for ethylene oxide. A chloromethyl group ($ClCH_2$—) is attached to position 2 of the ring in 2-(chloromethyl)oxirane.

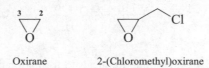

Oxirane   2-(Chloromethyl)oxirane

This compound is more commonly known as epichlorohydrin.

(c) Epoxides may be named by adding the prefix *epoxy* to the IUPAC name of a parent compound, specifying by number both atoms to which the oxygen is attached.

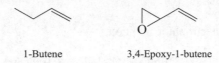

1-Butene   3,4-Epoxy-1-butene

17.2 1,2-Epoxybutane and tetrahydrofuran both have the molecular formula C_4H_8O (that is, they are constitutional isomers), and so it is appropriate to compare their heats of combustion directly. Angle strain from the three-membered ring of 1,2-epoxybutane causes it to have more internal energy than tetrahydrofuran, and its combustion is more exothermic.

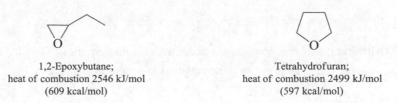

1,2-Epoxybutane;   Tetrahydrofuran;
heat of combustion 2546 kJ/mol   heat of combustion 2499 kJ/mol
(609 kcal/mol)   (597 kcal/mol)

17.3 Ethers can form hydrogen bonds to water, alkanes cannot; therefore, the compound that is more soluble in water is the ether.

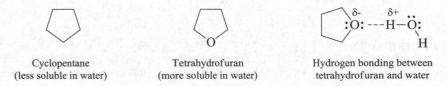

Cyclopentane   Tetrahydrofuran   Hydrogen bonding between
(less soluble in water)   (more soluble in water)   tetrahydrofuran and water

17.4 The compound is 1,4-dioxane; it has a six-membered ring and two oxygens separated by H_2C—CH_2 units.

1,4-Dioxane
("6-crown-2")

17.5 Protonation of the carbon–carbon double bond leads to the more stable carbocation.

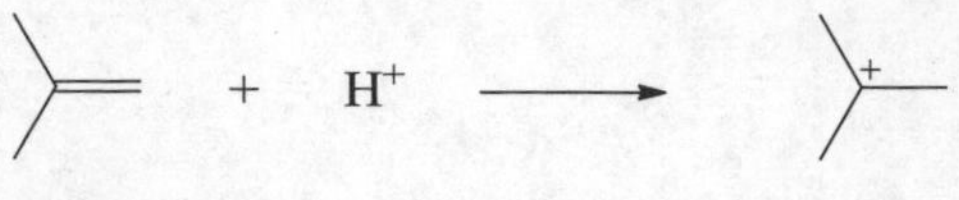

2-Methylpropene   *tert*-Butyl cation

Methanol acts as a nucleophile to capture *tert*-butyl cation.

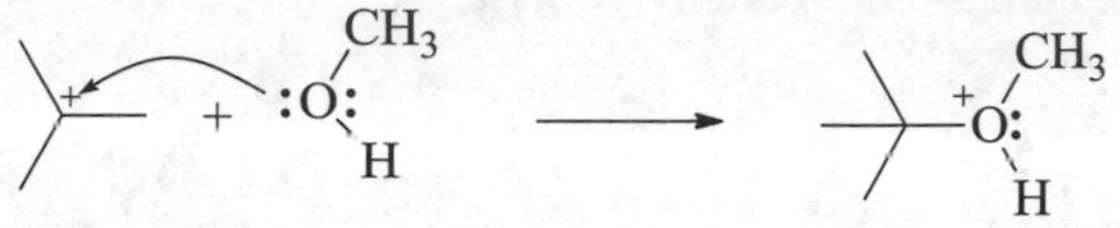

Deprotonation of the alkyloxonium ion leads to formation of *tert*-butyl methyl ether.

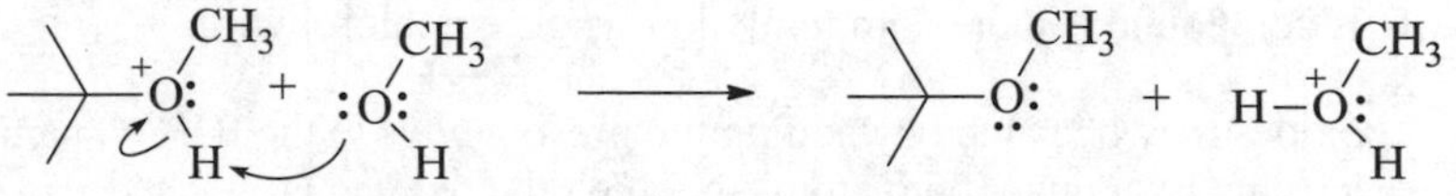

tert-Butyl methyl ether

The mechanism is electrophilic addition.

17.6 (*b*) Retrosynthetic analysis might suggest the two routes, A and B:

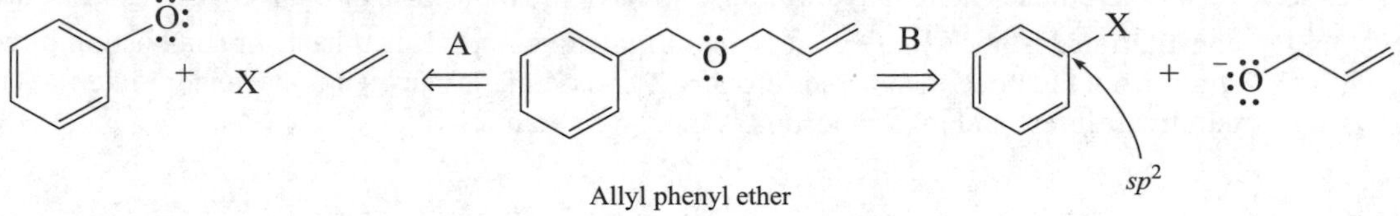

Allyl phenyl ether

However, S_N2 reactions cannot occur with leaving groups (X) bonded to sp^2-hybridized carbons, so path A is the only viable option. The equation to accomplish path a is

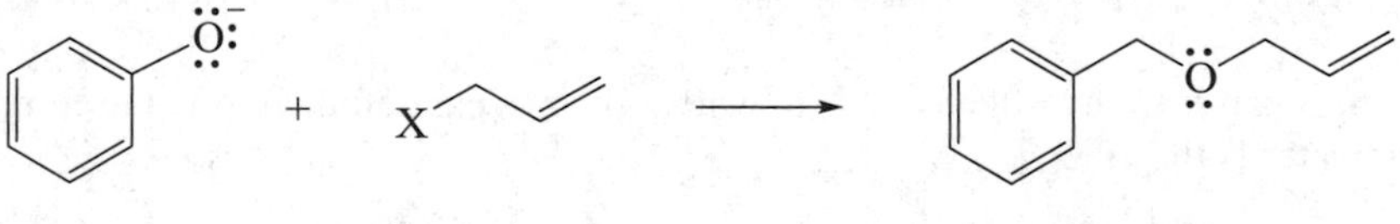

Allyl phenyl ether

17.7 (*b*) A primary carbon and a secondary carbon are attached to the ether oxygen. The secondary carbon can only be derived from the alkoxide, because secondary alkyl halides preferentially undergo elimination. The only effective method uses an allyl halide and sodium isopropoxide.

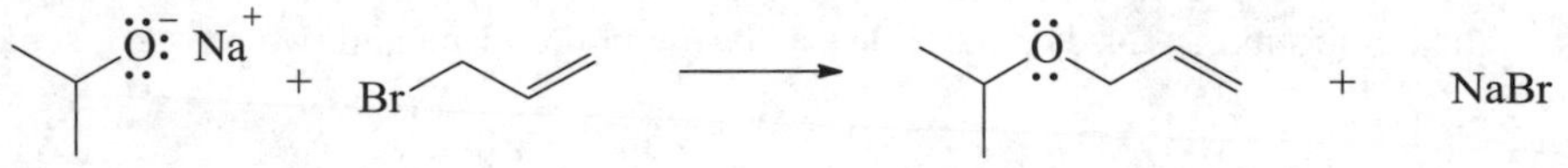

Sodium isopropoxide   Allyl bromide   Allyl isopropyl ether   Sodium bromide

(*c*) Here the ether is a mixed primary–tertiary one. The better combination uses the primary alkyl halide.

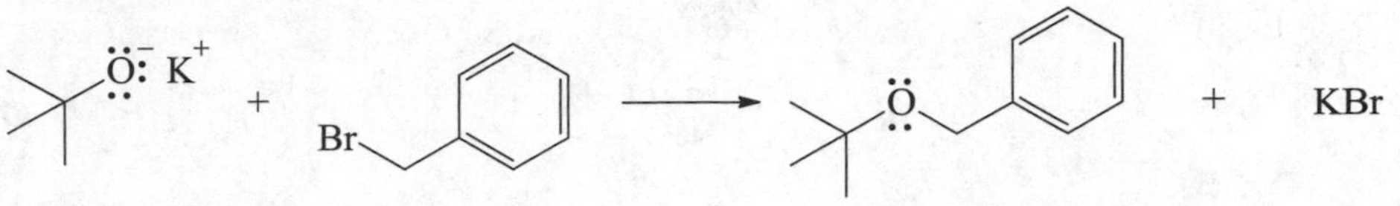

Potassium *tert*-butoxide | Benzyl bromide | Benzyl *tert*-butyl ether | Potassium bromide

The reaction between $(CH_3)_3CBr$ and $C_6H_5CH_2O^-$ would be elimination, not substitution.

17.8 The approach:

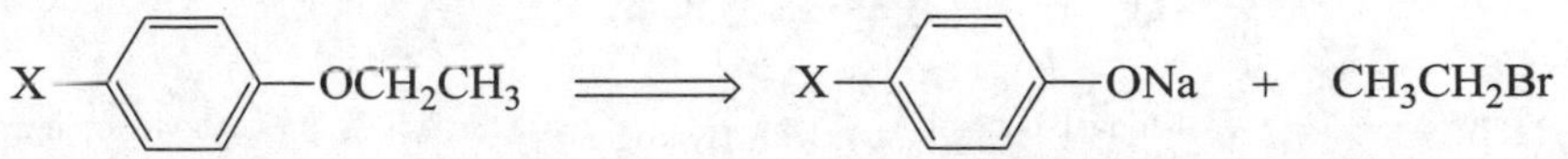

will give the desired S_N2 reaction involving ethyl bromide for both X = methyl and X = nitro. The other:

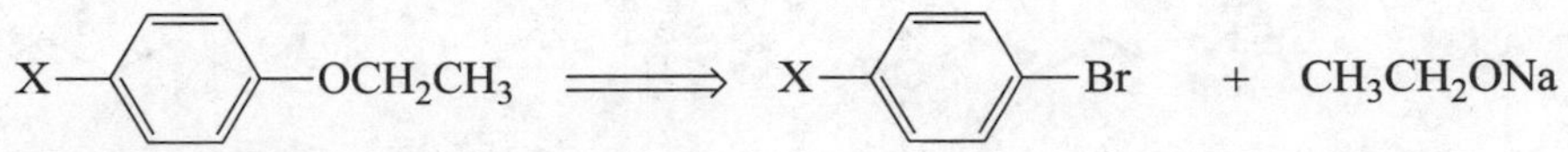

involves nucleophilic aromatic substitution and requires a strong electron-withdrawing group on the aromatic ring. It will work for X = nitro, but not for X = methyl.

17.9 (*b*) If benzyl bromide is the only organic product from reaction of a dialkyl ether with hydrogen bromide, then both alkyl groups attached to oxygen must be benzyl.

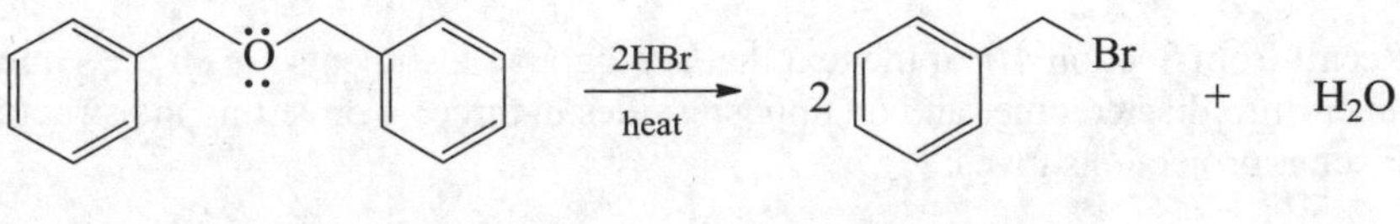

Dibenzyl ether | Benzyl bromide | Water

(*c*) Because *1 mole of a dihalide*, rather than 2 moles of a monohalide, is produced per mole of ether, the ether must be cyclic.

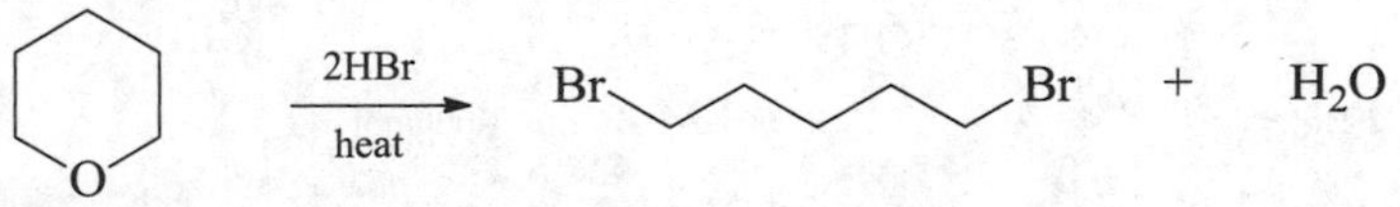

Tetrahydropyran | 1,5-Dibromopentane | Water

17.10 As outlined in Mechanism 17.1, the first step is protonation of the ether oxygen to give a dialkyloxonium ion.

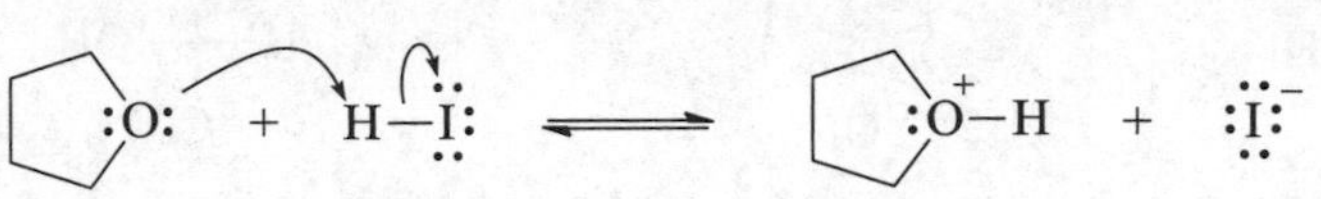

Tetrahydrofuran (THF) | Hydrogen iodide | Dialkyloxonium ion | Iodide ion

In the second step, nucleophilic attack of the halide ion on carbon of the dialkyloxonium ion gives 4-iodo-1-butanol.

| Iodide ion | Dialkyloxonium ion | | 4-Iodo-1-butanol |

The remaining two steps of the mechanism correspond to those in which an alcohol is converted to an alkyl halide, as discussed in Chapter 5.

| 4-Iodo-1-butanol | Hydrogen iodide | Alkyloxonium ion | Iodide ion |

| Alkyloxonium ion | Iodide ion | 1,4-Diiodobutane | Water |

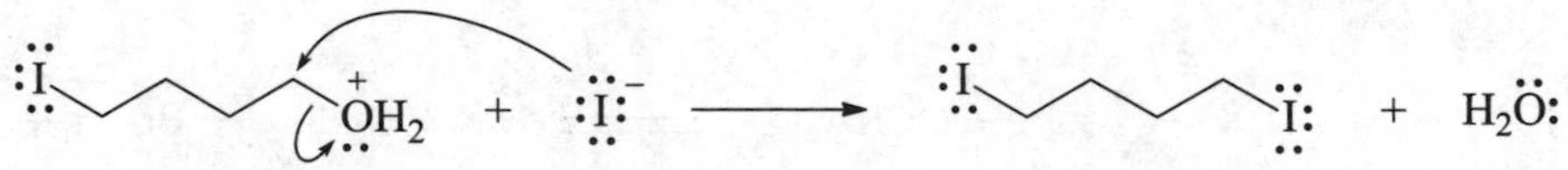

17.11 Inasmuch as Sharpless epoxidation using diethyl (2*R*,3*R*)-tartrate yields (2*S*,3*S*)-2,3-epoxy-1-hexanol from *trans*-2-hexen-1-ol, the corresponding (2*S*,3*S*) tartrate will yield the mirror image epoxide. The product is (2*R*,3*R*)-2,3-epoxy-1-hexanol.

17.12 Recall from Section 4.7 of the text that analogous substituents are on the same side of a Fischer projection in an erythro diastereomer and on opposite sides in threo. Converting the structural formulas given in the text to Fischer projections gives:

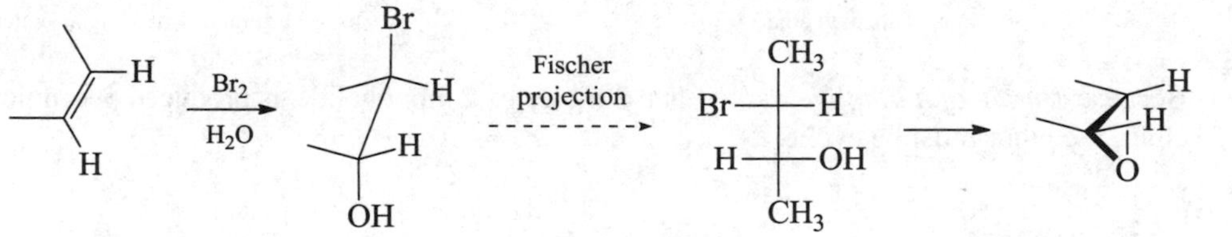

| *cis*-2-Butene | *threo*-3-Bromo-2-butanol | *cis*-2,3-Epoxybutane |

| *trans*-2-Butene | *erythro*-3-Bromo-2-butanol | *trans*-2,3-Epoxybutane |

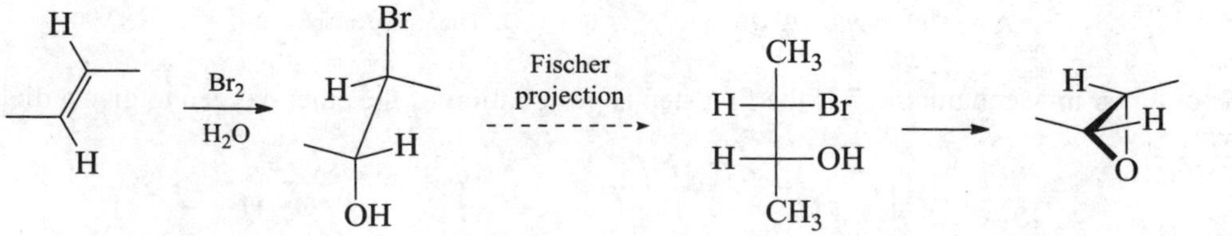

Thus, *cis*-2-butene gives the threo bromohydrin; *trans*-2-butene gives erythro. Both bromohydrins are chiral, but each is obtained as a mixture containing equal amounts of enantiomers by this method. Thus, neither of the product mixtures is optically active nor are the epoxides obtained from them. The *trans*-epoxide is chiral; the *cis* is meso.

17.13 (*b*) Azide ion [$\ddot{:}\overset{-}{N}=\overset{+}{N}=\overset{..}{N}\overset{-}{:}$] is a good nucleophile, reacting readily with ethylene oxide to yield 2-azidoethanol.

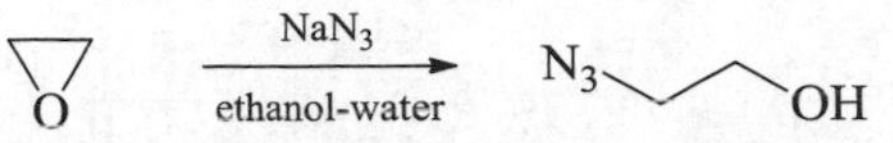

Ethylene oxide   2-Azidoethanol

(*c*) Ethylene oxide is hydrolyzed to ethylene glycol in the presence of aqueous base.

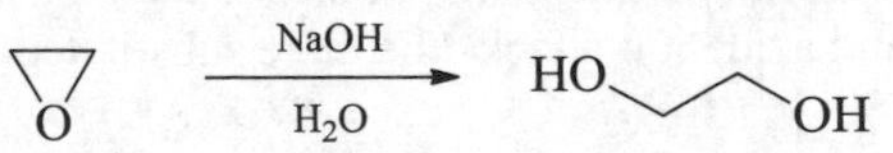

Ethylene oxide   Ethylene glycol

(*d*) Phenyllithium reacts with ethylene oxide in a manner similar to that of a Grignard reagent.

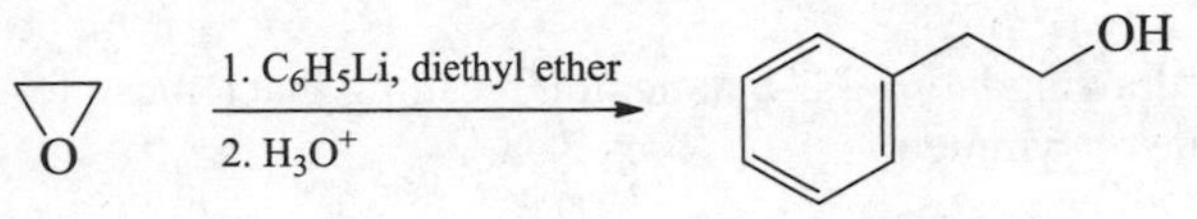

Ethylene oxide   2-Phenylethanol

(*e*) The nucleophilic species here is the acetylenic anion $CH_3CH_2C\equiv C\ddot{:}^-$, which attacks a carbon atom of ethylene oxide to give 3-hexyn-1-ol.

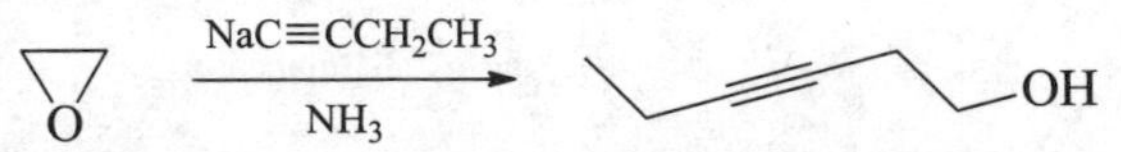

Ethylene oxide   3-Hexyn-1-ol (48%)

17.14 Because the reactant is not a meso form, both chirality centers have the same configuration. Therefore, we need to determine only one of them; the other will be the same. First, reorient the molecule so that the hydrogen at the rear-most hydrogen points away from you.

trans-2,3-Epoxybutane

ranking at C(2): O > C(1) > CH₃
is clockwise; therefore *R*

When the hydrogen at C-2 points away from you, the order of decreasing precedence traces a clockwise path. The configuration C-2 is *R*, and that at C-1 is the same.

Now assume ammonia bonds to C-2 from the side opposite the bond to oxygen.

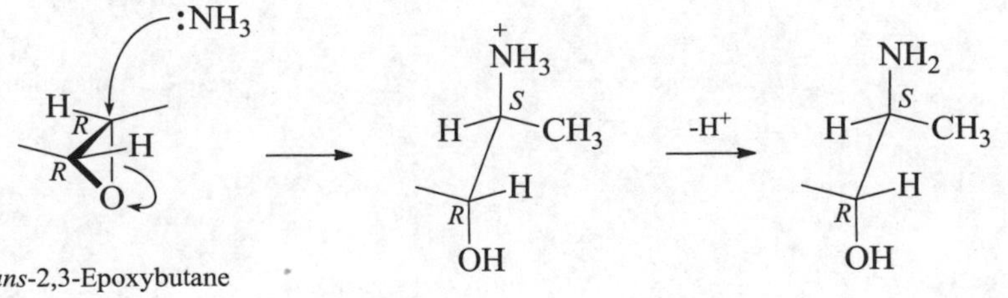

trans-2,3-Epoxybutane

17.15 At a pH of 9.5, the epoxide exists in its neutral form; that is, its oxygen is not protonated. Azide ion attacks the secondary (less-hindered) carbon in an S_N2 fashion.

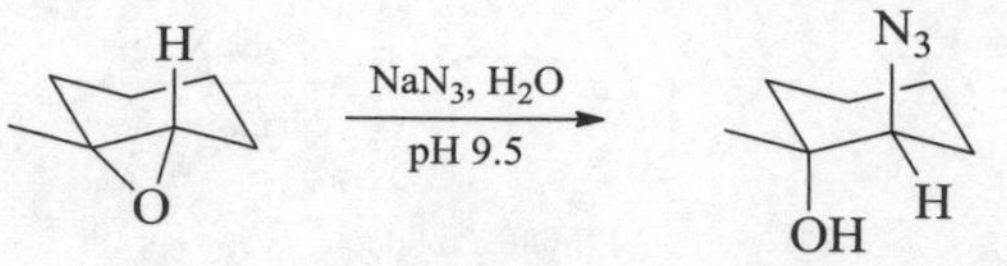

A pH of 4.2 corresponds to the pK_a of HN_3, so the concentrations of azide ion and its conjugate acid are equal. The protonated epoxide is more reactive than the neutral form, carbocation character develops in the transition state and azide ion attacks the more substituted carbon with inversion of configuration.

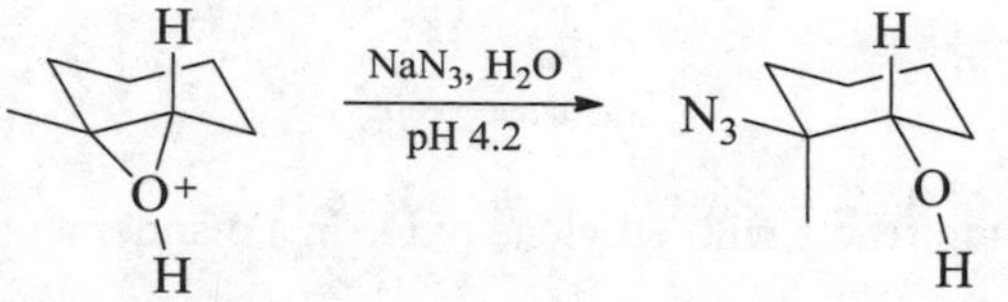

17.16 Begin by drawing *meso*-2,3-butanediol, recalling that a meso form is achiral. The eclipsed conformation has a plane of symmetry.

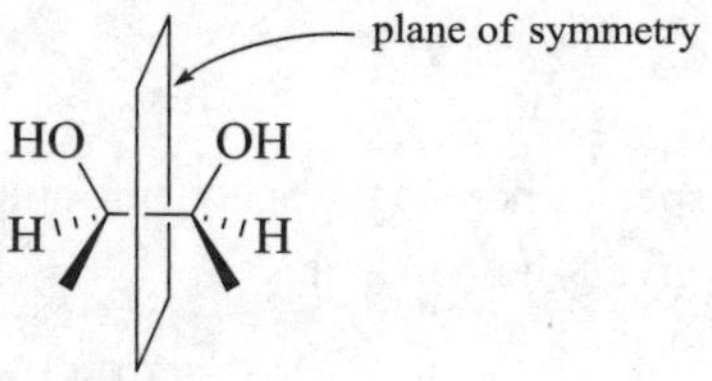

meso-2,3-Butanediol

Epoxidation followed by acid-catalyzed hydrolysis results in anti addition of hydroxyl groups to the double bond. *trans*-2-Butene is the required starting material.

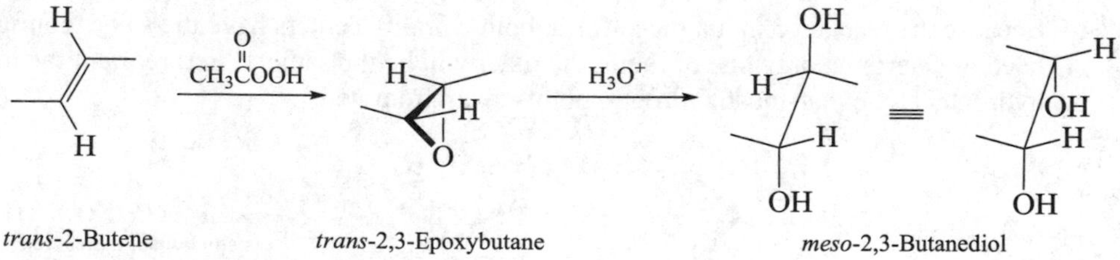

trans-2-Butene   *trans*-2,3-Epoxybutane   *meso*-2,3-Butanediol

Osmium tetraoxide dihydroxylation is a method of achieving syn dihydroxylation. The necessary starting material is *cis*-2-butene.

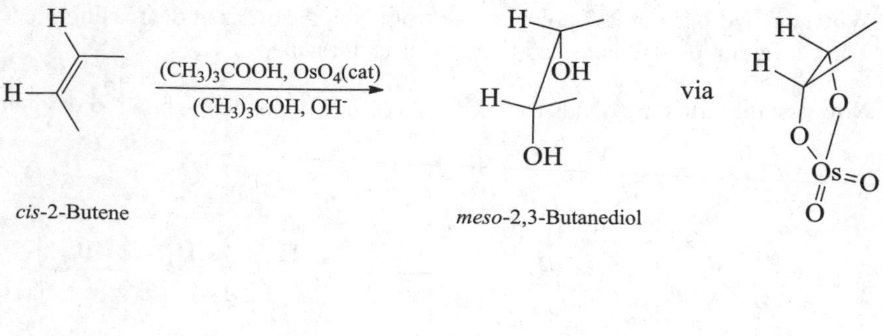

cis-2-Butene   *meso*-2,3-Butanediol

17.17 Reaction of (*R*)-2-octanol with *p*-toluenesulfonyl chloride yields a *p*-toluenesulfonate (tosylate) having the same configuration; the chirality center is not involved in this step. Reaction of the tosylate with a nucleophile proceeds by inversion of configuration in an S$_N$2 process. The product has the *S* configuration.

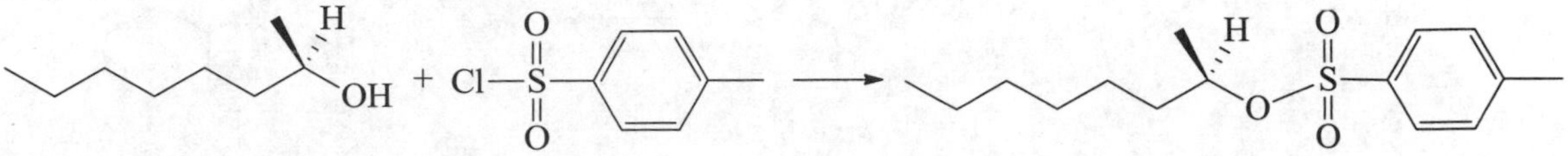

(*R*)-2-Octanol	*p*-Toluenesulfonyl chloride	(*R*)-1-Methylheptyl tosylate

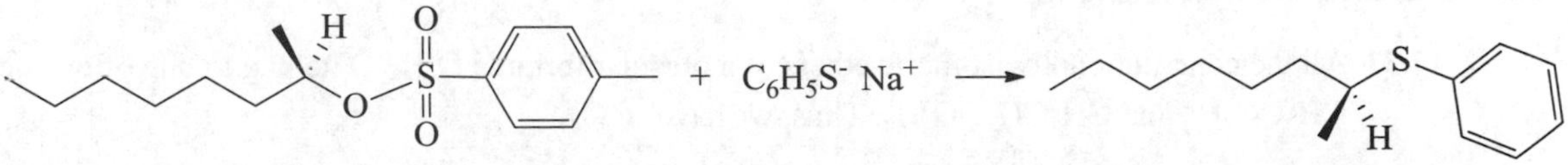

(*R*)-1-Methylheptyl tosylate	Sodium benzenethiolate	(*S*)-1-Methylheptyl phenyl sulfide

17.18 None of the carbons in omeprazole is a chirality center. The only chirality center is sulfur. In order of decreasing precedence, the groups attached to sulfur are: O > C=N > CH$_2$. Consider sulfur's unshared electron pair as its fourth and lowest-ranked "group." Use the structural formula given in the text and assume the unshared pair of sulfur is in the plane of the paper as shown for sulfoxides in Section 17.15. Oxygen which is also in the same plane, points down as shown in the structure at the left. The order of decreasing precedence traces a clockwise path making this the (*S*)-enantiomer (Nexium). To convert it to the (*R*)-enantiomer, swap the oxygen and electron pair positions as shown in the right structure.

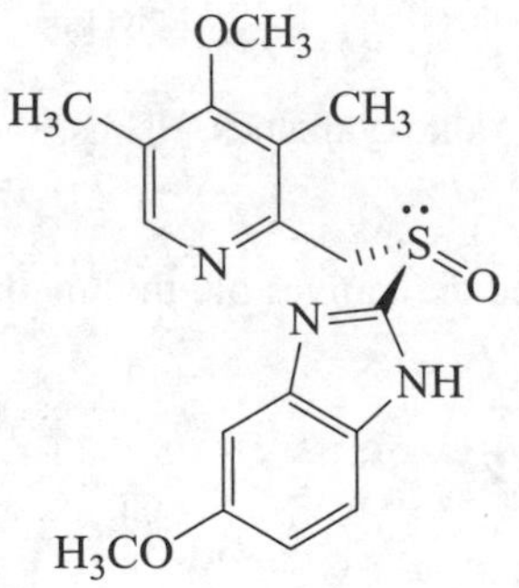

(*S*)-Omeprazole	(*R*)-Omeprazole

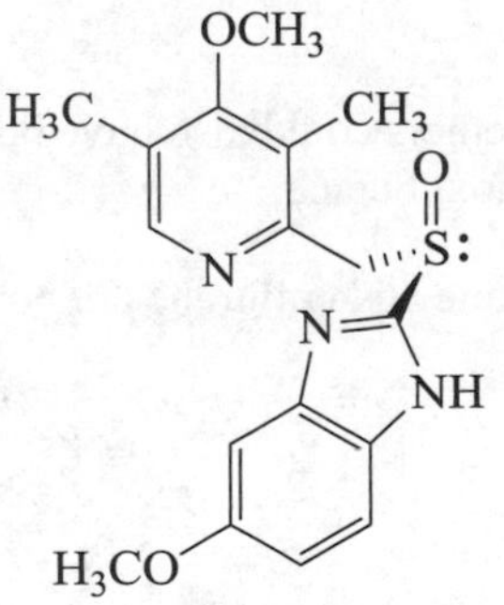

17.19 As shown in the text, dodecyldimethylsulfonium iodide may be prepared by reaction of dodecyl methyl sulfide with methyl iodide. An alternative method is the reaction of dodecyl iodide with dimethyl sulfide.

$$(CH_3)_2S \ + \ CH_3(CH_2)_{10}CH_2I \longrightarrow CH_3(CH_2)_{10}CH_2\overset{+}{S}(CH_3)_2 \ I^-$$

Dimethyl sulfide	Dodecyl iodide	Dodecyldimethylsulfonium iodide

The reaction of a sulfide with an alkyl halide is an S$_N$2 process. The faster reaction will be the one that uses the less sterically hindered alkyl halide. The method presented in the text will proceed faster.

17.20 The molecular ion from *sec*-butyl ethyl ether can also fragment by cleavage of a carbon–carbon bond in its ethyl group to give an oxygen-stabilized cation of *m/z* 87.

$$\text{(structure)} \longrightarrow \text{H}_3\text{C·} + \text{H}_2\text{C}=\overset{+}{\text{O}}\text{(structure)}$$

m/z 87

End of Chapter Problems

Structure and Nomenclature

17.21 All the constitutionally isomeric ethers of molecular formula $C_5H_{12}O$ belong to one of two general groups: $CH_3OC_4H_9$ and $CH_3CH_2OC_3H_7$. Thus, we have

Butyl methyl ether	*sec*-Butyl methyl ether	Isobutyl methyl ether	*tert*-Butyl methyl ether

and

Ethyl propyl ether	Ethyl isopropyl ether

These ethers could also have been named as "alkoxyalkanes." Thus, *sec*-butyl methyl ether would become 2-methoxybutane.

17.22 Isoflurane and enflurane are both halogenated derivatives of ethyl methyl ether.

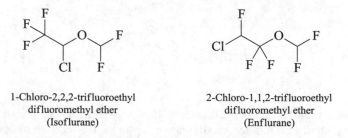

1-Chloro-2,2,2-trifluoroethyl difluoromethyl ether (Isoflurane)	2-Chloro-1,1,2-trifluoroethyl difluoromethyl ether (Enflurane)

17.23 (*a*) The parent compound is cyclopropane. It has a three-membered epoxide function, and thus a reasonable name is epoxycyclopropane. Numbers locating positions of attachment (as in "1,2-epoxycyclopropane") are not necessary, because no other structures (1,3 or 2,3) are possible here.

Epoxycyclopropane

(b) The longest continuous carbon chain has seven carbons, and so the compound is named as a derivative of heptane. The epoxy function bridges C-2 and C-4. Therefore

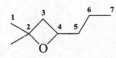

is 2,4-epoxy-2-methylheptane.

(c) The oxygen atom bridges the C-1 and C-4 atoms of a cyclohexane ring.

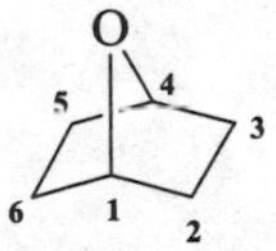

1,4-Epoxycyclohexane

(d) Eight carbon atoms are continuously linked and bridged by an oxygen. We name the compound as an epoxy derivative of cyclooctane.

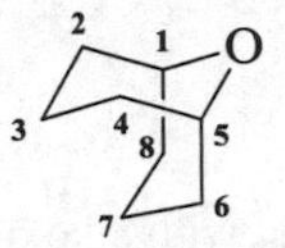

1,5-Epoxycyclooctane

17.24 (a) There are three methyl-substituted thianes, two of which are chiral.

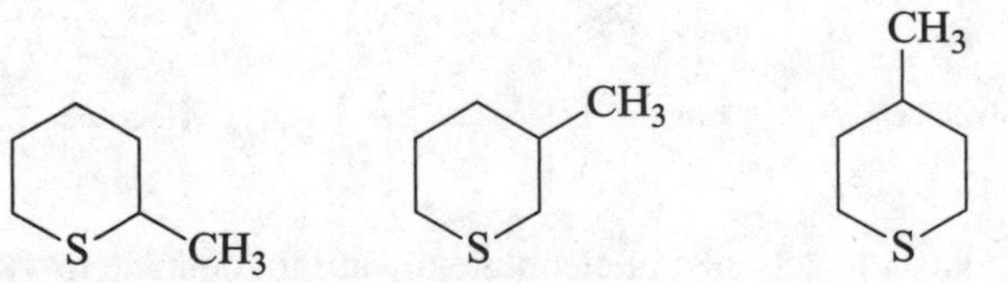

2-Methylthiane
(chiral)

3-Methylthiane
(chiral)

4-Methylthiane
(achiral)

(b) The locants in the name indicate the positions of the sulfur atoms in 1,4-dithiane and 1,3,5-trithiane.

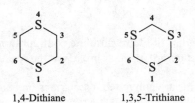

1,4-Dithiane

1,3,5-Trithiane

(c) Disulfides possess two adjacent sulfur atoms. 1,2-Dithiane is a disulfide.

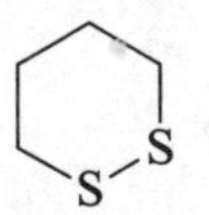

1,2-Dithiane

(*d*) Two chair conformations of the sulfoxide derived from thiane are possible; the oxygen atom may be either equatorial or axial.

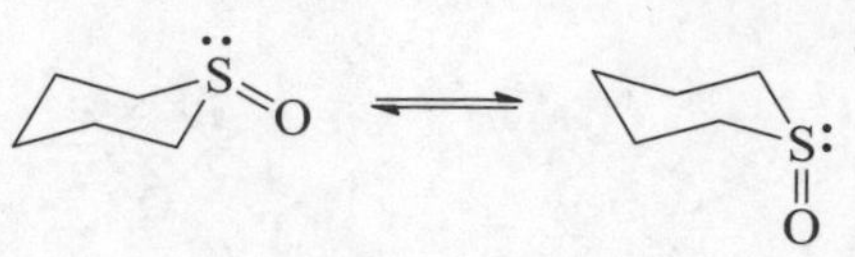

17.25 Of the two chair conformations of *cis*-3-hydroxythiane 1-oxide the more stable conformation has both groups axial due to hydrogen bonding.

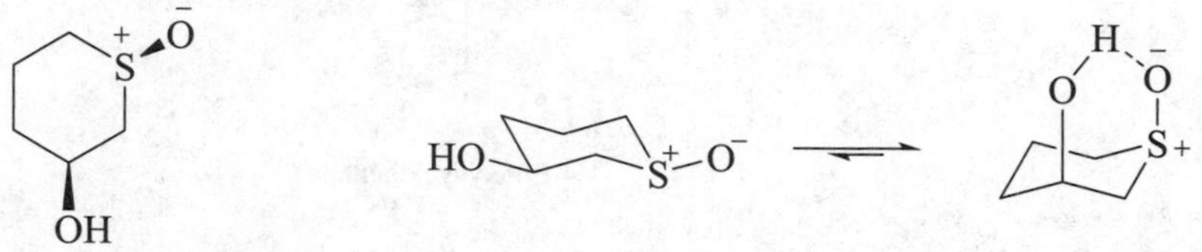

cis-3-Hydroxythiane 1-oxide   more stable
                                                               chair conformation

Reactions

17.26 (*a*) Secondary alkyl halides react with alkoxide bases by E2 elimination as the major pathway. The Williamson ether synthesis is not a useful reaction with secondary alkyl halides.

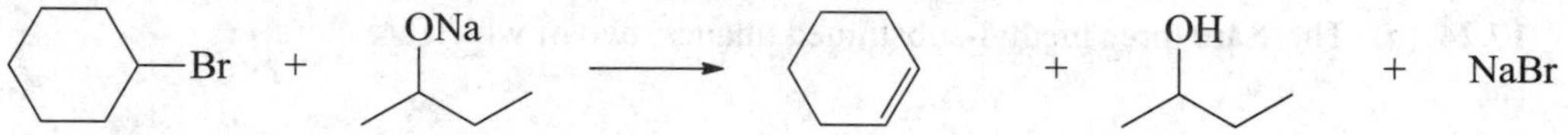

Bromocyclohexane   Sodium 2-butanolate   Cyclohexene   2-Butanol   Sodium
                                                                                  bromide

(*b*) Sodium alkoxide acts as a nucleophile toward iodoethane to yield an alkyl ethyl ether.

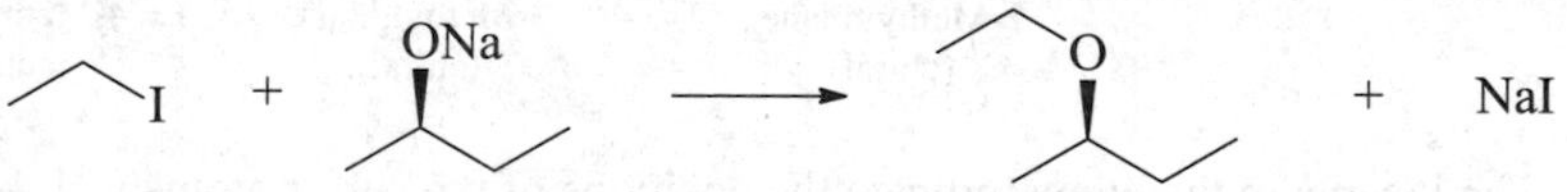

(*R*)-2-Ethoxybutane   Sodium
                                iodide

The ether product has the same absolute configuration as the starting alkoxide because no bonds to the chirality center are made or broken in the reaction.

(*c*) Vicinal halohydrins are converted to epoxides on being treated with base.

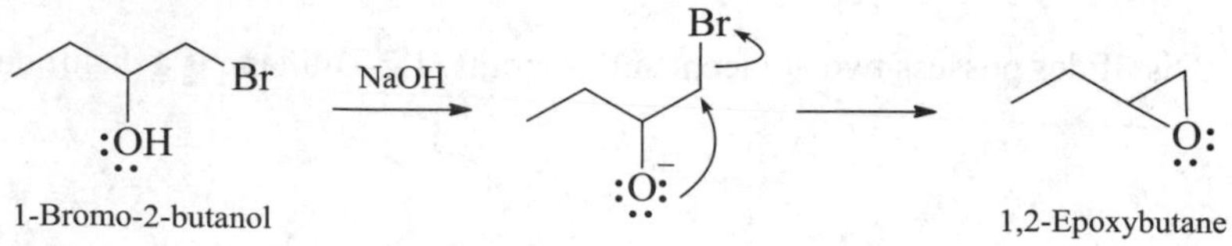

1-Bromo-2-butanol   1,2-Epoxybutane (*d*) The reactants, an alkene plus a peroxy acid, are customary ones for epoxide preparation. The reaction is a stereospecific syn addition of oxygen to the double bond.

| (*Z*)-1-Phenylpropene | Peroxybenzoic acid | *cis*-2-Methyl-3-phenyloxirane | Benzoic acid |

(*e*) Azide ion is a good nucleophile in S_N2 reactions with epoxides. Substitution occurs at carbon with inversion of configuration. The product is *trans*-2-azidocyclohexanol.

1,2-Epoxycyclohexane
(meso)

$\xrightarrow[\text{dioxane-water}]{\text{NaN}_3}$

trans-2-
Azidocyclohexanol (61%)
(racemic)

(*f*) Ammonia is a nucleophile capable of reacting with epoxides. It attacks the less hindered carbon of the epoxide.

$\xrightarrow[\text{methanol}]{\text{NH}_3}$

2-(2-Bromophenyl)-2-methyloxirane

1-Amino-2-(2-bromophenyl)-2-propanol

Aryl halides do not react with nucleophiles under these conditions, and so the bromine substituent on the ring is unaffected.

(*g*) Methoxide ion attacks the less substituted carbon of the epoxide ring with inversion of configuration.

$\xrightarrow{\text{CH}_3\text{OH}}$

1-Benzyl-1,2-epoxycyclohexane

1-Benzyl-*trans*-2-methoxycyclohexanol (98%)

(*h*) Under acidic conditions, substitution is favored at the carbon that can better support a positive charge. Aryl substituents stabilize carbocations, making the benzylic position the one that is attacked in an aryl substituted epoxide.

$\xrightarrow[\text{CHCl}_3]{\text{HCl}}$

2-Phenyloxirane

2-Chloro-2-phenylethanol (71%)

(*i*) Lithium aluminum hydride reduces epoxides to alcohols, and hydride is transferred to the less substituted carbon of the epoxide ring. The alkene double bond is not reduced by $LiAlH_4$.

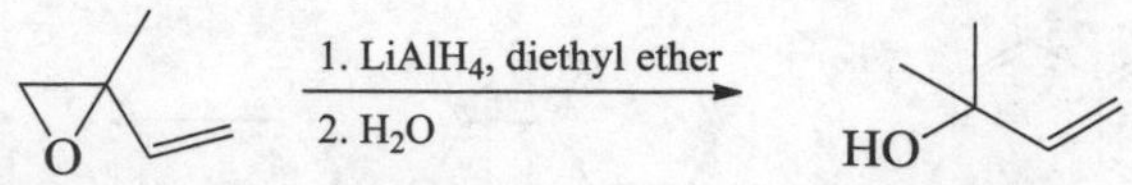

3,4-Epoxy-3-methyl-1-butene   2-Methyl-3-buten-2-ol

(*j*) Tosylates undergo substitution with nucleophiles such as sodium butanethiolate.

$$CH_3(CH_2)_{16}CH_2OTs \quad + \quad CH_3CH_2CH_2CH_2SNa \longrightarrow CH_3CH_2CH_2CH_2SCH_2(CH_2)_{16}CH_3$$

Octadecyl tosylate   Sodium butanethiolate   Butyl octadecyl sulfide

(*k*) Nucleophilic substitution proceeds with inversion of configuration.

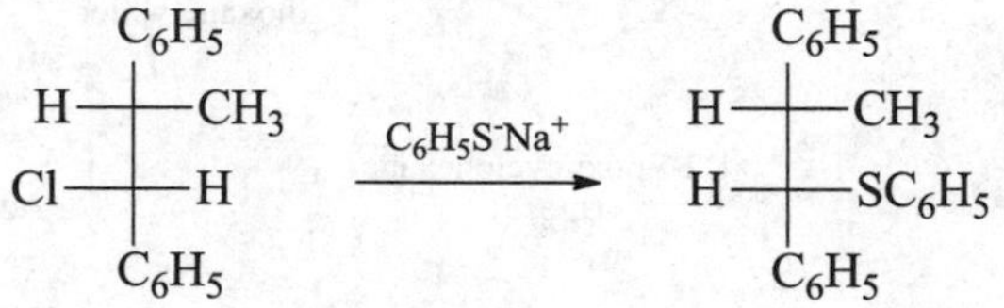

17.27 Compound A can be converted into a toluenesulfonate ester, compound B, with toluenesulfonyl chloride and pyridine. Although there are three alcohol groups, the primary one reacts selectively because it is the least hindered.

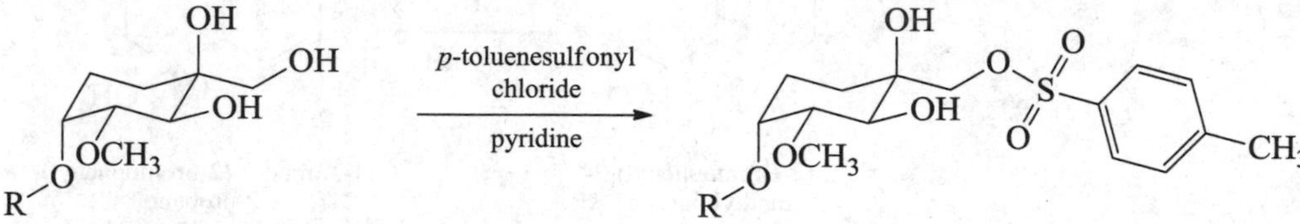

Compound A
(R = $(CH_3CH_2)_3Si$-

Compound B,
a toluene sulfonate ester
(R = $(CH_3CH_2)_3Si$-

In the presence of base, compound B undergoes an intramolecular S_N2 reaction to displace toluenesulfonate and form the epoxide.

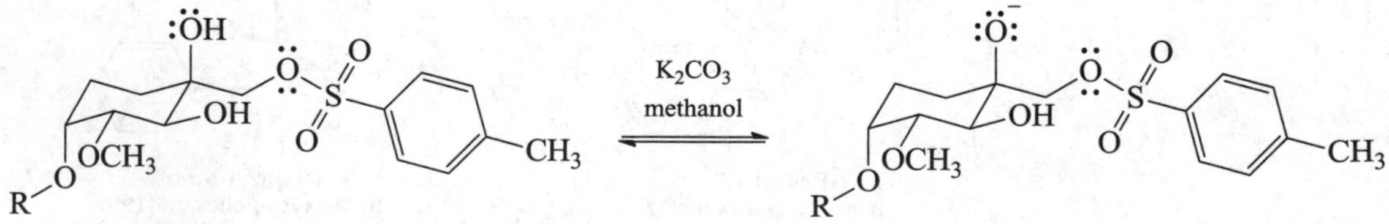

Compound B

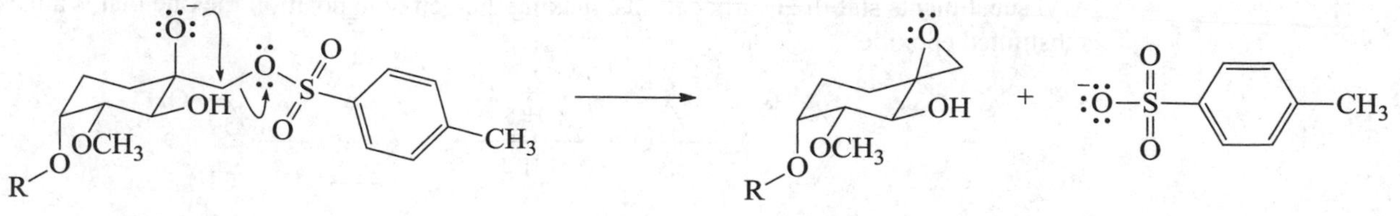

Compound C

17.28 Cineole contains no double or triple bonds and therefore must be bicyclic, on the basis of its molecular formula ($C_{10}H_{18}O$, index of hydrogen deficiency = 2). When cineole reacts with hydrogen chloride, one of the rings is broken and water is formed.

Cineole   +   2HCl $\longrightarrow$ [structure] + H_2O

($C_{10}H_{18}O$)   ($C_{10}H_{18}Cl_2$)

The reaction that takes place is hydrogen halide–promoted ether cleavage. In such a reaction with excess hydrogen halide, the C—O—C unit is cleaved and two carbon–halogen bonds are formed. This suggests that cineole is a cyclic ether because the product contains both newly formed carbon–halogen bonds. A reasonable structure consistent with these facts is

Cineole

17.29 Recall that *p*-toluenesulfonate (tosylate) is a good leaving group in nucleophilic substitution reactions. The nucleophile that displaces tosylate from carbon is the alkoxide ion derived from the hydroxyl group within the molecule. The product is a cyclic ether, and the nature of the union of the two rings is that they are spirocyclic.

($C_{15}H_{20}O$)

17.30 Lithium aluminum deuteride (LiAlD$_4$) will transfer a deuterium from the side of the ring opposite the oxygen of the epoxide. The *tert*-butyl group has the greatest preference for an equatorial orientation, so the deuterium will be axial in the most stable conformation of the product.

17.31 Oxidation of 4-*tert*-butylthiane yields two sulfoxides that are diastereomers of each other.

4-*tert*-Butylthiane

Oxidation of both stereoisomeric sulfoxides yields the same sulfone.

17.32 (*a*) The first step is a standard Grignard synthesis of a primary alcohol using formaldehyde. Compound A is 3-buten-1-ol.

H_2C=CHCH$_2$Br H_2C=CHCH$_2$CH$_2$OH

Allyl bromide   1. Mg   3-Buten-1-ol
                       2. H_2C=O   (compound A)
                       3. H_3O^+

Addition of bromine to the carbon–carbon double bond of 3-buten-1-ol takes place readily to yield the vicinal dibromide.

H_2C=CHCH$_2$CH$_2$OH $\xrightarrow{Br_2}$ H_2C—CHCH$_2$CH$_2$OH

3-Buten-1-ol   3,4-Dibromo-1-butanol
                            (compound B)

When compound B is treated with potassium hydroxide, it loses the elements of HBr to give compound C. Because further treatment of compound C with potassium hydroxide converts it to D by a second dehydrobromination, a reasonable candidate for C is 3-bromotetrahydrofuran.

3,4-Dibromo-1-butanol   3-Bromotetrahydrofuran   Compound D
(compound B)   (compound C)

Ring closure occurs by an intramolecular Williamson reaction.

3,4-Dibromo-1-butanol   3-Bromotetrahydrofuran
(compound B)   (compound C)

Dehydrohalogenation of compound C converts it to the final product, D.

The alternative series of events, in which double-bond formation precedes ring closure, is unlikely, because it requires nucleophilic attack by the alkoxide on a vinyl bromide.

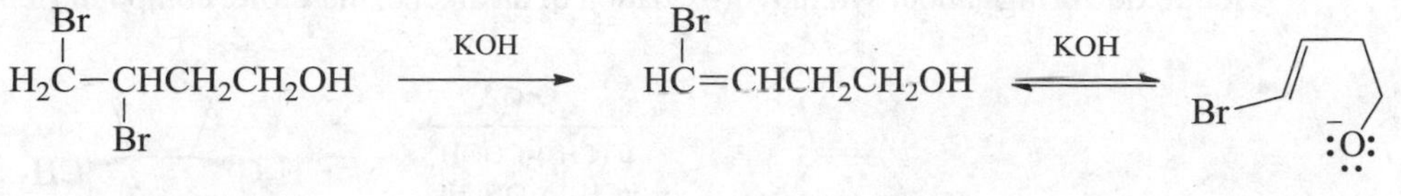

3,4-Dibromo-1-butanol
(compound B)

(Cyclization of this
intermediate
does not occur.)

(b) Lithium aluminum hydride reduces the carboxylic acid to the corresponding primary alcohol, compound E. Treatment of the vicinal chlorohydrin with base results in formation of an epoxide, compound F.

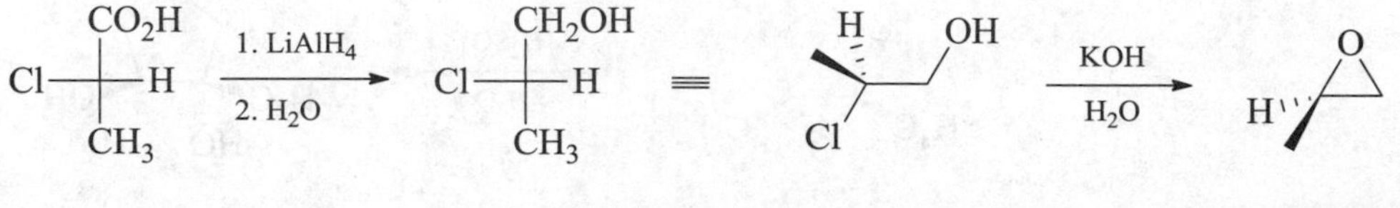

(S)-2-Chloro-1-propanol
(compound E)

(R)-1,2-Epoxypropane
(compound F)

As actually carried out, the first step proceeded in 56–58% yield, the second step in 65–70% yield.

(c) Treatment of the vicinal chlorohydrin with base results in ring closure to form an epoxide (compound G). Recall that attack occurs on the side opposite that of the carbon–chlorine bond. Compound G undergoes ring opening on reaction with sodium methanethiolate to give compound H.

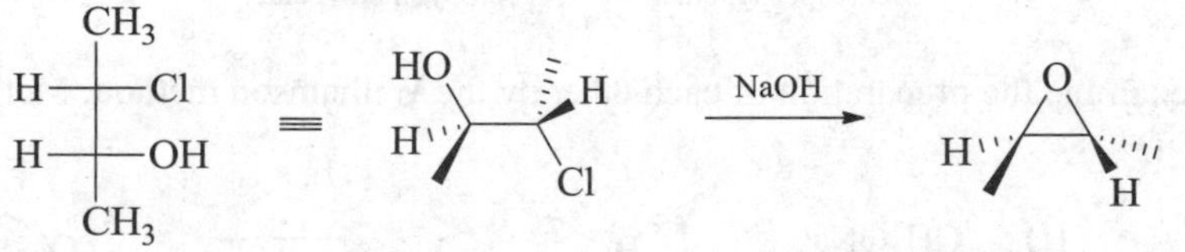

(2R,3S)-3-Chloro-2-butanol

trans-2,3-Epoxybutane
(compound G)

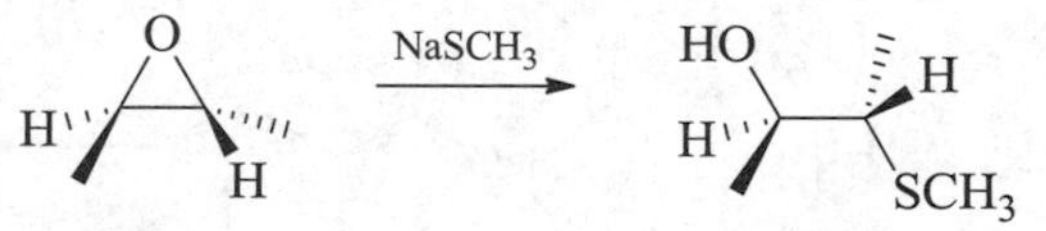

trans-2,3-Epoxybutane
(compound G)

Compound H

(d) Because it gives an epoxide on treatment with a peroxy acid, compound I must be an alkene; more specifically, it is 1,2-dimethylcyclopentene.

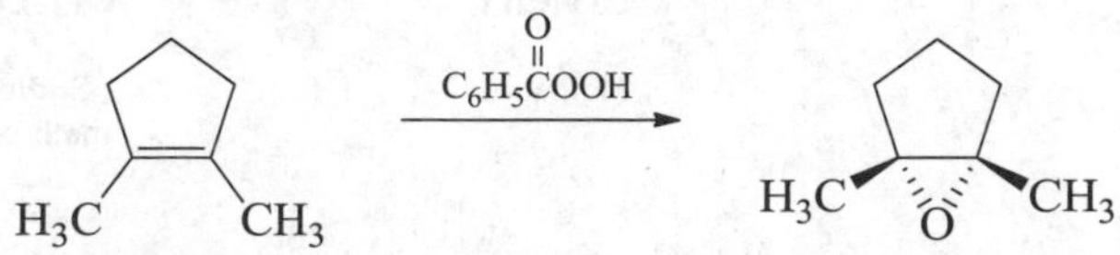

1,2-Dimethylcyclopentene
(compound I)

1,2-Epoxy-1,2-dimethylcyclopentane
(compound K)

Compounds J and L have the same molecular formula, $C_7H_{14}O_2$, but J is a liquid and L is a crystalline solid. Their molecular formulas correspond to the addition of two OH groups to compound I. Osmium tetraoxide brings about syn dihydroxylation of an alkene; therefore, compound J must be the cis diol.

1,2-Dimethylcyclopentene
(compound I)

cis-1,2-Dimethylcyclopentane-1,2-diol
(compound J)

Acid-catalyzed hydrolysis of an epoxide yields a trans diol (compound L):

1,2-Epoxy-1,2-dimethylcyclopentane
(compound K)

trans-1,2-Dimethylcyclopentane-1,2-diol
(compound L)

17.33 The ethers that are to be prepared are

Methyl propyl ether   Isopropyl methyl ether   Diethyl ether

First examine the preparation of each ether by the Williamson method. Methyl propyl ether can be prepared in two ways:

(I) CH_3ONa +   →   +   NaBr

Sodium   1-Bromopropane   Methyl propyl ether   Sodium
methoxide   bromide

(II) CH_3Br +   →   +   NaBr

Methyl   Sodium propoxide   Methyl propyl ether   Sodium
bromide   bromide

Either combination is satisfactory but the second is preferred, because there is no possibility of elimination. The necessary reagents are prepared as shown.

$$CH_3OH \xrightarrow{\text{Na}} CH_3ONa$$

Methanol   Sodium
methoxide

1-Propanol   1-Bromopropane

$$CH_3OH \xrightarrow[\text{(or HBr)}]{PBr_3} CH_3Br$$

Methanol   Methyl
            bromide

$$HO\diagdown \xrightarrow{Na} NaO\diagdown$$

1-Propanol   Sodium propoxide

Isopropyl methyl ether is best prepared by the reaction

$$CH_3Br \quad + \quad NaO\diagup\diagdown \longrightarrow \diagdown O\diagup\diagdown \quad + \quad NaBr$$

Methyl   Sodium isopropoxide   Isopropyl methyl ether   Sodium
bromide   bromide

The reaction of sodium methoxide with isopropyl bromide will proceed mainly by elimination. Methyl bromide is prepared as shown previously; sodium isopropoxide can be prepared by adding sodium to isopropyl alcohol.

Diethyl ether may be prepared as outlined:

$$HO\diagdown \xrightarrow{Na} NaO\diagdown$$

Ethanol   Sodium ethoxide

$$HO\diagdown \xrightarrow[\text{(or HBr)}]{PBr_3} Br\diagdown$$

Ethanol   Ethyl bromide

$$Br\diagdown \quad + \quad NaO\diagdown \longrightarrow \diagdown O\diagdown \quad + \quad NaBr$$

Ethyl bromide   Sodium ethoxide   Diethyl ether   Sodium
                                                          bromide

17.34   The best approach to this problem is to first write the equations in full stereochemical detail.

(a)

(R)-1,2-Epoxypropane   (R)-1,2-Propanediol

It now becomes clear that the arrangement of groups around the chirality center remains unchanged in going from starting materials to products. Therefore, choose conditions such that the nucleophile attacks the CH$_2$ group of the epoxide rather than the chirality center. Base-catalyzed hydrolysis is required; aqueous sodium hydroxide is appropriate.

The nucleophile (hydroxide ion) attacks the less hindered carbon of the epoxide ring.

(*b*)

(*S*)-1,2-Propanediol

Inversion of configuration at the chirality center is required. The nucleophile must therefore attack the chirality center, and acid-catalyzed hydrolysis should be chosen. Dilute sulfuric acid would be satisfactory.

The nucleophile (a water molecule) attacks that carbon atom of the ring that can better support a positive charge. Carbocation character develops at the transition state and is better supported by the carbon atom that is more highly substituted.

17.35 The simplest approach to this problem is to recognize that no bonds to the chirality center are made or broken beginning with (*S*)-glycidol. The bond to the chirality center projects outward from the paper in all the intermediates in the synthesis. The propranolol produced has the (*S*)-configuration.

(*S*)-Glycidol   (*S*)-Propranolol

17.36 (*a*) All the methods that we have so far discussed for the preparation of epoxides are based on alkenes as starting materials. This leads us to consider the partial retrosynthesis shown.

Target molecule   Key intermediate

The key intermediate, 1-phenylcyclohexene, is both a proper precursor to the desired epoxide and readily available from the given starting materials. A reasonable synthesis is

Preparation of the required tertiary alcohol, 1-phenylcyclohexanol, completes the synthesis.

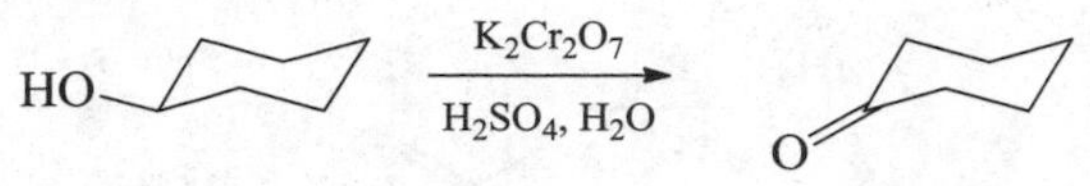

Cyclohexanol → Cyclohexanone

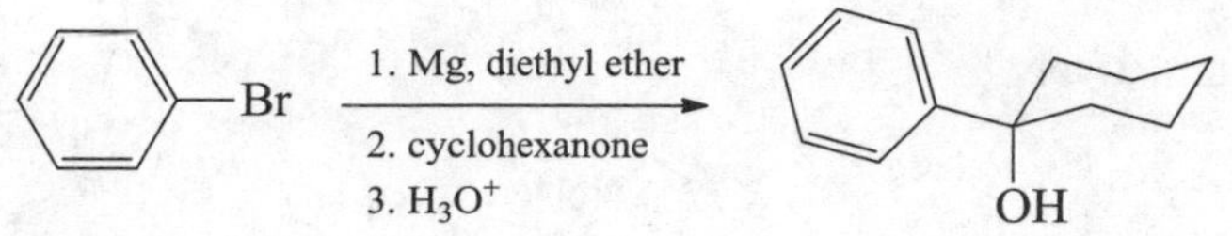

(b) The necessary carbon skeleton can be assembled through the reaction of a Grignard reagent with 1,2-epoxypropane.

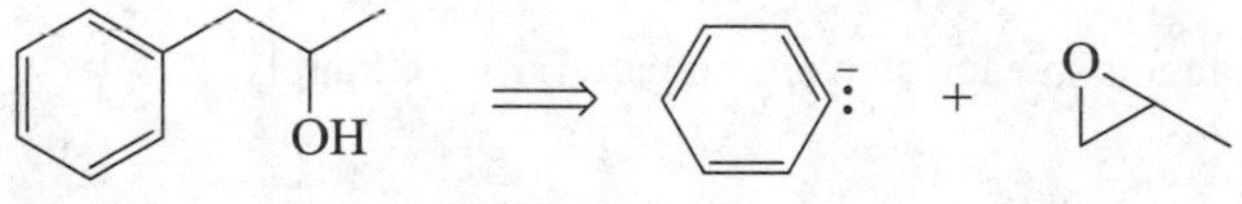

The reaction sequence is therefore

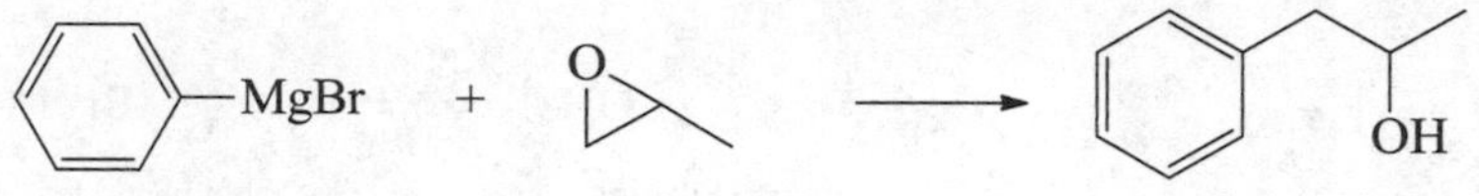

Phenylmagnesium bromide (from bromobenzene and magnesium) + 1,2-Epoxypropane → 1-Phenyl-2-propanol

The epoxide required in the first step, 1,2-epoxypropane, is prepared as follows from isopropyl alcohol:

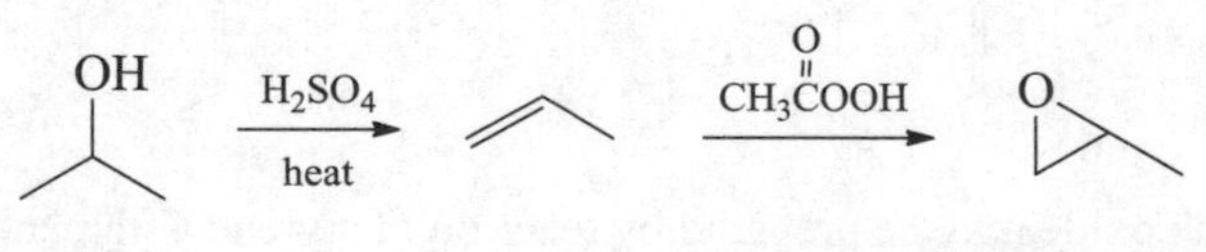

2-Propanol (isopropyl alcohol) → Propene → 1,2-Epoxypropane

(c) Because the target molecule is an ether, it ultimately derives from two alcohols.

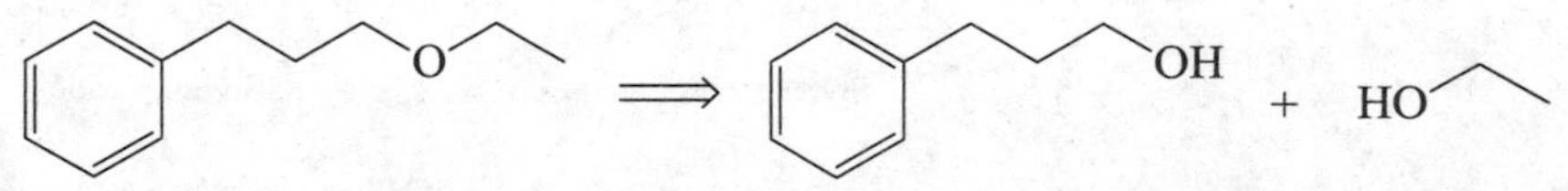

Our first task is to assemble 3-phenyl-1-propanol from the designated starting material benzyl alcohol. This requires formation of a primary alcohol with the original carbon chain extended by two carbons. The standard method for this transformation involves reaction of a Grignard reagent with ethylene oxide.

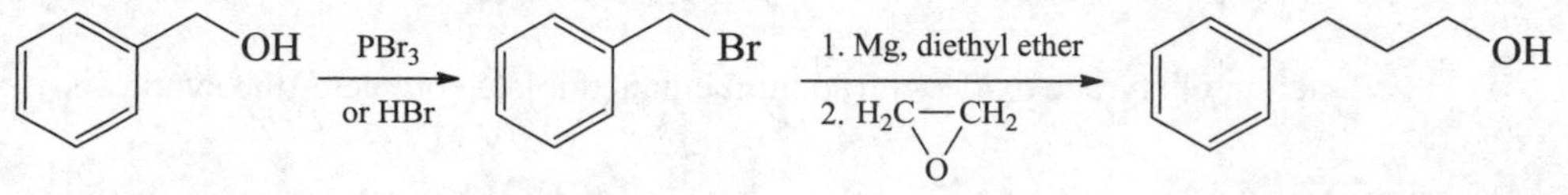

Benzyl alcohol → Benzyl bromide → 3-Phenyl-1-propanol

After 3-phenyl-1-propanol has been prepared, its conversion to the corresponding ethyl ether can be accomplished in either of two ways:

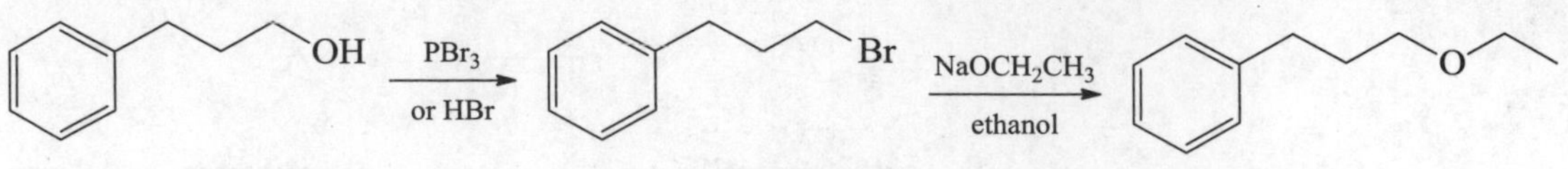

3-Phenyl-1-propanol　　　　　　　1-Bromo-3-phenylpropane　　　　　　　Ethyl 3-phenylpropyl ether

or alternatively

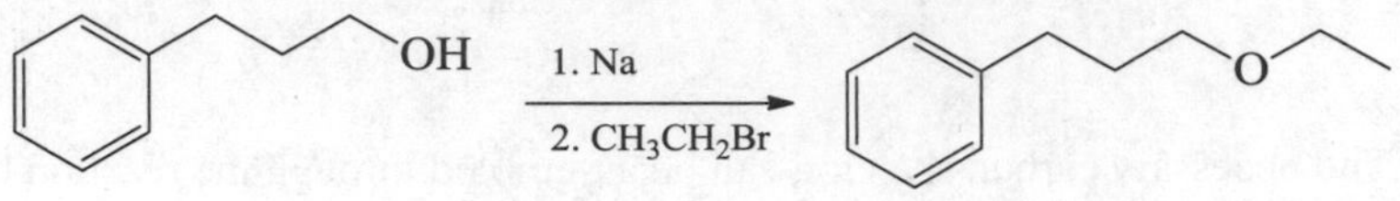

3-Phenyl-1-propanol　　　　　　　　　　　　Ethyl 3-phenylpropyl ether

The reagents in each step are prepared from ethanol.

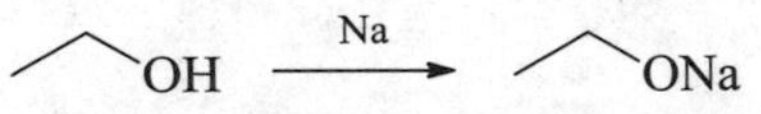

Ethanol　　　　　　　　　　　　　　Sodium ethoxide

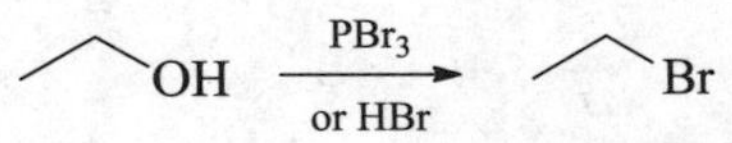

Ethanol　　　　　　　　　　　　　Ethyl bromide

(*d*) Retrosynthetic analysis reveals that the desired target molecule may be prepared by reaction of an epoxide with an ethanethiolate ion.

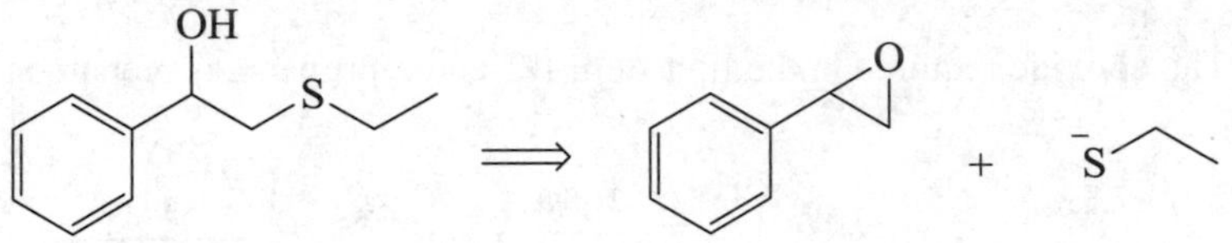

Styrene oxide may be prepared by reaction of styrene with peroxyacetic acid.

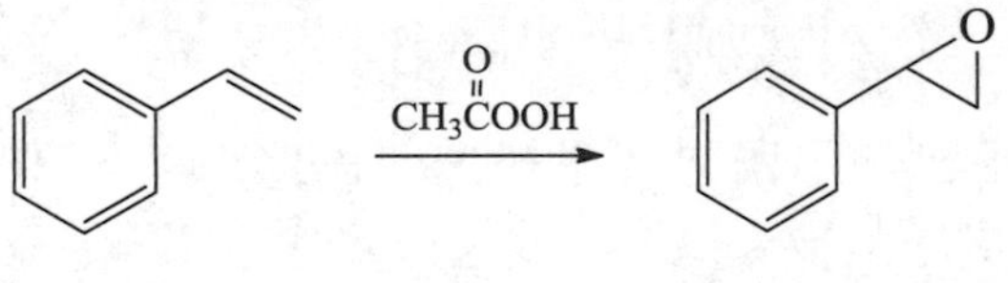

Styrene　　　　　　　　　　　　　　　Styrene oxide

The necessary thiolate anion is prepared from ethanol by way of the corresponding thiol.

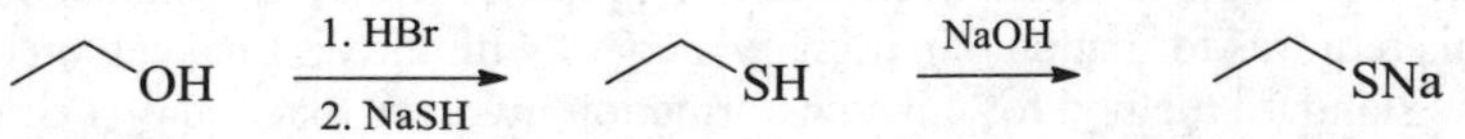

Ethanol　　　　　　　　　Ethanethiol　　　　　　Sodium ethanethiolate

Reaction of styrene oxide with sodium ethanethiolate completes the synthesis.

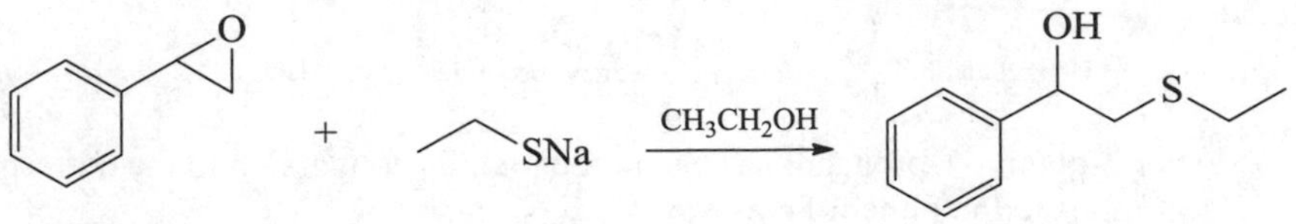

Styrene oxide　　　　Sodium ethanethiolate

Mechanisms

17.37 Protonation of oxygen to form an alkyloxonium ion is followed by loss of water. The resulting carbocation has a plane of symmetry and is achiral. Capture of the carbocation by methanol yields both enantiomers of 2-methoxy-2-phenylbutane. The product is racemic.

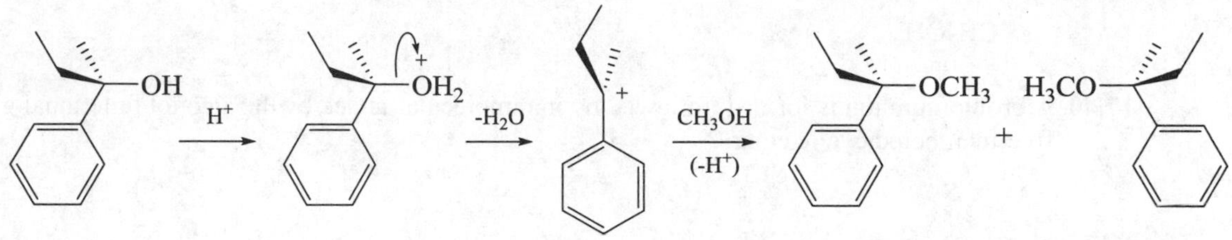

(*R*)-(+)-2-Phenyl-2-butanol   (Achiral carbocation)   2-Methoxy-2-phenylbutane (racemic)

17.38 Organolithium reagents react readily with the carbonyl group of ketones. The intermediate alkoxide ion can then undergo an intramolecular displacement of chloride to give the observed epoxide product. Although normally an S_N2 reaction would not occur with a tertiary halide, the fact that it is intramolecular allows the reaction to proceed.

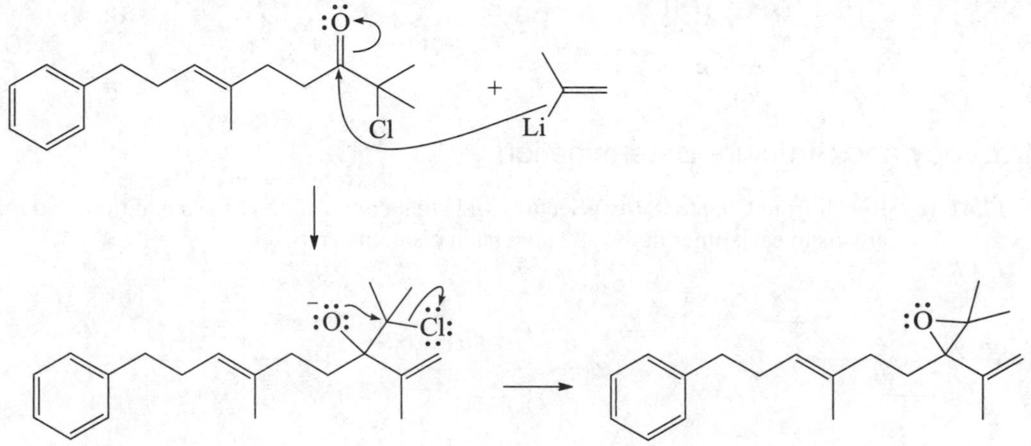

17.39 The fact that 1,2-dibromohexane is not converted to 1-bromo-2-methoxyhexane rules out the possibility of an S_N1 reaction from the dibromide product:

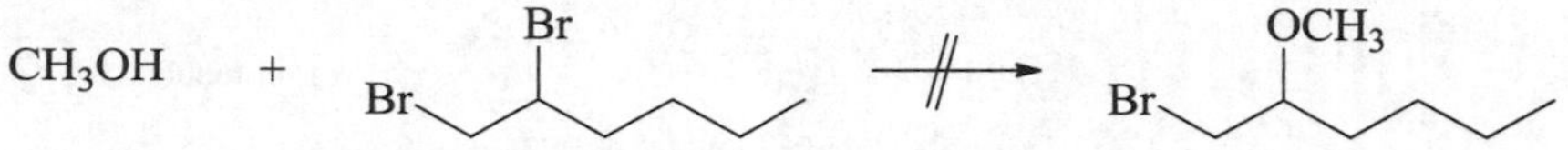

The mechanism is similar to the mechanism of halohydrin formation that is described in Chapter 8. A bromonium ion intermediate is formed in the first step of the mechanism.

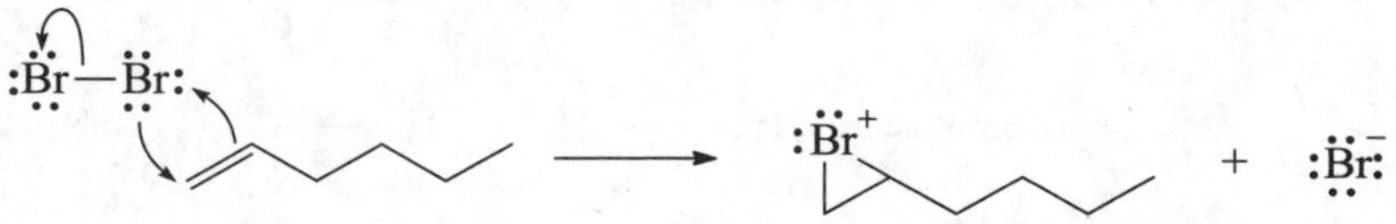

This reactive bromonium ion is then attacked by methanol. So instead of using H_2O as done in halohydrin formation, methanol attacks the bromonium ion intermediate to give the ether product.

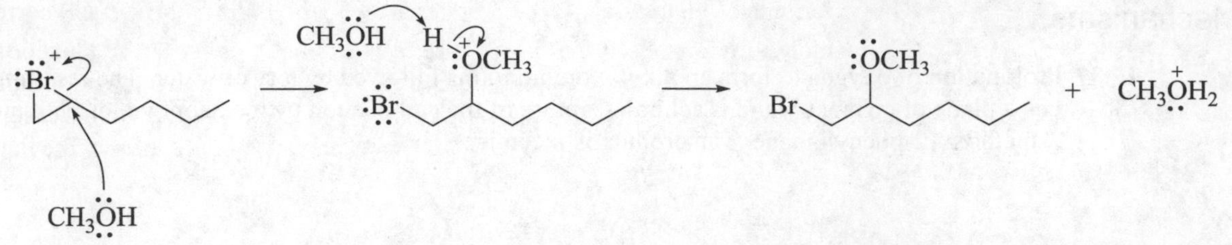

17.40 A bromonium ion is formed, followed by intramolecular attack by the alcohol functional group to form the five-membered cyclic ether.

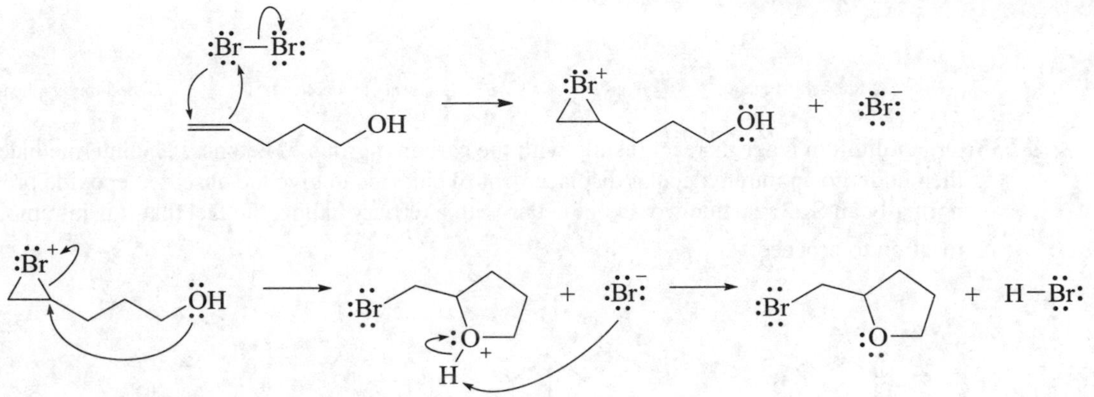

Spectroscopy and Structure Determination

17.41 (*a*) Recall from Chapter 8 (text Section 8.11) that epoxidation is a syn addition, and that substituents that are cis to each other in the alkene remain cis in the epoxide.

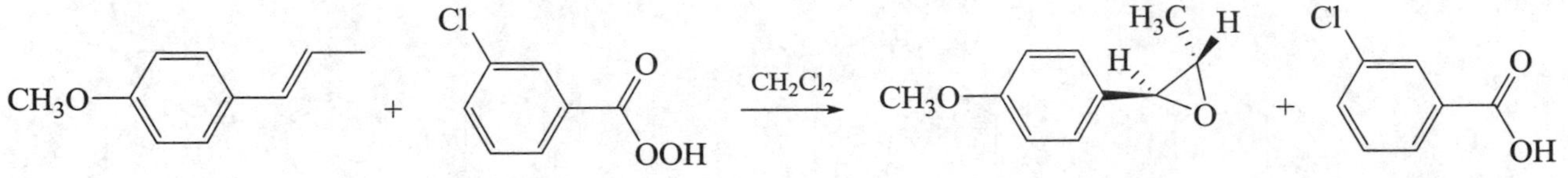

(*E*)-1-(*p*-Methoxyphenyl)propene   *m*-Chloroperoxybenzoic acid   *trans*-1,2-Epoxy-1-   *m*-Chlorobenzoic acid
                                                                                  (*p*-methoxyphenyl)propane

(*b*) The splitting patterns allow assignment of the protons on the epoxide ring; the proton on C-1 is a doublet, whereas the proton on C-2 is split both by the proton on C-1 and the protons of the methyl group. The higher-field ring protons are those ortho to the electron-donating methoxy group.

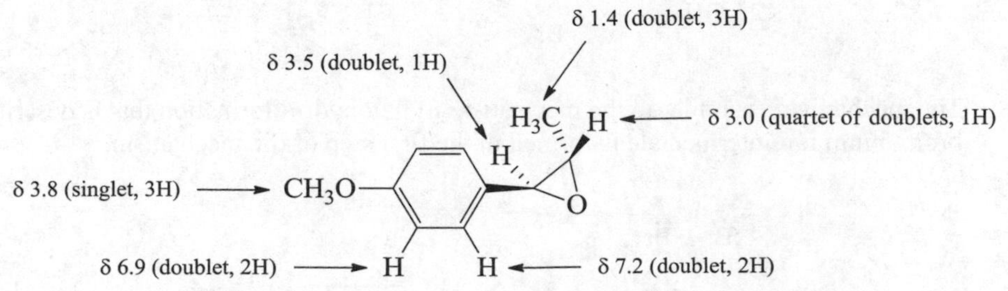

(*c*) From Table 14.2 in the text, you can see that the chemical shift range δ 50–65 in a ^{13}C NMR spectrum is typical for carbon atoms adjacent to an oxygen. The three signals in this region of the ^{13}C NMR spectrum of the epoxide arise from the two carbon atoms of the epoxide ring and the carbon of the methoxy group.

(*d*) The molecular formula of the product formed under acidic conditions ($C_{16}H_{17}O_4Cl$) corresponds to addition of *m*-chlorobenzoic acid to the epoxide. Recall from text Section 17.12 that epoxides undergo acid-catalyzed ring opening. Because carbocation character develops at the transition state, substitution is favored at the carbon that can better support a developing positive charge. In this case, substitution occurs at the benzylic carbon. For clarity, the *p*-methoxyphenyl group has been abbreviated as Ar in the following mechanism.

17.42 Given the information specified in the problem (molecular formula and method of preparation), the only possible alcohols are

1-Phenyl-1-propanol 1-Phenyl-2-propanol 2-Phenyl-1-propanol

The H—C—O signals in both spectra are in the δ 4–5 range and integrate for one proton. Therefore, we can eliminate 2-phenyl-1-propanol from consideration because its CH_2O group requires two protons.

The signal for the CH_3 group in compound A appears as a doublet at δ 1.2. This is consistent with compound A being 1-phenyl-2-propanol. Likewise, the signal for its H—C—O proton is consistent with the multiplet at δ 4 because this proton is coupled to five vicinal protons.

This leaves 1-phenyl-1-propanol as compound B, which is consistent with the NMR spectrum. The CH_3 signal is a triplet indicating that CH_3 is adjacent to CH_2. The same applies to the triplet for the H—C—O proton at δ 4.5.

To give 1-phenyl-2-propanol (compound A) as the product formed by the reaction of phenyllithium with 1,2-epoxypropane, phenyllithium must have attacked the CH_2 group of the epoxide.

1-Phenyl-2-propanol
(compound A)

The same is true for the formation of 1-phenyl-1-propanol (compound B) from the reaction of methyllithium with styrene oxide. Methyllithium must have attacked the CH_2 group of the epoxide.

1-Phenyl-1-propanol
(compound B)

The generalization we can draw from these experiments is that when organolithium reagents react with unsymmetrical epoxides, they attack the less crowded carbon of the ring. In this respect, the reactions resemble S_N2 processes.

17.43 A good way to address this problem is to consider the dibromide derived by treatment of compound A with hydrogen bromide. The presence of an NMR signal equivalent to four protons in the aromatic region at δ 7.3 indicates that this dibromide contains a disubstituted aromatic ring. The four remaining protons appear as a sharp singlet at δ 4.7 and are most reasonably contained in two equivalent methylene groups of the type $ArCH_2Br$. Because the dibromide contains all the carbons and hydrogens of the starting material and is derived from it by treatment with hydrogen bromide, it is likely that compound A is a cyclic ether in which a CH_2OCH_2 unit spans two of the carbons of a benzene ring. This can occur only when the positions involved are ortho to each other. Therefore

17.44 The molecular formula ($C_{10}H_{13}BrO$) indicates an index of hydrogen deficiency of 4. One of the products obtained on treatment of the compound with HBr is benzyl bromide ($C_6H_5CH_2Br$), which accounts for seven of its ten carbons and all the double bonds and rings. Thus, the compound is a benzyl ether having the formula $C_6H_5CH_2OC_3H_6Br$. The ^1H NMR spectrum includes a five-proton signal at δ 7.4 for a monosubstituted benzene ring and a two-proton singlet at δ 4.6 for the benzylic protons. This singlet appears at low field because the benzylic protons are bonded to oxygen.

$$C_6H_5CH_2OC_3H_6Br \xrightarrow[\text{heat}]{\text{HBr}} C_6H_5CH_2Br + C_3H_6Br_2$$

The six remaining protons appear as two overlapping two-proton triplets at δ 3.6 and 3.7, along with a two-proton pentet at δ 2.2, consistent with the unit $-OCH_2CH_2CH_2Br$. The compound is $C_6H_5CH_2OCH_2CH_2CH_2Br$.

17.45 The high index of hydrogen deficiency (5) of the unknown compound $C_9H_{10}O$ and the presence of six signals in the δ 120–140 region of the ^{13}C NMR spectrum suggests the presence of an aromatic ring. The problem states that the compound is a cyclic ether; thus, the oxygen atom is contained in a second ring fused to the benzene ring. As oxidation yields 1,2-benzenedicarboxylic acid, the second ring must be attached to the benzene ring by carbon atoms.

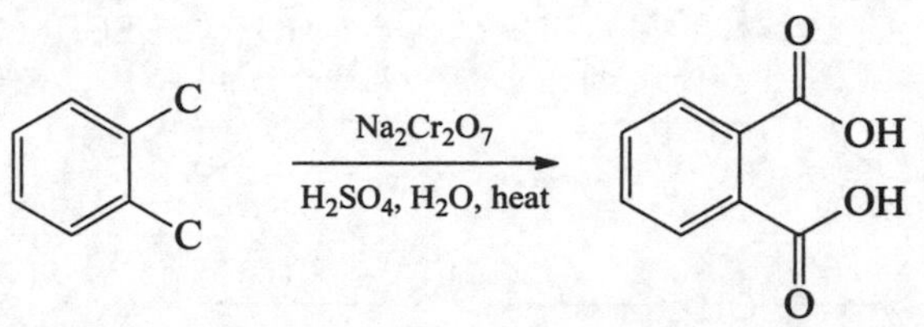

(C$_9$H$_{10}$O)   1,2-Benzenedicarboxylic acid

Two structures are possible with this information; however, only one of them is consistent with the presence of three CH$_2$ groups in the ^{13}C NMR spectrum. The compound is

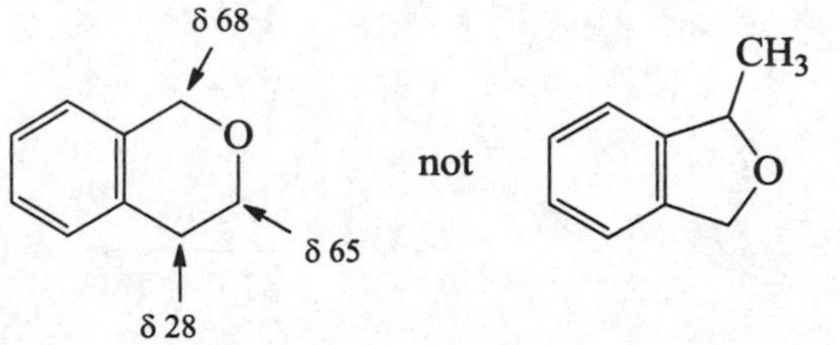

Answers to Interpretive Problems

17.46 C;   17.47 B;   17.48 B;   17.49 A;   17.50 C;   17.51 D

SELF-TEST

1. Write the structures and give a correct name for all the isomeric ethers of molecular formula C$_4$H$_{10}$O.

2. Give the structure of the product obtained from each of the following reactions. Show stereochemistry where it is important.

(a)

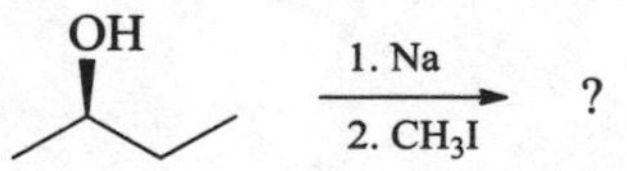

(d)

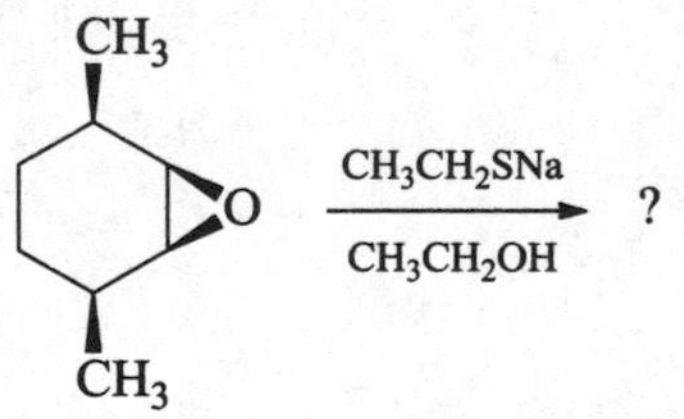

(b)

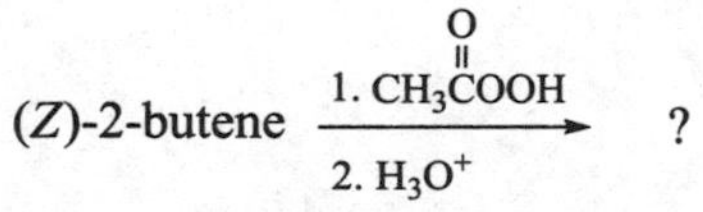

(Z)-2-butene $\xrightarrow[\text{2. H}_3\text{O}^+]{\text{1. CH}_3\overset{\text{O}}{\overset{\|}{\text{C}}}\text{OOH}}$?

(e)

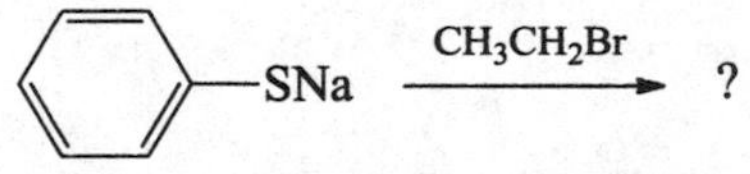

(c)

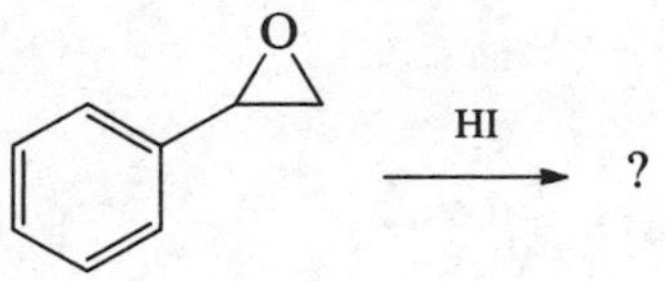

(f)

Product of part (e) $\xrightarrow{\text{NaIO}_4}$?

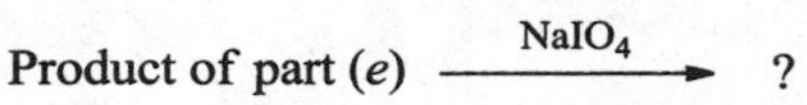

3. Outline a scheme for the preparation of cyclohexyl ethyl ether using the Williamson method.

4. Outline a synthesis of 2-ethoxyethanol, CH$_3$CH$_2$OCH$_2$CH$_2$OH, using ethanol as the source of all the carbon atoms.

5. Provide the reagents necessary to complete each of the following conversions. In each case, give the structure of the intermediate product.

(*a*)

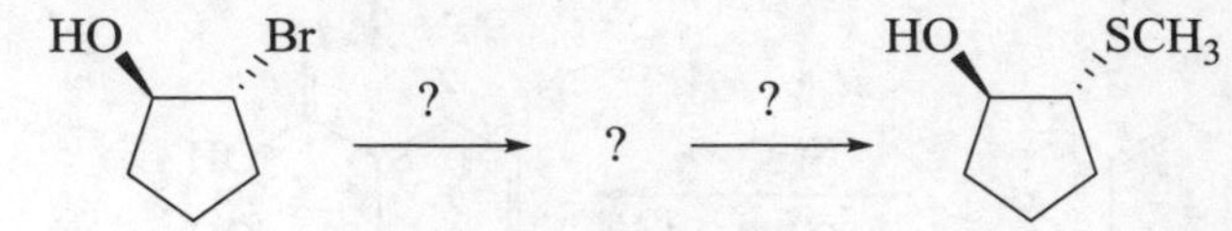
HO
Br
?
?
?
HO
SCH3

(*b*)

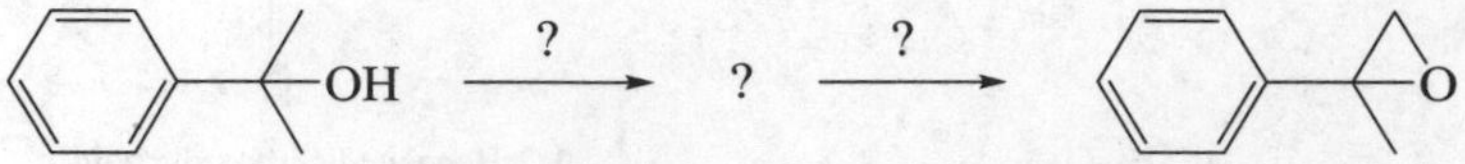
OH
?
?
?
O

6. Provide structures for compounds A and B in the following reaction scheme:

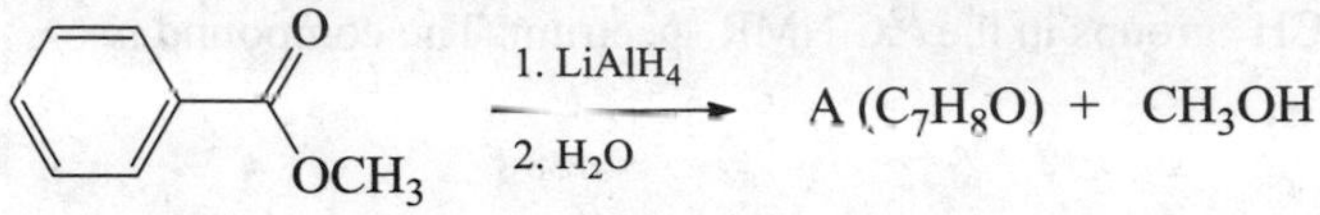
O
OCH3
1. LiAlH4
2. H2O
A (C7H8O) + CH3OH

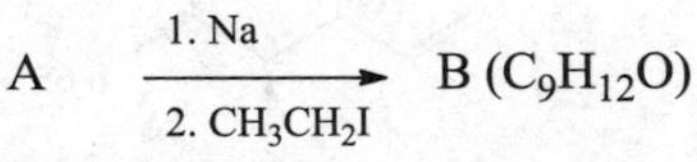
A
1. Na
2. CH3CH2I
B (C9H12O)

7. Using any necessary organic or inorganic reagents, provide the steps to carry out the following synthetic conversion:

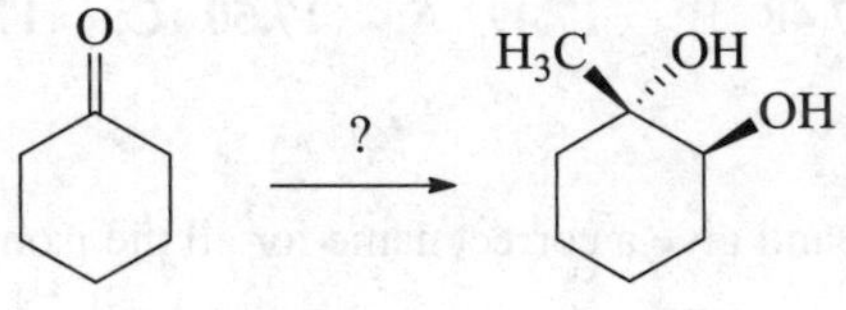
O
?
H3C
OH
OH

8. Give the final product, including stereochemistry, of the following reaction sequence:

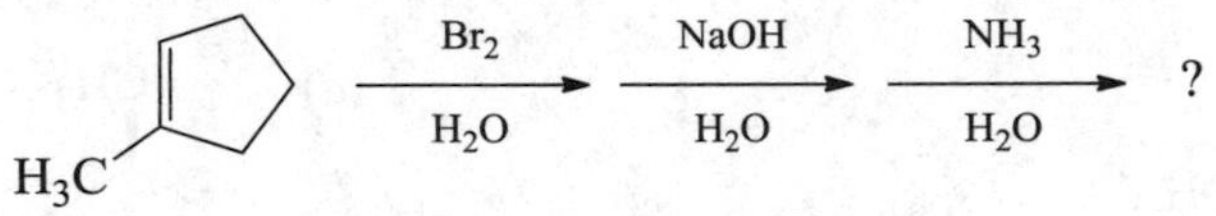
H3C
Br2
H2O
NaOH
H2O
NH3
H2O
?

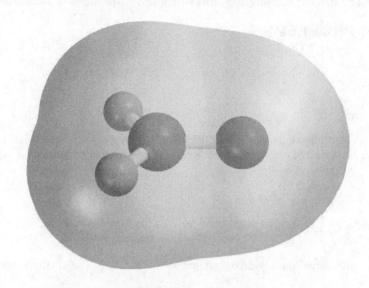

CHAPTER 18

Aldehydes and Ketones: Nucleophilic Addition to the Carbonyl Group

Table of Contents

SOLUTIONS TO TEXT PROBLEMS

In Chapter Problems

18.1 (*b*) The longest continuous chain in glutaraldehyde has five carbons and terminates in aldehyde functions at both ends. *Pentanedial* is an acceptable IUPAC name for this compound.

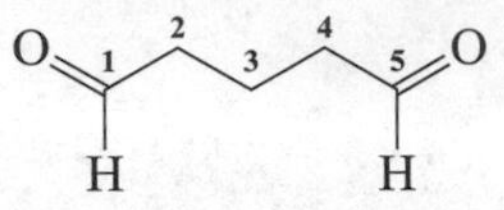

Pentanedial (glutaraldehyde)

(*c*) The three-carbon chain has hydroxy groups on C-2 and C-3. The aldehyde function is at C-1 and is not numbered.

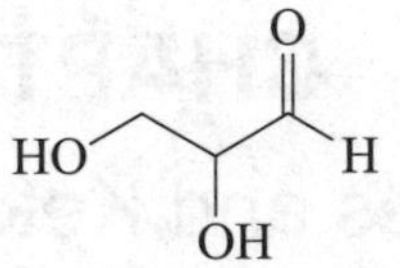

2,3-Dihydroxypropanal
(glyceraldehyde)

(*d*) Vanillin can be named as a derivative of benzaldehyde. Remember to cite the remaining substituents in alphabetical order.

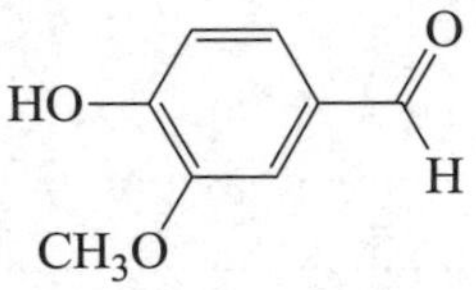

4-Hydroxy-3-methoxybenzaldehyde
(vanillin)

18.2 (*b*) First write the structural formula from the name given. Ethyl isopropyl ketone has an ethyl group and an isopropyl group bonded to a carbonyl group.

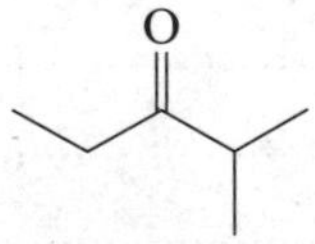

The substitutive name of ethyl isopropyl ketone is 2-methyl-3-pentanone. Its longest continuous chain has five carbons. The carbonyl carbon is C-3 irrespective of the direction in which the chain is numbered, and so we choose the direction that gives the lower number to the position that bears the methyl group. Alternatively, the ketone can be named 2-methylpentan-3-one.

(*c*) Methyl 2,2-dimethylpropyl ketone has a methyl group and a 2,2-dimethylpropyl group bonded to a carbonyl group.

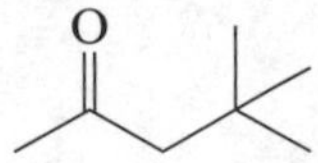

The longest continuous chain has five carbons, and the carbonyl carbon is C-2. Thus, the substitutive name of methyl 2,2-dimethylpropyl ketone is 4,4-dimethyl-2-pentanone or 4,4-dimethylpentan-2-one.

(*d*) The structure corresponding to allyl methyl ketone is

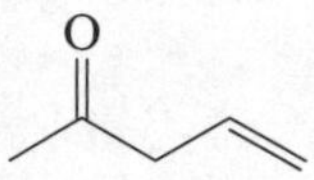

Because the carbonyl group is given the lowest possible number in the chain, the substitutive name is 4-penten-2-one, *not* 1-penten-4-one. An alternative name is pent-4-en-2-one.

18.3 The carbonyl oxygen of an aldehyde and a proton of water can form a hydrogen bond. The hydrogen bond between benzaldehyde and water can be represented as

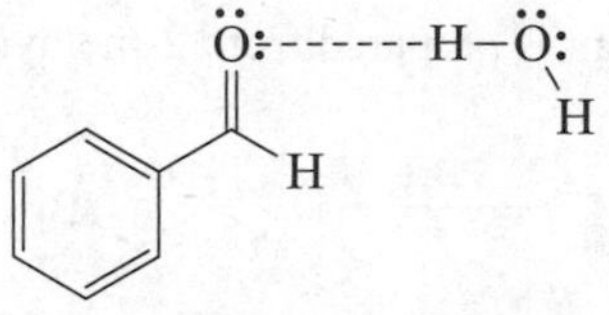

Benzaldehyde   Water

18.4 Catalytic hydrogenation cannot reduce carboxylic acids to alcohols. Lithium aluminum hydride is the only reagent we have discussed that is capable of reducing carboxylic acids (text Section 16.3).

18.5 (*a*) The target molecule, 2-butanone, contains four carbon atoms. The problem states that all of the carbons originate in acetic acid, which has two carbon atoms. This suggests the following disconnections:

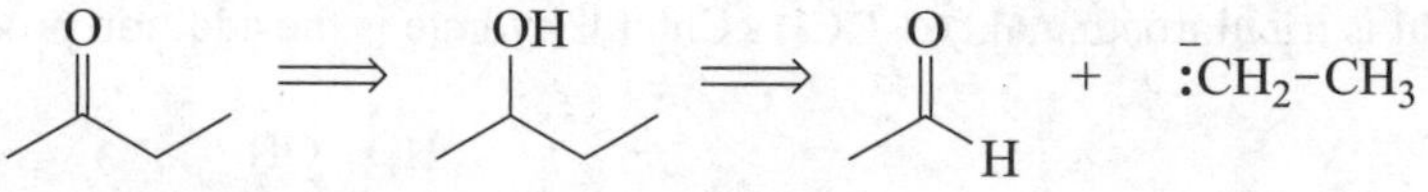

2-Butanone

The necessary aldehyde (acetaldehyde) is prepared from acetic acid by reduction followed by oxidation in an anhydrous medium.

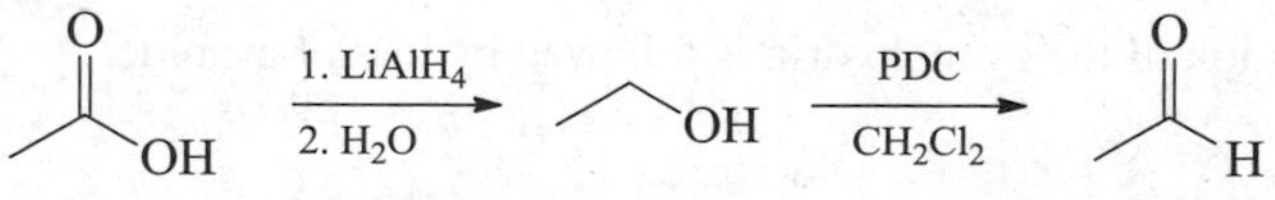

Acetic acid   Ethanol   Acetaldehyde

Ethylmagnesium bromide may be obtained from acetic acid by the following sequence:

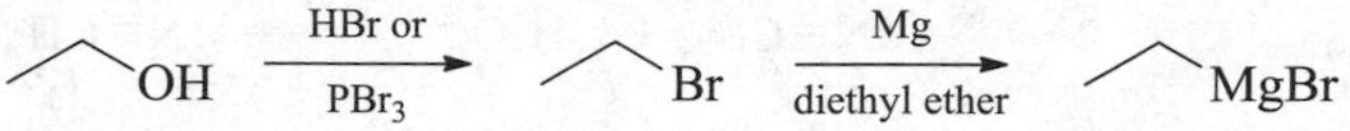

Ethanol   Ethyl bromide   Ethylmagnesium
(prepared as   bromide
previously)

The preparation of 2-butanone is completed as follows:

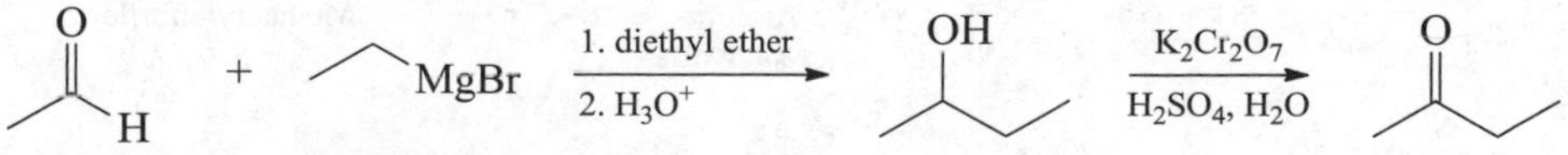

Acetaldehyde   Ethylmagnesium   2-Butanol   2-Butanone
                        bromide

(*b*) The target compound has eight carbon atoms. The problem states that it is prepared from four carbon compounds. This suggests the following disconnections:

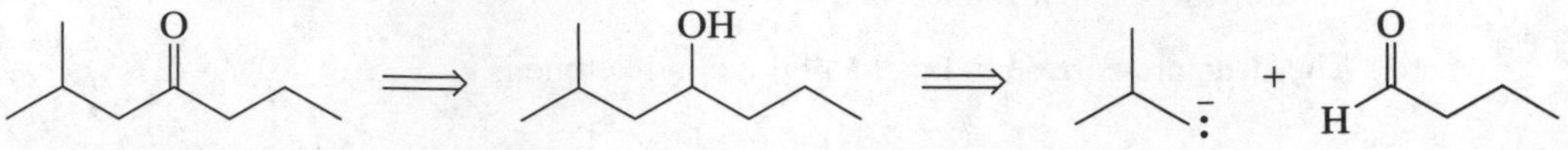

2-Methyl-4-heptanone

The necessary aldehyde (butanal) is prepared by oxidation from 1-butanol.

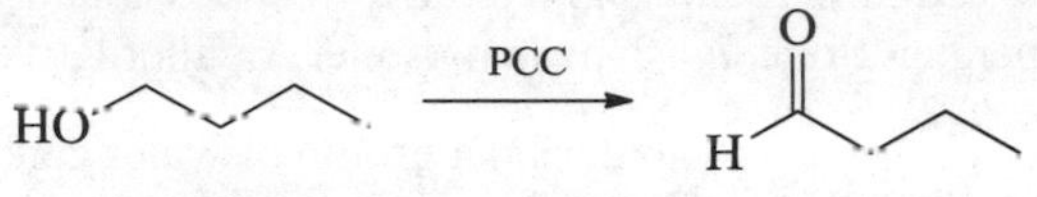

1-Butanol

The required Grignard reagent is prepared from 2-methyl-1-propanol by the following sequence:

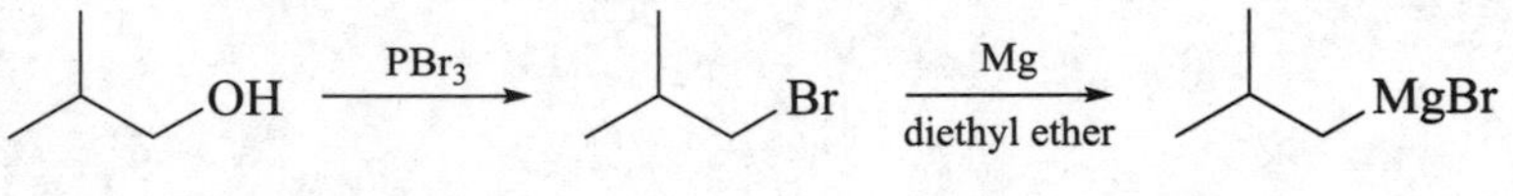

2-Methyl-1-propanol

The preparation of 2-methyl-4-heptanone is completed as follows:

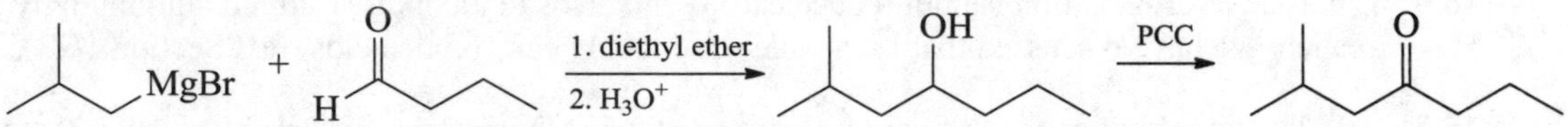

18.6   Chloral is trichloroethanal, Cl_3CCH. Chloral hydrate is the addition product of chloral and water.

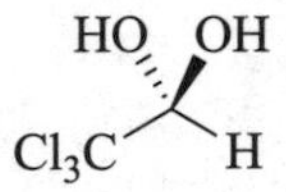

Chloral hydrate

18.7   Deprotonation of the cyanohydrin is followed by loss of cyanide.

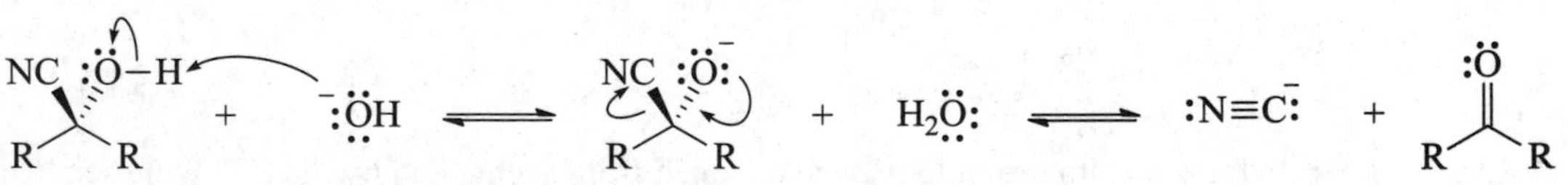

18.8   Methacrylonitrile is formed by the dehydration of acetone cyanohydrin and thus has the structure shown:

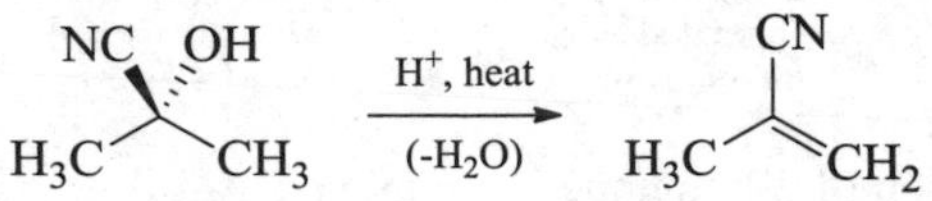

Acetone   Methacrylonitrile
cyanohydrin

18.9 The cyanohydrin and the ketone that it is derived from are revealed by the disconnections shown:

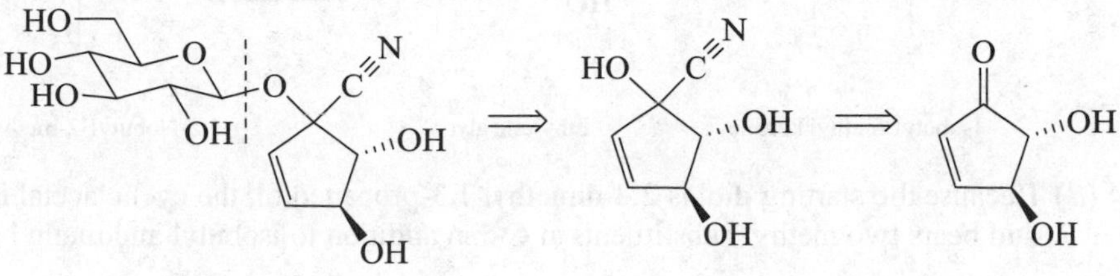

Gynocardin

18.10 The overall reaction is

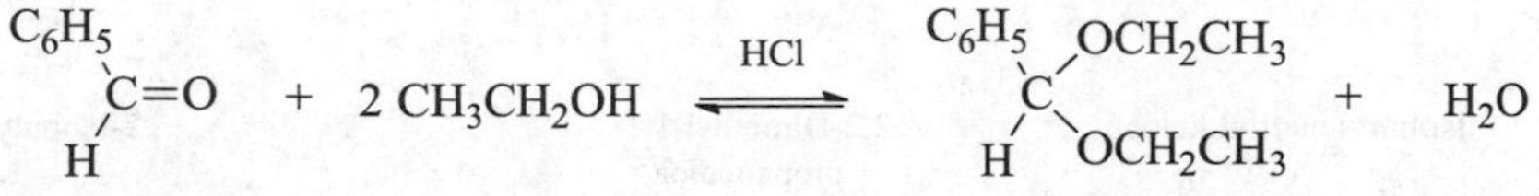

$$C_6H_5\text{—}CH\text{=}O + 2\ CH_3CH_2OH \rightleftharpoons C_6H_5\text{—}CH(OCH_2CH_3)_2 + H_2O$$

| Benzaldehyde | Ethanol | Benzaldehyde diethyl acetal | Water |

HCl is a strong acid and, when dissolved in ethanol, transfers a proton to ethanol to give ethyloxonium ion. Thus, we can represent the acid catalyst as the conjugate acid of ethanol.

The first three steps correspond to acid-catalyzed addition of ethanol to the carbonyl group to yield a hemiacetal.

Step 1:

Step 2:

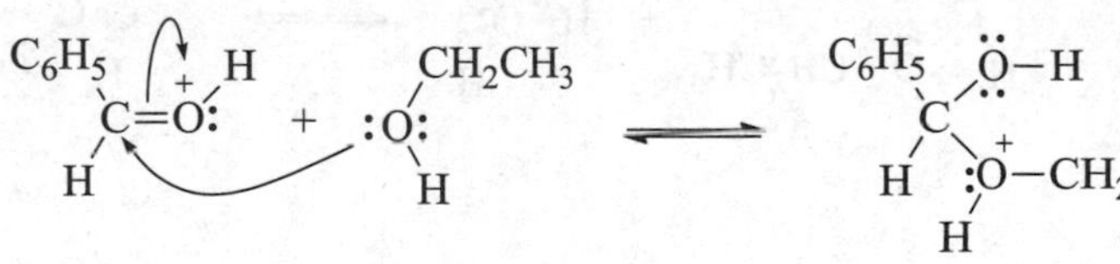

Step 3:

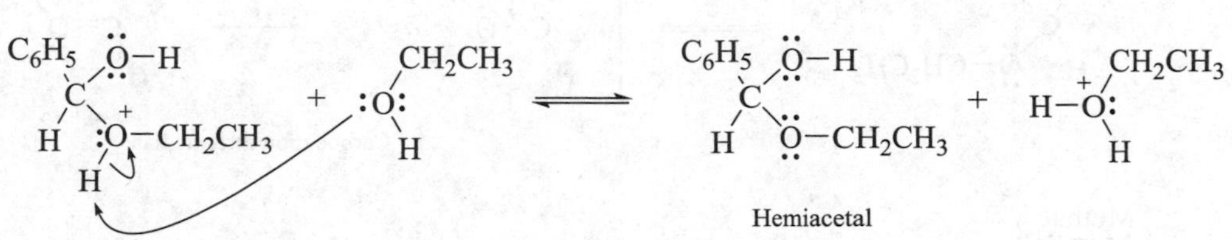

Hemiacetal

18.11 (b) 1,3-Propanediol forms acetals that contain a six-membered 1,3-dioxane ring.

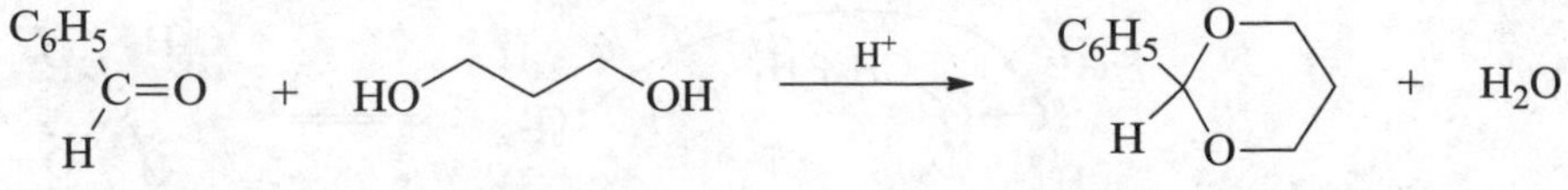

$$C_6H_5\text{—}CH\text{=}O + HO\text{—}CH_2CH_2CH_2\text{—}OH \xrightarrow{H^+} \text{2-Phenyl-1,3-dioxane} + H_2O$$

| Benzaldehyde | 1,3-Propanediol | 2-Phenyl-1,3-dioxane | Water |

(c) The cyclic acetal derived from isobutyl methyl ketone and ethylene glycol bears an isobutyl group and a methyl group at C-2 of a 1,3-dioxolane ring.

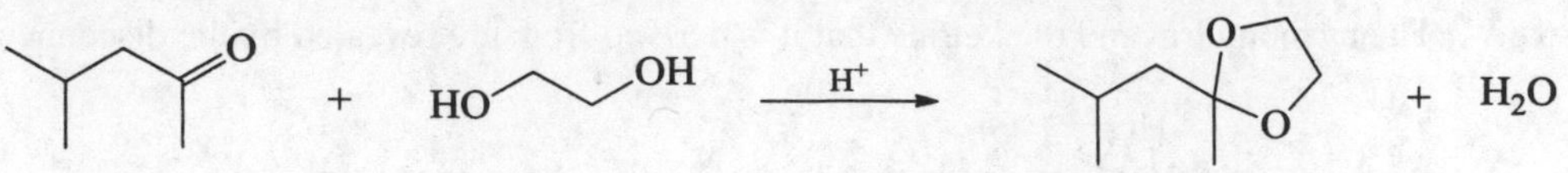

| Isobutyl methyl ketone | Ethylene glycol | 2-Isobutyl-2-methyl-1,3-dioxolane | Water |

(d) Because the starting diol is 2,2-dimethyl-1,3-propanediol, the cyclic acetal is six-membered and bears two methyl substituents at C-5 in addition to isobutyl and methyl groups at C-2.

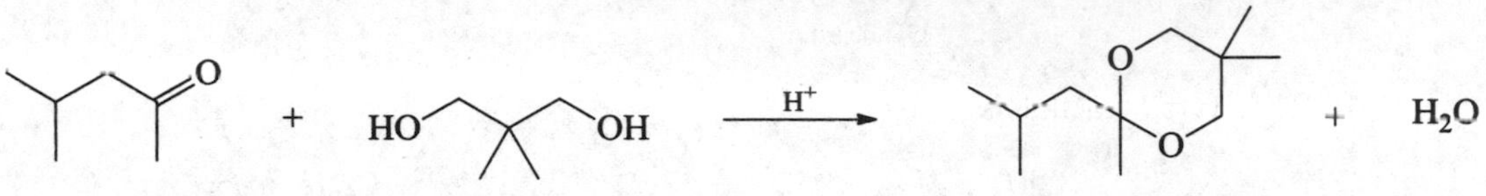

| Isobutyl methyl ketone | 2,2-Dimethyl-1,3-propanediol | 2-Isobutyl-2,5,5-trimethyl-1,3-dioxane | Water |

18.12 The overall reaction is

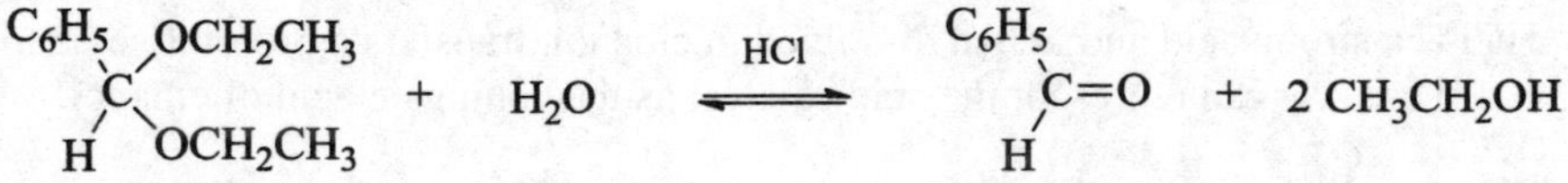

| Benzaldehyde diethyl acetal | Water | Benzaldehyde | Ethanol |

The mechanism of acetal hydrolysis is the reverse of acetal formation. The first four steps convert the acetal to the hemiacetal.

Step 1:

Step 2:

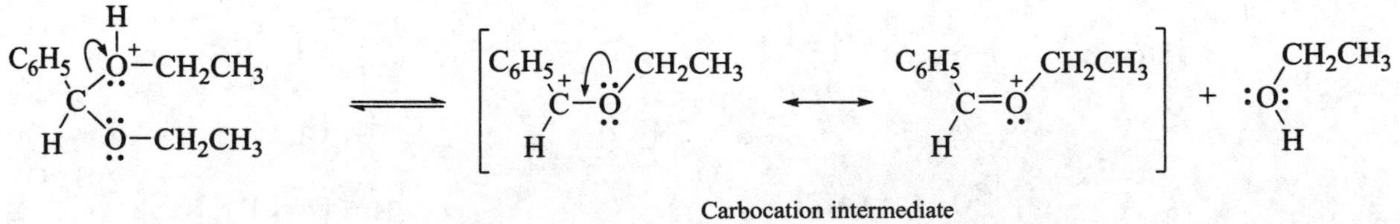

Carbocation intermediate

Step 3:

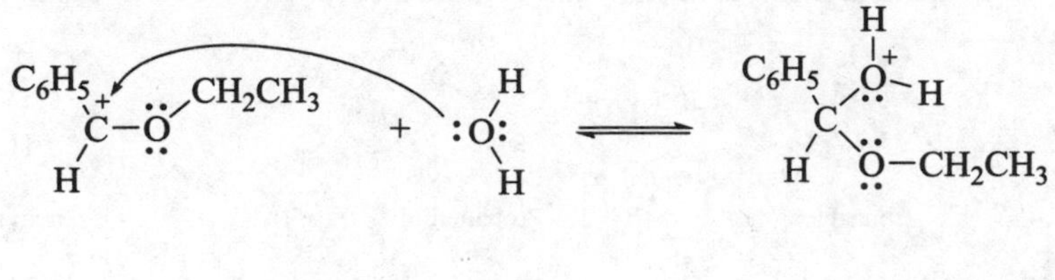

Step 4:

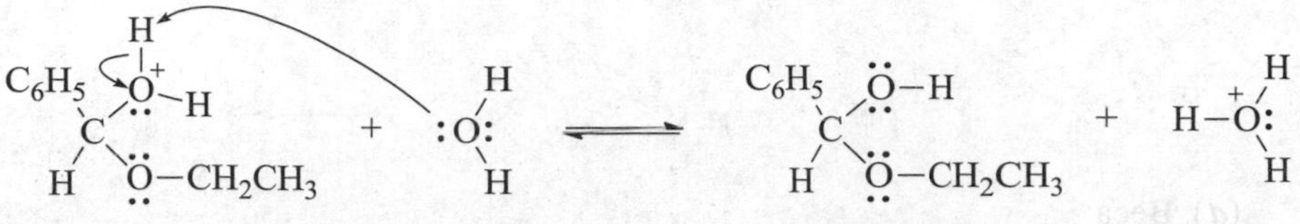

Hemiacetal

Step 5:

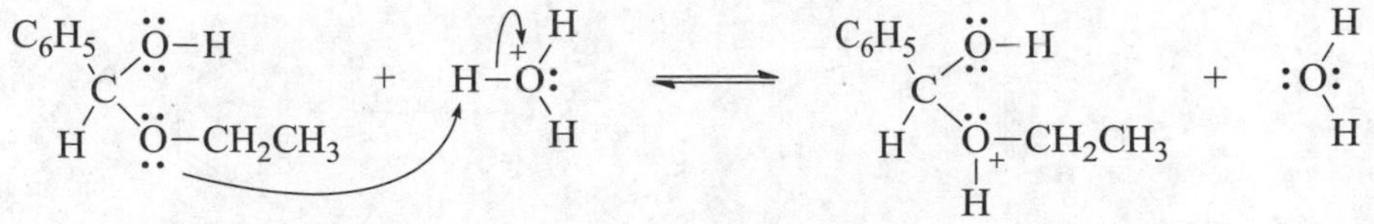

Hemiacetal

Step 6:

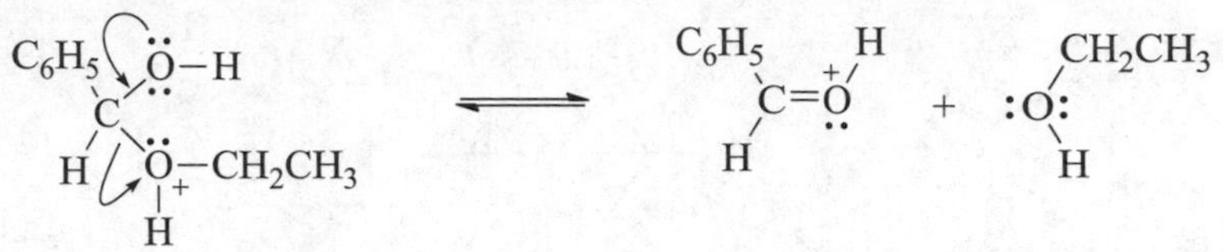

Step 7:

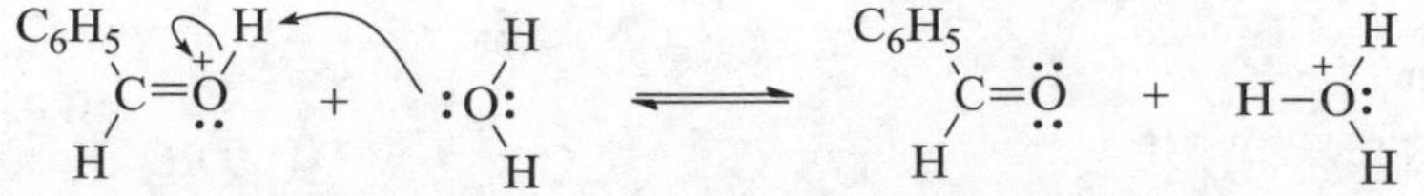

18.13 The conversion requires reduction; however, the conditions necessary ($LiAlH_4$) would also reduce the ketone carbonyl. The ketone functionality is therefore protected as the cyclic acetal.

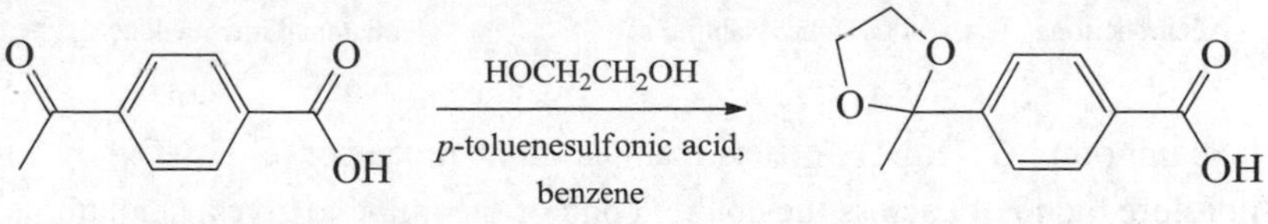

4-Acetylbenzoic acid

$HOCH_2CH_2OH$

p-toluenesulfonic acid, benzene

Reduction of the carboxylic acid may now be carried out.

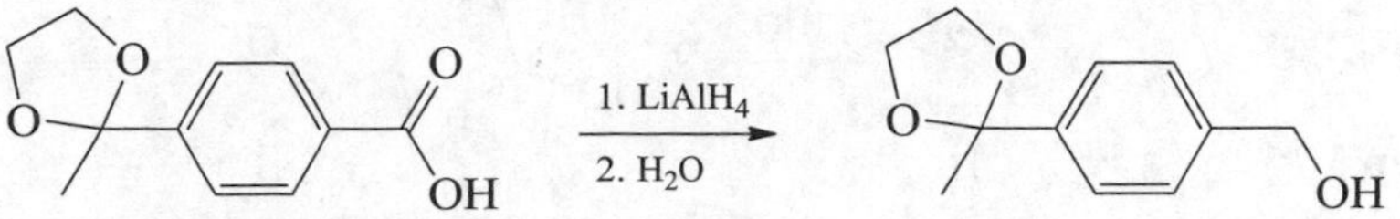

1. $LiAlH_4$
2. H_2O

Hydrolysis to remove the protecting group completes the synthesis.

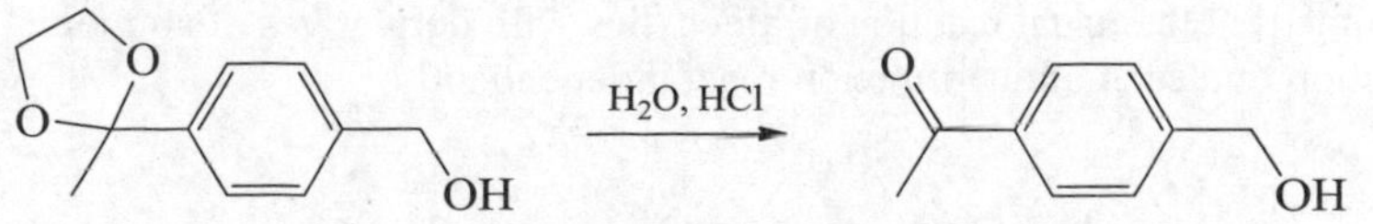

H_2O, HCl

4-Acetylbenzyl alcohol

18.14 (*b*)   Nucleophilic addition of butylamine to benzaldehyde gives the hemiaminal.

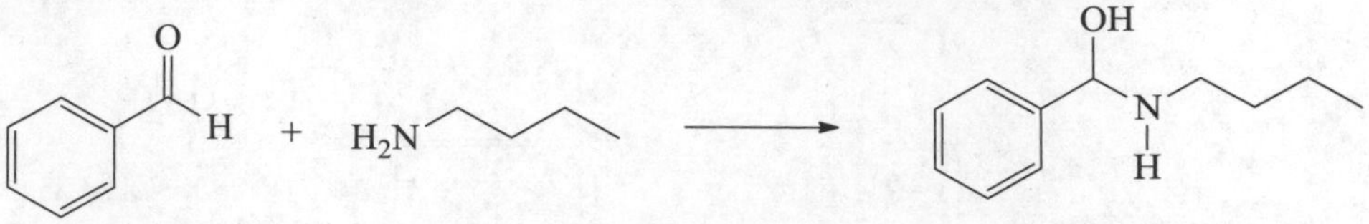

Benzaldehyde   Butylamine   Hemiaminal intermediate

Dehydration of the hemiaminal produces the imine.

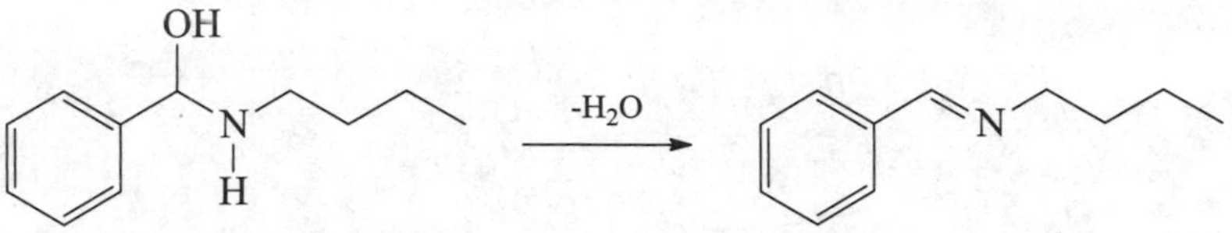

N-Benzylidenebutylamine

(*c*)   Cyclohexanone and *tert*-butylamine react according to the equation

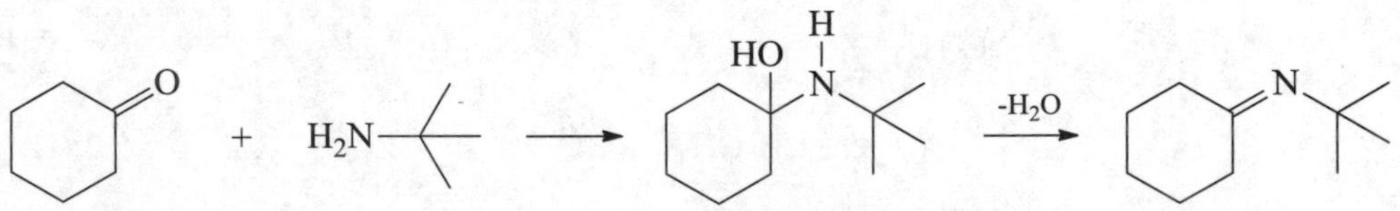

Cyclohexanone   *tert*-Butylamine   Hemiaminal intermediate   *N*-Cyclohexylidene *tert*-butylamine

(*d*)

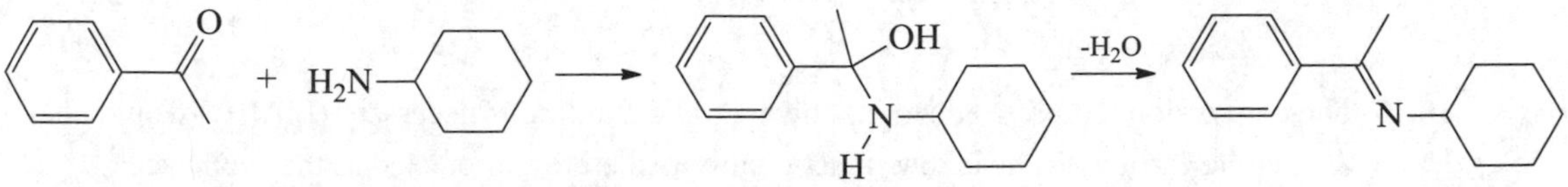

Acetophenone   Cyclohexylamine   Hemiaminal intermediate   *N*-(1-Phenylethylidene)-cyclohexylamine

18.15   The compound $C_5H_9NO_3$ differs from its imine precursor ($C_5H_7NO_2$) by a molecule of H_2O. Therefore, add H_2O across the double bond of the imine to give a hemiaminal as one of the answers. Recognize that the hemiaminal is in equilibrium with a noncyclic compound that contains both an aldehyde and a primary amine function to deduce the other answer.

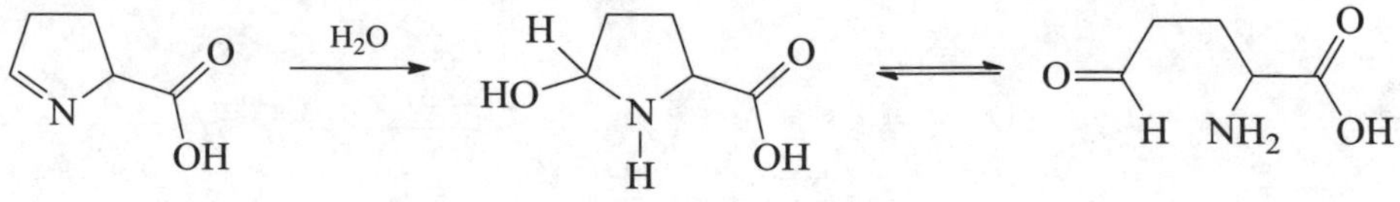

18.16   Assume that the usual reaction of aldehydes with derivatives of ammonia occurs and that each aldehyde function and each amine function react independently.

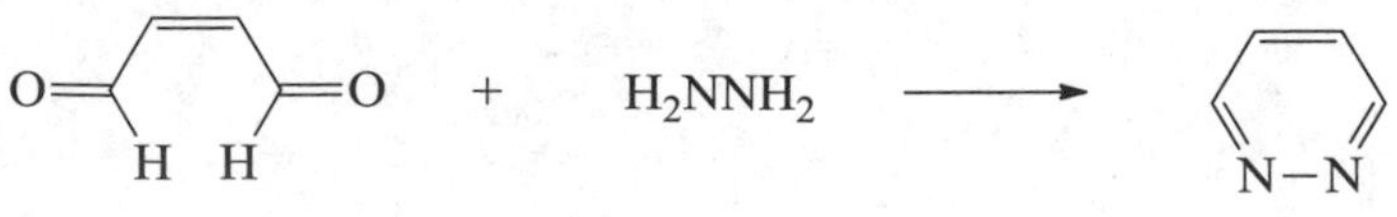

cis-2-Butenedial   Hydrazine   1,2-Diazine

18.17 (*b*) Pyrrolidine, a secondary amine, adds to 3-pentanone to give a hemiaminal.

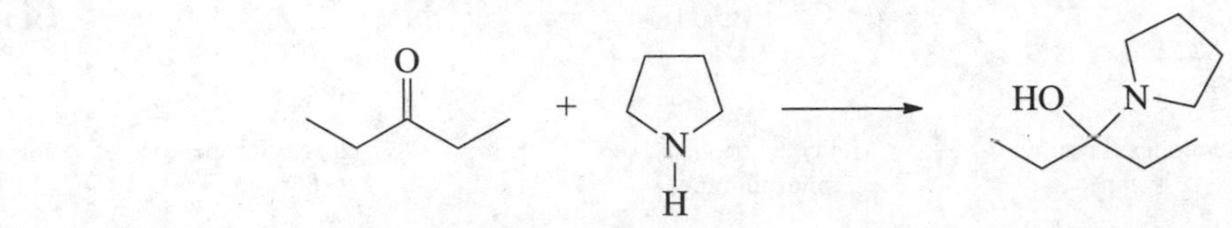

3-Pentanone   Pyrrolidine   Hemiaminal
                                              intermediate

Dehydration produces the enamine.

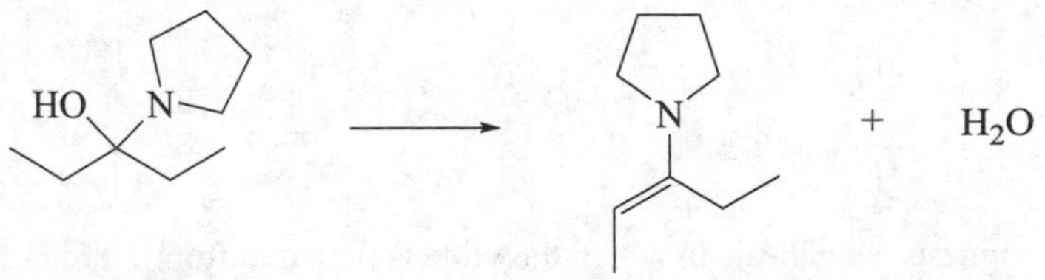

Hemiaminal   3-Pyrrolidino-2-pentene
intermediate

(*c*)

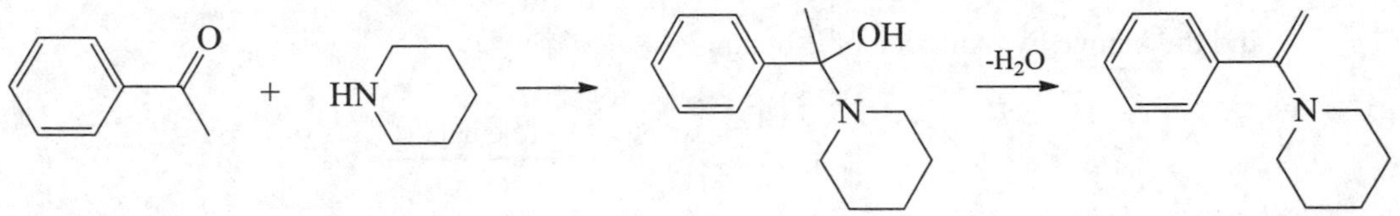

Acetophenone   Piperidine   Hemiaminal intermediate   1-Phenyl-1-piperidinoethene

18.18 The alternative disconnection and overall reaction are:

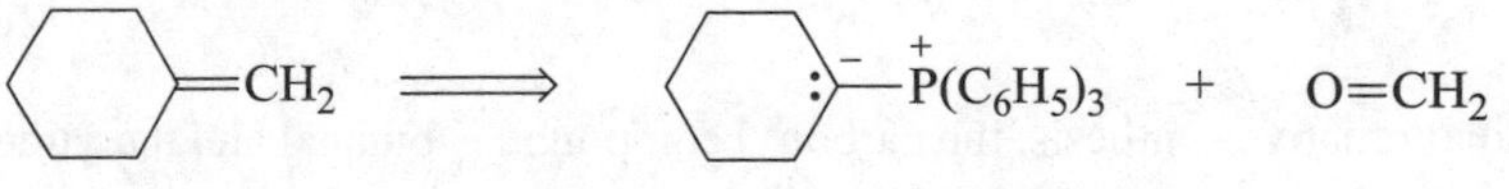

Methylenecyclohexane   Cyclohexylidenetriphenyl-   Formaldehyde
                                      phosphorane

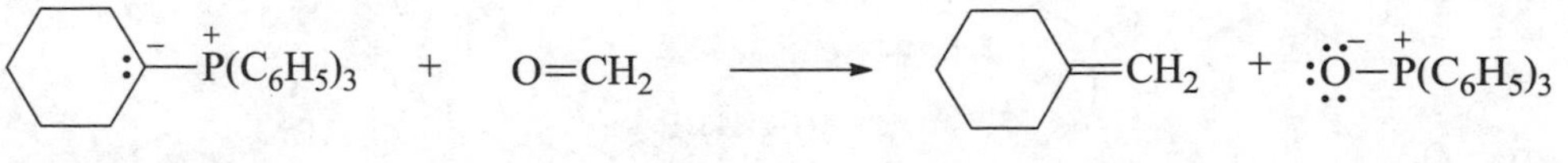

Cyclohexylidenetriphenyl-   Formaldehyde
phosphorane   Methylenecyclohexane   Triphenylphosphine
                                                                              oxide

18.19 (*b*) Here we see an example of the Wittig reaction applied to diene synthesis by use of an ylide containing a carbon–carbon double bond.

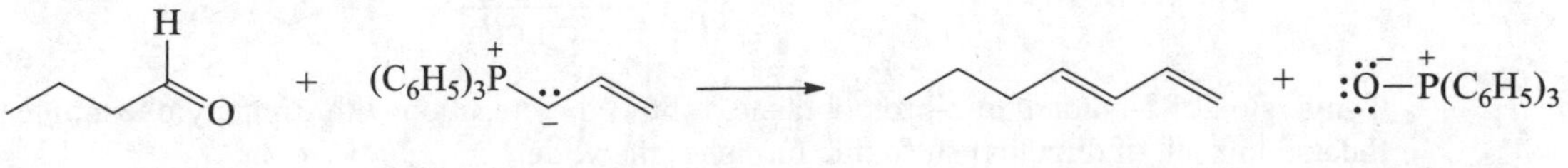

Butanal   Allylidenetriphenyl-   1,3-Heptadiene (52%)   Triphenylphosphine
                 phosphorane   oxide

(*c*) Methylene transfer from methylenetriphenylphosphorane is one of the most commonly used Wittig reactions.

$$\text{Cyclohexyl methyl ketone} + H_2\overset{-}{\ddot{C}}-\overset{+}{P}(C_6H_5)_3 \longrightarrow \text{2-Cyclohexylpropene} + :\overset{-}{\ddot{O}}-\overset{+}{P}(C_6H_5)_3$$

| Cyclohexyl methyl ketone | Methylenetriphenyl-phosphorane | 2-Cyclohexylpropene (66%) | Triphenylphosphine oxide |

18.20 One retrosynthesis is:

which suggests a synthesis in which the ylide is prepared from 1-bromobutane:

$$\text{OH} \xrightarrow[\text{heat}]{\text{HBr}} \text{Br} \xrightarrow{(C_6H_5)_3P} \overset{+}{P}(C_6H_5)_3 \ \overset{-}{Br} \xrightarrow{NaCH_2S(O)CH_3} \overset{-}{\ddot{\ }}\overset{+}{P}(C_6H_5)_3$$

and the ketone by oxidation of 2-butanol:

$$\text{HO} \xrightarrow[\text{CH}_2\text{Cl}_2]{\text{PDC or PCC}} \text{O}$$

Combining the two gives the desired alkene.

$$\overset{-}{\ddot{\ }}\overset{+}{P}(C_6H_5)_3 + \text{O} \longrightarrow$$

In an alternative synthesis, the carbonyl compound is butanal and the ylide is prepared from 2-butanol.

Butanal is prepared by oxidation of 1-butanol.

$$\text{OH} \xrightarrow[\text{CH}_2\text{Cl}_2]{\text{PDC or PCC}} \text{O}$$

Conversion of 2-butanol to 2-bromobutane, followed by reaction with triphenylphosphine the treatment with the sodium salt of dimethyl sulfoxide furnishes the ylide.

$$\text{HO} \xrightarrow[\text{heat}]{\text{HBr}} \text{Br} \xrightarrow{(C_6H_5)_3P} (C_6H_5)_3\overset{+}{P} \ \overset{-}{Br} \xrightarrow{NaCH_2S(O)CH_3} (C_6H_5)_3\overset{+}{P}\overset{-}{\ddot{\ }}$$

Reaction between the two gives 3-methyl-3-heptene.

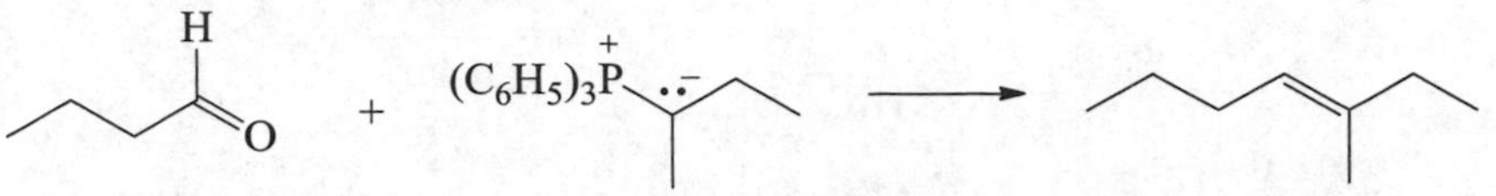

18.21 (*b*)

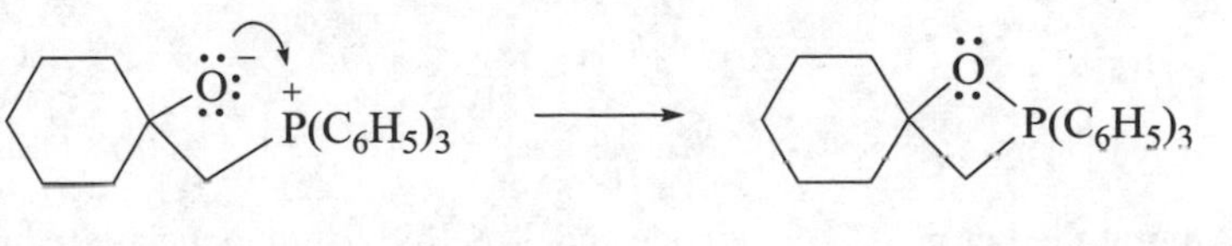

Betaine intermediate   Oxaphosphetane

(*c*)

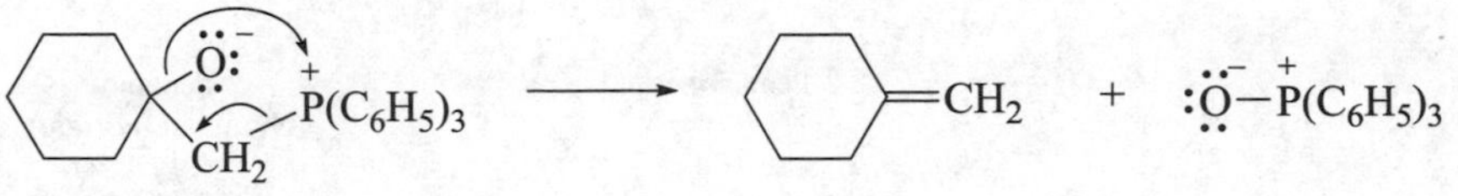

Betaine intermediate   Methylenecyclohexane   Triphenylphosphine oxide

18.22 The two products are stereoisomers of each other. However, they are not mirror images and thus are not enantiomers. They are diastereomers, that is, stereoisomers that are not enantiomers. The reaction is diastereoselective.

End of Chapter Problems

Structure and Nomenclature

18.23 (*a*) First consider all the isomeric aldehydes of molecular formula $C_5H_{10}O$.

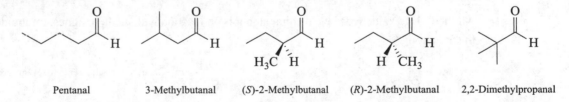

| Pentanal | 3-Methylbutanal | (*S*)-2-Methylbutanal | (*R*)-2-Methylbutanal | 2,2-Dimethylpropanal |

There are three isomeric ketones:

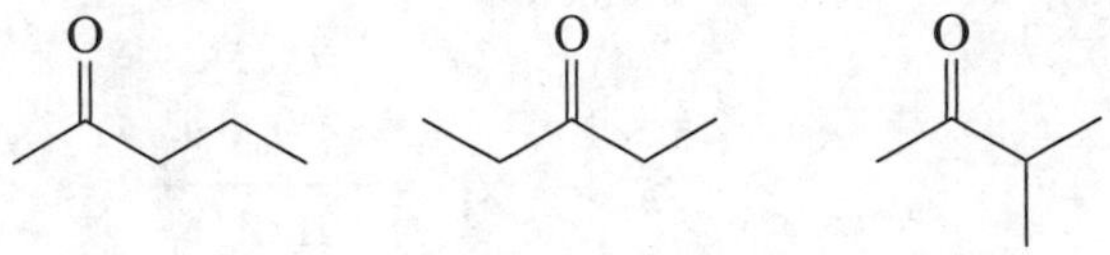

2-Pentanone   3-Pentanone   3-Methyl-2-butanone

(*b*) Reduction of an aldehyde to a primary alcohol does not introduce a chirality center into the molecule. The only aldehydes that yield chiral alcohols on reduction are therefore those that already contain a chirality center.

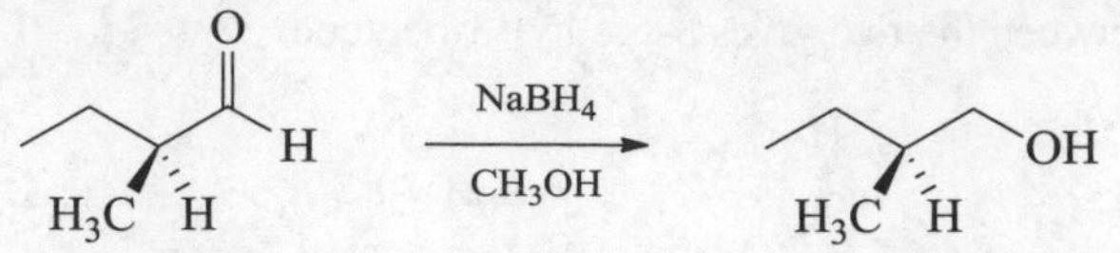

(*S*)-2-Methylbutanal → (*S*)-2-Methyl-1-butanol

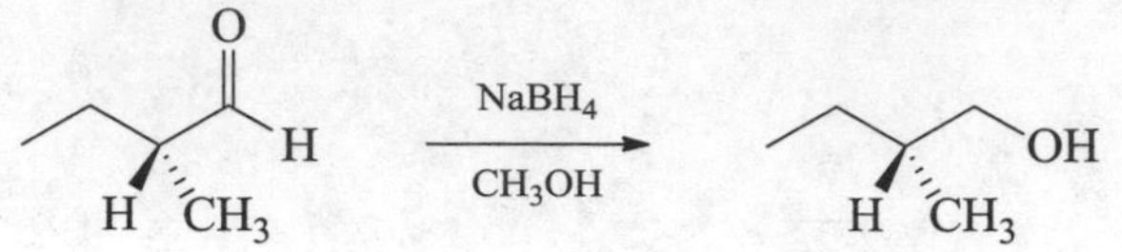

(*R*)-2-Methylbutanal → (*R*)-2-Methyl-1-butanol

Among the ketones, 2-pentanone and 3-methyl-2-butanone are reduced to chiral alcohols.

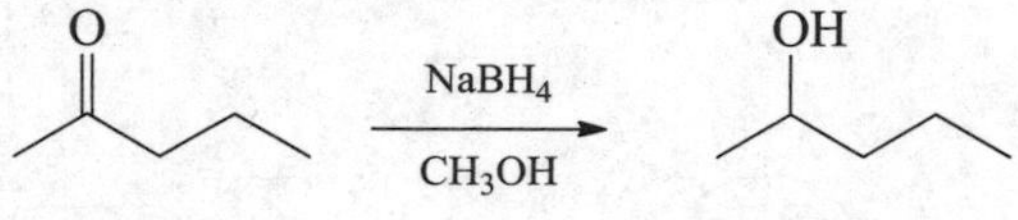

2-Pentanone → 2-Pentanol
(chiral but racemic)

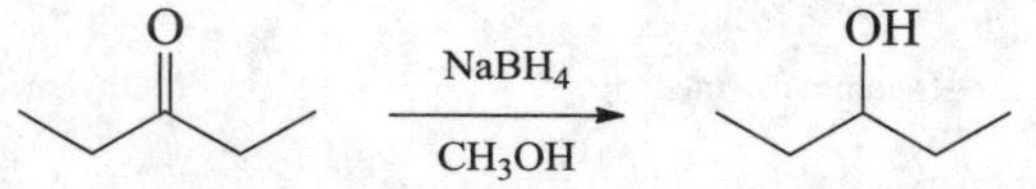

3-Pentanone → 3-Pentanol
(achiral)

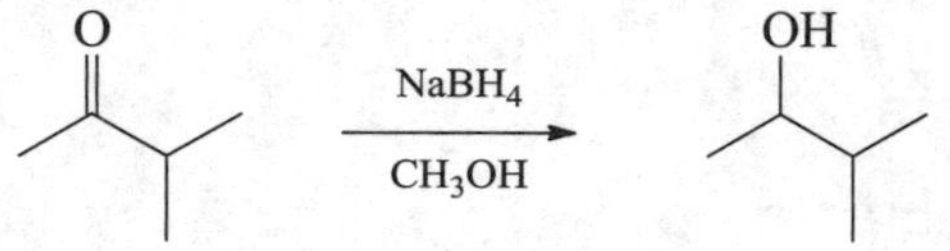

3-Methyl-2-butanone → 3-Methyl-2-butanol
(chiral but racemic)

(*c*) All the $C_5H_{10}O$ aldehydes yield chiral alcohols on reaction with methylmagnesium iodide according to the equation

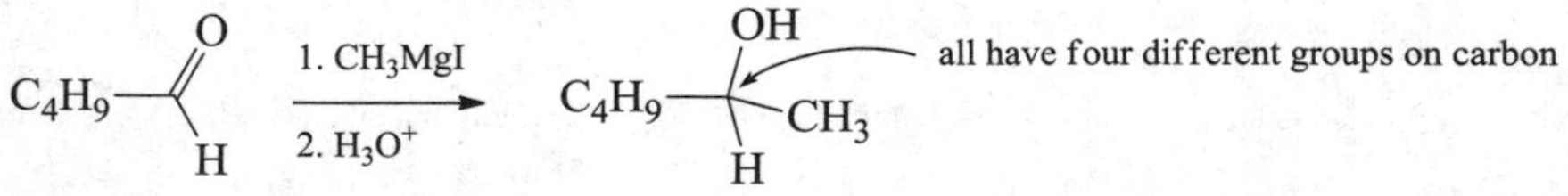

all have four different groups on carbon

None of the ketones yield chiral alcohols.

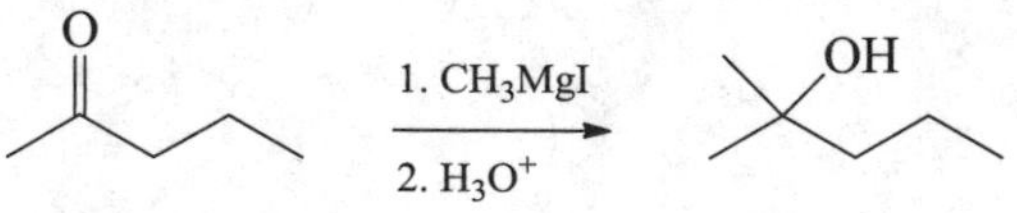

2-Pentanone → 2-Methyl-2-pentanol
(achiral)

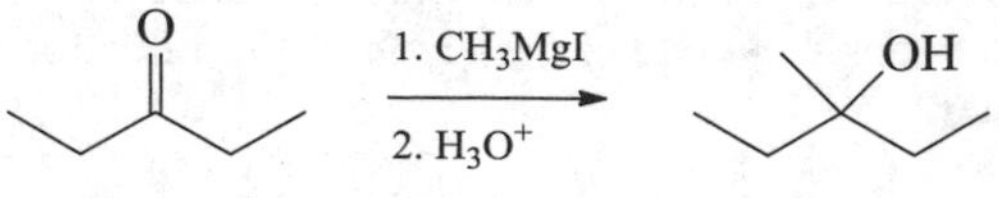

3-Pentanone → 3-Methyl-3-pentanol
(achiral)

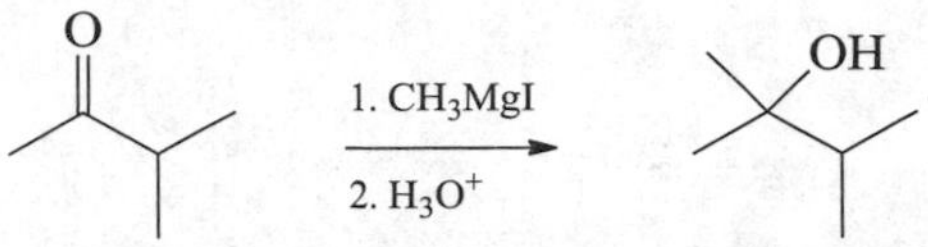

3-Methyl-2-butanone   2,3-Dimethyl-2-butanol
                                    (achiral)

18.24 (*a*) Chloral is the trichloro derivative of ethanal (acetaldehyde).

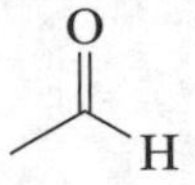

Ethanal 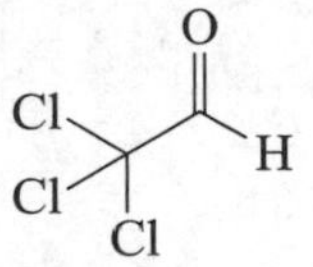   Trichloroethanal
                            (chloral)

(*b*) Pivaldehyde has two methyl groups attached to C-2 of propanal.

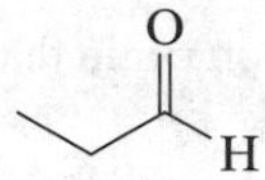

Propanal 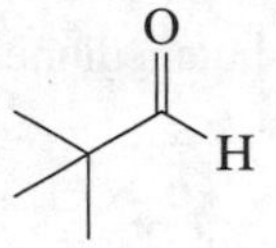   2,2-Dimethylpropanal
                           (pivaldehyde)

(*c*) Acrolein has a double bond between C-2 and C-3 of a three-carbon aldehyde.

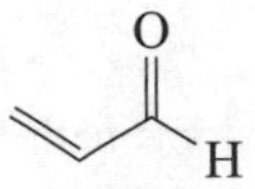

2-Propenal (acrolein)

(*d*) Crotonaldehyde has a trans double bond between C-2 and C-3 of a four-carbon aldehyde.

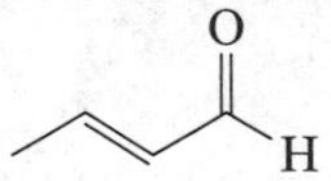

(*E*)-2-Butenal
(crotonaldehyde)

(*e*) Citral has two double bonds: one between C-2 and C-3 and the other between C-6 and C-7. The one at C-2 has the *E* configuration. There are methyl substituents at C-3 and C-7.

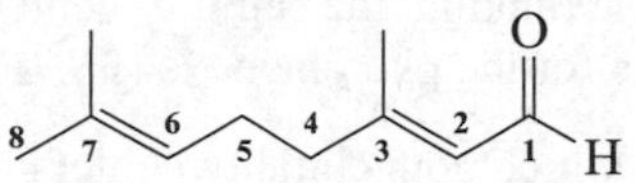

(*E*)-3,7-Dimethyl-2,6-octadienal
(citral)

(*f*) Diacetone alcohol is

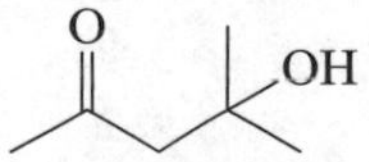

4-Hydroxy-4-methyl-
2-pentanone

(*g*) The parent ketone is 2-cyclohexenone.

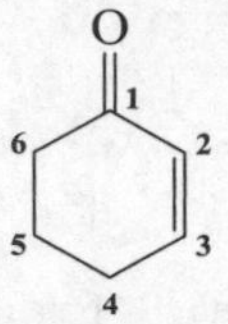

2-Cyclohexenone

Carvone has an isopropenyl group at C-5 and a methyl group at C-2.

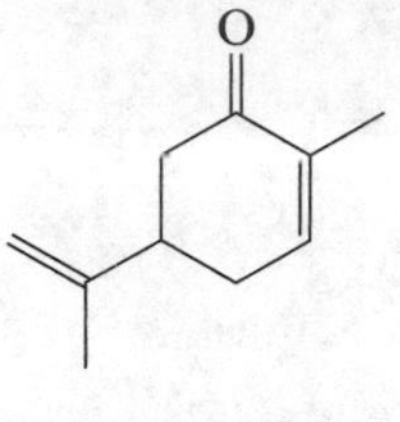

5-Isopropenyl-2-methyl-2-
cyclohexenone (carvone)

(*h*) Biacetyl is 2,3-butanedione. It has a four-carbon chain that incorporates ketone carbonyls as C-2 and C-3.

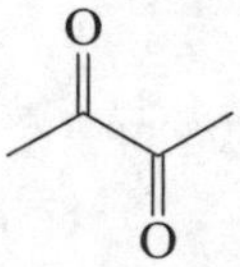

2,3-Butanedione
(biacetyl)

18.25 (*a*) The longest chain of carbon atoms in the crocodile pheromone is numbered from the end closest to the ketone carbonyl group.

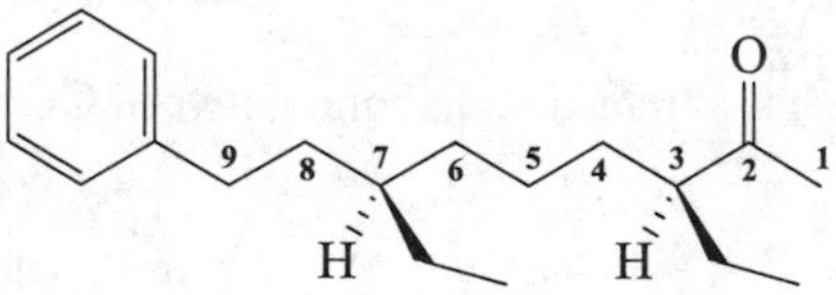

The absolute stereochemistry of the chirality centers can be determined by assigning the Cahn-Ingold-Prelog rankings to the four groups at each center.

$$\text{C-3:} \quad \overset{\overset{\displaystyle O}{\|}}{C} > CH_2CH_2 > CH_2CH_3 > H \qquad \text{C-7:} \quad CH_2CH_2C_6H_5 > CH_2CH_2CH_2 > CH_2CH_3 > H$$

In order of decreasing ranking, the chirality center at C-3 is *S*; the chirality center at C-7 is *R*. The systematic name for the pheromone is (3*S*,7*R*)-3,7-diethyl-9-phenylnonan-2-one.

(*b*) The stereoisomer in which both chirality centers have the *S* configuration (3*S*,7*S*) is a diastereomer of the (3*S*,7*R*) stereoisomer shown in the problem. The enantiomer of the (3*S*,7*R*) stereoisomer would be (3*R*,7*S*).

18.26 (*a*) The *Z* stereoisomer of CH₃CH=NCH₃ has its higher-ranked substituents on the same side of the double bond,

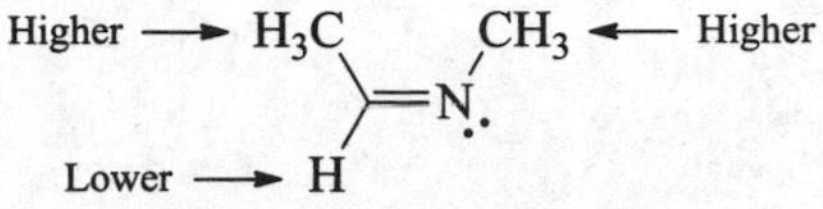

Higher ⟶ H_3C CH_3 ⟵ Higher

N

Lower ⟶ H

(*Z*)-*N*-Ethylidenemethylamine

The unshared electron pair of nitrogen is lower in rank than any other substituent.

(*b*) Higher-ranked groups are on opposite sides of the carbon–nitrogen double bond in the *E* oxime of acetaldehyde.

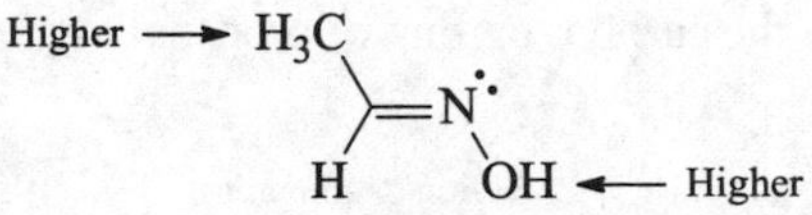

Higher ⟶ H_3C

N

H   OH ⟵ Higher

(*E*)-Acetaldehyde
oxime

(*c*) (*Z*)-2-Butanone hydrazone is

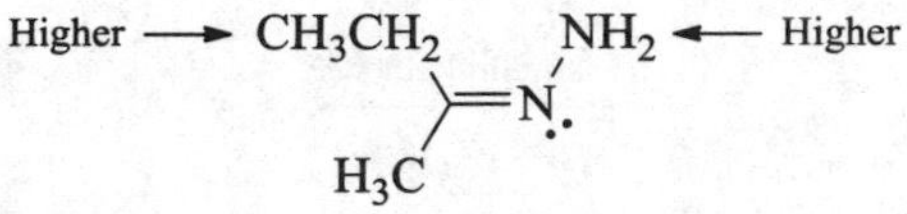

Higher ⟶ CH_3CH_2 NH_2 ⟵ Higher

N

H_3C

(*d*) (*E*)-Acetophenone semicarbazone is

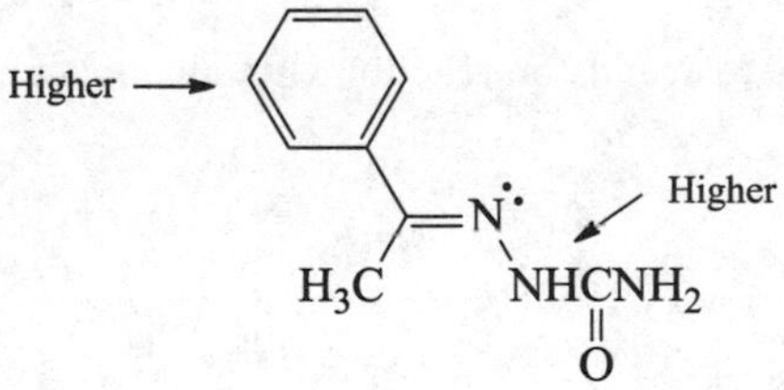

Higher ⟶

Higher

N

H_3C $NHCNH_2$

O

Reactions

18.27 (*a*) Lithium aluminum hydride reduces aldehydes to primary alcohols.

$$\text{Propanal} \quad \xrightarrow[\text{2. H}_2\text{O}]{\text{1. LiAlH}_4} \quad \text{1-Propanol}$$

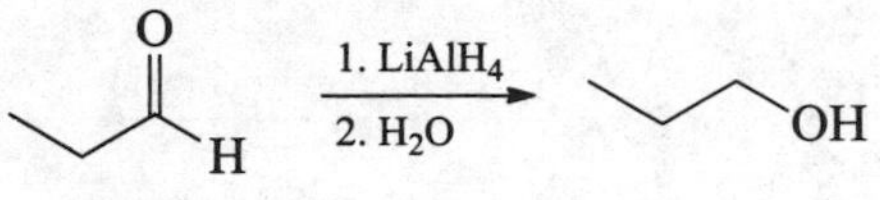

Propanal   1-Propanol

(*b*) Sodium borohydride reduces aldehydes to primary alcohols.

$$\text{Propanal} \quad \xrightarrow[\text{CH}_3\text{OH}]{\text{NaBH}_4} \quad \text{1-Propanol}$$

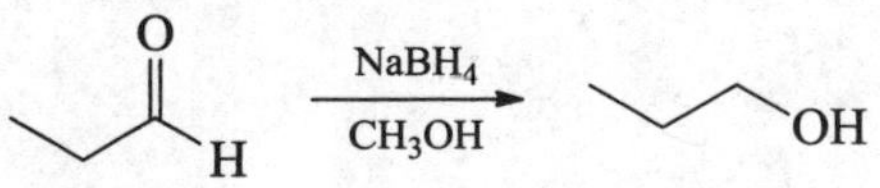

Propanal   1-Propanol

(*c*) Aldehydes can be reduced to primary alcohols by catalytic hydrogenation.

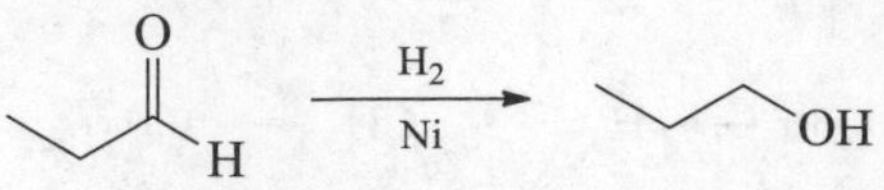

Propanal   1-Propanol

(*d*) Aldehydes react with Grignard reagents to form secondary alcohols.

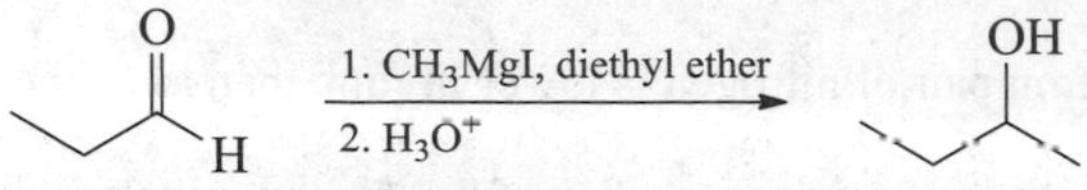

Propanal   2-Butanol

(*e*) Sodium acetylide adds to the carbonyl group of propanal to give an acetylenic alcohol.

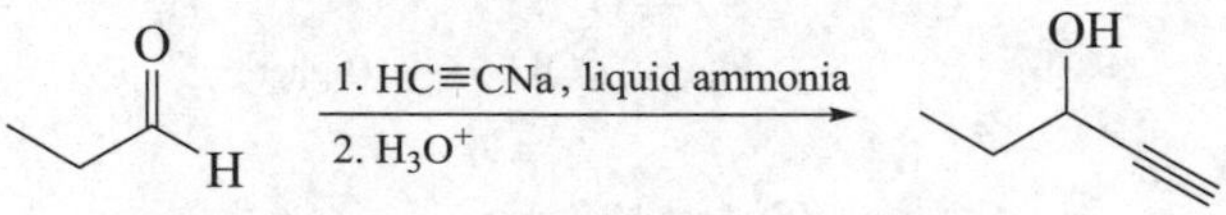

Propanal   1-Pentyn-3-ol

(*f*) Alkyl- or aryllithium reagents react with aldehydes in much the same way that Grignard reagents do.

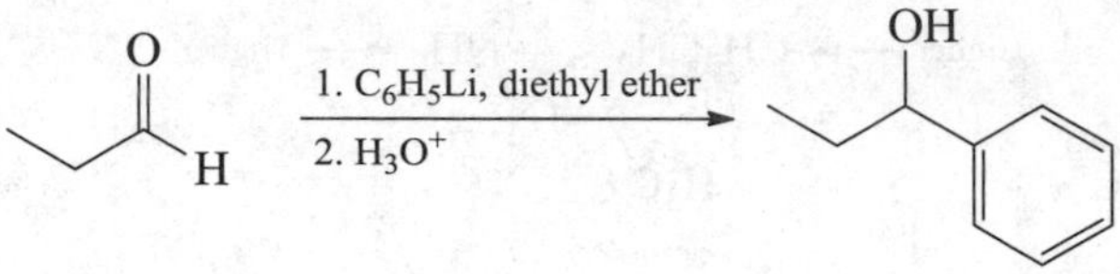

Propanal   1-Phenyl-1-propanol

(*g*) Aldehydes are converted to acetals on reaction with alcohols in the presence of an acid catalyst.

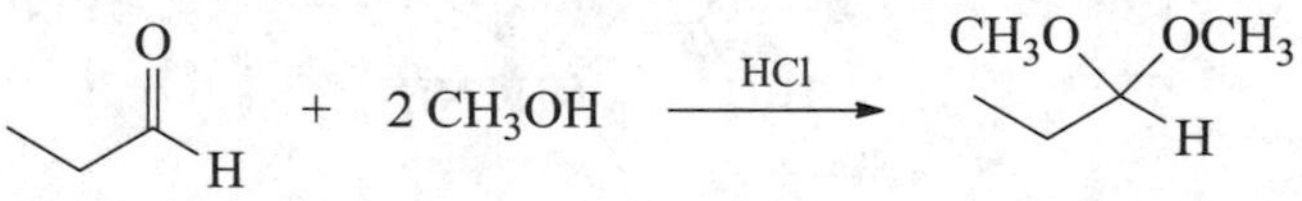

Propanal   Methanol   Propanal dimethyl acetal

(*h*) Cyclic acetal formation occurs when aldehydes react with ethylene glycol.

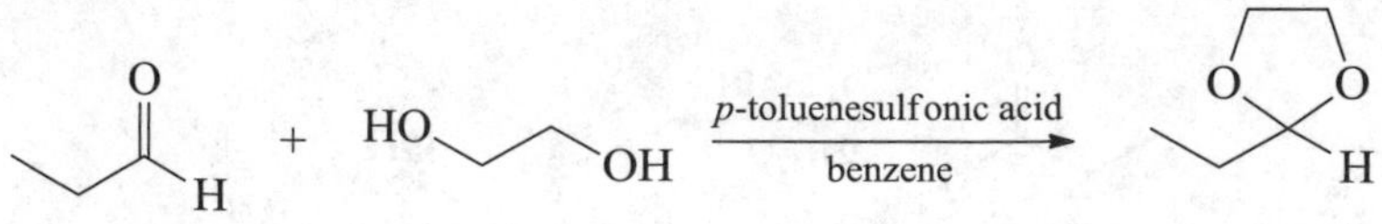

Propanal   Ethylene glycol   2-Ethyl-1,3-dioxolane

(*i*) Aldehydes react with primary amines to yield imines.

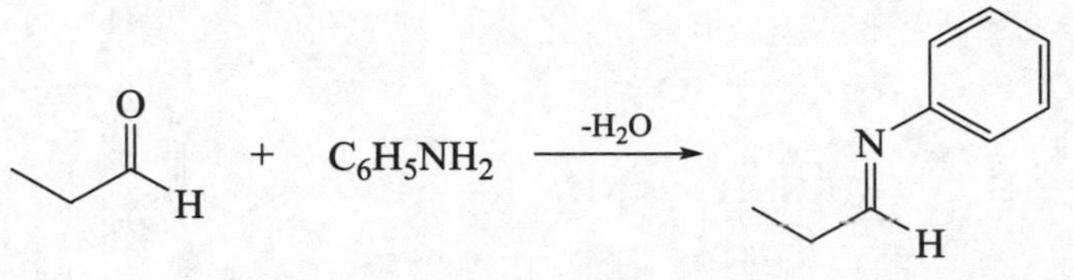

Propanal   Aniline   *N*-Propylideneaniline

(*j*) Secondary amines combine with aldehydes to yield enamines.

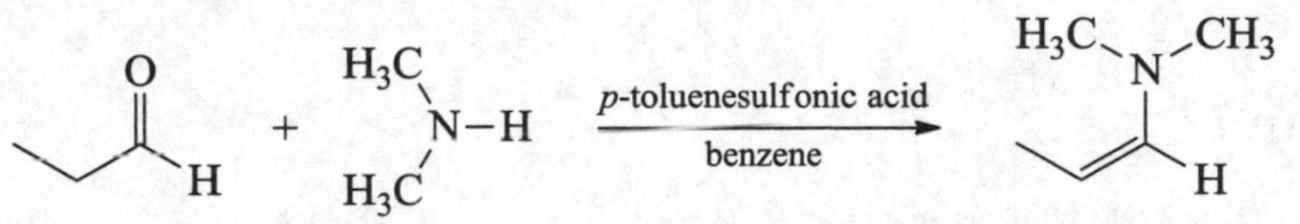

Propanal   Dimethylamine   1-(Dimethylamino)propene

(*k*) Oximes are formed on reaction of hydroxylamine with aldehydes.

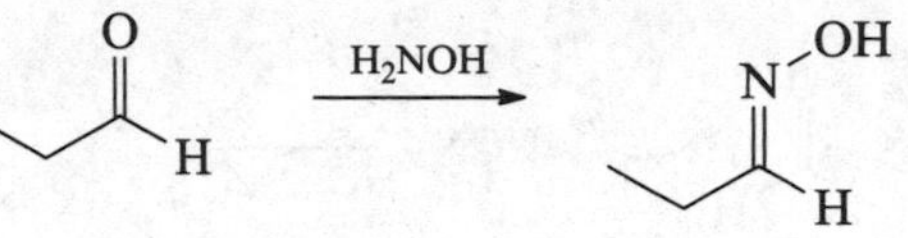

Propanal   Propanal oxime

(*l*) Hydrazine reacts with aldehydes to form hydrazones.

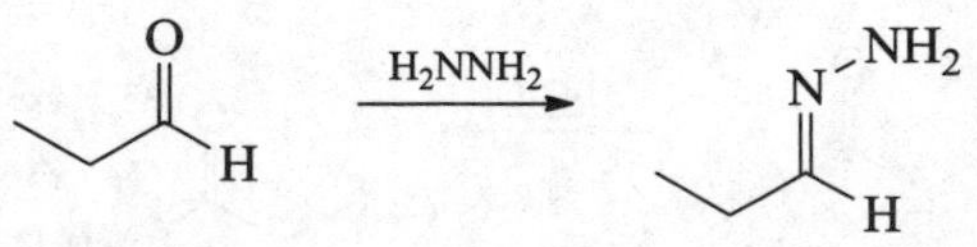

Propanal   Propanal hydrazone

(*m*) Hydrazone formation is the first step in the Wolff–Kishner reduction (text Section 13.8).

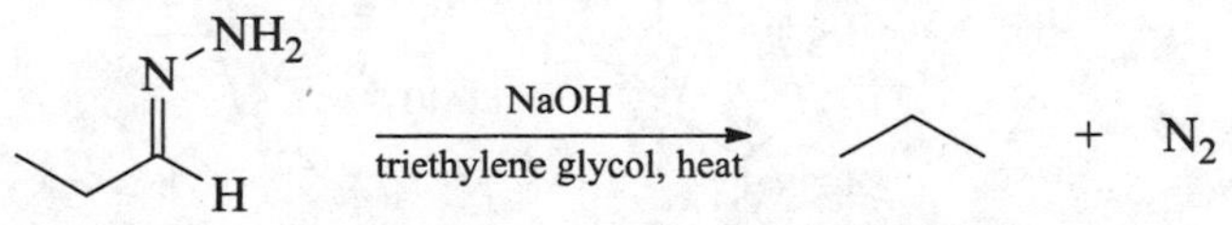

Propanal hydrazone   Propane

(*n*) The reaction of an aldehyde with *p*-nitrophenylhydrazine is analogous to that with hydrazine.

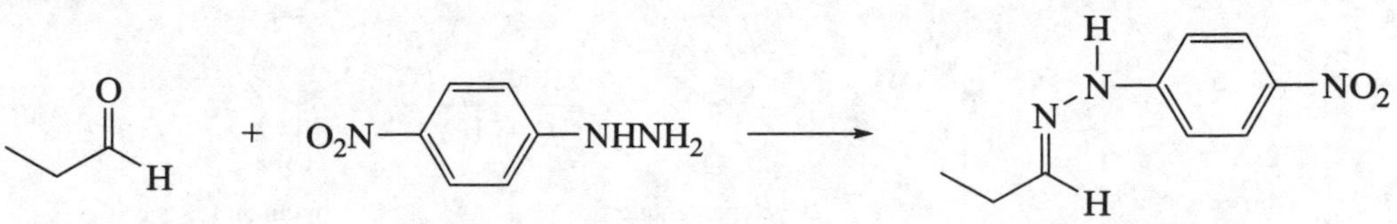

Propanal   *p*-Nitrophenylhydrazine   Propanal *p*-nitrophenylhydrazone

(*o*) Semicarbazide converts aldehydes to the corresponding semicarbazone.

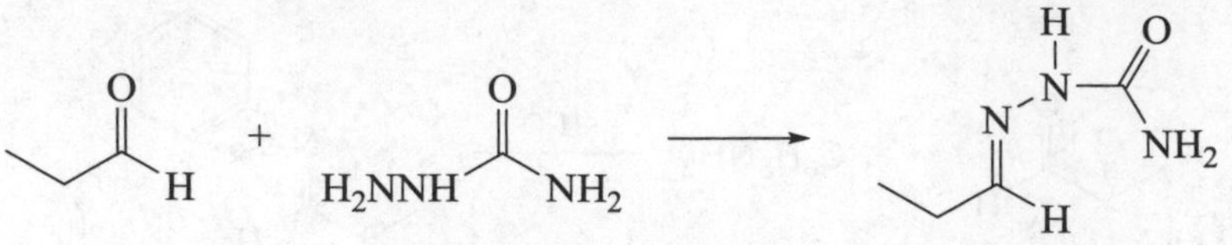

Propanal + Semicarbazide → Propanal semicarbazone

(*p*) Phosphorus ylides convert aldehydes to alkenes by a Wittig reaction.

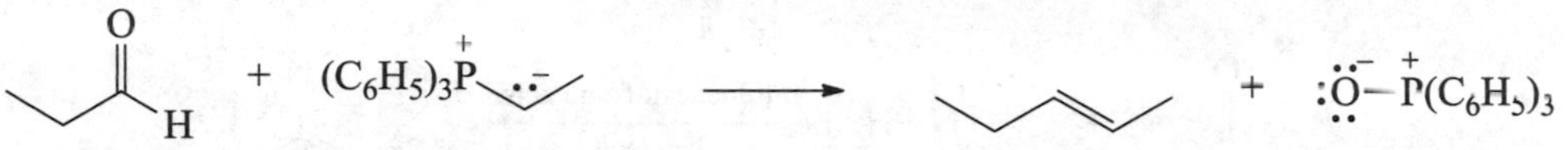

Propanal + Ethylidenetriphenyl-phosphorane → 2-Pentene + Triphenylphosphine oxide

(*q*) Acidification of solutions of sodium cyanide generates HCN, which reacts with aldehydes to form cyanohydrins.

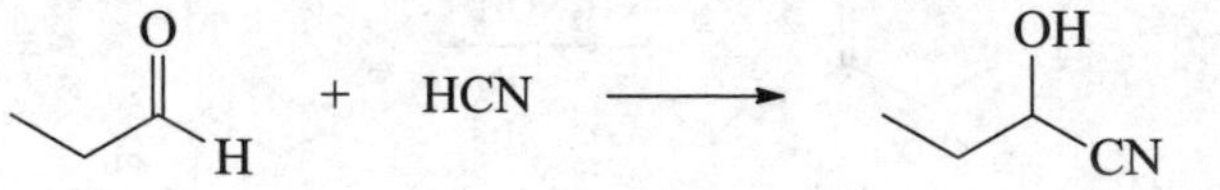

Propanal + Hydrogen cyanide → Propanal cyanohydrin

(*r*) Chromic acid oxidizes aldehydes to carboxylic acids.

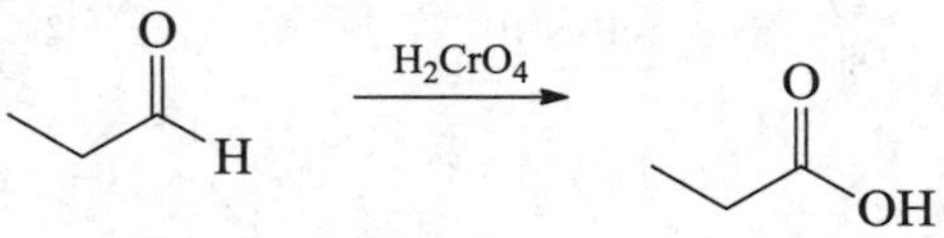

Propanal $\xrightarrow{H_2CrO_4}$ Propanoic acid

18.28 (*a*) Lithium aluminum hydride reduces ketones to secondary alcohols.

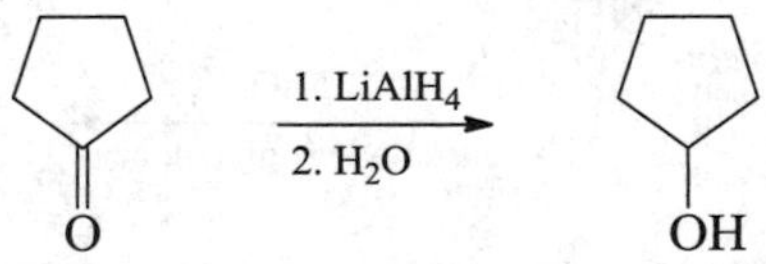

Cyclopentanone $\xrightarrow[\text{2. } H_2O]{\text{1. LiAlH}_4}$ Cyclopentanol

(*b*) Sodium borohydride converts ketones to secondary alcohols.

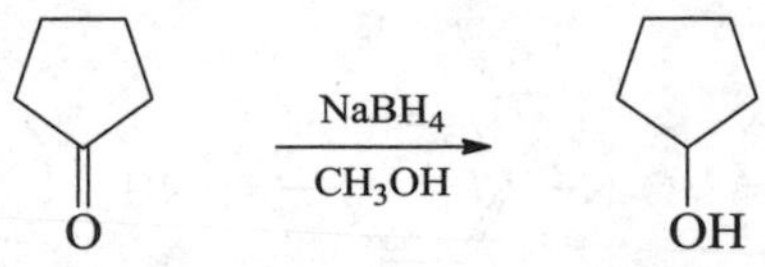

Cyclopentanone $\xrightarrow[\text{CH}_3\text{OH}]{\text{NaBH}_4}$ Cyclopentanol

(*c*) Catalytic hydrogenation of ketones yields secondary alcohols.

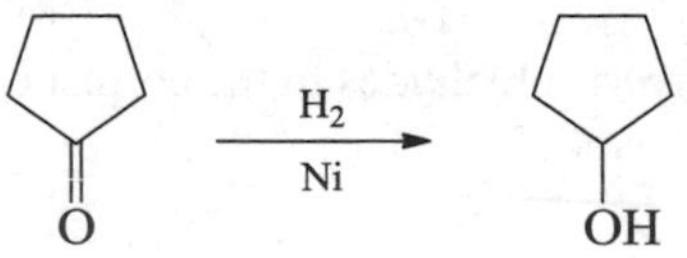

Cyclopentanone $\xrightarrow[\text{Ni}]{H_2}$ Cyclopentanol

(*d*) Grignard reagents react with ketones to form tertiary alcohols.

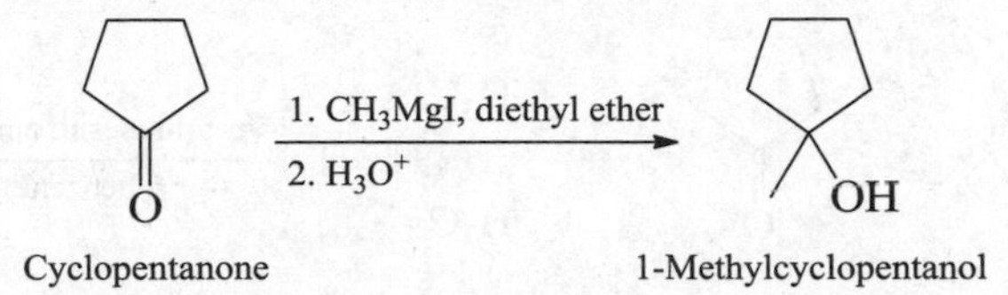

Cyclopentanone   1-Methylcyclopentanol

(*e*) Addition of sodium acetylide to cyclopentanone yields a tertiary acetylenic alcohol.

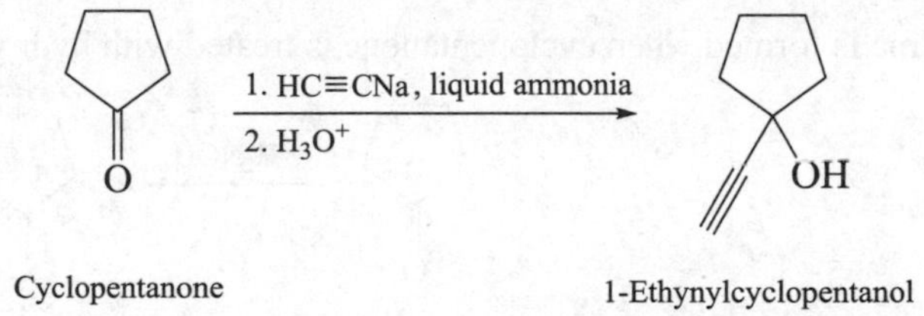

Cyclopentanone   1-Ethynylcyclopentanol

(*f*) Phenyllithium adds to the carbonyl group of cyclopentanone to yield 1-phenylcyclopentanol.

Cyclopentanone   1-Phenylcyclopentanol

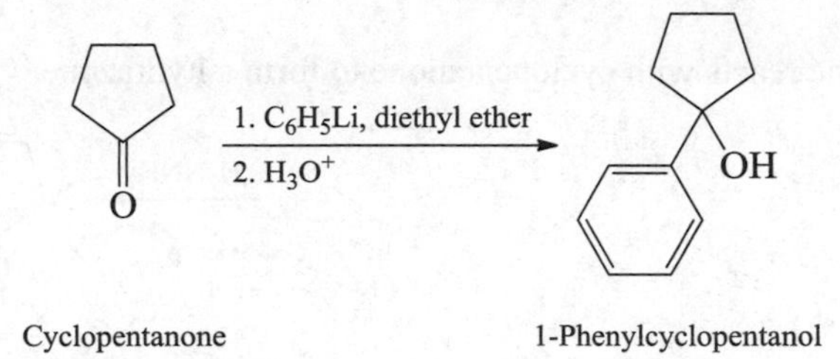

(*g*) The equilibrium constant for ketal formation from ketones is generally unfavorable.

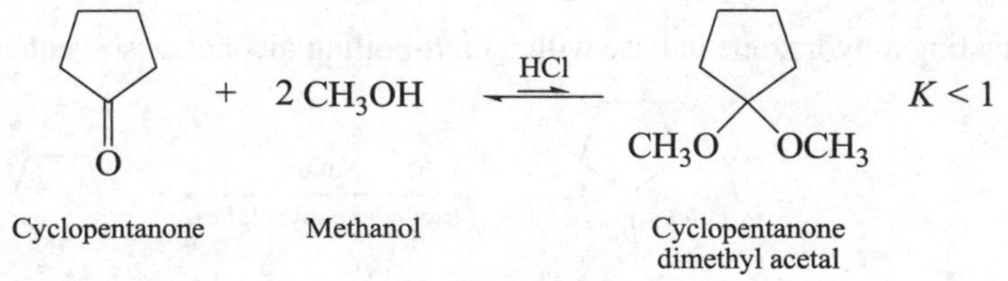

Cyclopentanone   Methanol   Cyclopentanone
                                     dimethyl acetal

(*h*) Cyclic ketal formation is favored even for ketones.

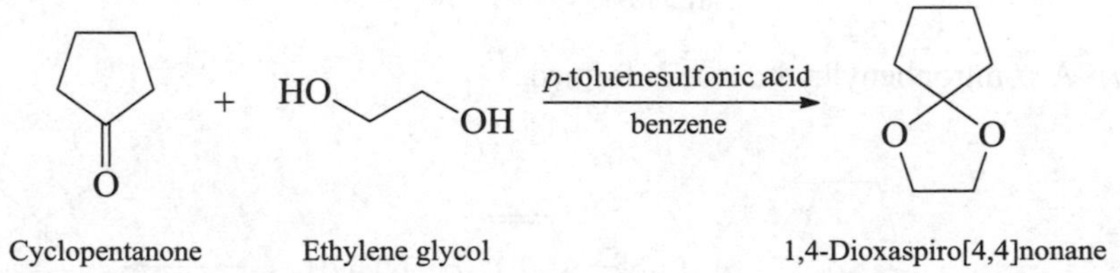

Cyclopentanone   Ethylene glycol   1,4-Dioxaspiro[4,4]nonane

(*i*) Ketones react with primary amines to form imines.

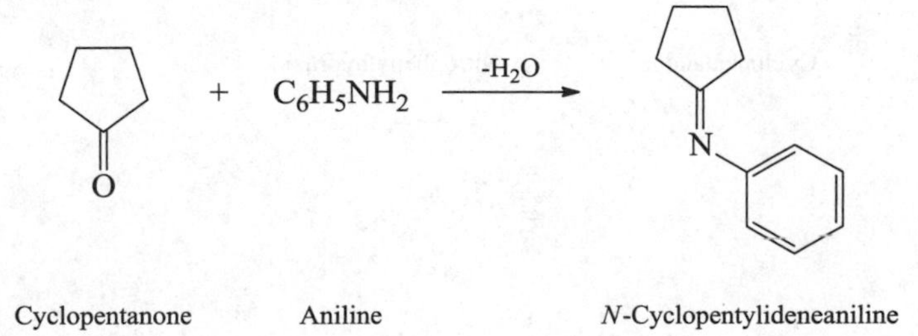

Cyclopentanone   Aniline   *N*-Cyclopentylideneaniline (*j*) Dimethylamine reacts with cyclopentanone to yield an enamine.

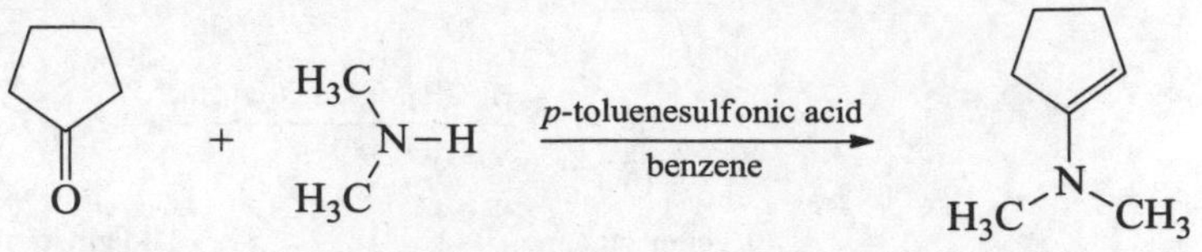

Cyclopentanone Dimethylamine   1-(Dimethylamino)-
cyclopentene

(*k*) An oxime is formed when cyclopentanone is treated with hydroxylamine.

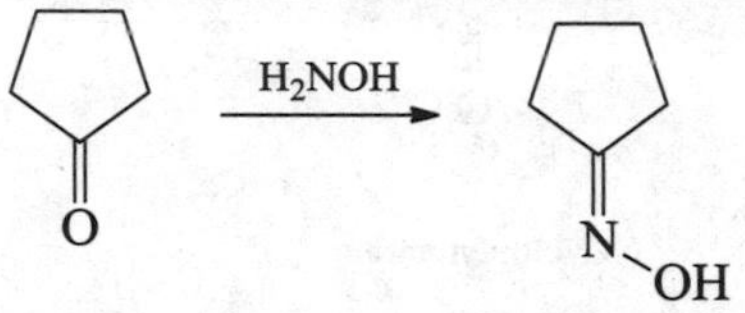

Cyclopentanone   Cyclopentanone
oxime

(*l*) Hydrazine reacts with cyclopentanone to form a hydrazone.

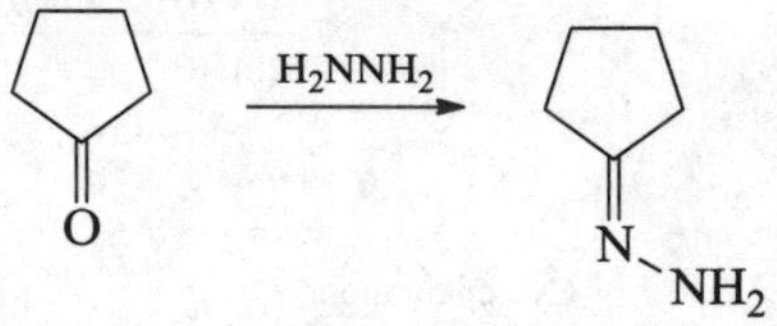

Cyclopentanone   Cyclopentanone
hydrazone

(*m*) Heating a hydrazone in base with a high-boiling alcohol as solvent converts it to an alkane.

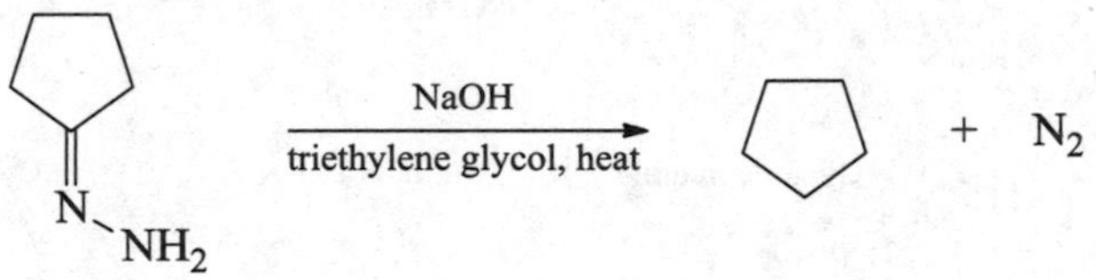

Cyclopentanone
hydrazone   Cyclopentane

(*n*) A *p*-nitrophenylhydrazone is formed.

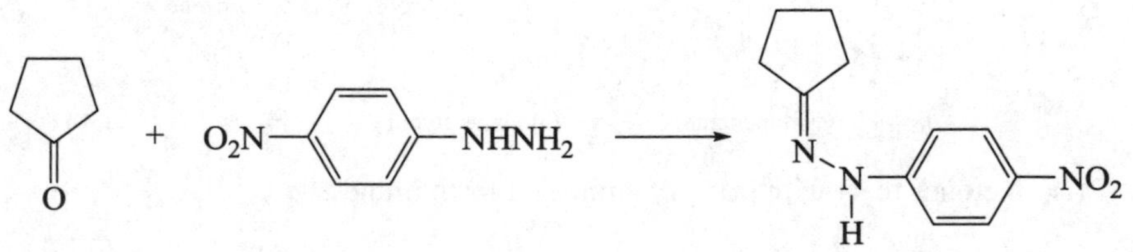

Cyclopentanone *p*-Nitrophenylhydrazine   Cyclopentanone
p-nitrophenylhydrazone

(*o*) Cyclopentanone is converted to a semicarbazone on reaction with semicarbazide.

Cyclopentanone   Semicarbazide   Cyclopentanone
                                                semicarbazone

(*p*) A Wittig reaction takes place, forming ethylidenecyclopentane.

Cyclopentanone   Ethylidenetriphenyl-   Ethylidenecyclo-   Triphenylphosphine
                  phosphorane   pentane   oxide

(*q*) Cyanohydrin formation takes place.

Cyclopentanone   Cyclopentanone
                        cyanohydrin

(*r*) Cyclopentanone is not oxidized by chromic acid.

18.29 (*a*) The first step in analyzing this problem is to write the structure of the starting ketone in stereochemical detail.

(*S*)-3-Phenyl-2-butanone   (2*R*,3*S*)-3-Phenyl-2-butanol   (2*S*,3*S*)-3-Phenyl-2-butanol

Reduction of the ketone introduces a new chirality center, which may have either the *R* or the *S* configuration; the configuration of the original chirality center is unaffected. In practice, the 2*R*,3*S* diastereomer is observed to form in greater amounts than the 2*S*,3*S* (ratio 2.5:1 for LiAlH$_4$ reduction).

(*b*) Reduction of the ketone can yield either *cis*- or *trans*-4-*tert*-butylcyclohexanol.

4-*tert*-Butylcyclohexanone   *cis*-4-*tert*-Butylcyclohexanol   *trans*-4-*tert*-Butylcyclohexanol

The major product obtained on reduction with either lithium aluminum hydride or sodium borohydride is the trans alcohol (trans/cis ≈ 9:1).

(c) The two reduction products are the exo and endo alcohols.

Bicyclo[2.2.1]heptan-2-one   exo-Bicyclo[2.2.1]heptan-2-ol   endo-Bicyclo[2.2.1]heptan-2-ol

The major product is the endo alcohol (endo/exo 9:1) for reduction with $NaBH_4$ or $LiAlH_4$.

The stereoselectivity observed in this reaction is due to decreased steric hindrance to attack of the hydride reagent from the exo face of the molecule, giving rise to the endo alcohol.

(d) The hydroxyl group may be on the same side as the double bond or on the opposite side.

Bicyclo[2.2.1]hept-2-en-7-one   syn-Bicyclo[2.2.1]hept-2-en-7-ol   anti-Bicyclo[2.2.1]hept-2-en-7-ol

The anti alcohol is formed in greater amounts (85:15) on reduction of the ketone with $LiAlH_4$.

Steric factors governing attack of the hydride reagent again explain the major product observed.

18.30 (a) Aldehydes undergo nucleophilic addition faster than ketones. Steric crowding in the rate-determining step of the ketone reaction raises the energy of the transition state, giving rise to a slower rate of reaction. Thus, benzaldehyde is reduced by sodium borohydride more rapidly than is acetophenone. The measured relative rates are

(b) The presence of an electronegative substituent on the α-carbon atom causes a dramatic increase in the K_{hydr} equilibrium constant. Trichloroethanal (chloral) is almost completely converted to its geminal diol (chloral hydrate) in aqueous solution.

Trichloroethanal   2,2,2-Trichloro-1,1-ethanediol
(chloral)   (chloral hydrate)

Electron-withdrawing groups such as Cl_3C destabilize carbonyl groups to which they are attached and make the energy change favoring the products of nucleophilic addition more favorable.

(c) Recall that the equilibrium constants for nucleophilic addition to carbonyl groups are governed by a combination of electronic effects and steric effects. Electronically, there is little difference between acetone and 3,3-dimethyl-2-butanone, but sterically, there is a significant difference. The cyanohydrin products are more crowded than the starting ketones, and so the bulkier the alkyl groups attached to the carbonyl, the more strained and less stable the cyanohydrin.

Ketone   Hydrogen   Cyanohydrin
                cyanide   [less strained for R = CH$_3$
                                  than for R = C(CH$_3$)$_3$]

$$K_{rel} = \frac{\underset{H_3C}{\overset{O}{\|}}\underset{CH_3}{}}{\underset{H_3C}{\overset{O}{\|}}} = 40$$

(d) Steric effects influence the rate of nucleophilic addition to these two ketones. Carbon is on its way from sp^2 to sp^3 at the transition state, and alkyl groups are forced closer together than they are in the ketone.

Transition state

The transition state is of lower energy, and the reaction is faster, when R is smaller. Acetone (R = methyl) is reduced faster than 3,3-dimethyl-2-butanone (R = *tert*-butyl).

$$k_{rel} = \frac{\underset{H_3C}{\overset{O}{\|}}\underset{CH_3}{}}{\underset{H_3C}{\overset{O}{\|}}} = 12$$

(e) In this problem we examine the rate of hydrolysis of acetals to the corresponding ketone or aldehyde. The rate-determining step is carbocation formation.

Carbocation

Hybridization at carbon changes from sp^3 to sp^2; crowding at this carbon is relieved as the carbocation is formed. The more crowded acetal (R = CH$_3$) forms a carbocation faster than the less crowded one (R = H). Another factor of even greater importance is the extent of stabilization of the carbocation intermediate; the more stable carbocation (R = CH$_3$) is formed faster than the less stable one (R = H).

$$k_{rel} = \dfrac{\begin{array}{c} H_3C_{,,,} \quad OCH_2CH_3 \\ H_3C \diagdown \\ OCH_2CH_3 \end{array}}{\begin{array}{c} H_{,,} \quad OCH_2CH_3 \\ H \diagdown \\ OCH_2CH_3 \end{array}} = 1.8 \times 10^7$$

18.31 (*a*) The reaction as written is the reverse of cyanohydrin formation, and the principles that govern equilibria in nucleophilic addition to carbonyl groups apply in reverse order to the dissociation of cyanohydrins to aldehydes and ketones. Cyanohydrins of ketones dissociate more at equilibrium than do cyanohydrins of aldehydes. More strain due to crowding is relieved when a ketone cyanohydrin dissociates and a more stabilized carbonyl group is formed. The equilibrium constant K_{diss} is larger for

| Acetone cyanohydrin | Acetone | Hydrogen cyanide |

than it is for

| Propanal cyanohydrin | Propanal | Hydrogen cyanide |

(*b*) Cyanohydrins of ketones have a more favorable equilibrium constant for dissociation than do cyanohydrins of aldehydes. Crowding is relieved to a greater extent when a ketone cyanohydrin dissociates and a more stable carbonyl group is formed. The measured dissociation constants are

$K = 4.7 \times 10^{-3}$

| Benzaldehyde cyanohydrin | Benzaldehyde | Hydrogen cyanide |

$K = 1.3$

| Acetophenone cyanohydrin | Acetophenone | Hydrogen cyanide |

(*c*) Ring strain in cyclopentanone would be greater than that of cyclohexanone because the $120°$ C-C-C bond angle for sp^2 hybridization in the carbonyl carbon would be substantially less in the five membered ring vs cyclohexanone. This would result in an equilibrium constant K_{diss} that is smaller for

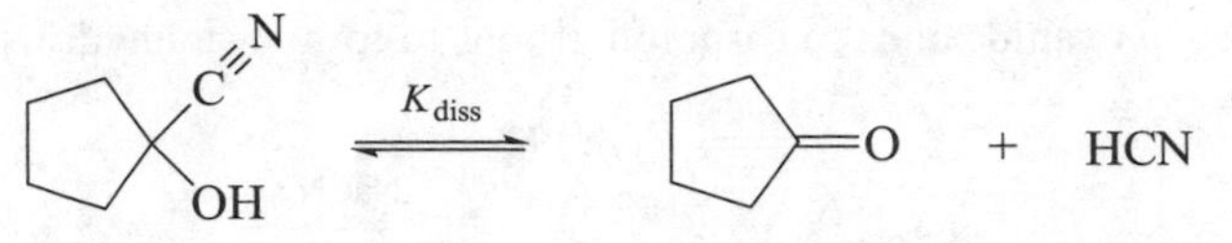

| Cyclopentanone cyanohydrin | | Cyclopentanone | Hydrogen cyanide |

than it is for

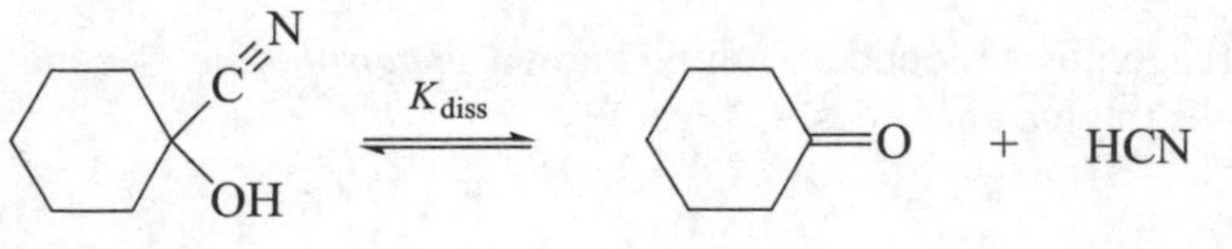

| Cyclohexanone cyanohydrin | | Cyclohexanone | Hydrogen cyanide |

18.32 (*a*) The reaction of an aldehyde with 1,3-propanediol in the presence of *p*-toluenesulfonic acid forms a cyclic acetal.

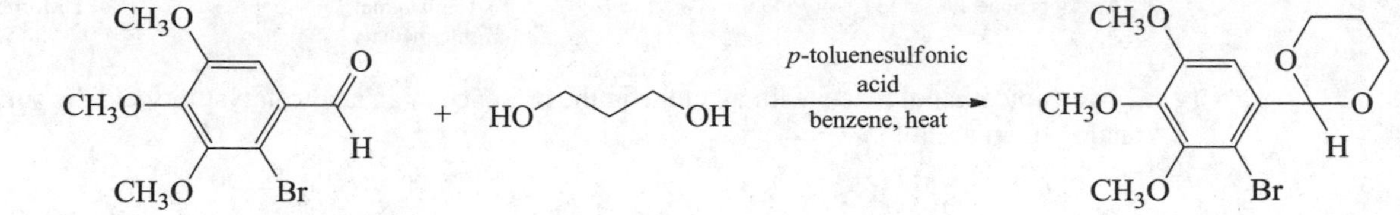

| 2-Bromo-3,4,5-trimethoxybenzaldehyde | 1,3-Propanediol | 2-(2'-Bromo-3',4',5'-trimethoxyphenyl)-1,3-dioxane (81%) |

(*b*) The reagent CH_3NH_2 is called *O*-methylhydroxylamine, and it reacts with aldehydes in a manner similar to hydroxylamine.

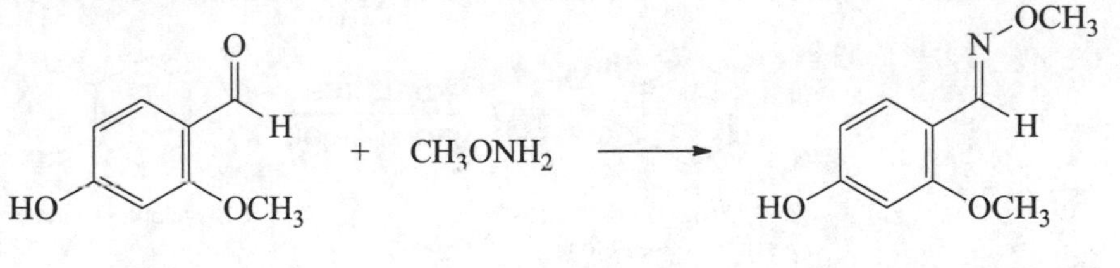

| 4-Hydroxy-2-methoxybenzaldehyde | *O*-Methyl hydroxylamine | 4-Hydroxy-2-methoxybenzaldehyde *O*-methyloxime |

(*c*) Propanal reacts with 1,1-dimethylhydrazine to yield the corresponding hydrazone.

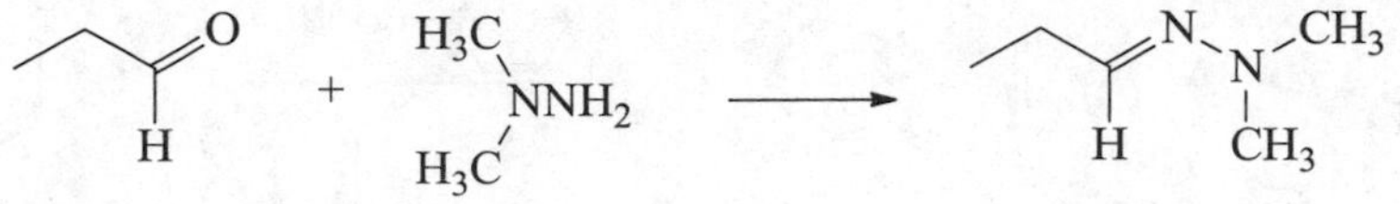

| Propanal | 1,1-Dimethylhydrazine | Propanal dimethylhydrazone |

(*d*) Acid-catalyzed hydrolysis of the acetal gives the aldehyde in 87% yield.

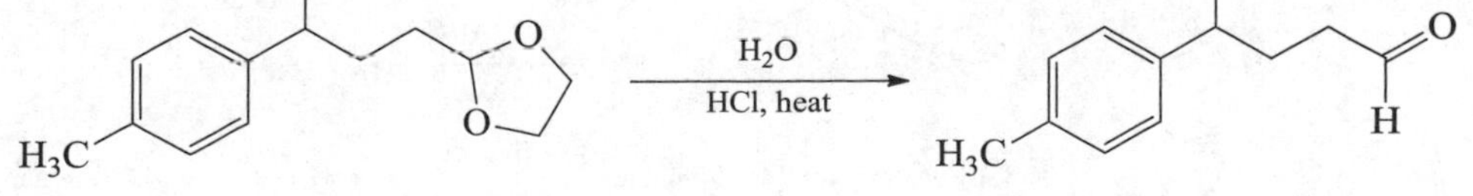

4-(*p*-Methylphenyl)pentanal

(*e*) Hydrogen cyanide adds to carbonyl groups to form cyanohydrins.

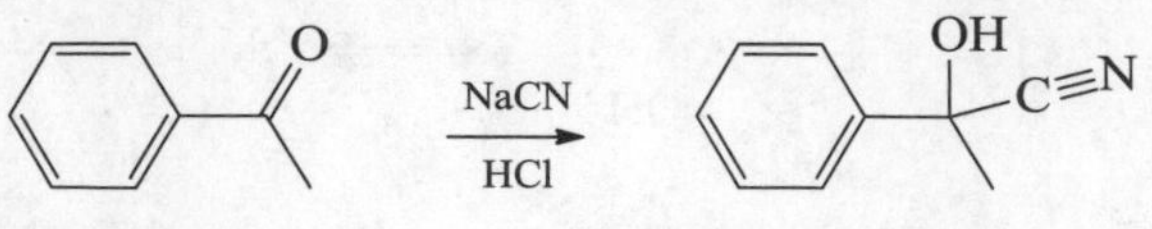

Acetophenone   Acetophenone cyanohydrin

(*f*) The reagent is a secondary amine known as *morpholine*. Secondary amines react with ketones to give enamines.

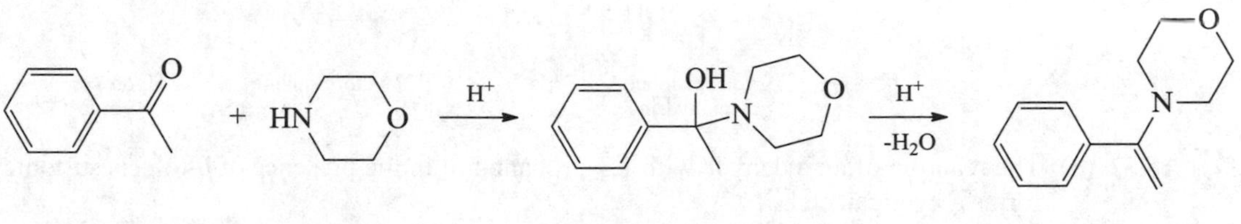

Acetophenone   Morpholine   Hemiaminal intermediate   1-Morpholinostyrene (57-64%)

(*g*) 2-Mercaptoethanol reacts with a ketone in the presence of an acid catalyst to yield the sulfur analog of an acetal.

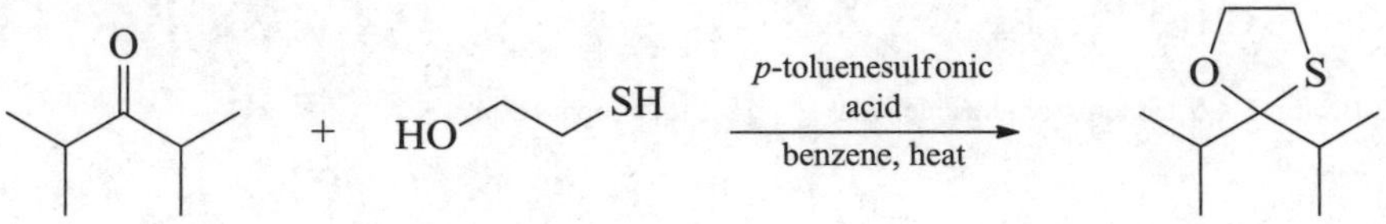

Diisopropyl ketone   2-Mercaptoethanol   2,2-Diisopropyl-1,3-oxathiolane

18.33 Wolff–Kishner reduction converts a carbonyl group ($C=O$) to a methylene group (CH_2).

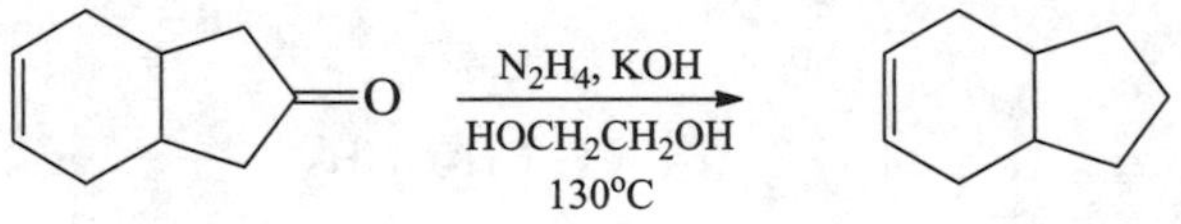

Bicyclo[4.3.0]non-3-en-8-one   Bicyclo[4.3.0]non-3-ene (compound A, 90%)

Treatment of the alkene with *m*-chloroperoxybenzoic acid produces an epoxide, compound B.

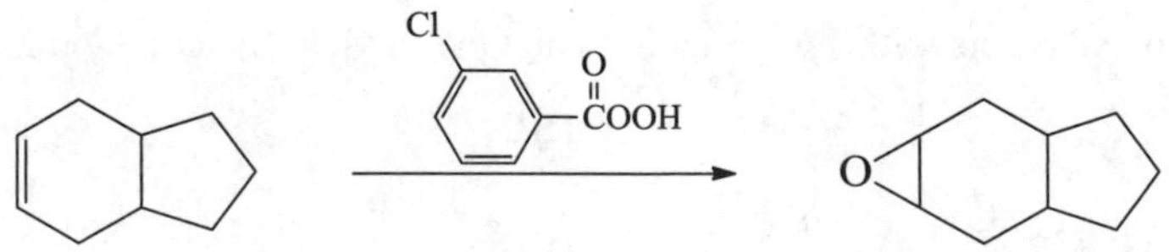

Bicyclo[4.3.0]non-3-ene   3.4-Epoxybicyclo[4.3.0]nonane (compound B, 92%)

Epoxides undergo reduction with lithium aluminum hydride to form alcohols (text Section 17.11).

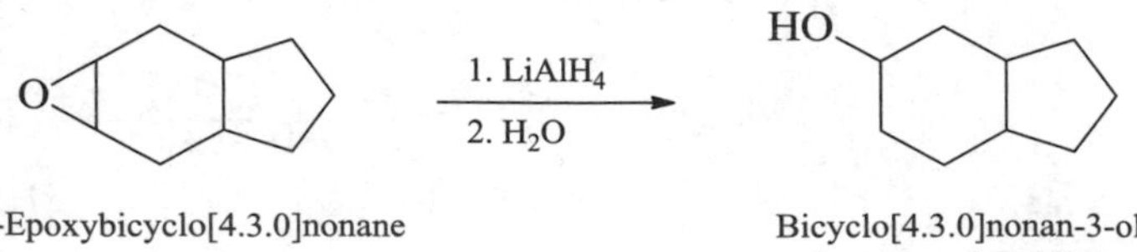

3.4-Epoxybicyclo[4.3.0]nonane   Bicyclo[4.3.0]nonan-3-ol (compound C, 90%)

Chromic acid oxidizes the alcohol to a ketone.

Bicyclo[4.3.0]nonan-3-ol   Bicyclo[4.3.0]nonan-3-one
(compound D, 75%)

18.34 Hydration of formaldehyde by $H_2{}^{17}O$ produces a *gem*-diol in which the labeled and unlabeled hydroxyl groups are equivalent. When this *gem*-diol reverts to formaldehyde, loss of either of the hydroxyl groups is equally likely and leads to eventual replacement of the mass-16 isotope of oxygen by ^{17}O.

This reaction has been monitored by ^{17}O NMR spectroscopy; ^{17}O gives an NMR signal, but ^{16}O does not.

18.35 First write out the chemical equation for the reaction that takes place. Vicinal diols (1,2-diols) react with aldehydes to give cyclic acetals.

Benzaldehyde   1,2-Octanediol   4-Hexyl-2-phenyl-1,3-dioxolane

Notice that the phenyl and hexyl substituents may be either cis or trans to each other. The two products are the cis and trans stereoisomers. Since the 1,2-octanediol is racemic, both dioxolanes are also racemic.

cis-4-Hexyl-2-phenyl-1,3-dioxolane   *trans*-4-Hexyl-2-phenyl-1,3-dioxolane

18.36 Cyclic hemiacetals are formed by intramolecular nucleophilic addition of a hydroxyl group to a carbonyl.

Cyclic hemiacetal

The ring oxygen is derived from the hydroxyl group; the carbonyl oxygen becomes the hydroxyl oxygen of the hemiacetal.

(*a*) This compound is the cyclic hemiacetal of 5-hydroxypentanal.

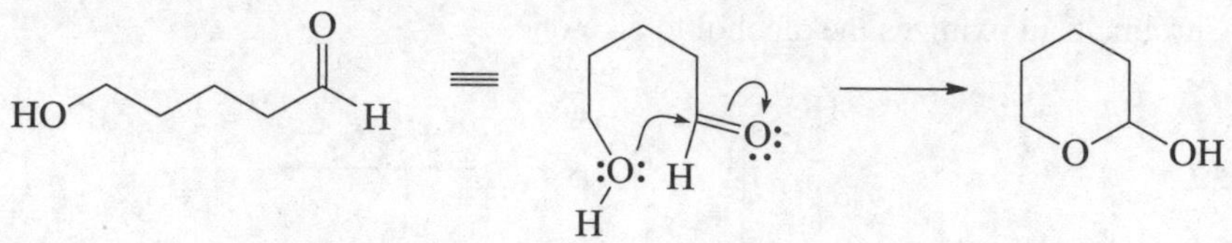

Indeed, 5-hydroxypentanal exists entirely as the cyclic hemiacetal. Its IR spectrum lacks absorption in the carbonyl region.

(b) The carbon connected to two oxygens is the one that is derived from the carbonyl group. Using retrosynthetic symbolism, disconnect the ring oxygen from this carbon.

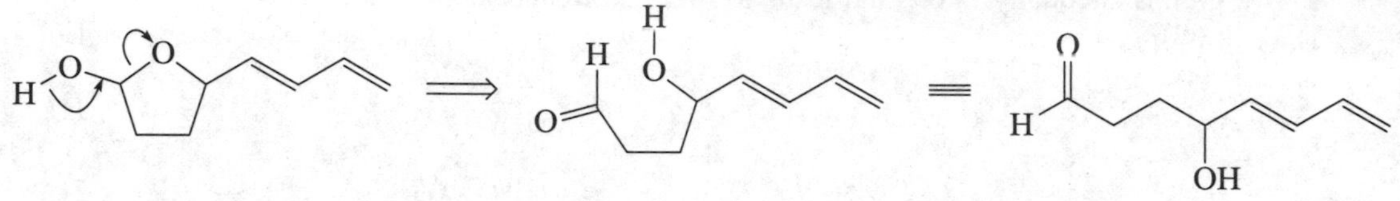

4-Hydroxy-5,7-octadienal

The next two compounds are cyclic acetals. The original carbonyl group is identifiable as the one that bears two oxygen substituents, which originate as hydroxyl oxygens of a diol.

(c)

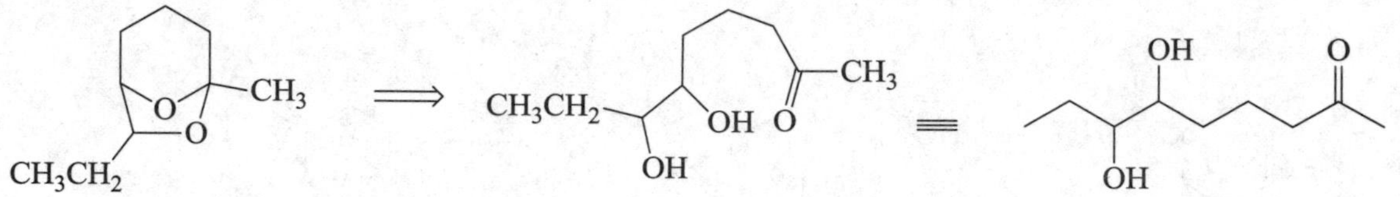

Brevicomin

6,7-Dihydroxy-2-nonanone

(d)

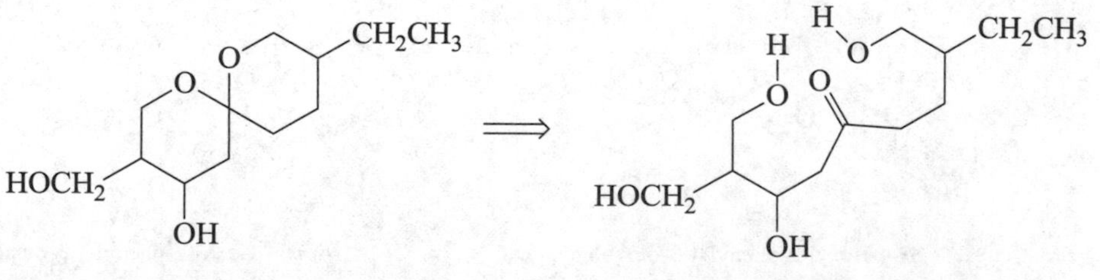

Talaromycin A

2,8-Di(hydroxymethyl)-1,3-dihydroxy-5-decanone

18.37 Benzaldehyde in anhydrous acid is required to make the acetal product.

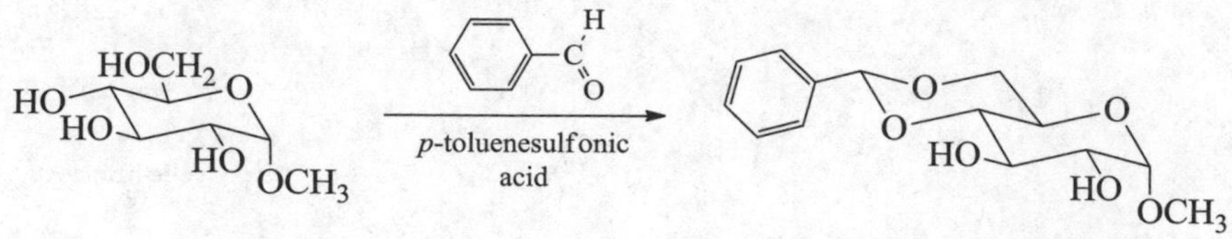

Methyl α-D-glucopyranoside

p-toluenesulfonic acid

Methyl 4,6-O-benzylidene-α-D-glucopyranoside

Synthesis

18.38 (*a*) Friedel–Crafts acylation of benzene with benzoyl chloride is a direct route to benzophenone.

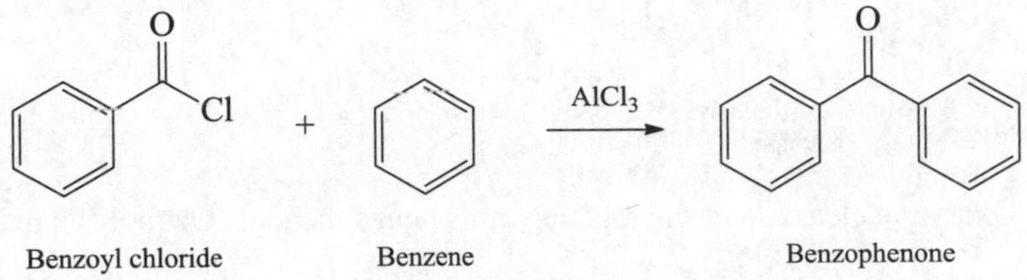

Benzoyl chloride   Benzene   Benzophenone

(*b*) On analyzing the overall transformation retrosynthetically, we see that the target molecule may be prepared by a Grignard synthesis followed by oxidation of the alcohol formed.

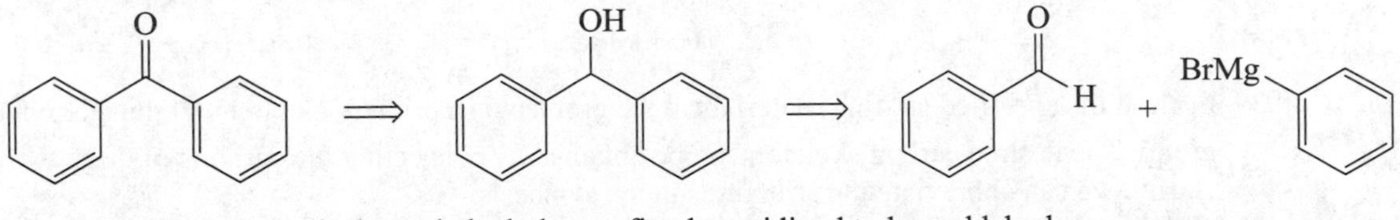

In the desired synthesis, benzyl alcohol must first be oxidized to benzaldehyde.

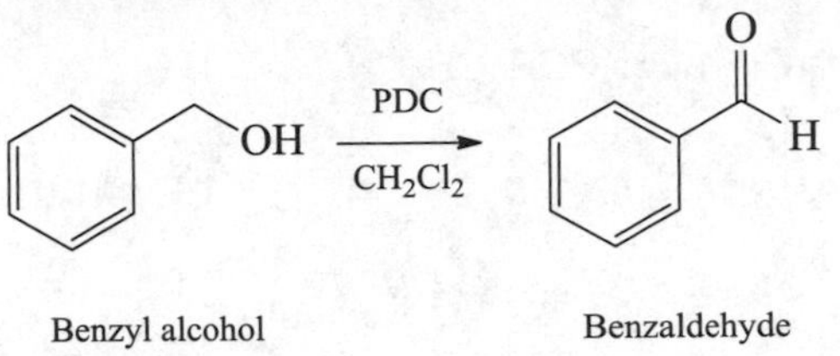

Benzyl alcohol   Benzaldehyde

Reaction of benzaldehyde with the Grignard reagent of bromobenzene followed by oxidation of the resulting secondary alcohol gives benzophenone.

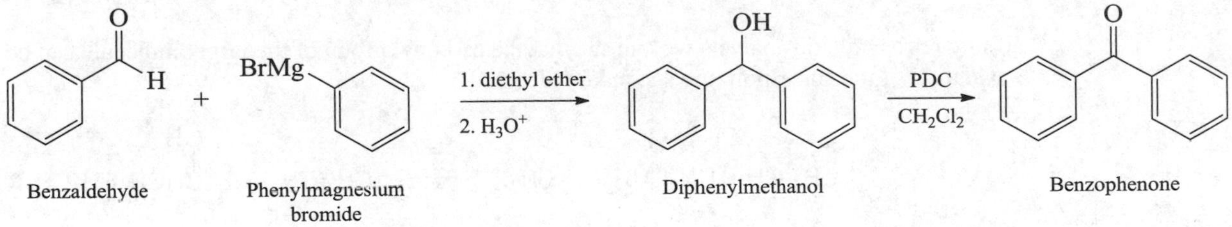

Benzaldehyde   Phenylmagnesium bromide   Diphenylmethanol   Benzophenone

(*c*) Hydrolysis of bromodiphenylmethane yields the corresponding alcohol, which can be oxidized to benzophenone as in part (*b*).

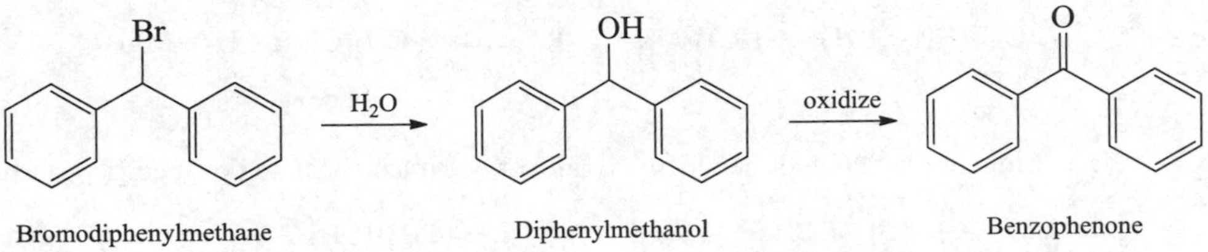

Bromodiphenylmethane   Diphenylmethanol   Benzophenone

(*d*) The starting material is the dimethyl acetal of benzophenone. All that is required is acid-catalyzed hydrolysis.

$$\text{Dimethoxydiphenylmethane} + 2H_2O \xrightarrow{H^+} \text{Benzophenone} + 2CH_3OH$$

| Dimethoxydiphenyl-methane | Water | | Benzophenone | Methanol |

(*e*) Oxidative cleavage of the alkene yields benzophenone. Ozonolysis may be used.

$$\text{1,1,2,2-Tetraphenylethene} \xrightarrow[\text{2. } H_2O, Zn]{\text{1. } O_3} 2 \text{ Benzophenone}$$

18.39 Each of the specified starting materials in the problem is a primary alcohol and thus terminates in a CH_2OH group. The desired carbon skeleton can be obtained by connecting the carbons of these two groups. For clarity, we can abbreviate the structures in the synthesis:

$$CH_3(CH_2)_8CH_2\overset{\overset{\displaystyle O}{\|}}{C}CH_2CH_2CH_2\underset{H}{\overset{H}{\underset{\displaystyle}{C}}}{=}\underset{CH_2CH_2CH_2CH=CH_2}{C} \quad \text{becomes}$$

$$CH_3(CH_2)_8CH_2\overset{\overset{\displaystyle O}{\|}}{C}CH_2CH_2CH_2R \quad \text{where R equals} \quad \underset{H}{\overset{}{C}}{=}\underset{CH_2CH_2CH_2CH=CH_2}{\overset{H}{C}}$$

Using retrosynthetic analysis we can see that the carbonyl group of the target molecule can be obtained by oxidation of the corresponding secondary alcohol.

$$CH_3(CH_2)_8CH_2\overset{\overset{\displaystyle O}{\|}}{C}CH_2CH_2CH_2R \implies CH_3(CH_2)_8CH_2\overset{\overset{\displaystyle OH}{|}}{C}HCH_2CH_2CH_2R$$

The secondary alcohol can be made by a Grignard reaction.

Disconnect here

$$CH_3(CH_2)_8CH_2{-}\overset{\overset{\displaystyle OH}{|}}{C}HCH_2CH_2CH_2R \implies CH_3(CH_2)_8CH_2{-}MgBr + \overset{\overset{\displaystyle O}{\|}}{H}CCH_2CH_2CH_2R$$

The Grignard reagent and the aldehyde can both be traced back to the specified starting materials.

$$CH_3(CH_2)_8CH_2{-}MgBr \implies CH_3(CH_2)_8CH_2{-}Br \implies CH_3(CH_2)_8CH_2{-}OH$$

$$\overset{\overset{\displaystyle O}{\|}}{H}CCH_2CH_2CH_2R \implies HO{-}CH_2CH_2CH_2CH_2R$$

This retrosynthetic analysis suggests the following synthesis.

First, prepare the aldehyde and the Grignard reagent.

$$HO-CH_2CH_2CH_2CH_2R \xrightarrow[CH_2Cl_2]{PCC} \overset{\overset{O}{\parallel}}{H}CCH_2CH_2CH_2R$$

(E)-5,10-Undecadien-1-ol   (E)-5,10-Undecadienal

$$CH_3(CH_2)_8CH_2-OH \xrightarrow[heat]{HBr} CH_3(CH_2)_8CH_2-Br \xrightarrow[diethyl\ ether]{Mg} CH_3(CH_2)_8CH_2-MgBr$$

1-Decanol   1-Bromodecane   Decylmagnesium bromide

Add the aldehyde to the Grignard reagent, then add aqueous acid to "work up" the reaction. The product is a secondary alcohol having the required carbon skeleton.

$$CH_3(CH_2)_8CH_2-MgBr + \overset{\overset{O}{\parallel}}{H}CCH_2CH_2CH_2R \xrightarrow[2.\ H_3O^+]{1.\ diethyl\ ether} CH_3(CH_2)_8CH_2-\overset{\overset{OH}{|}}{C}HCH_2CH_2CH_2R$$

(E)-1,6-Henicosadiene-11-ol

Complete the synthesis by oxidizing the secondary alcohol to the target ketone.

$$CH_3(CH_2)_8CH_2-\overset{\overset{OH}{|}}{C}HCH_2CH_2CH_2R \xrightarrow[CH_2Cl_2]{PCC} CH_3(CH_2)_8CH_2\overset{\overset{O}{\parallel}}{C}CH_2CH_2CH_2$$

$$\overset{H}{\underset{H}{}}C=C\overset{H}{\underset{CH_2CH_2CH_2CH=CH_2}{}}$$

(E)-1,6-Henicosadiene-11-one

18.40 The two alcohols given as starting materials contain all the carbon atoms of the desired product.

$$CH_3(CH_2)_8CH=CHCH_2CH=CHCH_2CH \vdots CHCH=CH_2 \Longrightarrow$$

$$CH_3(CH_2)_8CH=CHCH_2CH=CHCH_2CH_2-OH \quad and \quad HO-CH_2CH=CH_2$$

3,6-Hexadecadien-1-ol   Allyl alcohol

What is needed is to attach the two groups together so that the two primary alcohol carbons become doubly bonded to each other. This can be accomplished by using a Wittig reaction as the key step.

$$CH_3(CH_2)_8CH=CHCH_2CH=CHCH_2CH_2-OH \xrightarrow[CH_2Cl_2]{PCC} CH_3(CH_2)_8CH=CHCH_2CH=CHCH_2\overset{\overset{O}{\parallel}}{C}H$$

3,6-Hexadecadien-1-ol   3,6-Hexadecadienal

$$HO-CH_2CH=CH_2 \xrightarrow{PBr_3} Br-CH_2CH=CH_2 \xrightarrow{(C_6H_5)_3P} (C_6H_5)_3\overset{+}{P}-CH_2CH=CH_2\ Br^-$$

Allyl alcohol   Allyl bromide   Allyltriphenylphosphonium bromide

$$\downarrow \overset{CH_3CH_2CH_2CH_2Li}{\underset{THF}{}}$$

$$(C_6H_5)_3\overset{+}{P}-\overset{..}{C}HCH=CH_2$$

Allylidenetriphenylphosphorane

$$CH_3(CH_2)_8CH=CHCH_2CH=CHCH_2\overset{\overset{\displaystyle O}{\|}}{C}H \quad + \quad (C_6H_5)_3\overset{+}{P}-\overset{..}{\overset{-}{C}}HCH=CH_2 \longrightarrow$$

3.6-Hexadecadienal

Allylidenetriphenylphosphorane

$$CH_3(CH_2)_8CH=CHCH_2CH=CHCH_2CH=CHCH=CH_2$$

1,3,6,9-Nonadecatetraene

Alternatively, allyl alcohol could be oxidized to $H_2C=CHCHO$ for subsequent reaction with the ylide derived from $CH_3(CH_2)_8CH=CHCH_2CH=CHCH_2CH_2OH$ via its bromide and triphenylphosphonium salt.

18.41 Compound A is in equilibrium with its open-chain hydroxy aldehyde form, which undergoes nucleophilic addition with the phosphorous ylide. The alkene obtained, compound B, is mainly the *E*-alkene because the ylide is stabilized.

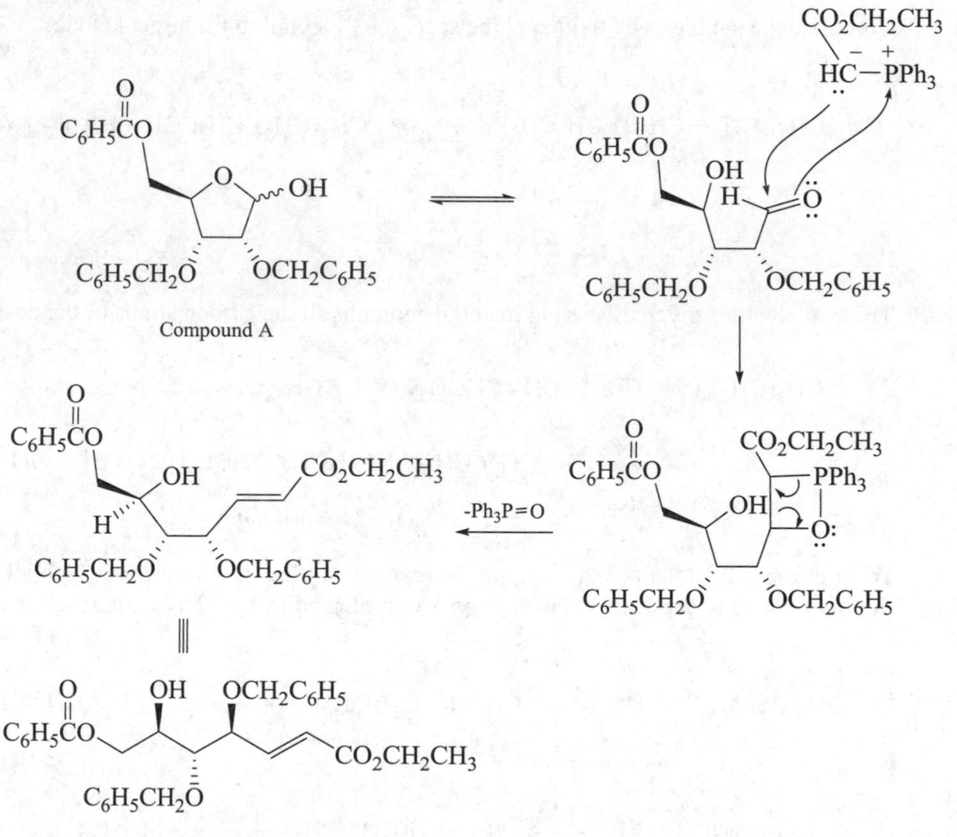

Compound A

Compound B

18.42 (*a*) Recalling that alkanes may be prepared by hydrogenation of the appropriate alkene, a synthesis of the desired product becomes apparent. What is needed is to convert —C=O into —C=CH$_2$; a Wittig reaction is appropriate.

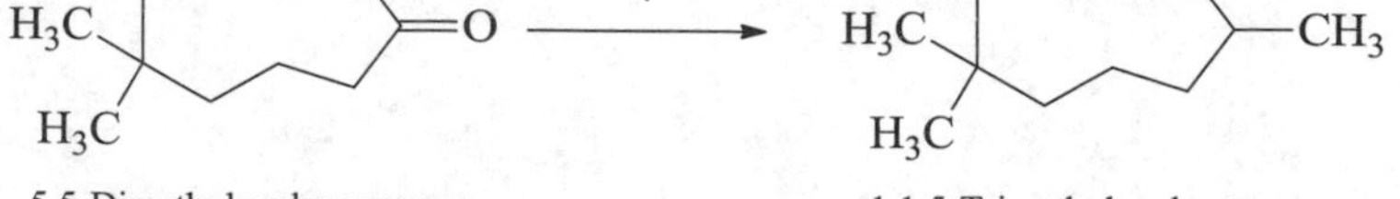

5,5-Dimethylcyclononanone

1,1,5-Trimethylcyclononanone

The two-step procedure that was followed used a Wittig reaction to form the carbon–carbon bond, then catalytic hydrogenation of the resulting alkene.

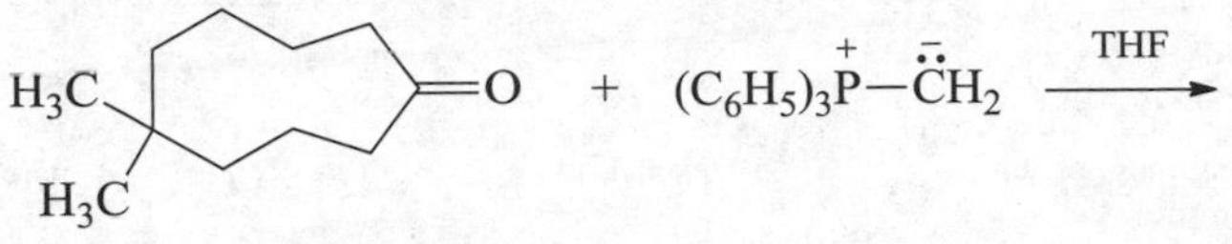

5,5-Dimethylcyclononanone

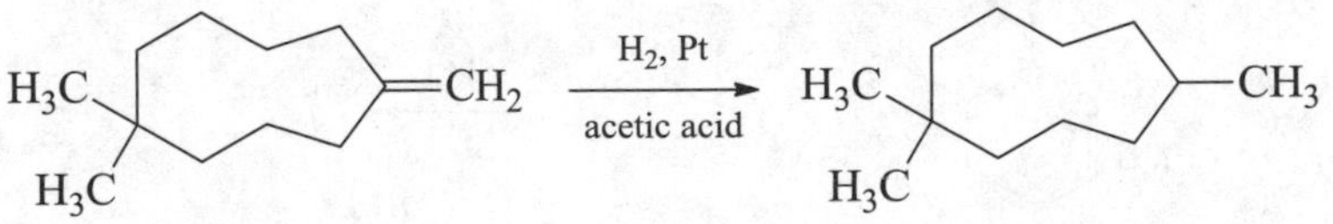

5,5-Dimethyl-1-
methylenecyclononane (59%)

1,1,5-Trimethylcyclononanone
(73%)

(b) In putting together the carbon skeleton of the target molecule, a methyl group has to be added to the original carbonyl carbon.

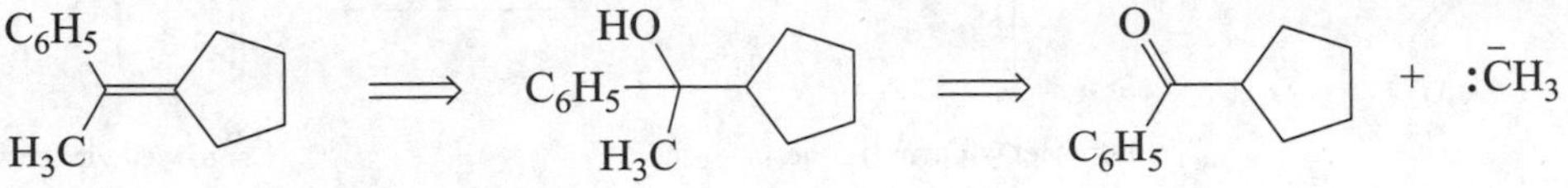

The logical way to do this is by way of a Grignard reagent.

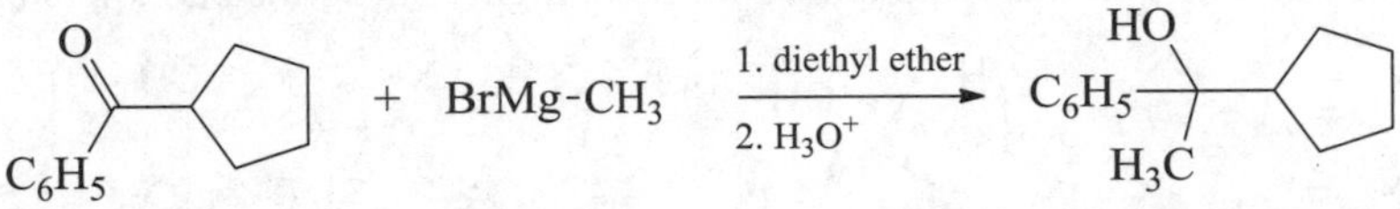

| Cyclopentyl phenyl ketone | Methylmagnesium iodide | 1-Cyclopentyl-1-phenylethanol |

Acid-catalyzed dehydration yields the more highly substituted alkene, the desired product, in accordance with the Zaitsev rule.

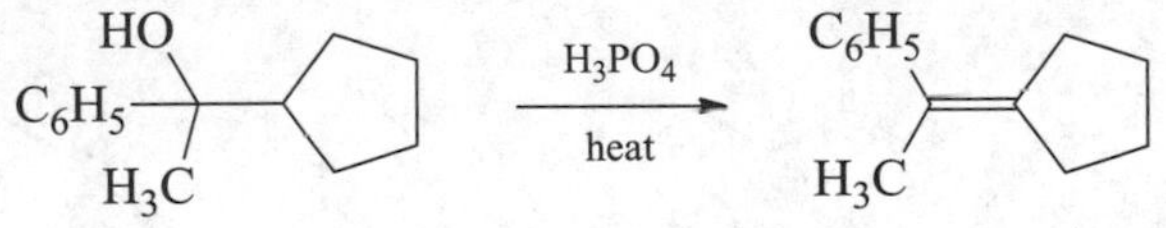

1-Cyclopentyl-1-phenylethanol   (1-Phenylethylidene)cyclopentane

(c) Analyzing the transformation retrosynthetically, keeping in mind the starting materials stated in the problem, we see that the carbon skeleton may be constructed in a straightforward manner.

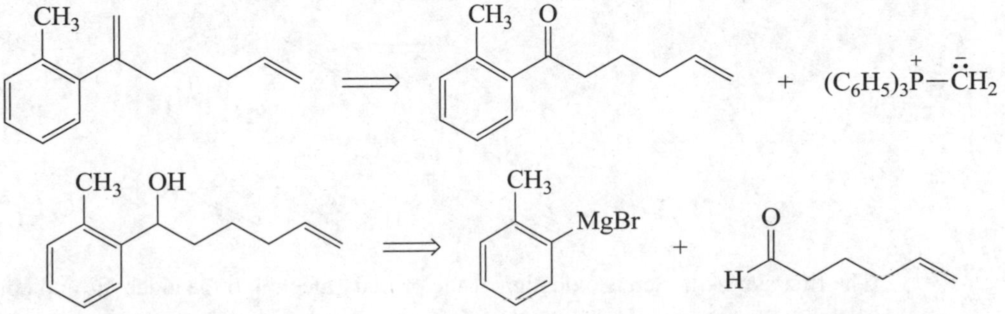

Proceeding with the synthesis in the forward direction, reaction between the Grignard reagent of *o*-bromotoluene and 5-hexenal produces most of the desired carbon skeleton.

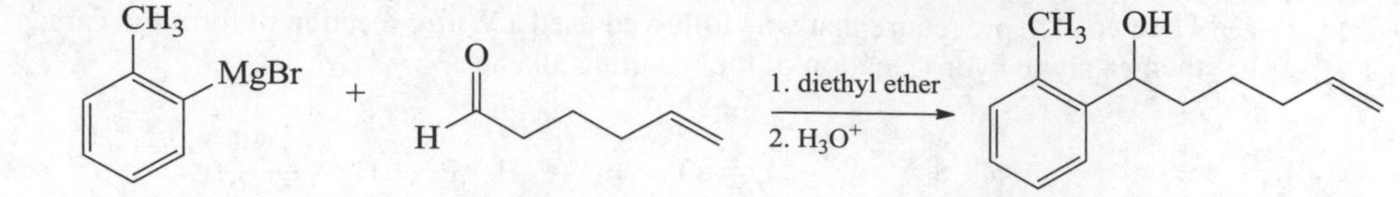

o-Methylphenylmagnesium
bromide

5-Hexenal

1-(o-Methylphenyl)-5-hexen-1-ol

Oxidation of the resulting alcohol to the ketone followed by a Wittig reaction leads to the final product.

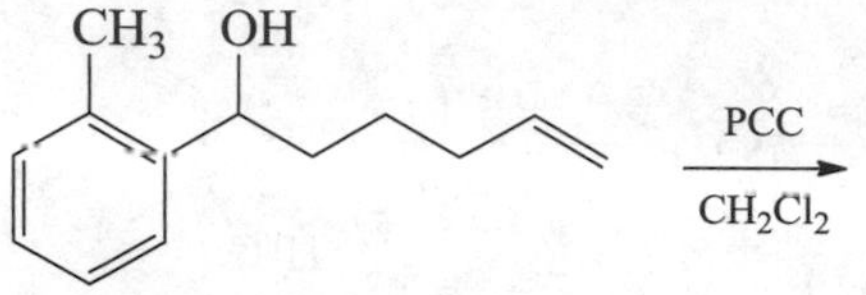

1-(o-Methylphenyl)-5-hexen-1-ol

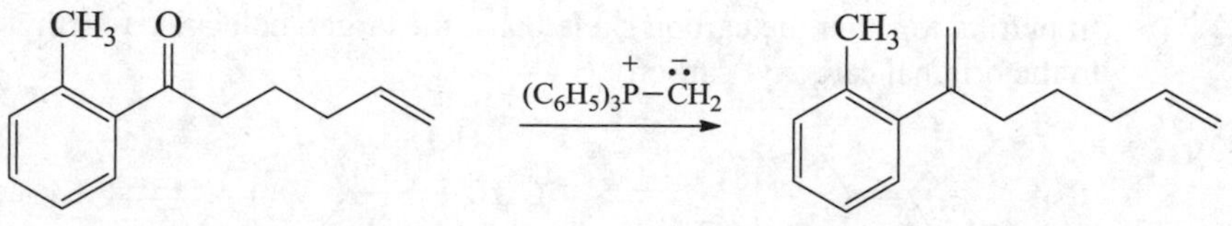

1-(o-Methylphenyl)-5-hexen-1-one

2-(o-Methylphenyl)-1,6-heptadiene

Acid-catalyzed dehydration of the corresponding tertiary alcohol would *not* be suitable because the major elimination product would have the more highly substituted double bond.

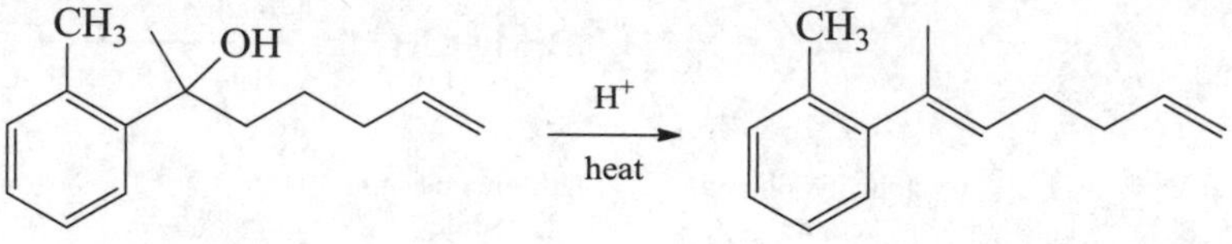

2-(o-Methylphenyl)-6-hepten-2-ol

6-(o-Methylphenyl)-1,5-heptadiene

(*d*) Remember that terminal alkynes can serve as sources of methyl ketones by hydration.

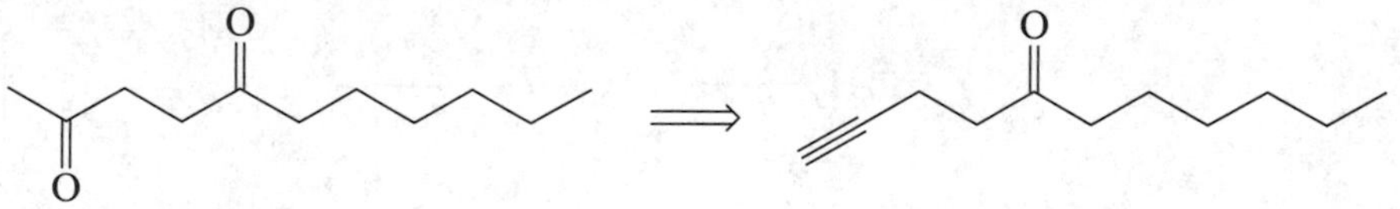

This gives us a clue as to how to proceed, because the acetylenic ketone may be prepared from the starting acetylenic alcohol.

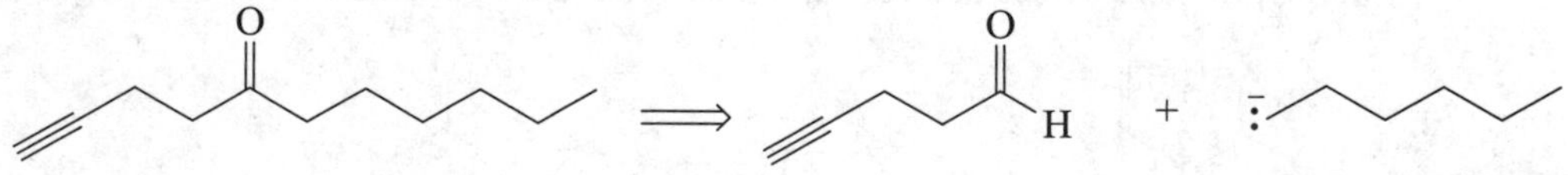

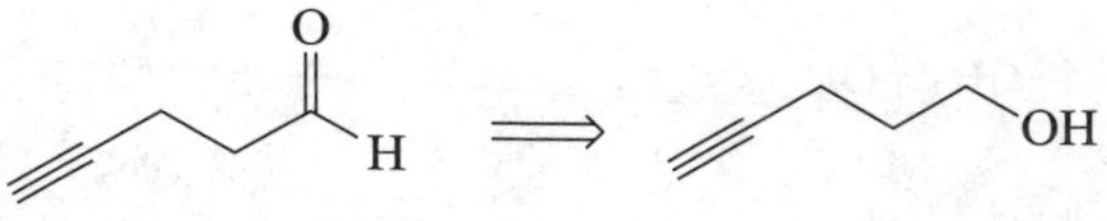

The first synthetic step is oxidation of the primary alcohol to the aldehyde and construction of the carbon skeleton by a Grignard reaction.

4-Pentyn-1-ol → 4-Pentynal → 1-Undecyn-5-ol

Oxidation of the secondary alcohol to a ketone and hydration of the terminal triple bond complete the synthesis.

1-Undecyn-5-ol

1-Undecyn-5-one → 2,5-Undecanedione

(e) The desired product is a benzylic ether. To prepare it, the aldehyde must first be reduced to the corresponding primary alcohol. Sodium borohydride was used in the preparation described in the literature, but lithium aluminum hydride or catalytic hydrogenation would also be possible. Once the alcohol is prepared, it can be converted to its alkoxide ion and this alkoxide ion treated with methyl iodide.

Alternatively, the alcohol could be treated with hydrogen bromide or with phosphorus tribromide to give the benzylic bromide and the bromide then allowed to react with sodium methoxide.

(f) Two operations need to be carried out: reduction of C=O to CH_2, and Friedel–Crafts acylation. A successful synthesis depends on the order in which the operations are accomplished.

By reducing C=O to CH_2 first, a compound is formed in which both positions on the benzene ring that are available for Friedel–Crafts acylation are equivalent and are activated toward eletrophilic aromatic substitution.

If Friedel–Crafts acylation were attempted first, it would probably fail because the C=O group in the starting material deactivates the aromatic ring toward electrophilic aromatic substitution. Furthermore, even if Friedel–Crafts acylation were successful, the product of that reaction would contain two carbonyl groups and there is no reason to believe that one could be reduced to CH₂ and not the other.

(*g*) Retrosynthetically, the desired product is available by a Diels–Alder cycloaddition.

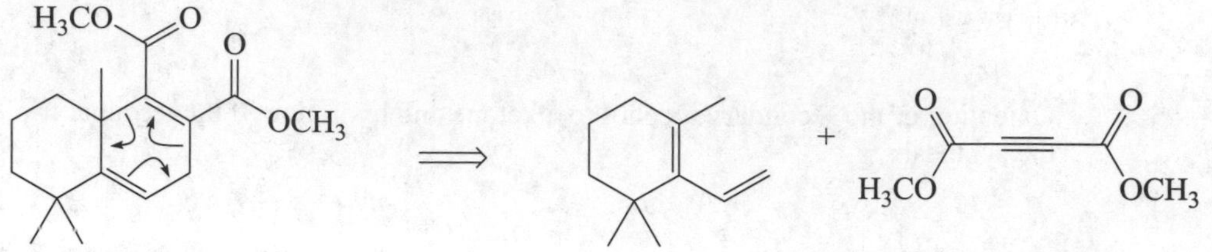

This suggests the following synthesis. The first reaction is the formation of the conjugated diene unit via a Wittig reaction.

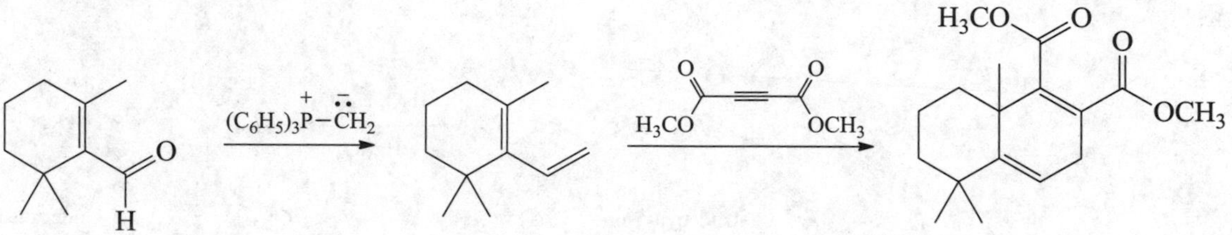

Mechanism

18.43 The best way to answer this problem is to work through the step-by-step mechanism.

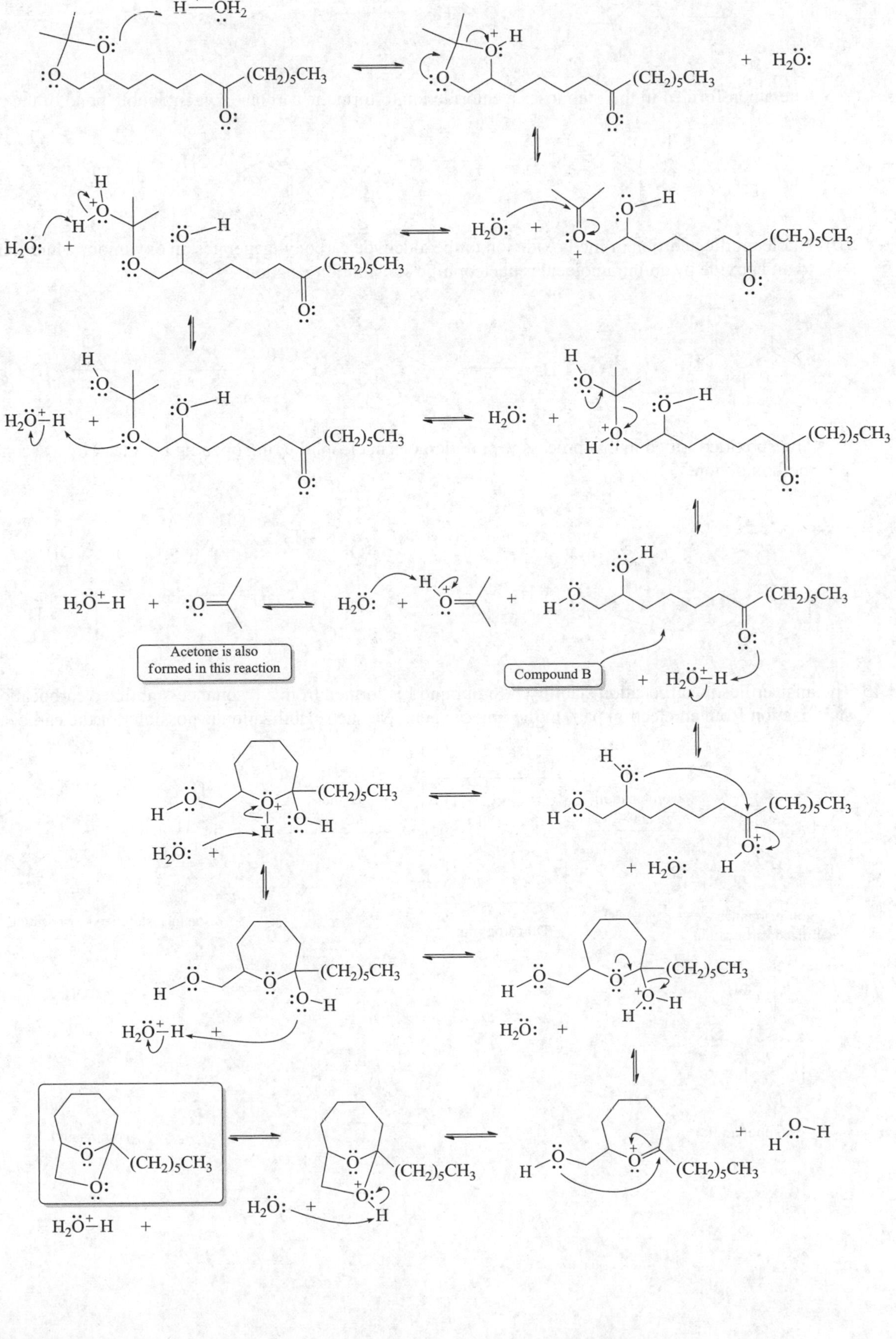

Acetone is also
formed in this reaction

Compound B

18.44 (*a*) Nucleophilic ring opening of the epoxide occurs by attack of methoxide at the less hindered carbon.

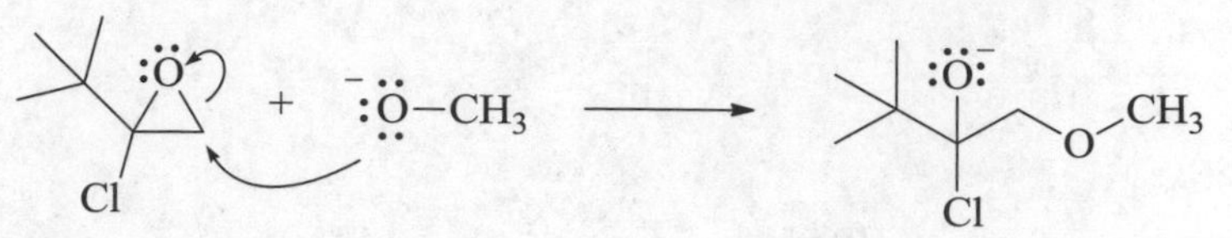

The anion formed in this step loses a chloride ion to form the carbon–oxygen double bond of the product.

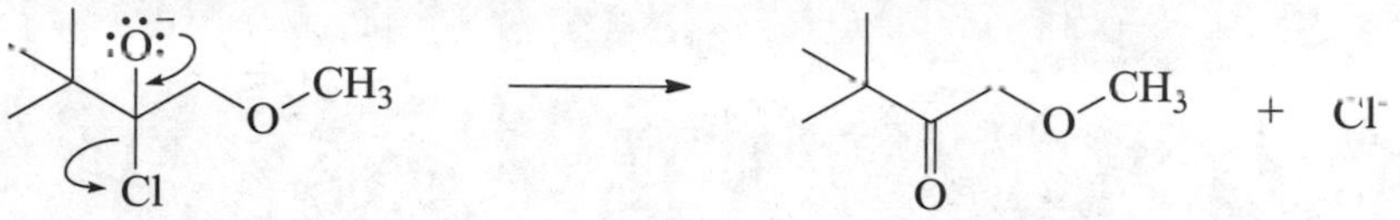

(*b*) Nucleophilic addition of methoxide ion to the aldehyde carbonyl generates an oxyanion, which can close to an epoxide by an intramolecular nucleophilic substitution reaction.

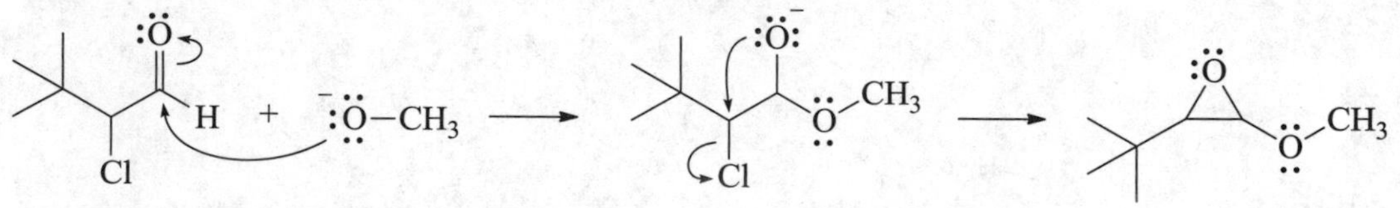

The epoxide formed in this process then undergoes nucleophilic ring opening on attack by a second methoxide ion.

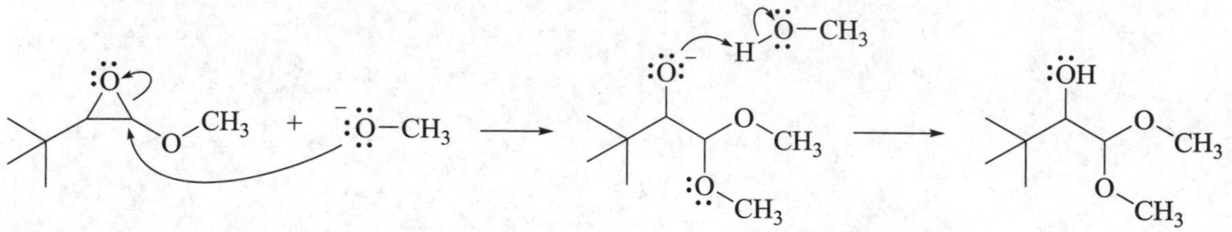

18.45 The answer lies in carbocation stability. Compound I is formed from a resonance-stabilized carbocation, with stabilization from an electron pair on the ring oxygen. No such stabilization is possible for the carbocation leading to II.

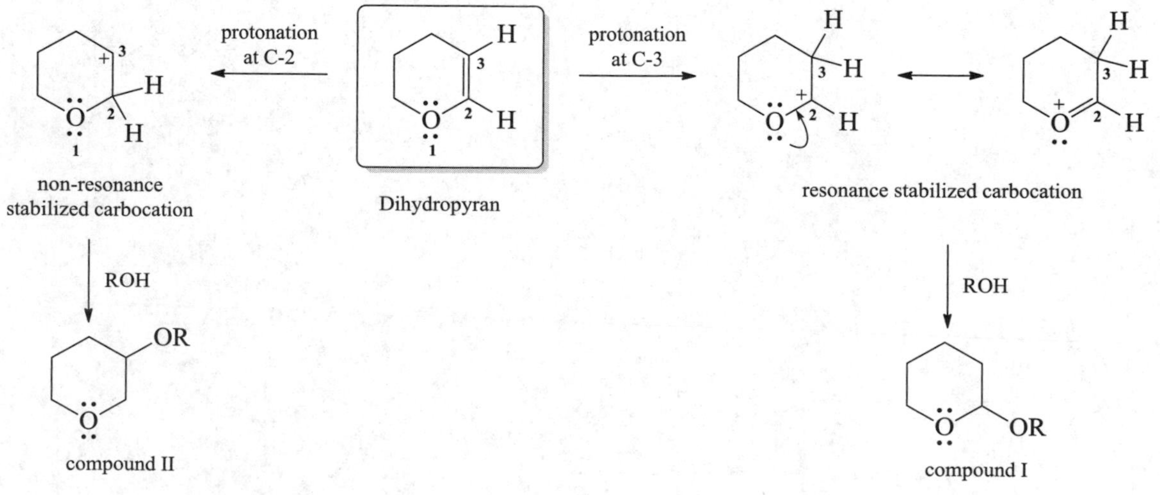

Spectroscopy

18.46 Given that the molecular formula is C_4H_8O and that the compound contains a carbonyl group, it cannot contain either a ring or another multiple bond. Its ^1H NMR spectrum is composed of four signals with the following chemical shifts, multiplicities, and relative areas.

0.9 ppm	triplet	3
1.7 ppm	sextet	2
2.4 ppm	triplet of doublets	2
9.8 ppm	triplet	1

The three-proton triplet at δ 0.9 ppm signals the presence a CH_3 group bonded to CH_2. This CH_2 group is responsible for the sextet at 1.7 ppm. The presence of a double bond along with a one-proton triplet at δ 9.8 tells us that the compound is an aldehyde and that C-2 is a CH_2 group. The signal for the two protons at C-2 is the triplet of doublets at δ 2.4 on the basis of its chemical shift and its splitting pattern. Thus, we have the structural units CH_3CH_2 and $CH_2CH=O$. The compound is butanal ($CH_3CH_2CH_2CH=O$).

18.47 A carbonyl group is evident from the strong IR absorption at 1710 cm^{-1}. Because all the ^1H NMR signals are singlets, there are no nonequivalent hydrogens in a vicinal or "three-bond" relationship. The three-proton signal at δ 2.1 and the two-proton signal at δ 2.3 can be understood as arising from a

$$CH_2\overset{\overset{\displaystyle O}{\|}}{C}CH_3$$

unit. The intense nine-proton singlet at δ 1.0 is due to the three equivalent methyl groups of a $(CH_3)_3C$ unit. The compound is 4,4-dimethyl-2-pentanone.

$$CH_3\overset{\overset{\displaystyle O}{\|}}{C}CH_2C(CH_3)_3$$

δ 2.1   δ 2.3   δ 1.0
singlet   singlet   singlet

4,4-Dimethyl-2-pentanone

18.48 The molecular formula of compounds A and B ($C_6H_{10}O_2$) indicates an index of hydrogen deficiency of 2. Because we are told the compounds are diketones, the two carbonyl groups account for all the unsaturations.

The ^1H NMR spectrum of compound A has only two peaks, both singlets, at δ 2.2 and 2.8. Their intensity ratio (6:4) is consistent with two equivalent methyl groups and two equivalent methylene groups. The chemical shifts are appropriate for

$$CH_3\overset{\overset{\displaystyle O}{\|}}{C} \quad \text{and} \quad CH_2\overset{\overset{\displaystyle O}{\|}}{C}$$

The simplicity of the spectrum can be understood if we are dealing with a symmetric diketone. The correct structure is

$$\underset{CH_3}{} \overset{O}{\underset{\|}{C}} CH_2 CH_2 \overset{O}{\underset{\|}{C}} CH_3$$

Equivalent methylene
groups do not split
each other.

2,5-Hexanedione (compound A)

Compound B is an isomer of compound A. The triplet–quartet pattern in the ^{1}H NMR spectrum is consistent with an ethyl group and, because the triplet is equivalent to six protons and the quartet to four, it is likely that two equivalent ethyl groups are present. The two ethyl groups account for four carbons, and because the problem stipulates that the molecule is a diketone, all the carbons are accounted for. The only $C_6H_{10}O_2$ diketone with two equivalent ethyl groups is 3,4-hexanedione.

$$CH_3CH_2 \overset{O}{\underset{\|}{C}} - \overset{O}{\underset{\|}{C}} CH_2CH_3$$

δ 1.3   δ 2.8
triplet   quartet

3,4-Hexanedione (compound B)

18.49 From its molecular formula ($C_{11}H_{14}O$), the compound has a total of five double bonds and/or rings. The presence of a strong peak near 1700 cm^{-1} in the IR spectrum indicates the presence of a carbonyl group, accounting for the remaining element of unsaturation. The presence of signals in the region δ 7 to 8 suggests an aromatic ring is present, accounting for four of the elements of unsaturation. The highest field peak in the NMR spectrum is a three-proton triplet, corresponding to the methyl group of a CH_3CH_2 unit. The two-proton signal at δ 3.0 corresponds to a CH_2 unit adjacent to the carbonyl group and, because it is a triplet, suggests the grouping $CH_2CH_2C=O$. The compound is butyl phenyl ketone (1-phenyl-1-pentanone).

δ 3.0   Multiplets at   δ 1.0 triplet
triplet   δ 1.3 and 1.7

Butyl phenyl ketone

18.50 With a molecular formula of $C_7H_{14}O$, the compound has an index of hydrogen deficiency of 1. We are told that it is a ketone, so it has no rings or double bonds other than the one belonging to its C=O group. The peak at δ 211 in the ^{13}C NMR spectrum corresponds to the carbonyl carbon. Only three other signals occur in the spectrum, and so there are only three types of carbons other than the carbonyl carbon. This suggests that the compound is the symmetrical ketone 4-heptanone.

4-Heptanone
(all chemical shifts in ppm)

18.51 Compounds A and B are isomers and have an index of hydrogen deficiency of 5. Signals in the region 125–140 in their ^{13}C NMR spectra suggest an aromatic ring, and a peak at δ 200 indicates a carbonyl group. An aromatic ring contributes one ring and three double bonds, and a carbonyl group contributes one double bond, and so the index of hydrogen deficiency of 5 is satisfied by a benzene ring and a carbonyl group. The carbonyl group is attached directly to the benzene ring, as evidenced by the presence of a peak at m/z 105 in the mass spectra of compounds A and B.

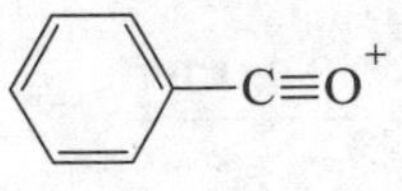

$$\text{C}_6\text{H}_5\text{—C}{\equiv}\text{O}^+$$

m/z 105

Each ^{13}C NMR spectrum shows four aromatic signals, and so the rings are monosubstituted.

Compound A has three unique carbons in addition to $C_6H_5C{=}O$ and so must be 1-phenyl-1-butanone. Compound B has only two additional signals and so must be 2-methyl-1-phenyl-1-propanone.

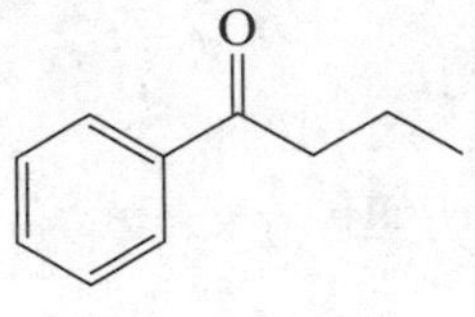

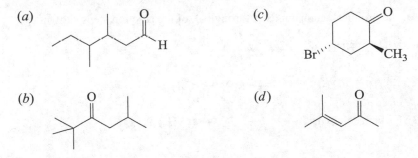

Compound A 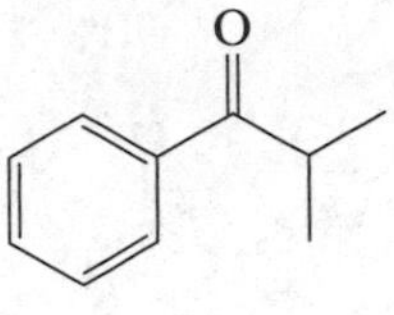   Compound B

Answers to Interpretive Problems 18

18.52 D; **18.53** A; **18.54** A; **18.55** C; **18.56** A; **18.57** C; **18.58** C

SELF-TEST

1. Give an acceptable IUPAC name for each of the following:

(*a*)   (*c*)

(*b*)   (*d*)

2. Write the structural formulas for

(*a*) (*E*)-3-Hexen-2-one

(*b*) 3-Cyclopropyl-2,4-pentanedione

(*c*) 3-Ethyl-4-phenylpentanal

3. For each of the following reactions, supply the structure of the missing reactant, reagent, or product:

(a)

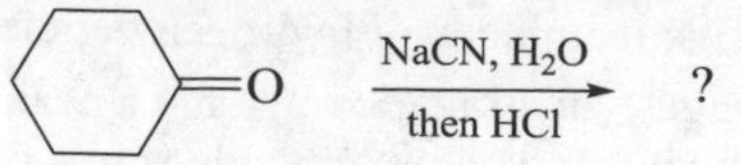

cyclohexanone $\xrightarrow[\text{then HCl}]{\text{NaCN, H}_2\text{O}}$?

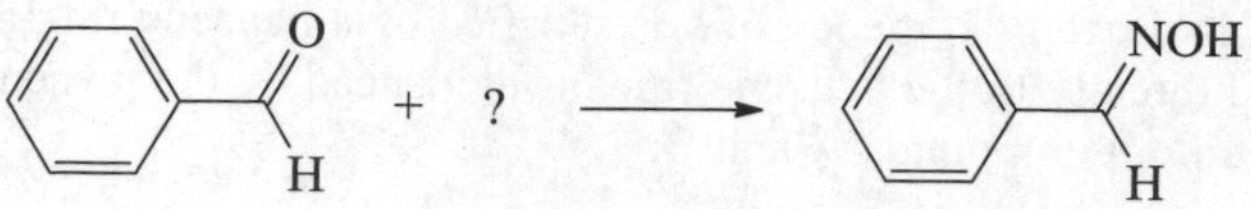

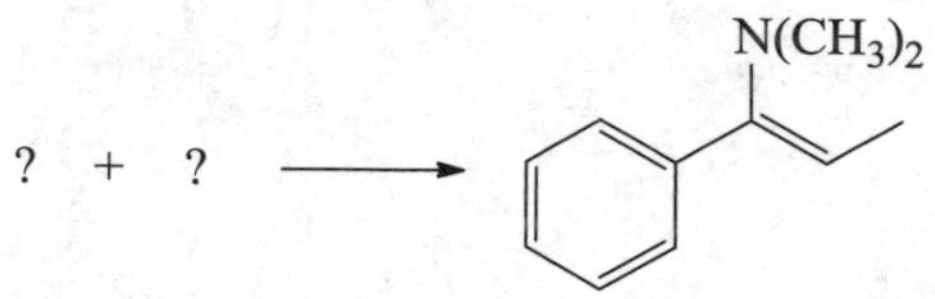

(b)

benzaldehyde + ? $\longrightarrow$ (benzaldoxime, NOH)

(c)

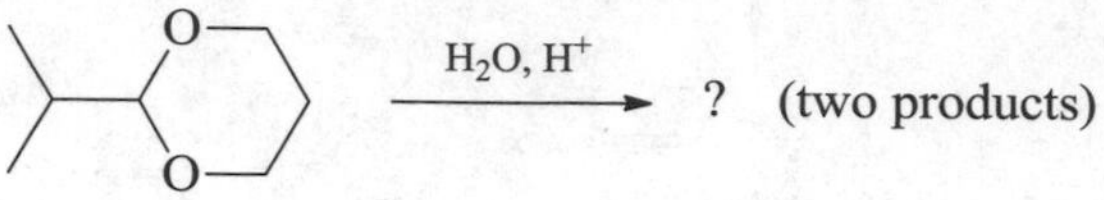

2-isopropyl-1,3-dioxane $\xrightarrow{\text{H}_2\text{O, H}^+}$? (two products)

(d)

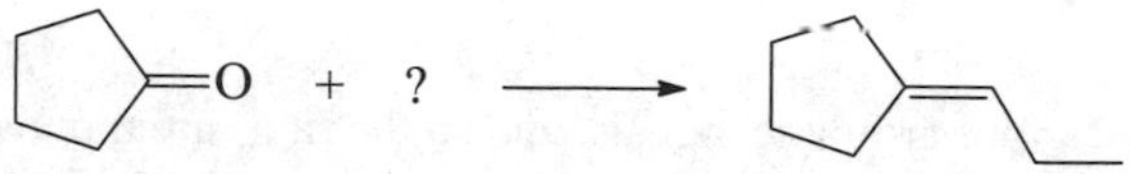

cyclopentanone =O + ? $\longrightarrow$ (ethylidenecyclopentane)

(e)

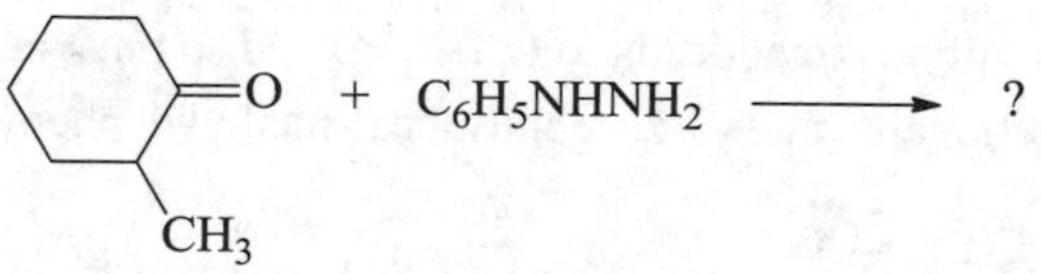

2-methylcyclohexanone + $C_6H_5NHNH_2$ $\longrightarrow$?

(f)

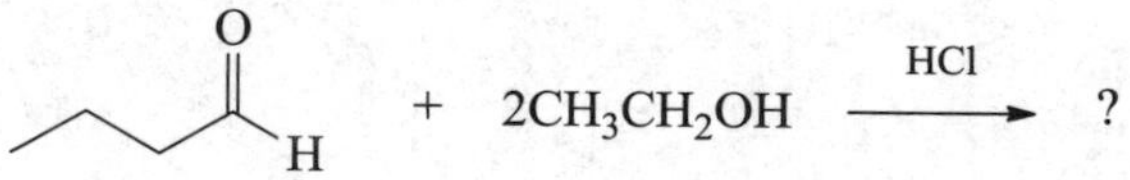

butanal + $2CH_3CH_2OH$ $\xrightarrow{\text{HCl}}$?

(g)

? + ? $\longrightarrow$ (1-phenyl-1-(dimethylamino)propene, $N(CH_3)_2$)

(h)

butanal $\xrightarrow[\text{H}^+, \text{H}_2\text{O}]{\text{Na}_2\text{Cr}_2\text{O}_7}$?

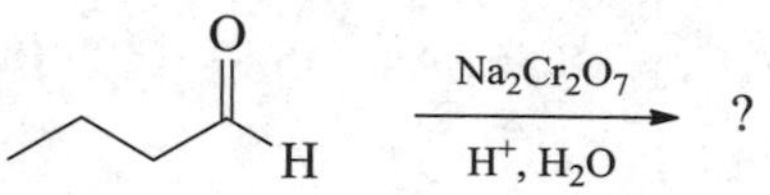

4.   Write the structures of the products, compounds A through D, of the reaction steps shown.

(a) $(C_6H_5)_3P$ + $(CH_3)_2CHCH_2Br$ $\longrightarrow$ A

A + $CH_3CH_2CH_2CH_2Li$ $\longrightarrow$ B + C_4H_{10}

B + benzaldehyde $\longrightarrow$ C + $(C_6H_5)_3\overset{+}{P}-\overset{-}{O}$

(b)

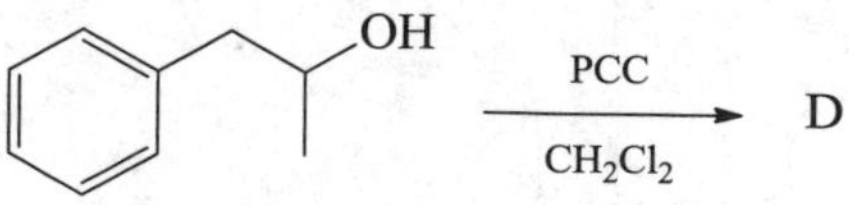

1-phenyl-2-propanol (OH) $\xrightarrow[\text{CH}_2\text{Cl}_2]{\text{PCC}}$ D

5.   Give the reagents necessary to convert cyclohexanone into each of the following compounds. More than one step may be necessary.

(a)   1-methylcyclohexene (CH_3)   (b)   methylenecyclohexane (CH_2)   (c)   1,4-dioxaspiro compound (O, O)

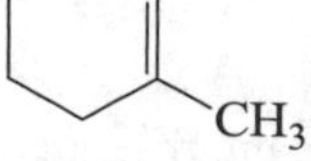

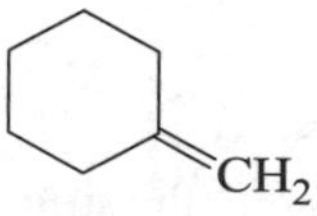

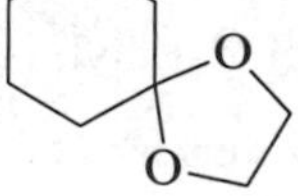

6. (*a*) What two organic compounds react together (in the presence of an acid catalyst) to give the compound shown, plus a molecule of water?

(*b*) Draw the structure of the open-chain form of the following cyclic acetal:

7. Outline reaction schemes to carry out each of the following interconversions, using any necessary organic or inorganic reagents.

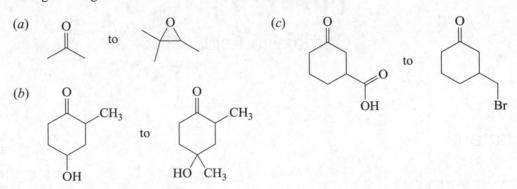

8. Write a stepwise mechanism for the formation of $CH_3CH(OCH_3)_2$ from acetaldehyde and methanol under conditions of acid catalysis.

9. Suggest a structure for an unknown compound, $C_9H_{10}O$, that exhibits a strong infrared absorption at 1710 cm^{-1} and has a ^1H NMR spectrum that consists of three singlets at δ 2.1 (3H), 3.7 (2H), and 7.2 (5H).

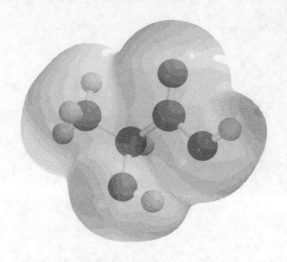

CHAPTER 19
Carboxylic Acids

Table of Contents

SOLUTIONS TO TEXT PROBLEMS

In Chapter Problems

19.1 (b) The four carbon atoms of crotonic acid form a continuous chain. Because there is a double bond between C-2 and C-3, crotonic acid is one of the stereoisomers of 2-butenoic acid. The stereochemistry of the double bond is E.

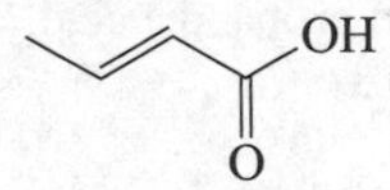

(E)-2-Butenoic acid or
(E)-But-2-enoic acid
(crotonic acid)

(c) Oxalic acid is a dicarboxylic acid that contains two carbons. It is *ethanedioic* acid.

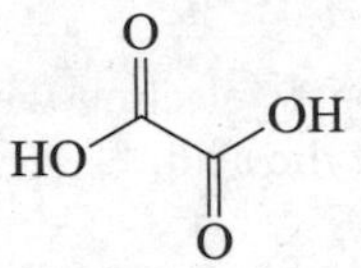

Ethanedioic acid
(oxalic acid)

(d) The name given to $C_6H_5CO_2H$ is benzoic acid. Because it has a methyl group at the para position, the compound shown is *p*-methylbenzoic acid, or 4-methylbenzoic acid.

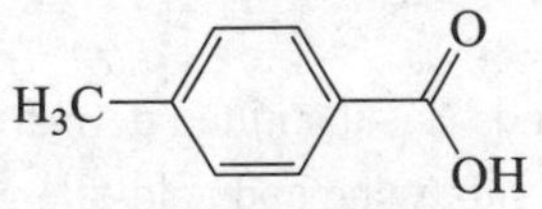

p-Methylbenzoic acid or
4-methylbenzoic acid
(*p*-toluic acid)

19.2 (b) As in part (a) of this problem we can calculate the conjugate base/acid ratio using the Henderson–Hasselbalch relationship.

$$\frac{[\text{conjugate base}]}{[\text{acid}]} = 10^{(\text{pH}-\text{p}K\text{a})}$$

$$\frac{[\text{lactate}]}{[\text{lactic acid}]} = 10^{(2.5-3.9)} = 10^{-1.4} = 0.04$$

At pH = 2.5, 25 times as much lactic acid is present as lactate ion.

19.3 (*b*) Propanoic acid is similar to acetic acid in its acidity. A hydroxyl group at C-2 is electron-withdrawing and stabilizes the carboxylate ion of lactic acid by a combination of inductive and field effects.

Hydroxyl group stabilizes
negative charge by attracting electrons.

Lactic acid is more acidic than propanoic acid. The measured pK_a's are

Lactic acid
pK_a = 3.8

Propanoic acid
pK_a = 4.9

(*c*) A carbonyl group is more strongly electron-withdrawing than a carbon–carbon double bond. Pyruvic acid is a stronger acid than acrylic acid.

Pyruvic acid
pK_a = 3.3

Acrylic acid
pK_a = 4.3

(*d*) Viewing the two compounds as substituted derivatives of acetic acid, RCH_2CO_2H, we judge CH_3SO_2— to be strongly electron-withdrawing and acid-strengthening, whereas an ethyl group has only a small effect.

Methanesulfonylacetic acid
pK_a = 2.4

Butanoic acid
pK_a = 4.7

19.4 The compound can only be a carboxylic acid; no other class containing only carbon, hydrogen, and oxygen is more acidic. A reasonable choice is $HC{\equiv}CCO_2H$; C-2 is *sp*-hybridized and therefore electron-withdrawing and acid-strengthening. This is borne out by its measured pK_a of 1.8.

19.5 Recall from Chapter 1 (text Section 1.14) that an acid–base equilibrium favors formation of the weaker acid and base. Also remember that the weaker acid forms the stronger conjugate base, and vice versa.

(*b*) The acid–base reaction between acetic acid and *tert*-butoxide ion is represented by the equation

| Acetic acid | + | tert-Butoxide | ⇌ | Acetate ion | + | tert-Butyl alcohol |
| (stronger acid) | | (stronger base) | | (weaker base) | | (weaker acid) |

Alcohols (pK_a = 16-18) are weaker acids than acetic acid (pK_a = 4.7) the equilibrium lies to the right. The equilibrium constant K is $10^{(18-4.7)} = 10^{13.3}$.

(c) Bromide ion is the conjugate base of hydrogen bromide, a strong acid.

| Acetic acid | + | Bromide ion | ⇌ | Acetate ion | + | Hydrogen bromide |
| (weaker acid) | | (weaker base) | | (stronger base) | | (stronger acid) |

In this case, the position of equilibrium favors the starting materials, because acetic acid is a much weaker acid than hydrogen bromide. The equilibrium constant K is $10^{(-5.8-4.7)} = 10^{-10.5}$.

(d) Acetylide ion is a rather strong base, and acetylene (pK_a = 26) is a much weaker acid than acetic acid. The position of equilibrium favors the formation of products. The equilibrium constant K is $10^{(26-4.7)} = 10^{21.3}$.

| Acetic acid | + | Acetylide ion | ⇌ | Acetate ion | + | Acetylene |
| (stronger acid) | | (stronger base) | | (weaker base) | | (weaker acid) |

(e) Nitrate ion is a very weak base; it is the conjugate base of nitric acid. The position of equilibrium lies far to the left. The equilibrium constant K is $10^{(-1.4-4.7)} = 10^{-6.1}$.

| Acetic acid | + | Nitrate ion | ⇌ | Acetate ion | + | Nitric acid |
| (weaker acid) | | (weaker base) | | (stronger base) | | (stronger acid) |

(f) Amide ion is a very strong base; it is the conjugate base of ammonia (pK_a = 36). The position of equilibrium lies far to the right. The equilibrium constant K is $10^{(36-4.7)} = 10^{31.3}$.

| Acetic acid | + | Amide ion | ⇌ | Acetate ion | + | Ammonia |
| (stronger acid) | | (stronger base) | | (weaker base) | | (weaker acid) |

19.6 For carbonic acid, the "true K_1" is given by

$$\text{True } K_1 = \frac{[H^+][HCO_3^-]}{[H_2CO_3]}$$

The "observed K" is given by the expression

$$4.3 \times 10^{-7} = \frac{[H^+][HCO_3^-]}{[CO_2]}$$

which can be rearranged to

$$[H^+][HCO_3^-] = (4.3 \times 10^{-7})[CO_2]$$

and therefore

$$\text{True } K_1 = \frac{(4.3 \times 10^{-7})[CO_2]}{[H_2CO_3]}$$

$$= \frac{(4.3 \times 10^{-7})(99.7)}{0.3}$$

$$= 1.4 \times 10^{-4}$$

Thus, when corrected for the small degree to which carbon dioxide is hydrated, it can be seen that carbonic acid is actually a stronger acid than acetic acid. Carboxylic acids dissolve in sodium bicarbonate solution because the equilibrium that leads to carbon dioxide formation is favorable, not because carboxylic acids are stronger acids than carbonic acid.

19.7   The order of steps is important. A Grignard reagent cannot be prepared from the starting diol because of the presence of the hydroxyl groups. The starting diol must first be converted to its dimethyl ether. This can be carried out with a Williamson ether synthesis:

2-Bromo-1,3-dimethoxybenzene

Formation of the Grignard reagent, followed by reaction with carbon dioxide and acidification gives the desired product.

2,6-Dimethoxybenzoic acid

19.8   (*b*) 2-Chloroethanol has been converted to 3-hydroxypropanoic acid by way of the corresponding nitrile.

2-Chloroethanol   2-Cyanoethanol   3-Hydroxypropanoic acid

The presence of the hydroxyl group in 2-chloroethanol precludes the preparation of a Grignard reagent from this material, and so any attempt at the preparation of 3-hydroxypropanoic acid via the Grignard reagent of 2-chloroethanol is certain to fail.

(c) Grignard reagents can be prepared from tertiary halides and react in the expected manner with carbon dioxide. The procedure shown is entirely satisfactory.

tert-Butyl chloride   *tert*-Butylmagnesium   2,2-Dimethylpropanoic
                              chloride   acid (61-70%)

Preparation by way of the nitrile will not be feasible. Rather than react with sodium cyanide by substitution, *tert*-butyl chloride will undergo elimination exclusively. The S_N2 reaction with cyanide ion is limited to primary and secondary alkyl halides.

19.9 Incorporation of ^{18}O into benzoic acid proceeds by a mechanism analogous to that of esterification. The nucleophile that adds to the protonated form of benzoic acid is ^{18}O-enriched water (the ^{18}O atom is represented by the shaded letter ⊙ in the following equations).

Benzoic acid   Tetrahedral
                                                                                intermediate

The three hydroxyl groups of the tetrahedral intermediate are equivalent except that one of them is labeled with ^{18}O. Any one of these three hydroxyl groups may be lost in the dehydration step; when the hydroxyl group that is lost is unlabeled, an ^{18}O label is retained in the benzoic acid.

Tetrahedral ^{18}O-enriched benzoic acid
intermediate

19.10 (b) The 16-membered ring of 15-pentadecanolide is formed from 15-hydroxypentadecanoic acid.

Disconnect this bond.

15-Pentadecanolide   15-Hydroxypentadecanoic
                                            acid

(c) Vernolepin has two lactone rings, which can be related to two hydroxy acid combinations.

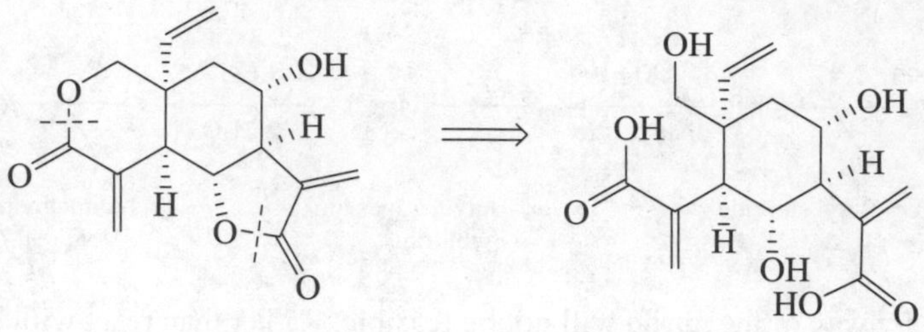

Be sure to keep the relative stereochemistry unchanged. Remember, the carbon–oxygen bond of an alcohol remains intact when the alcohol reacts with a carboxylic acid to give an ester.

19.11 (b) The starting material is a derivative of malonic acid. It undergoes efficient thermal decarboxylation in the manner shown.

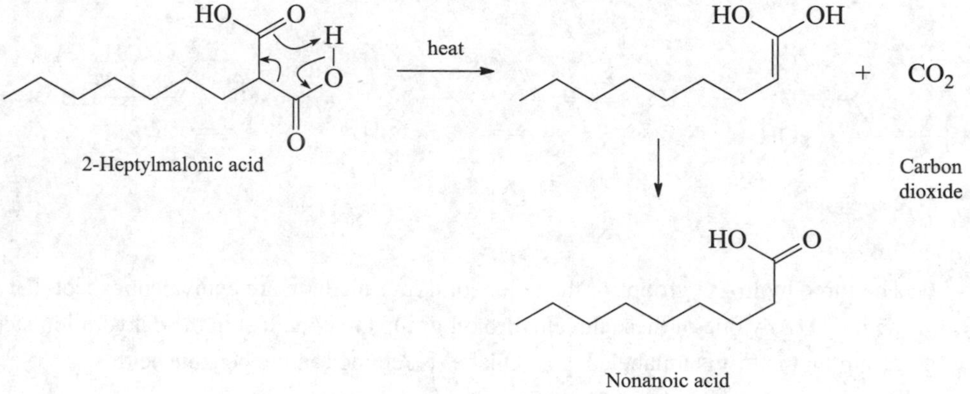

2-Heptylmalonic acid

Carbon dioxide

Nonanoic acid

(c) The phenyl and methyl substituents attached to C-2 of malonic acid play no role in the decarboxylation process.

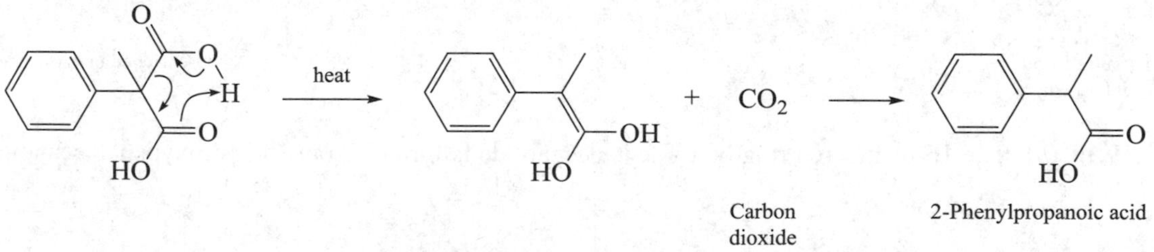

Carbon dioxide

2-Phenylpropanoic acid

19.12 (b) The thermal decarboxylation of β-keto acids resembles that of substituted malonic acids. The structure of 2,2-dimethylacetoacetic acid and the equation representing its decarboxylation were given in the text. The overall process involves the bonding changes shown.

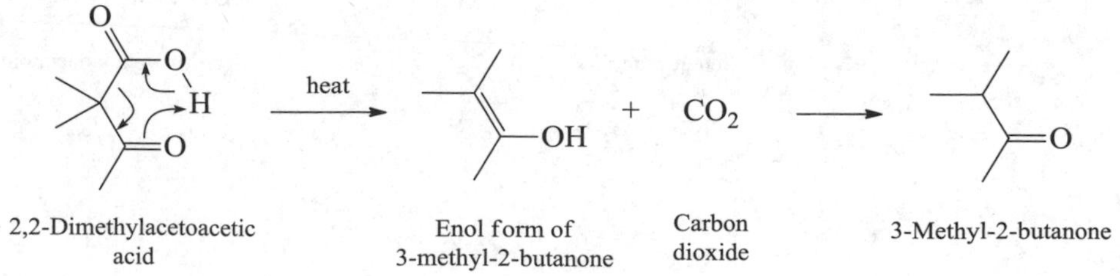

2,2-Dimethylacetoacetic acid

Enol form of 3-methyl-2-butanone

Carbon dioxide

3-Methyl-2-butanone

End of Chapter Problems

Structure and Nomenclature

19.13 (*a*) Lactic acid (2-hydroxypropanoic acid) is a three-carbon carboxylic acid that bears a hydroxyl group at C-2.

2-Hydroxypropanoic acid

(*b*) The parent name *ethanoic acid* tells us that the chain that includes the carboxylic acid function contains only two carbons. A hydroxyl group and a phenyl substituent are present at C-2.

2-Hydroxy-2-phenylethanoic acid
(mandelic acid)

(*c*) The parent alkane is *tetradecane,* which has an unbranched chain of 14 carbons. The terminal methyl group is transformed to a carboxyl function in tetradecanoic acid.

Tetradecanoic acid
(myristic acid)

(*d*) Undecane is the unbranched alkane with 11 carbon atoms, undecanoic acid is the corresponding carboxylic acid, and *undecenoic acid* is an 11-carbon carboxylic acid that contains a double bond. Because the carbon chain is numbered beginning with the carboxyl group, 10-undecenoic acid has its double bond at the opposite end of the chain from the carboxyl group.

10-Undecenoic acid
(undecylenic acid)

(*e*) Mevalonic acid has a five-carbon chain with hydroxyl groups at C-3 and C-5, along with a methyl group at C-3.

3,5-Dihydroxy-3-methylpentanoic acid
(mevalonic acid)

(*f*) The constitution represented by the systematic name 2-methyl-2-butenoic acid gives rise to two stereoisomers. Tiglic acid is the *E* isomer, and the *Z* isomer is known as *angelic acid*. The higher-ranked substituents, methyl and carboxyl, are placed on opposite sides of the double bond in tiglic acid and on the same side in angelic acid.

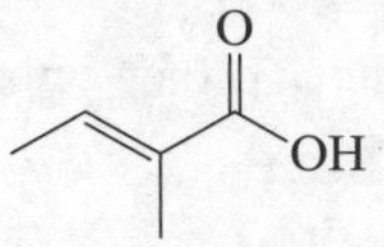

(E)-2-Methyl-2-butenoic acid
(tiglic acid)

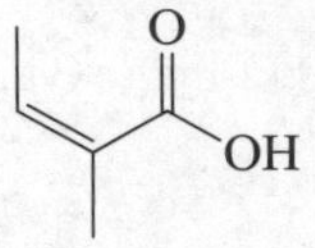

(Z)-2-Methyl-2-butenoic acid
(angelic acid)

(g) Butanedioic acid is a four-carbon chain in which both terminal carbons are carboxylic acid groups. Malic acid has a hydroxyl group at C-2.

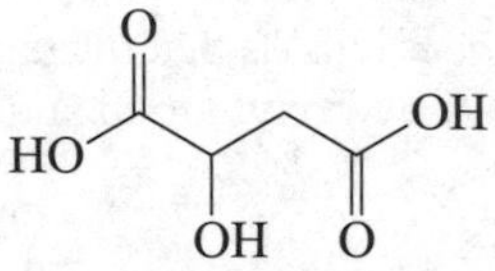

2-Hydroxybutanedioic acid
(malic acid)

(h) Each of the carbon atoms of propane bears a carboxyl group as a substituent in 1,2,3-propanetricarboxylic acid. In citric acid C-2 also bears a hydroxyl group.

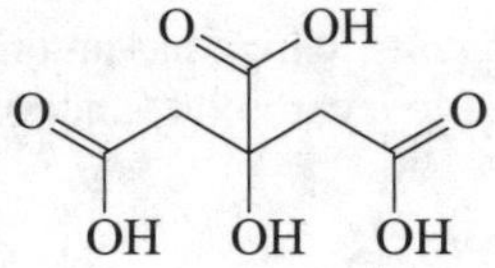

2-Hydroxy-1,2,3-propanetricarboxylic acid
(citric acid)

(i) There is an aryl substituent at C-2 of propanoic acid in ibuprofen. This aryl substituent is a benzene ring bearing an isobutyl group at the para position.

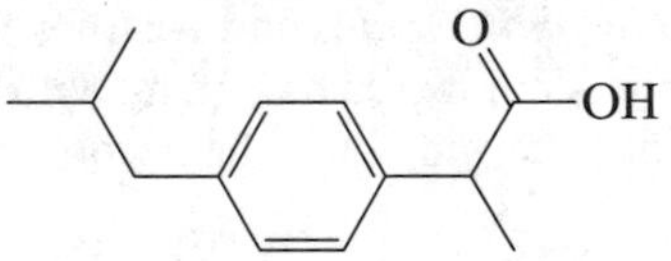

2-(p-Isobutylphenyl)propanoic acid

(j) Benzenecarboxylic acid is the systematic name for benzoic acid. *Salicylic acid* is a derivative of benzoic acid bearing a hydroxyl group at the position ortho to the carboxyl.

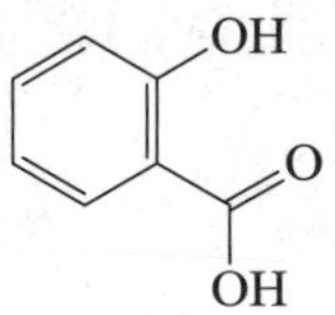

o-Hydroxybenzenecarbocylic acid
(salicylic acid)

19.14 (a) The carboxylic acid contains a linear chain of eight carbon atoms. The parent alkane is *octane,* and so the systematic name of $CH_3(CH_2)_6CO_2H$ is *octanoic acid.*

(b) The compound shown is the potassium salt of octanoic acid. It is *potassium octanoate.*

(c) The presence of a double bond in $H_2C=CH(CH_2)_5CO_2H$ is indicated by the ending *-enoic acid*. Numbering of the chain begins with the carboxylic acid, and so the double bond is between C-7 and C-8. The compound is named 7-*octenoic acid* or oct-7-enoic acid.

(d) Stereochemistry is systematically described by the *E–Z* notation. Here, the double bond between C-6 and C-7 in octenoic acid has the *Z* configuration; the higher-ranked substituents are on the same side.

(Z)-6-Octenoic acid or
(Z)-Oct-6-enoic acid

(e) A dicarboxylic acid is named as a *dioic acid*. The carboxyl functions are the terminal carbons of an eight-carbon chain; $HO_2C(CH_2)_6CO_2H$ is *octanedioic acid*. It is not necessary to identify the carboxylic acid locations by number because they can only be at the ends of the chain when the *-dioic acid* name is used.

(f) Pick the longest continuous chain that includes both carboxyl groups and name the compound as a *-dioic acid*. This chain contains only three carbons and bears a pentyl group as a substituent at C-2. It is not necessary to specify the position of the pentyl group, because it can only be attached to C-2.

Pentylpropanedioic acid

Malonic acid is an acceptable synonym for propanedioic acid; this compound may also be named *pentylmalonic acid*.

(g) A carboxylic acid function is attached as a substituent on a seven-membered ring. The compound is *cycloheptanecarboxylic acid*.

(h) The aromatic ring is named as a substituent attached to the eight-carbon carboxylic acid. Numbering of the chain begins with the carboxyl group.

6-Phenyloctanoic acid

Synthesis

19.15 (a) The conversion of 1-butanol to butanoic acid is simply the oxidation of a primary alcohol to a carboxylic acid. Chromic acid is a suitable oxidizing agent.

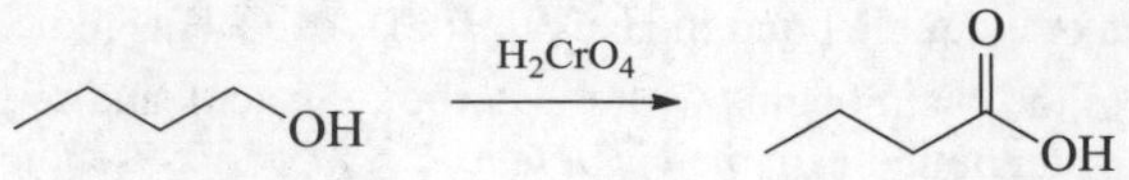

1-Butanol   Butanoic acid

(b) Aldehydes may be oxidized to carboxylic acids by any of the oxidizing agents that convert primary alcohols to carboxylic acids.

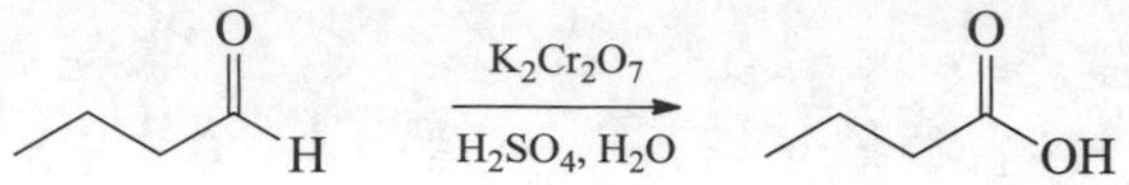

Butanal   Butanoic acid

(c) The starting material has the same number of carbon atoms as does butanoic acid, and so all that is required is a series of functional group transformations. Carboxylic acids may be obtained by oxidation of the corresponding primary alcohol. The alcohol is available from the designated starting material, 1-butene.

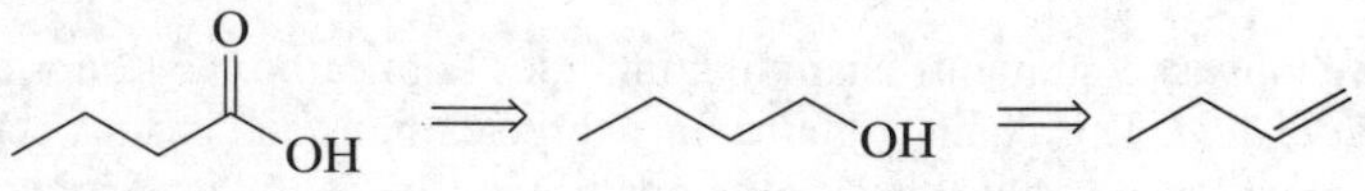

Hydroboration–oxidation of 1-butene yields 1-butanol, which can then be oxidized to butanoic acid as in part (a).

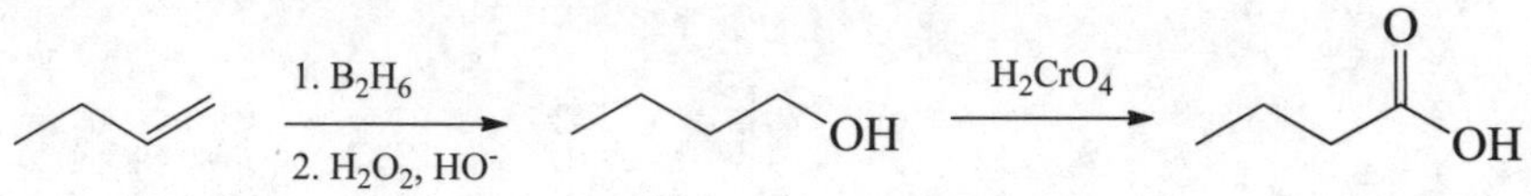

1-Butene   1-Butanol   Butanoic acid

(d) Converting 1-propanol to butanoic acid requires the carbon chain to be extended by one atom. Both methods for achieving this conversion, carboxylation of a Grignard reagent and formation and hydrolysis of a nitrile, begin with alkyl halides. Alkyl halides in turn are prepared from alcohols.

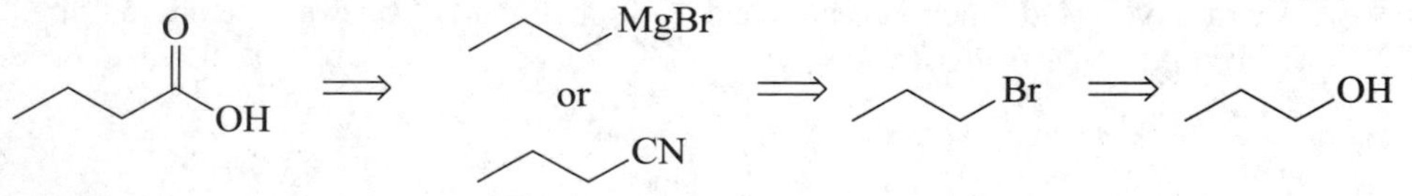

Either of the two following procedures is satisfactory:

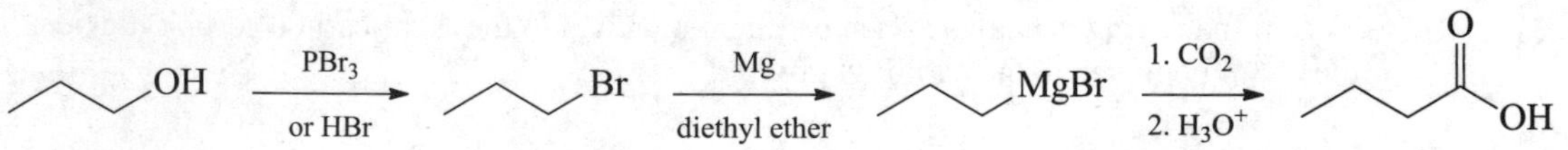

1-Propanol   1-Bromopropane   Butanoic acid

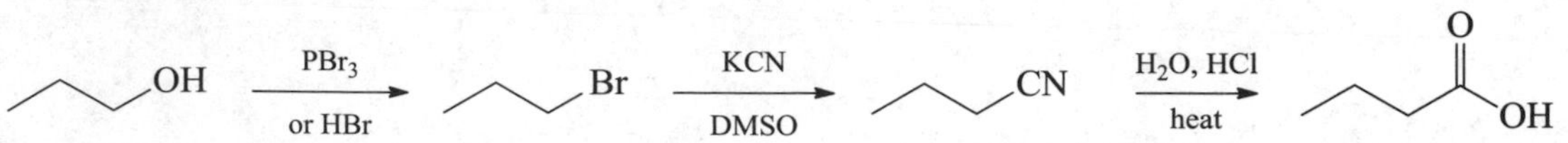

1-Propanol   1-Bromopropane   Butanenitrile   Butanoic acid

(e) Dehydration of 2-propanol to propene followed by free-radical addition of hydrogen bromide affords 1-bromopropane.

2-Propanol Propene 1-Bromopropane

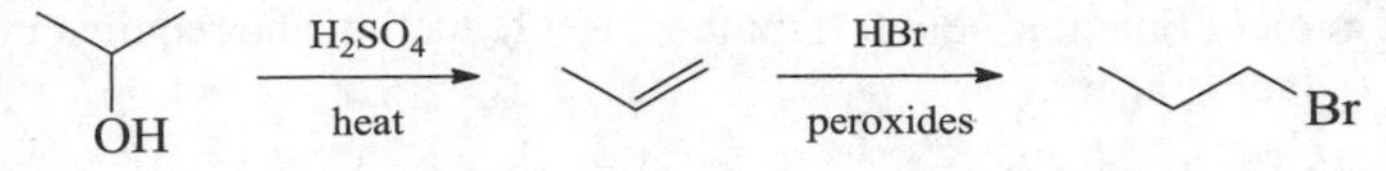

Once 1-bromopropane has been prepared, it is converted to butanoic acid as in part (*d*).

(*f*) Ethylmalonic acid belongs to the class of substituted malonic acids that undergo ready thermal decarboxylation. Decarboxylation yields butanoic acid.

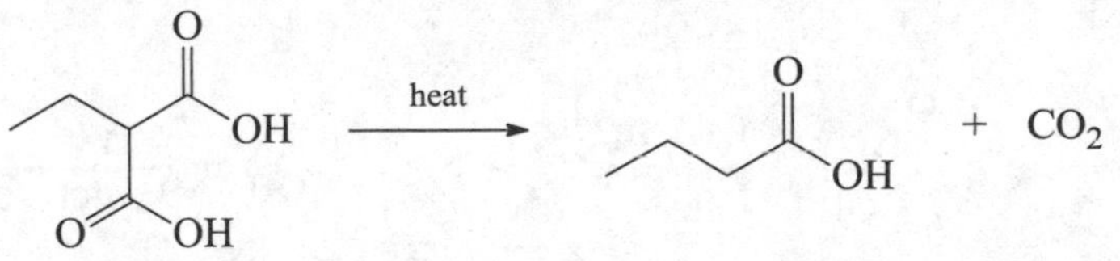

Ethylmalonic acid Butanoic acid Carbon dioxide

19.16 (*a*) The Friedel–Crafts alkylation of benzene by methyl chloride can be used to prepare ^{14}C-labeled toluene (C* = ^{14}C). Once prepared, toluene could be oxidized to benzoic acid.

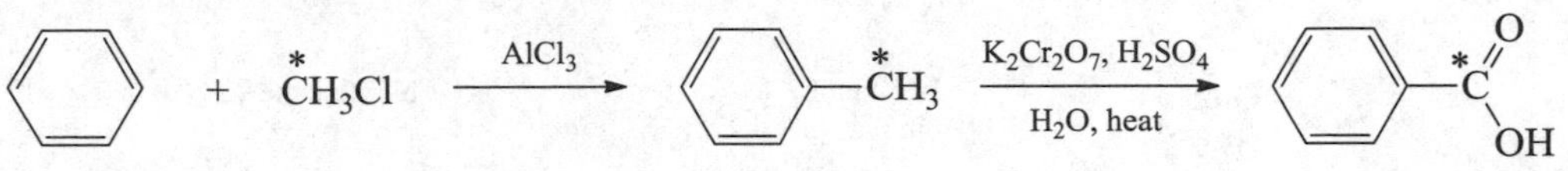

Benzene Methyl chloride, (^{14}C labeled) Toluene (^{14}C labeled) Benzoic acid (^{14}C labeled)

(*b*) Formaldehyde can serve as a one-carbon source if it is attacked by the Grignard reagent derived from bromobenzene.

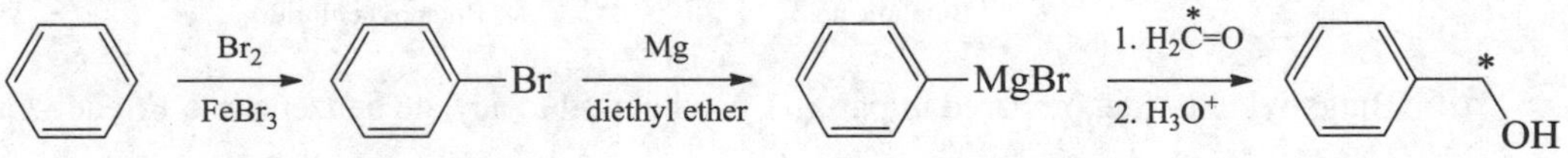

Benzene Bromobenzene Phenylmagnesium bromide Benzyl alcohol (^{14}C labeled)

This sequence yields ^{14}C-labeled benzyl alcohol, which can be oxidized to ^{14}C-labeled benzoic acid.

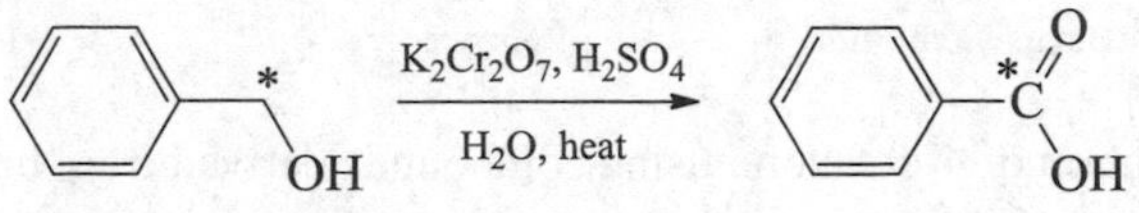

Benzyl alcohol (^{14}C labeled) Benzoic acid (^{14}C labeled)

(*c*) A direct route to ^{14}C-labeled benzoic acid utilizes a Grignard synthesis employing ^{14}C-labeled carbon dioxide.

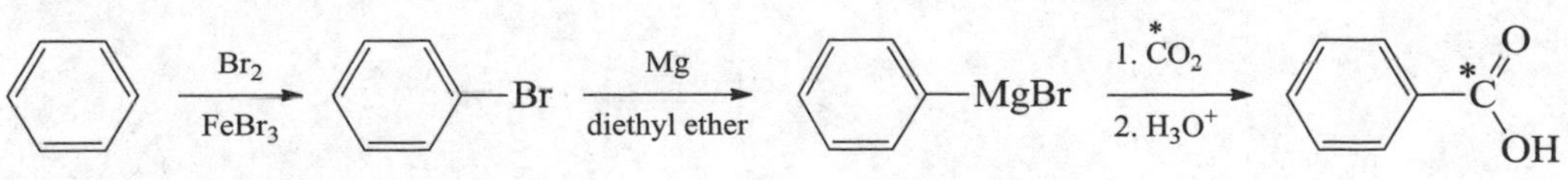

Benzene Bromobenzene Phenylmagnesium bromide Benzoic acid (^{14}C labeled)

19.17 (*a*) Conversion of butanoic acid to 1-butanol is a reduction and requires lithium aluminum hydride as the reducing agent.

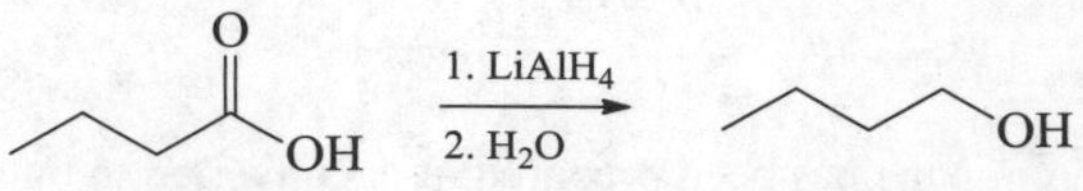

Butanoic acid   1-Butanol

(*b*) Carboxylic acids cannot be reduced directly to aldehydes. The following two-step procedure may be used:

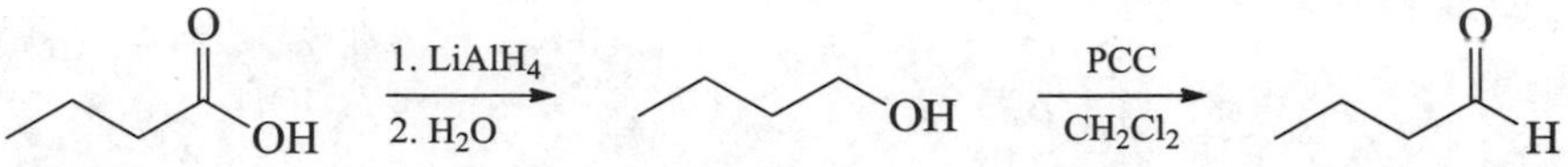

Butanoic acid   1-Butanol   Butanal

(*c*) Remember that alkyl halides are usually prepared from alcohols. 1-Butanol is therefore needed in order to prepare 1-chlorobutane.

1-Butanol   1-Chlorobutane
[from part (*a*)]

(*d*) Carboxylic acids are converted to their corresponding acyl chlorides with thionyl chloride.

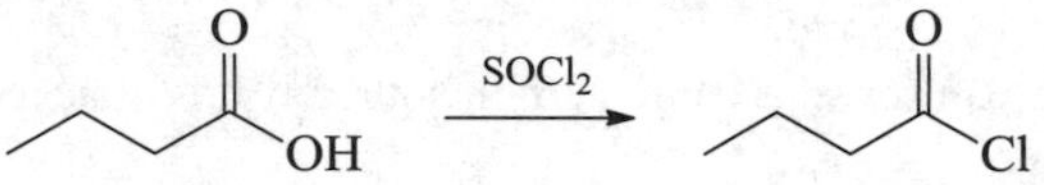

Butanoic acid   Butanoyl chloride

(*e*) Butanoyl chloride, prepared in part (*d*), can be used to acylate benzene in a Friedel–Crafts reaction.

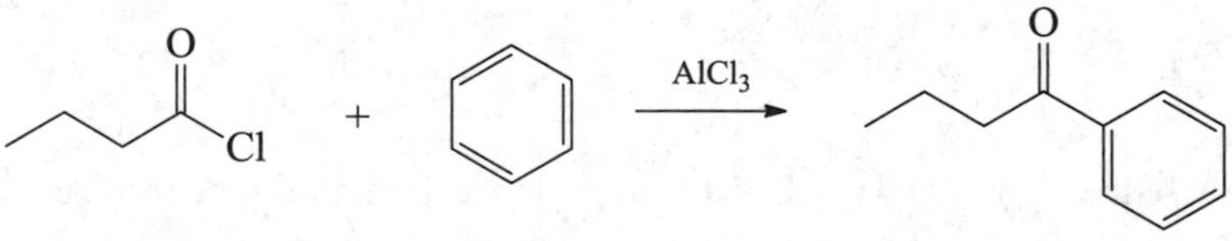

Butanoyl chloride   Benzene   Phenyl propyl ketone

(*f*) The preparation of 4-octanone using compounds derived from butanoic acid may be seen by using disconnections in a retrosynthetic analysis.

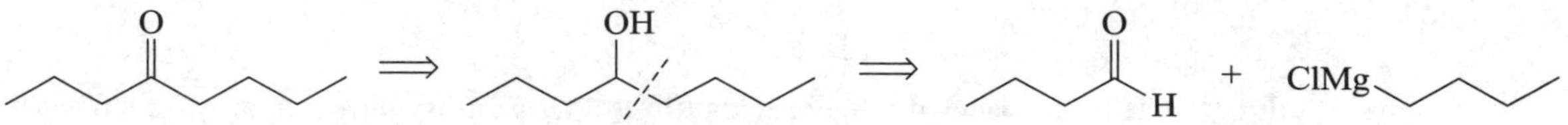

The reaction scheme that may be used is

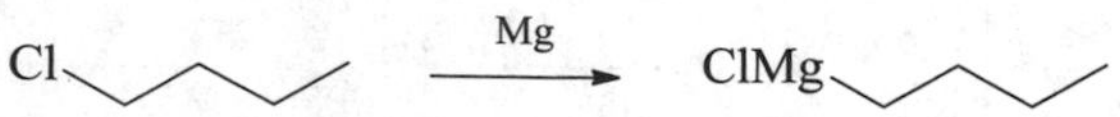

1-Chlorobutane   Butylmagnesium chloride
[from part (*c*)]

Butanal
[from part (*b*)] + Butylmagnesium chloride →[1. diethyl ether / 2. H₃O⁺] 4-Octanol →[H₂CrO₄] 4-Octanone

19.18 (*a*) The desired product and the starting material have the same carbon skeleton, and so all that is required is a series of functional group transformations. Recall that, as seen in Problem 19.15, a carboxylic acid may be prepared by oxidation of the corresponding primary alcohol. The needed alcohol is available from the appropriate alkene.

tert-Butyl alcohol →[H⁺ / heat] 2-Methylpropene →[1. B₂H₆ / 2. H₂O₂, OH⁻] 2-Methyl-1-propanol →[K₂Cr₂O₇ / H₂SO₄, H₂O] 2-Methylpropanoic acid

(*b*) The target molecule contains one more carbon than the starting material, and so a carbon–carbon bond-forming step is indicated. Two approaches are reasonable; one proceeds by way of nitrile formation and hydrolysis, the other by carboxylation of a Grignard reagent. In either case, the key intermediate is 1-bromo-2-methylpropane.

3-Methylbutanoic acid ⟹ 1-Bromo-2-methylpropane

The desired alkyl bromide may be prepared by free-radical addition of hydrogen bromide to 2-methylpropene.

tert-Butyl alcohol →[H⁺ / heat] 2-Methylpropene →[HBr / peroxides] 1-Bromo-2-methylpropane

Another route to the alkyl bromide utilizes the alcohol prepared in part (*a*).

2-Methyl-1-propanol →[PBr₃ or HBr] 1-Bromo-2-methylpropane

Conversion of the alkyl bromide to the carboxylic acid is then carried out as follows:

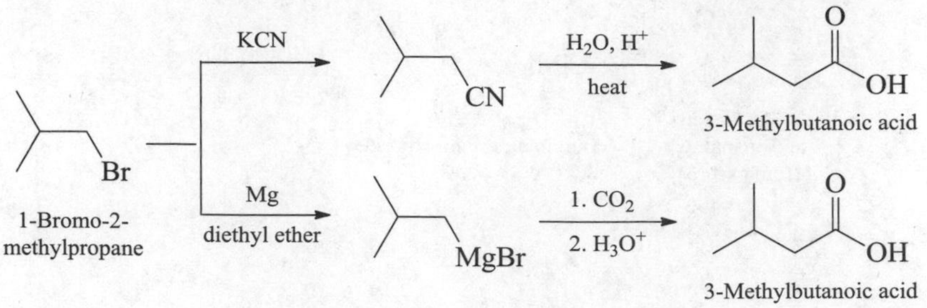

3-Methylbutanoic acid

3-Methylbutanoic acid

(c) Examining the target molecule reveals that it contains two more carbon atoms than the indicated starting material, suggesting use of ethylene oxide in a two carbon chain-extension process.

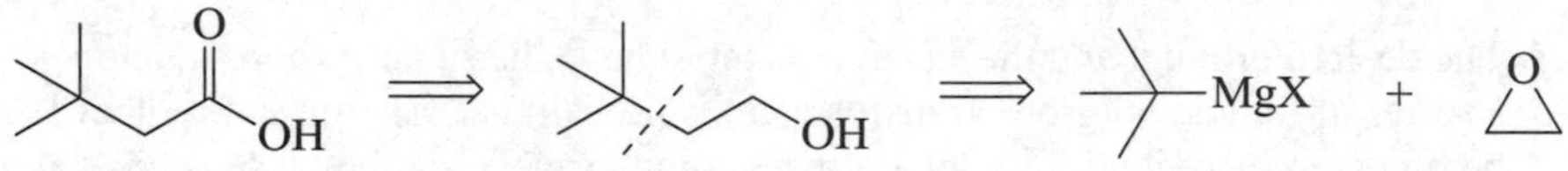

This suggests the following sequence of steps:

| *tert*-Butyl alcohol | 2-Bromo-2-methylpropane | *tert*-Butylmagnesium bromide | 3,3-Dimethyl-1-butanol |

3,3-Dimethylbutanoic acid

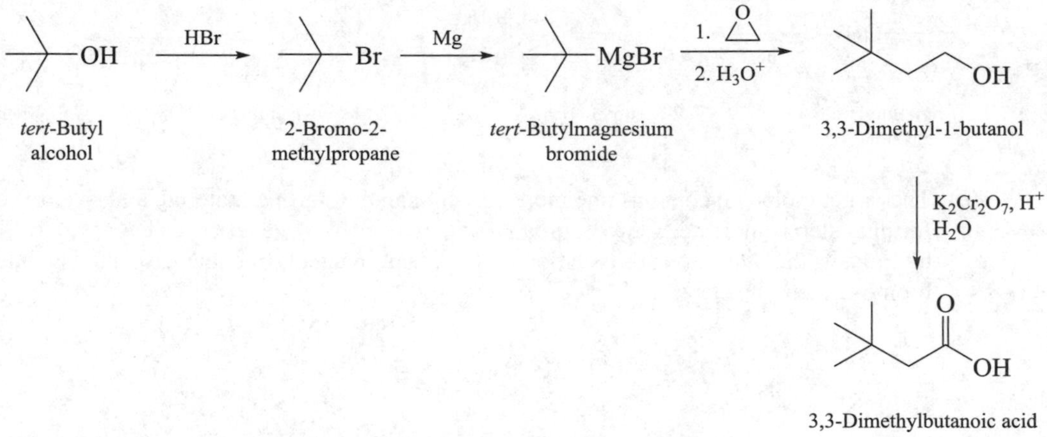

(d) This synthesis requires extending a carbon chain by two carbon atoms. One way to form dicarboxylic acids is by hydrolysis of dinitriles.

$$HO_2C(CH_2)_5CO_2H \implies NC(CH_2)_5CN \implies Br(CH_2)_5Br$$

This suggests the following sequence of steps:

$$HO_2C(CH_2)_3CO_2H \xrightarrow[\text{2. } H_2O]{\text{1. LiAlH}_4} HOCH_2(CH_2)_3CH_2OH \xrightarrow[\text{or HBr}]{PBr_3} BrCH_2(CH_2)_3CH_2Br$$

Pentanedioic acid   1,5-Pentanediol   1,5-Dibromopentane

$$HO_2CCH_2(CH_2)_3CH_2CO_2H \xleftarrow[\text{heat}]{H_2O, \, H^+} NCCH_2(CH_2)_3CH_2CN$$

Heptanedioic acid   1,5-Dicyanopentane (*e*) The desired alcohol cannot be prepared directly from the nitrile. It is available, however, by lithium aluminum hydride reduction of the carboxylic acid obtained by hydrolysis of the nitrile.

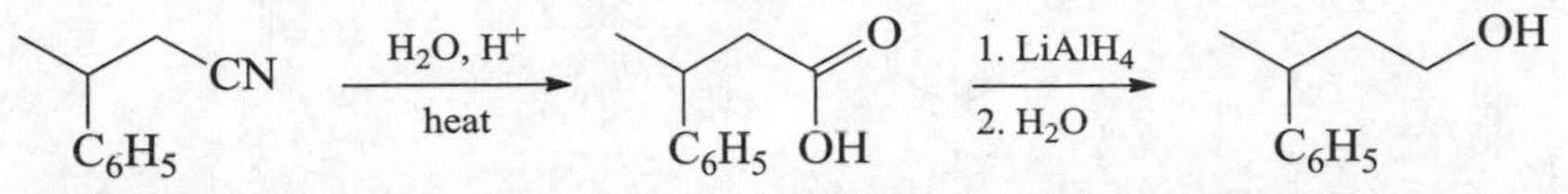

| 3-Phenylbutanenitrile | 3-Phenylbutanoic acid | 3-Phenyl-1-butanol |

(*f*) In this case, the halogen substituent is present at the β carbon rather than the α-carbon atom of the carboxylic acid. The starting material, a β-chloro unsaturated acid, can lead to the desired carbon skeleton by a Diels–Alder reaction.

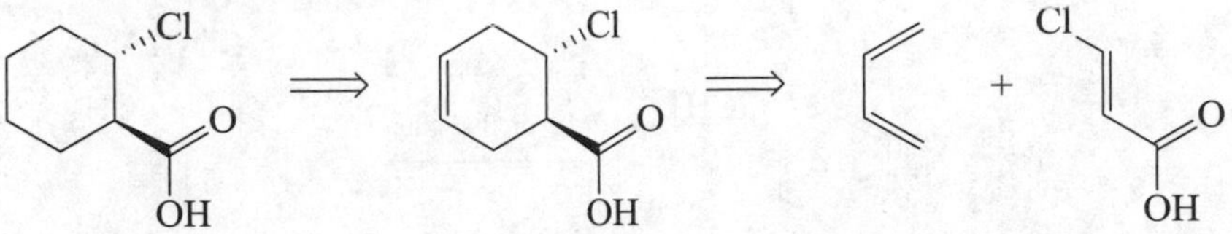

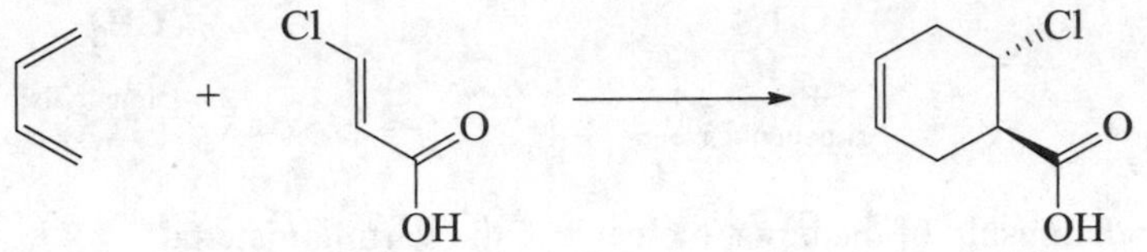

1,3-Butadiene   (*E*)-3-Chloropropenoic acid   *trans*-2-Chloro-4-cyclohexenecarboxylic acid

The required trans stereochemistry is a consequence of the stereospecificity of the Diels–Alder reaction.

Hydrogenation of the double bond of the Diels–Alder adduct gives the required product.

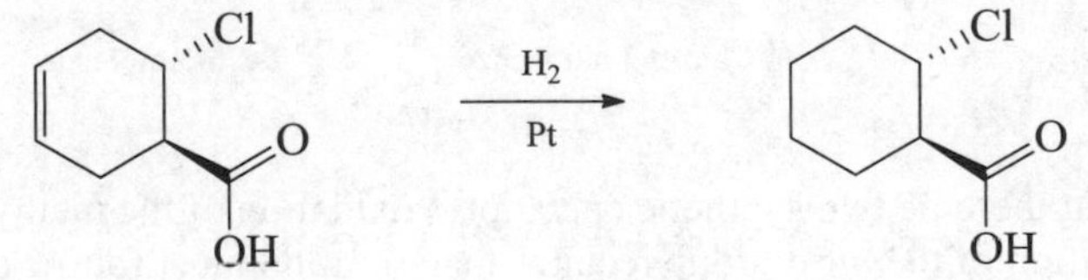

trans-2-Chloro-4-cyclohexenecarboxylic acid   *trans*-2-Chlorocyclohexanecarboxylic acid

(*g*) The target molecule can be related to the starting material by the retrosynthesis.

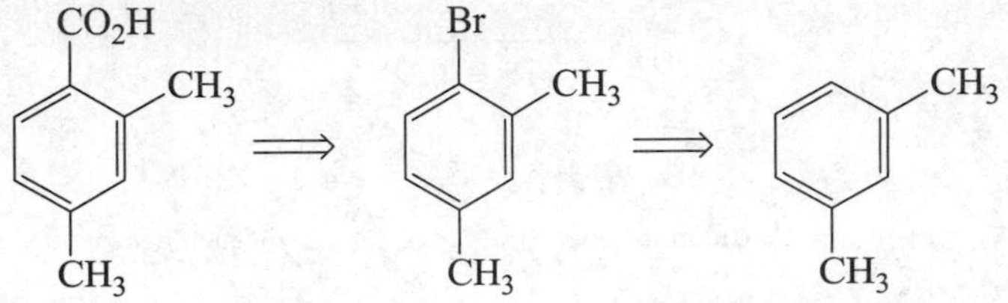

2,4-Dimethylbenzoic acid   *m*-Xylene

Bromine can be introduced by electrophilic substitution in the activated aromatic ring of *m*-xylene.

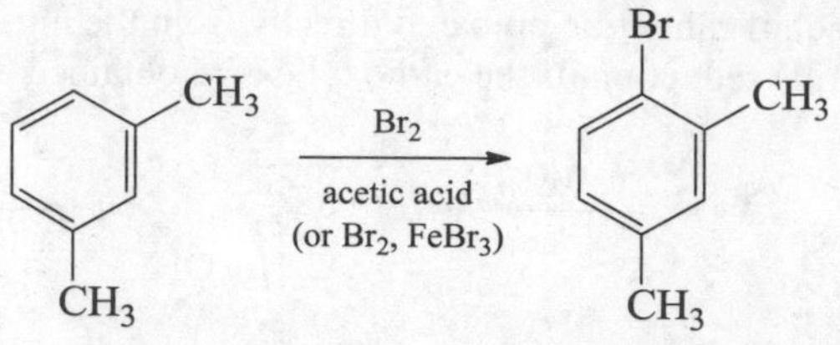

m-Xylene   1-Bromo-2,4-
                                              dimethylbenzene

The aryl bromide cannot be converted to a carboxylic acid by way of the corresponding nitrile, because aryl bromides are not reactive toward nucleophilic substitution. The Grignard route is necessary.

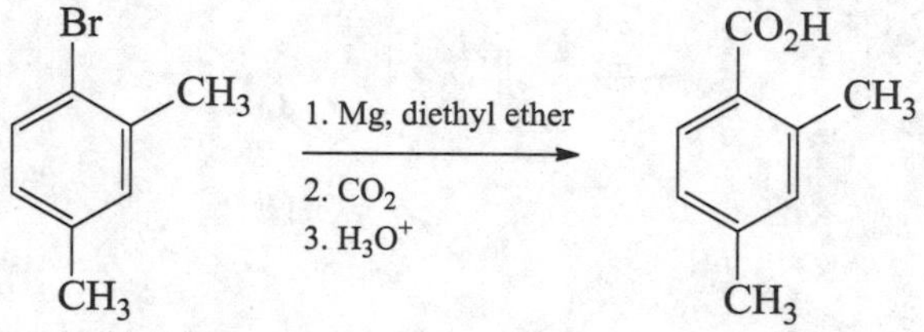

1-Bromo-2,4-   2,4-Dimethylbenzoic
dimethylbenzene   acid

(h) The relationship of the target molecule to the starting material

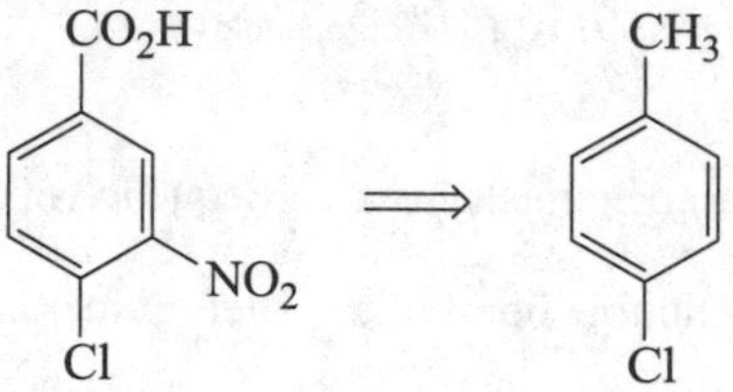

4-Chloro-3-nitrobenzoic   p-Chlorotoluene
        acid

requires that there be two synthetic operations: oxidation of the methyl group and nitration of the ring. The orientation of the nitro group requires that nitration must follow oxidation of the methyl group of the starting material.

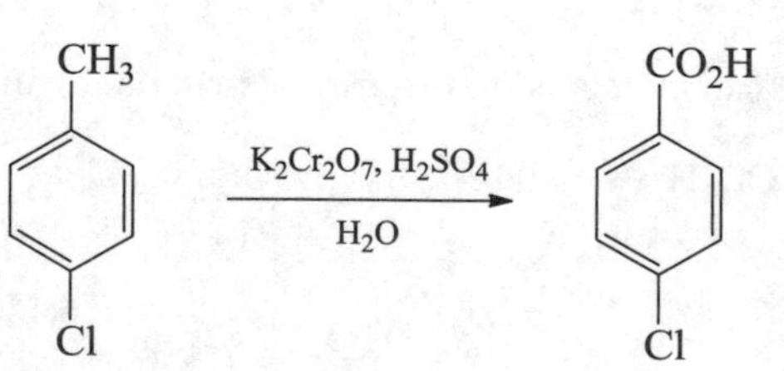

p-Chlorotoluene   p-Chlorobenzoic acid

Nitration of p-chlorobenzoic acid gives the desired product, because the directing effects of the chlorine (ortho, para) and the carboxyl (meta) groups reinforce each other.

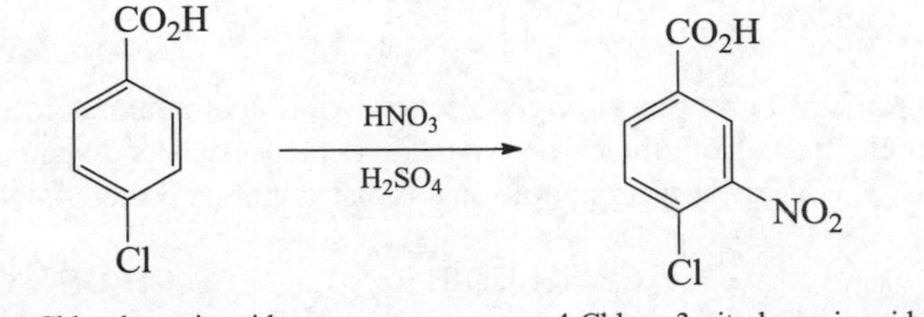

p-Chlorobenzoic acid   4-Chloro-3-nitrobenzoic acid

(*i*) The desired synthetic route becomes apparent when it is recognized that the *Z* alkene stereoisomer may be obtained from an alkyne, which, in turn, is available by carboxylation of the anion derived from the starting material.

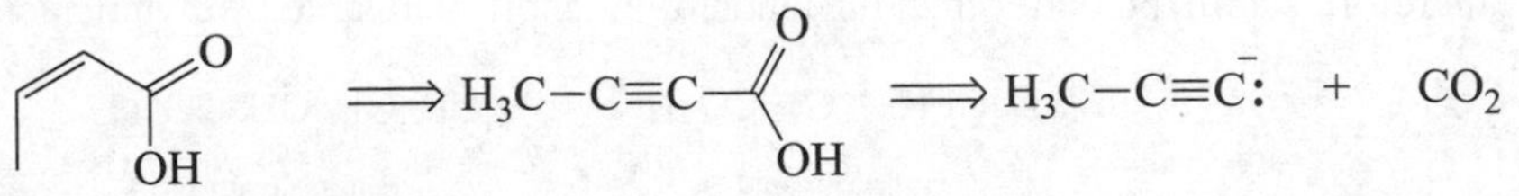

The desired reaction sequence is

$$H_3C-C\equiv CH \xrightarrow[NH_3]{NaNH_2} H_3C-C\equiv CNa \xrightarrow[2.\ H_3O^+]{1.\ CO_2} H_3C-C\equiv C-CO_2H$$

Propyne   Propynylsodium   2-Butynoic acid

Hydrogenation of the carbon–carbon triple bond of 2-butynoic acid over the Lindlar catalyst converts this compound to the *Z* isomer of 2-butenoic acid.

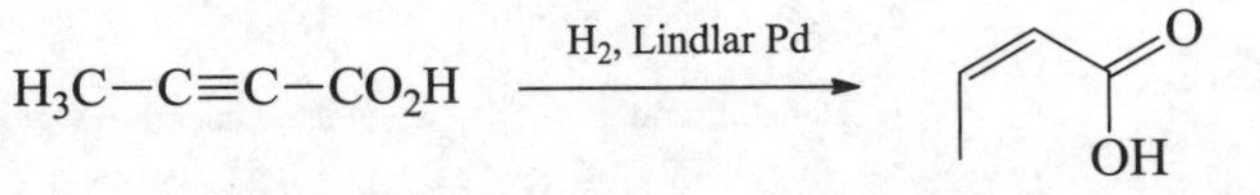

2-Butynoic acid   (*Z*)-2-Butenoic acid

19.19 Compound A is a δ-lactone. To determine its precursor, disconnect the ester linkage to a hydroxy acid.

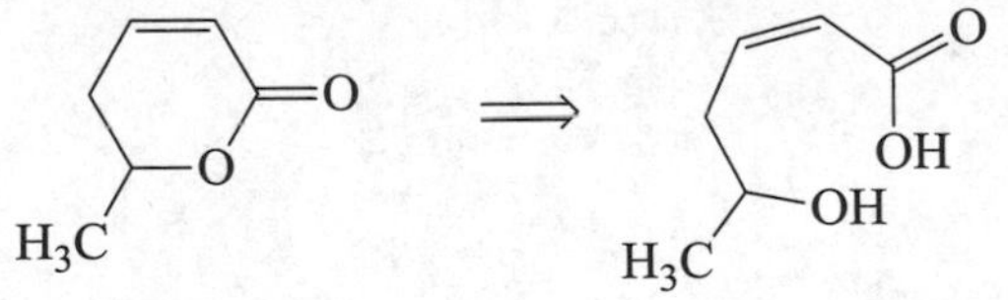

Compound A

The precursor has the same carbon skeleton as the designated starting material. All that is necessary is to hydrogenate the double bond of the alkynoic acid to the cis alkene. This can be done by using the Lindlar catalyst. Cyclization of the hydroxy acid to the lactone is spontaneous.

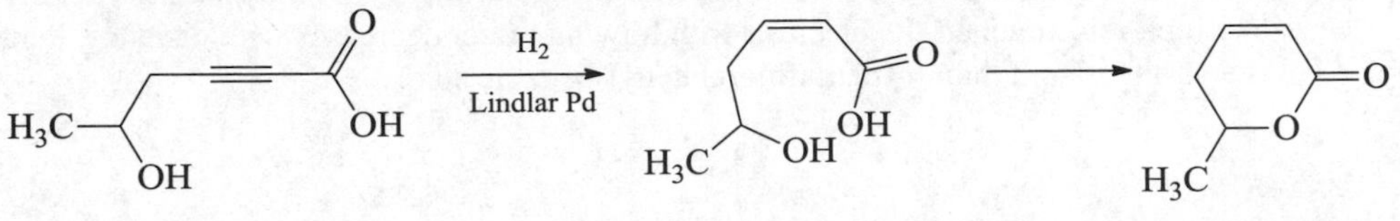

5-Hydroxy-2-hexynoic acid   (Not isolated)   Compound A

Reactions

19.20 (*a*) A trifluoromethyl group is strongly electron-withdrawing and acid-strengthening. Its ability to attract electrons from the carboxylate ion decreases as its distance down the chain increases. 3,3,3-Trifluoropropanoic acid is a stronger acid than 4,4,4-trifluorobutanoic acid.

$CF_3CH_2CO_2H$ $CF_3CH_2CH_2CO_2H$

3,3,3-Trifluoropropanoic acid   4,4,4-Trifluorobutanoic acid
$pK_a = 3.0$ $pK_a = 4.2$

(*b*) The carbon that bears the carboxyl group in 2-butynoic acid is *sp*-hybridized and is, therefore, more electron-withdrawing than the sp^3-hybridized α carbon of butanoic acid. The anion of 2-butynoic acid is therefore stabilized better than the anion of butanoic acid, and 2-butynoic acid is a stronger acid.

$H_3C-C\equiv C-CO_2H$ $CH_3CH_2CH_2CO_2H$

Butynoic acid   Butanoic acid
$pK_a = 2.6$ $pK_a = 4.8$

(*c*) Cyclohexanecarboxylic acid is a typical aliphatic carboxylic acid and is expected to be similar to acetic acid in acidity. The greater electronegativity of the sp^2-hybridized carbon attached to the carboxyl group in benzoic acid stabilizes benzoate anion better than the corresponding sp^3-hybridized carbon stabilizes cyclohexanecarboxylate. Benzoic acid is a stronger acid.

Benzoic acid   Cyclohexanecarboxylic acid
$pK_a = 4.2$ $pK_a = 4.9$

(*d*) Its five fluorine substituents make the pentafluorophenyl group more electron-withdrawing than an unsubstituted phenyl group. Thus, pentafluorobenzoic acid is a stronger acid than benzoic acid.

Pentafluorobenzoic acid   Benzoic acid
$pK_a = 3.4$ $pK_a = 4.2$

(*e*) The pentafluorophenyl substituent is electron-withdrawing and increases the acidity of a carboxyl group to which it is attached. Its electron-withdrawing effect decreases with distance. Pentafluorobenzoic acid is a stronger acid than *p*-(pentafluorophenyl)benzoic acid.

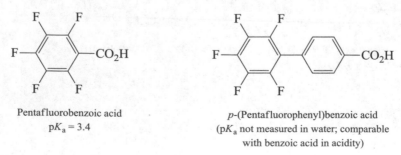

Pentafluorobenzoic acid   *p*-(Pentafluorophenyl)benzoic acid
$pK_a = 3.4$ (pK_a not measured in water; comparable
                                       with benzoic acid in acidity)

(*f*) The oxygen of the ring exercises an acidifying effect on the carboxyl group. This effect is largest when the oxygen is attached directly to the carbon that bears the carboxyl group. Furan-2-carboxylic acid is thus a stronger acid than furan-3-carboxylic acid.

Furan-2-carboxylic acid
$pK_a = 3.2$

Furan-3-carboxylic acid
$pK_a = 3.9$

(*g*) Furan-2-carboxylic acid has an oxygen attached to the carbon that bears the carboxyl group, whereas pyrrole-2-carboxylic acid has a nitrogen in that position. Oxygen is more electronegative than nitrogen and so stabilizes the carboxylate anion better. Furan-2-carboxylic acid is a stronger acid than pyrrole-2-carboxylic acid.

Furan-2-carboxylic acid
$pK_a = 3.2$

Pyrrole-2-carboxylic acid
$pK_a = 4.4$

19.21 (*a*) Carboxylic acids are the most acidic class of organic compounds containing only the elements C, H, and O. The order of decreasing acidity is

		pK_a
Acetic acid	CH_3CO_2H	4.7
Ethanol	CH_3CH_2OH	16
Ethane	CH_3CH_3	≈ 62

(*b*) Here again, the carboxylic acid is the strongest acid and the hydrocarbon the weakest.

		pK_a
Benzoic acid	$C_6H_5CO_2H$	4.2
Benzyl alcohol	$C_6H_5CH_2OH$	16–18
Benzene	C_6H_6	≈ 43

(*c*) Propanedioic acid is a stronger acid than propanoic acid because the electron-withdrawing effect of one carboxyl group enhances the ionization of the other.

		pK_a
Propanedioic acid	$HO_2CCH_2CO_2H$	2.9
Propanoic acid	$CH_3CH_2CO_2H$	4.9
1,3-Propanediol	$HOCH_2CH_2CH_2OH$	≈ 16

(*d*) Trifluoromethanesulfonic acid is by far the strongest acid in the group. It is structurally related to sulfuric acid, but its three fluorine substituents make it much stronger. Fluorine substituents increase the acidity of carboxylic acids and alcohols relative to their nonfluorinated analogs, but not enough to make fluorinated alcohols as acidic as carboxylic acids.

		pK_a
Trifluoromethanesulfonic acid	CF_3SO_2OH	−6
Trifluoroacetic acid	CF_3CO_2H	0.2
Acetic acid	CH_3CO_2H	4.7
2,2,2-Trifluoroethanol	CF_3CH_2OH	12.4
Ethanol	CH_3CH_2OH	≈16

19.22 (*a*) An acid–base reaction takes place when pentanoic acid is combined with sodium hydroxide.

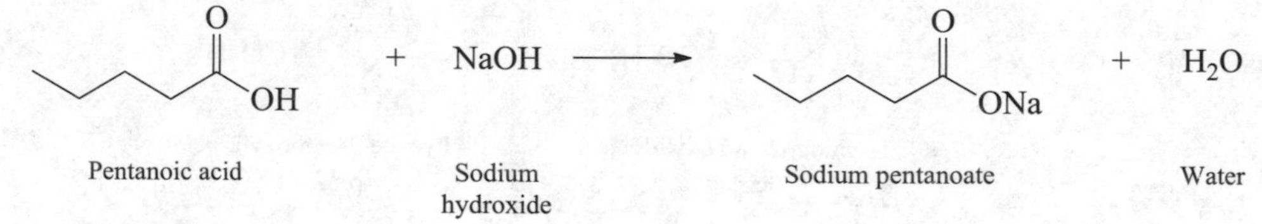

Pentanoic acid   Sodium hydroxide   Sodium pentanoate   Water

(*b*) Carboxylic acids react with sodium bicarbonate to give carbonic acid, which dissociates to carbon dioxide and water, so that the actual reaction that takes place is

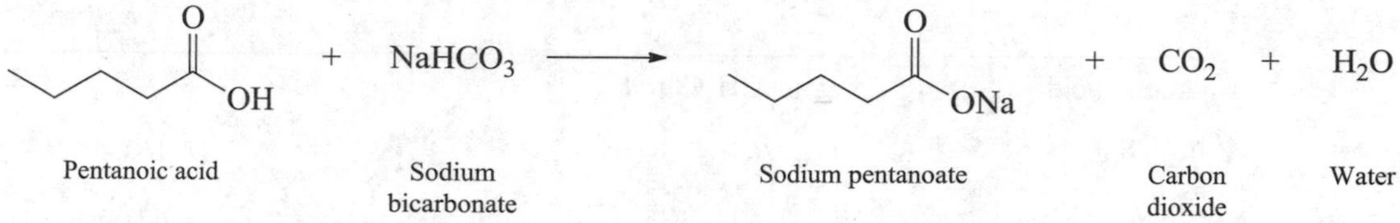

Pentanoic acid   Sodium bicarbonate   Sodium pentanoate   Carbon dioxide   Water

(*c*) Thionyl chloride is a reagent that converts carboxylic acids to the corresponding acyl chlorides.

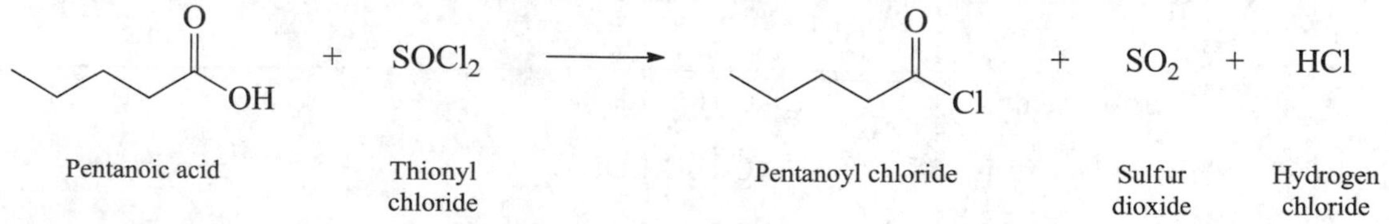

Pentanoic acid   Thionyl chloride   Pentanoyl chloride   Sulfur dioxide   Hydrogen chloride

(*d*) Phosphorus tribromide is used to convert carboxylic acids to their acyl bromides.

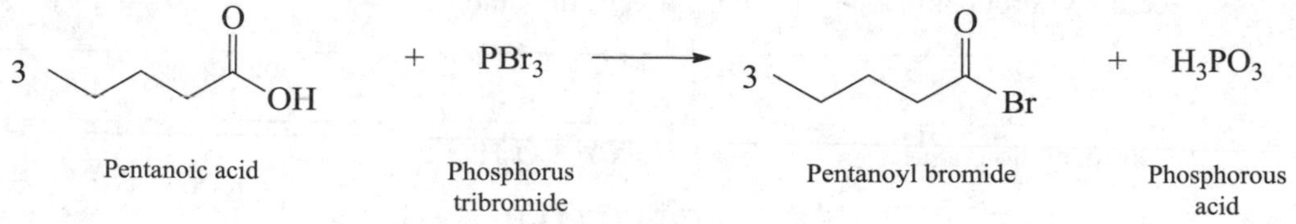

Pentanoic acid   Phosphorus tribromide   Pentanoyl bromide   Phosphorous acid

(e) Carboxylic acids react with alcohols in the presence of acid catalysts to give esters.

Pentanoic acid   Benzyl alchohol   Benzyl pentanoate   Water

(f) Lithium aluminum hydride is a powerful reducing agent and reduces carboxylic acids to primary alcohols.

Pentanoic acid   1-Pentanol

(g) Phenylmagnesium bromide acts as a base to abstract the carboxylic acid proton because Grignard reagents are not compatible with carboxylic acids. Proton transfer converts the Grignard reagent to the corresponding hydrocarbon.

Pentanoic acid   Phenylmagnesium   Bromomagnesium pentanoate   Benzene
                         bromide

19.23 (a) Only the cis stereoisomer of 4-hydroxycyclohexanecarboxylic acid is capable of forming a lactone. The most stable conformation of the starting hydroxy acid is a chair conformation; however, in the lactone, the cyclohexane ring must adopt a boat conformation.

cis-4-Hydroxycyclohexane-   cis-4-Hydroxycyclohexane-   Lactone
carboxylic acid   carboxylic acid
(chair conformation)   (boat conformation)

(b) As in part (a), lactone formation is possible only when the hydroxyl and carboxyl groups are cis.

cis-3-Hydroxycyclohexane-   Lactone
carboxylic acid

Although the most stable conformation of cis-3-hydroxycyclohexanecarboxylic acid has both substituents equatorial and is unable to close to a lactone, the diaxial orientation is accessible and is capable of lactone formation.

Neither conformation of trans-3-hydroxycyclohexanecarboxylic acid has the substituents close enough to each other to form an unstrained lactone.

trans-3-Hydroxycyclohexanecarboxylic acid: lactone formation impossible

19.24 Dicarboxylic acids in which both carboxyl groups are attached to the same carbon undergo ready thermal decarboxylation to produce the enol form of an acid.

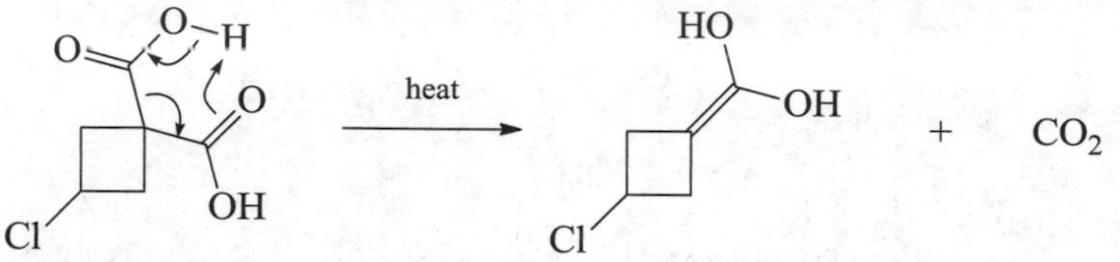

This enol yields a mixture of *cis*- and *trans*-3-chlorocyclobutanecarboxylic acid. The two products are stereoisomers.

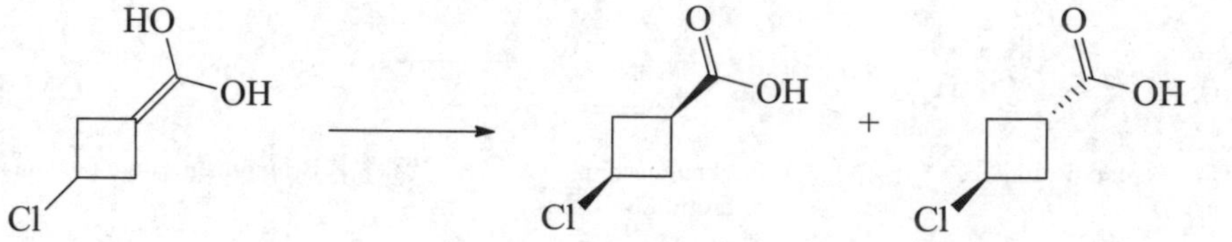

cis-3-Chlorocyclobutane-
carboxylic acid

trans-3-Chlorocyclobutane-
carboxylic acid

19.25 (*a*) Carboxylic acids are converted to ethyl esters when they are allowed to stand in ethanol in the presence of an acid catalyst.

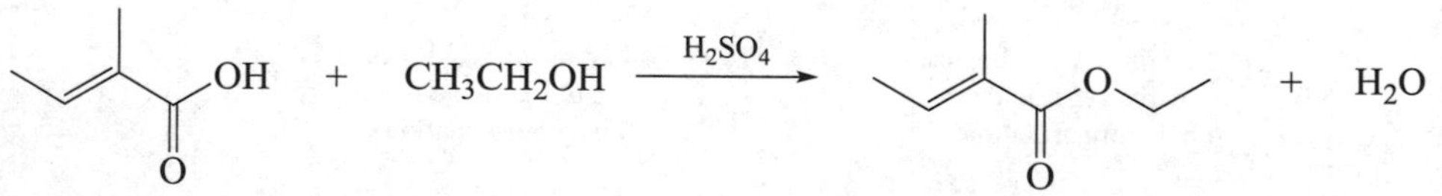

(*E*)-2-Methyl-2-butenoic acid   Ethanol   Ethyl (*E*)-2-methyl-2-butenoate   Water
                                        (74-80%)

(*b*) Lithium aluminum hydride, $LiAlH_4$, reduces carboxylic acids to primary alcohols. When $LiAlD_4$ is used, deuterium is transferred to the carbonyl carbon.

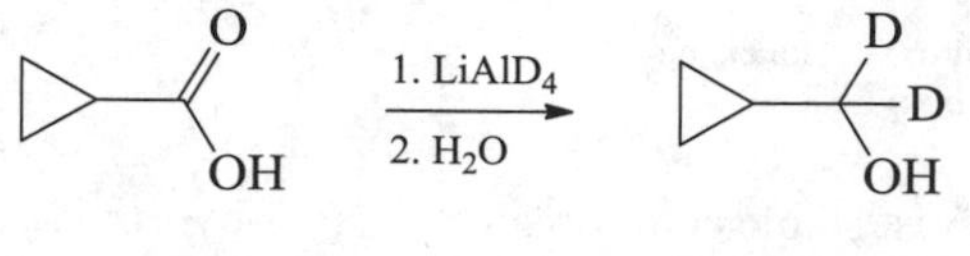

Cyclopropanecarboxylic
acid

1-Cyclopropyl-1,1-
dideuteriomethanol
(75%)

Notice that deuterium is bonded only to carbon. The hydroxyl proton is derived from water, not from the reducing agent.

(*c*) An alkyl or aryl bromide is much more readily converted to the corresponding Grignard reagent than the corresponding fluoride.

m-Bromo(trifluoromethyl)-
benzene

m (Trifluoromethyl)-
benzoic acid

(*d*) Cyano substituents are hydrolyzed to carboxyl groups in the presence of acid catalysts.

m-Chlorobenzyl
cyanide

m-Chlorophenylacetic
acid (61%)

(*e*) The carboxylic acid function plays no part in this reaction; free-radical addition of hydrogen bromide to the carbon–carbon double bond occurs.

$$H_2C{=}CH(CH_2)_8CO_2H \xrightarrow[\text{benzoyl peroxide}]{\text{HBr}} BrCH_2CH_2(CH_2)_8CO_2H$$

10-Undecenoic acid

11-Bromoundecanoic acid
(66-70%)

Recall that hydrogen bromide adds to alkenes in the presence of peroxides with a regioselectivity opposite to that of Markovnikov's rule.

19.26 The series of reactions given in this problem were used to convert the aldehyde starting material to a primary alcohol having one more carbon atom.

Reaction 1: The starting material is an aldehyde and is reduced by lithium aluminum hydride to the corresponding primary alcohol.

Reaction 2: Thionyl chloride converts alcohols to alkyl chlorides.

Reaction 3: The reactant in this step is a primary, benzylic chloride. It undergoes an S_N2 reaction with sodium cyanide to give a nitrile.

Reaction 4: The nitrile is converted to a carboxylic acid (as its potassium salt) on reaction with potassium hydroxide. Acidification of the reaction mixture gives the carboxylic acid.

Reaction 5: Lithium aluminum hydride reduces carboxylic acids to primary alcohols.

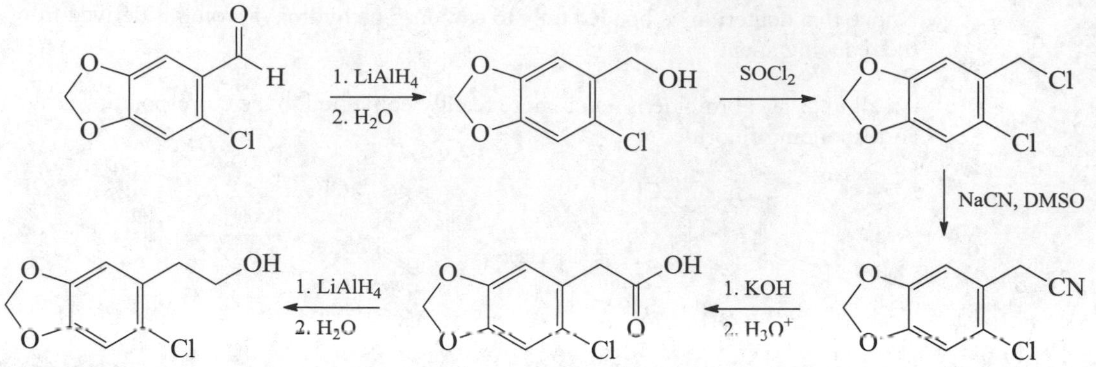

19.27 Examination of the molecular formula $C_{14}H_{26}O_2$ reveals that the compound has an index of hydrogen deficiency of 2. Because we are told that the compound is a carboxylic acid, one of these elements of unsaturation must be a carbon–oxygen double bond. The other must be a carbon–carbon double bond because the compound undergoes cleavage on ozonolysis. Examining the products of ozonolysis serves to locate the position of the double bond.

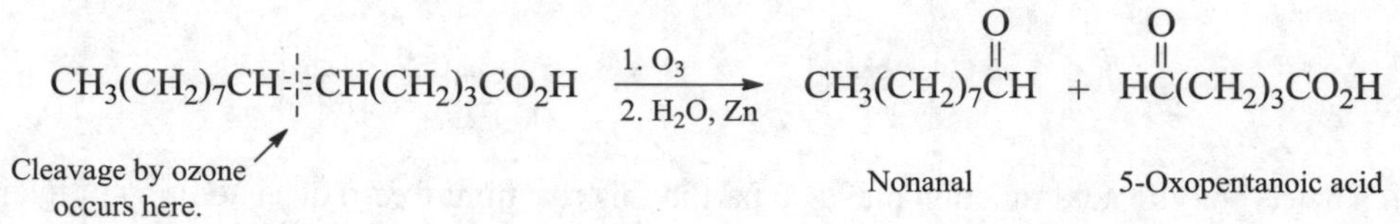

The starting acid must be 5-tetradecenoic acid. The stereochemistry of the double bond is not revealed by these experiments.

19.28 Hydrogenation of the starting material is expected to result in reduction of the ketone carbonyl while leaving the carboxyl group unaffected. Because the isolated product lacks a hydroxyl group, however, that group must react in some way. The most reasonable reaction is intramolecular esterification to form a γ-lactone.

Levulinic acid — 4-Hydroxypentanoic acid (not isolated) — 4-Pentanolide ($C_5H_8O_2$)

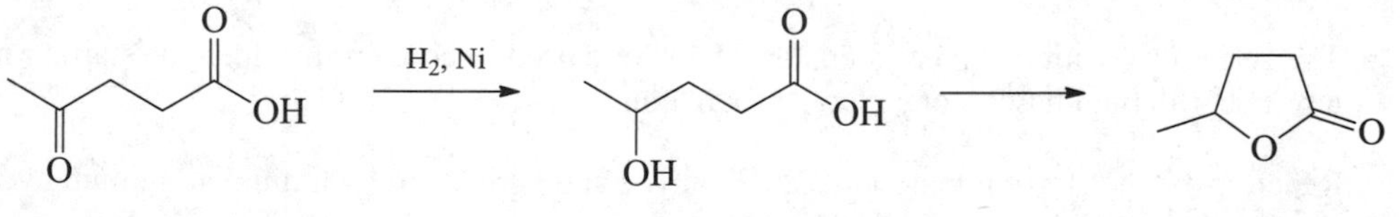

19.29 Compound A is a cyclic acetal and undergoes hydrolysis in aqueous acid to produce acetaldehyde, along with a dihydroxy carboxylic acid.

Compound A — 3,5-Dihydroxy-3-methylpentanoic acid — Acetaldehyde

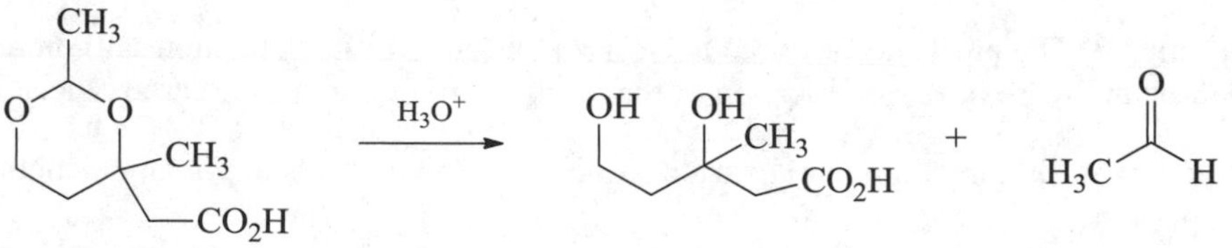

The dihydroxy acid that is formed in this step cyclizes to the δ-lactone mevalonolactone.

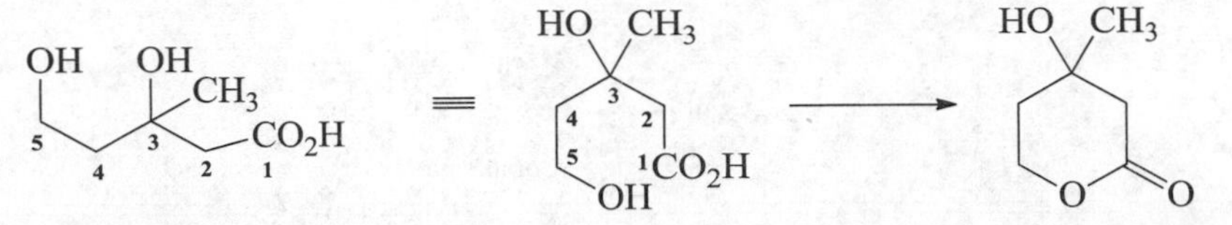

3,5-Dihydroxy-3-
methylpentanoic acid

Mevalonolactone

19.30 Hydration of the double bond can occur in two different directions:

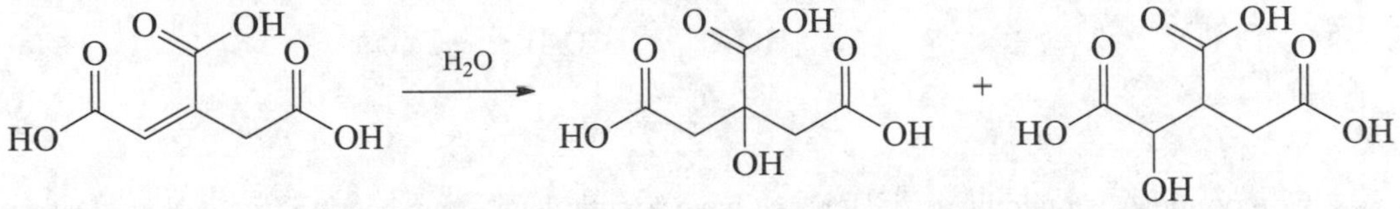

(a) The achiral isomer is citric acid.

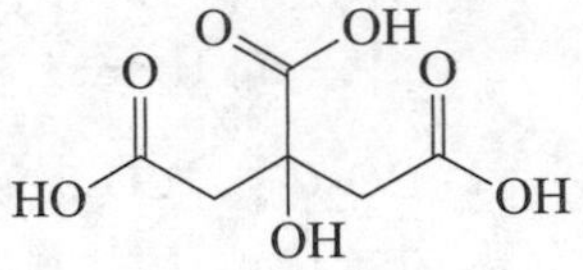

Citric acid has no chirality centers.

(b) The other isomer, isocitric acid, has two chirality centers (marked with an asterisk*). Isocitric acid has the constitution

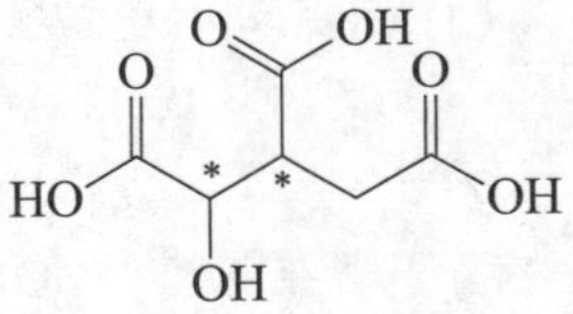

Isocitric acid

With two chirality centers, there are 2^2, or four, stereoisomers represented by this constitution. The one that is actually formed in this enzyme-catalyzed reaction is the $2R,3S$ isomer.

Spectroscopy

19.31 Carboxylic acid protons give signals in the range δ 10–12. A signal in this region suggests the presence of a carboxyl group but tells little about its environment. Thus, in assigning structures to compounds A, B, and C, the most useful data are the chemical shifts of the protons other than the carboxyl protons. Compare the three structures:

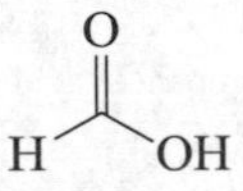

Formic acid 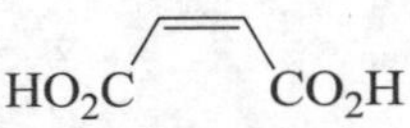   Maleic acid    Malonic acid

The proton that is diagnostic of structure in formic acid is bonded to a carbonyl group; it is an aldehyde proton. Typical chemical shifts of aldehyde protons are δ 8–10, and therefore formic acid is compound C.

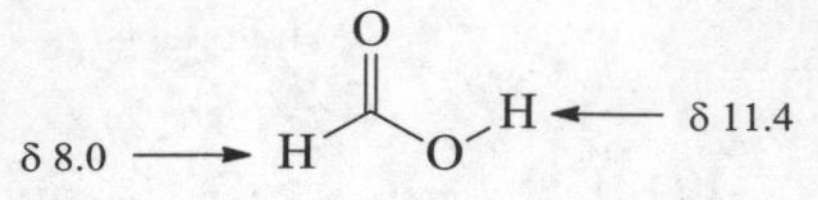

Compound C

The critical signal in maleic acid is that of the vinyl protons, which normally is found in the range δ 5–7. Maleic acid is compound B.

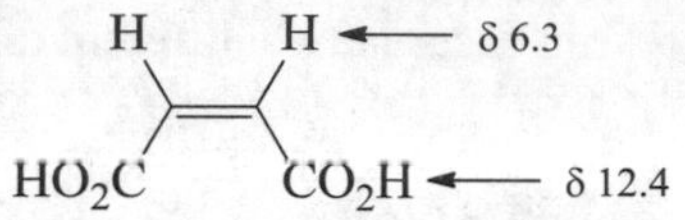

Compound B

Compound A is malonic acid. Here we have a methylene group bearing two carbonyl substituents. These methylene protons are more shielded than the aldehyde proton of formic acid or the vinyl protons of maleic acid.

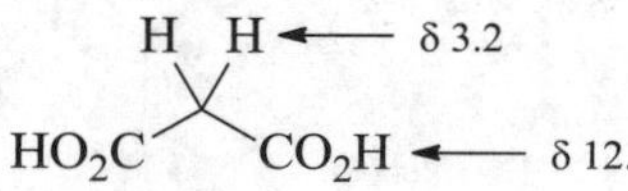

Compound A

19.32 Compounds A and B both exhibit ^{1}H NMR absorptions in the region δ 10–12 characteristic of carboxylic acids. The formula $C_4H_8O_3$ suggests an index of hydrogen deficiency of 1, accounted for by the carbonyl of the carboxyl group. Compound A has the triplet–quartet splitting indicative of an ethyl group, and compound B has two triplets, suggesting —CH_2CH_2—.

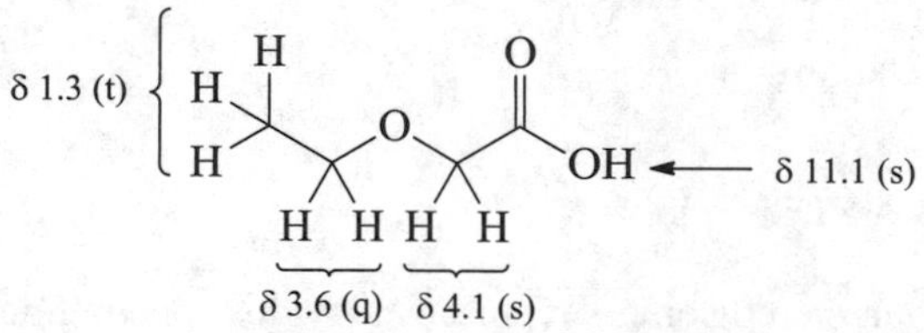

Compound A 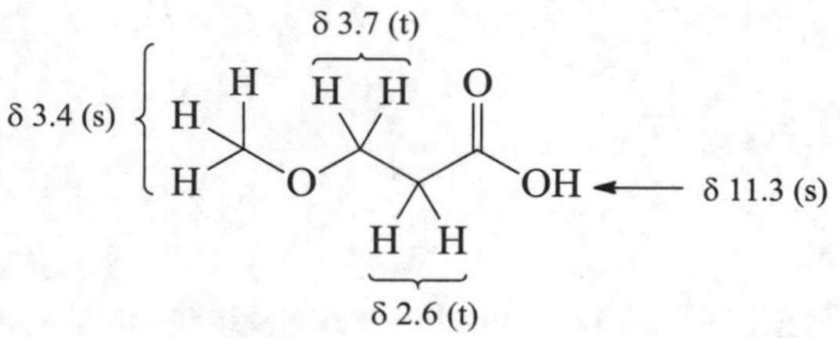   Compound B

19.33 (*a*) The formula of compound A ($C_3H_5ClO_2$) has an index of hydrogen deficiency of 1—the carboxyl group. Only two structures are possible:

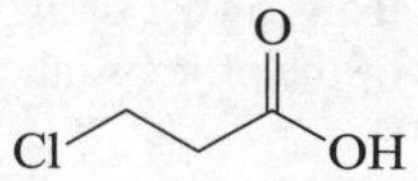

and

3-Chloropropanoic acid   2-Chloropropanoic acid

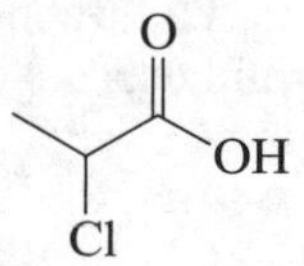

Compound A is determined to be 3-chloropropanoic acid on the basis of its ^{1}H NMR spectrum, which shows two triplets at δ 2.9 and δ 3.8.

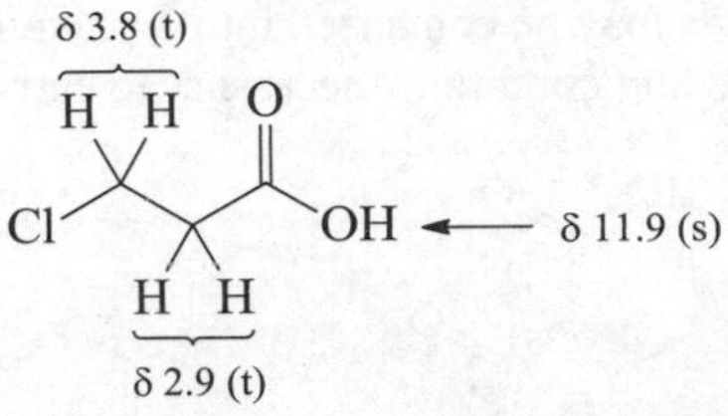

δ 3.8 (t)

H H   O

Cl   OH ◄—— δ 11.9 (s)

H H

δ 2.9 (t)

Compound A

Compound A cannot be 2-chloropropanoic acid, because that compound's ^1H NMR spectrum would show a three-proton doublet for the methyl group and a one-proton quartet for the methine proton.

(b) The formula of compound B ($C_9H_9NO_4$) corresponds to an index of hydrogen deficiency of 6.

The presence of an aromatic ring, as evidenced by the ^1H NMR absorptions at δ 7.5 and 8.3, accounts for four of the unsaturations. The appearance of the aromatic protons as a pair of doublets with a total area of 4 suggests a para-disubstituted ring.

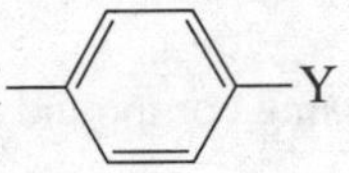

X———◯———Y

That compound B is a carboxylic acid is evidenced by the singlet (area = 1) at δ 11.9. The remaining ^1H NMR signals [a quartet at δ 3.8 (1H) and a doublet at δ 1.6 (3H)] suggest the fragment CH—CH_3. All that remains of the molecular formula is —NO_2. Combining this information identifies compound B as 2-(4-nitrophenyl)propanoic acid.

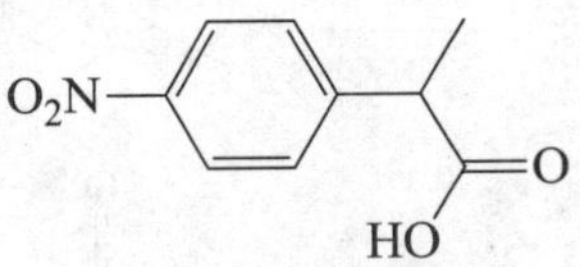

O_2N———◯———

HO

2-(4-Nitrophenyl)propanoic acid
(compound B)

Answers to Interpretive Problems 19

19.34 C; **19.35** B; **19.36** C; **19.37** B

SELF-TEST

1. Provide an acceptable IUPAC name for each of the following:

(a) CH_3

$C_6H_5CHCHCH_2CH_2CO_2H$

CH_3

(b) CO_2H

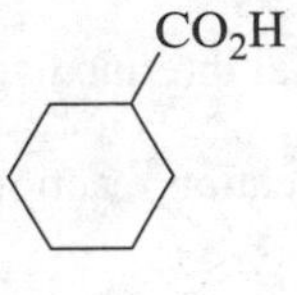

◯

(c)   Br   O

OH

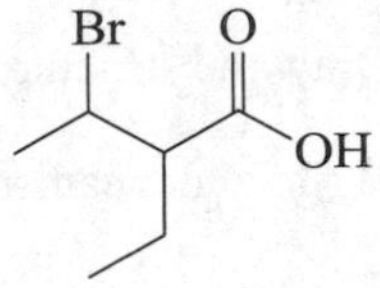

2. Both of the following compounds may be converted into 4-phenylbutanoic acid by one or more reaction steps. Give the reagents and conditions necessary to carry out these conversions.

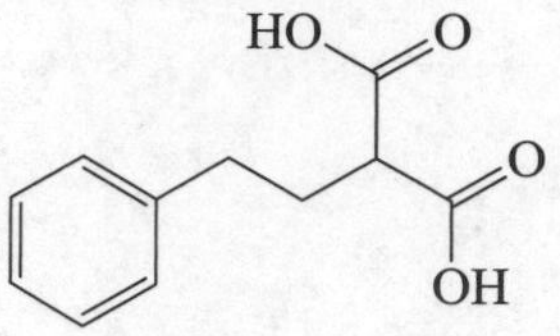

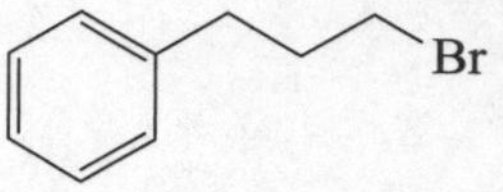

(Two methods)

3. The species whose structure is shown is an intermediate in an esterification reaction. Write the complete, balanced equation for this process.

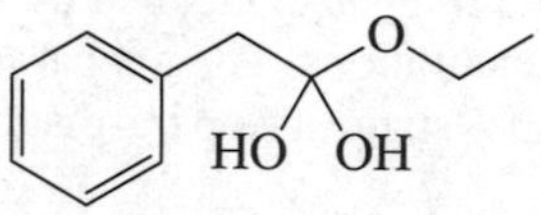

4. Give the missing reagent(s) and the missing compound in each of the following:

(a)

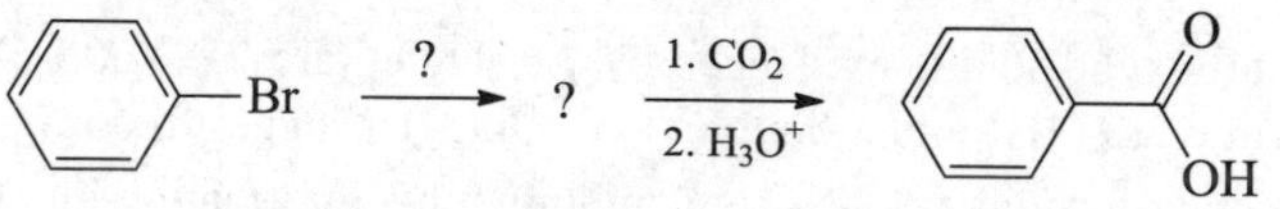

$$-Br \xrightarrow{?} ? \xrightarrow[\text{2. } H_3O^+]{\text{1. } CO_2} $$

(b)

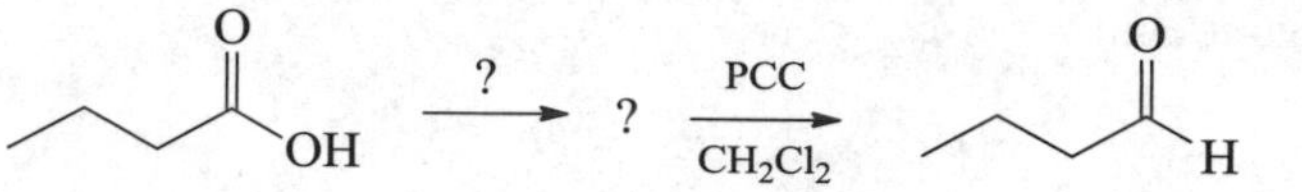

$$\xrightarrow{?} ? \xrightarrow[CH_2Cl_2]{PCC} $$

(c)

$$\xrightarrow[HCl]{NaCN} ? \xrightarrow{?} $$

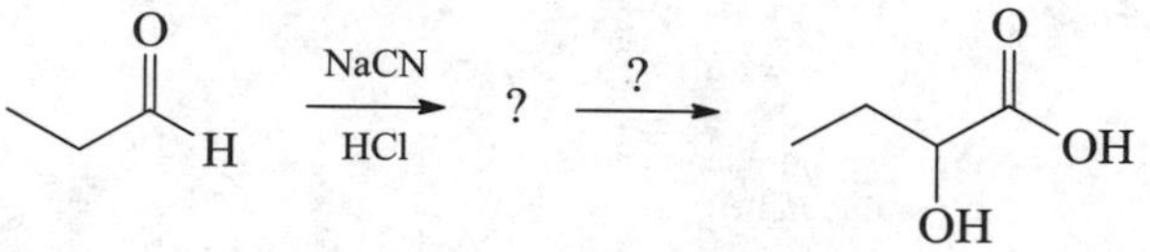

5. Identify the carboxylic acid ($C_4H_7BrO_2$) having the ^1H NMR spectrum consisting of

δ 1.1, 3H (triplet)
δ 2.0, 2H (pentet)
δ 4.2, 1H (triplet)
δ 12.1, 1H (singlet)

6. Draw the structure of the tetrahedral intermediate in the esterification of formic acid with 1-butanol.

7. Write a mechanism for the esterification reaction shown.

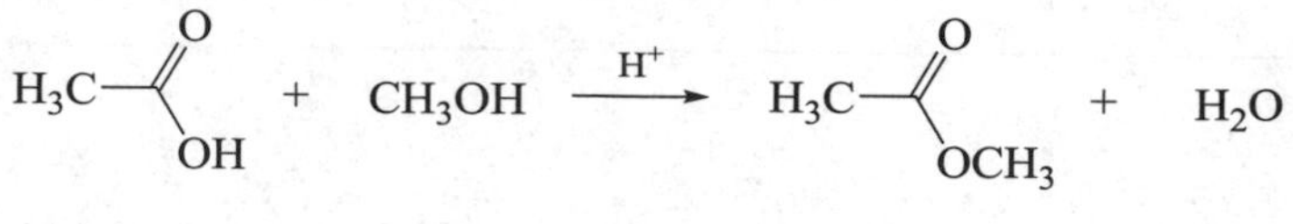

$$H_3C-\overset{O}{\underset{OH}{\|}} + CH_3OH \xrightarrow{H^+} H_3C-\overset{O}{\underset{OCH_3}{\|}} + H_2O$$

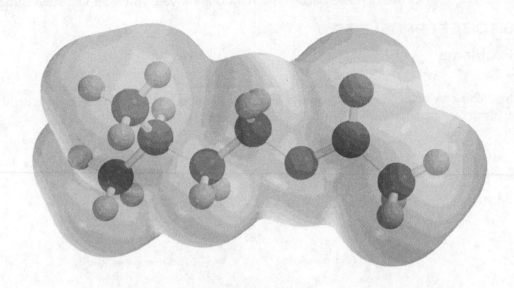

CHAPTER 20

Carboxylic Acid Derivatives:
Nucleophilic Acyl Substitution

Table of Contents

SOLUTIONS TO TEXT PROBLEMS

In Chapter Problems

20.1 (*b*) Acid anhydrides bear two acyl groups on oxygen, as in $\overset{\displaystyle O \quad O}{\underset{\displaystyle \| \quad \|}{RCOCR}}$. They are named as derivatives of carboxylic acids.

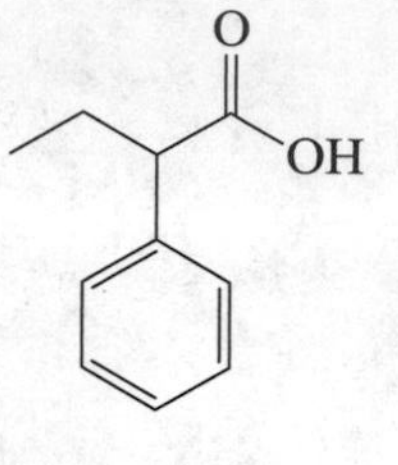

2-Phenybutanoic acid 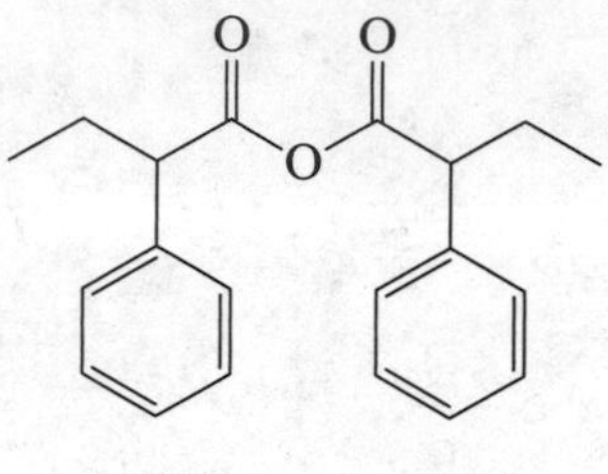   2-Phenybutanoic anyhydride

(*c*) Butyl 2-phenylbutanoate is the butyl ester of 2-phenylbutanoic acid.

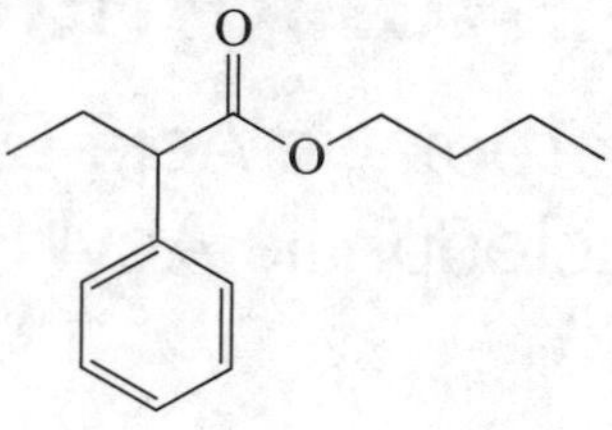

Butyl 2-phenylbutanoate

(*d*) In 2-phenylbutyl butanoate, the 2-phenylbutyl group is an alkyl group bonded to oxygen of the ester. It is not part of the acyl group of the molecule.

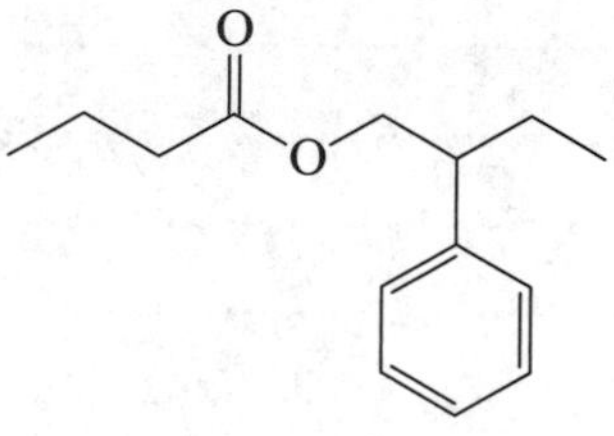

2-Phenylbutyl butanoate

(*e*) The ending -*amide* reveals this to be a compound of the type $\overset{\displaystyle O}{\underset{\displaystyle \|}{RCNH_2}}$.

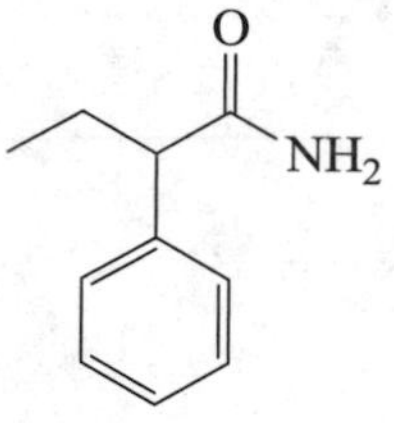

2-Phenylbutanamide

(*f*) This compound differs from 2-phenylbutanamide in part (*e*) only in that it bears an ethyl substituent on nitrogen.

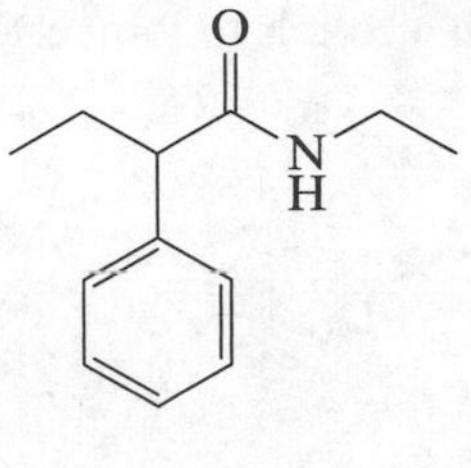

N-Ethyl-2-phenylbutanamide

(*g*) The -*nitrile* ending signifies a compound of the type RC≡N containing the same number of carbons as the alkane RCH_3. Alternatively, the compound may be named as an alkyl cyanide.

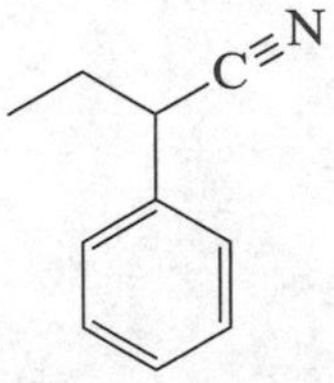

2-Phenylbutanenitrile or
1-phenylpropyl cyanide

20.2 The methyl groups in *N,N*-dimethylformamide are nonequivalent; one is cis to oxygen, the other is trans. The two methyl groups have different chemical shifts. The third signal is due to the carbonyl carbon.

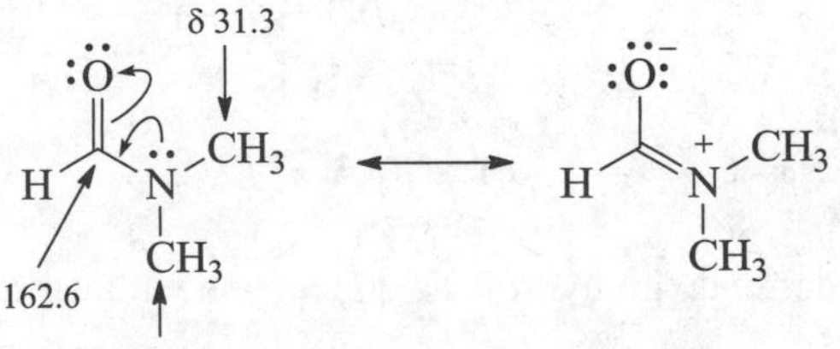

Rotation about the carbon–nitrogen bond is required to average the environments of the two methyl groups, but this rotation is slow enough in amides that the two distinct methyl groups can be detected. This is the result of the double-bond character imparted to the carbon–nitrogen bond, as shown by the resonance contributor on the right.

20.3 Electron release from the *p*-methoxy group stabilizes the acylium ion, which lowers the activation energy for its formation, and increases the reaction rate.

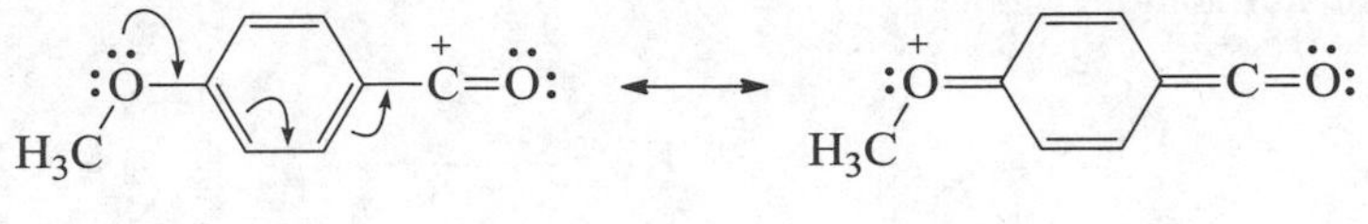

20.4 (*b*) Benzoyl chloride reacts with benzoic acid to give benzoic anhydride.

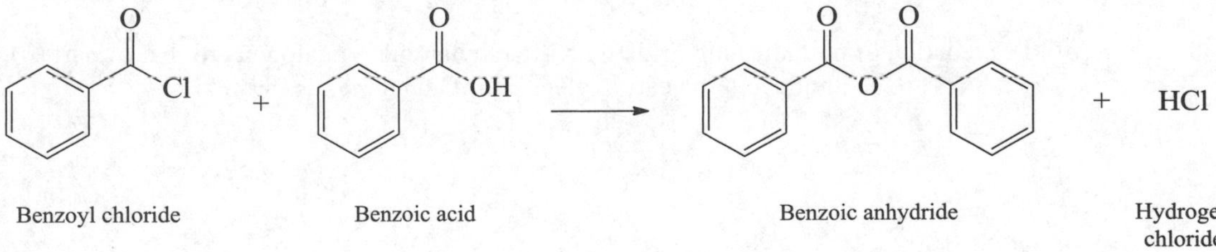

| Benzoyl chloride | Benzoic acid | | Benzoic anhydride | Hydrogen chloride |

(*c*) Acyl chlorides react with alcohols to form esters. The organic product is the ethyl ester of benzoic acid, ethyl benzoate.

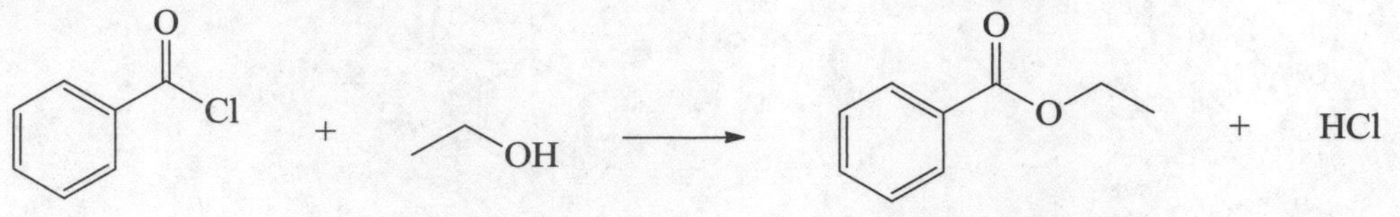

     Benzoyl chloride   Ethanol   Ethyl benzoate   Hydrogen chloride

(*d*) Acyl transfer from benzoyl chloride to the nitrogen of methylamine yields the amide *N*-methylbenzamide.

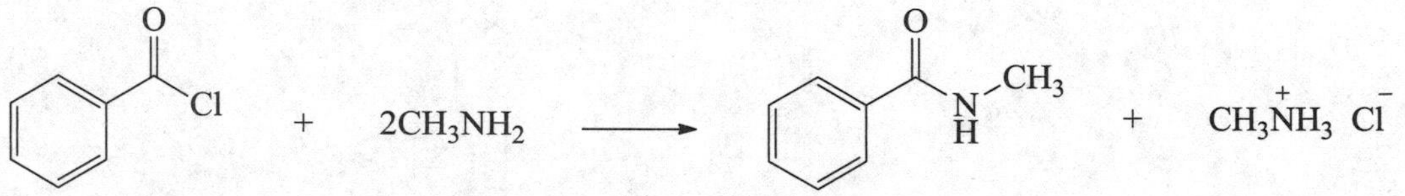

     Benzoyl chloride   Methylamine   *N*-Methylbenzamide

(*e*) As in part (*d*), an amide is formed. In this case, the amide has two methyl groups on nitrogen.

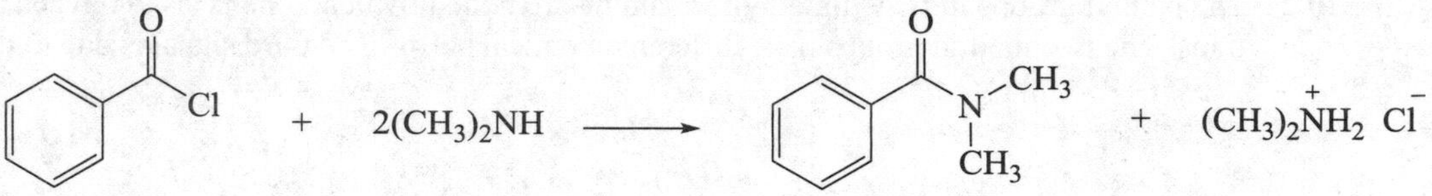

     Benzoyl chloride   Dimethylamine   *N,N*-Dimethylbenzamide

(*f*) Acyl chlorides undergo hydrolysis on reaction with water. The product is a carboxylic acid.

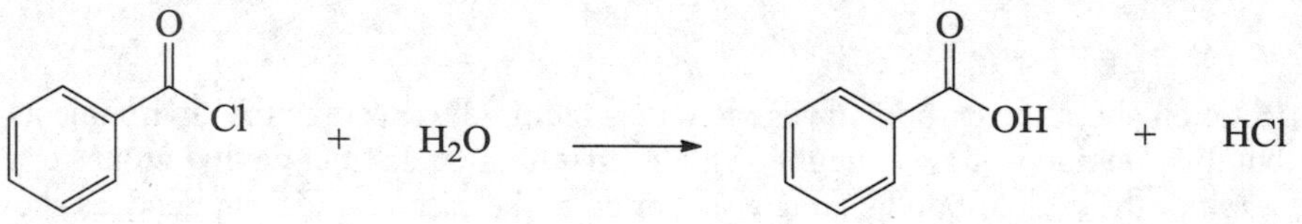

     Benzoyl chloride   Water   Benzoic acid   Hydrogen chloride

20.5 (*b*) Acyl transfer from an acid anhydride to ammonia yields an amide. The organic products are acetamide and ammonium acetate.

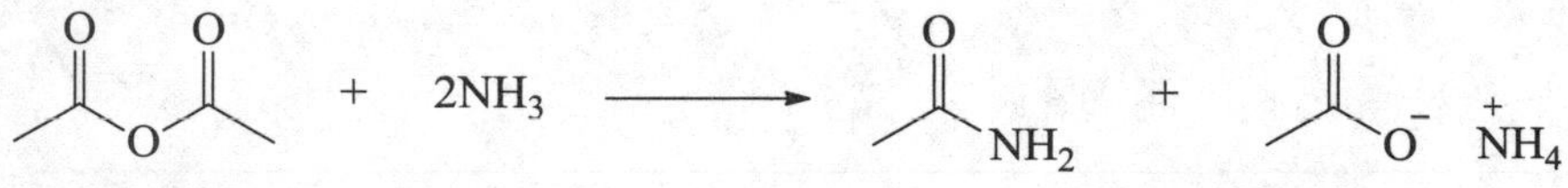

     Acetic anhydride   Ammonia   Acetamide   Ammonium acetate

(*c*) The reaction of phthalic anhydride with dimethylamine is analogous to that of part (*b*). The organic products are an amide and the carboxylate salt of an amine.

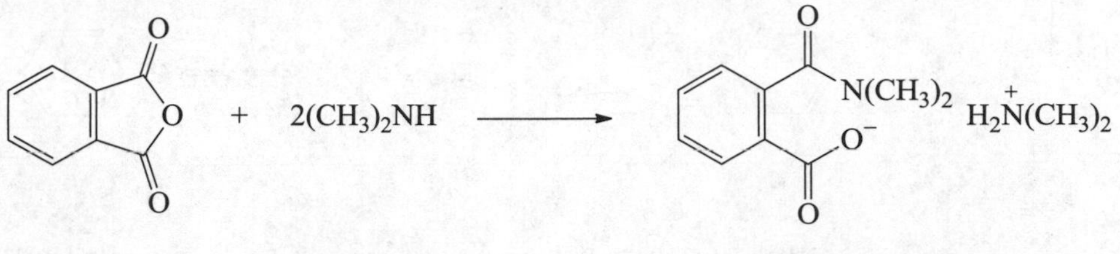

Phthalic anhydride   Dimethylamine   Product is an amine salt and
                                                   contains an amide functional group

In this case, both the amide function and the ammonium carboxylate salt are incorporated into the same molecule.

(*d*) The disodium salt of phthalic acid is the product of hydrolysis of phthalic acid in excess sodium hydroxide.

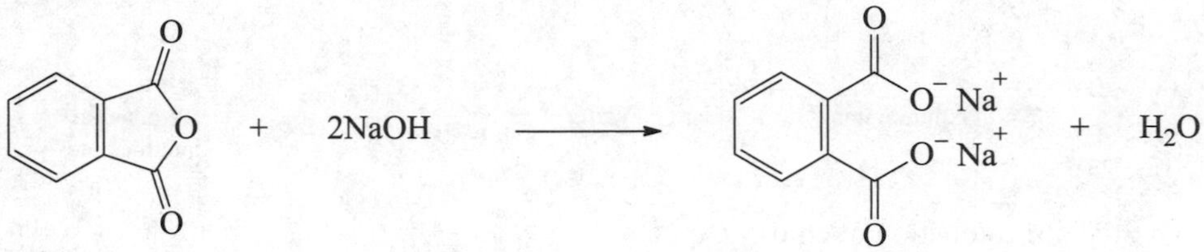

Phthalic anhydride   Sodium hydroxide   Sodium phthalate   Water

20.6 The starting material contains three acetate ester functions. All three undergo hydrolysis in aqueous sulfuric acid.

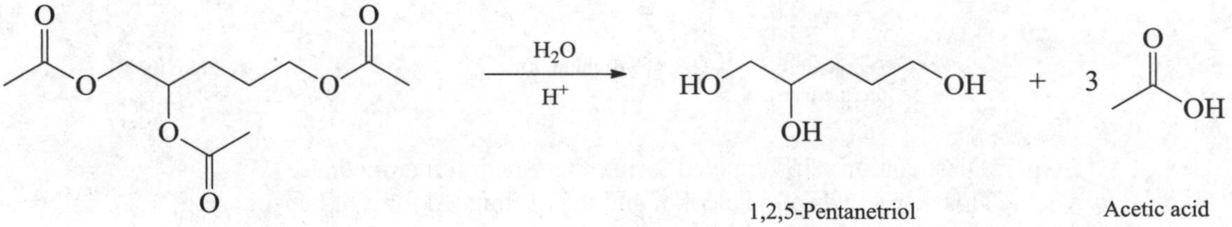

1,2,5-Pentanetriol   Acetic acid

The product is 1,2,5-pentanetriol. Also formed in the hydrolysis of the starting triacetate are three molecules of acetic acid.

20.7 Step 1: Protonation of the carbonyl oxygen

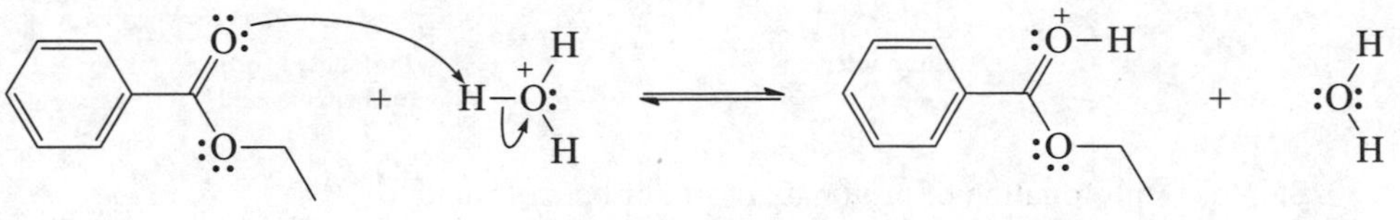

Ethyl benzoate   Hydronium   Protonated form of ester   Water
                        ion

Step 2: Nucleophilic addition of water

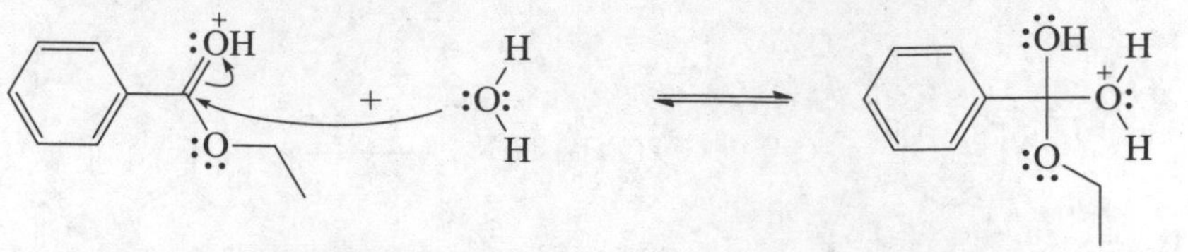

Protonated form of ester   Water   Oxonium ion

Step 3: Deprotonation of oxonium ion to give neutral form of tetrahedral intermediate

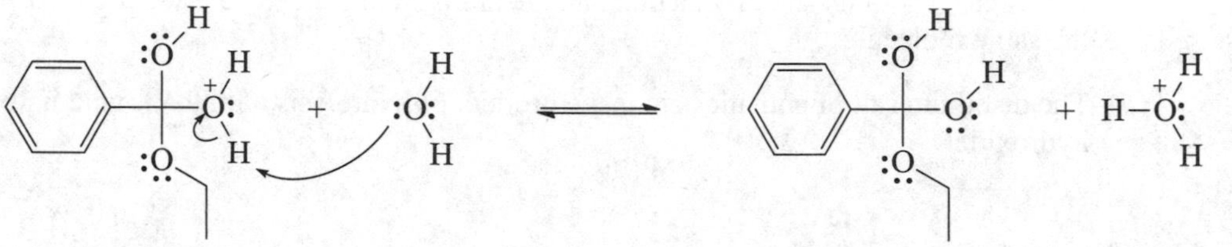

Oxonium ion   Water   Tetrahedral   Hydronium ion
                                                   intermediate

Step 4: Protonation of ethoxy oxygen

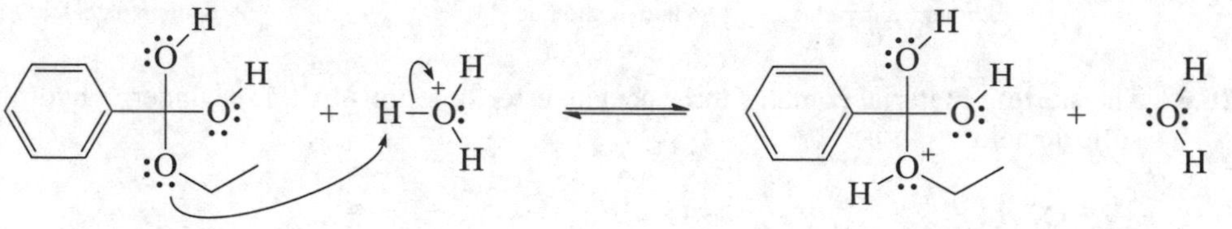

Tetrahedral   Hydronium ion   Oxonium ion   Water
intermediate

Step 5: Dissociation of protonated form of tetrahedral intermediate
This step yields ethyl alcohol and the protonated form of benzoic acid.

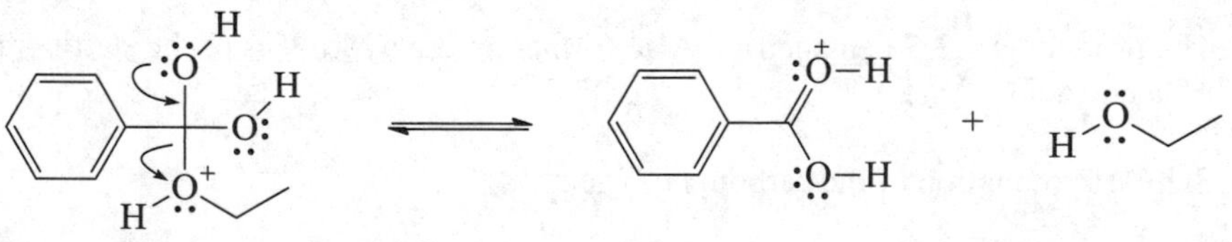

Oxonium ion   Protonated form   Ethyl alcohol
                                   of benzoic acid

Step 6: Deprotonation of protonated form of benzoic acid

Protonated form   Water   Protonated form   Hydronium ion
of benzoic acid   of benzoic acid

20.8 To determine which oxygen of 4-butanolide becomes labeled with ^{18}O, trace the path of ^{18}O-labeled water (● = ^{18}O) as it undergoes nucleophilic addition to the carbonyl group to form the tetrahedral intermediate.

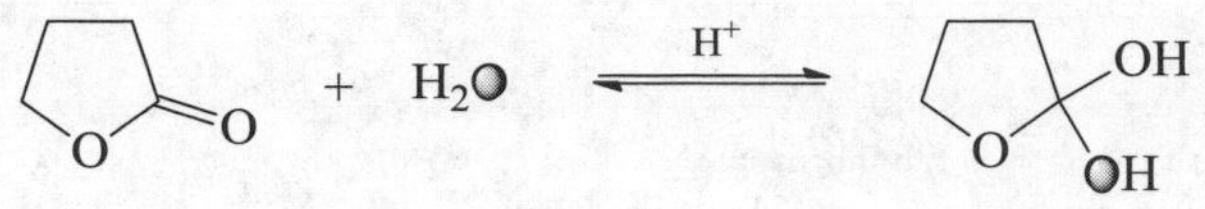

4-Butanolide ^{18}O-labeled water   Tetrahedral intermediate

The tetrahedral intermediate can revert to unlabeled 4-butanolide by loss of ^{18}O-labeled water. Alternatively it can lose ordinary water to give ^{18}O-labeled lactone.

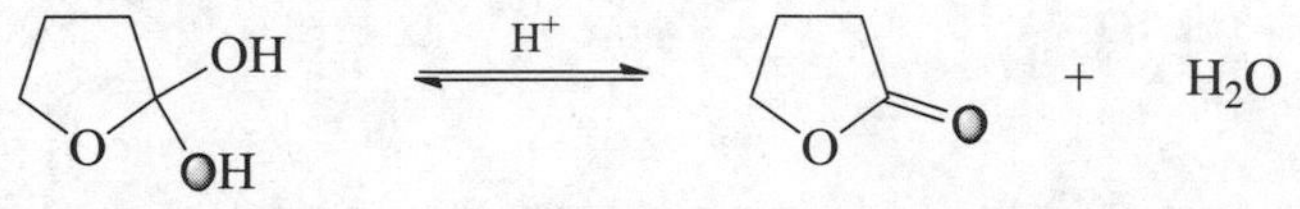

Tetrahedral intermediate ^{18}O-labeled 4-Butanolide   Water

The carbonyl oxygen is the one that is isotopically labeled in the ^{18}O-enriched 4-butanolide.

20.9 On the basis of trimyristin's molecular formula $C_{45}H_{86}O_6$ and of the fact that its hydrolysis gives only glycerol and tetradecanoic acid $CH_3(CH_2)_{12}CO_2H$, it must have the structure shown.

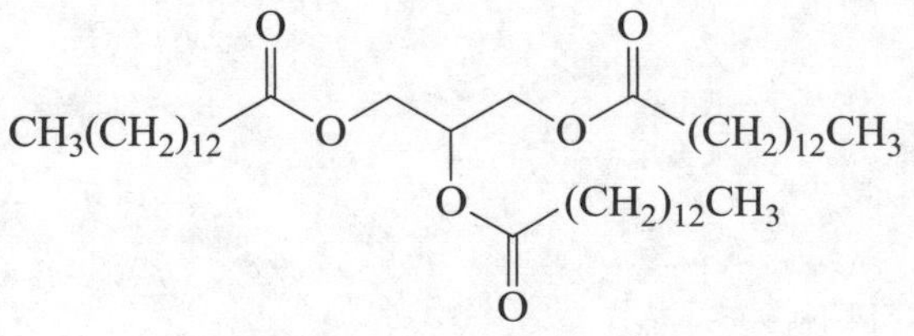

Trimyristin
($C_{45}H_{86}O_6$)

20.10 Because ester hydrolysis in base proceeds by acyl–oxygen cleavage, the ^{18}O label becomes incorporated into acetate ion (● = ^{18}O).

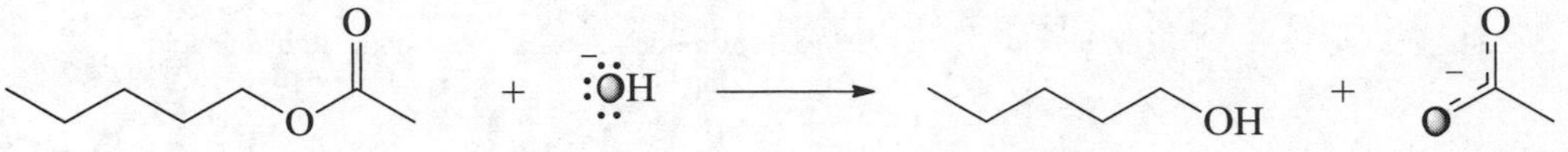

Pentyl acetate   Hydroxide ion   1-Pentanol   Acetate ion

20.11 Step 1: Nucleophilic addition of hydroxide ion to the carbonyl group

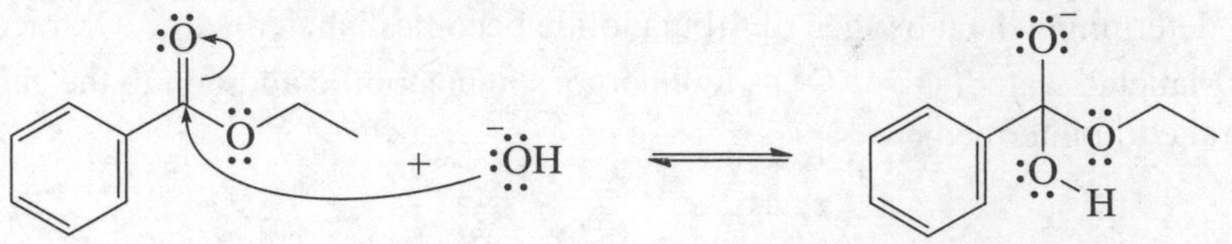

Ethyl benzoate   Hydroxide ion   Anionic form of tetrahedral intermediate

Step 2: Proton transfer from water to give neutral form of tetrahedral intermediate

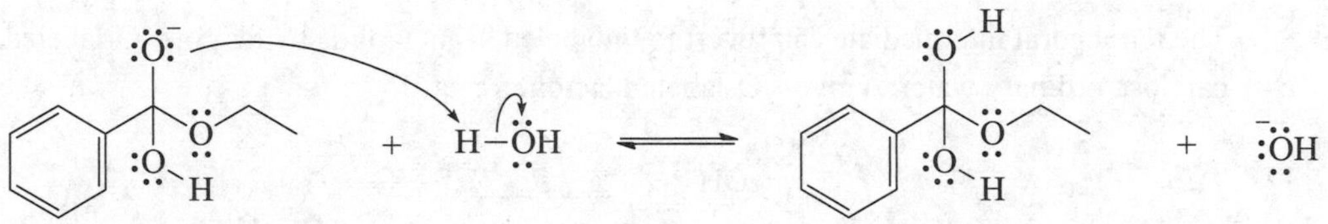

Anionic form of tetrahedral intermediate   Water   Tetrahedral intermediate   Hydroxide ion

Step 3: Dissociation of tetrahedral intermediate

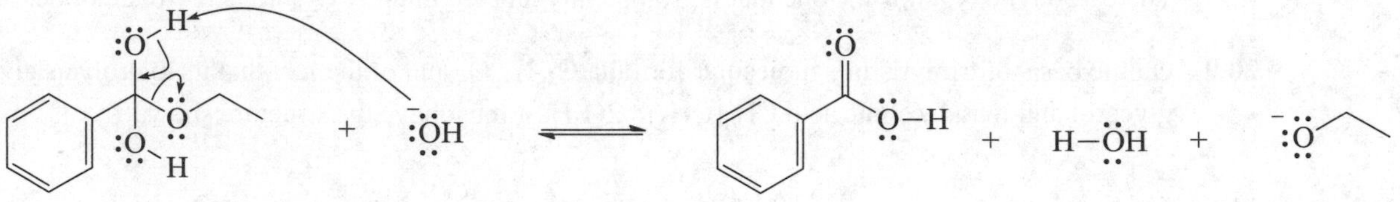

Tetrahedral intermediate   Hydroxide ion   Benzoic acid   Water   Ethoxide ion

Step 4: Proton transfer processes

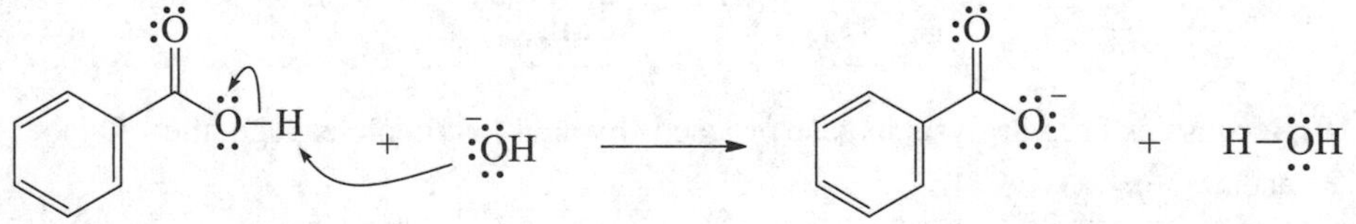

Benzoic acid   Hydroxide ion   Benzoic acid   Water

Ethoxide ion   Water   Ethanol   Hydroxide ion

20.12 (b) Stabilization of the ester carbonyl group is greater in the ester of cyclohexanol than in the ester of phenol, because of resonance involving the aromatic ring and oxygen in the latter. No such resonance is possible for the ester of cyclohexanol. Hydrolysis of the less stable ester of phenol occurs more rapidly.

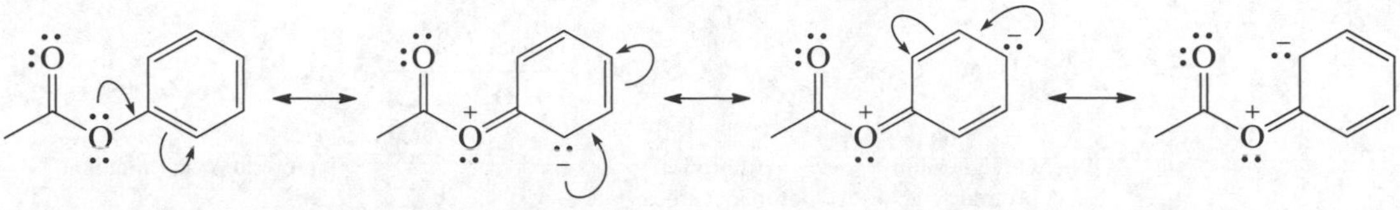

(*c*) The electron-withdrawing trifluoromethyl group destabilizes the ester carbonyl in the second compound by an inductive effect, so it hydrolyzes more rapidly.

(*d*) The large *tert*-butyl group destabilizes due to steric hindrance in the tetrahedral intermediate in saponification of the second compound, resulting in a larger energy of activation and slower hydrolysis. (Note that this applies to saponification, and not to hydrolysis in acid.)

20.13 The starting material is a lactone, a cyclic ester. The ester function is converted to an amide by nucleophilic acyl substitution.

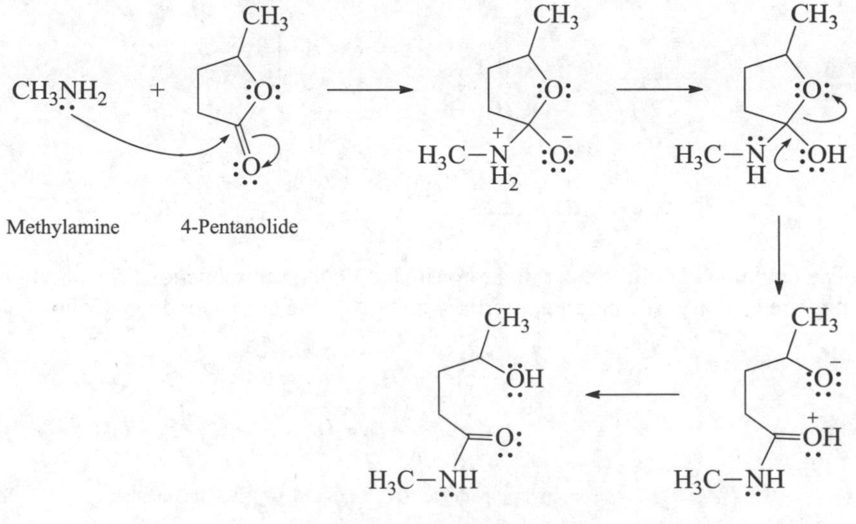

Methylamine   4-Pentanolide

4-Hydroxy-*N*-methylpentanamide

20.14 Nucleophilic attack by the amine on the ester carbonyl is the rate determining step that forms a tetrahedral intermediate.

slow

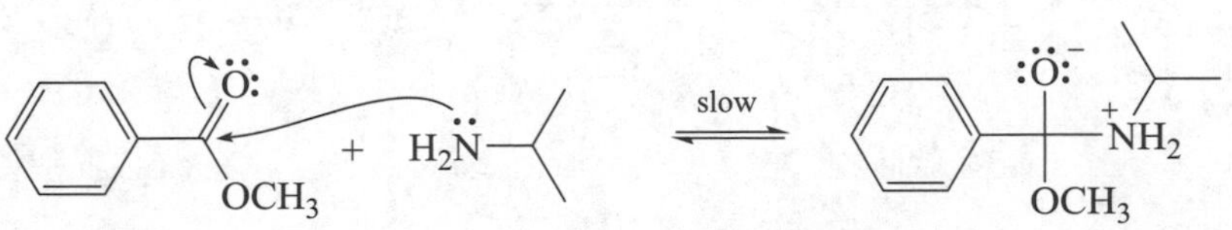

20.15 (*b*) Recall that the two identical groups bonded to the hydroxyl-bearing carbon of the alcohol arose from the Grignard reagent. That leads to the following retrosynthetic analysis:

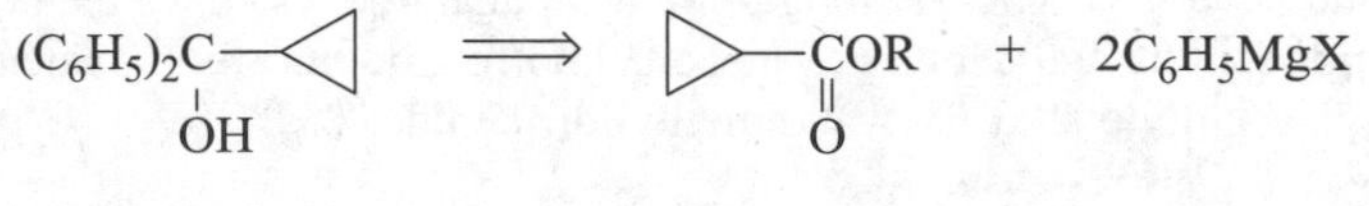

$(C_6H_5)_2C$—⊲ $\Longrightarrow$ ⊲—COR + $2C_6H_5MgX$
|
OH

O

Thus, the two phenyl substituents arise by addition of a phenyl Grignard reagent to an ester of cyclopropanecarboxylic acid.

$$2C_6H_5MgX \; + \; \triangle\text{--COCH}_3 \quad \xrightarrow[\text{2. } H_3O^+]{\text{1. diethyl ether}} \quad (C_6H_5)_2C\text{--}\triangle \; + \; CH_3OH$$

Phenylmagnesium bromide   Methyl cyclopropanecarboxylate   Cyclopropyldiphenylmethanol   Methanol

20.16 Ethyl benzoate is reduced by lithium aluminum hydride to give benzyl alcohol and ethanol.

$$C_6H_5\text{--C}(\text{=O})\text{--OCH}_2CH_3 \quad \xrightarrow[\text{2. } H_2O]{\text{1. LiAlH}_4} \quad C_6H_5\text{--CH}_2\text{OH} \; + \; H\ddot{O}CH_2CH_3$$

The aldehyde is benzaldehyde, formed by collapse of the tetrahedral intermediate. The benzaldehyde that is produced in the reaction is further reduced to benzyl alcohol.

Tetrahedral intermediate   Benzaldehyde

20.17 The acyl portion of the ester gives a primary alcohol on reduction. The alkyl group bonded to oxygen may be primary, secondary, or tertiary and gives the corresponding alcohol.

Isopropyl propanoate   1-Propanol   2-Propanol

20.18 The difference in the compounds lies in the number of N—H bonds in each.

Propanamide   *N*-Methylacetamide   *N,N*-Dimethylformide

2 N-H bonds   1 N-H bond   0 N-H bonds

The compound with the greatest number of N—H bonds will engage in the most intermolecular hydrogen bonds and have the highest boiling point. Propanamide will have the highest boiling point; *N,N*-dimethylformamide will have the lowest. The actual boiling points are: propanamide: 222°C; *N*-methylacetamide: 206°C; *N,N*-dimethylformamide: 153°C.

20.19 Recall (from text Section 20.2) that amide nitrogens are planar. Thus structure B is a planar monocyclic array of sp^2- hybridized atoms. Structure B also has six π electrons in the ring, two from each nitrogen and two from the double bond, and thus satisfies Hückel's rule.

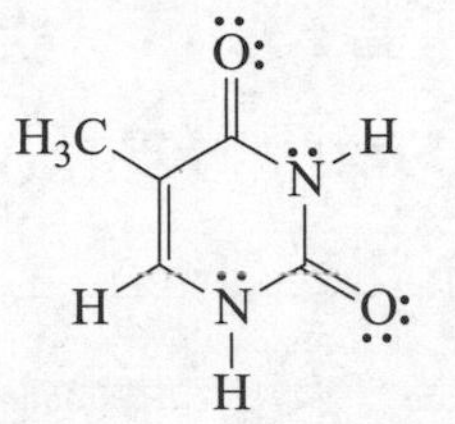

A resonance structure of B can be drawn with a ring analogous to benzene:

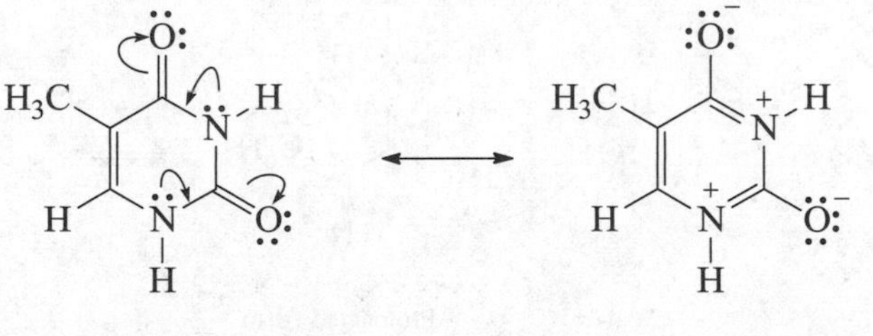

20.20 (*b*) Acetic anhydride is the acid anhydride that must be used; it transfers an acetyl group to suitable nucleophiles. The nucleophile in this case is methylamine.

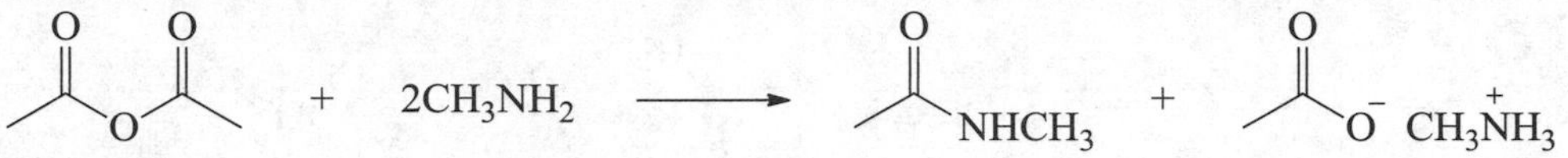

Acetic anhydride   Methylamine   *N*-Methylacetamide   Methylammonium acetate

(*c*) The acyl group is $HC-$. Because the problem specifies that the acyl transfer agent is a methyl ester, methyl formate is one of the starting materials.

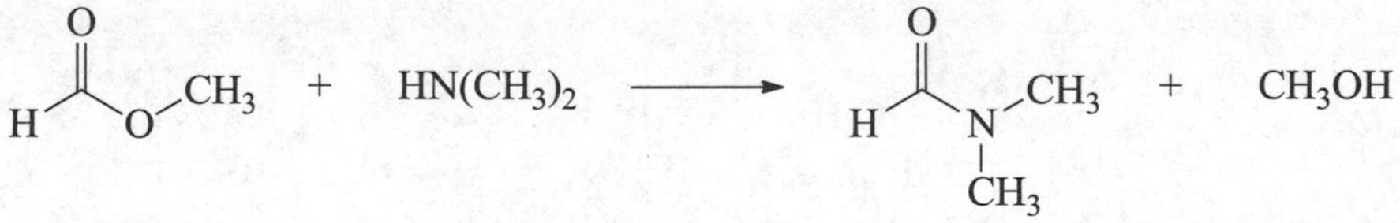

Methyl formate   Dimethylamine   *N,N*-Dimethylformamide   Methyl alcohol

20.21 The reaction that occurs between an amine and a carboxylic acid is proton transfer from the acid to the amine (acid-base).

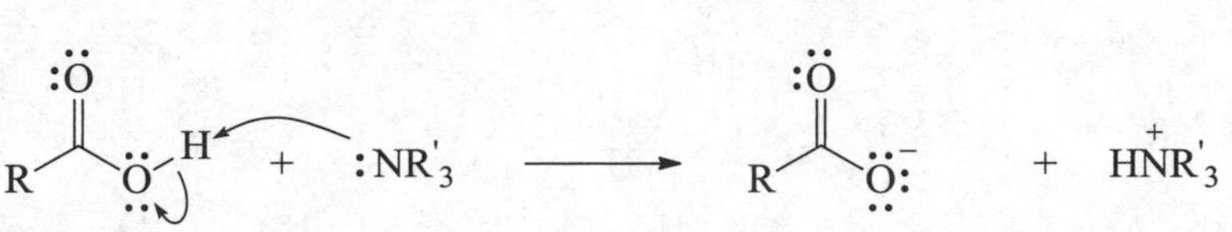

The carboxylate carbonyl is highly stabilized by electron delocalization and does not undergo nucleophilic acyl substitution and an ammonium ion is not a nucleophile; the nitrogen atom has no lone electron pairs.

20.22 Step 1: Protonation of the carbonyl oxygen

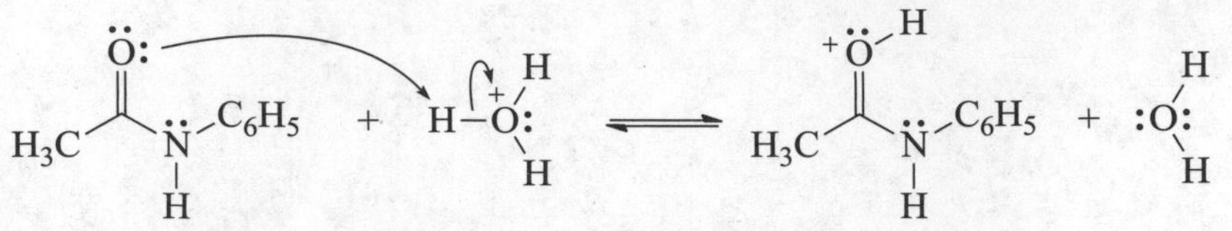

Acetanilide + Hydronium ion ⇌ Protonated form of amide + Water

Step 2: Nucleophilic addition of water

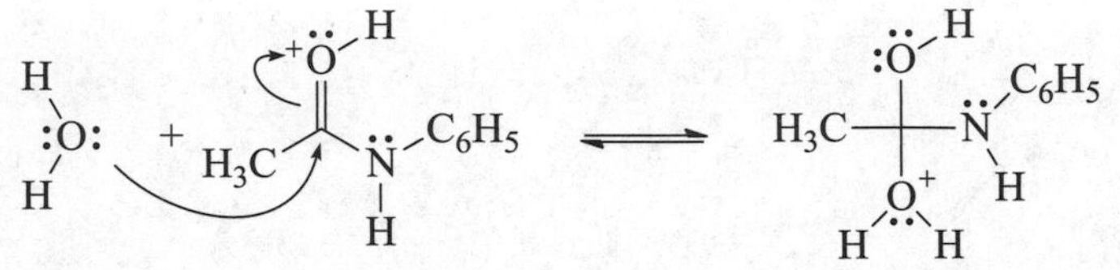

Water + Protonated form of amide ⇌ Oxonium ion

Step 3: Deprotonation of oxonium ion to give neutral form of tetrahedral intermediate

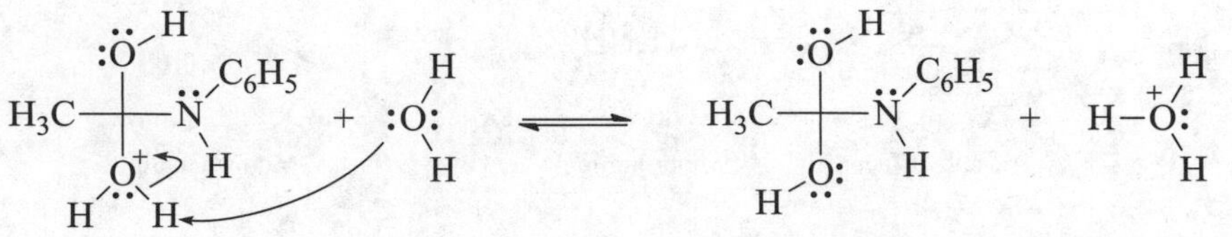

Oxonium ion + Water ⇌ Tetrahedral intermediate + Hydronium ion

Step 4: Protonation of amino group of tetrahedral intermediate

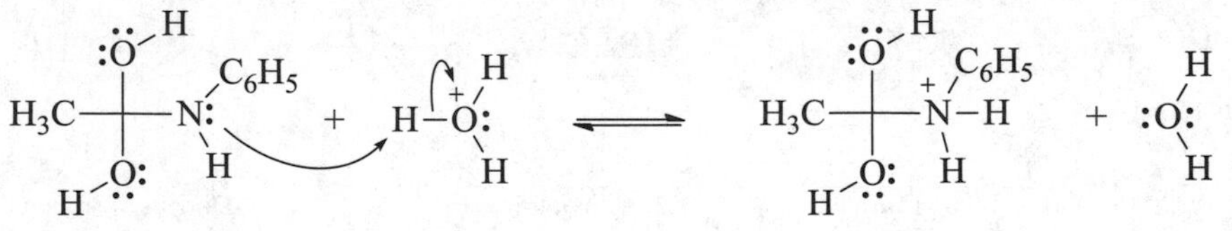

Tetrahedral intermediate + Hydronium ion ⇌ N-Protonated form of tetrahedral intermediate + Water

Step 5: Dissociation of N-protonated form of tetrahedral intermediate

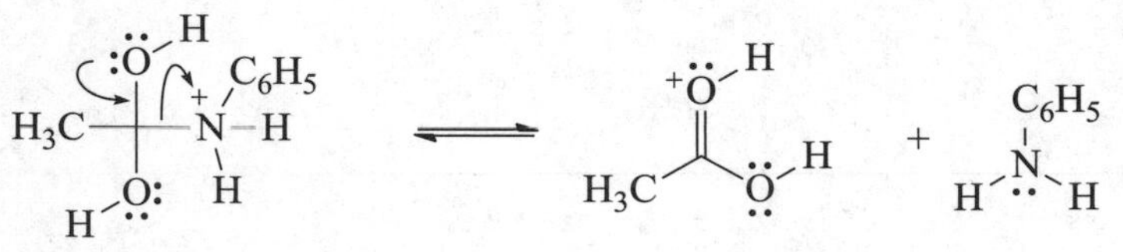

N-Protonated form of tetrahedral intermediate ⇌ Protonated form of acetic acid + Aniline

Step 6: Proton-transfer processes

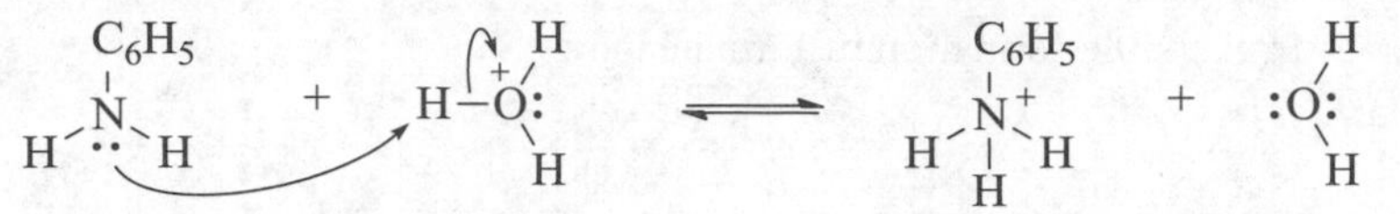

Aniline   Hydronium ion   Anilinium ion   Water

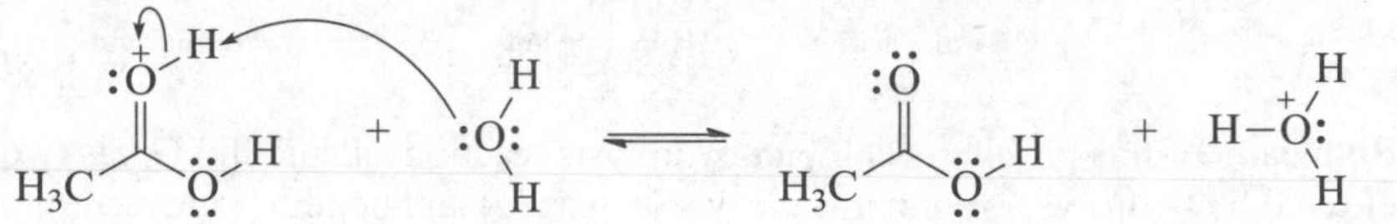

Protonated form
of acetic acid   Water   Acetic acid   Hydronium ion

20.23 Step 1: Nucleophilic addition of hydroxide ion to the carbonyl group

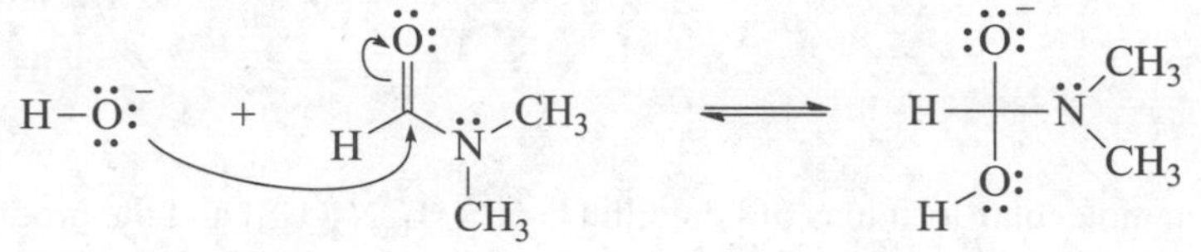

Hydroxide ion   *N,N*-Dimethylformamide   Anionic form of tetrahedral
intermediate

Step 2: Proton transfer to give neutral form of tetrahedral intermediate

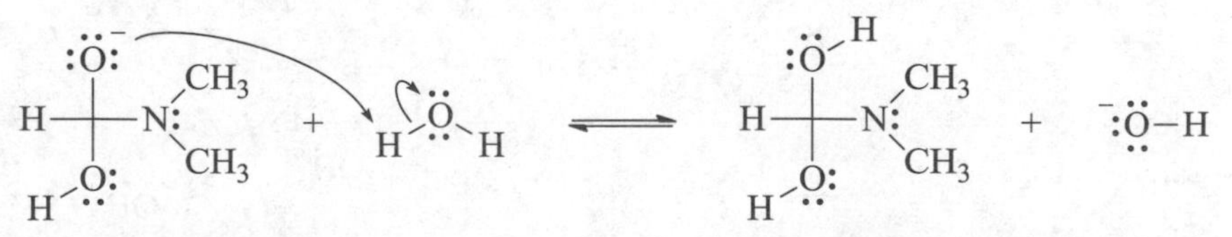

Anionic form of tetrahedral
intermediate   Water   Tetrahedral
intermediate   Hydroxide ion

Step 3: Proton transfer from water to nitrogen of tetrahedral intermediate

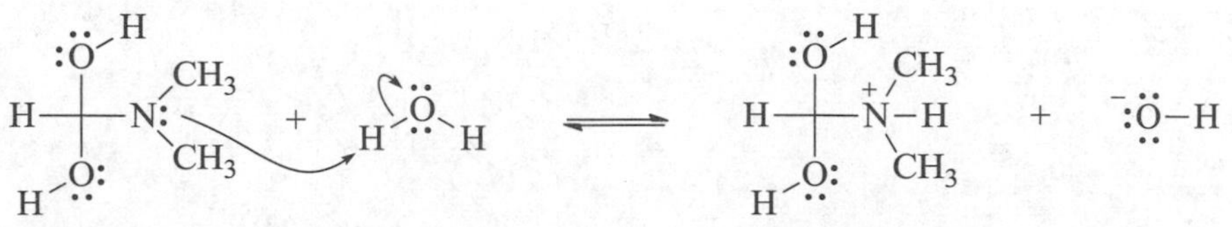

Tetrahedral
intermediate   Water   *N*-Protonated form of
tetrahedral intermediate   Hydroxide ion

Step 4: Dissociation of *N*-protonated form of tetrahedral intermediate

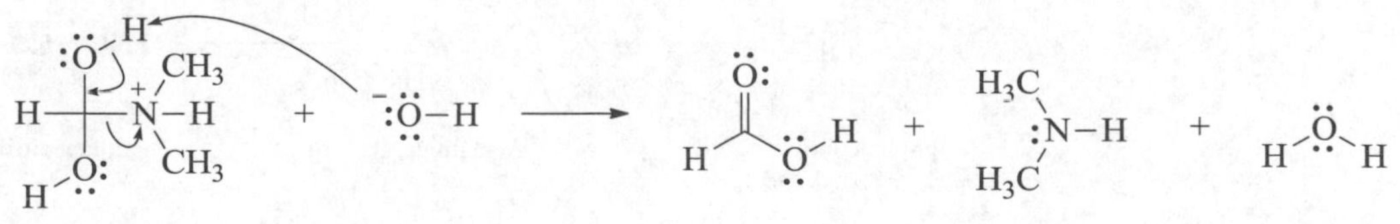

N-Protonated form of
tetrahedral intermediate   Hydroxide ion   Formic acid   Dimethylamine   Water

Step 5: Irreversible formation of formate ion

Formic acid Hydroxide ion Formic acid Water

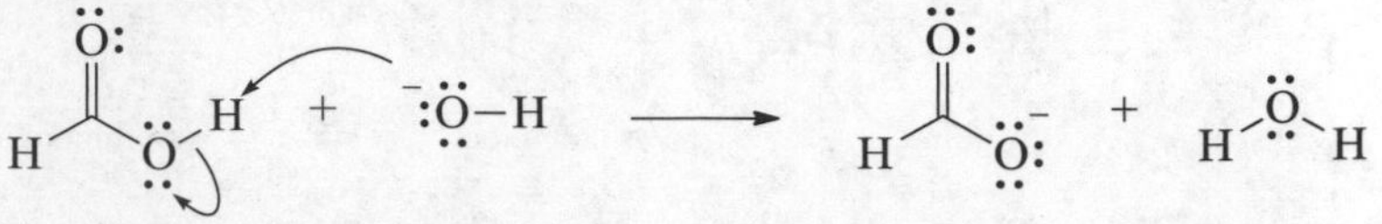

20.24 Both parts of this problem deal with hydrolysis reactions. Penicillin G has two amide bonds capable of being cleaved by hydrolysis. Bond *a* is cleaved in part (*a*) and bond *b* is cleaved in part (*b*).

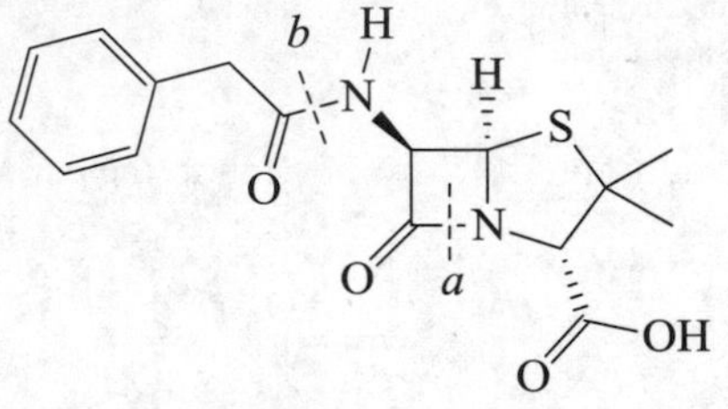

(*a*) The molecular formulas of penicillin G ($C_{16}H_{18}N_2O_4S$) and the product of β-lactamase-catalyzed hydrolysis ($C_{16}H_{20}N_2O_5S$) differ by H_2O. Therefore, all the atoms in penicillin G are retained on hydrolysis which is consistent only with cleaving the β-lactam ring. Therefore, the product is

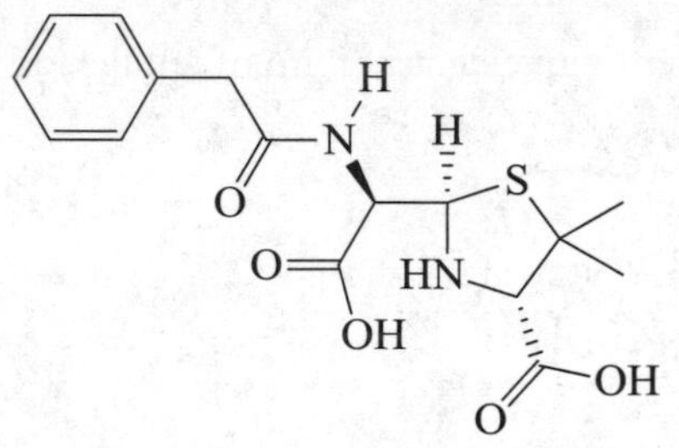

(*b*) The product of *penicillin acyl transferase*-catalyzed hydrolysis has a molecular formula ($C_8H_{12}N_2O_3S$) with eight fewer carbons than penicillin G. Therefore, the acyl group attached to the nitrogen of the side chain is cleaved from the molecule. The product is

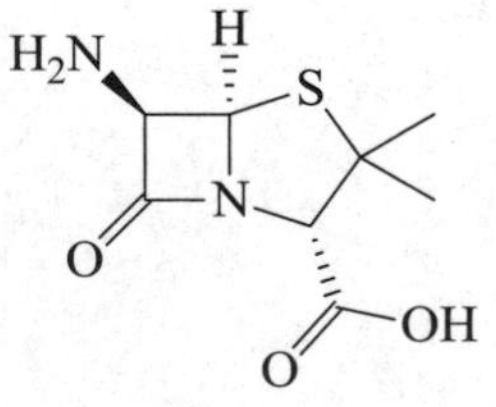

20.25 (*a*) Ethanenitrile has the same number of carbon atoms as ethyl alcohol. This suggests a reaction scheme proceeding via an amide.

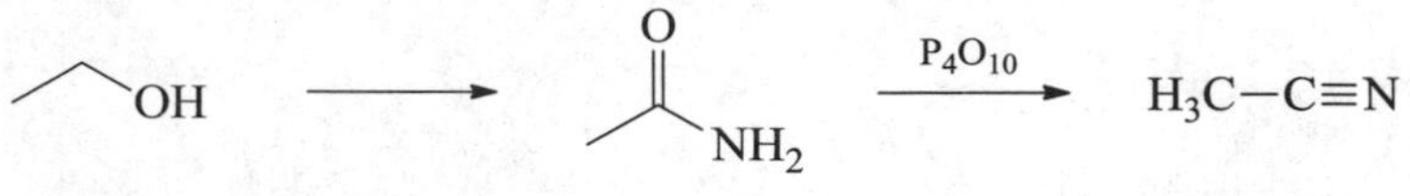

Ethyl alcohol Acetamide Ethanenitrile

The necessary amide is prepared from ethanol.

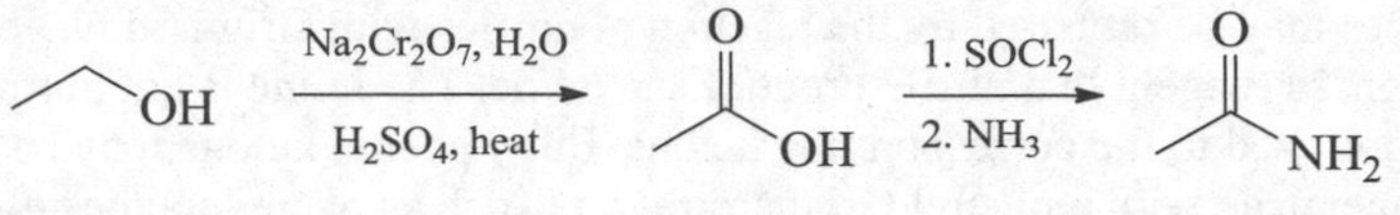

Ethyl alcohol   Acetic acid   Acetamide

(*b*) Propanenitrile may be prepared from ethyl alcohol by way of a nucleophilic substitution reaction of the corresponding bromide.

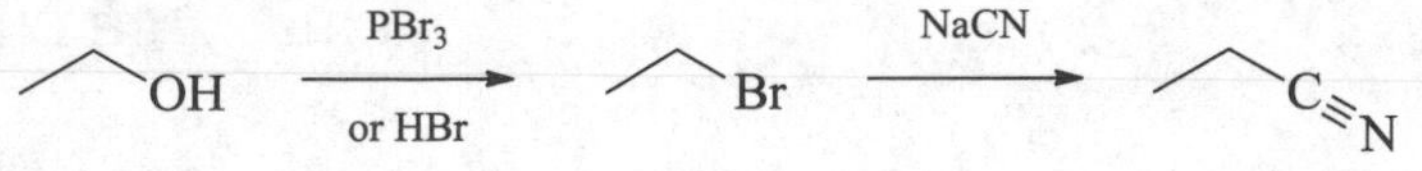

Ethyl alcohol   Ethyl bromide   Propanenitrile

20.26 Step 1: Protonation of the nitrile

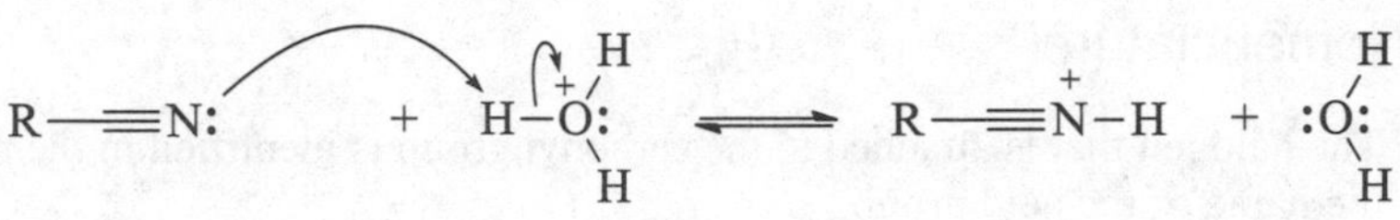

Nitrile   Hydronium ion   Protonated form   Water
                                                           of nitrile

Step 2: Nucleophilic addition of water

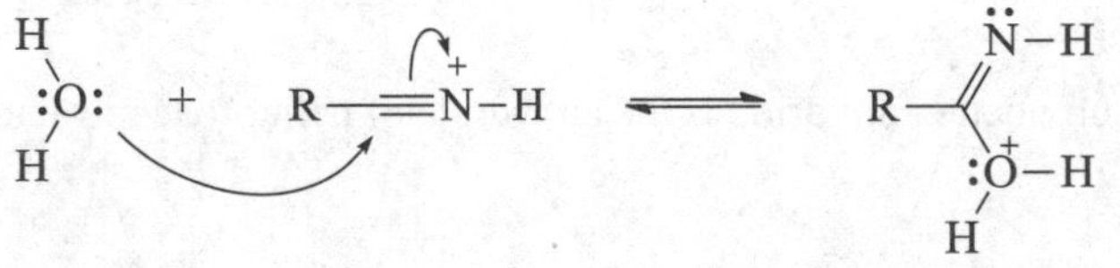

Water   Protonated form   Protonated form
                         of nitrile   imino acid

Step 3: Deprotonation of imino acid

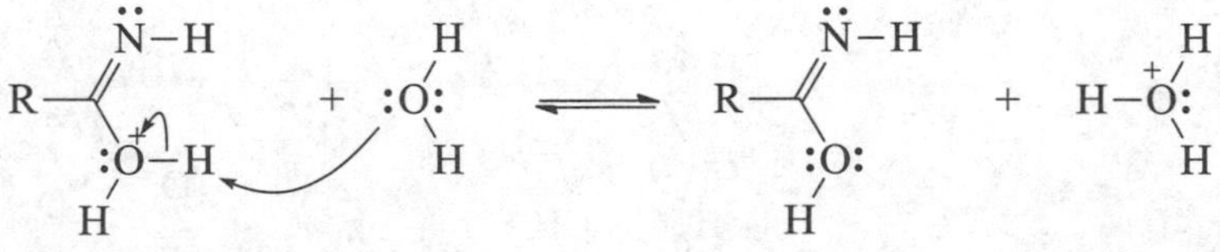

Protonated form   Water   Imino acid   Hydronium ion
imino acid

Steps 4 and 5: Proton transfers to give an amide

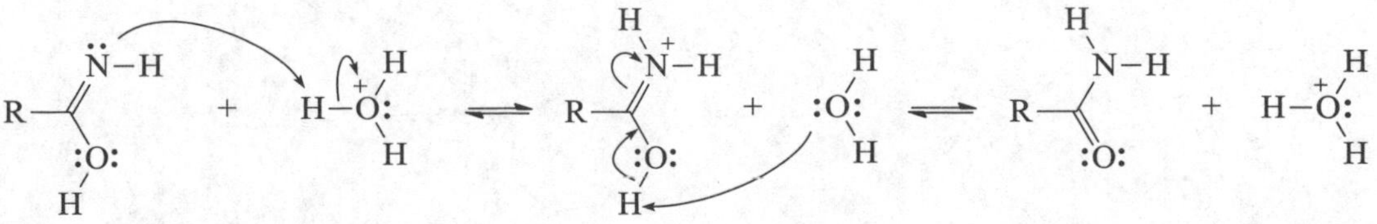

Imino acid   Hydronium ion   Conjugate acid   Water   Amide   Hydronium ion
                                          of amide

20.27 Ketones may be prepared by the reaction of nitriles with Grignard reagents. Nucleophilic addition of a Grignard reagent to a nitrile produces an imine. The imine is not normally isolated, however, but is hydrolyzed to the corresponding ketone. Ethyl phenyl ketone may be prepared by the reaction of propanenitrile with a phenyl Grignard reagent such as phenylmagnesium bromide, followed by hydrolysis of the imine.

$$CH_3CH_2\!-\!C\!\equiv\!N\colon \;+\; C_6H_5MgBr \xrightarrow{\text{diethyl ether}} \left[\substack{NH \\ \| \\ C_6H_5\diagup\diagdown CH_2CH_3} \right] \xrightarrow[\text{heat}]{H_2O,\ H^+} \substack{O \\ \| \\ C_6H_5\diagup\diagdown CH_2CH_3}$$

| Propanenitrile | Phenylmagnesium bromide | Imine (not isolated) | Ethyl phenyl ketone |

End of Chapter Problems

Structure and Nomenclature

20.28 (*a*) The halogen that is attached to the carbonyl group is identified in the name as a separate word following the name of the acyl group.

m-Chlorobenzoyl chloride

(*b*) Trifluoroacetic anhydride is the anhydride of trifluoroacetic acid. Notice that it contains six fluorines.

Trifluoroacetic anhydride

(*c*) This compound is the cyclic anhydride of *cis*-1,2-cyclopropanedicarboxylic acid.

| *cis*-1,2-Cyclopropanedicarboxylic acid | *cis*-1,2-Cyclopropanedicarboxylic anhydride |

(*d*) Ethyl cycloheptanecarboxylate is the ethyl ester of cycloheptanecarboxylic acid.

Ethyl cycloheptanecarboxylate

(*e*) 1-Phenylethyl acetate is the ester of 1-phenylethanol and acetic acid.

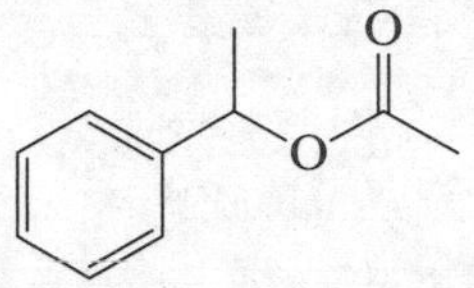

1-Phenylethyl acetate

(*f*) 2-Phenylethyl acetate is the ester of 2-phenylethanol and acetic acid.

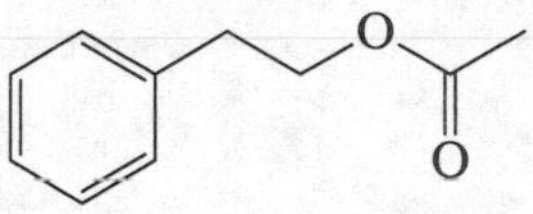

1-Phenylethyl acetate

(*g*) The parent compound in this case is benzamide. *p*-Ethylbenzamide has an ethyl substituent at the ring position para to the carbonyl group.

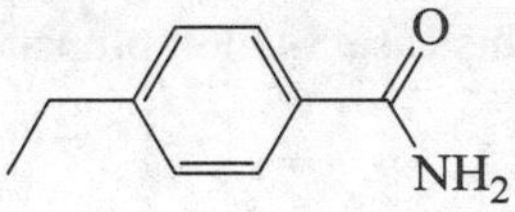

p-Ethylbenzamide

(*h*) The parent compound is benzamide. In *N*-ethylbenzamide, the ethyl substituent is bonded to nitrogen.

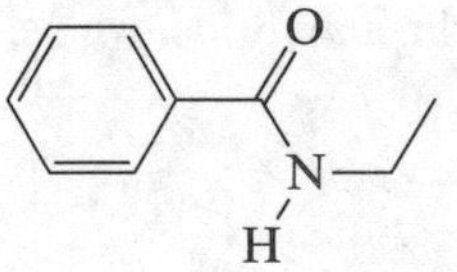

N-Ethylbenzamide

(*i*) Nitriles are named by adding the suffix *-nitrile* to the name of the alkane having the same number of carbons. Numbering begins at the nitrile carbon.

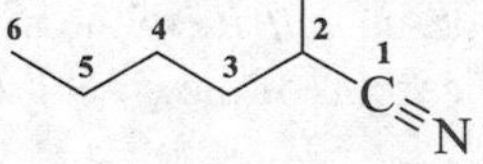

2-Methylhexanenitrile

20.29 (*a*) This compound, with a chlorine attached to its carbonyl group, is named as an acyl chloride. It is 3-chlorobutanoyl chloride.

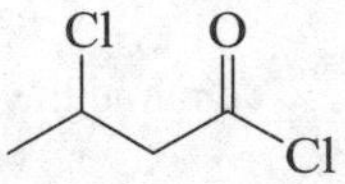

3-Chlorobutanoyl chloride

(*b*) The group attached to oxygen, in this case *benzyl,* is identified first in the name of the ester. This compound is the benzyl ester of acetic acid.

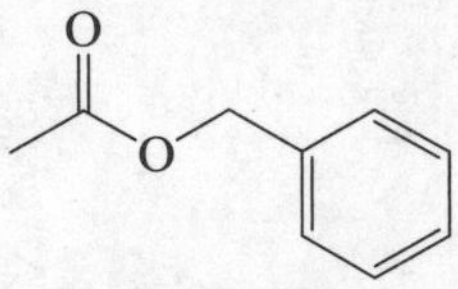

Benzyl acetate

(*c*) The group attached to oxygen is methyl; this compound is the methyl ester of phenylacetic acid.

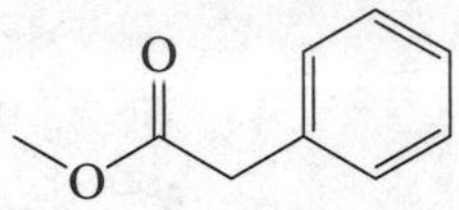

Methyl phenylacetate

(*d*) This compound contains the functional group —$\overset{\overset{O}{\|}}{C}O\overset{\overset{O}{\|}}{C}$— and thus is an anhydride of a carboxylic acid. We name the acid, in this case 3-chloropropanoic acid, drop the *acid* part of the name, and replace it by *anhydride*.

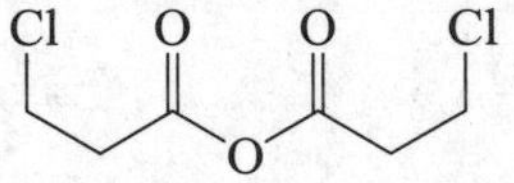

3-Chloropropanoic anhydride

(*e*) This compound is a cyclic anhydride, whose parent acid is 3,3-dimethylpentanedioic acid.

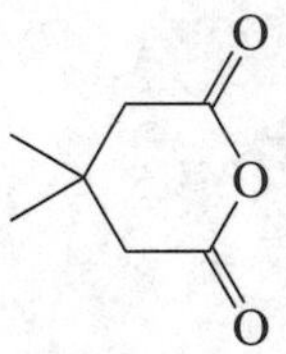

3,3-Dimethylpentanedioic
anhydride

(*f*) Nitriles are named by adding *-nitrile* to the name of the alkane having the same number of carbons. Remember to count the carbon of the C≡N group.

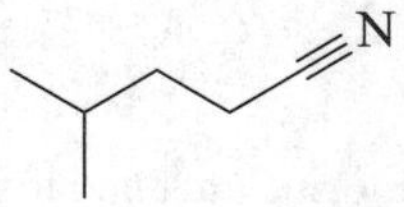

4-Methylpentanenitrile

(*g*) This compound is an amide. We name the corresponding acid and then replace the *-oic acid* suffix by *-amide*.

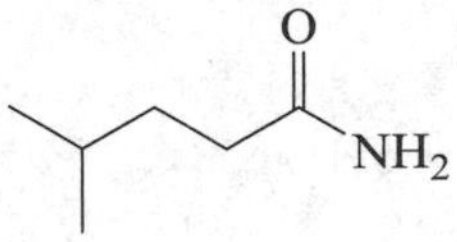

4-Methylpentanamide

(*h*) This compound is the *N*-methyl derivative of 4-methylpentanamide.

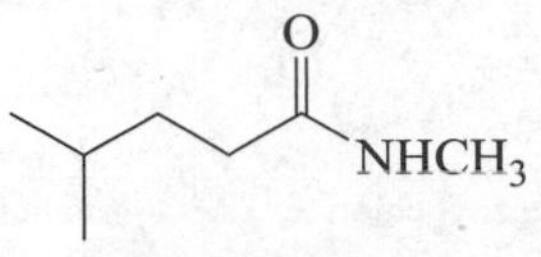

N,4-Dimethylpentanamide

(*i*) The amide nitrogen bears two methyl groups. We designate this as an *N*,*N*-dimethyl amide.

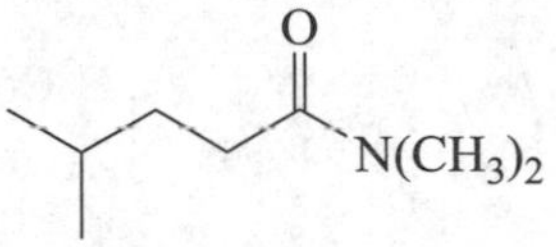

N,*N*,4-Trimethylpentanamide

20.30 The ester is formed from (*S*)-2-methylbutanoic acid and the hydroxyl group on the left-hand ring of mevinolin. The lactone can be formed from the open-chain hydroxy acid.

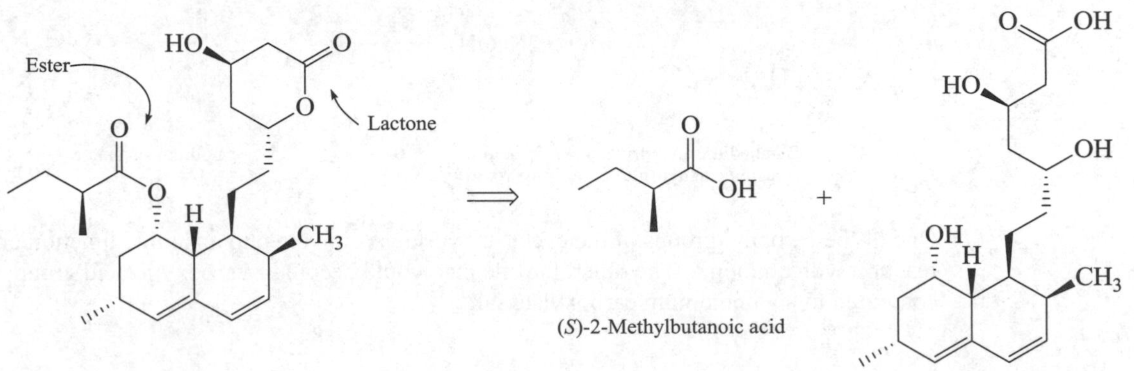

(*S*)-2-Methylbutanoic acid

Reactions

20.31 (*a*) Sodium propanoate acts as a nucleophile toward propanoyl chloride. The product is propanoic anhydride.

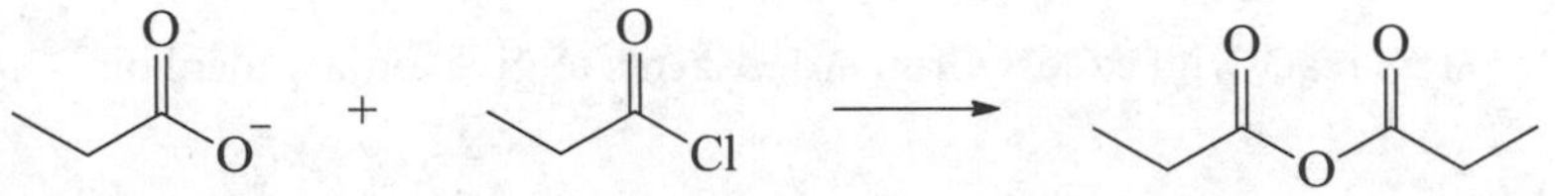

Propanoate anion   Propanoyl chloride   Propanoic anhydride

(*b*) Acyl chlorides convert alcohols to esters.

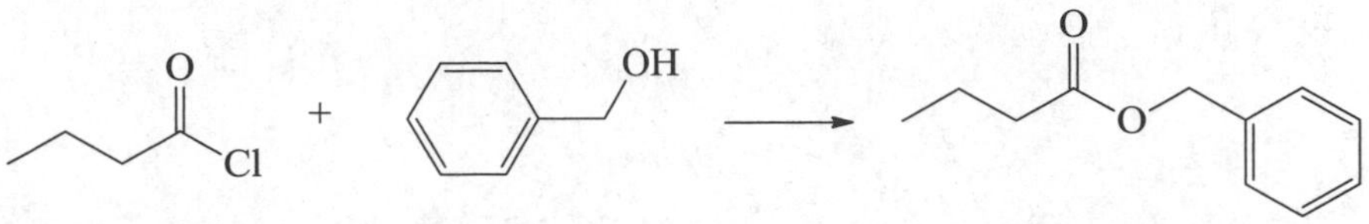

Butanoyl chloride   Benzyl alcohol   Benzyl butanoate

(*c*) Acyl chlorides react with ammonia to yield amides.

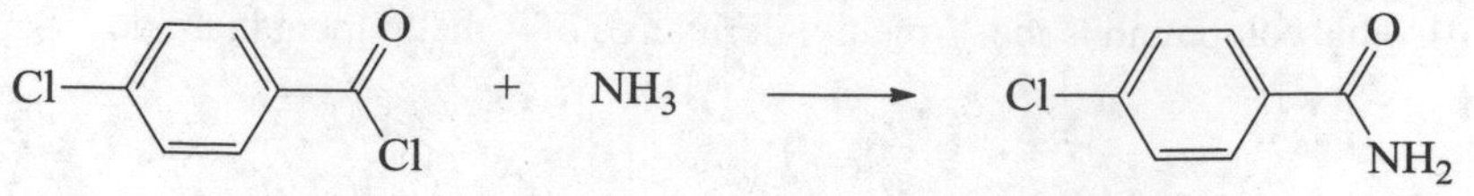

p-Chlorobenzoyl chloride   Ammonia   p-Chlorobenzamide

(d) The starting material is a cyclic anhydride. Acid anhydrides react with water to yield two carboxylic acid functions; when the anhydride is cyclic, a dicarboxylic acid results.

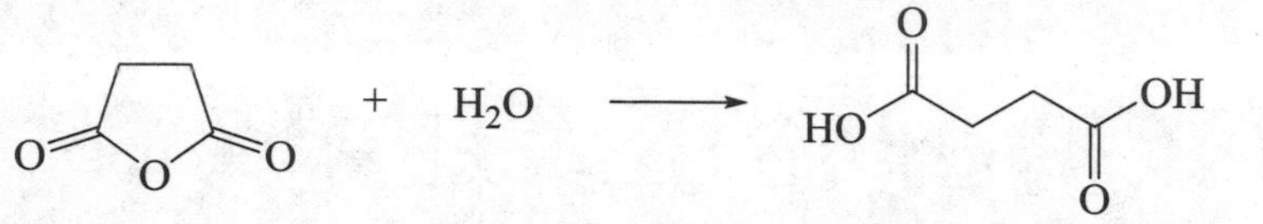

Butanedioic anhydride   Water   Butanedioic acid
(succinic anhydride)   (succinic acid)

(e) In dilute sodium hydroxide, the anhydride is converted to the disodium salt of the diacid.

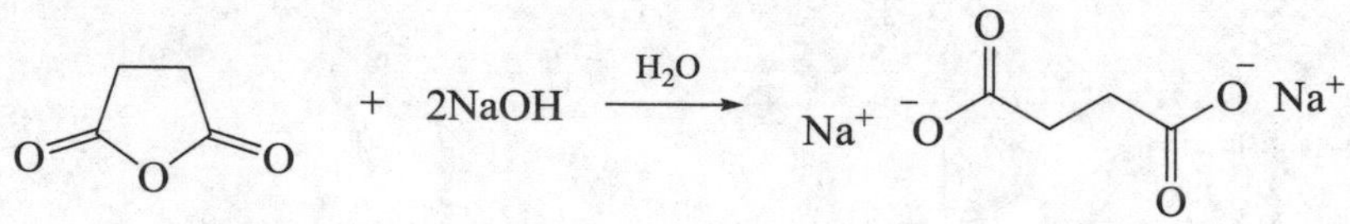

Butanedioic anhydride   Sodium   Sodium succinate
(succinic anhydride)   hydroxide

(f) One of the carbonyl groups of the cyclic anhydride is converted to an amide function on reaction with ammonia. The other, the one that would become a carboxylic acid group, is converted to an ammonium carboxylate salt.

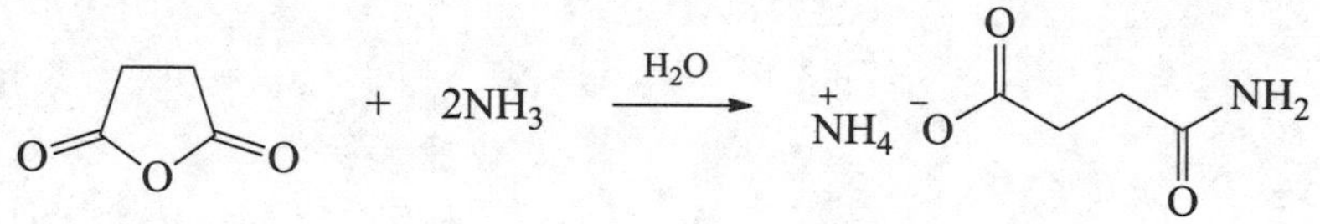

Succinic anhydride   Ammonia   Ammonium succinamate

(g) Esters react with excess Grignard reagents to give tertiary alcohols.

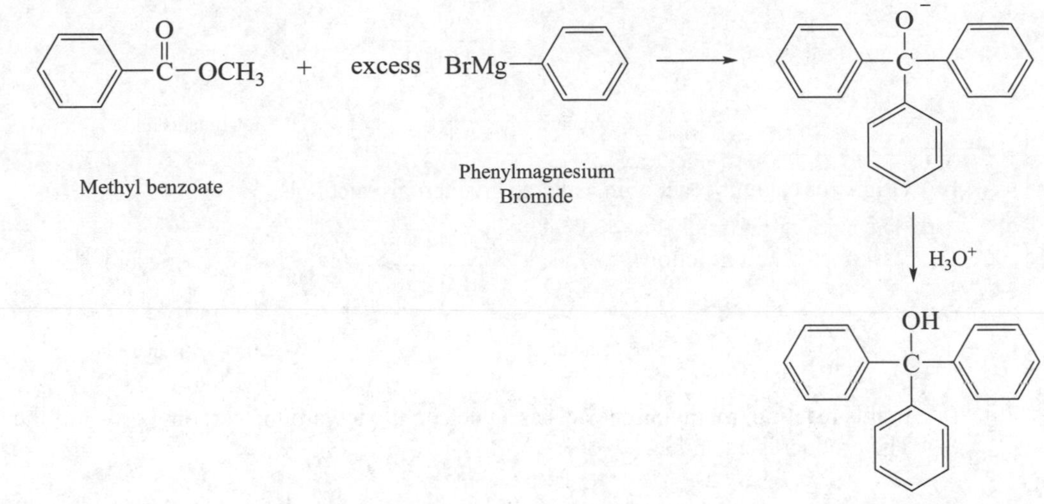

Methyl benzoate   +   excess   BrMg—⟨phenyl⟩   ⟶

Phenylmagnesium
Bromide

↓ H_3O^+

Triphenylmethanol

(*h*) Acid anhydrides react with alcohols to give an ester and a carboxylic acid.

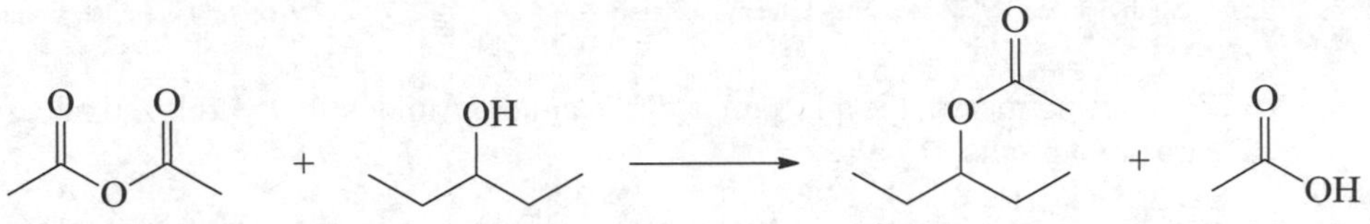

Acetic anhydride   +   3-Pentanol   ⟶   1-Ethylpropyl acetate   +   Acetic acid

(*i*) Esters are reduced by $LiAlH_4$ followed by acid to give alcohols.

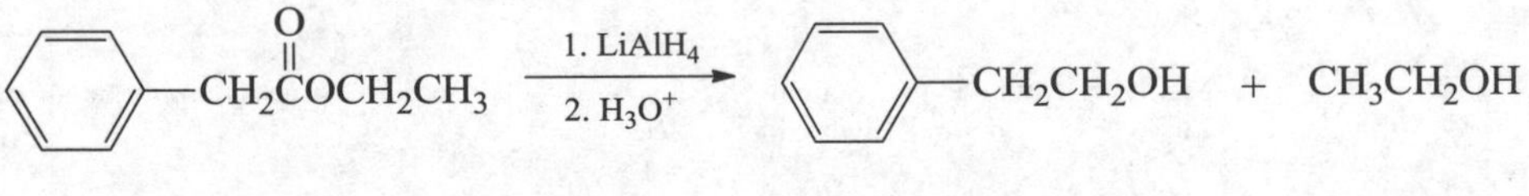

$—CH_2COCH_2CH_3$ $\xrightarrow[\text{2. } H_3O^+]{\text{1. } LiAlH_4}$ $—CH_2CH_2OH$ + CH_3CH_2OH

Ethyl phenylacetate   2-Phenylethanol   Ethanol

(*j*) The starting material is a cyclic ester, a lactone. Esters undergo saponification in aqueous base to give an alcohol and a carboxylate salt.

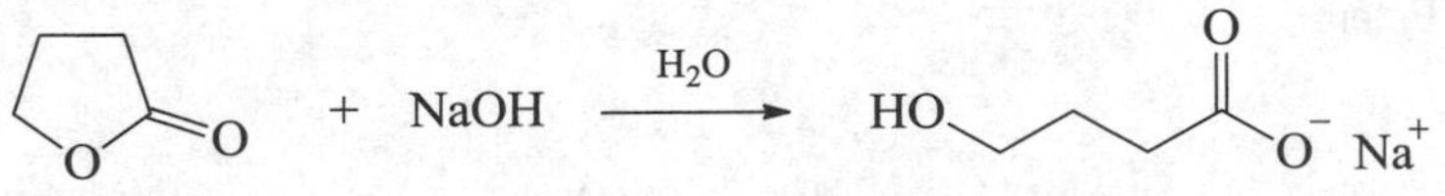

4-Butanolide   +   NaOH $\xrightarrow{H_2O}$ HO—⟶ $O^- Na^+$

Sodium
hydroxide

Sodium 4-hydroxybutanoate

(*k*) Ammonia reacts with esters to give an amide and an alcohol.

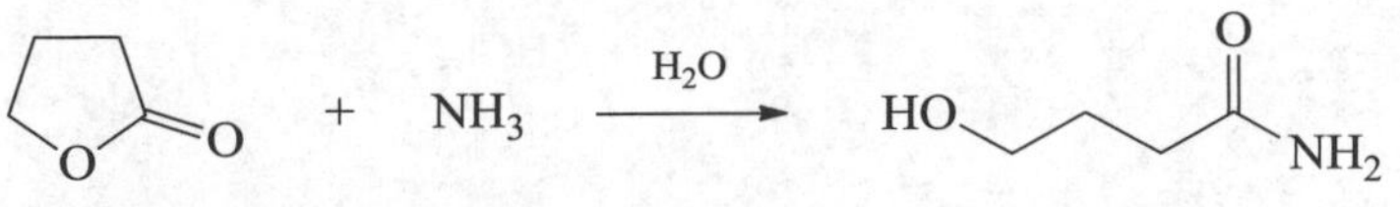

4-Butanolide   + NH_3 $\xrightarrow{H_2O}$ HO—⟶ NH_2

Ammonia

4-Hydroxybutanamide

(*l*) Lithium aluminum hydride reduces esters to two alcohols; the one derived from the acyl group is a primary alcohol. Reduction of a cyclic ester gives a diol.

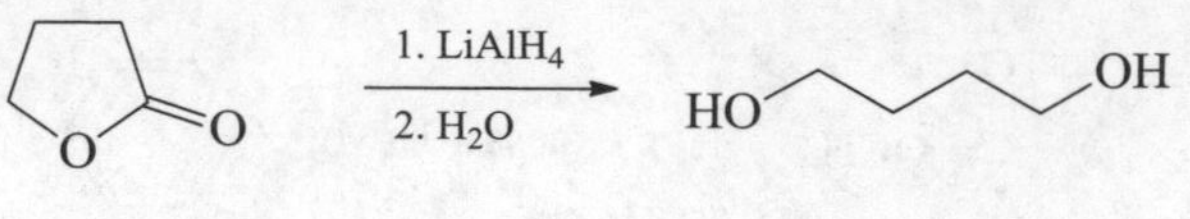

4-Butanolide → 1,4-Butanediol

(*m*) Grignard reagents react with esters to give tertiary alcohols.

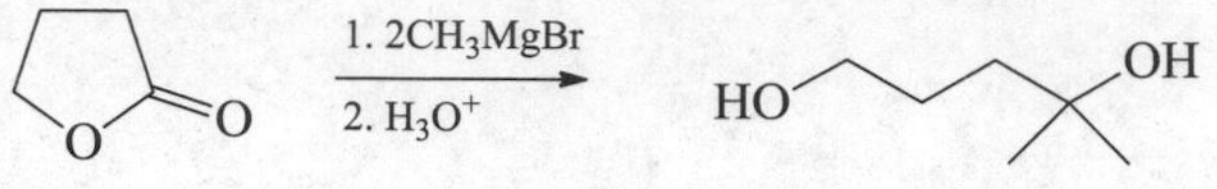

4-Butanolide → 4-Methyl-1,4-pentanediol

(*n*) In this reaction, methylamine acts as a nucleophile toward the carbonyl group of the ester. The product is an amide.

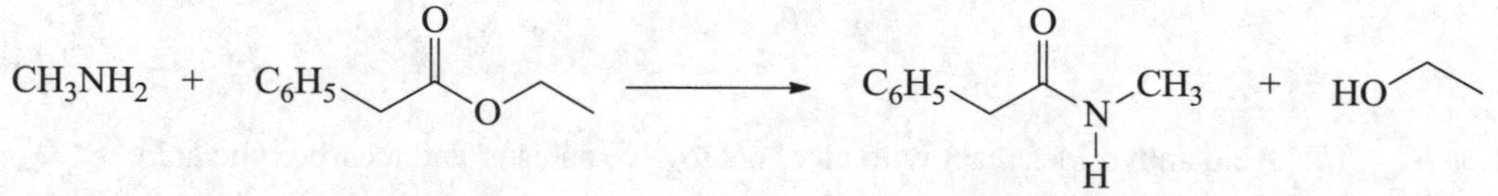

CH_3NH_2 + C_6H_5 (Ethyl phenylacetate) → N-Methylphenylacetamide + Ethyl alcohol

Methylamine Ethyl phenylacetate N-Methylphenylacetamide Ethyl alcohol

(*o*) The starting material is a lactam, a cyclic amide. Amides are hydrolyzed in base to amines and carboxylate salts.

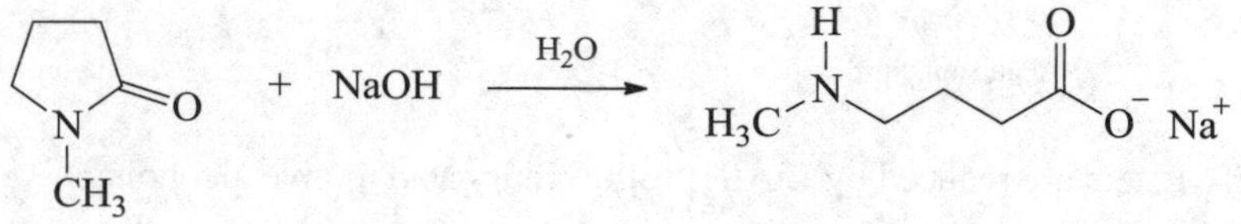

N-Methylpyrrolidone Sodium hydroxide → Sodium 4-(methylamino)butanoate

(*p*) In acid solution, amides yield carboxylic acids and ammonium salts.

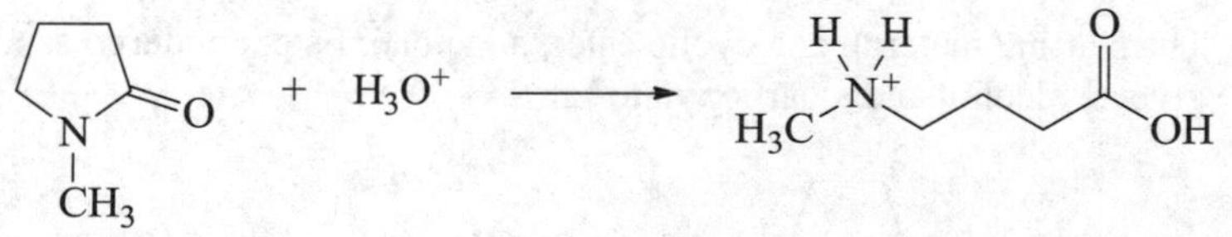

N-Methylpyrrolidone Hydronium ion → 4-(Methylammonio)butanoic acid

(*q*) Acetanilide is hydrolyzed in acid to acetic acid and the conjugate acid of aniline.

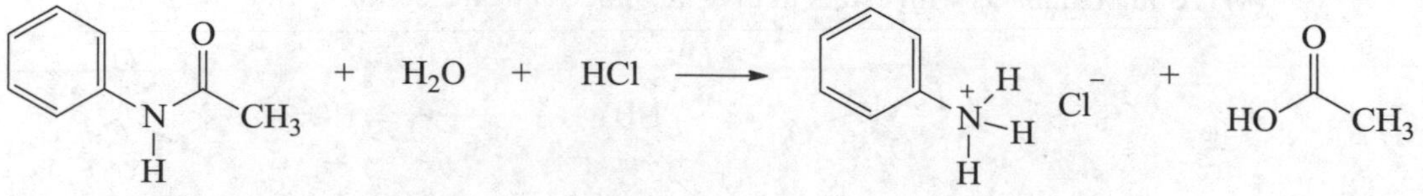

Acetanilide Water Hydrogen chloride → Anilinium chloride + Acetic acid

(*r*) This is another example of amide hydrolysis.

N-Methylbenzamide + H_2O + H_2SO_4 ⟶ Benzoic acid + $CH_3\overset{+}{N}H_3$ HSO_4^-

N-Methylbenzamide   Water   Sulfuric acid   Benzoic acid   Methylammonium
hydrogen sulfate

(s) One way to prepare nitriles is by dehydration of amides.

Cyclopentanecarboxamide $\xrightarrow{P_4O_{10}}$ Cyclopentyl cyanide + H_2O

Cyclopentanecarboxamide   Cyclopentyl cyanide   Water

(t) Nitriles are hydrolyzed to carboxylic acids in acidic media.

3-Methylbutanenitrile $\xrightarrow[\text{heat}]{\text{HCl, } H_2O}$ 3-Methylbutanoic acid

3-Methylbutanenitrile   3-Methylbutanoic acid

(u) Nitriles are hydrolyzed in aqueous base to salts of carboxylic acids.

p-Methoxybenzonitrile $\xrightarrow[\text{heat}]{\text{NaOH, } H_2O}$ Sodium p-methoxybenzoate + NH_3

p-Methoxybenzonitrile   Sodium p-methoxybenzoate   Ammonia

(v) Grignard reagents react with nitriles to yield ketones after addition of aqueous acid.

Propanenitrile $\xrightarrow[\text{2. } H_3O^+]{\text{1. } CH_3MgBr}$ 2-Butanone

Propanenitrile   2-Butanone

20.32 (a) By working through the sequence of reactions that occur when ethyl formate reacts with a Grignard reagent, we can see that this combination leads to *secondary alcohols*.

RMgX + Ethyl formate ⟶ Aldehyde + OMgX $\xrightarrow[\text{2. } H_3O^+]{\text{1. RMgX, diethyl ether}}$ Secondary alcohol

Grignard   Ethyl formate   Aldehyde   Secondary
reagent   alcohol

This is simply because the substituent on the carbonyl carbon of the ester, in this case a hydrogen, is carried through and becomes a substituent on the hydroxyl-bearing carbon of the alcohol.

(b) Diethyl carbonate has the potential to react with 3 moles of a Grignard reagent.

The tertiary alcohols that are formed by the reaction of diethyl carbonate with Grignard reagents have three identical R groups attached to the carbon that bears the hydroxyl substituent.

20.33 The enzyme-catalyzed hydrolysis of Penicillin G gives the following two compounds.

$C_8H_8O_2$ $C_8H_{12}N_2O_3S$

20.34 Compound A contains an amine, amide, and ester functional groups. The ester function is hydrolyzed in aqueous acid to give a carboxylic acid and an alcohol (ethanol). The amide is hydrolyzed in aqueous acid to give a carboxylic acid and an amine. This amine and the amine functional group present in compound A are protonated to give the hydrochloride salts.

Compound A

20.35 Acylation of the amino group is observed.

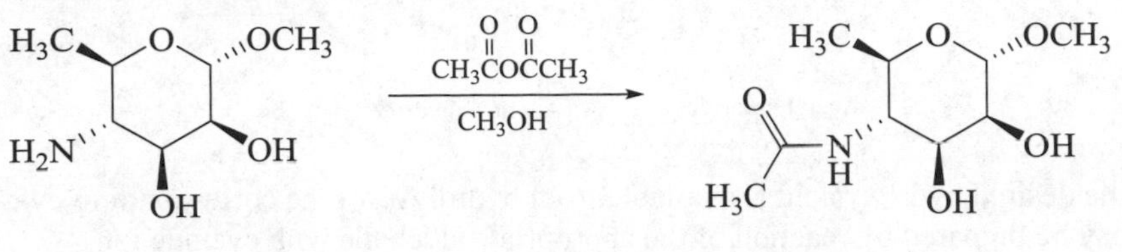

Compound A   Compound B

Synthesis

20.36 (*a*) Acetyl chloride is prepared by reaction of acetic acid with thionyl chloride. The first task then is to prepare acetic acid by oxidation of ethanol.

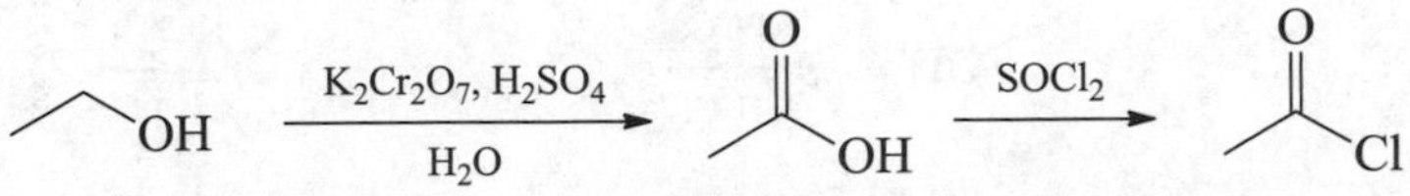

Ethanol   Acetic acid   Acetyl chloride

(*b*) Acetic acid and acetyl chloride, available from part (*a*), can be combined to form acetic anhydride.

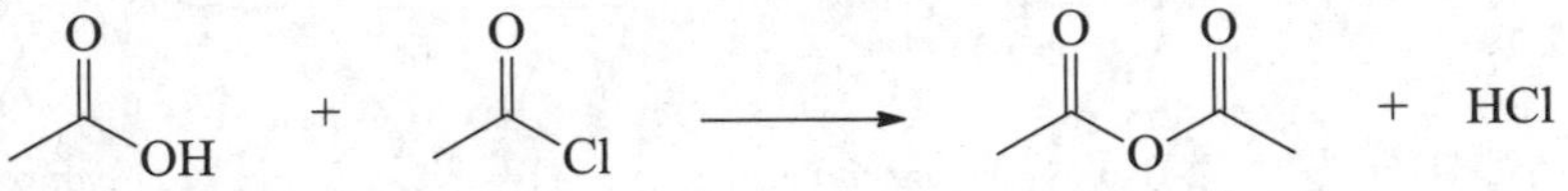

Acetic acid   Acetyl chloride   Acetic anhydride   Hydrogen chloride

(*c*) Ethanol can be converted to ethyl acetate by reaction with acetic acid, acetyl chloride, or acetic anhydride from parts (*a*) and (*b*).

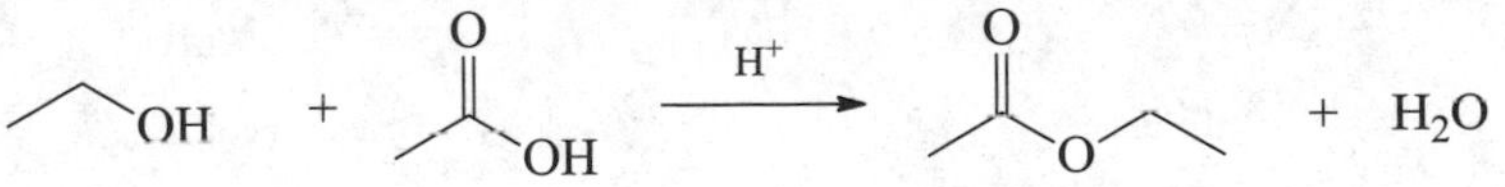

Ethanol   Acetic acid   Ethyl acetate   Water

or

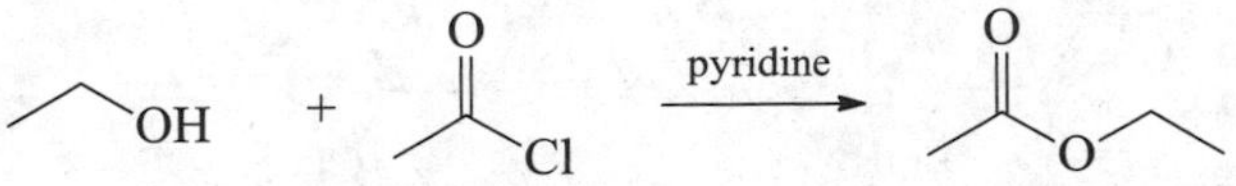

Ethanol   Acetyl chloride   Ethyl acetate

or

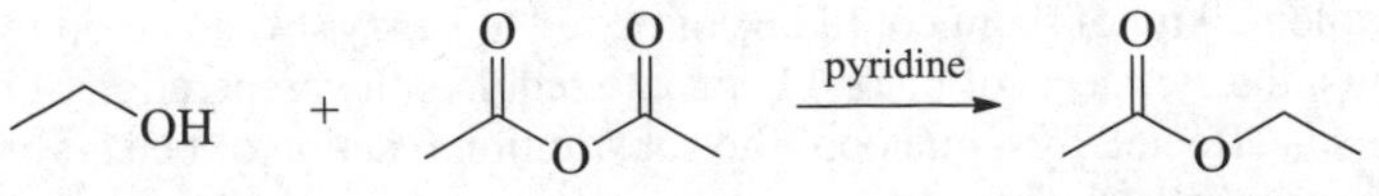

Ethanol   Acetic anhydride   Ethyl acetate

(*d*) Reaction of acetyl chloride, prepared in part (*a*), or acetic anhydride, from part (*b*), with ammonia gives acetamide.

Acetyl chloride   or   Acetic anhydride $\xrightarrow{NH_3}$ Acetamide

(e) The desired hydroxy acid is available from hydrolysis of the corresponding cyanohydrin, which may be prepared by reaction of the appropriate aldehyde with cyanide ion.

In this synthesis, the cyanohydrin is prepared from ethanol by way of acetaldehyde.

Ethanol $\xrightarrow[CH_2Cl_2]{PCC}$ Acetaldehyde $\xrightarrow[H^+]{KCN}$ 2-Hydroxypropanenitrile

2-Hydroxypropanenitrile

$$\xrightarrow[\substack{\text{or} \\ 1.\ HO^-,\ H_2O,\ heat \\ 2.\ H^+}]{H_2O,\ H^+,\ heat}$$

2-Hydroxypropanoeic acid

20.37 (a) Benzoyl chloride is made from benzoic acid. Oxidize toluene to benzoic acid, and then treat with thionyl chloride.

Toluene $\xrightarrow[H_2O,\ heat]{K_2Cr_2O_7,\ H_2SO_4}$ Benzoic acid $\xrightarrow{SOCl_2}$ Benzoyl chloride

(b) Benzoyl chloride and benzoic acid, both prepared from toluene in part (a), react with each other to give benzoic anhydride.

Benzoic acid   +   Benzoyl chloride $\longrightarrow$ Benzoic anhydride

(c) Benzoic acid, benzoyl chloride, and benzoic anhydride were prepared in parts (a) and (b) of this problem. Any of them could be converted to benzyl benzoate on reaction with benzyl alcohol. Thus, the synthesis of benzyl benzoate requires the preparation of benzyl alcohol from toluene. This is effected by a nucleophilic substitution reaction of benzyl bromide, in turn prepared by halogenation of toluene.

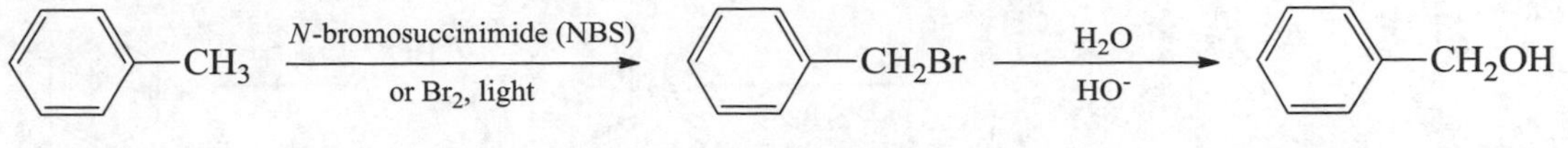

Toluene → Benzyl bromide → Benzyl alcohol

Alternatively, recall that primary alcohols may be obtained by reduction of the corresponding carboxylic acid.

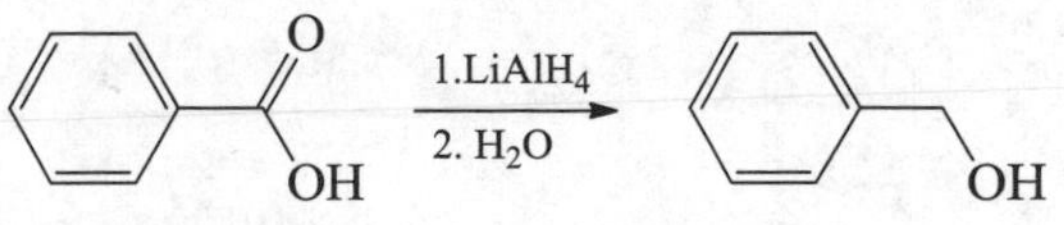

Benzoic acid → Benzyl alcohol

Then

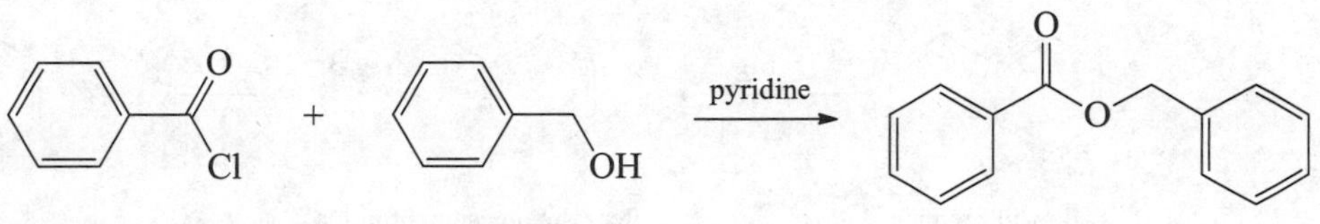

Benzoyl chloride + Benzyl alcohol → Benzyl benzoate

(d) Benzamide is prepared by reaction of ammonia with either benzoyl chloride from part (a) or benzoic anhydride from part (b).

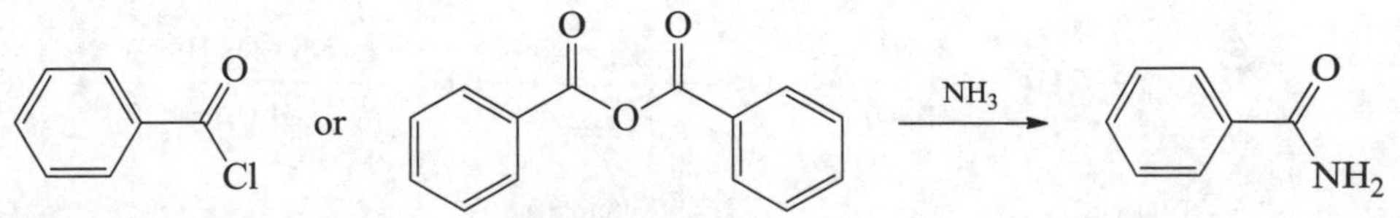

Benzoyl chloride   or   Benzoic anhydride → Benzamide

(e) Benzonitrile may be prepared by dehydration of benzamide.

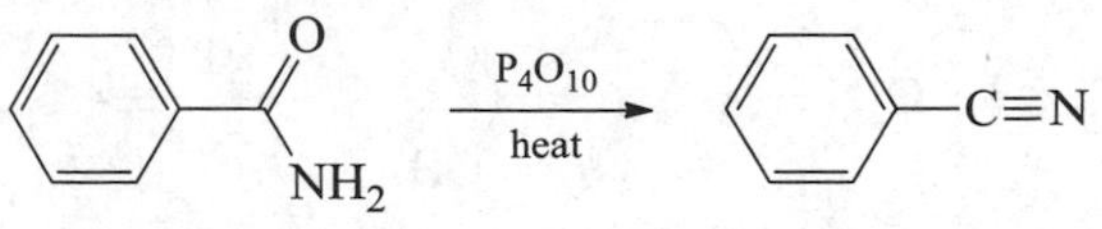

Benzamide → Benzonitrile

(f) Benzyl cyanide is the product of nucleophilic substitution by cyanide ion on benzyl bromide or benzyl chloride. The benzyl halides are prepared by free-radical halogenation of the toluene side chain.

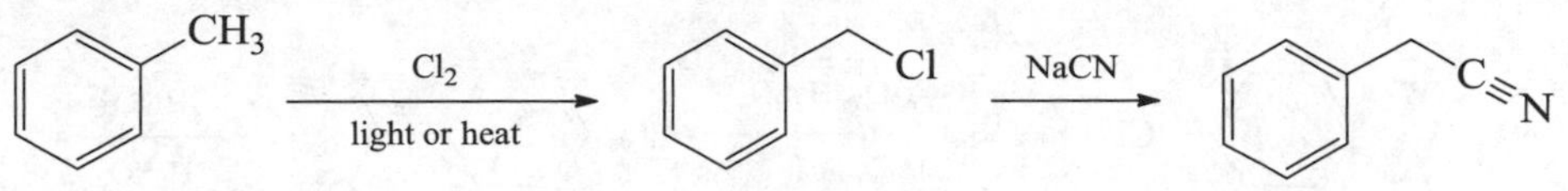

Toluene → Benzyl chloride → Benzyl cyanide

or

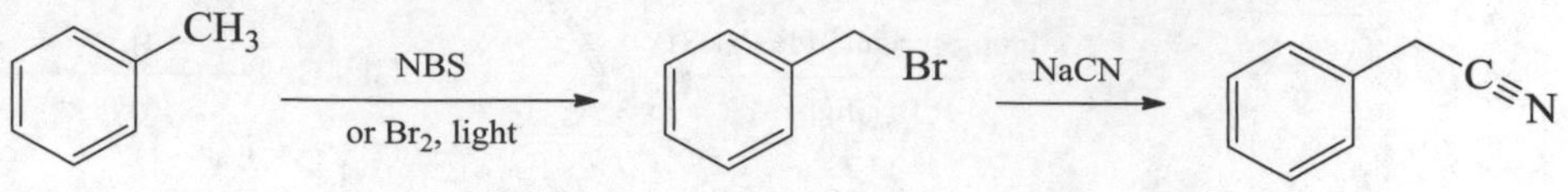

Toluene → Benzyl bromide → Benzyl cyanide

(g) Hydrolysis of benzyl cyanide yields phenylacetic acid.

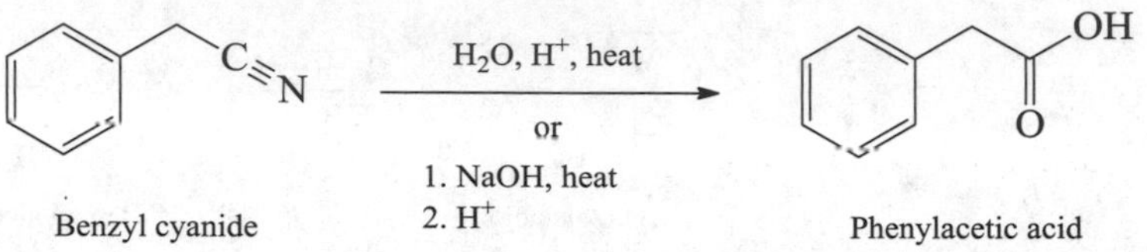

Benzyl cyanide → Phenylacetic acid

Alternatively, the Grignard reagent derived from benzyl bromide may be carboxylated.

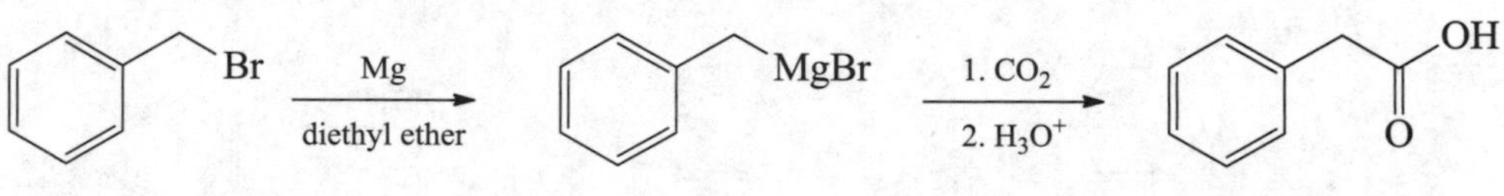

Benzyl bromide → Benzylmagnesium bromide → Phenylacetic acid

(h) The first goal is to synthesize *p*-nitrobenzoic acid because this may be readily converted to the desired acyl chloride. First convert toluene to *p*-nitrotoluene; then oxidize. Nitration must precede oxidation of the side chain in order to achieve the desired para orientation.

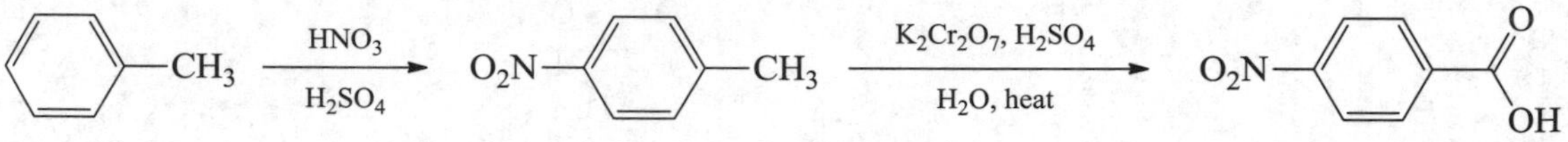

Toluene → *p*-Nitrotoluene (separate from ortho isomer) → *p*-Nitrobenzoic acid

Treatment of *p*-nitrobenzoic acid with thionyl chloride yields *p*-nitrobenzoyl chloride.

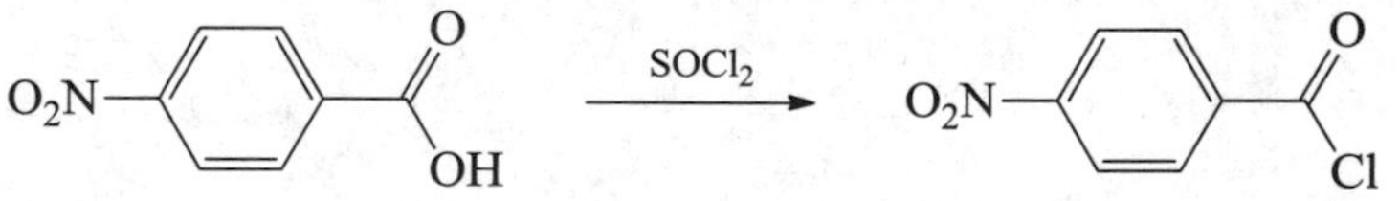

p-Nitrobenzoic acid → *p*-Nitrobenzoyl chloride

(i) To achieve the correct orientation in *m*-nitrobenzoyl chloride, oxidation of the methyl group must precede nitration.

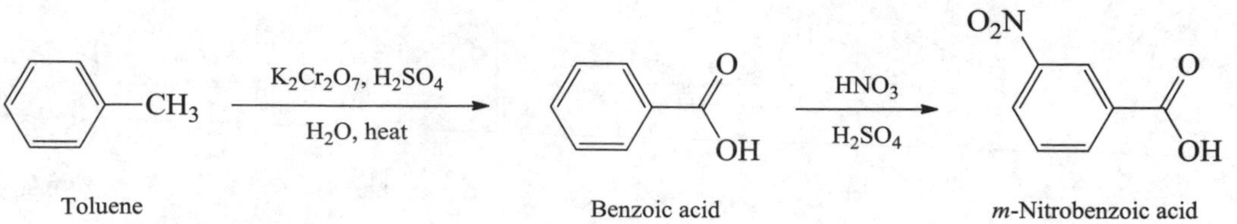

Toluene → Benzoic acid → *m*-Nitrobenzoic acid

Once *m*-nitrobenzoic acid has been prepared, it may be converted to the corresponding acyl chloride.

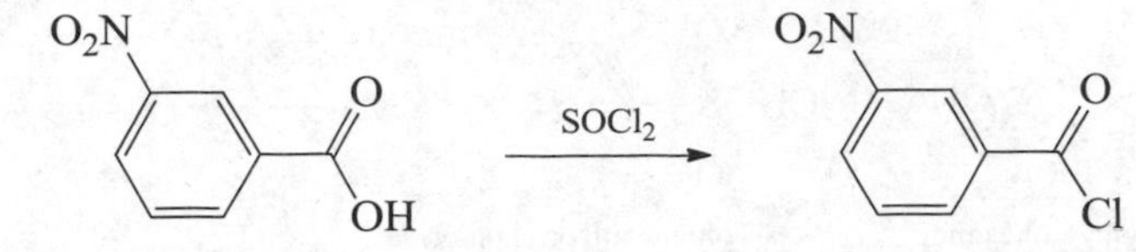

m-Nitrobenzoic acid   m-Nitrobenzoyl chloride

20.38 The synthetic format for the retrosynthesis starts with a Diels-Alder reaction, followed by reaction with 2 equivalents of methylmagnesium bromide to give the alcohol.

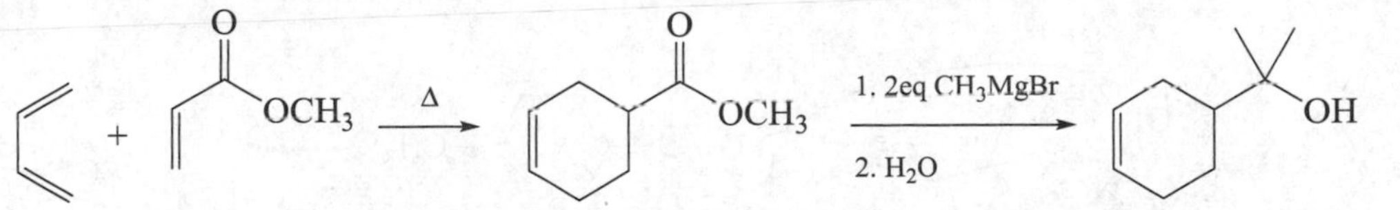

This is followed by a dehydration reaction to give the desired diene.

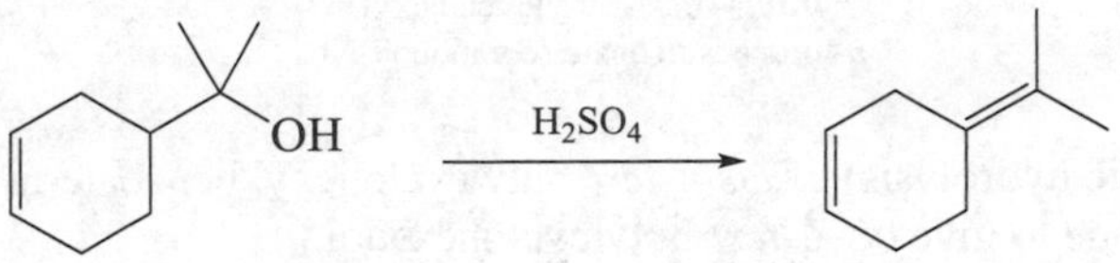

20.39 The problem specifies that $CH_3CH_2\overset{O}{\overset{\|}{C}}OCH_2CH_3$ is to be prepared from ^{18}O-labeled ethyl alcohol ($\bullet = {}^{18}O$).

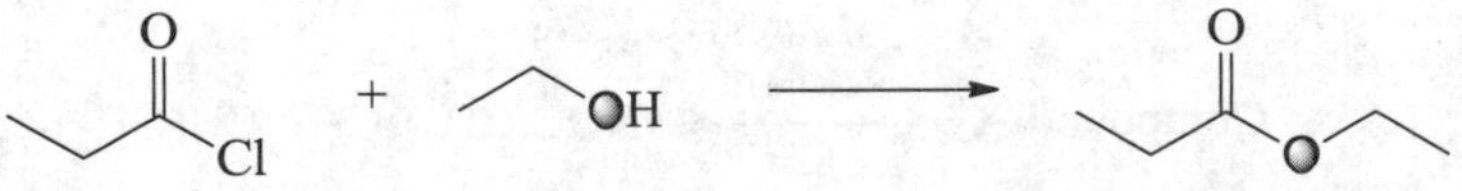

Propanoyl chloride   Ethyl alcohol   Ethyl propanoate

Thus, we need to prepare ^{18}O-labeled ethyl alcohol from the other designated starting materials, acetaldehyde and ^{18}O-enriched water. First, replace the oxygen of acetaldehyde with ^{18}O by the hydration–dehydration equilibrium in the presence of ^{18}O-enriched water.

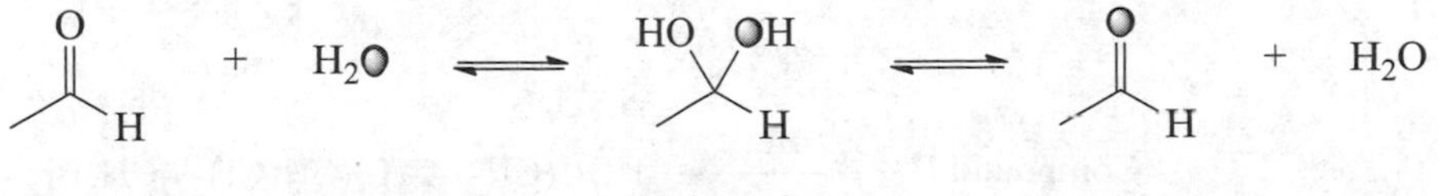

Acetaldehyde ^{18}O-enriched   Hydrate of ^{18}O-enriched   Water
                  water   acetaldehyde   acetaldehyde

Once ^{18}O-enriched acetaldehyde has been obtained, it can be reduced to ^{18}O-enriched ethanol.

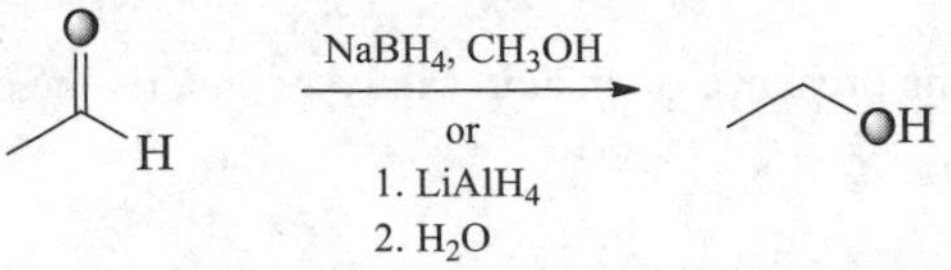

20.40 Compound A is the p-toluenesulfonate (tosylate) of trans-4-tert-butylcyclohexanol. The oxygen atom of the alcohol attacks the sulfur of p-toluenesulfonyl chloride, and so the reaction proceeds with retention of configuration.

trans-4-*tert*-Butylcyclohexanol   *p*-Toluenesulfonyl chloride   *trans*-4-*tert*-Butylcyclohexyl *p*-toluenesulfonate (compound A)

The second step is a nucleophilic substitution in which benzoate ion displaces *p*-toluenesulfonate with inversion of configuration.

Benzoate ion   *trans*-4-*tert*-Butylcyclohexyl *p*-toluenesulfonate (compound A)   *cis*-4-*tert*-Butylcyclohexyl benzoate (compound B)

Saponification (basic hydrolysis) of *cis*-4-*tert*-butylcyclohexyl benzoate in step 3 proceeds with acyl–oxygen cleavage to give *cis*-4-*tert*-butylcyclohexanol.

20.41 The first step is acid hydrolysis of an acetal protecting group.

Step 1:

$$\text{Compound A} \xrightarrow[\text{heat}]{\text{H}_2\text{O, H}^+} \underset{\substack{\quad\quad\quad\;\;| \quad\; | \\ \quad\quad\quad\;\text{HO} \quad \text{OH}}}{\overset{\overset{\displaystyle O}{\|}}{\text{HOC}}(\text{CH}_2)_5\text{CH}-\text{CH}(\text{CH}_2)_7\text{CH}_2\text{OH}}$$

Compound B
($\text{C}_{16}\text{H}_{32}\text{O}_5$)

All three alcohol functions are converted to bromide by reaction with hydrogen bromide in step 2.

Step 2:

$$\text{Compound B} \xrightarrow{\text{HBr}} \underset{\substack{\quad\quad\quad\;\;| \quad\;\; | \\ \quad\quad\quad\;\text{Br} \quad \text{Br}}}{\overset{\overset{\displaystyle O}{\|}}{\text{HOC}}(\text{CH}_2)_5\text{CH}-\text{CH}(\text{CH}_2)_7\text{CH}_2\text{Br}}$$

Compound C
($\text{C}_{16}\text{H}_{29}\text{Br}_3\text{O}_2$)

Reaction with ethanol in the presence of an acid catalyst converts the carboxylic acid to its ethyl ester in step 3.

Step 3:

$$\text{Compound C} \xrightarrow[\text{H}_2\text{SO}_4]{\text{ethanol}} \underset{\substack{\quad\quad\quad\quad\quad\quad\;\;| \quad\;\; | \\ \quad\quad\quad\quad\quad\quad\;\text{Br} \quad \text{Br}}}{\text{CH}_3\text{CH}_2\text{O}\overset{\overset{\displaystyle O}{\|}}{\text{C}}(\text{CH}_2)_5\text{CH}-\text{CH}(\text{CH}_2)_7\text{CH}_2\text{Br}}$$

Compound D
($\text{C}_{18}\text{H}_{33}\text{Br}_3\text{O}_2$)

The problem hint points out that zinc converts vicinal dibromides to alkenes. Of the three bromine substituents in compound D, two of them are vicinal. Step 4 is a dehalogenation reaction.

Step 4:

$$\text{Compound D} \xrightarrow[\text{ethanol}]{\text{Zn}} \text{CH}_3\text{CH}_2\text{OC(CH}_2)_5\text{CH}=\text{CH(CH}_2)_7\text{CH}_2\text{Br}$$

Compound E
($C_{18}H_{33}BrO_2$)

Step 5 is a nucleophilic substitution of the S_N2 type. Acetate ion is the nucleophile and displaces bromide from the primary carbon.

Step 5:

$$\text{Compound E} \xrightarrow[\text{CH}_3\text{CO}_2\text{H}]{\text{NaOCCH}_3} \text{CH}_3\text{CH}_2\text{OC(CH}_2)_5\text{CH}=\text{CH(CH}_2)_7\text{CH}_2\text{OCCH}_3$$

Compound F
($C_{20}H_{36}O_4$)

Step 6 is ester hydrolysis. It yields a 16-carbon chain having a carboxylic acid function at one end and an alcohol at the other.

Step 6:

$$\text{Compound F} \xrightarrow[\text{2. H}^+]{\text{1. KOH, ethanol}} \text{HOC(CH}_2)_5\text{CH}=\text{CH(CH}_2)_7\text{CH}_2\text{OH}$$

Compound G
($C_{16}H_{30}O_3$)

In step 7, compound G cyclizes to ambrettolide on heating.

Step 7:

Compound G   Ambrettolide

20.42 Reaction of ethyl trifluoroacetate with ammonia yields the corresponding amide, compound A. Compound A undergoes dehydration on heating with P_4O_{10} to give trifluoroacetonitrile, compound B. Grignard reagents react with nitriles to form ketones. *tert*-Butyl trifluoromethyl ketone is formed from trifluoroacetonitrile by treatment with *tert*-butylmagnesium chloride followed by aqueous hydrolysis.

$$\underset{\substack{\text{Ethyl trifluoroacetate}}}{\text{CF}_3\text{COCH}_2\text{CH}_3} \xrightarrow{\text{NH}_3} \underset{\substack{\text{Trifluoroacetamide}\\\text{(Compound A)}}}{\text{CF}_3\text{CNH}_2} \xrightarrow[\text{heat}]{\text{P}_4\text{O}_{10}} \underset{\substack{\text{Trifluoroacetonitrile}\\\text{(Compound B)}}}{\text{CF}_3\text{C}\equiv\text{N}}$$

$$CF_3C \equiv N \quad + \quad (CH_3)_3CMgCl \quad \xrightarrow[\text{2. } H_3O^+]{\text{1. diethyl ether}} \quad CF_3\overset{O}{\overset{\|}{C}}C(CH_3)_3$$

Compound B   *tert*-Butylmagnesium   *tert*-Butyl trifluoromethyl
                              chloride   ketone

20.43 (*a*) This step requires the oxidation of a primary alcohol to an aldehyde. As reported in the literature, pyridinium dichromate in dichloromethane was used to give the desired aldehyde in 84% yield.

$$HOCH_2CH = CH(CH_2)_7CO_2CH_3 \quad \xrightarrow[\text{CH}_2\text{Cl}_2]{\text{PDC}} \quad H\overset{O}{\overset{\|}{C}}CH = CH(CH_2)_7CO_2CH_3$$

Compound A (*E* isomer)   Compound B

(*b*) Conversion of $-\overset{O}{\overset{\|}{C}}H$ to $-CH=CH_2$ is a typical case in which a Wittig reaction is appropriate.

$$H\overset{O}{\overset{\|}{C}}CH = CH(CH_2)_7CO_2CH_3 \quad \xrightarrow{(C_6H_5)_3\overset{+}{P} - \overset{..}{C}H_2} \quad H_2C = CHCH = CH(CH_2)_7CO_2CH_3$$

Compound B   Compound C
                                      (observed yield, 53%)

(*c*) Lithium aluminum hydride was used to reduce the ester to a primary alcohol in 81% yield.

$$H_2C = CHCH = CH(CH_2)_7CO_2CH_3 \quad \xrightarrow[\text{2. } H_2O]{\text{1. LiAlH}_4} \quad H_2C = CHCH = CH(CH_2)_7CH_2OH$$

Compound C   Compound D

(*d*) The desired sex pheromone is the acetate ester of compound D. Compound D was treated with acetic anhydride to give the acetate ester in 99% yield.

$$H_2C = CHCH = CH(CH_2)_7CH_2OH \quad \xrightarrow[\text{pyridine}]{CH_3\overset{O}{\overset{\|}{C}}O\overset{O}{\overset{\|}{C}}CH_3} \quad H_2C = CHCH = CH(CH_2)_7CH_2O\overset{O}{\overset{\|}{C}}CH_3$$

Compound D   (*E*)-9,11-Dodecadien-1-yl acetate

Acetyl chloride could have been used in this step instead of acetic anhydride.

Kinetics and Mechanism

20.44 Incorporation of ^{18}O into the *tert*-butyl acetate product suggests that a carbocation was involved. A reasonable mechanism begins with protonation of the ester carbonyl ($\bullet = {}^{18}O$).

tert-Butyl acetate ^{18}O-labeled   Conjugate acid of ^{18}O-labeled
                              hydronium ion   *tert*-butyl acetate   water

Protonation of the ester carbonyl generates a good leaving group on the tertiary carbon and sets the stage for an S_N1 mechanism.

Conjugate acid of *tert*-butyl acetate   *tert*-Butyl cation   Acetic acid

tert-Butyl cation ^{18}O-labeled water   *tert*-Butyloxonium ion ^{18}O-labeled *tert*-butyl alcohol

20.45 (*a*) The rate-determining step in basic ester hydrolysis is nucleophilic addition of hydroxide ion to the carbonyl group. The intermediate formed in this step is negatively charged.

Ethyl acetate   Hydroxide ion   Rate-determining intermediate

The electron-withdrawing effect of a CF_3 group stabilizes the intermediate formed in the rate-determining step of ethyl trifluoroacetate saponification.

Ethyl trifluoroacetate   Hydroxide ion   Rate-determining intermediate

Because the intermediate is more stable, it is formed faster than the one from ethyl acetate.

(*b*) Crowding is increased as the transition state for nucleophilic addition to the carbonyl group is approached. The carbonyl carbon undergoes a change in hybridization from sp^2 to sp^3.

Ethyl 2,2-dimethylpropanoate   Hydroxide ion   Rate-determining intermediate; crowded

The *tert*-butyl group of ethyl 2,2-dimethylpropanoate causes more crowding than the methyl group of ethyl acetate; the tetrahedral intermediate is less stable and is formed more slowly.

(*c*) We see here another example of a steric effect of a *tert*-butyl group. The tetrahedral intermediate formed when hydroxide ion adds to the carbonyl group of *tert*-butyl acetate is more crowded and less stable than the tetrahedral intermediate formed from methyl acetate.

tert-Butyl acetate + HO⁻ → Rate-determining intermediate; more crowded

| *tert*-Butyl acetate | Hydroxide ion | | Rate-determining intermediate; more crowded |

Methyl acetate + HO⁻ → Rate-determining intermediate; less crowded than intermediate from *tert*-butyl acetate

(*d*) Here, as in part (*a*), we have an electron-withdrawing substituent increasing the rate of ester saponification. It does so by stabilizing the negatively charged intermediate formed in the rate-determining step.

more stable than

Rate-determining intermediate from methyl *m*-nitrobenzoate

Rate-determining intermediate from methyl benzoate

(*e*) Addition of hydroxide to 4-butanolide introduces torsional strain in the intermediate because of eclipsed bonds. The corresponding intermediate from 5-butanolide is more stable because the bonds are staggered in a six-membered ring.

Eclipsed bonds

Staggered bonds

Less stable; formed more slowly

More stable; formed faster

(*f*) Steric crowding increases more when hydroxide adds to the axial carbonyl group.

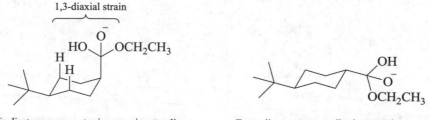

Cis diasteromer: greater increase in crowding when carbon changes from *sp*² to *sp*³; formed more slowly

Trans diasteromer: smaller increase in crowding when carbon changes from *sp*² to *sp*³; formed more rapidly

20.46 (*a*) The reaction given in the problem is between a lactone (cyclic ester) and a difunctional Grignard reagent. Esters usually react with 2 moles of a Grignard reagent; in this instance both Grignard functions of the reagent attack the lactone. The second attack is intramolecular, giving rise to the cyclopentanol ring of the product.

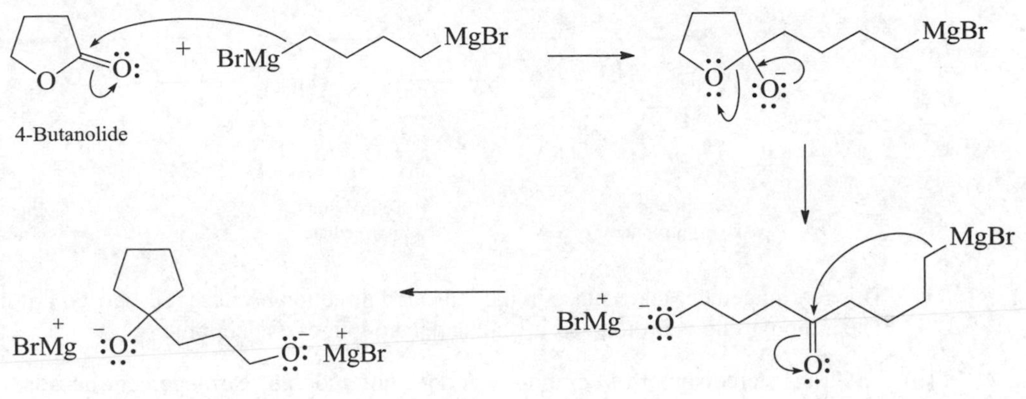

4-Butanolide

Protonation of both oxygens completes the reaction.

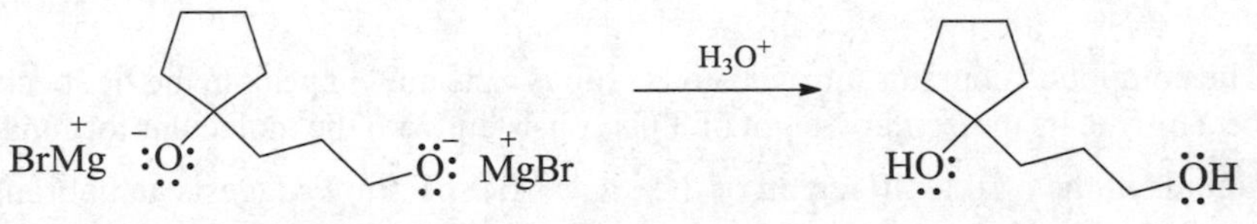

1-(3-Hydroxypropyl-cyclopentanol (88%)

(*b*) An intramolecular acyl transfer process takes place in this reaction. The amine group in the thiolactone starting material replaces sulfur on the acyl group to form a lactam (cyclic amide).

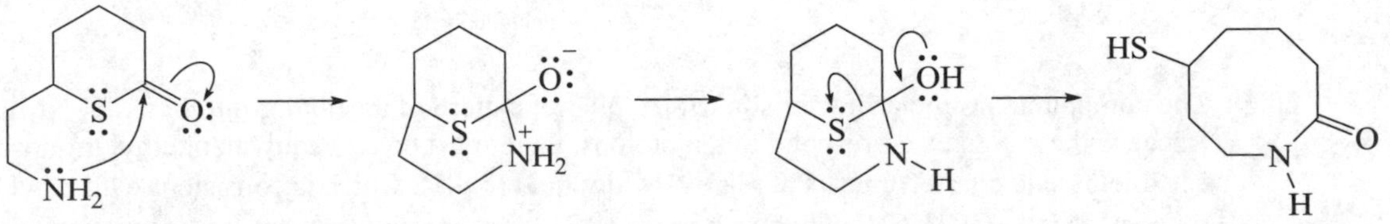

A thiolactone   Tetrahedral intermediate   A lactam

20.47 Compound A is an ester but has within it an amine function. Acyl transfer from oxygen to nitrogen converts the ester to a more stable amide, compound B.

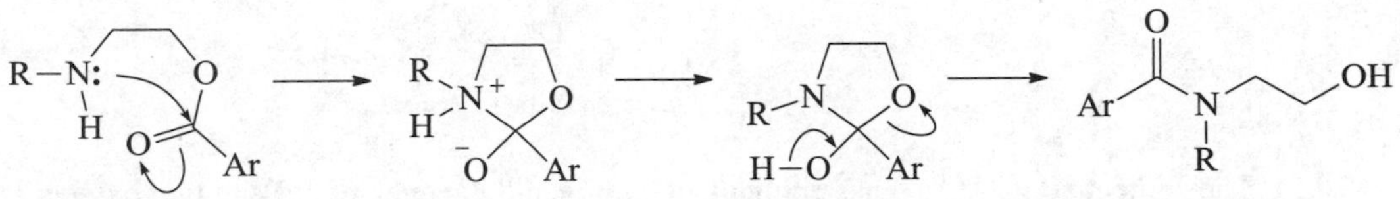

Compound A   Tetrahedral   Compound B
(Ar = *p*-nitrophenyl)   intermediate   (Ar = *p*-nitrophenyl)

The tetrahedral intermediate is the key intermediate in the reaction.

20.48 (*a*) The rearrangement in this problem is an acyl transfer from nitrogen to oxygen.

Compound A — Tetrahedral — Compound B
(Ar = *p*-nitrophenyl) — intermediate — (Ar = *p*-nitrophenyl)

This rearrangement takes place in the indicated direction because it is carried out in acid solution. The amino group is protonated in acid and is no longer nucleophilic.

(*b*) The trans stereoisomer of compound A does not undergo rearrangement because when the oxygen and nitrogen atoms on the five-membered ring are trans, the necessary tetrahedral intermediate cannot form.

Spectroscopy

20.49 The compound contains nitrogen and exhibits a prominent peak in the IR spectrum at 2270 cm^{-1}; it is likely to be a nitrile. Its molecular weight of 83 is consistent with the molecular formula C_5H_9N. The presence of four signals in the δ 10 to 30 region of the ^{13}C NMR spectrum suggests an unbranched carbon skeleton. This is confirmed by the presence of two triplets in the ^{1}H NMR spectrum at δ 1.0 (CH_3 coupled with adjacent CH_2) and at δ 2.3 (CH_2CN coupled with adjacent CH_2). The compound is pentanenitrile.

Pentanenitrile

20.50 The compound has the characteristic triplet–quartet pattern of an ethyl group in its ^{1}H NMR spectrum. Because these signals correspond to ten protons, there must be two equivalent ethyl groups in the molecule. The methylene quartet appears at relatively low field (δ 4.1), which is consistent with ethyl groups bonded to oxygen, as in —OCH_2CH_3. There is a peak at 1730 cm^{-1} in the IR spectrum, suggesting that these ethoxy groups reside in ester functions. The molecular formula $C_8H_{14}O_4$ reveals that if two ester groups are present, there can be no rings or double bonds. The remaining four hydrogens are equivalent in the ^{1}H NMR spectrum, and so two equivalent CH_2 groups are present. The compound is the diethyl ester of succinic acid.

Diethyl succinate

20.51 The compound ($C_4H_6O_2$) has an index of hydrogen deficiency of 2. With two oxygen atoms and a peak in the infrared at 1760 cm^{-1}, it is likely that one of the elements of unsaturation is the carbon–oxygen double bond of an ester. The ^{1}H NMR spectrum contains a three-proton singlet at δ 2.1, which is consistent with a CH_3C=O unit. It is likely that the compound is an acetate ester.

The ^{13}C NMR spectrum reveals that the four carbon atoms of the molecule are contained in one each of the fragments CH_3, CH_2, and CH, along with the carbonyl carbon. In addition to the two carbons of the acetate group, the remaining two carbons are the CH_2 and CH carbons of a vinyl group, CH=CH_2. The compound is vinyl acetate.

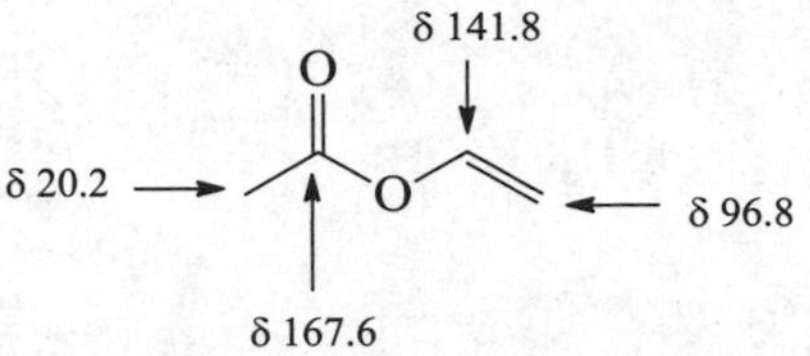

$$\delta\ 20.2 \longrightarrow \underset{\underset{\delta\ 167.6}{\uparrow}}{\overset{\overset{O}{\|}}{C}} \underset{}{O} \overset{\delta\ 141.8}{\underset{}{}} \longleftarrow \delta\ 96.8$$

Each vinyl proton is coupled to two other vinyl protons; each appears as a doublet of doublets in the ^1H NMR spectrum.

Answers to Interpretive Problems 20

20.52 C; **20.53** D; **20.54** B; **20.55** A; **20.56** A; **20.57** B; **20.58** D

SELF-TEST

1. Give the IUPAC name for each of the following acid derivatives:

(a)

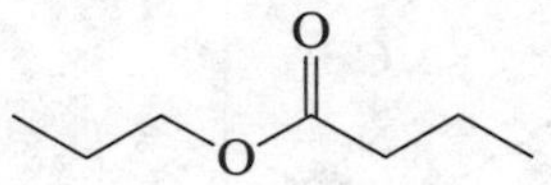

(c)

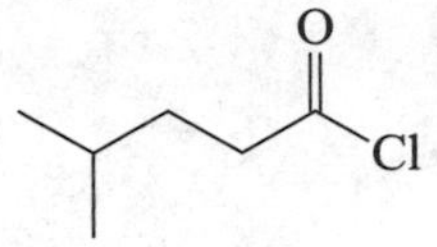

(b)

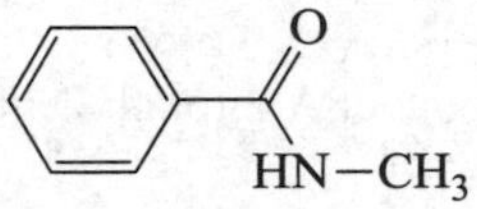

2. Provide the structure of

   (a) Benzoic anhydride

   (b) *N*-(1-Methylpropyl)acetamide

   (c) Phenyl benzoate

3. What reagents are needed to carry out each of the following conversions?

(a)

$$C_6H_5CH_2CO_2H \xrightarrow{\ ?\ } C_6H_5CH_2\overset{\overset{O}{\|}}{C}Cl$$

(b)

$$(CH_3)_2CHCH_2NH_2 \xrightarrow{\ ?\ } C_6H_5\overset{\overset{O}{\|}}{C}NHCH_2CH(CH_3)_2 \ + \ CH_3OH$$

(c)

$$C_6H_5CH_2\overset{\overset{O}{\|}}{C}OCH_3 \xrightarrow[2.\ H_3O^+]{1.\ ?} (CH_3)_2CH\underset{\underset{CH_2C_6H_5}{|}}{\overset{\overset{OH}{|}}{C}}CH(CH_3)_2 \ + \ CH_3OH$$

4. Write the structure of the product of each of the following reactions:

(a)

$$\text{Cyclohexyl acetate} \xrightarrow[2.\ H^+]{1.\ NaOH,\ H_2O} ?\ \text{(two products)}$$

(b)

$$\text{Cyclopentanol} \ + \ \text{benzoyl chloride} \xrightarrow{\text{pyridine}} ?$$

(c)

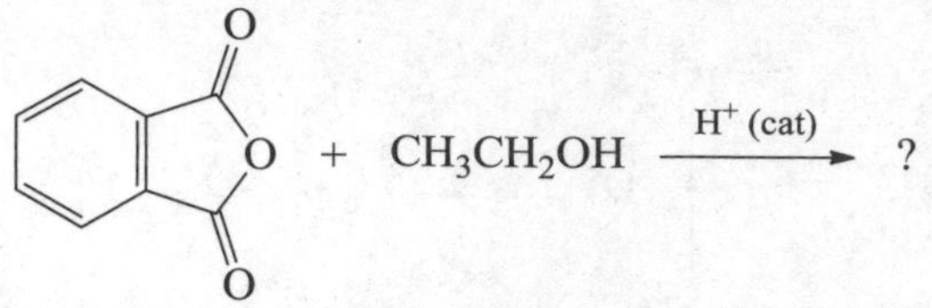

$+\ CH_3CH_2OH\ \xrightarrow{H^+\ (cat)}\ ?$

(d)   Ethyl propanoate + dimethylamine ⟶ ? (two products)

(e)

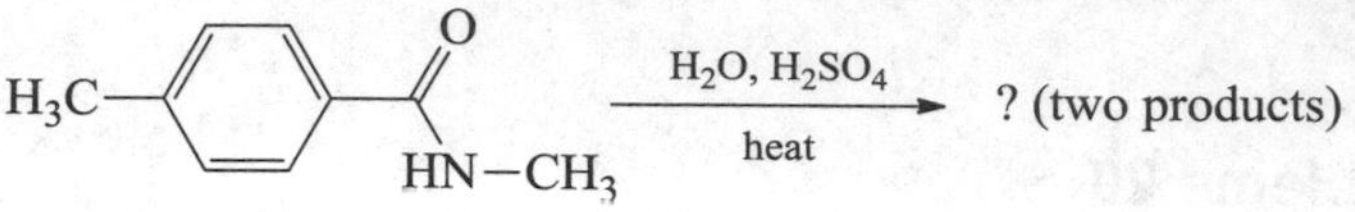

$\xrightarrow[\text{heat}]{H_2O,\ H_2SO_4}$? (two products)

(f)

$\xrightarrow[\text{2. H}_3\text{O}^+]{\text{1. 2C}_6\text{H}_5\text{MgBr}}$?

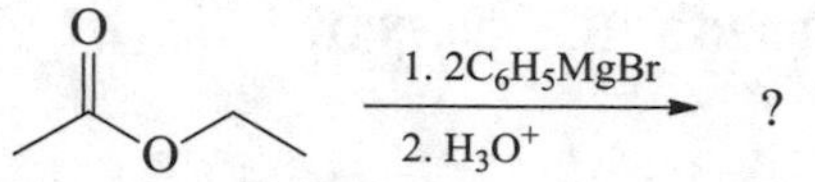

5. The following reaction occurs when the reactant is allowed to stand in pentane. Write the structure of the key neutral uncharged intermediate in this process.

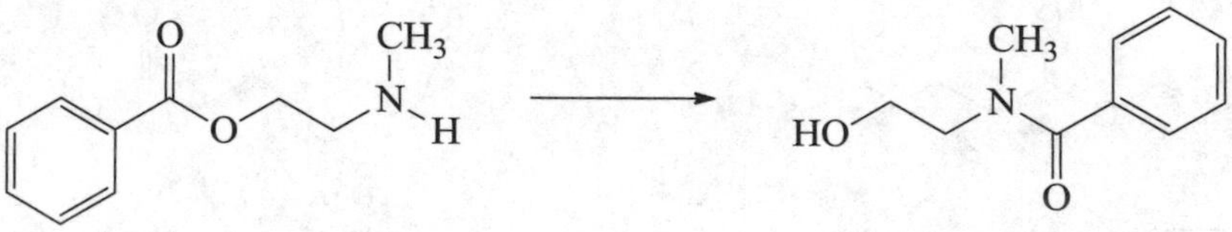

6. Give the structures, clearly showing stereochemistry, of each compound, A through D, in the following sequence of reactions:

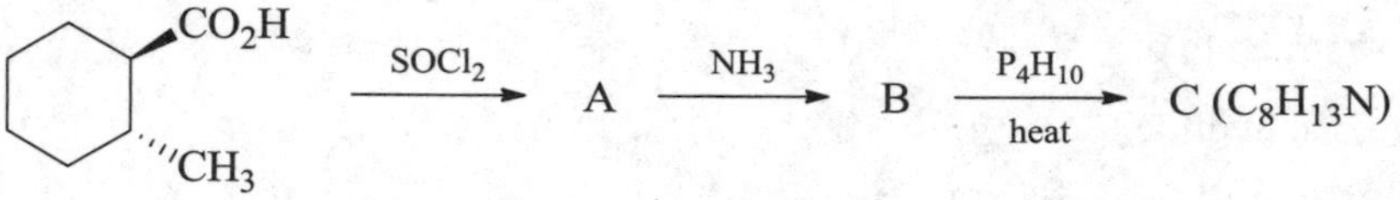

$\xrightarrow{SOCl_2}$ A $\xrightarrow{NH_3}$ B $\xrightarrow[\text{heat}]{P_4H_{10}}$ C $(C_8H_{13}N)$

7. Write the structure of the neutral uncharged form of the tetrahedral intermediate in the

   (a) Acid-catalyzed hydrolysis of methyl acetate

   (b) Reaction of ammonia with acetic anhydride

8. Write the steps necessary to prepare

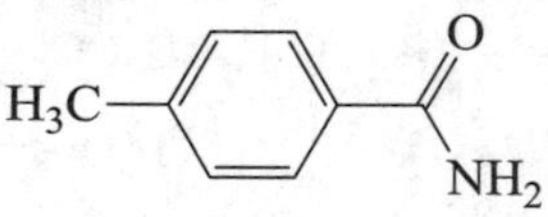

from

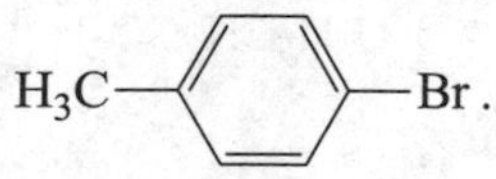

.

9. Outline a synthesis of benzyl benzoate using toluene as the source of all the carbon atoms.

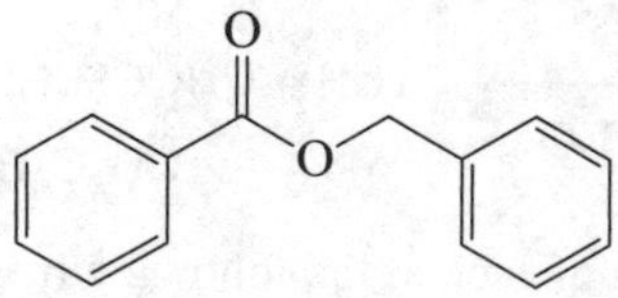

Benzyl benzoate

10. The IR spectrum of a compound (C_3H_6ClNO) has an intense peak at 1680 cm^{-1}. Its ^{1}H NMR spectrum consists of a doublet (3H, δ 1.5), a quartet (1H, δ 4.1), and a broad singlet (2H, δ 6.5). What is the structure of the compound? How would you prepare it from 2-chloropropanoic acid?

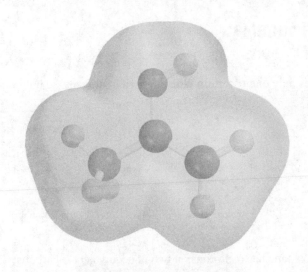

CHAPTER 21

Enols and Enolates

Table of Contents

SOLUTIONS TO TEXT PROBLEMS

In Chapter Problems

21.1 (*b*) Acetophenone has only one α carbon with attached hydrogens, so only one enol is possible. There are no *E* and *Z* isomers.

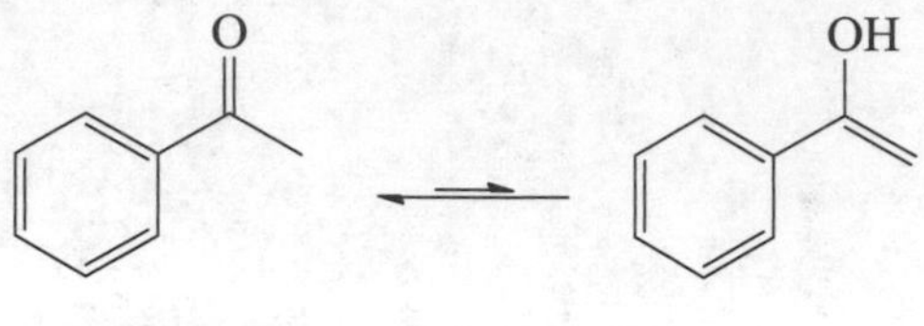

Acetophenone (keto form)   1-Phenylethenol (enol form)

(*c*) 2-Methylcyclohexanone has α protons on two nonequivalent α carbons. Because it is a cyclic six-membered ketone, only one isomer (*E*) is possible. The two enols are constitutional isomers.

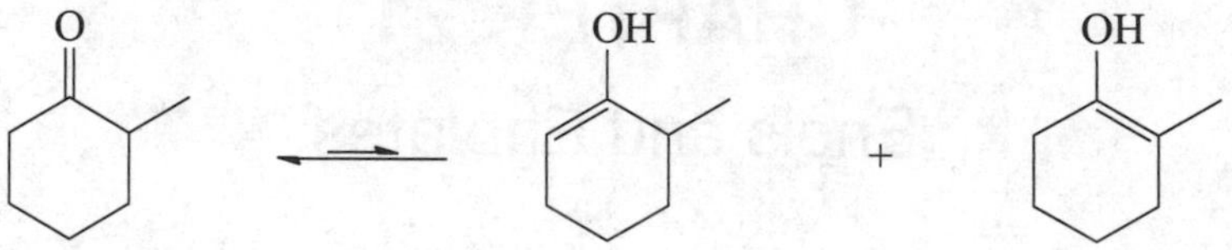

2-Methylcyclohexanone (keto form)   6-Methyl-1-cyclohexenol (enol form)   2-Methyl-1-cyclohexenol (enol form)

(*d*) Methyl vinyl ketone has protons on both α carbons, but one of them is attached to an sp^2 carbon and does not participate in enol formation. (The enol formed at this carbon would result in an unstable cumulated diene.) One enol is possible, involving the methyl group α protons.

3-Buten-2-one (keto form)   1,3-Butadien-2-ol (enol form)

21.2 (*a*) The two most stable enol forms have the carbon–carbon double bond conjugated with the remaining carbonyl group. The enols are further stabilized by an intramolecular hydrogen bond, which is only possible in the *Z* isomers. The dashed line signifies hydrogen bonding.

(*b*) The intramolecular hydrogen bond stabilizes the *Z* isomers in each of the two enols.

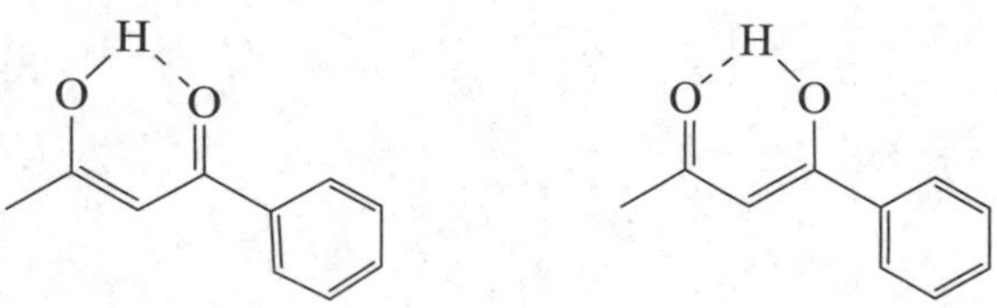

(*c*) The most stable enols have the carbon–carbon bond conjugated with the carbonyl group, and an intramolecular hydrogen bond.

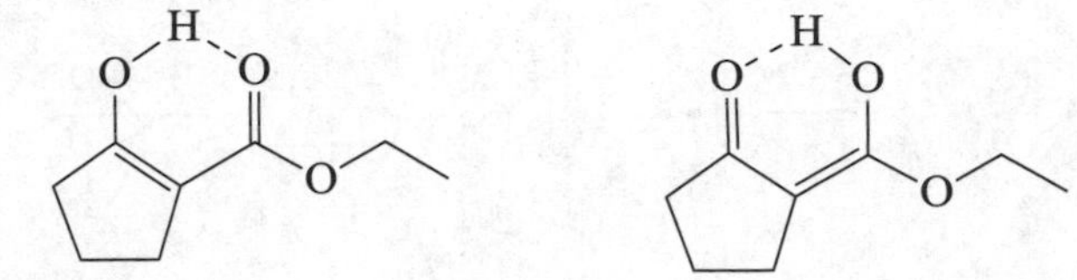

21.3 The enol form is aromatic.

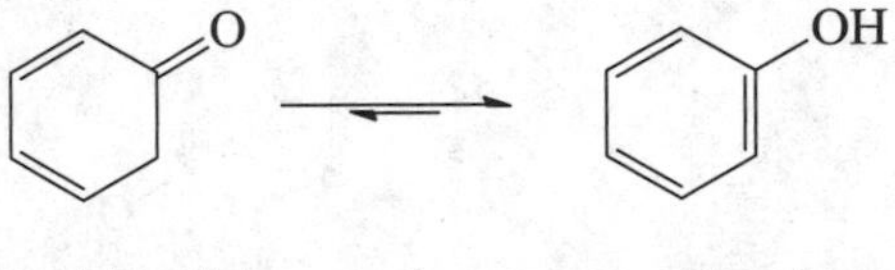

2,4-Cyclohexadienone   Phenol

21.4 (*a*) As shown in the general equation and the examples in the text, halogen substitution is specific for the α-carbon atom. The ketone 2-butanone has two nonequivalent α carbons, and so substitution is possible at either position. Both 1-chloro-2-butanone and 3-chloro-2-butanone are formed in the reaction.

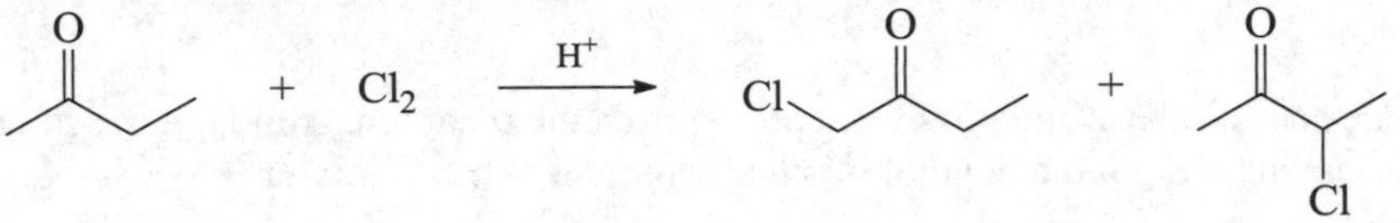

2-Butanone   Chlorine   1-Chloro-2-butanone   3-Chloro-2-butanone

(*b*)   The enol intermediates that react with Cl$_2$ are

1-Buten-2-ol   2-Buten-2-ol
                            (*E* and *Z* isomers)

(*c*)   A pair of electrons from the double bond in each enol is used to form the product C—Cl bonds.

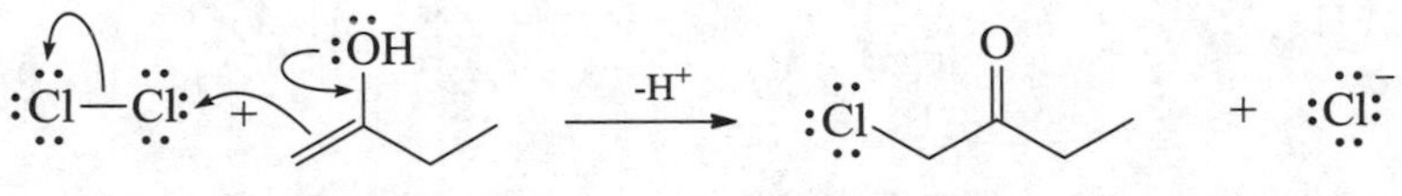

Chlorine   1-Buten-2-ol   1-Chloro-2-butanone   Chloride
                                                                   ion

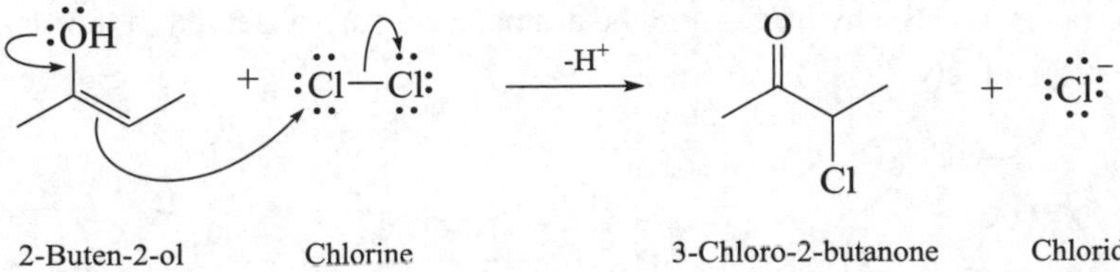

2-Buten-2-ol   Chlorine   3-Chloro-2-butanone   Chloride
                                                                     ion

21.5 Bromination occurs at the α carbon, which is then substituted with the amino group by treatment with ammonia.

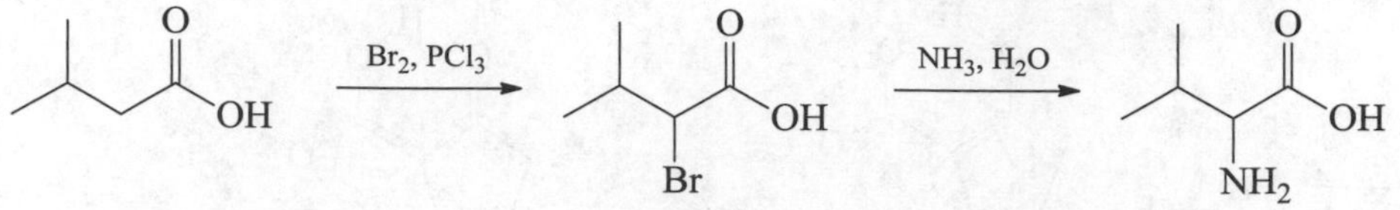

3-Methylbutanoic acid   2-Bromo-3-methylbutanoic acid   Valine

Valine primarily exists in the zwitterionic form (internal salt) over a wide pH range.

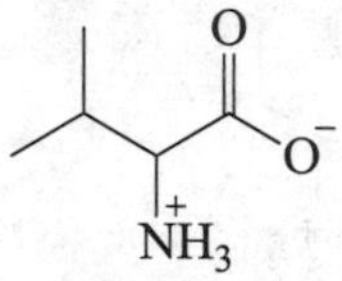

Valine
(zwitterionic form)

The optical rotation of valine prepared by the Hell–Volhard–Zelinsky route is zero, because the product is racemic.

21.6 (b) Hydroxide acts as a base and removes a proton from the α carbon. Although the structure of the enolate is shown as having a Z double bond, it is actually present as a E/Z mixture.

| Hydroxide | 3-Methylbutanal | Water | Enolate |
| (base) | (acid) | (conjugate acid) | (conjugate base) |

(c) Hydroxide acts as a base and removes a proton from the α carbon. As in part (b) of this question, both E and Z isomers of the enolate will be present.

| Hydroxide | Methyl propanoate | Water | Enolate |
| (base) | (acid) | (conjugate acid) | (conjugate base) |

21.7 Each of the five α hydrogens has been replaced by deuterium by base-catalyzed enolization. Only the OCH_3 hydrogens and the hydrogens on the aromatic ring are observed in the ^1H NMR spectrum at δ 3.9 and δ 6.7–6.9, respectively.

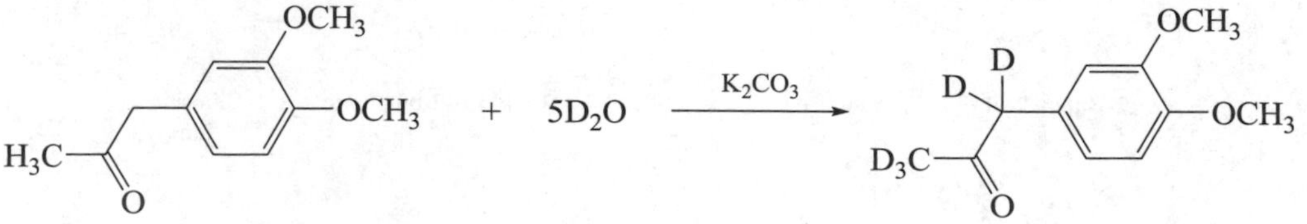

21.8 (a) The equation for the reaction of ethoxide with ethyl acetoacetate is:

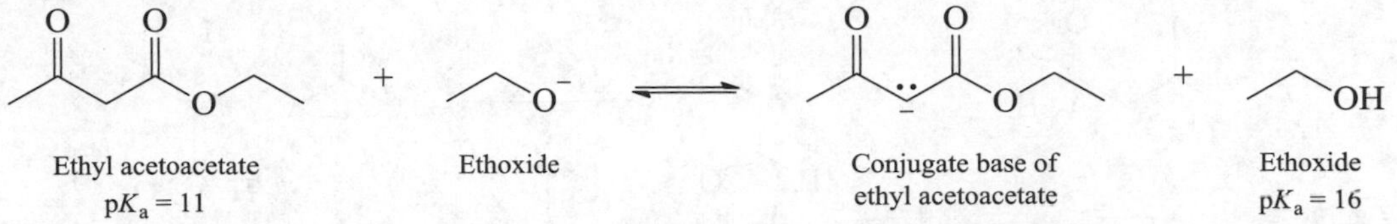

Ethyl acetoacetate
$pK_a = 11$

Ethoxide

Conjugate base of
ethyl acetoacetate

Ethoxide
$pK_a = 16$

Subtracting the pK_a of the product from the pK_a of the reactant gives -5 for the pK of the reaction. $pK = -\log K$ and $K = 10^5$.

(b) An analogous calculation for diethyl malonate ($pK_a = 13$) gives $K = 10^3$.

21.9 (b) Approaching this problem mechanistically in the same way as part (a), write the structure of the enolate ion from 2-methylbutanal.

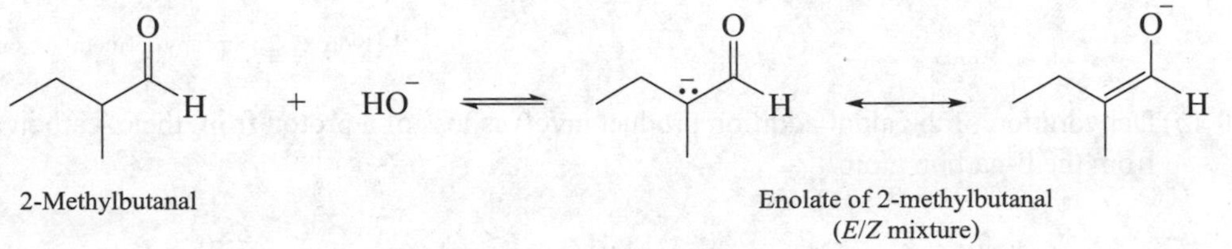

2-Methylbutanal + HO⁻

Enolate of 2-methylbutanal
(E/Z mixture)

This enolate adds to the carbonyl group of the aldehyde.

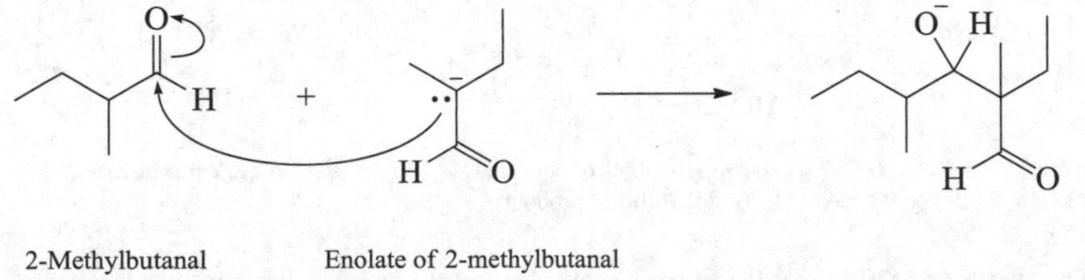

2-Methylbutanal   Enolate of 2-methylbutanal

A proton transfer from solvent yields the product of aldol addition.

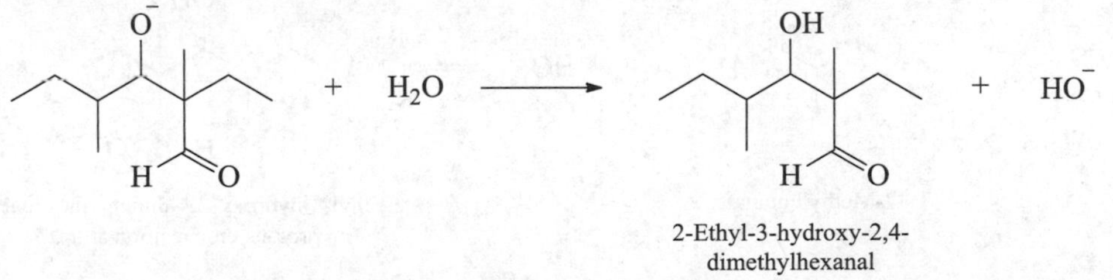

+ H₂O

+ HO⁻

2-Ethyl-3-hydroxy-2,4-
dimethylhexanal

(c) The aldol addition product of 3-methylbutanal can be identified through the same mechanistic approach.

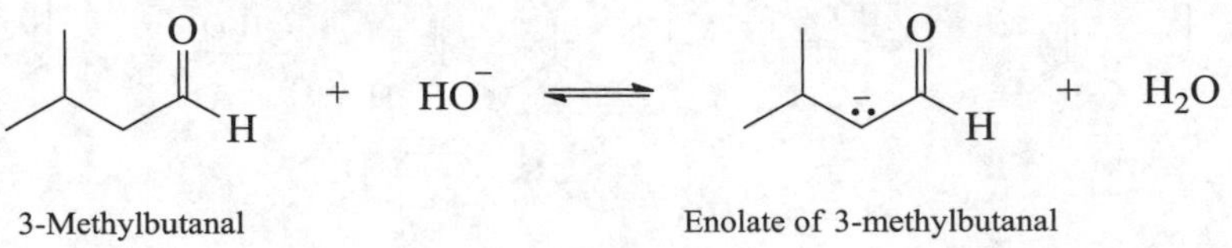

3-Methylbutanal + HO⁻

Enolate of 3-methylbutanal + H₂O

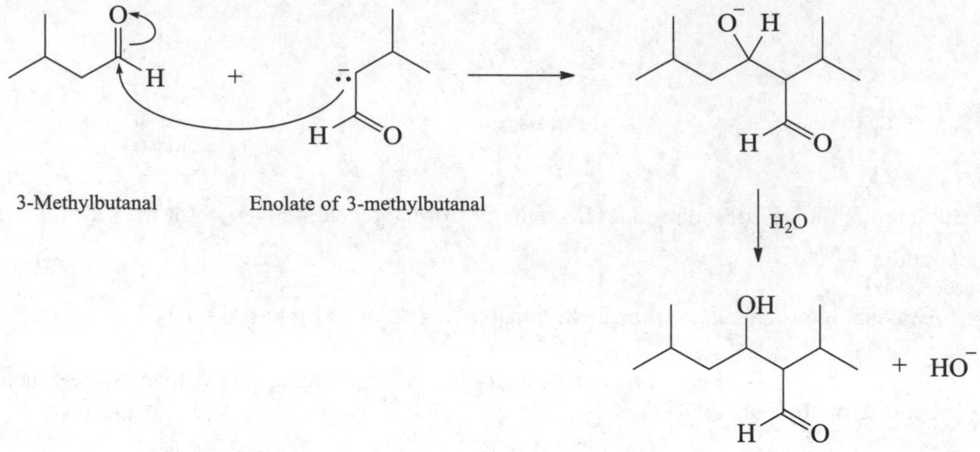

3-Methylbutanal   Enolate of 3-methylbutanal

3-Hydroxy-2-isopropyl-5-methylhexanal

21.10 (*a*) Dehydration of the aldol addition product involves loss of a proton from the α-carbon atom and hydroxide from the β-carbon atom.

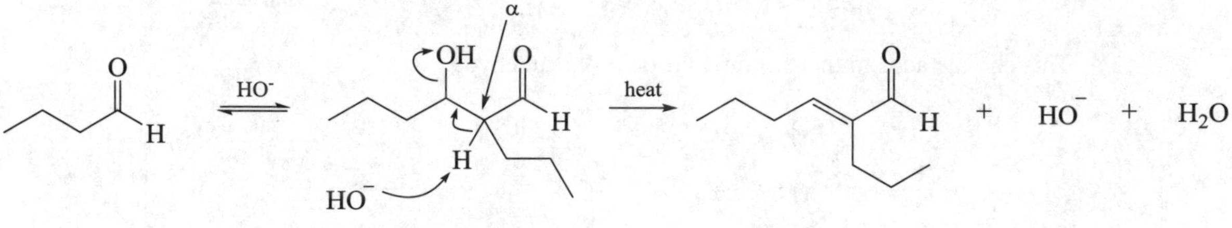

Pentanal   2-Propyl-3-hydroxyhexanal   (*E/Z* mixture)
                                (aldol addition product)

(*b*) The product of aldol addition of 2-methylbutanal has no α hydrogens. It cannot dehydrate to an aldol condensation product.

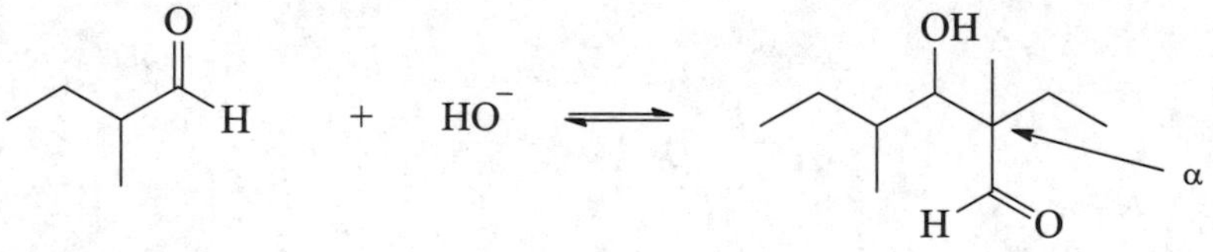

2-Methylbutanal   2-Ethyl-3-hydroxy-2,4-dimethylhexanal
                                         (no protons on α-carbon atom)

(*c*) Aldol condensation is possible with 3-methylbutanal.

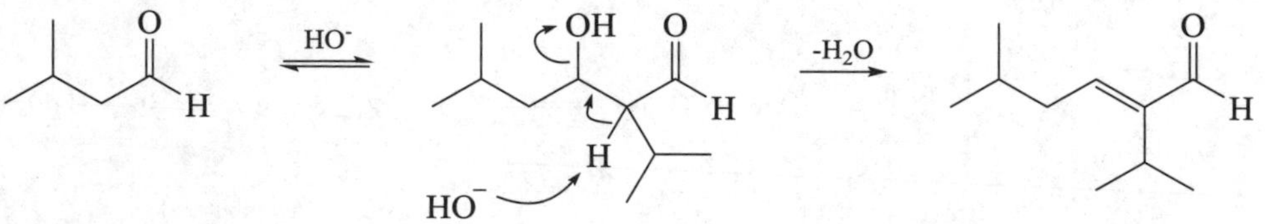

3-Methylbutanal   3-Hydroxy-2-isopropyl-5-methylhexanal   2-Isopropyl-5-methyl-2-hexenal
                                                                           (*E/Z* mixture)

21.11 (*b*) The last step for the aldol condensation product is formation of a carbon-carbon double bond of the α, β unsaturated ketone. Removing this double bond and placing a carbonyl on the β carbon gives the starting reactant.

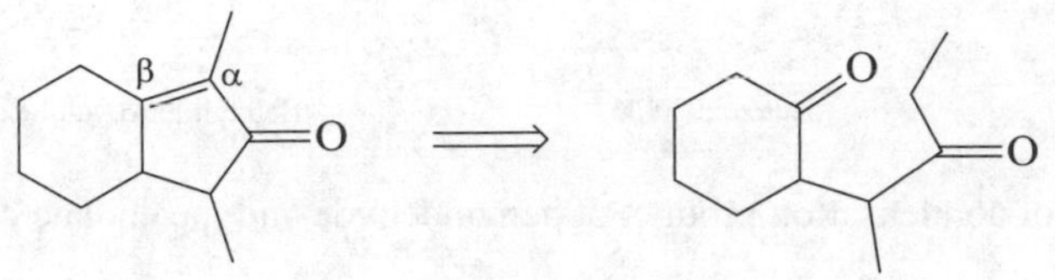

(*c*) The solution for this problem is similar to part (*b*)

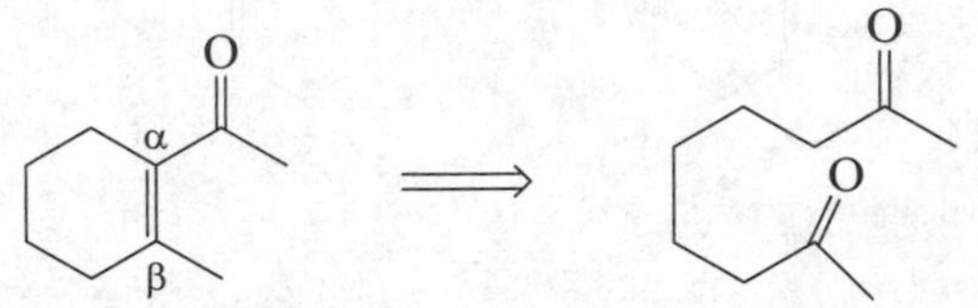

20.12 (*b*) The only enolate that can be formed from *tert*-butyl methyl ketone arises by proton abstraction from the methyl group attached to the carbonyl.

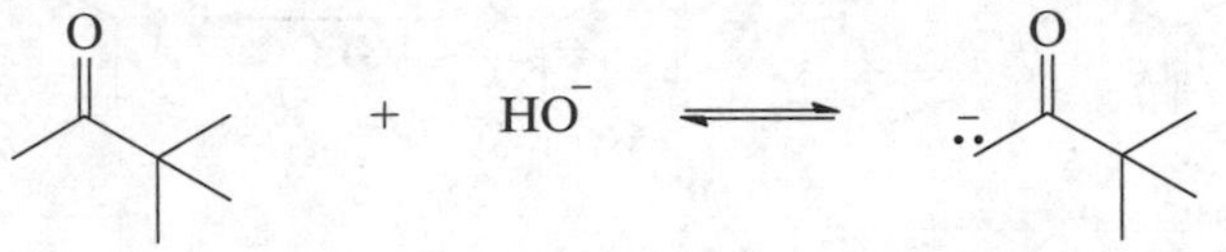

tert-Butyl methyl ketone

Enolate of
tert-butyl methyl ketone

This enolate adds to the carbonyl group of benzaldehyde to give the mixed aldol addition product, which then dehydrates under the reaction conditions.

Enolate of
tert-butyl methyl ketone

H₂O

Product of mixed aldol addition

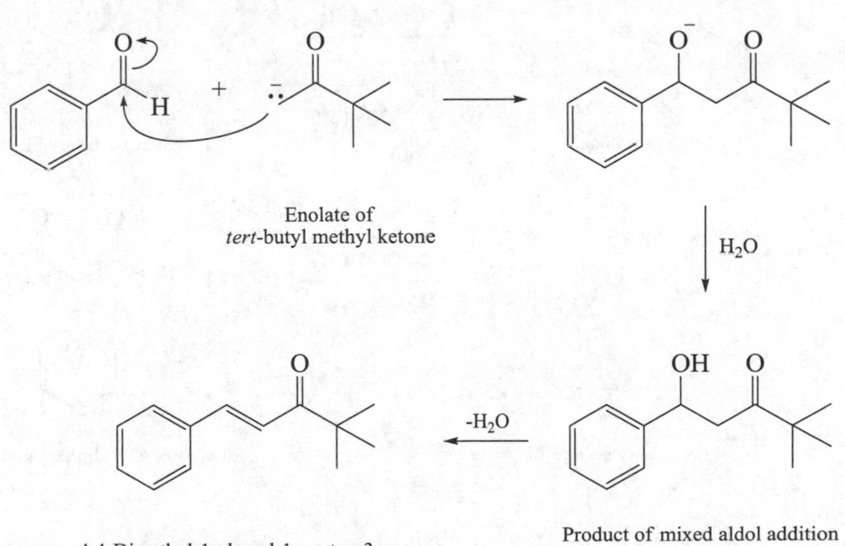

4,4-Dimethyl-1-phenyl-1-penten-3-one
(product of mixed aldol condensation)

-H₂O

(*c*) The enolate of cyclohexanone adds to benzaldehyde. Dehydration of the mixed aldol addition product takes place under the reaction conditions to give the following mixed aldol condensation product.

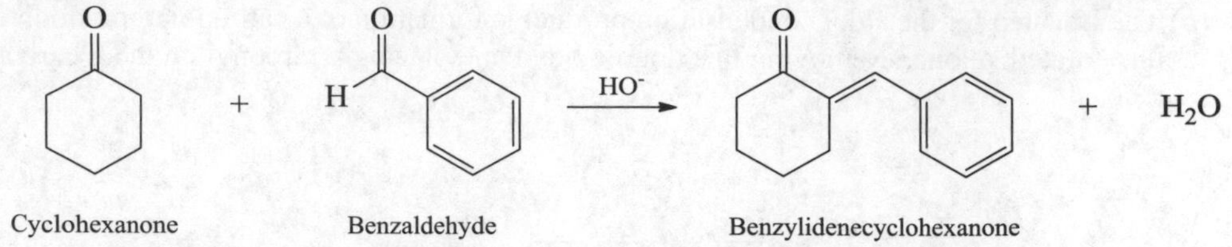

Cyclohexanone   Benzaldehyde   Benzylidenecyclohexanone

21.13 Chalcone is the aldol condensation product of benzaldehyde and the enolate of acetophenone.

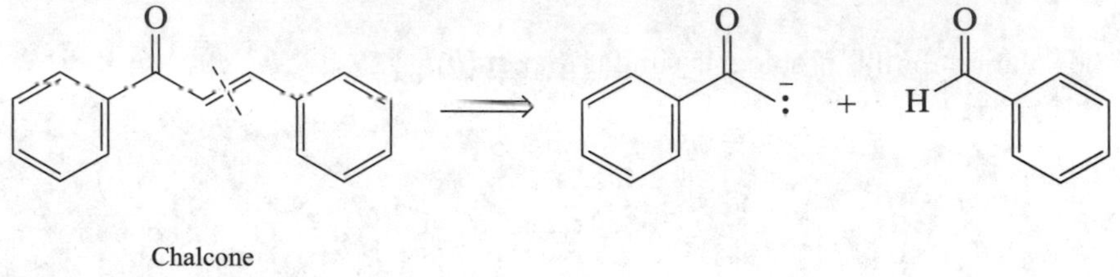

Chalcone

A reasonable synthesis is:

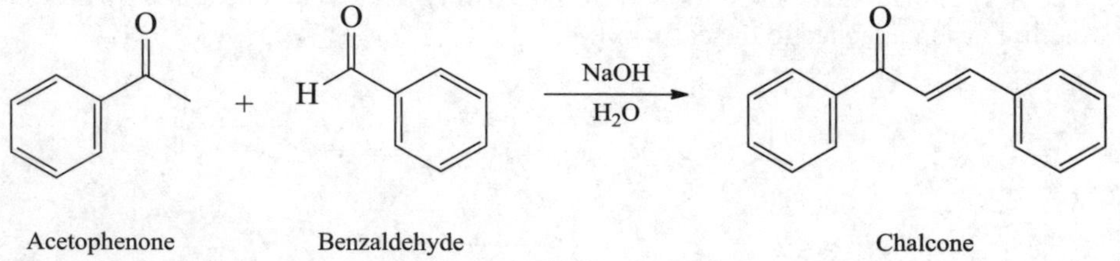

Acetophenone   Benzaldehyde   Chalcone

21.14 Ethyl benzoate cannot undergo the Claisen condensation, because it has no protons on its α-carbon atom and so cannot form an enolate. Ethyl pentanoate and ethyl phenylacetate can undergo the Claisen condensation.

Ethyl pentanoate   Ethyl 3-oxo-2-propylheptanoate

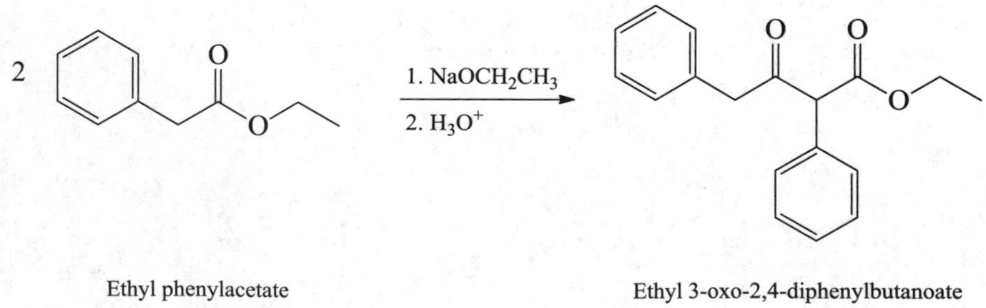

Ethyl phenylacetate   Ethyl 3-oxo-2,4-diphenylbutanoate

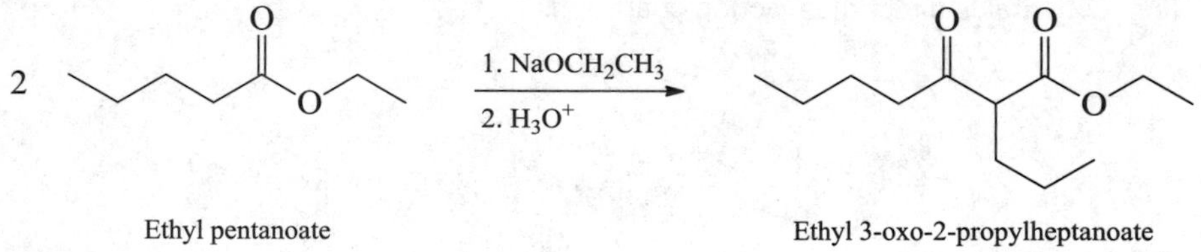

21.15 (*b*) The enolate formed by proton abstraction from the α-carbon atom of diethyl 4-methylheptanedioate cyclizes to form a six-membered β-keto ester.

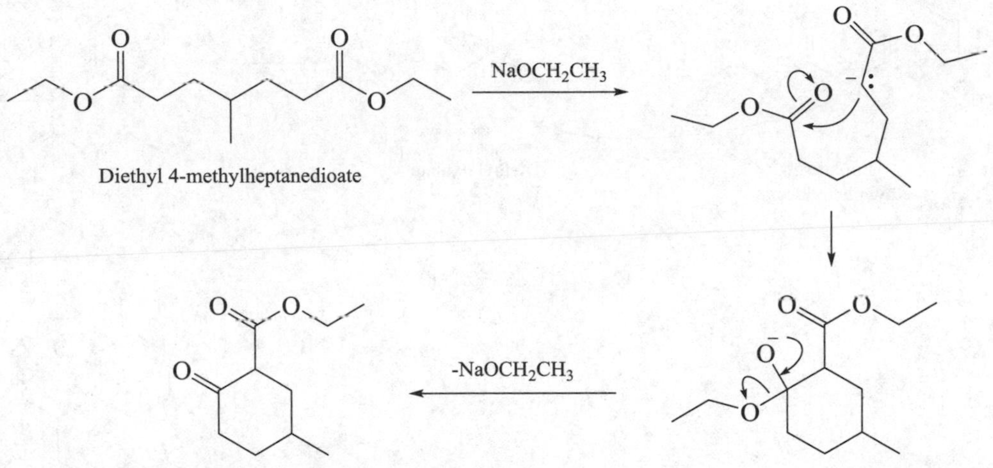

Diethyl 4-methylheptanedioate

Ethyl 5-methyl-2-oxocyclohexanecarboxylate

(*c*) The two α carbons of this diester are not equivalent. Cyclization by attack of the enolate at C-2 gives

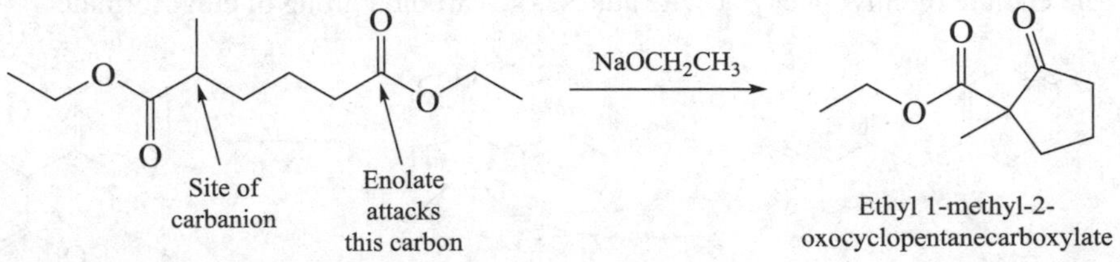

Site of carbanion

Enolate attacks this carbon

Ethyl 1-methyl-2-oxocyclopentanecarboxylate

This β-keto ester cannot form a stable enolate by deprotonation. It is present in only small amounts at equilibrium. The major product is formed by way of the other enolate.

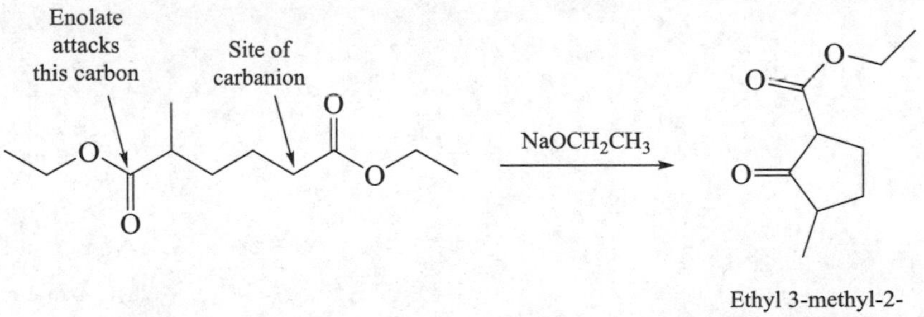

Enolate attacks this carbon

Site of carbanion

Ethyl 3-methyl-2-oxocyclopentanecarboxylate

This β-keto ester is converted to a stable enolate on deprotonation, causing the equilibrium to shift in its favor.

21.16 (*b*) Both carbonyl groups of diethyl oxalate are equivalent. The enolate of ethyl phenylacetate attacks one of them.

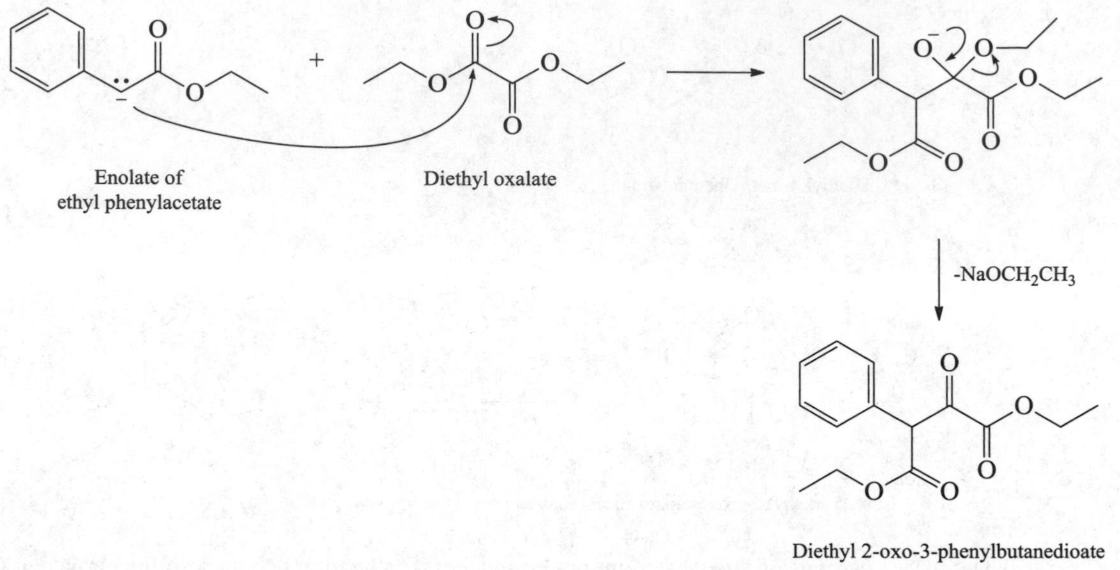

Enolate of
ethyl phenylacetate   Diethyl oxalate

-NaOCH₂CH₃

Diethyl 2-oxo-3-phenylbutanedioate

(*c*) The enolate of ethyl phenylacetate attacks the carbonyl group of ethyl formate.

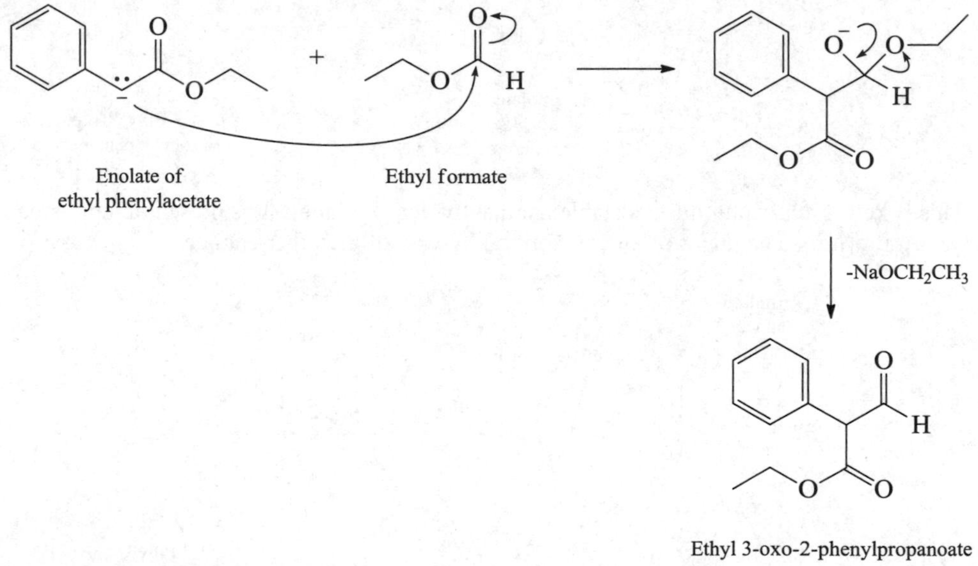

Enolate of
ethyl phenylacetate   Ethyl formate

-NaOCH₂CH₃

Ethyl 3-oxo-2-phenylpropanoate

21.17 For a five-membered ring to be formed, C-5 must be the carbanionic site that attacks the ester carbonyl.

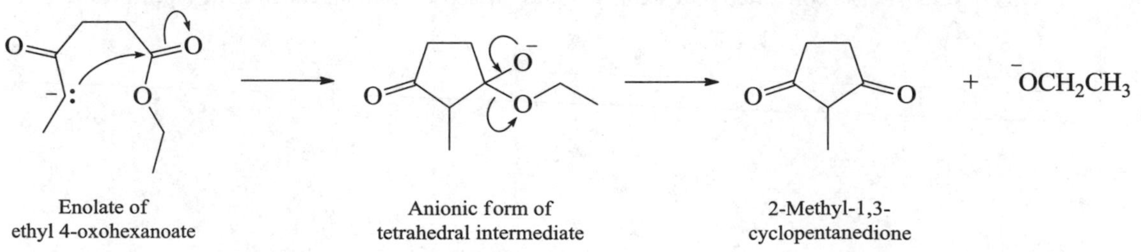

Enolate of   Anionic form of   2-Methyl-1,3-
ethyl 4-oxohexanoate   tetrahedral intermediate   cyclopentanedione   +   ⁻OCH₂CH₃

21.18 The resonance structures for enolates of ethyl acetoacetate and diethyl malonate are shown below. Resonance with both carbonyls of each structure is important.

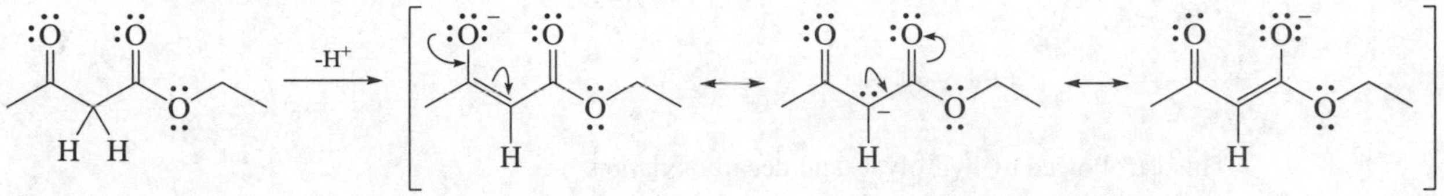

Ethyl acetoacetate   Enolate resonance structures for ethyl acetoacetate

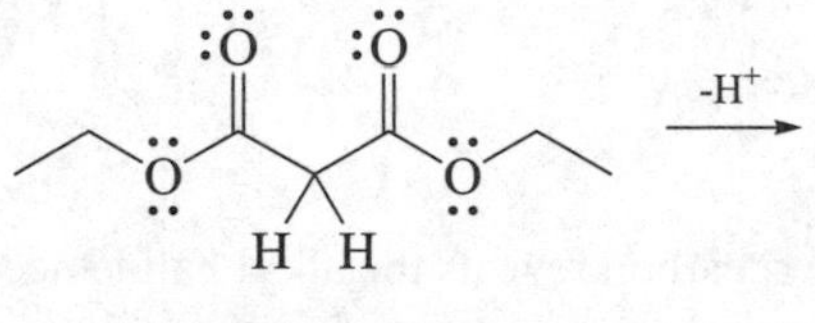

Diethyl malonate

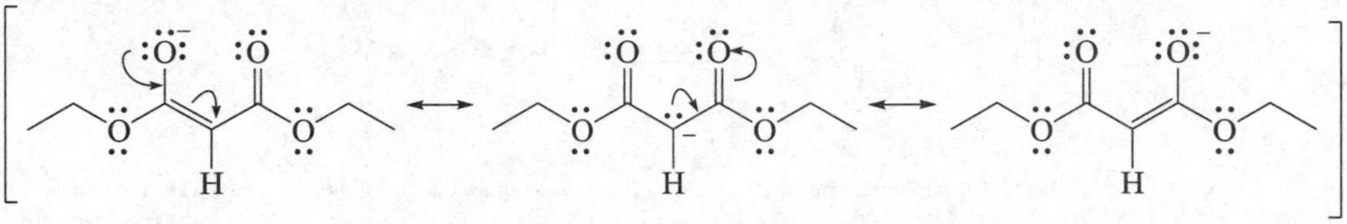

Enolate resonance structures for diethyl malonate

21.19 (*a*) The acetoacetic ester protocol including hydrolysis and decarboxylation gives the product. First, alkylation forms the substituted ethyl acetoacetate.

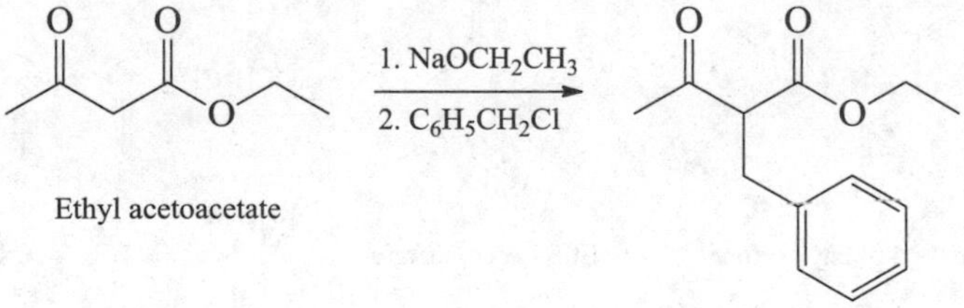

Ethyl acetoacetate

This is followed by hydrolysis and decarboxylation.

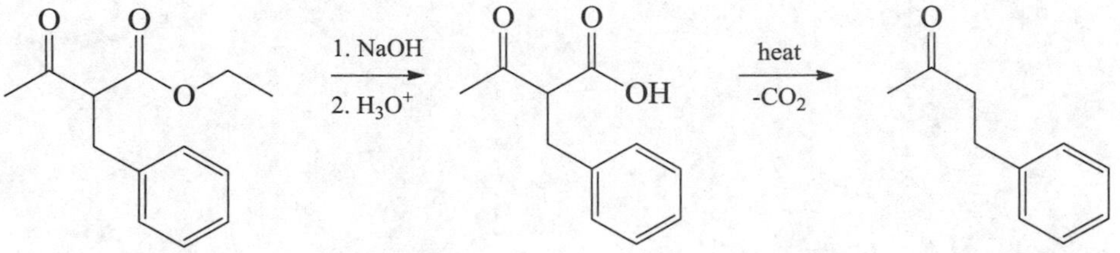

(*b*) The malonic ester protocol including hydrolysis and decarboxylation gives the product. First, alkylation forms the substituted diethyl malonate.

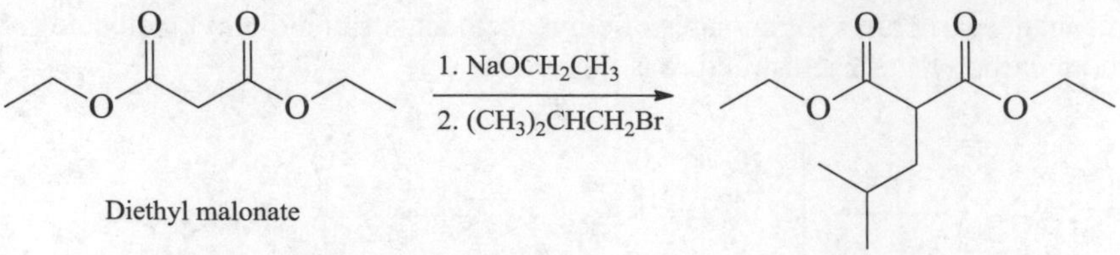

Diethyl malonate

This is followed by hydrolysis and decarboxylation.

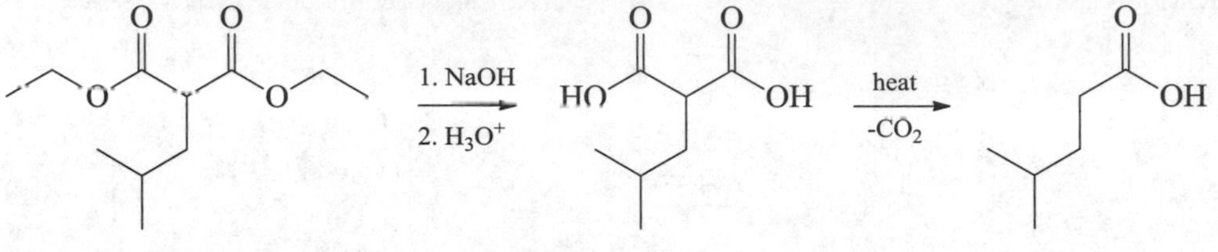

21.20 (*b*) Disconnection of the target molecule adjacent to the α carbon reveals the alkyl halide needed to react with the enolate derived from ethyl acetoacetate.

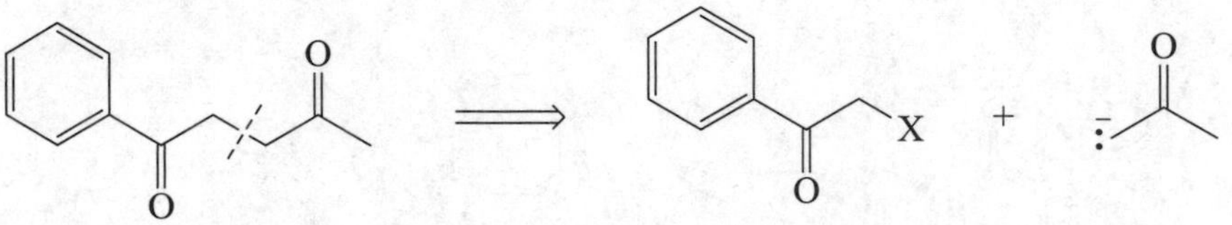

1-Phenyl-1,4-pentanedione   Required alkyl halide   Derived from
                                                                               ethyl acetoacetate

Analyzing the target molecule in this way reveals that the required alkyl halide is an α ketone. Thus, a suitable starting material would be bromomethyl phenyl ketone.

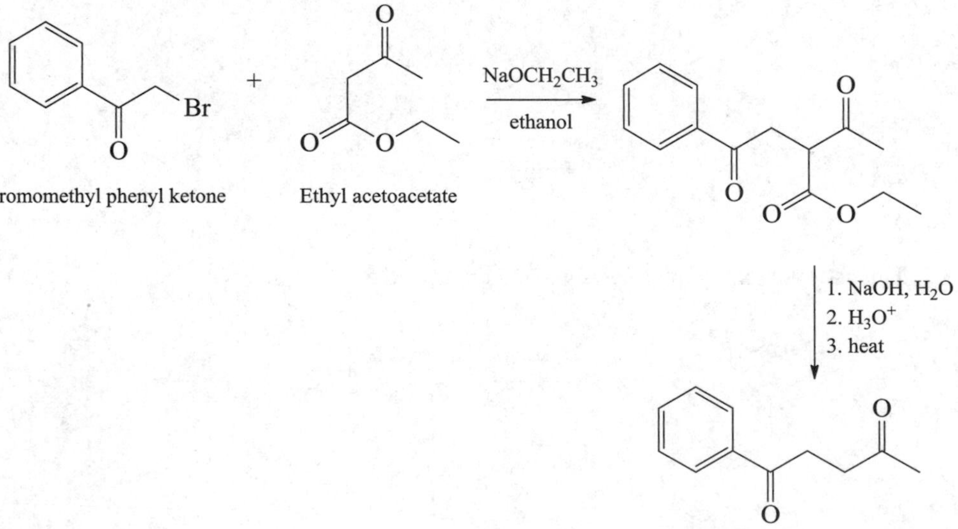

Bromomethyl phenyl ketone   Ethyl acetoacetate

1. NaOH, H$_2$O
2. H$_3$O$^+$
3. heat

1-Phenyl-1,4-pentanedione

(*c*) Disconnection of the target molecule adjacent to the α carbon reveals the alkyl halide needed to react with the enolate derived from diethyl malonate.

OH
X
+
OH

4-Methylhexanoic acid Required alkyl halide Derived from diethyl malonate

The necessary alkyl halide in this synthesis is 1-bromo-2-methylbutane.

Br
+
$NaOCH_2CH_3$
ethanol
1. NaOH, H_2O
2. H_3O^+
3. heat
OH

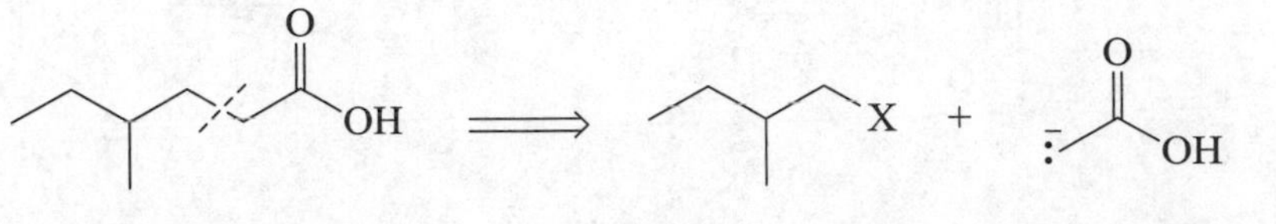

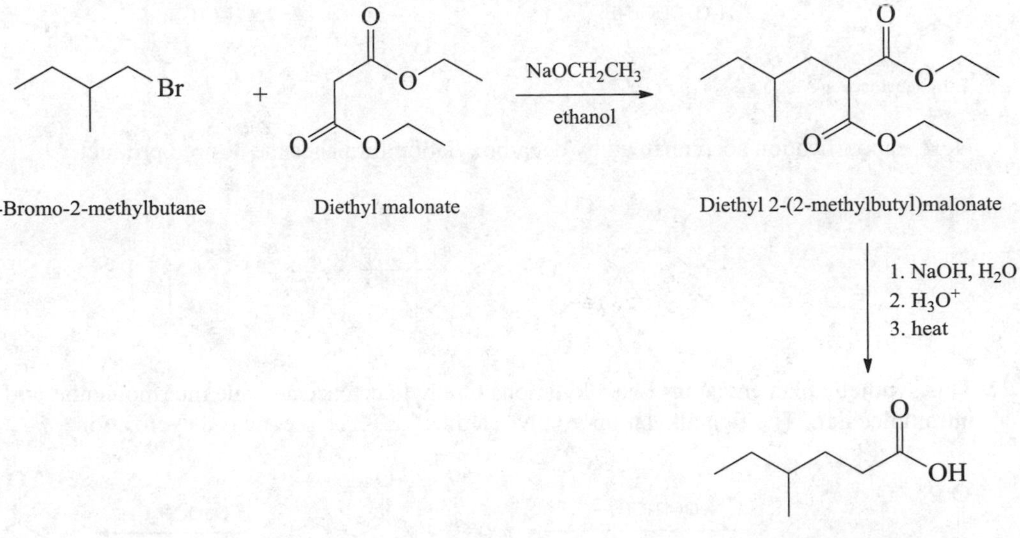

1-Bromo-2-methylbutane Diethyl malonate Diethyl 2-(2-methylbutyl)malonate

4-Methylhexanoic acid

(*d*) The disconnection approach to retrosynthetic analysis reveals that the preparation of 5-hexen-2-one by the acetoacetic ester synthesis requires an allylic halide.

X
+

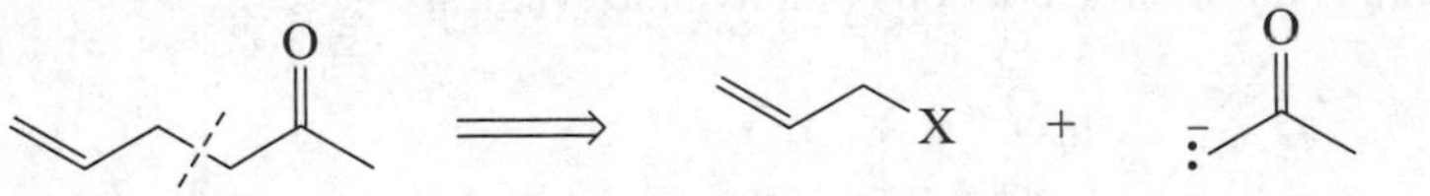

5-Hexen-2-one Required alkyl halide Derived from ethyl acetoacetate

The necessary alkyl halide in this synthesis is allyl bromide.

Br
+
$NaOCH_2CH_3$
ethanol
1. HO^-, H_2O
2. H_3O^+
3. heat

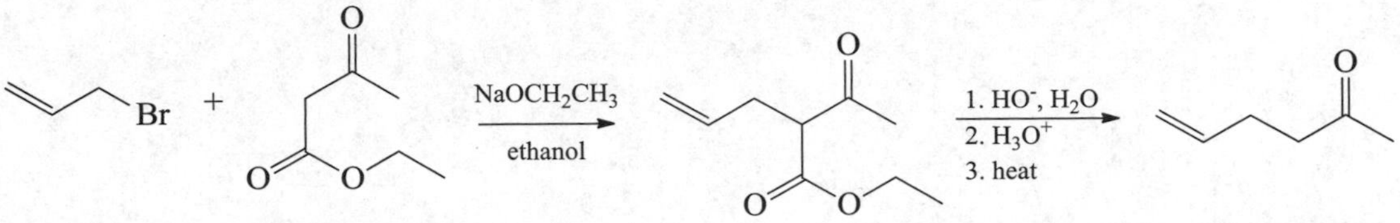

Allyl bromide Ethyl acetoacetate 5-Hexen-2-one

21.21 The structure of the target compound indicates it can be considered a dialkylated derivative of acetone and suggests an acetoacetic ester synthesis.

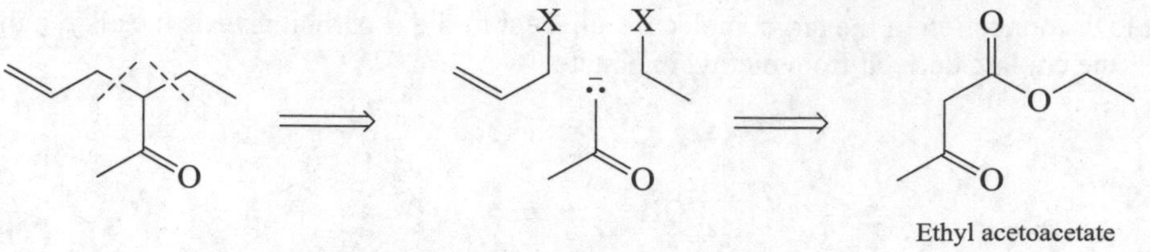

Ethyl acetoacetate

First, ethyl acetoacetate is subjected to two alkylations.

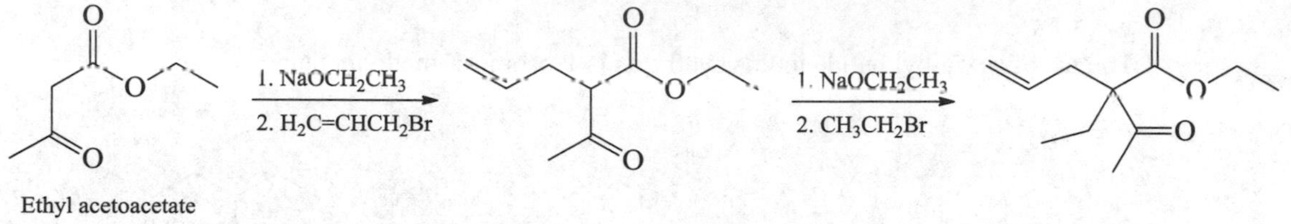

Ethyl acetoacetate

Next, saponification accompanied by decarboxylation furnishes the desired product.

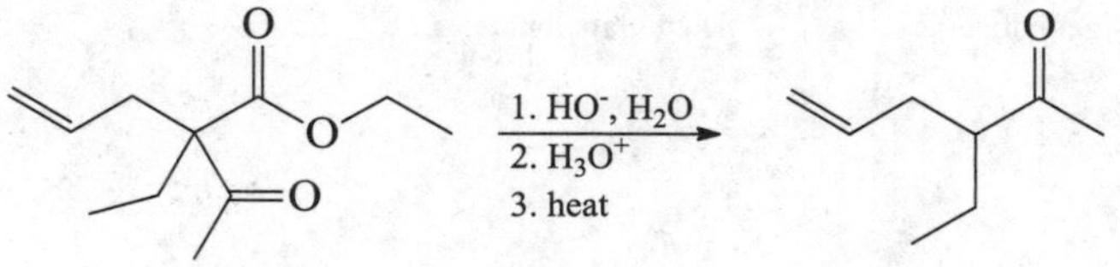

21.22 This synthetic plan envisions two alkylations of ethyl acetoacetate, one intermolecular and the other intramolecular. The first alkylation involves a dihalide. The second is a cyclization.

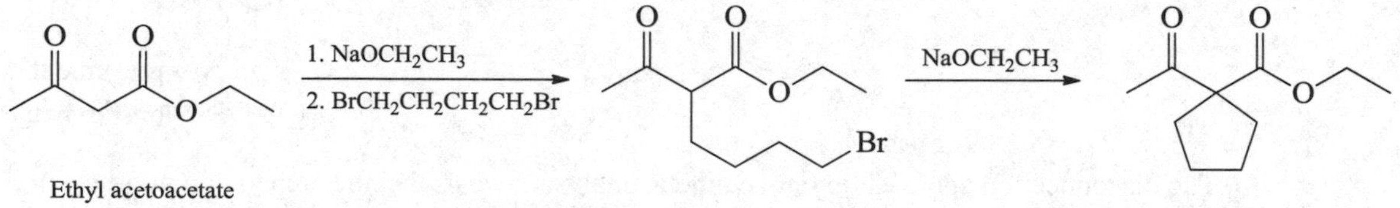

Ethyl acetoacetate

Saponification and decarboxylation complete the synthesis.

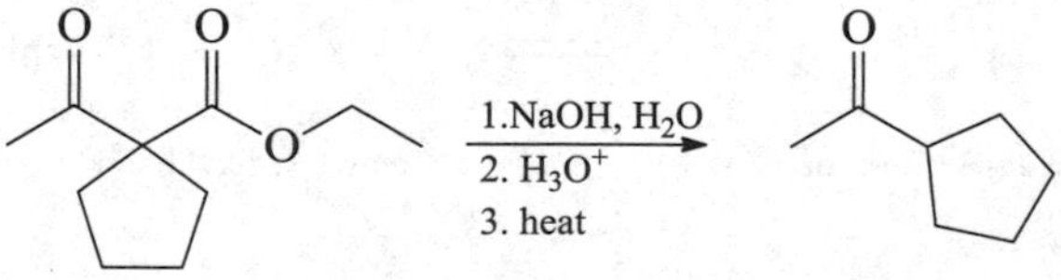

21.23 The highest yields of carboxylic acids are obtained from methyl ketones that can enolize in only one direction. With ketones that can enolize in two directions, mixtures are obtained. Thus, the most suitable compound, (4-isopropylphenyl) methyl ketone, best fits this criteria. It is a methyl ketone that can enolize only toward the methyl group. The carboxylic acid obtained is 4-isopropylbenzoic acid.

Benzyl methyl ketone   (4-Isopropylphenyl)   *tert*-Butyl phenyl ketone   Ethyl phenyl ketone
                        methyl ketone

Benzyl methyl ketone can enolize in both directions and would give a mixture of products containing a lower yield of a single carboxylic acid. The *tert*-butyl phenyl ketone cannot enolize because it has no α protons and therefore cannot undergo the haloform reaction. Ethyl phenyl ketone is not a methyl ketone so it cannot form a trihalomethyl ketone.

21.24 Mesityl oxide is an α,β-unsaturated ketone. It is prepared by an aldol condensation with acetone. Traces of acids or bases can catalyze its isomerization so that some of the less stable β,γ-unsaturated isomer is present.

Mesityl oxide;   4-Methyl-4-penten-2-one
4-methyl-3-penten-2-one   (less stable)
(more stable)

20.25 Because sodium azide, NaN_3, does not react to give a product with propanal, one can presume that the reaction with acrolein should be a different reaction than the addition to a carbonyl. The relationship between the molecular formula of acrolein (C_3H_4O) and the product ($C_3H_5N_3O$) corresponds to an addition of HN_3. Because a C=C is present, it is probable that 1,4 addition takes place.

Acrolein   Sodium   3-Azidopropanal
                azide

21.26 Michael addition of the enolate of diethyl malonate and methyl vinyl ketone forms the carbon structure required to solve this problem.

Methyl vinyl ketone   Diethyl malonate

Hydrolysis followed by heat gives the final product.

21.27 The enolate of dibenzyl ketone adds to methyl vinyl ketone in the conjugate addition step.

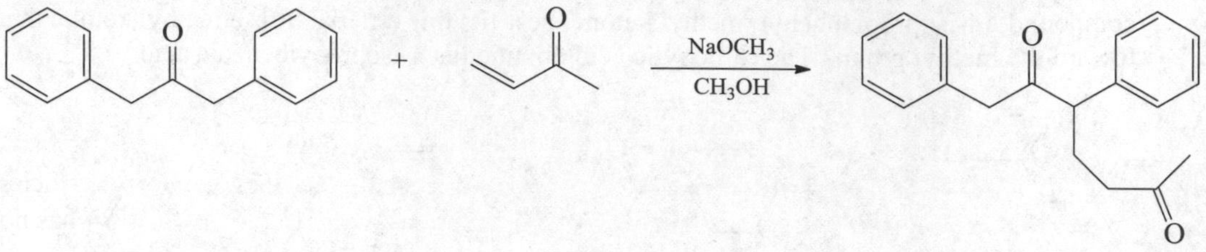

Dibenzyl ketone   Methyl vinyl ketone   1,3-Diphenyl-2,6-heptanedione

via

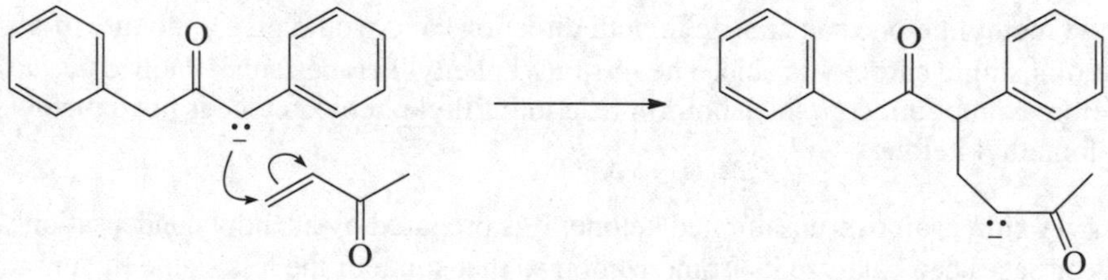

The intramolecular aldol condensation that gives the observed product is

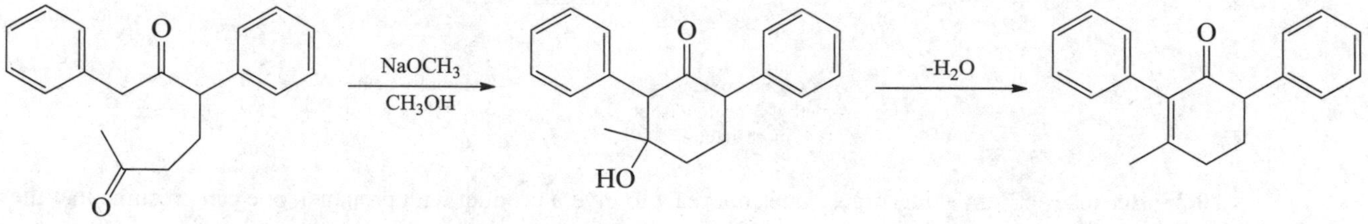

1,3-Diphenyl-2,6-heptanedione   3-Methyl-2,6-diphenyl-2-cyclohexenone

21.28 A second solution to the synthesis of 4-methyl-2-octanone by conjugate addition of a lithium dialkylcuprate reagent to an α,β-unsaturated ketone is revealed by the disconnection shown.

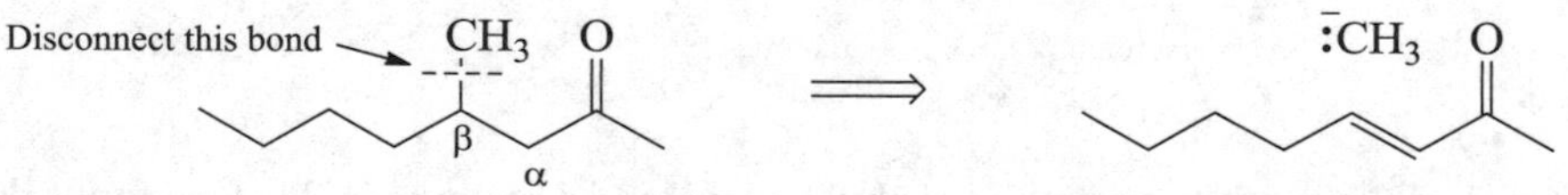

According to this disconnection, the methyl group is derived from lithium dimethylcuprate.

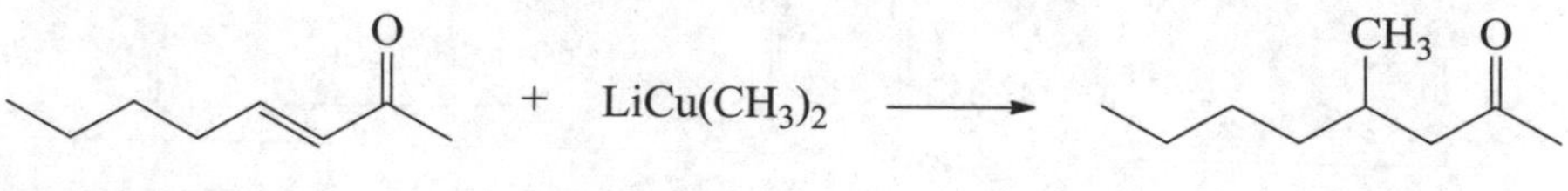

3-Octen-2-one   Lithium dimethylcuprate   3-Methyl-2-octanone

End of Chapter Problems

Enols and Enolization

21.29 (a) 2-Methylpropanal has the greater enol content.

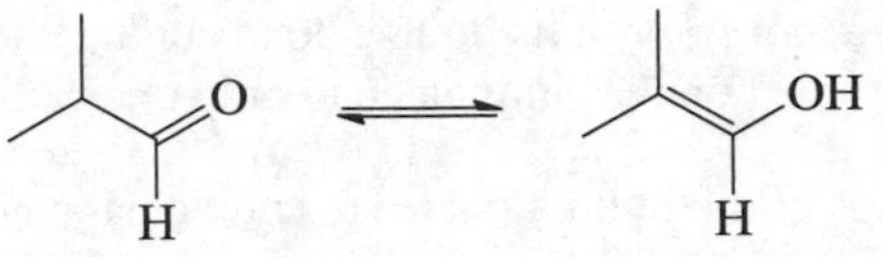

2-Methylpropanal   Enol form

Although the enol content of 2-methylpropanal is quite small, the compound is nevertheless capable of enolization, whereas the other compound, 2,2-dimethylpropanal, cannot enolize—it has no α hydrogens.

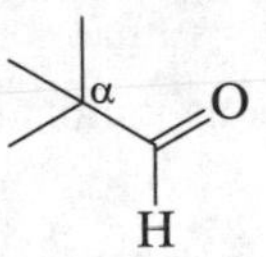

(Enolization is impossible)

(*b*) The ketone enolizes in a straightforward manner. The lactone has no hydrogens on its α carbon and cannot enolize.

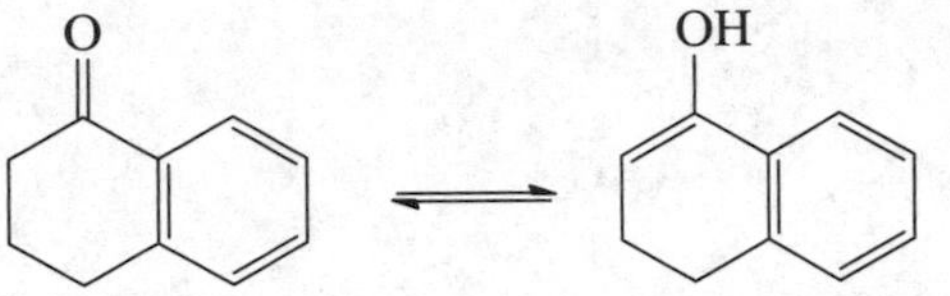

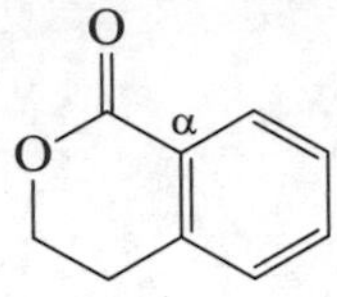

(*c*) The enol content is greater if the double bond is conjugated.

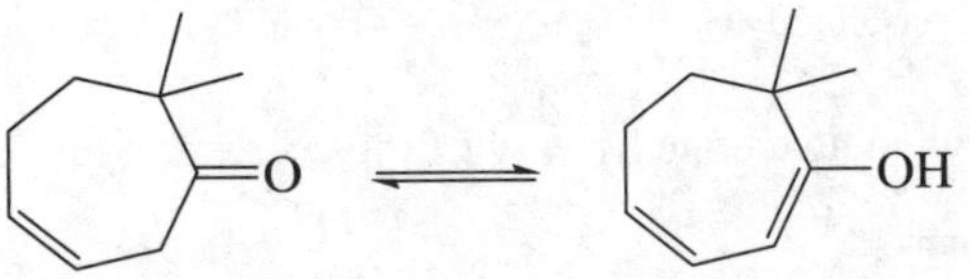

Double bonds   Double bonds
are conjugated.   are not conjugated.

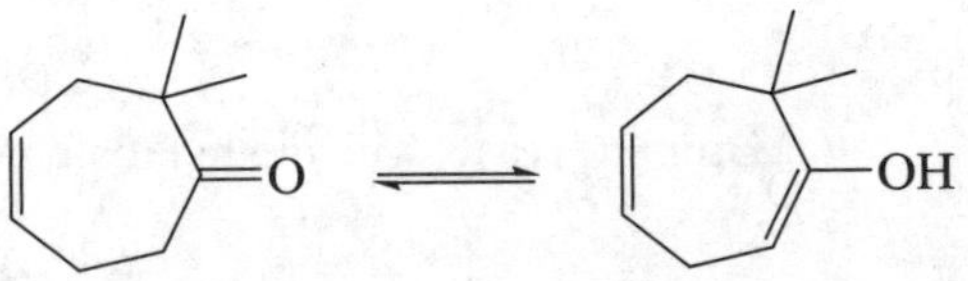

(*d*) Here, we are comparing a simple ketone, dibenzyl ketone, with a β-diketone. The β-diketone enolizes to a much greater extent than the simple ketone because its enol form is stabilized by conjugation of the double bond with the remaining carbonyl group and by intramolecular hydrogen bonding.

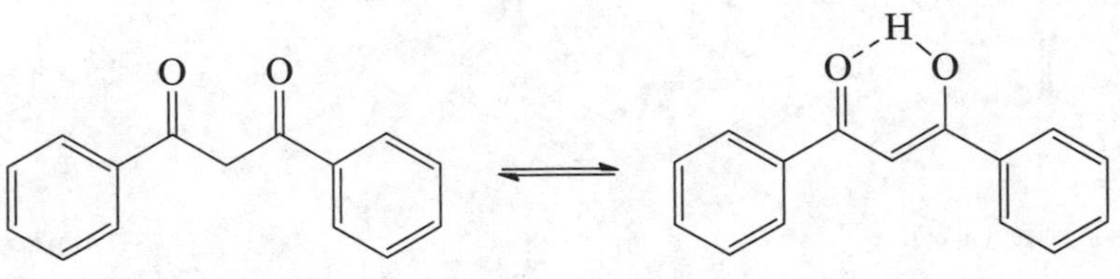

1,3-Diphenyl-1,3-propanedione   Enol form

21.30 The carbon atom formed by two carbonyl groups is the one involved in the enolization of terreic acid.

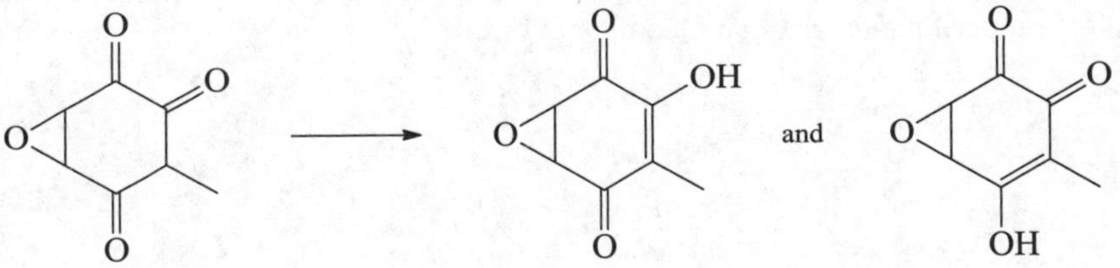

Terreic acid   Enol A   and   Enol B

Of these two structures, enol A, with its double bond conjugated to two carbonyl groups, is more stable than enol B, in which the double bond is conjugated to only one carbonyl.

21.31 (*a*) Both carbonyl groups of diethyl malonate are equivalent, and so enolization can occur in either direction.

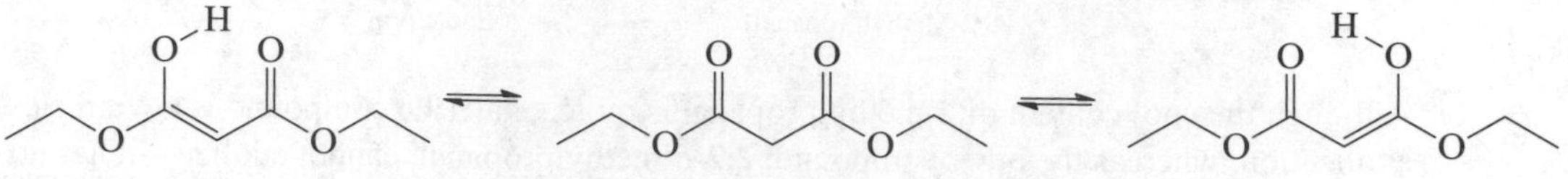

Diethyl malonate

(*b*) Ethyl acetoacetate can give three constitutionally isomeric enols:

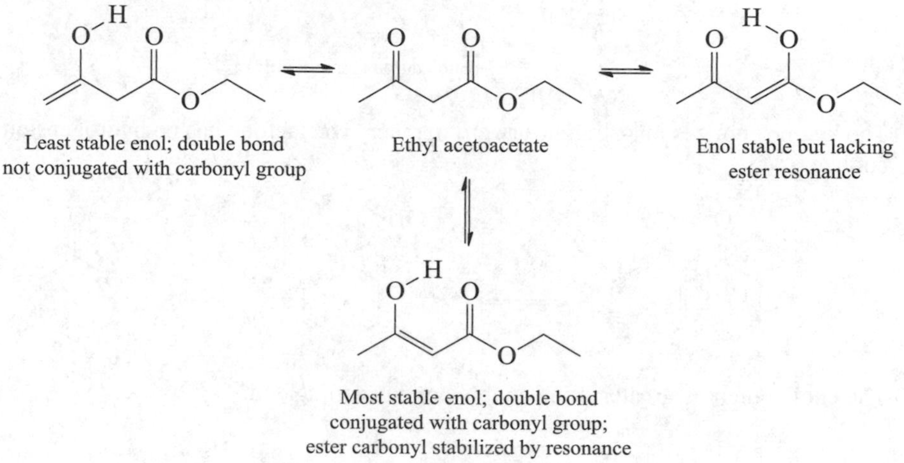

Least stable enol; double bond   Ethyl acetoacetate   Enol stable but lacking
not conjugated with carbonyl group   ester resonance

Most stable enol; double bond
conjugated with carbonyl group;
ester carbonyl stabilized by resonance

(*c*) Bromine reacts with diethyl malonate and ethyl acetoacetate by way of the corresponding enols:

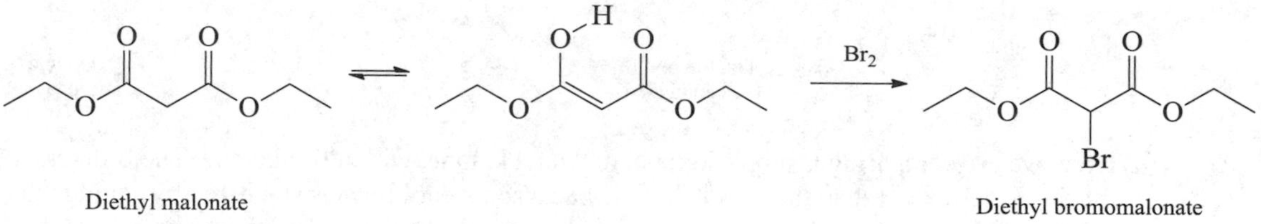

Diethyl malonate   Diethyl bromomalonate

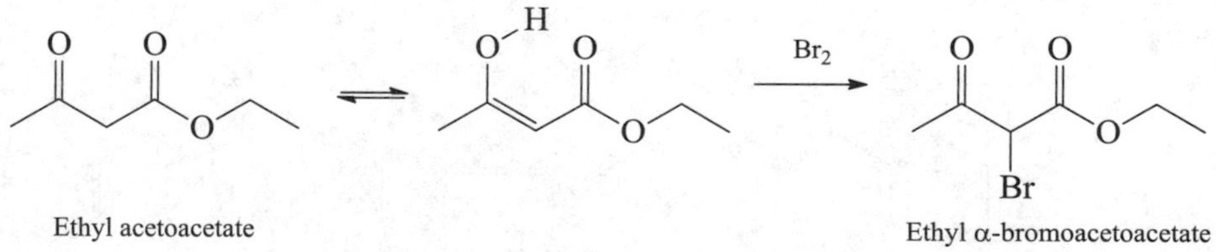

Ethyl acetoacetate   Ethyl α-bromoacetoacetate

Reactions

21.32 (*a*) The chirality center in (*R*) piperitone is located at a carbon that also has an α hydrogen. In the presence of base, piperitone is converted to its enolate, which is achiral. From this enolate, conversion to (*R*)- and (*S*)-piperitone results in a racemic mixture.

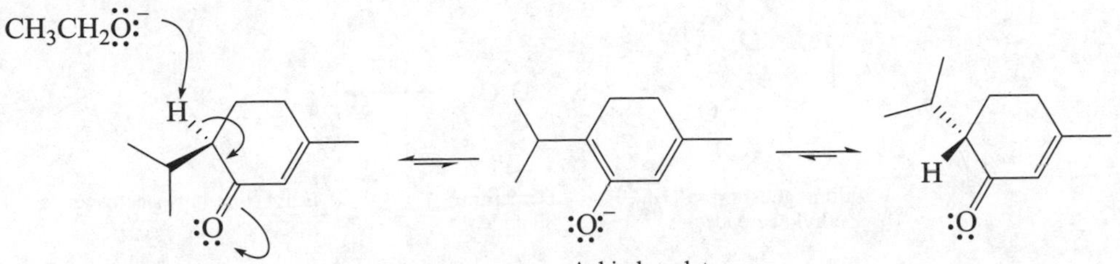

(*R*)-Piperitone Achiral enolate + $CH_3CH_2\ddot{O}H$ (*S*)-Piperitone

(*b*) The enol formed from menthone can revert to either menthone or isomenthone.

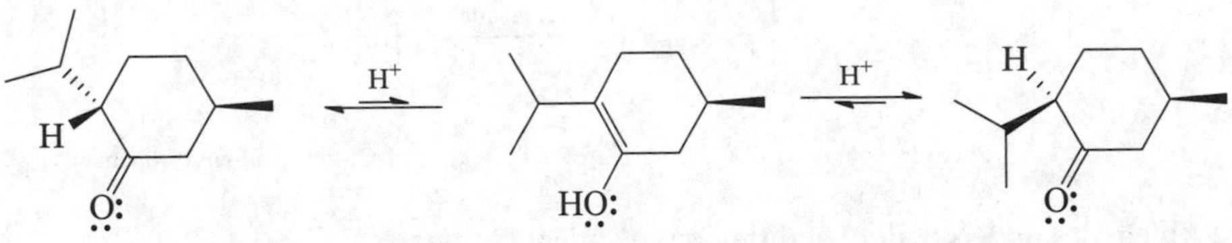

Menthone Enol form Isomenthone

Only the stereochemistry at the α-carbon atom is affected by enolization. The other chirality center in menthone (the one bearing the methyl group) is not affected.

21.33 (*a*) Recall that aldehydes and ketones are in equilibrium with their *hydrates* in aqueous solution (text Section 18.6). Thus, the principal substance present when $(C_6H_5)_2CHCH{=}O$ is dissolved in aqueous acid is $(C_6H_5)_2CHCH(OH)_2$ (81%).

(*b*) The problem states that the major species present in aqueous base is *not* $(C_6H_5)_2CHCH{=}O$, its enol, or its hydrate. The most reasonable species is the *enolate ion:*

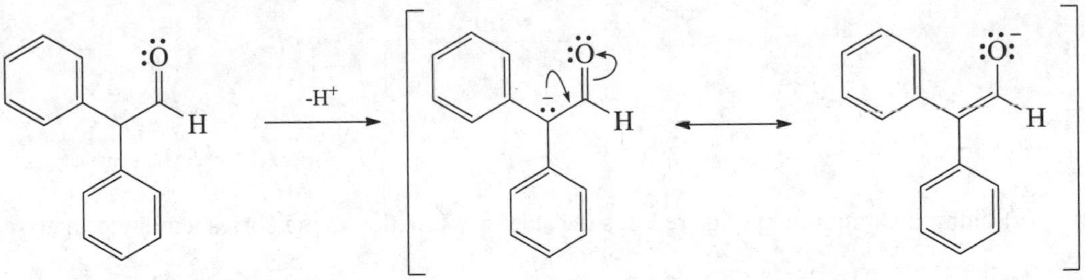

21.34 (*a*) Recall that Grignard reagents are destroyed by reaction with proton donors. Ethyl acetoacetate is a stronger acid than water; it transfers a proton to a Grignard reagent.

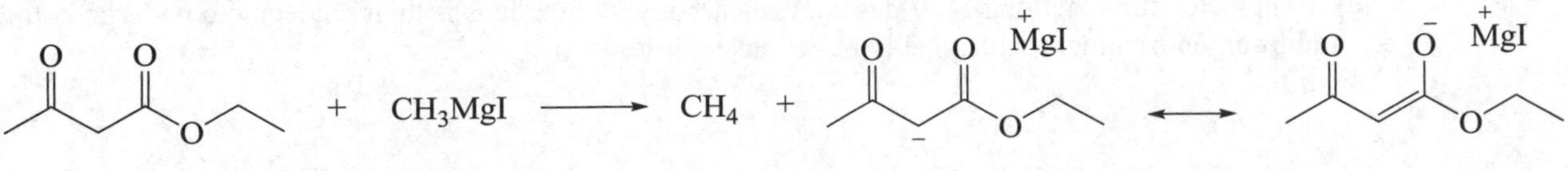

Ethyl acetoacetate Methylmagnesium iodide Methane Iodomagnesium salt of ethyl acetoacetate

(*b*) Adding D_2O and DCl to the reaction mixture leads to D^+ transfer to the α-carbon atom of ethyl acetoacetate.

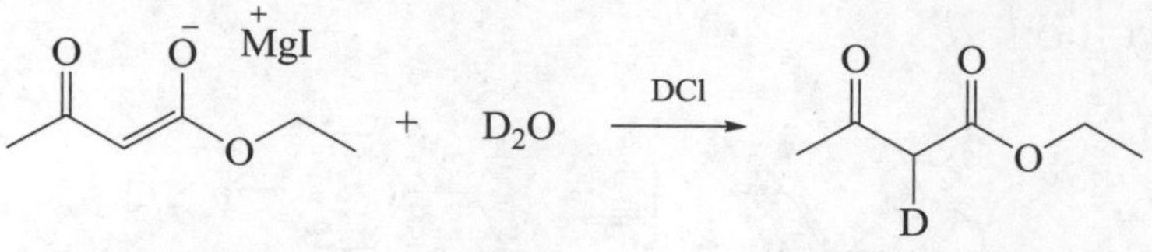

Iodomagnesium salt of   Deuterium   Ethyl α-deuterioacetoacetate
ethyl acetoacetate   oxide

21.35 (*a*) Chlorination of 3-phenylpropanal under conditions of acid catalysis occurs via the enol form and yields the α-chloro derivative.

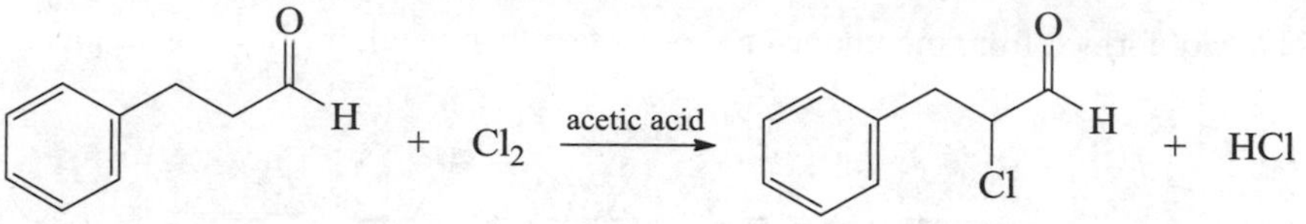

3-Phenylpropanal   2-Chloro-3-phenylpropanal

(*b*) Aldehydes undergo aldol addition on treatment with base.

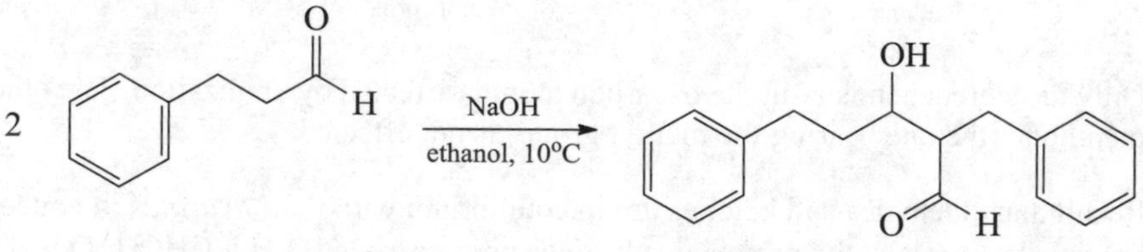

3-Phenylpropanal   2-Benzyl-3-hydroxy-5-phenylpentanal

(*c*) Dehydration of the aldol addition product occurs when the reaction is carried out at elevated temperature.

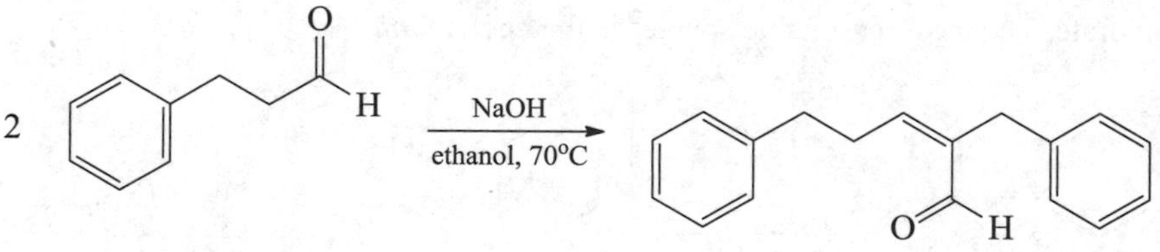

3-Phenylpropanal   2-Benzyl-5-phenyl-2-pentenal

(*d*) Lithium aluminum hydride reduces the aldehyde function to the corresponding primary alcohol.

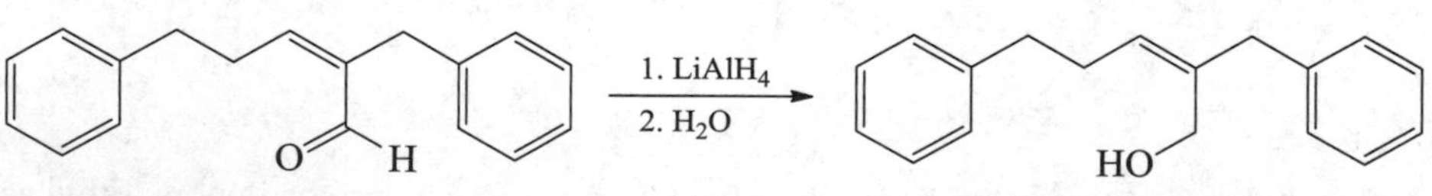

2-Benzyl-5-phenyl-2-pentenal   2-Benzyl-5-phenyl-2-penten-1-ol

(*e*) A characteristic reaction of α,β-unsaturated carbonyl compounds is their tendency to undergo conjugate addition on treatment with weakly basic nucleophiles.

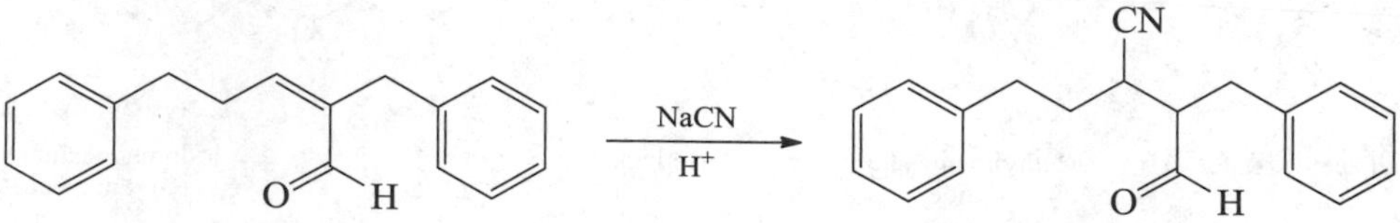

2-Benzyl-5-phenyl-2-pentenal   2-Benzyl-3-cyano-5-phenylpentanal

21.36 (*a*) Chlorination can only occur at the one α carbon containing hydrogens.

1-(*o*-Chlorophenyl)-1-propanone

2-Chloro-1-(*o*-chlorophenyl)-1-propanone

(*b*) The combination of $C_6H_5CH_2SH$ and NaOH yields $C_6H_5CH_2S^-$ (as its sodium salt), which is a good nucleophile and adds to α,β-unsaturated ketones by conjugate addition.

2-Isopropylidene-5-
methylcyclohexanone

2-(1-Benzylthio-1-methylethyl)-5-
methylcyclohexanone (89-90%)

(*c*) The aldehyde given as the starting material is called *furfural* and is based on a furan unit as an aromatic ring. Furfural cannot form an enolate. It reacts with the enolate of acetone in a manner much as benzaldehyde would.

Furfural   Acetone   (Not isolated)   4-Furyl-3-buten-2-one
                                                                                                    (60-66%)

(*d*) Lithium dialkylcuprates transfer an alkyl group to the β-carbon atom of α,β-unsaturated ketones.

2,4,4-Trimethyl-2-cyclohexenone   2,3,4,4-Tetramethylcyclohexanone

A mixture of stereoisomers was obtained in 67% yield in this reaction.

(*e*) Two nonequivalent α-carbon atoms occur in the starting ketone. Although enolate formation is possible at either position, only reaction at the methylene carbon leads to an intermediate that can undergo dehydration.

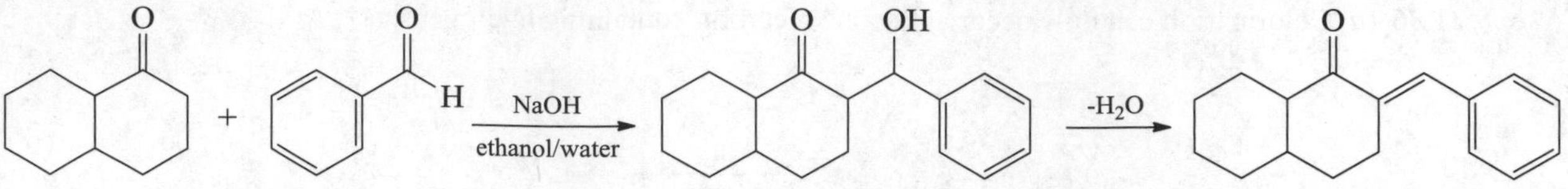

Reaction at the other α position gives an intermediate that cannot dehydrate.

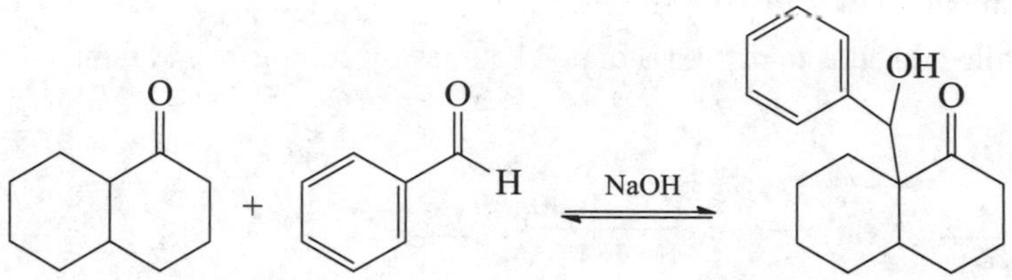

(Cannot dehydrate; reverts to starting material)

(*f*) β-Diketones readily undergo alkylation by primary halides at the most acidic position, on the carbon between the carbonyls.

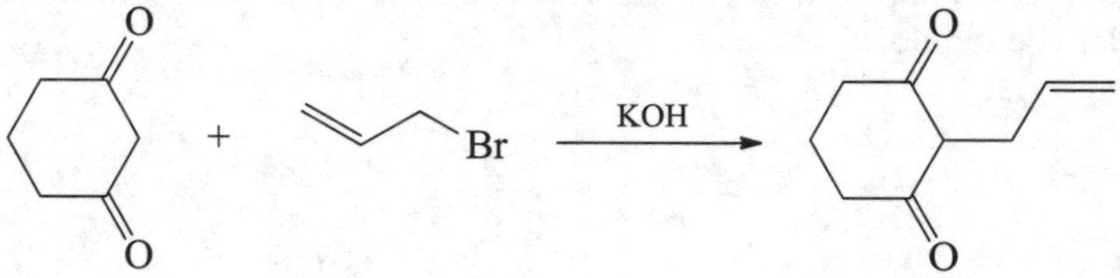

1,3-Cyclohexanedione   Allyl bromide   2-Allyl-1,3-cyclohexanedione
                                                        (75%)

21.37 Bromination of ketones replaces only protons α to the carbonyl group.

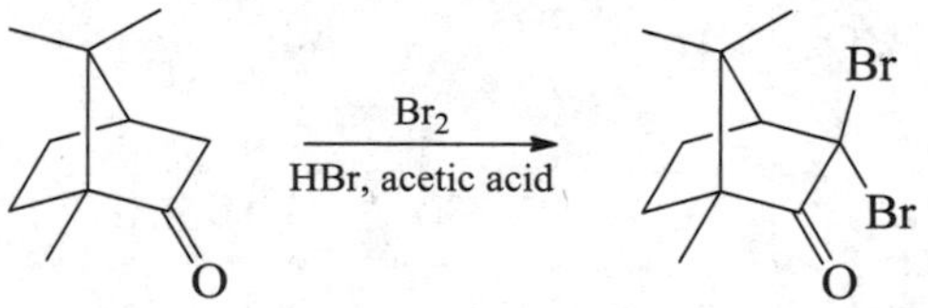

21.38 Bromination can occur at either of the two α-carbon atoms.

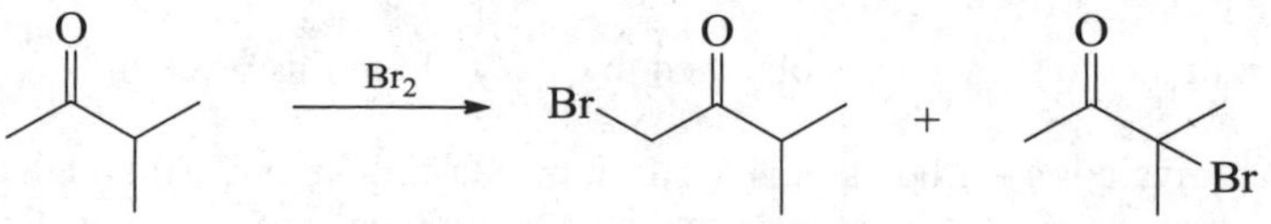

1-Bromo-3-methyl-   3-Bromo-3-methyl-
2-butanone   2-butanone

The ^{1}H NMR spectrum of the major product, compound A, is consistent with the structure of 1-bromo-3-methyl-2-butanone. The minor product B is 3-bromo-3-methyl-2-butanone.

O

Br 　←── δ 1.2 doublet

←── δ 3.0 septet

←── δ 1.2 doublet

δ 4.1 singlet

Compound A

O

←── δ 1.9 singlet

Br

δ 2.5 singlet

←── δ 1.9 singlet

Compound B

21.39 Three dibromination products are possible from a halogenation of 2-butanone.

O

Br

Br

1,1-Dibromo-2-butanone

O

Br

Br

1,3-Dibromo-2-butanone

O

Br

Br

3,3-Dibromo-2-butanone

The product is 1,3-dibromo-2-butanone, on the basis of its ^{1}H NMR spectrum which showed two signals at low field. One is a two-proton singlet as δ 4.6 assignable to CH_2Br and the other a one-proton quartet at δ 5.2 assignable to CHBr.

21.40 (*a*) On treatment with base, ethyl acetoacetate is converted to its enolate, which reacts as a nucleophile toward 1-bromobutane.

O   O

O

+

Br

$\xrightarrow[\text{ethanol}]{NaOCH_2CH_3}$

O   O

O

Ethyl acetoacetate

1-Bromobutane

Ethyl 2-acetylhexanoate

(*b*) Saponification and decarboxylation of the product from part (a) completes the preparation of 2-heptanone by the acetoacetic ester synthesis.

O   O

O

$\xrightarrow{\begin{array}{l}\text{1.NaOH, H}_2\text{O}\\\text{2. H}_3\text{O}^+\\\text{3. heat}\end{array}}$

O

Ethyl 2-acetylhexanoate

2-Heptanone

(*c*) The enolate of acetophenone attacks the carbonyl group of diethyl carbonate.

O

+

O

O   O

$\xrightarrow{\begin{array}{l}\text{1. NaOCH}_2\text{CH}_3\\\text{2. H}_3\text{O}^+\end{array}}$

O   O

O

Acetophenone

Diethyl carbonate

3-Oxo-3-phenylpropanoate

(*d*) Diethyl oxalate acts as an acylating agent toward the enolate of acetone.

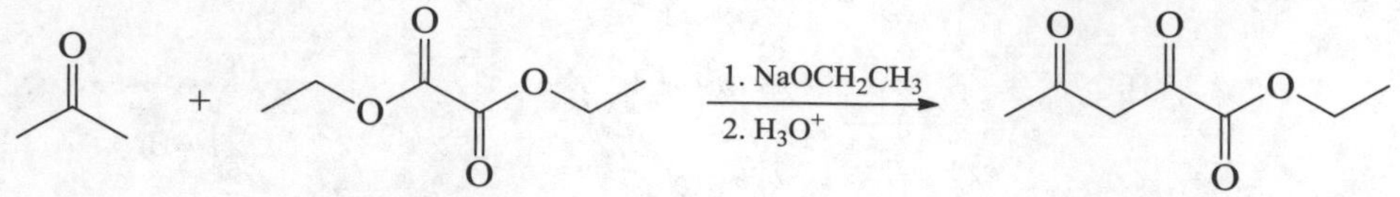

Acetone   Diethyl oxalate   Ethyl 2,4-dioxopentanoate

(*e*) The first stage of the malonic ester synthesis is the alkylation of diethyl malonate with an alkyl halide.

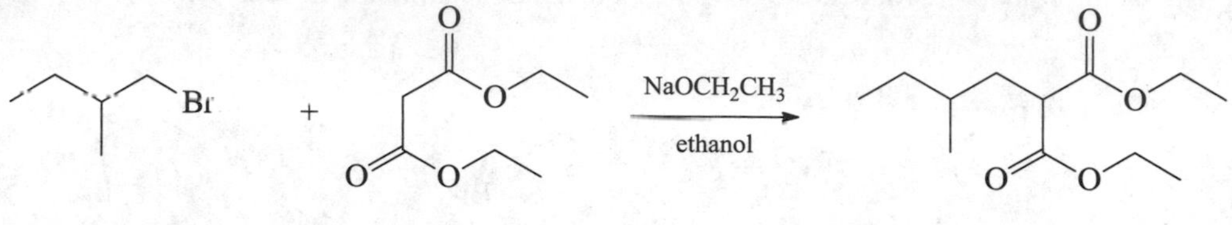

1-Bromo-2-methylbutane   Diethyl malonate   Diethyl 3-methylpentane-1,1-dicarboxylate

(*f*) Alkylation of diethyl malonate is followed by saponification and decarboxylation to give a carboxylic acid.

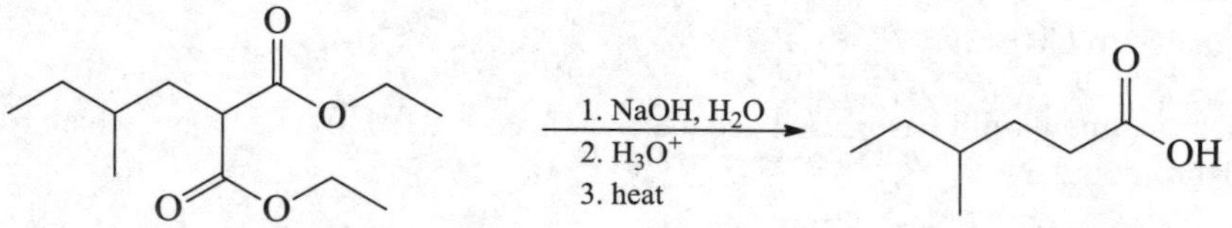

Diethyl 3-methylpentane-1,1-dicarboxylate   4-Methylhexanoic acid

(*g*) The anion of diethyl malonate undergoes Michael addition to 6-methyl-2-cyclohexenone.

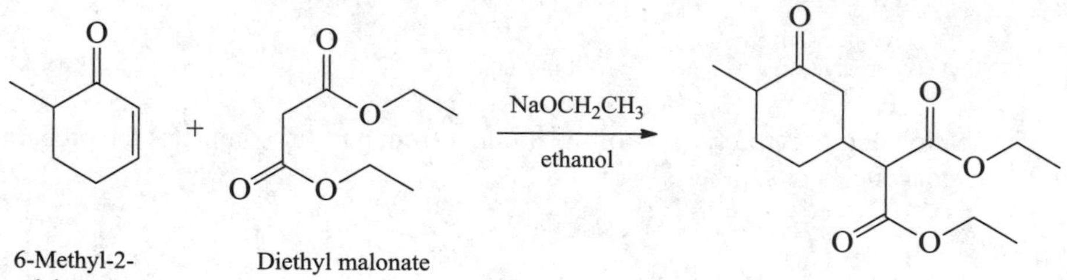

6-Methyl-2-
cyclohexenone   Diethyl malonate

Diethyl 2-(4-methyl-3-oxocyclohexyl)malonate

(*h*) Acid hydrolysis converts the diester in part (*g*) to a malonic acid derivative, which then undergoes decarboxylation.

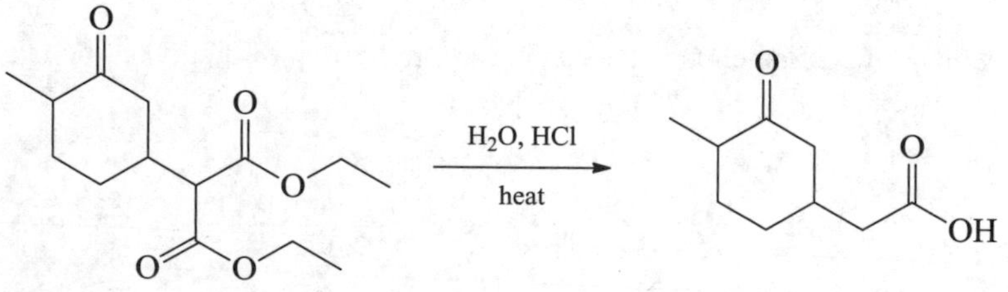

Diethyl 2-(4-methyl-3-oxocyclohexyl)malonate   (4-Methyl-3-oxocyclohexyl)acetic acid

(*i*) Lithium diisopropylamide (LDA) is used to convert esters quantitatively to their enolate ions. In this reaction, the enolate of *tert*-butyl acetate adds to benzaldehyde.

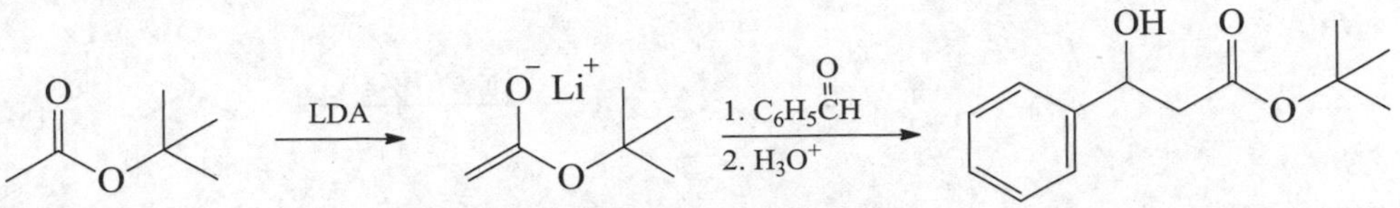

tert-Butyl acetate   Lithium enolate of   *tert*-Butyl 3-hydroxy-3-
                      tert-Butyl acetate   phenylpropanoate

21.41 (*a*) Ethyl acetoacetate is converted to its enolate ion with sodium ethoxide; this anion then acts as a nucleophile toward 1-bromopentane.

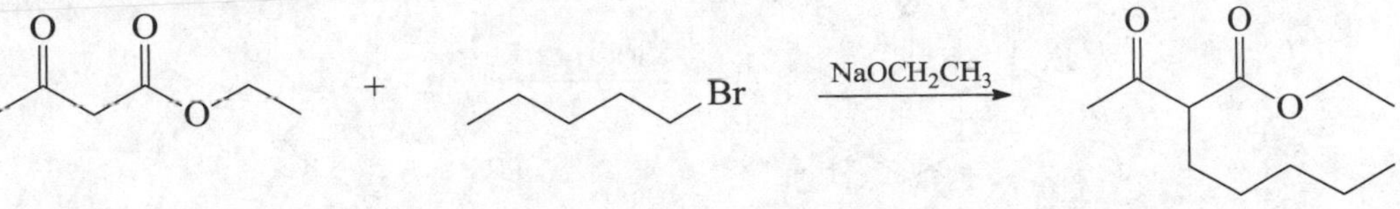

Ethyl acetoacetate   1-Bromopentane   Ethyl 2-acetylheptanoate

(*b*) Saponification and decarboxylation of the product in part (*a*) yields 2-octanone.

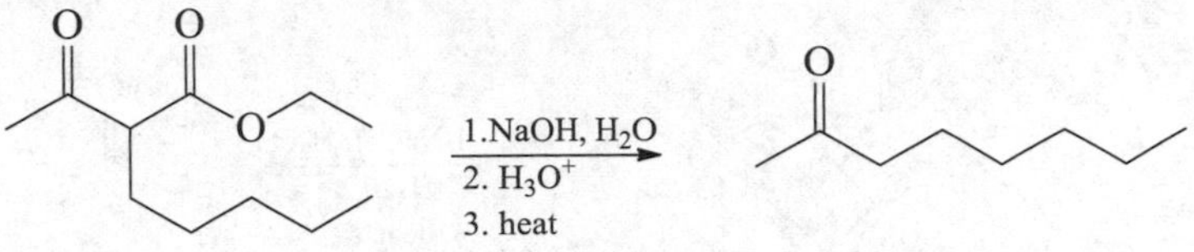

Ethyl 2-acetylheptanoate   2-Octanone

(*c*) The product derived from the reaction in part (*a*) can be alkylated again:

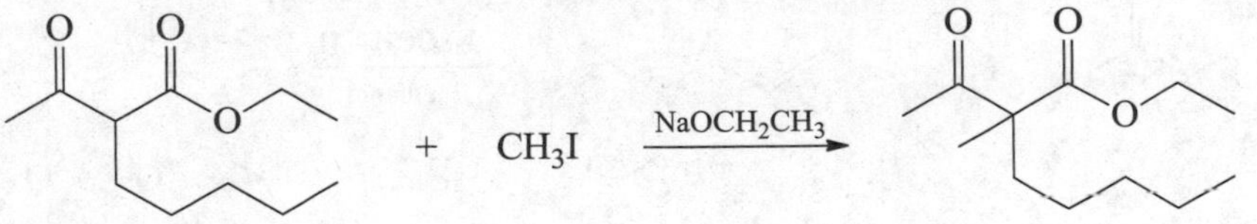

Ethyl 2-acetylheptanoate   Ethyl 2-acetyl-2-methylheptanoate

(*d*) The dialkylated derivative of acetoacetic ester formed in part (*c*) can be converted to a ketone by saponification and decarboxylation.

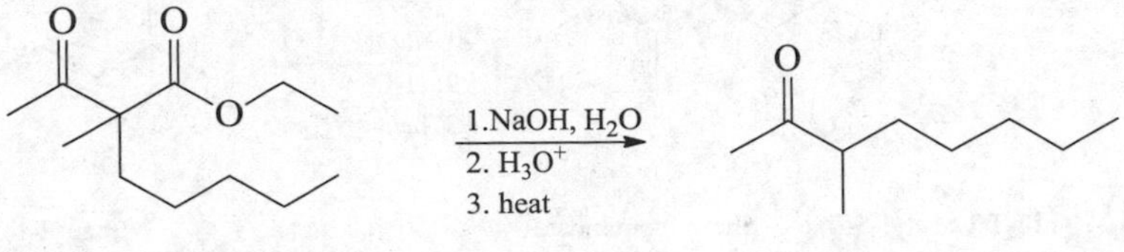

Ethyl 2-acetyl-2-methylheptanoate   3-Methyl-2-octanone

(*e*) The anion of ethyl acetoacetate acts as a nucleophile toward 1-bromo-3-chloropropane. Bromide is a better leaving group than chloride and is displaced preferentially.

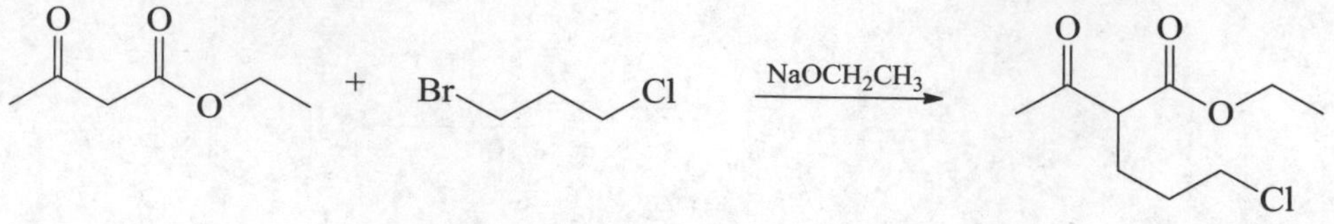

Ethyl acetoacetate   1-Bromo-3-chloropropane   Ethyl 2-acetyl-5-chloropentanoate

(*f*) Treatment of the product of part (*e*) with sodium ethoxide gives an enolate ion that cyclizes by intramolecular nucleophilic substitution of chloride.

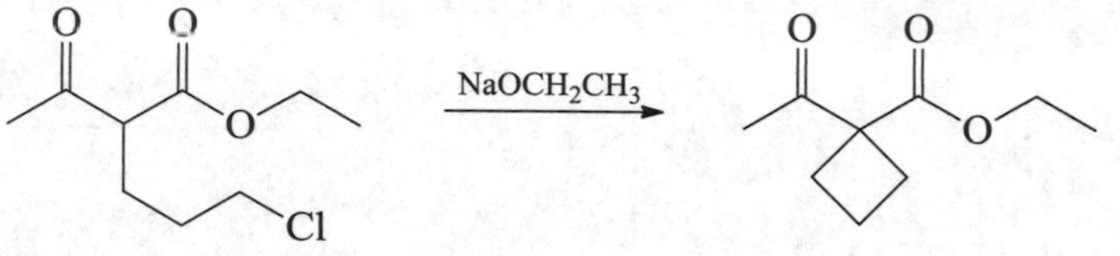

Ethyl 2-acetyl-5-chloropentanoate   Ethyl 1-acetylcyclobutanecarboxylate

(*g*) Cyclobutyl methyl ketone is formed by saponification and decarboxylation of the product in part (*f*).

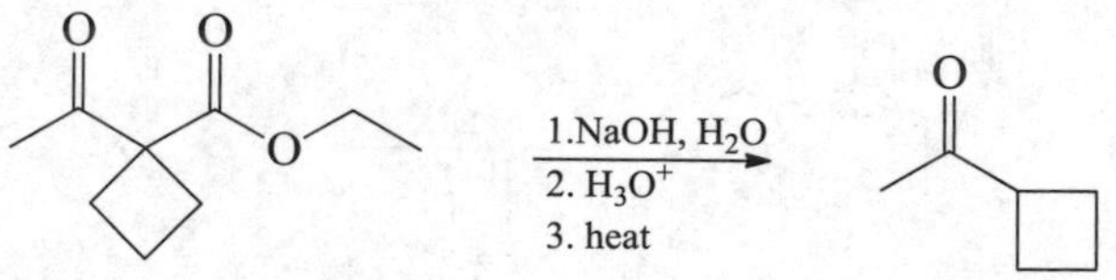

Ethyl 1-acetylcyclobutanecarboxylate   Cyclobutyl methyl ketone

(*h*) Ethyl acetoacetate undergoes Michael addition to phenyl vinyl ketone in the presence of base.

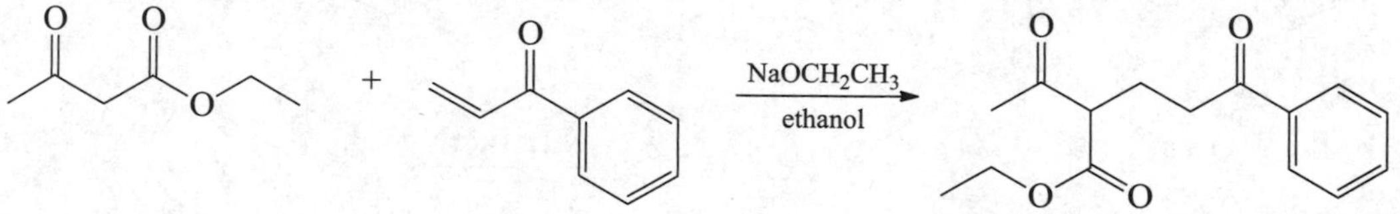

Ethyl acetoacetate   Phenyl vinyl ketone   Ethyl 2-acetyl-5-oxo-5-phenylpentanoate

(*i*) A diketone results from saponification and decarboxylation of the Michael adduct.

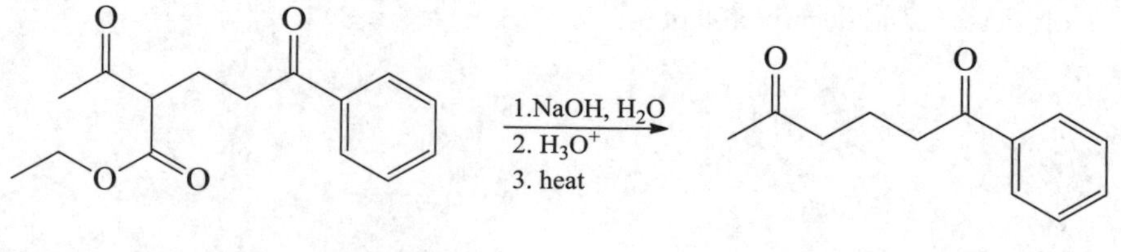

Ethyl 2-acetyl-5-oxo-5-phenylpentanoate   1-Phenyl-1,5-hexanedione

21.42 Diethyl malonate reacts with the reagents given in the preceding problem in a manner analogous to that of ethyl acetoacetate.

(*a*)

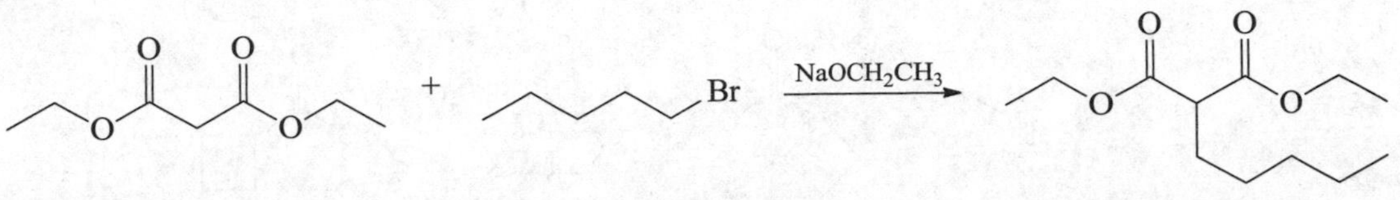

Diethyl malonate   1-Bromopentane   Diethyl 1,1-hexanedicarboxylate

(*b*)

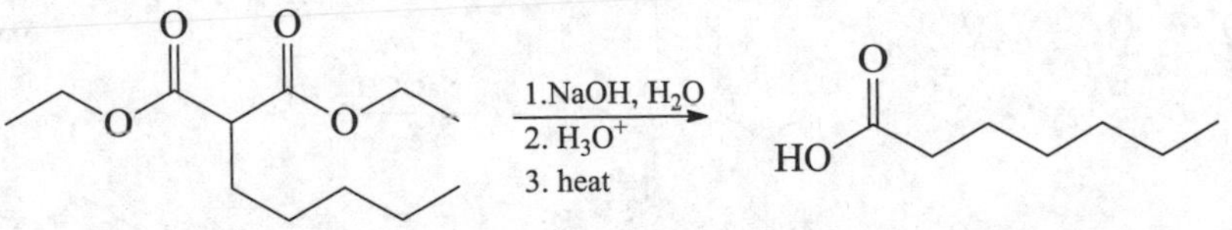

Diethyl 1,1-hexanedicarboxylate   Heptanoic acid

(*c*)

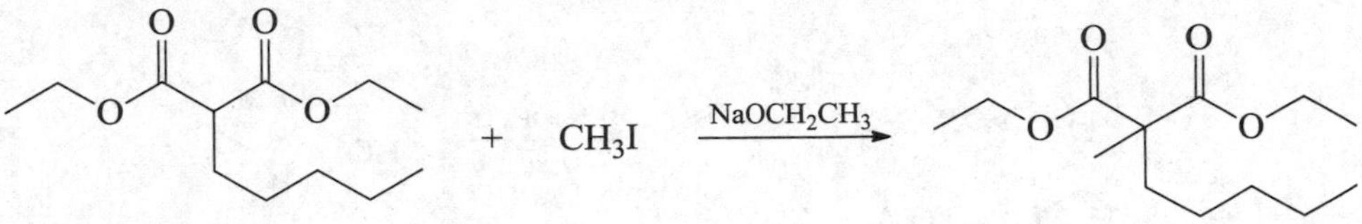

Diethyl 1,1-hexanedicarboxylate   Diethyl 2,2-heptanedicarboxylate

(*d*)

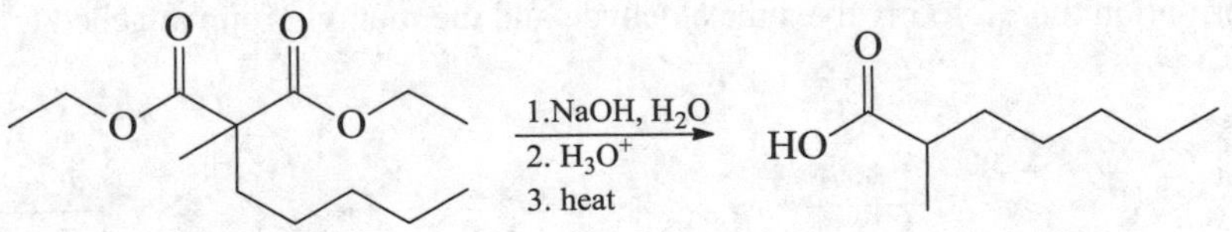

Diethyl 2,2-heptanedicarboxylate   2-Methylheptanoic acid

(*e*)

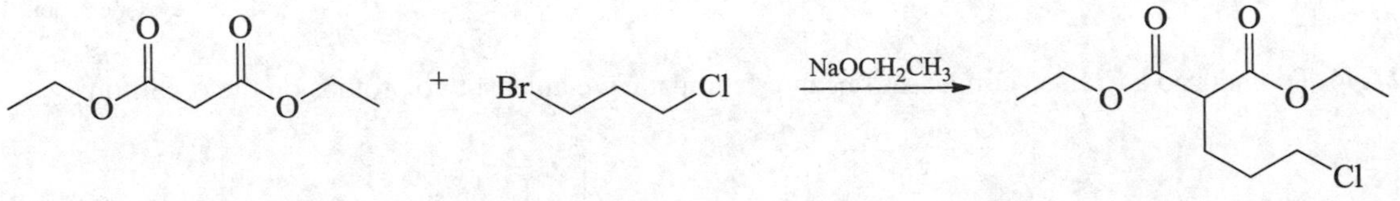

Diethyl malonate   1-Bromo-3-chloropropane   Diethyl 4-chloro-1,1-butanedicarboxylate

(*f*)

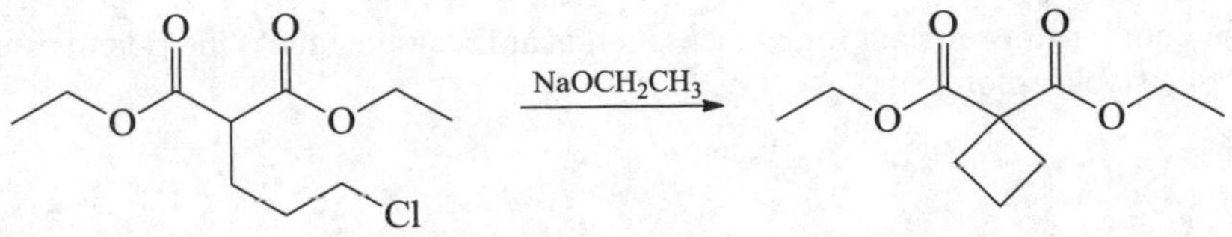

Diethyl 4-chloro-1,1-butanedicarboxylate   Diethyl cyclobutane-1,1-dicarboxylate

(*g*)

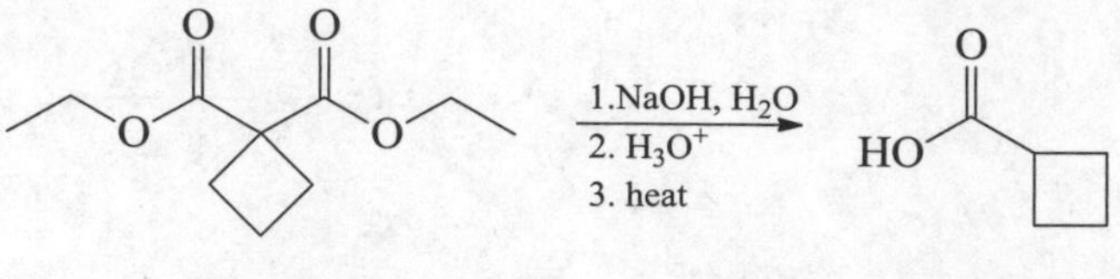
1.NaOH, H_2O
2. H_3O^+
3. heat
HO

Diethyl cyclobutane-1,1-dicarboxylate　　Cyclobutyl methyl ketone

(*h*)

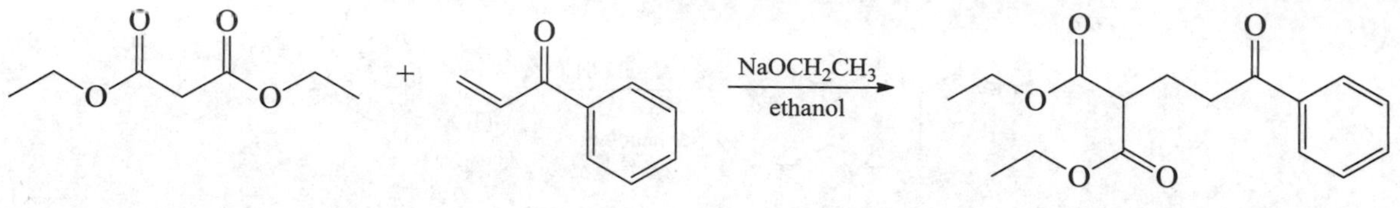
$NaOCH_2CH_3$
ethanol

Diethyl malonate　　Phenyl vinyl ketone　　Diethyl 4-oxo-4-phenylbutane-1,1-dicarboxylate

(*i*)

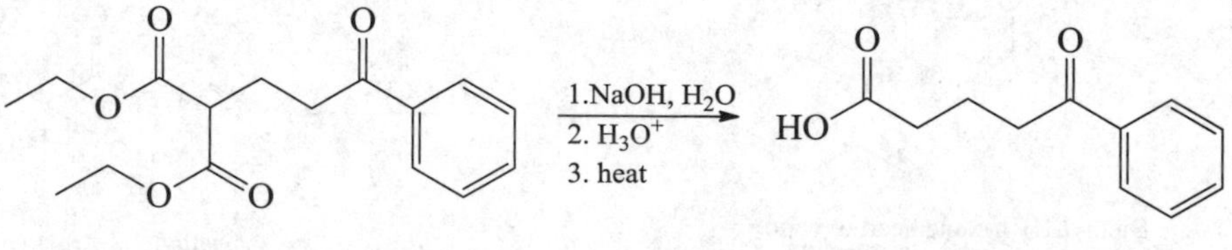
1.NaOH, H_2O
2. H_3O^+
3. heat
HO

Ethyl 2-acetyl-5-oxo-5-phenylpentanoate　　5-Oxo-5-phenylpentanoic acid

21.43 Intramolecular aldol condensations occur best when a five- or six-membered ring is formed. Carbon-carbon bond formation therefore involves the aldehyde and the methyl group attached to the ketone carbonyl.

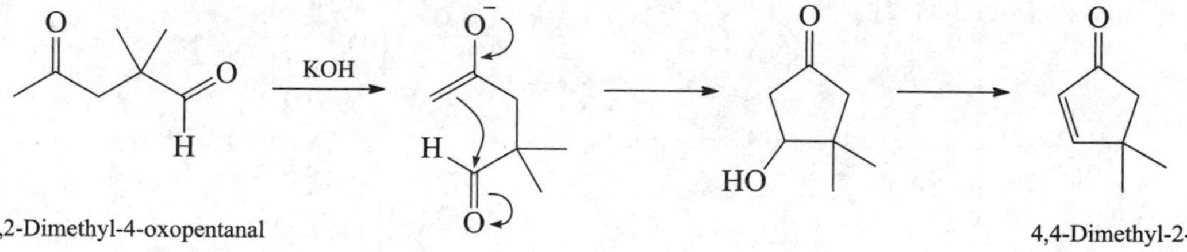
KOH
H
HO

2,2-Dimethyl-4-oxopentanal　　4,4-Dimethyl-2-cyclopentenone (63%)

21.44 To undergo a Claisen condensation, an ester must have at least two protons on the α carbon:

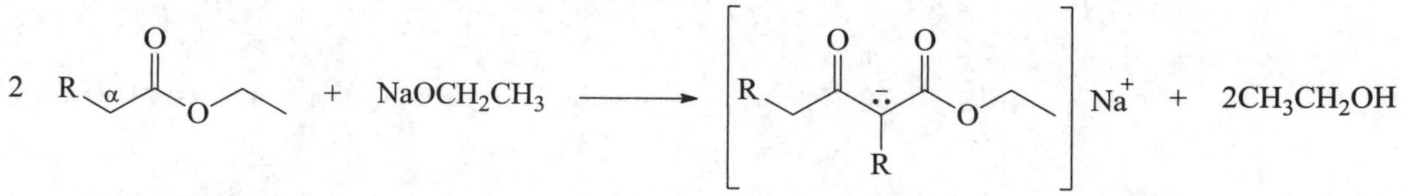
2
R
α
+ $NaOCH_2CH_3$
R
R
Na^+
+ $2CH_3CH_2OH$

The equilibrium constant for condensation is unfavorable unless the β-keto ester can be deprotonated to form a stable anion.

(*a*) Among the esters given, ethyl pentanoate and ethyl 3-methylbutanoate undergo the Claisen condensation.

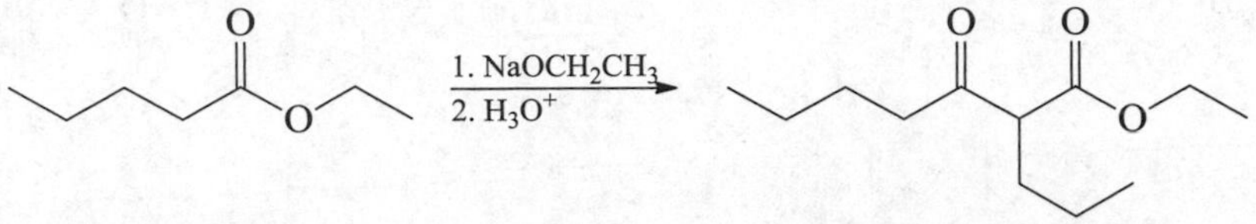

Ethyl pentanoate   Ethyl 3-oxo-2-propylheptanoate

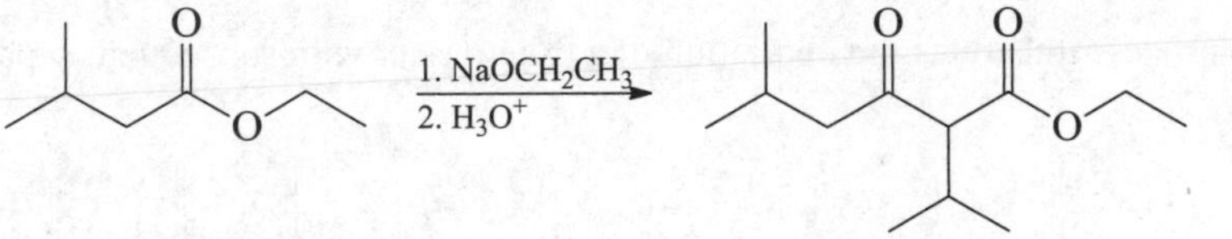

Ethyl 3-methylbutanoate   Ethyl 2-isopropyl-5-methyl-
                                                3-oxohexanoate

(*b*) The Claisen condensation product of ethyl 2-methylbutanoate cannot be deprotonated; the equilibrium constant for its formation is less than 1.

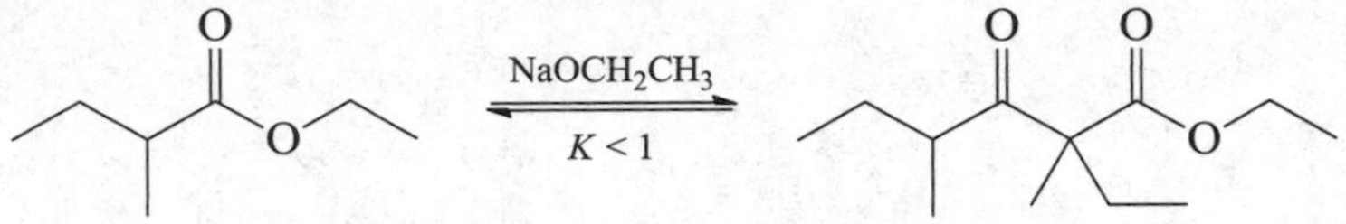

Ethyl 2-methylbutanoate   No protons on α-carbon atom; cannot
                                             form stabilized enolate by deprotonation

(*c*) Ethyl 2,2-dimethylpropanoate has no protons on its α carbon; it cannot form the ester enolate required in the first step of the Claisen condensation.

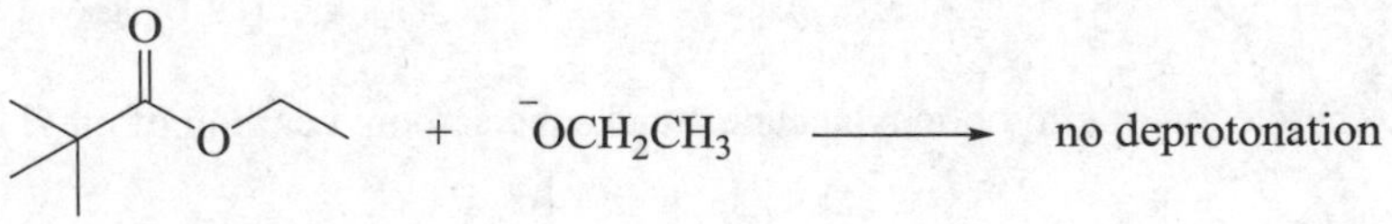

Ethyl 2,2-dimethylpropanoate

21.45 (*a*) The Claisen condensation of ethyl phenylacetate is given by the equation

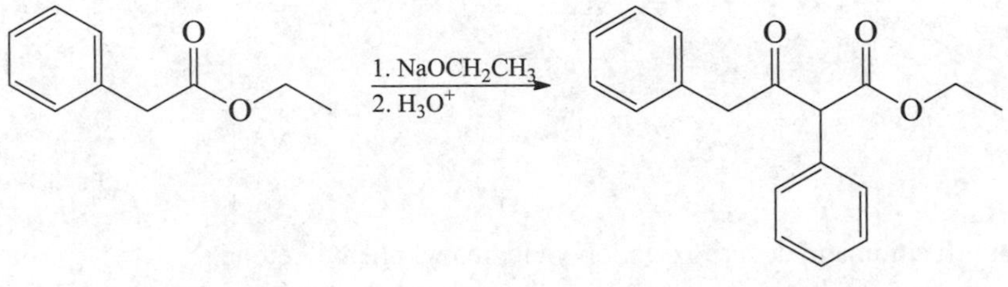

Ethyl phenylacetate   Ethyl 3-oxo-2,4-diphenylbutanoate

(*b*) Saponification and decarboxylation of this β-keto ester gives dibenzyl ketone.

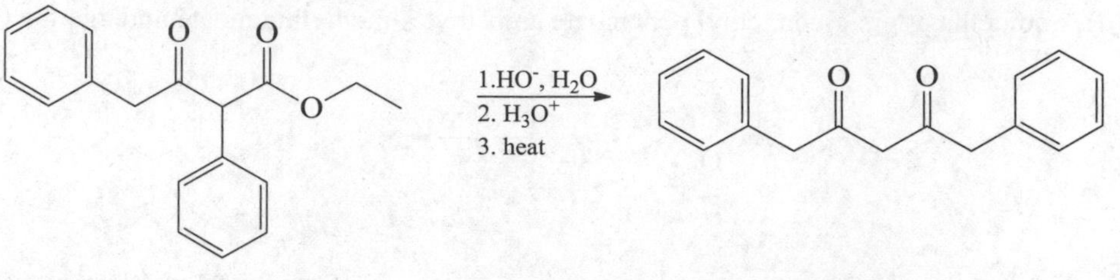

Ethyl 3-oxo-2,4-diphenylbutanoate   Dibenzyl ketone

(*c*) This process illustrates the alkylation of a β-keto ester with subsequent saponification and decarboxylation.

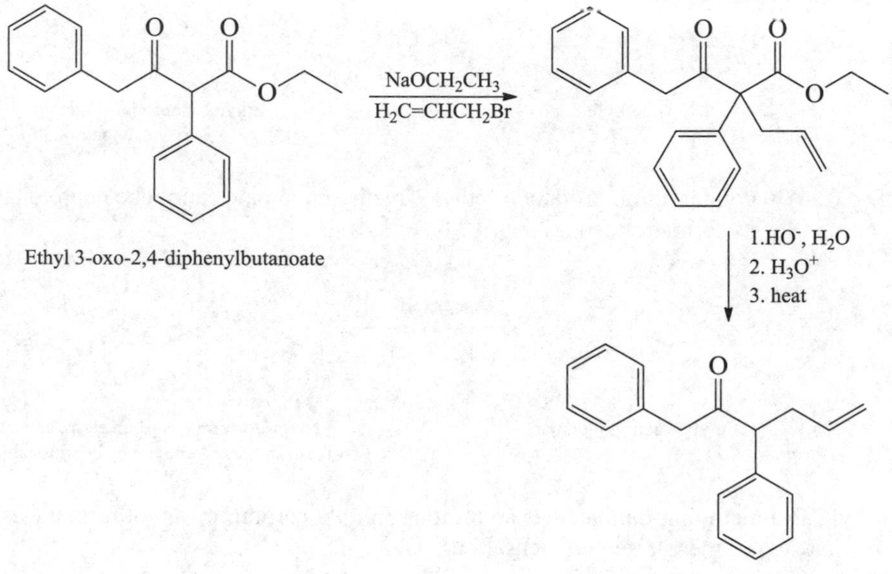

Ethyl 3-oxo-2,4-diphenylbutanoate

1,3-Diphenyl-5-hexen-2-one

(*d*) The enolate ion of ethyl phenylacetate attacks the carbonyl carbon of ethyl benzoate.

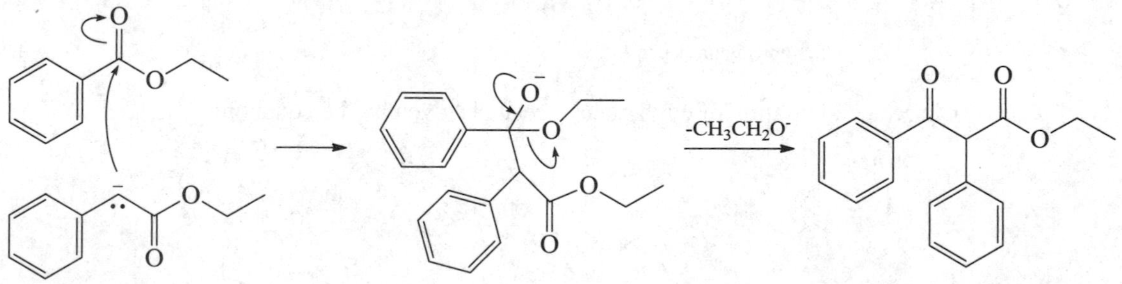

Ethyl 3-oxo-2,3-diphenylpropanoate

(*e*) Saponification and decarboxylation yield benzyl phenyl ketone.

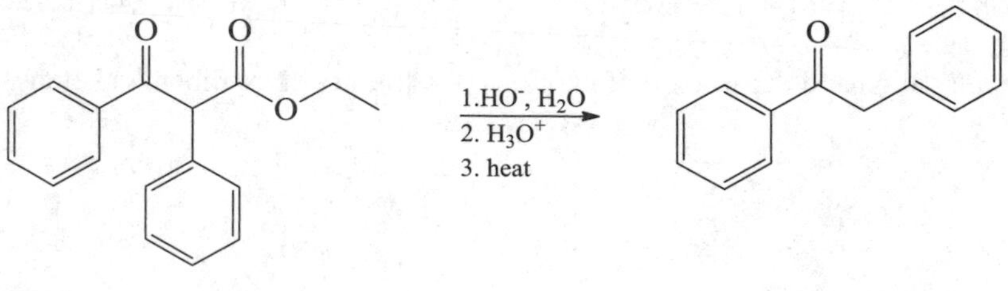

Ethyl 3-oxo-2,3-diphenylpropanoate   Benzyl phenyl ketone (*f*) This sequence is analogous to that of part (*c*).

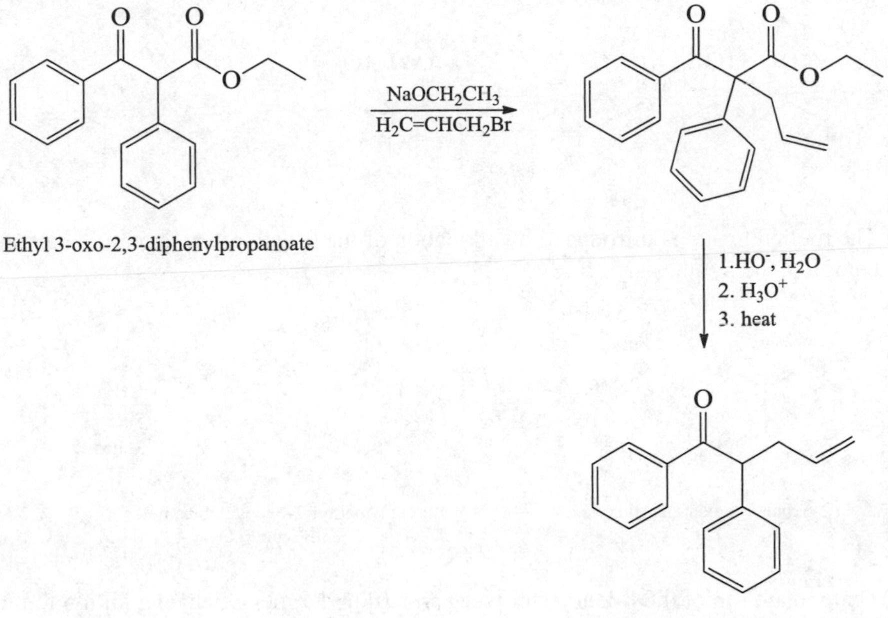

Ethyl 3-oxo-2,3-diphenylpropanoate

1. HO⁻, H₂O
2. H₃O⁺
3. heat

1,2-Diphenyl-4-penten-1-one

21.46 (*a*) The Dieckmann cyclization is the intramolecular version of the Claisen condensation. It employs a diester as starting material.

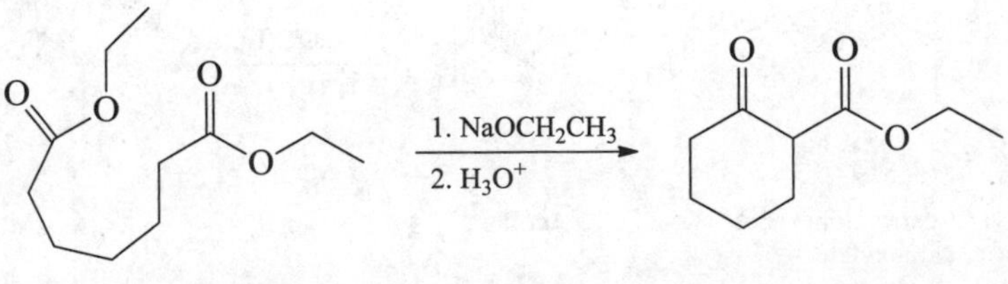

Diethyl heptanedioate   Ethyl (2-oxocyclohexane)carboxylate

(*b*) Acylation of cyclohexanone with diethyl carbonate yields the same β-keto ester formed in part (*a*).

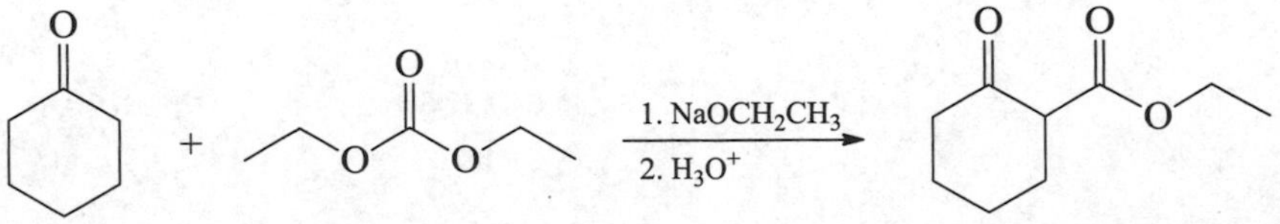

Cyclohexanone   Diethyl carbonate   Ethyl (2-oxocyclohexane)carboxylate

(*c*) The two most stable enol forms are those that involve the proton on the carbon flanked by the two carbonyl groups.

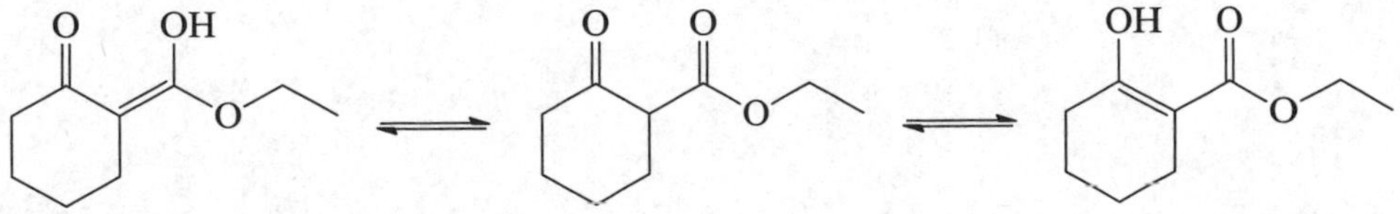

(*d*) Deprotonation of the β-keto ester involves the acidic proton at the carbon flanked by the two carbonyl groups.

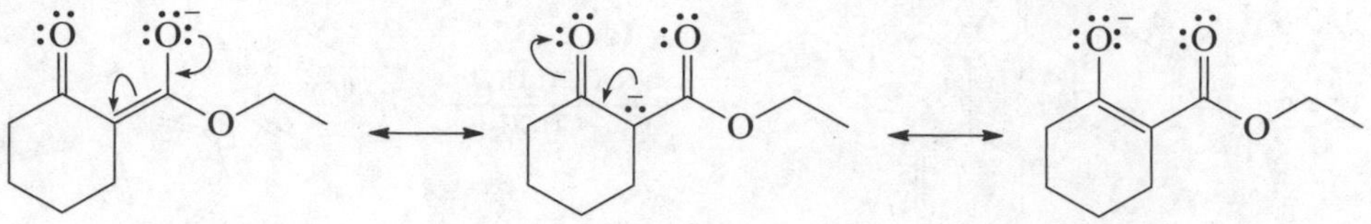

(*e*) The methyl group is introduced by alkylation of the β-keto ester. Saponification and decarboxylation complete the synthesis.

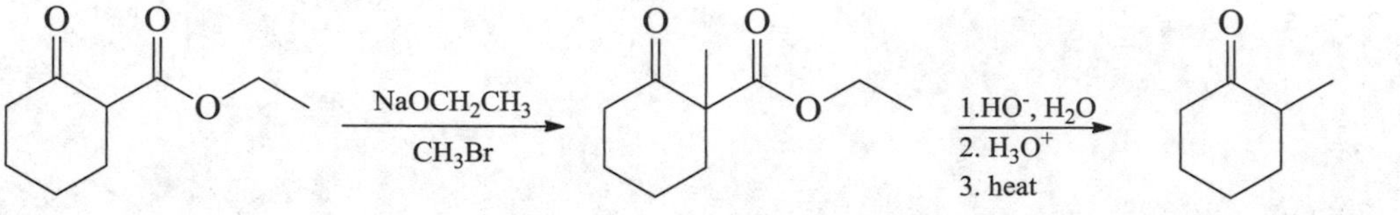

Ethyl (2-oxocyclohexane)carboxylate   Ethyl (1-methyl-2-oxocyclohexane)-   2-Methylcyclohexanone
                                                         carboxylate

(*f*) The enolate ion of the β-keto ester [see part (*d*)] undergoes Michael addition to the carbon–carbon double bond of acrolein.

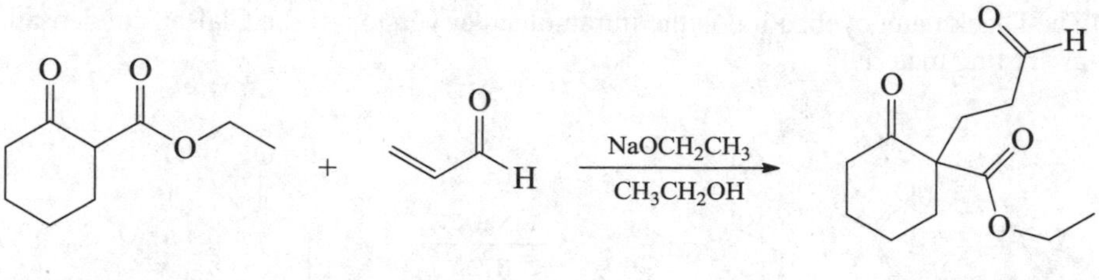

Ethyl (2-oxocyclohexane)-   Acrolein   Michael adduct
carboxylate

21.47 (*a*) Both ester functions in this molecule are β to a ketone carbonyl. Hydrolysis is followed by decarboxylation.

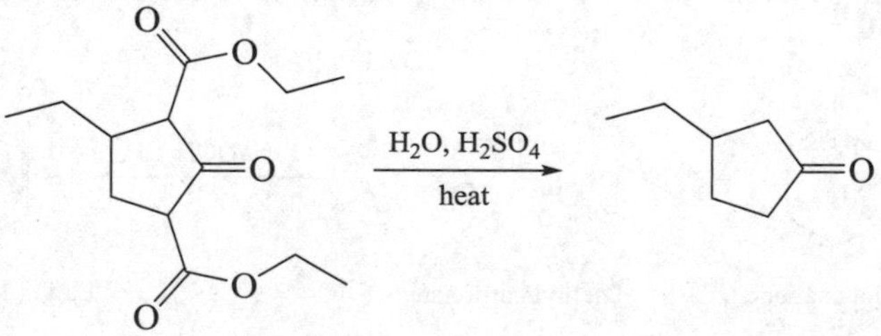

Diethyl   3-Ethylcyclopentanone
3-ethylcyclopentanone-2,5-dicarboxylate   ($C_7H_{12}O$)

(*b*) A Dieckmann cyclization occurs, giving a five-membered ring fused to the original three-membered ring.

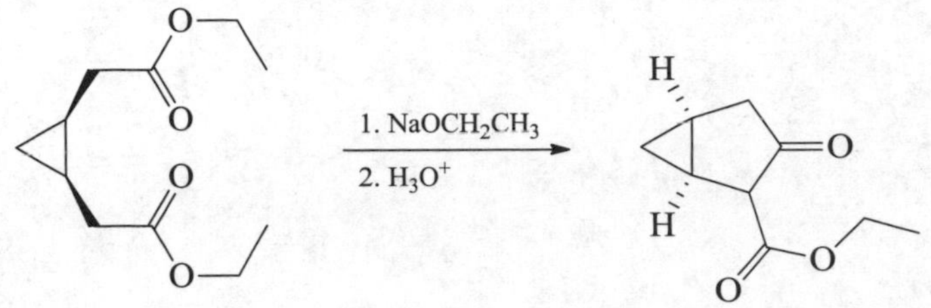

Diethyl *cis*-1,2-
cyclopropanediacetate

1. NaOCH$_2$CH$_3$
2. H$_3$O$^+$

Diethyl bicyclo[3.1.0]-
hexan-3-one-2-carboxylate
(C$_9$H$_{12}$O$_3$, 79%)

(*c*) The compound given in the problem contains three functionalities that can undergo acid-catalyzed hydrolysis: an acetal and two equivalent ester groups. Hydrolysis yields 3-oxo-1,1-cyclobutanedicarboxylic acid and 2 moles each of methanol and 2-propanol. The hydrolysis product is a malonic acid derivative that decarboxylates on heating. The final product of the reaction is 3-oxocyclobutanecarboxylic acid (C$_5$H$_6$O$_3$).

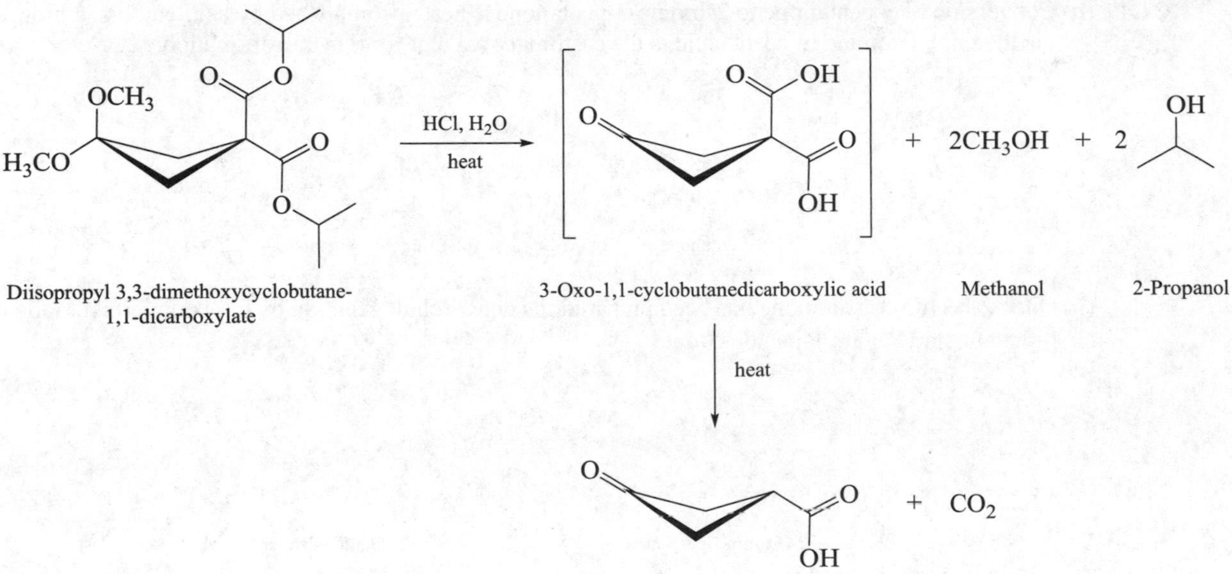

Diisopropyl 3,3-dimethoxycyclobutane-
1,1-dicarboxylate

HCl, H$_2$O
heat

3-Oxo-1,1-cyclobutanedicarboxylic acid

+ 2CH$_3$OH + 2

Methanol   2-Propanol

heat

3-Oxocyclobutanedicarboxylic acid

+ CO$_2$

Carbon
dioxide

Synthetic Applications

21.48 Styrene oxide will be attacked by the anion of diethyl malonate at its less hindered ring position.

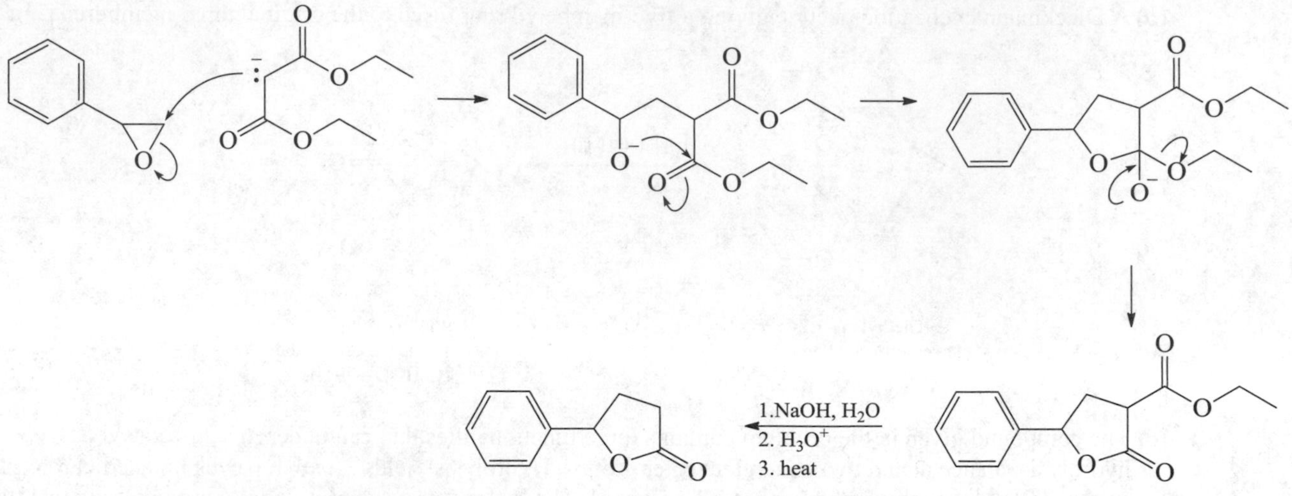

The product is 4-phenylbutanolide. It has been prepared in 72% yield by this procedure.

21.49 (*a*) Conversion of 3-pentanone to 2-bromo-3-pentanone is best accomplished by acid-catalyzed bromination via the enol. Bromine in acetic acid is the customary reagent for this transformation.

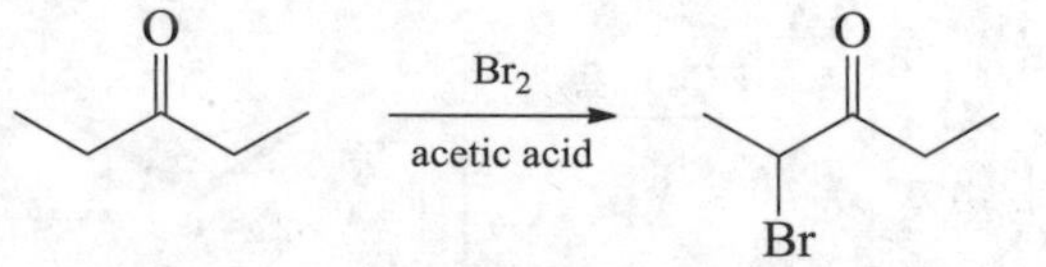

3-Pentanone $\xrightarrow[\text{acetic acid}]{Br_2}$ 2-Bromo-3-pentanone

(*b*) Once 2-bromo-3-pentanone has been prepared, its dehydrohalogenation by base converts it to the desired α, β-unsaturated ketone 1-penten-3-one.

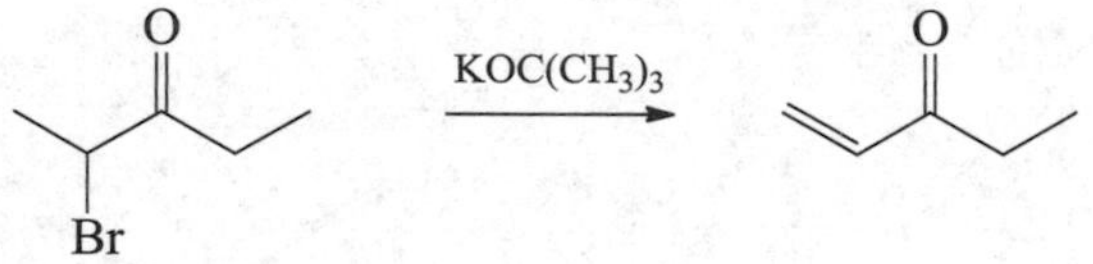

2-Bromo-3-pentanone $\xrightarrow{KOC(CH_3)_3}$ 1-Penten-3-one

Potassium *tert*-butoxide is a good base for bringing about elimination reactions of secondary alkyl halides; suitable solvents include *tert*-butyl alcohol and dimethyl sulfoxide.

(*c*) Reduction of the carbonyl group of 1-penten-3-one converts it to the desired alcohol.

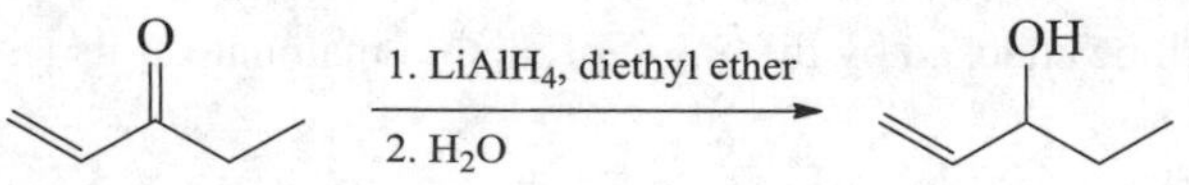

1-Penten-3-one $\xrightarrow[\text{2. } H_2O]{\text{1. LiAlH}_4, \text{ diethyl ether}}$ 1-Penten-3-ol

Sodium borohydride would also be suitable; catalytic hydrogenation would not because reduction of the carbon-carbon double bond would accompany reduction of the carbonyl group.

(*d*) Conversion of 3-pentanone to 3-hexanone requires addition of a methyl group to the β-carbon atom.

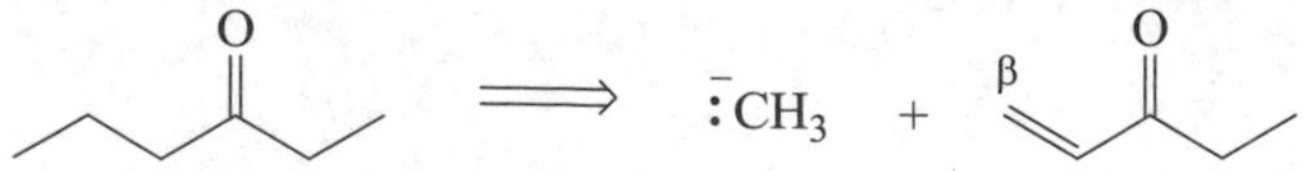

$\Longrightarrow$ $^-:CH_3$ +   β

The best way to add an alkyl group to the β carbon of a ketone is via conjugate addition of a dialkylcuprate reagent to an α,β-unsaturated ketone.

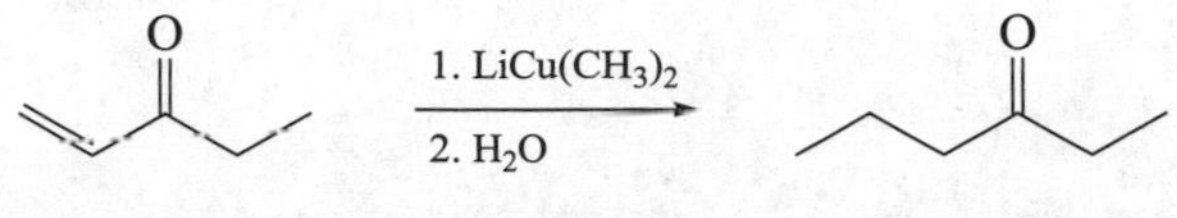

1-Penten-3-one
[prepared as described in part (*b*)]

3-Hexanone

(*e*) The compound to be prepared is the mixed aldol condensation product of 3-pentanone and benzaldehyde.

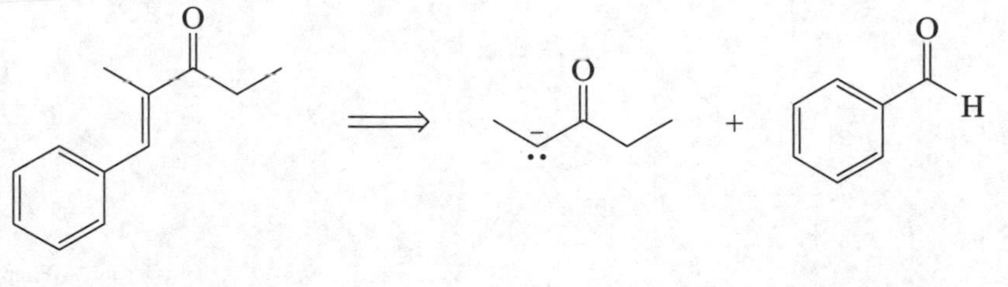

2-Methyl-1-phenyl-
1-penten-3-one

The desired reaction sequence is

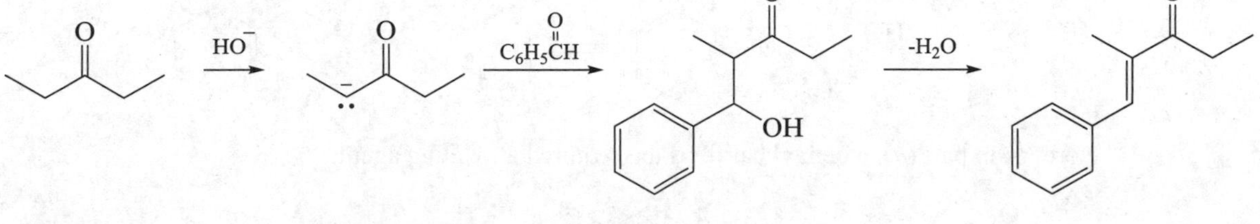

3-Pentanone

Enolate of 3-pentanone

Aldol addition product
(not isolated; dehydration
occurs under conditions
of its formation)

2-Methyl-1-phenyl-
1-penten-3-one

21.50 (*a*) First write out the structure of 4-phenyl-2-butanone and identify the synthon that is derived from ethyl acetoacetate.

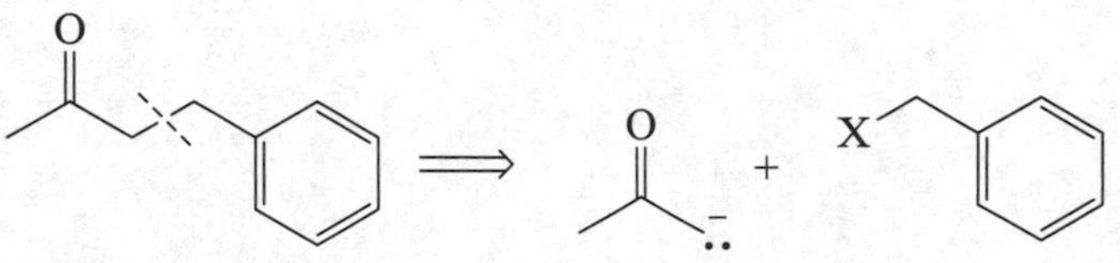

Therefore, carry out the acetoacetic ester synthesis using a benzyl halide as the alkylating agent.

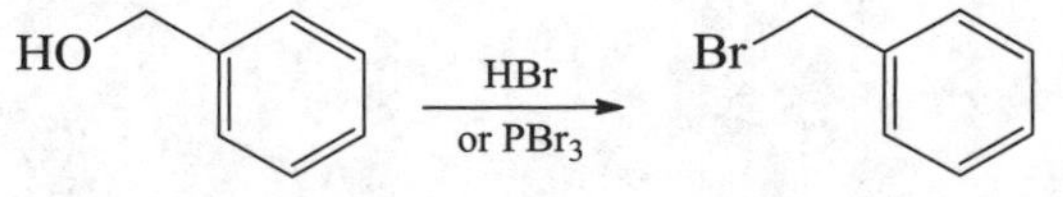

Benzyl alcohol

Benzyl bromide

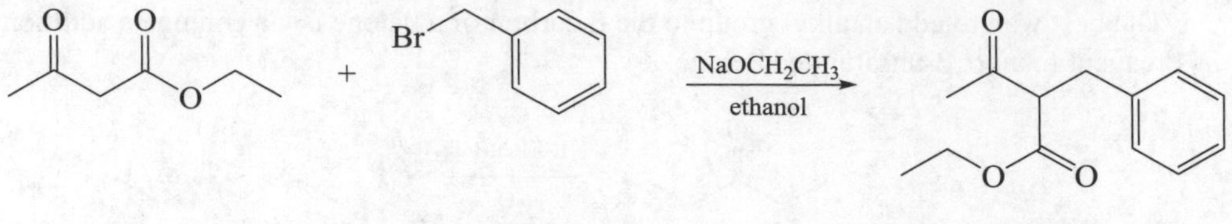

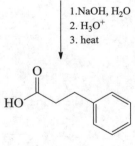

Ethyl acetoacetate + Benzyl bromide → (NaOCH₂CH₃ / ethanol) → Ethyl 2-benzyl-3-oxobutanoate

$$\text{Ethyl acetoacetate} + \text{Benzyl bromide} \xrightarrow[\text{ethanol}]{\text{NaOCH}_2\text{CH}_3} \text{Ethyl 2-benzyl-3-oxobutanoate}$$

1. NaOH, H_2O
2. H_3O^+
3. heat

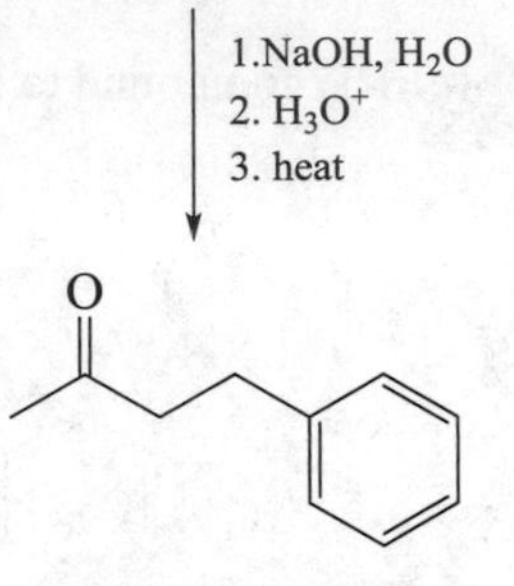

4-Phenyl-2-butanone

(b) Identify the synthon in 3-phenylpropanoic acid that is derived from malonic ester by disconnecting the molecule at its α-carbon atom.

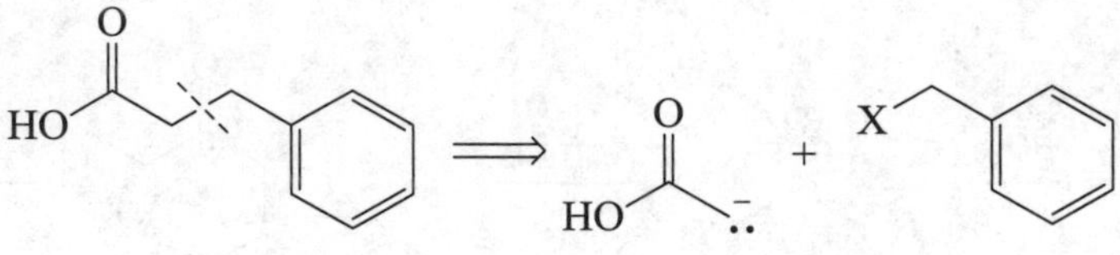

Here, as in part (a), a benzyl halide is the required alkylating agent.

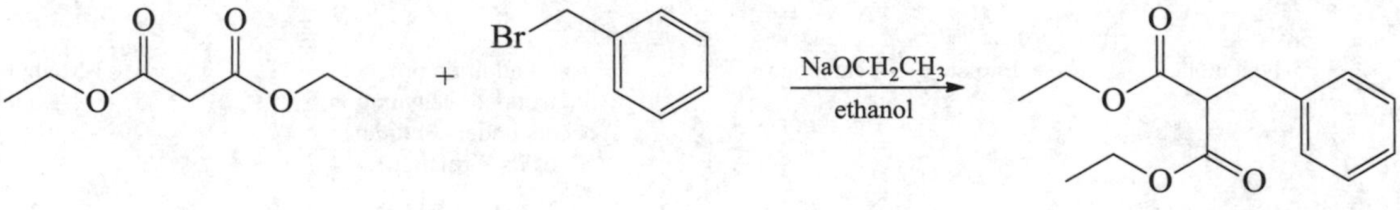

$$\text{Diethyl malonate} + \text{Benzyl bromide} \xrightarrow[\text{ethanol}]{\text{NaOCH}_2\text{CH}_3} \text{Diethyl benzylmalonate}$$

Diethyl malonate + Benzyl bromide → Diethyl benzylmalonate

1. NaOH, H_2O
2. H_3O^+
3. heat

3-Phenylpropanoic acid

(c) In this synthesis, the desired 1,3-diol function can be derived by reduction of a malonic ester derivative. First, propene must be converted to an allyl halide for use as an alkylating agent.

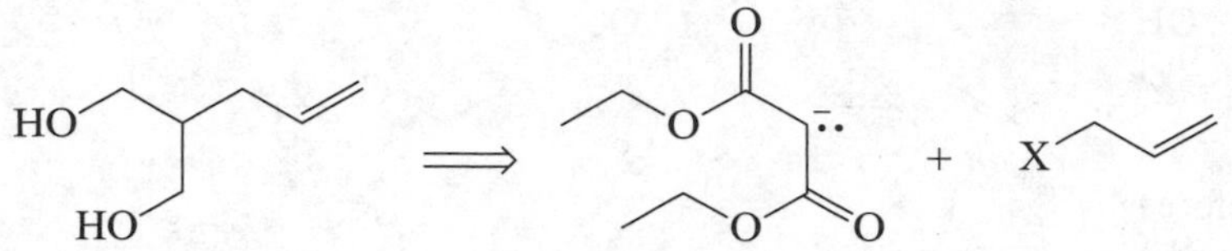

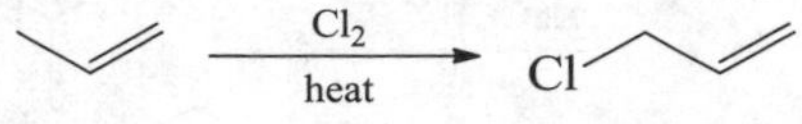

Propene   Allyl chloride

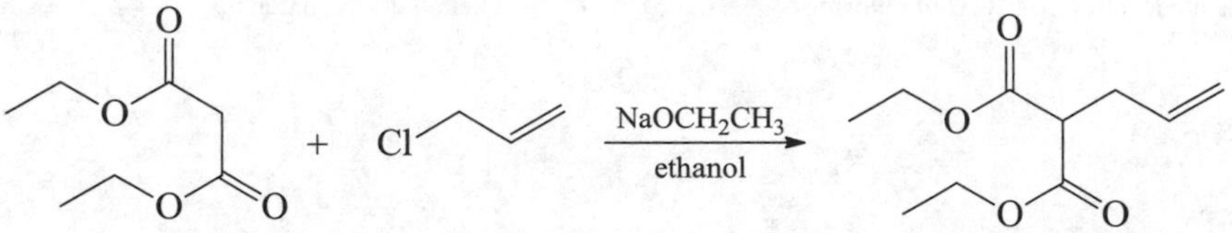

Diethyl malonate   Allyl chloride   Diethyl 2-allylmalonate

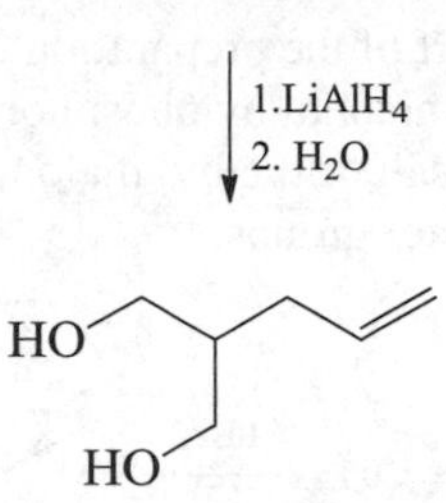

2-Allyl-1,3-propanediol

(*d*) The desired primary alcohol may be prepared by reduction of the corresponding carboxylic acid, which in turn is available from the malonic ester synthesis using allyl chloride, including saponification and decarboxylation of the diester [prepared in part (*c*)].

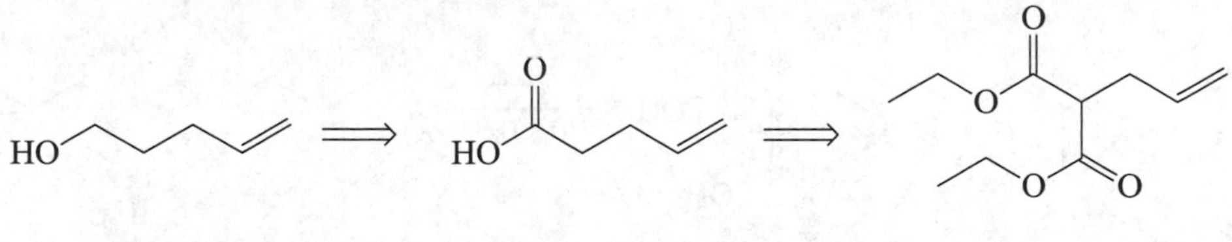

4-Penten-1-ol

The correct sequence of reactions is

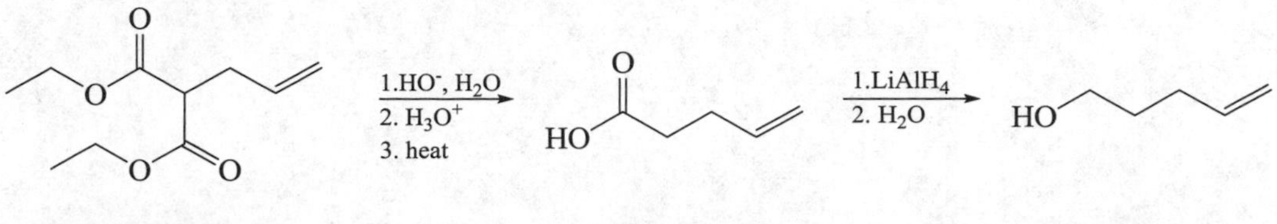

Diethyl 2-allylmalonate   4-Pentenoic acid   4-Penten-1-ol
[prepared as in part (*c*)]

(*e*) The desired product is an alcohol. It can be prepared by reduction of a ketone, which in turn can be prepared by the acetoacetic ester synthesis.

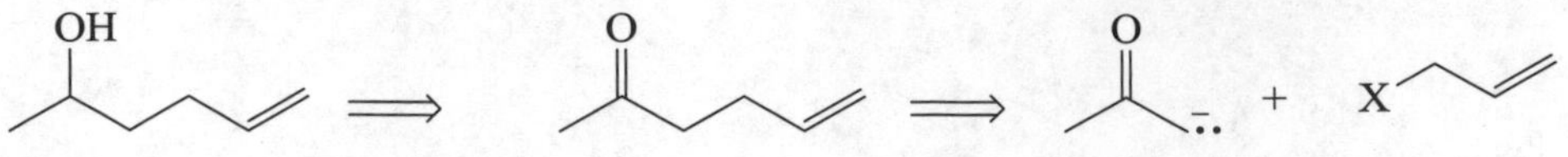

Therefore

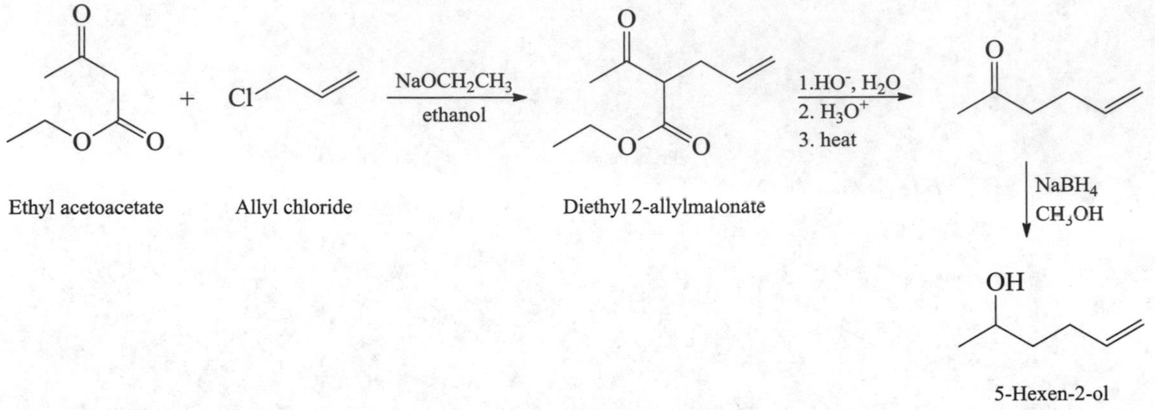

Ethyl acetoacetate   Allyl chloride $\xrightarrow[\text{ethanol}]{\text{NaOCH}_2\text{CH}_3}$ Diethyl 2-allylmalonate

1. HO⁻, H₂O
2. H₃O⁺
3. heat

NaBH₄
CH₃OH

5-Hexen-2-ol

21.51 The heart of the preparation of capsaicin is a malonic ester synthesis. The first step is bromination of the primary alcohol by phosphorous tribromide. The resulting primary alkyl bromide is used to alkylate the sodium salt of diethyl malonate. A substituted malonic acid derivative is obtained following basic hydrolysis of the ester groups.

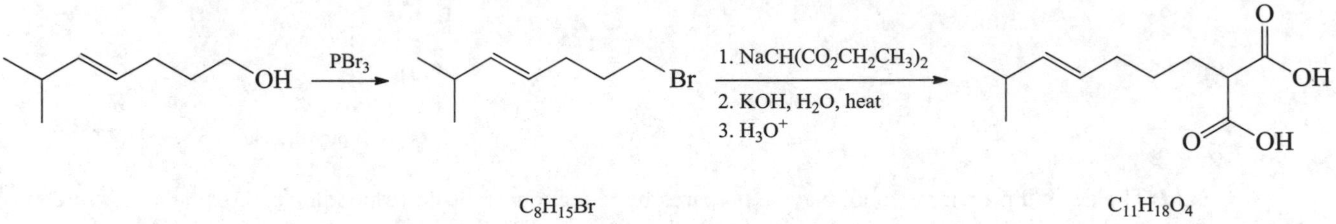

$\xrightarrow{\text{PBr}_3}$

$C_8H_{15}Br$

1. NaCH(CO₂CH₂CH₃)₂
2. KOH, H₂O, heat
3. H₃O⁺

$C_{11}H_{18}O_4$

Malonic acid derivatives undergo decarboxylation on heating.

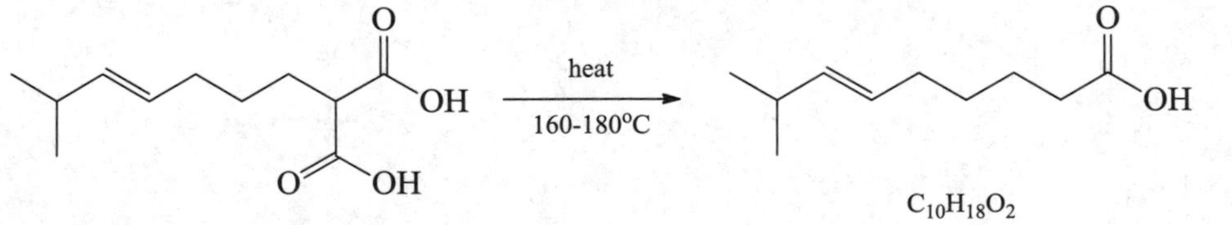

$\xrightarrow[\text{160-180°C}]{\text{heat}}$

$C_{10}H_{18}O_2$

Formation of the amide completes the synthesis of capsaicin.

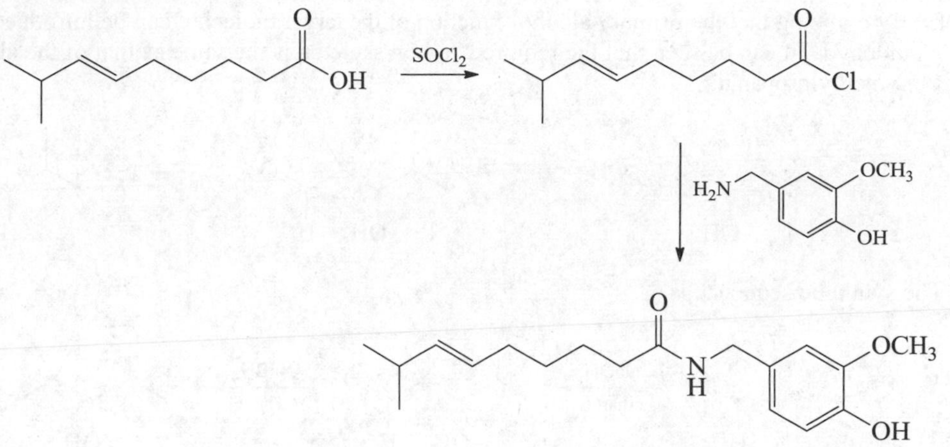

Capsaicin ($C_{18}H_{27}NO_3$)

21.52 We can solve this by retrosynthetic analysis. First, put the water that was lost to form (i) back in (ii), then disconnect the indicated α-carbon–(carbonyl-carbon) bond (iii).

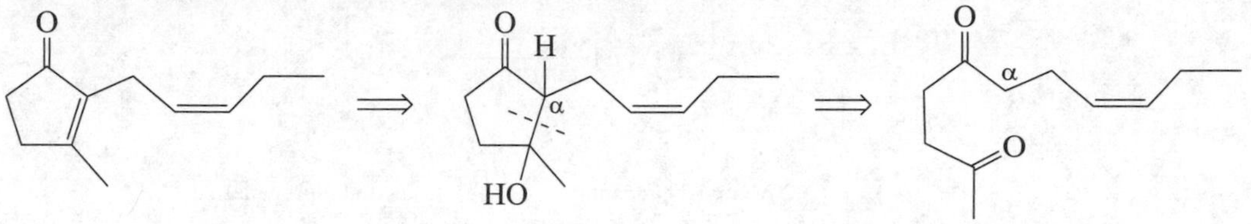

i   ⇒   ii   ⇒   iii

We might expect to see the following product from the enolate formed at the methyl group carbon.

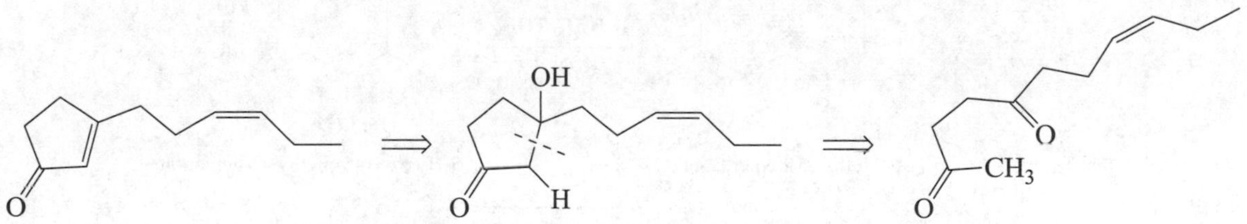

21.53 (*a*) Claisen condensation of ethyl pentanoate gives a β-keto ester:

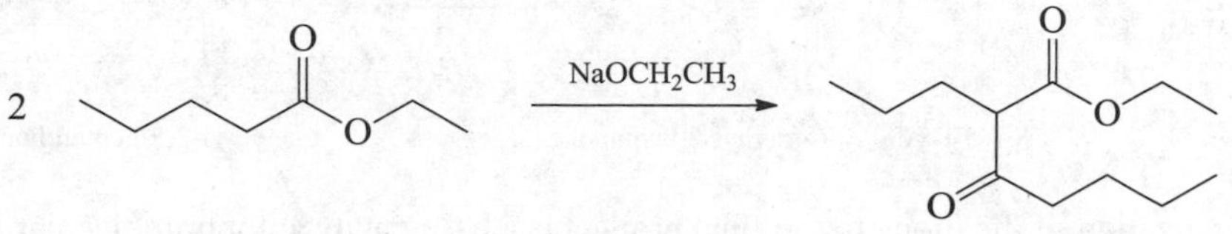

Saponification of the ester, acidification, and decarboxylation gives 5-nonanone.

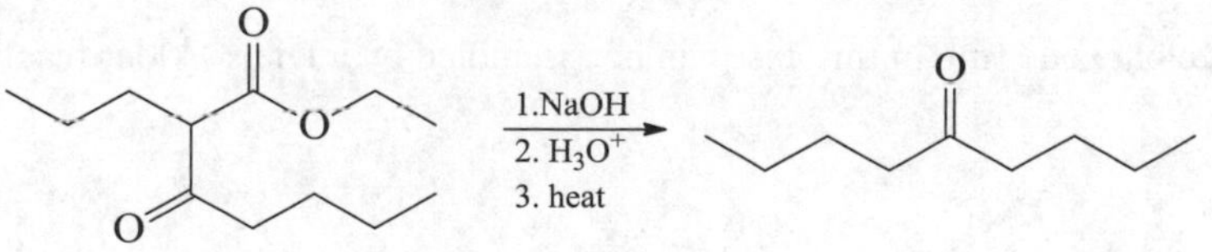

(*b*) By realizing that the primary alcohol function of the target molecule can be introduced by reduction of an aldehyde, it can be seen that the required carbon skeleton is the same as that of the aldol addition product of 2-methylpropanal.

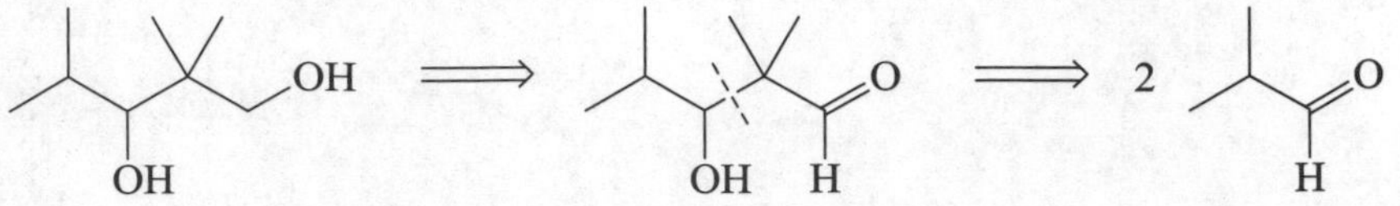

The synthetic sequence is

2-Methylpropanal — 3-Hydroxy-2,2,4-trimethylpentanal — 2,2,4-Trimethyl-1,3-pentanediol

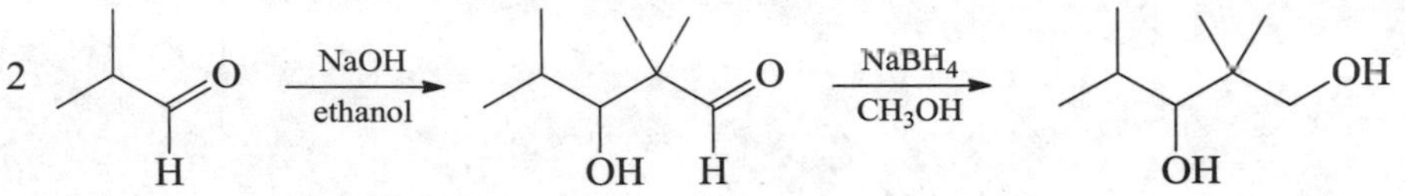

The starting aldehyde is prepared by oxidation of 2-methyl-1-propanol.

2-Methyl-1-propanol — 2-Methylpropanal

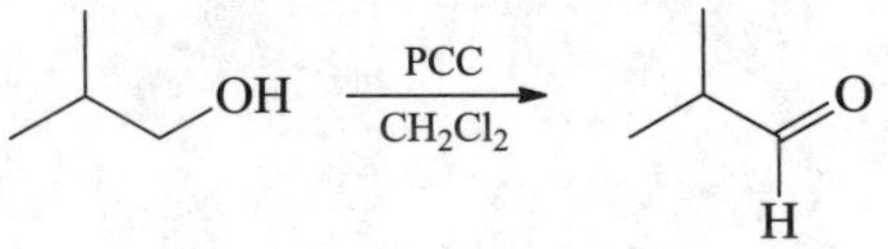

(*c*) The first step in this synthesis is the hydration of the alkene function to an alcohol. Notice that this hydration must take place with a regioselectivity opposite to that of Markovnikov's rule and therefore requires a hydroboration-oxidation sequence.

6-Methyl-5-hepten-2-one — 5-Hydroxy-6-methyl-2-heptanone

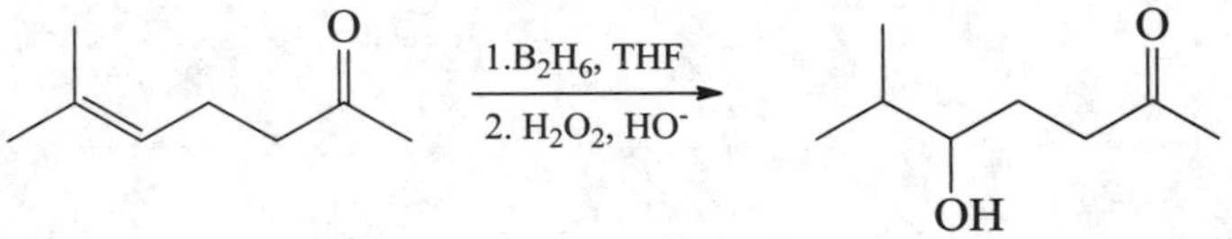

Conversion of the secondary alcohol function to a carbonyl group can be achieved with any of a number of oxidizing agents.

5-Hydroxy-6-methyl-2-heptanone — 6-Methyl-2,5-heptandione

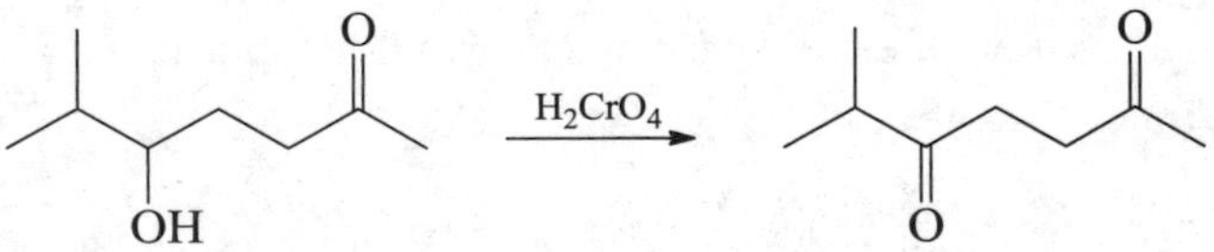

Cyclization of the dione to the final product is a base-catalyzed intramolecular aldol condensation and was accomplished in 71% yield by treatment of the dione with a 2% solution of sodium hydroxide in aqueous ethanol.

(*d*) The cyclohexene ring in this case can be assembled by a Diels–Alder reaction.

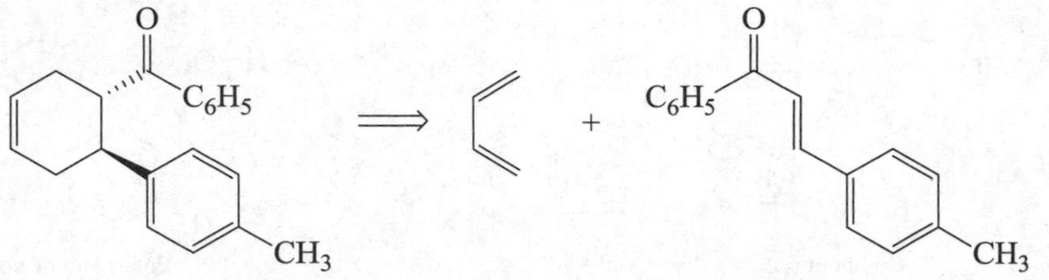

1,3-Butadiene is one of the given starting materials; the α,β-unsaturated ketone is the mixed aldol condensation product of 4-methylbenzaldehyde and acetophenone.

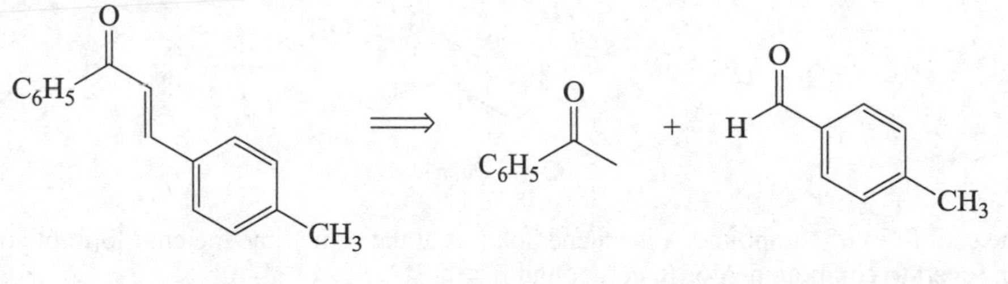

The complete synthetic sequence is

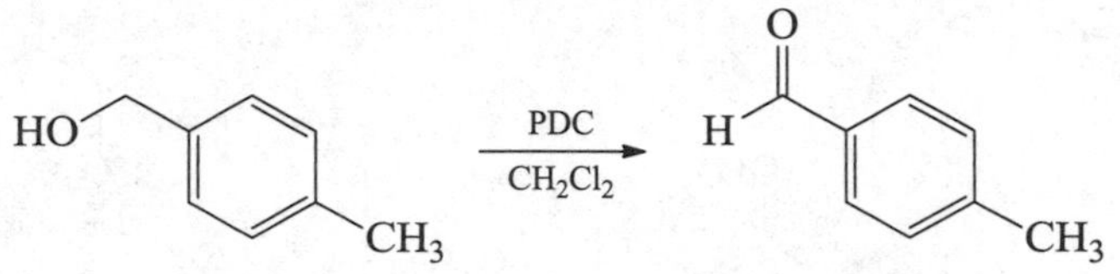

4-Methylbenzyl alcohol   4-Methylbenzaldehyde

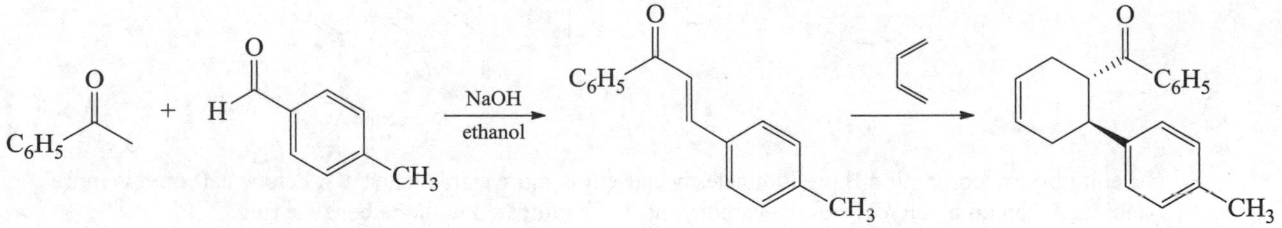

Acetophenone   4-Methylbenzaldehyde

trans-4-Benzoyl-5-
(4-methylphenyl)cyclohexene

α,β-Unsaturated ketones are good dienophiles in Diels–Alder reactions.

Mechanisms

21.54 At first glance this transformation is an internal oxidation–reduction reaction. An aldehyde function is reduced to a primary alcohol, and a secondary alcohol is oxidized to a ketone.

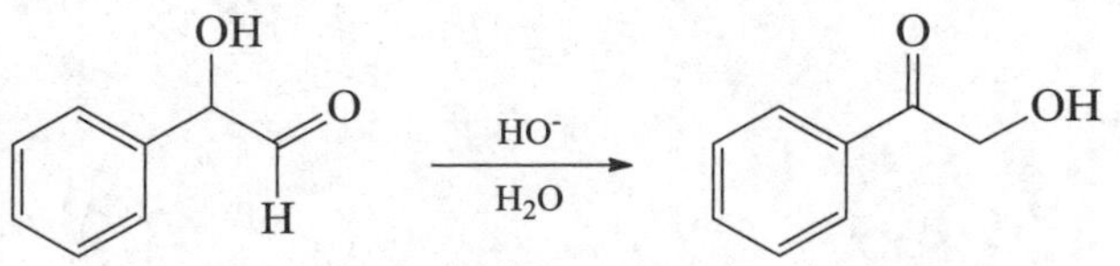

Compound A   Compound B

Once one realizes that enolization can occur, however, a mechanism involving only proton-transfer reactions, emerges.

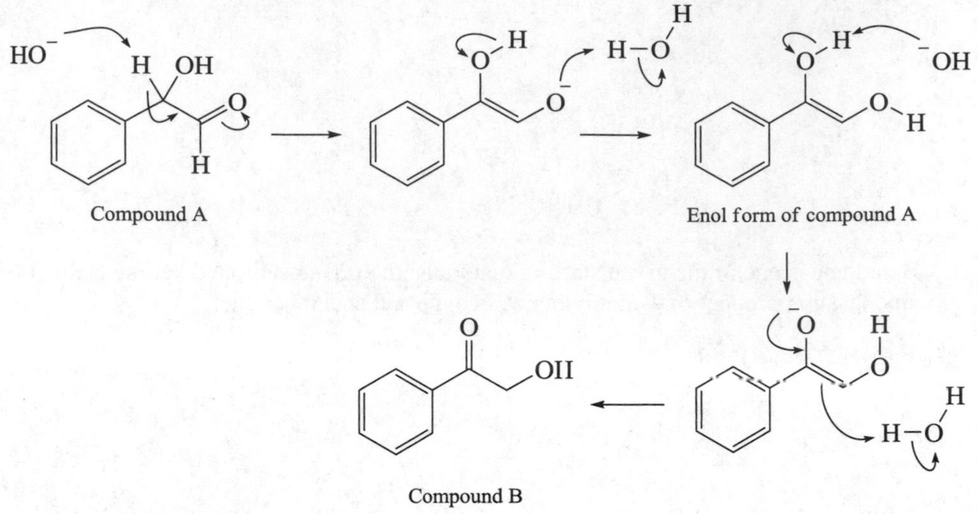

Compound A   Enol form of compound A

Compound B

The enol form of compound A is an enediol; it is at the same time the enol form of compound B. The enediol can revert to compound A or to compound B.

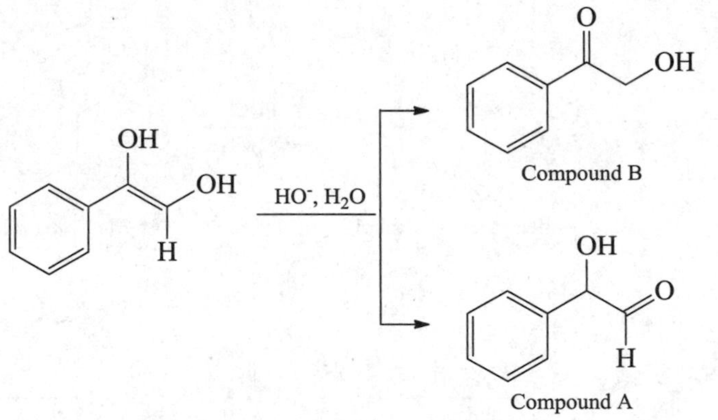

Compound B

Compound A

At equilibrium, compound B predominates because it is more stable than A. A ketone carbonyl is more stabilized than an aldehyde, and the carbonyl in B is conjugated with the benzene ring.

21.55 The mechanism for the key step is as follows:

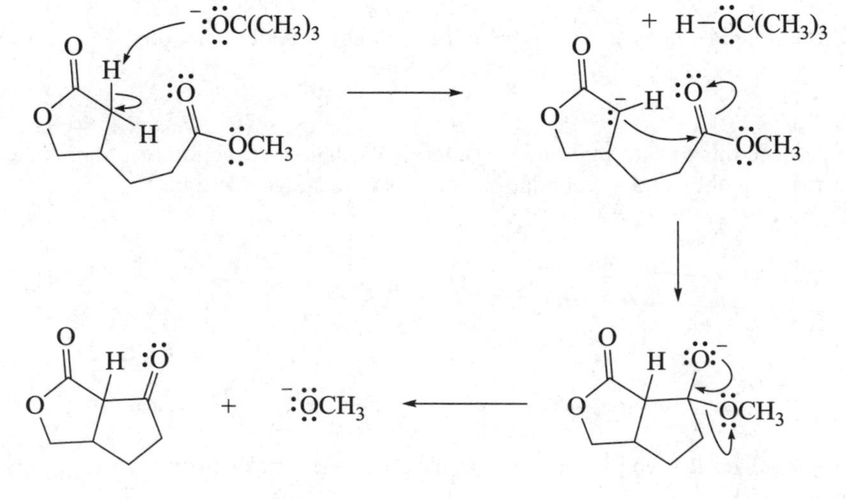

21.56 The first step is the conversion of the thioester to its enolate. LDA is a strong base and removes a proton from the α carbon.

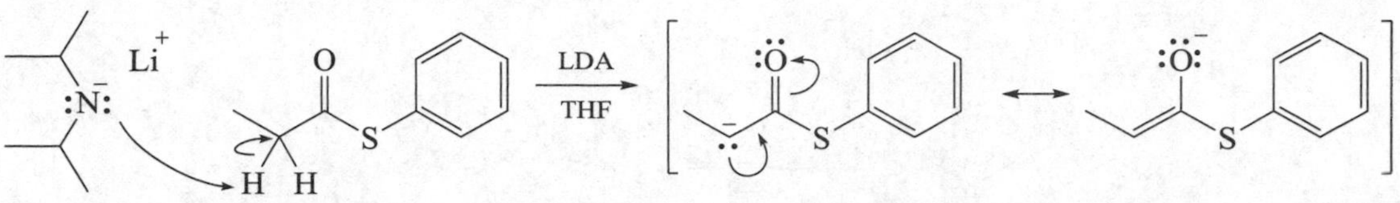

The enolate adds to the carbonyl group of cyclohexanone, and the species formed in this step undergoes intramolecular displacement of thiophenoxide from thioester carbonyl.

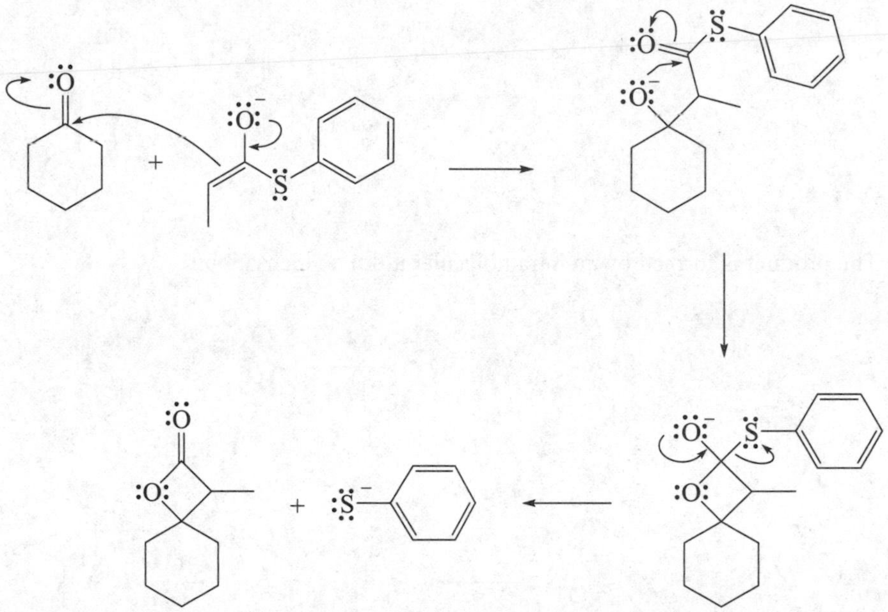

21.57 (*a*) This reaction is an intramolecular alkylation of a ketone. Although alkylation of a ketone with a separate alkyl halide molecule is usually difficult, *intramolecular* alkylation reactions can be carried out effectively. The enolate formed by proton abstraction from the α-carbon atom carries out a nucleophilic attack on the carbon that bears the leaving group.

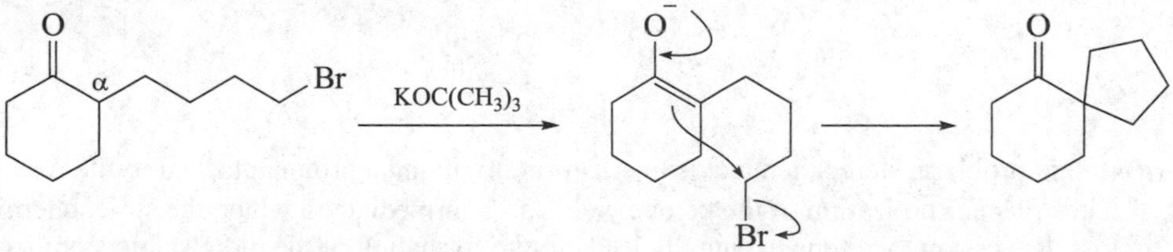

(*b*) The starting material, known as *citral,* is converted to the two products by a reversal of an aldol condensation. The first step is conjugate addition of hydroxide.

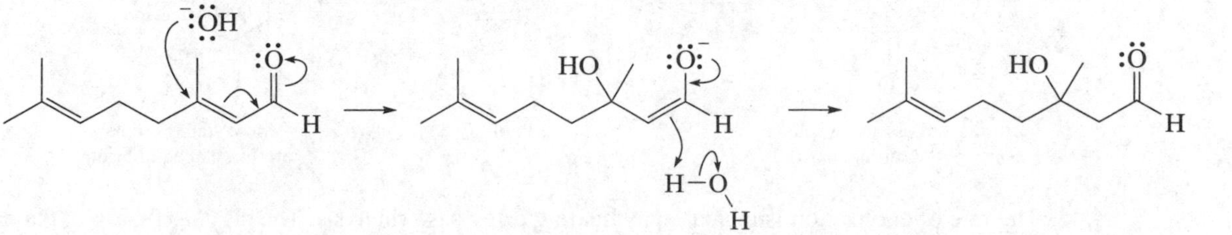

The product of this conjugate addition is a β-hydroxy aldehyde. It undergoes base-catalyzed cleavage to the observed products.

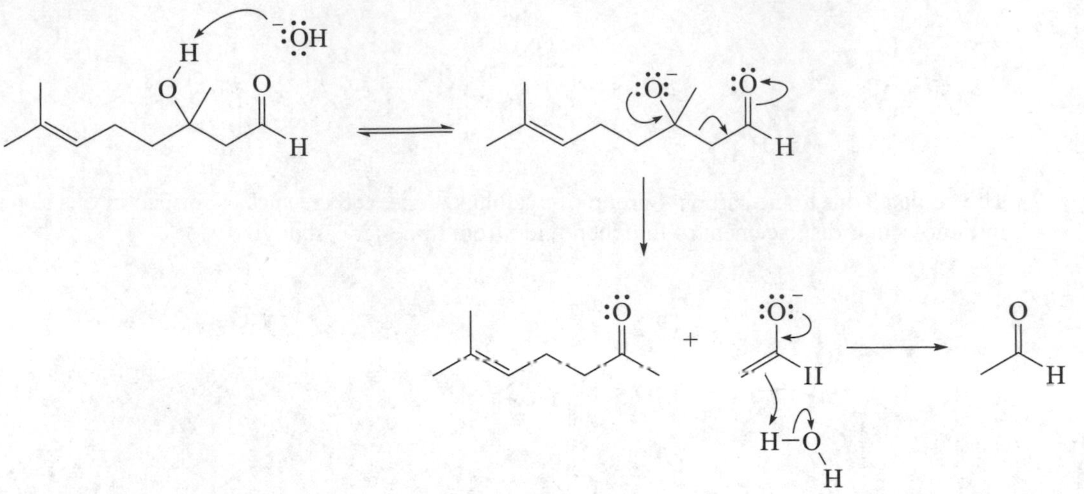

(*c*) The product is formed by an intramolecular aldol condensation.

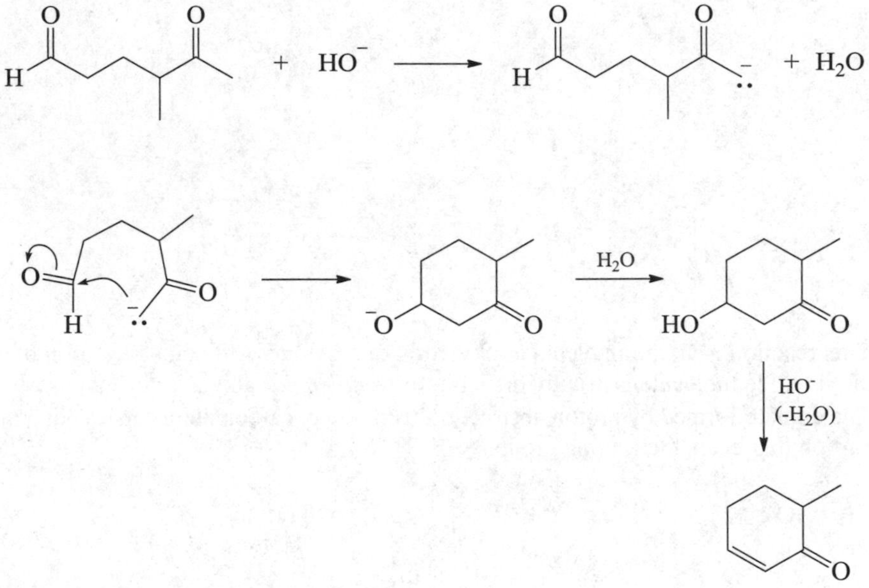

(*d*) In this problem, stereochemical isomerization involving a proton attached to the α-carbon atom of a ketone takes place. Enolization of the ketone yields an intermediate in which the stereochemistry of the α carbon is lost. Reversion to ketone eventually leads to the formation of the more stable stereoisomer at equilibrium.

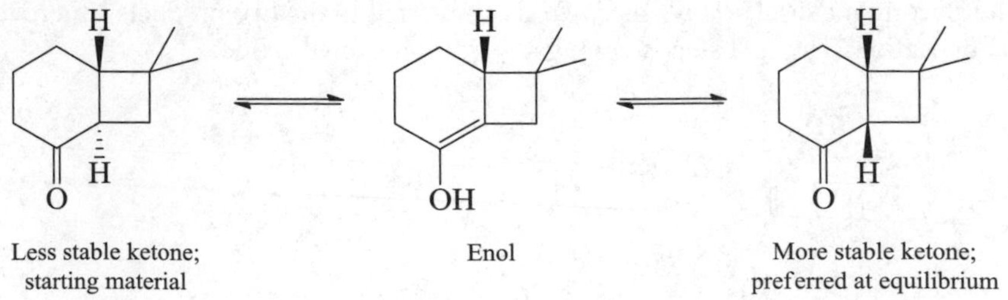

| Less stable ketone; starting material | Enol | More stable ketone; preferred at equilibrium |

The rate of enolization is increased by heating or by base catalysis. The cis ring fusion in the product is more stable than the trans because there are not enough atoms in the six-membered ring to span *trans*-1,2 positions in the four-membered ring without excessive strain.

Answers to Interpretive Problems 21

    21.58 D;   **21.59** C;   **21.60** A;   **21.61** C;   **21.62** A;   **21.63** B

SELF-TEST

1. Write the correct structure(s) for each of the following:

   (*a*)   The two enol forms of 2-butanone

   (*b*)   The most stable enolate derived from reaction of 1,3-cyclohexanedione with sodium methoxide

   (*c*)   The carbonyl form of the following enol

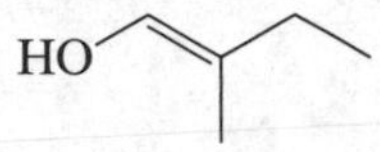

2. Give the correct structures for compounds A and B in the following reaction schemes:

   (*a*)

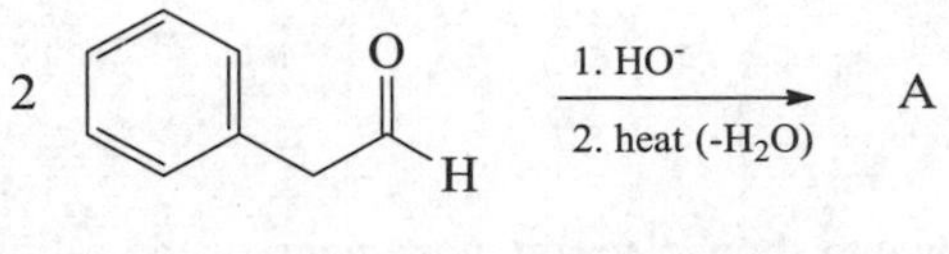

   (*b*)

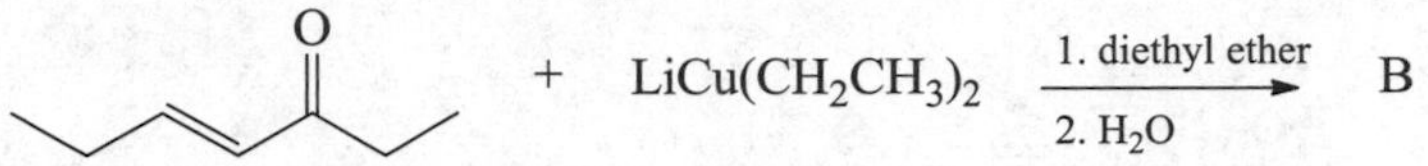

3. Write the structures of all the possible aldol addition products that may be obtained by reaction of a mixture of propanal and 2-methylpropanal with base.

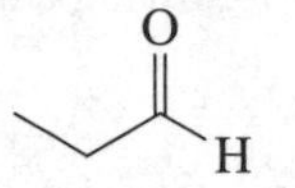

        Propanal 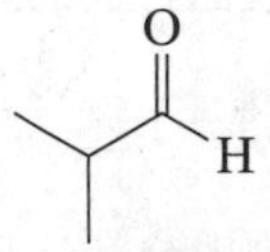   2-Methylpropanal

4. Using any necessary organic or inorganic reagents, outline a synthesis of 1,3-butanediol from ethanol as the only source of carbons.

5. Outline a series of reaction steps that will allow the preparation of compound B from 1,3-cyclopentanedione, compound A.

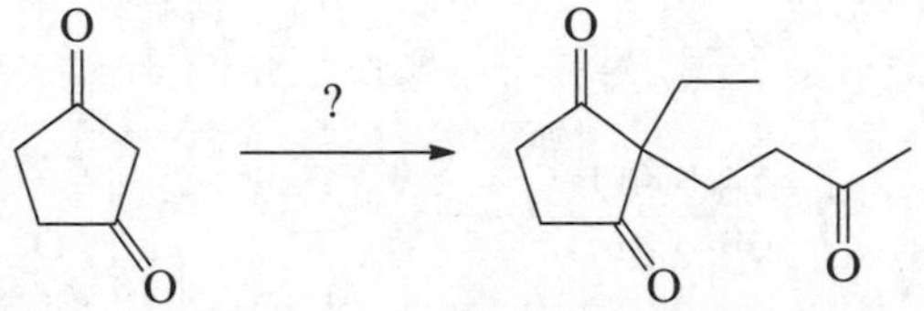

        A   B

6. Give the structure of the product formed in each of the following reactions:

   (*a*)

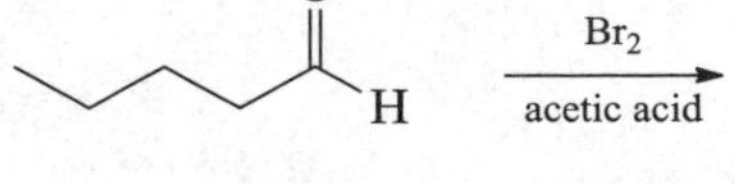

(b)

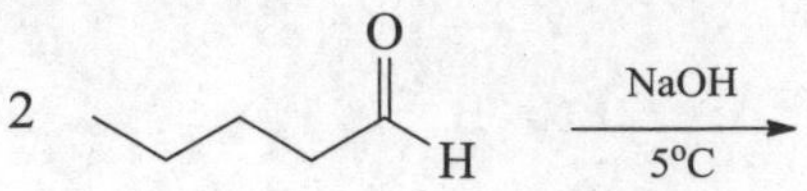

$$2 \quad \text{(pentanal)} \quad \xrightarrow[\text{5°C}]{\text{NaOH}}$$

(c)

$$\xrightarrow[(-H_2O)]{\text{NaOH, heat}}$$

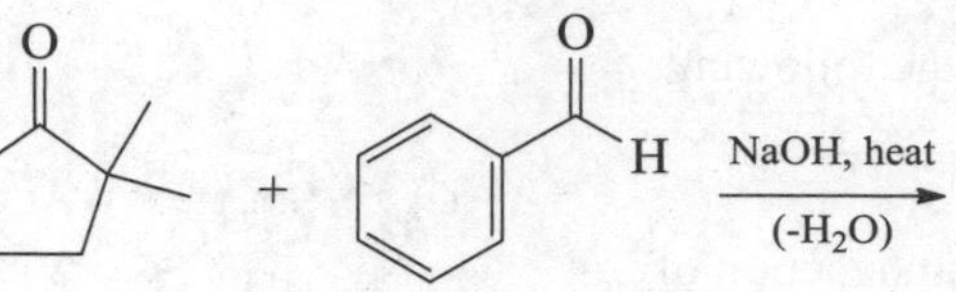

(d)

$$\xrightarrow[\text{ethanol}]{\text{NaSCH}_3}$$

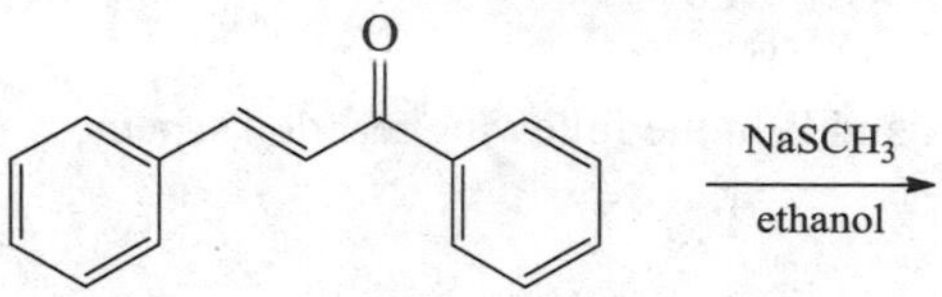

7. Write out the mechanism, using curved arrows to show electron movement, of the following aldol addition reaction.

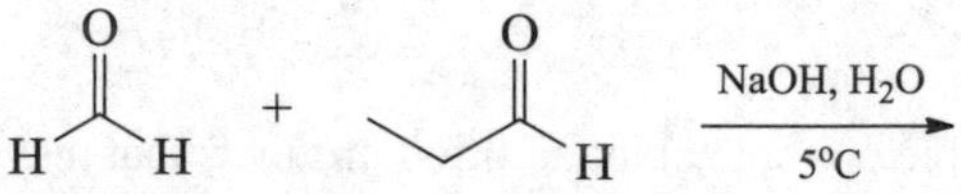

$$\xrightarrow[\text{5°C}]{\text{NaOH, H}_2\text{O}}$$

8. Identify the two starting materials needed to make the following compound by a mixed aldol condensation.

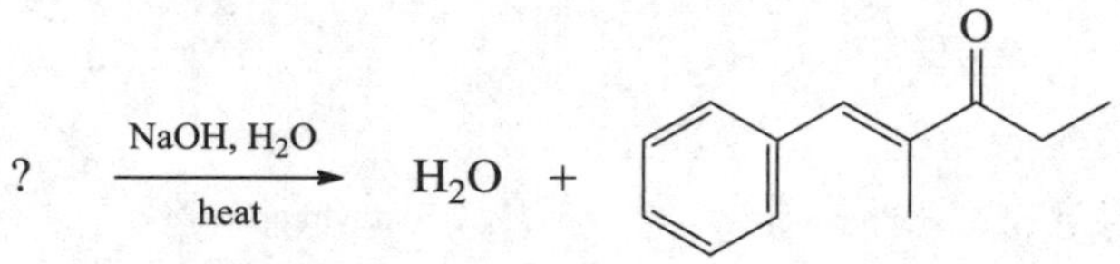

$$? \quad \xrightarrow[\text{heat}]{\text{NaOH, H}_2\text{O}} \quad \text{H}_2\text{O} \quad +$$

9. Give the structure of the reactant, reagent, or product omitted from each of the following:

(a)

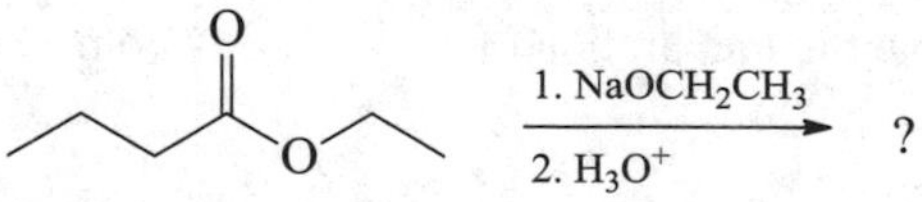

1. NaOCH$_2$CH$_3$
2. H$_3$O$^+$?

(b)

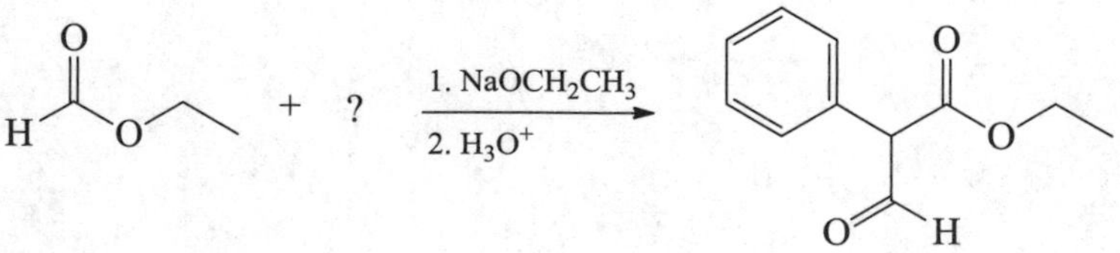

$$+ \quad ? \quad \xrightarrow[\text{2. H}_3\text{O}^+]{\text{1. NaOCH}_2\text{CH}_3}$$

(c)

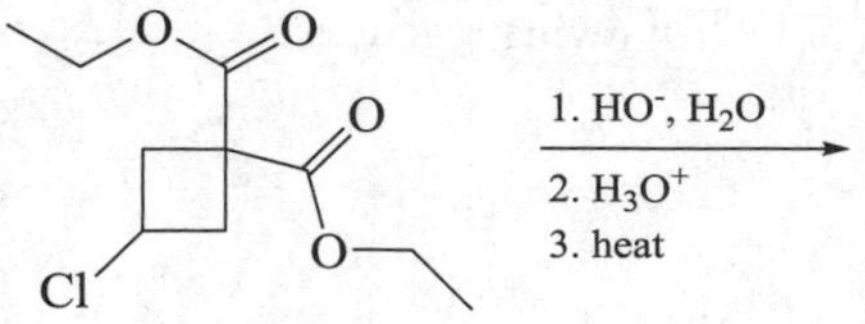

1. HO$^-$, H$_2$O
2. H$_3$O$^+$? (two stereoisomeric products; C$_5$H$_7$ClO$_2$)
3. heat

(d)

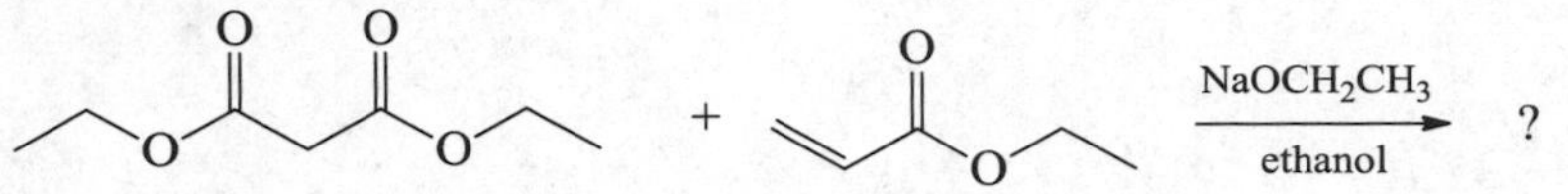

NaOCH₂CH₃ / ethanol → ?

(e)

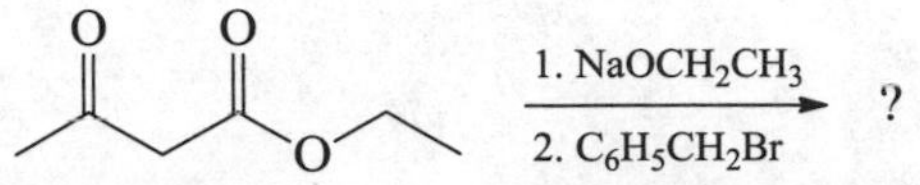

1. NaOCH₂CH₃
2. C₆H₅CH₂Br → ?

(f)

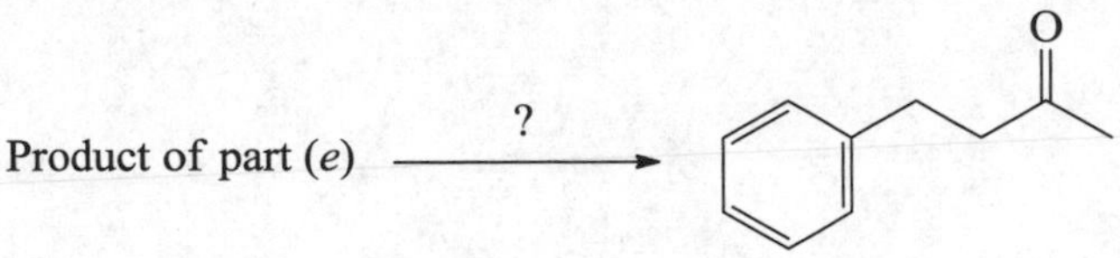

Product of part (e) ⟶ ?

(g)

heat ⟶ CO₂ + ?

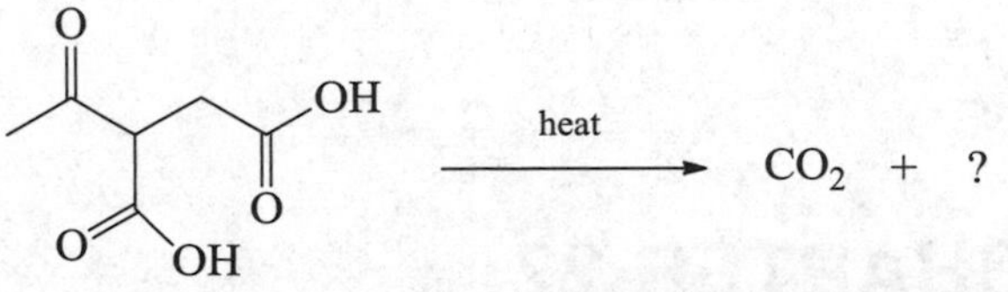

10. Provide the correct structures of compounds A through E in the following reaction sequences:

(a)

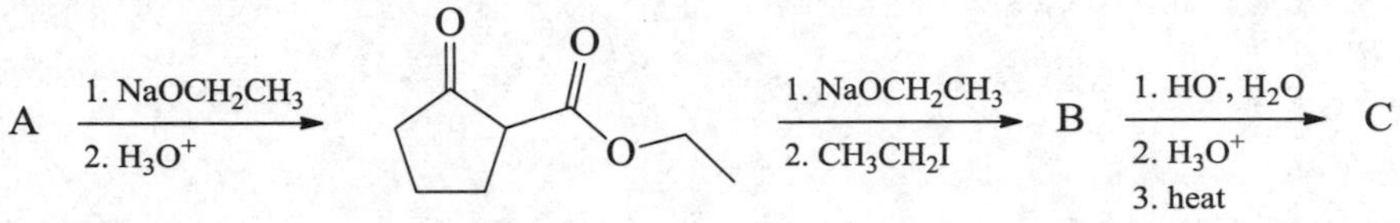

A   1. NaOCH₂CH₃ / 2. H₃O⁺ →

1. NaOCH₂CH₃ / 2. CH₃CH₂I → B   1. HO⁻, H₂O / 2. H₃O⁺ / 3. heat → C

(b)

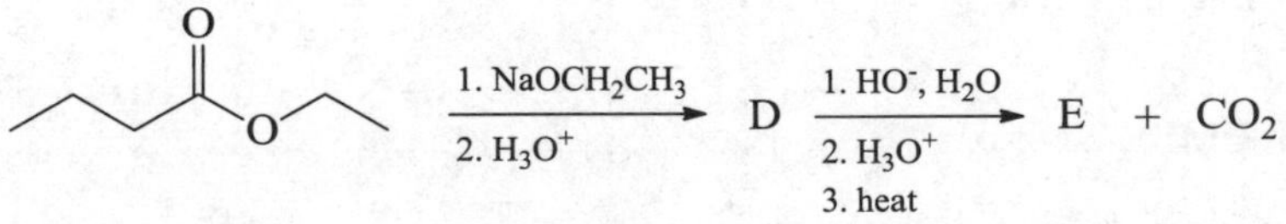

1. NaOCH₂CH₃ / 2. H₃O⁺ → D   1. HO⁻, H₂O / 2. H₃O⁺ / 3. heat → E + CO₂

11. Give a series of steps that will enable preparation of each of the following compounds from the starting material(s) given and any other necessary reagents.

(a)

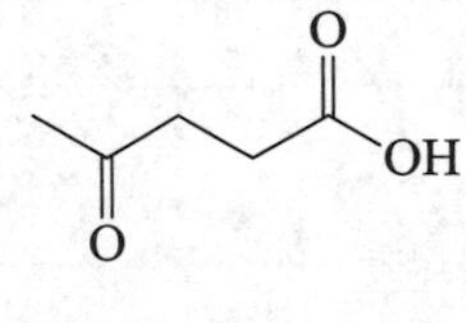

from ethyl acetoacetate

(b)

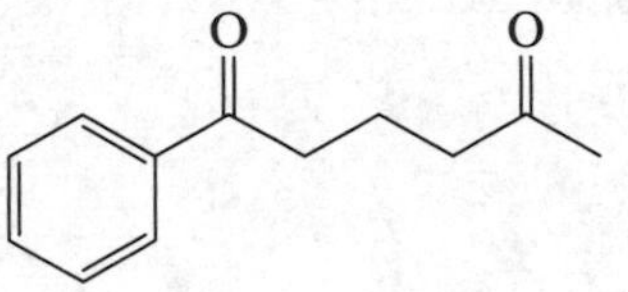

from

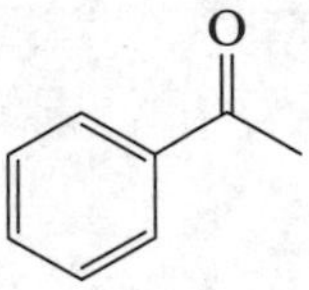

and diethyl carbonate

12. Write a stepwise mechanism for the reaction of ethyl propanoate with sodium ethoxide in ethanol.

13. Ethyl 2-methylbutanoate does not undergo a Claisen condensation, whereas ethyl 3-methylbutanoate does. Provide a mechanistic explanation for this observation.

CHAPTER 22

Amines

Table of Contents

SOLUTIONS TO TEXT PROBLEMS

In Chapter Problems

22.1 (*b*) The aminc is at the C-1 position of the cyclohexane ring.

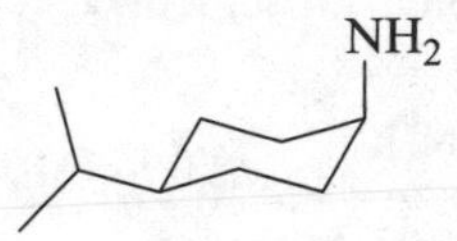

cis-4-Isopropylcyclohexylamine, or
cis-4-Isopropylcyclohexanamine

(*c*)

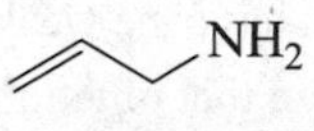

Allylamine, or
2-Propen-1-amine

22.2 *N,N*-Dimethylcycloheptylamine may also be named as a dimethyl derivative of cycloheptanamine.

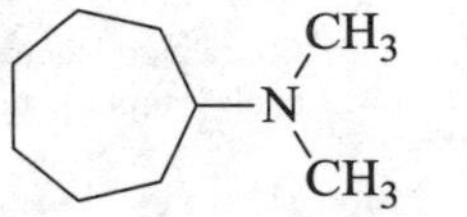

N,N-Dimethylcycloheptanamine

22.3 Three substituents are attached to the nitrogen atom; the amine is tertiary. In alphabetical order, the substituents present on the aniline nucleus are ethyl, isopropyl, and methyl. Their positions are specified as *N*-ethyl, 4-isopropyl, and *N*-methyl.

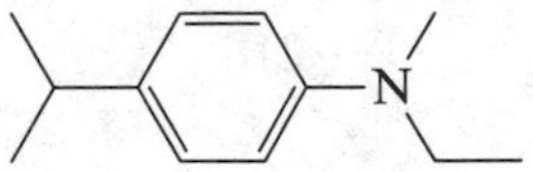

N-Ethyl-4-isopropyl-*N*-methylaniline

22.4 The electron-donating amino group and the electron-withdrawing nitro group are both conjugated in *p*-nitroaniline. The planar geometry of *p*-nitroaniline suggests that the delocalized resonance form shown is a major contributor to the structure of the compound.

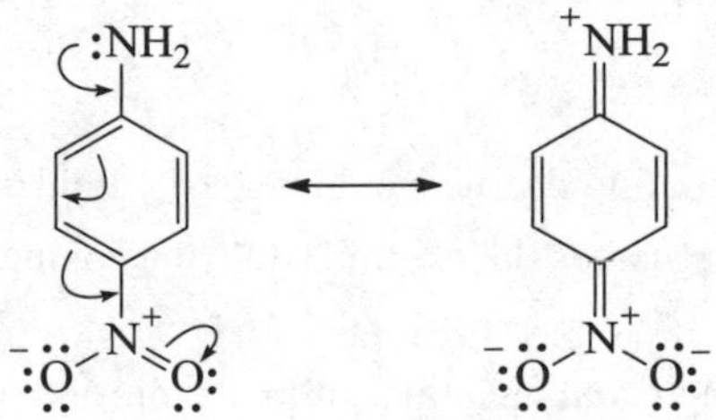

22.5 The Henderson–Hasselbalch equation described in text Section 19.4 can be applied to bases such as amines, as well as carboxylic acids. The ratio $[CH_3NH_3^+]/[CH_3NH_2]$ is given by

$$\frac{[CH_3NH_3^+]}{[CH_3NH_2]} = 10^{(pK_a - pH)}$$

The pK_a of methylammonium ion is given in the text as 10.7. Therefore

$$\frac{[CH_3NH_3^+]}{[CII_3NH_2]} = 10^{(10.7-7)} = 10^{3.7} = 5 \times 10^3$$

22.6 Tetrahydroisoquinoline is an alkylamine because the nitrogen atom is attached only to sp^3-hybridized carbons. Tetrahydroquinoline, on the other hand, is an arylamine because nitrogen is directly attached to the aromatic ring. Arylamines are less basic than alkylamines because the nitrogen unshared-electron pair is delocalized into the π system of the aromatic ring. Thus, tetrahydroisoquinoline is the stronger base. Recalling that the stronger base will have the weaker conjugate acid, the pK_a of the conjugate acid of tetrahydroisoquinoline is smaller than that of tetrahydroquinoline.

Tetrahydroisoquinoline (an alkylamine): more basic	Tetrahydroquinoline (an arylamine): less basic

pK_a (of conjugate acid):　　　　　9.4　　　　　　　　　　4.0

22.7 (b) An acetyl group attached directly to nitrogen as in acetanilide delocalizes the nitrogen lone pair into the carbonyl group. Amides are weaker bases than amines.

(c) An acetyl group in a position para to an amine function is conjugated to it and delocalizes the nitrogen lone pair.

22.8 (b) An aqueous solution of imidazole would be basic. Imidazole is more basic than water; the pK_a of its conjugate acid is 7, whereas the pK_a of H_3O^+ (the conjugate acid of water) is –1.7.

(c) A solution containing equal quantities of imidazole and its conjugate acid would be neutral, with a pH = 7. Recall the Henderson–Hasselbalch equation from text Section 19.4:

$$pH = pK_a + \log \frac{[\text{conjugate base}]}{[\text{conjugate acid}]}$$

When the concentration of imidazole (the conjugate base) and imidazolium ion (the conjugate acid) are equal, the pH of the solution equals pK_a because $\log (1) = 0$.

22.9 Nicotine contains two different nitrogen atoms. The more basic of the two is the nitrogen of the pyrrolidine ring.

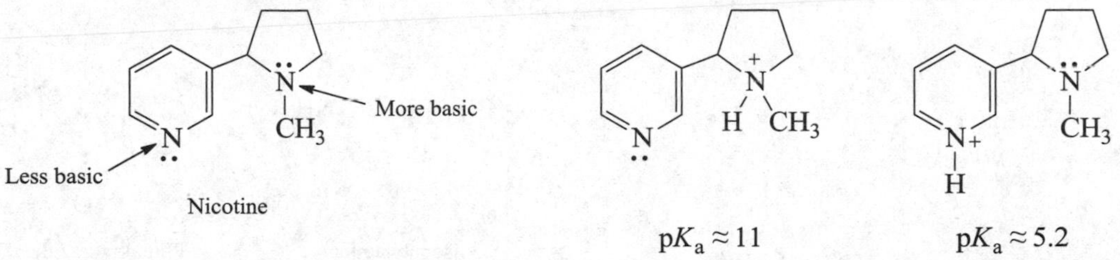

The pK_a of the protonated pyrrolidine unit is expected to be about 11, similar to that of piperidinium ion found in Table 1.8. The pK_a of protonated nicotine's pyridinium moiety is approximately 5.2.

22.10 The reaction that leads to allylamine is nucleophilic substitution by ammonia on allyl chloride.

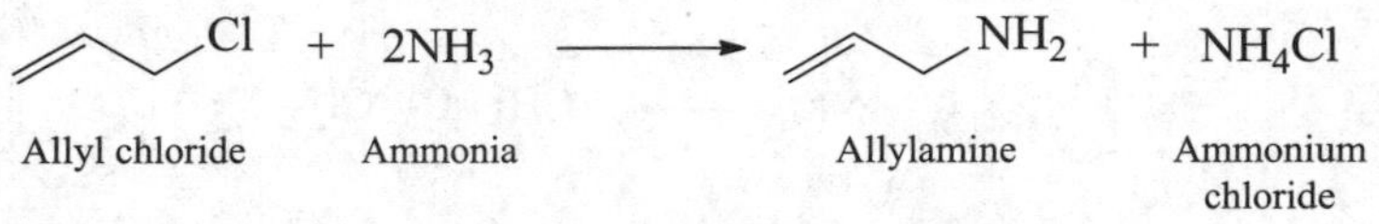

Allyl chloride is prepared by free-radical chlorination of propene.

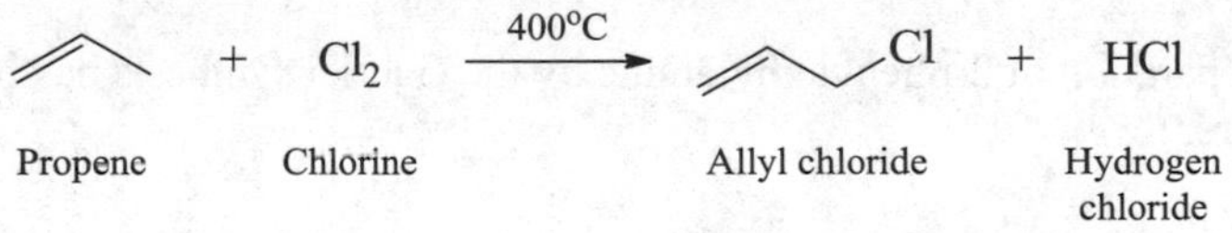

22.11 (*b*) Isobutylamine is $(CH_3)_2CHCH_2NH_2$. It is a primary amine of the type RCH_2NH_2 and can be prepared from a primary alkyl halide by the Gabriel synthesis.

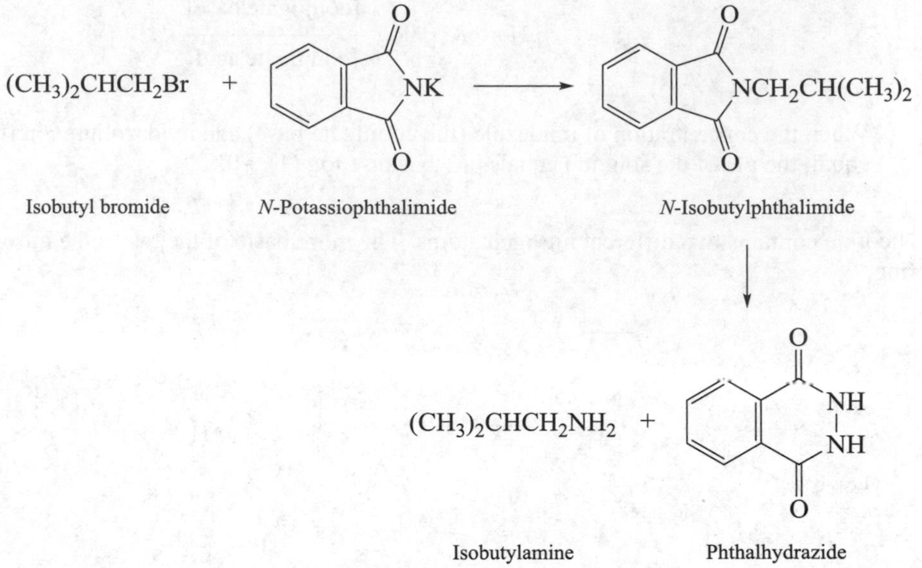

(CH3)2CHCH2Br
NK
NCH2CH(CH3)2
Isobutyl bromide
N-Potassiophthalimide
N-Isobutylphthalimide
(CH3)2CHCH2NH2
NH
NH
Isobutylamine
Phthalhydrazide

(*c*) Although *tert*-butylamine $(CH_3)_3CNH_2$ is a primary amine, it cannot be prepared by the Gabriel method, because it would require an S_N2 reaction on a tertiary alkyl halide in the first step. Elimination occurs instead.

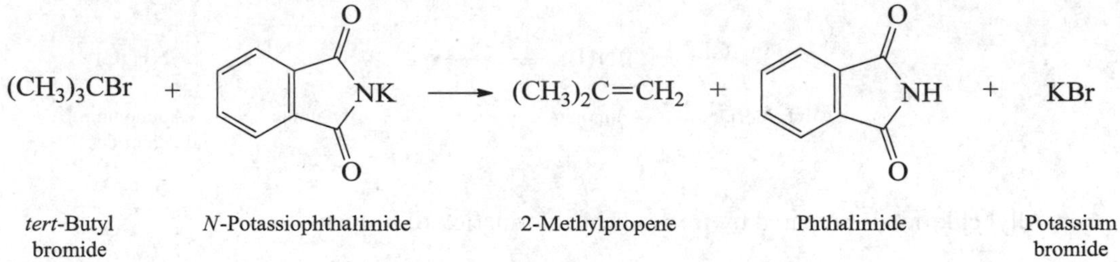

(CH3)3CBr
NK
(CH3)2C=CH2
NH
KBr
tert-Butyl bromide
N-Potassiophthalimide
2-Methylpropene
Phthalimide
Potassium bromide

(*d*) The preparation of 2-phenylethylamine by the Gabriel synthesis has been described in the chemical literature.

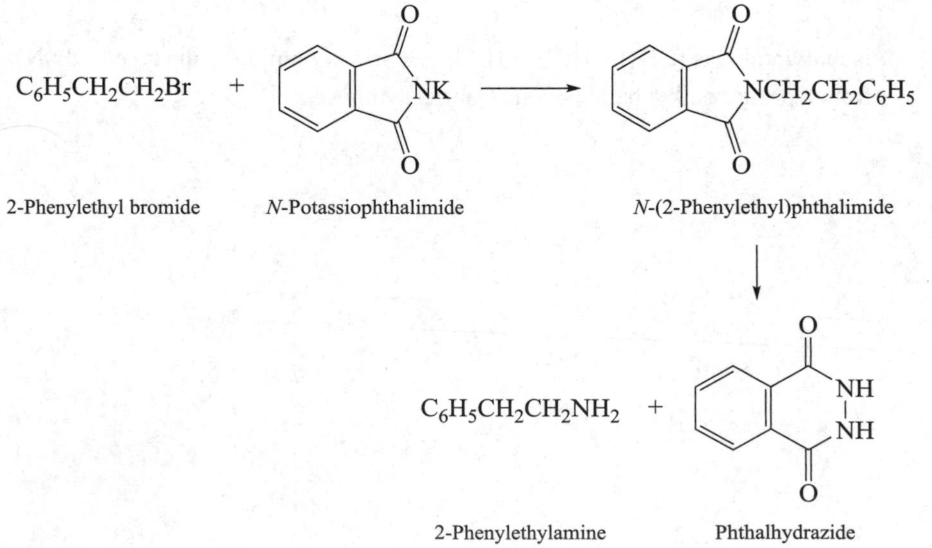

C6H5CH2CH2Br
NK
NCH2CH2C6H5
2-Phenylethyl bromide
N-Potassiophthalimide
N-(2-Phenylethyl)phthalimide
C6H5CH2CH2NH2
NH
NH
2-Phenylethylamine
Phthalhydrazide

(e) The Gabriel synthesis leads to primary amines; *N*-methylbenzylamine is a secondary amine and cannot be prepared by this method.

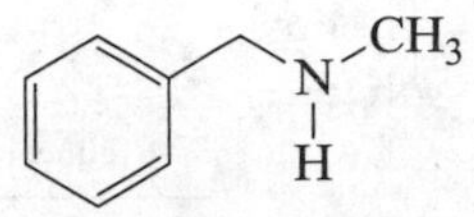

N-Methylbenzylamine
(two carbon substituents on
nitrogen; a secondary amine)

(f) Aniline cannot be prepared by the Gabriel method. Aryl halides do not undergo nucleophilic substitution under these conditions.

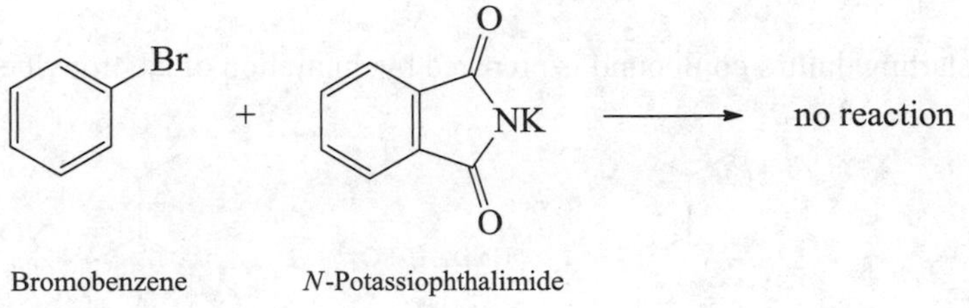

Bromobenzene   *N*-Potassiophthalimide

22.12 For each part of this problem, keep in mind that aromatic amines are derived by reduction of the corresponding aromatic nitro compound. Each synthesis should be approached from the standpoint of how best to prepare the necessary nitroaromatic compound.

$$Ar-NH_2 \implies Ar-NO_2 \implies Ar-H$$

(Ar = substituted aromatic ring)

(b) The para isomer of isopropylaniline may be prepared by a procedure analogous to that used for its ortho isomer in part (a).

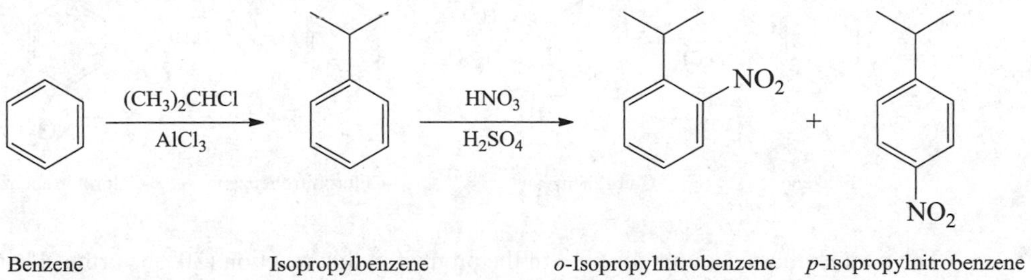

Benzene   Isopropylbenzene   *o*-Isopropylnitrobenzene   *p*-Isopropylnitrobenzene

After separating the ortho, para mixture by distillation, the nitro group of *p*-isopropylnitrobenzene is reduced to yield the desired *p*-isopropylaniline.

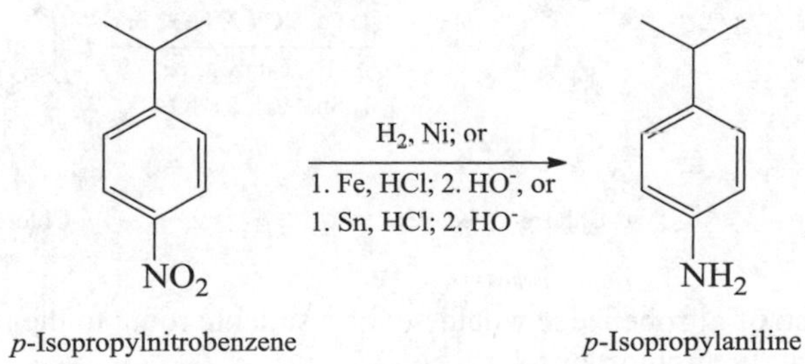

p-Isopropylnitrobenzene   *p*-Isopropylaniline

(c) The target compound is the reduction product of 1-isopropyl-2,4-dinitrobenzene.

1-Isopropyl-2,4-dinitrobenzene   4-Isopropyl-1,3-benzenediamine

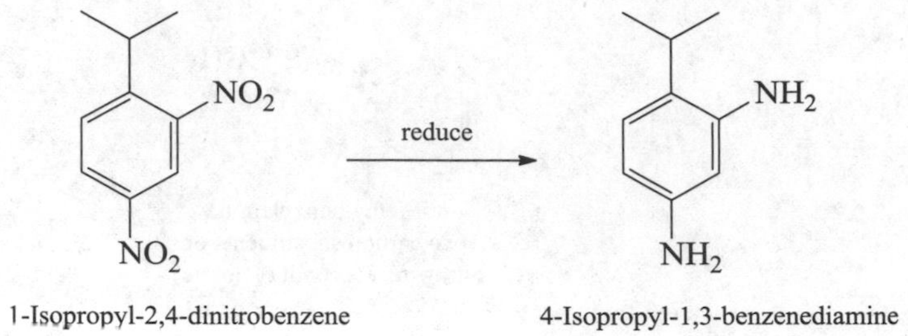

This reduction is carried out in the same way as reduction of an arene that contains only a single nitro group. In this case, hydrogenation over a nickel catalyst gave the desired product in 90% yield.

The starting dinitro compound is prepared by dinitration of isopropylbenzene.

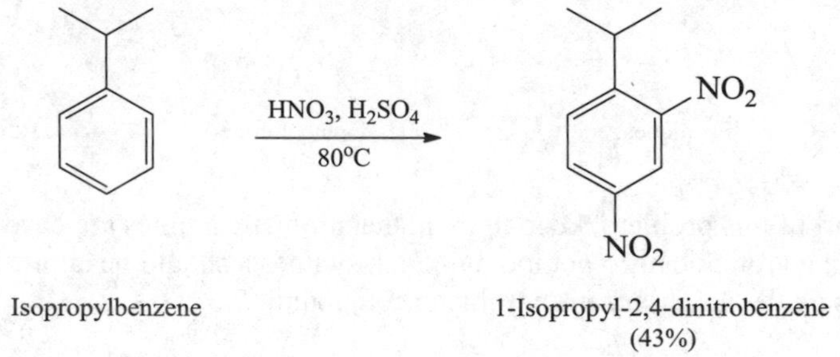

Isopropylbenzene   1-Isopropyl-2,4-dinitrobenzene
                                                                    (43%)

(d) The conversion of *p*-chloronitrobenzene to *p*-chloroaniline was cited as an example in the text to illustrate reduction of aromatic nitro compounds to arylamines. *p*-Chloronitrobenzene is prepared by nitration of chlorobenzene.

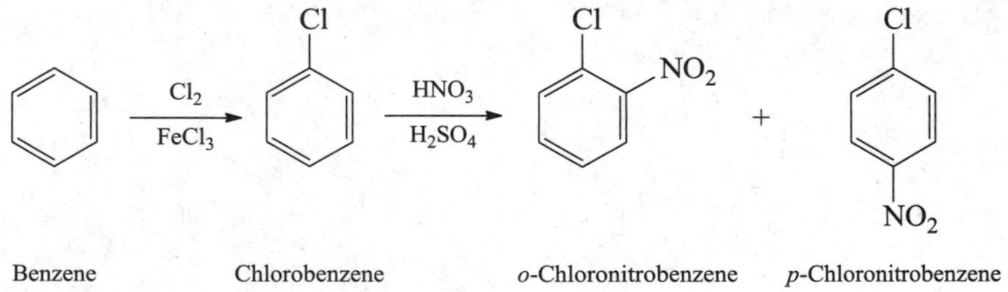

Benzene   Chlorobenzene   *o*-Chloronitrobenzene   *p*-Chloronitrobenzene

The para isomer accounts for 69% of the product in this reaction (30% is ortho, 1% meta). Separation of *p*-chloronitrobenzene and its reduction completes the synthesis.

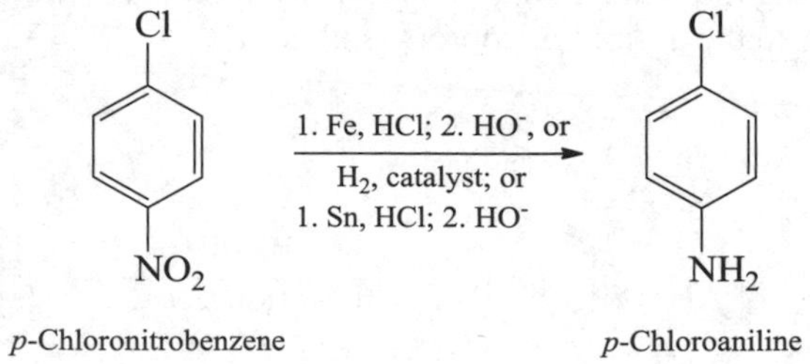

p-Chloronitrobenzene   *p*-Chloroaniline

Chlorination of nitrobenzene would not be a suitable route to the required intermediate, because it would produce mainly *m*-chloronitrobenzene.

(*e*) The synthesis of *m*-aminoacetophenone may be carried out by the scheme shown:

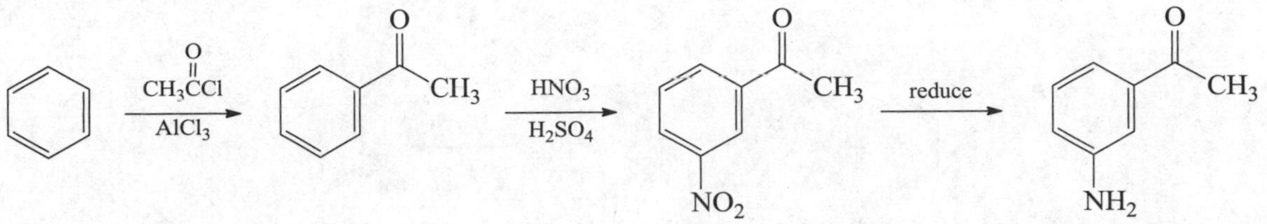

Benzene   Acetophenone   *m*-Nitroacetophenone   *m*-Aminoacetophenone

The acetyl group is attached to the ring by Friedel–Crafts acylation. It is a meta director, and its nitration gives the proper orientation of substituents. The order of the first two steps cannot be reversed, because Friedel–Crafts acylation of nitrobenzene is not possible (text Section 13.16). Once prepared, *m*-nitroacetophenone can be reduced to *m*-nitroaniline by any of a number of reagents. Indeed, all three reducing combinations described in the text have been employed for this transformation.

	Reducing agent	Yield (%)
m-Nitroacetophenone	H_2, Pt	94
↓	Fe, HCl	84
m-Aminoacetophenone	Sn, HCl	82

22.13 (*b*) Dibenzylamine is a secondary amine and can be prepared by reductive amination of benzaldehyde with benzylamine.

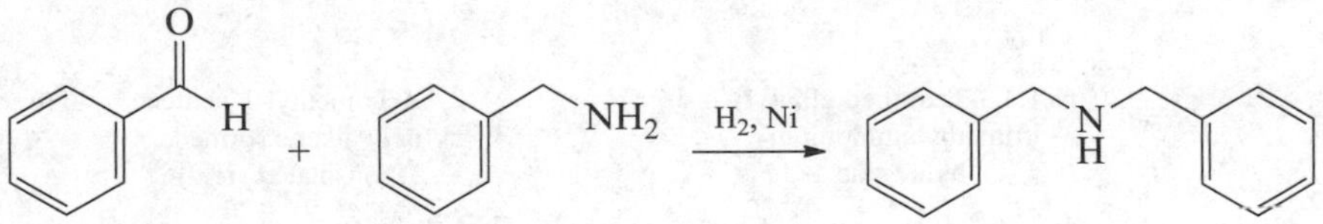

Benzaldehyde   Benzylamine   Dibenzylamine

(*c*) *N,N*-Dimethylbenzylamine is a tertiary amine. Its preparation from benzaldehyde requires dimethylamine, a secondary amine.

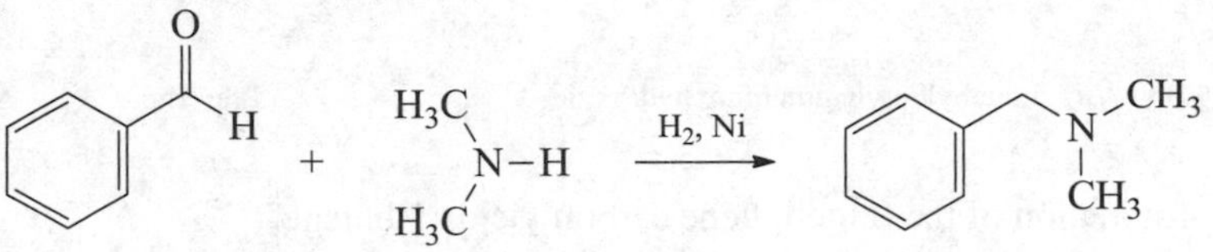

Benzaldehyde   Dimethylamine   *N,N*-Dimethylbenzylamine (*d*) The preparation of *N*-butylpiperidine by reductive amination is described in text Section 22.10. An analogous procedure is used to prepare *N*-benzylpiperidine.

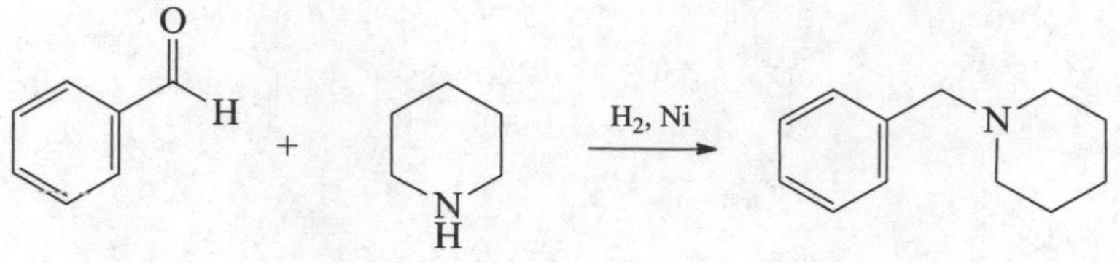

Benzaldehyde   Piperidine   *N*-Benzylpiperidine

22.14 (*b*) First identify the available β hydrogens. Elimination must involve a proton from the carbon atom adjacent to the one that bears the nitrogen.

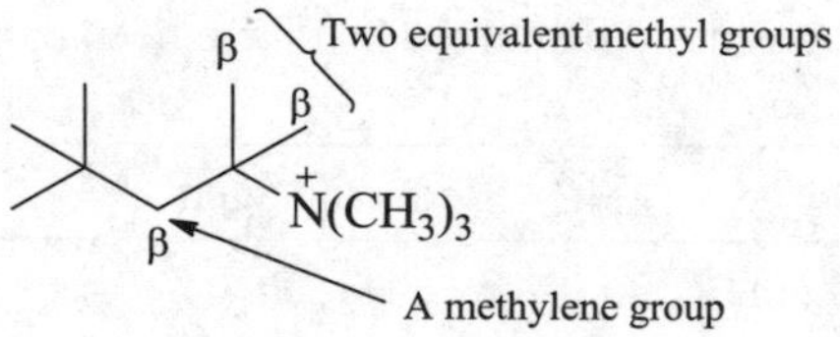

Two equivalent methyl groups

A methylene group

It is a proton from one of the methyl groups, rather than from the more sterically hindered methylene, that is lost on elimination.

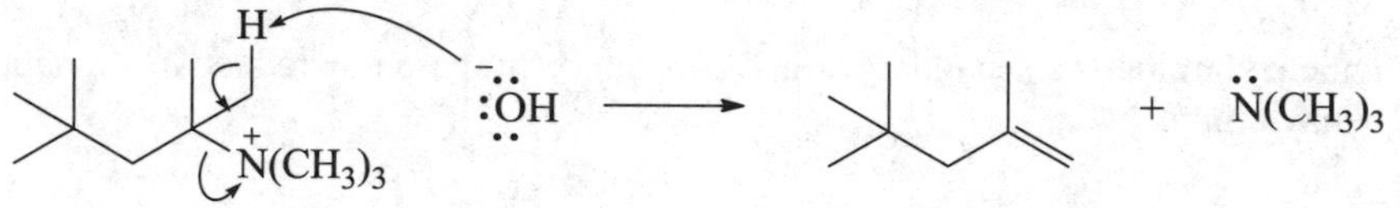

(1,1,3,3-Tetramethylbutyl)- trimethylammonium hydroxide

2,4,4-Trimethyl-1-pentene (only alkene formed, 70% isolated yield)   Trimethylamine

(*c*) The base may abstract a proton from either of two β carbons. Deprotonation of the β methyl carbon yields ethylene.

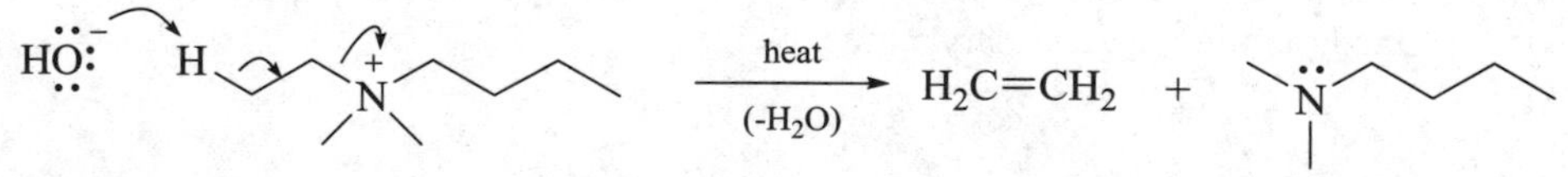

N-Ethyl-*N*,*N*-dimethylbutylammonium hydroxide   Ethylene   *N*,*N*-Dimethylbutylamine

Deprotonation of the β methylene carbon yields 1-butene.

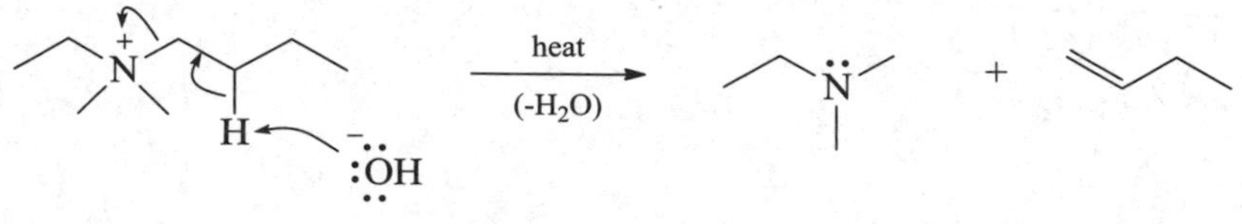

N-Ethyl-*N*,*N*-dimethylbutylammonium hydroxide   *N*,*N*-Dimethylethylamine   1-Butene

The preferred order of proton removal in Hofmann elimination reactions is β CH$_3$ > β CH$_2$ > β CH. Ethylene is the major alkene formed, the observed ratio of ethylene to 1-butene being 98:2.

22.15 (*b*) The pattern of substituents in 2,4-dinitroaniline suggests that they can be introduced by dinitration. Because nitration of aniline itself is not practical, the amino group must be protected by conversion to its *N*-acetyl derivative.

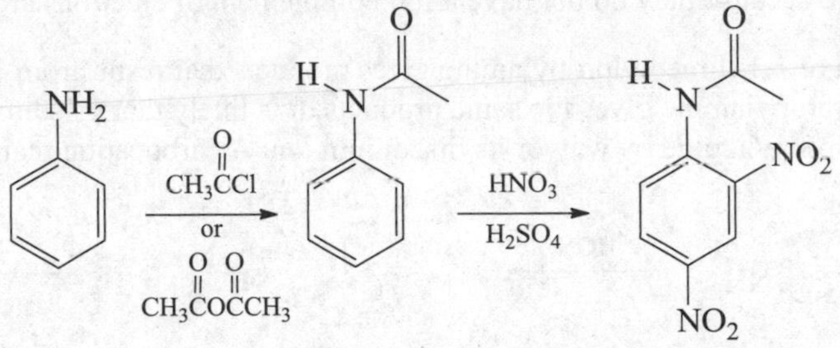

| Aniline | Acetanilide | 2,4-Dinitroacetanilide |

Hydrolysis of the amide bond in 2,4-dinitroacetanilide furnishes the desired 2,4-dinitroaniline.

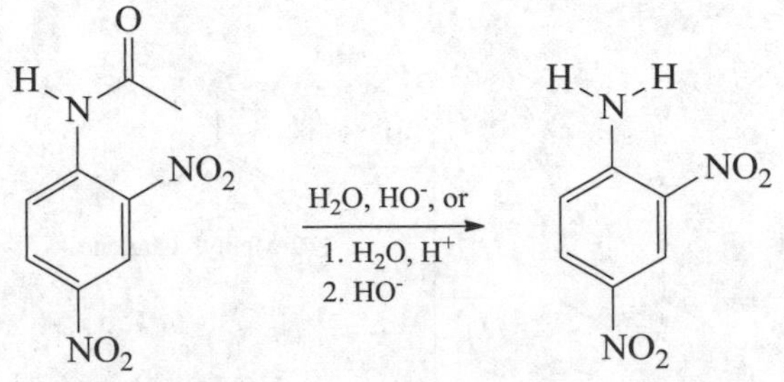

| 2,4-Dinitroacetanilide | 2,4-Dinitroaniline |

(*c*) Retrosynthetically, *p*-aminoacetanilide may be derived from *p*-nitroacetanilide.

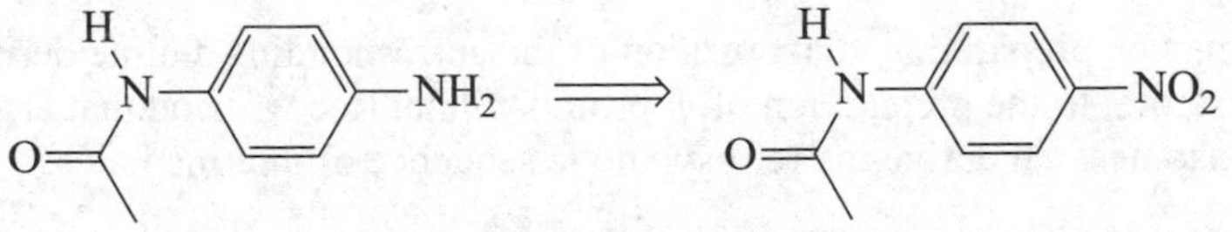

| *p*-Aminoacetanilide | *p*-Nitroacetanilide |

This suggests the sequence

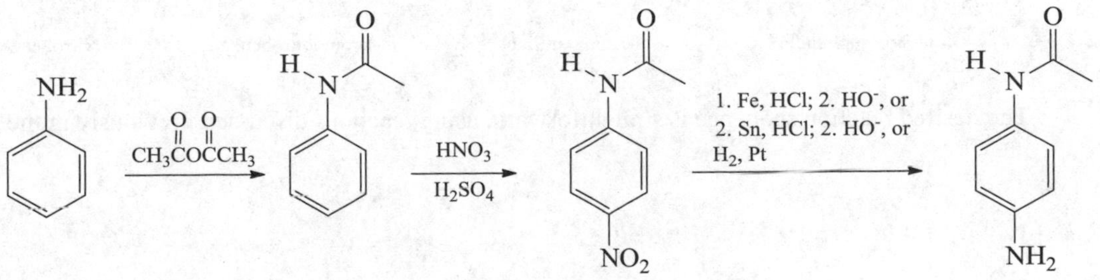

| Aniline | Acetanilide | *p*-Nitroacetanilide (separate from ortho isomer) | *p*-Aminoacetanilide |

22.16 The principal resonance contributors of *N*-nitrosodimethylamine are

All atoms (except hydrogen) have octets of electrons in each of these structures. Other resonance contributors are less stable because they do not have a full complement of electrons around each atom.

22.17 Deamination of 1,1-dimethylpropylamine gives products that result from 1,1-dimethylpropyl cation. Because 2,2-dimethylpropylamine gives the same products, it is likely that 1,1-dimethylpropyl cation is formed from 2,2-dimethylpropylamine by way of its diazonium ion. A carbocation rearrangement is indicated.

2,2-Dimethylpropylamine   2,2-Dimethylpropyldiazonium ion   1,1-Dimethylpropyl cation

Once formed, 1,1-dimethylpropyl cation loses a proton to form an alkene or is captured by water to give an alcohol.

2-Methyl-1-butene   +   2-Methyl-2-butene

1,1-Dimethylpropyl cation

2-Methyl-2-butanol

22.18 Phenols may be prepared by diazotization of the corresponding aniline derivative. The problem simplifies itself, therefore, to the preparation of *m*-bromoaniline. Recognizing that arylamines are ultimately derived from nitroarenes, we derive the retrosynthetic sequence of intermediates:

m-Bromophenol   *m*-Bromoaniline   *m*-Bromonitrobenzene   Nitrobenzene

The desired reaction sequence is straightforward, using reactions discussed previously in the text.

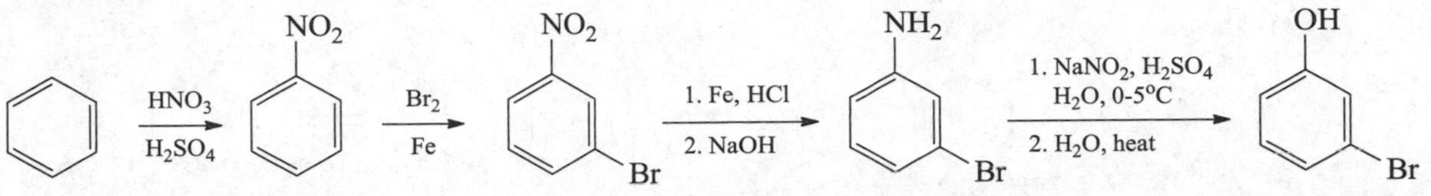

Benzene → Nitrobenzene → *m*-Bromonitrobenzene → *m*-Bromoaniline → *m*-Bromophenol

22.19 The key to this problem is to recognize that the iodine substituent in *m*-bromoiodobenzene is derived from an arylamine by diazotization.

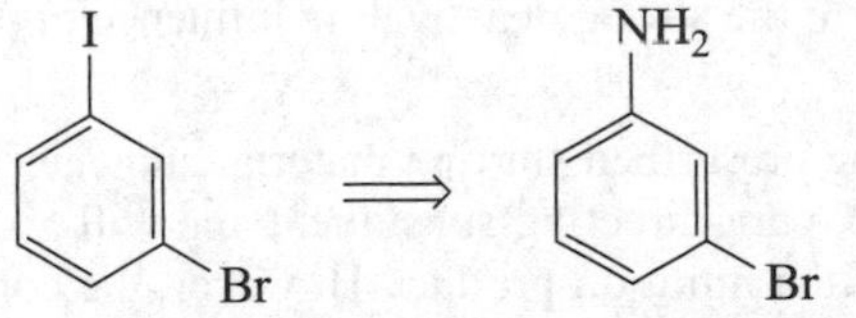

m-Bromoiodobenzene   *m*-Bromoaniline

The preparation of *m*-bromoaniline from benzene has been described in Problem 22.18. All that remains is to write the equation for its conversion to *m*-bromoiodobenzene.

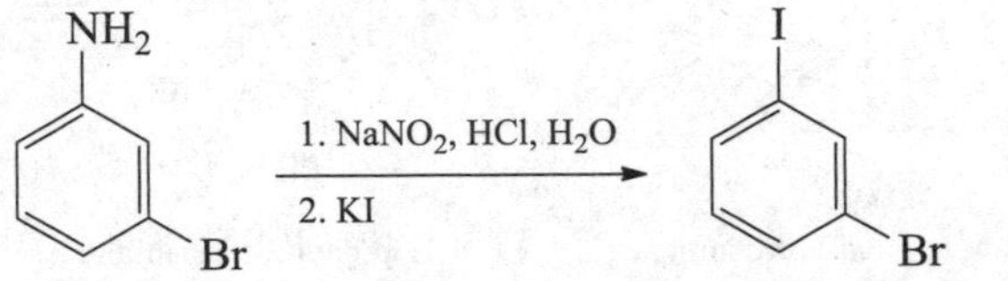

m-Bromoaniline   *m*-Bromoiodobenzene

22.20 The final step in the preparation of ethyl *m*-fluorophenyl ketone is shown in the text example immediately preceding this problem; therefore all that is necessary is to describe the preparation of *m*-aminophenyl ethyl ketone.

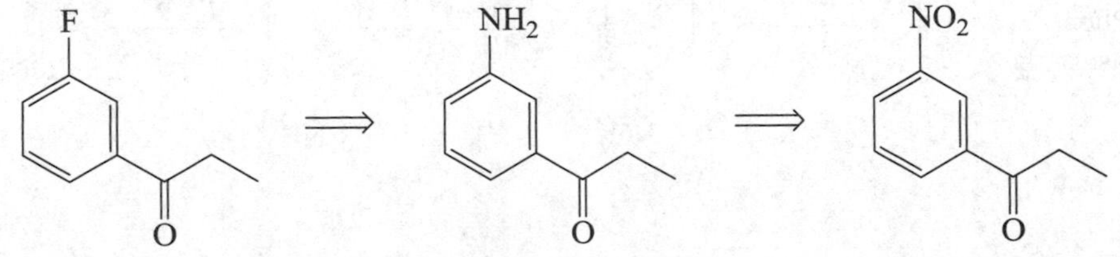

Ethyl *m*-fluorophenyl ketone   *m*-Aminophenyl ethyl ketone   Ethyl *m*-nitrophenyl ketone

Recalling that arylamines are normally prepared by reduction of nitroarenes, we see that ethyl *m*-nitrophenyl ketone is a pivotal synthetic intermediate. It is prepared by nitration of ethyl phenyl ketone, which is analogous to nitration of acetophenone, shown in text Section 13.16.

The preparation of ethyl phenyl ketone by Friedel–Crafts acylation of benzene is shown in text Section 13.7.

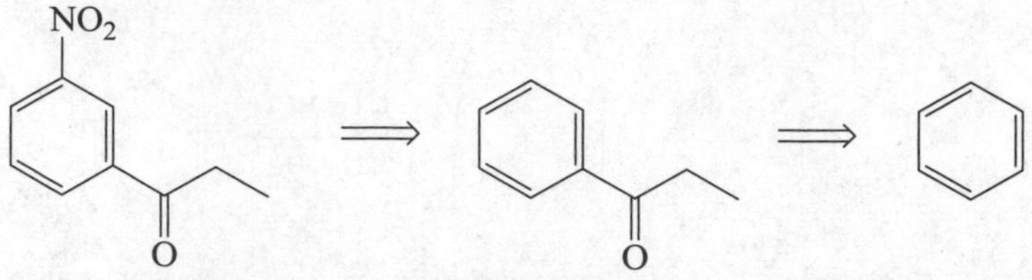

Ethyl *m*-nitrophenyl ketone   Ethyl phenyl ketone   Benzene

Reversing the order of introduction of the nitro and acyl groups will not lead to the desired product. It is possible to nitrate ethyl phenyl ketone, but not possible to carry out a Friedel–Crafts acylation on nitrobenzene, owing to the strong deactivating influence of the nitro group.

22.21 Direct nitration of the prescribed starting material cumene (isopropylbenzene) is not suitable, because isopropyl is an ortho, para-directing substituent and will give the target molecule *m*-nitrocumene as only a minor component of the nitration product. However, the conversion of 4-isopropyl-2-nitroaniline to *m*-isopropylnitrobenzene, which was used to illustrate reductive deamination of arylamines in the text, establishes the last step in the synthesis.

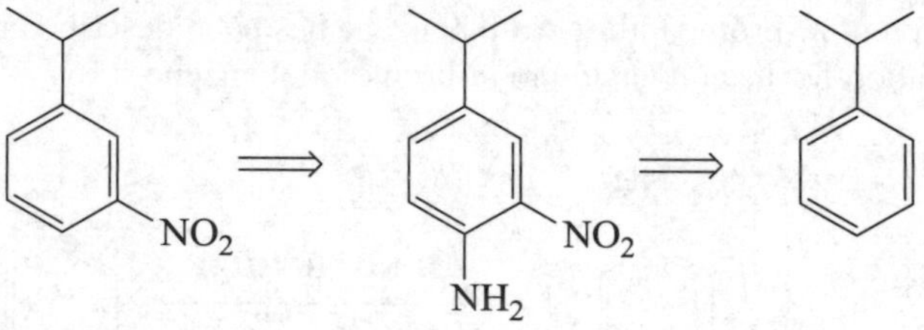

m-Nitrocumene   4-Isopropyl-2-nitroaniline   Cumene

Our task simplifies itself to the preparation of 4-isopropyl-2-nitroaniline from cumene. The following procedure is a straightforward extension of the reactions and principles developed in this chapter.

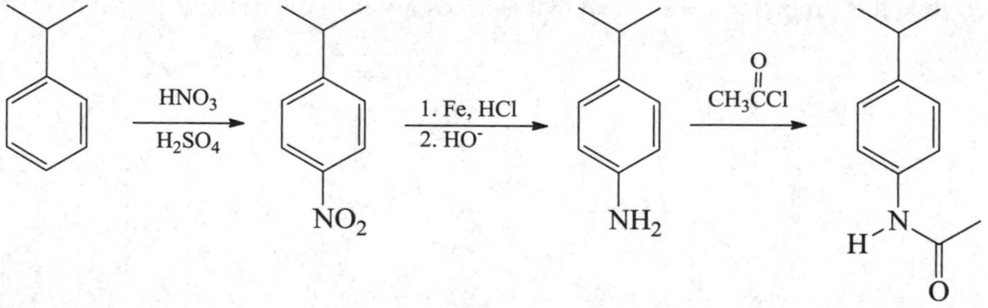

Cumene   *p*-Nitrocumene   *p*-Isopropylaniline   *p*-Isopropylacetanilide

Reductive dcamination of 4-isopropyl-2-nitroaniline by diazotization in the presence of ethanol or hypophosphorous acid yields *m*-nitrocumene and completes the synthesis.

22.22 Chrysoidine can be prepared from the reaction of benzenediazonium ion and *m*-benzenediamine.

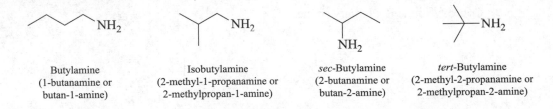

Chrysoidine   Benzenediazonium   *m*-Benzenediamine
ion

End of Chapter Problems

Structure and Nomenclature

22.23 Amines may be primary, secondary, or tertiary. The $C_4H_{11}N$ primary amines, compounds of the type $C_4H_9NH_2$, and their systematic names are

Butylamine
(1-butanamine or
butan-1-amine)

Isobutylamine
(2-methyl-1-propanamine or
2-methylpropan-1-amine)

sec-Butylamine
(2-butanamine or
butan-2-amine)

tert-Butylamine
(2-methyl-2-propanamine or
2-methylpropan-2-amine)

Secondary amines have the general formula R_2NH. Those of molecular formula $C_4H_{11}N$ are

Diethylamine
(*N*-ethylethanamine)

N-Methylpropylamine
(*N*-methyl-1-propanamine or
N-methylpropan-1-amine)

N-Methylisopropylamine
(*N*-methyl-2-propanamine or
N-methylpropan-2-amine)

There is only one tertiary amine (R_3N) of molecular formula $C_4H_{11}N$.

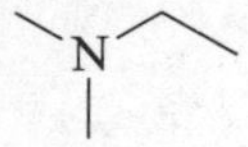

N,N-Dimethylethylamine
(*N,N*-dimethylethanamine)

22.24 (*a*) The name 2-ethyl-1-butanamine designates a four-carbon chain terminating in an amino group and bearing an ethyl group at C-2.

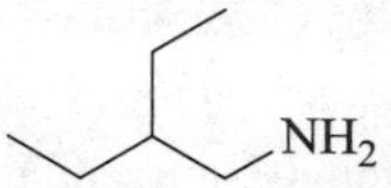

2-Ethyl-1-butanamine

(*b*) The prefix *N*- in *N*-ethyl-1-butanamine identifies the ethyl group as a substituent on nitrogen in a secondary amine.

N-Ethyl-1-butanamine

(*c*) Dibenzylamine is a secondary amine. It bears two benzyl groups on nitrogen.

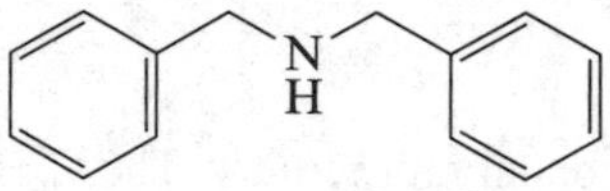

Dibenzylamine

(*d*) Tribenzylamine is a tertiary amine.

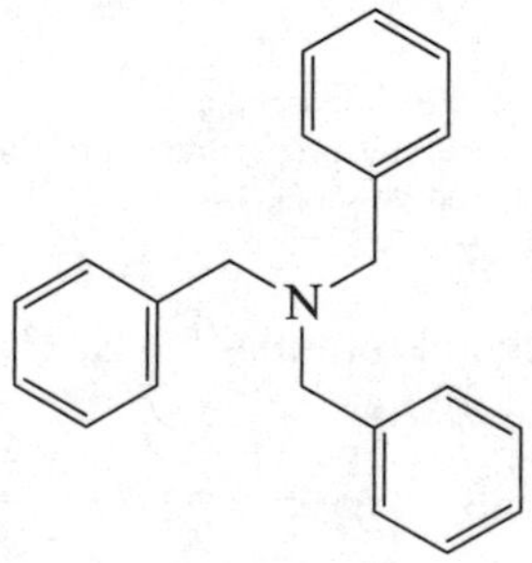

Tribenzylamine

(*e*) Tetraethylammonium hydroxide contains a quaternary ammonium ion.

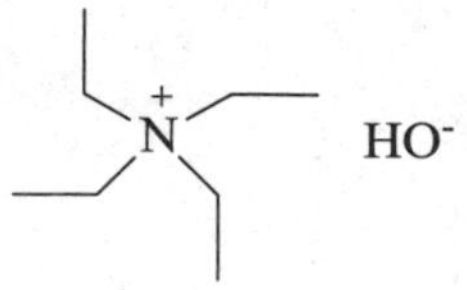

Tetraethylammonium hydroxide

(*f*) This compound is a secondary amine; it bears an allyl substituent on the nitrogen of cyclohexylamine.

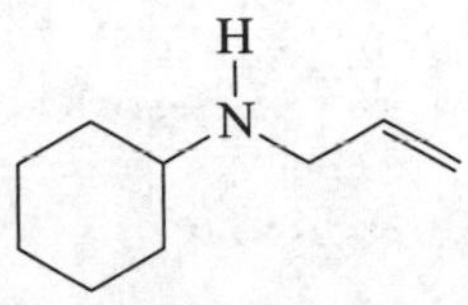

N-Allylcyclohexylamine

(*g*) Piperidine is a cyclic secondary amine that contains nitrogen in a six-membered ring. *N*-Allylpiperidine is a tertiary amine.

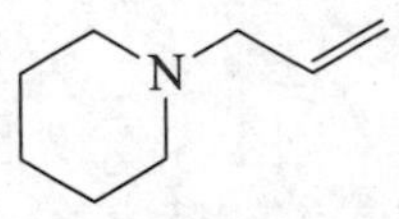

N-Allylpiperidine

(*h*) The compound is the benzyl ester of 2-aminopropanoic acid.

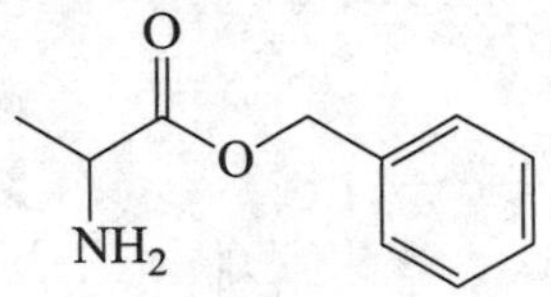

Benzyl 2-aminopropanoate

(*i*) The parent compound is cyclohexanone. The substituent $(CH_3)_2N$ is attached to C-4.

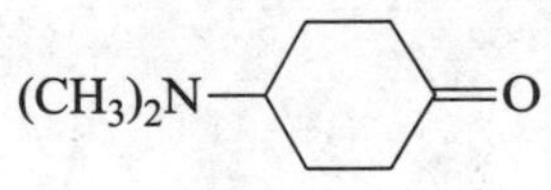

$(CH_3)_2N$—⟨ ⟩=O

4-(*N*,*N*-Dimethylamino)cyclohexanone

(*j*) The suffix *-diamine* reveals the presence of two amino groups, one at either end of a three-carbon chain that bears two methyl groups at C-2.

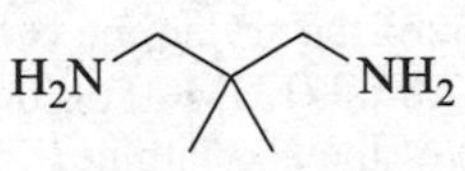

H_2N NH$_2$

2,2-Dimethyl-1,3-propanediamine

22.25 (*a*) A phenyl group and an amino group are trans to each other on a three-membered ring in this compound.

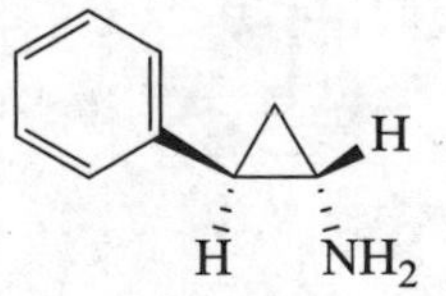

trans-2-Phenylcyclopropylamine
(tranylcypromine)

(*b*) This compound is a tertiary amine. It bears a benzyl group, a methyl group, and a 2-propynyl group on nitrogen.

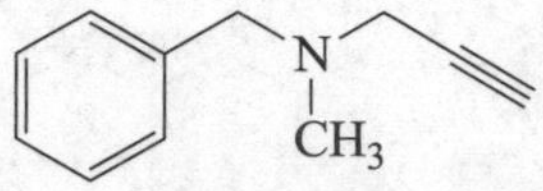

N-Benzyl-*N*-methyl-2-propynylamine
(pargyline)

(*c*) The amino group is at C-2 of a three-carbon chain that bears a phenyl substituent at its terminus.

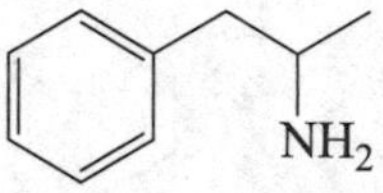

1-Phenyl-2-propanamine
(amphetamine)

(*d*) Phenylephrine is named systematically as an ethanol derivative.

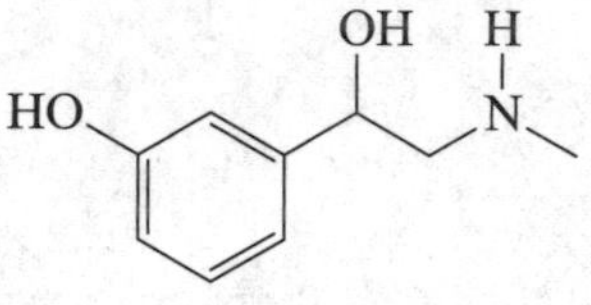

1-(*m*-Hydroxyphenyl)-2-(*N*-methylamino)ethanol

22.26 For inversion to occur, both aryl- and alkylamines must go through a trigonal planar (sp^2) transition state:

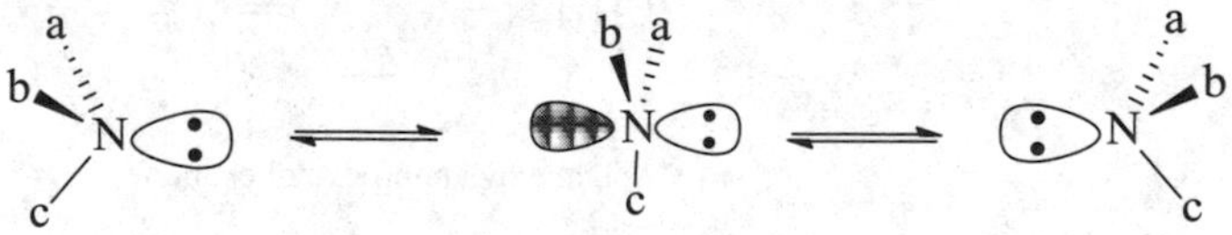

Transition state
(trigonal planar sp^2)

The transition state for inversion of the arylamine is stabilized by the interaction of the filled *p*-orbital with the lowest occupied molecular orbital (LUMO) of the aromatic system and is therefore lower in energy than the transition state for inversion of the alkylamine.

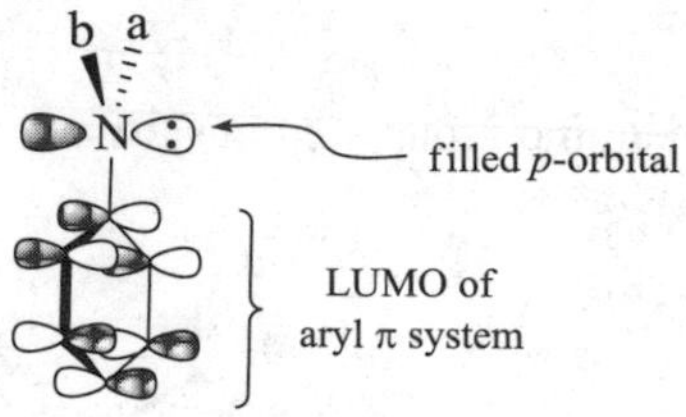

filled *p*-orbital

LUMO of
aryl π system

Reactions

22.27 (*a*) There are five isomers of C_7H_9N that contain a benzene ring.

Benzylamine   N-Methylaniline   o-Methylaniline   m-Methylaniline   p-Methylaniline

(b) Benzylamine is the strongest base because its amine group is bonded to an sp^3-hybridized carbon. Benzylamine is a typical alkylamine; the pK_a of its conjugate acid is 9.3. All the other isomers are arylamines; the pK_a's of their conjugate acids are about 4.

(c) The formation of N-nitrosoamines on reaction with sodium nitrite and hydrochloric acid is a characteristic reaction of secondary amines. The only C_7H_9N isomer in this problem that is a secondary amine is N-methylaniline.

N-Methylaniline   N-Methyl-N-nitrosoaniline

(d) Ring nitrosation is a characteristic reaction of tertiary arylamines.

Tertiary arylamine   p-Nitroso-N,N-dialkylaniline

None of the C_7H_9N isomers in this problem is a tertiary amine; hence, none will undergo ring nitrosation.

22.28 (a) Basicity decreases in proceeding left to right across a row in the periodic table. The increased nuclear charge as one progresses from carbon to nitrogen to oxygen to fluorine causes the electrons to be bound more strongly to the atom and thus less readily shared.

$$H_3C\overset{..}{:}^- \; > \; H_2\overset{..}{N}\overset{..}{:}^- \; > \; H\overset{..}{O}\overset{..}{:}^- \; > \; \overset{..}{:}\overset{..}{F}\overset{..}{:}^-$$

	Strongest base			Weakest base
pK_a of conjugate acid	60	36	16	3.4

(b) The strongest base in this group is amide ion, H_2N^-, and the weakest base is water, H_2O. Ammonia is a weaker base than hydroxide ion; the equilibrium lies to the left.

$$:NH_3 \; + \; H_2O \; \rightleftharpoons \; \overset{+}{N}H_4 \; + \; HO^-$$

| Weaker base | Weaker acid | Stronger acid | Stronger base |

The correct order is

$$H_2\ddot{\overset{..}{N}}{}^- \; > \; H\ddot{\overset{..}{O}}{}^- \; > \; :NH_3 \; > \; H_2\ddot{O}:$$

| Strongest base | | | Weakest base |

(c) These anions can be ranked according to their basicity by considering the respective acidities of their conjugate acids.

Base	Conjugate acid	pK_a of conjugate acid
H_2N^-	NH_3	36
HO^-	H_2O	16
$:C{\equiv}N:^-$	$HC{\equiv}N:$	9.1
$:\ddot{O}{-}\overset{\ddot{O}:}{\underset{:\ddot{O}:^-}{N^+}}$	$HO{-}\overset{\ddot{O}:}{\underset{:\ddot{O}:^-}{N^+}}$	−1.4

The order of basicities is the opposite of the order of acidities of their conjugate acids.

$$H_2\ddot{\overset{..}{N}}{}^- \; > \; H\ddot{\overset{..}{O}}{}^- \; > \; :C{\equiv}N:^- \; > \; :\ddot{O}{-}\overset{\ddot{O}:}{\underset{:\ddot{O}:^-}{N^+}}$$

| Strongest base | | | Weakest base |

(d) A carbonyl group attached to nitrogen stabilizes its negative charge. The strongest base is the anion that has no carbonyl groups on nitrogen; the weakest base is phthalimide anion, which has two carbonyl groups.

Strongest base   Weakest base

22.29 In evaluating base strengths in each part of this problem, be sure to recall that the strongest base has the weakest conjugate acid (largest pK_a).

(*a*) An alkyl substituent on nitrogen is electron-releasing and base-strengthening; thus, methylamine is a stronger base than ammonia. An aryl substituent is electron-withdrawing and base-weakening, and so aniline is a weaker base than ammonia.

$$CH_3-\ddot{N}H_2 \quad > \quad :NH_3 \quad > \quad C_6H_5-\ddot{N}H_2$$

	Methylamine, strongest base:	Ammonia:	Aniline, weakest base:
pK_a of conjugate acid	11.6	9.3	4.6

(*b*) An acetyl group is an electron-withdrawing and base-weakening substituent, especially when bonded directly to nitrogen. Amides are weaker bases than amines; thus, acetanilide is a weaker base than aniline. Alkyl groups are electron-releasing; *N*-methylaniline is a slightly stronger base than aniline.

	N-Methylaniline, strongest base:	Aniline	Acetanilide, weakest base:
pK_a of conjugate acid	4.9	4.6	-1.0

(*c*) Chlorine substituents are slightly electron-withdrawing, and methyl groups are slightly electron-releasing. 2,4-Dimethylaniline is therefore a stronger base than 2,4-dichloroaniline. Nitro groups are strongly electron-withdrawing, their base-weakening effect being especially pronounced when a nitro group is ortho or para to an amino group because the two groups are then directly conjugated.

	2,4-Dimethylaniline, strongest base:	2,4-Dichloroaniline:	2,4-Dinitroaniline, weakest base:
pK_a of conjugate acid	4.9	2.0	-4.5

(*d*) Nitro groups are more electron-withdrawing than chlorine, and the base-weakening effect of a nitro substituent is greater when it is ortho or para to an amino group than when it is meta to it.

	3,4-Dichloroaniline, strongest base:	4-Chloro-3-nitroaniline:	4-Chloro-2-nitroaniline, weakest base:
pK_a of conjugate acid	≈ 3	1.9	-1.0

(*e*) According to the principle applied in part (*a*) (alkyl groups increase basicity, aryl groups decrease it), the order of decreasing basicity is as shown:

	Dimethylamine, strongest base:	*N*-Methylaniline:	Diphenylamine, weakest base:
pK_a of conjugate acid	10.7	4.9	0.8

22.30 There is no obvious reason why the dimethylamino group in 4-(*N*,*N*-dimethylamino)pyridine should be appreciably more basic than it is in *N*,*N*-dimethylaniline; it is the ring nitrogen of 4-(*N*,*N*-dimethylamino)pyridine that is more basic. Note that protonation of the ring nitrogen permits delocalization of the dimethylamino lone pair and dispersal of the positive charge.

Most stable protonated form of
4-(*N*,*N*-dimethylamino)pyridine

22.31 Only the unshared electron pair on nitrogen that is not part of the π electron cloud of the aromatic system will be available for protonation. Treatment of 5-methyl-γ-carboline with acid will give the salt shown.

5-Methyl-γ-carboline

22.32 The imidazole nitrogen is the other basic site (in addition to the terminal amino group) in carnosine and is protonated in the conjugate acid. The amide nitrogen is not a basic site (its two electrons are conjugated with the carbonyl) and is thus not protonated.

Carnosine   Conjugate acid of carnosine

22.33 Nitrogen ④ is the most basic and the most nucleophilic of the three nitrogen atoms of physostigmine and is the one that reacts with methyl iodide.

Physostigmine   Methyl   "Physostigmine methiodide"
iodide

The nitrogen that reacts is the one that is a tertiary alkylamine. Of the other two nitrogens, ⓑ is attached to an aromatic ring and is much less basic and less nucleophilic. The third nitrogen, ©, is an amide nitrogen; amides are less nucleophilic than amines.

22.34 A mixture of 9-aminofluorene and fluorene can be separated by taking advantage of the amine group in 9-aminofluorene. Because this substituent is basic and can be protonated to make a salt that is soluble in water, the following extraction process can be applied.

Step 1: Add ether to the mixture. From this solution, extract the 9-aminofluorene into the aqueous phase with aqueous hydrochloric acid. This process converts 9-aminofluorene to its HCl salt, which will be soluble in the aqueous phase. The neutral fluorene remains in the ether phase.

9-Aminofluorene   HCl salt of 9-aminofluorene
(water soluble)

Step 2: After separating the organic phase from the aqueous phase, the aqueous phase is treated with base to convert the HCl salt back to 9-aminofluorene. The neutral 9-aminofluorene can then be extracted into ether.

HCl salt of 9-aminofluorene   9-Aminofluorene
(water soluble)   (ether soluble)

Step 3: Concentration of the separate ether solutions will provide 9-aminofluorene and fluorene.

22.35 (*a*) Amines are basic and are protonated by hydrogen halides.

Benzylamine   Hydrogen   Benzylammonium
bromide   bromide

(b) Equimolar amounts of benzylamine and sulfuric acid yield benzylammonium hydrogen sulfate as the product.

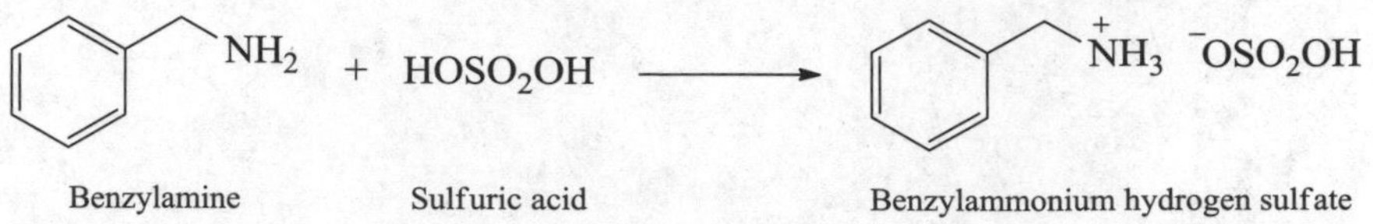

Benzylamine   +   Sulfuric acid   →   Benzylammonium hydrogen sulfate

(c) Acetic acid transfers a proton to benzylamine.

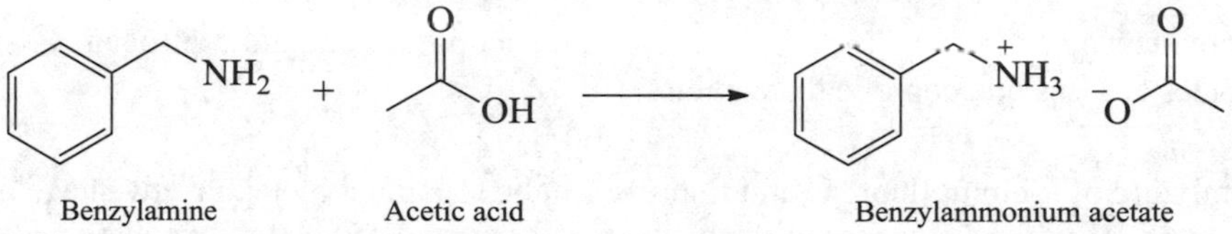

Benzylamine   +   Acetic acid   →   Benzylammonium acetate

(d) Acetyl chloride reacts with benzylamine to form an amide.

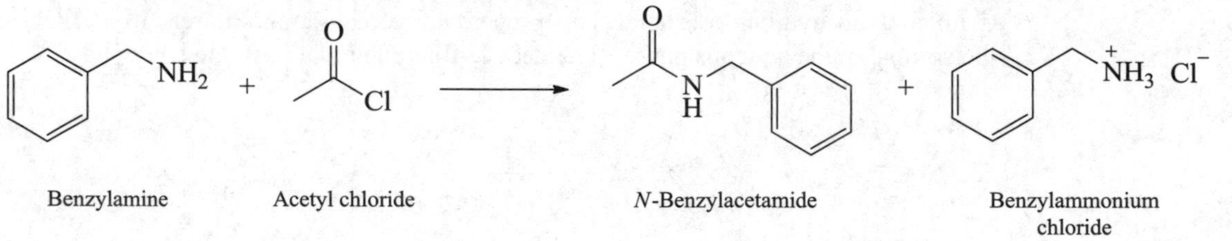

Benzylamine   +   Acetyl chloride   →   N-Benzylacetamide   +   Benzylammonium chloride

(e) Acetic anhydride also gives an amide with benzylamine.

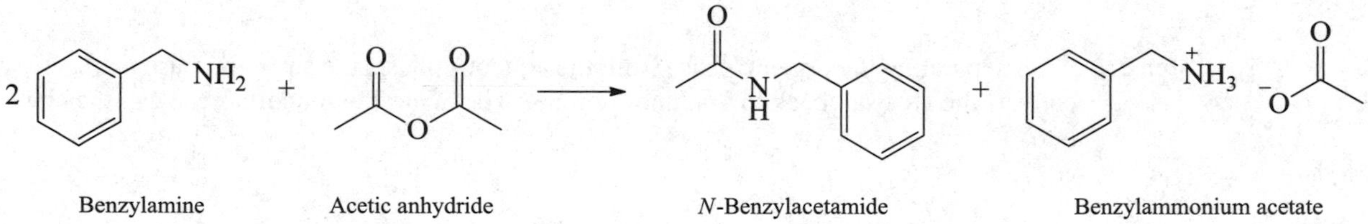

Benzylamine   +   Acetic anhydride   →   N-Benzylacetamide   +   Benzylammonium acetate

(f) Primary amines react with ketones to give imines.

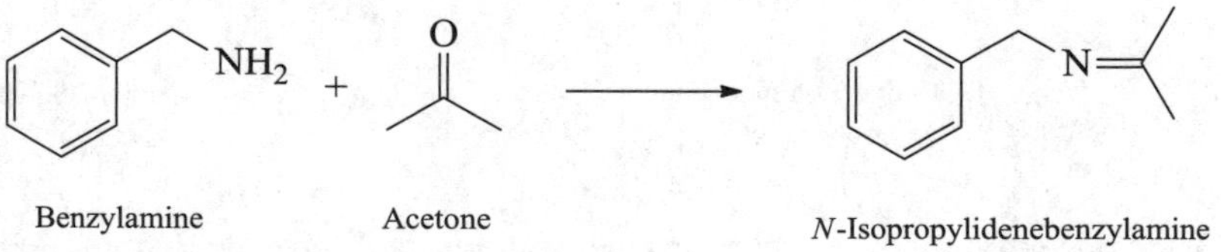

Benzylamine   +   Acetone   →   N-Isopropylidenebenzylamine

(g) These reaction conditions lead to reduction of the imine formed in part (f). The overall reaction is reductive amination.

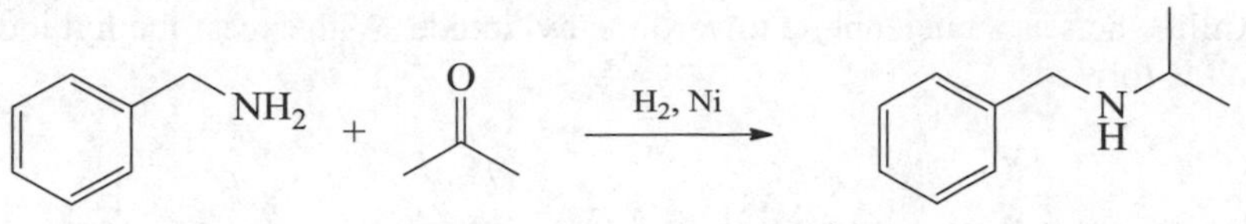

Benzylamine + Acetone →(H₂, Ni) N-Isopropylbenzylamine

(*h*) Amines are nucleophilic and bring about the opening of epoxide rings.

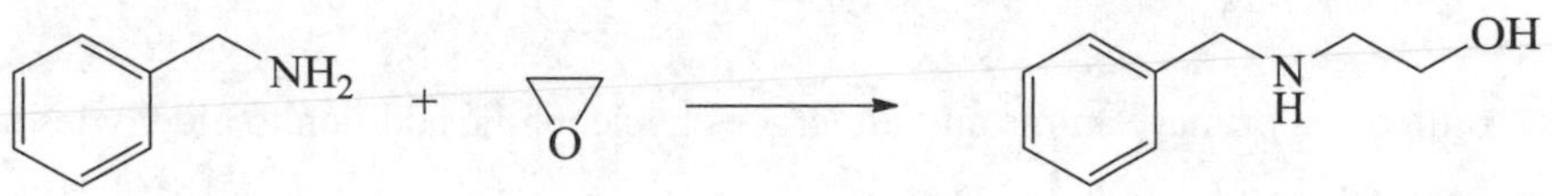

Benzylamine + Ethylene oxide → 2-(N-Benzylamino)ethanol

(*i*) In these nucleophilic ring-opening reactions, the amine attacks the less sterically hindered carbon of the ring.

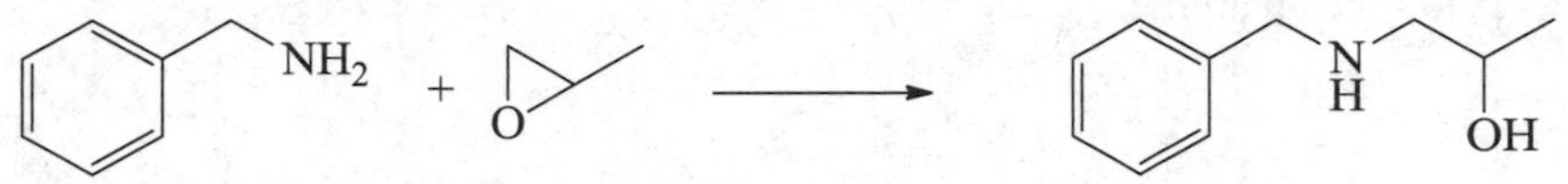

Benzylamine + 1,2-Epoxypropane → 1-(N-Benzylamino)-2-propanol

(*j*) With excess methyl iodide, amines are converted to quaternary ammonium iodides.

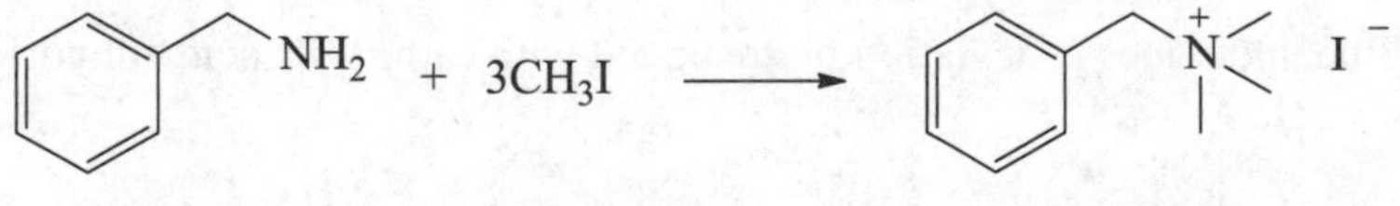

Benzylamine + 3CH₃I → Benzyltrimethylammonium iodide

(*k*) Nitrous acid forms from sodium nitrite in dilute hydrochloric acid. Nitrosation of benzylamine in water gives benzyl alcohol via a diazonium ion intermediate.

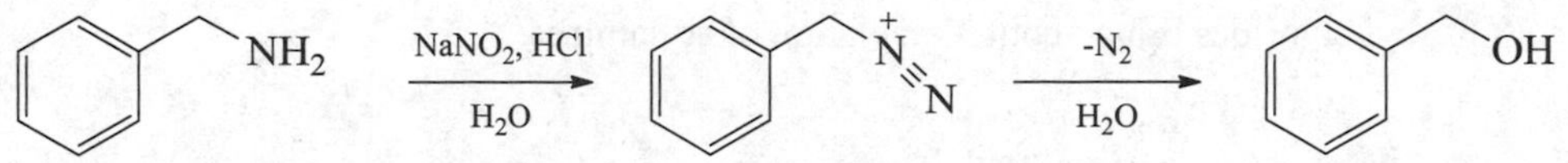

Benzylamine →(NaNO₂, HCl / H₂O) Benzyldiazonium ion →(-N₂ / H₂O) Benzyl alcohol

22.36 (*a*) Aniline is a weak base and yields a salt on reaction with hydrogen bromide.

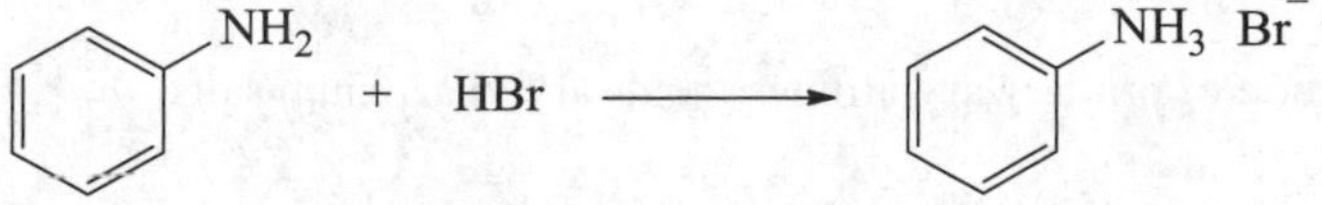

Aniline + Hydrogen bromide → Anilinium bromide

(*b*) Aniline acts as a nucleophile toward methyl iodide. With excess methyl iodide, a quaternary ammonium salt is formed.

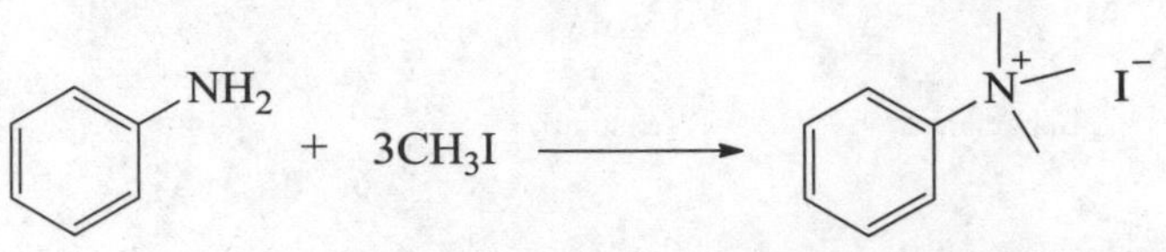

Aniline   Methyl iodide   *N,N,N*-Trimethylanilinium iodide

(*c*) Aniline is a primary amine and undergoes nucleophilic addition to aldehydes and ketones to form imines.

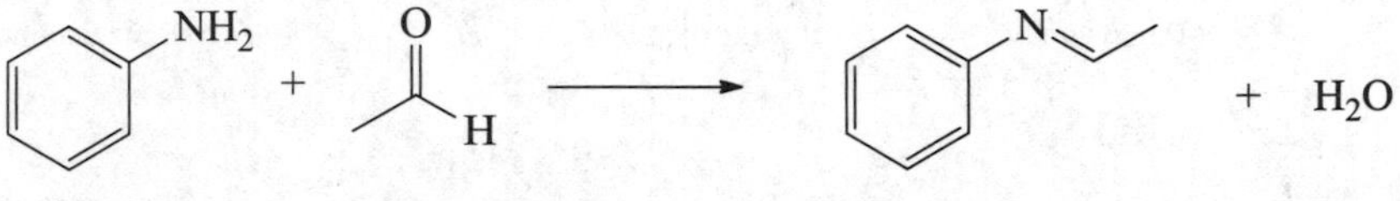

Aniline   Acetaldehyde   *N*-Phenylacetaldimine   Water

(*d*) When an imine is formed in the presence of hydrogen and a suitable catalyst, reductive amination occurs to give an amine.

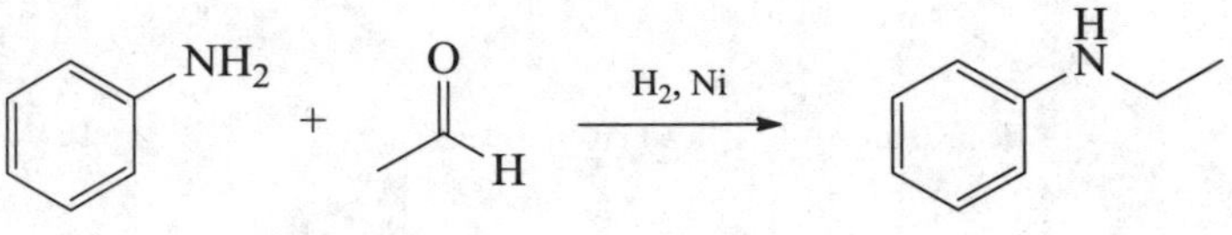

Aniline   Acetaldehyde   *N*-Ethylaniline

(*e*) Aniline undergoes *N*-acylation on treatment with carboxylic acid anhydrides.

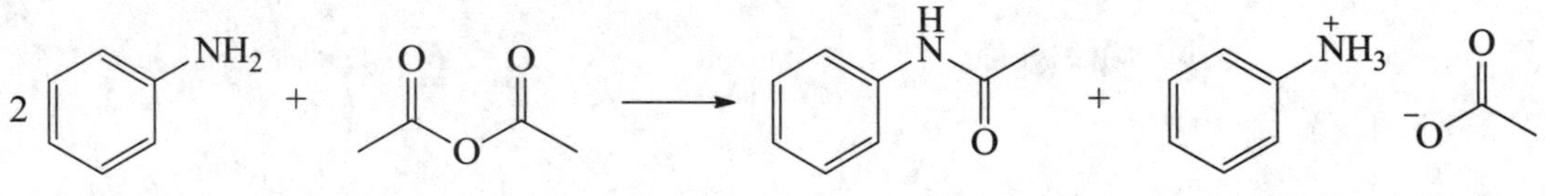

Aniline   Acetic anhydride   Acetanilide   Anilinium acetate

(*f*) Acyl chlorides bring about *N*-acylation of arylamines.

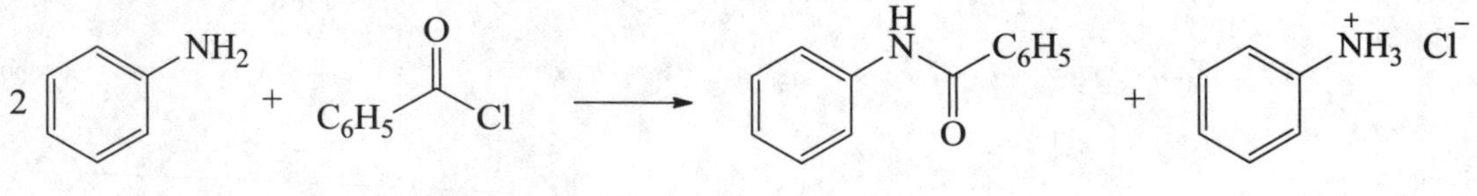

Aniline   Benzoyl chloride   *N*-Benzylacetamide   Anilinium chloride

(*g*) Nitrosation of primary arylamines yields aryl diazonium salts.

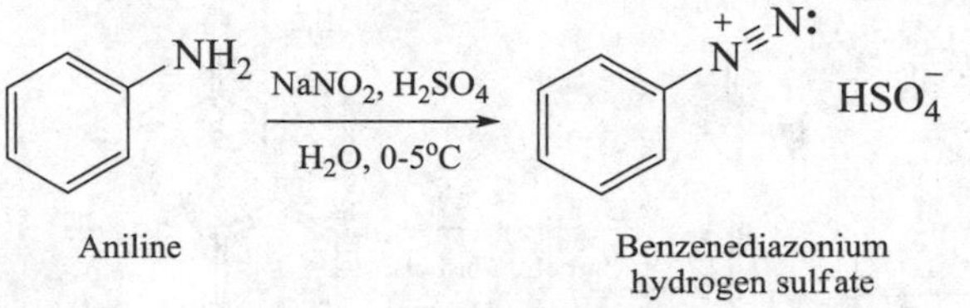

Aniline → Benzenediazonium hydrogen sulfate

$NaNO_2, H_2SO_4$ / $H_2O, 0\text{-}5°C$

22.37 (*a*) Amides are reduced to amines by lithium aluminum hydride.

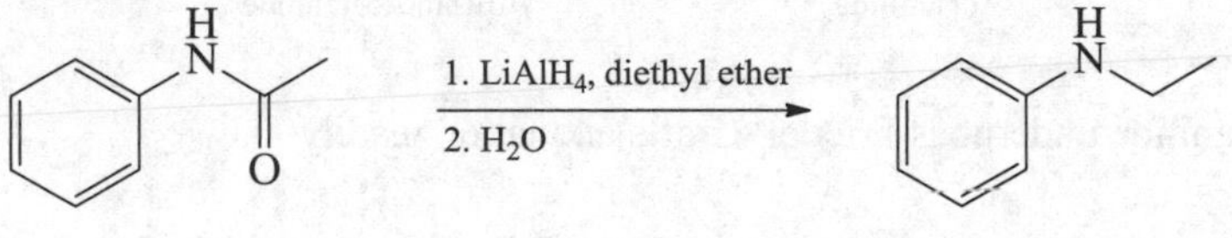

1. LiAlH₄, diethyl ether
2. H₂O

Acetanilide → N-Ethylaniline

(*b*) Acetanilide is strongly activated toward electrophilic aromatic substitution. An acetamido group is ortho, para-directing.

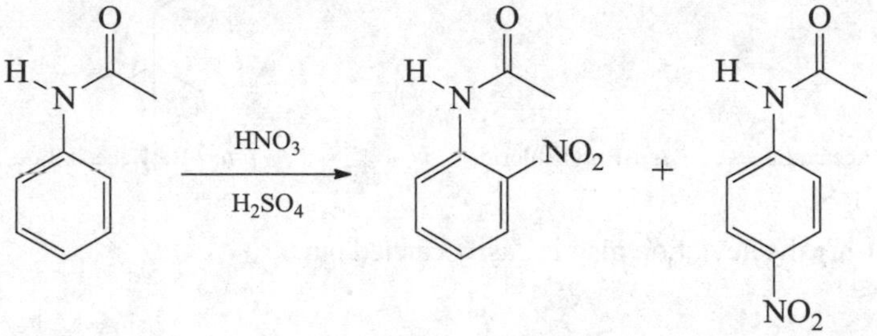

HNO_3 / H_2SO_4

Acetanilide → o-Nitroacetanilide + p-Nitroacetanilide

(*c*) Sulfonation of the ring occurs.

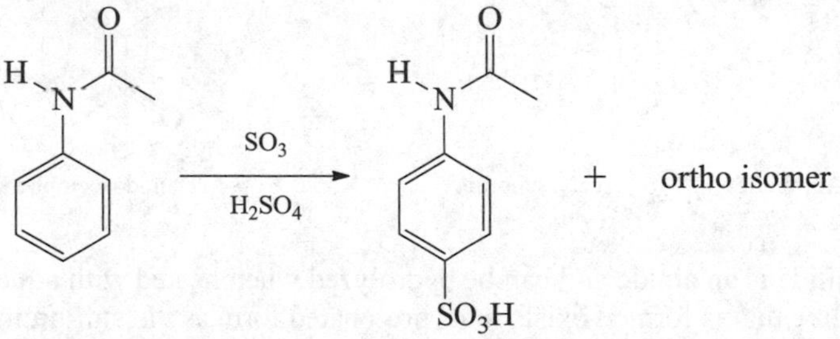

SO_3 / H_2SO_4

Acetanilide → p-Acetamidobenzenesulfonic acid + ortho isomer

(*d*) Bromination of the ring takes place.

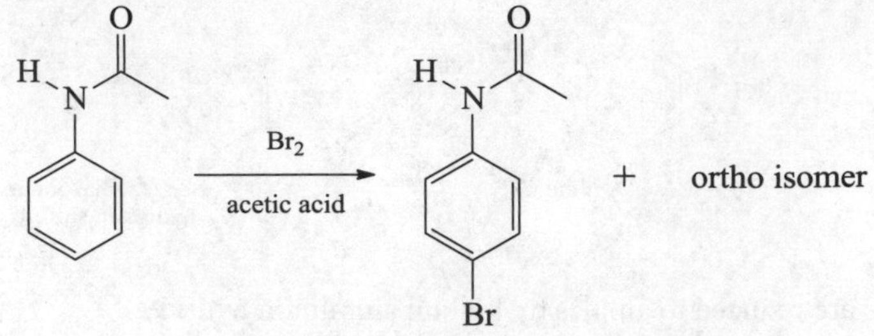

Acetanilide   *p*-Bromoacetanilide

(*e*) Acetanilide undergoes Friedel–Crafts alkylation readily.

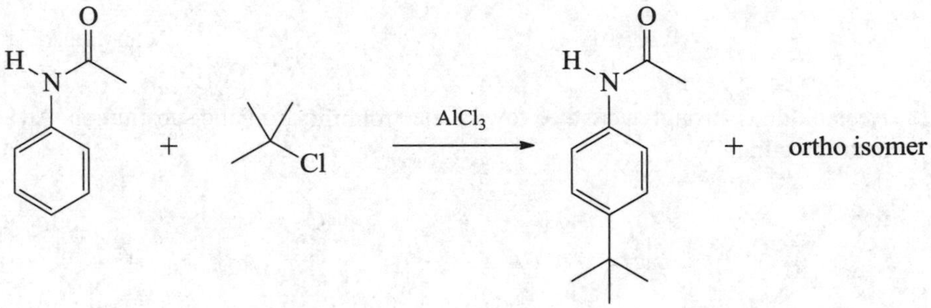

Acetanilide   *tert*-Butyl chloride   *p*-*tert*-Butylacetanilide

(*f*) Friedel–Crafts acylation also is easily carried out.

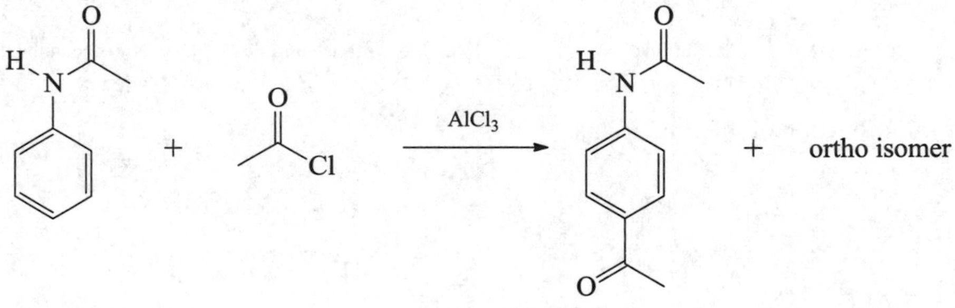

Acetanilide   Acetyl chloride   *p*-Acetamidoacetophenone

(*g*) Acetanilide is an amide and can be hydrolyzed when heated with aqueous acid. Under acidic conditions, the aniline that is formed exists in its protonated form as the anilinium cation.

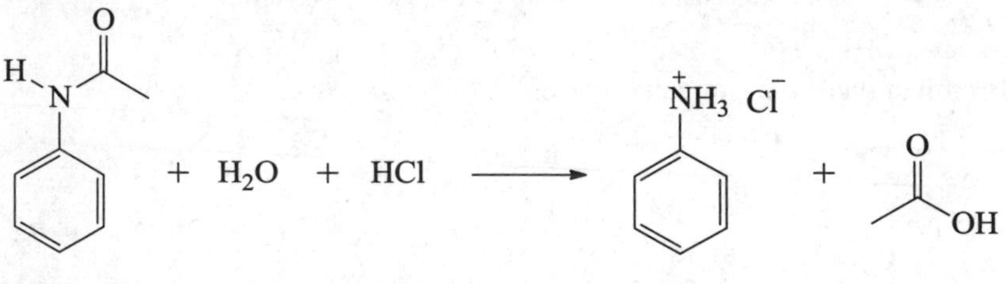

Acetanilide   Water   Hydrogen   Anilinium   Acetic acid
                          chloride   chloride

(*h*) Amides are hydrolyzed in base.

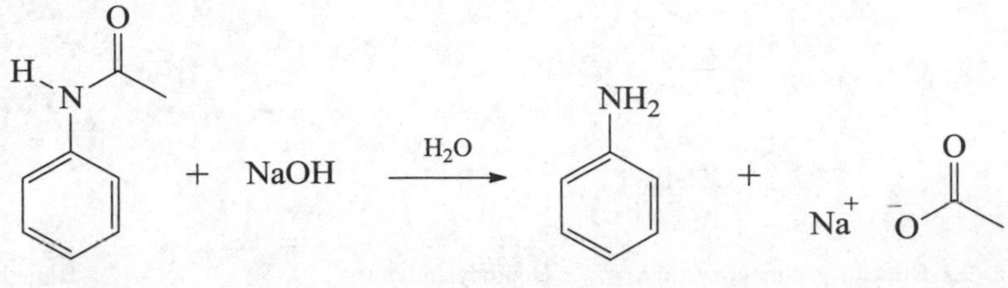

| Acetanilide | Sodium hydroxide | | Aniline | | Sodium acetate |

22.38 (*a*) The reaction illustrates the preparation of a secondary amine by reductive amination.

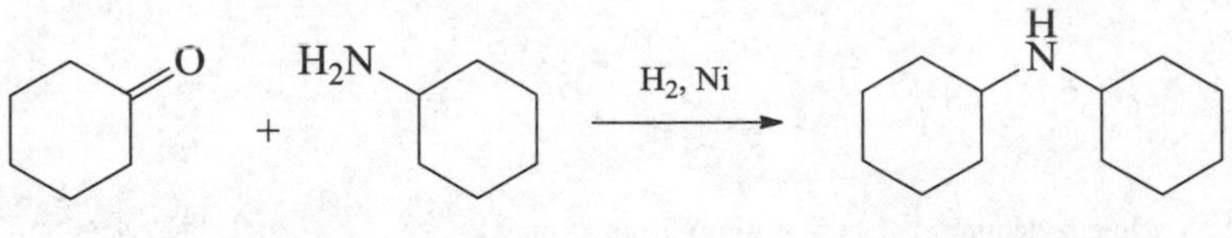

| Cyclohexanone | Cyclohexylamine | | Dicyclohexylamine (70%) |

(*b*) Amides are reduced to amines by lithium aluminum hydride.

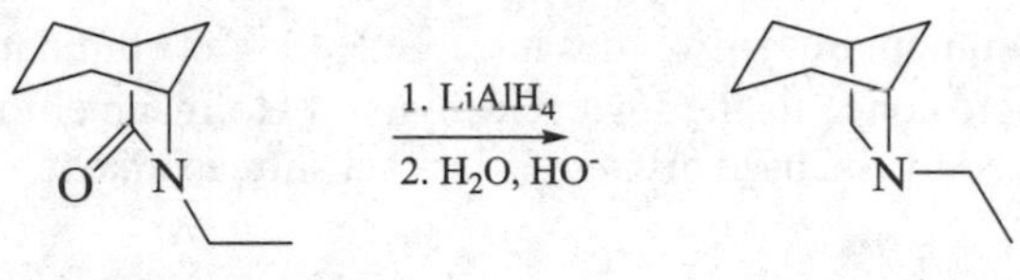

| 6-Ethyl-6-azabicyclo[3.2.1]octan-7-one | | 6-Ethyl-6-azabicyclo[3.2.1]octane |

(*c*) Treatment of alcohols with *p*-toluenesulfonyl chloride converts them to *p*-toluenesulfonates.

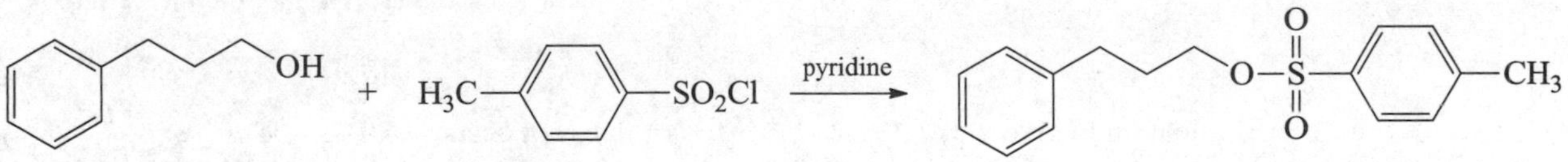

| 3-Phenyl-1-propanol | *p*-Toluenesulfonyl chloride | | 3-Phenylpropyl *p*-toluenesulfonate |

p-Toluenesulfonate is an excellent leaving group in nucleophilic substitution reactions. Dimethylamine is the nucleophile.

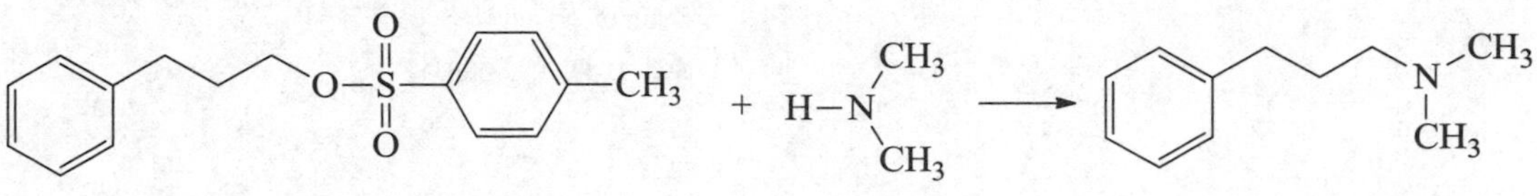

| 3-Phenylpropyl *p*-toluenesulfonate | Dimethylamine | *N*,*N*-Dimethyl-3-phenyl-1-propanamine (86%) |

(*d*) Amines are sufficiently nucleophilic to react with epoxides. Attack occurs at the less substituted carbon of the epoxide.

2-(2,5-Dimethoxyphenyl)oxirane   Isopropylamine   1-(2,5-Dimethoxyphenyl)-2-
                                                     (*N*-isopropylamino)ethanol (67%)

(*e*) α-Halo ketones are reactive substrates in nucleophilic substitution reactions. Dibenzylamine is the nucleophile.

$(C_6H_5CH_2)_2NH$ + 1-Chloro-2-propanone ⟶ $N(CH_2C_6H_5)_2$

Dibenzylamine   1-Chloro-2-propanone   1-(Dibenzylamino)-2-propanone (87%)

Because the reaction liberates hydrogen chloride, it is carried out in the presence of added base—in this case, triethylamine—so as to avoid converting the dibenzylamine to its hydrochloride salt.

(*f*) Quaternary ammonium hydroxides undergo Hofmann elimination when they are heated. A point to be considered here concerns the regioselectivity of Hofmann eliminations: it is the less hindered β proton that is removed by the base giving the less substituted alkene.

       $+\overset{+}{N}(CH_3)_3$ $\xrightarrow{-H_2O}$ +   :$N(CH_3)_3$

                                       trans-1-Isopropenyl-4-   Trimethylamine
                                       methylcyclohexane (98%)

Elimination to give   does not occur.

(*g*) The combination of sodium nitrite and aqueous acid is a nitrosating agent. Secondary alkylamines react with nitrosating agents to give *N*-nitroso amines as the isolated products.

          $\xrightarrow[\text{HCl, }H_2O]{NaNO_2}$

Diisopropylamine   *N*-Nitrosodiisopropylamine (91%)

22.39 (*a*) Catalytic hydrogenation reduces nitro groups to amino groups.

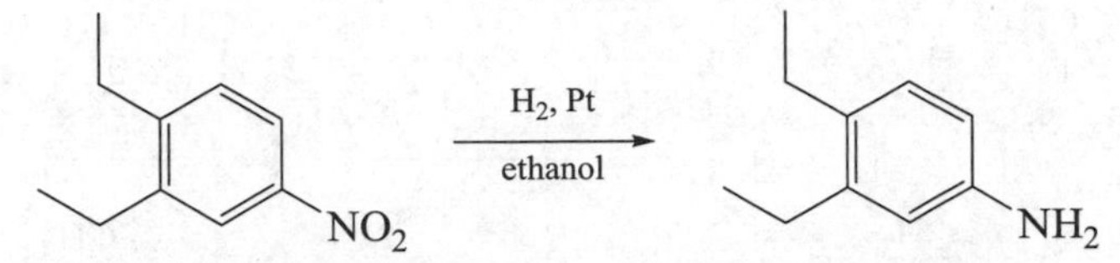

1,2-Diethyl-4-nitrobenzene   3,4-Diethylaniline (93-99%)

(b) Nitro groups are readily reduced by tin(II) chloride.

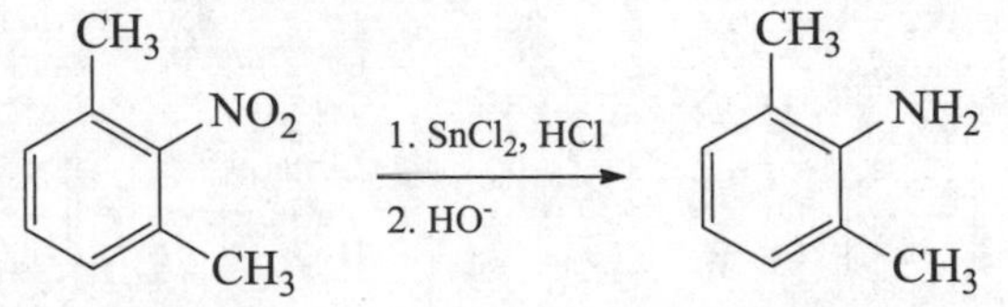

1,3-Dimethyl-2-nitrobenzene   2,6-Dimethylaniline

This reaction is the first step in a synthesis of the drug *lidocaine*.

(c) The amino group of arylamines is nucleophilic and undergoes acylation on reaction with chloroacetyl chloride. Chloroacetyl chloride is a difunctional compound—it is both an acyl chloride and an alkyl chloride. Acyl chlorides react with nucleophiles faster than do alkyl chlorides. Therefore, acylation of the amine nitrogen occurs rather than alkylation.

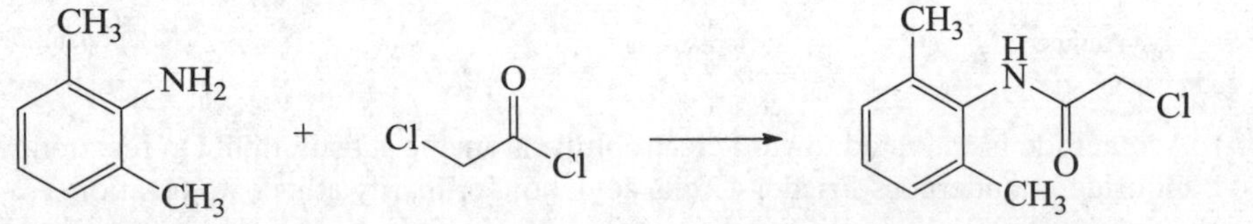

2,6-Dimethylaniline   Chloroacetyl chloride   *N*-(Chloroacetyl)-2,6-dimethylaniline

(d) The final step in the synthesis of lidocaine is displacement of the chloride by diethylamine from the α-halo amide formed in part (c) in a nucleophilic substitution reaction.

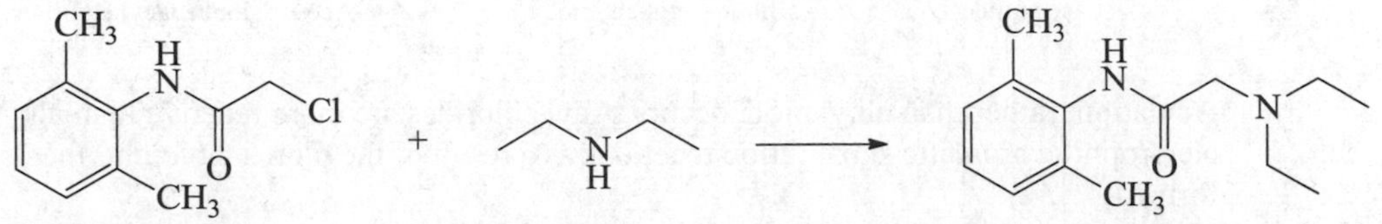

N-(Chloroacetyl)-2,6-dimethylaniline   Diethylamine   Lidocaine

The reaction is carried out with excess diethylamine, which acts as a base to neutralize the hydrogen chloride formed.

(e) For use as an anesthetic, lidocaine is made available as its hydrochloride salt. Of the two nitrogens in lidocaine, the amine nitrogen is more basic than the amide.

Lidocaine   Lidocaine hydrochloride

(*f*) Lithium aluminum hydride reduction of amides is one of the best methods for the preparation of amines, including arylamines.

N-Phenylbutanamide   *N*-Butylaniline (92%)

(*g*) Arylamines react with aldehydes and ketones in the presence of hydrogen and nickel to give the product of reductive amination.

Aniline   Heptanal   *N*-Heptylaniline (65%)

(*h*) Acetanilide is activated toward electrophilic aromatic substitution. On reaction with chloroacetyl chloride, it undergoes Friedel–Crafts acylation, primarily at its para position.

Acetanilide   Chloroacetyl chloride   *p*-Acetamidophenacyl chloride (79-83%)

Acylation, rather than alkylation, occurs. Acyl chlorides are more reactive than alkyl chlorides toward electrophilic aromatic substitution reactions as a result of the more stable intermediate (acylium ion) formed.

(*i*) Reduction with iron in hydrochloric acid is one of the most common methods for converting nitroarenes to arylamines.

4-Bromo-4'-nitrobiphenyl   4-Amino-4'-bromobiphenyl (94%)

(*j*) Primary arylamines are converted to aryl diazonium salts on treatment with sodium nitrite in aqueous acid. When the aqueous acidic solution containing the diazonium salt is heated, a phenol is formed.

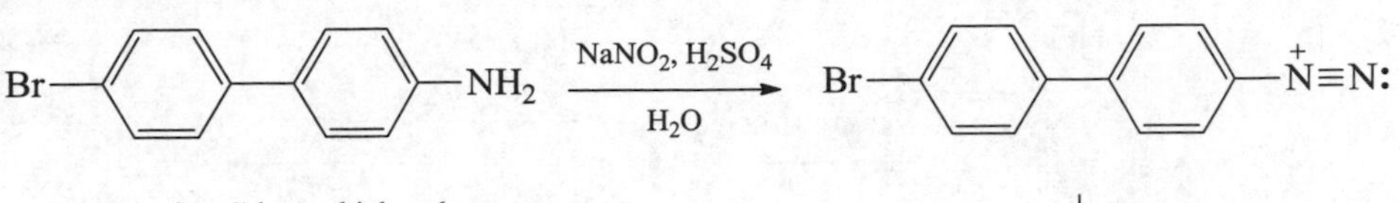

4-Amino-4'-bromobiphenyl

4-Bromo-4'-hydroxybiphenyl (85%)

(k) This problem illustrates the conversion of an arylamine to an aryl chloride by the Sandmeyer reaction.

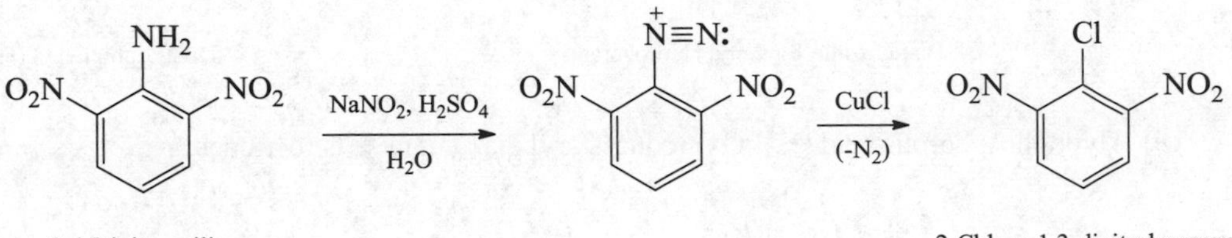

2,6-Dinitroaniline

2-Chloro-1,3-dinitrobenzene
(71-74%)

(l) Diazotization of primary arylamines followed by treatment with copper(I) bromide converts them to aryl bromides.

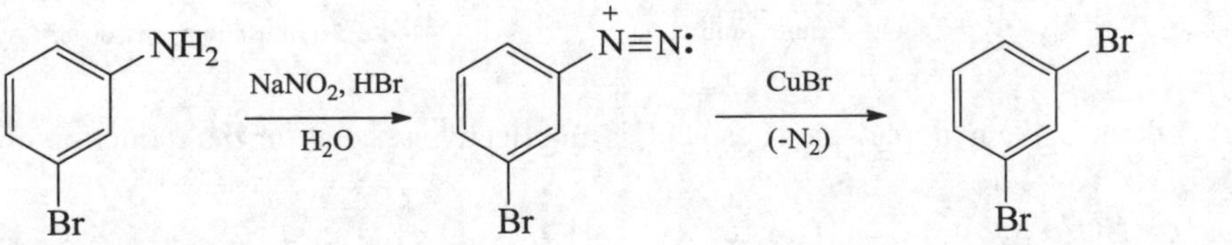

m-Bromoaniline

m-Dibromobenzcne (80-87%)

(m) Nitriles are formed when aryl diazonium salts react with copper(I) cyanide.

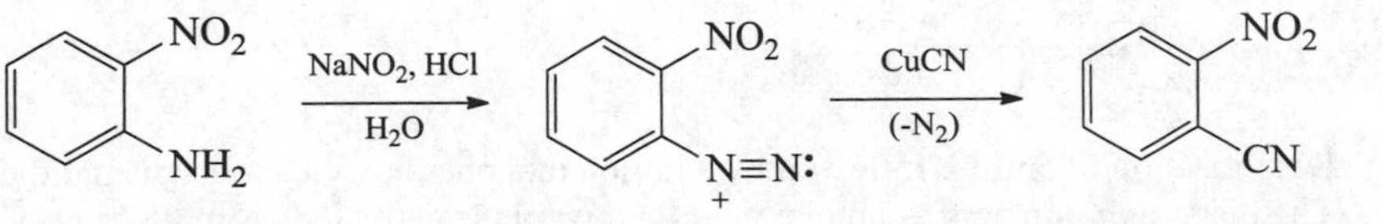

o-Nitroaniline

o-Nitrobenzonitrile (87%)

(n) An aryl diazonium salt is converted to an aryl iodide on reaction with potassium iodide.

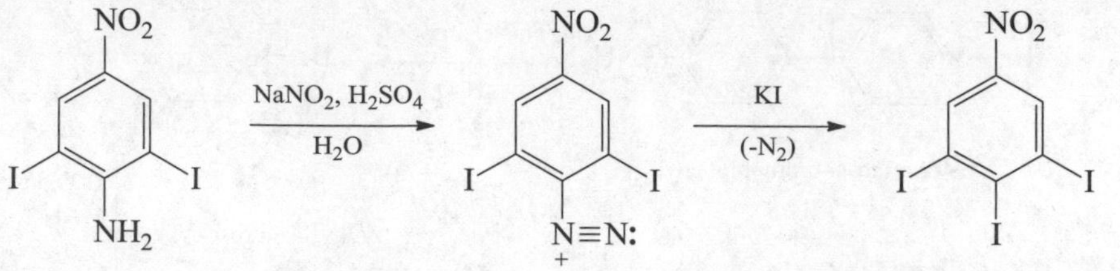

2,6-Diiodo-4-nitroaniline

NaNO$_2$, H$_2$SO$_4$
H$_2$O

KI
(-N$_2$)

1,2,3-Triiodo-5-nitrobenzene
(94-95%)

(*o*) Aryl diazonium fluoroborates are converted to aryl fluorides when heated. Both diazonium salt functions in the starting material undergo this reaction.

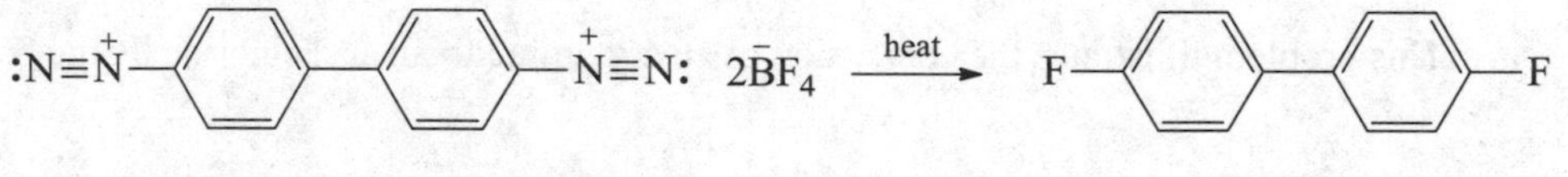

4,4'-Bis(diazonio)biphenyl fluoroborate

heat

4,4'-Difluorobiphenyl (82%)

(*p*) Hypophosphorous acid (H$_3$PO$_2$) reduces aryl diazonium salts to arenes.

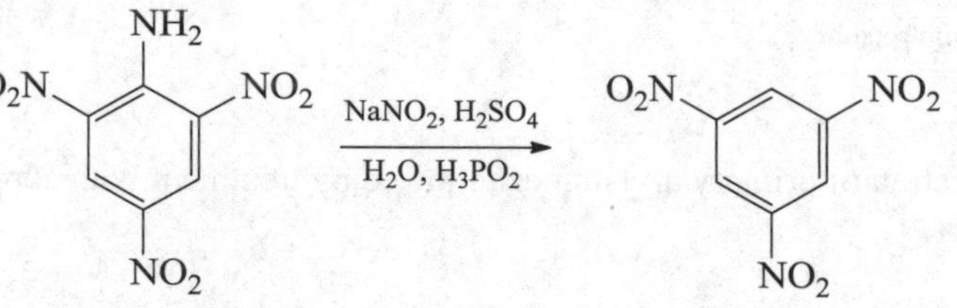

2,4,6-Trinitroaniline

NaNO$_2$, H$_2$SO$_4$
H$_2$O, H$_3$PO$_2$

1,3,5-Trinitrobenzene (60-65%)

(*q*) Ethanol, like hypophosphorous acid, is an effective reagent for the reduction of aryl diazonium salts.

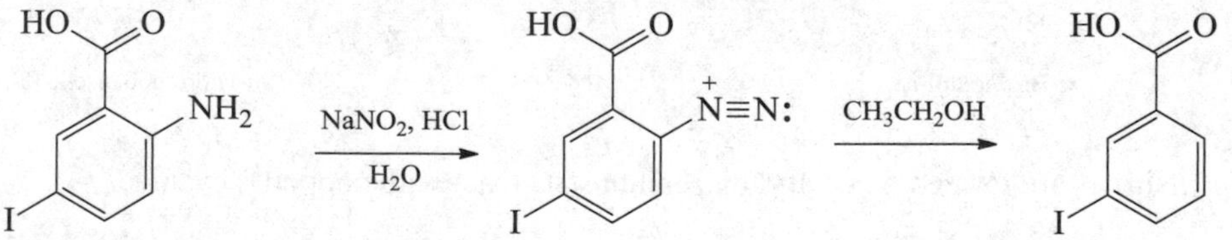

2-Amino-5-iodobenzoic acid

NaNO$_2$, HCl
H$_2$O

CH$_3$CH$_2$OH

m-Iodobenzoic acid (86-93%)

(*r*) Diazotization of aniline followed by addition of a phenol yields a bright-red diazo-substituted phenol. The diazonium ion acts as an electrophile toward the activated aromatic ring of the phenol.

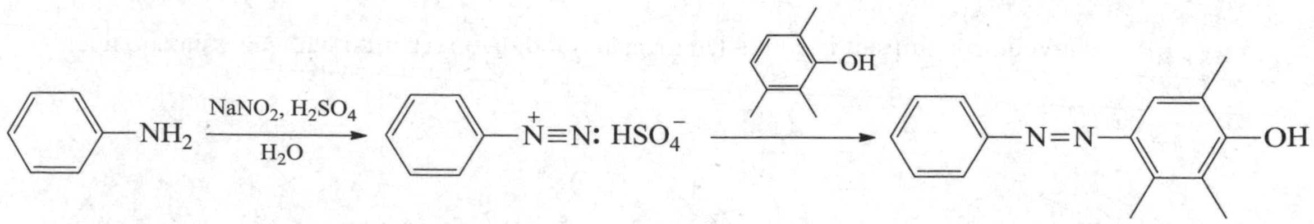

Aniline

NaNO$_2$, H$_2$SO$_4$
H$_2$O

Benzenediazonium hydrogen sulfate

2,3,6-Trimethyl-4-(phenylazo)phenol (98%)

(*s*) Nitrosation of *N,N*-dialkylarylamines takes place on the ring at the position para to the dialkylamino group.

N,N-Dimethyl-*m*-toluidine   3-Methyl-4-nitroso-*N,N*-dimethylaniline (83%)

22.40 (*a*) 4-Methylpiperidine can participate in intermolecular hydrogen bonding in the liquid phase.

These hydrogen bonds must be broken for individual 4-methylpiperidine molecules to escape into the gas phase. *N*-Methylpiperidine lacks a proton bonded to nitrogen and so cannot engage in intermolecular hydrogen bonding. Less energy is required to transfer a molecule of *N*-methylpiperidine to the gaseous state, and therefore it has a lower boiling point than 4-methylpiperidine.

N-Methylpiperidine;
no hydrogen bonding possible to
other *N*-methylpiperidine molecules

(*b*) The two products are diastereomeric quaternary ammonium chlorides that differ in the configuration at the nitrogen atom.

4-*tert*-Butyl-*N*-methylpiperidine

(*c*) Tetramethylammonium hydroxide cannot undergo Hofmann elimination. The only reaction that can take place is nucleophilic substitution.

Tetramethylammonium hydroxide   Trimethylamine   Methanol

(*d*) The key intermediate in the reaction of an amine with nitrous acid is the corresponding diazonium ion.

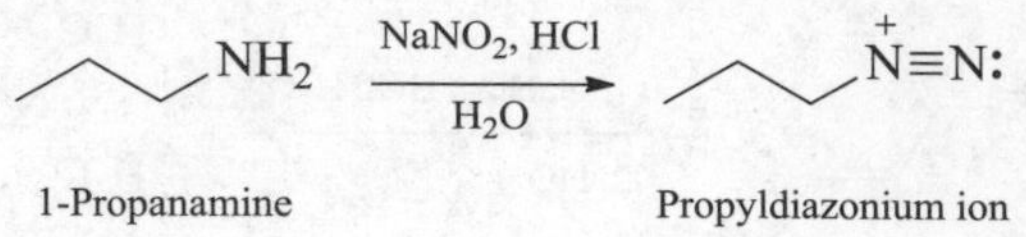

1-Propanamine   Propyldiazonium ion

Loss of nitrogen from this diazonium ion is accompanied by a hydride shift to form a secondary carbocation.

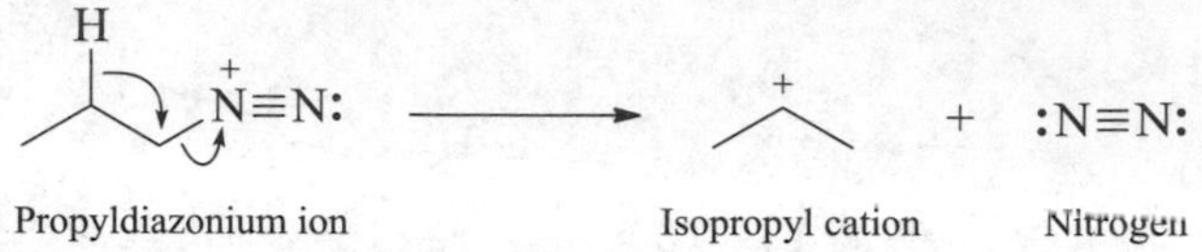

Propyldiazonium ion   Isopropyl cation   Nitrogen

Capture of isopropyl cation by water yields the major product of the reaction, 2-propanol.

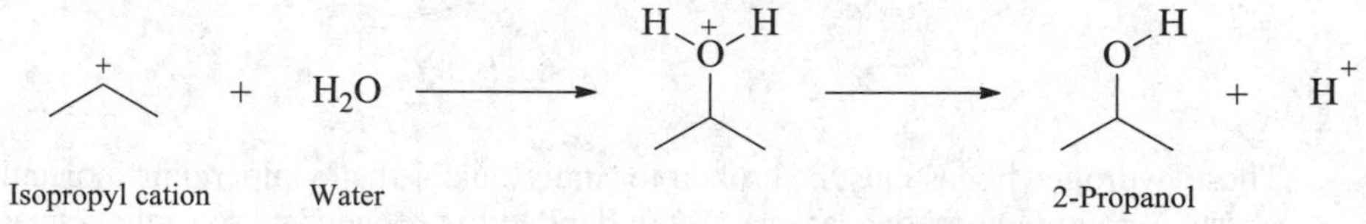

Isopropyl cation   Water   2-Propanol

Synthesis

22.41 (*a*) Looking at the problem retrosynthetically, it can be seen that a variety of procedures are available for preparing ethylamine from ethanol. The methods by which a primary amine may be prepared include

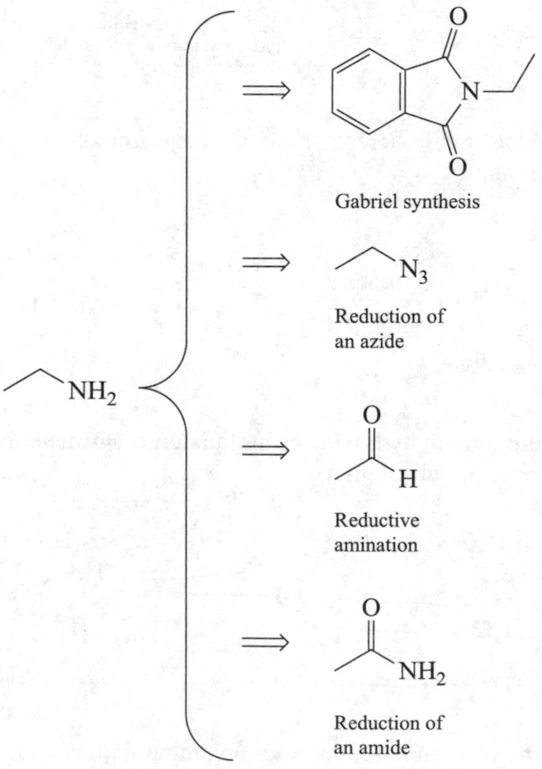

Gabriel synthesis

Reduction of
an azide

Reductive
amination

Reduction of
an amide

Two of these methods, the Gabriel synthesis and the preparation and reduction of the corresponding azide, begin with ethyl bromide.

Ethanol Ethyl bromide

Ethyl bromide *N*-Potassiophthalimide *N*-Ethylphthalimide Ethylamine

Ethyl bromide Ethyl azide Ethylamine

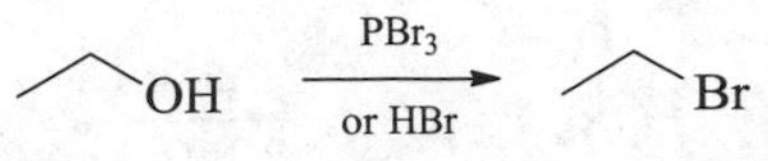

To use reductive amination, we must begin with oxidation of ethanol to acetaldehyde.

Ethanol Acetaldehyde

Acetaldehyde Ethylamine

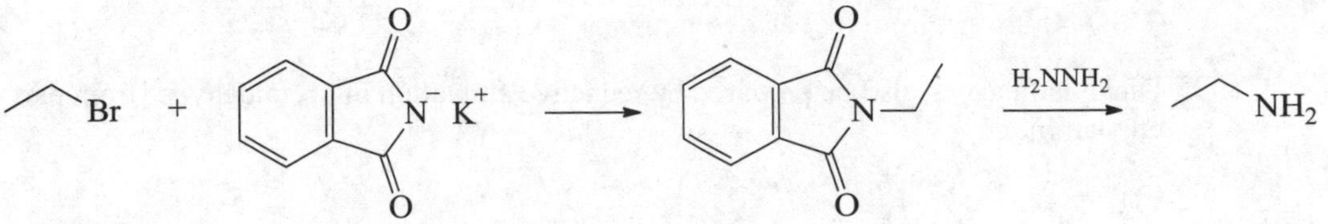

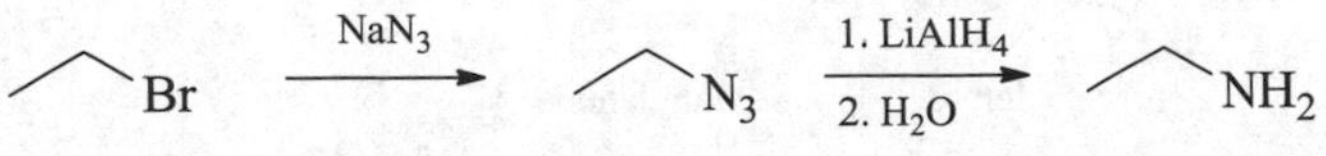

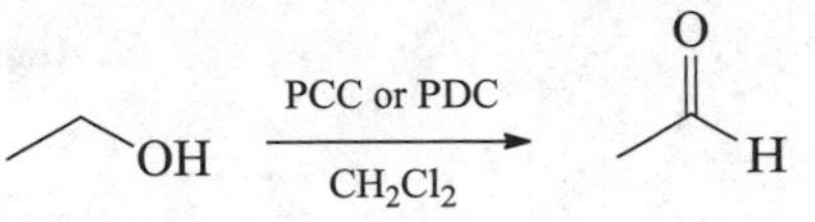

Another possibility is reduction of acetamide. This requires an initial oxidation of ethanol to acetic acid.

Ethanol Acetic acid Acetamide Ethylamine

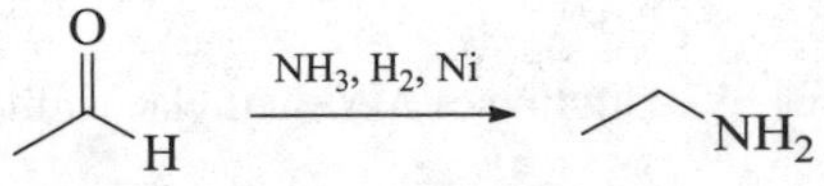

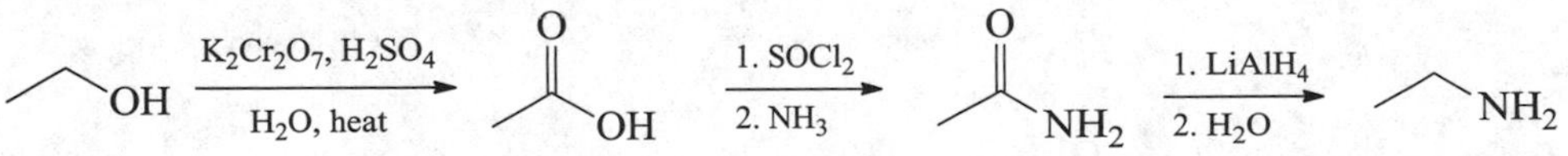

(*b*) Acylation of ethylamine with acetyl chloride, prepared in part (*a*), gives the desired amide.

Acetyl chloride Ethylamine *N*-Ethylacetamide Ethylammonium chloride

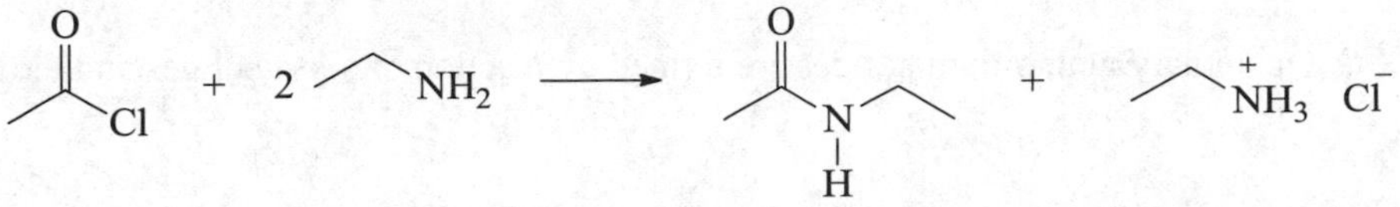

Excess ethylamine can be allowed to react with the hydrogen chloride formed in the acylation reaction. Alternatively, equimolar amounts of acyl chloride and amine can be used in the presence of aqueous hydroxide as the base.

(*c*) Reduction of the *N*-ethylacetamide prepared in part (*b*) yields diethylamine.

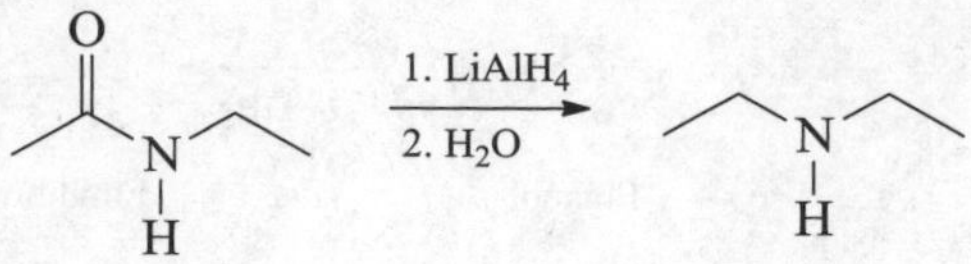

N-Ethylacetamide   Diethylamine

Diethylamine can also be prepared by reductive amination of acetaldehyde [from part (*a*)] with ethylamine.

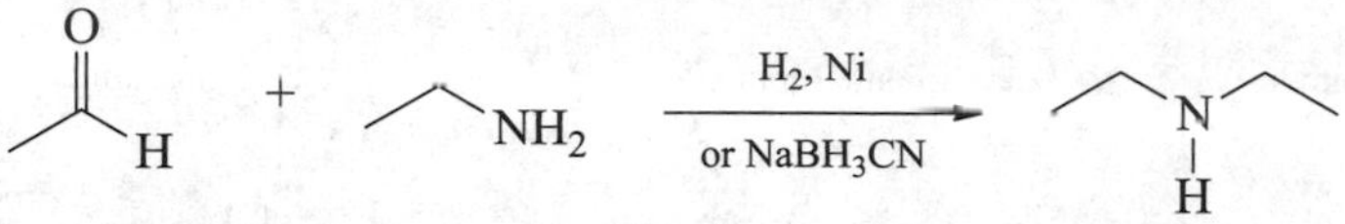

Acetaldehyde   Ethylamine   Diethylamine

(*d*) The preparation of *N*,*N*-diethylacetamide is a standard acylation reaction. The reactants, acetyl chloride and diethylamine, have been prepared in previous parts of this problem.

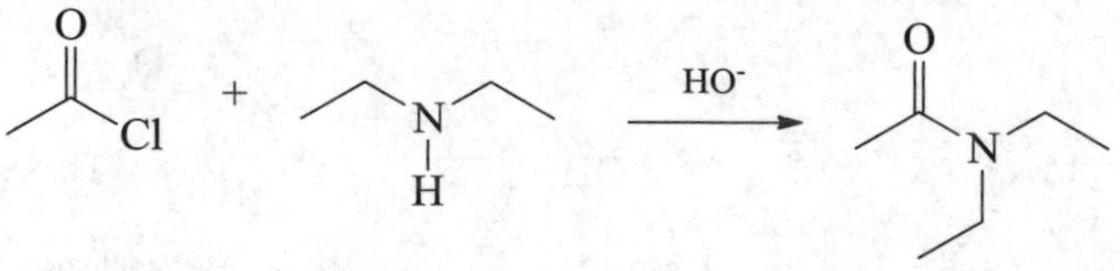

Acetyl chloride   Diethylamine   *N*,*N*-Diethylacetamide

(*e*) Triethylamine arises by reduction of *N*,*N*-diethylacetamide or by reductive amination.

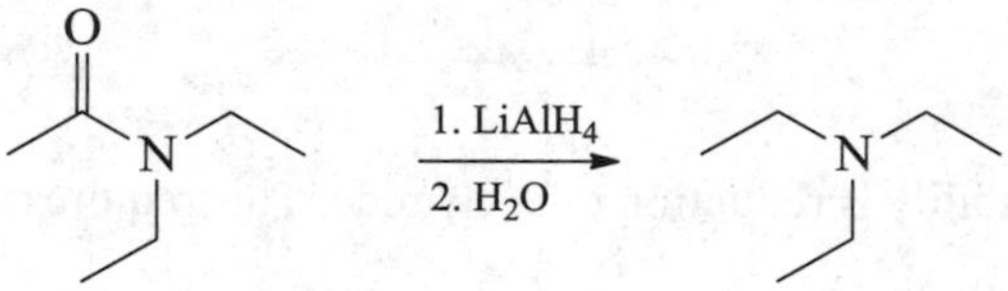

N,*N*-Diethylacetamide   Triethylamine

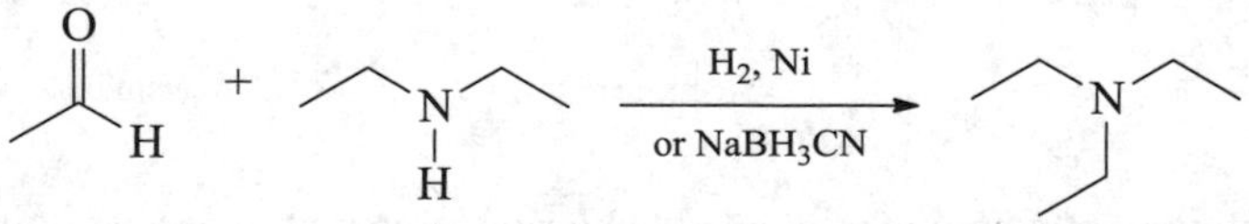

Acetaldehyde   Diethylamine   Triethylamine

(*f*) Quaternary ammonium halides are formed by reaction of alkyl halides and tertiary amines.

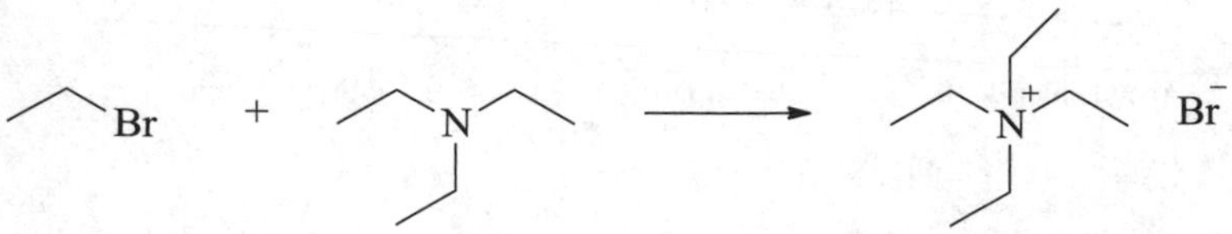

Ethyl bromide   Triethylamine   Tetraethylammonium bromide

22.42 (*a*) In this problem, a primary alkanamine must be prepared with a carbon chain extended by one carbon. This can be accomplished by way of a nitrile.

$$RCH_2NH_2 \implies RCN \implies RBr \implies ROH$$

$$(R = CH_3CH_2CH_2CH_2\text{—})$$

The desired reaction sequence is therefore

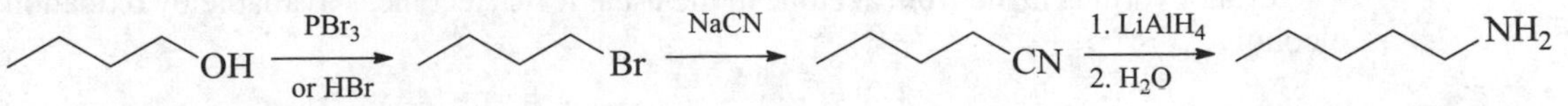

1-Butanol   Butyl bromide   Pentanenitrile   1-Pentanamine

(*b*) The carbon chain of *tert*-butyl chloride cannot be extended by a nucleophilic substitution reaction; the S_N2 reaction that would be required on the tertiary halide would not work. The sequence employed in part (*a*) is therefore not effective in this case. The best route is carboxylation of the Grignard reagent and subsequent conversion of the corresponding amide to the desired primary amine product.

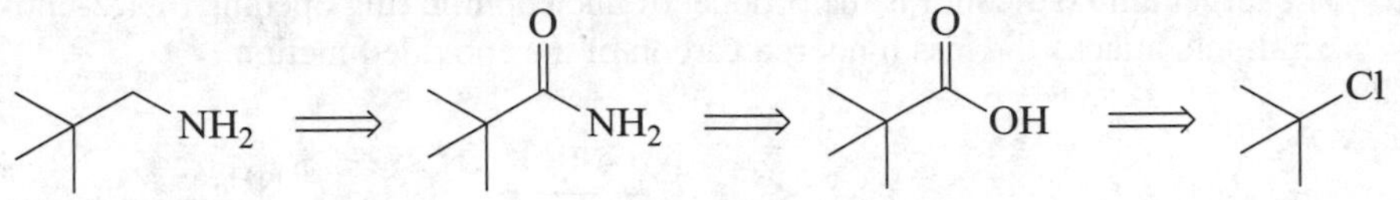

The reaction sequence to be used is

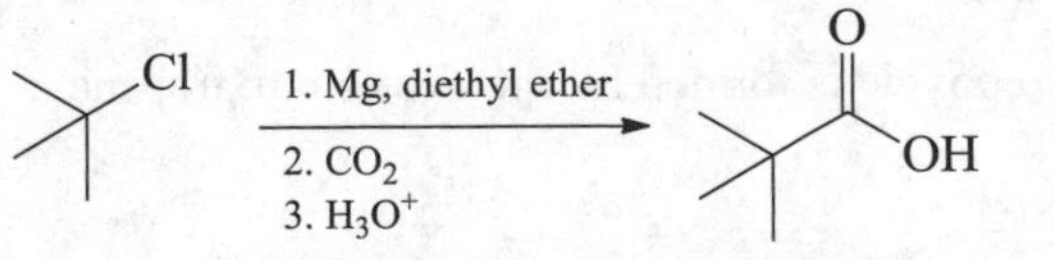

tert-Butyl   2,2-Dimethylpropanoic
chloride   acid

Once the carboxylic acid has been obtained, it is converted to the desired amine by reduction of the corresponding amide.

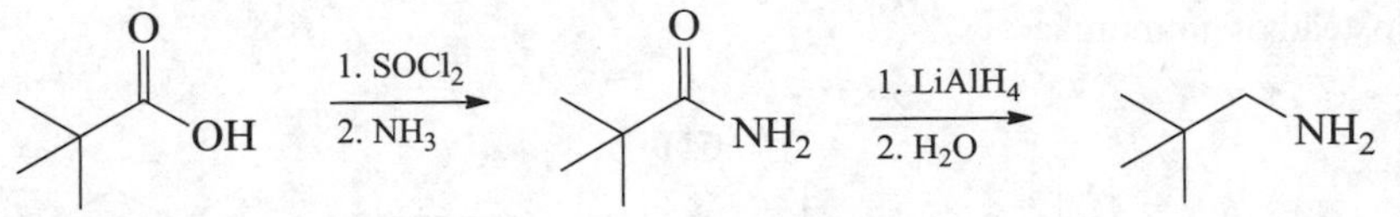

2,2-Dimethylpropanoic   2,2-Dimethylpropanamide   2,2-Dimethyl-1-propanamine
acid

(*c*) Oxidation of cyclohexanol to cyclohexanone gives a compound suitable for reductive amination.

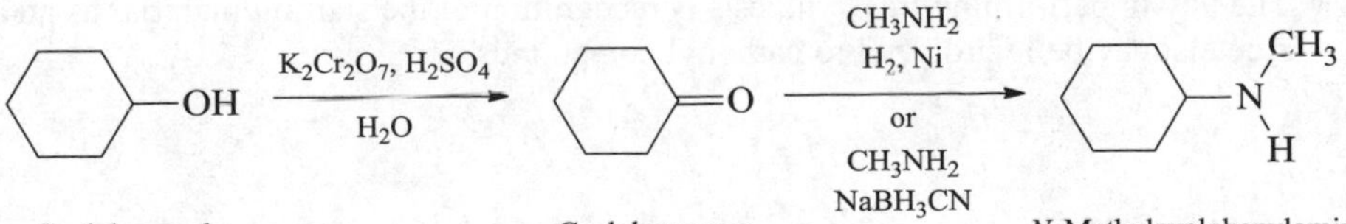

Cyclohexanol   Cyclohexanone   *N*-Methylcyclohexylamine (*d*) The desired product is the reduction product of the cyanohydrin of acetone.

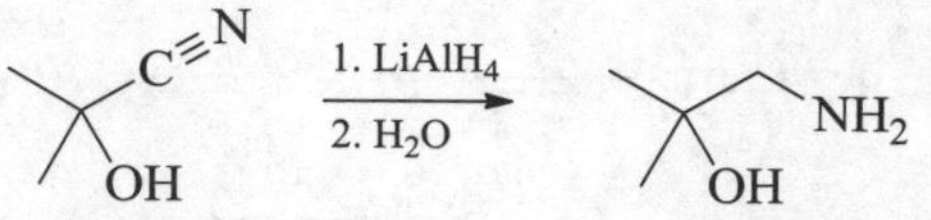

Acetone
cyanohydrin

1-Amino-2-methyl-
2-propanol

The cyanohydrin is made from acetone in the usual way. Acetone is available by oxidation of isopropyl alcohol.

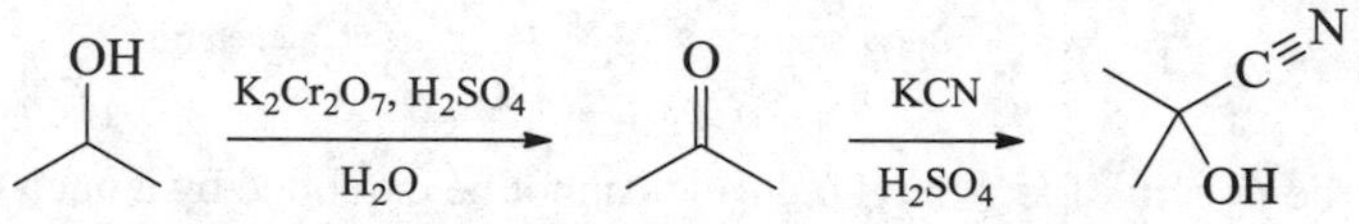

Isopropyl
alcohol

Acetone

Acetone
cyanohydrin

(*e*) The target amino alcohol is the product of nucleophilic ring opening of 1,2-epoxypropane by ammonia. Ammonia attacks the less hindered carbon of the epoxide function.

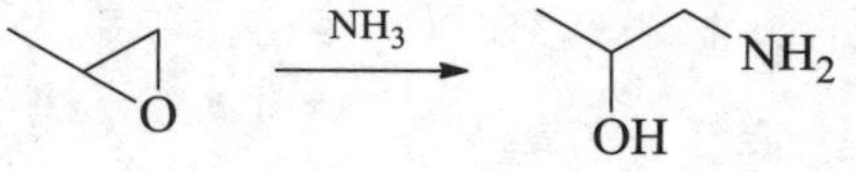

1,2-Epoxypropane

1-Amino-2-propanol

The necessary epoxide is formed by epoxidation of propene.

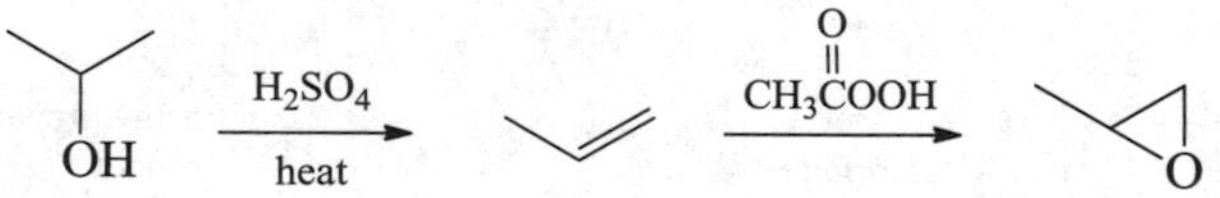

Isopropyl
alcohol

Propene

1,2-Epoxypropane

(*f*) The reaction sequence is the same as in part (*e*) except that dimethylamine is used as the nucleophile instead of ammonia.

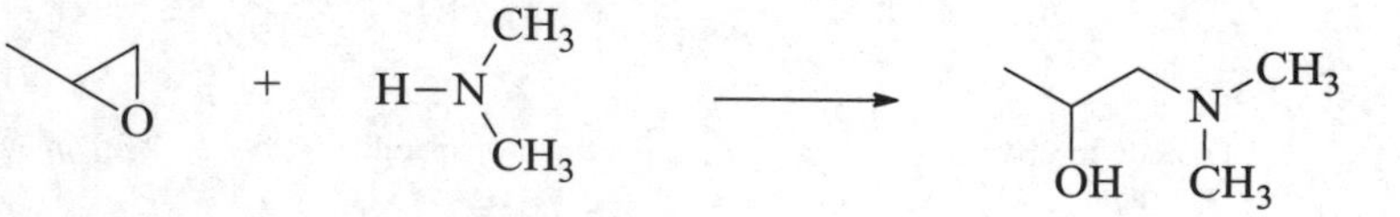

1,2-Epoxypropane
[prepared as in part (*e*)]

Dimethylamine

1-(*N*,*N*-Dimethylamino)-2-propanol

(*g*) The key to performing this synthesis is recognition of the starting material as an acetal of acetophenone. Acetals may be hydrolyzed to carbonyl compounds.

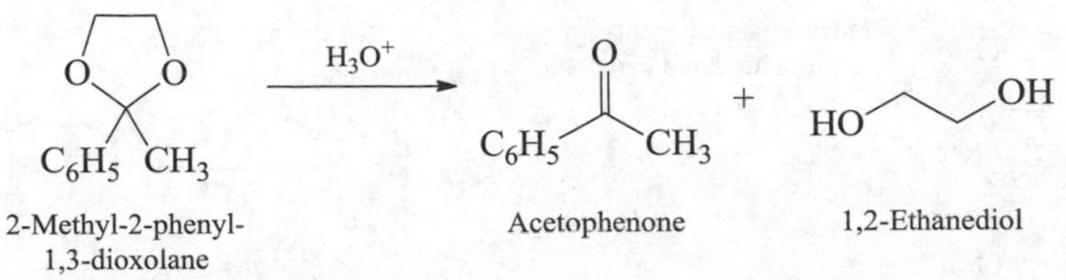

2-Methyl-2-phenyl-
1,3-dioxolane | Acetophenone | 1,2-Ethanediol

Once acetophenone has been obtained, it may be converted to the required product by reductive amination.

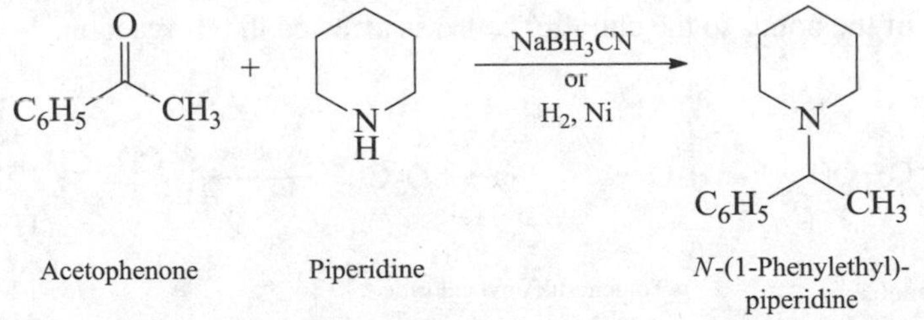

Acetophenone   Piperidine   N-(1-Phenylethyl)-
piperidine

22.43 (a) The reaction of alkyl halides with N-potassiophthalimide (the first step in the Gabriel synthesis of amines) is a nucleophilic substitution reaction. Alkyl bromides are more reactive than alkyl fluorides; that is, bromide is a better leaving group than fluoride.

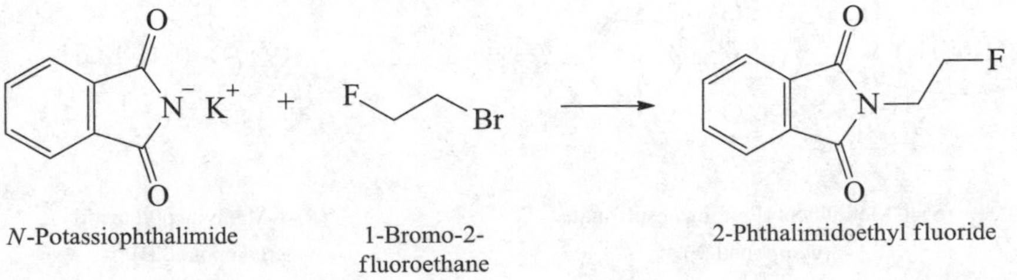

N-Potassiophthalimide   1-Bromo-2-
fluoroethane   2-Phthalimidoethyl fluoride

(b) In this example, one bromine is attached to a primary and the other to a secondary carbon. Phthalimide anion is a good nucleophile and reacts with alkyl halides by the S_N2 mechanism. It attacks the less hindered primary carbon.

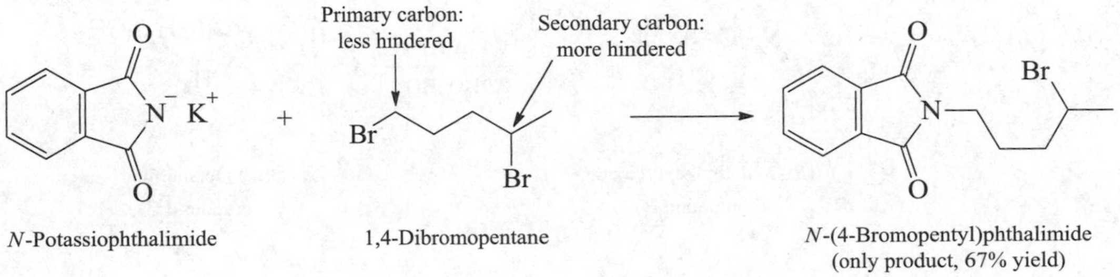

N-Potassiophthalimide   1,4-Dibromopentane   N-(4-Bromopentyl)phthalimide
(only product, 67% yield)

(c) Both bromines are bonded to primary carbons, but branching at the adjacent carbon hinders nucleophilic attack at one of them.

N-Potassiophthalimide   **1,4-Dibromopentane**   **N-(4-Bromo-3,3-dimethylbutyl)phthalimide**
                                                              (only product, 53% yield)

22.44 Alcohols are converted to *p*-toluenesulfonates by reaction with *p*-toluenesulfonyl chloride. None of the bonds to the chirality center is affected in this reaction.

(*S*)-2-Octanol   *p*-Toluenesulfonyl chloride   (*S*)-1-Methylheptyl *p*-toluenesulfonate
                                                               (compound A)

Displacement of the *p*-toluenesulfonate leaving group by sodium azide is an S_N2 process and proceeds with inversion of configuration.

(*S*)-1-Methylheptyl *p*-toluenesulfonate   (*R*)-1-Methylheptyl azide
(compound A)   (compound B)

Reduction of the azide yields a primary amine. A nitrogen–nitrogen bond is cleaved; all the bonds to the chirality center remain intact.

(*R*)-1-Methylheptyl azide   (*R*)-2-Octanamine
(compound B)   (compound C)

22.45 (*a*) The overall transformation can be expressed as RBr → RCH$_2$NH$_2$. In many cases, this can be carried out via a nitrile, as RBr → RCN → RCH$_2$NH$_2$. In this case, however, the substrate is 1-bromo-2,2-dimethylpropane, an alkyl halide that reacts very slowly in nucleophilic substitution.

Carbon–carbon bond formation with 1-bromo-2,2-dimethylpropane can be achieved more effectively by carboxylation of the corresponding Grignard reagent.

1-Bromo-2,2-dimethylpropane

1. Mg
2. CO_2
3. H_3O^+

3,3-Dimethylbutanoic acid (63%)

The carboxylic acid can then be converted to the desired amine by reduction of the derived amide.

3,3-Dimethylbutanoic acid

1. $SOCl_2$
2. NH_3

3,3-Dimethylbutanamide (51%)

1. $LiAlH_4$
2. H_2O

3,3-Dimethyl-1-butanamine (57%)

The yields listed in parentheses are those reported in the chemical literature for this synthesis.

(*b*) Consider the starting materials in relation to the desired product.

$H_2C=CH(CH_2)_8CH_2\text{-}N$

N-(10-Undecenyl)pyrrolidine

$\Longrightarrow$

$H_2C=CH(CH_2)_8\overset{O}{\overset{\|}{C}}-OH$ + $H-N$

10-Undecenoic acid   Pyrrolidine

The synthetic tasks are to form the necessary carbon–nitrogen bond and to reduce the carbonyl group to a methylene group. This has been accomplished by way of the amide as a key intermediate.

$H_2C=CH(CH_2)_8\overset{O}{\overset{\|}{C}}-OH$

10-Undecenoic acid

1. $SOCl_2$
2. pyrrolidine

$H_2C=CH(CH_2)_8\overset{O}{\overset{\|}{C}}-N$

N-(10-Undecenoyl)pyrrolidine (75%)

1. $LiAlH_4$
2. H_2O

$H_2C=CH(CH_2)_8CH_2-N$

N-(10-Undecenyl)pyrrolidine (66%)

A second approach converts the starting carboxylic acid to an aldehyde followed by reductive amination.

$H_2C=CH(CH_2)_8\overset{O}{\overset{\|}{C}}-OH$

10-Undecenoic acid

1. $LiAlH_4$
2. H_2O

$H_2C=CH(CH_2)_8CH_2-OH$

10-Undecen-1-ol

PCC or PDC
CH_2Cl_2

$H_2C=CH(CH_2)_8\overset{O}{\overset{\|}{C}}-H$

10-Undecenal

The reducing agent in the reductive amination process cannot be hydrogen, because that would result in hydrogenation of the double bond. Sodium cyanoborohydride is required.

$H_2C=CH(CH_2)_8\overset{O}{\overset{\|}{C}}-H$ + $H-N$

10-Undecenal   Pyrrolidine

$NaBH_3CN$

$H_2C=CH(CH_2)_8CH_2-N$

N-(10-Undecenyl)pyrrolidine

(*c*) It is stereochemistry that determines the choice of which synthetic method to employ in introducing the amine group. The carbon–nitrogen bond must be formed with inversion of configuration at the alcohol carbon. Conversion of the alcohol to its *p*-toluenesulfonate ensures that the leaving group is introduced with the same stereochemistry as the alcohol.

cis-2-Phenoxycyclopentanol   *p*-Toluenesulfonyl chloride   *cis*-2-Phenoxycyclopentyl
                                                                                                                       p-toluenesulfonate

Once the leaving group has been introduced with the proper stereochemistry, it can be displaced by a nitrogen nucleophile suitable for subsequent conversion to an amine.

cis-2-Phenoxycyclopentyl   *trans*-2-Phenoxycyclopentyl   *trans*-2-Phenoxy-
p-toluenesulfonate   azide   cyclopentylamine

(As actually reported, the azide was reduced by hydrogenation over a palladium catalyst, and the amine was isolated as its hydrochloride salt in 66% yield.)

(*d*) Recognition that the primary amine is derivable from the corresponding nitrile by reduction,

and that the necessary tertiary amine function can be introduced by a nucleophilic substitution reaction between the two given starting materials suggests the following synthesis.

N-Methylbenzylamine   4-Bromobutanenitrile   *N*-Benzyl-*N*-methyl-1,4-butanediamine

Alkylation of *N*-methylbenzylamine with 4-bromobutanenitrile has been achieved in 92% yield in the presence of potassium carbonate as a weak base to neutralize the hydrogen bromide produced. The nitrile may be reduced with lithium aluminum hydride, as shown in the equation, or by catalytic hydrogenation. Catalytic hydrogenation over platinum gave the desired diamine, isolated as its hydrochloride salt, in 90% yield.

(e) The overall transformation may be viewed retrosynthetically as follows:

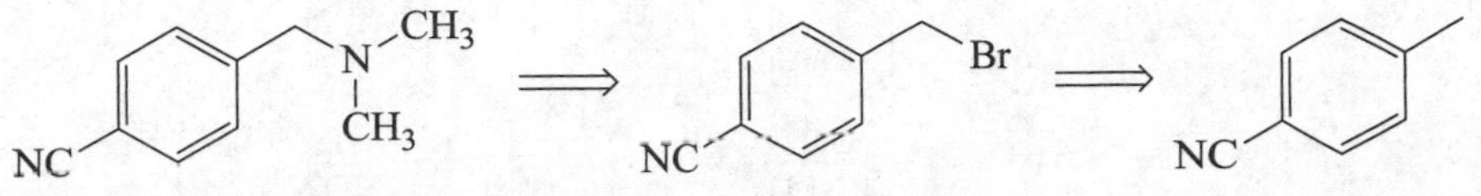

The sequence that presents itself begins with benzylic bromination with *N*-bromosuccinimide.

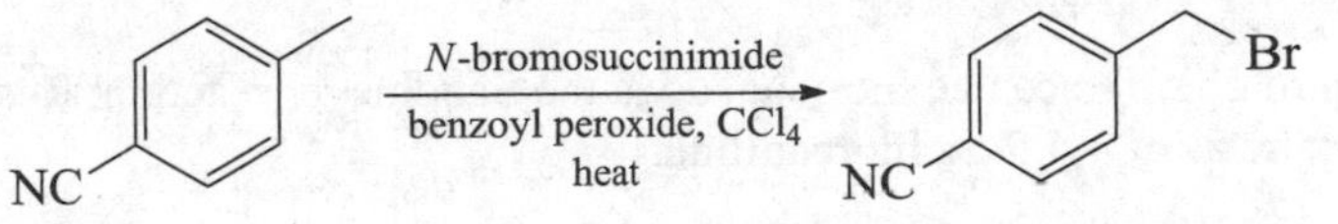

p-Cyanotoluene   p-Cyanobenzyl bromide

The reaction shown in the equation has been reported in the chemical literature and gave the benzylic bromide in 60% yield. Treatment of this bromide with dimethylamine gives the desired product. (The isolated yield was 83% by this method.)

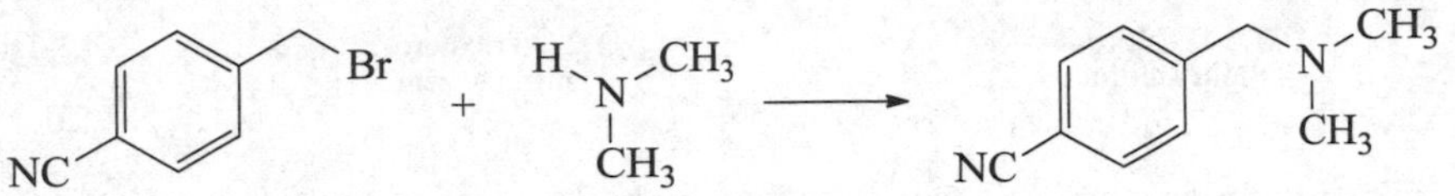

p-Cyanobenzyl bromide   Dimethylamine   p-Cyano-*N*,*N*-dimethylbenzylamine

22.46 (a) This problem illustrates the application of the Sandmeyer reaction to the preparation of aryl cyanides. Diazotization of *p*-nitroaniline followed by treatment with copper(I) cyanide converts it to *p*-nitrobenzonitrile.

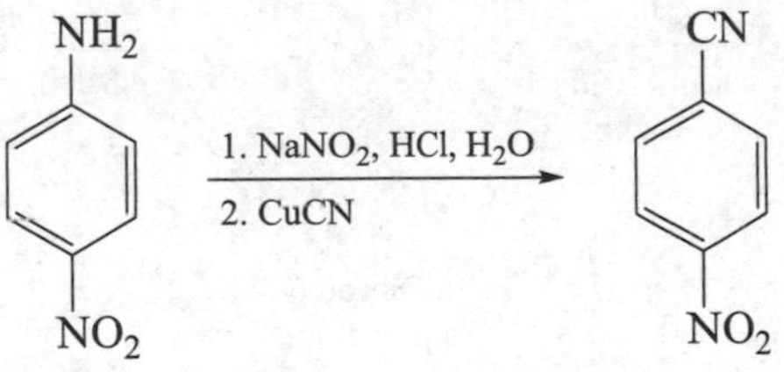

1. NaNO$_2$, HCl, H$_2$O
2. CuCN

p-Nitroaniline   p-Nitrobenzonitrile

(b) An acceptable pathway becomes apparent when it is realized that the amino group in the product is derived from the nitro group of the starting material. Two chlorines are introduced by electrophilic aromatic substitution, the third by a Sandmeyer reaction.

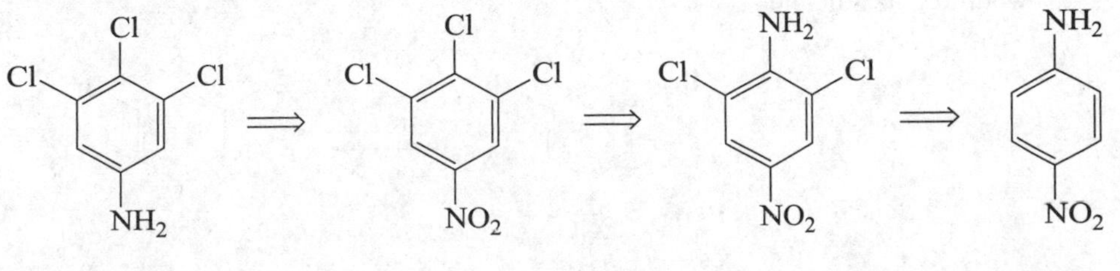

Two of the required chlorine atoms can be introduced by chlorination of the starting material, *p*-nitroaniline.

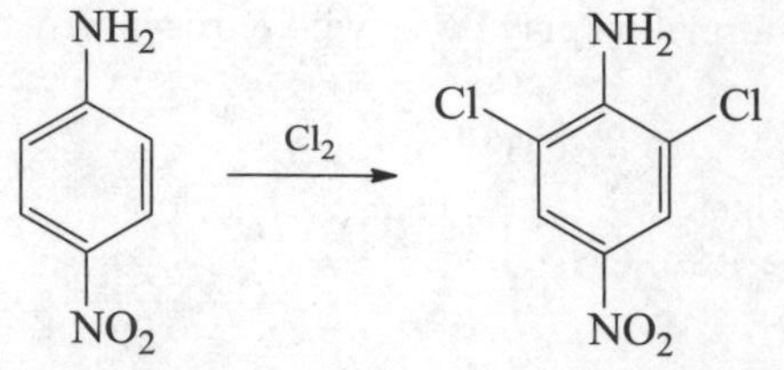

p-Nitroaniline   2,6-Dichloro-4-
nitroaniline

The third chlorine can be introduced via the Sandmeyer reaction. Reduction of the nitro group completes the synthesis of 3,4,5-trichloroaniline.

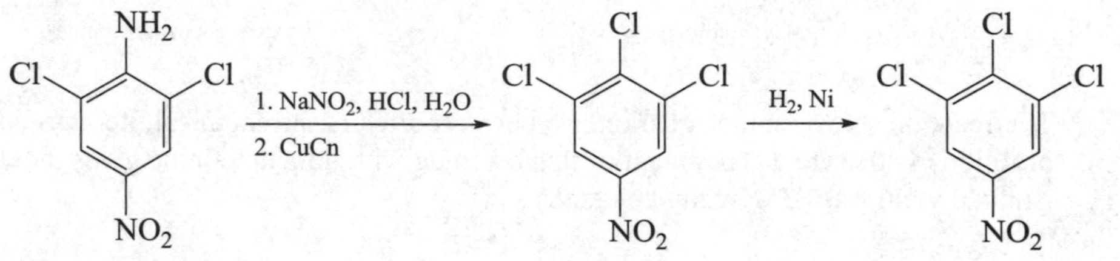

2,6-Dichloro-4-
nitroaniline   1,2,3-Trichloro-5-
nitrobenzene   3,4,5-Trichloroaniline

The reduction step has been carried out by hydrogenation with a nickel catalyst in 70% yield.

(*c*) The amino group that is present in the starting material facilitates the introduction of the bromine substituents and is then removed by reductive deamination.

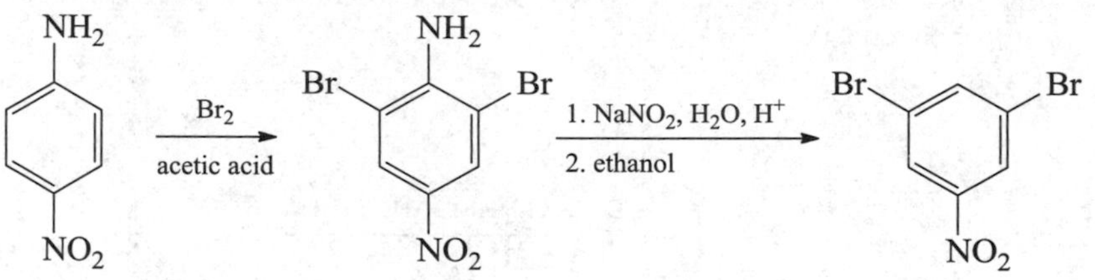

p-Nitroaniline   2,6-Dibromo-4-
nitroaniline   1,3-Dibromo-5-
nitrobenzene
(70%)

Hypophosphorous acid has also been used successfully in the reductive deamination step.

(*d*) Reduction of the nitro group of the 1,3-dibromo-5-nitrobenzene prepared in the preceding part (*c*) this problem gives the desired product. The customary reducing agents used for the reduction of nitroarenes would all be suitable.

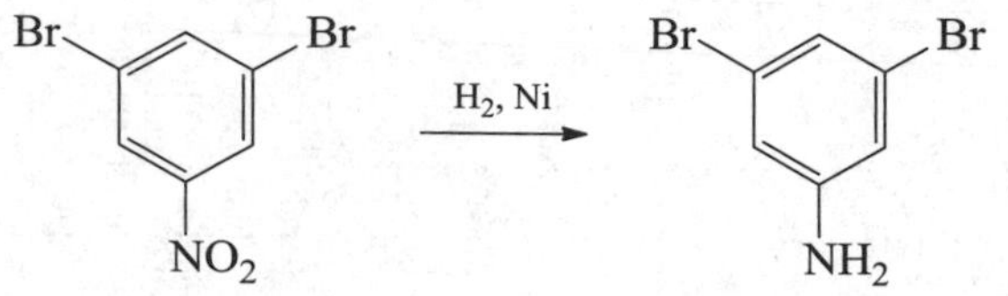

1,3-Dibromo-5-nitrobenzene   3,5-Dibromoaniline
[prepared from *p*-nitroaniline as in part (*c*)]   (80%)

(*e*) The synthetic objective is

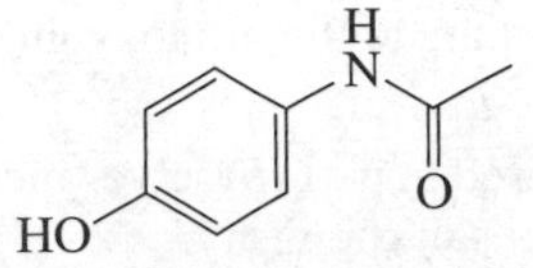

p-Acetamidophenol

This compound, known as *acetaminophen* and used as an analgesic to reduce fever and relieve minor pain, may be prepared from *p*-nitroaniline by way of *p*-nitrophenol.

p-Nitroaniline *p*-Nitrophenol *p*-Acetamidophenol

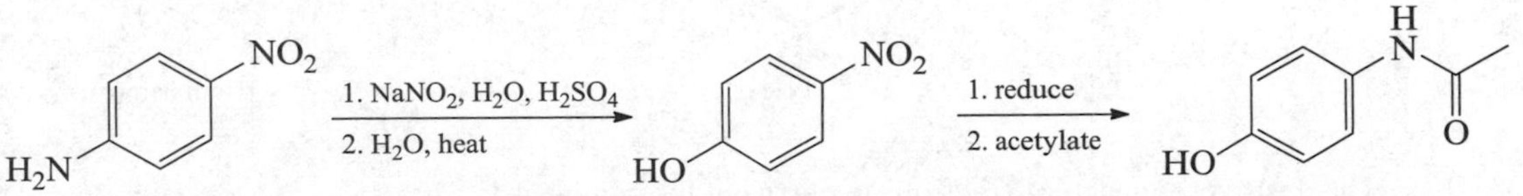

Any of the customary reducing agents suitable for converting aryl nitro groups to arylamines (Fe, HCl; Sn, HCl; H_2, Ni) may be used. Acetylation of *p*-aminophenol may be carried out with acetyl chloride or acetic anhydride. The amino group of *p*-aminophenol is more nucleophilic than the hydroxyl group and is acetylated preferentially.

22.47 (*a*) Replacement of an amino substituent by a bromine is readily achieved by the Sandmeyer reaction.

o-Anisidine *o*-Bromoanisole (88-93%)

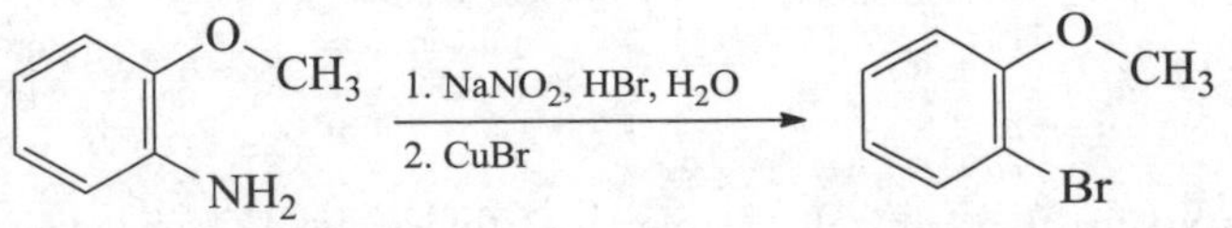

(*b*) This conversion demonstrates the replacement of an amino substituent by fluorine via the Schiemann reaction.

o-Anisidine *o*-Methoxybenzenediazonium fluoroborate (57%) *o*-Fluoroanisole (53%)

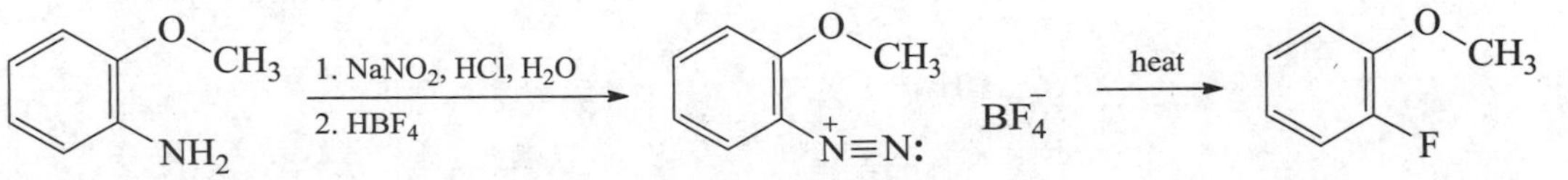

(*c*) We can use the *o*-fluoroanisole prepared in part (*b*) to prepare 3-fluoro-4-methoxyacetophenone by Friedel–Crafts acylation.

o-Anisidine *o*-Fluoroanisole 3-Fluoro-4-methoxyacetophenone (70-80%)

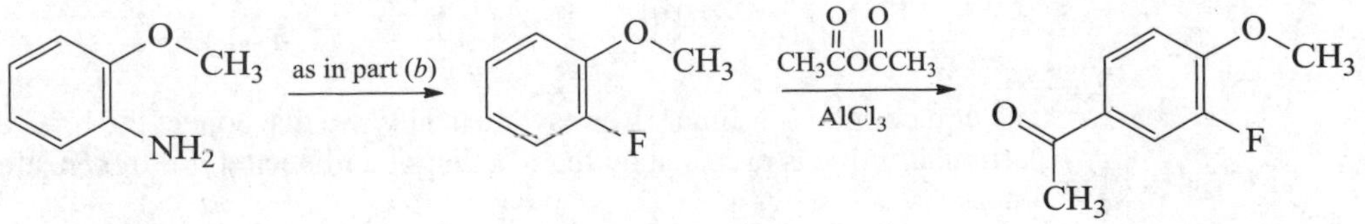

Remember from text Section 13.16 that it is the more activating substituent that determines the regioselectivity of electrophilic aromatic substitution when an arene bears two different substituents. Methoxy is a strongly activating substituent; fluorine is slightly deactivating. Friedel–Crafts acylation takes place at the position para to the methoxy group.

(d) The o-fluoroanisole prepared in part (b) serves nicely as a precursor to 3-fluoro-4-methoxy-benzonitrile via diazonium salt chemistry.

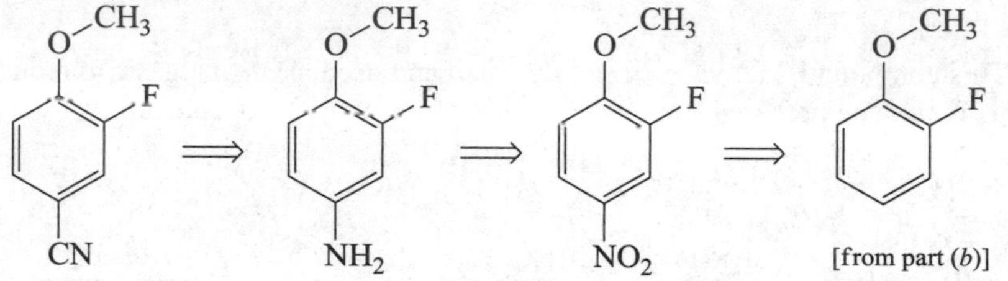

The desired sequence of reactions to carry out the synthesis is

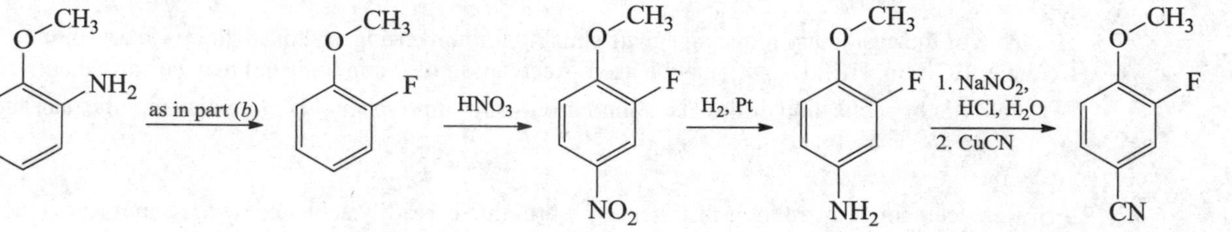

| o-Anisidine | o-Fluoroanisole | 2-Fluoro-4-nitroanisole (53%) | 4-Amino-2-fluoroanisole (85%) | 3-Fluoro-4-methoxybenzonitrile (46%) |

Conversion of o-fluoroanisole to 4-amino-2-fluoroanisole proceeds in the conventional way by preparation and reduction of a nitro derivative. Once the necessary arylamine is at hand, it is converted to the nitrile by a Sandmeyer reaction.

(e) Diazotization followed by hydrolysis of the 4-amino-2-fluoroanisole prepared as an intermediate in part (d) yields the desired phenol.

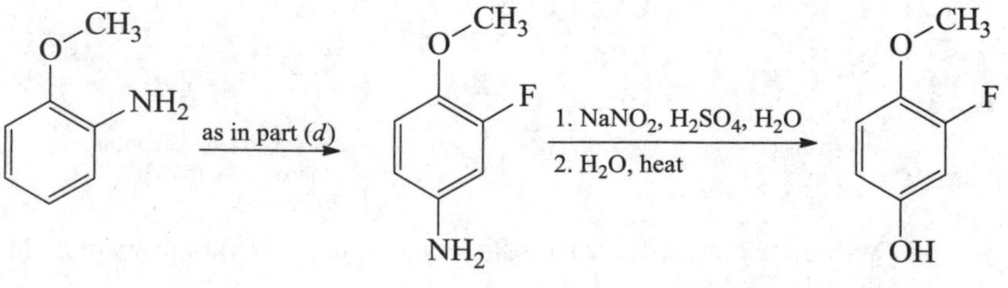

| o-Anisidine | 4-Amino-2-fluoroanisole | 3-Fluoro-4-methoxyphenol (70%) |

22.48 (a) The compound is an imine. Retrosynthetically, we disconnect the N=C double bond to see that this particular imine is prepared by the reaction of aniline and p-nitrobenzaldehyde.

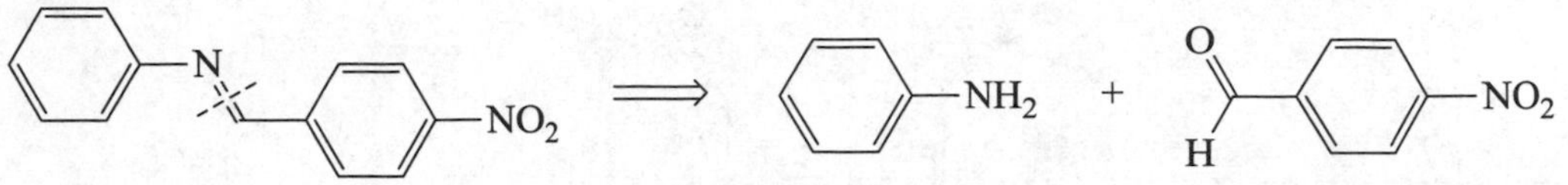

One of the reactants given in the problem is nitrobenzene, from which we can prepare aniline by reduction.

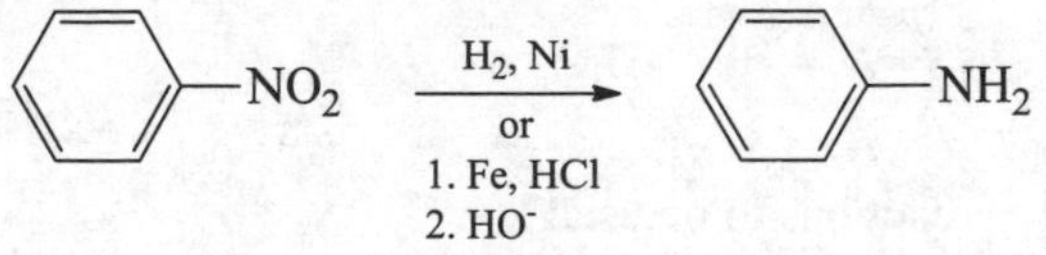

The other reactant given is *p*-nitrobenzyl alcohol, which can be oxidized to *p*-nitrobenzaldehyde with PCC or PDC in dichloromethane.

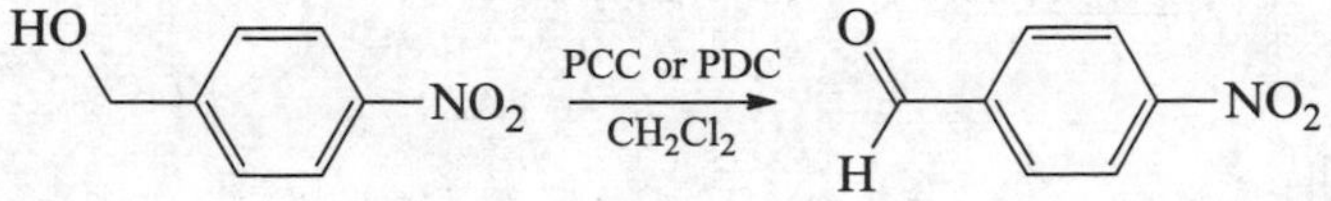

Combining aniline and *p*-nitrobenzaldehyde gives the desired imine. An acid catalyst such as a trace of HCl can be used to speed up the reaction.

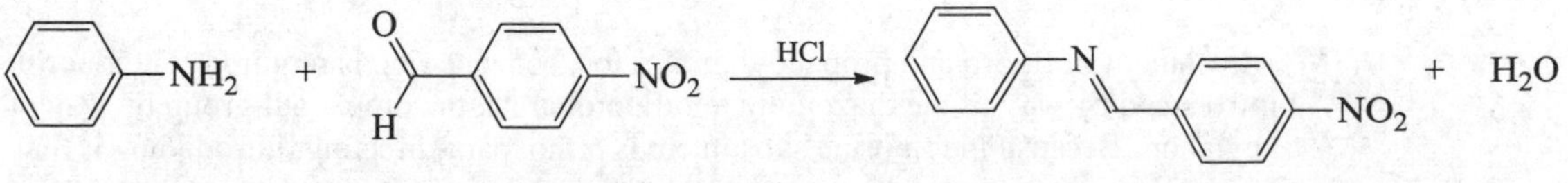

(*b*) The most direct way to approach this synthesis is to convert *p*-nitrotoluene to *p*-nitrobenzyl alcohol, then proceed as in part (*a*). Benzylic bromination with *N*-bromosuccinimide (NBS), followed by hydrolysis will work.

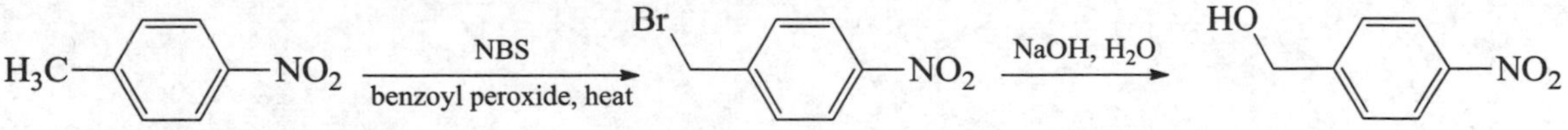

The alternative sequence

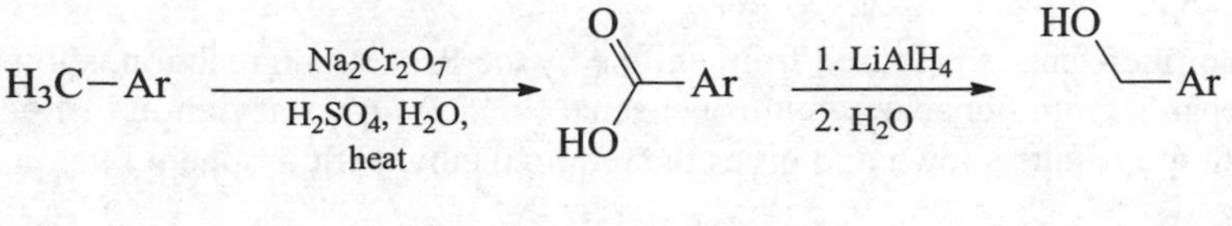

will not work in this synthesis because the nitro group will also be reduced by lithium aluminum hydride.

22.49 (*a*) The carboxyl group of *p*-aminobenzoic acid can be derived from the methyl group of *p*-methylaniline by oxidation. First, however, the nitrogen must be acylated so as to protect the ring from oxidation.

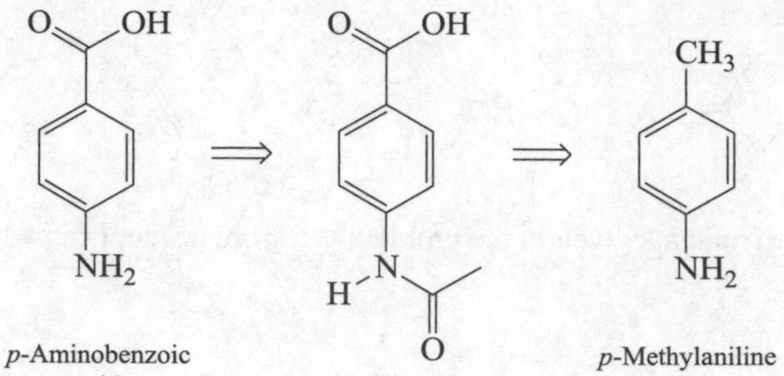

p-Aminobenzoic acid   *p*-Methylaniline

The sequence of reactions to be used is

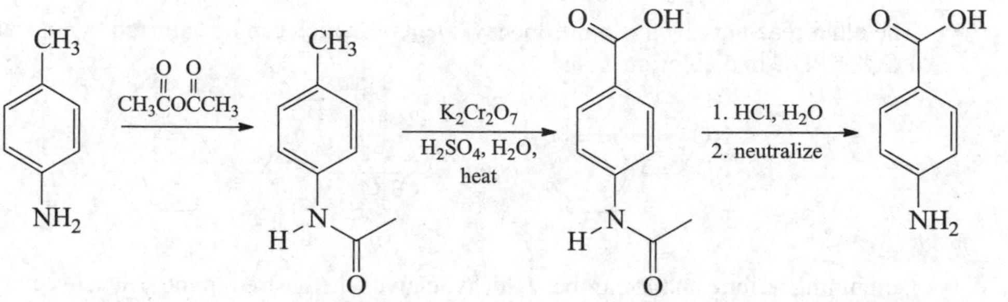

p-Methylaniline   *p*-Methylacetanilide   *p*-Acetamidobenzoic acid   *p*-Aminobenzoic acid

(*b*) Attachment of fluoro and propanoyl groups to a benzene ring is required. The fluorine substituent can be introduced by way of the diazonium tetrafluoroborate, the propanoyl group by way of a Friedel–Crafts acylation. Because the fluorine substituent is ortho, para-directing, introducing it first gives the proper orientation of substituents.

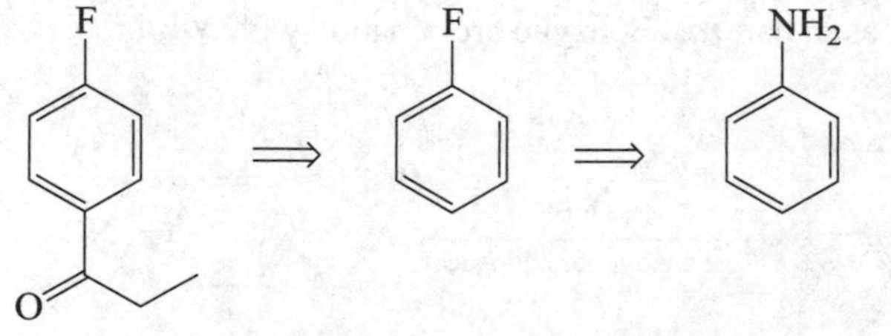

Ethyl *p*-fluorophenyl ketone   Fluorobenzene   Aniline

Fluorobenzene is prepared from aniline by the Schiemann reaction, shown in Section 22.17. Aniline is prepared from benzene via nitrobenzene. Friedel–Crafts acylation of fluorobenzene has been carried out with the results shown and gives the required ethyl *p*-fluorophenyl ketone as the major product.

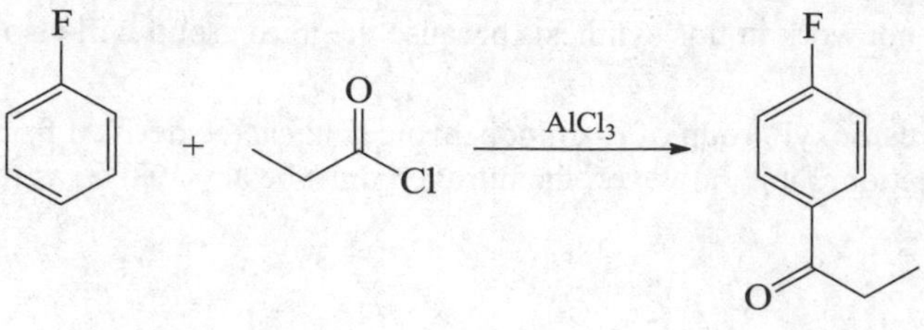

Fluorobenzene   Propanoyl chloride   Ethyl *p*-fluorophenyl ketone (86%)

(*c*) The synthetic plan is based on the essential step of forming the fluorine derivative from an amine by way of a diazonium salt.

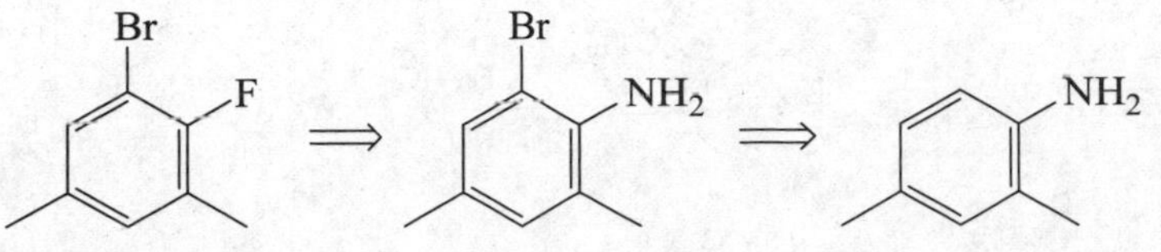
Br
F
Br
NH_2
NH_2

1-Bromo-2-fluoro-3,5-dimethylbenzene

2,4-Dimethylaniline

The required substituted aniline is derived from *m*-xylene by a standard synthetic sequence.

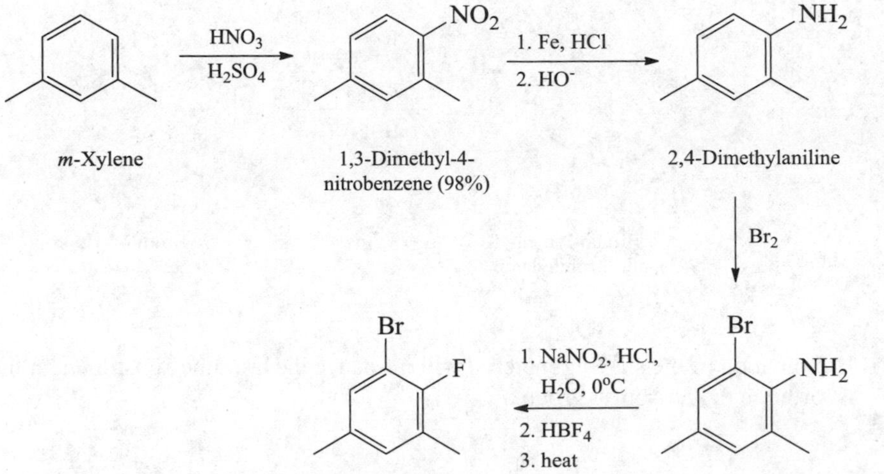
HNO_3
H_2SO_4
NO_2
1. Fe, HCl
2. HO^-
NH_2
Br_2
Br
NH_2
1. $NaNO_2$, HCl, H_2O, 0°C
2. HBF_4
3. heat
Br
F

m-Xylene

1,3-Dimethyl-4-nitrobenzene (98%)

2,4-Dimethylaniline

2-Bromo-4,6-dimethylaniline

2-Bromo-2-fluoro-3,5-dimethylbenzene (60%)

(*d*) In this problem, two nitrogen-containing groups of the starting material are each to be replaced by a halogen substituent. The task is sufficiently straightforward that it may be confronted directly.

Replace amino group by bromine:

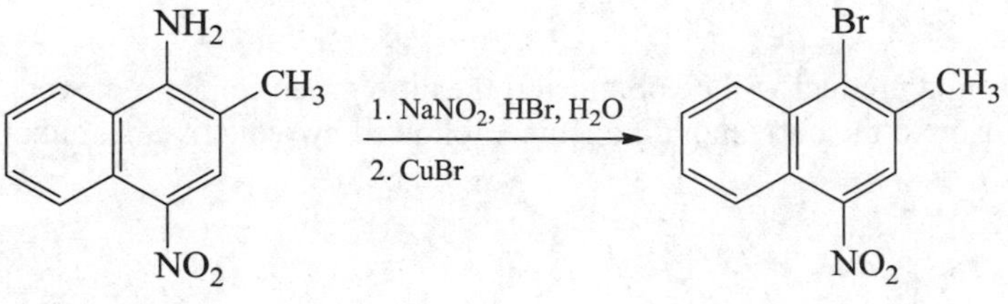
NH_2
CH_3
NO_2
1. $NaNO_2$, HBr, H_2O
2. CuBr
Br
CH_3
NO_2

2-Methyl-4-nitro-1-naphthylamine

1-Bromo-2-methyl-4-nitronaphthalene (82%)

Reduce nitro group to amine:

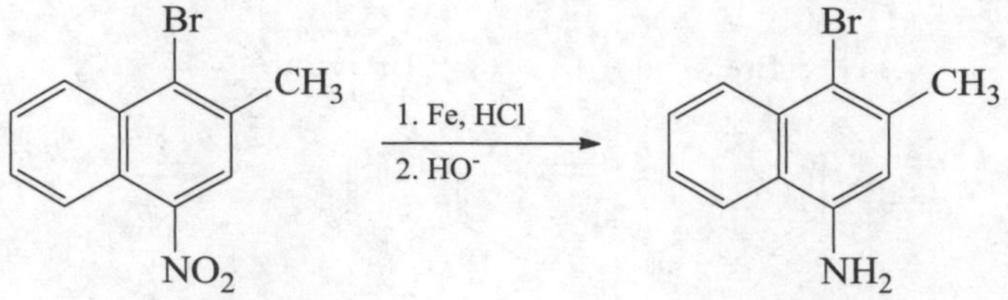
Br
CH3
1. Fe, HCl
2. HO-
NO2
Br
CH3
NH2

1-Bromo-2-methyl-4-nitronaphthalene (82%)

4-Bromo-3-methyl-1-nitronaphthylamine

Replace amino group by fluorine:

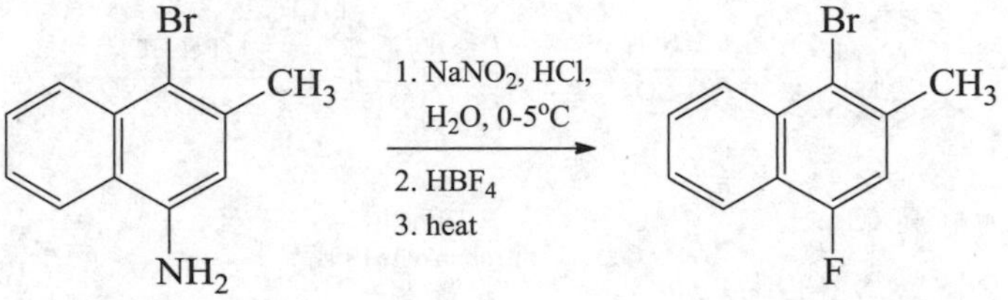
Br
CH3
1. NaNO2, HCl,
H2O, 0-5°C
2. HBF4
3. heat
NH2
Br
CH3
F

4-Bromo-3-methyl-1-nitronaphthylamine

1-Bromo-4-fluoro-2-methylnaphthalene (64%)

(*e*) Bromination of the starting material will introduce the bromine substituent at the correct position, that is, ortho to the *tert*-butyl group.

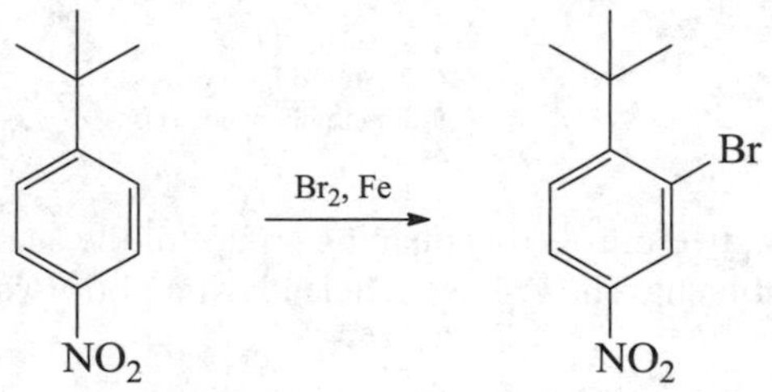
Br2, Fe
NO2
Br
NO2

p-tert-Butylnitrobenzene

2-Bromo-1-*tert*-butyl-4-nitrobenzene

The desired product will be obtained if the nitro group can be removed. This is achieved by its conversion to the corresponding amine, followed by reductive deamination.

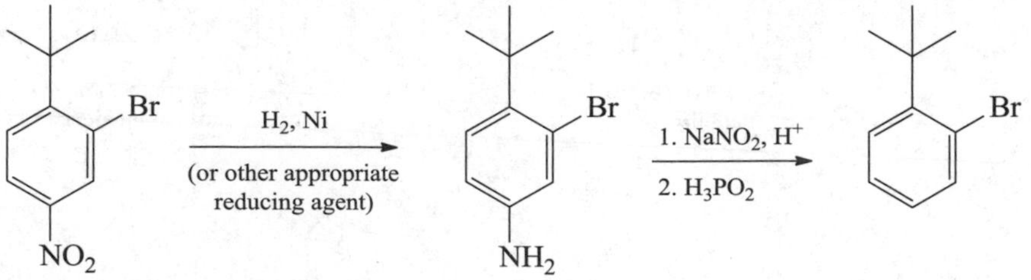
Br
NO2
H2, Ni
(or other appropriate
reducing agent)
Br
NH2
1. NaNO2, H+
2. H3PO2
Br

2-Bromo-1-*tert*-butyl-4-nitrobenzene

3-Bromo-4-*tert*-butylaniline

o-Bromo-*tert*-butylbenzene

(*f*) The proper orientation of the chlorine substituent can be achieved only if it is introduced after the nitro group is reduced.

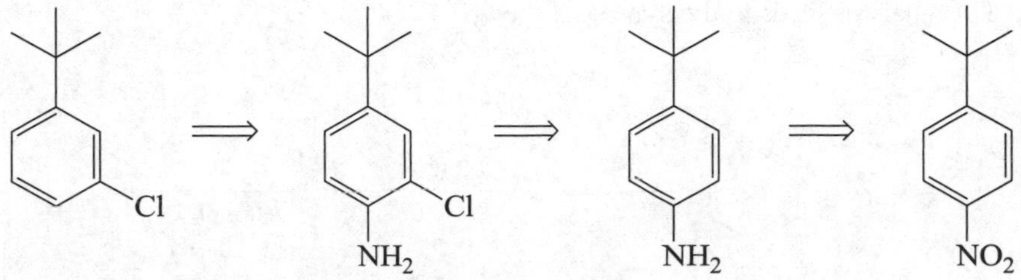

The correct sequence of reactions to carry out this synthesis is shown.

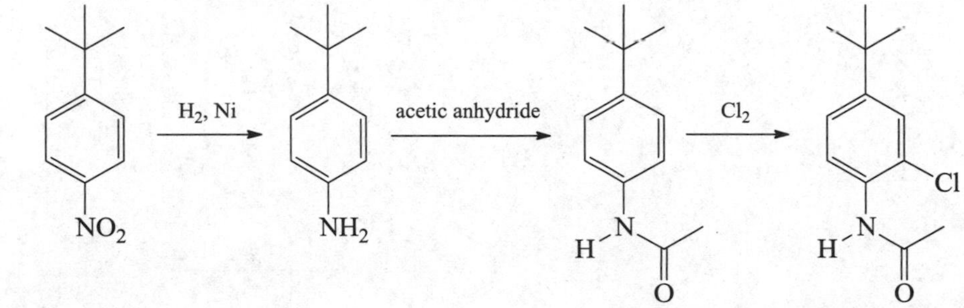

| *p-tert-* Butylnitrobenzene | *p-tert-* Butylaniline | *p-tert-* Butylacetanilide | 4-*tert*-Butyl-1- chloroacetanilide |

Hydrolysis to remove the acetyl group followed by reductive deamination completes the synthesis.

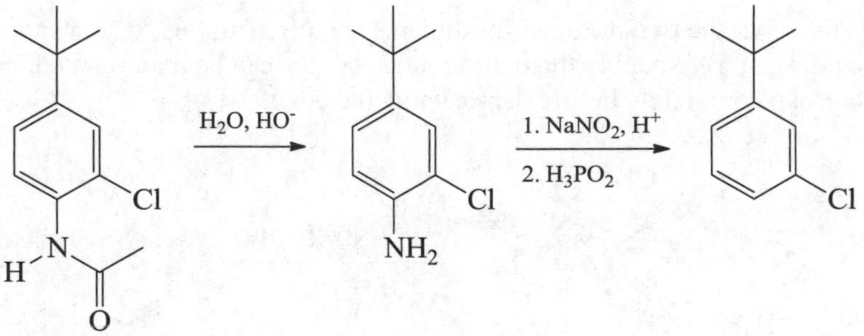

| 4-*tert*-Butyl-1- chloroacetanilide | 4-*tert*-Butyl-2- chloroaniline | *m-tert*-Butylchlorobenzene |

(g) The orientation of substituents in the target molecule can be achieved by using an amino group to control the regiochemistry of bromination, then removing it by reductive deamination.

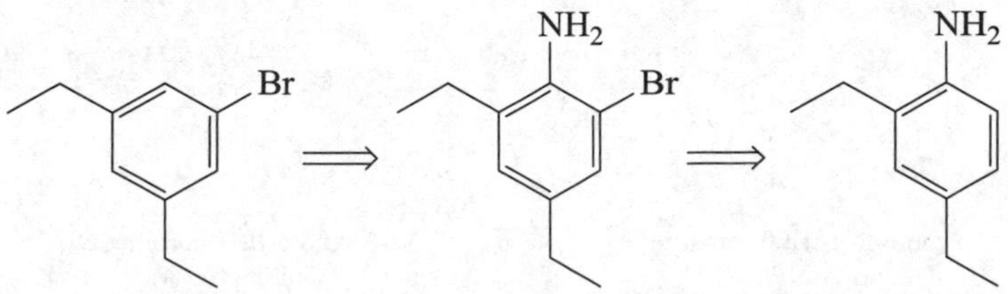

The amino group is introduced in the standard fashion by nitration of an arene followed by reduction.

This analysis leads to the synthesis shown.

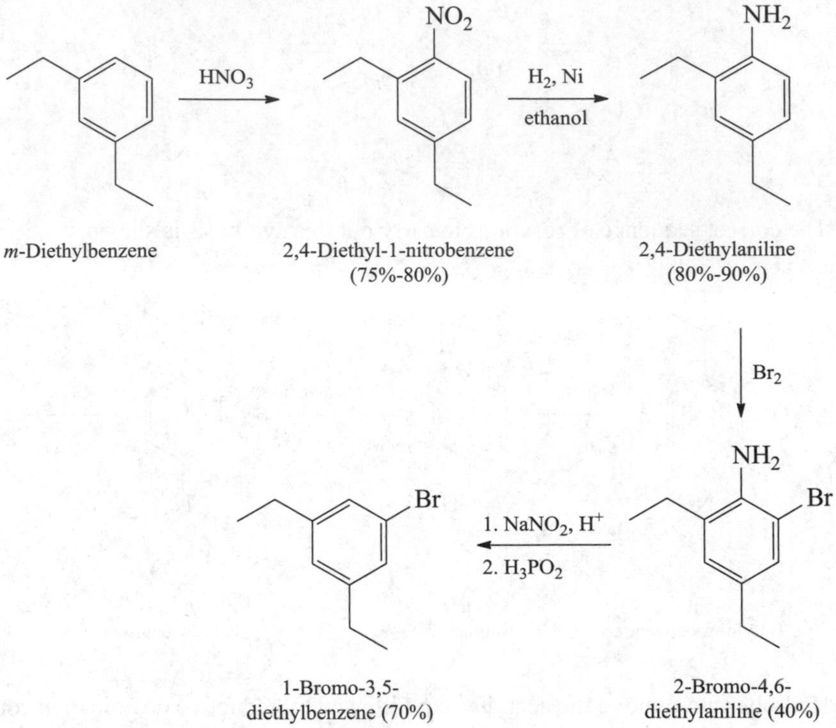

NO2
NH2
HNO3
H2, Ni
ethanol
m-Diethylbenzene
2,4-Diethyl-1-nitrobenzene (75%-80%)
2,4-Diethylaniline (80%-90%)
Br2
NH2
Br
Br
1. NaNO2, H+
2. H3PO2
1-Bromo-3,5-diethylbenzene (70%)
2-Bromo-4,6-diethylaniline (40%)

(*h*) In this exercise, the two nitrogen substituents are differentiated; one is an amino nitrogen, the other an amide nitrogen. By keeping them differentiated, they can be manipulated independently. Remove one amino group completely before deprotecting the other.

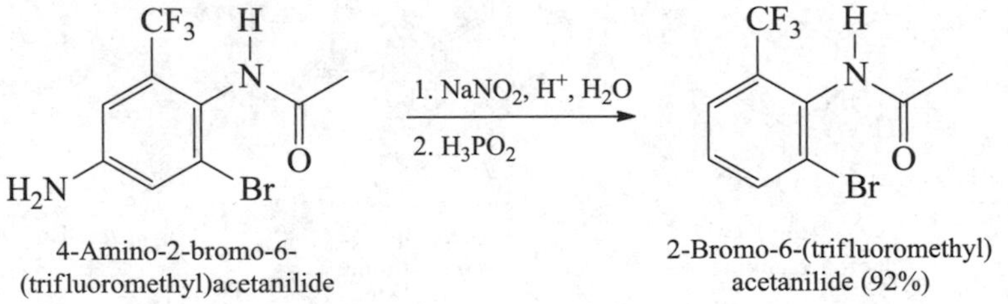

CF3
H
N
O
H2N
Br
1. NaNO2, H+, H2O
2. H3PO2
4-Amino-2-bromo-6-(trifluoromethyl)acetanilide
2-Bromo-6-(trifluoromethyl) acetanilide (92%)

Once the acetyl group has been removed by hydrolysis, the molecule is ready for introduction of the iodo substituent by way of a diazonium salt.

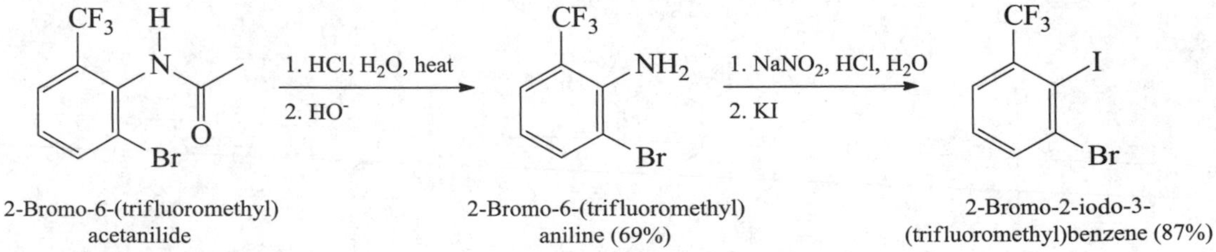

CF3
H
N
O
Br
1. HCl, H2O, heat
2. HO-
NH2
1. NaNO2, HCl, H2O
2. KI
I
2-Bromo-6-(trifluoromethyl) acetanilide
2-Bromo-6-(trifluoromethyl) aniline (69%)
2-Bromo-2-iodo-3-(trifluoromethyl)benzene (87%)

(*i*) To convert the designated starting material to the indicated product, both the nitro group and the ester function must be reduced and a carbon–nitrogen bond must be formed. Converting the starting material to an amide gives the necessary carbon–nitrogen bond and has the advantage that amides can be reduced to amines by lithium aluminum hydride. The amide can be formed intramolecularly by reducing the nitro group to an amine, then heating to cause cyclization.

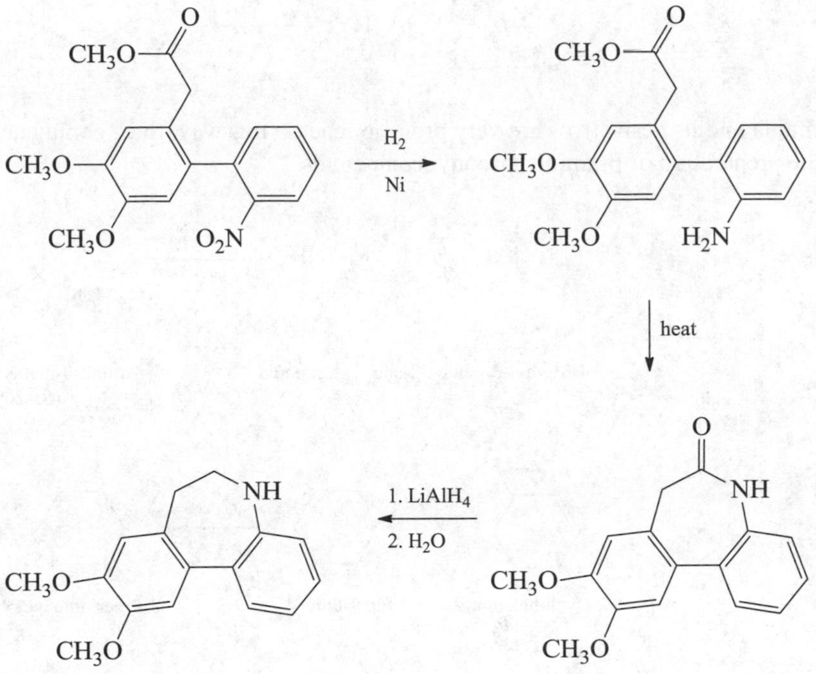

This synthesis is the one described in the chemical literature. Other routes are also possible, but the one shown is short and efficient.

22.50 The synthesis requires two steps. In the first step, *N*-potassiophthalimide reacts with the alkyl halide 1-bromo-2-(2-bromoethyl)benzene in an S_N2 reaction. The bromine on the aryl ring does not react, and over alkylation of the nitrogen does not occur in the Gabriel synthesis.

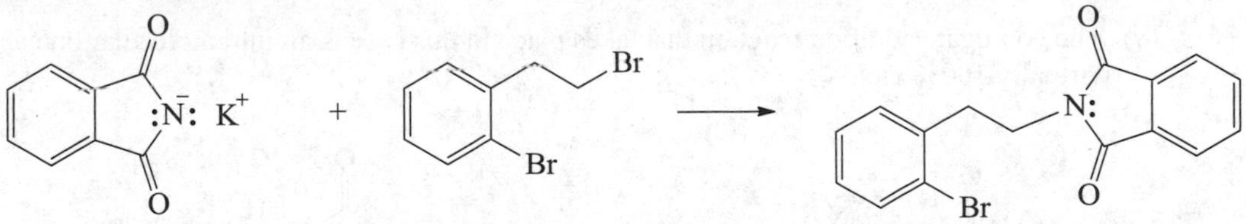

N-Potassiophthalimide

1-Bromo-2-(2-bromoethyl)benzene

Compound A

In the second step, the phthalimide group is cleaved by treatment with hydrazine to give the final product.

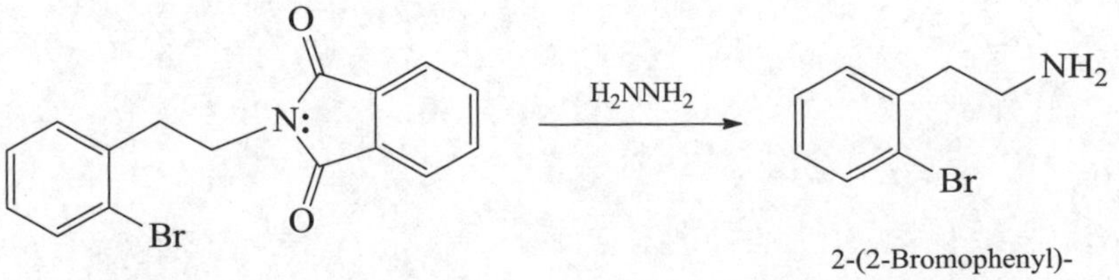

2-(2-Bromophenyl)-ethanamine

22.51 Weakly basic nucleophiles react with α,β-unsaturated carbonyl compounds by conjugate addition.

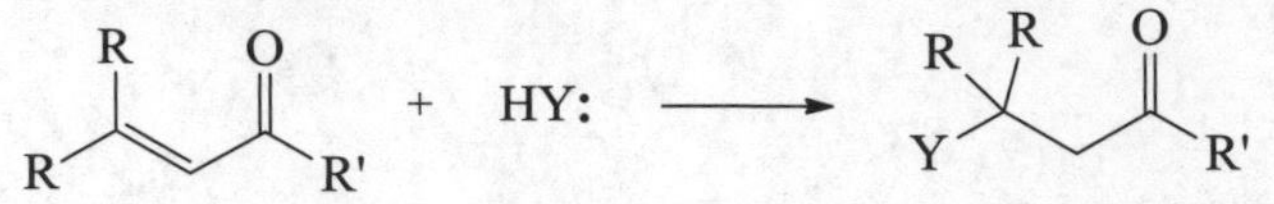

Ammonia and its derivatives are very prone to react in this way; thus, conjugate addition provides a method for the preparation of β-amino carbonyl compounds.

(*a*)

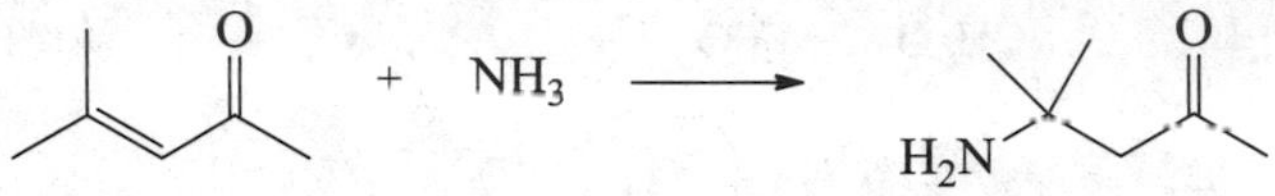

4-Methyl-3-penten-2-one   Ammonia   4-Amino-4-methyl-2-pentanone
                                                        (63–70%)

(*b*)

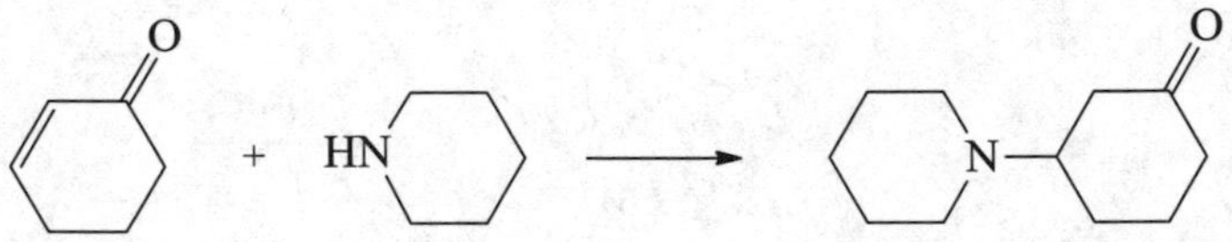

2-Cyclohexanone   Piperidine   3-Piperidinocyclohexanone (45%)

(*c*)

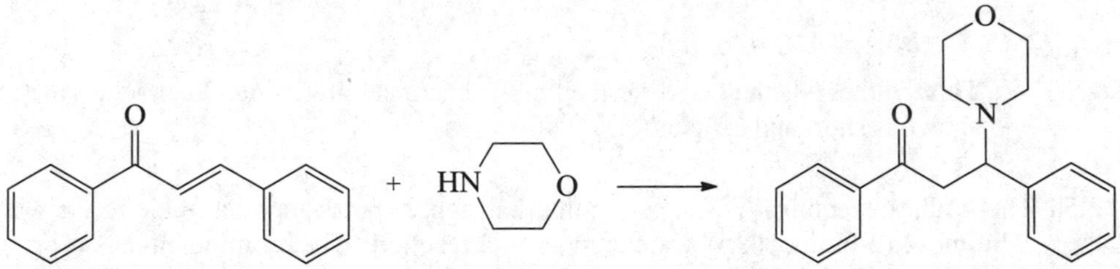

1,3-Diphenyl-2-propen-1-one   Morpholine   3-Morpholino-1,3-diphenyl-1-propanone (91%)

(*d*) The conjugate addition reaction that takes place in this case is an intramolecular one and occurs in virtually 100% yield.

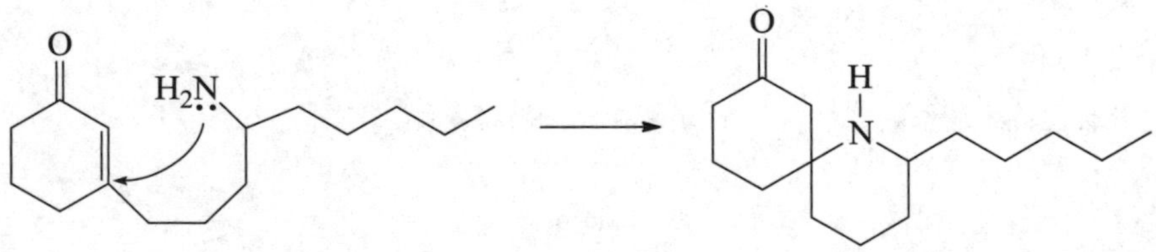

22.52 The first step in the synthesis is the conjugate addition of methylamine to ethyl acrylate. Two sequential Michael additions take place.

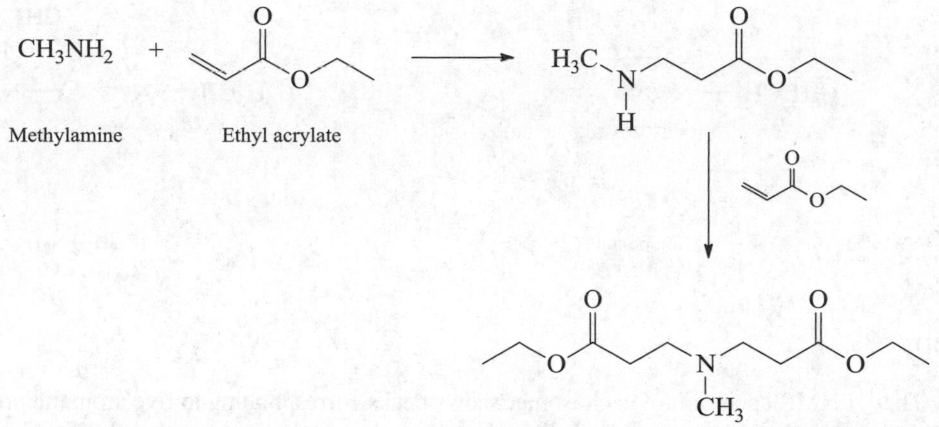

Methylamine   Ethyl acrylate

Conversion of this intermediate to the desired *N*-methyl-4-piperidone requires a Dieckmann cyclization followed by decarboxylation of the resulting β-keto ester.

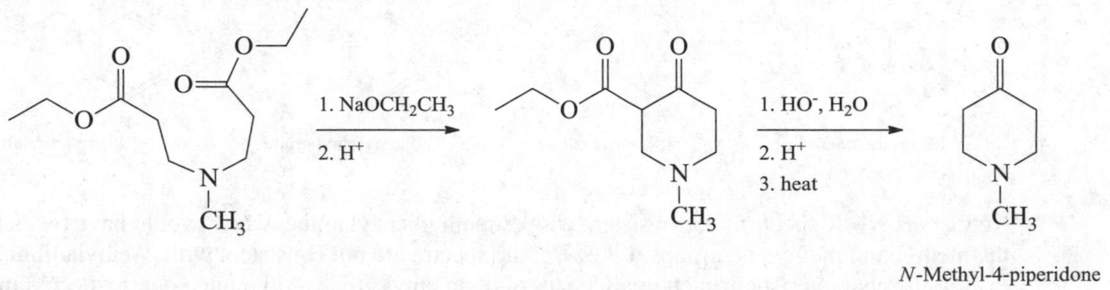

1. NaOCH$_2$CH$_3$
2. H$^+$

1. HO$^-$, H$_2$O
2. H$^+$
3. heat

N-Methyl-4-piperidone

Treatment of *N*-methyl-4-piperidone with the Grignard reagent derived from bromobenzene gives a tertiary alcohol that can be dehydrated to an alkene. Hydrogenation of the alkene completes the synthesis.

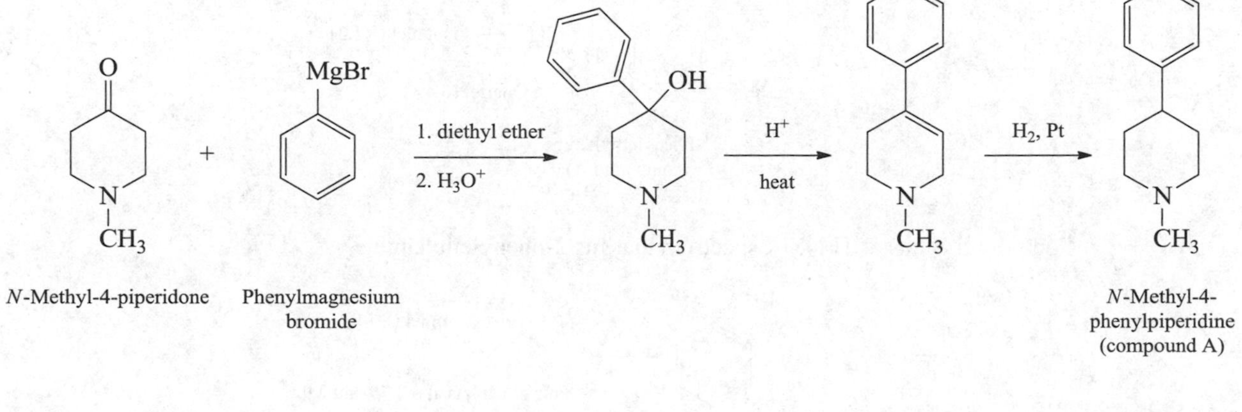

N-Methyl-4-piperidone   Phenylmagnesium bromide

1. diethyl ether
2. H$_3$O$^+$

H$^+$
heat

H$_2$, Pt

N-Methyl-4-phenylpiperidine (compound A)

22.53 An imine is formed in the first step. This is reduced to give an aniline derivative in the second step of the synthesis.

Compound A
$(C_{19}H_{21}FN_2)$

$(C_{19}H_{23}FN_2)$

Spectroscopy

22.54 The ^{1}H NMR spectrum of each isomer shows peaks corresponding to five aromatic protons, so compounds A and B each contain a monosubstituted benzene ring. Only four compounds of molecular formula $C_8H_{11}N$ meet this requirement.

N-Methylbenzylamine   *N*-Ethylaniline   1-Phenylethylamine   2-Phenylethylamine

Neither ^{1}H NMR spectrum is consistent with *N*-methylbenzylamine, which would have two singlets due to the methyl and methylene groups. Likewise, the spectra are not consistent with *N*-ethylaniline, which would exhibit the characteristic triplet–quartet pattern of an ethyl group. Although a quartet occurs in the spectrum of compound A, it corresponds to only one proton, not the two that an ethyl group requires. The one-proton quartet in compound A arises from an H—C—CH$_3$ unit. Compound A is 1-phenylethylamine.

NH$_2$ ← Singlet (δ 1.4)
CH$_3$ ← Doublet (δ 1.3)
H ← Quartet (δ 4.1)

1-Phenylethylamine
(compound A)

Compound B has a ^{1}H NMR spectrum that fits 2-phenylethylamine.

NH$_2$ ← Singlet (δ 1.0)
Pair of triplets at δ 2.75 and 2.92

2-Phenylethylamine
(compound B)

22.55 Write the structural formulas for the two possible compounds given in the problem and consider how their ^{13}C NMR spectra will differ from each other. Both will exhibit their CH$_3$ carbon signals at high field, but they differ in the positions of their CH_2 and quaternary carbons. A carbon bonded to nitrogen is more shielded than one bonded to oxygen, because nitrogen is less electronegative than oxygen.

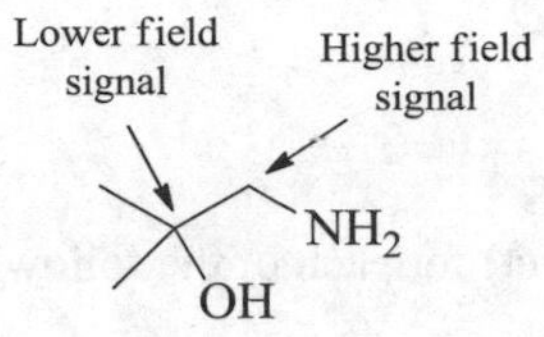

1-Amino-2-methyl-2-propanol 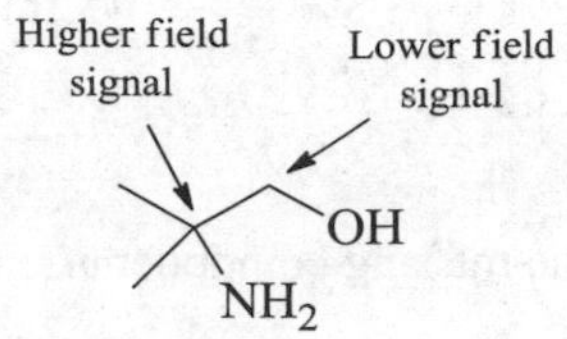   2-Amino-2-methyl-1-propanol

In one isomer, the lowest field signal is a quaternary carbon; in the other, it is a CH_2 group. The spectrum shown in text Figure 22.11 shows the lowest field signal as a CH_2 group. The compound is therefore 2-amino-2-methyl-1-propanol, $(CH_3)_2CCH_2OH$.
$$\underset{NH_2}{|}$$

This compound *cannot* be prepared by reaction of ammonia with an epoxide, because in basic solution nucleophiles attack epoxides at the less hindered carbon, and therefore epoxide ring opening will give 1-amino-2-methyl-2-propanol rather than 2-amino-2-methyl-1-propanol.

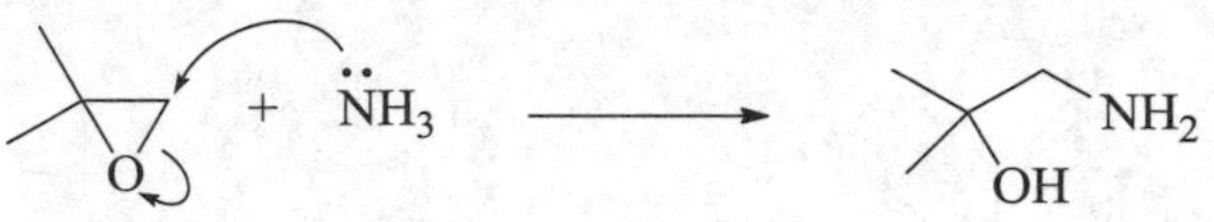

2,2-Dimethyloxirane   Ammonia   1-Amino-2-methyl-2-propanol

Answers to Interpretive Problems 22

22.56 A;   **22.57** B;   **22.58** D;   **22.59** A;   **22.60** C;   **22.61** B

SELF-TEST

1. Give an acceptable name for each of the following. Identify each compound as a primary, secondary, or tertiary amine.

(a)   (b) 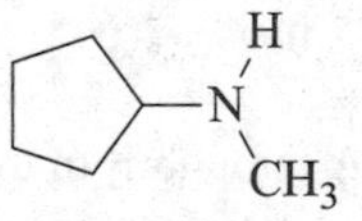   (c)

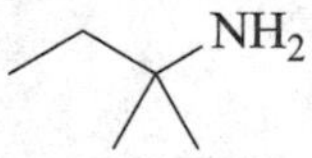

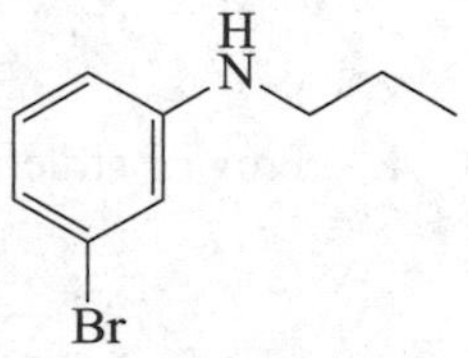

2. Provide the correct formula or structure of the reagent omitted from each of the following reactions:

(a)

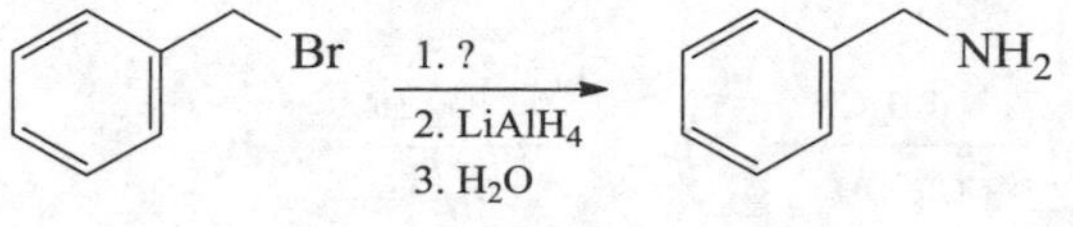

(b)

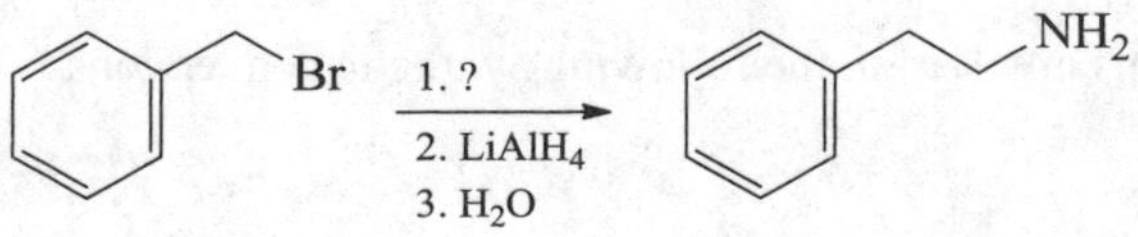

(c)

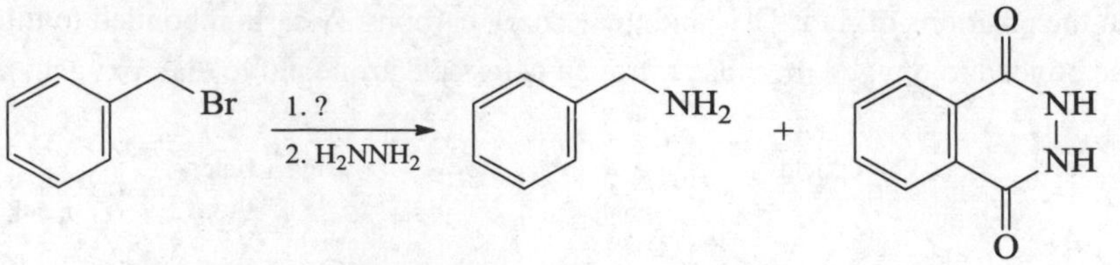

3. Provide the missing component (reagent or product) for each of the following:

(a)

(b)

Product of part (a) $\xrightarrow{\text{CuBr}}$?

(c)

Product of part (a) $\xrightarrow{?}$ toluene

(d)

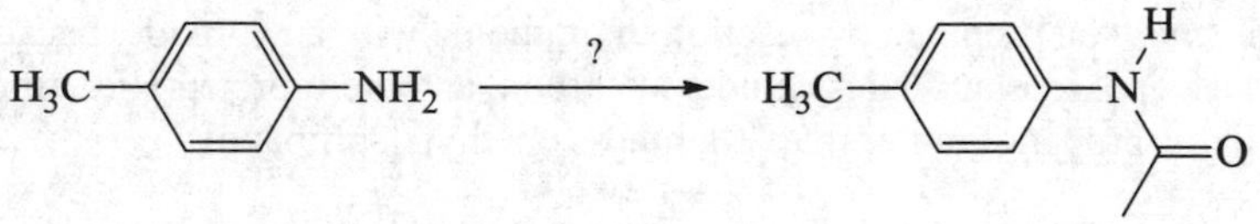

(e)

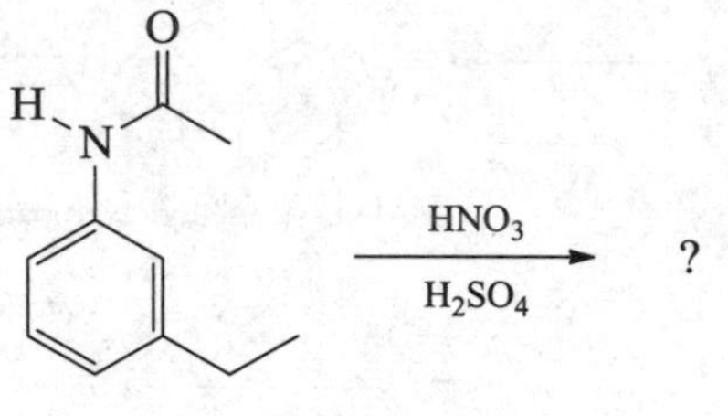

(f)

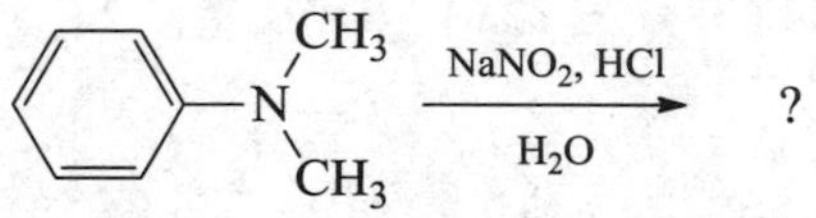

(g)

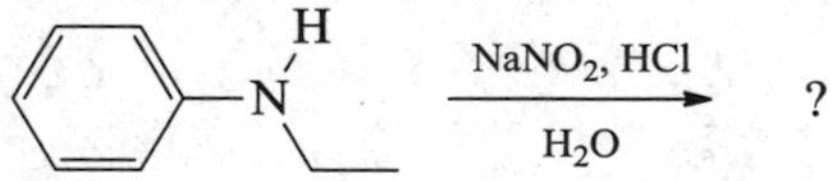

4. Provide structures for compounds A through E in the following reaction sequences:

(a)

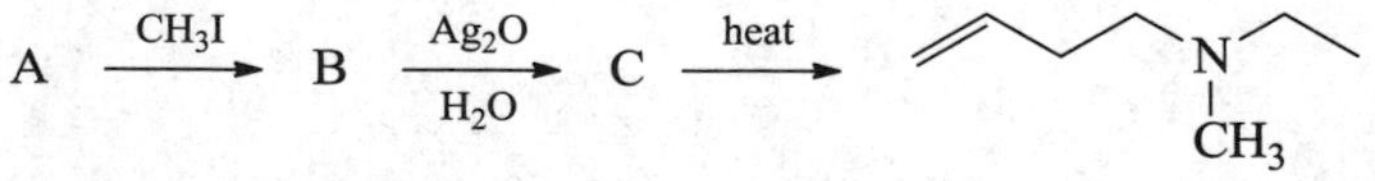

(b)

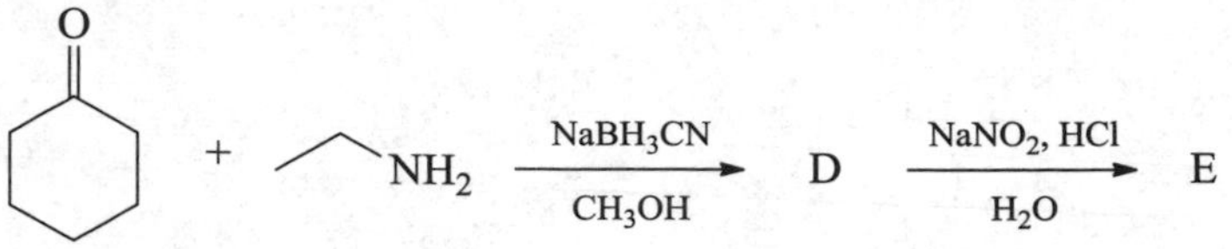

5. Give the series of reaction steps involved in the following synthetic conversions:

(*a*)

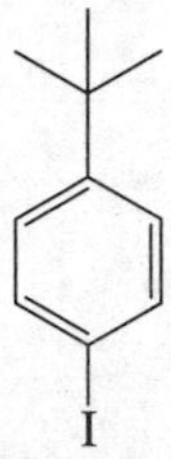

from benzene

(*b*)   *m*-Chloroaniline from benzene

(*c*)

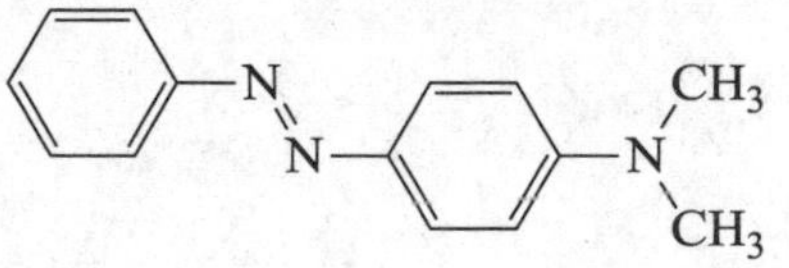

from aniline

6. *p*-Nitroaniline is less basic than *m*-nitroaniline. Using resonance structures, explain the reason for this difference.

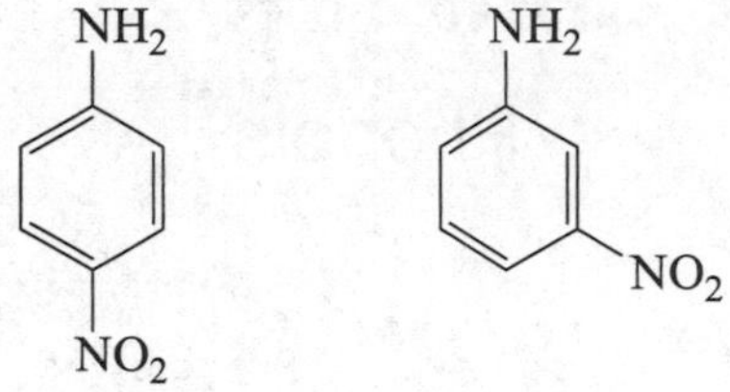

7. Identify the strongest and weakest bases among the following:

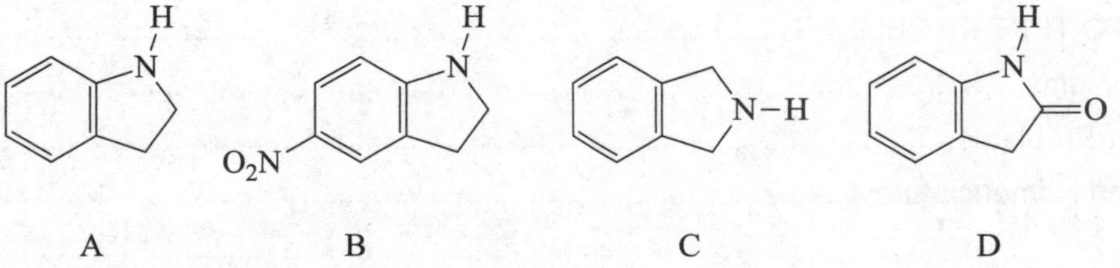

A   B   C   D

8. Write the structures of the compounds A–D formed in the following reaction sequence:

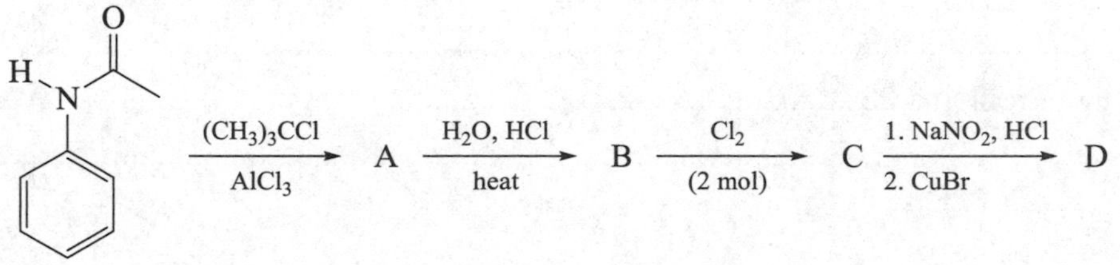

CHAPTER 23
Phenols

Table of Contents

SOLUTIONS TO TEXT PROBLEMS

In Chapter Problems

23.1 (*b*) A benzyl group (C$_6$H$_5$CH$_2$—) is ortho to the phenolic hydroxyl group in *o*-benzylphenol.

(*c*) Naphthalene is numbered as shown. 3-Nitro-1-naphthol has a hydroxyl group at C-1 and a nitro group at C-3.

Naphthalene   3-Nitro-1-naphthol

23.2 Intramolecular hydrogen bonding between the hydroxyl group and the ester carbonyl can occur when these groups are ortho to each other.

Methyl salicylate

Intramolecular hydrogen bonds form at the expense of intermolecular ones, and intramolecularly hydrogen-bonded phenols have lower boiling points than isomers in which only intermolecular hydrogen bonding is possible.

23.3 Since benzoic is a stronger acid than phenol, shaking an ether solution of benzoic acid and phenol with an aqueous solution of dilute sodium bicarbonate converts benzoic acid quantitatively into its sodium salt. Sodium benzoate is more soluble in water than ether so it is extracted into the aqueous phase. Phenol stays in the ether phase.

Phenol   Benzoic acid   Phenol   Sodium benzoate
(pK_a = 10)   (pK_a = 4.2)   (remains in ether phase)   (in aqueous phase)
(in ether phase)   (in ether phase)

Phenols are more acidic than alcohols so a mixture can be separated by shaking an ether solution with dilute aqueous sodium hydroxide.

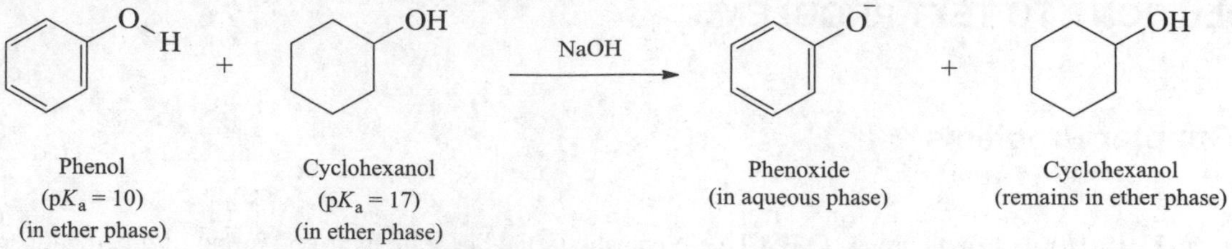

NaOH
Phenol
(pK_a = 10)
(in ether phase)
Cyclohexanol
(pK_a = 17)
(in ether phase)
Phenoxide
(in aqueous phase)
Cyclohexanol
(remains in ether phase)

23.4 (*b*) A cyano group withdraws electrons from the ring by resonance. A *p*-cyano substituent is conjugated directly with the negatively charged oxygen and stabilizes the anion more than does a *m*-cyano substituent.

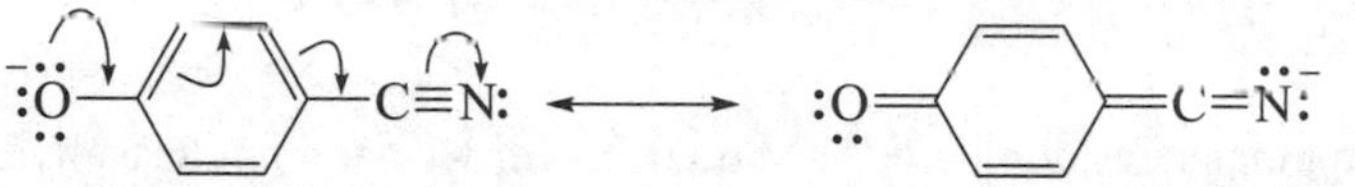

p-Cyanophenol is slightly more acidic than *m*-cyanophenol, the pK_a values being 8.0 and 8.5, respectively.

(*c*) The electron-withdrawing inductive effect of fluorine will be more pronounced at the ortho position than at the para. *o*-Fluorophenol (pK_a = 8.7) is a stronger acid than *p*-fluorophenol (pK_a = 9.9).

23.5 In order to carry out this conversion, an intermediate compound that can be converted to a phenol is needed.

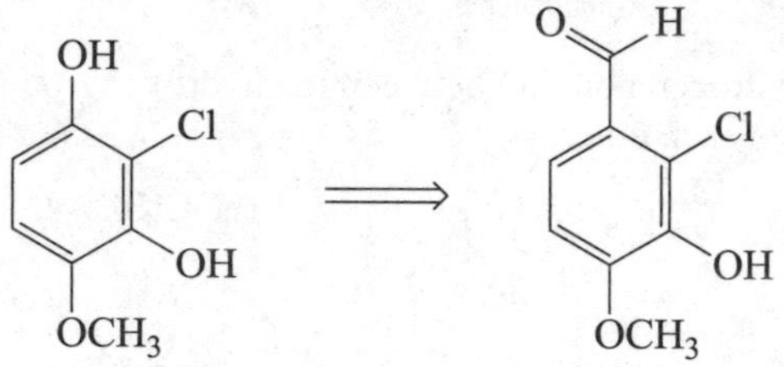

A Baeyer-Villiger oxidation (Descriptive Passage and Interpretive Problems 18) converts a methyl ketone to an acetate. The acetate is then hydrolyzed to the final product.

CF_3CO_3H
HCl

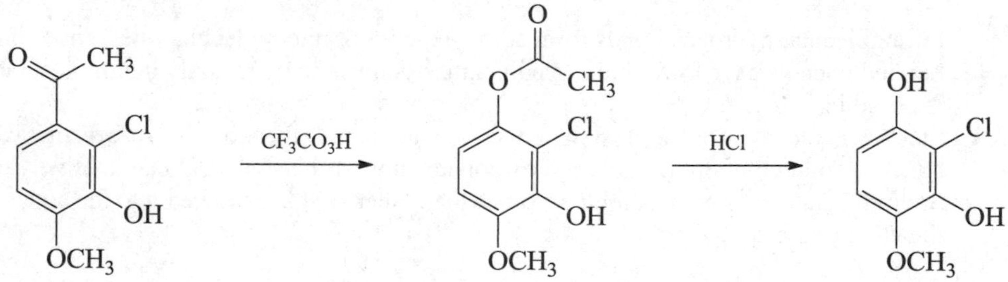

The methyl ketone can be formed by reaction with a Grignard reagent to give an alcohol followed by oxidation.

1. CH_3MgBr
2. H_2O
$K_2Cr_2O_7$
H_2SO_4, H_2O

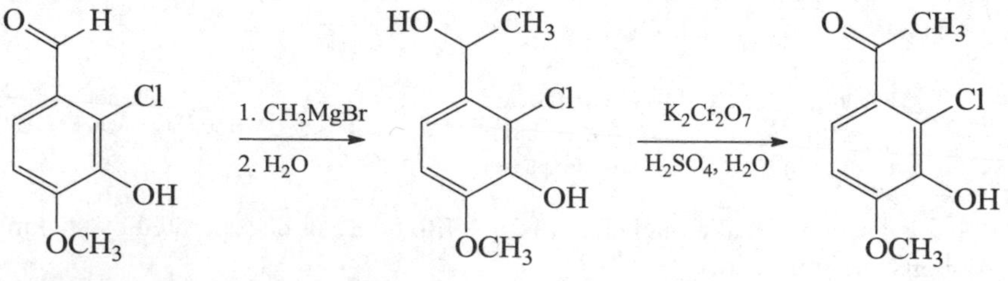

Although not covered in the text, aromatic aldehydes undergo Baeyer-Villiger oxidation to yield the corresponding aryl formate ester. Thus, an alternative synthesis is:

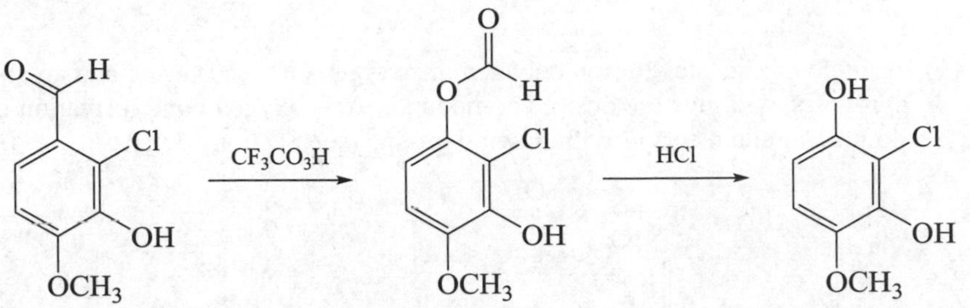

23.6 The Fries rearrangement of phenyl benzoate yields the two thermodynamically more stable benzophenones.

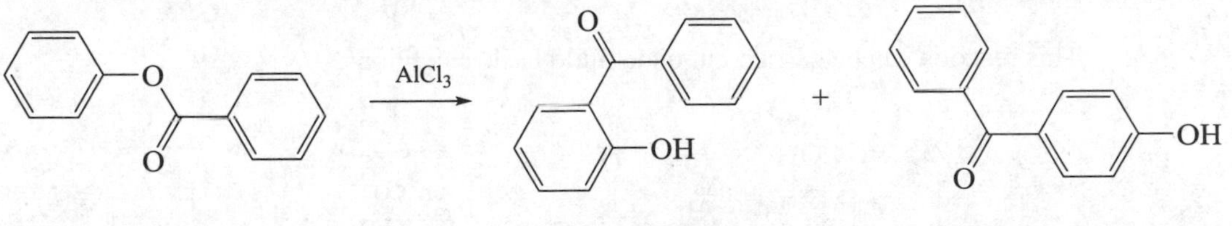

Phenyl benzoate

o-Hydroxybenzophenone
(9%)

p-Hydroxybenzophenone
(64%)

23.7 (*b*) The reaction is Friedel–Crafts alkylation. Proton transfer from sulfuric acid to 2-methylpropene gives *tert*-butyl cation. Because the position para to the hydroxyl substituent already bears a bromine, the *tert*-butyl cation attacks the ring at the position ortho to the hydroxyl.

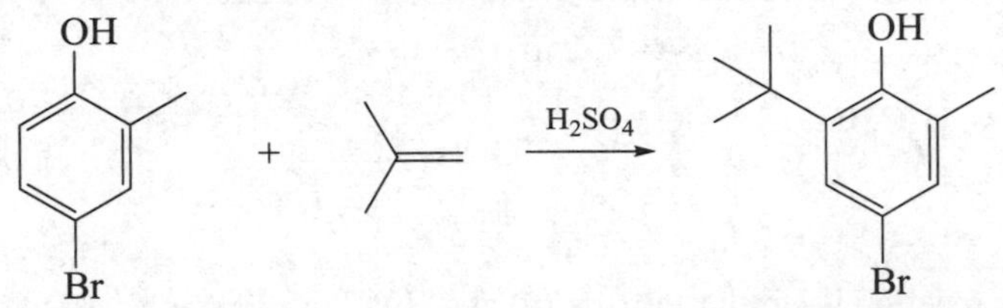

4-Bromo-2-methylphenol

2-Methylpropene

4-Bromo-2-*tert*-butyl-6-methylphenol
(isolated yield, 70%)

(*c*) Acidification of sodium nitrite produces nitrous acid, which nitrosates the strongly activated aromatic ring of phenols.

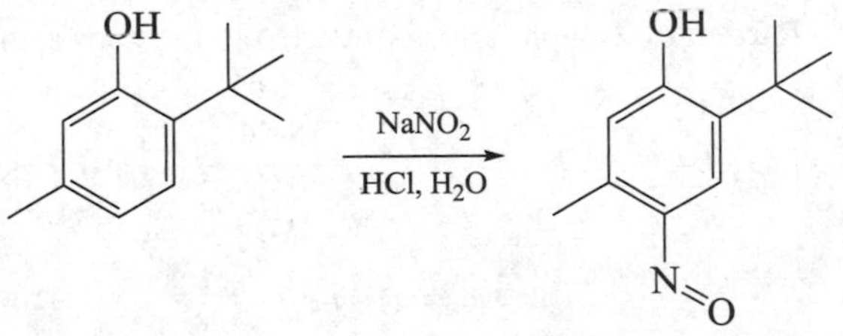

2-Isopropyl-5-methylphenol

2-Isopropyl-5-methyl-4-nitrosophenol
(isolated yield, 87%)

(*d*) Friedel–Crafts acylation occurs ortho to the hydroxyl group.

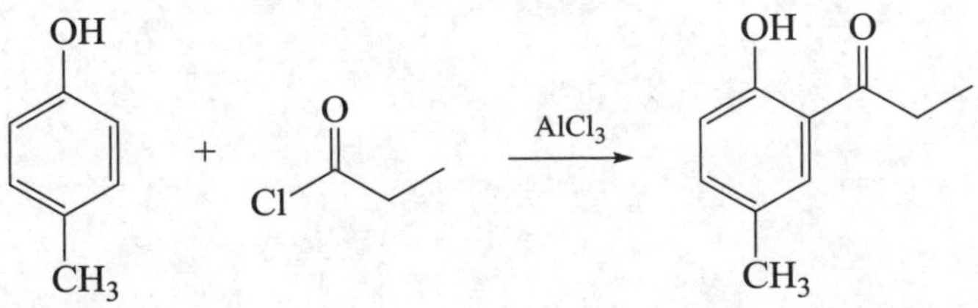

4-Methylphenol
(*p*-cresol)

1-(2-Hydroxy-5-methylphenyl)-1-propanone (isolated yield, 87%)

23.8 (*b*) A nucleophilic substitution between the oxygen on the aryl ring and an alkyl halide, a Williamson synthesis, will give the desired compound. Aryl oxygen bond formation is difficult for aryl halides that do not contain electron-withdrawing groups (see Section 13.19).

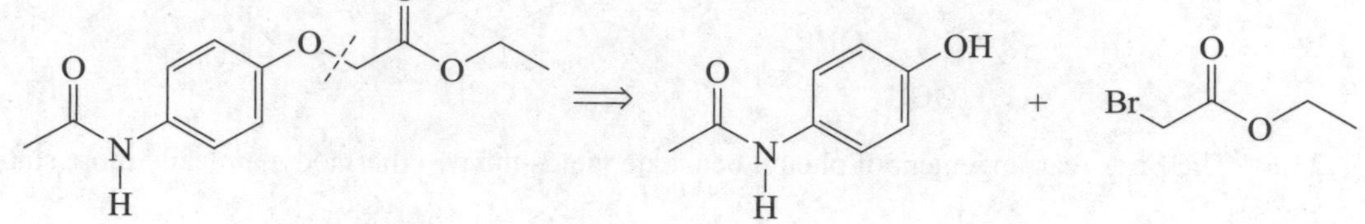

This reaction can be carried out under mild basic conditions.

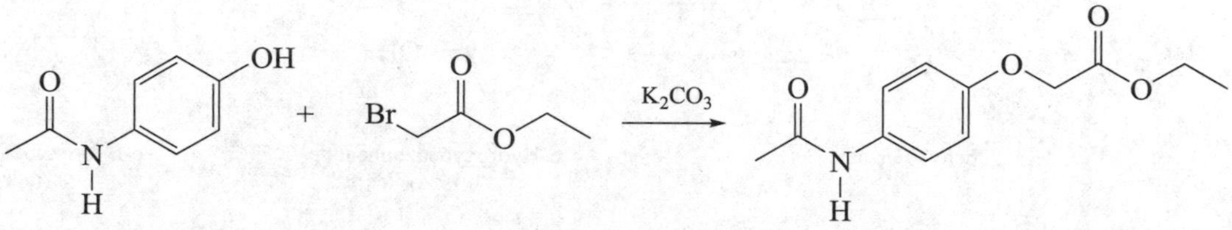

(*c*) A nucleophilic substitution between the oxygen on the aryl ring and an alkyl halide, a Williamson synthesis, will give the desired compound. Aryl oxygen bond formation is difficult for aryl halides that do not contain electron-withdrawing groups (see Section 13.19).

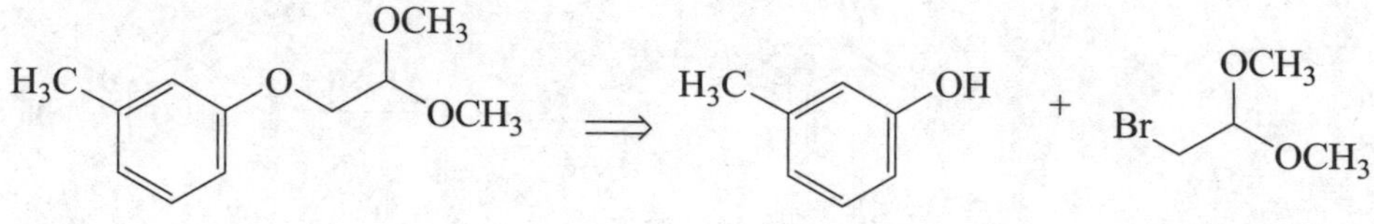

This reaction is similar to parts (*a*) and (*b*). It also can be carried out under mild basic conditions.

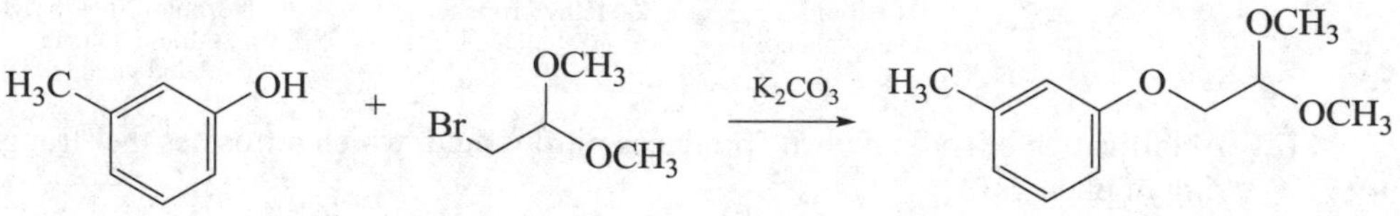

23.9 (*b*) The hydroxyl group of 2-naphthol is converted to the corresponding acetate ester.

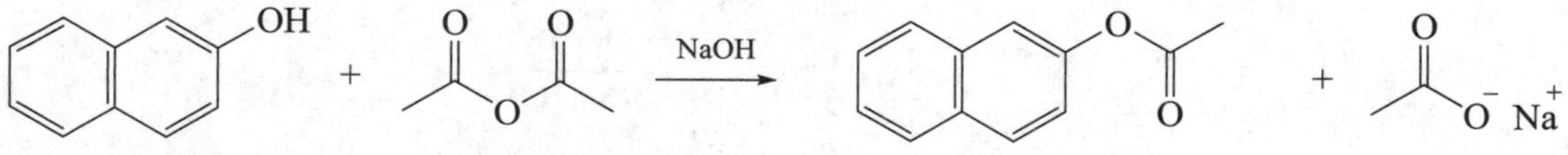

2-Naphthol   Acetic anhydride   2-Naphthyl acetate   Sodium acetate

(*c*) Benzoyl chloride acylates the hydroxyl group of phenol.

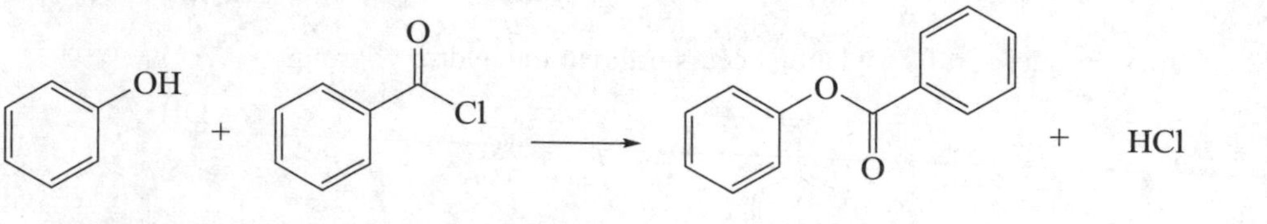

Phenol   Benzoyl chloride   Phenyl benzoate   Hydrogen chloride

23.10 A hydrogen atom is abstracted from the oxygen.

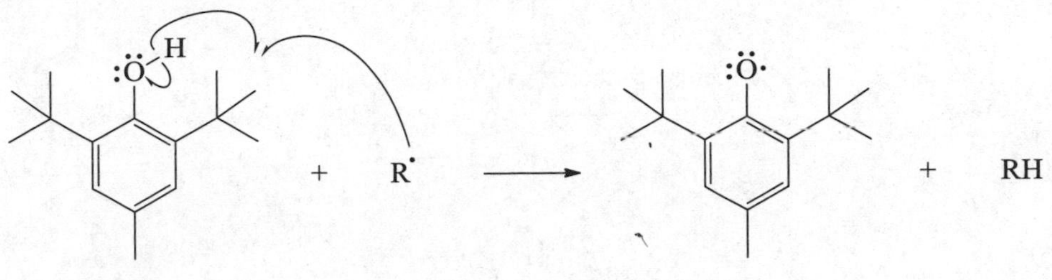

BHT

The radical is stable because of the steric hindrance provided by the adjacent *tert*-butyl groups and the resonance stabilization of the aromatic ring.

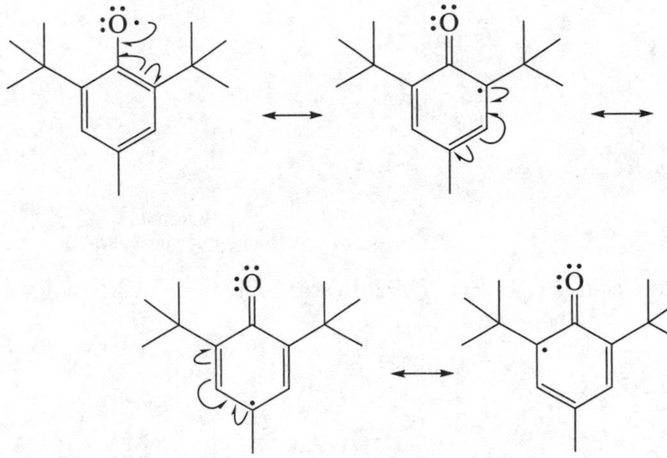

23.11 Substituted allyl aryl ethers undergo a Claisen rearrangement similar to the reaction described in text Section 23.12 for allyl phenyl ether. *E*-2-Butenyl phenyl ether rearranges on heating to give 2-(1-methyl-2-propenyl)phenol.

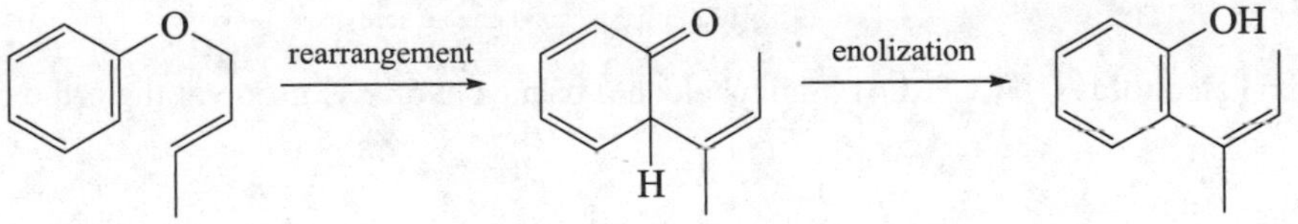

2-Butenyl phenyl ether   2-(1-Methyl-2-propenyl)phenol

End of Chapter Problems

Structure and Nomenclature

23.12 (*a*) The parent compound is benzaldehyde. Vanillin bears a methoxy group (CH₃O) at C-3 and a hydroxyl group (HO) at C-4.

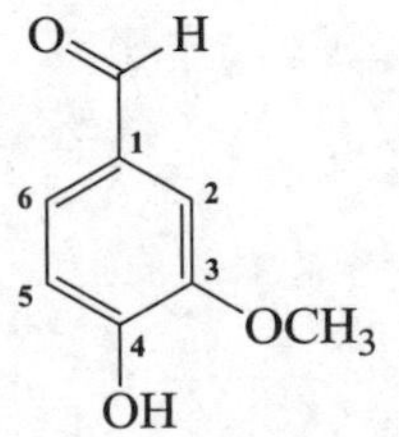

Vanillin
(4-hydroxy-3-methoxybenzaldehyde)

(*b, c*) Thymol and carvacrol differ with respect to the position of the hydroxyl group.

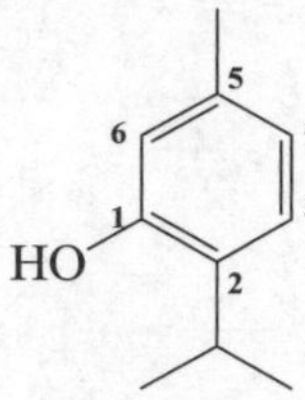

Thymol
(2-isopropyl-5-methylphenol)

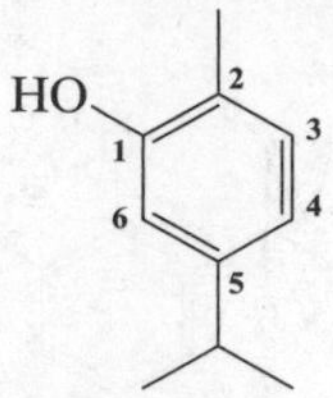

Carvacrol
(5-isopropyl-2-methylphenol)

(*d*) An allyl substituent is —CH$_2$CH=CH$_2$.

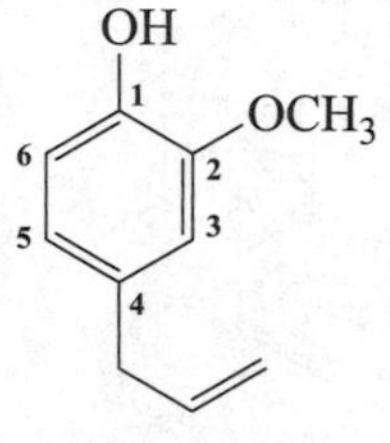

Eugenol
(4-allyl-2-methoxyphenol)

(*e*) Benzoic acid is C$_6$H$_5$CO$_2$H. Gallic acid bears three hydroxyl groups, located at C-3, C-4, and C-5.

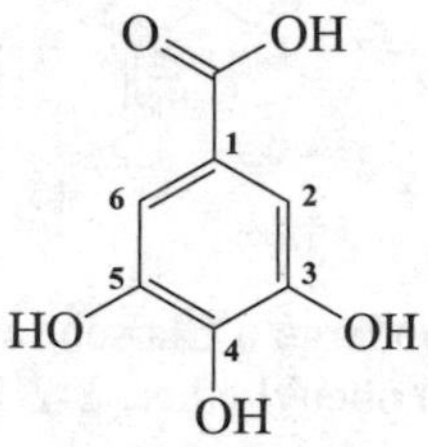

Gallic acid
(3,4,5-trihydroxybenzoic acid)

(*f*) Benzyl alcohol is C$_6$H$_5$CH$_2$OH. Salicyl alcohol bears a hydroxyl group at the ortho position.

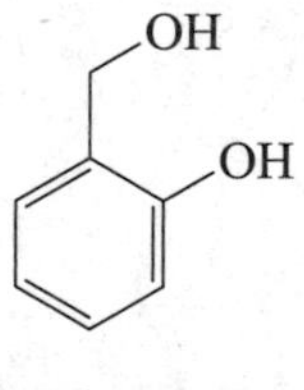

Salicyl alcohol
(*o*-hydroxybenzyl alcohol)

23.13 (*a*) The compound is named as a derivative of phenol. The substituents (ethyl and nitro) are cited in alphabetical order with numbers assigned in the direction that gives the lowest number at the first point of difference.

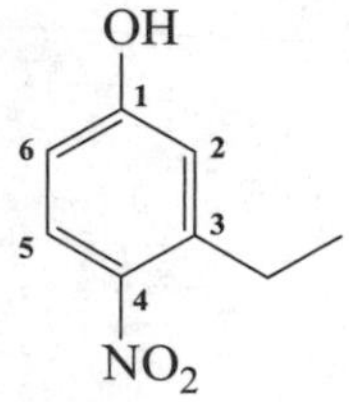

3-Ethyl-4-nitrophenol

(b) An isomer of the compound in part (a) is 4-ethyl-3-nitrophenol.

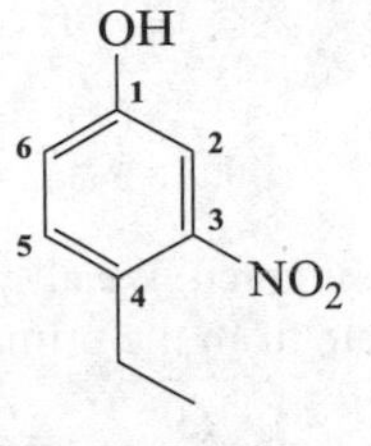

OH

4-Ethyl-3-nitrophenol

(c) The parent compound is phenol. It bears, in alphabetical order, a benzyl group at C-4 and a chlorine at C-2.

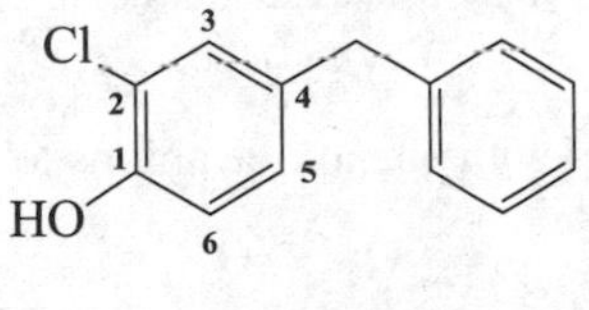

4-Benzyl-2-chlorophenol

(d) This compound is named as a derivative of anisole, $C_6H_5OCH_3$. Because multiplicative prefixes (di-, tri-, etc.) are not considered when alphabetizing substituents, isopropyl precedes dimethyl.

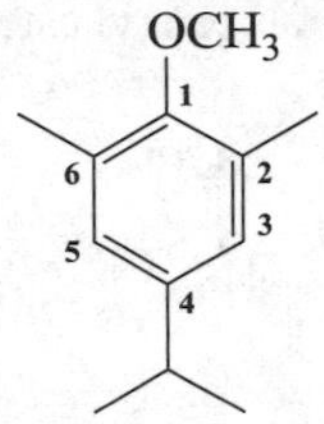

OCH_3

4-Isopropyl-2,6-dimethylanisole

(e) The compound is an aryl ester of trichloroacetic acid. The aryl group is 2,5-dichlorophenyl.

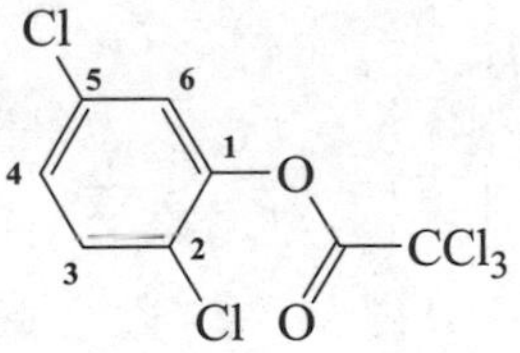

2,5-Dichlorophenyl trichloroacetate

Reactions

23.14 (a) The reaction is an acid–base reaction. Phenol is the acid; sodium hydroxide is the base.

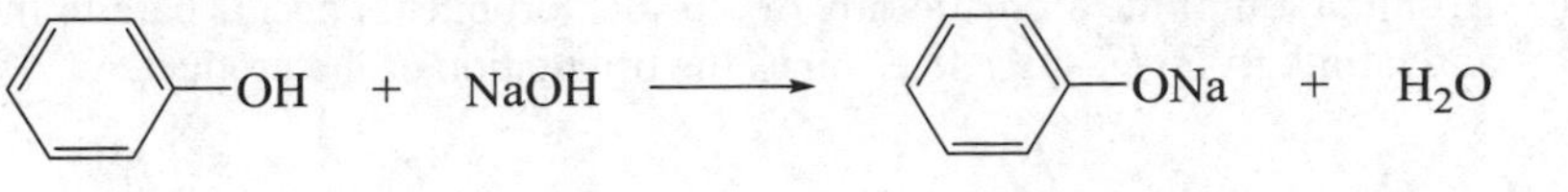

$$\text{OH} + \text{NaOH} \longrightarrow \text{ONa} + H_2O$$

| Phenol (stronger acid) | Sodium hydroxide (stronger base) | Sodium phenoxide (weaker base) | Water (weaker acid) |

(b) Sodium phenoxide reacts with ethyl bromide to yield ethyl phenyl ether in a Williamson reaction. Phenoxide ion acts as a nucleophile.

Sodium phenoxide   Ethyl bromide   Ethyl phenyl ether   Sodium bromide

(c) *p*-Toluenesulfonates behave much like alkyl halides in nucleophilic substitutions. Phenoxide ion displaces *p*-toluenesulfonate from the primary carbon.

Sodium phenoxide   Butyl *p*-toluenesulfonate   Butyl phenyl ether   Sodium *p*-toluenesulfonate

(d) Acid anhydrides react with phenoxide anions to yield aryl esters.

Sodium phenoxide   Acetic anhydride   Phenyl acetate   Sodium acetate

(e) Acyl chlorides convert phenols to aryl esters.

2-Methylphenol   Benzoyl chloride   2-Methylphenyl benzoate   Hydrogen chloride
(*o*-cresol)

(f) Phenols react as nucleophiles toward epoxides.

3-Methylphenol   Ethylene oxide   2-(3-Methylphenoxy)ethanol
(*m*-cresol)

The reaction as written conforms to the requirements of the problem that a balanced equation be written. Of course, the reaction will be much faster if catalyzed by acid or base, but the catalysts do not enter into the equation representing the overall process.

(g) Bromination of the aromatic ring of 2,6-dichlorophenol occurs para to the hydroxy group. The more activating group (—OH) determines the orientation of the product.

2,6-Dichlorophenol   Bromine   4-Bromo-2,6-   Hydrogen
                                                          dichlorophenol   bromide

(*h*) In aqueous solution, bromination occurs at all the open positions that are ortho and para to the hydroxyl group.

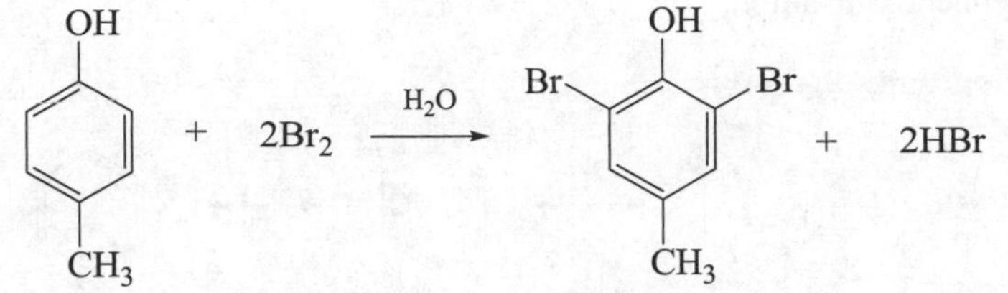

| 4-Methylphenol (*p*-cresol) | Bromine | 2,6-Dibromo-4-methylphenol | Hydrogen bromide |

(*i*) Hydrogen bromide cleaves ethers to give an alkyl halide and a phenol.

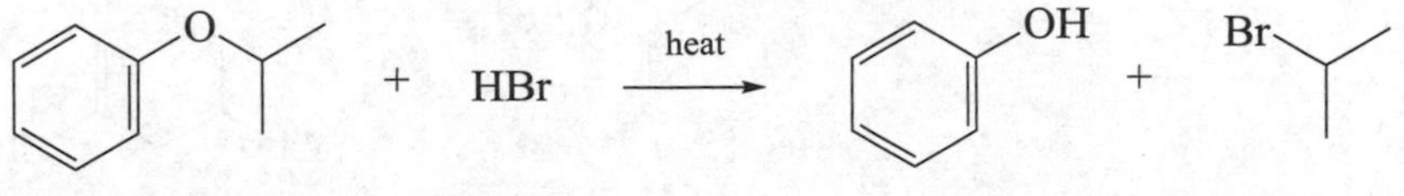

| Isopropyl phenyl ether | Hydrogen bromide | Phenol | Isopropyl bromide |

23.15 (*a*) Strongly electron-withdrawing groups, particularly those such as —NO$_2$, increase the acidity of phenols by resonance stabilization of the resulting phenoxide anion. Electron-releasing substituents such as —CH$_3$ exert a very small acid-weakening effect.

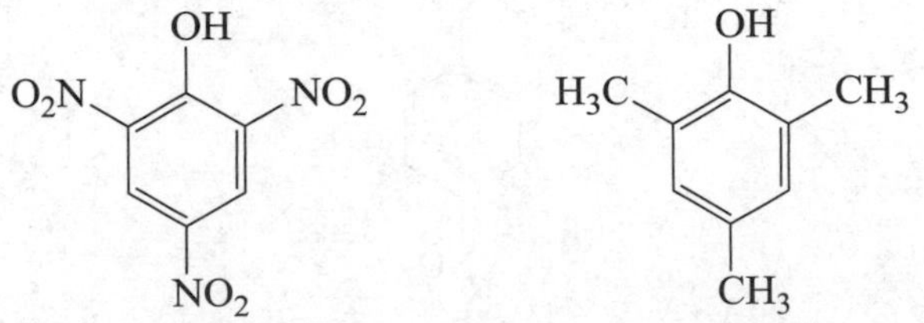

2,4,6-Trinitrophenol, more acidic (pK_a = 0.4)

2,4,6-Trimethylphenol, less acidic (pK_a = 10.9)

2,4,6-Trinitrophenol, also known as picric acid, is a stronger acid by far than 2,4,6-trimethylphenol. All three nitro groups participate in resonance stabilization of the picrate anion.

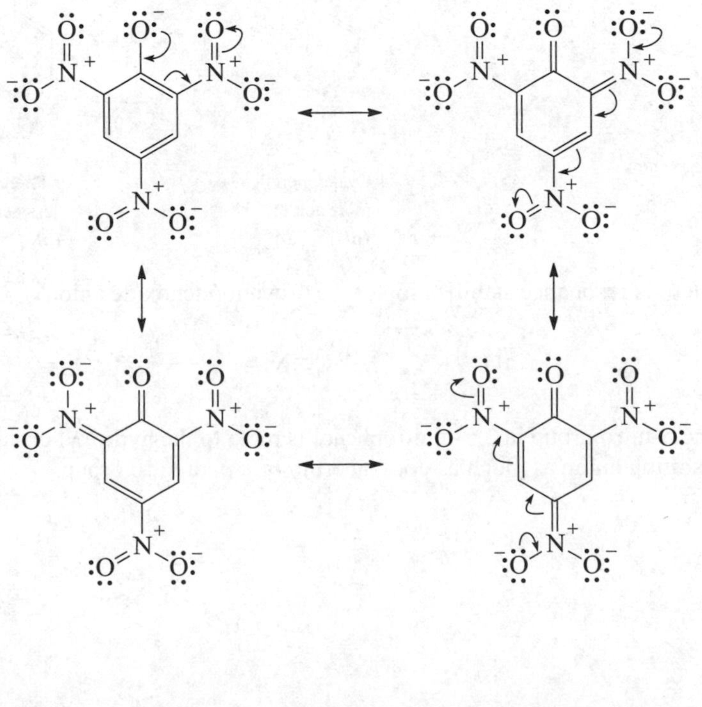

(b) Stabilization of a phenoxide anion is most effective when electron-withdrawing groups are present at the ortho and para positions because these carbons bear most of the negative charge in phenoxide anion.

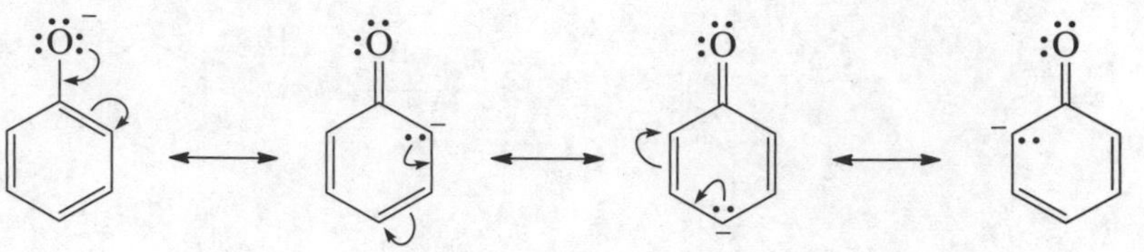

2,6-Dichlorophenol is therefore expected to be (and is) a stronger acid than 3,5-dichlorophenol.

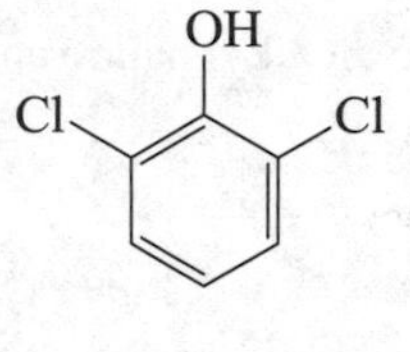

2,6-Dichlorophenol,
more acidic
($pK_a = 6.8$)

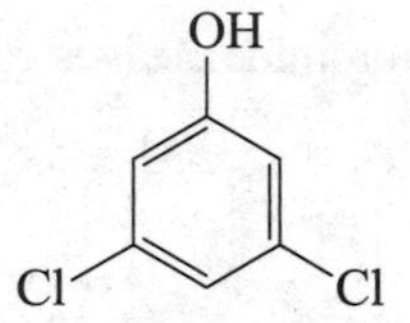

3,5-Dichlorophenol,
less acidic
($pK_a = 8.2$)

(c) The same principle is at work here as in part (b). A nitro group para to the phenol oxygen is directly conjugated to it and stabilizes the anion better than one at the meta position.

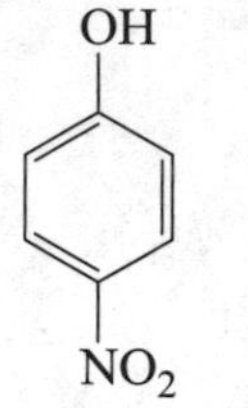

4-Nitrophenol,
stronger acid
($pK_a = 7.2$)

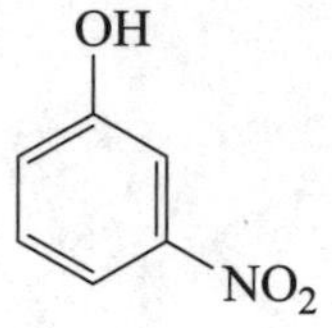

3-Nitrophenol,
weaker acid
($pK_a = 8.4$)

(d) A cyano group is strongly electron-withdrawing, and so 4-cyanophenol is a stronger acid than phenol.

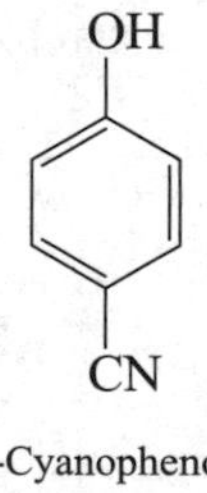

4-Cyanophenol,
more acidic
($pK_a = 8.0$)

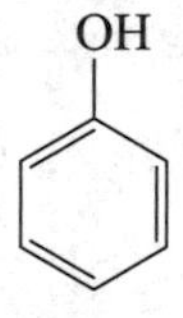

Phenol,
less acidic
($pK_a = 10$)

There is resonance stabilization of the 4-cyanophenoxide anion.

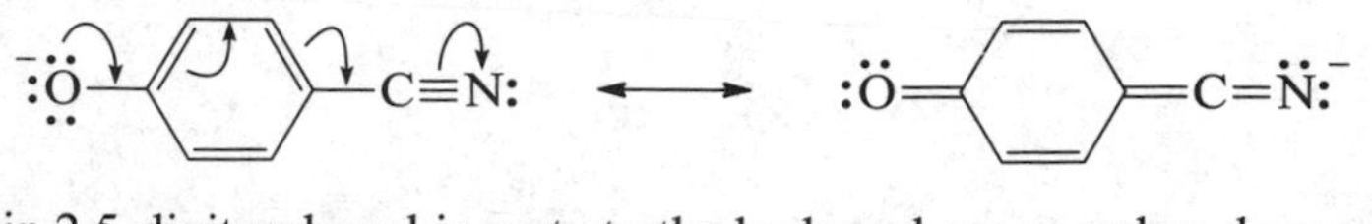

(e) The 5-nitro group in 2,5-dinitrophenol is meta to the hydroxyl group and so does not stabilize the resulting anion as much as does an ortho or a para nitro group.

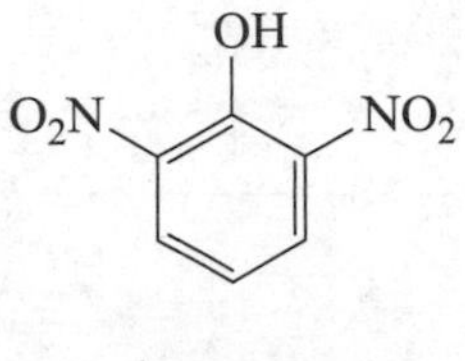

2,6-Dinitrophenol,
more acidic
($pK_a = 3.7$)

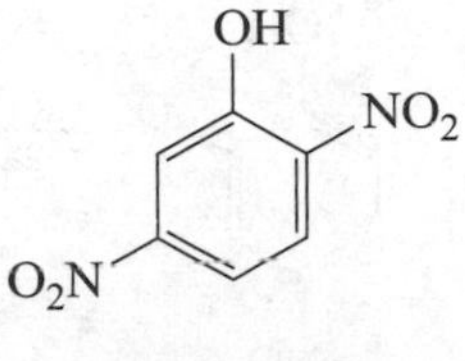

2,5-Dinitrophenol,
less acidic
($pK_a = 5.2$)

23.16 (*a*) Allyl bromide is a reactive alkylating agent and converts the free hydroxyl group of the aryl compound (a natural product known as *guaiacol*) to its corresponding allyl ether.

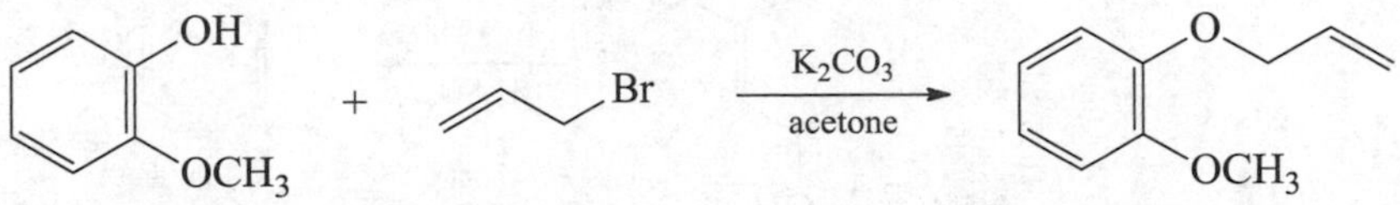

Guaiacol   Allyl bromide   2-Allyloxyanisole (80-90%)

    $\xrightarrow[\text{acetone}]{K_2CO_3}$

(*b*) Sodium phenoxide acts as a nucleophile in this reaction and is converted to an ether.

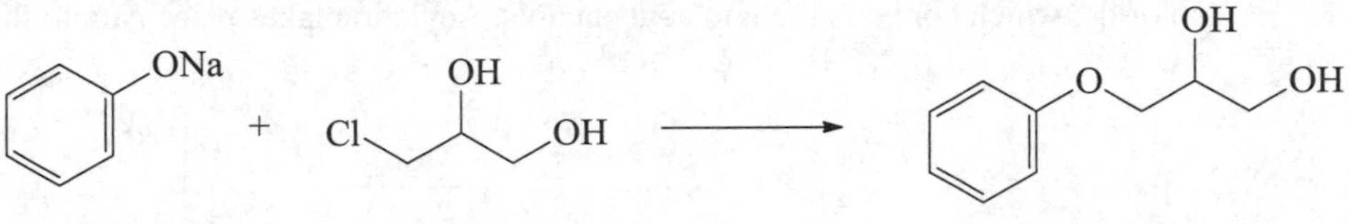

Sodium
phenoxide   3-Chloro-1,2-propanediol   3-Phenoxy-1,2-propanediol
(61-63%)

(*c*) Orientation in nitration is governed by the most activating substituent, in this case the hydroxyl group.

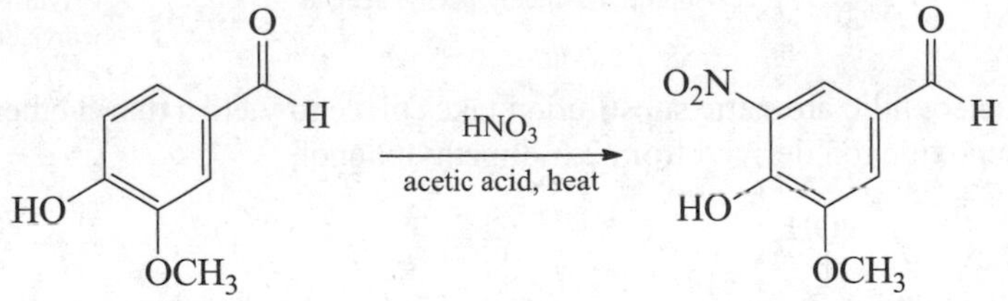

Vanillin $\xrightarrow[\text{acetic acid, heat}]{HNO_3}$ 4-Hydroxy-3-methoxy-5-
nitrobenzaldehyde (83%)

(*d*) Allyl aryl ethers undergo a Claisen rearrangement on heating. Heating *p*-acetamidophenyl allyl ether gave an 83% yield of 4-acetamido-2-allylphenol.

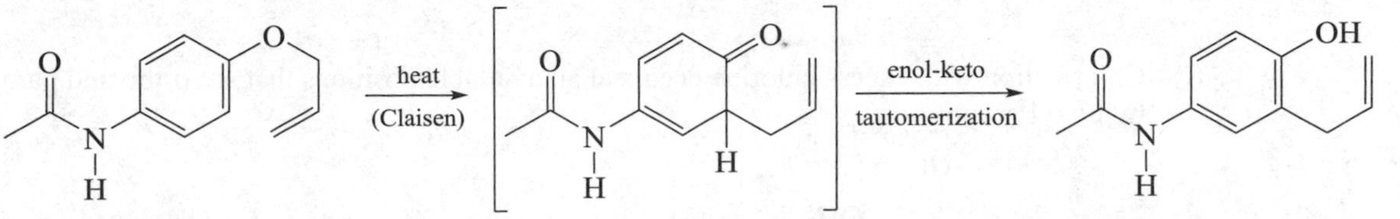

4-Acetamidophenyl allyl ether $\xrightarrow[\text{(Claisen)}]{\text{heat}}$ $\xrightarrow[\text{tautomerization}]{\text{enol-keto}}$ 4-Acetamidophenyl-2-allylphenol

(*e*) The hydroxyl group, as the most activating substituent, controls the orientation of electrophilic aromatic substitution. Bromination takes place ortho to the hydroxyl group.

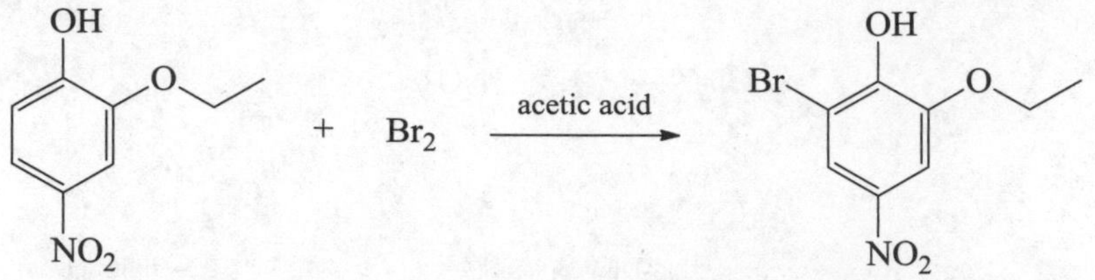

2-Ethoxy-4-nitrophenol   2-Bromo-6-ethoxy-4-nitrophenol (65%)

(*f*) Oxidation of hydroquinone derivatives (*p*-dihydroxybenzenes) with Cr(VI) reagents is a method for preparing quinones.

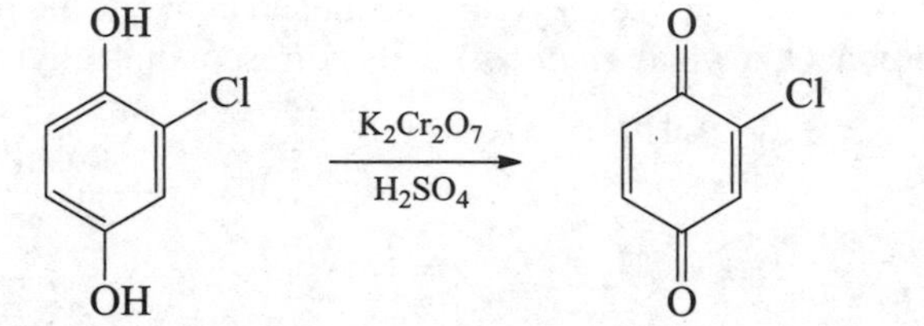

2-Chloro-1,4-benzenediol   2-Chloro-1,4-benzoquinone (88%)

(*g*) Aryl esters undergo a reaction known as the *Fries rearrangement* on being treated with aluminum chloride, which converts them to acyl phenols. Acylation takes place para to the hydroxyl in this case.

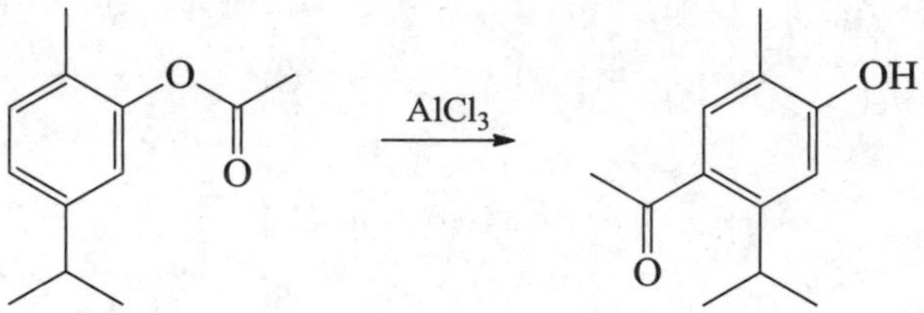

5-Isopropyl-2-methylphenyl acetate   4-Hydroxy-2-isopropyl-
5-methylacetophenone (90%)

(*h*) Nucleophilic aromatic substitution takes place to yield a diaryl ether. The nucleophile is the phenoxide ion derived from 2,6-dimethylphenol.

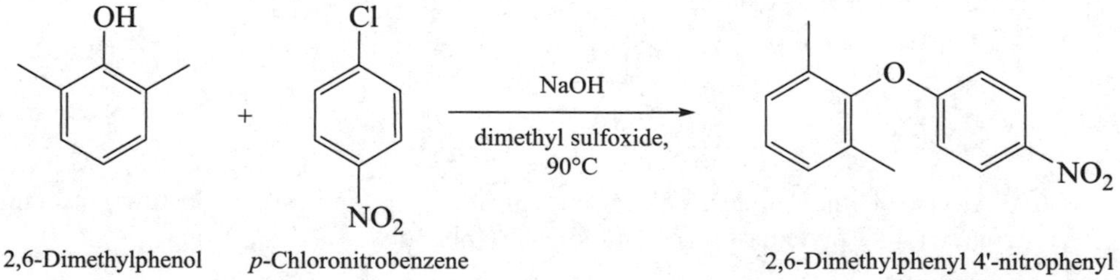

2,6-Dimethylphenol   *p*-Chloronitrobenzene   2,6-Dimethylphenyl 4'-nitrophenyl
ether (82%)

(*i*) Chlorination with excess chlorine occurs at all available positions that are ortho and para to the hydroxyl group.

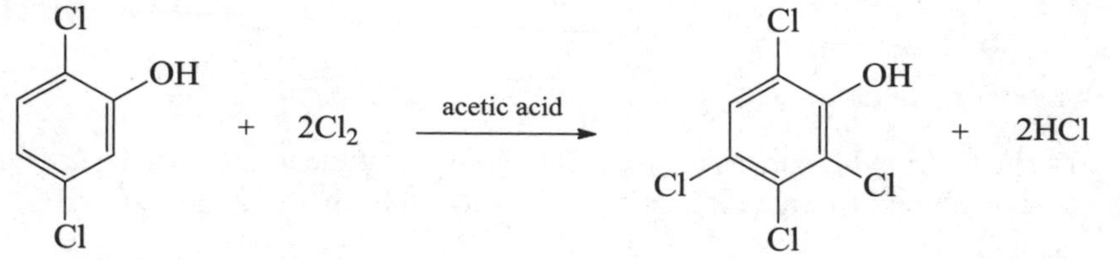

2,5-Dichlorophenol   Chlorine   2,3,4,6-Tetrachlorophenol   Hydrogen
(isolated yield, 100%)   chloride

(*j*) Amines react with esters to give amides. In the case of a phenyl ester, phenol is the leaving group.

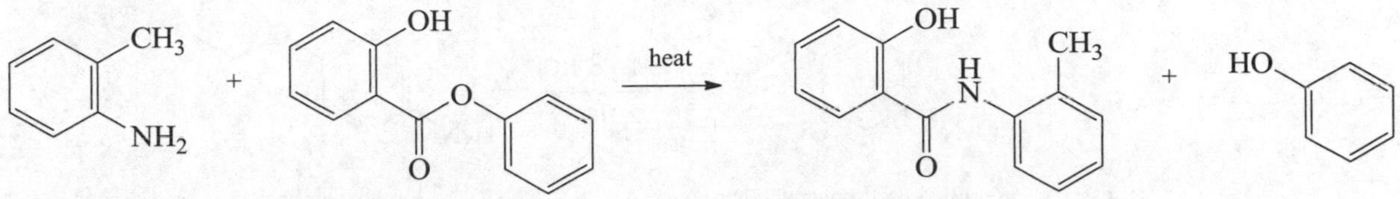

| *o*-Methylaniline | Phenyl salicylate | *N*-(*o*-Methylphenyl)salicylamide (isolated yield, 73-77%) | Phenol |

(*k*) Aryl diazonium salts attack electron-rich aromatic rings, such as those of phenols, to give the products of electrophilic aromatic substitution.

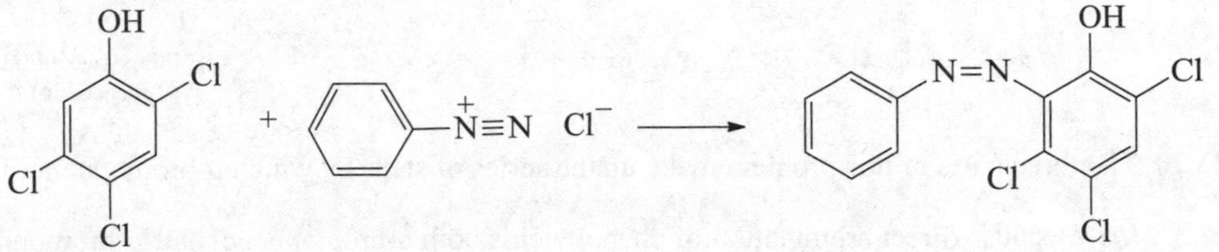

| 2,4,5-Trichlorophenol | Benzenediazonium chloride | 2-Benzeneazo-3,4,6-trichlorophenol (80%) |

23.17 Epoxides are sensitive to nucleophilic ring-opening reactions. Phenoxide ion attacks the less hindered carbon to yield 1-phenoxy-2-propanol.

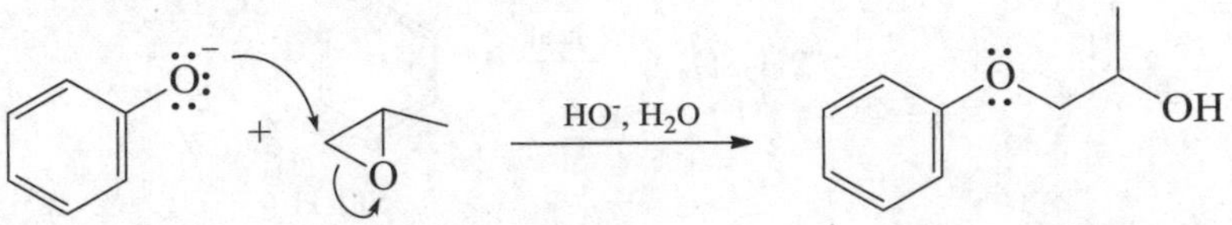

| Phenoxide ion | 1,2-Epoxypropane | 1-Phenoxy-2-propanol |

23.18 Compound A is the ester that is formed by *O*-acylation. It rearranges to the more stable C-acyl product in a process known as the Fries rearrangement.

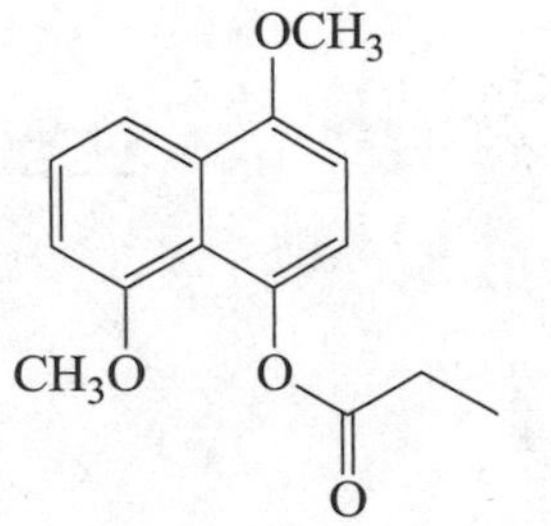

Compound A

Synthesis

23.19 In the first step, *p*-nitrophenol is alkylated on its phenolic oxygen with ethyl bromide.

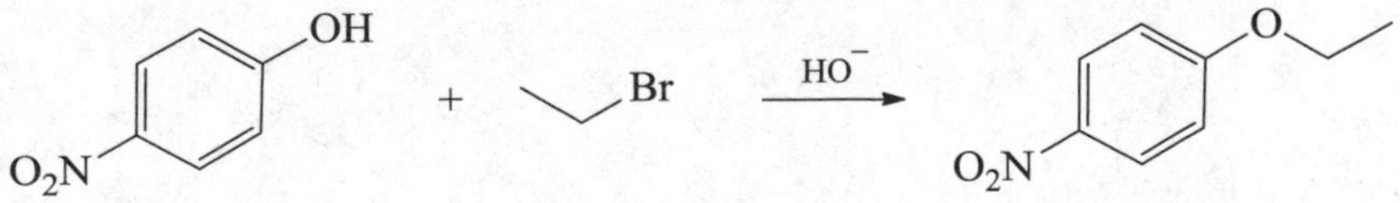

| *p*-Nitrophenol | Ethyl bromide | Ethyl *p*-nitrophenyl ether |

Reduction of the nitro group gives the corresponding arylamine.

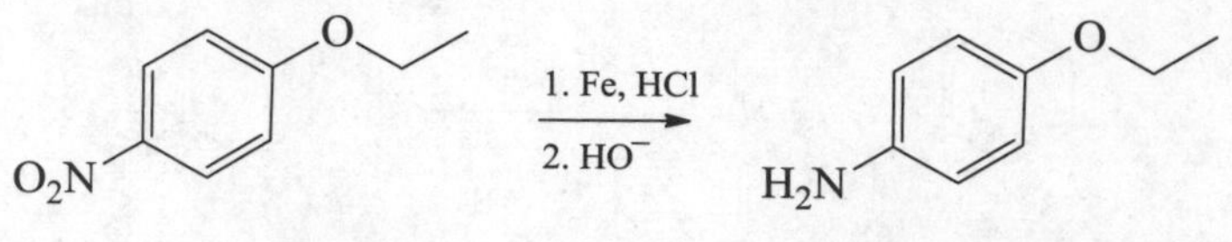

Ethyl *p*-nitrophenyl ether   *p*-Ethoxyaniline

Treatment of *p*-ethoxyaniline with acetic anhydride gives phenacetin.

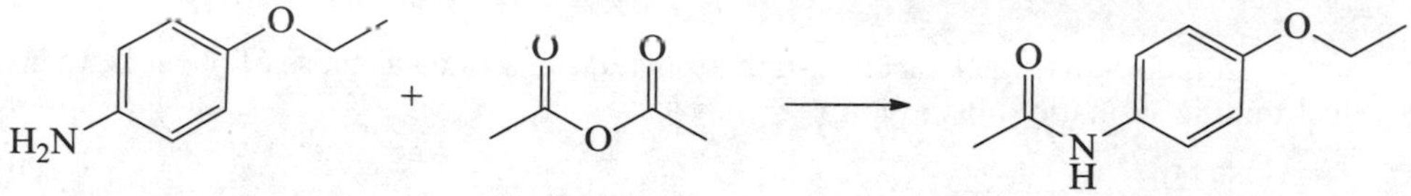

p-Ethoxyaniline   Acetic anhydride   *p*-Ethoxyacetanilide
                                                                (phenacetin)

23.20 The three parts of this problem make up the series of steps by which *o*-bromophenol is prepared.

(*a*) Because direct bromination of phenol yields both *o*-bromophenol and *p*-bromophenol, it is essential that the para position be blocked prior to the bromination step. In practice, what is done is to disulfonate phenol, which blocks the para and one of the ortho positions.

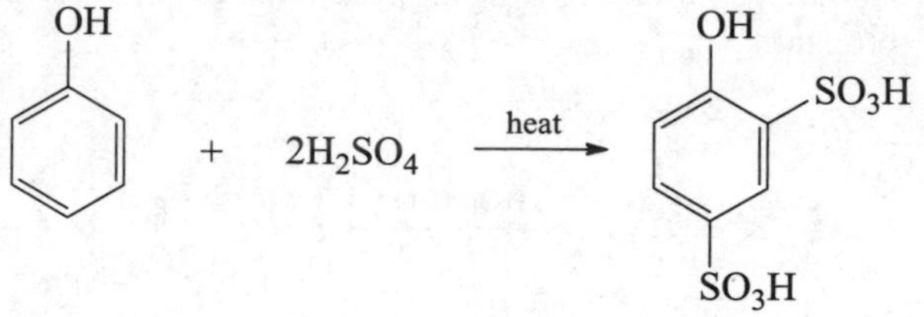

Phenol   4-Hydroxy-1,3-
                                                benzenedisulfonic
                                                acid (compound A)

(*b*) Bromination then can be accomplished cleanly at the open position ortho to the hydroxyl group.

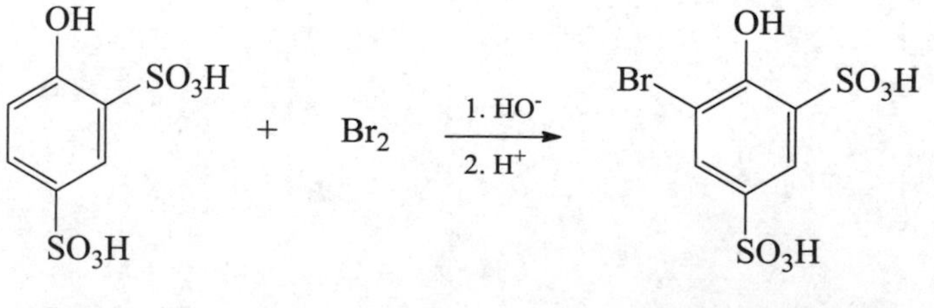

Compound A   5-Bromo-4-hydroxy-1,3-
                                        benzenedisulfonic
                                        acid (compound B)

(*c*) After bromination, the sulfonic acid groups are removed by acid-catalyzed hydrolysis.

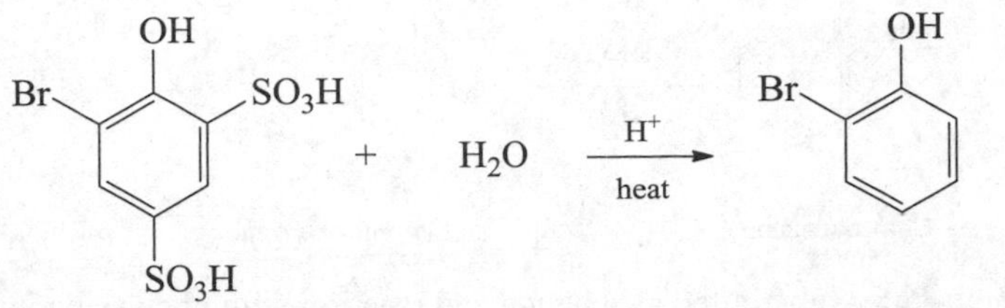

Compound B

o-Bromophenol
(compound C)

23.21 Nitration of 3,5-dimethylphenol gives a mixture of the 2-nitro and 4-nitro derivatives.

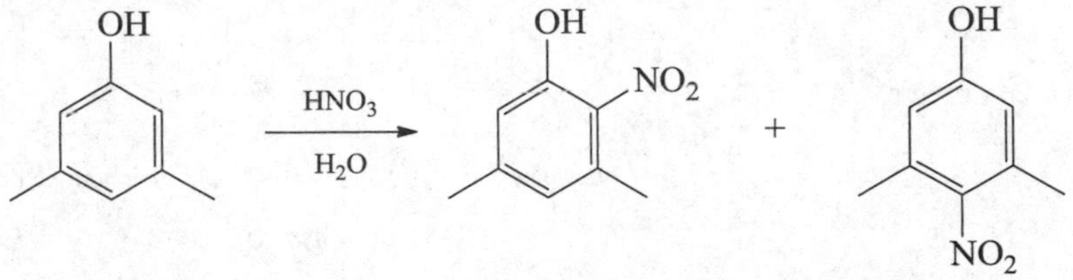

3,5-Dimethylphenol   3,5-Dimethyl-2-nitrophenol   3,5-Dimethyl-4-nitrophenol

The more volatile compound (compound A), isolated by steam distillation, is the 2-nitro derivative. Intramolecular hydrogen bonding is possible between the nitro group and the hydroxyl group.

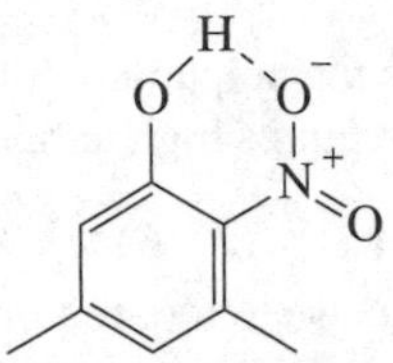

Intramolecular hydrogen bonding
in 3,5-dimethyl-2-nitrophenol

The 4-nitro derivative participates in intermolecular hydrogen bonds and has a much higher boiling point; it is compound B.

23.22 The relationship between the target molecule and the starting materials tells us that two processes are required, formation of a diaryl ether linkage and nitration of an aromatic ring. The proper order of carrying out these two separate processes is what needs to be considered.

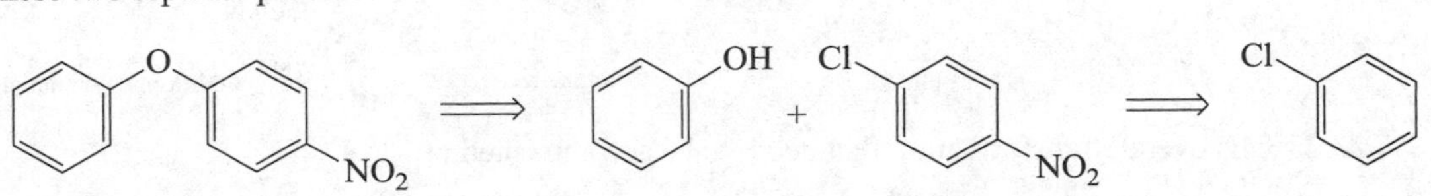

The critical step is ether formation, a step that is feasible for the reactants shown:

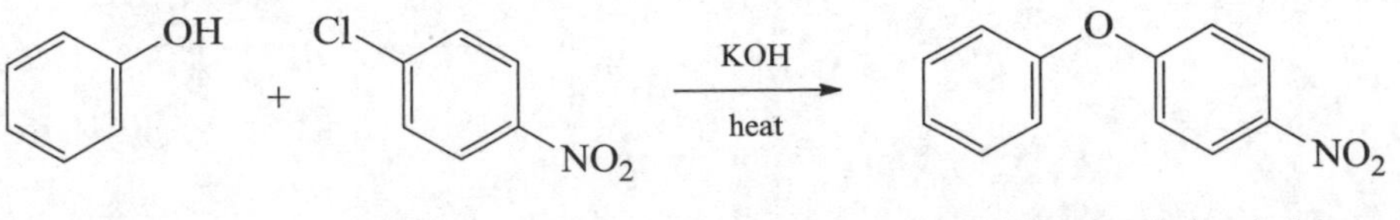

Phenol   p-Chloronitrobenzene   4-Nitrophenyl phenyl ether

The reason this reaction is suitable is that it involves nucleophilic aromatic substitution by the addition–elimination mechanism on a p-nitro-substituted aryl halide. Indeed, this reaction has been carried out and gives an 80–82% yield. A reasonable synthesis would therefore begin with the preparation of p-chloronitrobenzene.

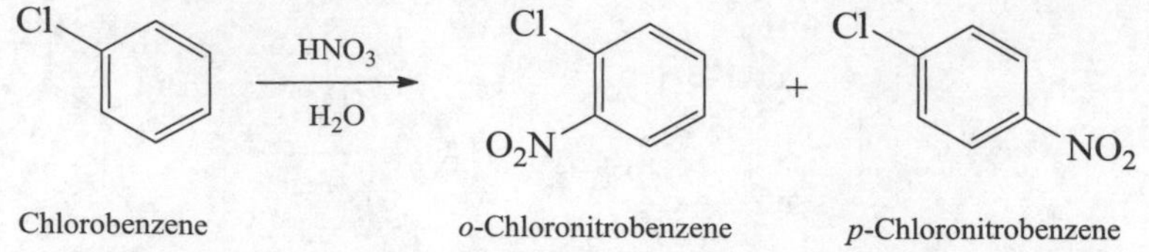

Chlorobenzene   *o*-Chloronitrobenzene   *p*-Chloronitrobenzene

Separation of the *p*-nitro-substituted aryl halide and reaction with phenoxide ion complete the synthesis.

The following alternative route is less satisfactory:

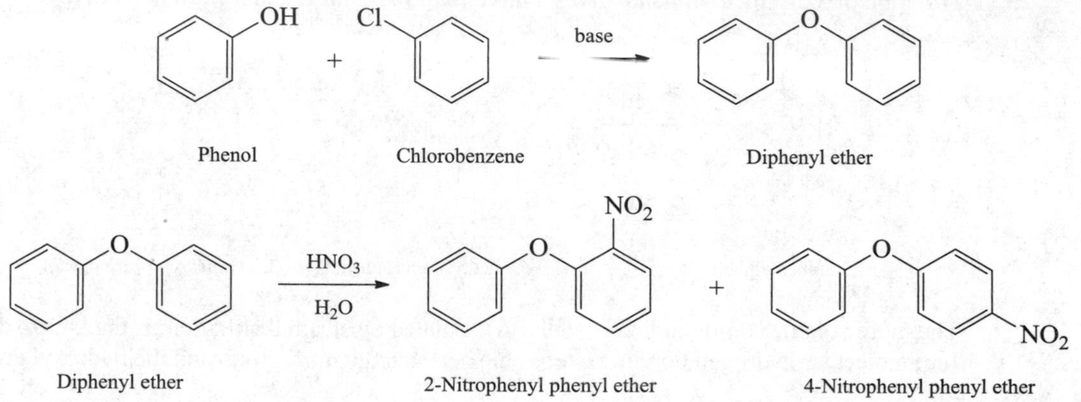

Phenol   Chlorobenzene   Diphenyl ether

Diphenyl ether   2-Nitrophenyl phenyl ether   4-Nitrophenyl phenyl ether

The difficulty with this route concerns the preparation of diphenyl ether. Direct reaction of phenoxide ion with chlorobenzene is very slow and requires high temperatures because chlorobenzene is a poor substrate for nucleophilic substitution.

A third route is also unsatisfactory because it, too, requires nucleophilic substitution on chlorobenzene.

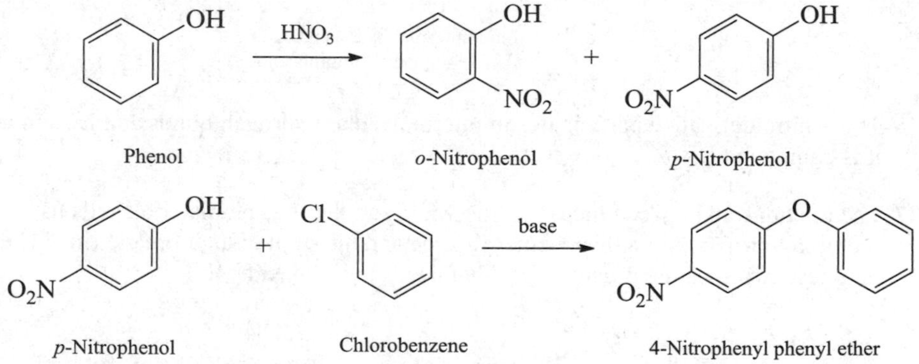

Phenol   *o*-Nitrophenol   *p*-Nitrophenol

p-Nitrophenol   Chlorobenzene   4-Nitrophenyl phenyl ether

23.23 The overall transformation that needs to be accomplished is

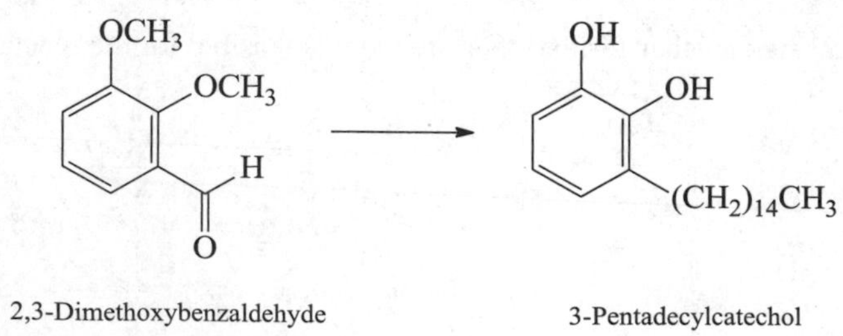

2,3-Dimethoxybenzaldehyde   3-Pentadecylcatechol

A reasonable place to begin is with the attachment of the side chain. The aldehyde function allows for chain extension by a Wittig reaction.

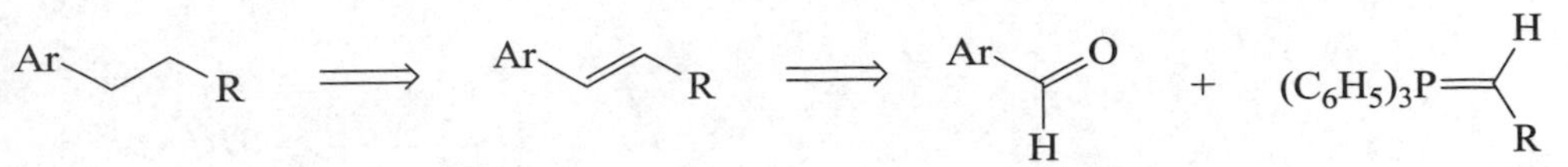

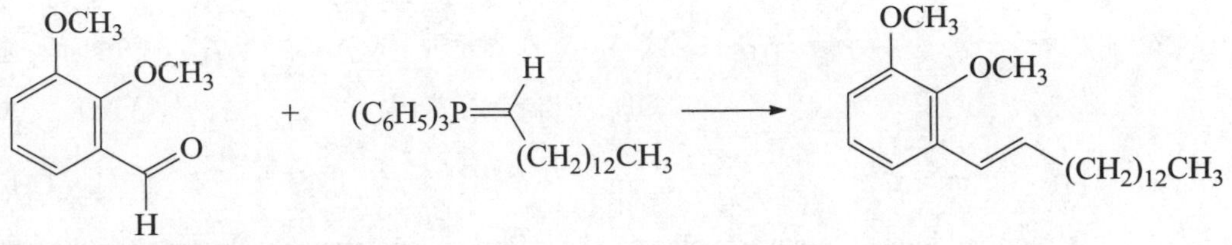

2,3-Dimethoxybenzaldehyde

Hydrogenation of the double bond and hydrogen halide cleavage of the ether functions complete the synthesis.

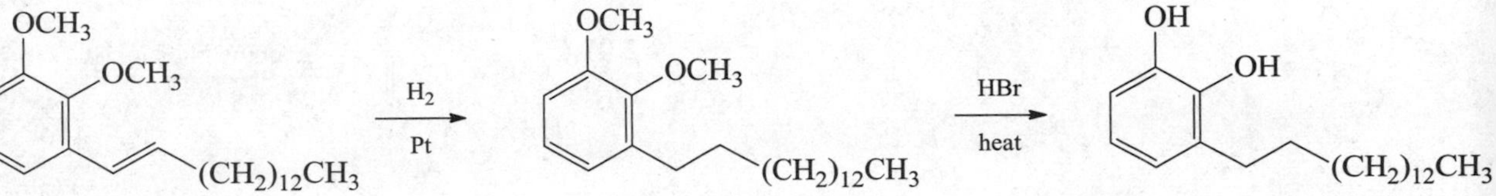

Other synthetic routes are of course possible. One of the earliest approaches used a Grignard reaction to attach the side chain.

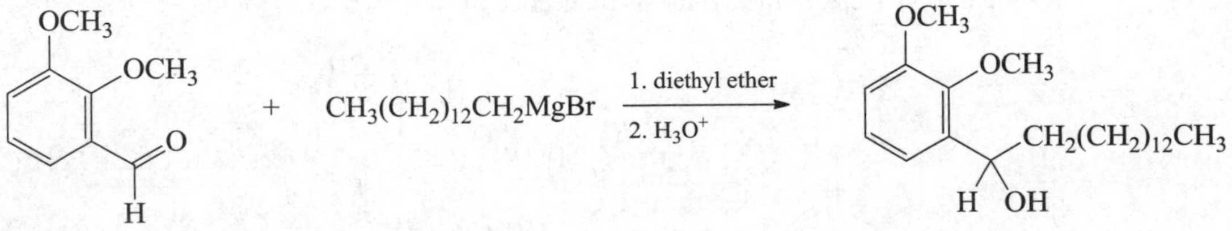

2,3-Dimethoxybenzaldehyde

The resulting secondary alcohol can then be dehydrated to the same alkene intermediate prepared in the preceding synthetic scheme.

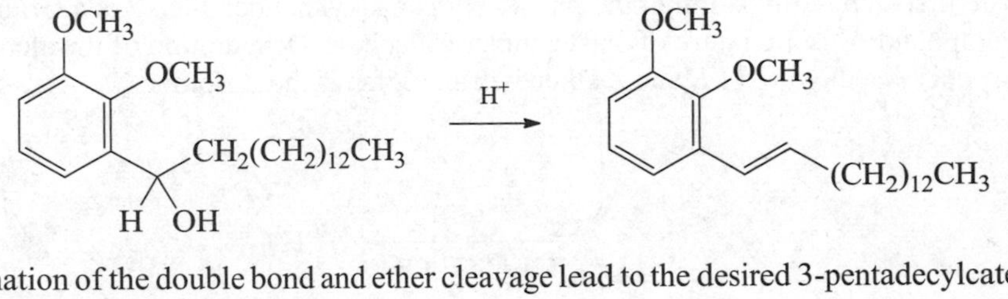

Again, hydrogenation of the double bond and ether cleavage lead to the desired 3-pentadecylcatechol.

23.24 Recall that the Claisen rearrangement converts an aryl allyl ether to an ortho-substituted allyl phenol. The presence of an allyl substituent in the product ortho to an aryl ether thus suggests the following retrosynthesis:

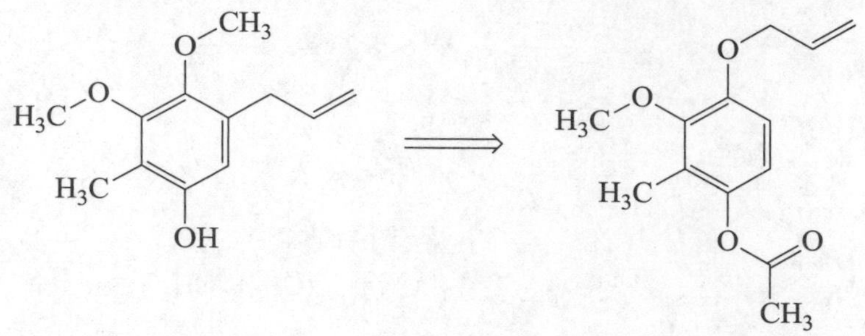

As reported in the literature synthesis, the starting phenol may be converted to the corresponding allyl ether by reaction with allyl bromide in the presence of base. This step was accomplished in 80% yield. Heating the allyl ether yields the *o*-allyl phenol.

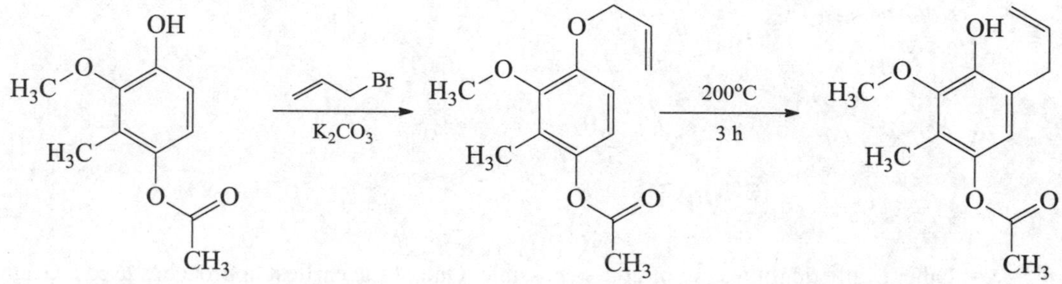

The synthesis is completed by methylation of the phenolic oxygen and saponification of the acetate ester. The final three steps of the synthesis proceeded in an 82% overall yield.

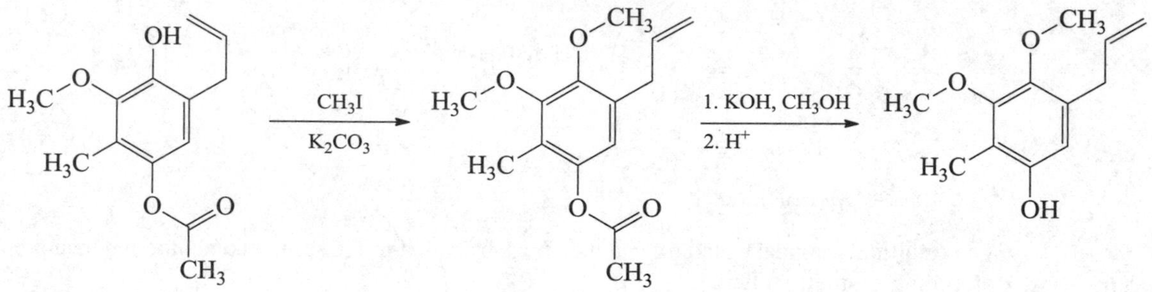

23.25 The first step is alkylation of the phenol with an alkyl halide. Step 2 is a Grignard reaction, and the product, compound A, is a mixture of diastereomeric alcohols. Dehydration of the alcohol gives Tamoxifen. The product is a mixture of *E* and *Z* alkenes. Tamoxifen is the *Z* isomer.

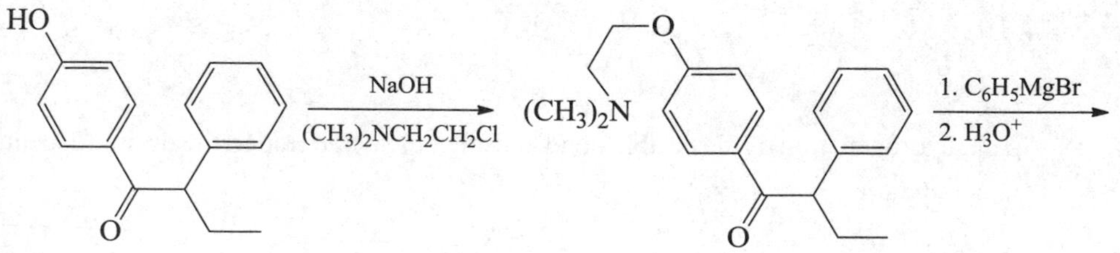

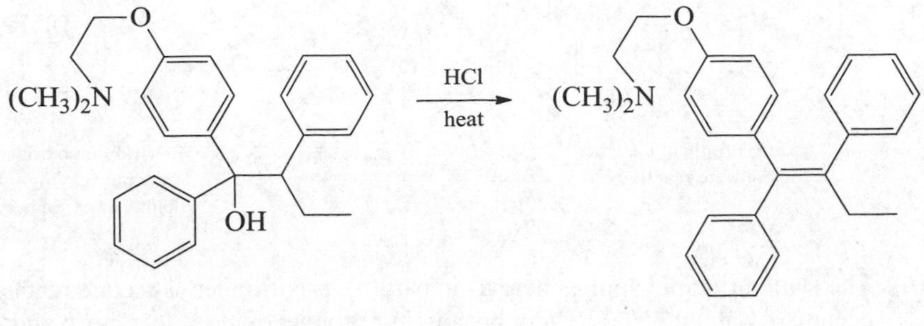

Compound A   Tamoxifen

23.26 The first step is a Baeyer–Villiger reaction using trifluoroacetic acid. Migration of the phenyl group occurs preferentially to give the ester of acetic acid. Hydrolysis of the ester gives *p*-fluorophenol.

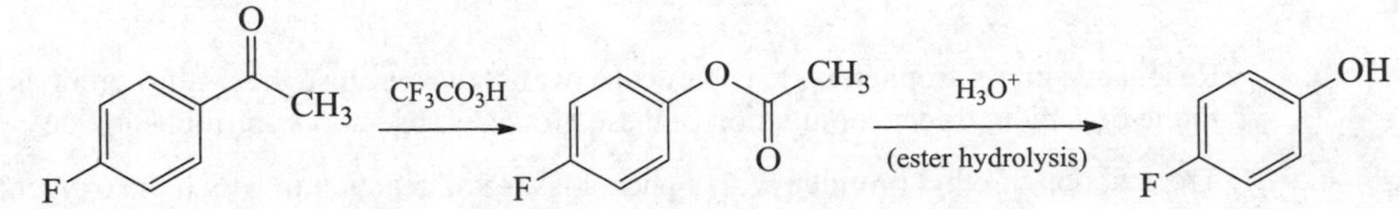

p-Fluorophenol

Note: Migration of the methyl group in the Baeyer–Villiger oxidation would give a different ester. This is formed only in trace amounts under the conditions that were used.

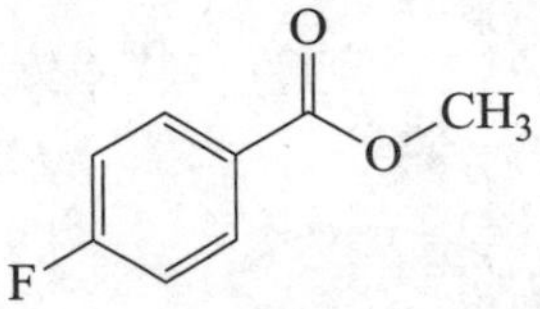

Rate and Mechanism

23.27 (*a*) The rate-determining step of ester hydrolysis in basic solution is formation of the tetrahedral intermediate.

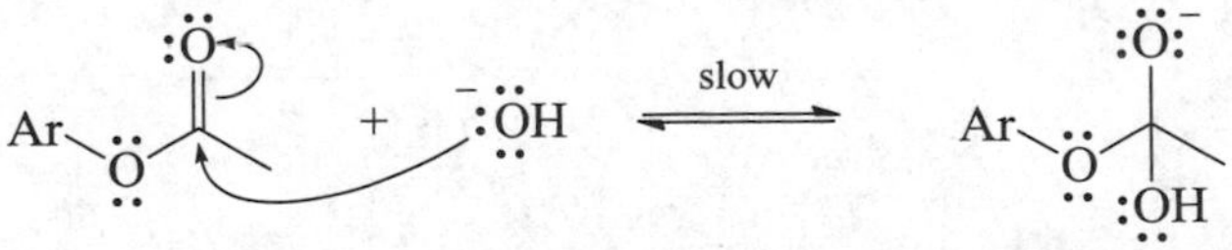

Because this intermediate is negatively charged, there will be a small effect favoring its formation when the aryl group bears an electron-withdrawing substituent. Furthermore, this intermediate can either return to starting materials or proceed to products.

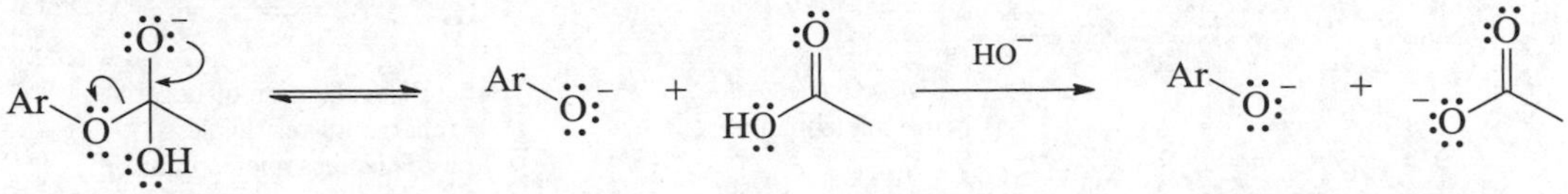

The proportion of the tetrahedral intermediate that goes on to products increases as the leaving group ArO$^-$ becomes less basic. This is strongly affected by substituents; electron-withdrawing groups stabilize ArO$^-$. The prediction is that *m*-nitrophenyl acetate undergoes hydrolysis in basic solution faster than phenol. Indeed, this is observed to be the case; *m*-nitrophenyl acetate reacts some ten times faster than does phenyl acetate at 25°C.

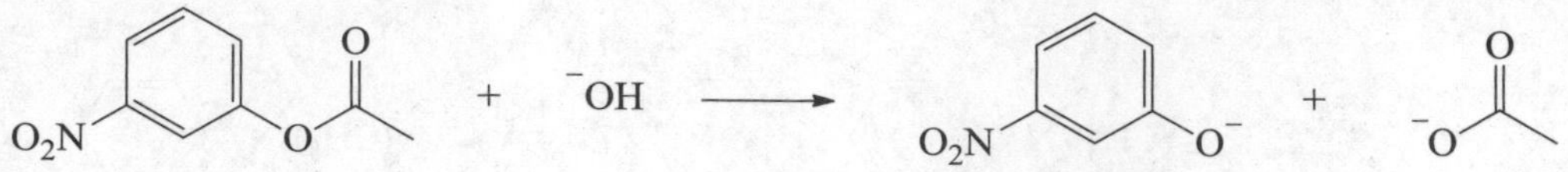

m-Nitrophenyl acetate
(more reactive)

m-Nitrophenoxide anion
(a better leaving group
than phenoxide because
it is less basic)

(*b*) The same principle applies here as in part (*a*). *p*-Nitrophenyl acetate reacts faster than *m*-nitrophenyl acetate (by about 45%) largely because *p*-nitrophenoxide is less basic and thus a better leaving group than *m*-nitrophenoxide.

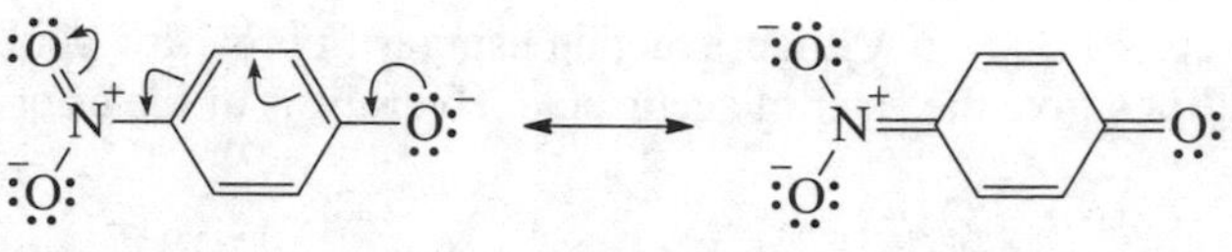

Resonance in *p*-nitrophenoxide is particularly effective because the *p*-nitro group is directly conjugated to the oxyanion; direct conjugation of these groups is absent in *m*-nitrophenoxide.

(*c*) The reaction of ethyl bromide with a phenol is an S_N2 reaction in which the oxygen of the phenol is the nucleophile. The reaction is much faster with sodium phenoxide than with phenol because an anion is more nucleophilic than a corresponding neutral molecule.

Faster reaction:

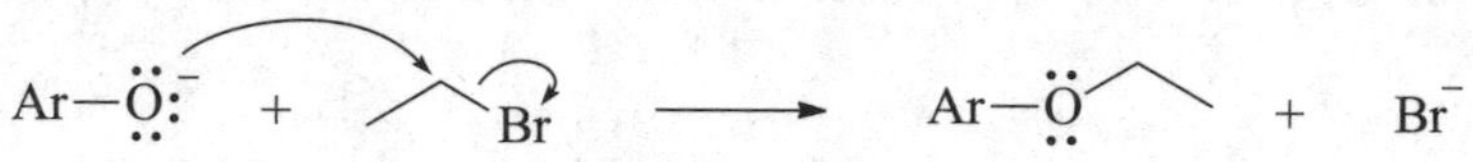

Slower reaction:

(*d*) The answer here also depends on the nucleophilicity of the attacking species, which is a phenoxide anion in both reactions.

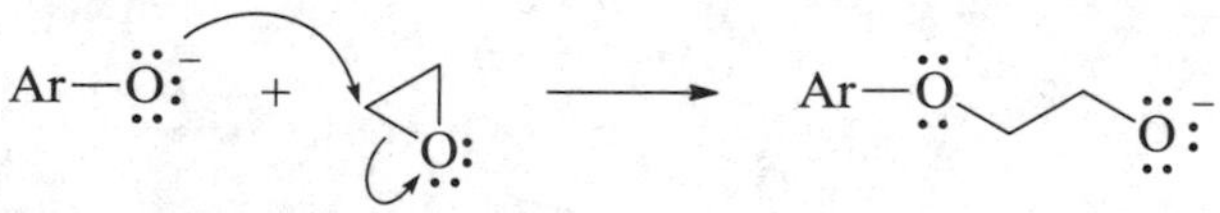

The more nucleophilic anion is phenoxide ion, because it is more basic than *p*-nitrophenoxide.

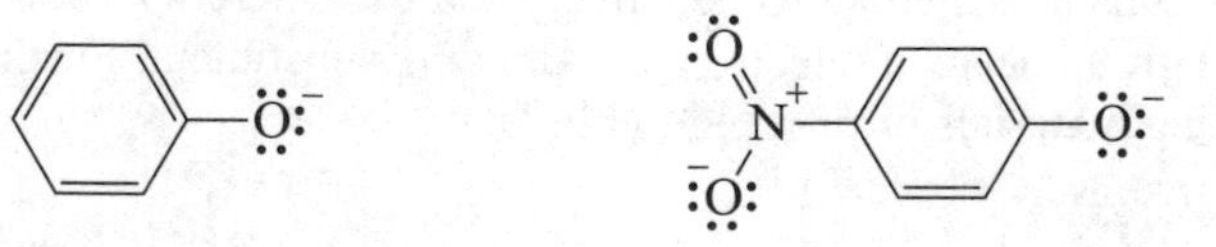

More basic
better nucleophile

Better delocalization of negative
charge makes this less
basic and less nucleophilic

Rate measurements reveal that sodium phenoxide reacts 17 times faster with ethylene oxide (in ethanol at 70°C) than does its *p*-nitro derivative.

(*e*) This reaction is electrophilic aromatic substitution. Because a hydroxy substituent is more activating than an acetate group, phenol undergoes bromination faster than does phenyl acetate.

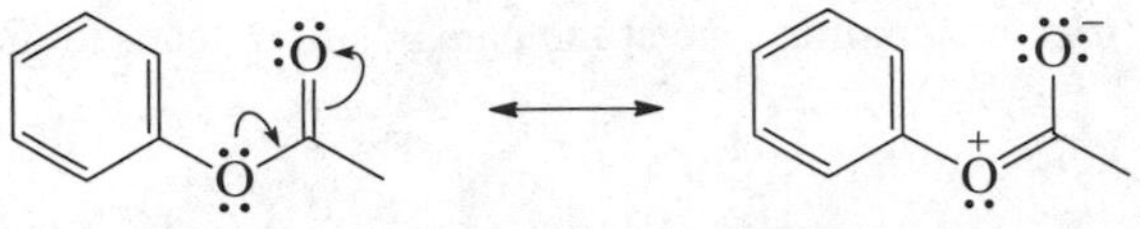

Resonance involving ester group reduces
tendency of oxygen to donate electrons to ring.

23.28 Nucleophilic aromatic substitution by the elimination–addition mechanism is impossible, owing to the absence of any protons that might be abstracted from the substrate. The addition–elimination pathway is available, however.

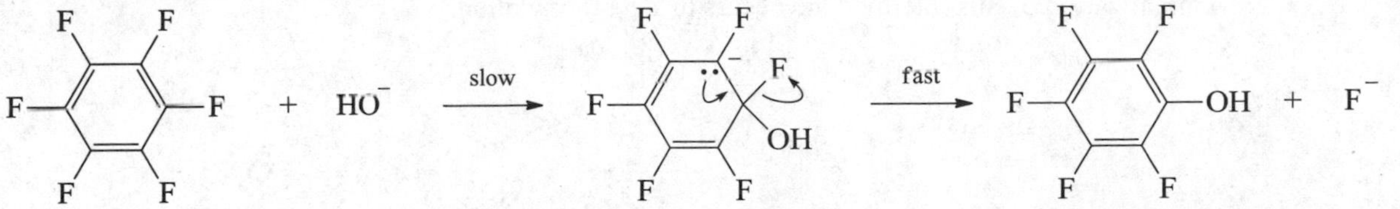

Hexafluorobenzene   Pentafluorophenol

This pathway is favorable because the cyclohexadienyl anion intermediate formed in the rate-determining step is stabilized by the electron-withdrawing inductive effect of its fluorine substituents.

23.29 The driving force for this reaction is the stabilization that results from formation of the aromatic ring. A reasonable series of steps begins with protonation of the carbonyl oxygen.

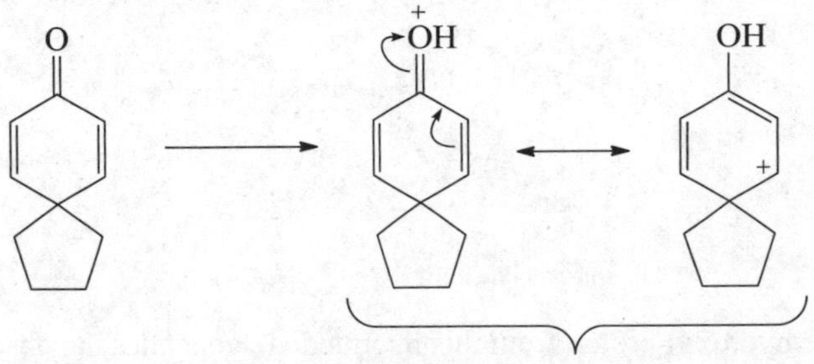

Resonance forms of protonated ketone

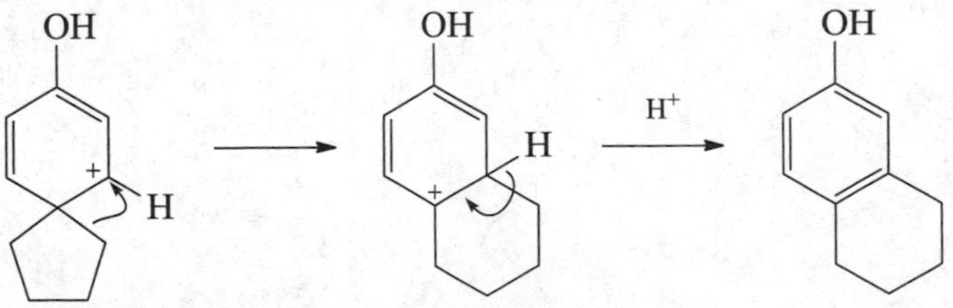

Protonated ketone
can rearrange by
alkyl migration

Aromatization of this
intermediate occurs
by loss of a proton

23.30 Bromination of *p*-hydroxybenzoic acid takes place in the normal fashion at both positions ortho to the hydroxy group.

p-Hydroxybenzoic acid   3,5-Dibromo-4-
hydroxybenzoic acid

A third bromination step, this time at the para position, leads to the intermediate shown.

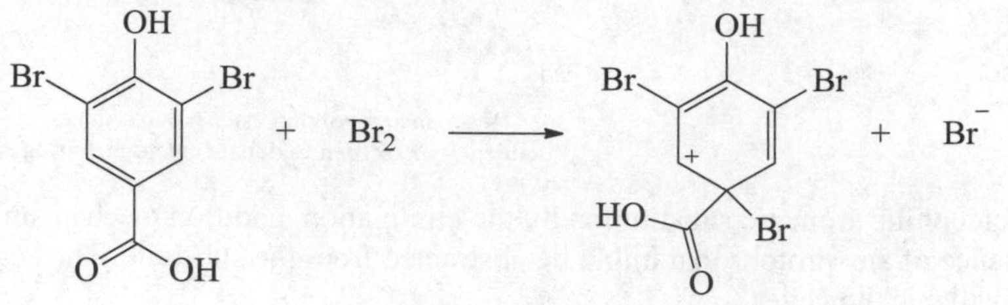

Aromatization of this intermediate occurs by decarboxylation.

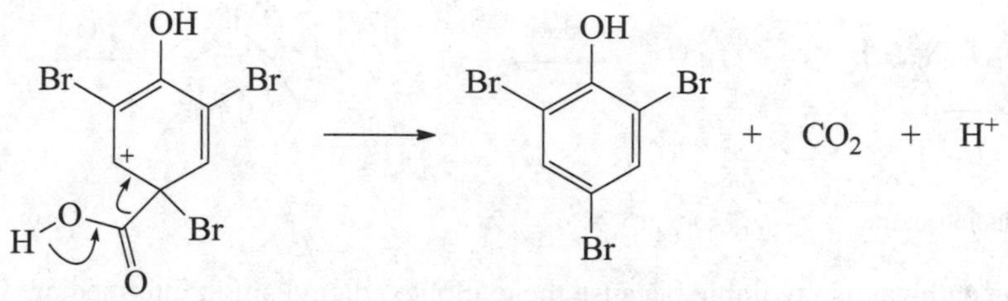

2,4,6-Tribromophenol

23.31 Electrophilic attack of bromine on 2,4,6-tribromophenol leads to a cationic intermediate.

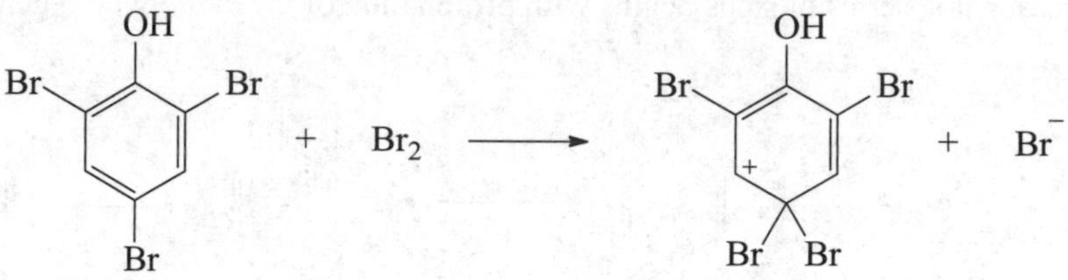

2,4,6-Tribromophenol

Loss of the hydroxyl proton from this intermediate generates the observed product.

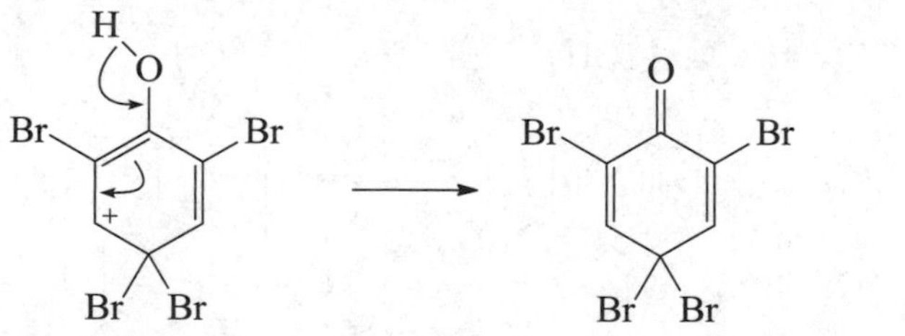

2,4,4,6-Tetrabromocyclohexadienone

23.32 A reasonable first step is protonation of the hydroxyl oxygen.

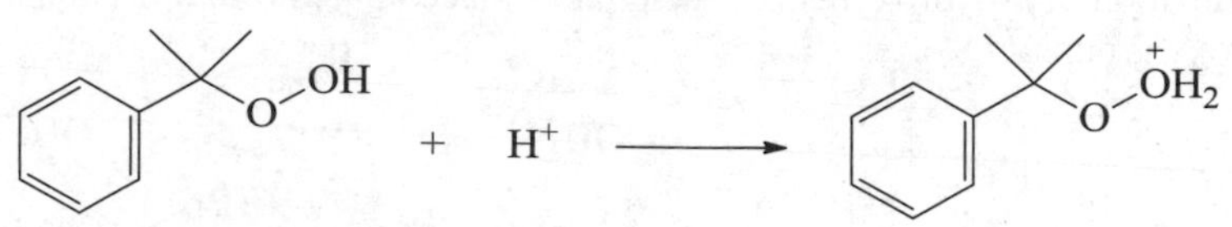

Cumene hydroperoxide

The weak oxygen–oxygen bond can now be cleaved, with loss of water as the leaving group.

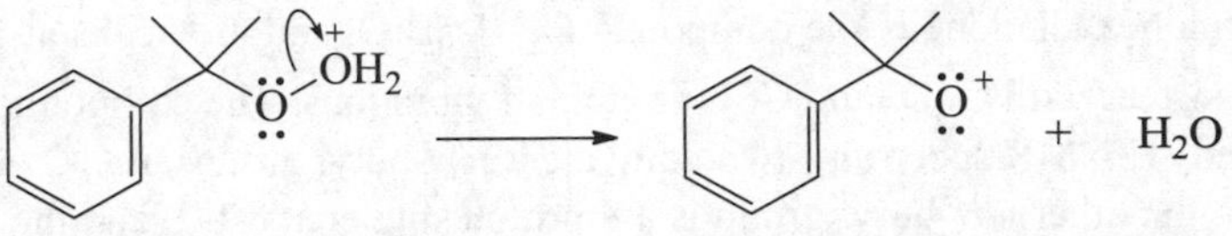

This intermediate bears a positively charged oxygen with only six electrons in its valence shell. Like a carbocation, such a species is highly electrophilic. The electrophilic oxygen attacks the π system of the neighboring aromatic ring to give an unstable intermediate.

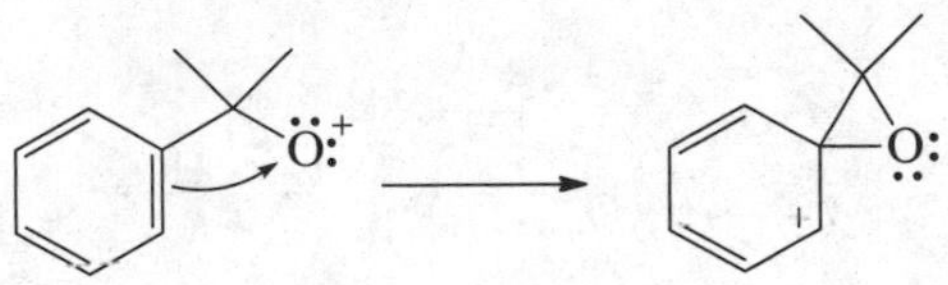

Ring opening of this intermediate is assisted by one of the lone pairs of oxygen and restores the aromaticity of the ring.

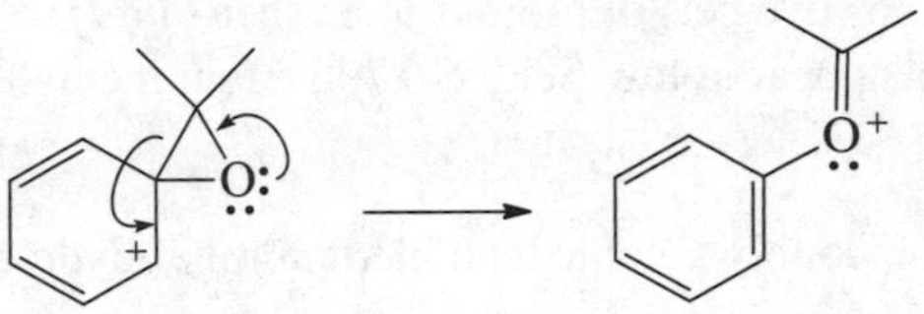

The cation formed by ring opening is captured by a water molecule to yield the hemiacetal product.

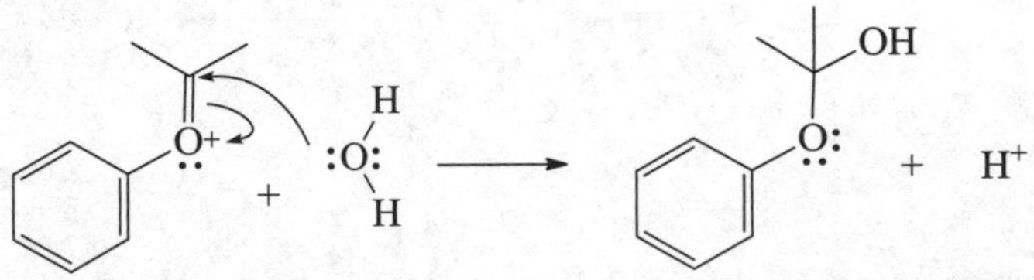

Spectroscopy

23.33 A good way to approach this problem is to assume that bromine attacks the aromatic ring of the phenol in the usual way, that is, para to the hydroxyl group.

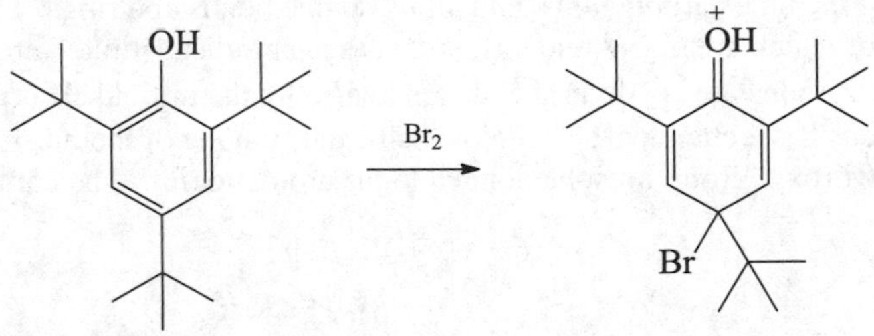

2,4,6-Tri-*tert*-butylphenol

This cation cannot yield the product of electrophilic aromatic substitution by loss of a proton from the ring but can lose a proton from oxygen to give a cyclohexadienone derivative.

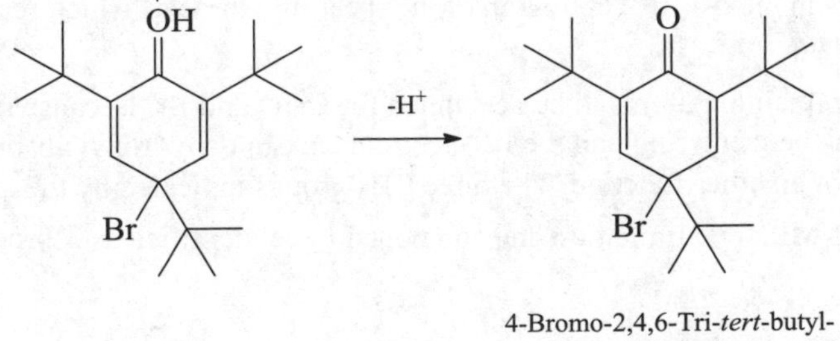

4-Bromo-2,4,6-Tri-*tert*-butyl-
2,5-cyclohexadienone

This cyclohexadienone is the compound $C_{18}H_{29}BrO$, and the peaks at 1655 and 1630 cm^{-1} in the infrared are consistent with C=O and C=C stretching vibrations. The compound's symmetry is consistent with the observed ^{1}H NMR spectrum; two equivalent *tert*-butyl groups at C-2 and C-6 appear as an 18-proton singlet at δ 1.3, the other *tert*-butyl group is a 9-proton singlet at δ 1.2, and the two equivalent vinyl protons of the ring appear as a singlet at δ 6.9.

23.34 Because the starting material is an acetal and the reaction conditions lead to hydrolysis with the production of 1,2-ethanediol, a reasonable reaction course is

Compound A   1,2-Ethanediol   Compound B

Indeed, dione B satisfies the spectroscopic criteria. Carbonyl bands are seen in the infrared spectrum, and compound B has two sets of protons to be seen in its ^{1}H NMR spectrum. The two vinyl protons are equivalent and appear at low field, δ 6.7; the four methylene protons are equivalent to each other and are seen at δ 2.9.

Compound B is the doubly ketonic tautomeric form of hydroquinone, compound C, to which it isomerizes on standing in water.

Compound B   Compound C
(hydroquinone)

23.35 (*a*) The molecular formula of the compound ($C_9H_{12}O$) tells us that it has a total of four double bonds and rings (index of hydrogen deficiency = 4). The prominent peak in the IR spectrum is the hydroxyl absorption of an alcohol or a phenol at 3300 cm^{-1}.

Peaks in the δ 110–160 region of the ^{13}C NMR spectrum suggest an aromatic ring, which accounts for six of the nine carbon atoms and all its double bonds and rings. The presence of four peaks in this region, two of which are C and two CH, indicates a para-disubstituted aromatic derivative. That the remaining three carbons are sp^3-hybridized is indicated by the upfield absorptions at δ 15, 26, and 38. None of these carbons has a chemical shift below δ 40, and so none of them can be bonded to the hydroxyl group. Thus, the hydroxyl group must be bonded to the aromatic ring. The compound is 4-propylphenol.

4-Propylphenol

(*b*) Once again the molecular formula ($C_9H_{11}BrO$) indicates a total of four double bonds and rings. The four peaks in the δ 110–160 region of the spectrum, three of which represent CH, suggest a monosubstituted aromatic ring.

The remaining atoms to be accounted for are O and Br. Because all the unsaturations are accounted for by the benzene ring and the IR spectrum lacks any hydroxyl absorption, the oxygen atom must be part of an ether function. The three CH$_2$ groups indicated by the absorptions at δ 32, 35, and 66 in the ^{13}C NMR spectrum allow the compound to be identified as 3-bromopropyl phenyl ether.

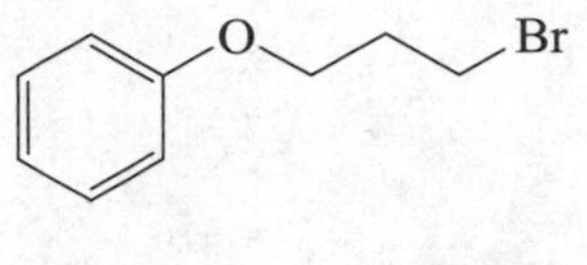

3-Bromopropyl phenyl ether

Answers to Interpretive Problems 23

23.36 A; **23.37** B; **23.38** A; **23.39** C

SELF TEST

1. Which is the stronger acid, *m*-hydroxybenzaldehyde or *p*-hydroxybenzaldehyde? Explain your answer, using resonance structures.

2. The cresols are methyl-substituted phenols. Predict the major products to be obtained from the reactions of *o*-, *m*-, and *p*-cresol with dilute nitric acid.

3. Give the structure of the product from the reaction of *p*-cresol with propanoyl chloride, $CH_3CH_2\overset{O}{\overset{\|}{C}}Cl$, in the presence of $AlCl_3$. What product is obtained in the absence of $AlCl_3$?

4. Provide the structure of the reactant, reagent, or product omitted from each of the following:

(a)

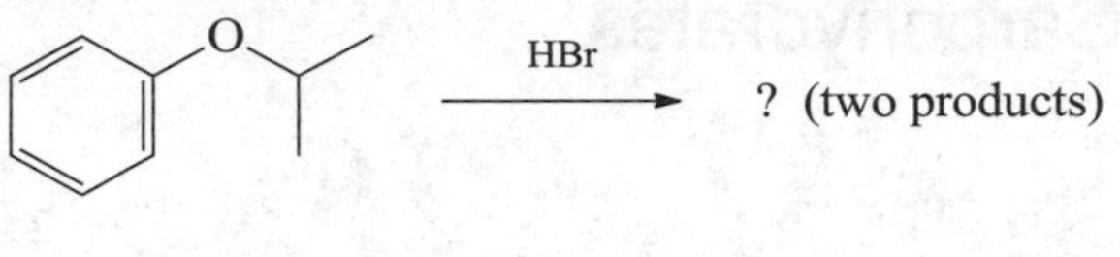

(b)

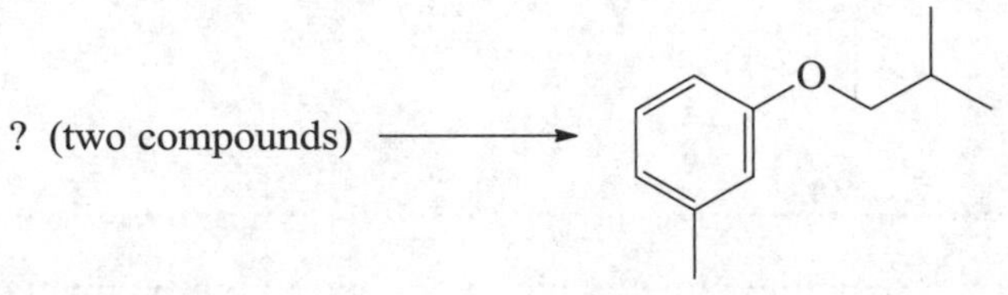

(c)

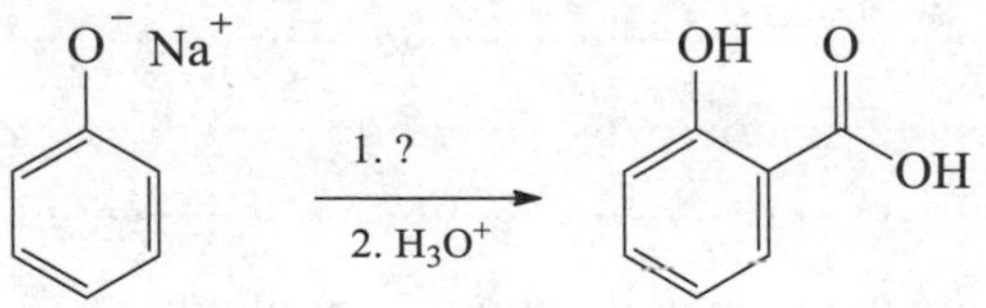

(d)

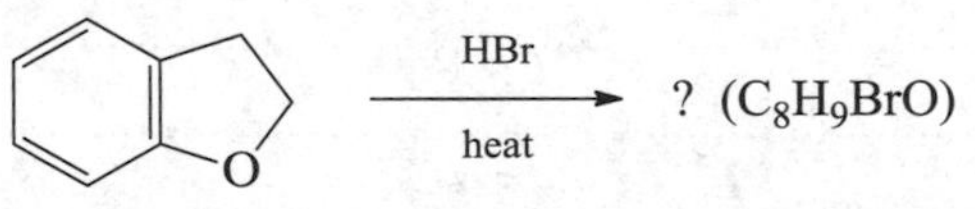

5. Provide the structures of compounds A and B in the following sequence of reactions:

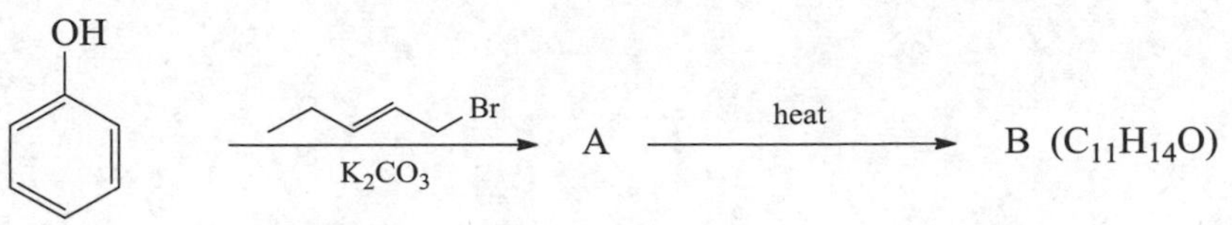

6. Prepare *p-tert*-butylphenol from *tert*-butylbenzene using any necessary organic or inorganic reagents.

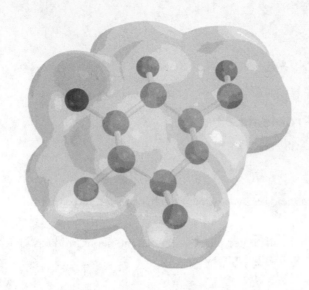

CHAPTER 24

Carbohydrates

Table of Contents

SOLUTIONS TO TEXT PROBLEMS

In Chapter Problems

24.1 (b) Reorient the three-dimensional representation, putting the aldehyde group at the top and the primary alcohol at the bottom.

$$\text{HOCH}_2\text{—C—CHO} \xrightarrow{\text{turn } 90^\circ} \text{H—C—OH}$$

What results is not equivalent to a proper Fischer projection, because the horizontal bonds are directed "back" when they should be "forward." The opposite is true for the vertical bonds. To make the drawing correspond to a proper Fischer projection, we need to rotate it 180° around a vertical axis.

$$\text{H—C—OH} \longrightarrow \text{HO—C—H} \quad \text{is equivalent to} \quad \text{HO——H}$$

rotate 180°

Now, having the molecule arranged properly, we see that it is L-glyceraldehyde.

(c) Look at the drawing from a perspective that permits you to see the carbon chain oriented vertically with the aldehyde at the top and the CH$_2$OH at the bottom. Both groups should point away from you. When examined from this perspective, the hydrogen is to the left and the hydroxyl to the right with both pointing toward you.

$$\text{HOCH}_2\text{—C—H} \quad \text{is equivalent to} \quad \text{H—C—OH}$$

The molecule is D-glyceraldehyde.

24.2 Begin by drawing a perspective view of the molecular model shown in the problem. To view the compound as a Fischer projection, redraw it in an eclipsed conformation.

Staggered conformation   Same molecule in eclipsed conformation

The eclipsed conformation shown, when oriented so that the aldehyde carbon is at the top, vertical bonds back, and horizontal bonds pointing outward from their chirality centers, is readily transformed into the Fischer projection of L-erythrose.

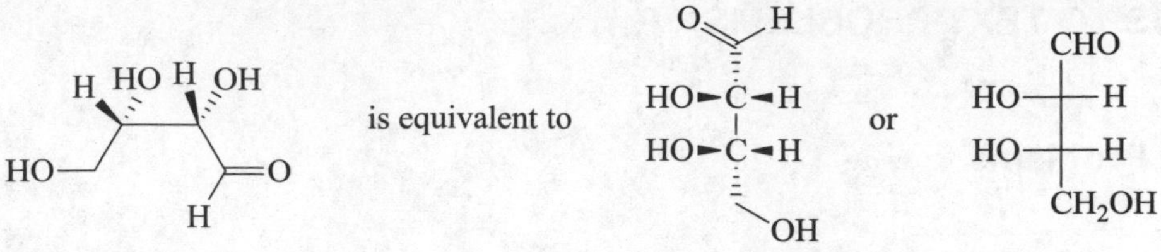

L-Erythrose

24.3   L-Arabinose is the mirror image of D-arabinose, the structure of which is given in text Figure 24.2. The configuration at *each* chirality center of D-arabinose must be inverted to transform it into L-arabinose.

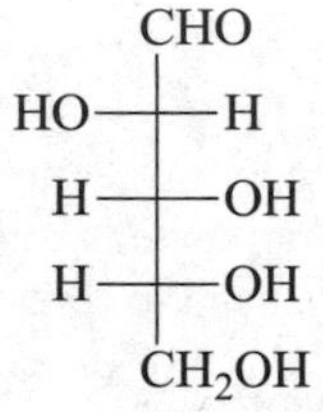

D-(−)-Arabinose 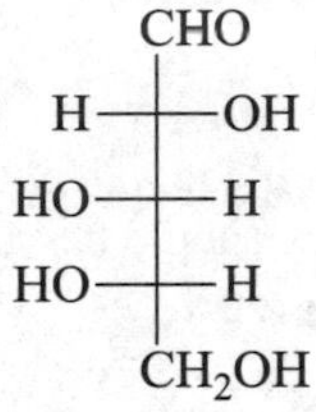   L-(+)-Arabinose

24.4   The configuration at C-5 is opposite to that of D-(+)-glyceraldehyde; it is *S*. This particular carbohydrate therefore belongs to the L series. Comparing it with the Fischer projection formulas of the eight D-aldohexoses reveals it to be in the mirror image of D-(+)-talose; it is L-(−)-talose.

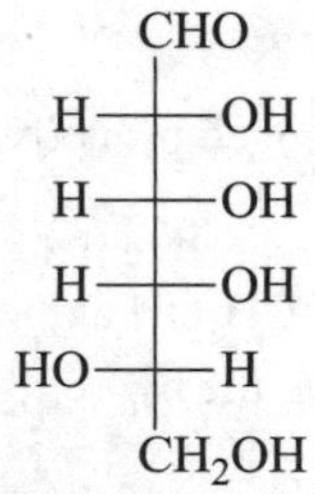

24.5   (*b*)   As in part (*a*), the —OH group on the highest-numbered chirality center (C-3) is up, and this stereoisomer belongs to the L series. The —OH group at the anomeric carbon is also up, making this the α-furanose form.

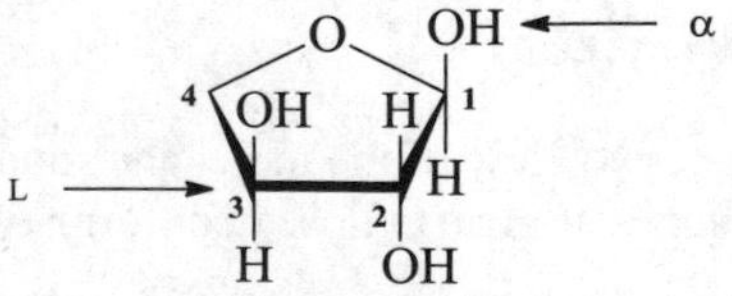

α-L-Threofuranose

(*c, d*)   The —OH group on the highest-numbered chirality center (C-3) is down in both of these structures, and these stereoisomers belong to the D series. For D series carbohydrates, the configuration of the anomeric carbon is α if its hydroxyl group is down and β if the hydroxyl group is up.

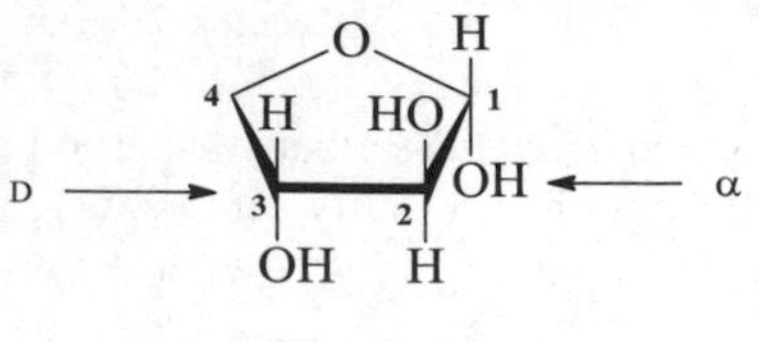

α-D-Threofuranose 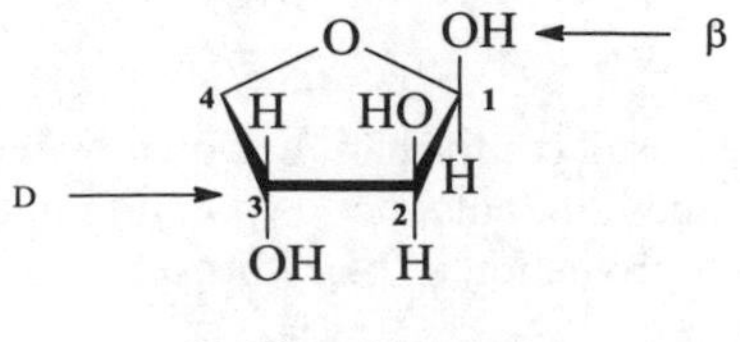   β-D-Threofuranose

24.6 (*b*) The Fischer projection of D-arabinose may be found in text Figure 24.2. The Fischer projection and the eclipsed conformation corresponding to it are

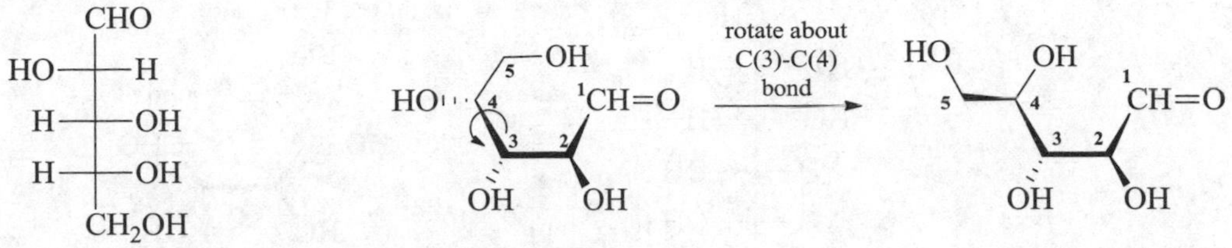

D-Arabinose · Eclipsed conformation of D-arabinose · rotate about C(3)-C(4) bond · Conformation suitable for furanose ring formation

Cyclic hemiacetal formation between the carbonyl group and the C-4 hydroxyl yields the α- and β-furanose forms of D-arabinose.

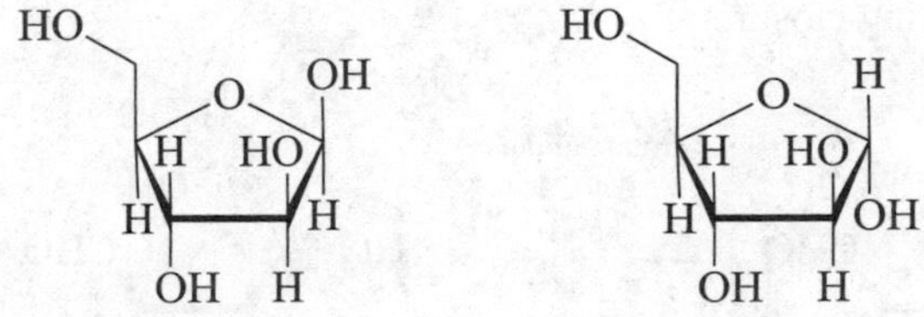

β-D-Arabinofuranose · α-D-Arabinofuranose

(*c*) The mirror image of D-arabinose [from part (*b*)] is L-arabinose.

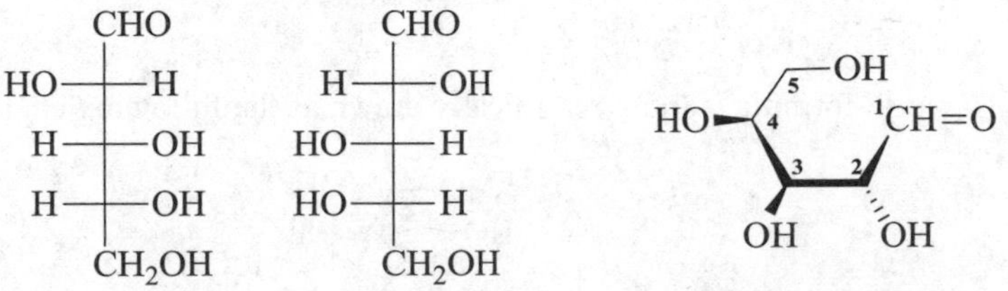

D-Arabinose · L-Arabinose · Eclipsed conformation of L-arabinose

The C-4 atom of the eclipsed conformation of L-arabinose must be rotated 120° in a counterclockwise sense so as to bring its hydroxyl group into the proper orientation for furanose ring formation.

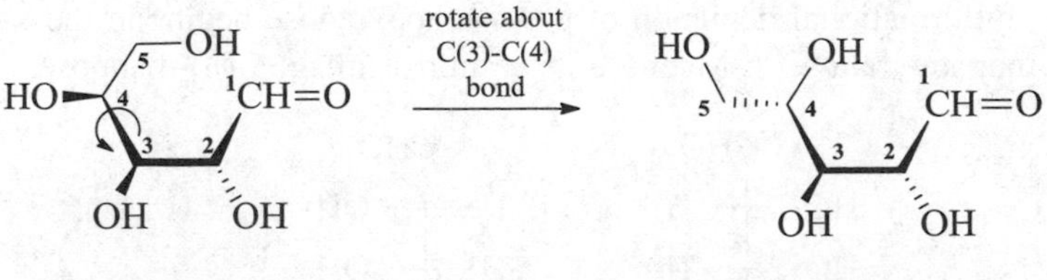

Eclipsed conformation of L-arabinose · rotate about C(3)-C(4) bond · Conformation suitable for furanose ring formation

Cyclization gives the α- and β-furanose forms of L-arabinose.

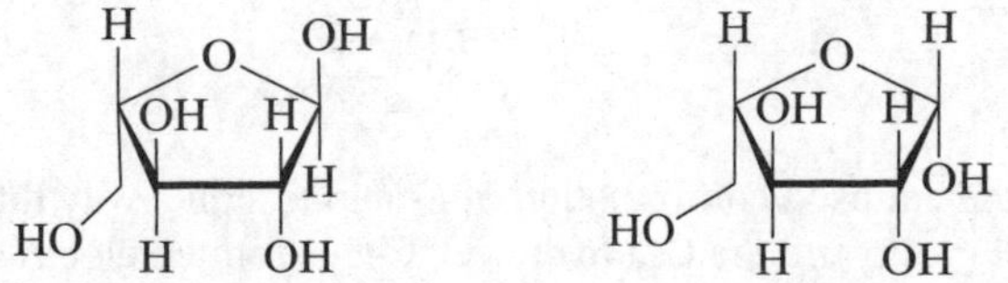

α-L-Arabinofuranose · β-L-Arabinofuranose

In the L series, the anomeric hydroxyl is up in the α isomer and down in the β isomer.

24.7 (*b*) To convert the Fischer projection to a Haworth formula for D-mannose first turn the Fischer projection on its side.

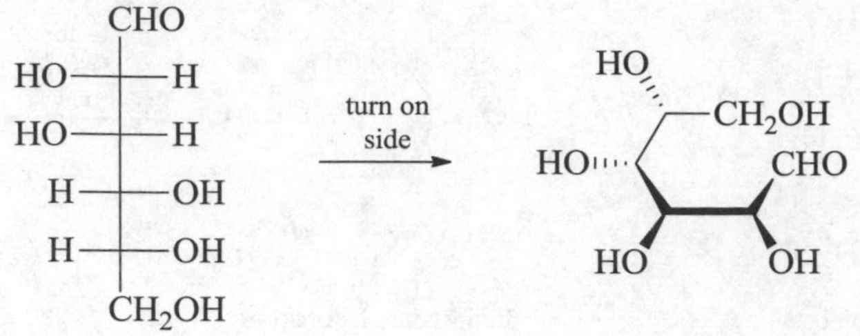

D-Mannose

D-Mannose eclipsed conformation

Rotate the C4-C5 bond in the eclipsed conformation to properly orient the hydroxy group on C5 and follow with ring closure.

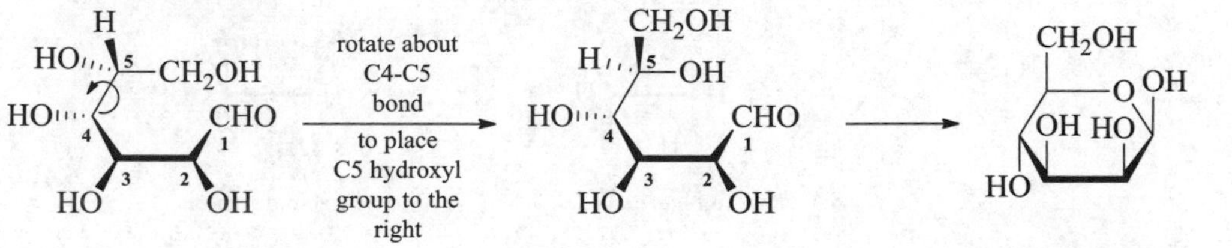

β-D-Mannopyranose (Haworth formula for pyranose form of D-mannose)

The Haworth formula is more realistically drawn as the following chair conformation:

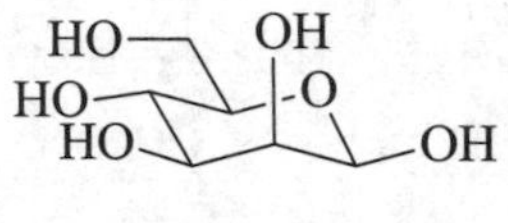

β-D-Mannopyranose

Mannose differs from glucose in configuration at C-2. All hydroxyl groups are equatorial in β-D-glucopyranose; the hydroxyl at C-2 is axial in β-D-mannopyranose.

(*c*) The conformational depiction of β-L-mannopyranose begins in the same way as that of β-D-mannopyranose. L-Mannose is the mirror image of D-mannose.

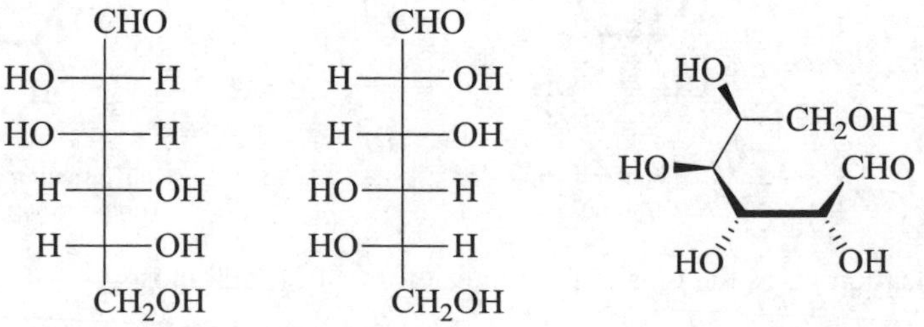

D-Mannose L-Mannose Eclipsed conformation of L-mannose

To rewrite the eclipsed conformation of L-mannose in a way that permits hemiacetal formation between the carbonyl group and the C-5 hydroxyl, C-5 is rotated clockwise 120°.

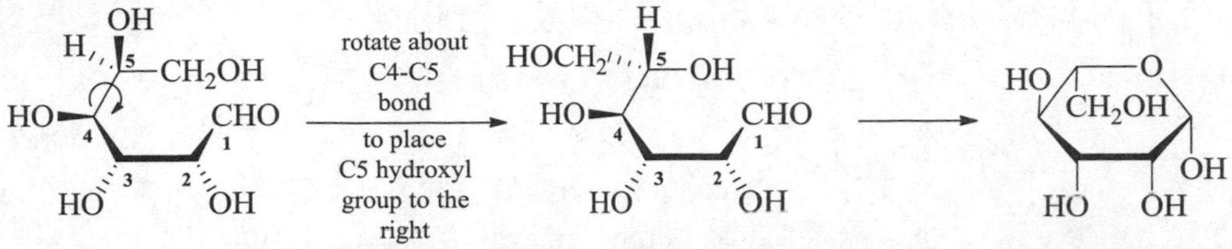

β-L-Mannopyranose
(remember, the anomeric
hydroxyl is down in L series)

Translating the Haworth formula into a proper conformational depiction requires that a choice be made between the two chair conformations shown.

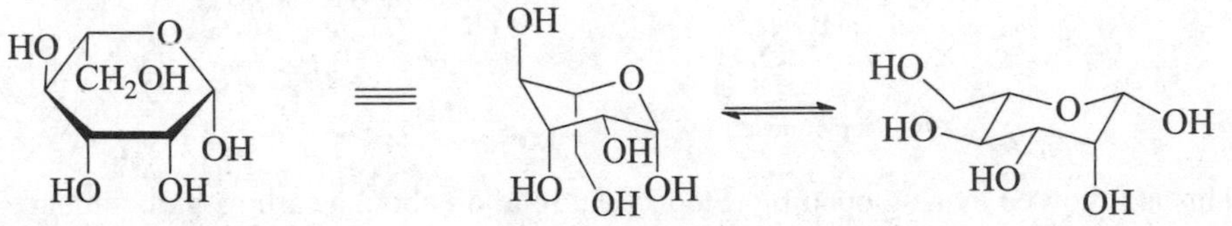

| Haworth formula of β-L-Mannopyranose | Less stable chair conformation; CH₂OH is axial | More stable chair conformation; CH₂OH is equatorial |

(*d*) The Fischer projection formula for L-ribose is the mirror image of that for D-ribose.

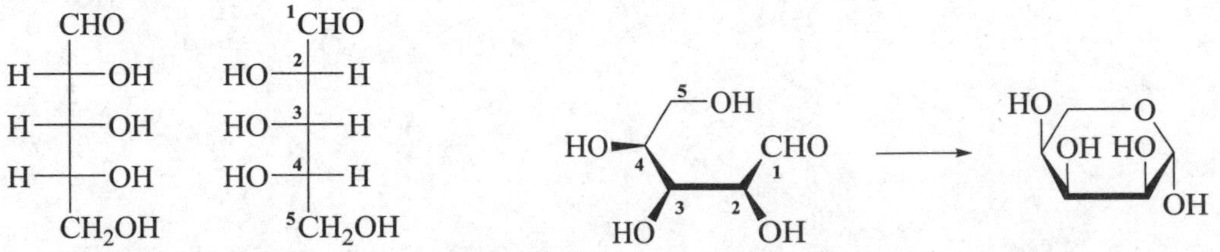

D-Ribose   L-Ribose   Eclipsed conformation of L-ribose is oriented properly for ring closure   Haworth formula of β-L-ribopyranose

Of the two chair conformations of β-L-ribose, the one with the greater number of equatorial substituents is more stable.

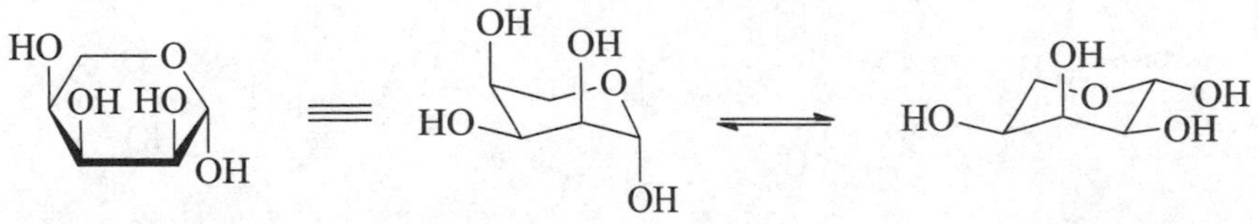

| | Less stable chair conformation of β-L-ribopyranose | More stable chair conformation of β-L-ribopyranose |

24.8 The equation describing the equilibrium is

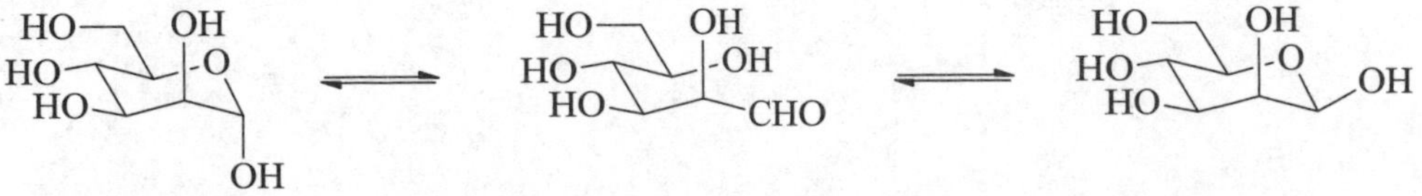

α-D-Mannopyranose   Open-chain form of D-mannose   β-D-Mannnopyranose

$[\alpha]_D^{20}$ +29.3° $[\alpha]_D^{20}$ -17.0°

Let A = percent α isomer; $100 - A$ = percent β isomer. Then

$$A(+29.3^{o})+(100\text{-}A)(-17.0^{o}) = 100(+14.2^{o})$$

$$A(46.3) = 3120$$

$$\text{Percent } \alpha \text{ isomer} = 67\%$$

$$\text{Percent } \beta \text{ isomer} = (100 - A) = 33\%$$

24.9 As shown in the text, Step 1 of this mutarotation is deprotonation of α-D-glucopyranose.

α-D-Glucopyranose

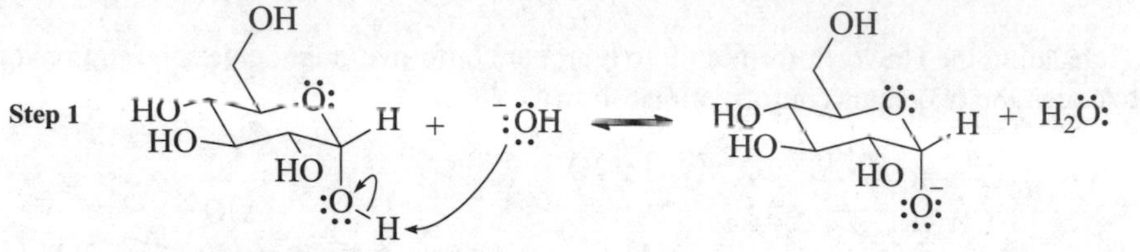

This is followed by ring opening (Step 2) and rotation about a carbon–carbon bond. Ring closing (Step 3) followed by protonation (Step 4) gives β-D-glucopyranose.

β-D-Glucopyranose

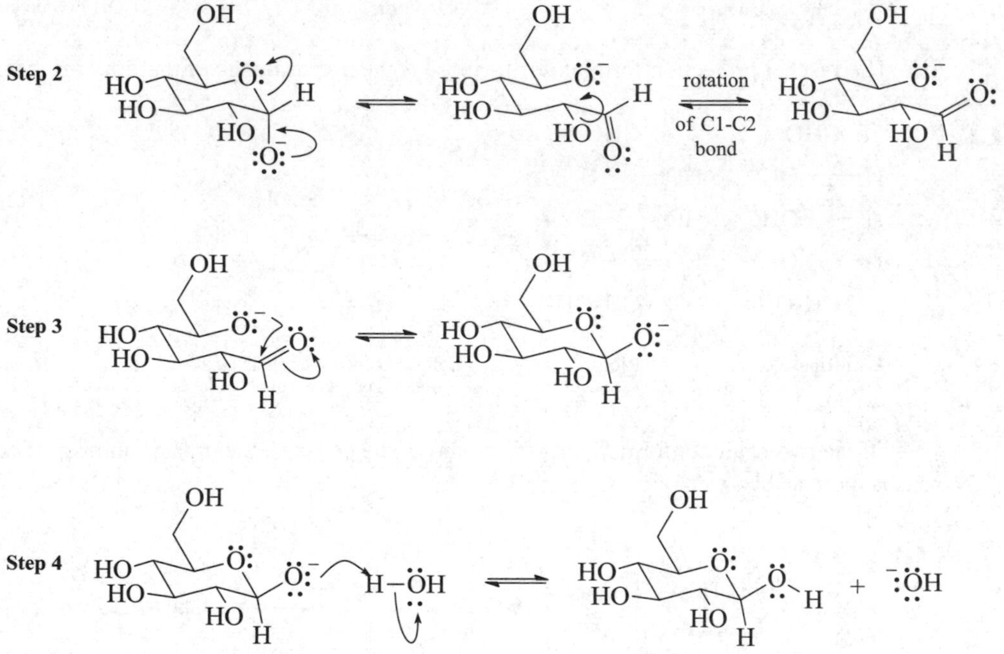

24.10 Based on the anomeric effect, the C–Cl bond and the O–CH_3 bond are gauche in the most stable conformation of $ClCH_2OCH_3$.

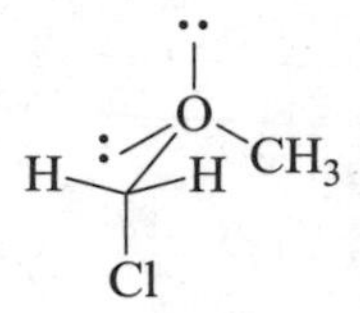

Sawhorse projection of chloromethyl methyl ether

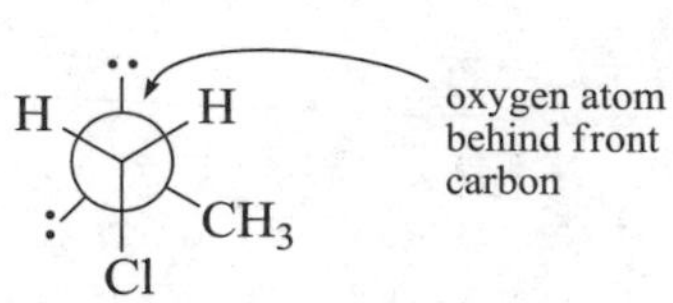

Newman projection of chloromethyl methyl ether

24.11 Review carbohydrate terminology by referring to text Table 23.1. A *ketotetrose* is a four-carbon ketose. Writing a Fischer projection for a four-carbon ketose reveals that only one chirality center is present, and thus there are only two ketotetroses. They are enantiomers of each other and are known as D- and L-erythrulose.

$$
\begin{array}{cc}
CH_2OH & CH_2OH \\
C=O & C=O \\
H\text{---OH} & HO\text{---H} \\
CH_2OH & CH_2OH \\
\text{D-Erythrulose} & \text{L-Erythrulose}
\end{array}
$$

24.12 To convert the Fischer projection to a Haworth formula for D-fructose, first turn the Fischer projection on its side.

D-Fructose   turn on side   D-Fructose eclipsed conformation

Rotate the C4-C5 bond in the eclipsed conformation to properly orient the hydroxy group on C5 and follow with ring closure.

D-Fructose eclipsed conformation   rotate about C4-C5 bond to place C5 hydroxyl group to the right   β-D-Fructose

24.13 (b) Because L-rhamnose is 6-deoxy-l-mannose, first write the Fischer projection formula of D-mannose, and then transform it to its mirror image, l-mannose. Transform the C-6 CH_2OH group to CH_3 to produce 6-deoxy-l-mannose.

$$
\begin{array}{ccc}
CHO & CHO & CHO \\
HO\text{---}H & H\text{---}OH & H\text{---}OH \\
HO\text{---}H & H\text{---}OH & H\text{---}OH \\
H\text{---}OH & HO\text{---}H & HO\text{---}H \\
H\text{---}OH & HO\text{---}H & HO\text{---}H \\
CH_2OH & CH_2OH & CH_3 \\
\end{array}
$$

D-Mannose   L-Mannose   6-Deoxy-L-mannose
(from text Figure 23.2)   (L-rhamnose)

24.14 Numbering of the nine-carbon chain begins at the carboxyl group in *N*-acetylneuraminic acid. The compound can be described as a ketose as C-2 is the hemiacetal of a ketone carbonyl. It is also a deoxy sugar as evidenced by the lack of a hydroxyl group on C-3. The highest-numbered chirality center is C-8. It is D; it has the same configuration as D-glyceraldehyde.

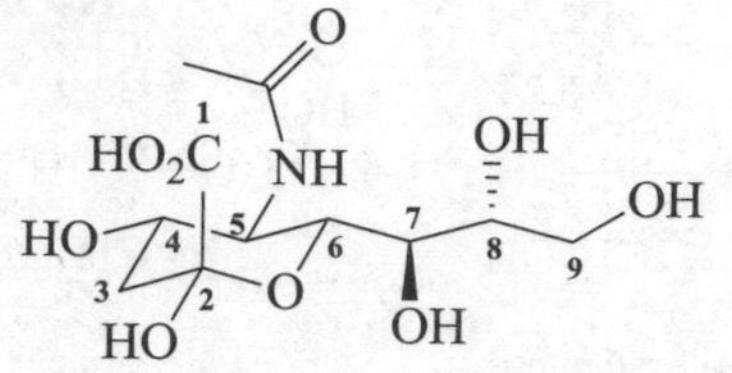

N-Acetylneuraminic acid

24.15 Reaction of D-galactose with methanol in the presence of HCl gives the α and β methyl acetal products. Because of the anomeric effect, the α-galactopyranoside is the major product.

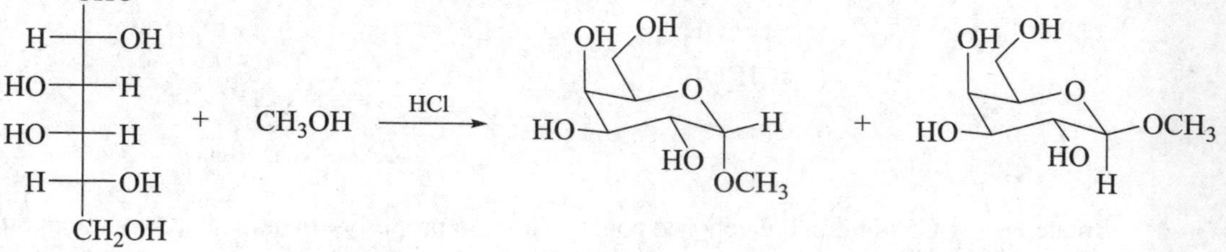

D-Galactose

Methyl α-D-galactopyranoside
(major product)

Methyl β-D-galactopyranoside

24.16 The curved arrows indicate the electron pair movements.

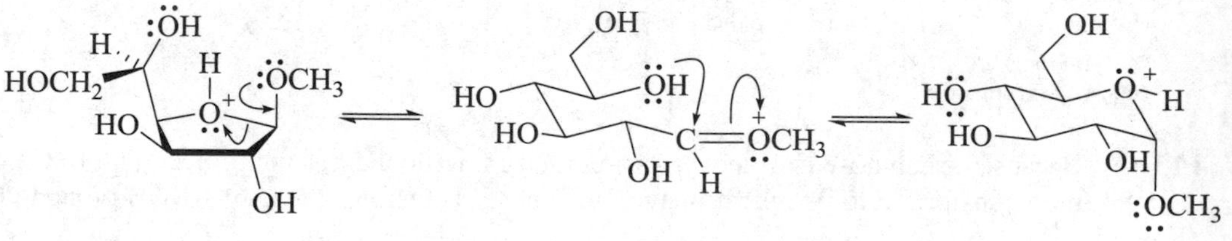

24.17 In the first step of this mechanism, loss of methanol gives an oxonium ion intermediate. The second step is the reverse of the first step except CD_3OH is used instead of CH_3OH.

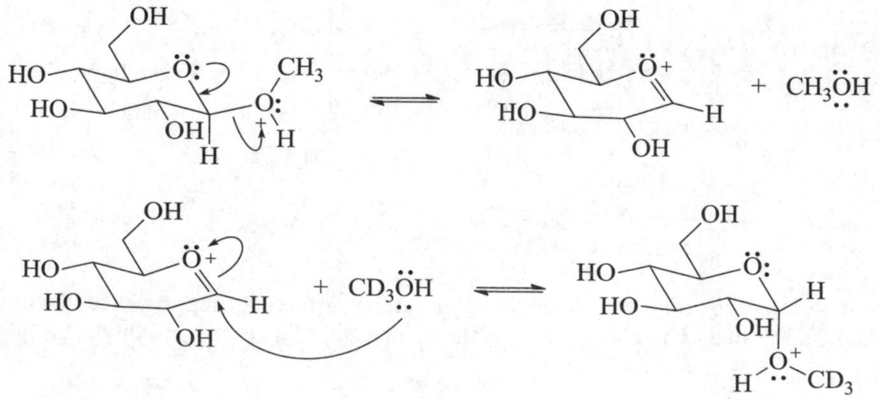

24.18 This problem can be solved by starting at the end of Mechanism 24.2 (step 7) and proceeding backwards.

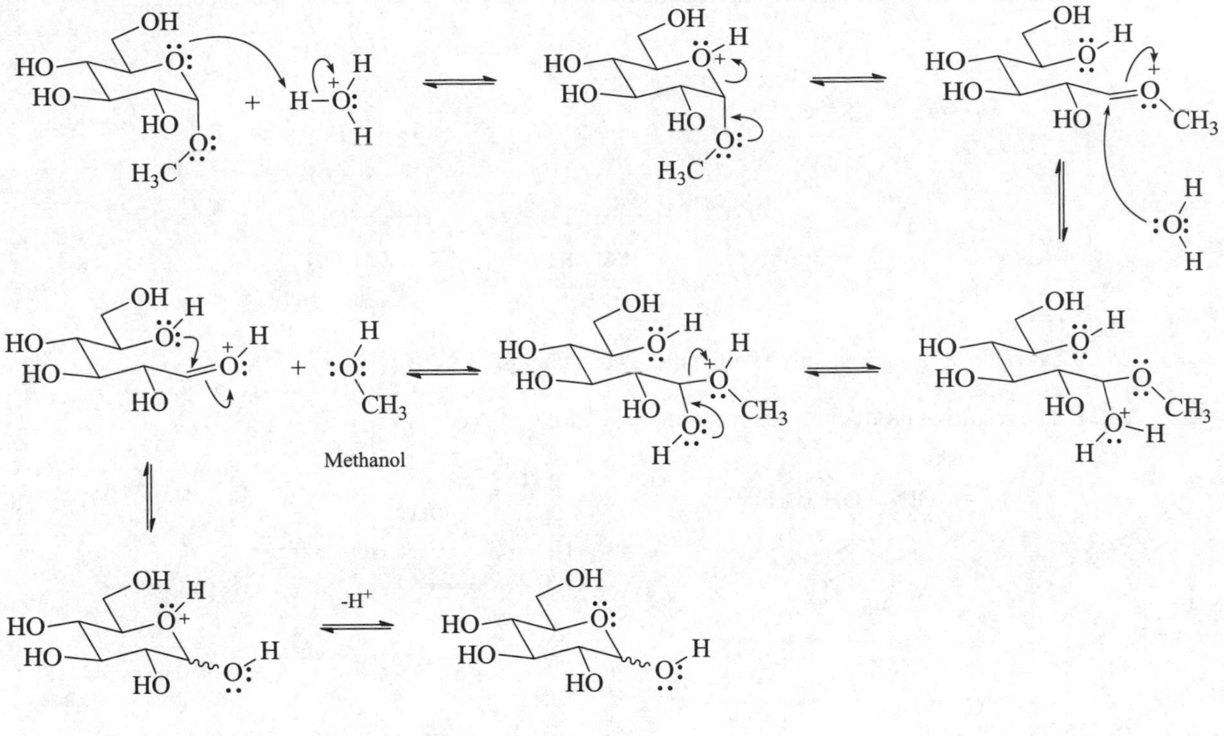

Methanol

D-Glucose

24.19 The hemiacetal opens to give an intermediate containing a free aldehyde function. Cyclization of this intermediate can produce either the α or the β configuration at this center. The axial and equatorial orientations of the anomeric hydroxyl can best be seen by drawing maltose with the pyranose rings in chair conformations.

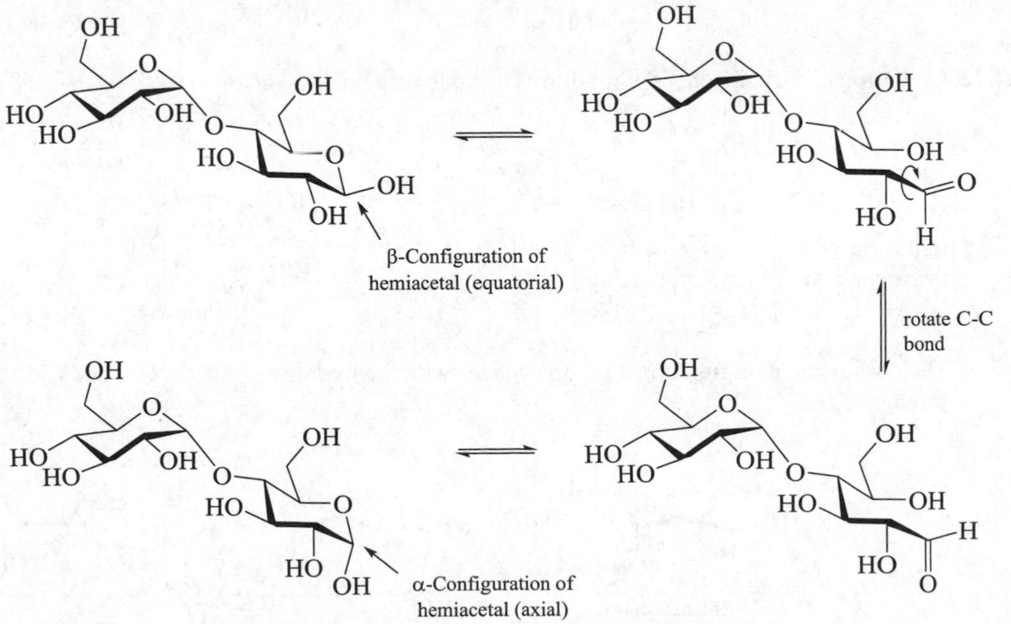

β-Configuration of
hemiacetal (equatorial)

rotate C-C
bond

α-Configuration of
hemiacetal (axial)

Only the configuration of the hemiacetal function is affected in this process. The α configuration of the glycosidic linkage remains unchanged.

24.20 D–Galactitol and D–mannitol are the alditols that are formed from reduction of D-fructose. The two alditols that are formed are stereoisomers called epimers; they differ only at one stereocenter.

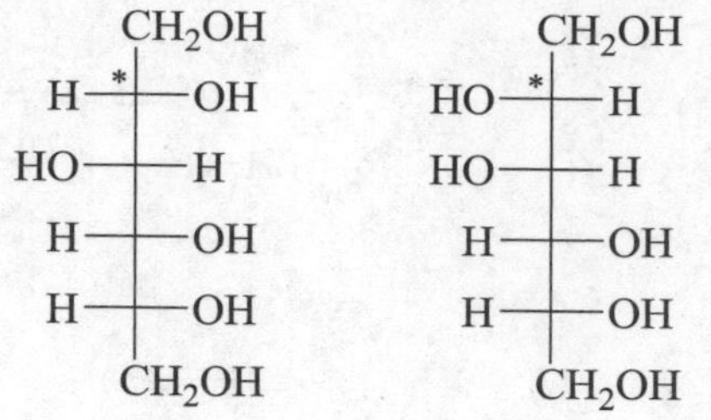

D-Galactitol   D-Mannitol

The aldose that gives only D-mannitol is D-mannose.

24.21 The resulting product of this reaction is the triol that cyclizes.

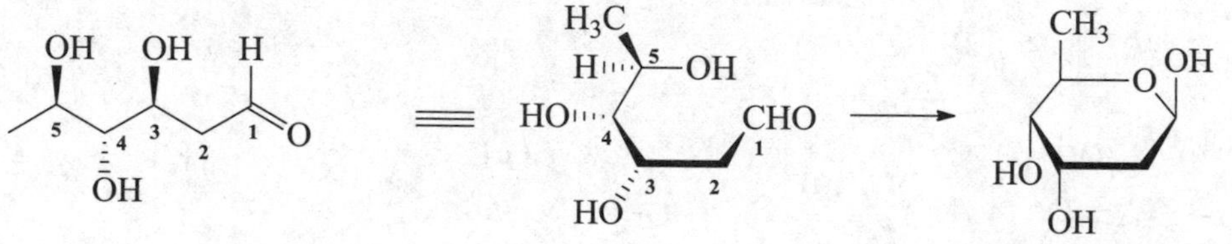

$C_6H_{12}O_4$

Haworth formula of
α-D-digitoxopyranose

The lower energy 6-membered ring is the conformation with the α configuration.

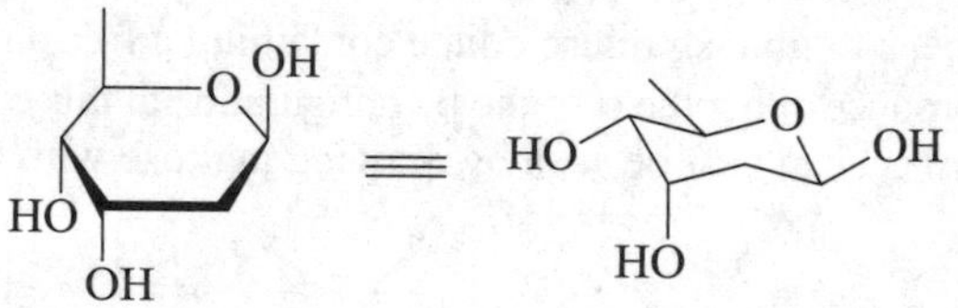

24.22 Oxidation of L-rhamnose initially forms the aldonic acid.

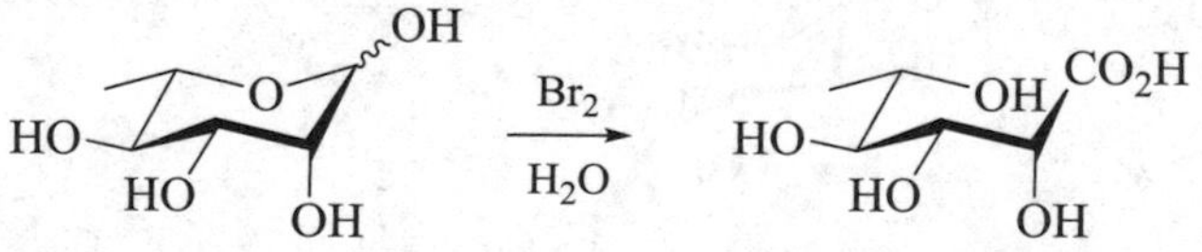

L-Rhamnose

Aldonic acid
of L-rhamnose

The carboxylic acid functional group cyclizes with the hydroxy groups at C-4 and C-5 to form the lactone products.

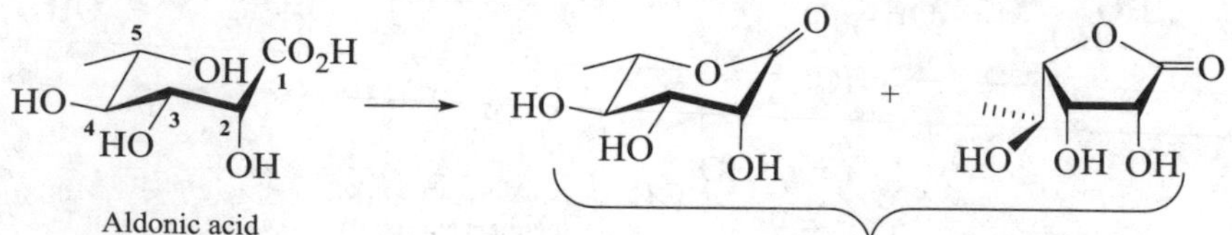

Aldonic acid
of L-rhamnose

L-Rhamnolactone (80%)

24.23 D-Glucuronic acid structure derives from the oxidation of the C-6 primary alcohol of D-glucose. Cyclization between the aldehyde and the hydroxy group of C-5 (as indicated on D-glucose), gives the β-pyranose structure.

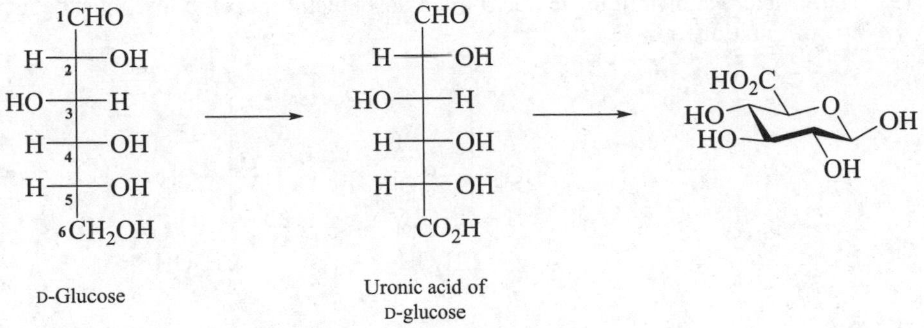

D-Glucose   Uronic acid of D-glucose

24.24 Because the groups at both ends of the carbohydrate chain are oxidized to carboxylic acid functions, two combinations of one CH_2OH with one CHO group are possible.

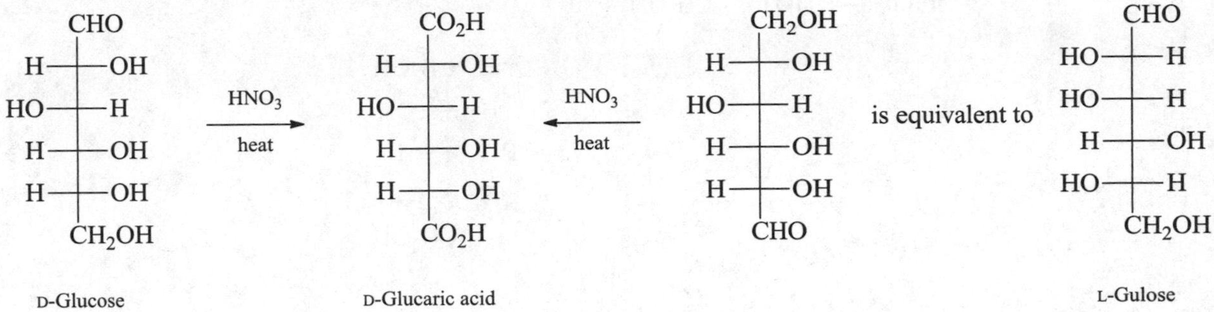

D-Glucose   D-Glucaric acid   is equivalent to   L-Gulose

L-Gulose yields the same aldaric acid on oxidation as does D-glucose.

24.25 The three glycosidic linkages are indicated below. The α and β designate the stereochemistry at the anomeric positions. The numbers specify the ring carbons involved.

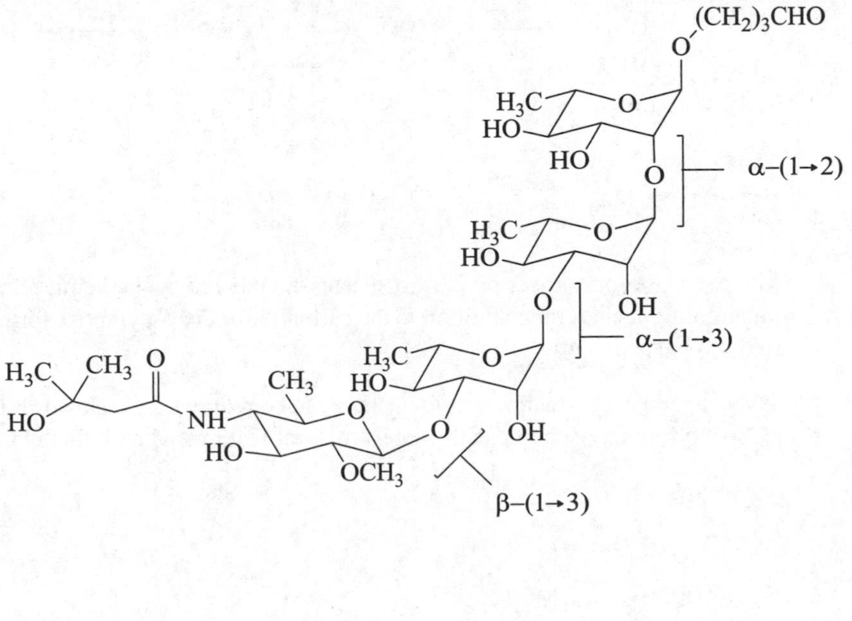

α–(1→2)

α–(1→3)

β–(1→3)

End of Chapter Problems

Structure

24.26 (*a*) The structure shown in the text is D-(+)-xylose; therefore (−)-xylose must be its mirror image and has the L-configuration at C-4.

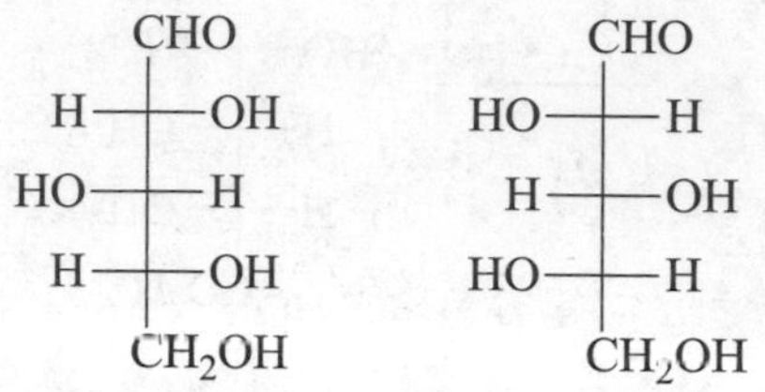

D-(+)-Xylose   L-(-)-Xylose

(*b*) Alditols are the reduction products of carbohydrates; xylitol is derived from xylose by conversion of the terminal —CHO to —CH$_2$OH.

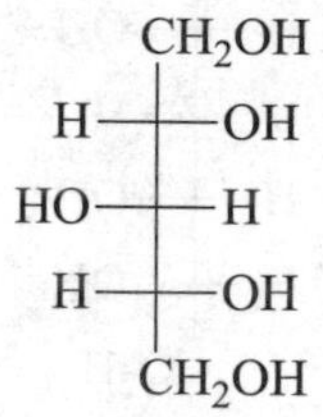

Xylitol

(*c*) Redraw the Fischer projection of D-xylose in its eclipsed conformation.

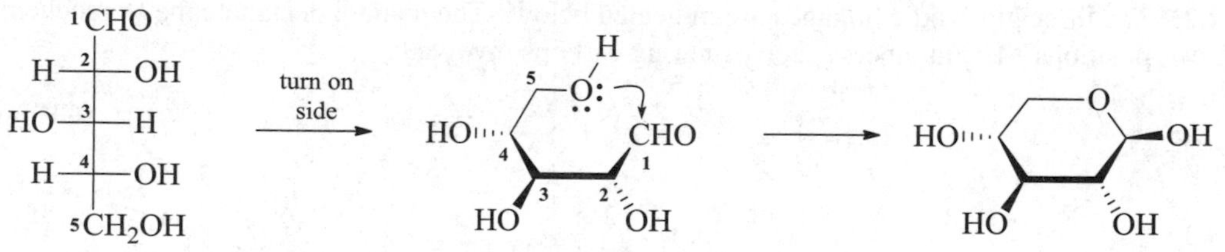

D-Xylose   Eclipsed conformation   Eclipsed conformation
                          of D-xylose   of β-D-xylopyranose

The pyranose form arises by closure to a six-membered cyclic hemiacetal, with the C-5 hydroxyl group undergoing nucleophilic addition to the carbonyl. In the β-pyranose form of D-xylose the anomeric hydroxyl group is up.

The eclipsed conformation representation can be redrawn by a Haworth formula. The chair conformation of β-D-xylopyranose represents more structural features with all the hydroxyl groups equatorial.

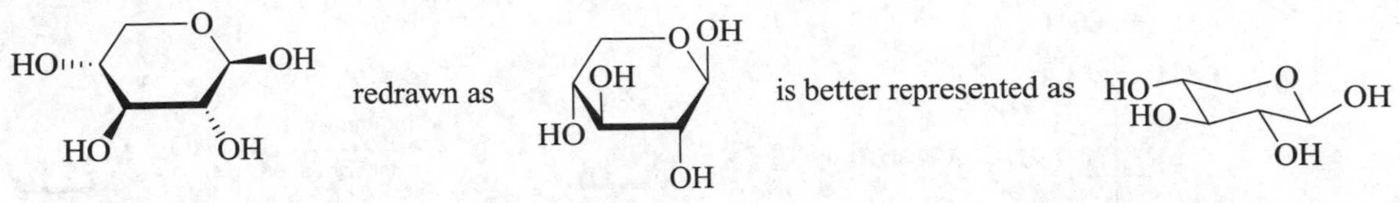

Eclipsed conformation
of β-D-xylopyranose

Haworth formula
of β-D-xylopyranose

Chair conformation of
of β-D-xylopyranose

(*d*) L-Xylose is the mirror image of D-xylose.

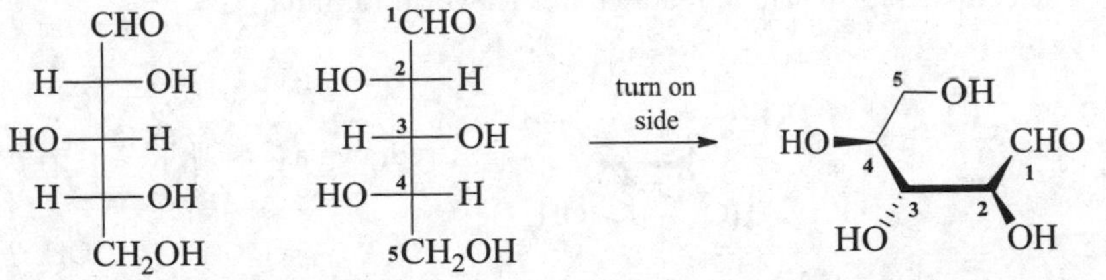

D-Xylose

L-Xylose

Eclipsed conformation
of L-xylose

To construct the furanose form of L-xylose, the hydroxyl at C-4 needs to be brought into the proper orientation to form a five-membered ring.

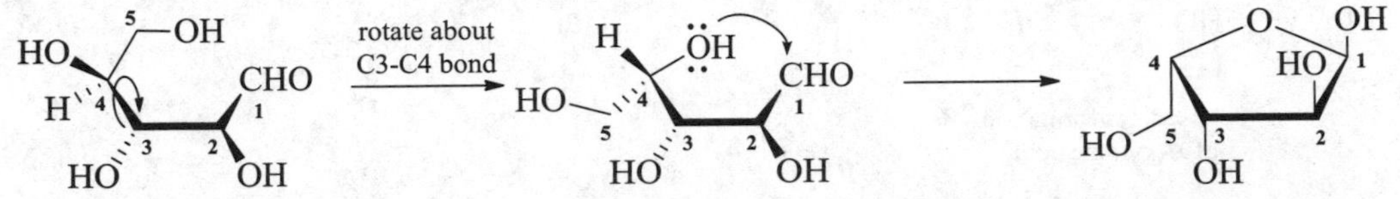

rotate about
C3-C4 bond

The α-anomeric hydroxyl group is up in the L series.

(*e*) Methyl α-L-xylofuranoside is the methyl glycoside corresponding to the structure in part (*d*).

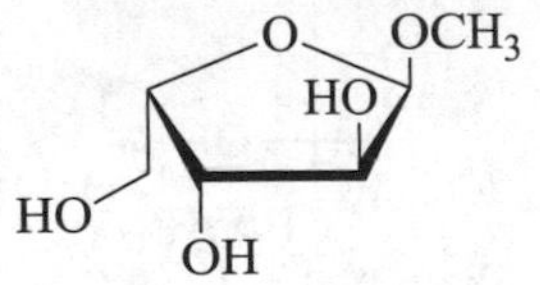

(*f*) Aldonic acids are derived from aldoses by oxidation of the terminal aldehyde to a carboxylic acid.

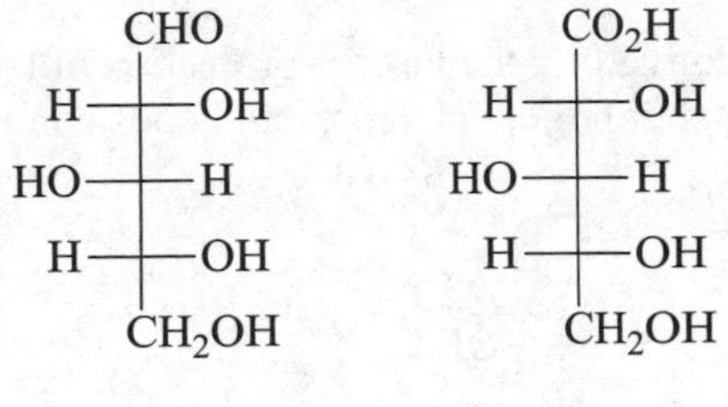

D-Xylose

D-Xylonic acid

(*g*) Aldonic acids tend to exist as lactones. A δ-lactone has a six-membered ring.

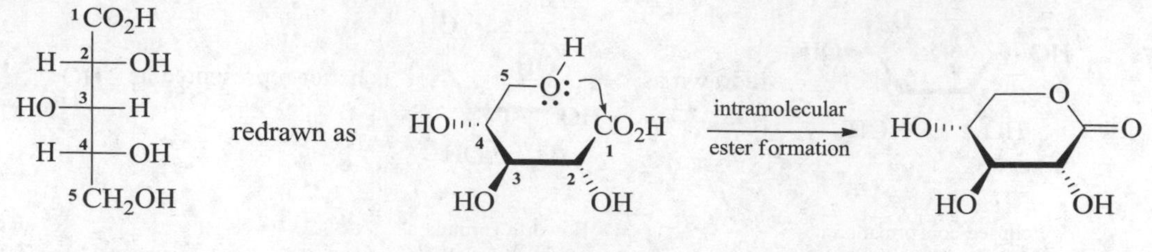

D-Xylonic acid   redrawn as   Eclipsed conformation
                                             of D-xylonic acid   intramolecular
                                                                        ester formation   δ-Lactone of D-xylonic acid

The eclipsed form can be redrawn in a Haworth formula.

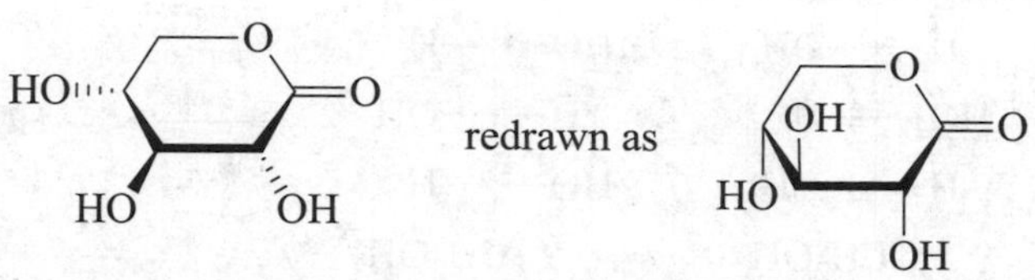

redrawn as

(*h*) A γ-lactone has a five-membered ring.

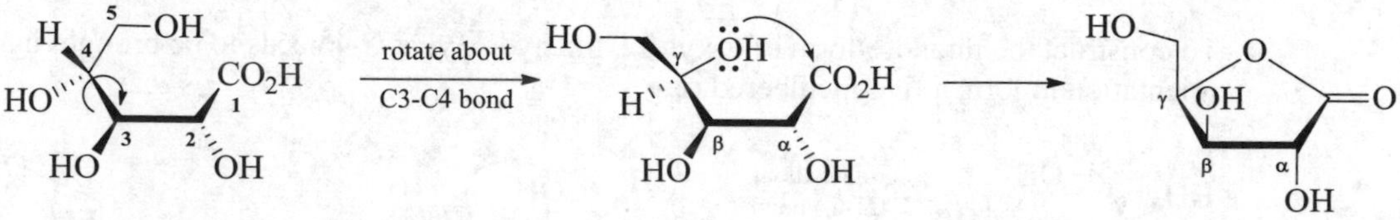

Eclipsed conformation   rotate about   γ-Lactone of
of D-xylonic acid   C3-C4 bond   of D-xylonic acid

(*i*) Aldaric acids have carboxylic acid groups at both ends of the chain.

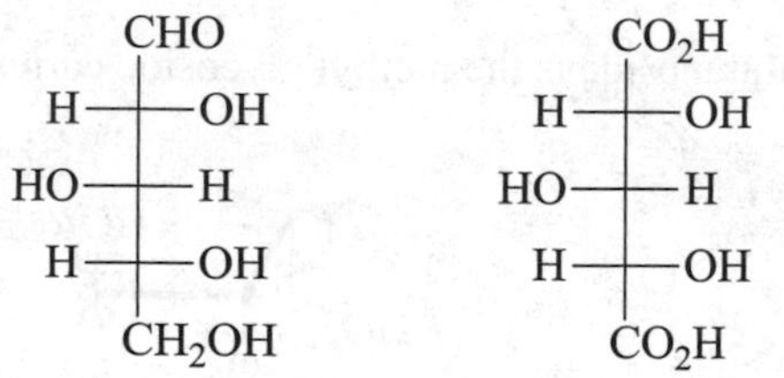

D-Xylose   Xylaric acid

24.27 Begin the problem by converting the Fischer projection of D-xylose to a perspective view. Remember that the horizontal lines of a Fischer projection represent bonds coming toward you, and the vertical lines are going away from you.

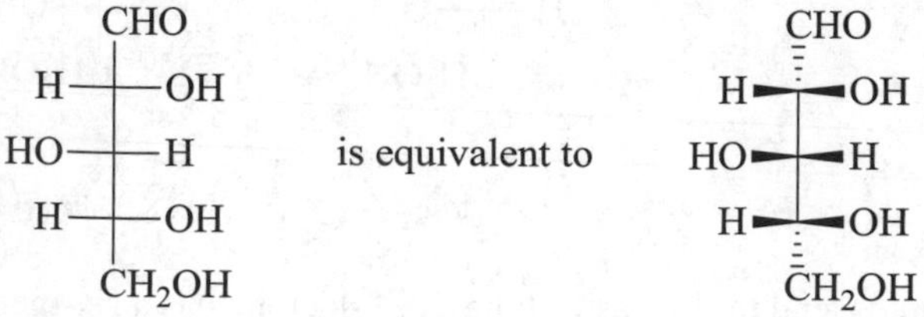

is equivalent to

D-Xylose

Rank the groups attached to each chirality center. Identify each chirality center as either *R* or *S* according to the methods described in Chapter 4. Remember that the proper orientation of the lowest-ranked group (usually H) is away from you. Two of the chirality centers in D-xylose have the *R* configuration. The IUPAC name of D-xylose is (2*R*,3*S*,4*R*)-2,3,4,5-tetrahydroxypentanal.

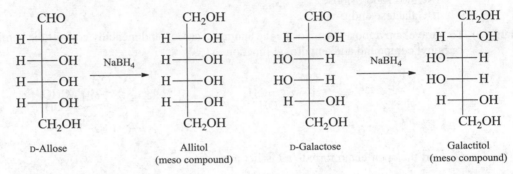

(The Fischer projection shows:)

CHO
H——OH   C2 is R
HO——H   C3 is S
H——OH   C4 is R
CH₂OH

24.28 (*a*) Reduction of aldoses with sodium borohydride yields polyhydroxy alcohols called *alditols*. Optically inactive alditols are those that have a plane of symmetry, that is, those that are meso forms. The D-aldohexoses that yield optically inactive alditols are D-allose and D-galactose.

CHO
H——OH
H——OH   NaBH₄
H——OH   ——→
H——OH
CH₂OH

D-Allose

CH₂OH
H——OH
H——OH
H——OH
H——OH
CH₂OH

Allitol
(meso compound)

CHO
H——OH
HO——H   NaBH₄
HO——H   ——→
H——OH
CH₂OH

D-Galactose

CH₂OH
H——OH
HO——H
HO——H
H——OH
CH₂OH

Galactitol
(meso compound)

(*b*) All the aldonic acids and their lactones obtained on oxidation of the aldohexoses with bromine are optically active. The presence of a carboxyl group at one end of the carbon chain and a CH₂OH at the other precludes the existence of meso forms.

(*c*) Nitric acid oxidation of aldoses converts them to aldaric acids. The same D-aldoses found to yield optically inactive alditols in part (*a*) yield optically inactive aldaric acids.

CHO
H——OH
H——OH   HNO₃
H——OH   ——→
H——OH
CH₂OH

D-Allose

CO₂H
H——OH
H——OH
H——OH
H——OH
CO₂H

Allaric acid
(meso compound)

CHO
H——OH
HO——H   HNO₃
HO——H   ——→
H——OH
CH₂OH

D-Galactose

CO₂H
H——OH
HO——H
HO——H
H——OH
CO₂H

Galactaric acid
(meso compound)

(*d*) Aldoses that differ in configuration only at C-2 enolize to the same enediol.

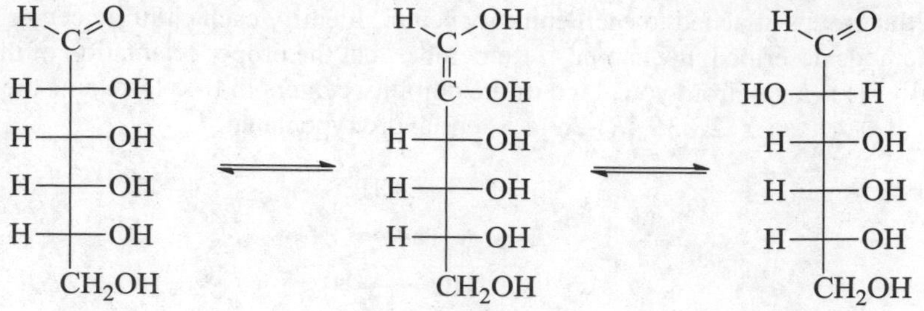

D-Allose   Enediol   D-Altrose

The chirality center at C-2 in the D-aldose becomes sp^2-hybridized in the enediol. The other pairs of D-aldohexoses that form the same enediols are

> D-Glucose and D-mannose
> D-Gulose and D-idose
> D-Galactose and D-talose

24.29 (*a*) To unravel a pyranose form, locate the anomeric carbon and mentally convert the hemiacetal linkage to a carbonyl compound and a hydroxyl function.

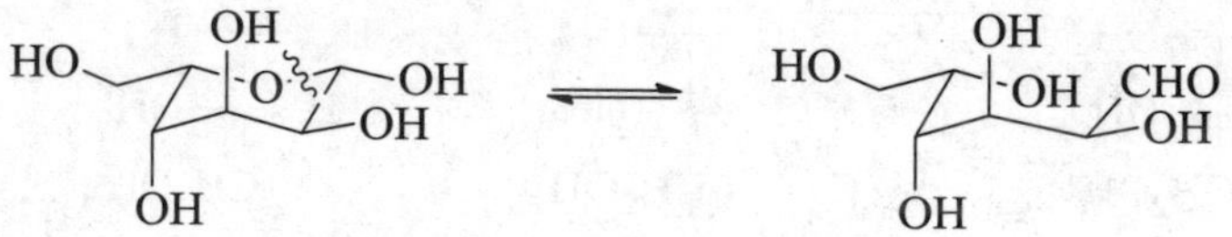

Convert the open-chain form to a Fischer projection.

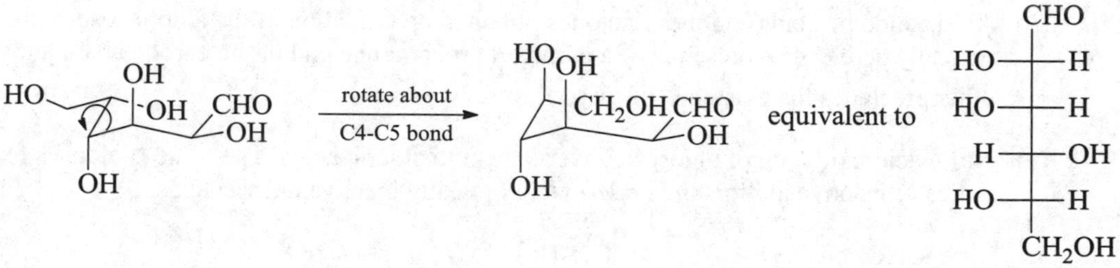

(*b*) Proceed in the same manner as in part (*a*) and unravel the furanose sugar by disconnecting the hemiacetal function.

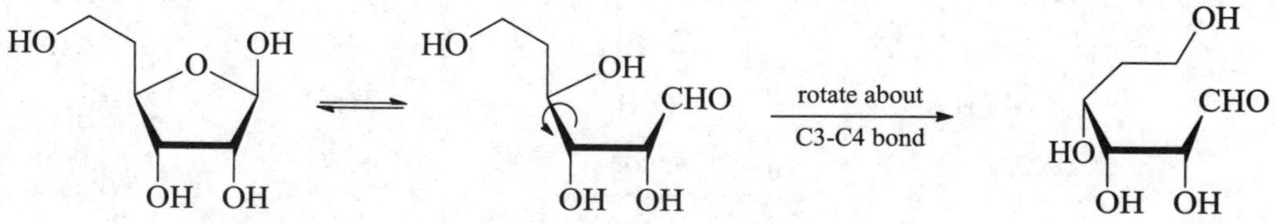

The Fischer projection is

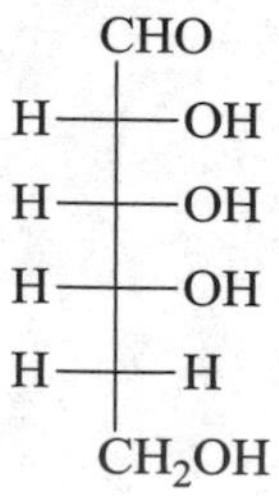

(c) By disconnecting and unraveling as before, the Fischer projection is revealed.

$$
\begin{array}{c}
\text{CHO} \\
\text{H}\!-\!\!-\!\text{OH} \\
\text{H}_3\text{C}\!-\!\!-\!\text{OH} \\
\text{HO}\!-\!\!-\!\text{H} \\
\text{CH}_2\text{OH}
\end{array}
$$

equivalent to

(d) Remember in disconnecting cyclic hemiacetals that the anomeric carbon is the one that bears two oxygen substituents.

rotate about
C3-C4 bond

$$
\begin{array}{c}
\text{CH}_2\text{OH} \\
=\!\!\text{O} \\
\text{H}\!-\!\!-\!\text{OH} \\
\text{H}\!-\!\!-\!\text{OH} \\
\text{H}\!-\!\!-\!\text{OH} \\
\text{H}\!-\!\!-\!\text{OH} \\
\text{CH}_2\text{OH}
\end{array}
$$

24.30 (a) The L sugars have the hydroxyl group to the left at the highest numbered chirality center in their Fischer projection. The L sugars are the ones in Problem 24.29a and c.

$$
\begin{array}{c}
\text{CHO} \\
\text{HO}\!-\!\!-\!\text{H} \\
\text{HO}\!-\!\!-\!\text{H} \\
\text{H}\!-\!\!-\!\text{OH} \\
\text{HO}\!-\!\!-\!\text{H} \\
\text{CH}_2\text{OH}
\end{array}
$$

Highest numbered
chirality center is L

$$
\begin{array}{c}
\text{CHO} \\
\text{H}\!-\!\!-\!\text{OH} \\
\text{H}_3\text{C}\!-\!\!-\!\text{OH} \\
\text{HO}\!-\!\!-\!\text{H} \\
\text{CH}_2\text{OH}
\end{array}
$$

Highest numbered
chirality center is L

(b) Deoxy sugars are those that lack an oxygen substituent on one of the carbons in the main chain. The carbohydrate in Problem 24.29b is a deoxy sugar.

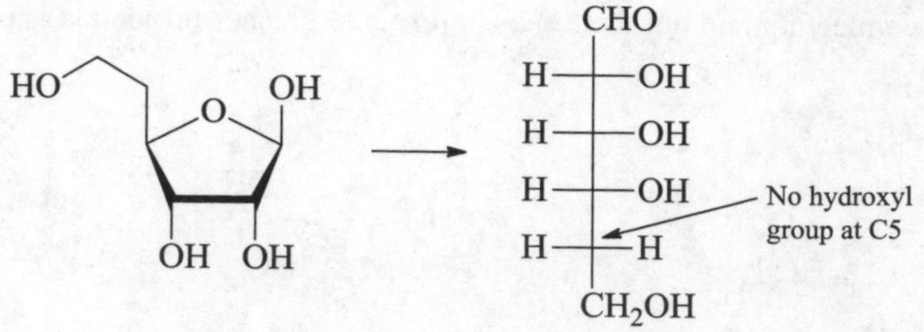

(c) Branched-chain sugars have a carbon substituent attached to the main chain; the carbohydrate in Problem 24.29c fits this description.

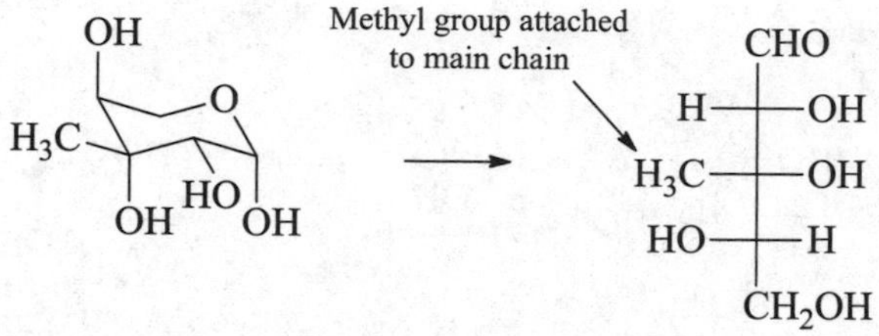

(d) Only the sugar in Problem 24.29d is a ketose.

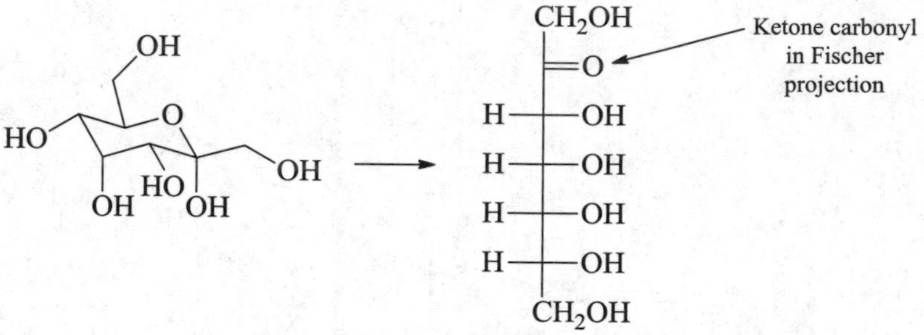

(e) A furanose ring is a five-membered cyclic hemiacetal. Only the compound in Problem 24.29b is a furanose form.

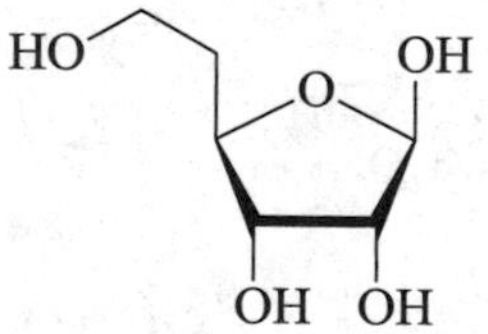

(f) In D sugars, the α configuration corresponds to the condition in which the hydroxyl group at the anomeric carbon is down. The α-D sugar is that in Problem 24.29d.

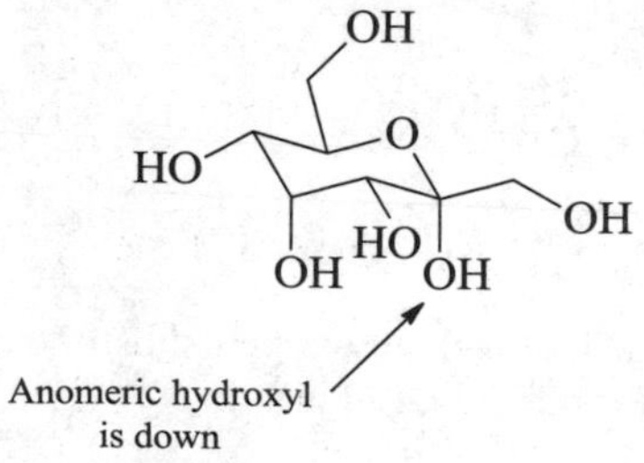

α-Pyranose form of a D-ketose

In the α-L series the anomeric hydroxyl is up. Neither of the L sugars—namely, those of Problem 24.29*a* and *c*—is α; both are β.

24.31 There are seven possible ketopentoses. The ketone carbonyl can be located at either C-2 or C-3. When the carbonyl group is at C-2, there are two chirality centers, giving rise to four stereoisomers (two pairs of enantiomers).

D-Ribulose L-Ribulose D-Xylulose L-Xylulose

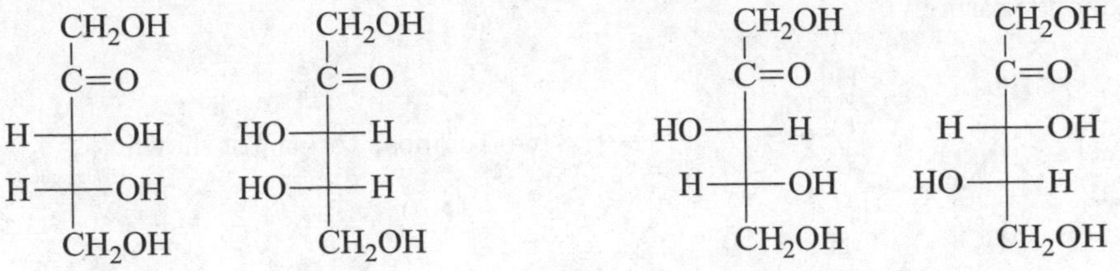

When the carbonyl group is located at C-3, there are only three stereoisomers, because one of them is a meso form and is superimposable on its mirror image.

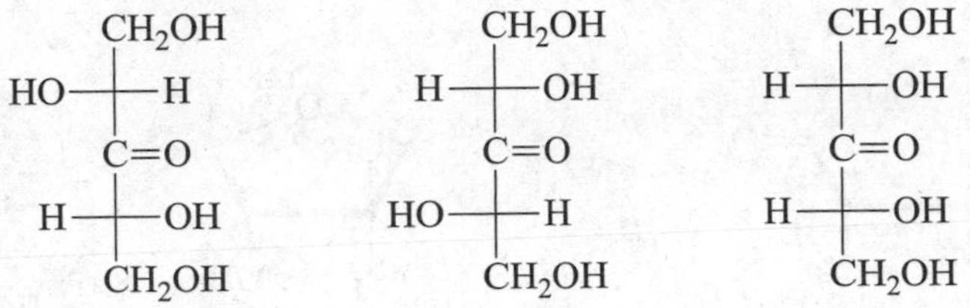

24.32 (*a*) Carbon-2 is the only chirality center in D-apiose.

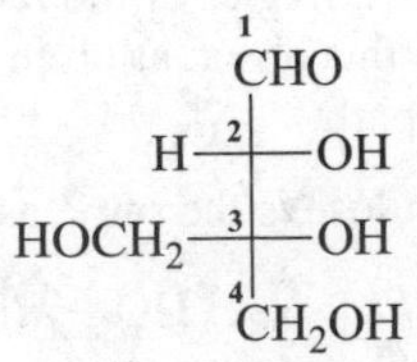

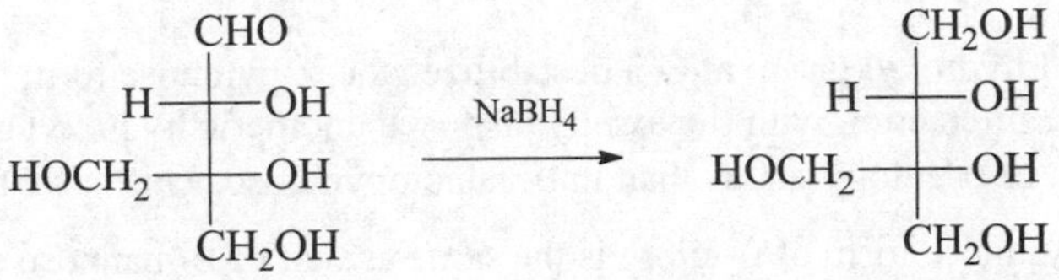

D-Apiose
(optically active)

Carbon-3 is not a chirality center; it bears two identical CH_2OH substituents.

(*b*) The alditol obtained on reduction of D-apiose retains the chirality center. It is chiral and optically active.

D-Apiose
(optically active)

D-Apiitol
(optically active)

(*c, d*) Cyclic hemiacetal formation in D-apiose involves addition of a CH_2OH hydroxyl group to the aldehyde carbonyl.

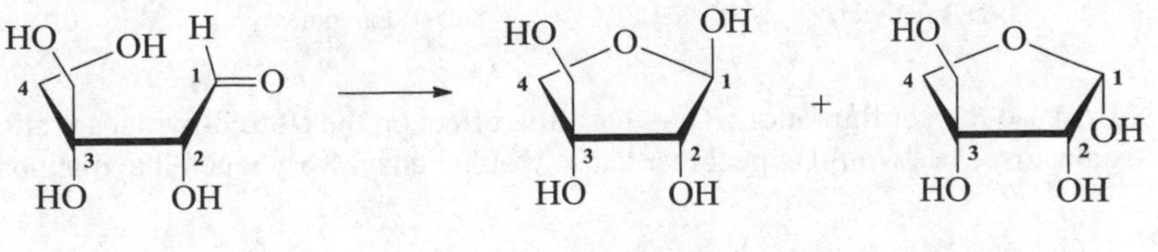

Three chirality centers occur in the furanose form, namely, the anomeric carbon C-1 and the original chirality center C-2, as well as a new chirality center at C-3.

In addition to the two furanose forms just shown, two more are possible. Instead of the reaction of the CH$_2$OH group that was shown to form the cyclic hemiacetal, the other CH$_2$OH group may add to the aldehyde carbonyl.

rotate C3
120° about the
C2-C3 bond

24.33 Comparing D-glucose, D-mannose, and D-galactose, it can be said that the configuration of C-2 has a substantial effect on the relative energies of the α- and β-pyranose forms, but that the configuration of C-4 has virtually no effect. With this observation in mind, write the structures of the pyranose forms of the carbohydrates given in each part.

(a) The β-pyranose form of D-gulose is the same as that of D-galactose except for the configuration at C-3.

β-D-Galactopyranose
(64% at equilibrium)

β-D-Gulopyranose

α-D-Gulopyranose
(1,3-diaxial repulsion
between hydroxy groups)

The axial hydroxyl group at C-3 destabilizes the α-pyranose form more than the β form because of its repulsive interaction with the axially disposed anomeric hydroxyl group. There should be an even higher β/α ratio in D-gulopyranose than in D-galactopyranose. This is so; the observed β/α ratio is 88:12.

(b) The β-pyranose form of D-talose is the same as that of D-mannose except for the configuration at C-4.

β-D-Talopyranose

α-D-Talopyranose

α-D-Mannopyranose
(68% at equilbrium)

Because the configuration at C-4 has little effect on the α- to β-pyranose ratio (compare D-glucose and D-galactose), we would expect that talose would behave very much like mannose and that the α-pyranose form would be preferred at equilibrium. This is indeed the case; the α-pyranose form predominates at equilibrium, the observed α/β ratio being 78:22.

(c) The pyranose form of D-xylose is just like that of D-glucose except that it lacks a CH_2OH group.

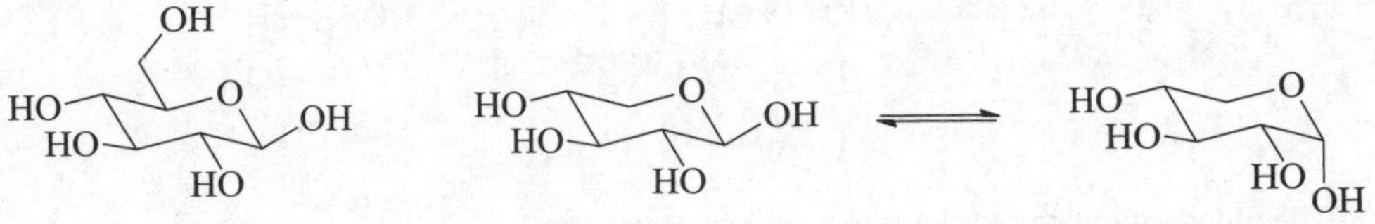

β-D-Glucopyranose   β-D-Xylopyranose   α-D-Xylopyranose
(64% and equilibrium)

We would expect the equilibrium between pyranose forms in D-xylose to be much like that in D-glucose and predict that the β-pyranose form would predominate. It is observed that the β/α ratio in D-xylose is 64:36, exactly the same as in D-glucose.

(d) The pyranose form of D-lyxose is like that of D-mannose except that it lacks a CH_2OH group. As in D-mannopyranose, the α form should predominate over the β.

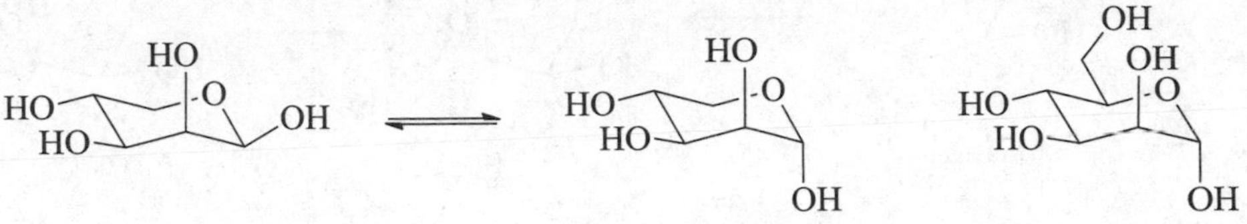

β-D-Lyxopyranose   α-D-Lyxopyranose   α-D-Mannopyranose
(68% at equilbrium)

The observed α/β distribution ratio in D-lyxopyranose is 73:27.

24.34 The arrows point to the acetal carbons of topiramate.

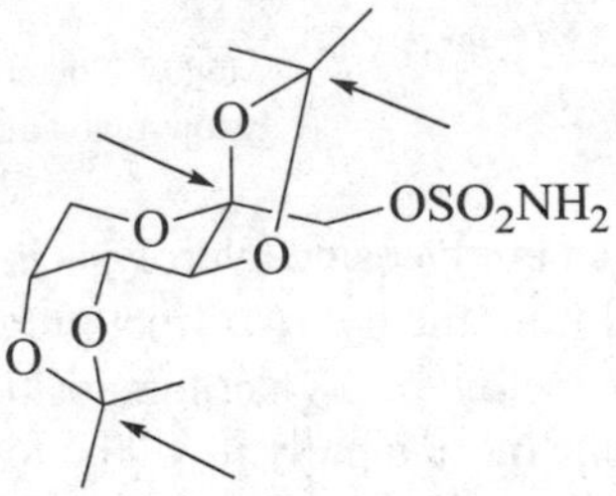

Topiramate

24.35 Unravel the lactone and orient the carbon chain to reveal its stereochemistry.

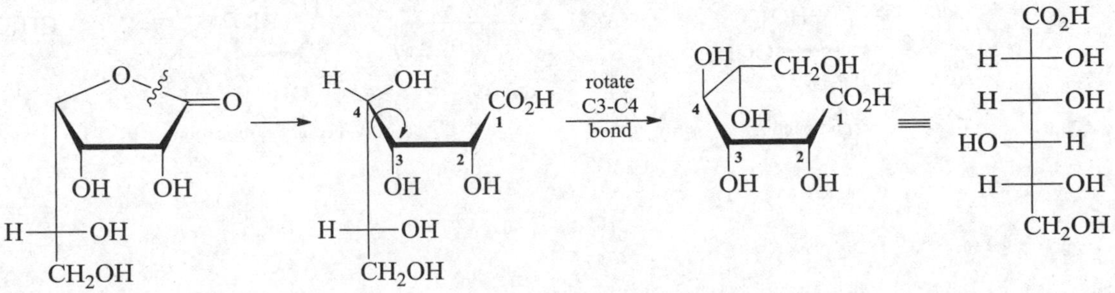

The chain-extension method adds a carboxyl group to the aldehyde end of an aldose. To determine the aldose, disconnect the carboxyl group and transform the next carbon to an aldehyde.

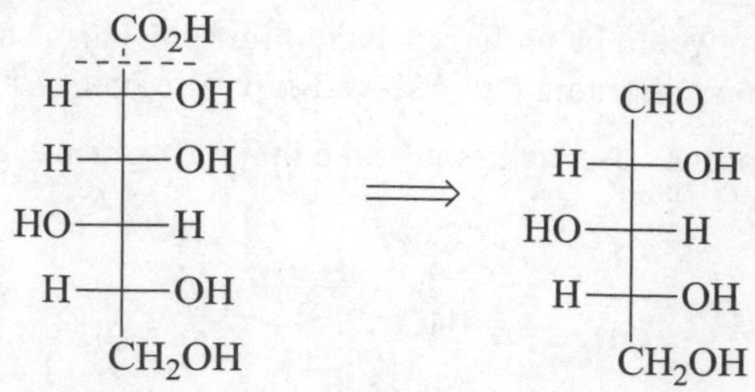

The aldose subjected to chain extension is D-xylose.

Mechanism

24.36 Acid-catalyzed addition of methanol to the glycal proceeds by regioselective protonation of the double bond in the direction that leads to the more stable carbocation. The more stable carbocation is the one stabilized by the ring oxygen.

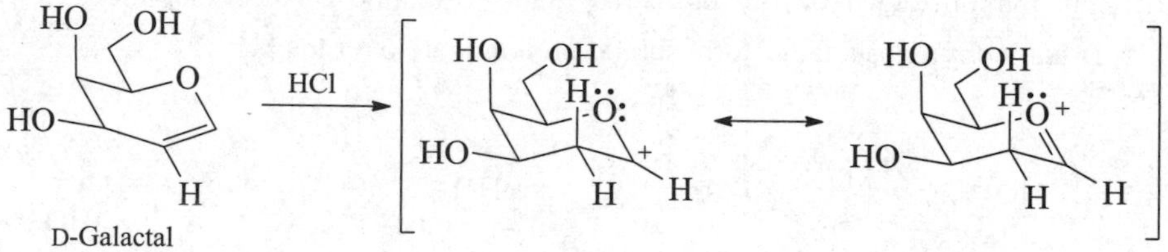

D-Galactal

Capture on either face of the carbocation by methanol yields the α- and β-methyl glycosides.

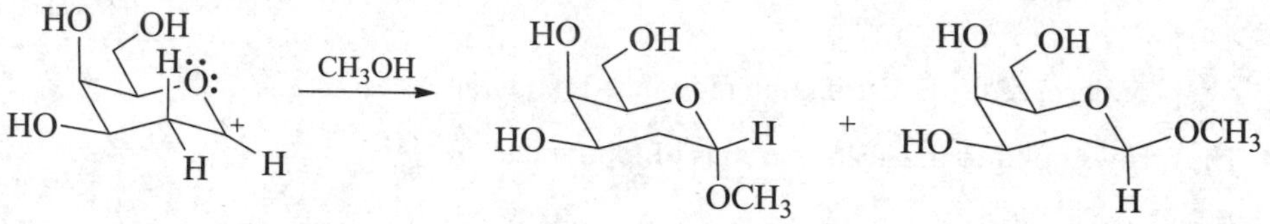

Methyl 2-deoxy-α-D-lyxopyranoside (38%)

Methyl 2-deoxy-β-D-lyxopyranoside (36%)

24.37 (*a*) The rate-determining step in glycoside hydrolysis is carbocation formation at the anomeric position. The carbocation formed from methyl α-D-fructofuranoside (compound A) is tertiary and therefore more stable than the one from methyl α-D-glucofuranoside (compound B), which is secondary. The more stable a carbocation is, the more rapidly it will be formed.

Faster:

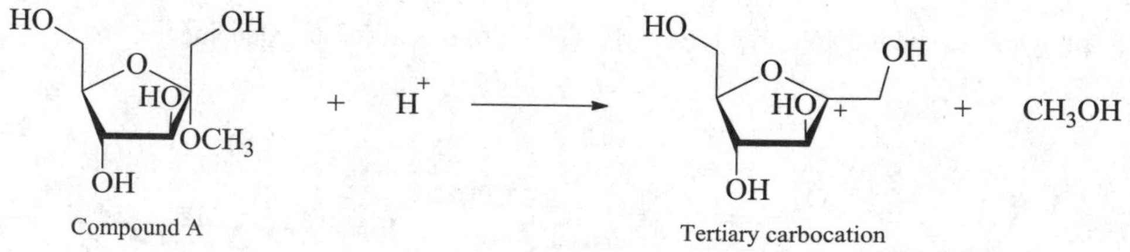

Compound A

Tertiary carbocation

Slower:

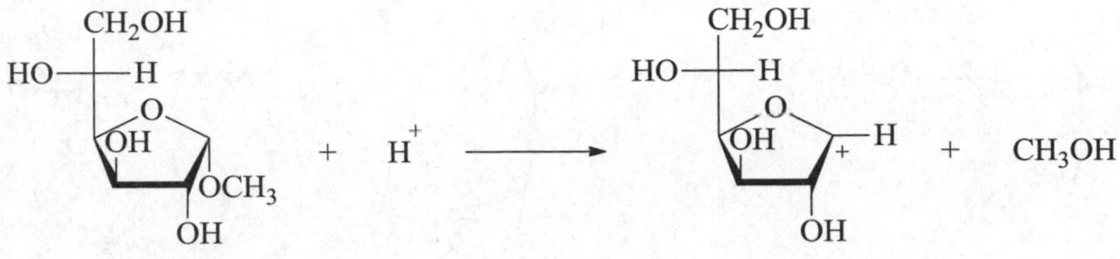

Compound B   Secondary carbocation

(b) The carbocation formed from methyl β-D-glucopyranoside (compound D) is less stable than the one from its 2-deoxy analog (compound C) and is formed more slowly. It is destabilized by the electron-withdrawing inductive effect of the hydroxyl group at C-2.

Faster:

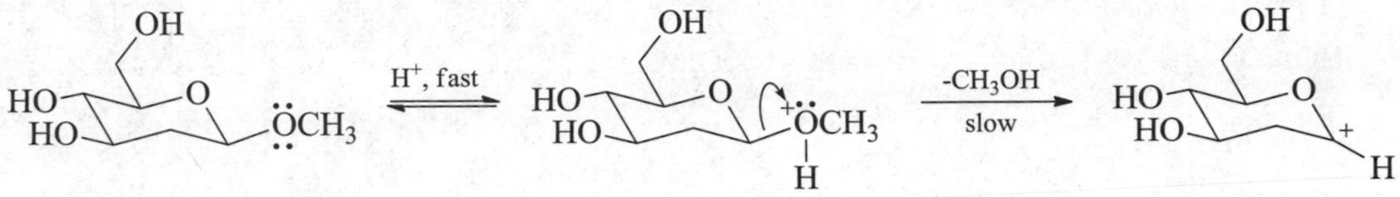

Compound C   More stable

Slower:

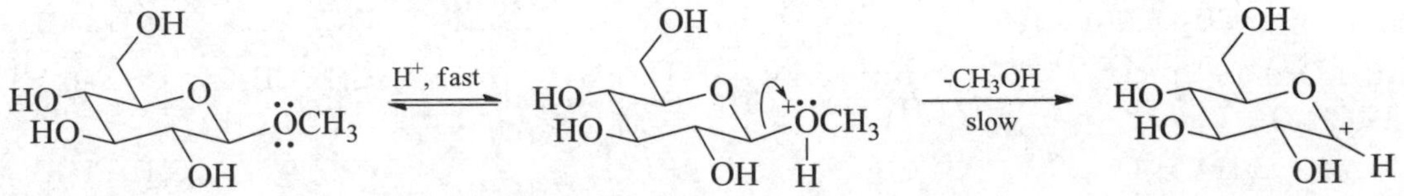

Compound D   Less stable

24.38 An acetal formation mechanism that is outlined in Mechanism 18.4, Chapter 18, is followed. The mechanism could also be written with the other secondary alcohol group reacting with protonated acetone.

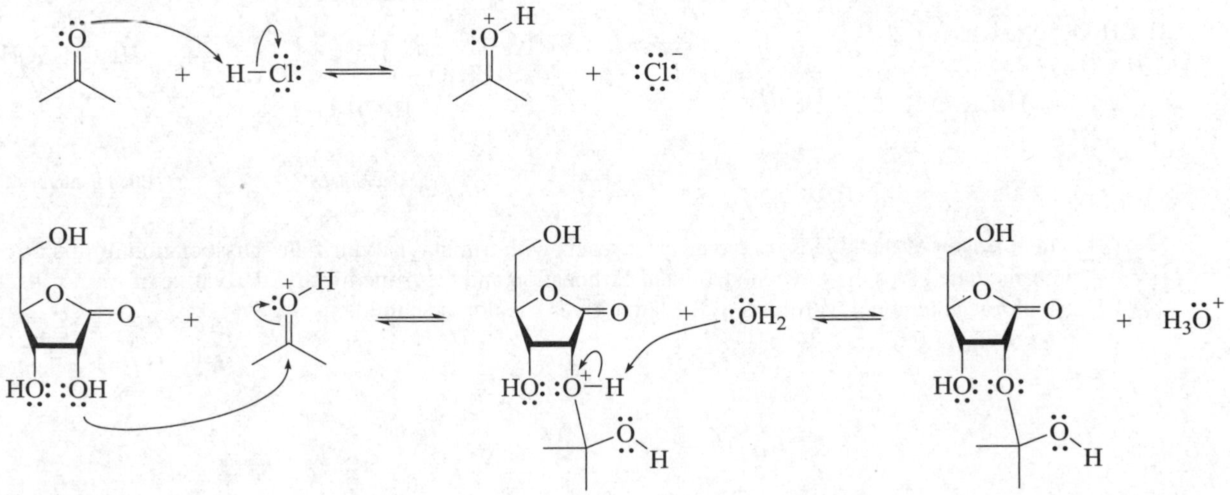

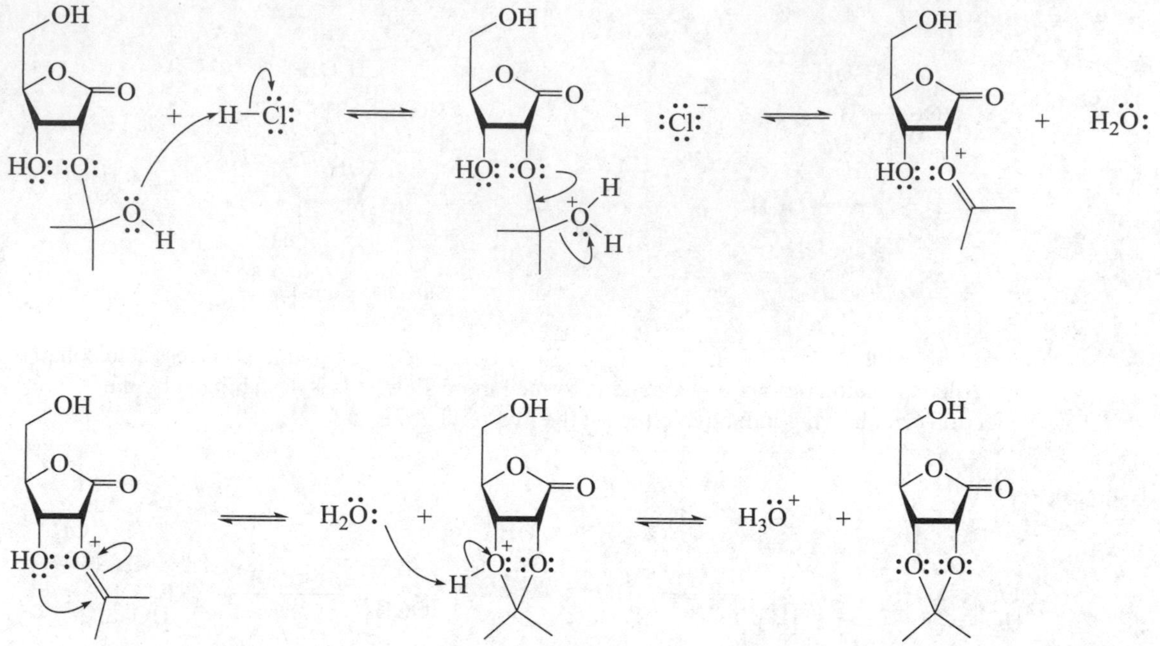

24.39 (*a*) Methylation of the thioglycoside occurs to give a sulfonium salt that dissociates to give the oxygen-stabilized carbocation and ethyl methyl sulfide. Alkylation of sulfides to give sulfonium salts is described in Section 17.17.

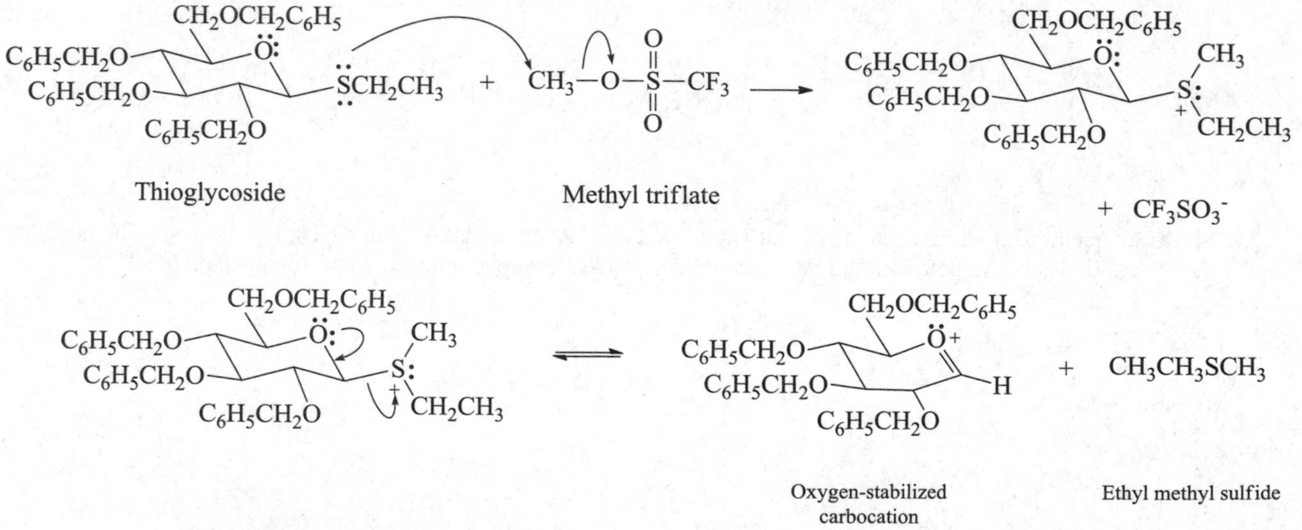

Thioglycoside   Methyl triflate

+ CF₃SO₃⁻

Oxygen-stabilized   Ethyl methyl sulfide
carbocation

(*b*) The nitrogen of the glycosyl imidate ester reacts with trimethylsilyl triflate. Dissociation of this intermediate gives the oxygen-stabilized carbocation and the trimethylsilyl derivative of trichloroacetamide. Hydrolysis of the latter gives trichloroacetamide.

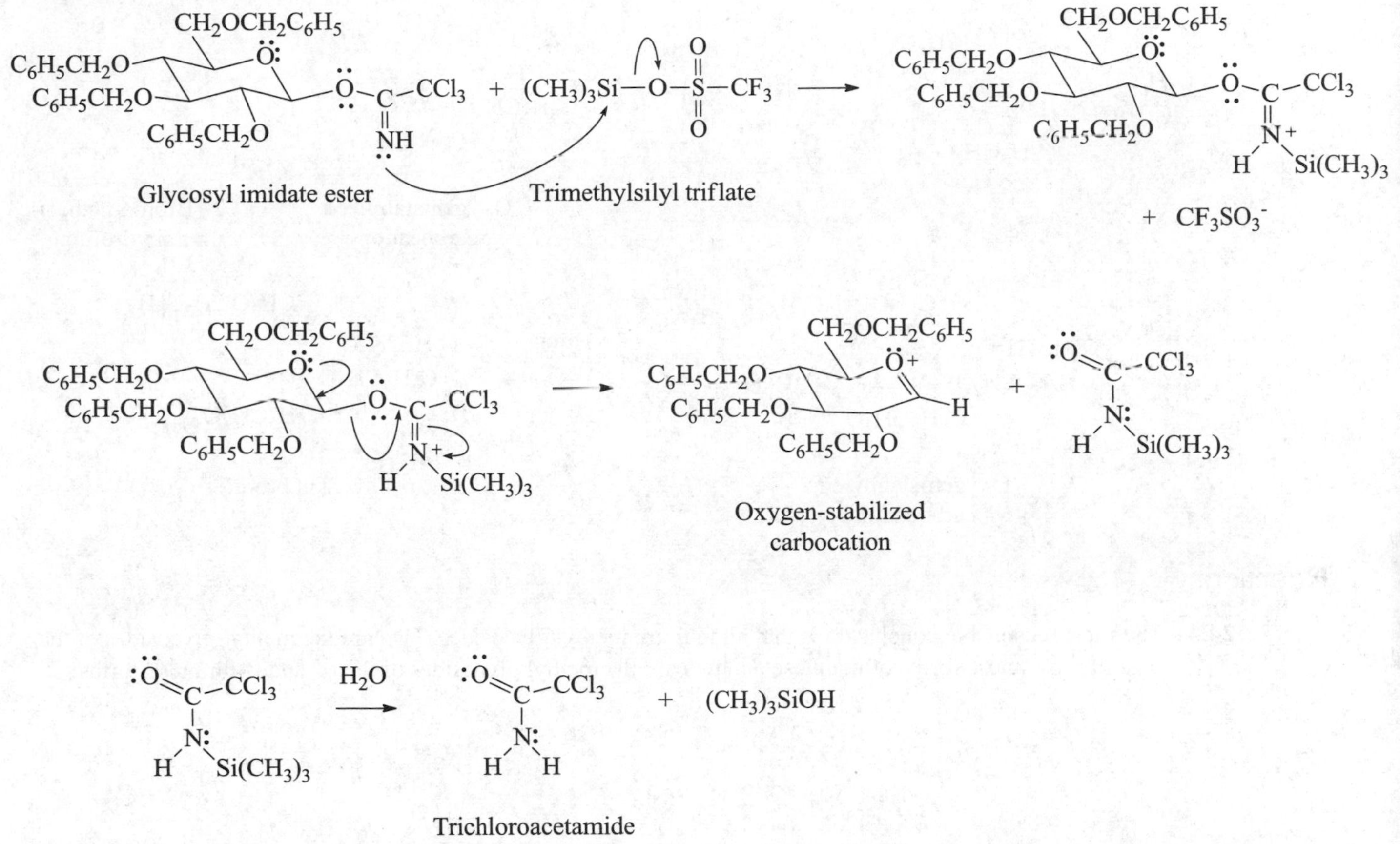

24.40 Bromination of the alkene (Step 1) gives a bromonium ion that can cyclize by reaction with the oxygen attached to the anomeric center to give an oxonium ion (Step 2). Dissociation of 2-(bromomethyl)-tetrahydrofuran (Step 3) gives the oxygen-stabilized carbocation. The oxygen-stabilized carbocation reacts with bromide (Step 4) to give the α anomer shown. The α anomer is favored over the β anomer by the anomeric effect.

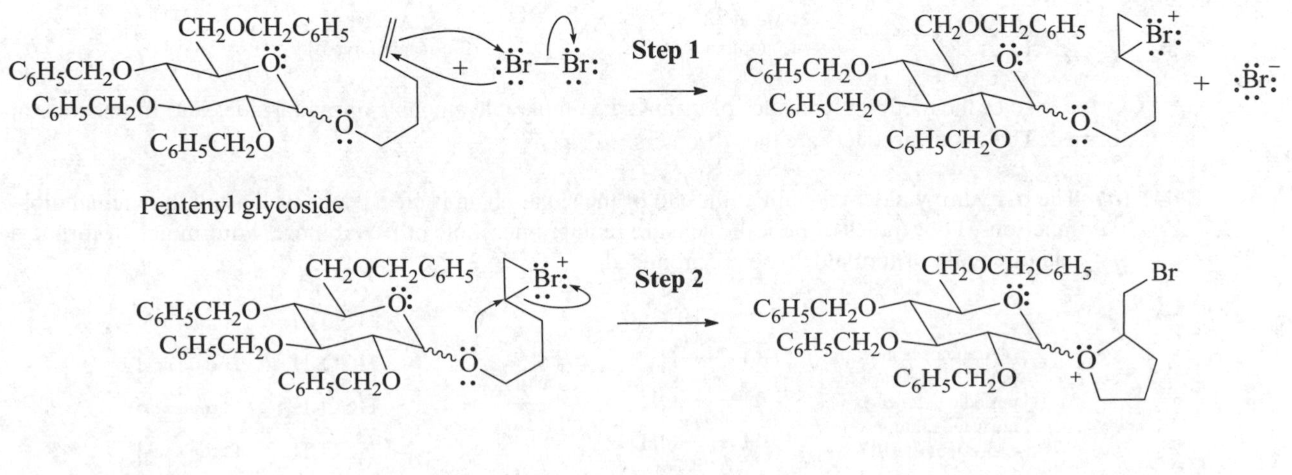

Pentenyl glycoside

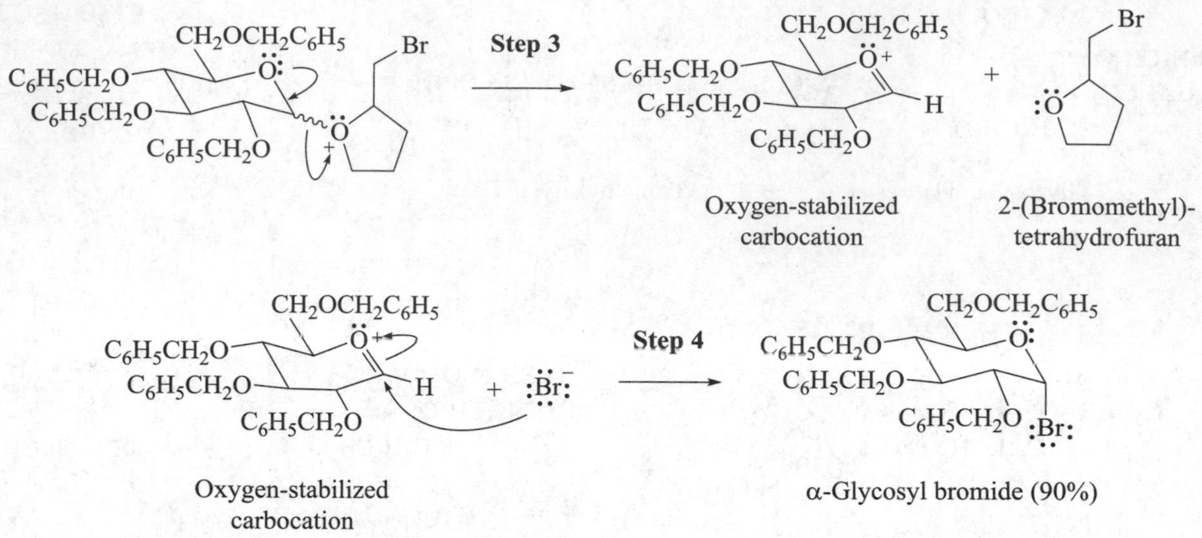

Oxygen-stabilized carbocation   2-(Bromomethyl)-tetrahydrofuran

Oxygen-stabilized carbocation   α-Glycosyl bromide (90%)

Reactions

24.41 The most reasonable conclusion is that all four are methyl glycosides. Two are the methyl glycosides of the α- and β-pyranose forms of mannose and two are the methyl glycosides of the α- and β-furanose forms.

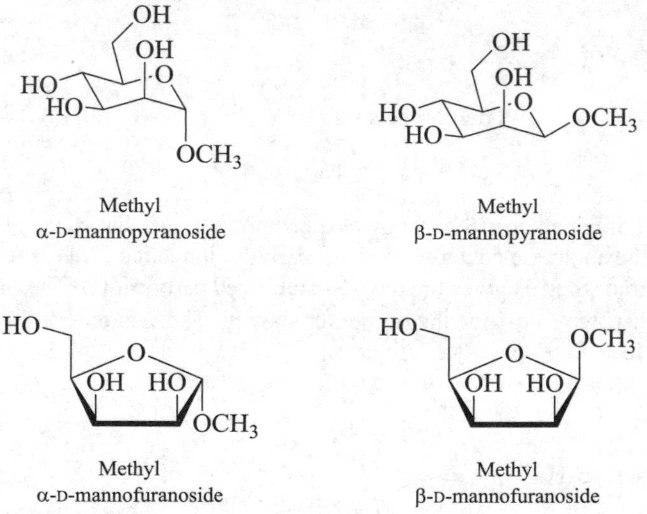

Methyl
α-D-mannopyranoside

Methyl
β-D-mannopyranoside

Methyl
α-D-mannofuranoside

Methyl
β-D-mannofuranoside

In the case of the methyl glycosides of mannose, comparable amounts of pyranosides and furanosides are formed. The major products are the α isomers.

24.42 (*a*) The α-hydroxy aldehyde unit at the end of the sugar chain is cleaved, as well as all the vicinal diol functions. Four moles of periodic acid are required per mole of D-arabinose. Four moles of formic acid and one mole of formaldehyde are produced.

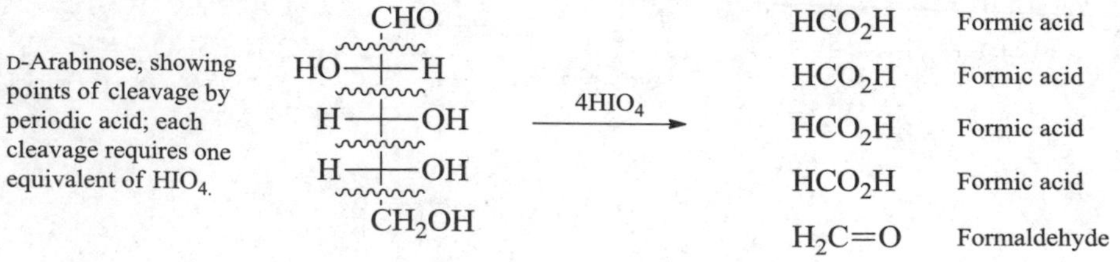

D-Arabinose, showing points of cleavage by periodic acid; each cleavage requires one equivalent of HIO_4.

HCO_2H Formic acid

HCO_2H Formic acid

HCO_2H Formic acid

HCO_2H Formic acid

$H_2C{=}O$ Formaldehyde

(*b*) The points of cleavage of D-ribose on treatment with periodic acid are as indicated.

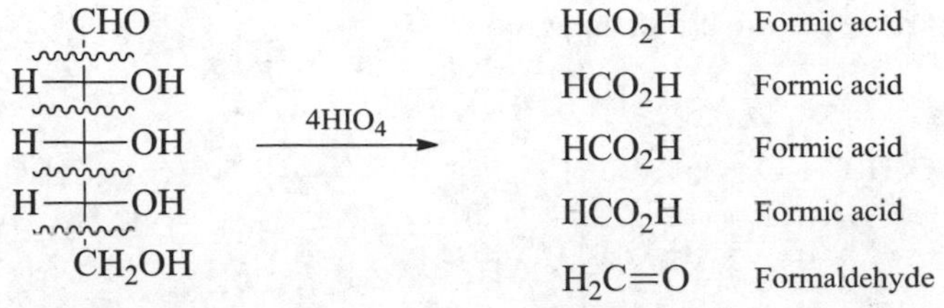

D-Ribose + 4 HIO$_4$ → 4 HCO$_2$H (Formic acid) + H$_2$C=O (Formaldehyde)

Four moles of periodic acid per mole of D-ribose is required. Four moles of formic acid and one mole of formaldehyde are produced.

(c) Write the structure of methyl β-D-glucopyranoside so as to identify the adjacent alcohol functions.

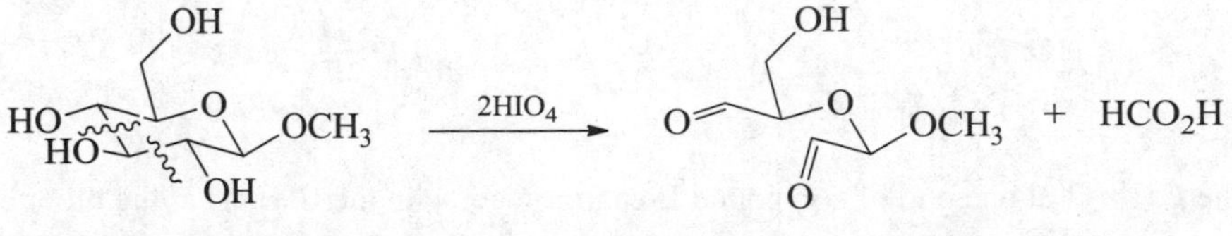

Methyl β-D-glucopyranoside + 2 HIO$_4$ → product + HCO$_2$H

Two moles of periodic acid per mole of glycoside is required. One mole of formic acid is produced.

(d) There are two independent vicinal diol functions in this glycoside. Two moles of periodic acid are required per mole of substrate.

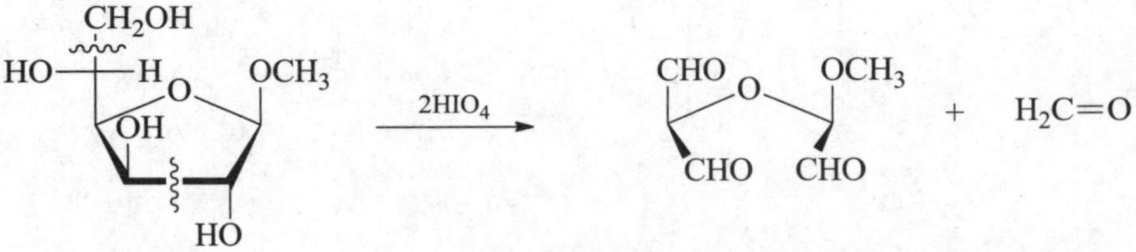

substrate + 2 HIO$_4$ → product + H$_2$C=O

24.43 Given the fact that triphenylmethyl chloride reacts preferentially with primary alcohols and there is only one primary alcohol in the starting material, and that this reaction is reversible with acid, this reagent can be utilized as a protecting group.

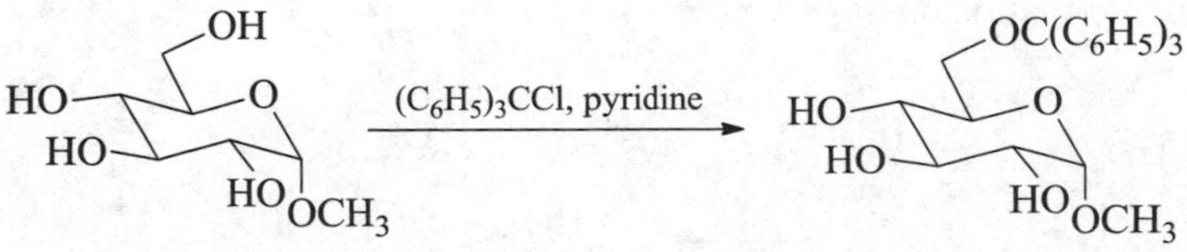

starting material + (C$_6$H$_5$)$_3$CCl, pyridine → product

Reaction of the secondary alcohol groups can be accomplished by treatment with benzyl chloride and base in an S$_N$2 reaction (Table 24.2).

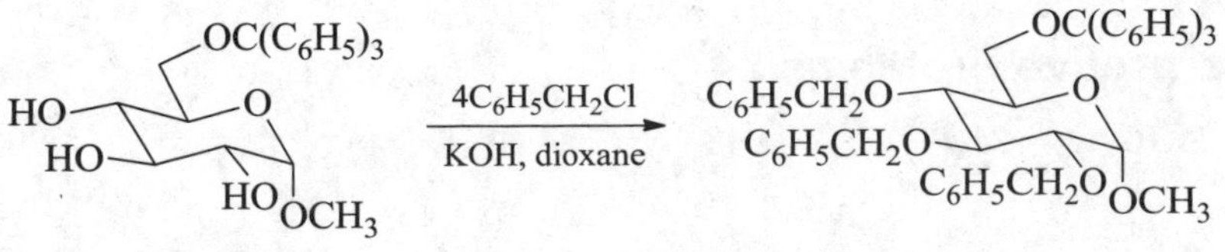

product + 4 C$_6$H$_5$CH$_2$Cl, KOH, dioxane → benzylated product

The last step removes the protecting group.

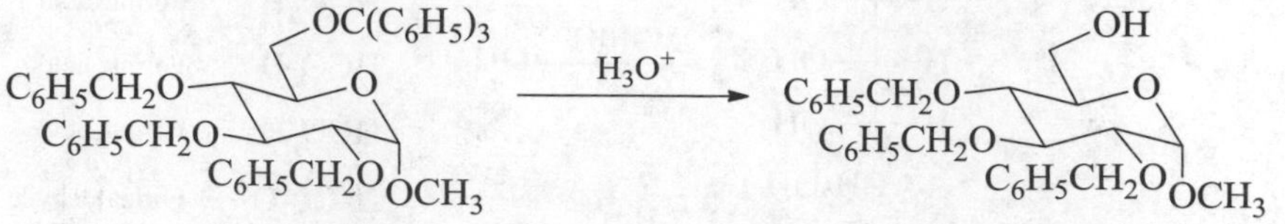

24.44 Periodic acid cleaves the vicinal diol function of compound A.

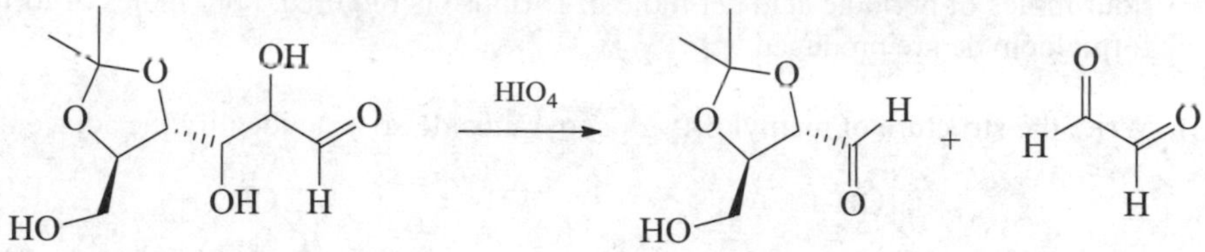

| Compound A | Compound B |

The CH—O at one end of compound B cannot react with the CH$_2$OH at the other end to give a cyclic hemiacetal because the structure produced would have a highly strained trans fusion between two five-membered rings.

Compound B is an acetal. Acid hydrolysis converts it to acetone and compound C.

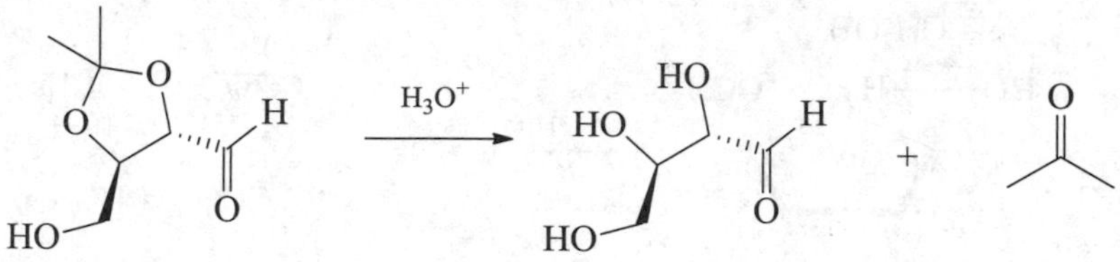

Compound B

Compound C
(C$_4$H$_8$O$_4$)

The acyclic structure of compound C is in equilibrium with the cyclic hemiacetal structure. Oxidation of compound C converts it to the γ-lactone (compound D). You can refer to Problem 24.22 and Section 24.18 for more information on this last step.

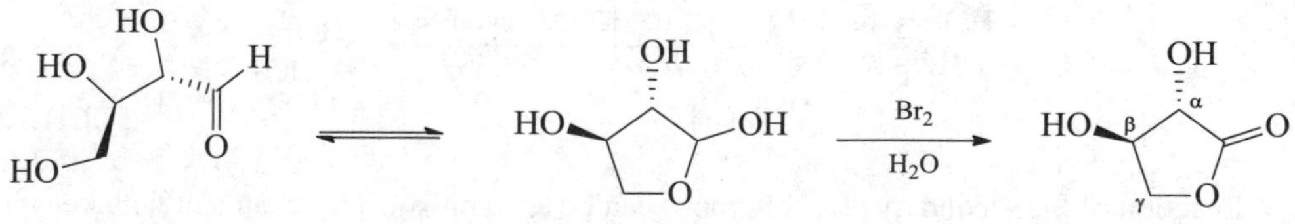

Compound C
(C$_4$H$_8$O$_4$)

Compound C
(hemiacetal form)

Compound D
(C$_4$H$_6$O$_4$)

Answers to Interpretive Problems 24

24.45 E;   **24.46** F;   **24.47** C;   **24.48** A;   **24.49** D

SELF-TEST

1. The structure of D-erythrose is shown. Draw the following:

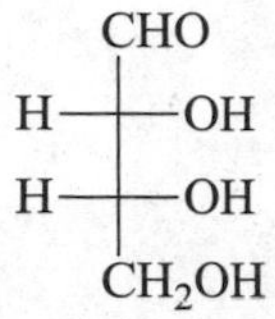

$$
\begin{array}{c}
\text{CHO} \\
\text{H}\!-\!\!-\!\text{OH} \\
\text{H}\!-\!\!-\!\text{OH} \\
\text{CH}_2\text{OH}
\end{array}
$$

(a) The enantiomer of D-erythrose

(b) A diastereomer of D-erythrose

(c) The α-furanose form of D-erythrose (use a Haworth formula)

(d) The anomer of the structure in part (c)

(e) Assign the configuration of each chirality center of D-erythrose as either R or S.

2. The structure of D-mannose is

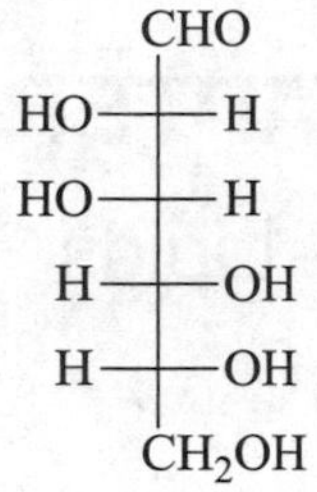

$$
\begin{array}{c}
\text{CHO} \\
\text{HO}\!-\!\!-\!\text{H} \\
\text{HO}\!-\!\!-\!\text{H} \\
\text{H}\!-\!\!-\!\text{OH} \\
\text{H}\!-\!\!-\!\text{OH} \\
\text{CH}_2\text{OH}
\end{array}
$$

D-Mannose

Using Fischer projections, draw the product of the reaction of D-mannose with

(a) $NaBH_4$ in H_2O

(b) Excess periodic acid

3. Referring to the structure of D-arabinose shown, draw the following:

(a) The α-pyranose form of D-arabinose

(b) The β-furanose form of D-arabinose

(c) The β-pyranose form of L-arabinose

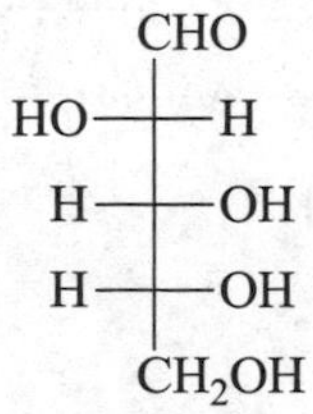

$$
\begin{array}{c}
\text{CHO} \\
\text{HO}\!-\!\!-\!\text{H} \\
\text{H}\!-\!\!-\!\text{OH} \\
\text{H}\!-\!\!-\!\text{OH} \\
\text{CH}_2\text{OH}
\end{array}
$$

D-Arabinose

4. Using text Figure 24.2, identify the following carbohydrate:

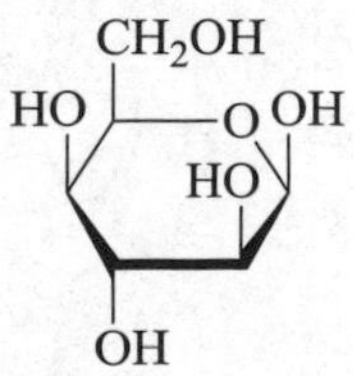

5. Write structural formulas for the α- and β-methyl pyranosides formed from the reaction of D-mannose (see Problem 2 for its structure) with methanol in the presence of hydrogen chloride. How are the two products related? Are they enantiomers? Diastereomers?

CHAPTER 25
Lipids

Table of Contents

SOLUTIONS TO TEXT PROBLEMS

In Chapter Problems

25.1 The triacylglycerol shown in text Figure 25.2*a*, with an oleyl group at C-2 of the glycerol unit and two stearyl groups at C-1 and C-3, yields stearic and oleic acids in a 2:1 molar ratio on hydrolysis. A constitutionally isomeric structure in which the oleyl group is attached to C-1 of glycerol would yield the same hydrolysis products.

Oleic acid Glycerol Stearic acid

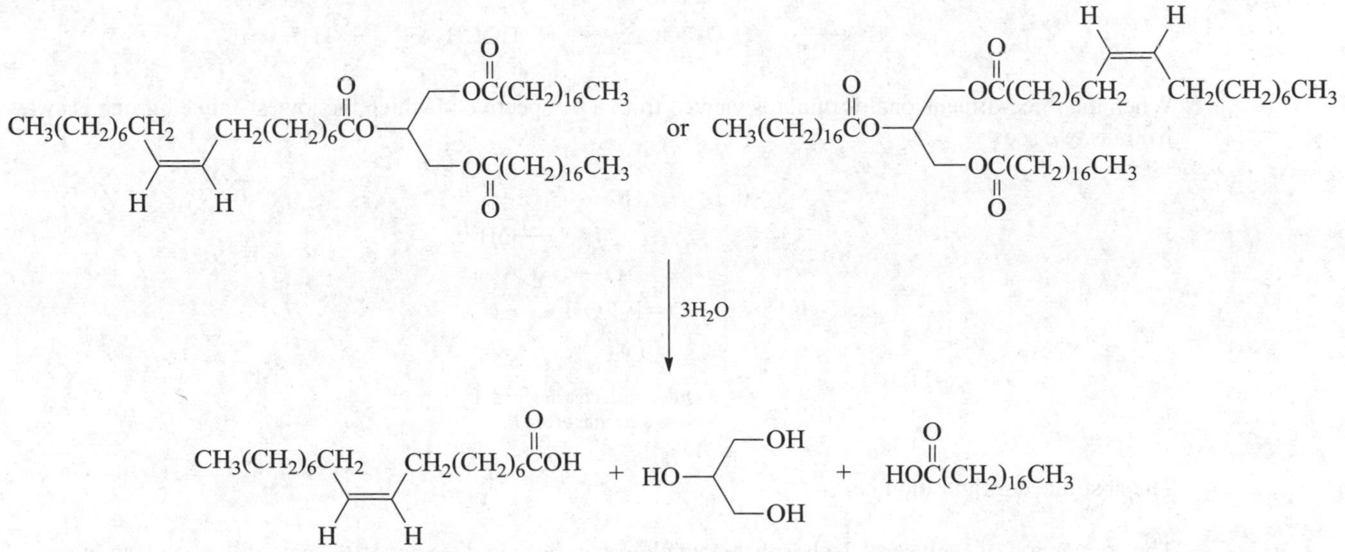

25.2 The first two cycles are shown in the text. The outline for these and the last two is:

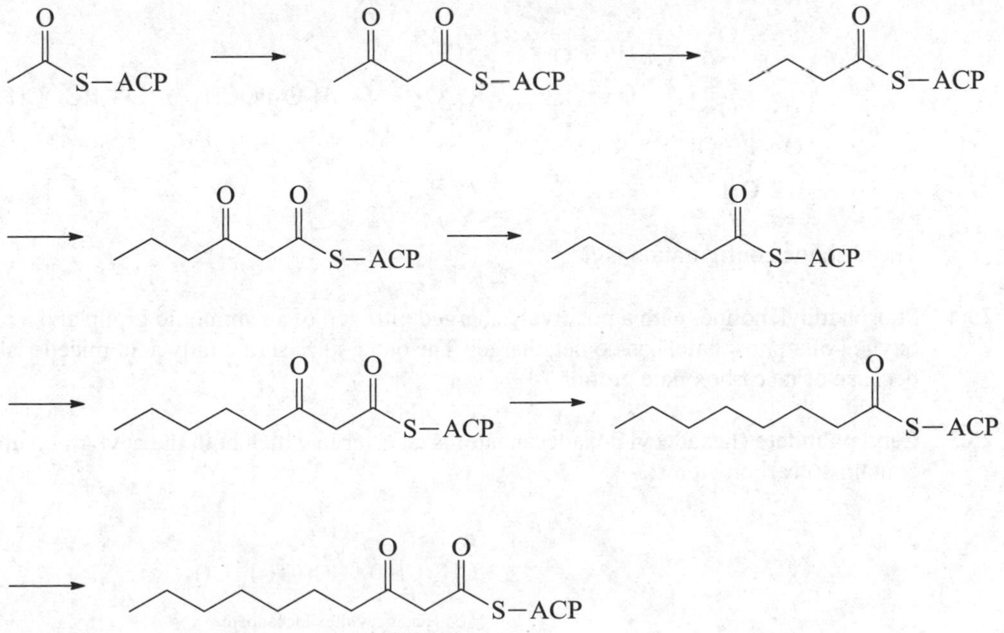

Keto acyl-ACP after four cycles

25.3 The structure of L-glycerol 3-phosphate is shown in dash-wedge notation. The hydrogen is understood to be projected behind the plane of the paper.

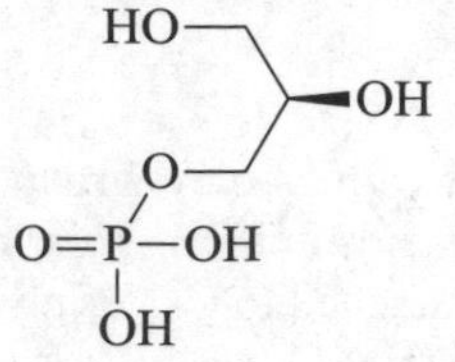

same as

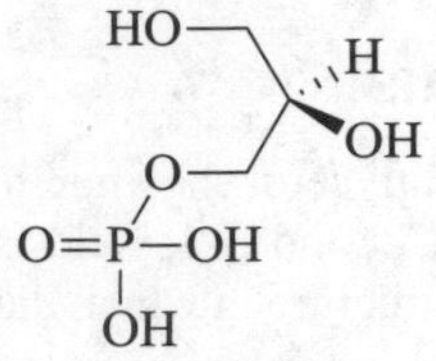

The order of decreasing sequence rule precedence is

$$HO- \quad > \quad H_2O_3POCH_2- \quad > \quad HOCH_2- \quad > \quad H-$$

When the three-dimensional formula is viewed from a perspective in which the lowest-ranked group is away from us, we see

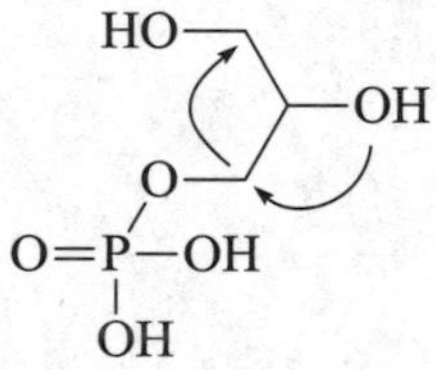

Order of decreasing rank is
clockwise, therefore *R*.

The absolute configuration is *R*.

The conversion of L-glycerol 3-phosphate to a phosphatidic acid does not affect any of the bonds to the chirality center, nor does it alter the sequence rule ranking of the groups.

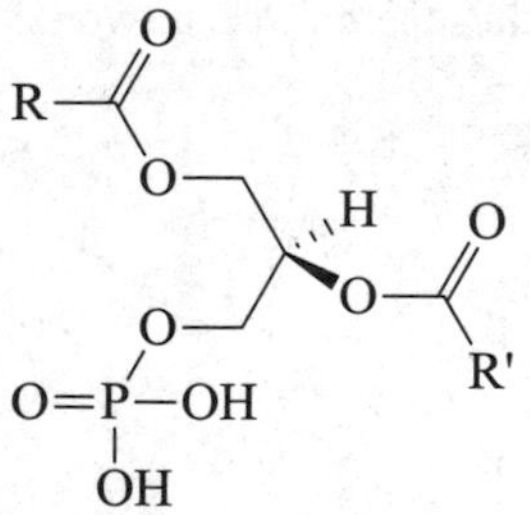

$$R'\overset{O}{\overset{\|}{C}}O- \quad > \quad H_2O_3POCH_2- \quad > \quad R\overset{O}{\overset{\|}{C}}OCH_2- \quad > \quad H-$$

The absolute configuration is *R*.

25.4 Phosphatidylcholine, with a positively charged nitrogen of an ammonio group and a negatively charged oxygen of a phosphate, has no net charge. The outer surface of a fatty acid micelle is negatively charged because of its carboxylate groups.

25.5 Cetyl palmitate (hexadecyl hexadecanoate) is an ester in which both the acyl group and the alkyl group contain 16 carbon atoms.

$$CH_3(CH_2)_{14}\overset{O}{\overset{\|}{C}}O(CH_2)_{15}CH_3$$

Hexadecyl hexadecanoate

25.6 The biosynthesis of PGE$_2$ is described in the text. The fatty acid precursor of PGE$_2$ is *cis,cis,cis,cis*-5,8,11,14-icosatetraenoic acid, better known by its common name arachidonic acid.

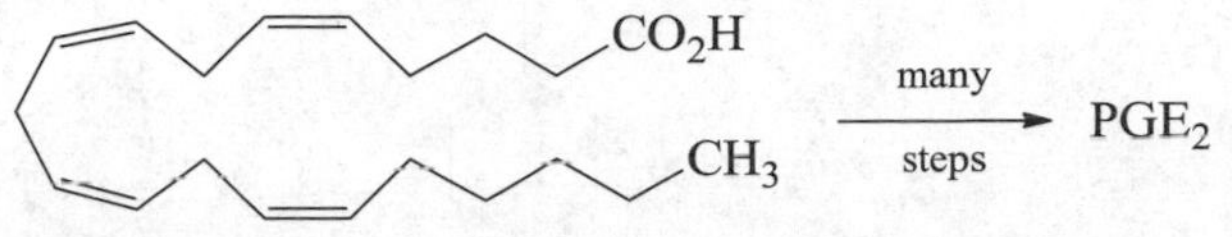

cis,cis,cis,cis-5,8,11,14-Icosatetraenoic
acid (arachidonic acid)

PGE$_1$, shown in text Figure 25.6, has one less double bond than PGE$_2$. By analogy, the fatty acid precursor to PGE$_1$ also has one less double bond.

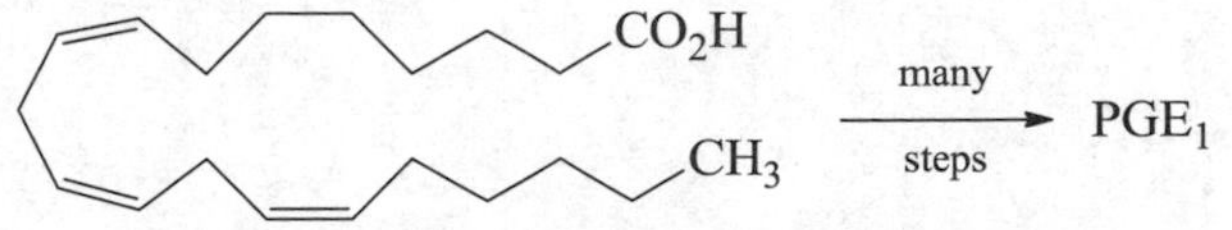

cis,cis,cis-8,11,14-Icosatrienoic acid

25.7 We are told that LTA$_4$ is an epoxide. The thiol group of glutathione (abbreviated GSH in the following reaction) acts as a nucleophile and opens the epoxide ring. In abbreviated form, the reaction can be viewed as

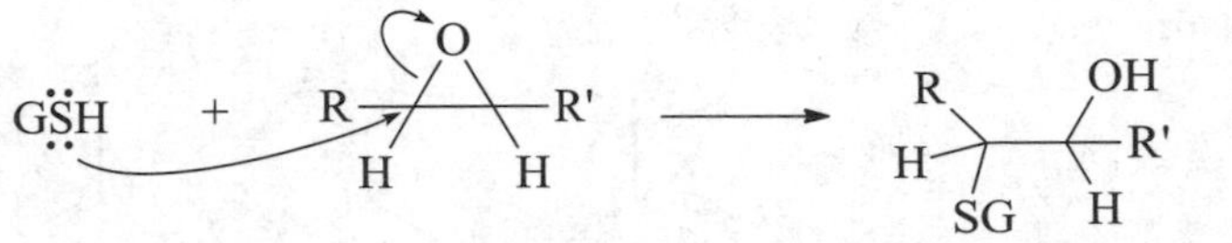

Reasoning backward from the structure of leukotriene C$_4$ (LTC$_4$) in the text, we can deduce the structure of the epoxide leukotriene A$_4$ (LTA$_4$).

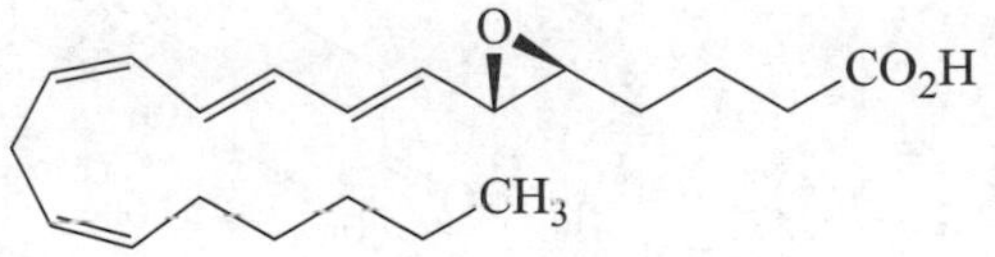

Leukotriene A$_4$ (LTA$_4$)

25.8 Isoprene units are 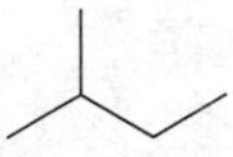 fragments in the carbon skeleton. Functional groups and multiple bonds are ignored when structures are examined for the presence of isoprene units.

α-Phellandrene (two equally correct answers):

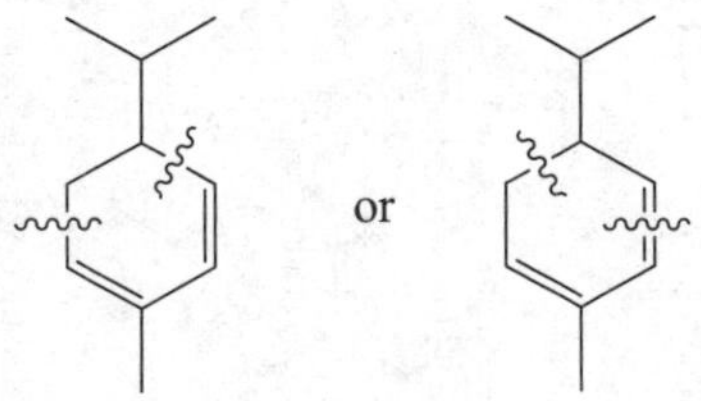

Menthol (same carbon skeleton as α-phellandrene but different functionality):

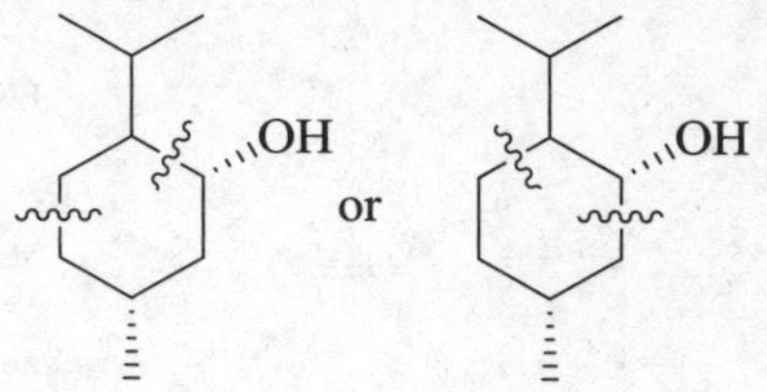

Citral:

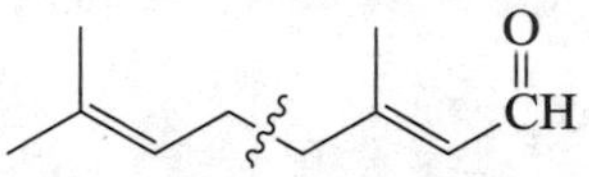

α-Selinene:

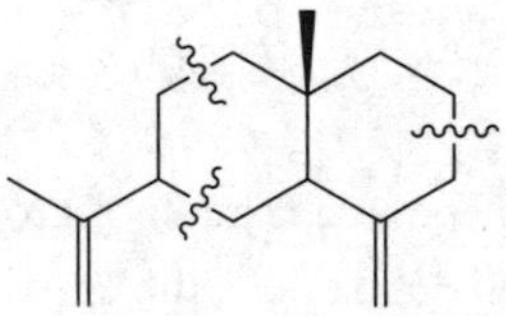

Farnesol:

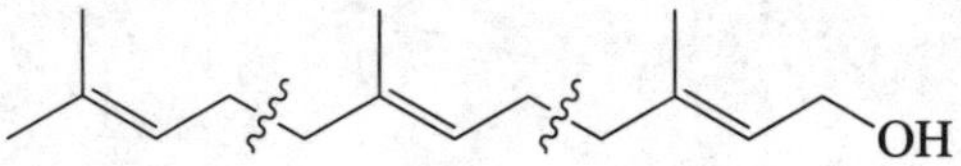

Abscisic acid:

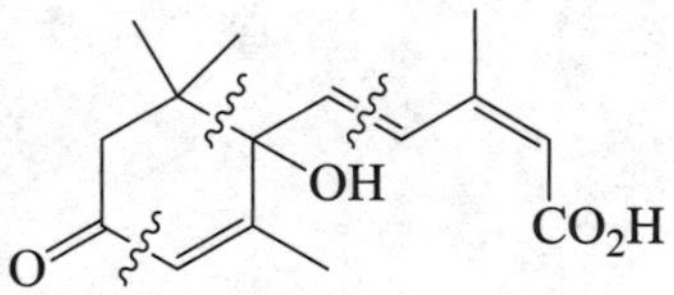

Cembrene (two equally correct answers):

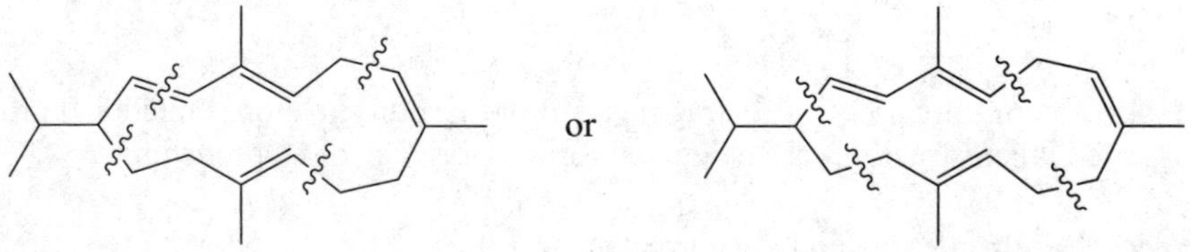

Vitamin A:

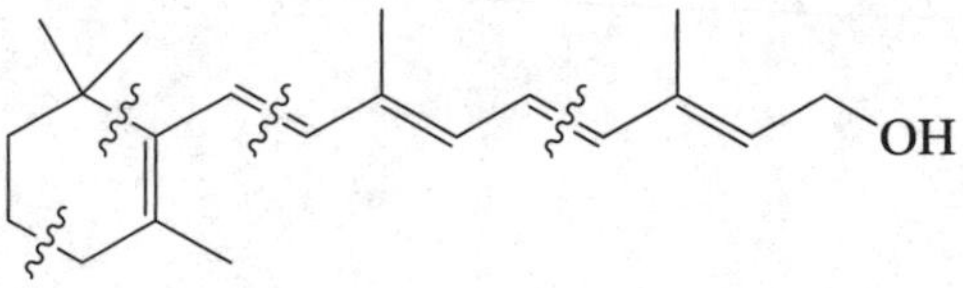

25.9 β-Carotene is a tetraterpene because it has 40 carbon atoms. The tail-to-tail linkage is at the midpoint of the molecule and connects two 20-carbon fragments.

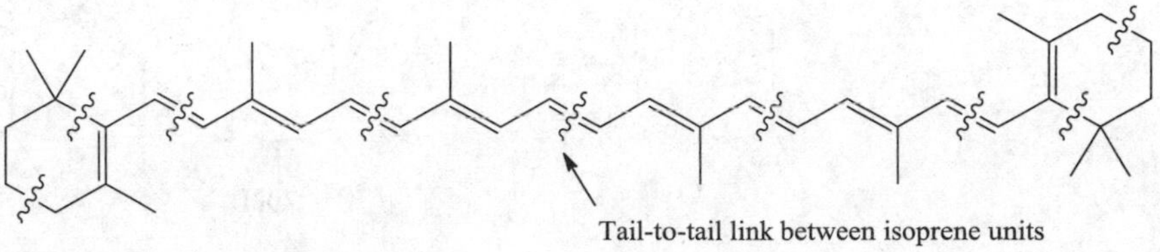

Tail-to-tail link between isoprene units

25.10 Isopentenyl diphosphate acts as an alkylating agent toward farnesyl diphosphate. Alkylation is followed by loss of a proton from the carbocation intermediate, giving geranylgeranyl diphosphate. Hydrolysis of the diphosphate yields geranylgeraniol.

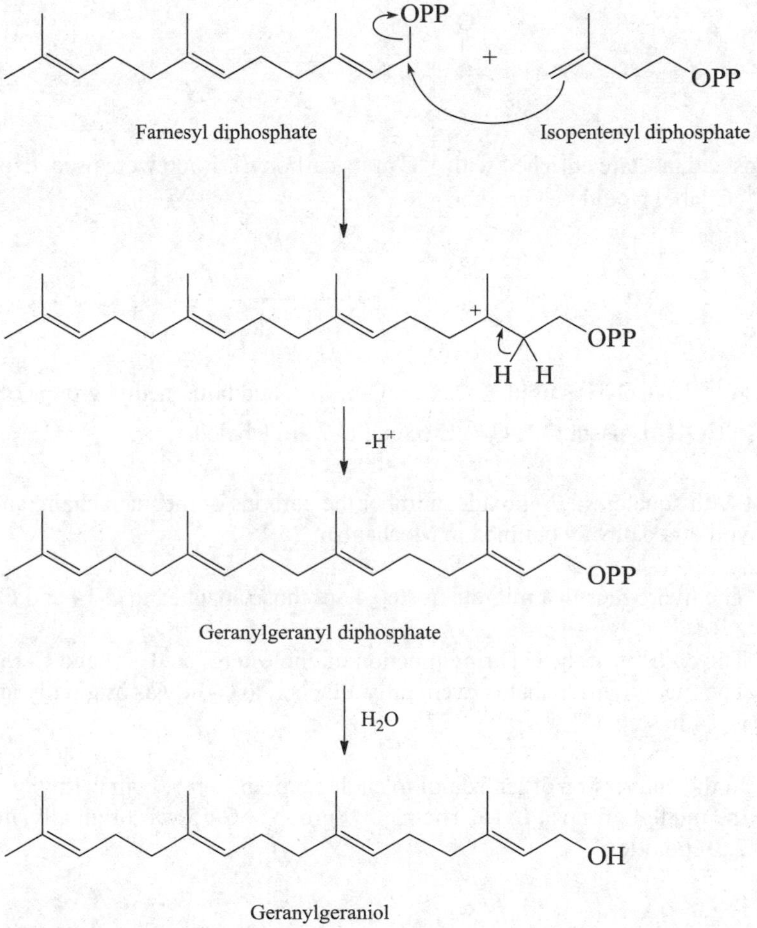

Farnesyl diphosphate   Isopentenyl diphosphate

Geranylgeranyl diphosphate

Geranylgeraniol

25.11 Borneol, the structure of which is given in text Figure 25.8, is a secondary alcohol. Oxidation of borneol converts it to the ketone camphor.

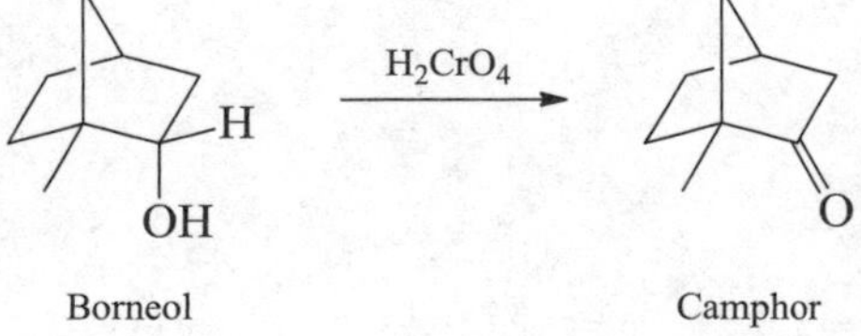

Borneol   Camphor

Reduction of camphor with sodium borohydride gives a mixture of stereoisomeric alcohols, of which one is borneol and the other isoborneol.

Camphor   Borneol   Isoborneol

25.12 Figure 25.9 in the text describes the distribution of ^{14}C (denoted by *) in citronellal biosynthesized from acetate enriched with ^{14}C in its methyl group.

If, instead, acetate enriched with ^{14}C at its carbonyl carbon were used, exactly the opposite distribution of the ^{14}C label would be observed.

When $^{14}CH_3CO_2H$ is used, C-2, C-4, C-6, C-8, and both methyl groups of citronellal are labeled. When $CH_3{}^{14}CO_2H$ is used, C-1, C-3, C-5, and C-7 are labeled.

25.13 Start with squalene 2,3-epoxide, number the carbons of the main chain, and track these carbons through the biosynthetic pathway outlined in Mechanism 25.1.

(*b*) The hydrogens that migrate in step 4 are those attached to C-14 and C-18 of squalene 2,3-epoxide.

(*c*) The carbons at the C,D ring junction of cholesterol are C-14 and C-15 of squalene 2,3-epoxide. The methyl group that is eventually attached to C-14 was originally attached to C-15 and migrated to C-14 in step 4.

(*d*) In the conversion of lanosterol to cholesterol in step 5, both methyl groups at C-2 are lost, as well as the methyl group at C-15. The methyl group at C-15 was originally attached to C-10 of squalene 2,3-epoxide.

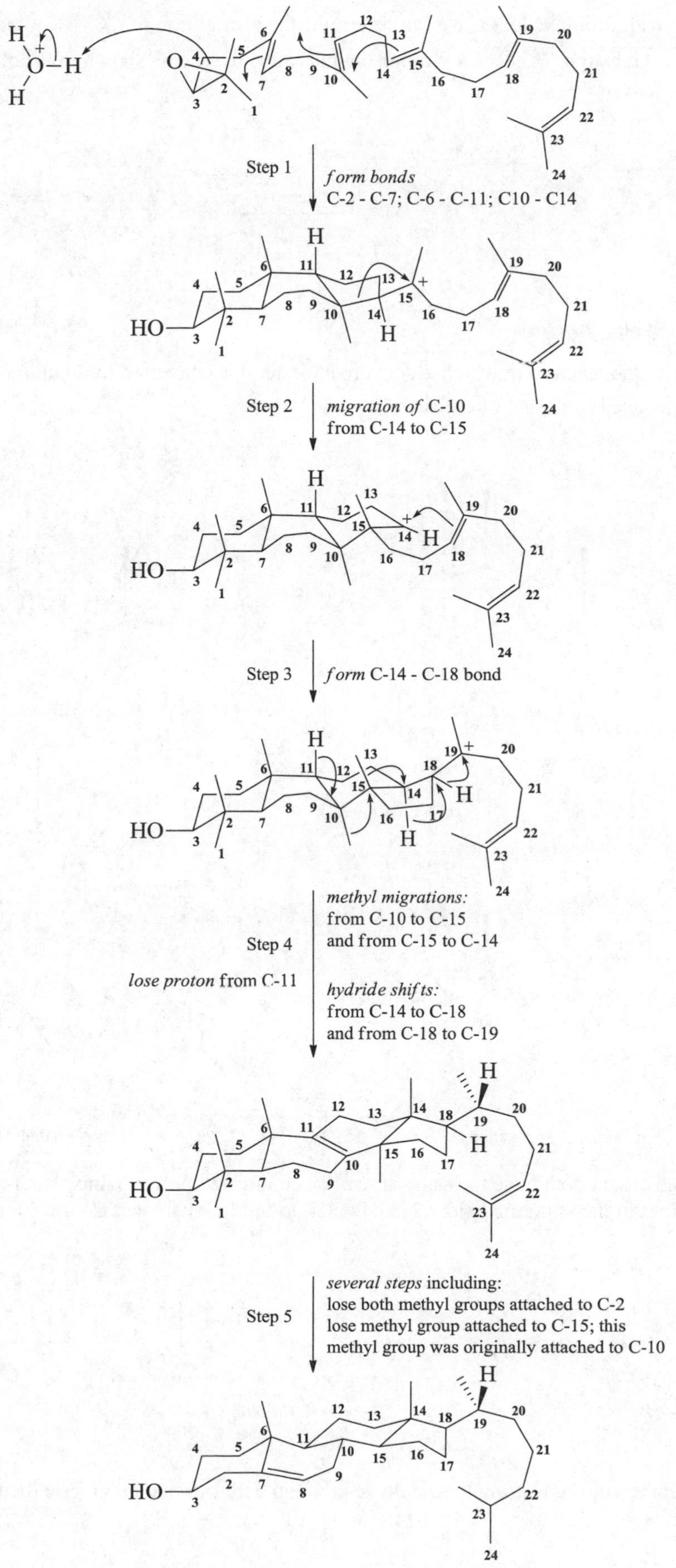

Step 1 | *form bonds*
C-2 - C-7; C-6 - C-11; C10 - C14

Step 2 | *migration of* C-10
from C-14 to C-15

Step 3 | *form* C-14 - C-18 bond

methyl migrations:
from C-10 to C-15
and from C-15 to C-14

Step 4

lose proton from C-11

hydride shifts:
from C-14 to C-18
and from C-18 to C-19

several steps including:
lose both methyl groups attached to C-2
lose methyl group attached to C-15; this
methyl group was originally attached to C-10

Step 5

25.14 By analogy with Problem 25.12, we can determine the distribution of ^{14}C (denoted by *) in squalene 2,3-epoxide. Track the ^{14}C labels through the biosynthetic scheme shown in Problem 25.13 to locate their position in cholesterol.

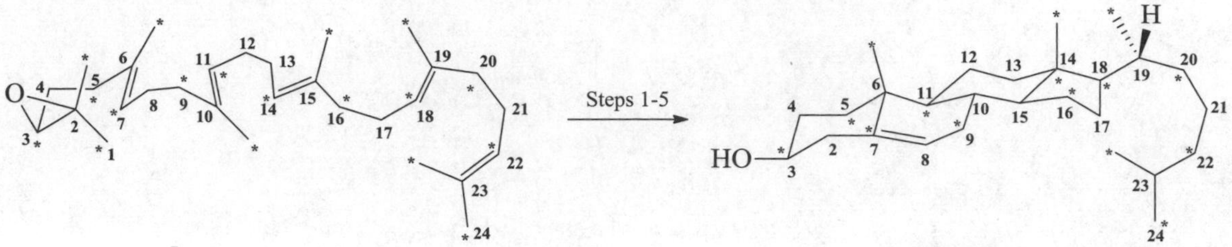

Squalene 2,3-epoxide   Cholesterol

25.15 By analogy to the reaction in which 7-dehydrocholesterol is converted to vitamin D_3, the structure of vitamin D_2 can be deduced from that of ergosterol.

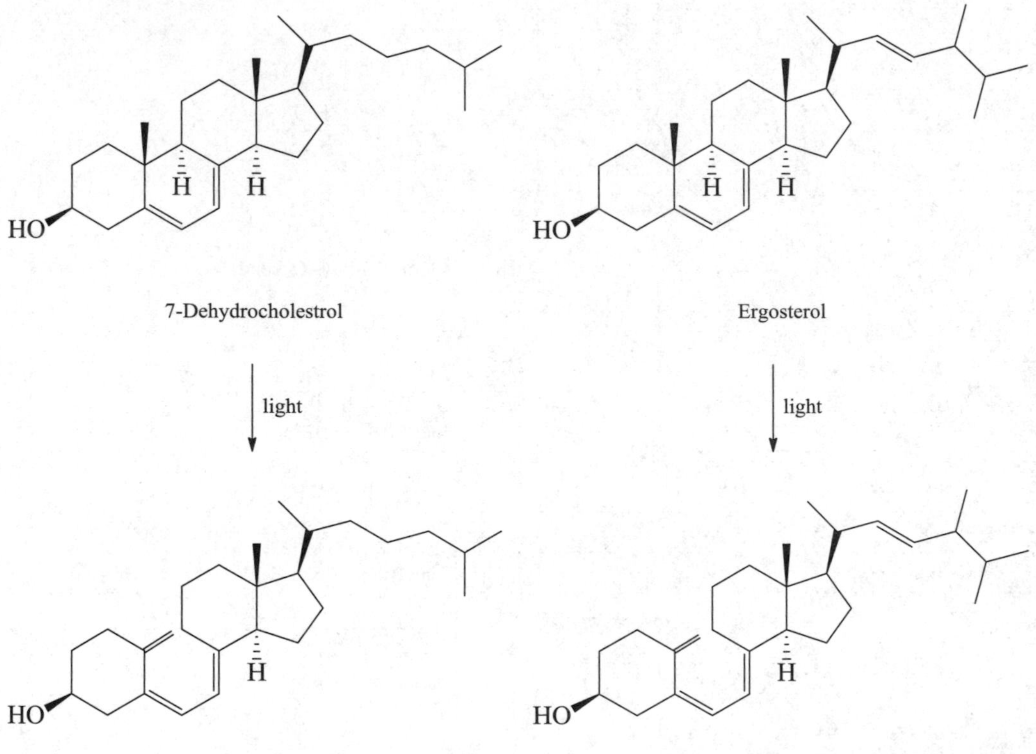

7-Dehydrocholestrol   Ergosterol

light   light

Vitamin D_3 Vitamin D_2

25.16 Crocetin and crocin both have the same 20-carbon conjugated polyene unit. The isoprene units are shown by the wavy lines in the structure below. Note the tail-to-tail link between isoprene units in the center.

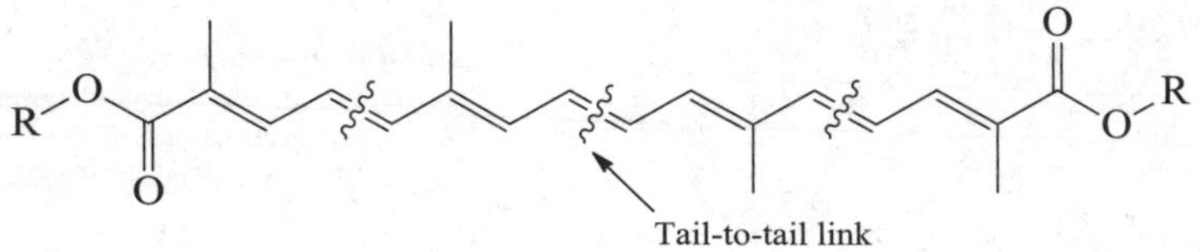

Tail-to-tail link

R = H: Crocetin
R = gentiobiose: Crocin

The six-membered ring bearing the aldehyde group in both safranal and picrocrocin is derived from two isoprene units.

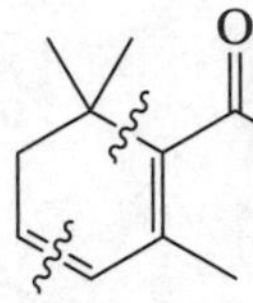

Safranal 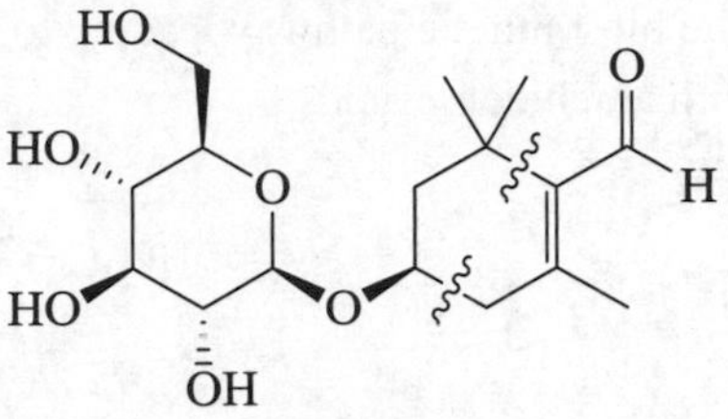   Picrocrocin

End of Chapter Problems

Structure

25.17 (*a*) Fatty acid biosynthesis proceeds by the joining of acetate units.

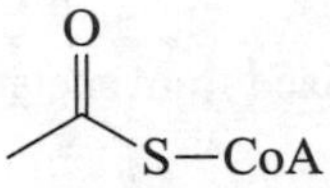

Acetyl coenzyme A 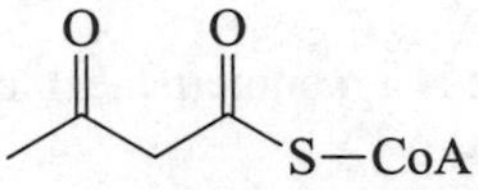   Acetoacetyl coenzyme A

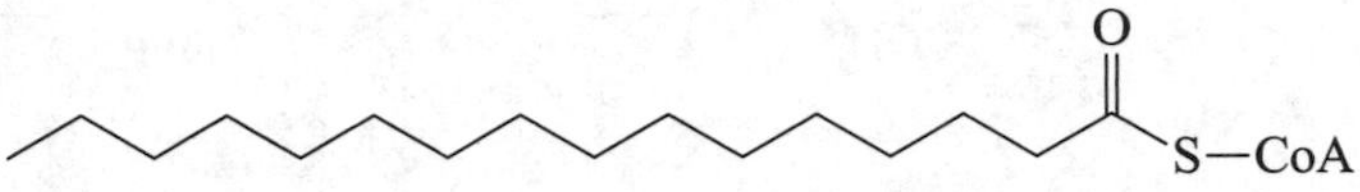

Palmitoyl coenzyme A

Thus, the even-numbered carbons will be labeled with ^{14}C when palmitic acid is biosynthesized from $^{14}CH_3CO_2H$ (^{14}C is represented as *).

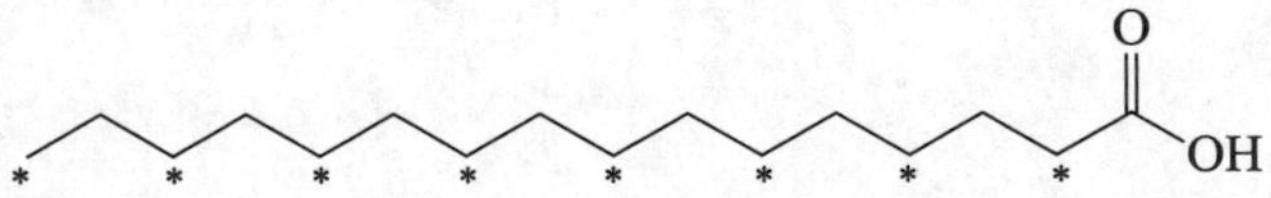

(*b*) As noted in text Section 25.6, arachidonic acid is the biosynthetic precursor of PGE_2. The distribution of the ^{14}C label in PGE_2 biosynthesized from $^{14}CH_3CO_2H$ reflects the fatty acid origin of the prostaglandins.

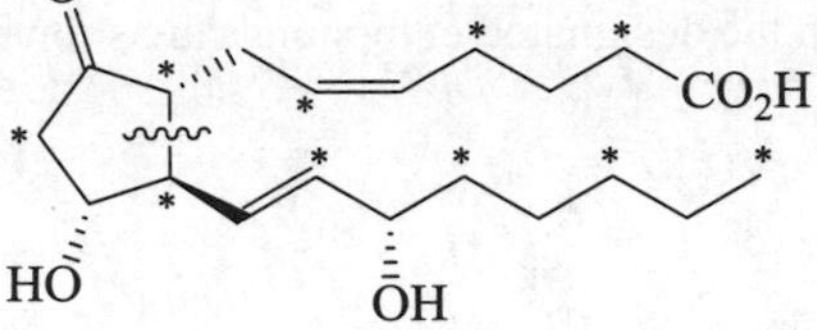

PGE_2

(*c*) The biosynthetic pathway leading to prostacyclin I₂ (PGI₂), like the pathway leading to PGE₂, begins with arachidonic acid.

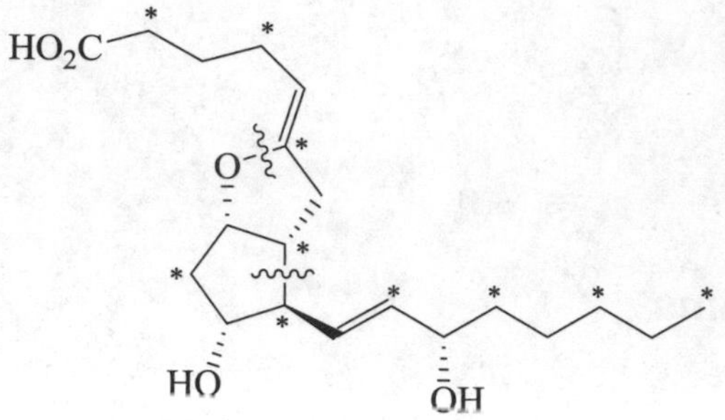

PGI₂

(*d*) Limonene is a monoterpene, biosynthesized from acetate by way of mevalonate and isopentenyl diphosphate.

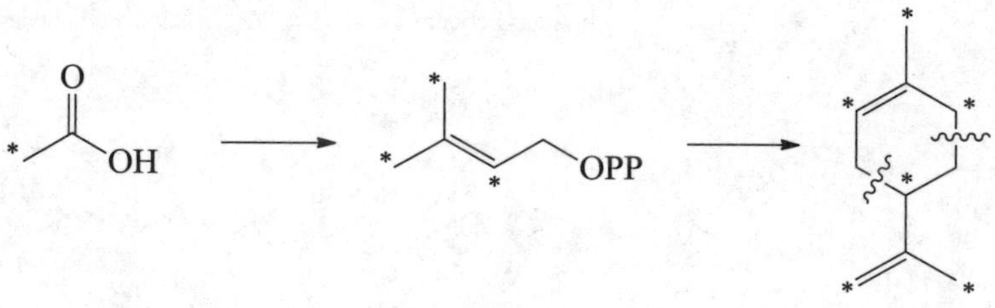

¹⁴C Labeled
acetic acid

Isopentenyl
diphosphate

Limonene

(*e*) The distribution of the ¹⁴C label in β carotene becomes evident once its isoprene units are identified.

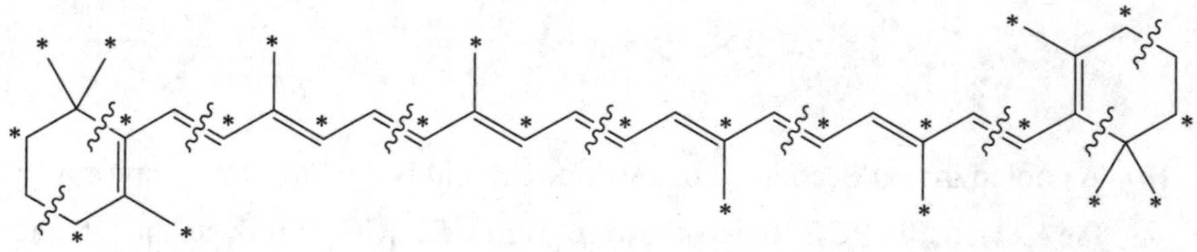

β Carotene

25.18 The isoprene units in the designated compounds are shown by disconnections in the structural formulas.

(*a*) Ascaridole:

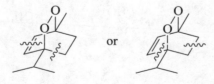

or

(*b*) Dendrolasin:

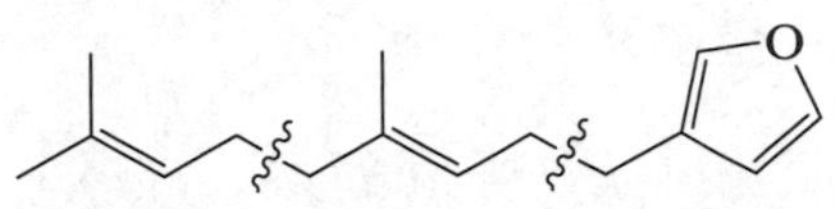

(*c*) γ Bisabolene:

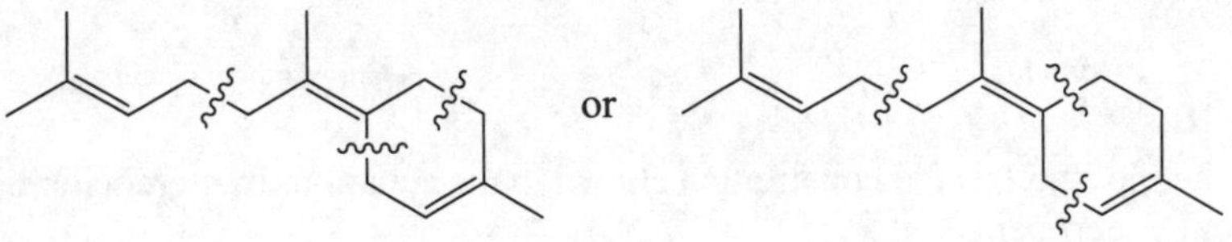

or

(*d*) α-Santonin:

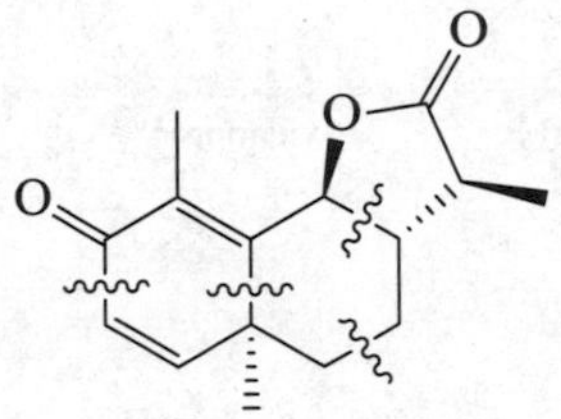

(*e*) Tetrahymanol

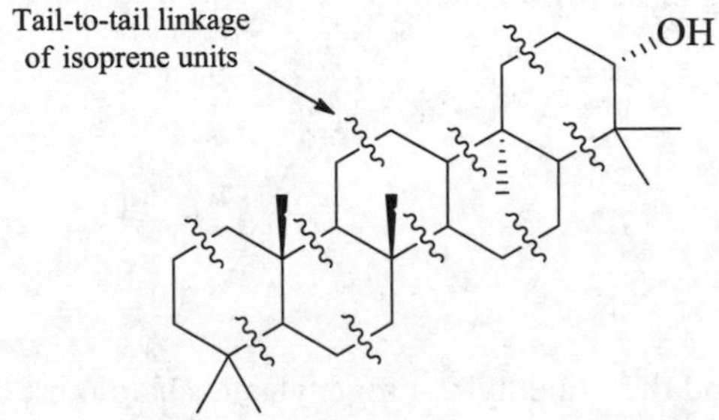

Tail-to-tail linkage
of isoprene units

25.19 Of the four isoprene units of cubitene, three of them are joined in the usual head-to-tail fashion, but the fourth one is joined in an irregular way.

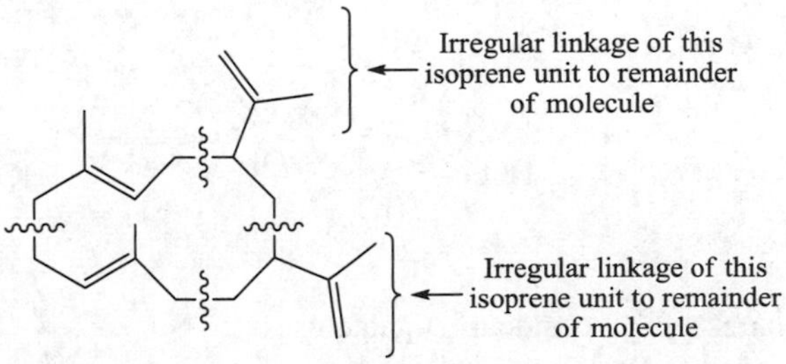

Irregular linkage of this
isoprene unit to remainder
of molecule

Irregular linkage of this
isoprene unit to remainder
of molecule

25.20 (*a*) Cinerin I is an ester, the acyl portion of which is composed of two isoprene units, as follows:

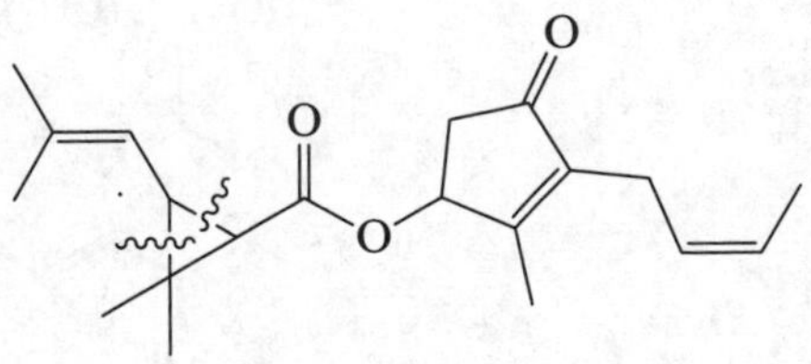

Cinerin I

(b) Hydrolysis of cinerin I involves cleavage of the ester unit.

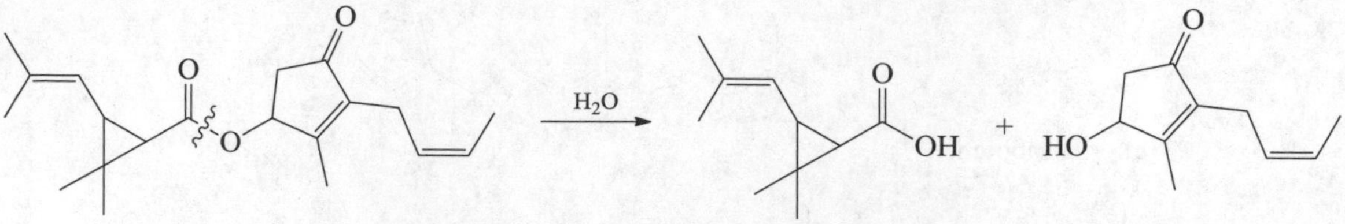

Cinerin I   (+)-Chrysanthemic acid

Chrysanthemic acid has the constitution shown in the equation. Its stereochemistry is revealed by subsequent experiments.

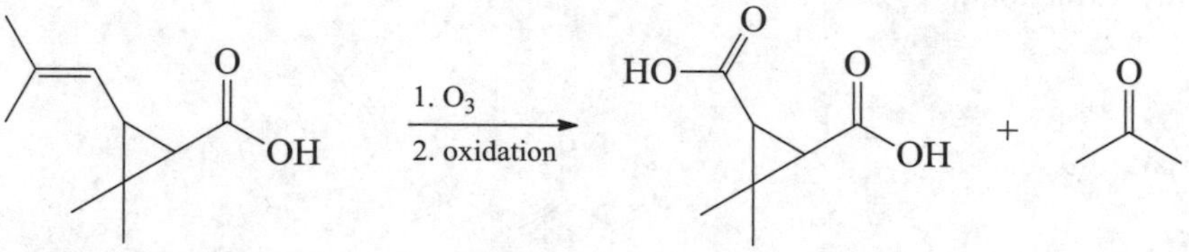

(+)-Chrysanthemic acid   (−)-Caronic acid   Acetone

Because caronic acid is optically active, its carboxyl groups must be trans to each other. (The cis stereoisomer is an optically inactive meso form.) The structure of (+)-chrysanthemic acid must therefore be either the following or its mirror image.

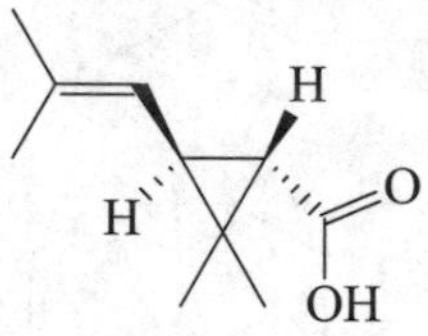

The carboxyl group and the 2-methyl-1-propenyl side chain must be trans to each other.

25.21 (a) Hydrolysis of phrenosine cleaves the glycosidic bond. The carbohydrate liberated by this hydrolysis is D-galactose.

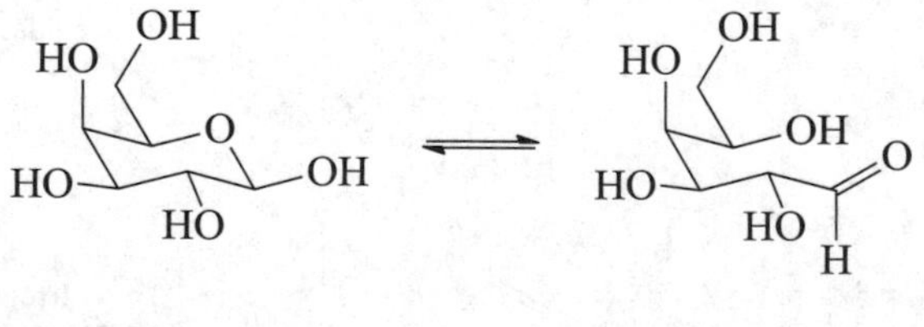

Phrenosine is a β-glycoside of D-galactose.

(b) The species that remains on cleavage of the galactose unit has the structure

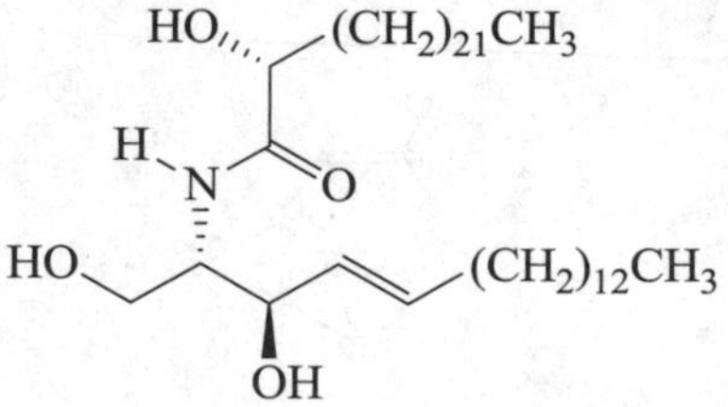

The two substances, sphingosine and cerebronic acid, that are formed along with D-galactose arise by hydrolysis of the amide bond.

Sphingosine   Cerebronic acid

Reactions and Mechanism

25.22 (*a*) Catalytic hydrogenation over Lindlar palladium converts alkynes to cis alkenes.

$$CH_3(CH_2)_7-C\equiv C-(CH_2)_7CO_2H \;+\; H_2 \xrightarrow{\text{Lindlar Pd}}$$

9-Octadecynoic acid   (Z)-9-Octadecenoic acid (74%)
(stearolic acid)   (oleic acid)

(*b*) Carbon–carbon triple bonds are converted to trans alkenes by reduction with lithium and ammonia.

$$CH_3(CH_2)_7-C\equiv C-(CH_2)_7CO_2H \xrightarrow[\text{2. H}^+]{\text{1. Li, NH}_3}$$

9-Octadecynoic acid   (E)-9-Octadecenoic acid (97%)
(stearolic acid)   (elaidic acid)

(*c*) The carbon–carbon double bond is hydrogenated readily over a platinum catalyst. Reduction of the ester function does not occur.

Ethyl (Z)-9-octadecenoate   Ethyl 9-octadecanoate (91%)
(ethyl oleate)   (ethyl stearate)

(*d*) Lithium aluminum hydride reduces the ester function but leaves the carbon–carbon double bond intact.

Methyl (Z)-12-hydroxy-9-octadecenoate   (Z)-9-octadecen-1,12-diol (52%)   Methanol
(methyl ricinoleate)

(*e*) Epoxidation of the double bond occurs when an alkene is treated with a peroxy acid. The reaction is stereospecific; substituents that are cis to each other in the alkene remain cis in the epoxide.

Oleic acid + Peroxybenzoic acid ⟶ *cis*-9,10-Epoxyoctadecanoic acid (62-67%) + Benzoic acid

(*f*) Acid-catalyzed hydrolysis of the epoxide yields a diol; its stereochemistry corresponds to net anti dihydroxylation of the double bond of the original alkene.

cis-9,10-Epoxyoctadecanoic acid $\xrightarrow{H_3O^+}$ 9,10-Dihydroxyoctadecanoic acid

The product is chiral but is formed as a racemic mixture containing equal amounts of the 9*R*,10*R* and 9*S*,10*S* stereoisomers when the starting epoxide is racemic.

(*g*) Dihydroxylation of carbon–carbon double bonds with osmium tetraoxide proceeds with syn addition of hydroxyl groups.

Oleic acid $\xrightarrow{\text{1. OsO}_4, (CH_3)_3COOH, HO^- \\ \text{2. } H^+}$ 9,10-Dihydroxyoctadecanoic acid (70%)

The product is chiral but is formed as a racemic mixture containing equal amounts of the 9*R*,10*S* and 9*S*,10*R* stereoisomers.

(*h*) Hydroboration–oxidation gives syn hydration of carbon–carbon double bonds with a regioselectivity opposite to Markovnikov's rule. The reagent attacks the less hindered face of the double bond of α-pinene.

Methyl group shields top face of double bond

$\xrightarrow{\text{1. } B_2H_6, \text{ diglyme} \\ \text{2. } H_2O_2, HO^-}$

B$_2$H$_6$ attacks from this direction

Isopinocampheol (79%)

(*i*) The starting alkene in this case is β-pinene. As in the preceding exercise with α-pinene, diborane adds to the bottom face of the double bond.

Methyl group shields top face of double bond

$\xrightarrow{\text{1. } B_2H_6, \text{ diglyme} \\ \text{2. } H_2O_2, HO^-}$

B$_2$H$_6$ attacks from this direction

cis-Mystanol (81%)

(*j*) The starting material is an acetal. It undergoes hydrolysis in dilute aqueous acid to give a ketone.

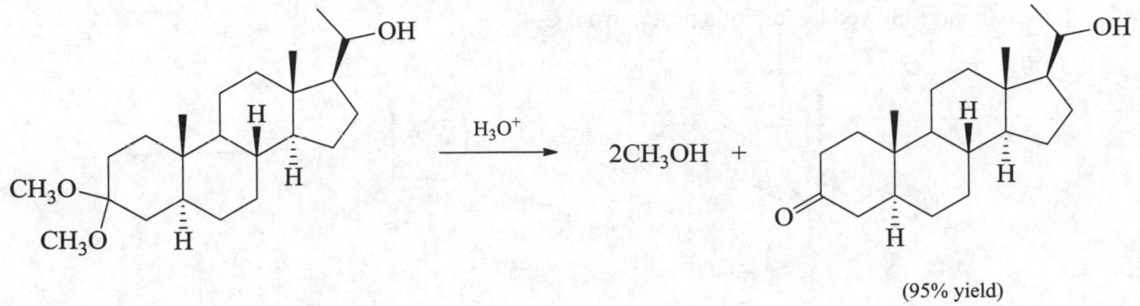
OH
H
H
H
CH_3O
CH_3O
H_3O^+
$2CH_3OH$ +
O
(95% yield)

25.23 The first step is a 1,4 addition of hydrogen bromide to the conjugated diene system of isoprene.

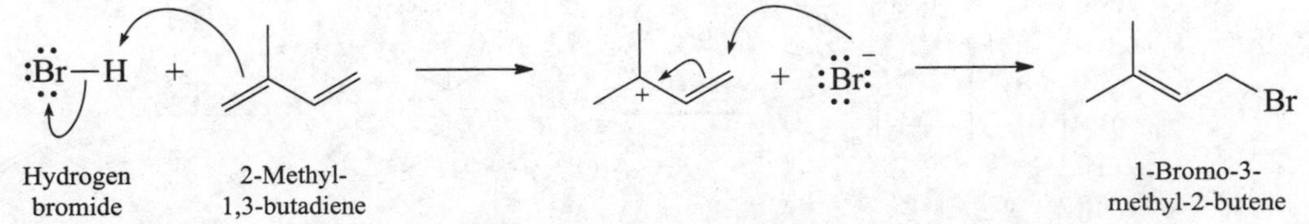
:Br—H +
+ :Br:
Br
Hydrogen bromide
2-Methyl-1,3-butadiene
1-Bromo-3-methyl-2-butene

This is followed by Markovnikov addition of hydrogen bromide to the remaining double bond.

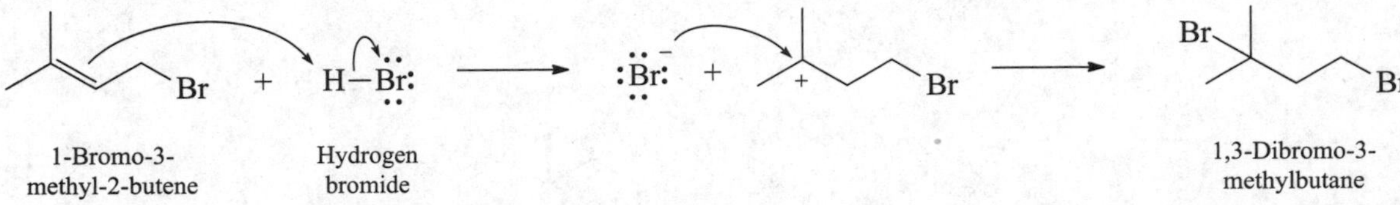
Br
+
H—Br:
:Br: +
Br
Br
Br
1-Bromo-3-methyl-2-butene
Hydrogen bromide
1,3-Dibromo-3-methylbutane

25.24 A reasonable mechanism is protonation of the isolated carbon–carbon double bond, followed by cyclization.

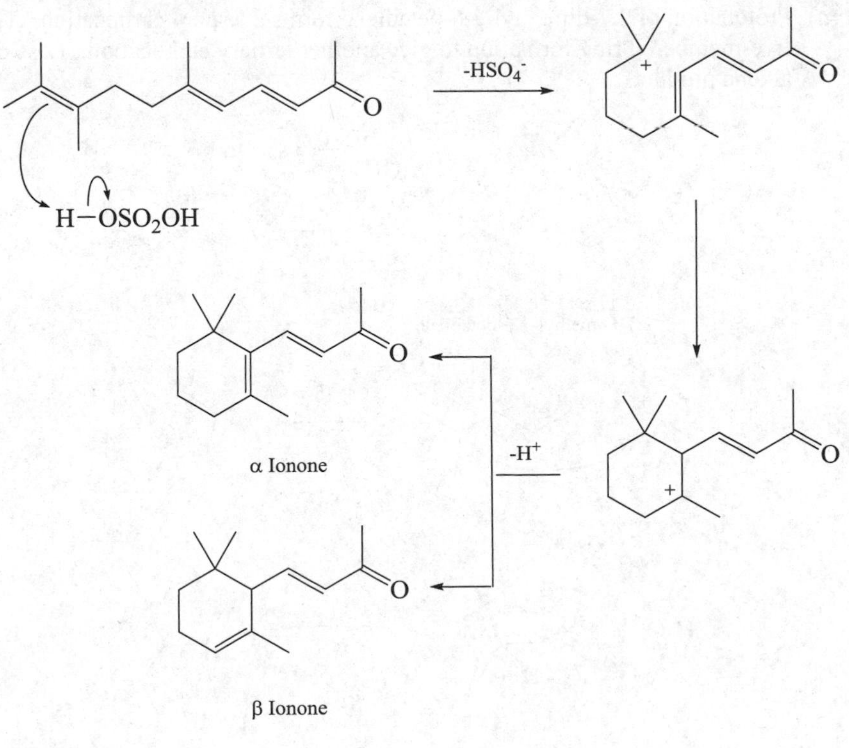
O
H—OSO₂OH
-HSO₄⁻
O
O
α Ionone
-H⁺
O
O
β Ionone

25.25 The double bond has a tendency to become conjugated with the carbonyl group. Two mechanisms are more likely than any others under conditions of acid catalysis. One of these involves protonation of the double bond followed by loss of a proton from C-4.

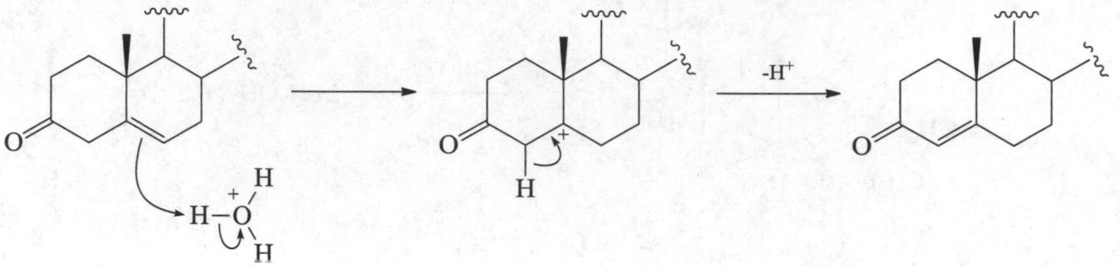

The other mechanism proceeds by enolization followed by proton-induced double-bond migration.

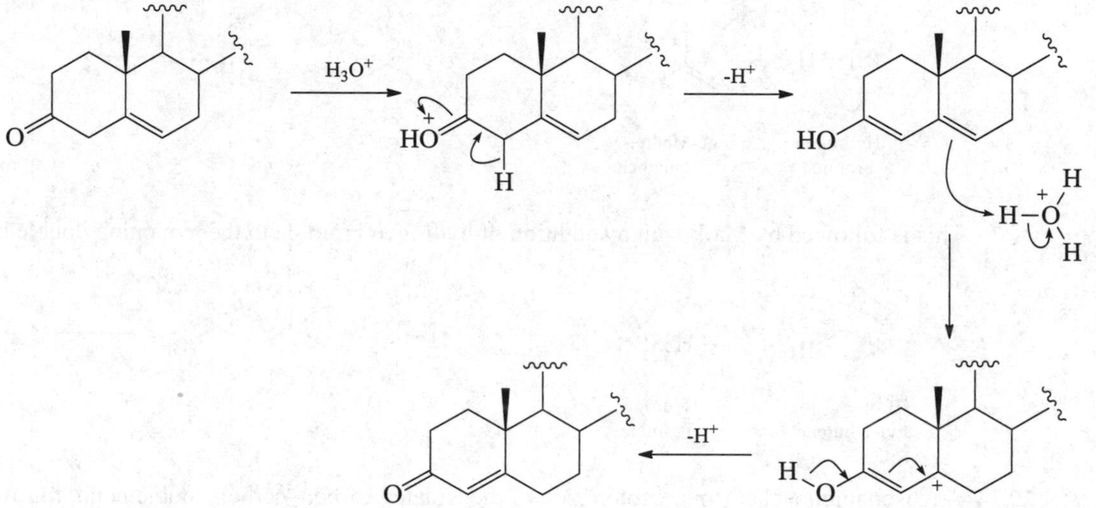

25.26 (*a*) Protonation of 2,7-dimethyl-2,6-octadiene forms a tertiary carbocation. This is followed by five-membered ring formation to give another tertiary carbocation. Loss of a proton then gives the alkene product.

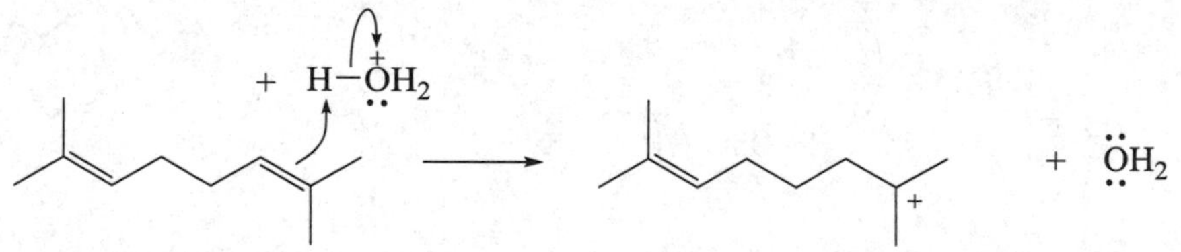

2,7-Dimethyl-2,6-octadiene

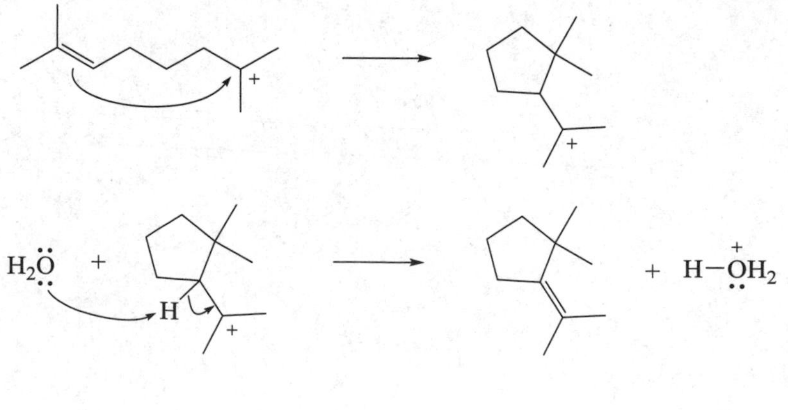

(b) The remaining two products are formed from the cyclic carbocation in the preceding mechanism. A carbocation rearrangement, a hydride shift, gives a second cyclic tertiary carbocation. Loss of a proton gives one of the two remaining products.

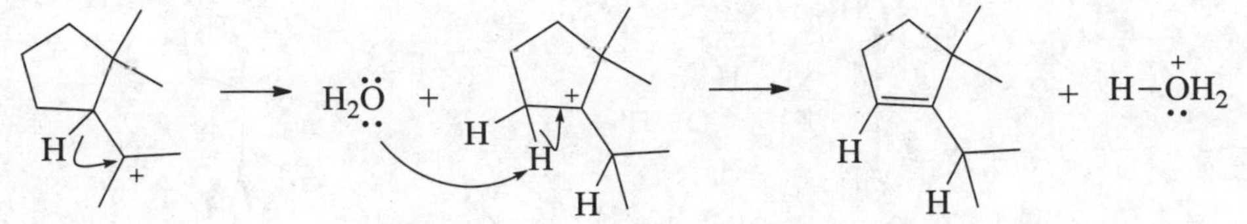

The other cyclic product is formed from the second cyclic tertiary carbocation by another carbocation rearrangement, a methyl shift, followed by loss of a proton as shown.

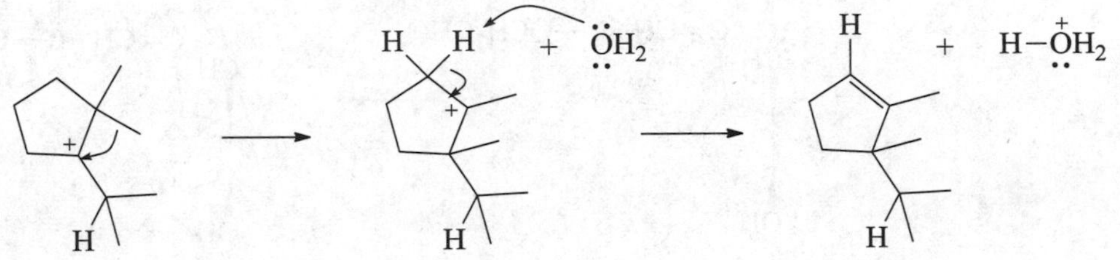

25.27 This mechanism involves steps that have been introduced in earlier chapters. Carbocation formation and tautomerism are both involved.

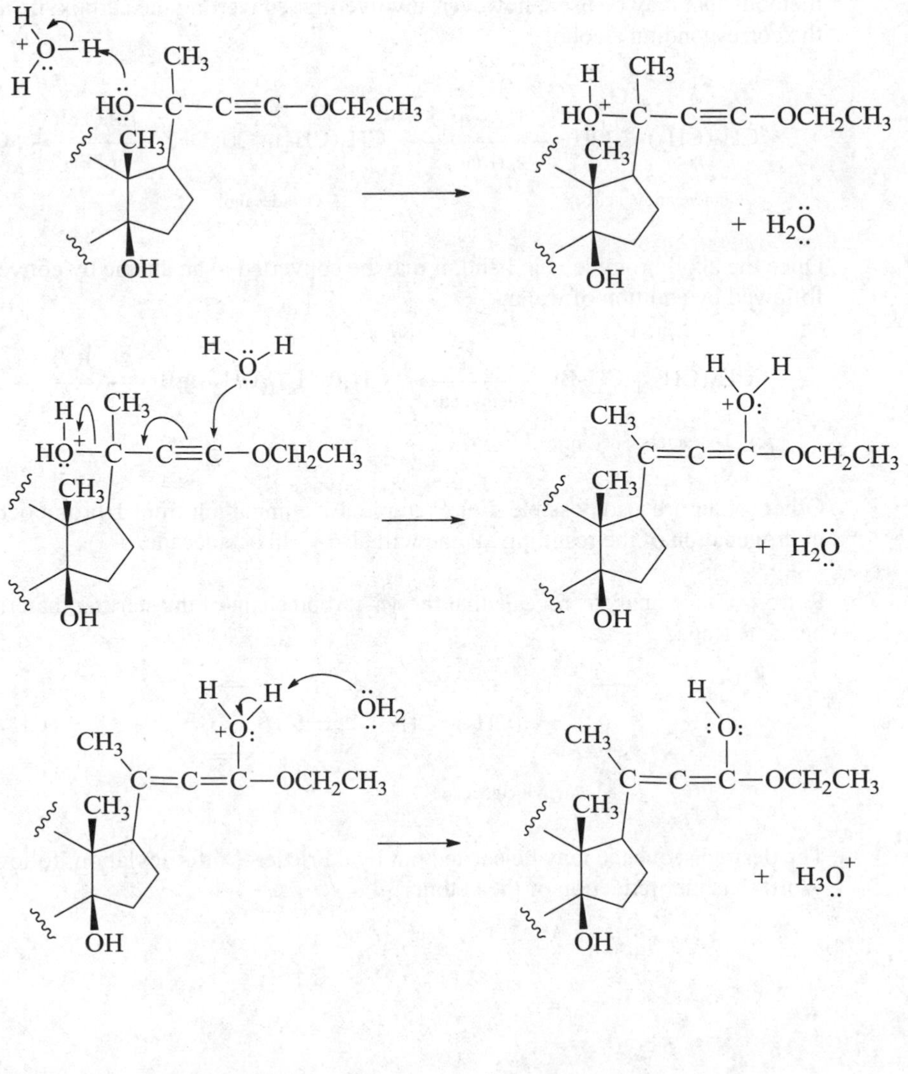

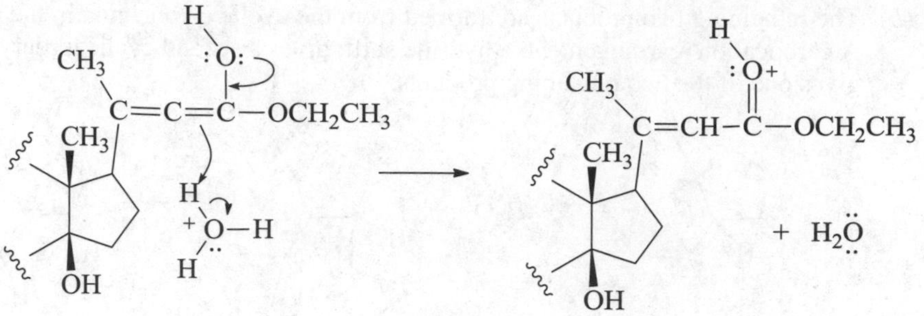

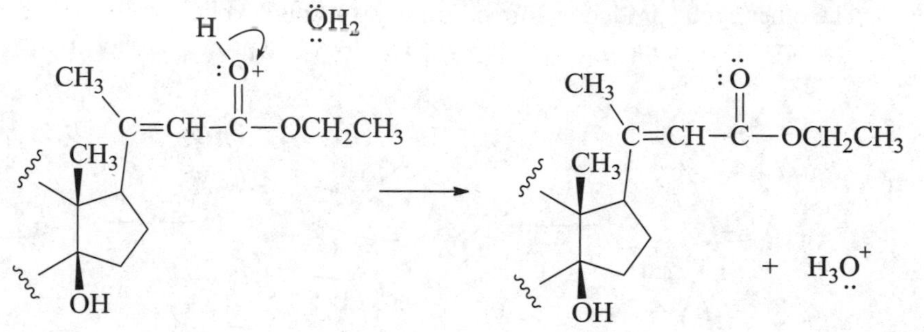

Synthesis

25.28 (*a*) There are no direct methods for the reduction of a carboxylic acid to an alkane. A number of indirect methods that may be used, however, involve first converting the carboxylic acid to an alkyl bromide via the corresponding alcohol.

$$CH_3(CH_2)_{16}\overset{\overset{\displaystyle O}{\|}}{C}OH \xrightarrow[\text{2. } H_2O]{\text{1. LiAlH}_4} CH_3(CH_2)_{16}CH_2OH \xrightarrow[\text{or PBr}_3]{\text{HBr, heat}} CH_3(CH_2)_{16}CH_2Br$$

Octadecanoic acid   1-Octadecanol   1-Bromooctadecane

Once the alkyl bromide is in hand, it may be converted to an alkane by conversion to a Grignard reagent followed by addition of water.

$$CH_3(CH_2)_{16}CH_2Br \xrightarrow[\text{diethyl ether}]{\text{Mg}} CH_3(CH_2)_{16}CH_2MgBr \xrightarrow{H_2O} CH_3(CH_2)_{16}CH_3$$

1-Bromooctadecane   Octadecane

Other routes are also possible. For example, E2 elimination from 1-bromooctadecane followed by hydrogenation of the resulting alkene will also yield octadecane.

(*b*) Retrosynthetic analysis reveals that the 18-carbon chain of the starting material must be attached to a benzene ring.

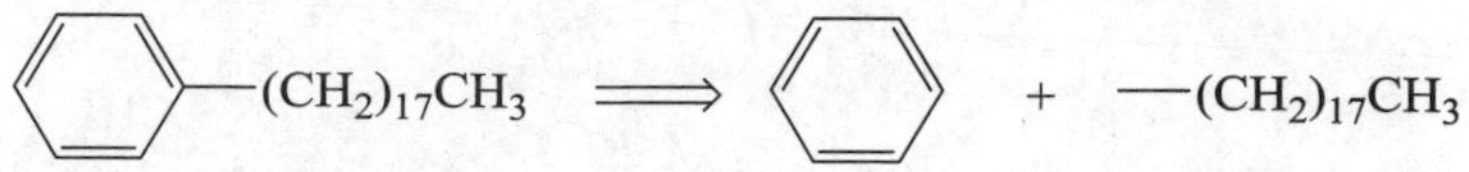

1-Phenyloctadecane

The desired sequence may be carried out by a Friedel–Crafts acylation, followed by Clemmensen or Wolff–Kishner reduction of the ketone.

$$CH_3(CH_2)_{16}\overset{\overset{\displaystyle O}{\|}}{C}OH \xrightarrow{\text{SOCl}_2} CH_3(CH_2)_{16}\overset{\overset{\displaystyle O}{\|}}{C}Cl \xrightarrow[\text{AlCl}_3]{\text{benzene}} \text{C}_6\text{H}_5-\overset{\overset{\displaystyle O}{\|}}{C}(CH_2)_{16}CH_3$$

Octadecanoic acid   Octadecanoic acid   1-Phenyl-1-octadecanone

$$\downarrow \text{Zn(Hg), HCl}$$

$$\text{C}_6\text{H}_5-CH_2(CH_2)_{16}CH_3$$

1-Phenyloctadecane

(*c*) First examine the structure of the target molecule 3-ethylicosane.

$$CH_3(CH_2)_{16}\underset{\underset{\displaystyle CH_2CH_3}{|}}{CH}CH_2CH_3$$

Retrosynthetic analysis reveals that two ethyl groups have been attached to a C_{18} unit.

$$CH_3(CH_2)_{16}\underset{\underset{\displaystyle CH_2CH_3}{|}}{CH}-CH_2CH_3 \implies CH_3(CH_2)_{16}\overset{|}{CH}- \quad + \quad 2CH_3CH_2-$$

The necessary carbon–carbon bonds can be assembled by the reaction of an ester with two moles of a Grignard reagent.

$$CH_3(CH_2)_{16}\overset{\overset{\displaystyle O}{\|}}{C}OCH_2CH_3 + 2CH_3CH_2MgBr \xrightarrow[\text{2. H}_3\text{O}^+]{\text{1.diethyl ether}} CH_3(CH_2)_{16}\overset{\overset{\displaystyle OH}{|}}{\underset{\underset{\displaystyle CH_2CH_3}{|}}{C}}-CH_2CH_3$$

Ethyl octadecanoate   Ethylmagnesium   3-Ethyl-3-icosanol
(from octadecanoic acid   bromide
and ethanol)

With the correct carbon skeleton in place, all that is needed is to convert the alcohol to the alkene. This can be accomplished by dehydration and reduction.

$$CH_3(CH_2)_{16}\overset{\overset{\displaystyle OH}{|}}{\underset{\underset{\displaystyle CH_2CH_3}{|}}{C}}-CH_2CH_3 \xrightarrow[\text{heat}]{\text{H}_2\text{SO}_4} CH_3(CH_2)_{16}\underset{\underset{\displaystyle CH_2CH_3}{|}}{C}=CHCH_3 \quad + \quad CH_3(CH_2)_{15}CH=\underset{\underset{\displaystyle CH_2CH_3}{|}}{C}CH_2CH_3$$

3-Ethyl-3-icosanol   3-Ethyl-2-icosene   3-Ethyl-3-icosene

$$\downarrow \text{H}_2, \text{Pt}$$

$$CH_3(CH_2)_{16}\underset{\underset{\displaystyle CH_2CH_3}{|}}{CH}-CH_2CH_3$$

3-Ethylicosane

(*d*) Icosanoic acid contains two more carbon atoms than octadecanoic acid.

$$CH_3(CH_2)_{18}\overset{\overset{\displaystyle O}{\|}}{C}OH \implies CH_3(CH_2)_{16}CH_2Br \ + \ \overset{..}{C}H_2\overset{\overset{\displaystyle O}{\|}}{C}OH$$

Icosanoic acid

A reasonable approach utilizes a malonic ester synthesis.

$$CH_3(CH_2)_{16}CH_2Br \ + \ CH_2(CO_2CH_2CH_3)_2 \ \xrightarrow{\ NaOCH_2CH_3\ } \ CH_3(CH_2)_{16}CH_2CH(CO_2CH_2CH_3)_2$$

1-Bromooctadecane   Diethyl malonate   Diethyl 2-octadecylmalonate
[from part (*a*)]

1. HO⁻, H₂O
2. H₃O⁺
3. heat

$$CH_3(CH_2)_{16}CH_2CH_2CO_2H$$

Icosanoic acid

(*e*) Lithium aluminum hydride reduction of octadecanamide gives the corresponding amine. The amide can be prepared from octadecanoic acid.

$$CH_3(CH_2)_{16}\overset{\overset{\displaystyle O}{\|}}{C}OH \ \xrightarrow[\text{2. NH}_3]{\text{1. SOCl}_2} \ CH_3(CH_2)_{16}\overset{\overset{\displaystyle O}{\|}}{C}NH_2 \ \xrightarrow[\text{2. H}_2O]{\text{1. LiAlH}_4} \ CH_3(CH_2)_{16}CH_2NH_2$$

Octadecanoic acid   Octadecanamide   1-Octadecamine

(*f*) Chain extension can be achieved via cyanide displacement of bromine from 1-bromooctadecane. Reduction of the cyano group completes the synthesis.

$$CH_3(CH_2)_{16}CH_2Br \ \xrightarrow{\ KCN\ } \ CH_3(CH_2)_{16}CH_2C{\equiv}N \ \xrightarrow[\text{2. H}_2O]{\text{1. LiAlH}_4} \ CH_3(CH_2)_{16}CH_2CH_2NH_2$$

1-Bromooctadecane   Nonadecanenitrile   1-Nonadecamine
[from part (*a*)]

25.29 First acylate the free hydroxyl group with an acyl chloride.

Treatment with aqueous acid brings about hydrolysis of the acetal function.

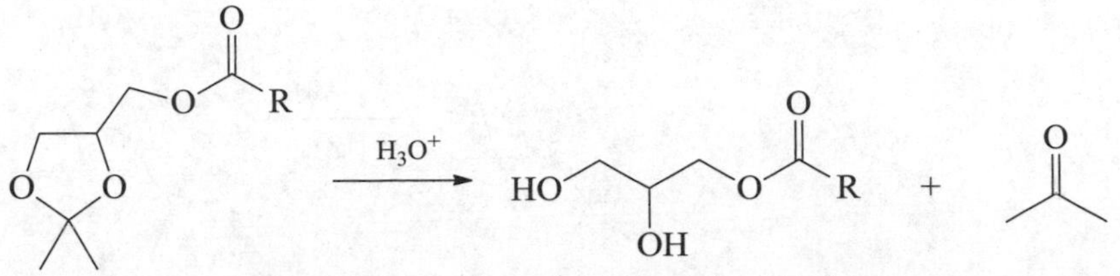

The two hydroxyl groups of the resulting diol are then esterified with 2 moles of the second acyl chloride.

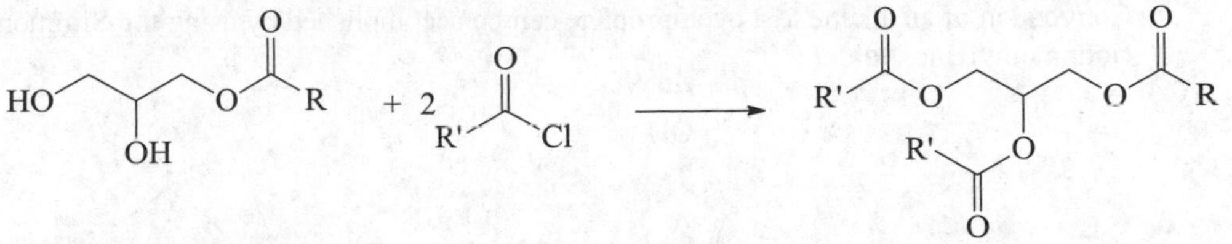

25.30 The overall transformation

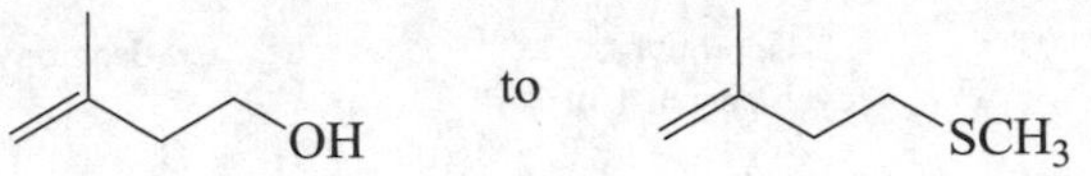

requires converting the alcohol function to some suitable leaving group, followed by substitution by an appropriate nucleophile.

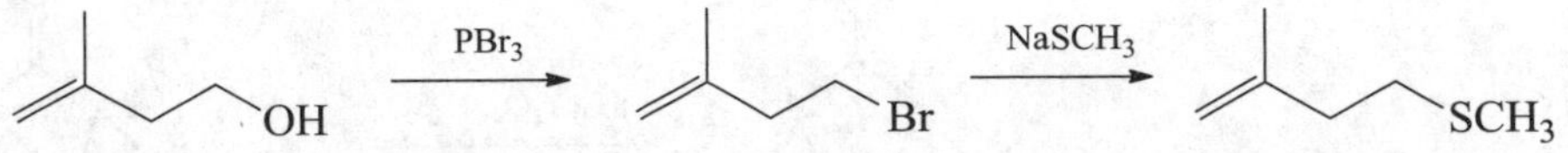

3-Methyl-3-buten-1-ol

4-Bromo-2-methyl-1-butene

3-Methyl-3-butenyl methyl sulfide

As reported in the literature, the alcohol was converted to its corresponding *p*-toluenesulfonate and this substance was then used as the substrate in the nucleophilic substitution step to produce the desired sulfide in 76% yield.

25.31 The first transformation is an intramolecular aldol condensation. This reaction was carried out under conditions of base catalysis.

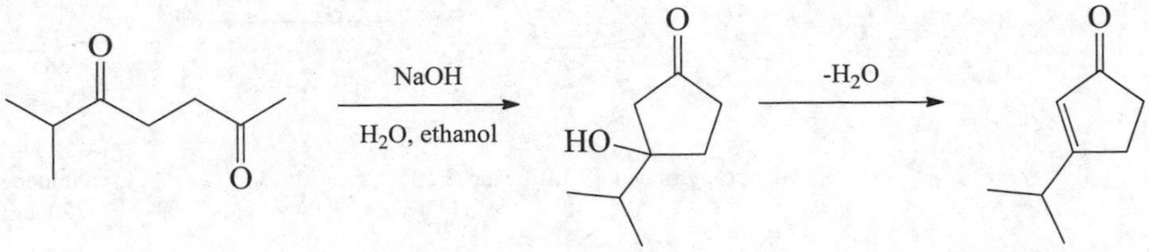

6-Methyl-2,5-heptanedione

(Not isolated)

3-Isopropyl-2-cyclopentenone (71%)

The next step is reduction of a ketone to a secondary alcohol. Lithium aluminum hydride is suitable; it reduces carbonyl groups but leaves the double bond intact.

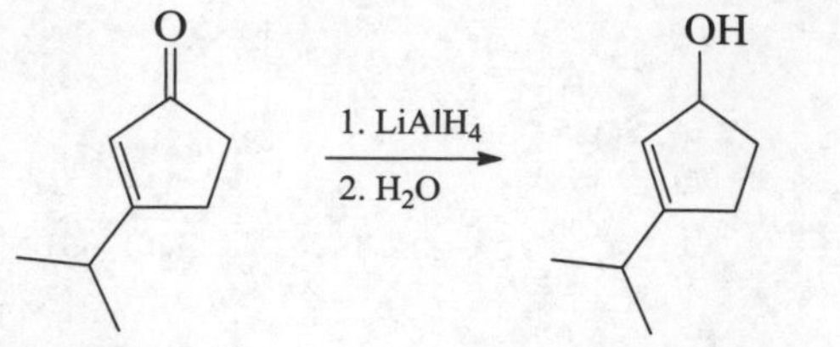

3-Isopropyl-2-
cyclopentenone

1. LiAlH$_4$
2. H$_2$O

3-Isopropyl-2-
cyclopenten-1-ol (97%)

Conversion of an alkene to a cyclopropane can be accomplished by using the Simmons–Smith reagent (iodomethylzinc iodide).

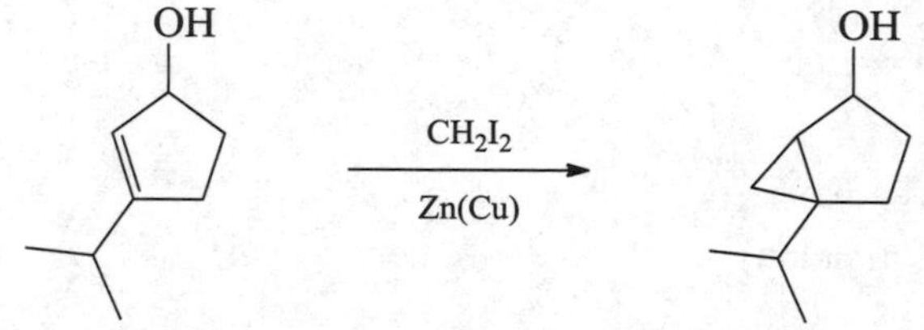

3-Isopropyl-2-
cyclopenten-1-ol

CH$_2$I$_2$

Zn(Cu)

5-Isopropylbicyclo[3.1.0]hexan-2-ol
(66%)

Oxidation of the secondary alcohol to the ketone can be accomplished with any of a number of oxidizing agents. The chemists who reported this synthesis used chromic acid.

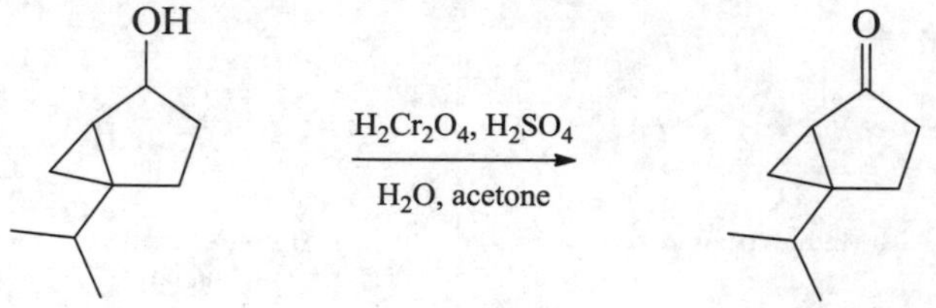

5-Isopropylbicyclo[3.1.0]hexan-2-ol

H$_2$Cr$_2$O$_4$, H$_2$SO$_4$

H$_2$O, acetone

5-Isopropylbicyclo[3.1.0]hexan-2-one
(89%)

A Wittig reaction converts the ketone to sabinene.

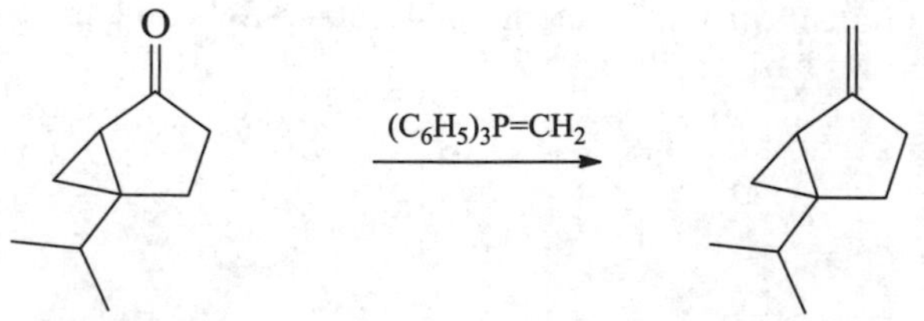

5-Isopropylbicyclo[3.1.0]hexan-2-one

(C$_6$H$_5$)$_3$P=CH$_2$

Sabinene
(70%)

25.32 This involves the conversion of a carboxylic acid to an amide. Treating the acid with thionyl chloride, followed by ethylamine to give the amide, will not work because thionyl chloride will also react with the hydroxyl groups.

The acid can be treated with methyl iodide in the presence of potassium carbonate in acetone. In this reaction, the acid is converted to a carboxylate anion, which then acts as a nucleophile in an S$_N$2 reaction (Table 6.1). Reaction of the ester with ethylamine gives the amide, *bimatoprost*.

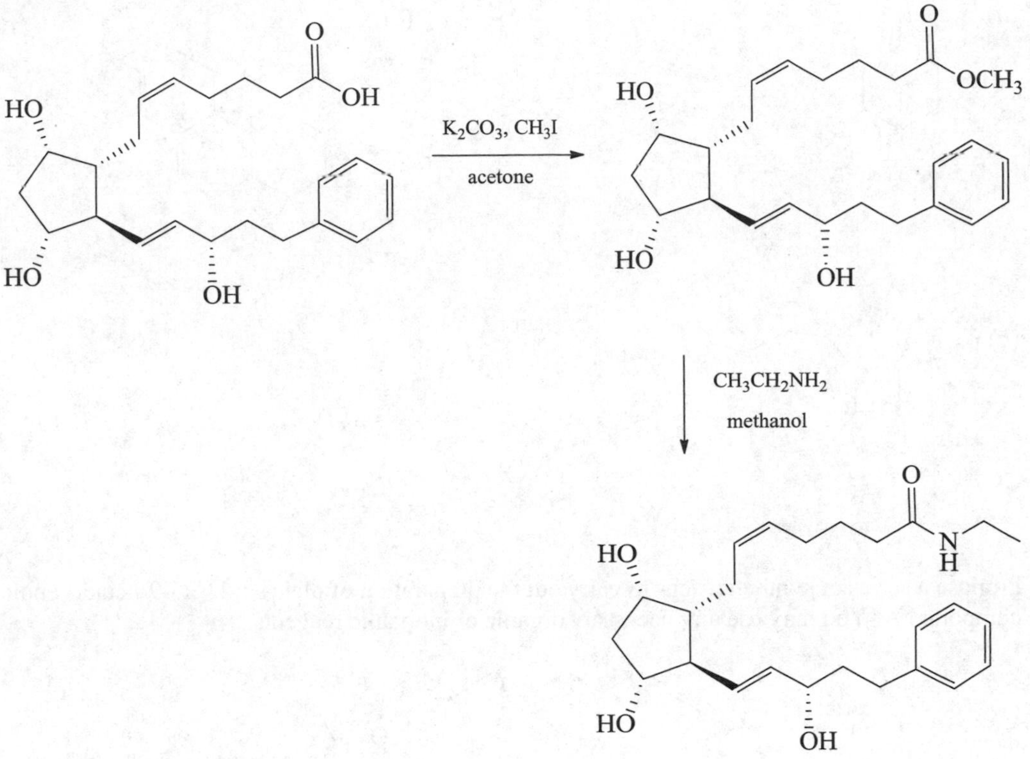

Answers to Interpretive Problems 25

25.33 A; **25.34** D; **25.35** B; **25.36** B; **25.37** C; **25.38** A

SELF-TEST

1. Write a balanced chemical equation for the basic hydrolysis of tristearin.

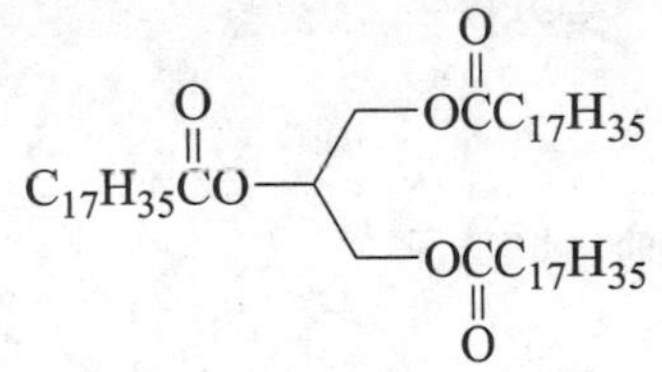

Tristearin

2. Both waxes and fats are lipids that contain the ester functional group. In what way do the structures of these lipids differ?

3. Classify each of the following isoprenoid compounds as a monoterpene, a diterpene, and so on. Indicate with dashed lines the isoprene units that make up each structure.

(a)

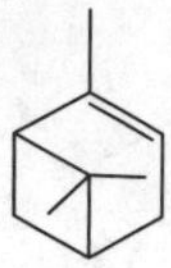

α Pinene

(b)

Caryophyllene

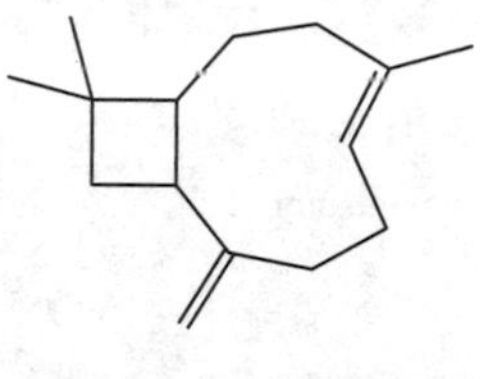

(c)

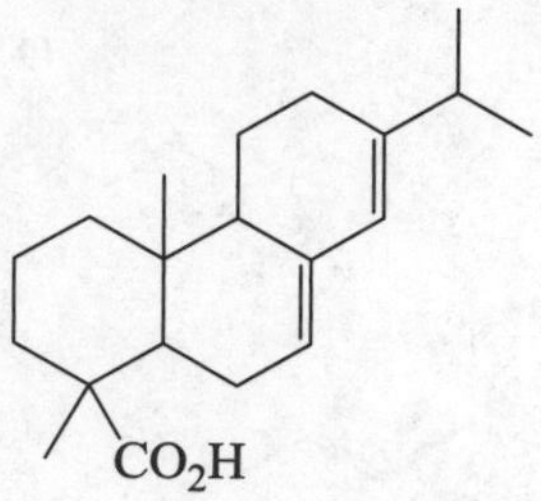

Adietic acid

4. Propose a series of synthetic steps to carry out the preparation of oleic acid [(Z)-9-octadecenoic acid] from compound A. You may use any necessary organic or inorganic reagents.

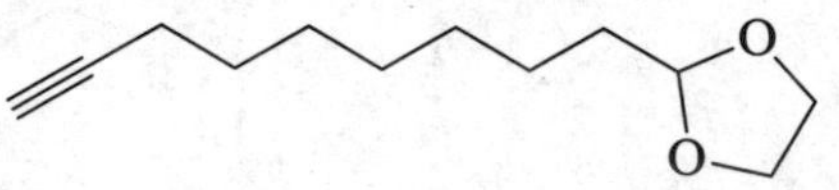

Compound A

5. Write a mechanism for the biosynthetic pathway by which limonene is formed from geranyl diphosphate.

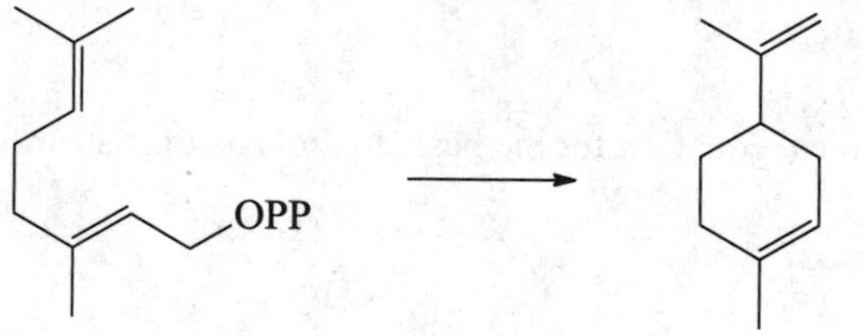

Geranyl diphosphate

Limonene

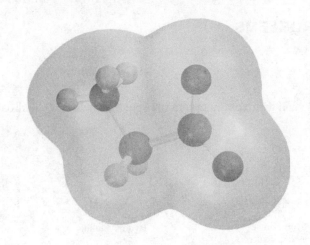

CHAPTER 26
Amino Acids, Peptides, and Proteins

Table of Contents

SOLUTIONS TO TEXT PROBLEMS

In Chapter Problems

26.1 The chirality centers and their configurations are specified below. Dehydroalanine does not have any chirality centers

Dehydroalanine
(no chirality centers)

Hydroxyproline

Selenocysteine

Pyrrolysine

The reason for selenocysteine's R configuration is that the $-CH_2SeH$ group outranks the $-CO_2^-$ group.

26.2 (*b*) L-Cysteine is the only amino acid in text Table 26.1 that has the R configuration at its chirality center.

L-Cysteine

The order of decreasing sequence rule precedence is

$$H_3\overset{+}{N}- \ > \ HSCH_2- \ > \ -CO_2^- \ > \ H-$$

When the molecule is oriented so that the lowest-ranked atom (H) is held away from us, the order of decreasing precedence traces a clockwise path.

Clockwise; therefore R

The reason why L-cysteine has the R configuration and all the other L-amino acids have the S configuration lies in the fact that the $-CH_2SH$ substituent is the only side chain that outranks $-CO_2^-$ according to the sequence rule. Remember, rank order is determined by atomic number at the first point of difference, and $-C-S$ outranks $-C-O$. In all the other amino acids, $-CO_2^-$ outranks the substituent at the chirality center. The reversal in the Cahn–Ingold–Prelog descriptor comes not from any change in the spatial arrangement of substituents at the chirality center but rather from a reversal in the relative ranks of the carboxylate group and the side chain.

(*c*) The order of decreasing sequence rule precedence in L-methionine is

$$\overset{+}{H_3N}— \;>\; —CO_2^- \;>\; —CH_2CH_2SCH_3 \;>\; H—$$

Sulfur is one atom farther removed from the chirality center, and so C—O outranks C—C—S.

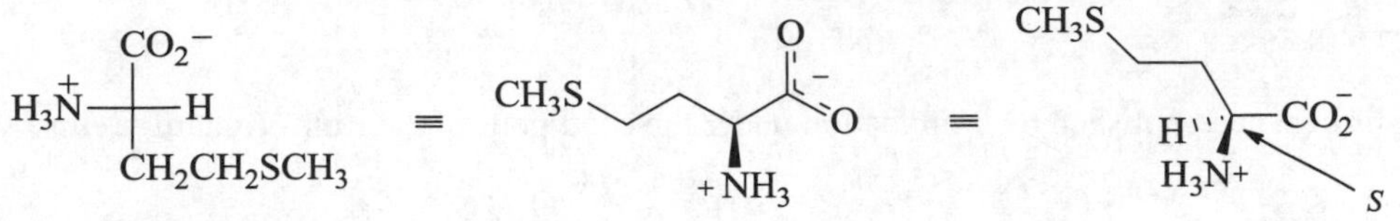

L-Methionine

The absolute configuration is *S*.

26.3 Fischer projections were discussed in Section 4.7 of the text. The following is the Fischer projection for (2*S*,3*R*)-2-amino-3-hydroxybutanoic acid.

(2*S*,3*R*)-2-Amino-3-hydroxybutanoic acid

26.4 The isoelectric point of an amino acid is midway between the pK_a value of the zwitterion and its conjugate acid. For cysteine, the equilibria are

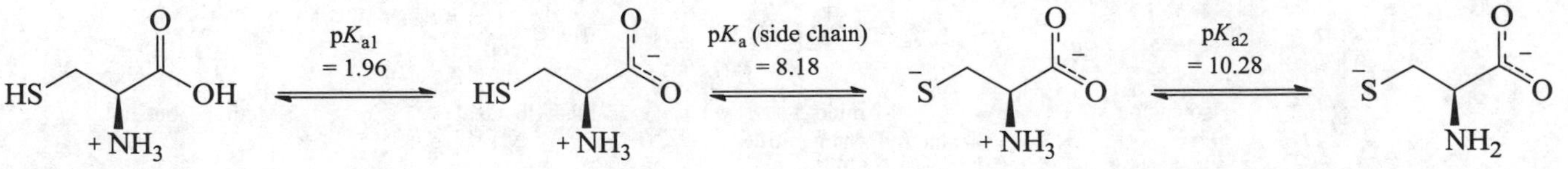

Conjugate acid of zwitterion

Zwitterion

The isoelectric point of cysteine is the average of the pK_{a1} and pK_a (side chain), or 5.07.

26.5 The zwitterion form of tyrosine is shown

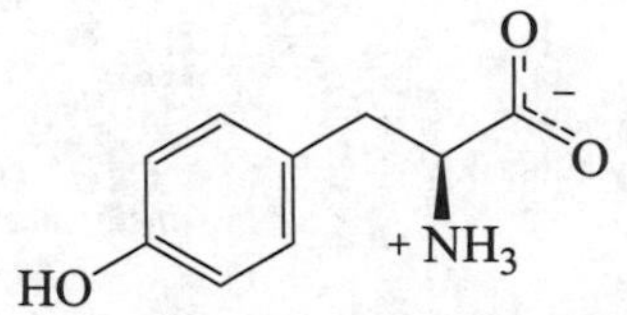

Zwitterion form of tyrosine

As the pH is raised, protons are lost from both the ammonium group and the phenolic hydroxyl group to form a dianion.

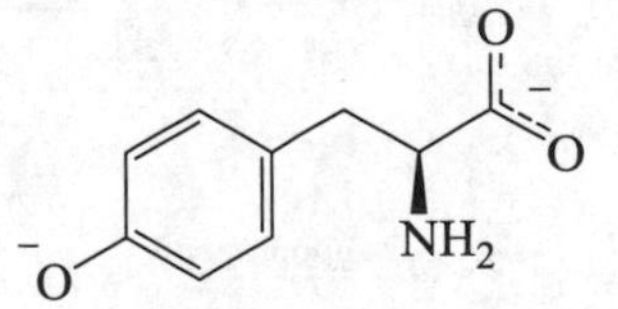

Dianion form of tyrosine

26.6 To convert 3-methylbutanoic acid to valine, a leaving group must be introduced at the α carbon prior to displacement by ammonia.

This is best accomplished by bromination under the conditions of the Hell–Volhard–Zelinsky reaction.

3-Methylbutanoic acid   2-Bromo-3-methylbutanoic acid   Valine

Valine has been prepared by this method. The Hell–Volhard–Zelinsky reaction was carried out in 88% yield, but reaction of the α-bromo acid with ammonia was not very efficient, valine being isolated in only 48% yield in this step.

26.7 The alcohol functional group in 2-methyl-1-butanol will allow further changes at that position. The methyl branch in 2-methyl-1-butanol matches the same branching in isoleucine. The retrosynthetic reaction sequence allows the extension of the carbon change by one using the Strecker synthesis.

Isoleucine   2-Amino-3-   2-Methylbutanal   2-Methyl-1-butanol
                      methylpentanenitrile

A reasonable synthesis for the transformation includes oxidation of the alcohol, followed by the Strecker synthesis and then hydrolysis. A racemic mixture of isoleucine enantiomers will be observed.

2-Methyl-1-butanol   2-Methylbutanal   2-Amino-3-   Isoleucine
                                                         methylpentanenitrile   (racemic mixture
                                                                                 of diasteromers)

26.8 The alkyl halide with which the anion of diethyl acetamidomalonate is treated is 2-bromopropane.

Diethyl acetamidomalonate   2-Bromopropane   Diethyl
                                                               acetamidoisopropylmalonate

This is the difficult step in the synthesis; it requires a nucleophilic substitution of the S_N2 type involving a secondary alkyl halide. Competition of elimination with substitution results in only a 37% observed yield of alkylated diethyl acetamidomalonate.

Hydrolysis and decarboxylation of the alkylated derivative are straightforward and proceed in 85% yield to give valine.

Diethyl   2-Aminoisopropylmalonic acid   Valine
acetamidoisopropylmalonate   (racemic)

The overall yield of valine (31%) is the product of 37% × 85%.

26.9 (a) Esterification conditions gives the corresponding ester.

(b) Reduction of the carboxylate yields an alcohol. The configuration of the stereocenter is retained since the reduction does not involve that carbon center.

(c) Oxidation of the sulfide functional group is accomplished with hydrogen peroxide. This can be accomplished stepwise to first give the sulfoxide that can be isolated. Further oxidation gives the sulfone.

$C_5H_{11}NO_3S$

26.10 The carbon that bears the amino group of 4-aminobutanoic acid corresponds to the α carbon of an α-amino acid, glutamic acid.

4-Aminobutanoic acid   Glutamic Acid

26.11 The conversion of 3,4-dihydroxyphenylalanine to dopamine is catalyzed by an amino acid decarboxylase.

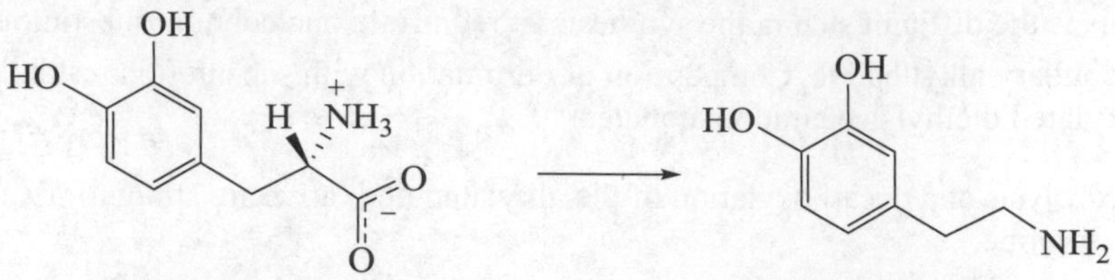

3,4-Dihydroxyphenylalanine
(L-Dopa)

Dopamine

26.12 In transamination, the ketone carbonyl of an α-keto acid and the α-amine group of an amino acid effectively trade places by the steps outlined in text Mechanism 26.2. Thus, the transamination reaction of α-ketoglutaric acid with L-aspartic acid yields L-glutamic acid and oxaloacetic acid.

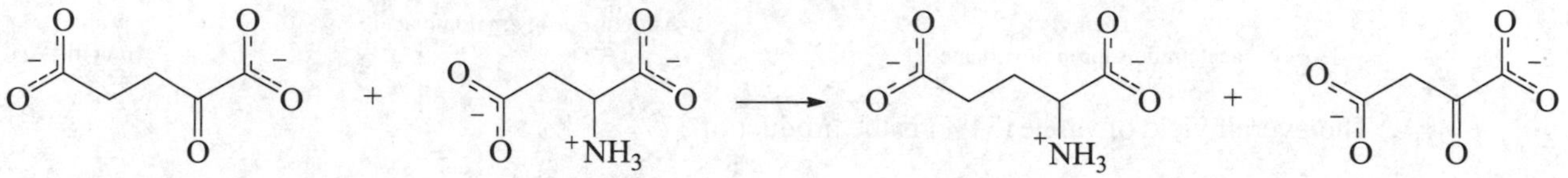

α-Ketoglutaric acid   L-Aspartic acid   L-Glutamic acid   Oxaloacetic acid

26.13 (*b*) Alanine is the N-terminal amino acid in Ala-Phe. Its carboxyl group is joined to the nitrogen of phenylalanine by a peptide bond.

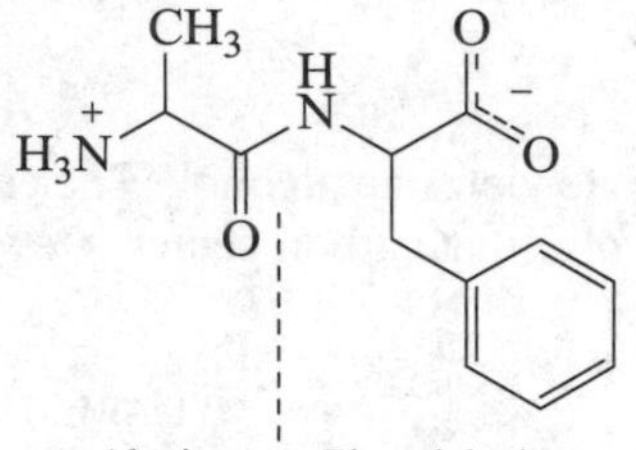

AF

Alanine ┊ Phenylalanine

(*c*) The positions of the amino acids are reversed in Phe-Ala. Phenylalanine is the N terminus and alanine is the C terminus.

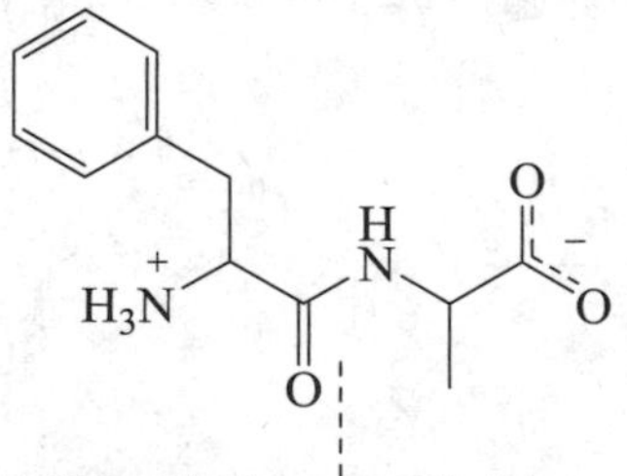

FA

Phenylalanine ┊ Alanine

(*d*) The carboxyl group of glycine is joined by a peptide bond to the amino group of glutamic acid.

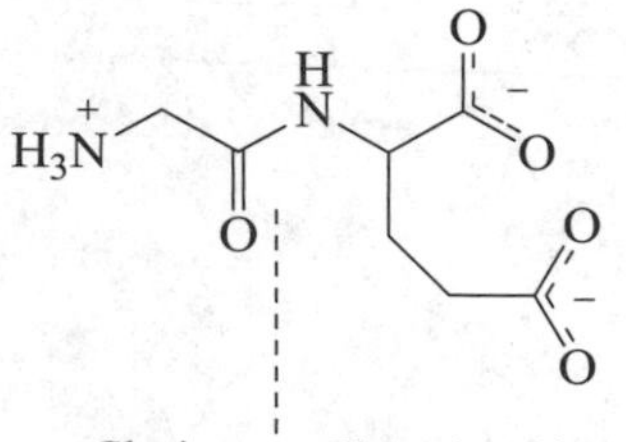

GE

Glycine ┊ Glutamic acid

The dipeptide is written in its anionic form because the carboxyl group of the side chain is ionized at pH 7. Alternatively, it could have been written as a neutral zwitterion with a $CH_2CH_2CO_2H$ side chain.

(e) The peptide bond in Lys-Gly is between the carboxyl group of lysine and the amino group of glycine.

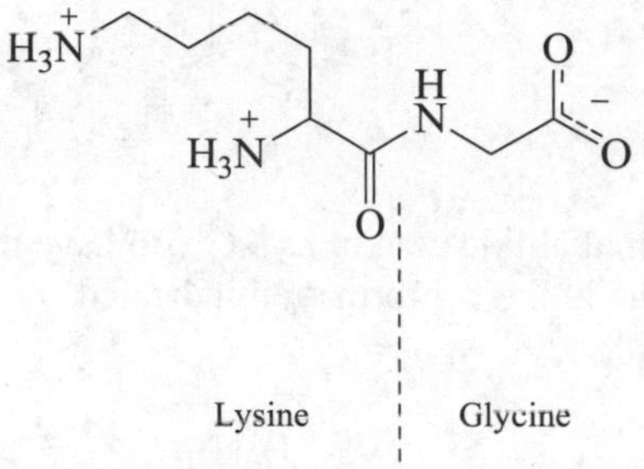

KG

Lysine   Glycine

The amino group of the lysine side chain is protonated at pH 7, and so the dipeptide is written here in its cationic form. It could have also been written as a neutral zwitterion with the side chain $H_2NCH_2CH_2CH_2CH_2$.

(f) Both amino acids are alanine in D-Ala-D-Ala. The fact that they have the D configuration has no effect on the constitution of the dipeptide.

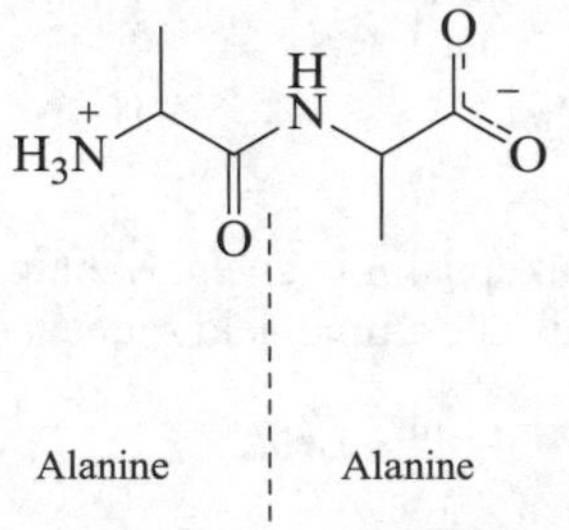

D-A-D-A

Alanine   Alanine

26.14 (b) When amino acid residues in a dipeptide are indicated without a prefix, it is assumed that the configuration at the α-carbon atom is L. For all amino acids except cysteine, the L configuration corresponds to S. The stereochemistry of Ala-Phe may therefore be indicated for the zigzag conformation as shown.

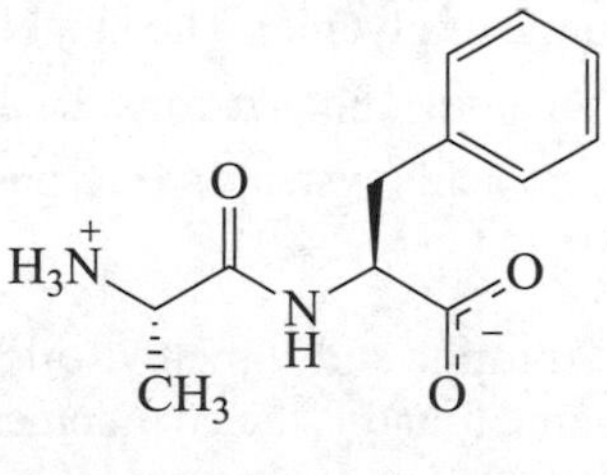

The L configuration corresponds to S for each of the chirality centers in Ala-Phe.

(c) Similarly, Phe-Ala has its substituent at the N-terminal amino acid directed away from us, whereas the C-terminal side chain is pointing toward us, and the L configuration corresponds to S for each chirality center.

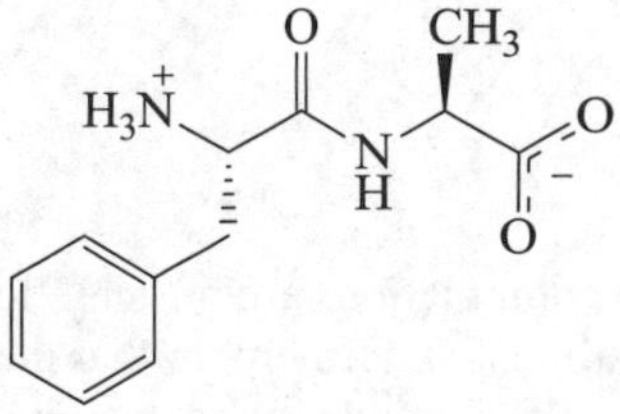

(*d*) There is only one chirality center in Gly-Glu. It has the L (or *S*) configuration.

(*e*) In order for the N-terminal amino acid in Lys-Gly to have the L (or *S*) configuration, its side chain must be directed away from us in the conformation indicated.

(*f*) The configuration at both α-carbon atoms in D-Ala-D-Ala is exactly the reverse of the configuration of the chirality centers in parts (*a*) through (*e*). Both chirality centers have the D (or *R*) configuration.

26.15 In the text the structure of leucine enkephalin is given. Methionine enkephalin differs from it only with respect to the C-terminal amino acid. The amino acid sequences of the two pentapeptides are

Tyr-Gly-Gly-Phe-Leu   Tyr-Gly-Gly-Phe-Met

Leucine enkephalin   Methionine enkephalin

The peptide sequence of a polypeptide can also be expressed using the one-letter abbreviations listed in text Table 26.1. Methionine enkephalin becomes YGGFM.

26.16 The structure of oxytocin is given in Figure 26.5. The amino acids present are Cys, Tyr, Ile, Gln, Asn, Pro, Leu, and Gly-NH$_2$ (the amide of glycine). The number of negative charges is zero because the C-terminal amino acid (Gly-NH$_2$) is an amide, not a carboxylic acid, and none of the other amino acids has an acidic side chain. The N-terminal amino acid (cysteine) has a positively charged nitrogen. Therefore, the net charge of oxytocin is +1.

26.17 The two amino acids are L-proline and *N*-methyl-D-leucine. The proline is derivatized as an amide of (*S*)-lactic acid. The D-leucine found is the enantiomer of L-leucine, and it is methylated at the nitrogen.

(*S*)-Lactic acid

26.18 Twenty-four tetrapeptide combinations are possible for the four amino acids alanine (A), glycine (G), phenylalanine (F), and valine (V). Remember that the order is important; AG is not the same peptide as GA. Using the one-letter abbreviations for each amino acid, the possibilities are

AGFV	AGVF	AFGV	AFVG	AVGF	AVFG
GAFV	GAVF	GFAV	GFVA	GVFA	GVAF
FAGV	FAVG	FVAG	FVGA	FGAV	FGVA
VAGF	VAFG	VGAF	VGFA	VFAG	VFGA

26.19 Chymotrypsin cleaves a peptide selectively at the carboxyl group of amino acids that have aromatic side chains. The side chain of phenylalanine is a benzyl group, $C_6H_5CH_2$—. If the dipeptide isolated after treatment with chymotrypsin contains valine (V) and phenylalanine (F), its sequence must be VF.

Valine | Phenylalanine | Valinylphenylalanine (VF)

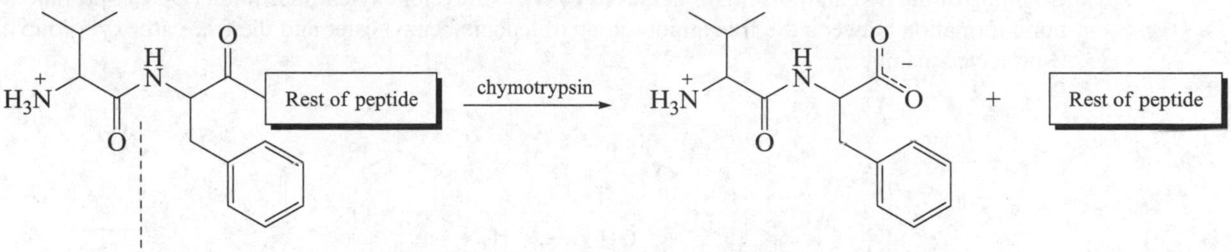

The possible sequences for the unknown tetrapeptide are VFAG and VFGA.

26.20 The Edman degradation removes the N-terminal amino acid, which is identified as a phenylthiohydantoin (PTH) derivative. The first Edman degradation of Val-Phe-Gly-Ala gives the PTH derivative of valine; the second gives the PTH derivative of phenylalanine.

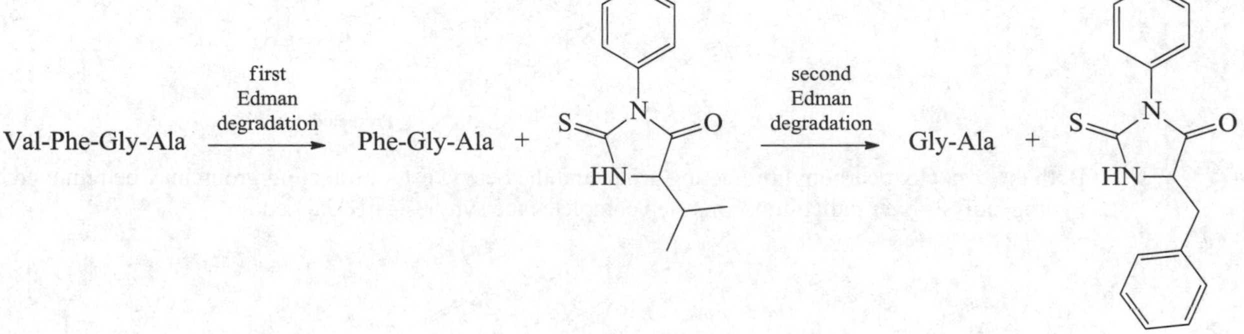

26.21 Leucine and isoleucine are isomers. Therefore, they have the same molecular weight and the same *m/z* value.

26.22 The peptide bond of Ala-Leu connects the carboxyl group of alanine and the amino group of leucine. We therefore need to protect the amino group of alanine and the carboxyl group of leucine.

Protect the amino group of alanine as its benzyloxycarbonyl derivative.

Alanine | Benzyloxycarbonyl chloride | *Z*-Protected alanine

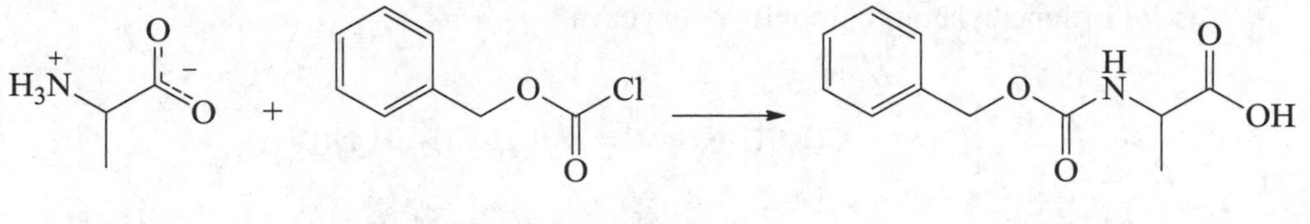

Protect the carboxyl group of leucine as its benzyl ester.

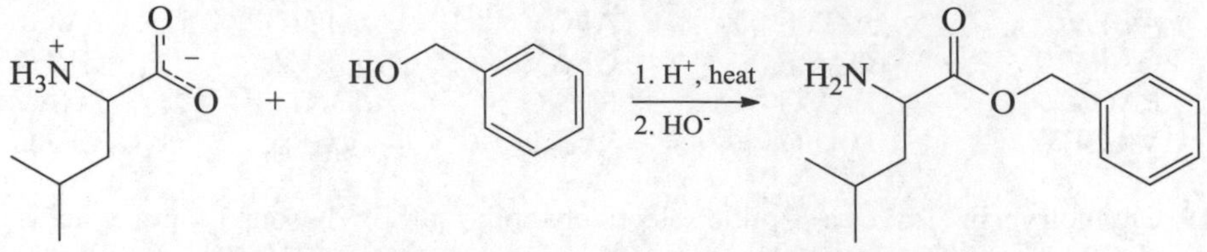

Leucine   Benzyl alcohol   Leucine benzyl ester

Coupling of the two amino acids is achieved by N,N'-dicyclohexylcarbodiimide (DCCI)-promoted amide bond formation between the free amino group of leucine benzyl ester and the free carboxyl group of Z-protected alanine.

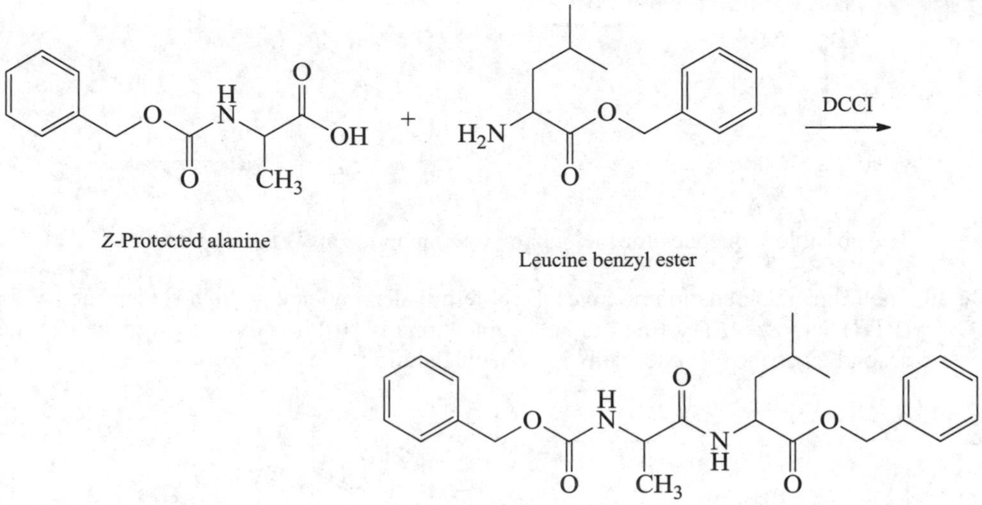

Z-Protected alanine

Leucine benzyl ester

Protected dipeptide

Both the benzyloxycarbonyl protecting group and the benzyl ester protecting group may be removed by hydrogenolysis over palladium. This step completes the synthesis of Ala-Leu.

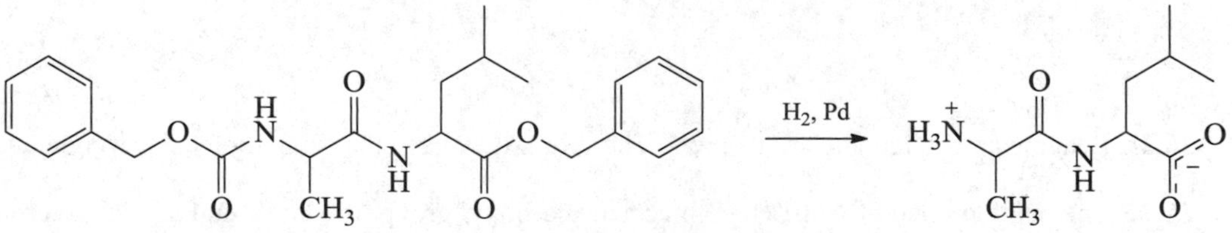

Protected dipeptide   Ala-Leu

26.23 The urea that is produced in the coupling reaction of a protected amino acid and a carboxylic acid with EDCI is N-(3-(dimethylamino)propyl)-N'-ethylurea.

$$CH_3CH_2NH-\overset{\overset{\textstyle O}{\|}}{C}-NHCH_2CH_2CH_2N(CH_3)_2$$

N-(3-(Dimethylamino)propyl)-N'-ethylurea

26.24 Amino acid residues are added by beginning at the C terminus in the Merrifield solid-phase approach to peptide synthesis. Thus, the synthesis of Phe-Gly requires glycine to be anchored to the solid support. Begin by protecting glycine as its *tert*-butoxycarbonyl (Boc) derivative.

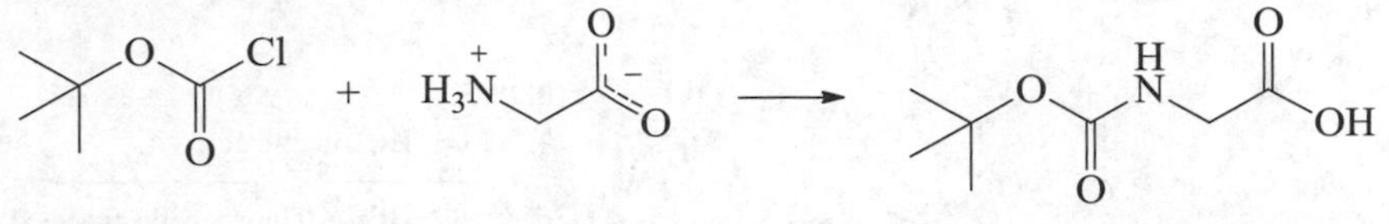

tert-Butyoxycarbonyl
chloride   Glycine   Boc-protected glycine

The protected glycine is attached via its carboxylate anion to the solid support.

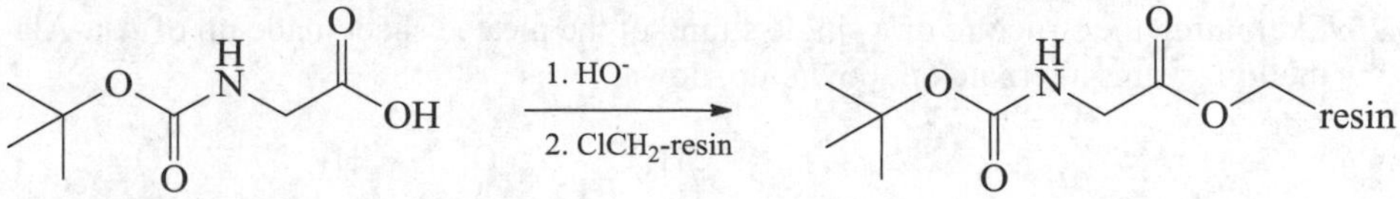

Boc-protected glycine   Boc-protected, resin-bound glycine

The amino group of glycine is then exposed by removal of the protecting group. Typical conditions for this step involve treatment with hydrogen chloride in acetic acid.

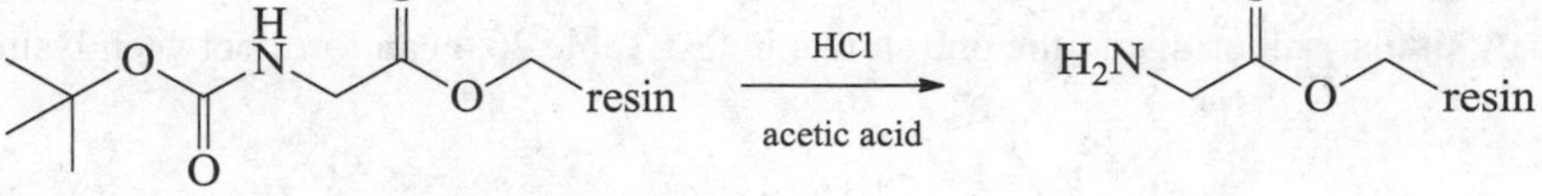

Boc-protected, resin-bound glycine   Resin-bound glycine

To attach phenylalanine to resin-bound glycine, we must first protect the amino group of phenylalanine. A Boc protecting group is appropriate.

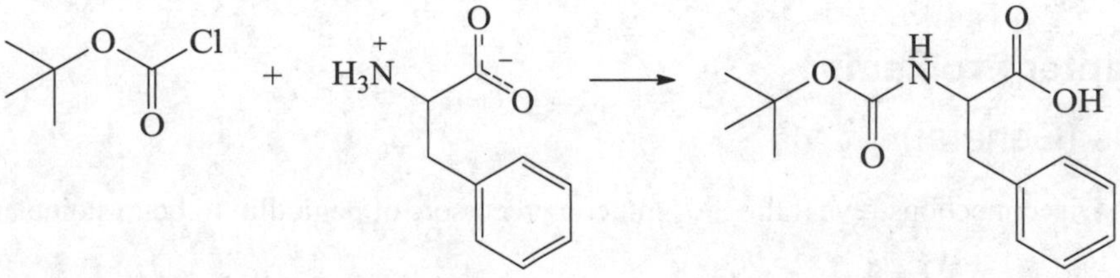

tert-Butyoxycarbonyl
chloride   Phenylalanine   Boc-protected phenylalanine

Peptide bond formation occurs when the resin-bound glycine and Boc-protected phenylalanine are combined in the presence of DCCI.

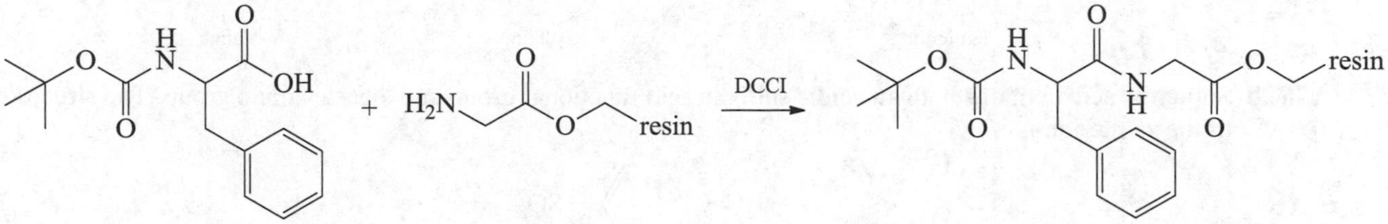

Boc-protected phenylalanine   Resin-bound glycine   Boc-protected, resin-bound Phe-Gly

Remove the Boc group with HCl and then treat with HBr in trifluoroacetic acid to cleave Phe-Gly from the solid support.

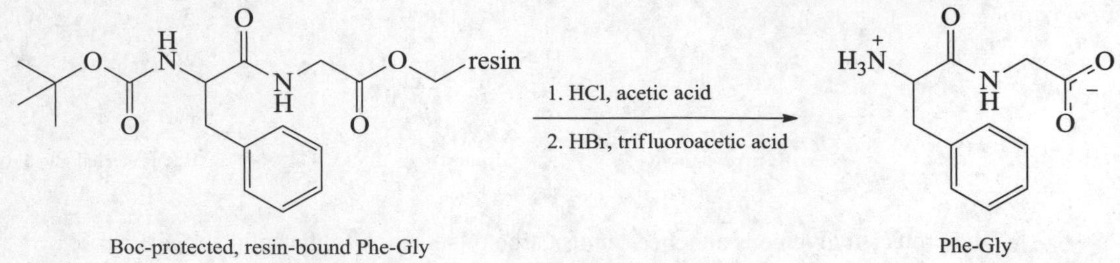

Boc-protected, resin-bound Phe-Gly → Phe-Gly

1. HCl, acetic acid
2. HBr, trifluoroacetic acid

26.25 Examining the structure of a single strand of the pleated sheet made up of Ala-Ala-Ala-Ala reveals that the methyl groups alternate up, down, up, down.

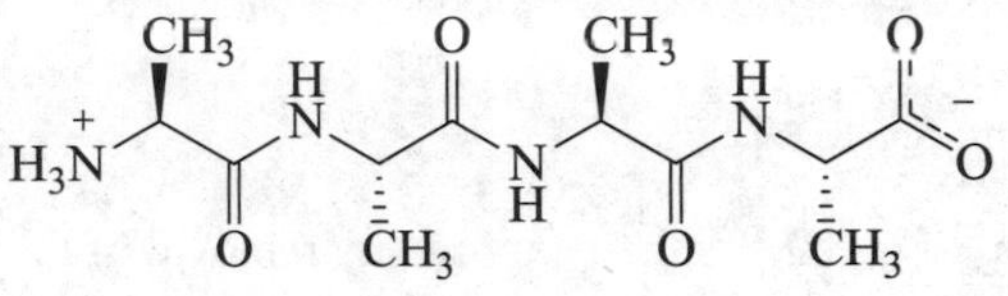

Ala-Ala-Ala-Ala

26.26 A similar salt bridge to the one shown in text Table 26.4 can form between lysine and glutamic acid.

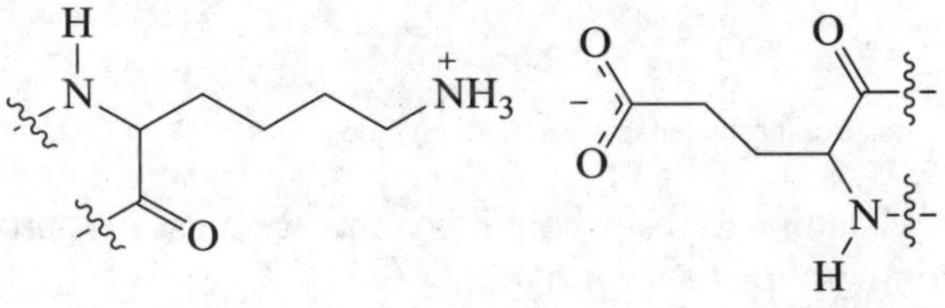

Lysine   Glutamic acid

End of Chapter Problems

Amino Acids (General)

26.27 Disconnections reveal the biosynthetic precursors of penicillin to be cysteine and valine.

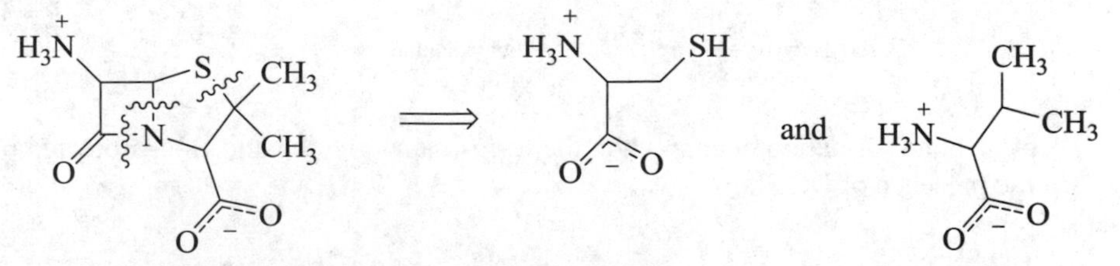

Penicillin   Cysteine   Valine

26.28 Sulfanilic acid contains both an acidic sulfonic acid functional group and a basic amino group. The structure of the zwitterion is

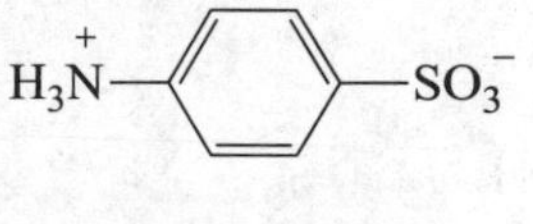

Zwitterion of
sulfanilic acid

26.29 Acid hydrolysis of the triester converts all its ester functions to free carboxyl groups and cleaves both amide bonds.

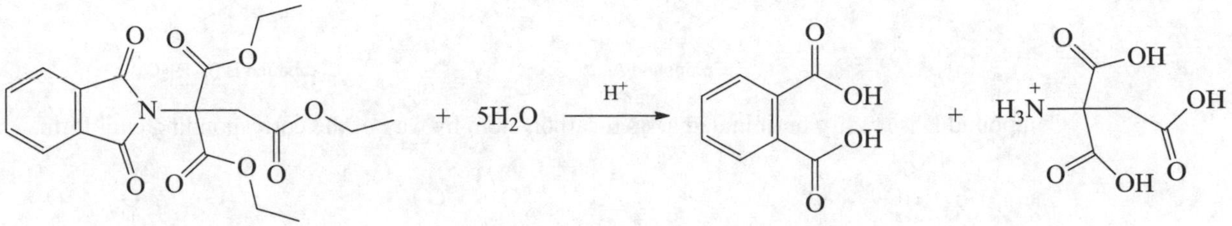

The hydrolysis product is a substituted derivative of malonic acid and undergoes decarboxylation on being heated. The product of this decarboxylation is aspartic acid (in its protonated form under conditions of acid hydrolysis).

Aspartic acid   Carbon dioxide

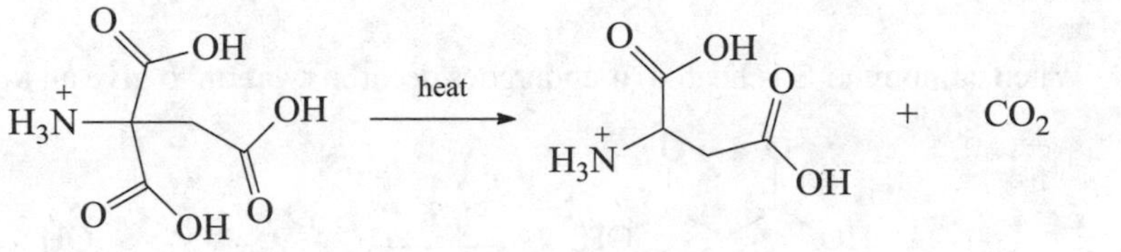

Aspartic acid is chiral but is formed as a racemic mixture, so the product of this reaction is not optically active. The starting triester is achiral and cannot give an optically active product when it reacts with optically inactive reagents.

26.30 The following outlines a synthesis of β alanine in which conjugate addition to acrylonitrile plays a key role.

Acrylonitrile   3-Aminopropanenitrile   β Alanine

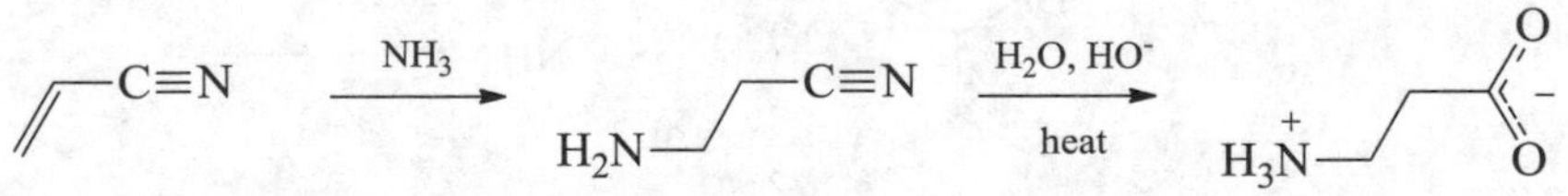

Addition of ammonia to acrylonitrile has been carried out in modest yield (31–33%). Hydrolysis of the nitrile group can be accomplished in the presence of either acids or bases. Hydrolysis in the presence of $Ba(OH)_2$ has been reported in the literature to give β alanine in 85–90% yield.

26.31 (*a*) The first step involves alkylation of diethyl malonate by 2-bromobutane.

2-Bromobutane   Anion of diethyl malonate   Compound A

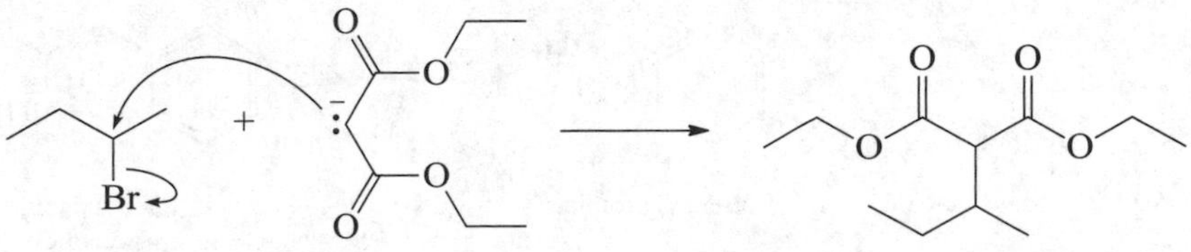

In the second step of the synthesis, compound A is subjected to ester saponification. Following acidification, the corresponding diacid (compound B) is isolated.

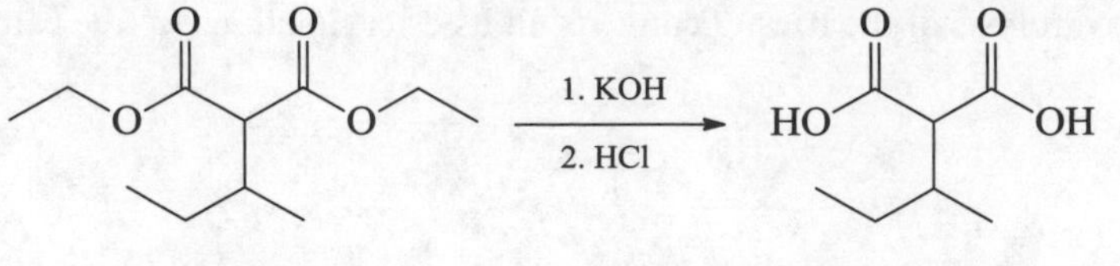

Compound A   Compound B ($C_7H_{12}O_4$)

Compound B is readily brominated at its α-carbon atom by way of the corresponding enol form.

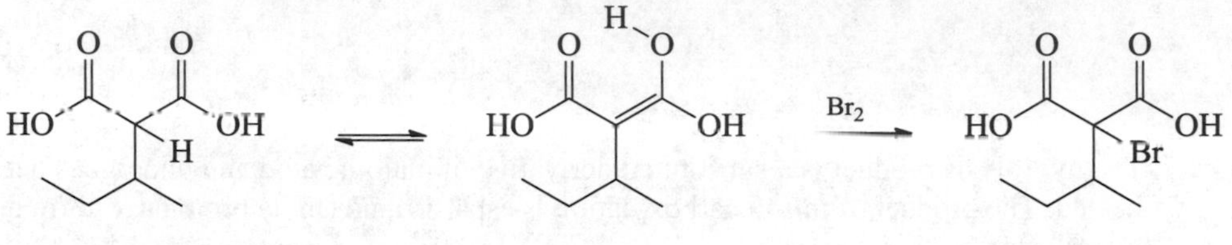

Compound B   Enol form   Compound C
                                                            ($C_7H_{11}BrO_4$)

When compound C is heated, it undergoes decarboxylation to give an α-bromo carboxylic acid.

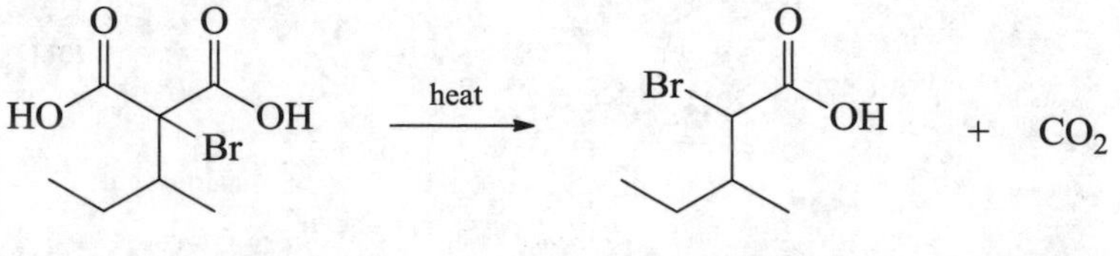

Compound C   Compound D   Carbon dioxide

Treatment of compound D with ammonia converts it to isoleucine by nucleophilic substitution.

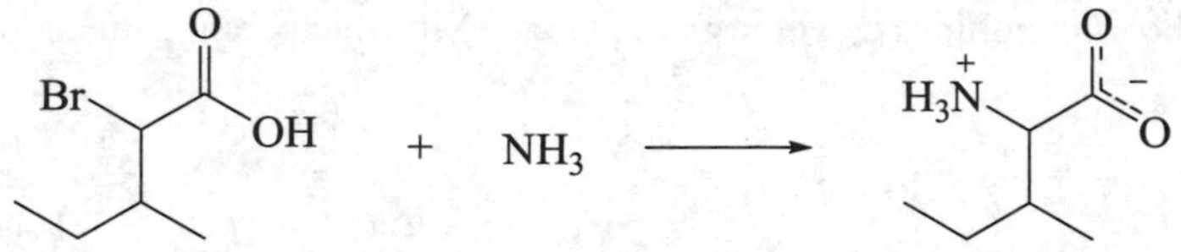

Compound D   Isoleucine
                                          (racemic mixture
                                          of diastereomers)

(b) The procedure just described can be adapted to the synthesis of other amino acids. The group attached to the α-carbon atom is derived from the alkyl halide used to alkylate diethyl malonate. Benzyl bromide (or chloride or iodide) would be appropriate for the preparation of phenylalanine.

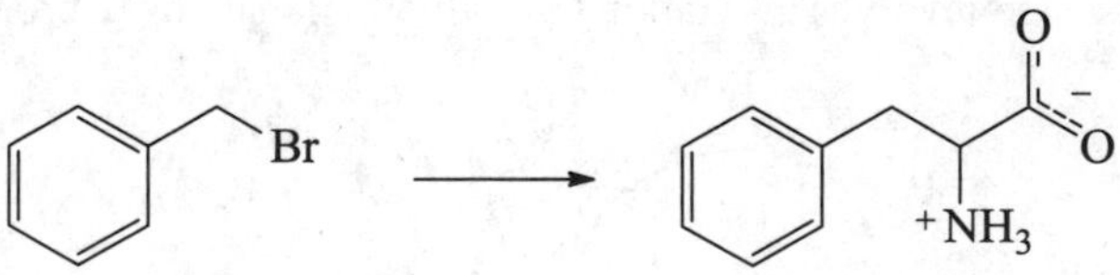

Benzyl bromide   Phenylalanine
                                      (racemic)

26.32 (*a*) Nitration occurs ortho to the ortho, para-directing, strongly activating OH group.

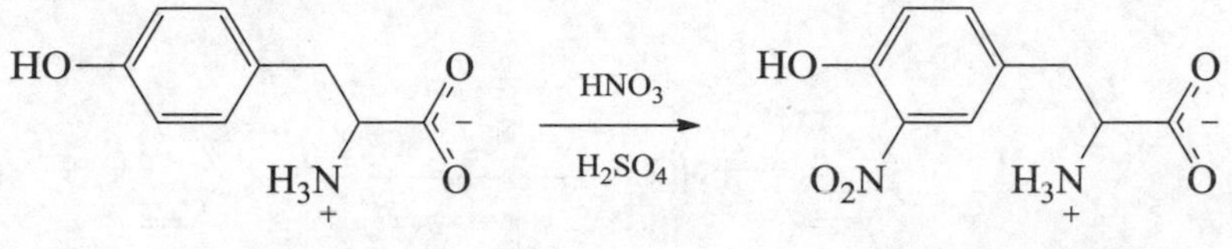

Tyrosine

3-Nitrotyrosine ($C_9H_{10}N_2O_5$)

(*b*) The phenolic OH is alkylated with allyl bromide to give an *O*-allyl ether. In the second step, a methyl ester is converted to an amide by reaction with ammonia. In the following structural formulas, the Boc protecting group is shown in detail.

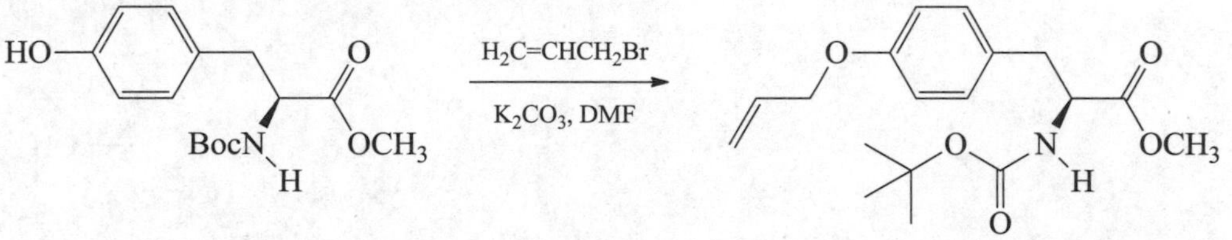

$C_{18}H_{25}NO_5$

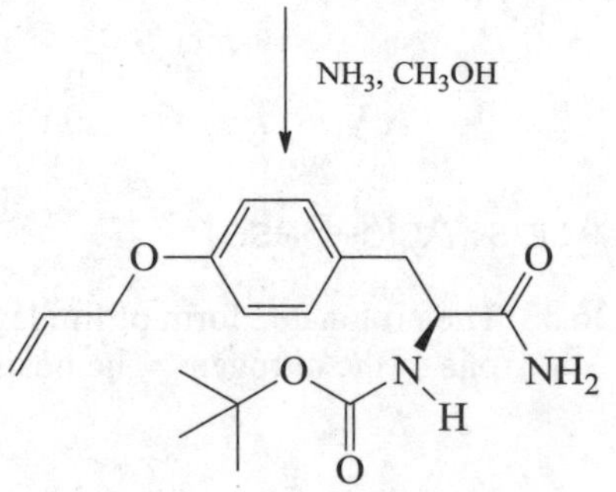

NH$_3$, CH$_3$OH

$C_{17}H_{24}N_2O_4$

(*c*) With both chlorine atoms in ortho-para relationships with three nitro groups, the aryl halide is especially reactive to nucleophilic aromatic substitution. The molecular formulas of the products indicate both halogens are replaced by glycines.

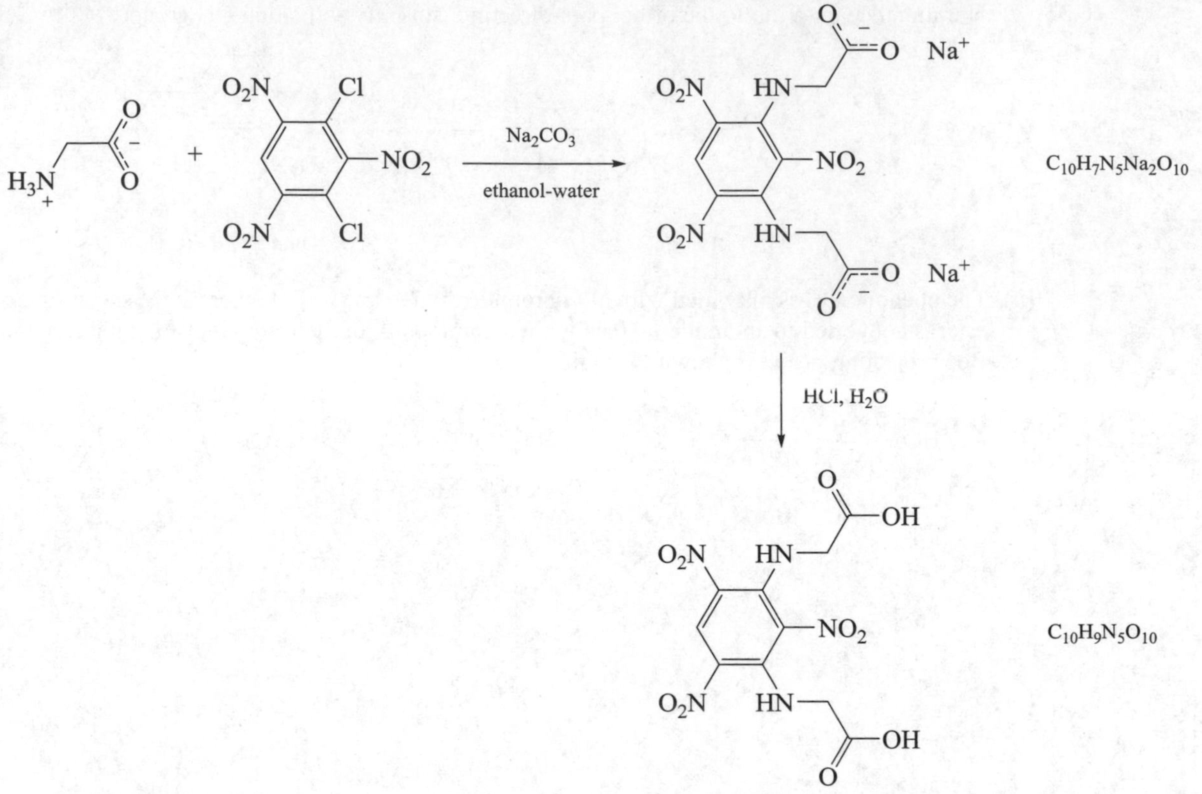

Amino Acids (Acid–Base)

26.33 The protonated form of imidazole represented by structure A is stabilized by delocalization of the lone pair of one of the nitrogens. The positive charge is shared by both nitrogens.

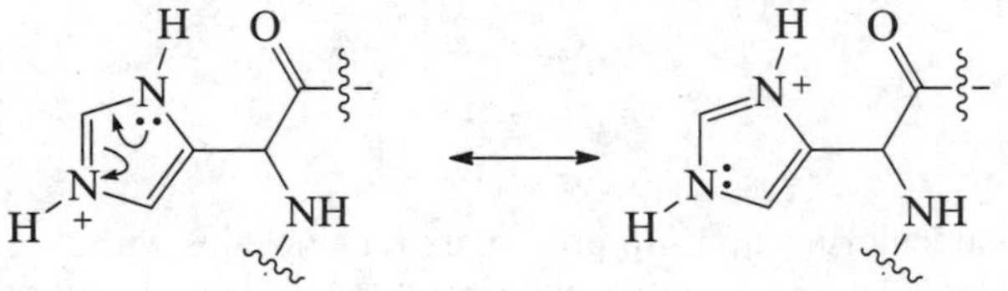

A

The positive charge in structure B is localized on a single nitrogen. Resonance stabilization of the type shown in structure A is not possible; in addition, the imidazole ring is no longer aromatic, thus, structure A is the more stable protonated form.

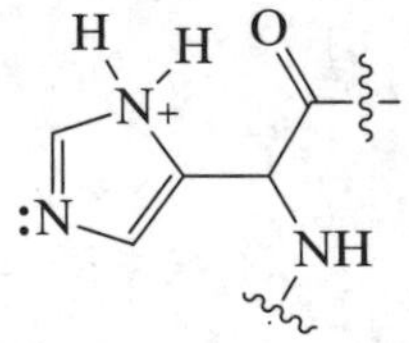

B

26.34 (*a*) The amino acid in the problem is valine [$R=CH(CH_3)_2$] A is the zwitterion of valine; B is its conjugate base. The pK_a for deprotonation of the ammonium ion, from text Table 26.2, is 9.62. We can use the

Henderson–Hasselbalch equation (text Section 19.4) to determine the ratio of the zwitterion (A) to the conjugate base (B).

$$pH = pK_a + \log \frac{[\text{conjugate base}]}{[\text{acid}]}$$

$$pH = pK_a + \log \frac{[B]}{[A]}$$

$$\log \frac{[B]}{[A]} = pH - pK_a$$

$$\log \frac{[B]}{[A]} = 7.00 - 9.62 = -2.62$$

$$\frac{[B]}{[A]} = 10^{-2.62} = 2.40 \times 10^{-3}$$

$$\frac{[A]}{[B]} = \frac{1}{2.40 \times 10^{-3}} = 417$$

(b) The maximum concentration of the zwitterion (A) is found at the isoelectric point, pI. The pH where the concentration of A is at a maximum is 5.96.

26.35 From the molecular formulas and net charges given in the text, the structures of citrulline, ornithine, and putrescine can be determined.

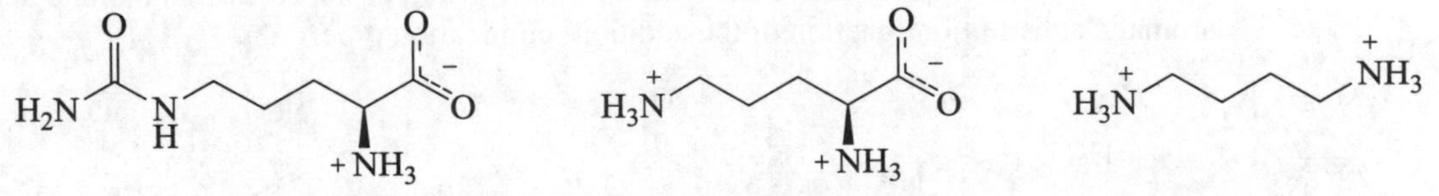

Citrulline   Ornithine   Putrescine

Peptides

26.36 Asparagine and glutamine each contain an amide function in their side chain. Under the conditions of peptide bond hydrolysis that characterize amino acid analysis, the side-chain amide is also hydrolyzed, giving ammonia.

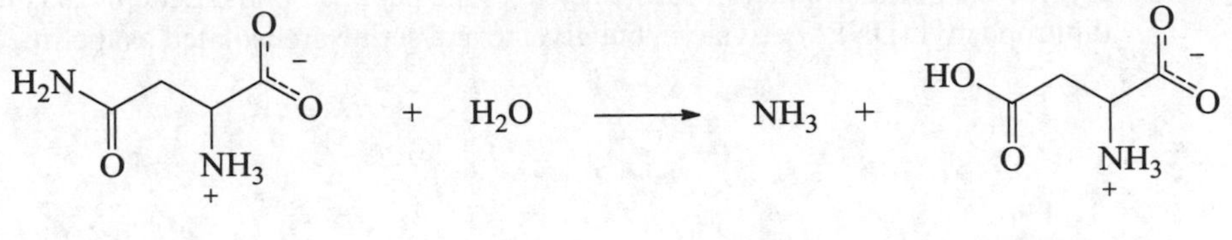

Asparagine   Water   Ammonia   Aspartic acid

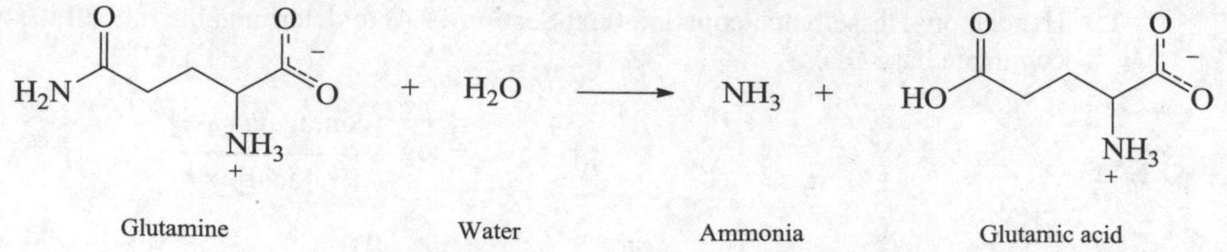

| Glutamine | Water | Ammonia | Glutamic acid |

26.37 The amino acids are valine (Val), asparagine (Asn), alanine (Ala), glutamic acid (Glu), and phenylalanine (Phe). There is modified leucine between the Asn and Ala residues that has the linker $CH(OH)CH_2CH(CH_3)$.

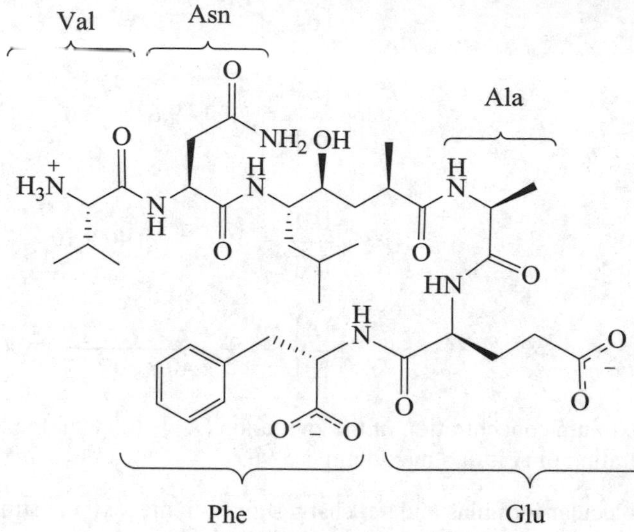

26.38 (*a*) 1-Fluoro-2,4-dinitrobenzene reacts with the amino group of the N-terminal amino acid in a nucleophilic aromatic substitution reaction of the addition–elimination type.

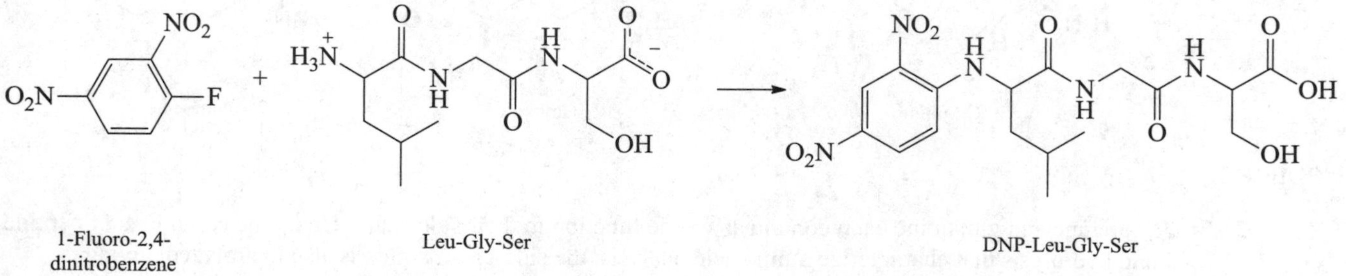

1-Fluoro-2,4-dinitrobenzene   Leu-Gly-Ser   DNP-Leu-Gly-Ser

(*b*) Hydrolysis of the product in part (*a*) cleaves the peptide bonds. Leucine is isolated as its 2,4-dinitrophenyl (DNP) derivative, but glycine and serine are isolated as the free amino acids.

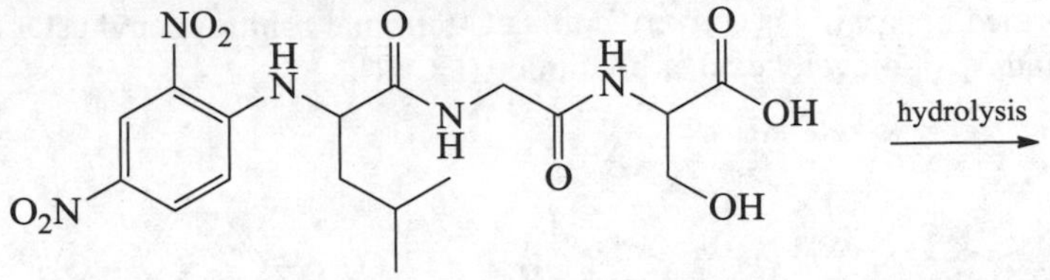

DNP-Leu-Gly-Ser

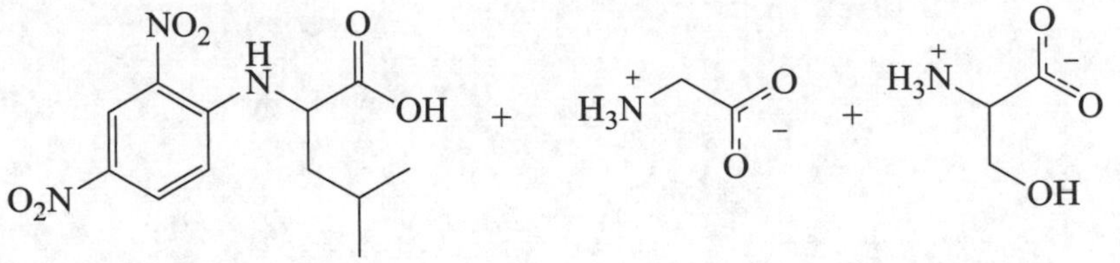

DNP-Leu   Gly   Ser

(*c*) Phenyl isothiocyanate is a reagent used to identify the N-terminal amino acid of a peptide by the Edman degradation. The N-terminal amino acid is cleaved as a phenylthiohydantoin (PTH) derivative, the remainder of the peptide remaining intact.

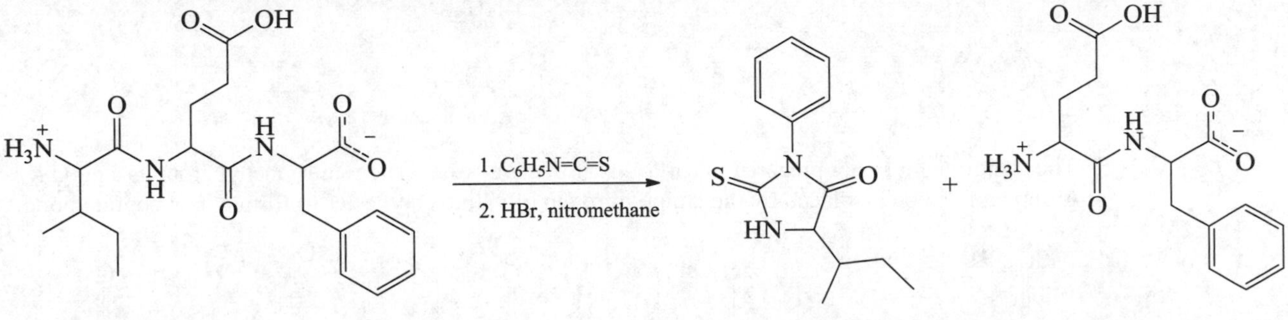

Ile-Glu-Phe   1. $C_6H_5N=C=S$ 2. HBr, nitromethane   PTH derivative of isoleucine   Glu-Phe

(*d*) Benzyloxycarbonyl chloride reacts with amino groups to convert them to amides. The only free amino group in Asn-Ser-Ala is the N terminus. The amide function of asparagine does not react with benzyloxycarbonyl chloride.

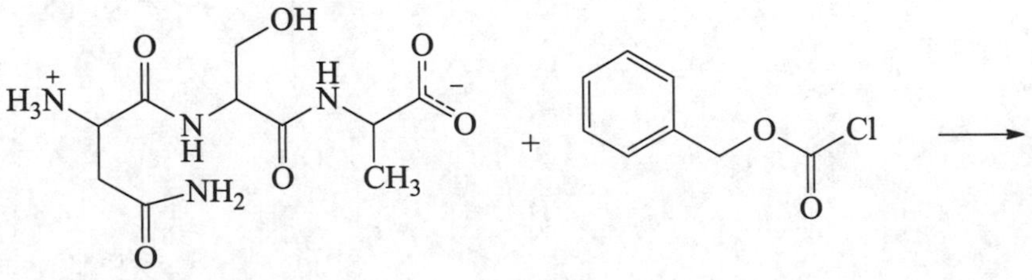

Asn-Ser-Ala

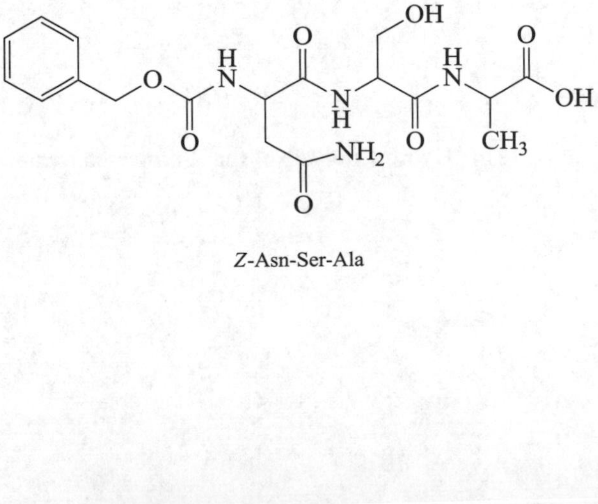

Z-Asn-Ser-Ala

(e) The Z-protected tripeptide formed in part (d) is converted to its C-terminal p-nitrophenyl ester on reaction with p-nitrophenol and N,N'-dicyclohexylcarbodiimide (DCCI).

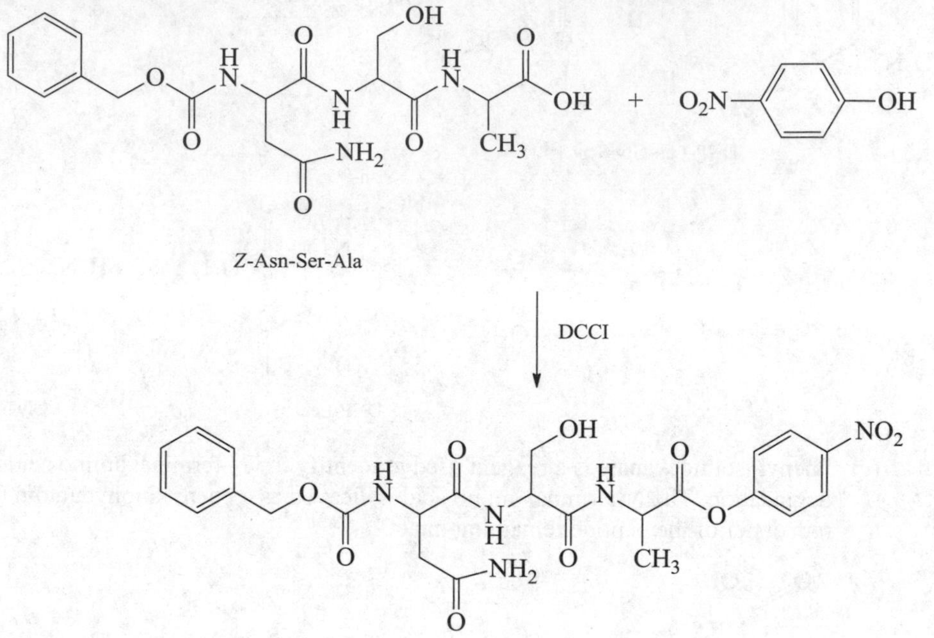

Z-Asn-Ser-Ala

DCCI

Z-Asn-Ser-Ala p-nitrophenyl ester

(f) The p-nitrophenyl ester prepared in part (e) is an "active" ester. The p-nitrophenyl group is a good leaving group and can be displaced by the amino nitrogen of valine ethyl ester to form a new peptide bond.

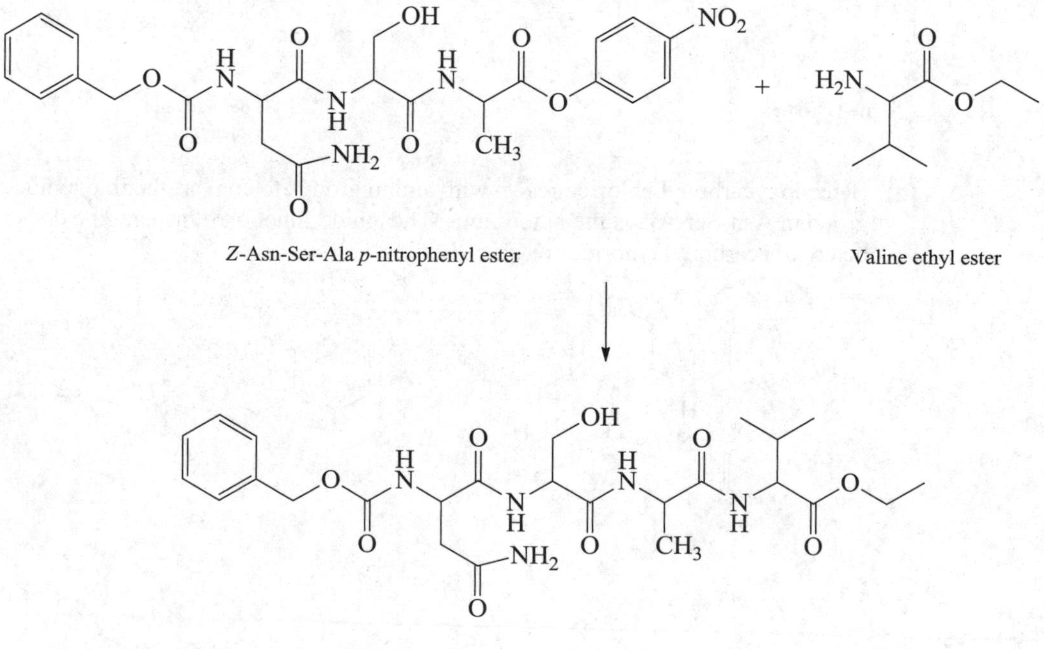

Z-Asn-Ser-Ala p-nitrophenyl ester   Valine ethyl ester

Z-Asn-Ser-Ala-Val ethyl ester

(g) Hydrogenolysis of the Z-protected tetrapeptide ester formed in part (f) removes the Z protecting group.

Z-Asn-Ser-Ala-Val ethyl ester

H_2, Pd

Asn-Ser-Ala-Val ethyl ester

26.39 (*a*) Trypsin catalyzes the hydrolysis of peptide bonds involving the carboxyl group of a lysine (K) or arginine (R) residue. The primary sequence of a peptide is written from the N terminus on the left to the C terminus on the right, so the cleavages are to the right of K and R.

AQDDYR –§– YIHFLTQYDAK –§– PKGR –§– NDEYCFNMMK

Trypsin-catalyzed cleavage occurs at the points indicated.

(*b*) Chymotrypsin-catalyzed cleavage is selective for peptide bonds involving the carboxyl groups (that is, to the right) of amino acids with aromatic side chains: phenylalanine (F), tyrosine (Y), and tryptophan (W).

AQDDY –§– RY –§– IHF –§– LTQHY –§– DAKPKGRNDEY –§– CF –§– NMMK

Chymotrypsin-catalyzed cleavage occurs at the points indicated.

26.40 Somatostatin is a tetradecapeptide and so is composed of 14 amino acids. The fact that Edman degradation gave the PTH derivative of alanine identifies this as the N-terminal amino acid. A major piece of information is the amino acid sequence of a hexapeptide obtained by partial hydrolysis:

Ala-Gly-Cys-Lys-Asn-Phe

Using this as a starting point and searching for overlaps with the other hydrolysis products gives the entire sequence.

Ala—Gly—Cys—Lys—Asn—Phe

Asn—Phe—Phe—Trp—Lys

Phe—Trp

Lys—Thr—Phe

Thr—Phe—Thr—Ser—Cys

Thr—Ser—Cys

Ala—Gly—Cys—Lys—Asn—Phe—Phe—Trp—Lys—Thr—Phe—Thr—Ser—Cys

| 1 | 2 | 3 | 4 | 5 | 6 | 7 | 8 | 9 | 10 | 11 | 12 | 13 | 14 |

The disulfide bridge in somatostatin is between cysteine 3 and cysteine 14. Thus, the primary structure is

Lys—Asn—Phe—Phe—Trp—Lys

Ala—Gly—Cys

S—S—Cys—Ser—Thr—Phe—Thr

26.41 The amino acids leucine, phenylalanine, and serine each have one chirality center.

Leucine: $R = (CH_3)_2CHCH_2$
Phenylalanine: $R = C_6H_5CH_2$
Serine: $R = HOCH_2$

When prepared by the Strecker synthesis, each of these amino acids is obtained as a racemic mixture containing 50% of the D enantiomer and 50% of the L enantiomer.

Chiral, but racemic

Thus, preparation of the tripeptide Leu-Phe-Ser will yield a mixture of 2^3 (eight) stereoisomers.

D-Leu-D-Phe-D-Ser   L-Leu-L-Phe-L-Ser
D-Leu-D-Phe-L-Ser   L-Leu-L-Phe-D-Ser
D-Leu-L-Phe-D-Ser   L-Leu-D-Phe-L-Ser
D-Leu-L-Phe-L-Ser   L-Leu-D-Phe-D-Ser

26.42 It is the C-terminal amino acid that is anchored to the solid support in the preparation of peptides by the Merrifield method. Refer to the structure of oxytocin in Figure 26.5 of the text and note that oxytocin, in fact, has no free carboxyl groups; all the acyl groups of oxytocin appear as amide functions. Thus, the carboxyl terminus of oxytocin has been modified by conversion to an amide. There are three amide functions of the type $-\overset{O}{\overset{\|}{C}}NH_2$, two of which belong to side chains of asparagine and glutamine, respectively. The third amide belongs to the C-terminal amino acid, glycine, $-NHCH_2\overset{O}{\overset{\|}{C}}OH$, which in oxytocin has been modified so that it appears as $-NHCH_2\overset{O}{\overset{\|}{C}}NH_2$. Therefore, attach glycine to the solid support in the first step of the Merrifield synthesis. The carboxyl group can be modified to the required amide after all the amino acid residues have been added and the completed peptide is removed from the solid support.

26.43 This seems like a long mechanism, but much of it is simply acyl transfer reactions that occur through tetrahedral intermediates, just like the ones in Chapter 20.

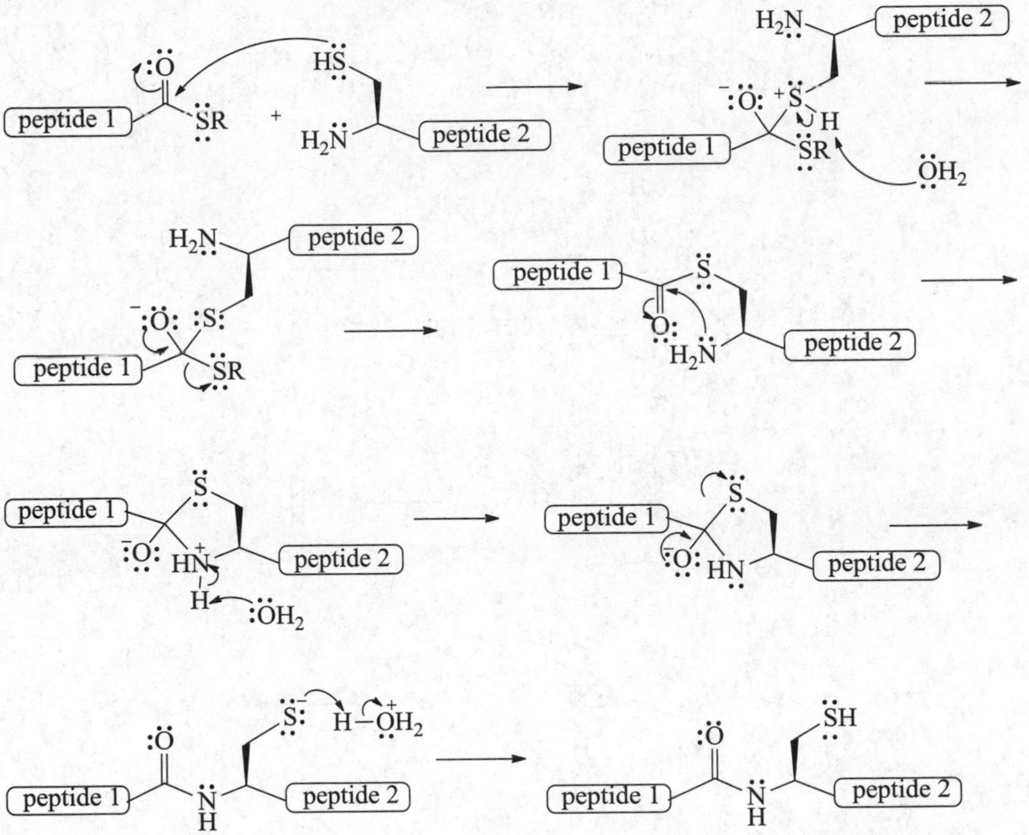

26.44 (*a*) The cyclization reaction is facilitated by the good leaving group, methyl thiocyanate.

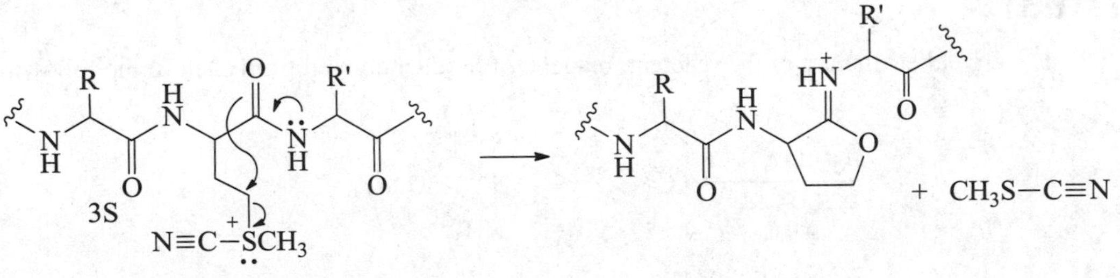

(*b*) The hydrolysis reaction occurs as shown.

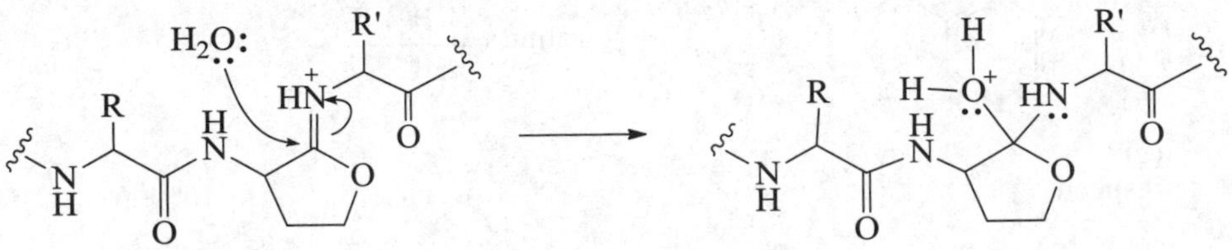

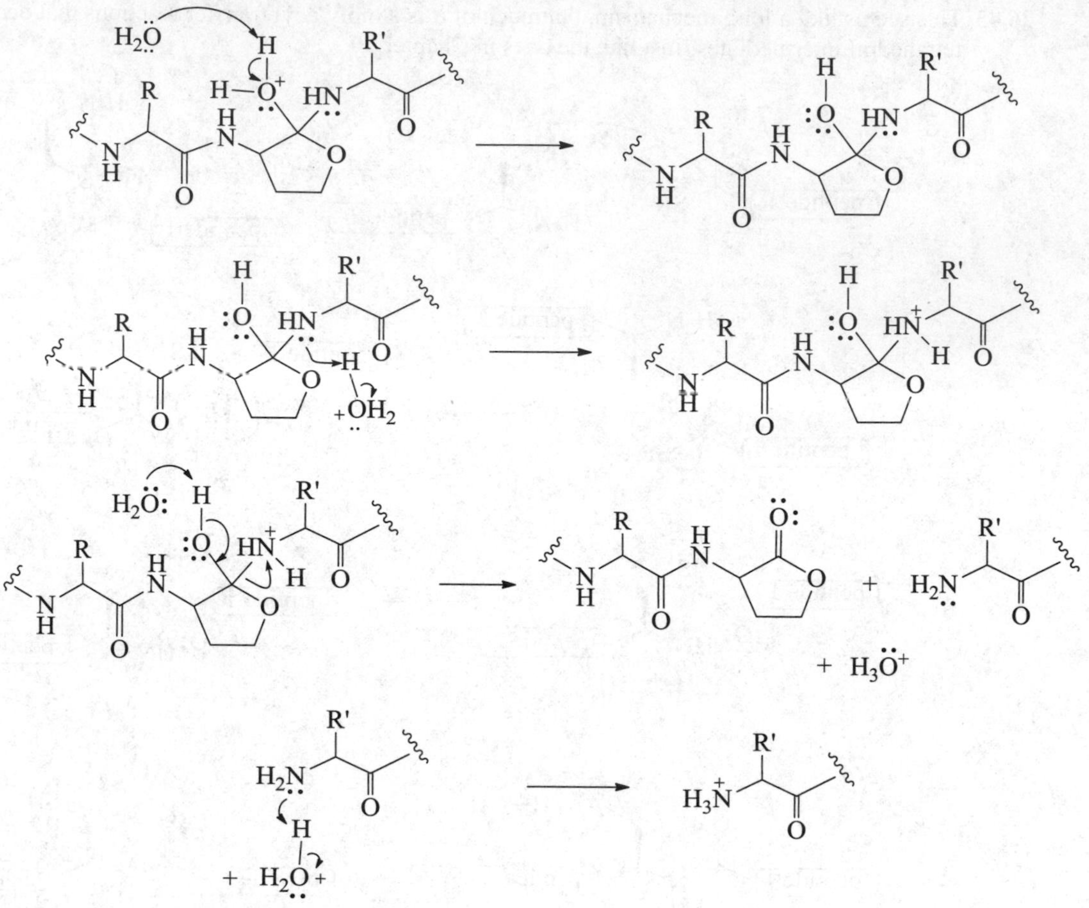

Answers to Interpretive Problems 26

26.45 A; 26.46 C; 26.47 B; 26.48 D; 26.49 C; 26.50 A

SELF-TEST

1. Give the structure of the reactant, reagent, or product omitted from each of the following:

(a)

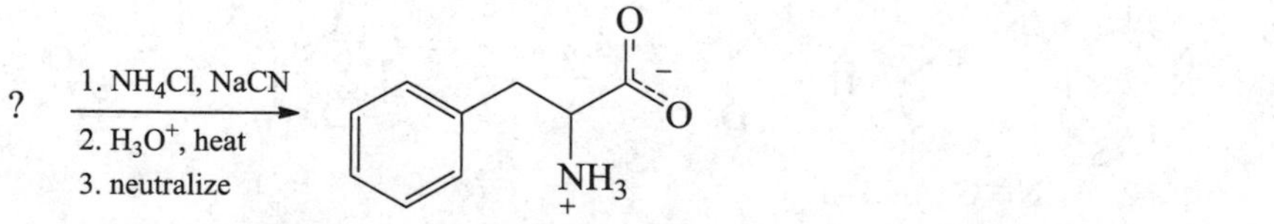

? →
1. NH$_4$Cl, NaCN
2. H$_3$O$^+$, heat
3. neutralize

(b)

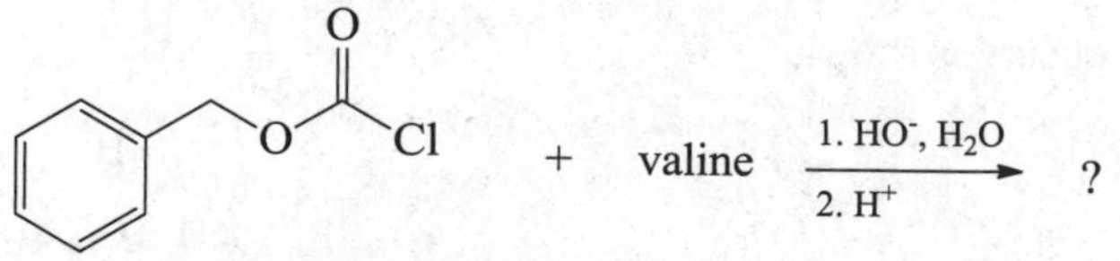

+ valine →
1. HO$^-$, H$_2$O
2. H$^+$
?

(c)

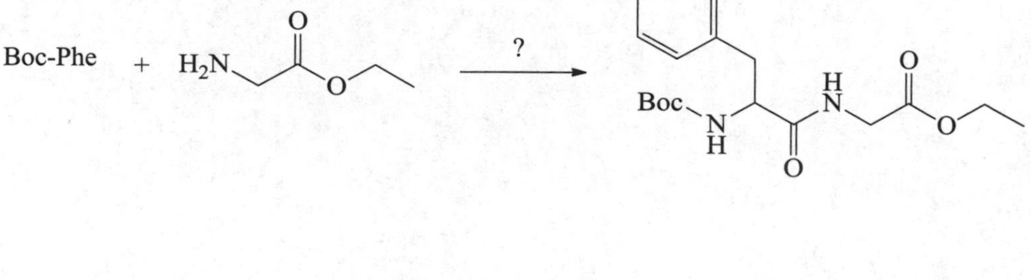

Boc-Phe + H$_2$N—CH$_2$—CO—O—CH$_2$CH$_3$ → ?

2. Give the structure of the derivative that would be obtained by treatment of Phe-Ala with Sanger's reagent (1-fluoro-2,4-dinitrobenzene) followed by hydrolysis.

3. Outline a sequence of steps that would allow the following synthetic conversions to be carried out:

    (*a*)

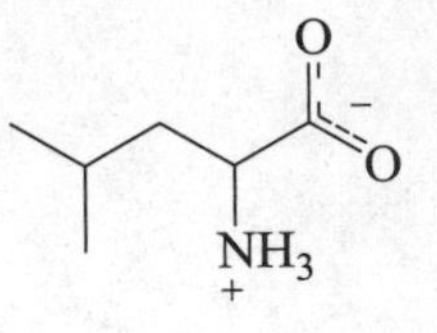

    (leucine) from

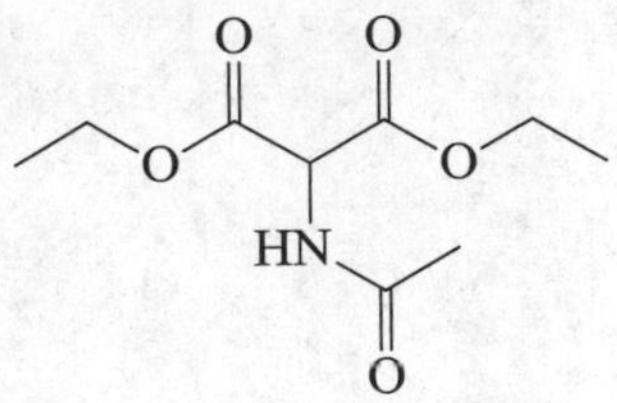

    (*b*)

    Leu-Val from leucine and (valine)

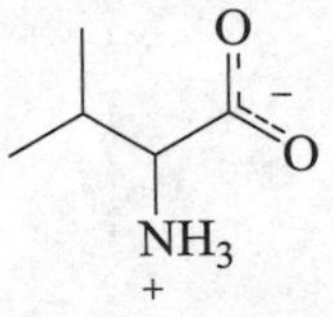

4. The carboxypeptidase-catalyzed hydrolysis of a pentapeptide yielded phenylalanine (Phe). One cycle of an Edman degradation gave a derivative of leucine (Leu). Partial hydrolysis yielded the fragments Leu-Val-Gly and Gly-Ala, among others. Deduce the structure of the peptide.

5. Consider the following compound:

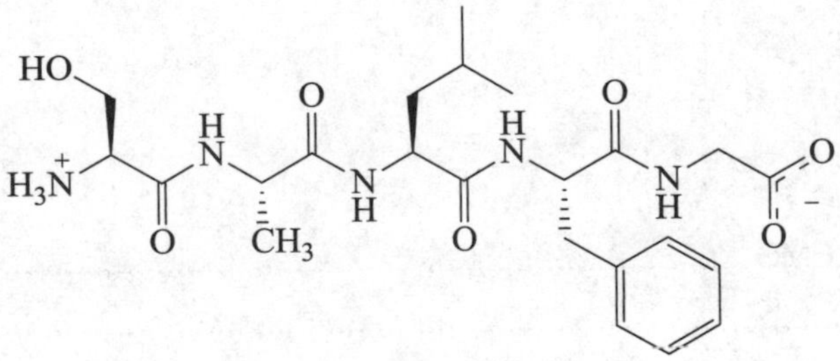

    (*a*) What kind of peptide does this structure represent? (For example, dipeptide)
    (*b*) How many peptide bonds are present?
    (*c*) Give the name for the N-terminal amino acid.
    (*d*) Give the name for the C-terminal amino acid.
    (*e*) Using three-letter abbreviations, write the sequence.

6. Consider the tetrapeptide Ala-Gly-Phe-Leu. What are the products obtained from each of the following? Be sure to account for all the amino acids of the peptide.

    (*a*) Treatment with 1-fluoro-2,4-dinitrobenzene followed by hydrolysis in concentrated HCl at 100°C
    (*b*) Treatment with chymotrypsin
    (*c*) Treatment with carboxypeptidase
    (*d*) Reaction with benzyloxycarbonyl chloride

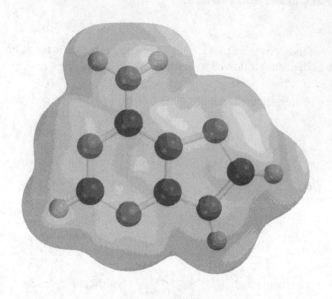

CHAPTER 27
Nucleosides, Nucleotides, and Nucleic Acids

Table of Contents

SOLUTIONS TO TEXT PROBLEMS

In Chapter Problems

27.1   The enol form of cytosine has a double bond between nitrogen at position 1 (see text Table 27.1 for the numbering scheme) in the ring and the carbon at position 2:

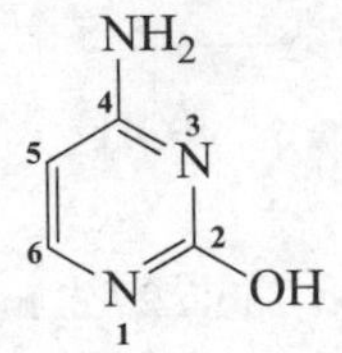

27.2 (*b*) Three zwitterionic resonance forms of uracil that that involve amide resonance can be formed as shown from the non-zwitterionic structure:

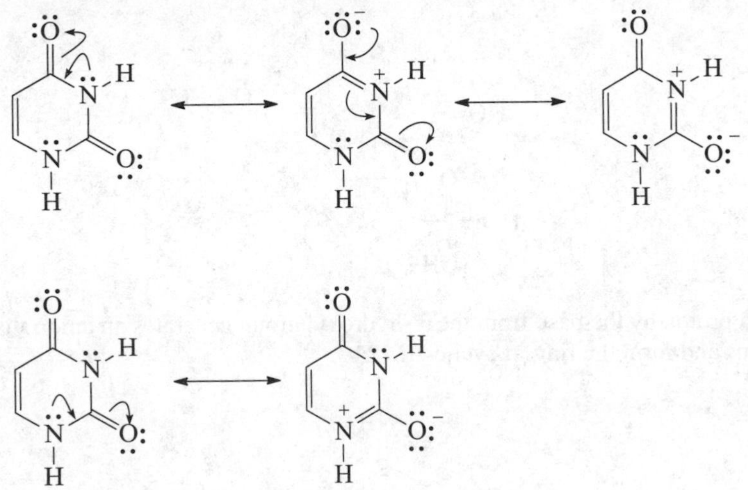

27.3   Caffeine and theobromine are both purines. Caffeine lacks H—N—C=O units so it cannot enolize. Two constitutionally isomeric enols are possible for theobromine.

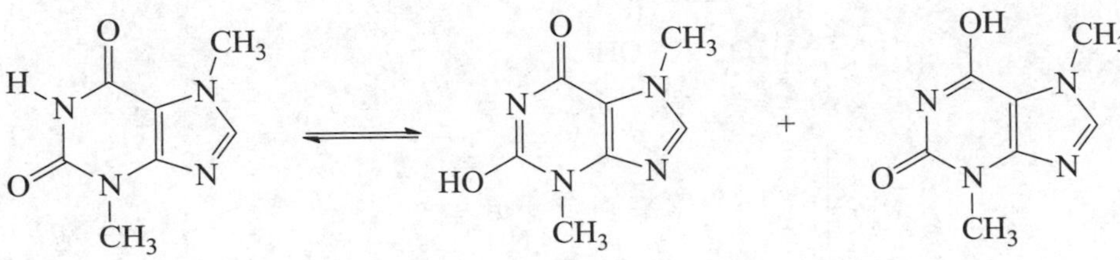

27.4 The structure of adenosine is shown in text Table 27.2. 3'-Deoxyadenosine (cordycepin) has an H replacing the OH group on C-3 of the ribose portion of the molecule.

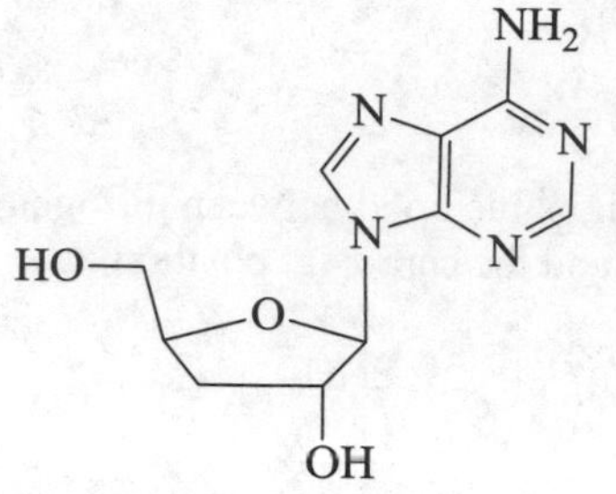

27.5 The structure of cytidine is shown in text Table 27.2. In 2'-deoxycytidine-3'-monophosphate, an H has replaced the OH at C-2 of the ribose portion of the molecule. In addition, C-3 of the ribose portion is bonded to a phosphate group.

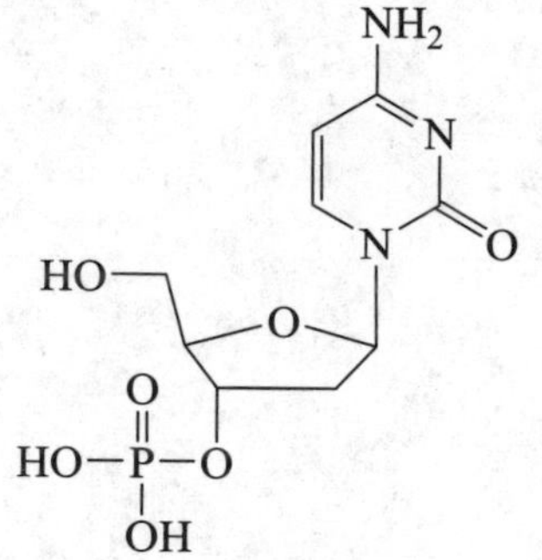

27.6 Removal of a proton by the base from the 3'-hydroxyl group generates an anion that can act as a nucleophile on phosphorus and form the ring in cyclic-AMP.

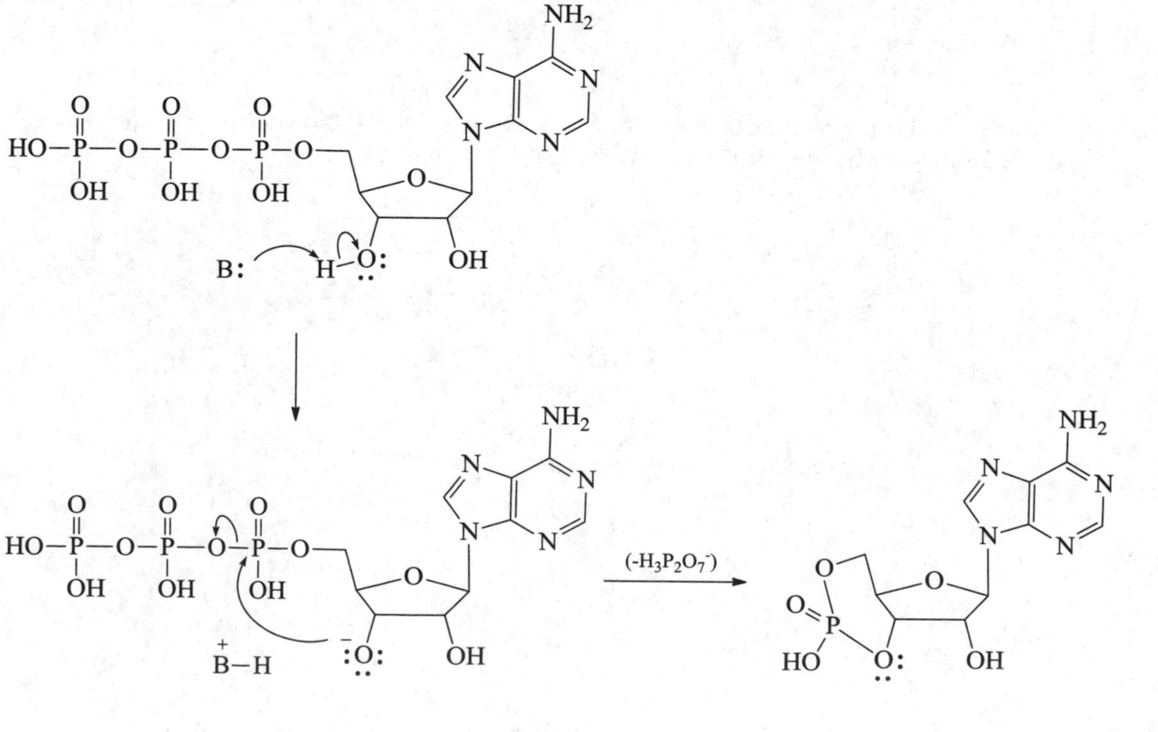

$(-H_3P_2O_7^-)$

27.7 When equations are added, species that appear on both the left and the right of the arrow are canceled. Thus, adding the hydrolysis of ATP to equation 1 does, indeed, produce equation 2.

Hydrolysis of ATP:

$$ATP + H_2O \longrightarrow ADP + HPO_4^{2-}$$

Equation 1:

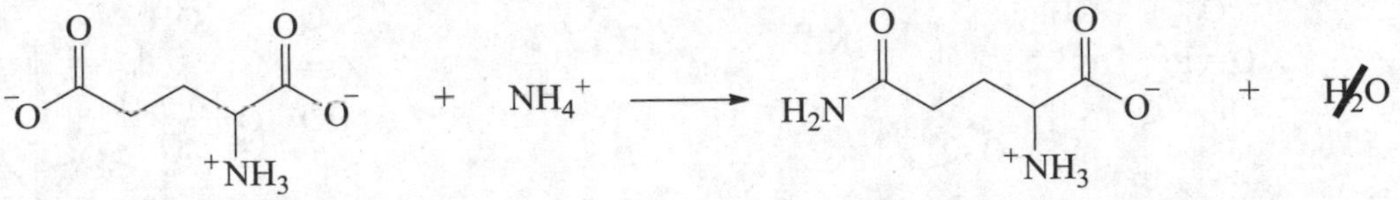

Equation 2:

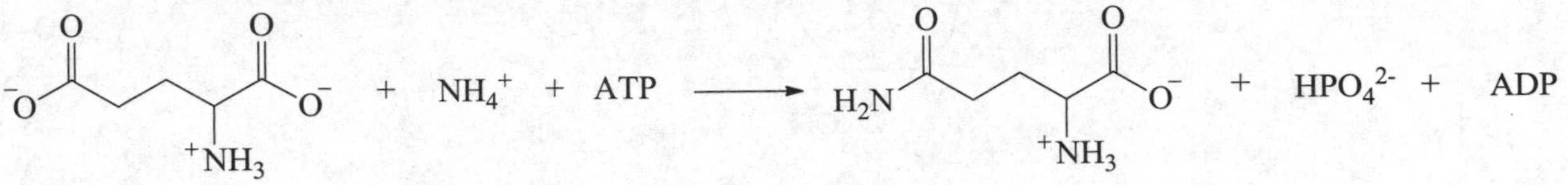

27.8 The reaction of ammonia with γ-glutamyl phosphate to give glutamine is analogous to the reaction of ammonia with an acid anhydride to yield an amide. As noted in the text, γ-glutamyl phosphate is a mixed anhydride of glutamic acid and phosphoric acid. Addition of ammonia to the carbonyl carbon is followed by elimination to give glutamine and phosphate ion.

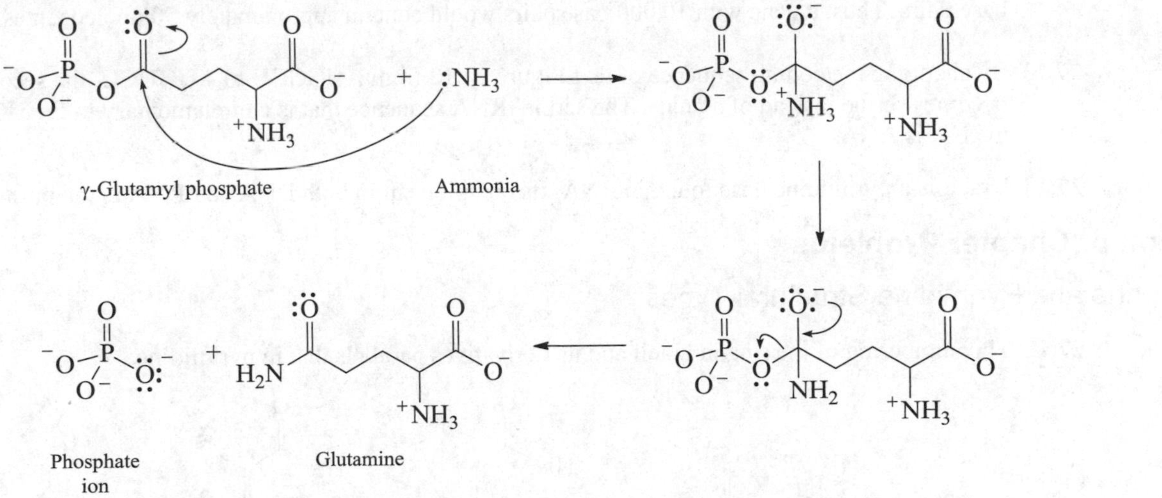

27.9 The free energy of hydrolysis for ATP is 31 kJ/mol versus 13.9 kJ/mol for glucose-6-phosphate, so the phosphorylation of glucose by ATP would be expected to have a large, positive equilibrium constant that would be much greater than 1.

27.10 Figure 27.1 in the text shows a trinucleotide of 2′-deoxy-D-ribose. If the pentose were D-ribose instead, an OH group would be found on C-2 of the ribose portion of each nucleotide segment. The arrows in the following structures point to the OH groups not found in nucleotides formed from 2′-deoxy-D-ribose.

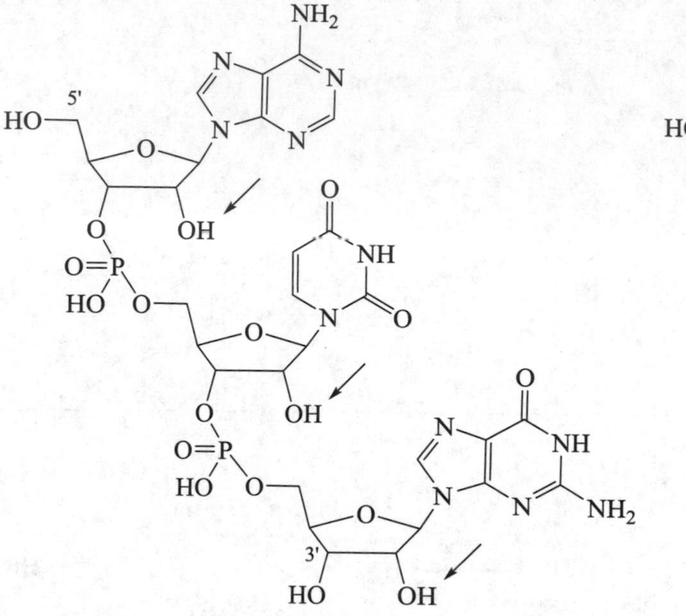

AUG 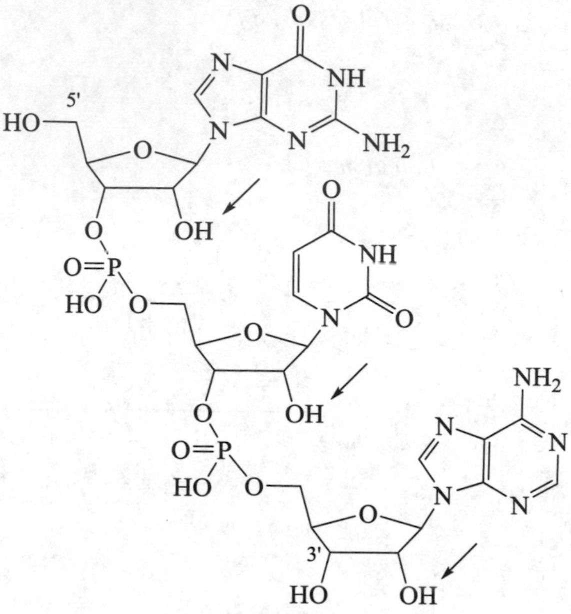   GUA

27.11 The ratio of the purine guanine to the pyrimidine cytosine in DNA is 1:1. Thus, the guanine content in turtle DNA is the same as the cytosine content, 21.3%.

27.12 Each nucleosome is 146 base pairs long and is separated from the next nucleosome by a linker of about 50 base pairs. Thus, a gene with 10,000 base pairs would contain approximately 50 nucleosomes.

27.13 By changing the coding sequence of text Figure 27.12 from AUGGCU to AUGUCU, the second amino acid becomes serine instead of alanine. The serine tRNA sequence that is complementary to the UCU sequence of mRNA is AGA.

27.14 Uracil is a pyrimidine base found in RNA. Its complement in both DNA and RNA is the purine adenine.

End of Chapter Problems

Purine and Pyrimidine Structural Types

27.15 The numbering of the ring in uracil and its derivatives parallels that in pyrimidine.

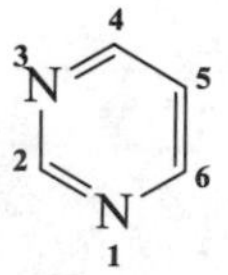

Pyrimidine 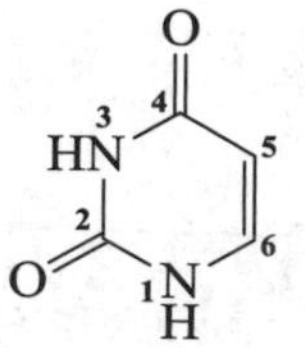   Uracil 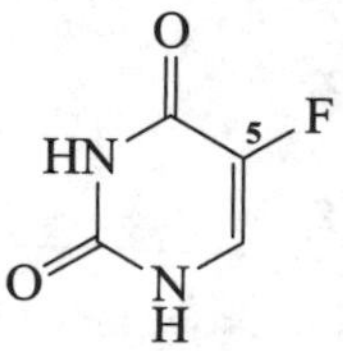   5-Fluorouracil

27.16 (*a, b*) Oxygen is more electronegative than nitrogen, and we would expect the OH proton to be more acidic than the NH proton. As the following scheme illustrates, the conjugate bases of the keto form and of the enol form are resonance contributors to the same anion.

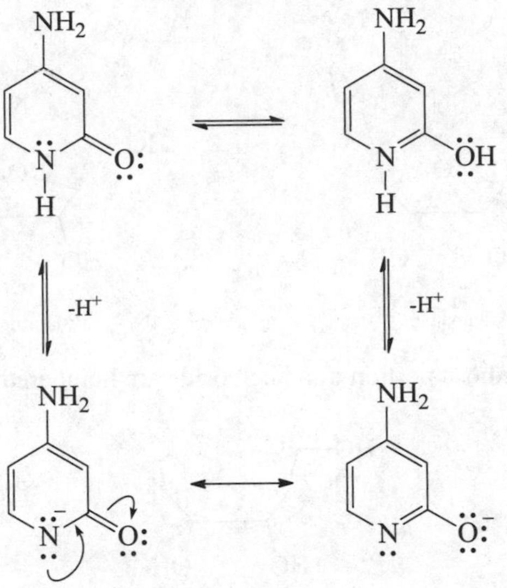

27.17 Purine has the molecular formula $C_5H_4N_4$. Because we are told that uric acid is a purine having the formula $C_5H_4N_4O_3$ and has no C—H bonds, it is reasonable to presume that three C=O groups have replaced the C—H groups of purine.

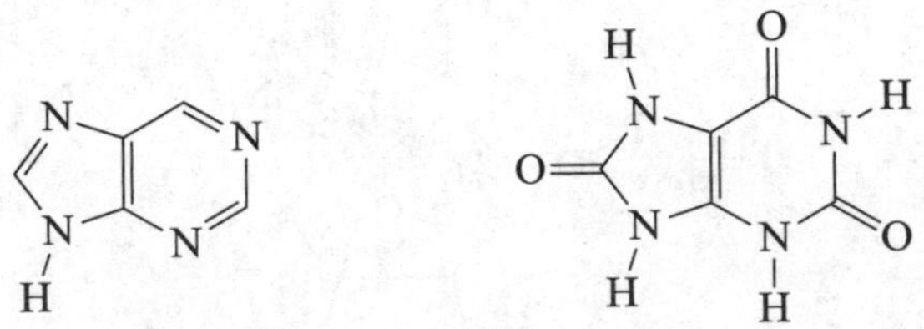

Purine   Uric acid

27.18 Purine and its numbering system are as shown:

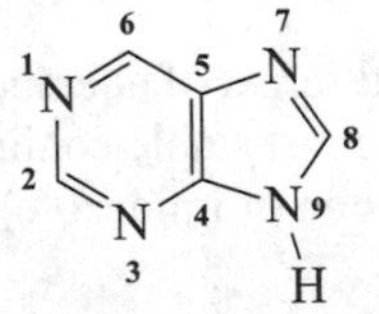

In nebularine, D-ribose in its furanose form is attached to position 9 of purine. The stereochemistry at the anomeric position is β.

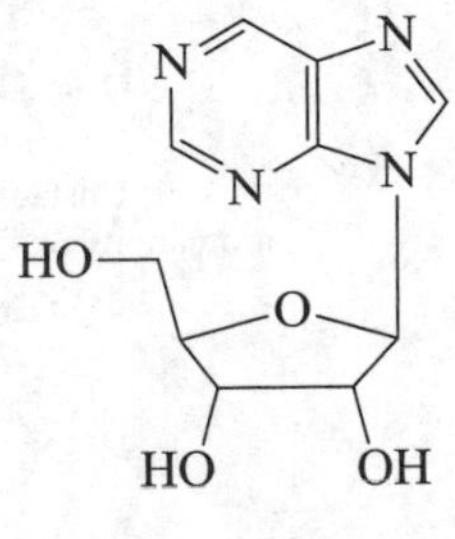

9-β-D-Ribofuranosylpurine
(nebularine)

27.19 The problem states that vidarabine is the arabinose analog of adenosine. Arabinose and ribose differ only in their configuration at C-2.

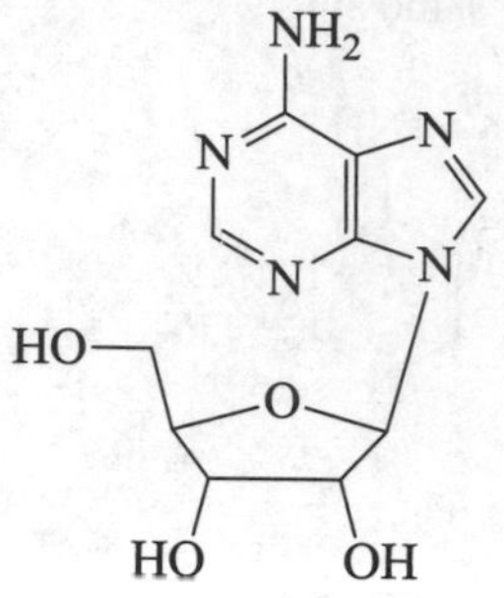

Adenosine 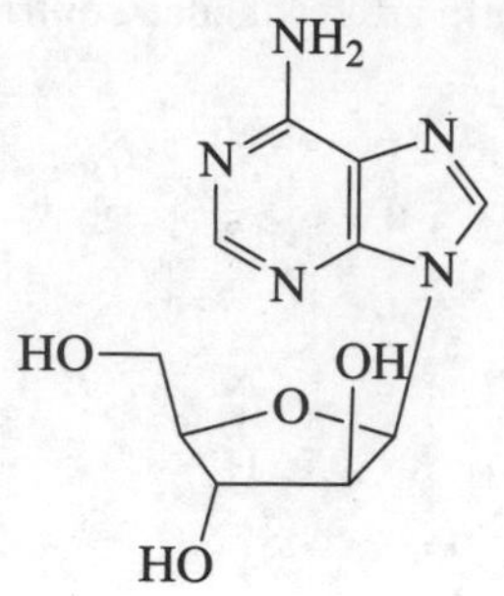   Vidarabine

27.20 The carbon atoms of the ribose portion of a nucleoside are numbered as follows:

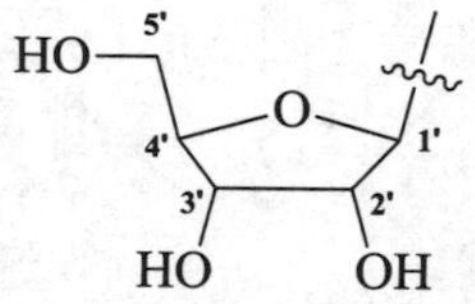

A 5'-nucleotide has a phosphate group attached to the C-5' hydroxyl.

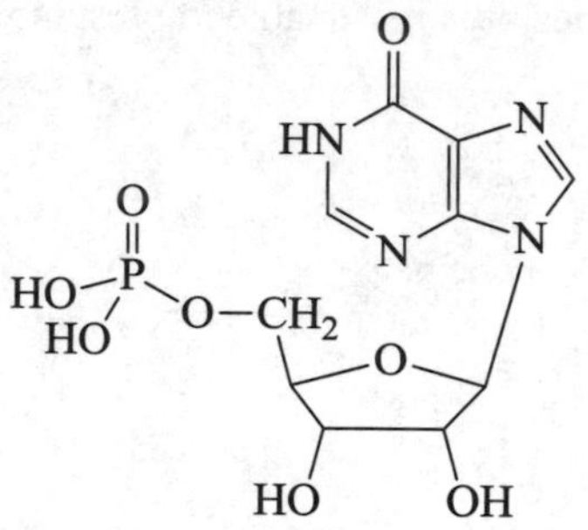

Inosinic acid

Reactions of Purines and Pyrimidines

27.21 Evaluating each of the three protonated forms of adenine will reveal which is the most stable. The most basic nitrogen will be the one that forms the most stable conjugate acid. *Hint:* You might find it helpful to review amine basicity (particularly the discussion of imidazole) in text Section 22.4.

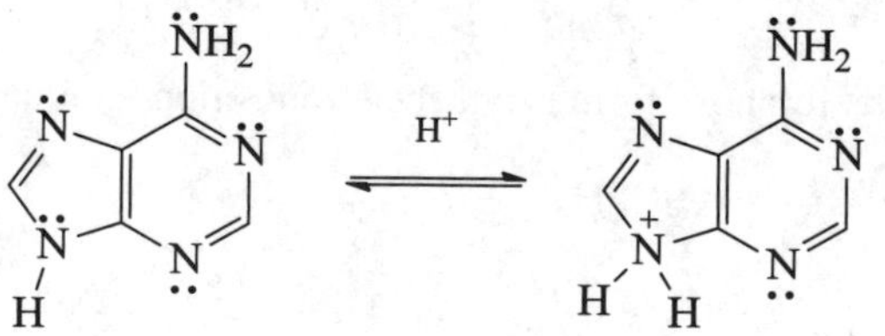

Unstable; disrupts
aromaticity of five-membered ring

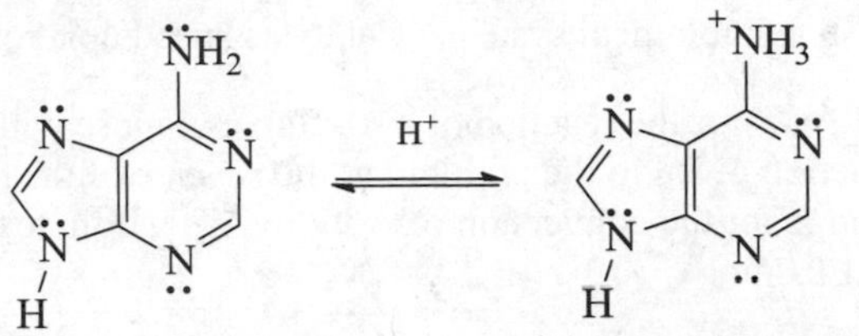

Not the most stable; disrupts
delocalization of arylamine
unshared electron pair

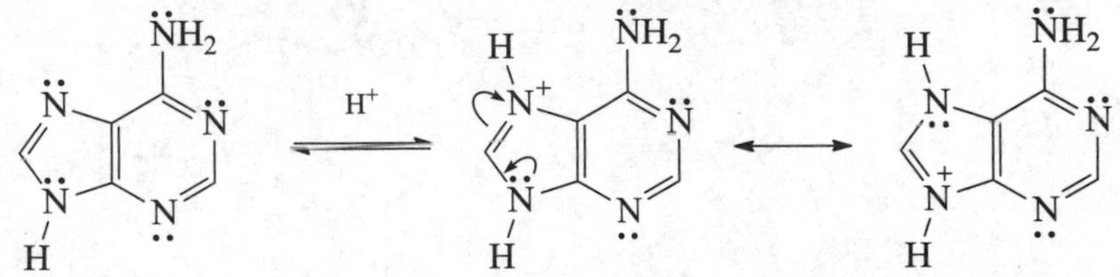

Stable; Positive charge shared by two
nitrogen atoms as shown by resonance structures

27.22 (*a*) The hydrogen with the lower pK_a of 9.5 is due to the resonance stabilization with both carbonyl groups. Of the two resonance contributors shown, the one on the left is more important.

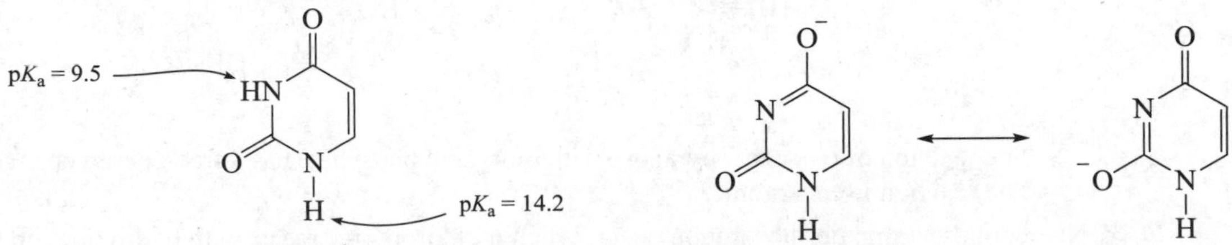

pK_a = 9.5

pK_a = 14.2

Uracil

Most stable contributor
for the monoanion.

Deprotonation to make the dianion gives a resonance structure that is aromatic, albeit a dianion.

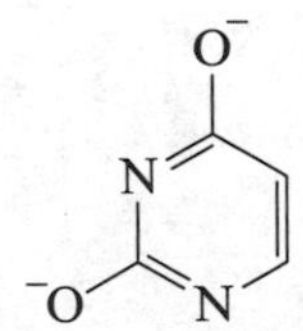

Most stable contributor
for the dianion.

(*b*) Triethylamine, with a pK_a of 10.4 for its conjugate acid, is not sufficiently basic to generate the dianion, because the pK_a of the less acidic hydrogen is higher, at 14.2. It is sufficient to generate the monoanion. The $K_{eq} = 10^{0.9} = 7.9$.

27.23 (*a*) The value of $\Delta G^{\circ\prime}$ for the phosphorylation of α-D-glucopyranose is −23 kJ/mol. Reactions with a negative change in free energy are exergonic.

(b) Enzymes are catalysts in biological systems. Catalysts speed up a reaction but do not affect $\Delta G^{\circ\prime}$.

(c) To find the value of $\Delta G^{\circ\prime}$ for the reaction of α-D-glucopyranose with inorganic phosphate, add the phosphorylation reaction given in the problem to the *reverse* of the hydrolysis of ATP to ADP. The free energy change for the conversion of ADP to ATP (from text Section 27.5) is $-(-31 \text{ kJ/mol}) = +31 \text{ kJ/mol}$.

Phosphorylation:

$\text{ATP} \quad + \quad \text{[glucopyranose]} \quad \longrightarrow \quad \text{[phosphorylated glucopyranose]} \quad + \quad \text{ADP} \qquad \Delta G^{\circ\prime} = -23 \text{ kJ/mol}$

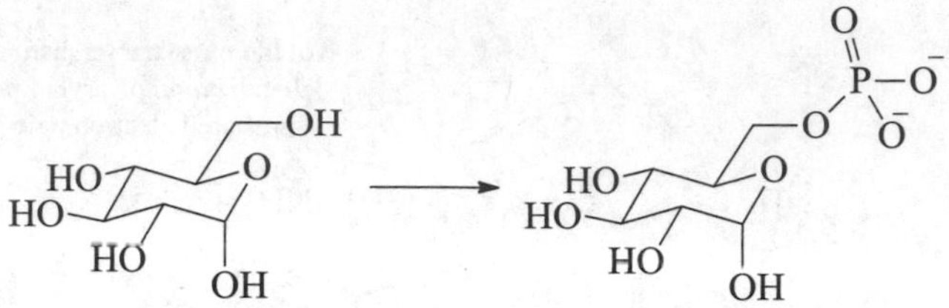

Reverse of ATP Hydrolysis:

$$\text{ADP} \quad + \quad \text{HPO}_4^{2-} \quad \longrightarrow \quad \text{ATP} \quad + \quad \text{H}_2\text{O} \qquad \Delta G^{\circ\prime} = +31 \text{ kJ/mol}$$

Sum:

$\text{HPO}_4^{2-} \quad + \quad \text{[glucopyranose]} \quad \longrightarrow \quad \text{[phosphorylated glucopyranose]} \quad + \quad \text{H}_2\text{O} \qquad \Delta G^{\circ\prime} = +8 \text{ kJ/mol}$

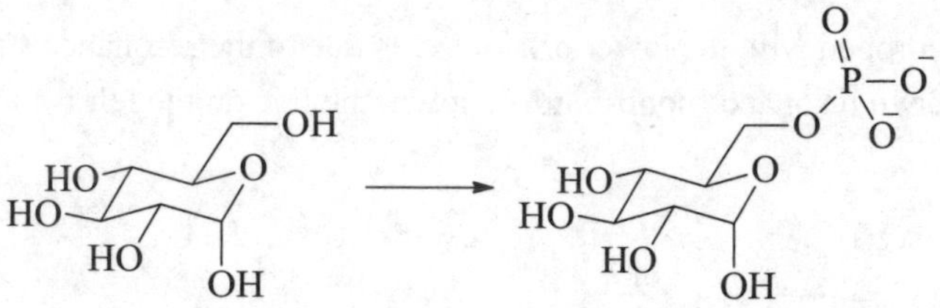

The reaction of α-D-glucopyranose with inorganic phosphate has a free-energy change, $\Delta G^{\circ\prime} = +8 \text{ kJ}$. The reaction is endergonic.

27.24 Nucleophilic aromatic substitution occurs when 6-chloropurine reacts with hydroxide ion by an addition–elimination pathway.

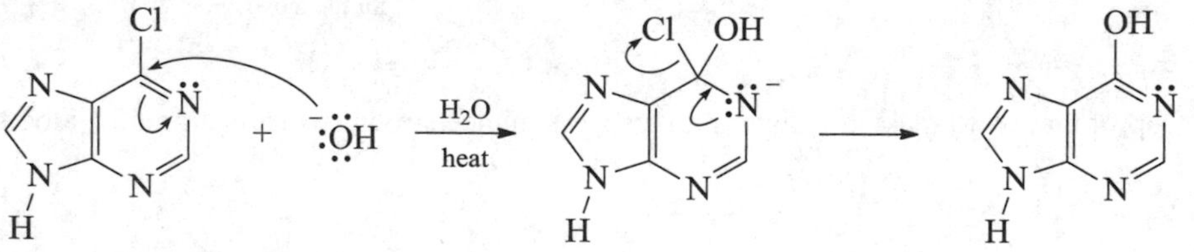

The enol tautomerizes to give hypoxanthine.

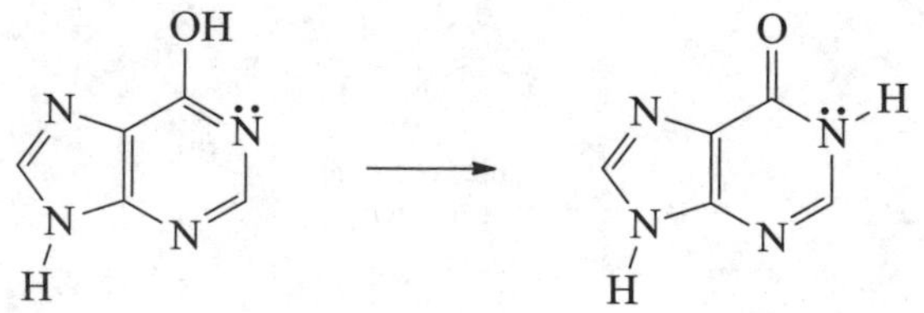

Hypoxanthrine

27.25 Nitrous acid reacts with aromatic primary amines to yield diazonium ions.

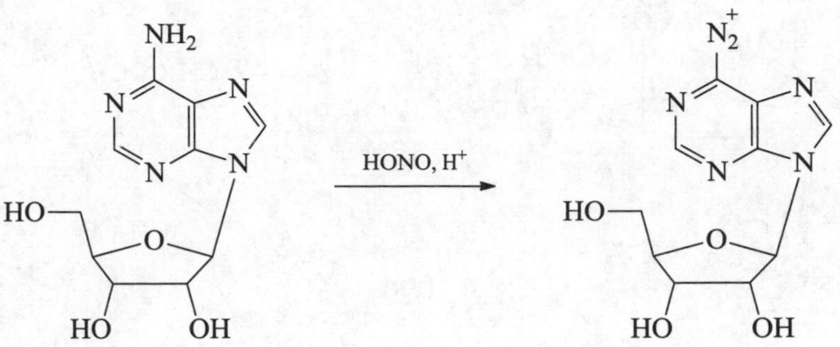

Adenosine

Treatment of the diazonium ion with water yields a phenol. Tautomerization gives inosine.

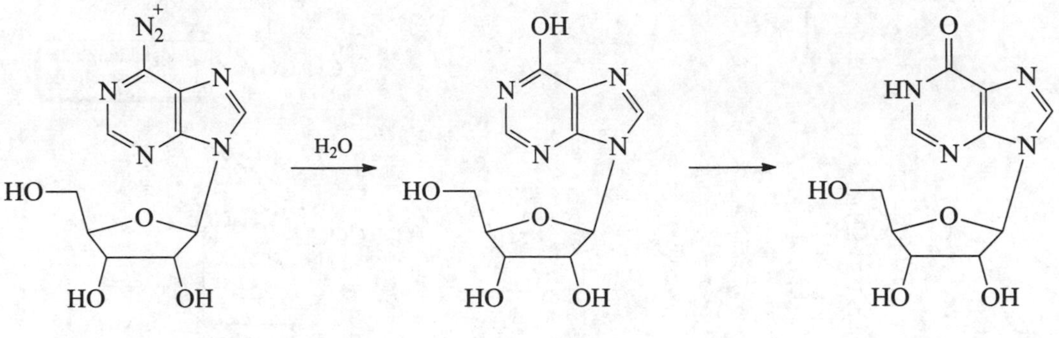

Inosine

Synthesis of Nucleosides and Nucleotides

27.26 Working through the mechanism, one finds that the sugar reacts with the acid catalyst to form a resonance-stabilized carbocation from the ester group at C-2. This participation by the ester group blocks one side of the ring, so attack occurs selectively from the less hindered side to give the α anomer.

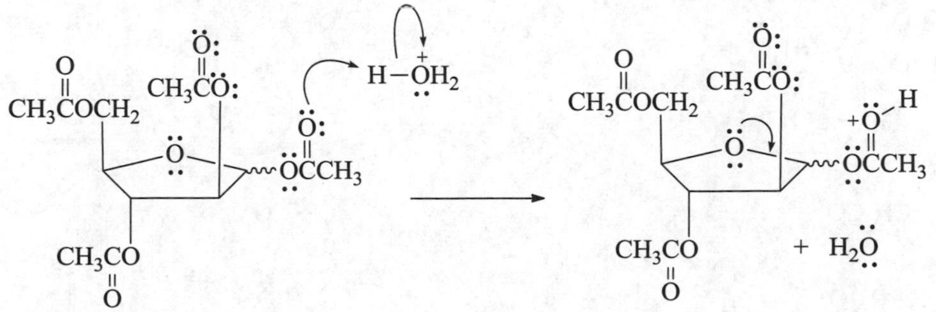

1,2,3,4-Tetra-*O*-acetyl-
α,β-arabinofuranose

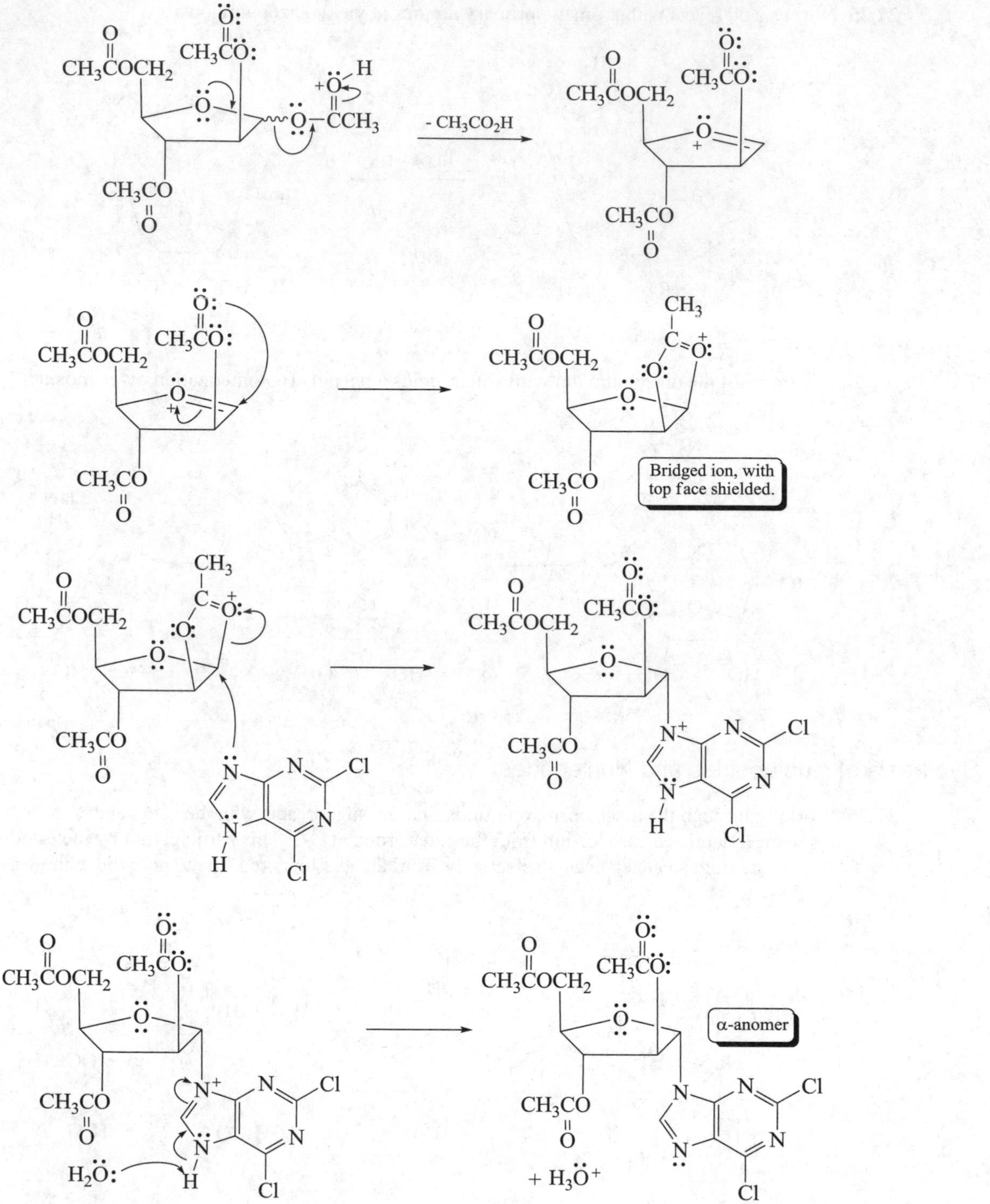

27.27 All the bases in the synthetic messenger RNA prepared by Nirenberg were U; therefore, the codon is UUU. By referring to the codons in text Table 27.4, we see that the UUU codes for phenylalanine. A polypeptide in which all the amino acid residues were phenylalanine was isolated in Nirenberg's experiment.

27.28 (*a*) The unprotected hydroxyl group of the glycosyl donor attacks the epoxide in a nucleophilic ring-opening reaction (see Section 17.12).

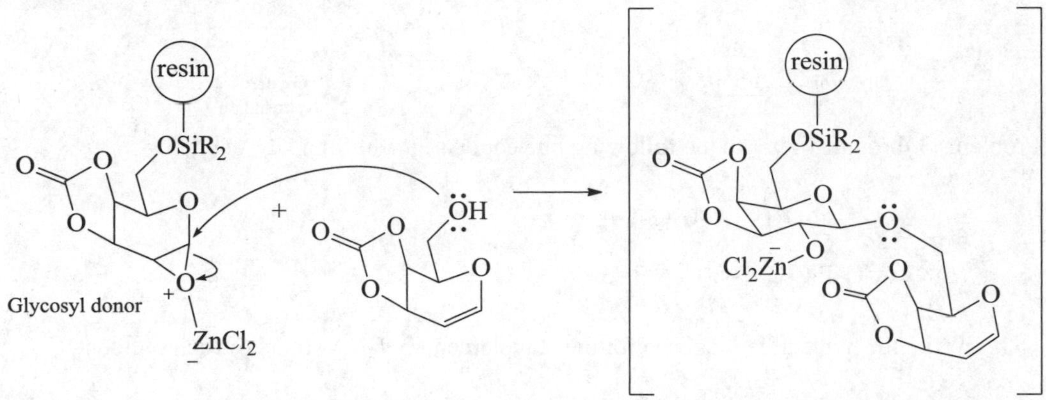

(*b*) The ring opening is assisted by complexation of the epoxide oxygen with zinc chloride. Carbocation character develops at the adjacent carbon, which favors nucleophilic attack there, similar to acid-catalyzed epoxide opening.

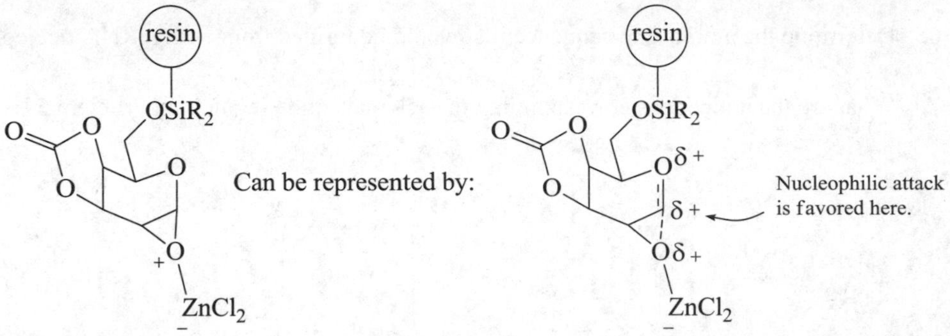

Answers to Interpretive Problems 27

27.29 C; **27.30** D; **27.31** B; **27.32** A; **27.33** B; **27.34** A

SELF-TEST

1. How does the nucleoside adenosine found in RNA differ from that found in DNA?

2. The standard free-energy change ($\Delta G°'$) for the reaction of glycerol with inorganic phosphate ion to give glycerol 1-phosphate is 9.2 kJ.

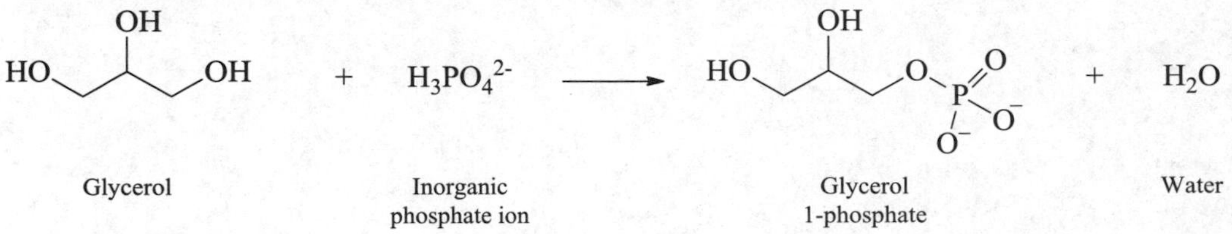

Glycerol	Inorganic phosphate ion	Glycerol 1-phosphate	Water

Using the free-energy change for the conversion of ATP to ADP (text Section 27.4), calculate the free-energy change for the reaction:

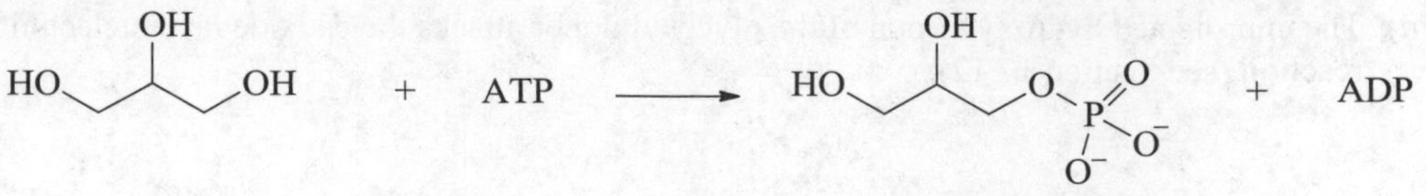

Glycerol

Glycerol
1-phosphate

Problems 3 through 7 refer to the following nucleotide segments of a DNA strand.

(*a*) A-A-A-G-G-T-C-C-C-G-T-A

(*b*) T-A-C-T-C-G-C-G-G-A-T-G

3. Write the nucleotide sequence for the complementary DNA strand of each segment.

4. Write the mRNA nucleotide sequence that would be produced by transcription of each DNA segment.

5. List the codons present in each mRNA segment.

6. Determine the amino acid sequence that would be formed from each mRNA nucleotide segment.

7. What are the anticodons corresponding to each nucleotide segment in Problem 5?

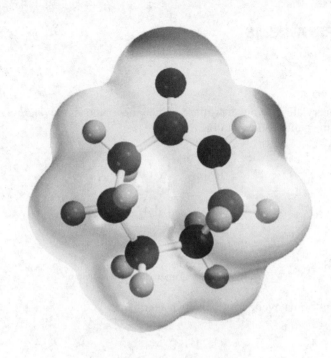

CHAPTER 28
Synthetic Polymers

Table of Contents

SOLUTIONS TO TEXT PROBLEMS

In Chapter Problems

28.1 (*b*) Methyl methacrylate is two words; thus, its polymer name is *poly(methyl methacrylate)*. The repeating unit follows the pattern shown in the text.

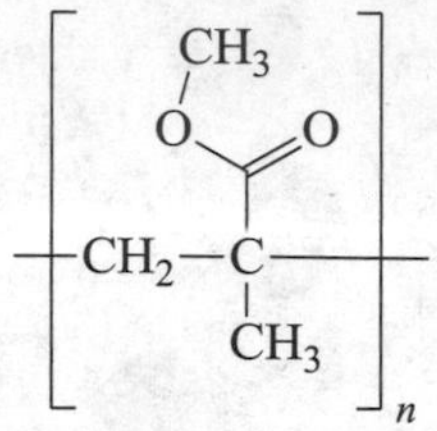

Poly(methyl methacrylate)

28.2 Formaldehyde can form a cyclic trimer, known as trioxane.

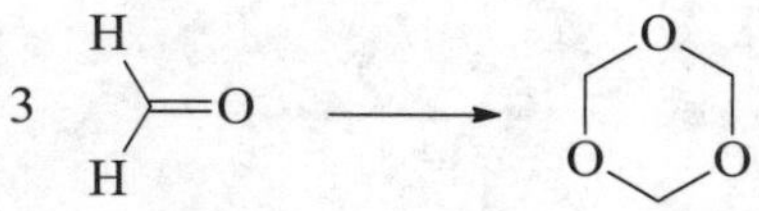

Formaldehyde   Trioxane

28.3 Nucleophilic acyl substitution follows an addition–elimination pathway (text Section 20.3). Nucleophilic addition by 1,4-benzenediamine to the carbonyl group of terephthaloyl chloride gives the tetrahedral intermediate shown.

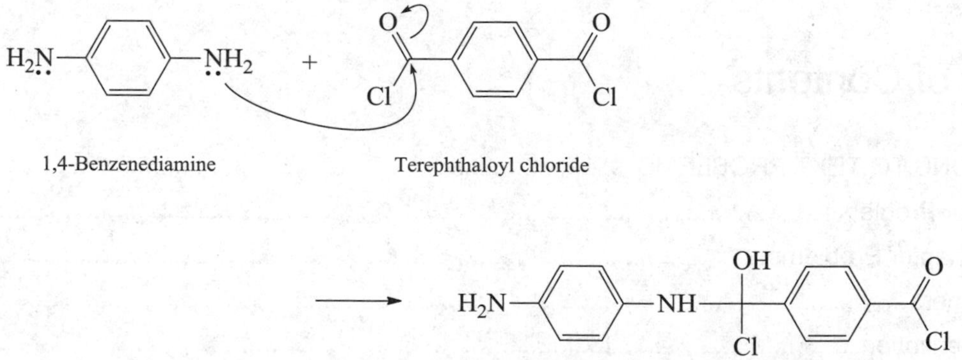

1,4-Benzenediamine   Terephthaloyl chloride

Tetrahedral intermediate

28.4 The product of the reaction contains all of the atoms of the starting materials. This is an example of an *addition* reaction.

28.5 (*b*) Azobisisobutyronitrile (AIBN) undergoes homolytic bond cleavage as shown in the text to give nitrogen and two 1-cyano-1-methylethyl radicals. These radicals initiate polymerization of styrene by addition to the vinyl group.

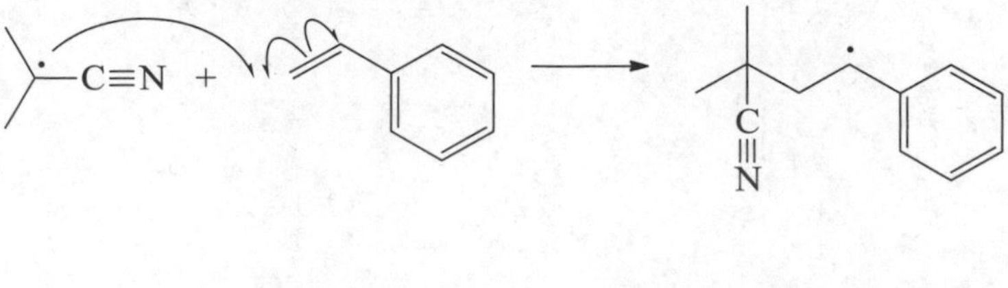

28.6 The terminal free radical of the growing polymer chain can react with the free-radical initiator, in this case an alkoxy radical.

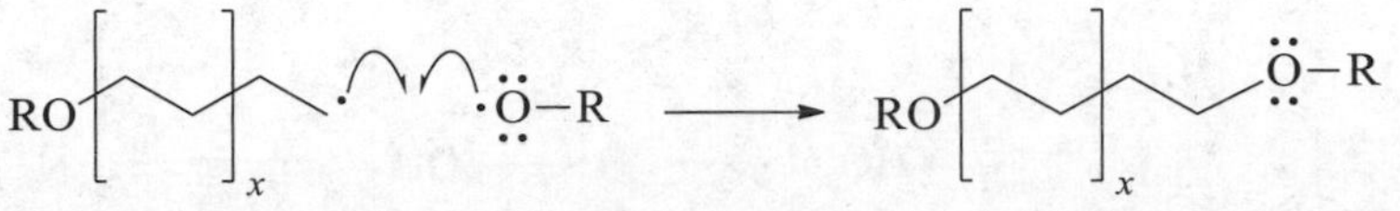

28.7 As shown in text Mechanism 28.1, four-carbon branches result from a strain-free six-membered chair cyclohexane-like transition state. Transition states leading to chains shorter than four carbons would have angle strain, which would increase ΔH of activation. Transition states leading to chains longer than four carbons require restricting the motion of more atoms and are more ordered. This would make ΔS of activation more negative.

28.8 As noted in the text, the number of chains is equal to the number of molecules of butyllithium used. Thus, if less butyllithium is present, there will be fewer chains and each one will be longer. In other words, the average chain length is a function of the ratio of moles of styrene to moles of butyllithium.

28.9 Hydroxide ion adds to methyl 2-cyanoacrylate to give an anion stabilized by delocalization of the negative charge to oxygen and nitrogen.

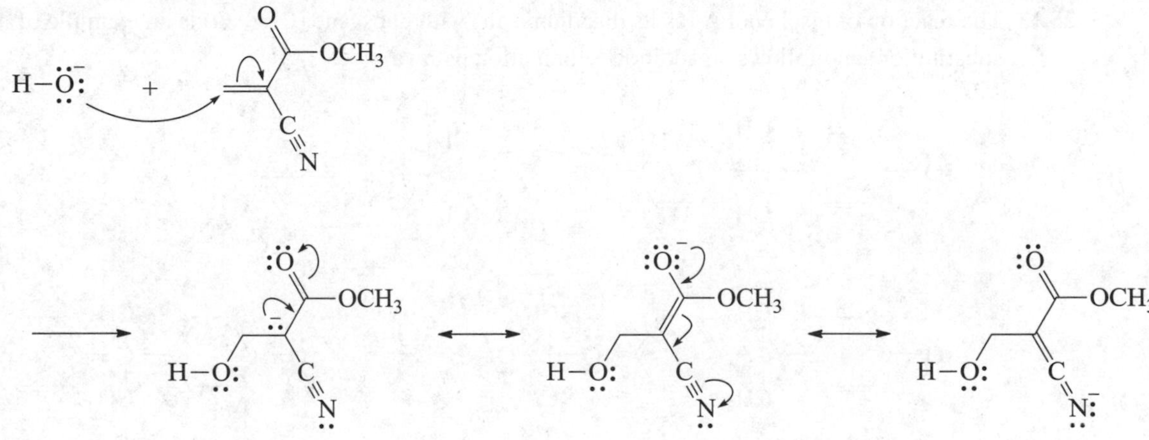

28.10 A lactam is a cyclic amide (text Section 20.14). ε-Caprolactam has a seven-membered ring.

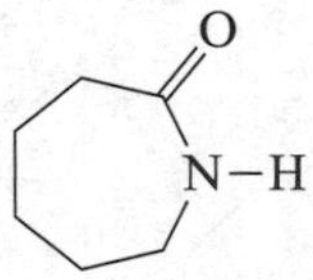

ε-Caprolactam

28.11 1,3-Benzenediamine reacts with 1,3-benzenedicarboxylic acid to give the repeating unit of Nomex.

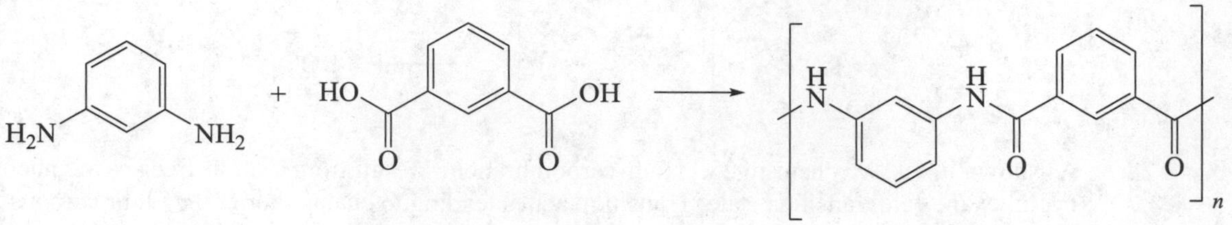

1,3-Benzenediamine   1,3-Benzenedicaroxylic acid   Repeating unit of Nomex

28.12 The polymer made from ε-caprolactone is a polyester.

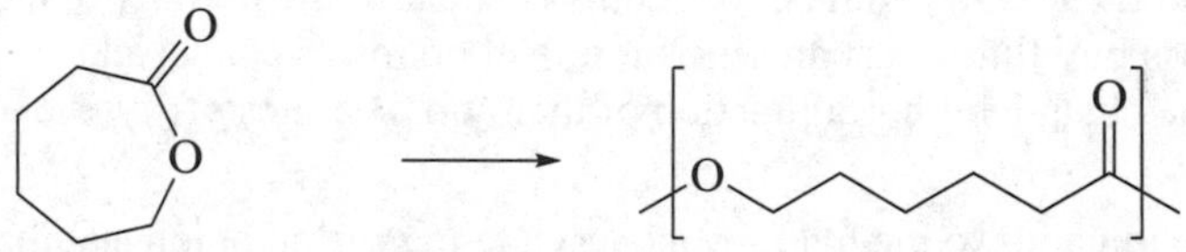

ε-Caprolactone   Repeating unit of poly(ε-caprolactone)

28.13 The reaction of bisphenol A (as its disodium salt) with phosgene ($Cl_2C=O$) is an example of nucleophilic acyl substitution, and follows an addition–elimination pathway.

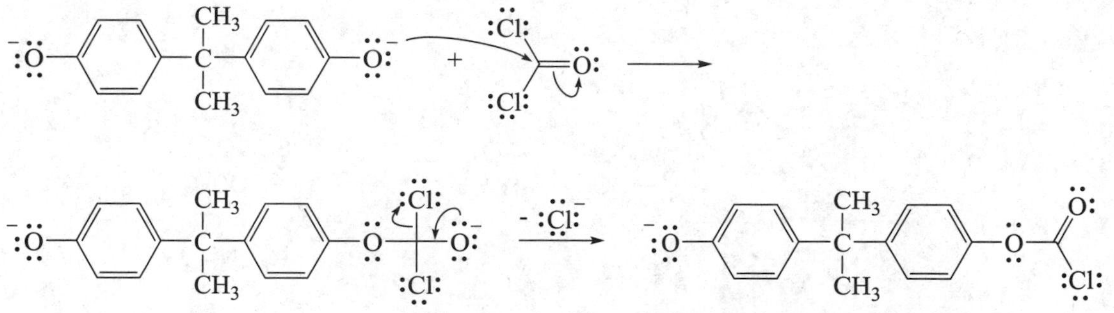

28.14 If the "polymeric diol" used in the polymer shown in the text were derived from 1,2-epoxypropane, its structure would be as shown:

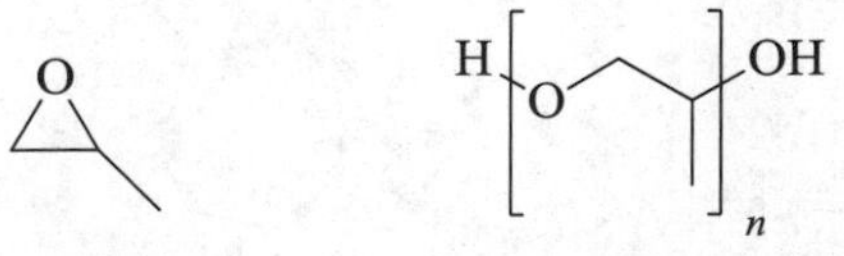

1,2-Epoxypropane   "Polymeric diol"

End of Chapter Problems

Monomers

28.15 Polyvinylene is a polymer of acetylene. Its source-based name is polyacetylene.

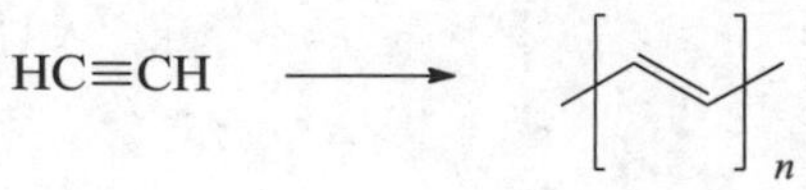

Acetylene   Polyvinylene

28.16 The lactone used to prepare the polymer shown is 3-propanolide, better known by its common name β-propiolactone.

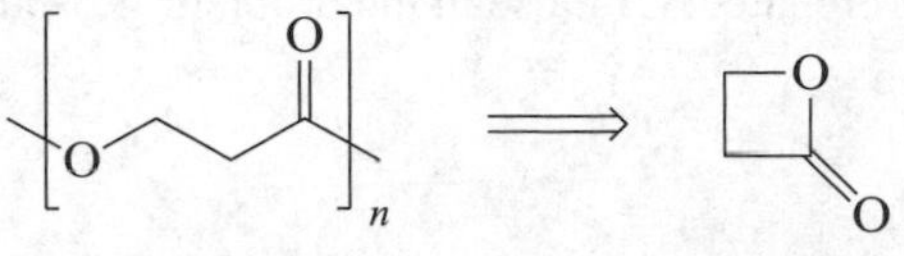

3-Propanolide
(β-propiolactone)

28.17 The monomers used to prepare the polymer in Kodel fibers are 1,4-benzenedicarboxylic acid (terephthalic acid) and cyclohexane-1,4-dimethanol.

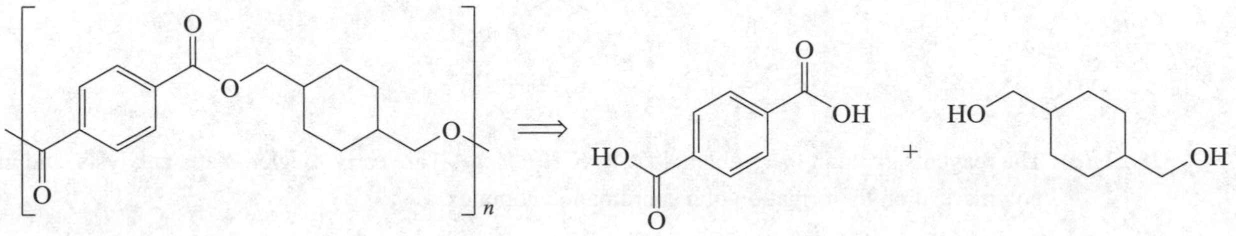

1,4-Benzenedicarboxylic acid   Cyclohexane-1,4-dimethanol
(terephthalic acid)

28.18 3-Hydroxynonanoic acid will polymerize to form a polyester with a hexyl group side chain in the repeating unit.

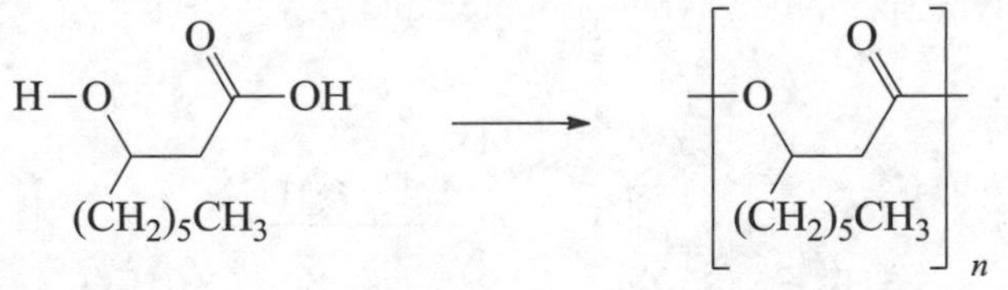

3-Hydroxynonanoic acid   Polyester repeating unit

28.19 Nylon 11 is a condensation polymer as water is released in the polymerization process, as shown.

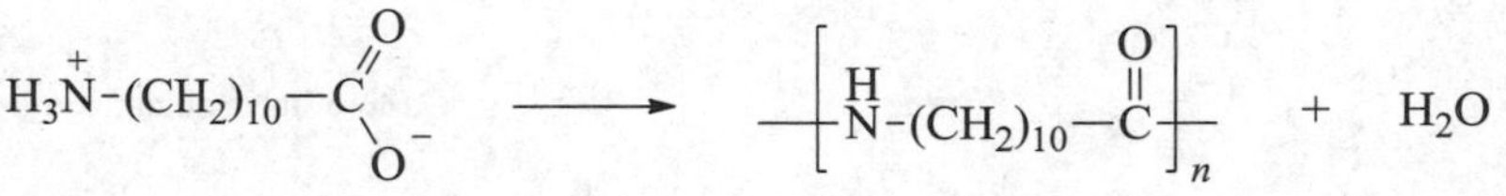

11-Aminoundecanoic acid   Nylon 11   Water

As the polymer grows, each chain has two growth points, a free amino group at one end and a free carboxyl at the other. The starting material reacts rapidly to form oligomers, which can then react together to form the polymer. Nylon 11 is a step-growth polymer.

28.20 The protein biosynthesis process described in text Figure 27.12 extends the protein by one amino acid at a time on the end of the chain. It is a chain-growth process. The process converts an ester to an amide, so an alcohol is formed as each additional amino acid is added to the growing chain. The protein is a condensation polymer.

Polymerization

28.21 The alkene that would be most suited for cationic polymerization is the one that forms the most stable carbocation when protonated. Of those in the problem, 1,3-butadiene is the best candidate.

1,3-Butadiene

28.22 An alkene suitable for anionic polymerization is one that forms a relatively stable anion. Thus, alkenes with electron-withdrawing groups attached to the double bond are the best candidates. Of the alkenes in the problem, acrylonitrile is most suited for anionic polymerization.

Acrylonitrile

28.23 (*a*) The reagents given in the problem [TiCl$_4$, (CH$_3$CH$_2$)$_3$Al] are early Ziegler–Natta catalysts, and initiate polymerization by formation of a coordination complex.

(*b*) Organic peroxides readily undergo homolytic bond cleavage of the O–O bond, and initiate polymerization by a free-radical mechanism.

(*c*) Boron trifluoride (BF$_3$) is a powerful Lewis acid and initiates polymerization by a cationic mechanism.

28.24 *p*-Methoxystyrene contains an electron-donating methoxy group, and thus the anion formed from this compound would be less stable, and formed more slowly, than that formed from styrene. As a result, the anionic polymerization of styrene is more rapid.

p-Methoxystyrene   Less stable anion; formed more slowly

Styrene   More stable anion; formed more rapidly

28.25 The monomer that forms the more stable anion will react faster and should be used first to form the living polymer. Thus, acrylonitrile should be polymerized first, and styrene added second to form the copolymer.

Acrylonitrile   Living polymer   Copolymer

28.26 The repeating unit of poly(vinyl butyral) is an acetal, formed from the reaction of the diol starting polymer and an aldehyde. Compound A is butanal.

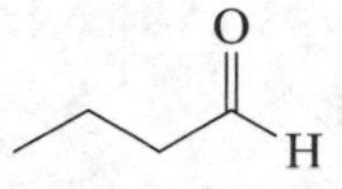

Butanal
(compound A)

28.27 Polymerization of ethylene under Ziegler–Natta conditions yields a linear polymer of methylene (CH_2) groups. Adding a small quantity of 1-hexene will result in introduction of butyl side chains into the polymer.

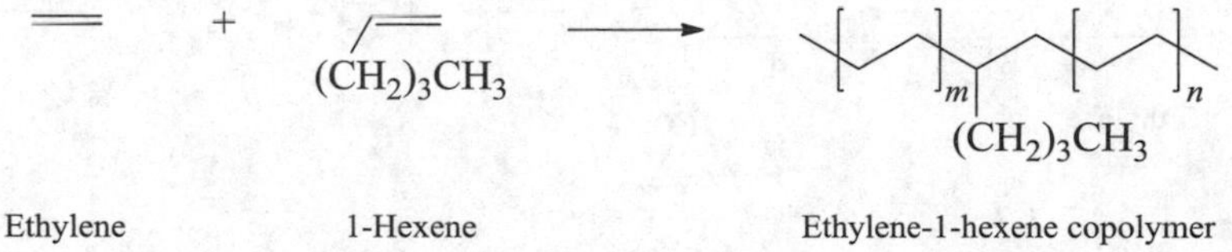

Ethylene   1-Hexene   Ethylene-1-hexene copolymer

Mechanisms

28.28 (*a*) The formation of bisphenol A begins by electrophilic aromatic substitution at the para position of phenol to form a tertiary alcohol. The electrophile is the protonated form of acetone [$(CH_3)_2C=O$].

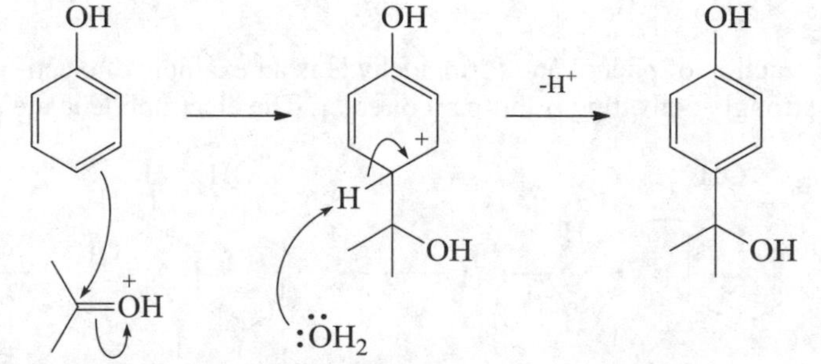

A second mole of phenol then reacts with this alcohol to form bisphenol A.

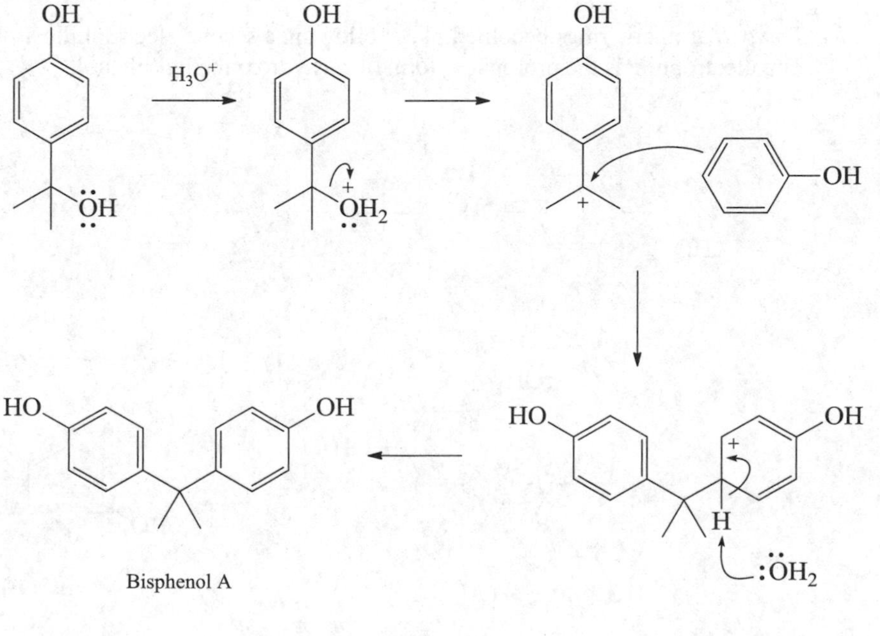

Bisphenol A

(b) Bisphenol B differs from bisphenol A in that one of the methyl groups has been replaced by an ethyl group.

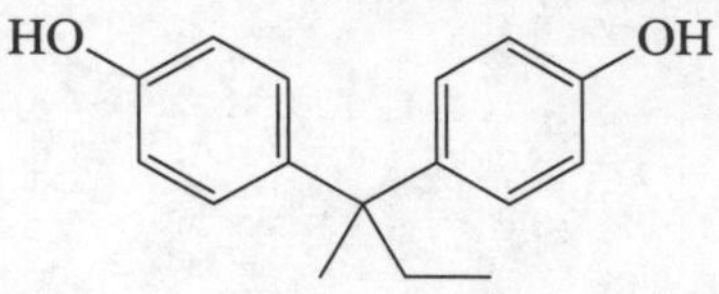

Bisphenol B

28.29 In the presence of an acid catalyst, the oxygen of ethylene oxide is protonated in the polymerization process.

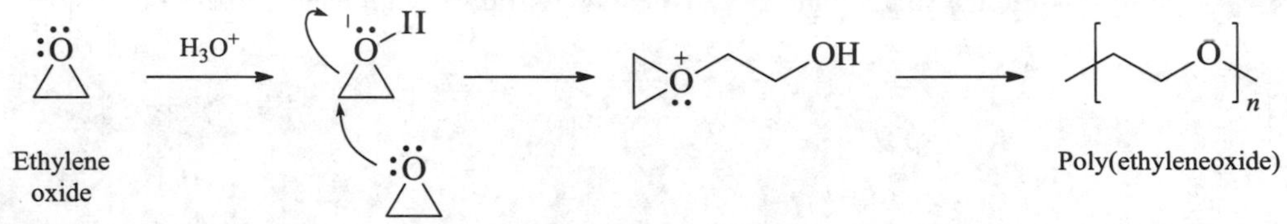

Ethylene
oxide

Poly(ethyleneoxide)

Under basic conditions, ethylene oxide is attacked directly by hydroxide ion.

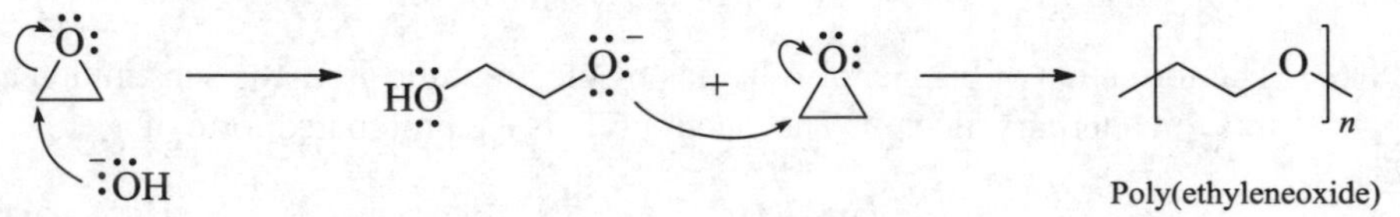

Poly(ethyleneoxide)

28.30 (a) The reaction of phenol and formaldehyde is an example of electrophilic aromatic substitution. Hydroxyl is a strongly activating ortho-para director. The electrophile is the protonated form of formaldehyde.

Phenol   Protonated form
                  of formaldehyde

o-(Hydroxymethyl)phenol

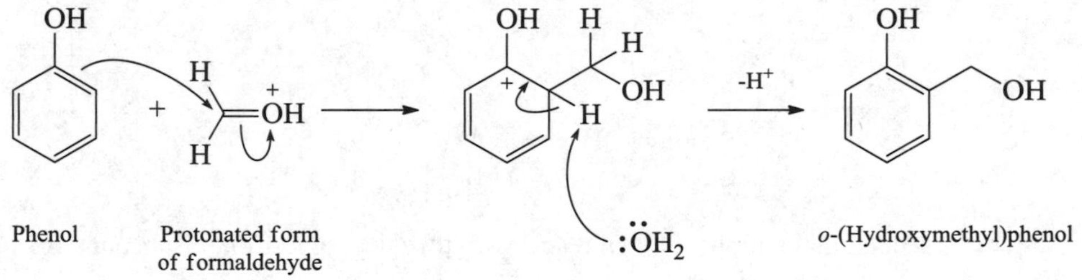

(b) The two aromatic rings become linked following a second electrophilic aromatic substitution. The electrophile is the protonated form of p-(hydroxymethyl)phenol.

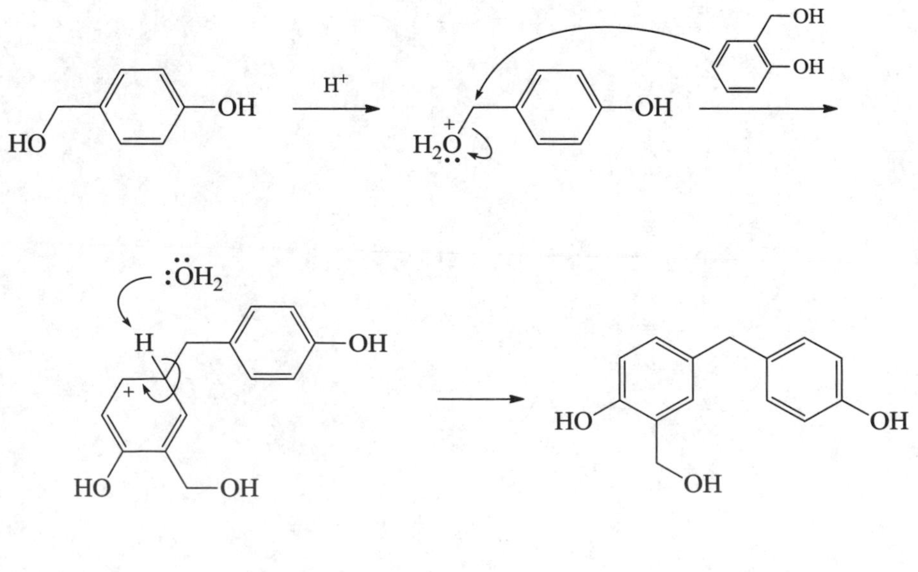

28.31 In the second polymerization step, a molecule of formaldehyde adds to the formaldehyde-boron trifluoride complex.

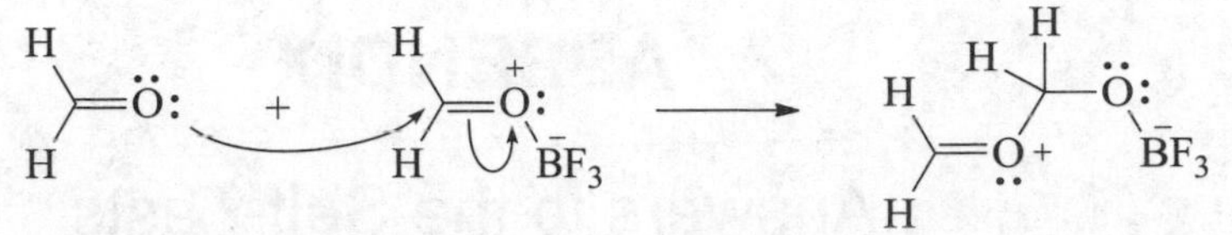

Answers to Interpretive Problems 28

28.32 C;   **28.33** B;   **28.34** A;   **28.35** D;   **28.36** B;   **28.37** B;   **28.38** C

SELF-TEST

1. Briefly describe the difference between a chain-growth and a step-growth polymer.

2. (*a*) What monomer of molecular formula C_2H_4O gives the polymer with the repeating unit shown?

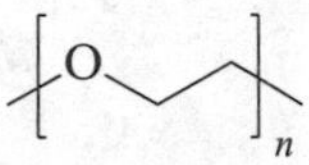

   (*b*) What is the source-based name of this polymer?

3. Bisphenol A is used to form the polycarbonate polymer known as *Lexan,* as shown in the following reaction. Is this polymer formed in an addition process, or a condensation?

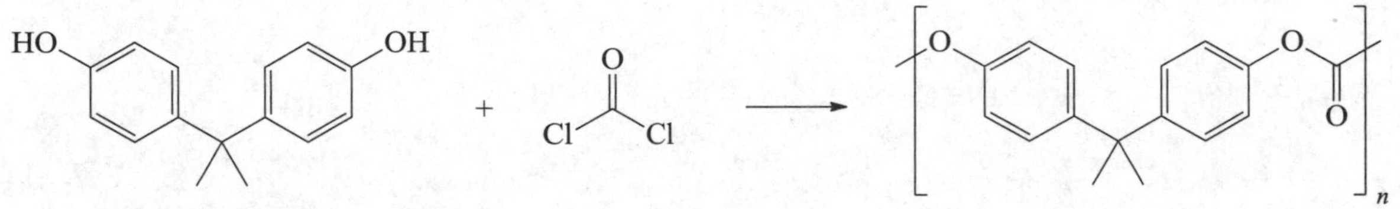

| Bisphenol A | Phosgene | Polycarbonate repeating unit |

4. What is the monomer used to prepare poly(vinyl acctate)?

5. Poly(vinyl acetate) is used to prepare poly(vinyl alcohol) by the reaction:

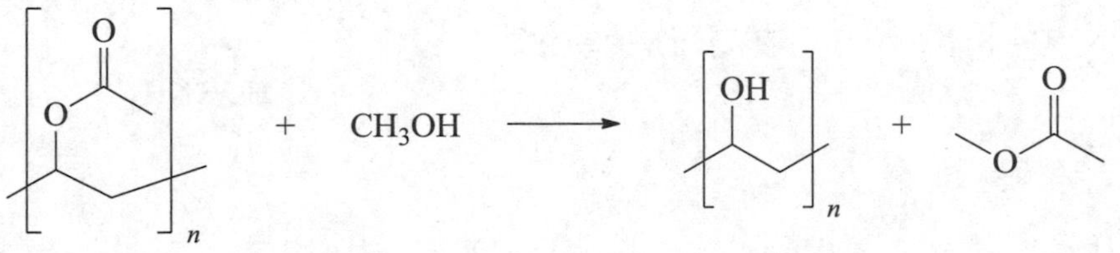

| Poly(vinyl acetate) | Methanol | Poly(vinyl alcohol) | Methyl acetate |

   Why is it impractical to make poly(vinyl alcohol) from vinyl alcohol?

APPENDIX

Answers to the Self-Tests

CHAPTER 1

1. (a) P; $1s^2 2s^2 2p^6 3s^2 3p^3$ (b) S^{2-}; $1s^2 2s^2 2p^6 3s^2 3p^6$

2. (a)

$:\ddot{N}=C=\ddot{S}:$

Formal charge: −1 0 0 Net charge: −1

(c)

$\overset{-1}{:\ddot{O}:}$
|
$HC=NH_2$

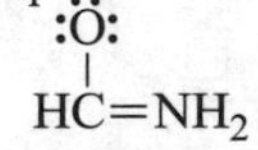

Formal charge: 0 +1 Net charge: 0

(b)

$:O\equiv N-\ddot{O}:$

Formal charge: +1 +1 −1 Net charge: +1

3. (a)

$:N\equiv C-\ddot{\underset{..}{S}}:$

Formal charge: 0 0 −1 Net charge: −1

(c)

$\overset{0}{:\ddot{O}}$
‖
$HC-\ddot{N}H_2$

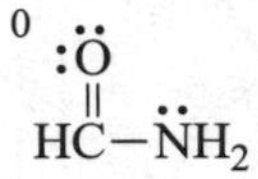

Formal charge: 0 0 Net charge: 0

(b)

$:\ddot{O}=N=\ddot{O}:$

Formal charge: 0 +1 0 Net charge: +1

The more stable Lewis structures are

(a) $^-:\ddot{N}=C=\ddot{S}:$

(b) $:\ddot{O}=\overset{+}{N}=\ddot{O}:$

(c) $:\ddot{O}$
‖
$HC-\ddot{N}H_2$

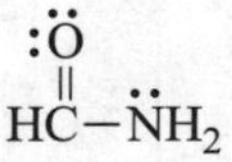

4. (a)

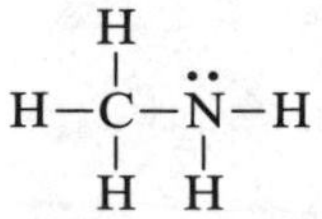

```
      H
      |
H — C — Ṅ — H
      |   |
      H   H
```

(b)

```
   H   H
   |   |
H — C — C = Ö:
   |
   H
```

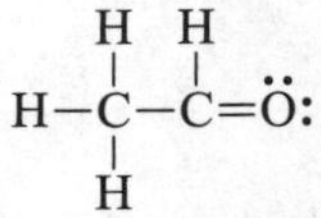

5. (*a*)

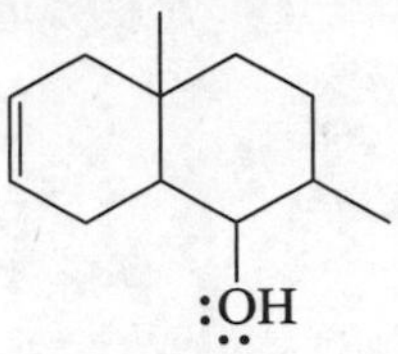

$C_{12}H_{20}O$

(*c*)

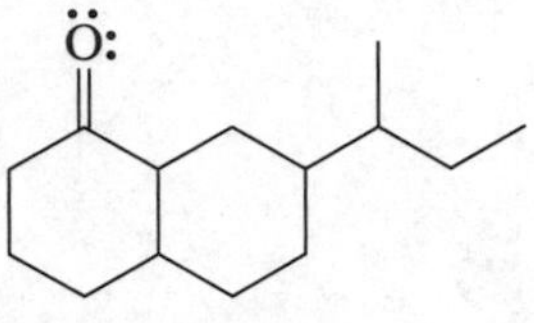

$C_{14}H_{24}O$

(*b*)

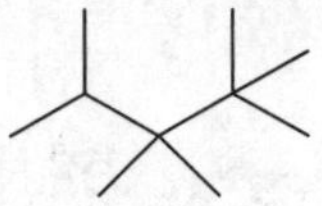

$C_{10}H_{22}$

(*d*)

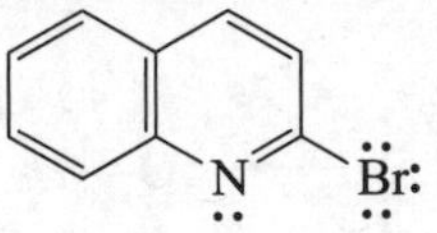

C_9H_6BrN

6.

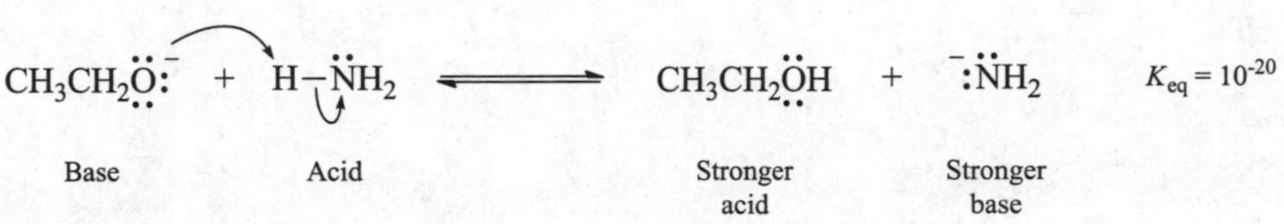

$$CH_3CH_2\ddot{O}:^- \ + \ H-\dot{N}H_2 \ \rightleftharpoons \ CH_3CH_2\ddot{O}H \ + \ ^-:\dot{N}H_2 \qquad K_{eq} = 10^{-20}$$

Base   Acid   Stronger   Stronger
                                     acid   base

7.

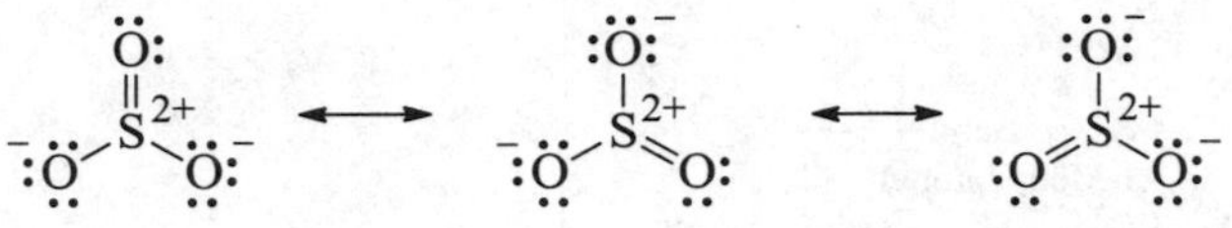

:Ö:
‖
⁻:Ö—S²⁺—Ö:⁻   ⟷   :Ö:⁻ / ⁻:Ö—S²⁺—Ö: ⟷   :Ö:⁻ / :Ö═S²⁺—Ö:⁻

8.   :Ö—C≡N:

Formal
charge: -1   0   0   Net charge: -1

9. (*a*)

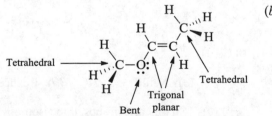

(*b*)

Pyramidal;   :N(Cl)(Cl)(Cl)   Yes, it is polar.

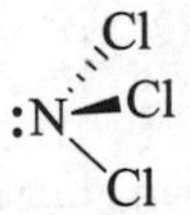

10. (*a*) Linear   (*b*) Linear   (*c*) Bent

11. (*a*) D   (*c*) None   (*e*) None   (*g*) A

(*b*) A, B   (*d*) B   (*f*) A, D   (*h*) C

12.

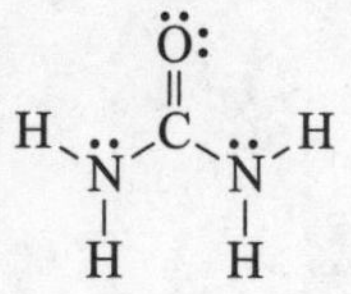

Urea

13. (*a*)

More
stable

(*b*)

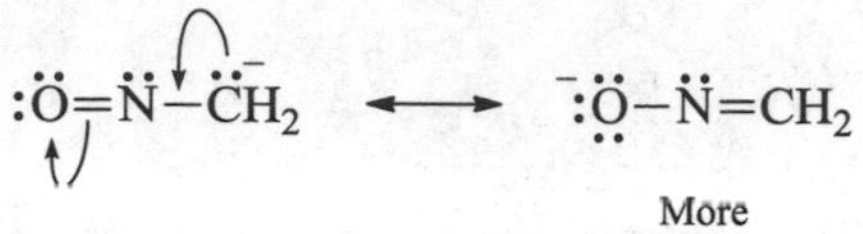

More
stable

CHAPTER 2

1.

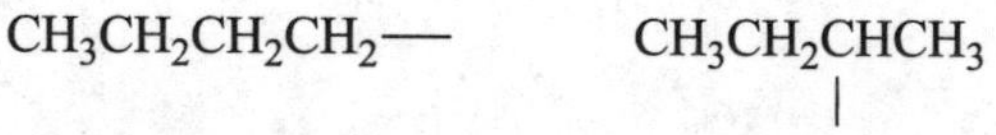

$$CH_3CH_2CH_2CH_2-\qquad CH_3CH_2CHCH_3$$
$$|$$

Common:	*n*-Butyl	*sec*-Butyl
Systematic:	Butyl	1-Methylpropyl

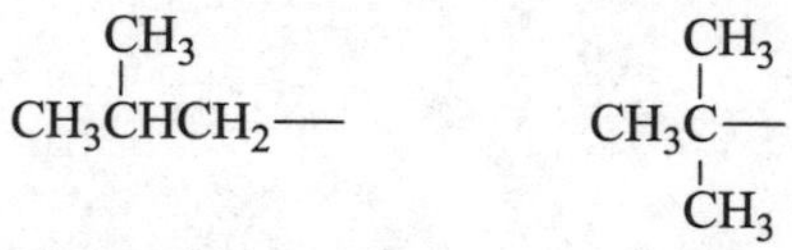

$$\begin{array}{c} CH_3 \\ | \\ CH_3CHCH_2- \end{array}\qquad \begin{array}{c} CH_3 \\ | \\ CH_3C- \\ | \\ CH_3 \end{array}$$

Common:	Isobutyl	*tert*-Butyl
Systematic:	2-Methylpropyl	1,1-Dimethylpropyl

2. (*a*) 28 (8 C—C; 20 C—H)   (*b*) 27 (9 C—C; 18 C—H)

3. (*a*) Oxidized   (*b*) Neither   (*c*) Neither   (*d*) Reduced

4. (*a*)

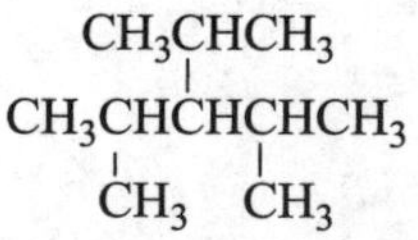

$$\begin{array}{c} CH_3CHCH_3 \\ | \\ CH_3CHCHCHCH_3 \\ |\quad\ | \\ CH_3\ \ CH_3 \end{array}$$

(*b*) Six methyl groups, three isopropyl groups

5. (*a*) 3,4-Dimethylheptane   (*b*) (1,2-Dimethylpropyl)cyclohexane

6.

	Primary	Secondary	Tertiary
(*a*)	4	3	2
(*b*)	3	5	3

7. (*a*) 1,3-Dimethylbutyl; secondary

(*b*) 1,1-Diethylpropyl; tertiary

(*c*) 2,2-Diethylbutyl; primary

8.
$$\begin{array}{c} CH_3 \\ | \\ CH_3CHCHCH_2CH_3 \\ | \\ CH_3 \end{array} ; \; C_7H_{16} \qquad C_7H_{16} + 11O_2 \longrightarrow 7CO_2 + 8H_2O$$

9.

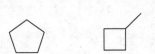

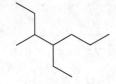

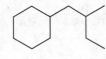

Cyclopentane

Methylcyclobutane

Ethylcyclopropane

1,1-Dimethylcyclopropane

1,2-Dimethylcyclopropane

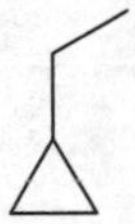

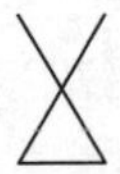

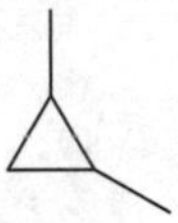

10. (*a*)

4-Ethyl-3-methylheptane

(*b*)

(2-Methylbutyl)cyclohexane

(*c*)

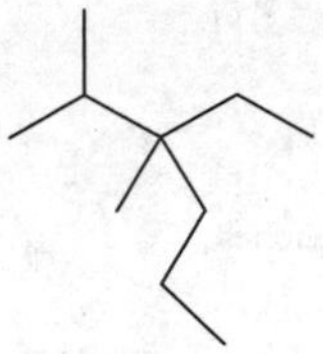

3-Ethyl-2,3-dimethylhexane

11. (*a*) $CH_3CH_2CH_2CH_2CH_2CH_2CH_2CH_3$

(*b*) $(CH_3)_3CC(CH_3)_3$

(*c*) $\begin{array}{c} (CH_3)_2CHCHCH(CH_3)_2 \\ | \\ CH_3 \end{array}$

(*d*) $(CH_3)_3CC(CH_3)_3$

12.

2,2-Dimethylpentane

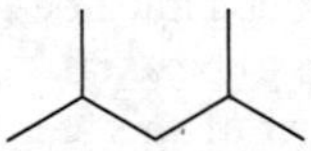

2,4-Dimethylpentane

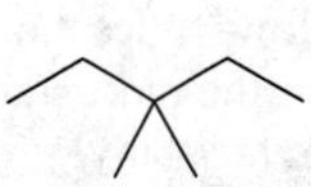

3,3-Dimethylpentane

2,3-Dimethylpentane

3-Ethylpentane

13. 10,049 kJ/mol

14. (*a*) 11σ; 1π (*b*) 9σ; 2π (*c*) 12σ; 4π (*d*) 13σ; 4π

15. (a) $CH_3CH=CHCH_3$

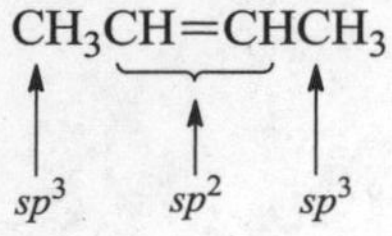

$sp^3 \quad sp^2 \quad sp^3$

(b) $HC{\equiv}CCH_2CH_3$

$sp \quad sp^3$

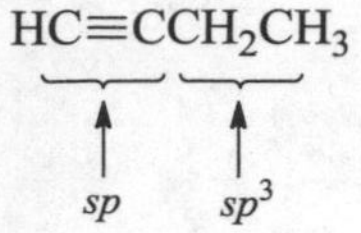

(c)

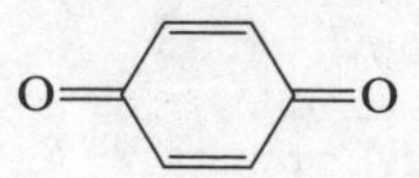

All carbons are sp^2

(d)

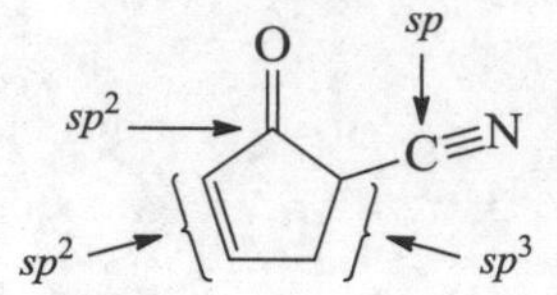

$sp^2 \quad sp$

$sp^2 \quad sp^3$

CHAPTER 3

1.

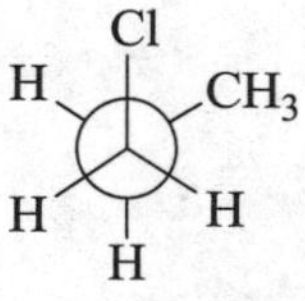

Gauche 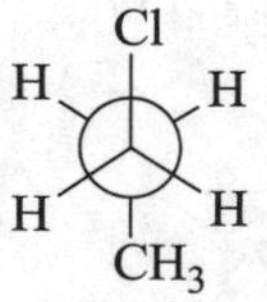   Anti

2. (a)

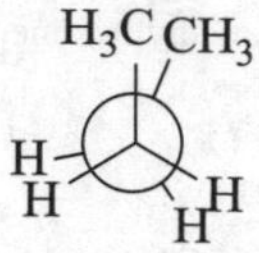

(Eclipsed)

(b)

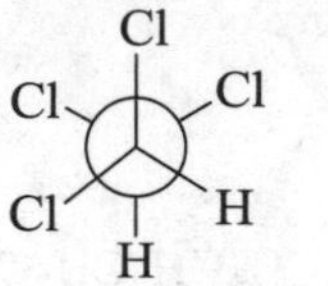

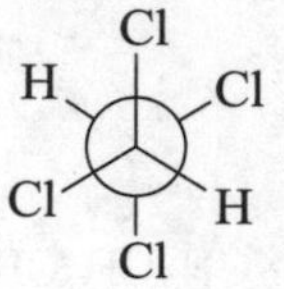

3. $(CH_3)_3CCH_2C(CH_3)_3 = $ 2,2,4,4-tetramethylpentane

4.

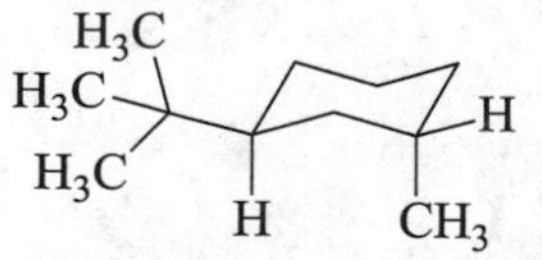

For the less stable stereoisomer, the methyl and *tert*-butyl groups must be trans. This allows at least one group to be axial.
For the most stable isomer of this stereoisomer, the *tert*-butyl group must be equatorial.

5.

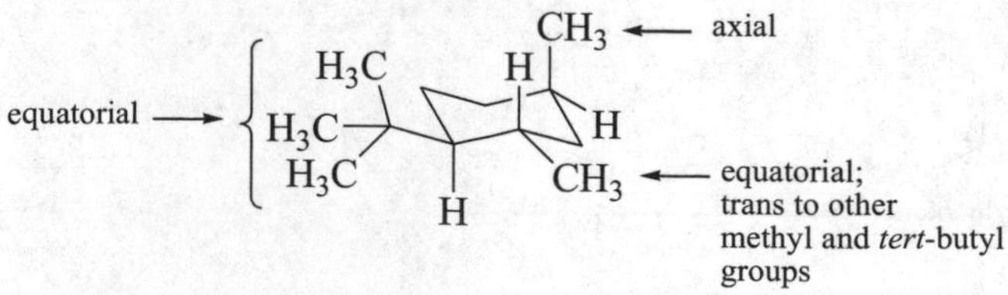

equatorial $\rightarrow$

CH$_3$ $\leftarrow$ axial

equatorial; trans to other methyl and *tert*-butyl groups

6.

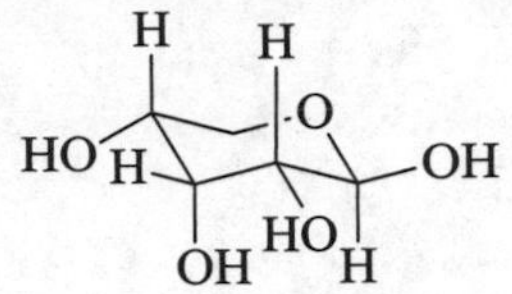

7. (*a*) C (*b*) A and B (*c*) D (*d*) A

8.

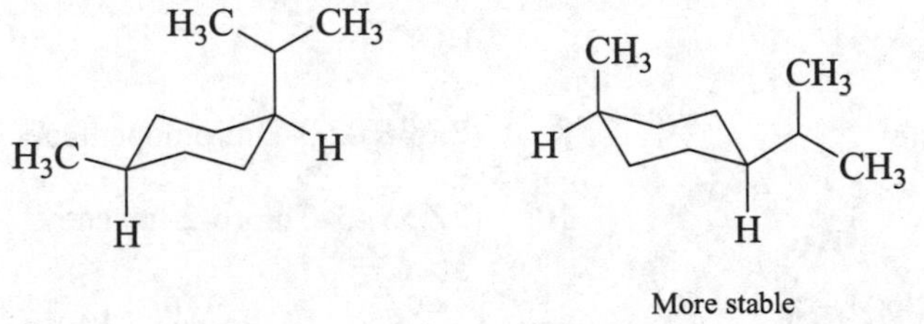

More stable

9. *cis*-1-Ethyl-3-methylcyclohexane has the lower heat of combustion.

10. Tricyclic; $C_{10}H_{16}$

11. The form of the curve more closely resembles ethane than butane.

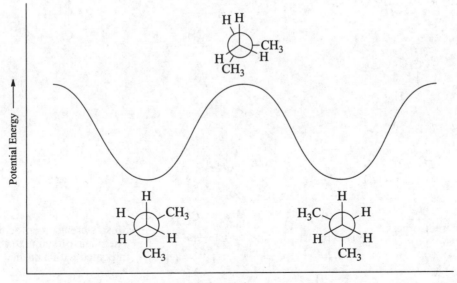

CHAPTER 4

1. (*a*) 1 and 2, both achiral; identical

   (*b*) 3 and 4, both chiral; enantiomers

(c)   5 chiral, 6 achiral (meso); diastereomers

(d)   7 and 8, both chiral; diastereomers

(e)   9 and 10, both chiral; diastereomers

2.   3:   (R)-2-Chlorobutane   4:   (S)-2-Chlorobutane

5:

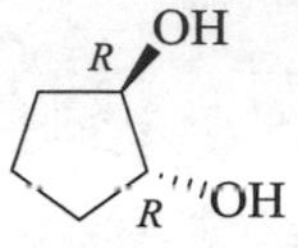

6:

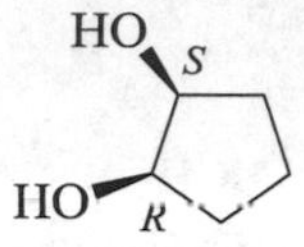

7:   (2S,3R)-2,3-Dibromopentane   8:   (2R,3R)-2,3-Dibromopentane

9:   (2E,5R)-5-Chloro-2-hexene   10:   (2Z,5S)-5-Chloro-2-hexene

3.   (a)   Three; meso form is possible   (c)   Four; no meso form possible

(b)   Eight; no meso form possible

4.   (a)

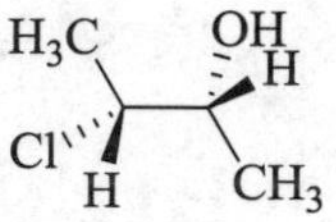

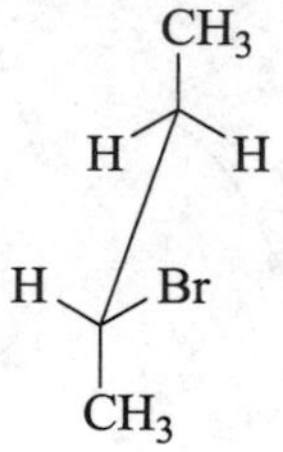

(c)

(b)

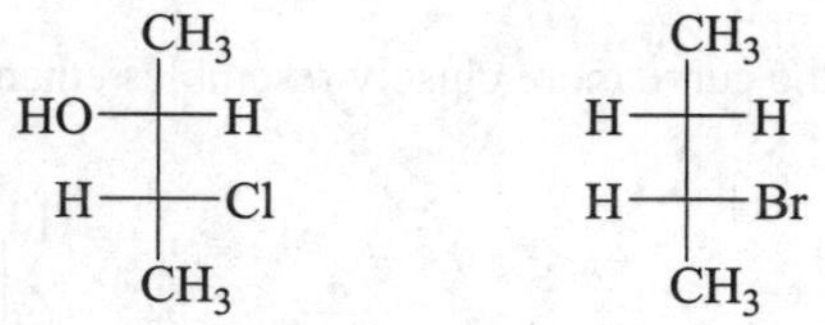

5.

Chiral stereoisomers:

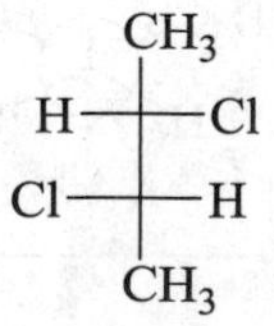

(2S,3S)-2,3-
Dichlorobutane

and

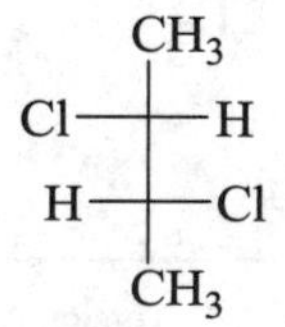

(2R,3R)-2,3-
Dichlorobutane

Meso stereoisomer (achiral):
The plane of symmetry is
indicated with a dashed line.

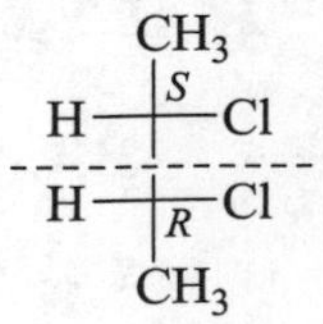

meso-2,3-
Dichlorobutane

6.   (a) $[\alpha] = -31.2°$ (b)   30% S

7.   a S; b S; c R; d R; e R

8.   (a)   (2S,3S)-1,3-Dibromo-2-chlorobutane

(b)   (R)-1-Ethylcyclohex-2-enol

9.   There are six chirality centers, counting nitrogen.

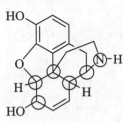

10. 2-methyl-3-pentanol has one chirality center. The can have two stereoisomers that are enantiomers.

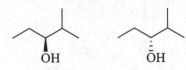

CHAPTER 5

1.   Alcohol, alkene, ester, ketone

2.   (*a*) *trans*-1-Bromo-3-methylcyclopentane

    (*b*) 2-Ethyl-4-methyl-1-hexanol

3.   (*a*)   (*b*)

4.   (*a*) **Functional class:**   1-Ethyl-3-methylbutyl alcohol
            Substitutive: 5-Methyl-3-hexanol

    (*b*) **Functional class:**   1,1,2-Trimethylbutyl chloride
            Substitutive: 2-Chloro-2,3-dimethylpentane

5.   (*a*) $CH_3CH_2CH_2Cl$ (*b*)

6.   (*a*)   (*b*)   (*c*)

7.   The alcohol that reacts fastest is 2,4-dimethylpentan-2-ol reacts because it is a tertiary alcohol that forms the more stable tertiary carbocation in an S_N1 reaction. The slowest reacting alcohol is 5-methylhexan-1-ol because it is primary. Primary alcohols do react in an S_N2 reaction but slower.

       2,4-Dimethylpentan-2-ol   5-Methylhexan-1-ol

8. (*a*)

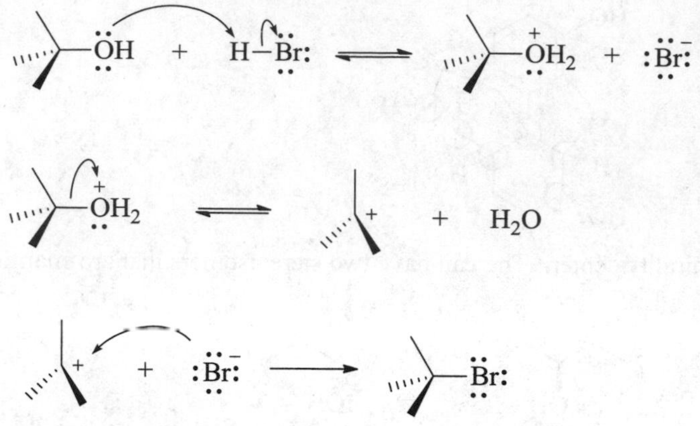

(*b*)

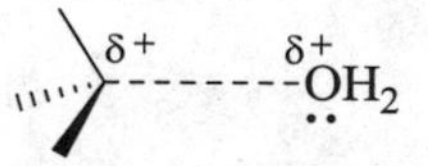

(*c*) Water is displaced directly from the oxonium ion of 1-butanol by bromide ion. A primary carbocation is not involved.

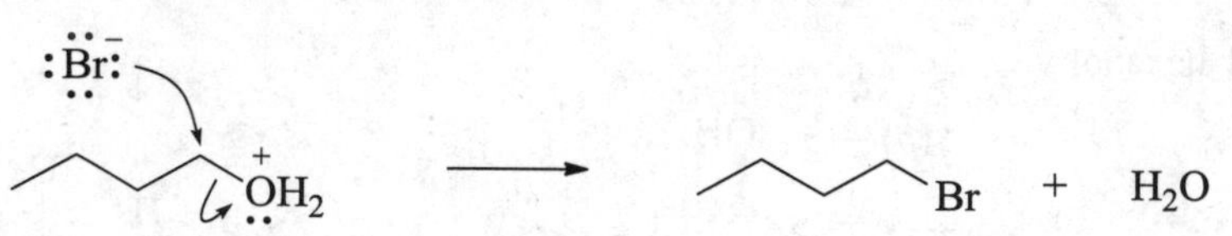

9. 3-Methyl-3-pentanol reacts faster because it is a tertiary alcohol that gives a tertiary carbocation in an SN1 reaction. The other compound, 2-methyl-3-pentanol is a secondary alcohol.

CHAPTER 6

1. (*a*)

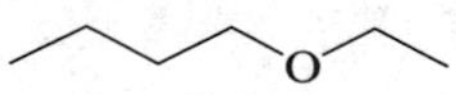

(*d*)

(*b*)

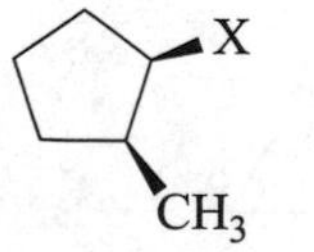

(X = OTs, Br, I)

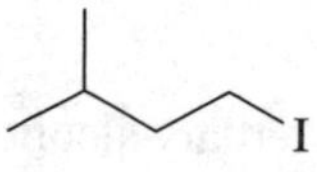

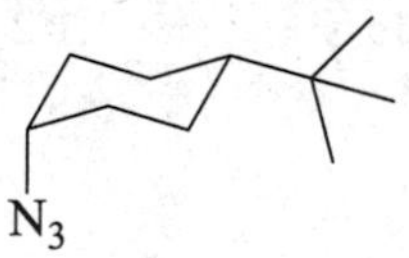

(*e*)

(*c*)

(*f*)

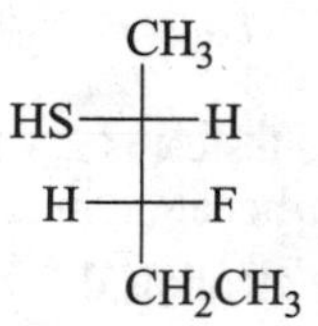

2. $(CH_3)_2CHO^- Na^+ + CH_3CH_2CH_2Br$

3. (a)

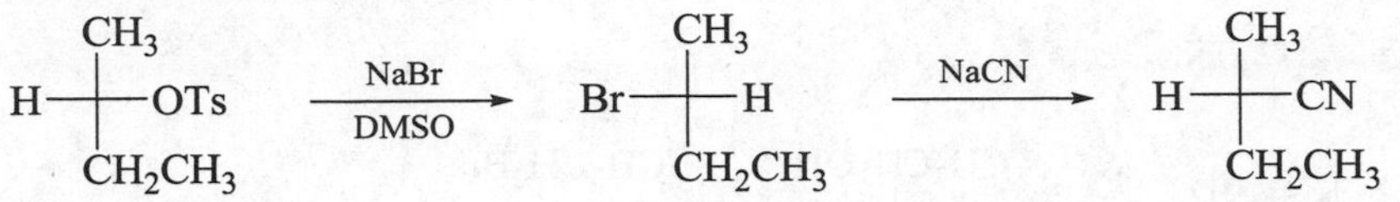

(b)

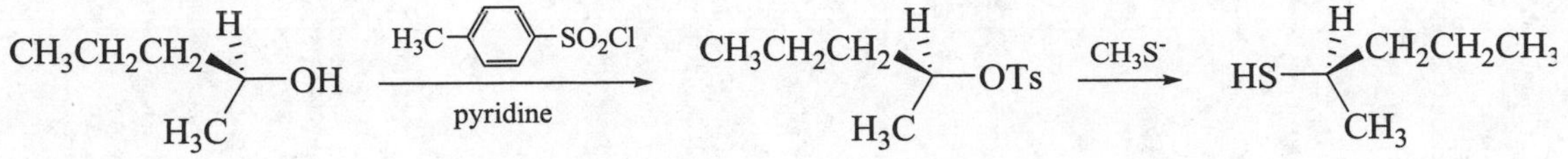

(S)-2-Pentanol

4. **Step 1:** Ionization to form a secondary carbocation

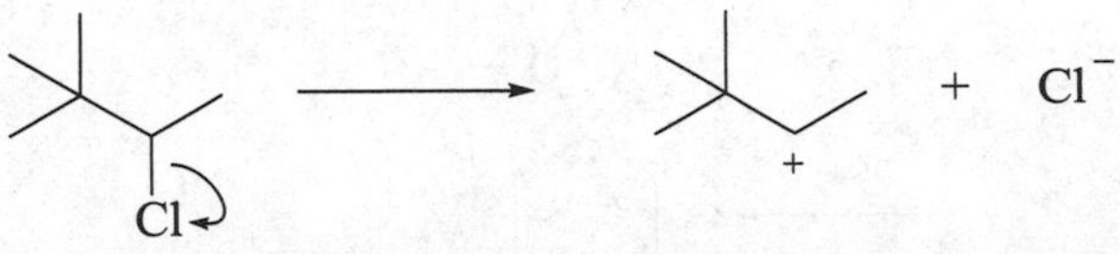

Step 2: Rearrangement by methyl migration to form a more stable tertiary carbocation

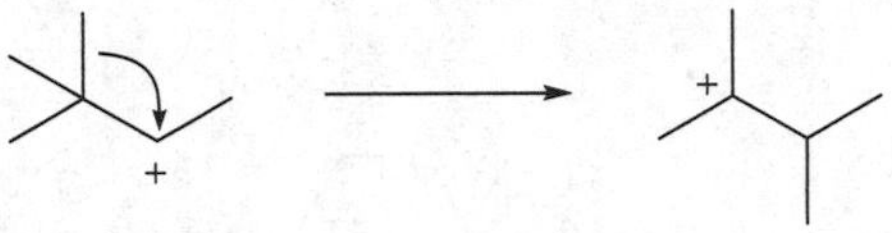

Step 3: Capture of the carbocation by water, followed by deprotonation

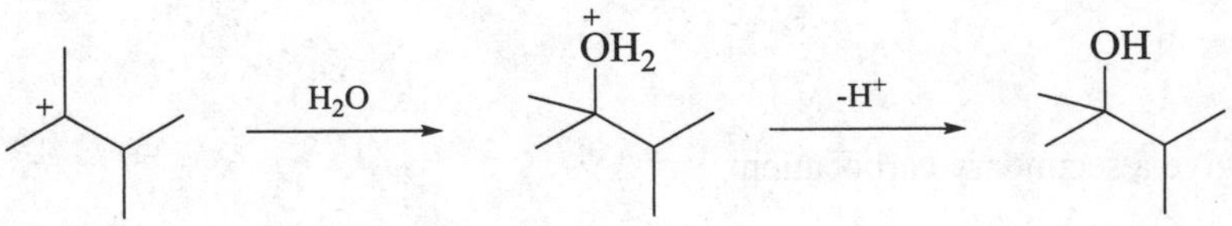

5. (a)

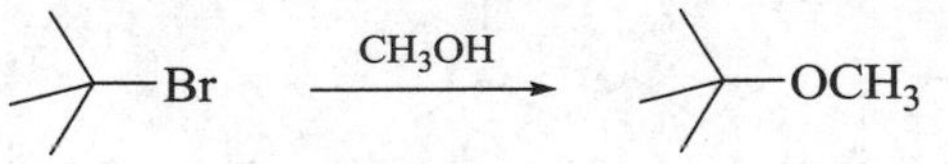

S_N1, unimolecular substitution; rate = $k[(CH_3)_3CBr]$

(b)

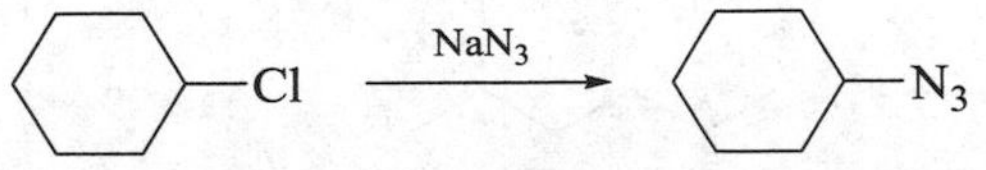

S_N2, bimolecular substitution; rate = $k[C_6H_{11}Cl][NaN_3]$

6. (a) Sodium iodide is soluble in acetone, whereas the byproduct of the reaction, sodium bromide, is not. According to Le Châtelier's principle, the reaction will shift in the direction that will replace the component removed from solution, in this case toward product.

(b) Protic solvents such as water form hydrogen bonds to anionic nucleophiles, thus stabilizing them and decreasing their nucleophilic strength. Aprotic solvents such as DMSO do not solvate anions very strongly, leaving them more able to express their nucleophilic character.

7.

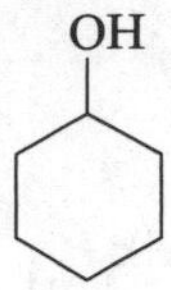

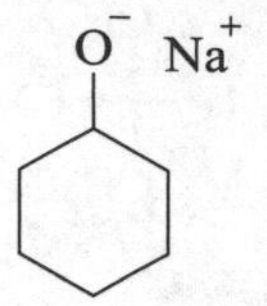

A   B CH_3CH_2OH CH_3CH_2Br

C   D

8.

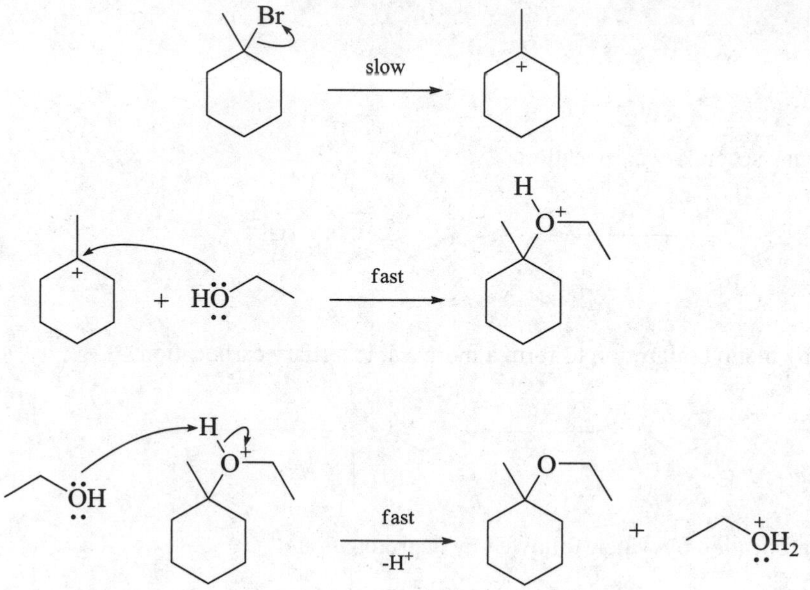

9. Dissociation to give a secondary carbocation

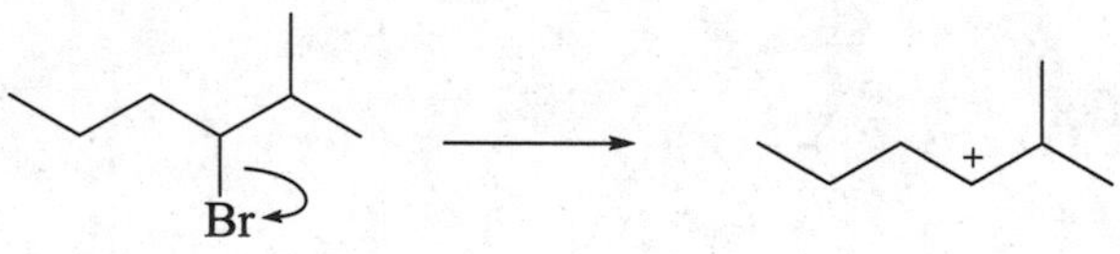

Rearrangement by hydride migration to give a tertiary carbocation

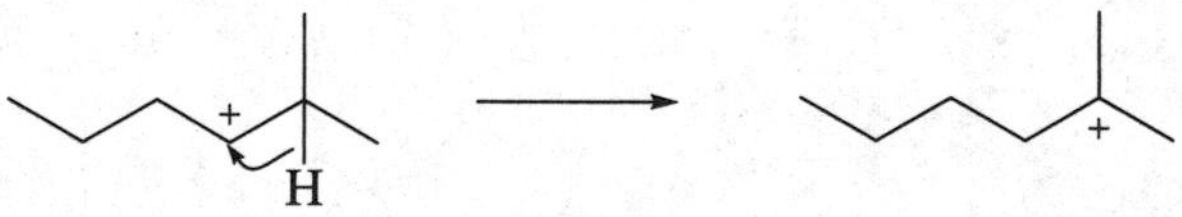

Capture of the carbocation by water and loss of a proton to give product

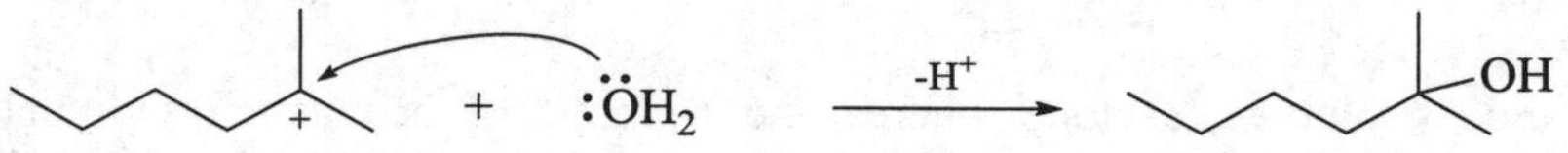

CHAPTER 7

1. (a) 2,4,4-Trimethyl-2-pentene or 2,4,4-trimethylpent-2-ene

(b) (*E*)-3,5-Dimethyl-4-octene or (*E*)-3,5-dimethyloct-4-ene

(c) (*E*)-2,7-Dibromo-3-(2-methylpropyl)-2-heptene or (*E*)-2,7-dibromo-3-(2-methylpropyl)hept-2-ene

(d) 5-Methyl-4-hexen-3-ol or 5-methylhex-4-en-3-ol

2. (*a*)

2,3-Dimethyl-2-pentene

(*c*)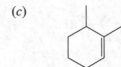

1,6-Dimethylcyclohexene

(*b*)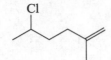

5-Chloro-2-methyl-1-hexene

(*d*)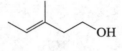

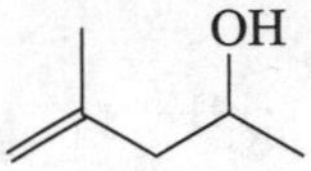

4-Methyl-4-penten-2-ol

3. (*a*)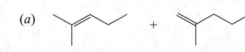

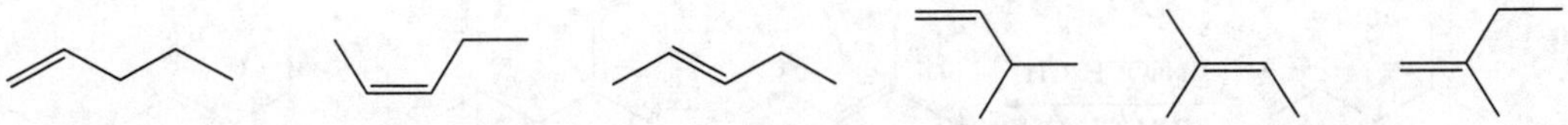

(*b*) Isomer 5   (*c*) Isomers 1 and 4   (*d*) Isomers 2 and 3

4. Two sp^2 C atoms; four sp^3 C atoms; three sp^2–sp^3 σ bonds

5. (*a*) 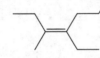 (*b*) 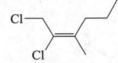 (*c*)

6.

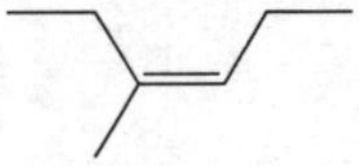

(*Z*)-3-Methyl-3-hexene 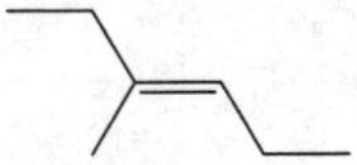   (*E*)-3-Methyl-3-hexene

7. (*a*) 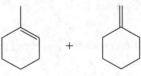 +

(major)

(*c*)

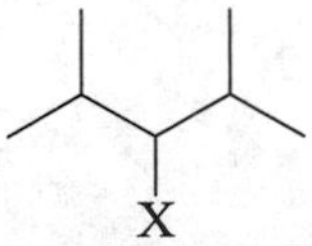

X

(X = Cl, Br, or I)

(*b*)

+

(major)

(*d*)

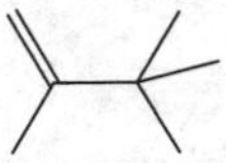

8.

9. **Step 1:** Protonation

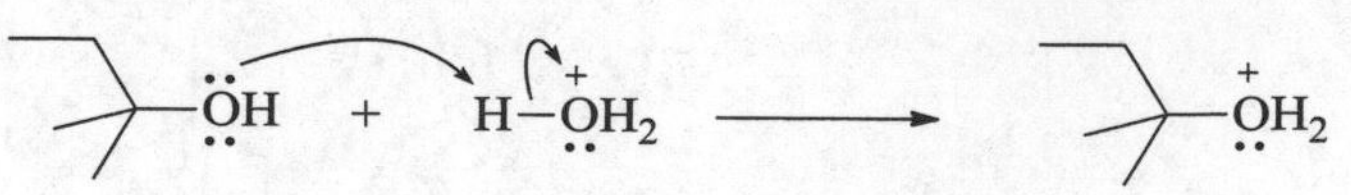

Step 2: Dissociation

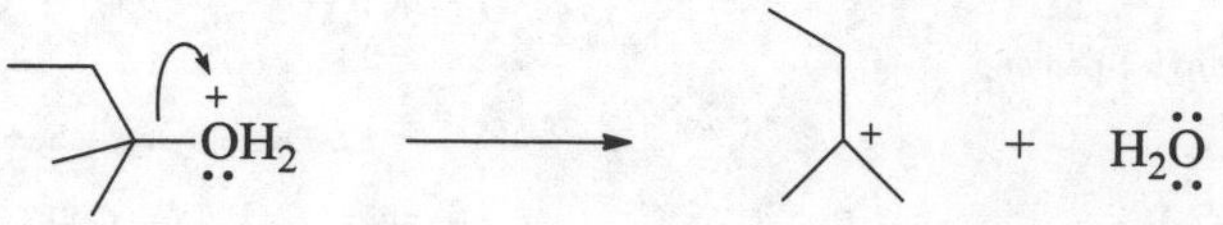

Step 3: Deprotonation

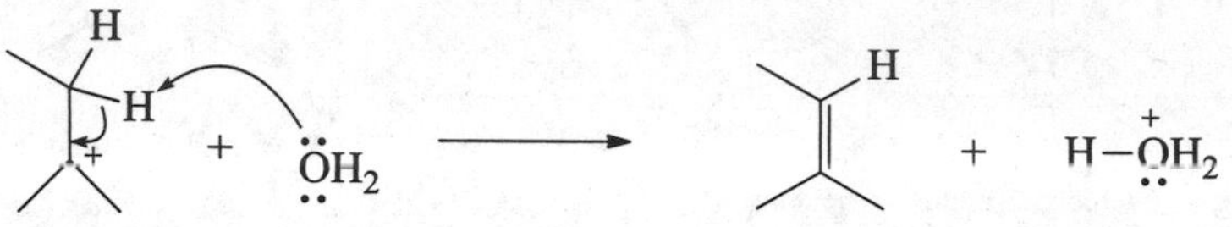

10.

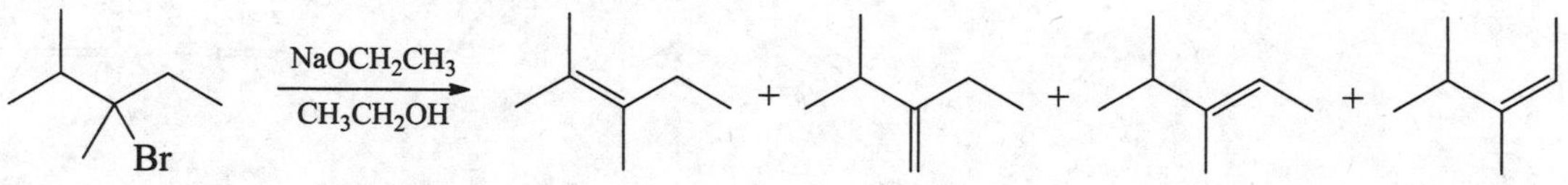

(major)

11.

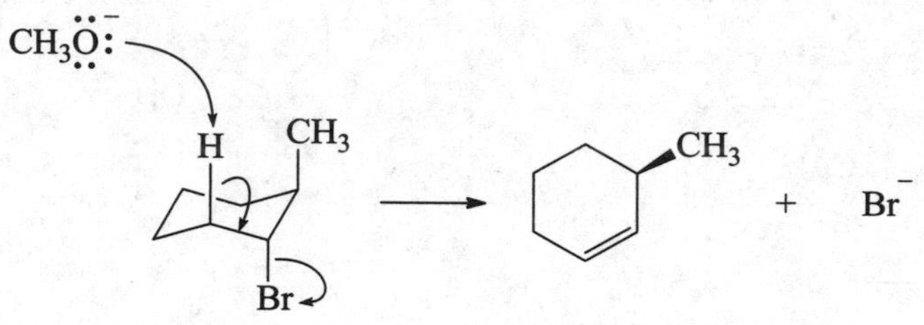

12. Cis isomer:

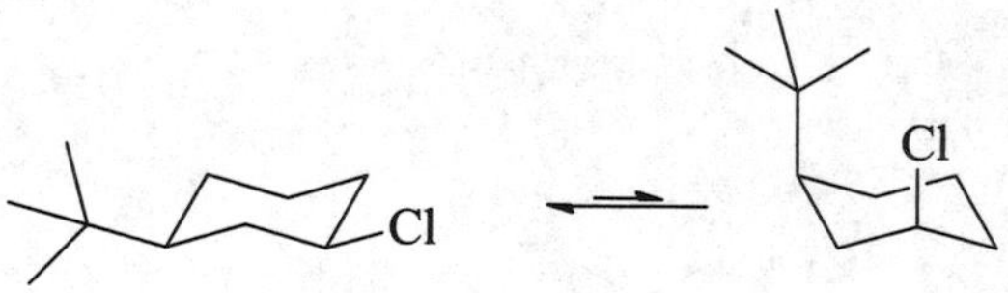

Trans isomer:

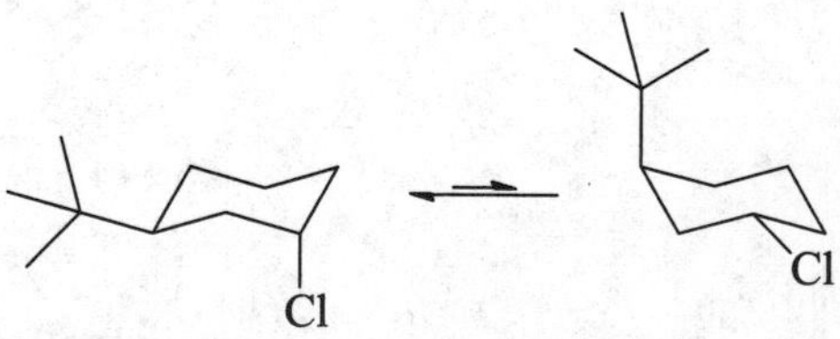

The trans isomer will react faster because its most stable conformation (with the isopropyl group equatorial) has an axial Cl able to undergo E2 elimination.

13. Rearrangement (hydride migration) occurs to form a more stable carbocation.

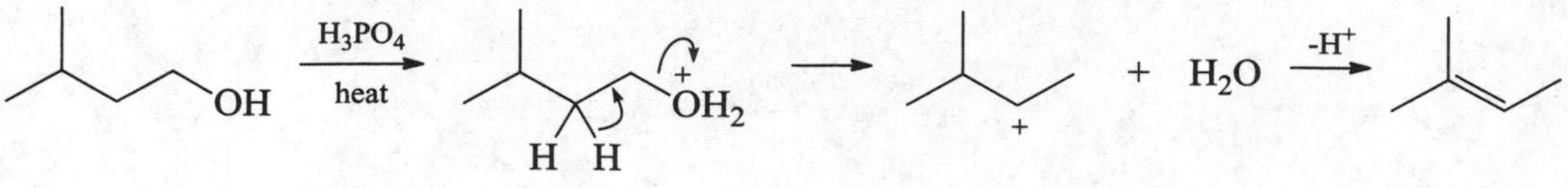

CHAPTER 8

1. Five different alkenes will give 2,3-dimethylpentane on catalytic hydrogenation:

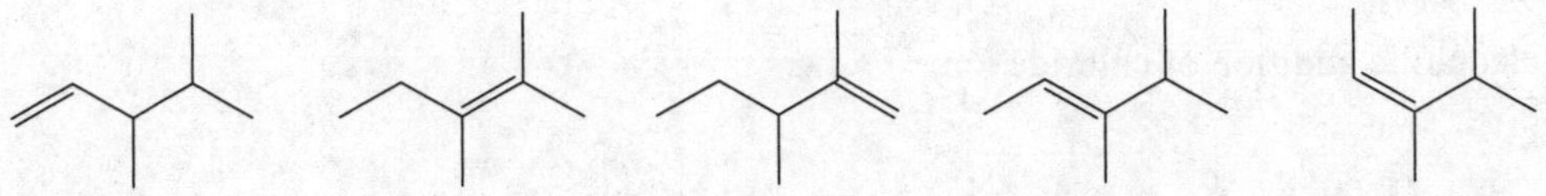

| 3,4-Dimethyl-
1-pentene | 2,3-Dimethyl-
2-pentene | 2,3-Dimethyl-
1-pentene | (*E*)-3,4-Dimethyl-
2-pentene | (*Z*)-3,4-Dimethyl-
2-pentene |

2. (*a*)

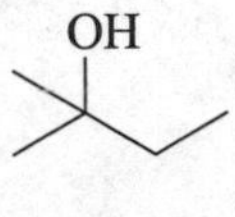

(*c*)

(*b*) 1. BH$_3$, THF

  2. H$_2$O$_2$, OH$^-$

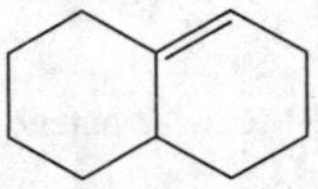

(*d*)

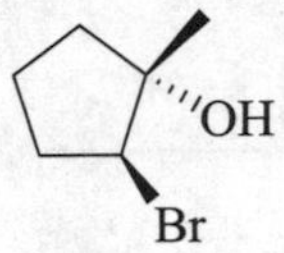

3. (*a*)

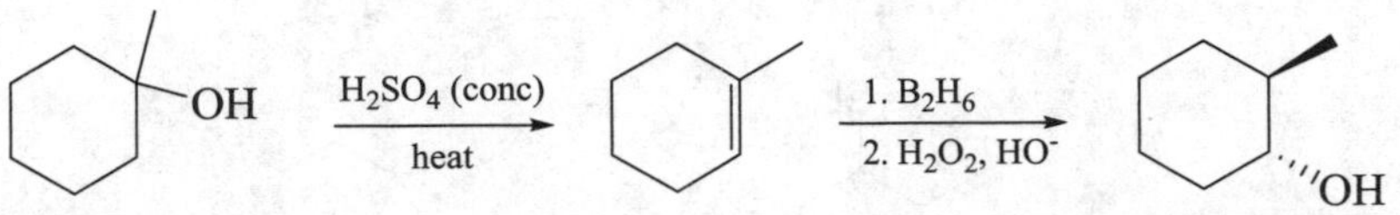

(*b*)

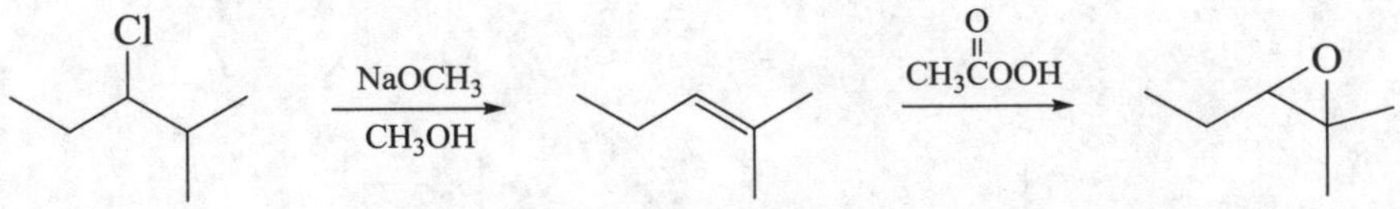

4.

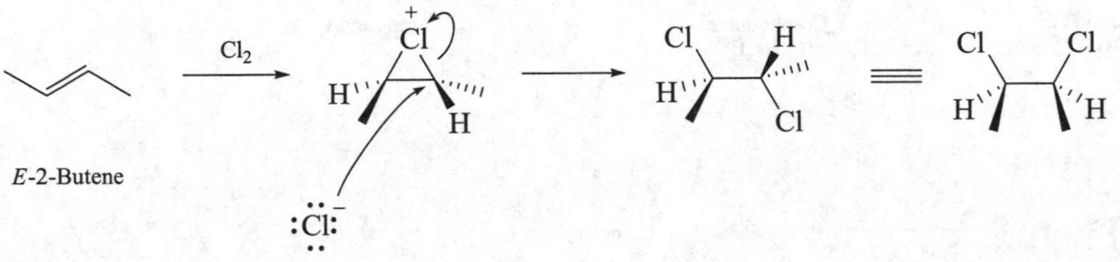

E-2-Butene

5.

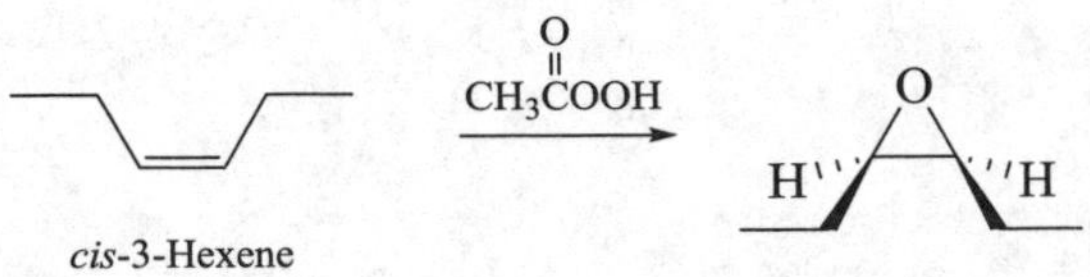

cis-3-Hexene

6. Step 1: Protonation to form a carbocation

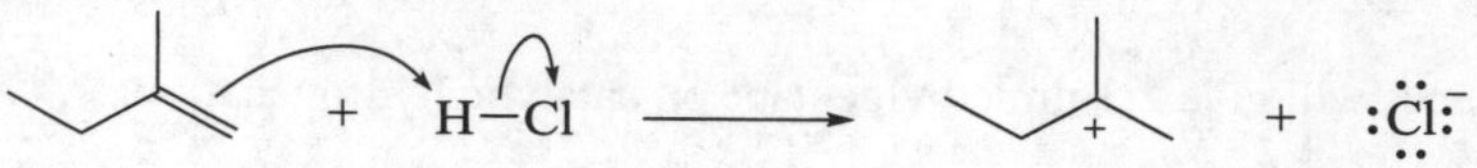

Step 2: Nucleophilic addition of chloride ion

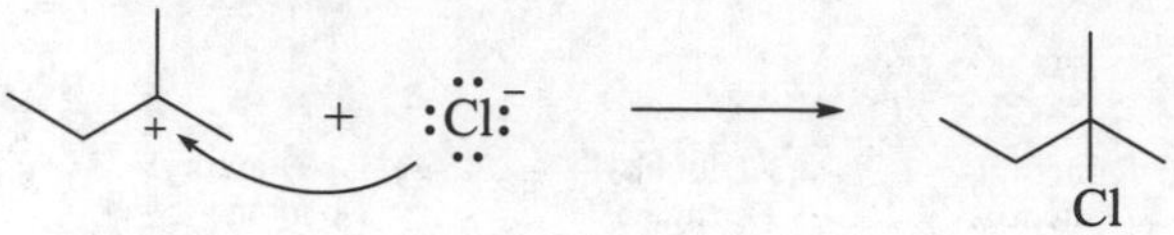

7.

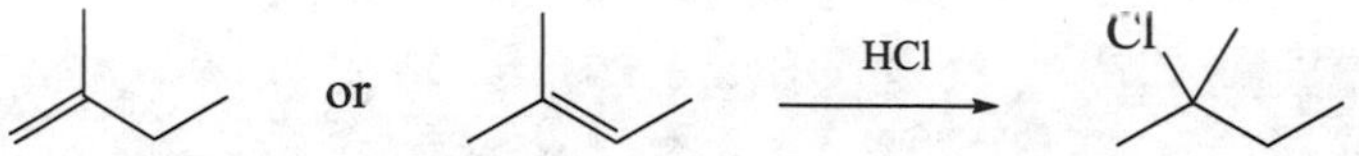

2-Methyl-1-butene   2-Methyl-2-butene   2-Chloro-2-methylbutane

8.

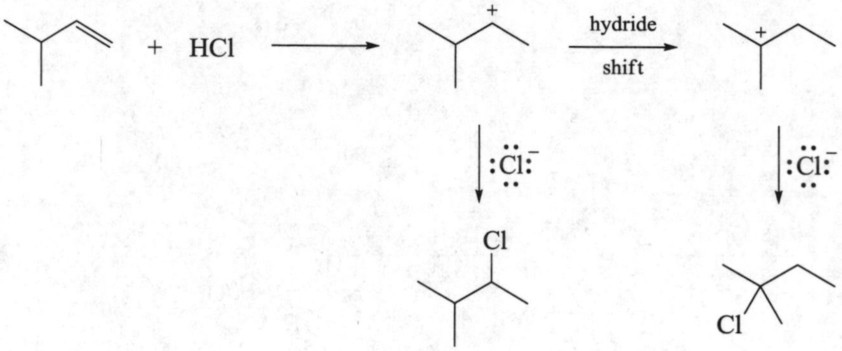

9.

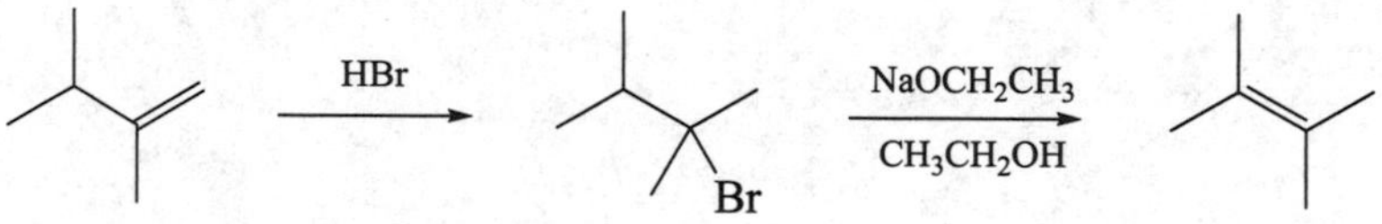

Compound A   Compound B   Compound C
(C$_6$H$_{12}$)   (C$_6$H$_{13}$Br)

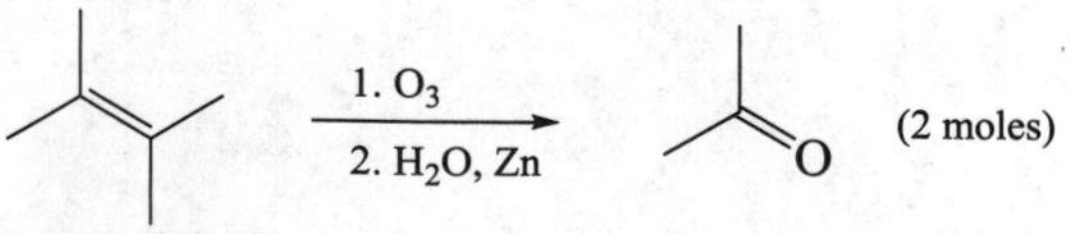

Compound C

CHAPTER 9

1. (*a*) 4,5-Dimethyl-2-hexyne   (*c*) 6,6-Dimethylcyclodecyne

　　(*b*) 4-Ethyl-3-propyl-1-heptyne

2. (a)

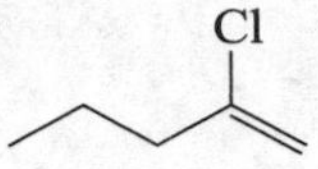

(e)

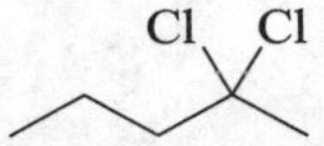

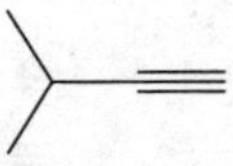

(b)

(f)   Na, NH$_3$ (l)

(c)   H$_2$O, H$_2$SO$_4$, HgSO$_4$

(g)

(d)

(cis)

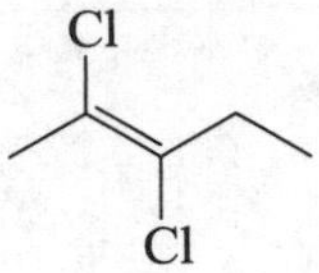

(h)

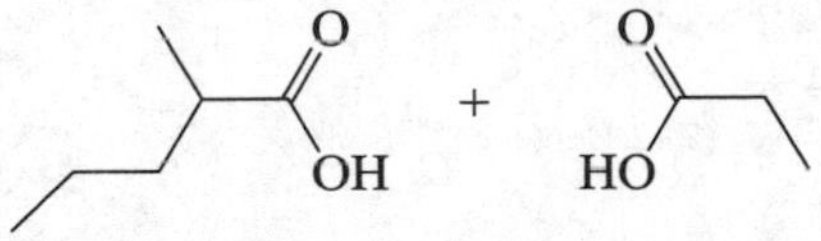

3.   Reaction (2) is effective; the desired product is formed by an S$_N$2 reaction.

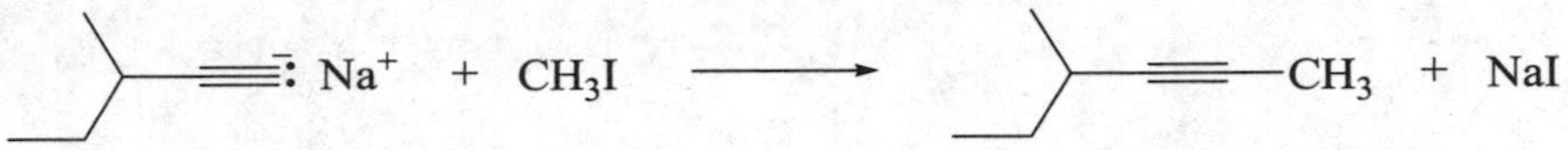

Reaction (1) is not effective, owing to E2 elimination from the secondary bromide.

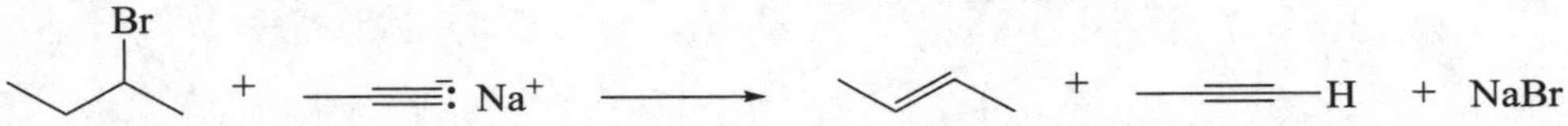

4.   (a)

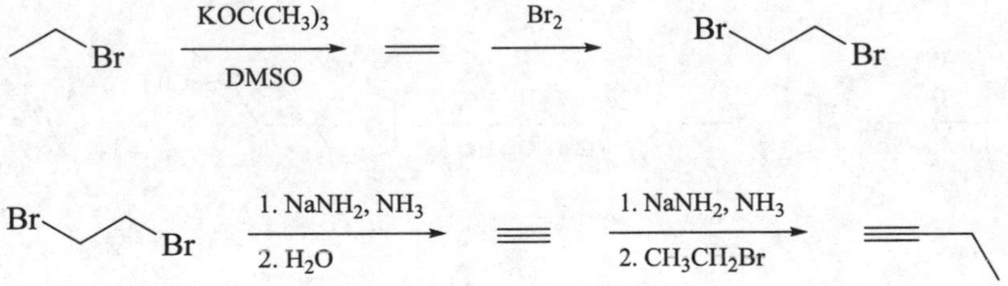

(b)

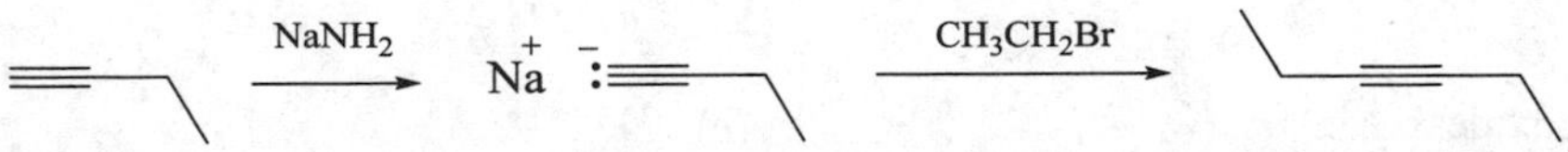

(c)

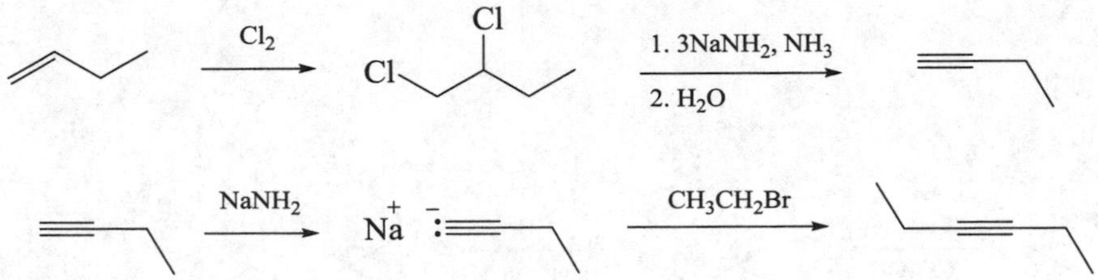

(d)

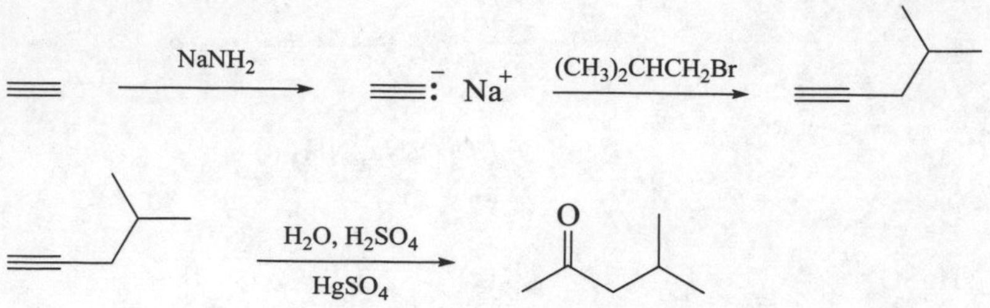

5.

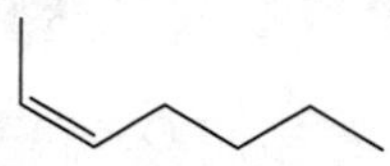

(Z)-2-Heptene

6.

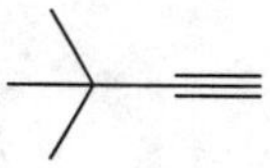

A 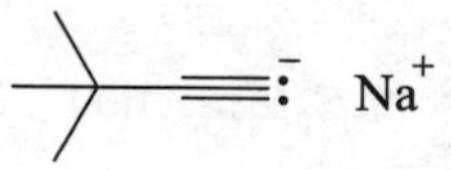   B 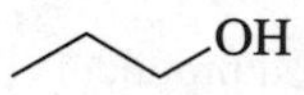   C 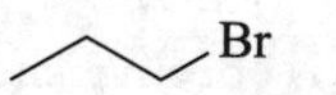   D

7.

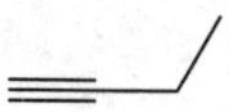

E 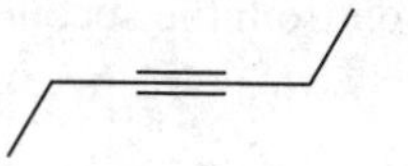   F

8.

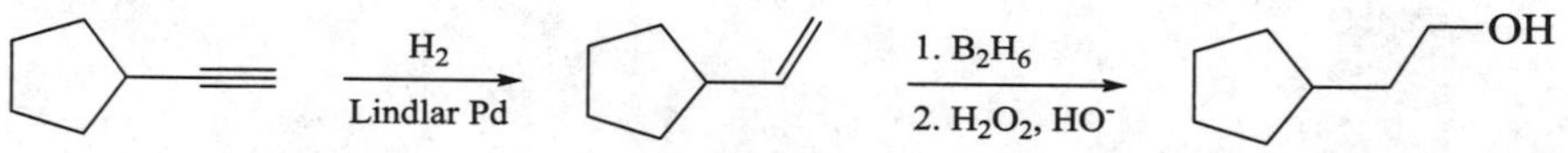

CHAPTER 10

1.   *(a)*   HBr, peroxides

(c)

(b)   Br$_2$, light

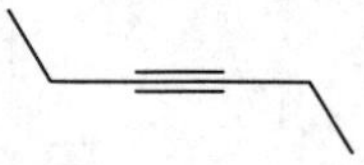

2.   *(a)*

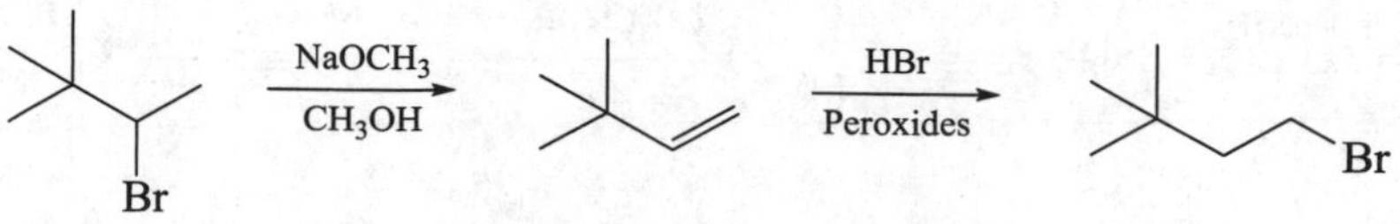

(b)

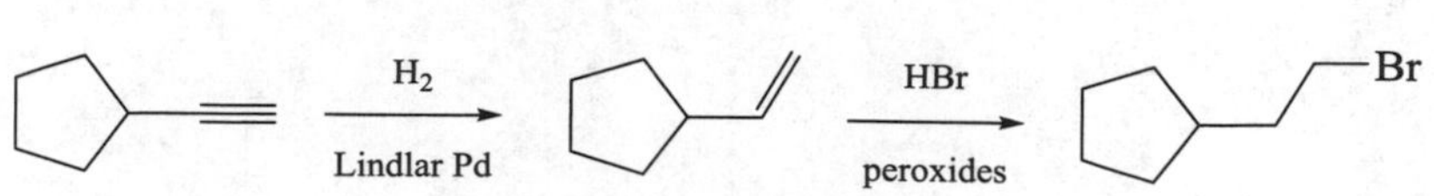

(c)

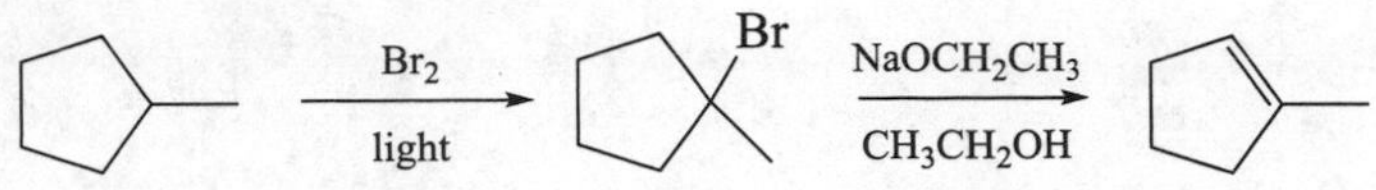

3.

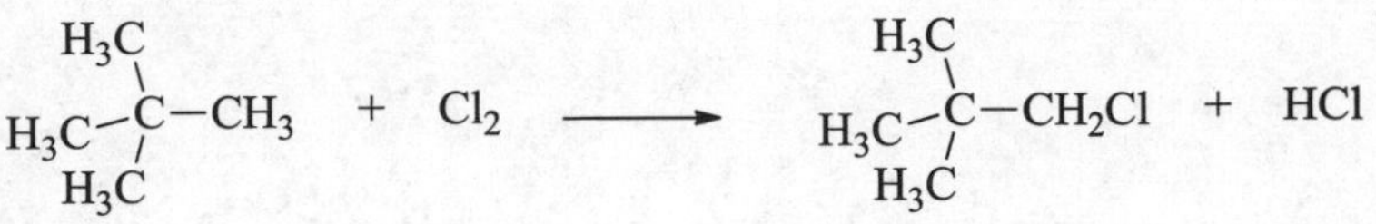

4. Initiation:

$$ROOR \xrightarrow[\text{or heat}]{\text{light}} 2RO\cdot$$

$$RO\cdot + HBr \longrightarrow ROH + Br\cdot$$

Propagation:

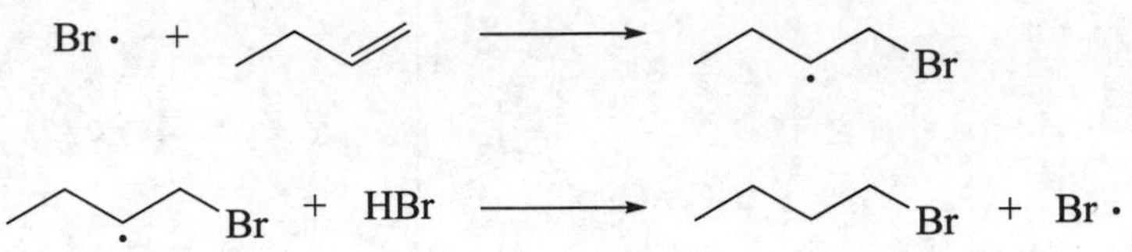

5.

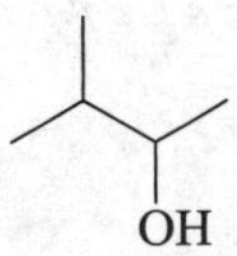

6.

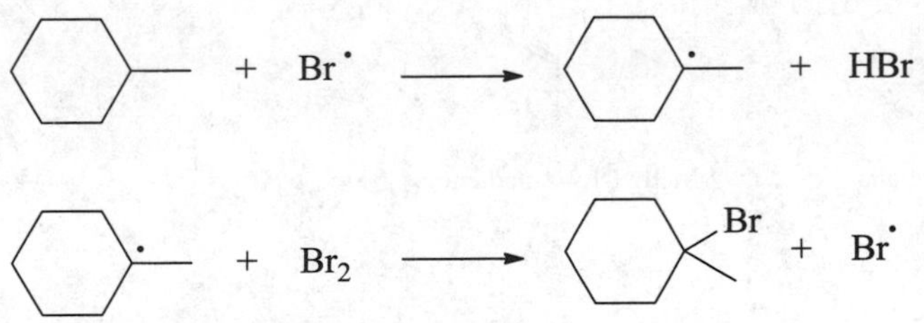

7. $\Delta H = -57$ kJ/mol (-13.5 kcal/mol)

8. (a) Ethyl radical, $CH_3\dot{C}H_2$ (b) Cl_2

9.

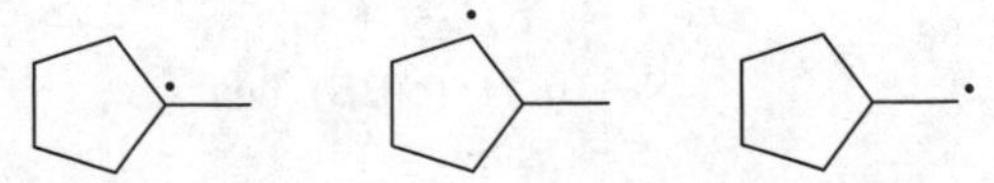

tertiary radical   primary radical
(most stable)   (least stable)

10.

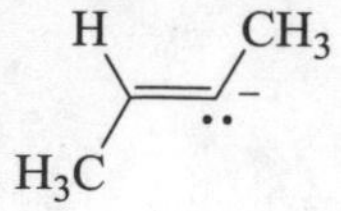

CHAPTER 11

1.

(Conjugated)
cis and trans are constitutionally identical

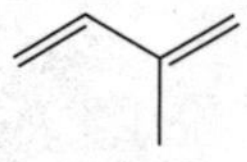

(Conjugated)

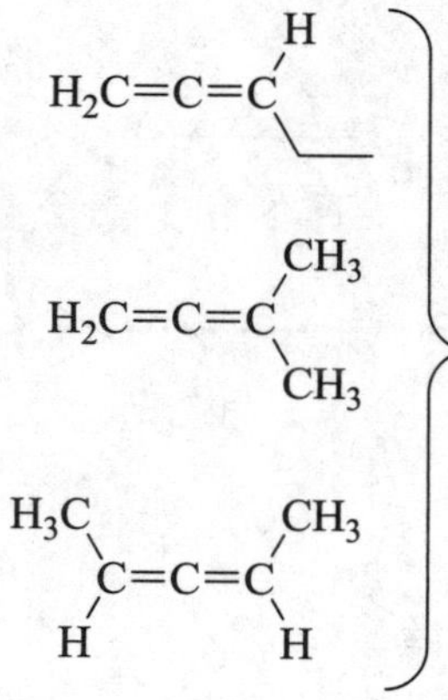

$H_2C=C=C$ ...

Allenes

2.

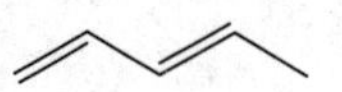

(3E)-1,3-Pentadiene   (3Z)-1,3-Pentadiene 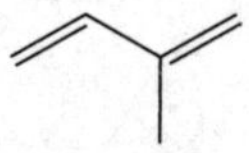   2-Methyl-1,3-butadiene

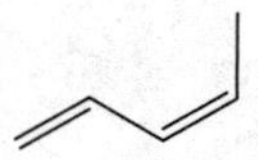

3.

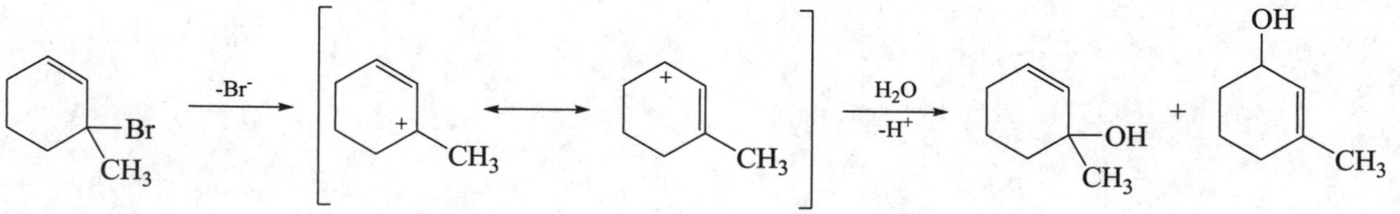

4. (a)

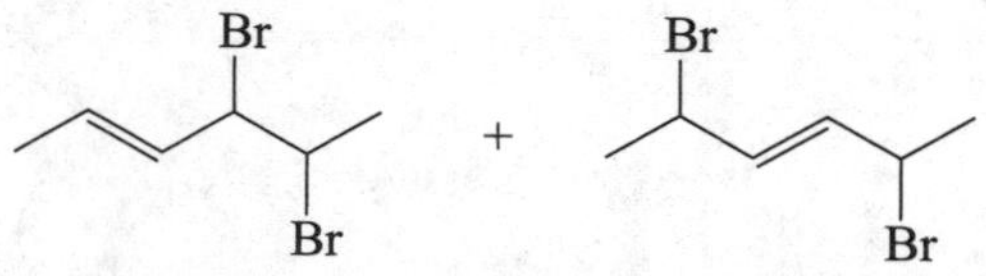

(1,2-Addition)   (1,4-Addition)

(d)

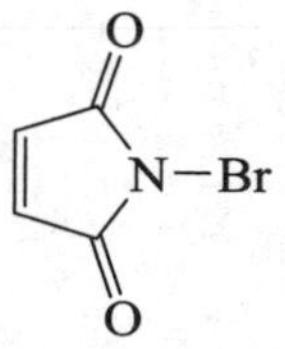

(NBS), heat

(b)

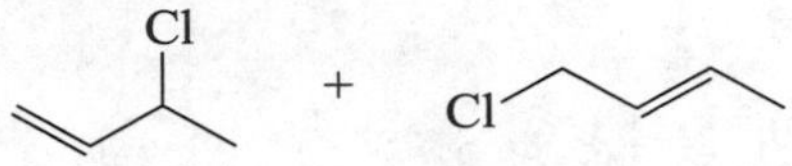

+

(e)

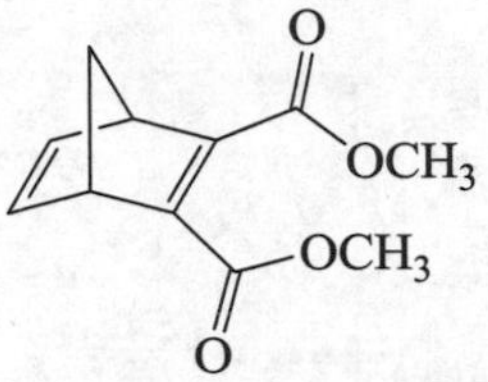

(c)

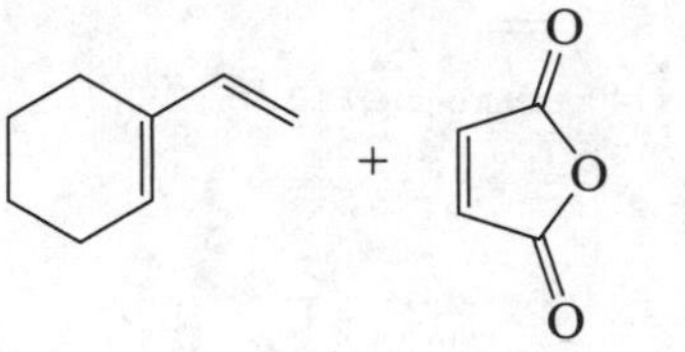

+

5.

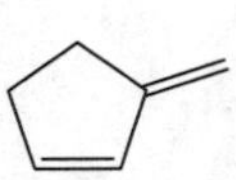

(cannot adopt the required *s*-cis conformation)

6.

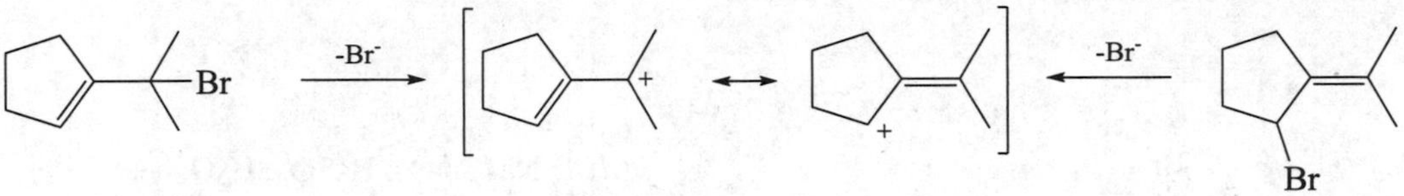

7.

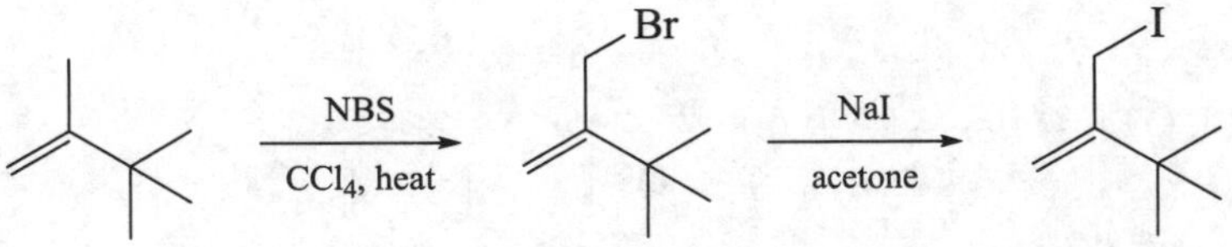

Compound A   Compound B

8.

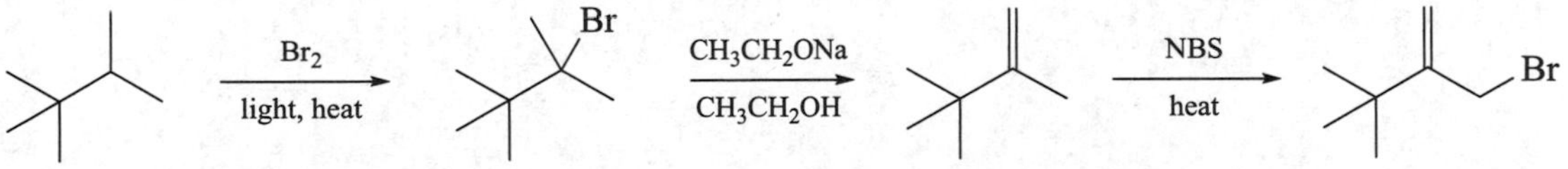

CHAPTER 12

1. (*a*) *m*-Bromotoluene   (*c*) *o*-Chloroacetophenone

(*b*) 2-Chloro-3-phenylbutane   (*d*) 2,4-Dinitrophenol

2. (*a*)

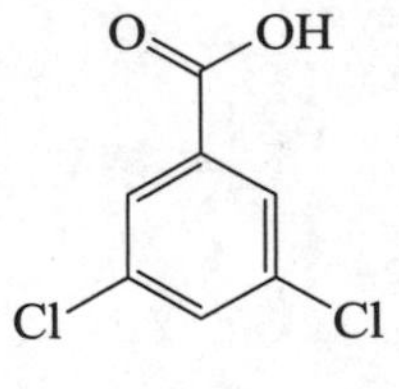

(*b*)

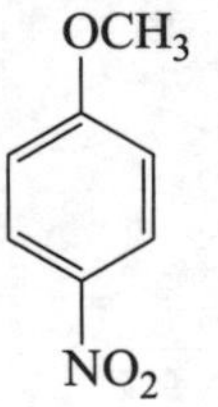

(*c*)

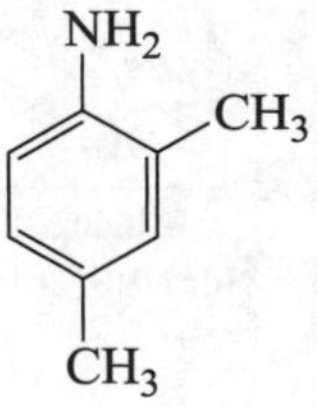

(*d*)

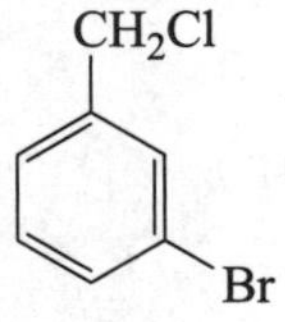

3. (*a*)

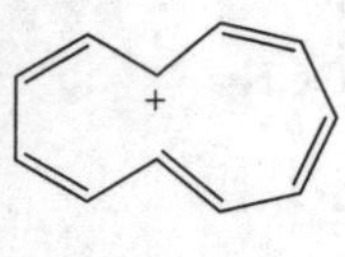

(10 π electrons)

(*b*)

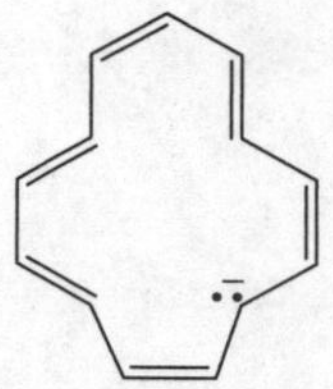

(14 π electrons)

4. (*a*) 8 π electrons. No, the substance is not aromatic.

(*b*) 6 π electrons. Yes, it is aromatic.

(*c*) 14 π electrons. Yes, it is aromatic.

5.

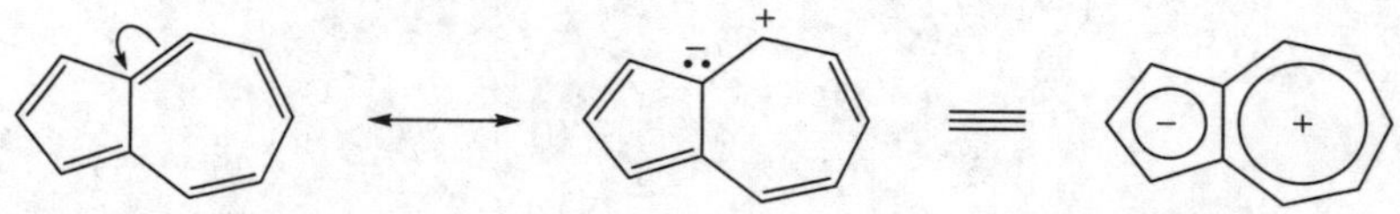

6. (*a*)

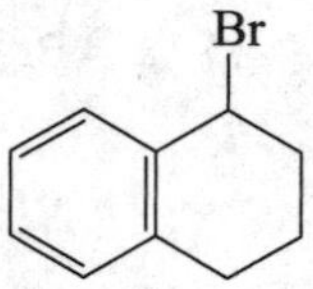

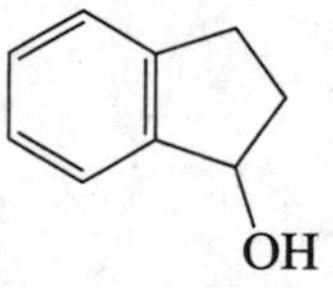

(*d*) $Na_2Cr_2O_7, H_2SO_4, H_2O$, heat

(*b*) $C_6H_5CH_2X$ (X = Cl, Br, I, OTs)

(*e*)

(*c*)

(*f*)

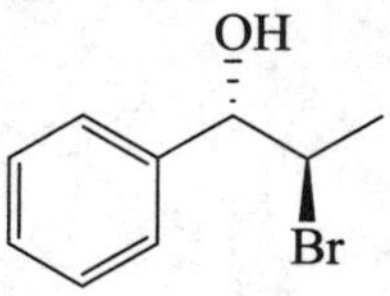

(racemic)

7.

(I)

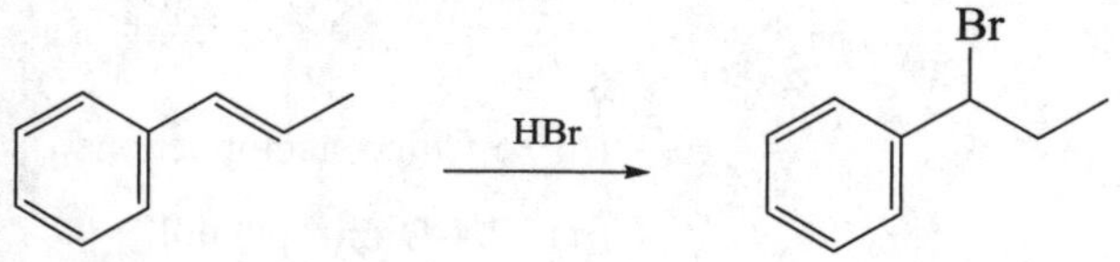

$$\xrightarrow{\text{HBr}}$$

(II)

$$\xrightarrow[\text{(or NBS, heat)}]{\underset{\text{light}}{Br_2}}$$

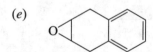

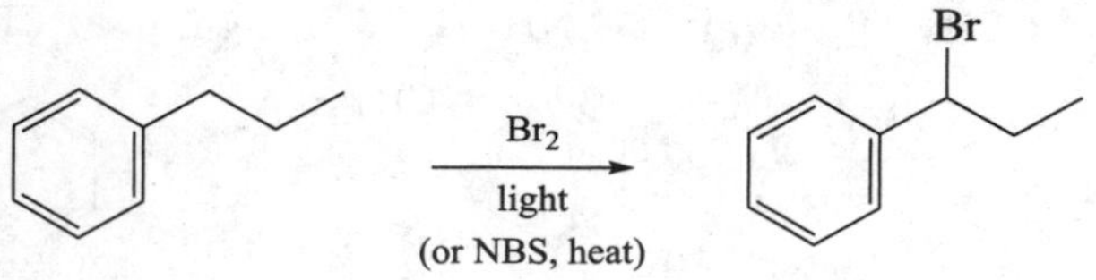

8.

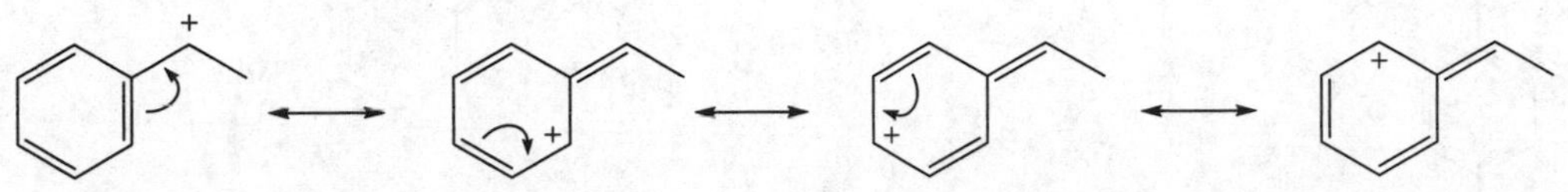

9.

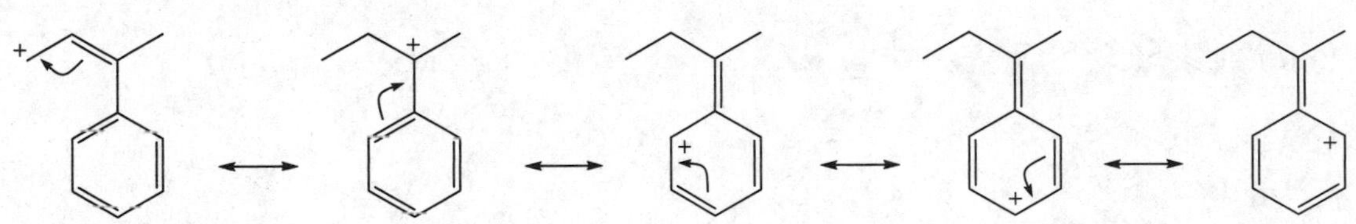

10.

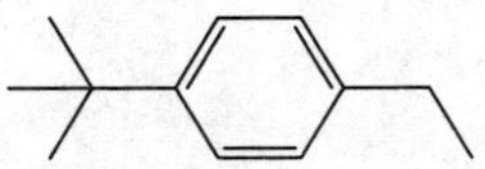

CHAPTER 13

1.

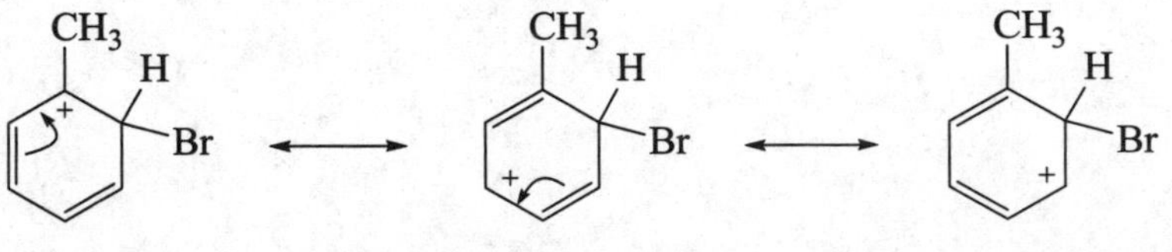

2. (*a*)

Slower

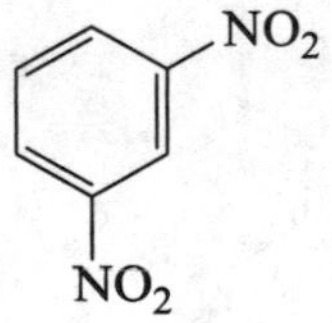

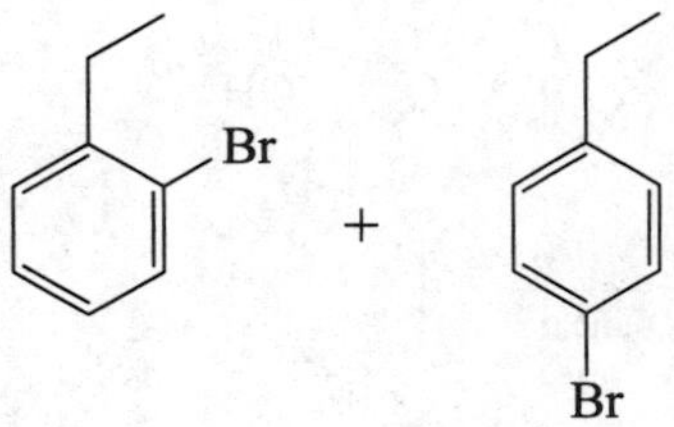

(*c*)

Slower

(*b*)

Faster

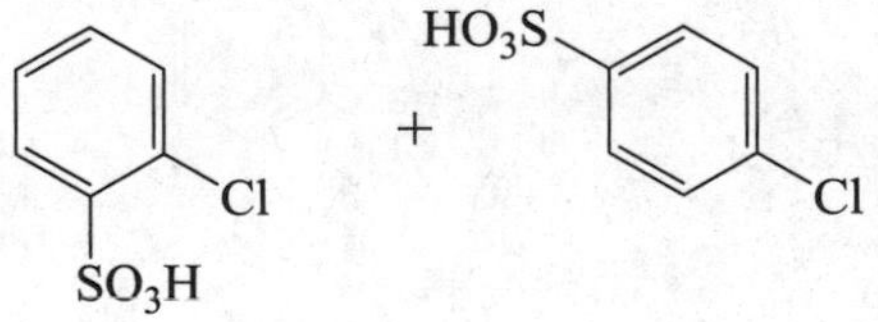

3. (*a*) NO_2^+ (*b*) $Br-\overset{+}{Br}-\overset{-}{Fe}Br_3$ (*c*) SO_3

4. (*a*)

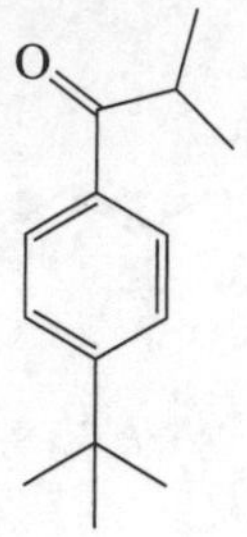

(*b*)

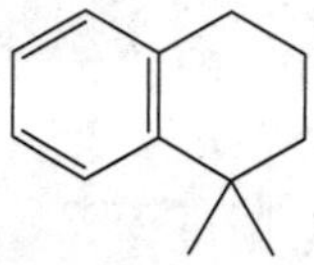

(*c*)

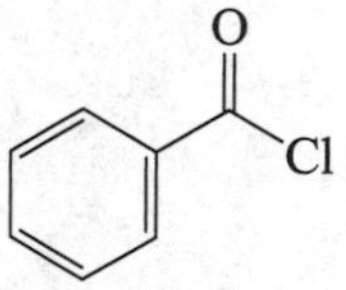

, AlCl$_3$ or

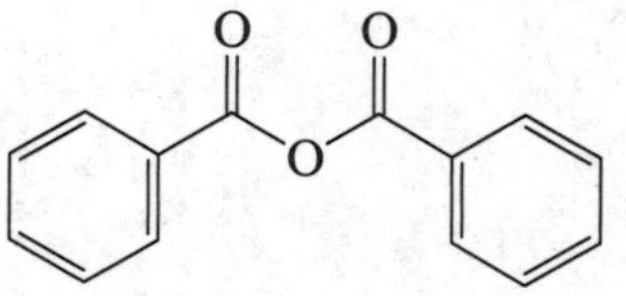

, AlCl$_3$

(*d*)

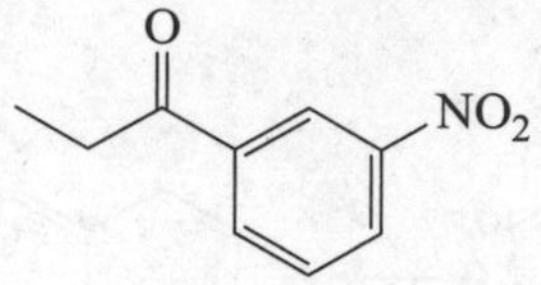

(*e*)

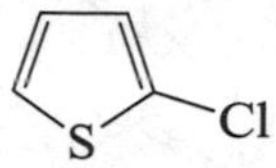

5. (*a*)

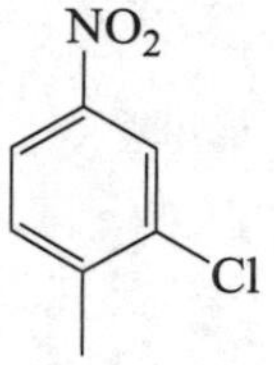

(*b*)

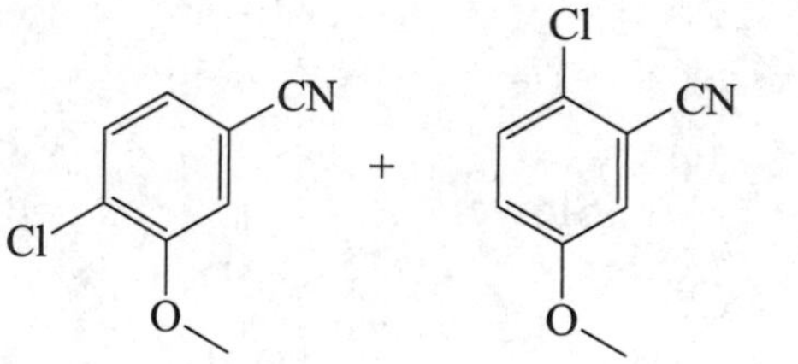

+

(*c*)

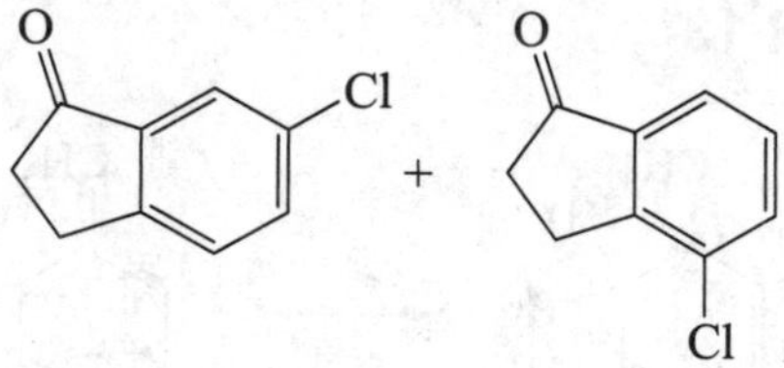

+

(*d*)

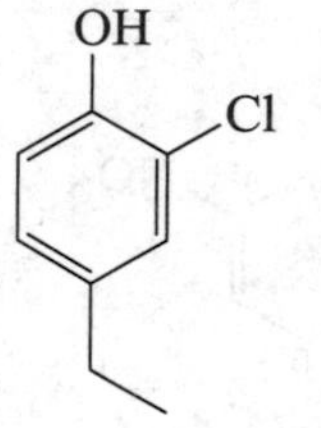

6. (*a*)

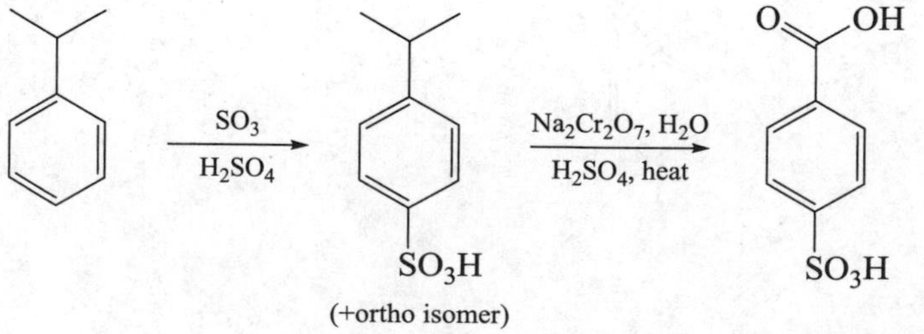

$$\xrightarrow[\text{H}_2\text{SO}_4]{\text{SO}_3}$$

(+ortho isomer)

$$\xrightarrow[\text{H}_2\text{SO}_4,\ \text{heat}]{\text{Na}_2\text{Cr}_2\text{O}_7,\ \text{H}_2\text{O}}$$

(*b*)

$$\xrightarrow[\text{AlCl}_3]{\text{C}_6\text{H}_5\text{CH}_2\overset{\text{O}}{\overset{\|}{\text{C}}}\text{Cl}}$$

$$\xrightarrow{\text{Cl}_2,\ \text{FeCl}_3}$$

$$\xrightarrow[\substack{\text{or N}_2\text{H}_4,\ \text{KOH} \\ \text{heat}}]{\text{Zn(Hg)},\ \text{HCl}}$$

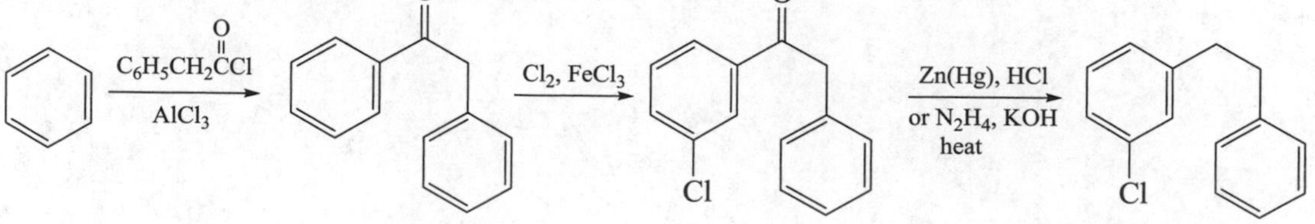

(c)

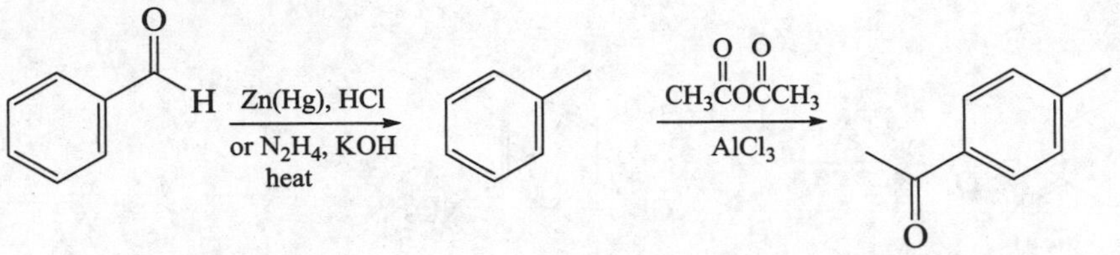

(d)

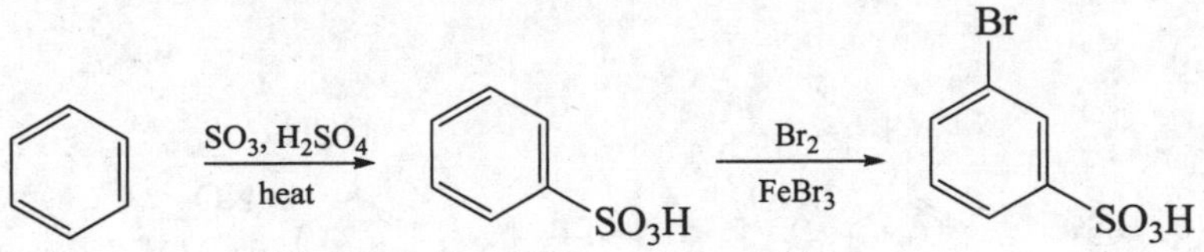

(e)

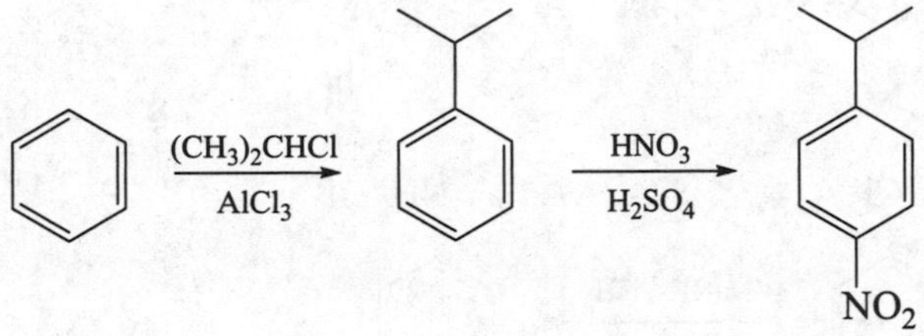

7. (a)

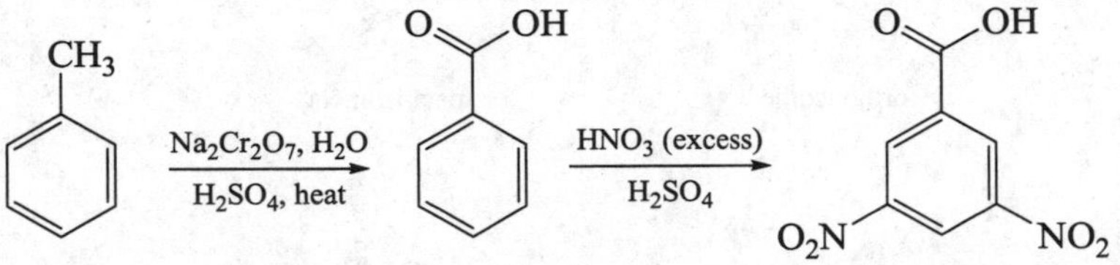

(b)

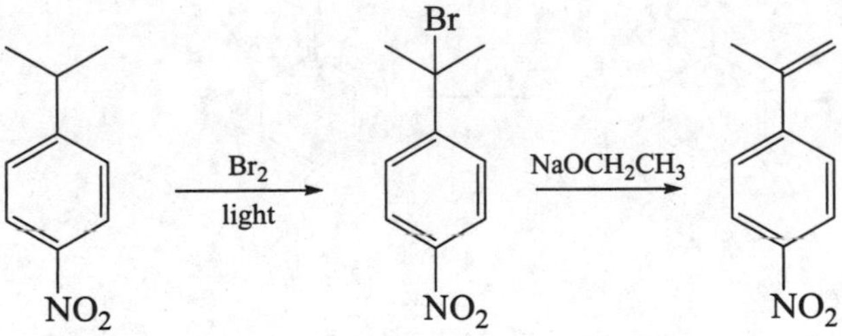

[Prepared from benzene as in
Problem 6(e)]

(c)

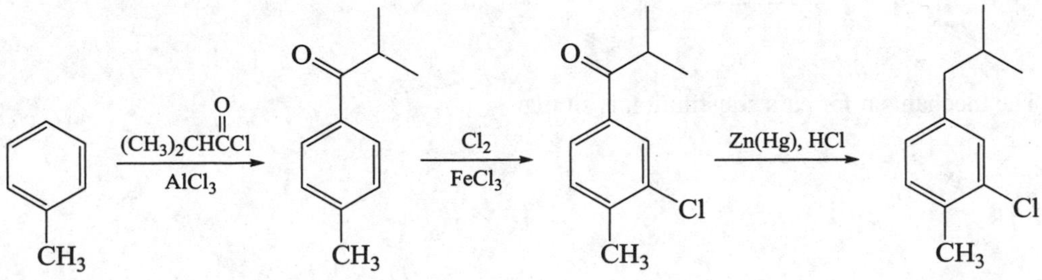

8.

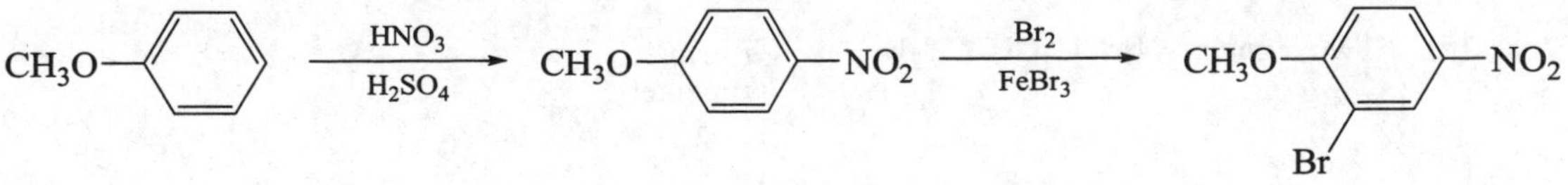

9.

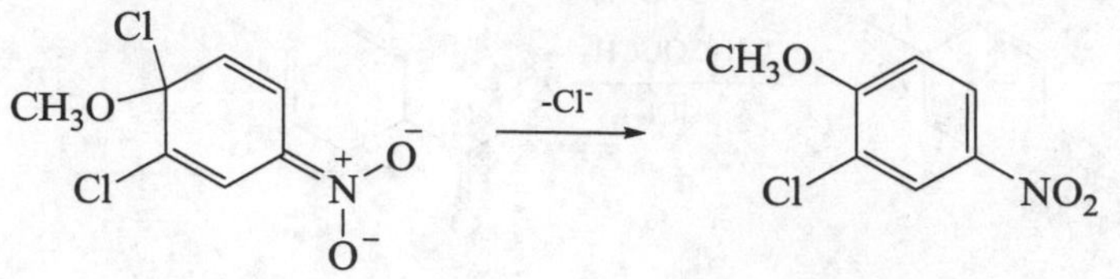

10. (a)

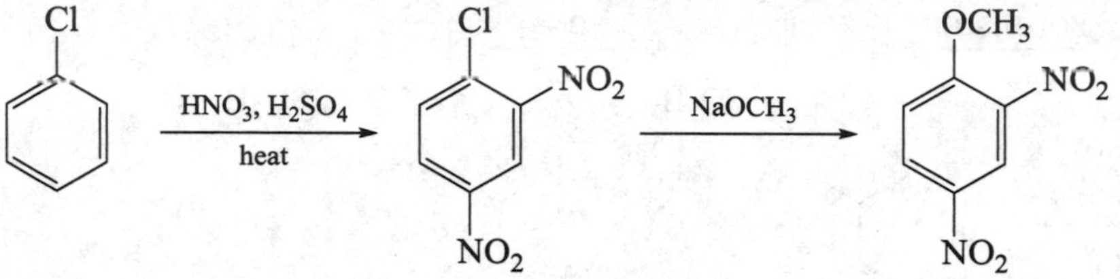

(b)

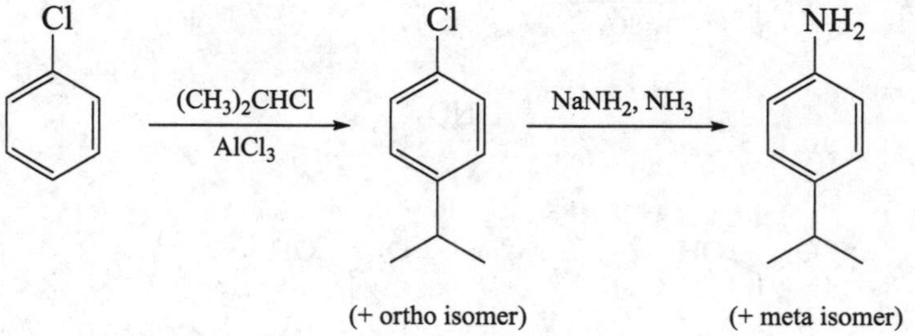

(+ ortho isomer)   (+ meta isomer)

11.

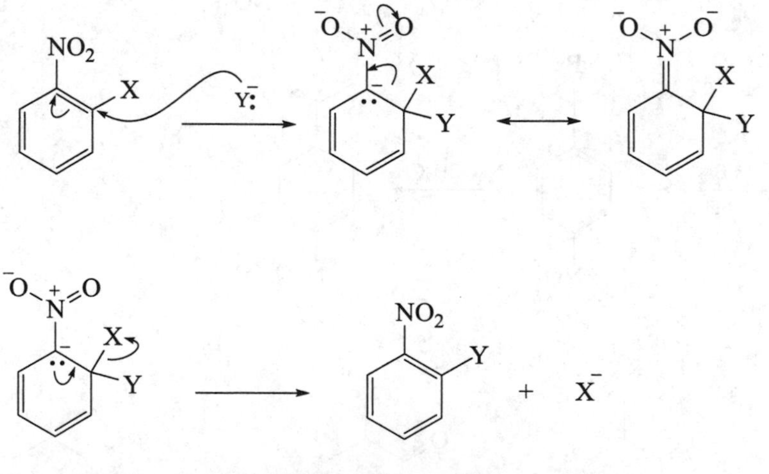

The mechanism for para substitution is similar.

CHAPTER 14

1.   1: 6.10 ppm   3: 200 MHz

2: 1305 Hz   4: 0.00 ppm

2.   (a)   Two signals $BrCH_2CH_2CH_2Br$ a: triplet
                                      a   b   a   b: pentet

(b)

Two signals $\underset{a \quad b \ | \ b \quad a}{CH_3CH_2\overset{\overset{\displaystyle Cl}{|}}{\underset{\displaystyle Cl}{C}}CH_2CH_3}$ a: triplet
b: quartet

(c)   Three signals, all singlets

3.

A: $CH_3\overset{\overset{\displaystyle O}{\|}}{C}OC(CH_3)_3$ B: $CH_3O\overset{\overset{\displaystyle O}{\|}}{C}C(CH_3)_3$

4.   (a)

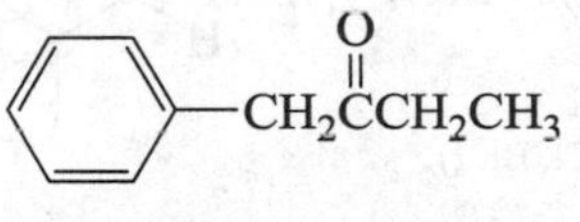

$\text{---}CH_2\overset{\overset{\displaystyle O}{\|}}{C}CH_2CH_3$

(c)

$HC(\overset{\overset{\displaystyle O}{\|}}{C}OCH_2CH_3)_3$

(b)

$(CH_3)_2\overset{\overset{\displaystyle HO}{|}}{C}\text{---}\overset{\overset{\displaystyle OH}{|}}{C}(CH_3)_2$

(d)

$(CH_3)_2\overset{\overset{\displaystyle OH}{|}}{C}\text{---}C{\equiv}N$

5.   Seven signals:

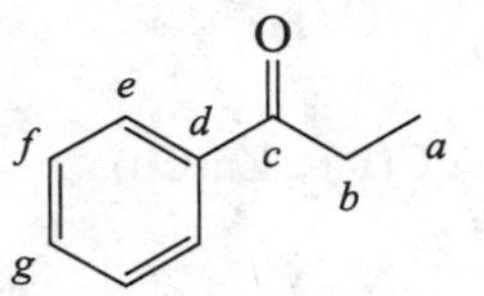

a: δ 10–30
b: δ 20–40
c: δ 190–220
d–g: δ 110–175

6.   Pentane: three signals; 2-methylbutane: four signals; 2,2-dimethylpropane: two signals

7.   2,3-Dimethylbutane: $(CH_3)_2CHCH(CH_3)_2$

CHAPTER 15

1.   (a)

+   2Li $\longrightarrow$ +   LiX

(X = Cl, Br, I)

(b)

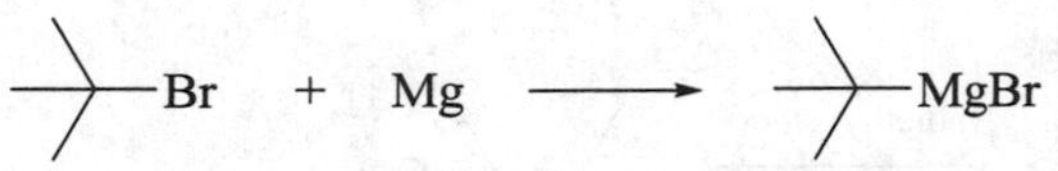

(c)

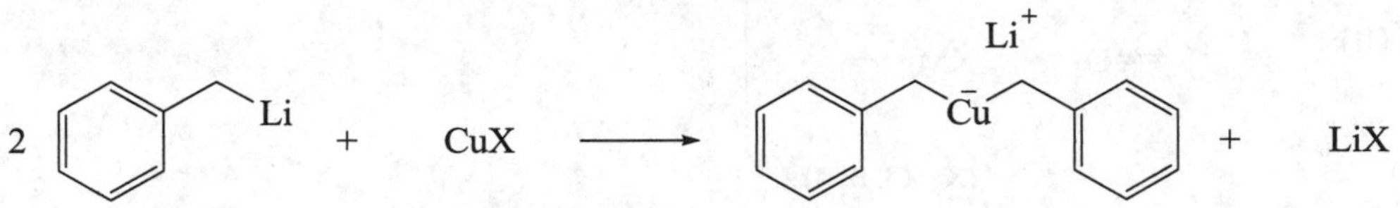

(X = Cl, Br, I)

2.   (*a*)

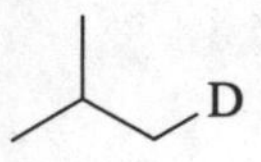

(*c*)

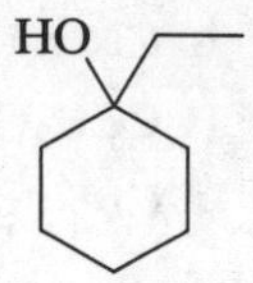

(*b*)

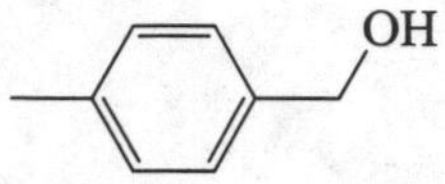

3.   (*a*)

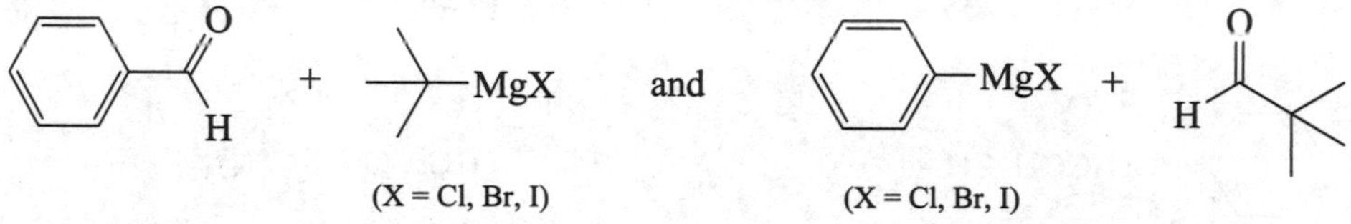

(X = Cl, Br, I)   (X = Cl, Br, I)

(*b*)

(X = Cl, Br, I)   (X = Cl, Br, I)

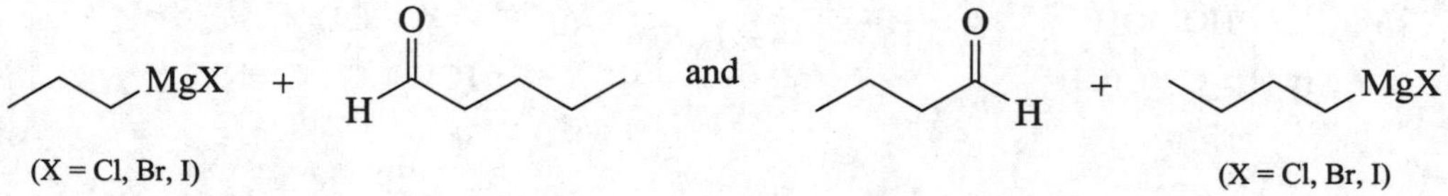

4.   (*a*) $(CH_3CH_2CH_2)_2CuLi$ (*b*) CH_2I_2, Zn(Cu)

5.   Solvents A, B, and E are suitable; they are all ethers. Solvents C and F have acidic hydrogens and will react with a Grignard reagent. Solvent D is an ester, which will react with a Grignard reagent.

6.

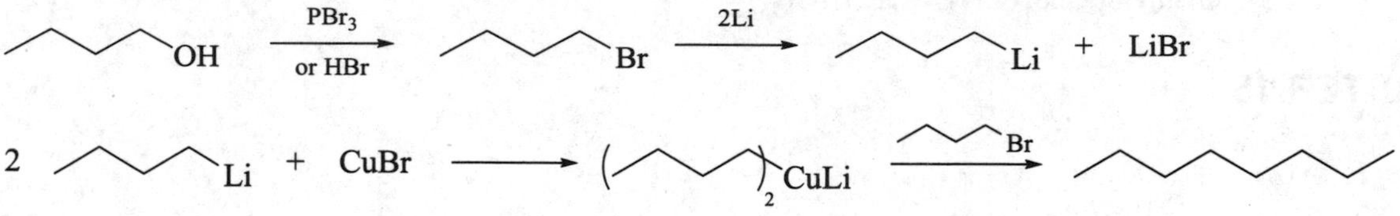

7.

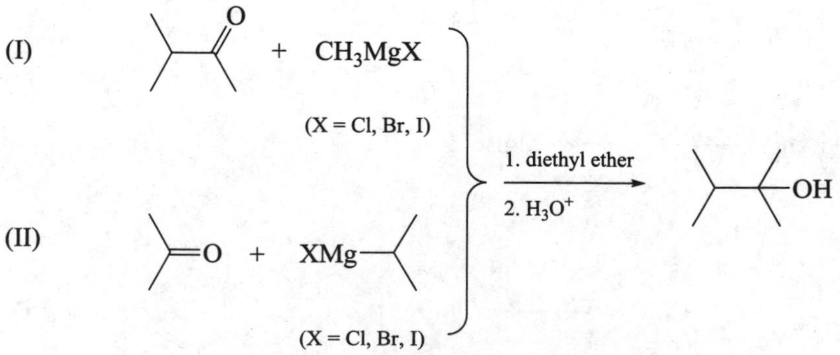

   (I)

8.

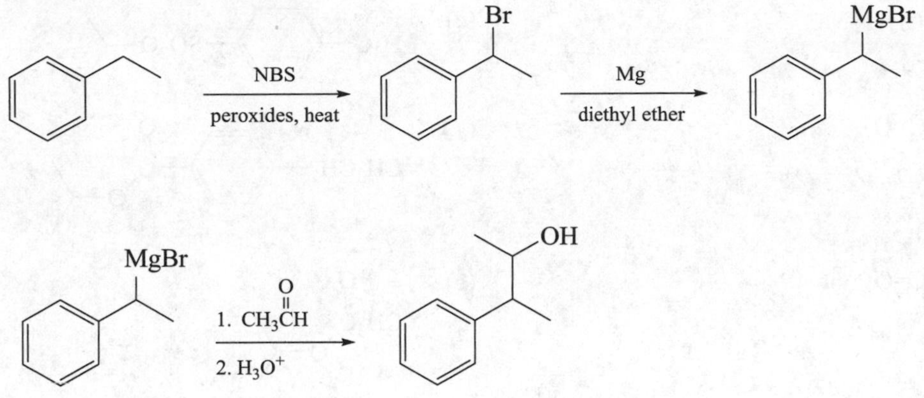

9. (a)

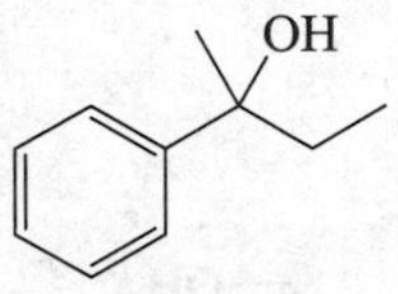

(b)

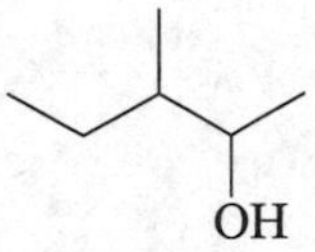

(c)

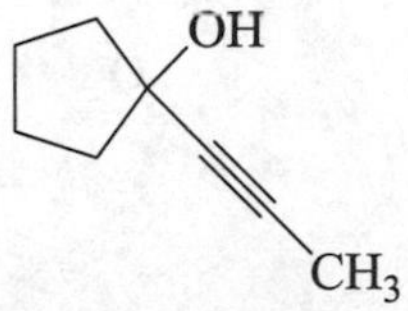

CHAPTER 16

1. (a)

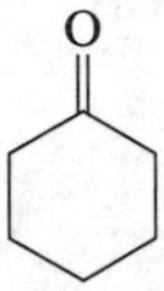

(c) OsO_4, $(CH_3)_3COOH$, $(CH_3)_3COH$, HO^-

(d) NaSH

(b) 1. B_2H_6; 2. H_2O_2, HO^-

2. (a)

(c) $(C_6H_5CH_2CH_2)_2O$

(d) $K_2Cr_2O_7$, H^+, H_2O, heat

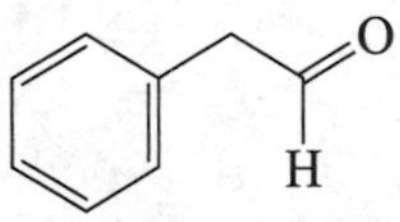

(b)

$CH_3\overset{\overset{O}{\|}}{C}Cl$, pyridine; or

$(CH_3\overset{\overset{O}{\|}}{C})_2O$, pyridine; or

CH_3CO_2H, H^+

(e)

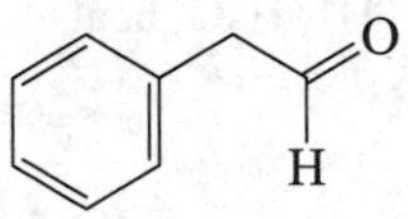

3. (*a*) $(CH_3)_2CHO^- Na^+$

(*b*) $(CH_3)_2C=O$

(*c*) $(CH_3)_2C=O$

(*d*)

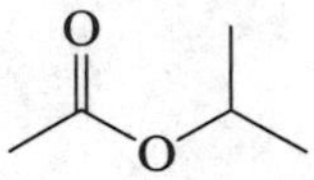

(*e*)

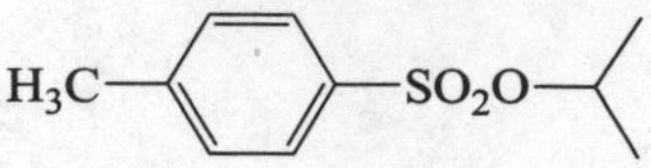

H_3C- benzene ring $-SO_2O-$ isopropyl

(*f*)

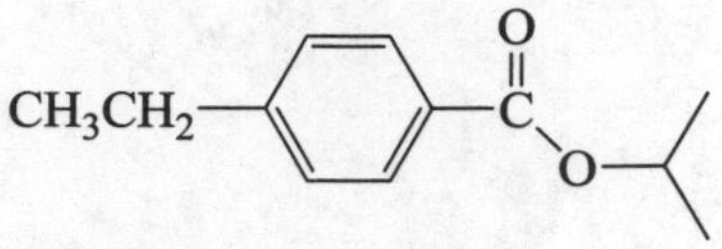

CH_3CH_2- benzene ring $-C(=O)O-$ isopropyl

(*g*)

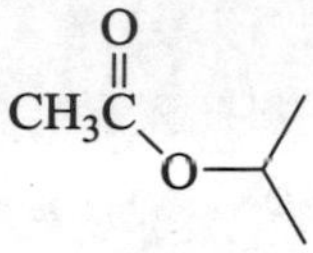

$CH_3C(=O)O-$ isopropyl

(*h*) $(CH_3)_2C=O$

4.

(I)

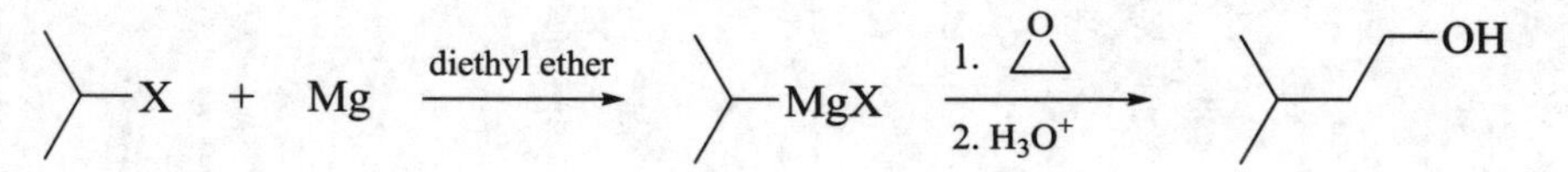

$\rangle-X + Mg \xrightarrow{\text{diethyl ether}} \rangle-MgX \xrightarrow[\text{2. H}_3\text{O}^+]{\text{1. } \triangle\text{O}} $ product —OH

(X = Cl, Br, I)

(II)

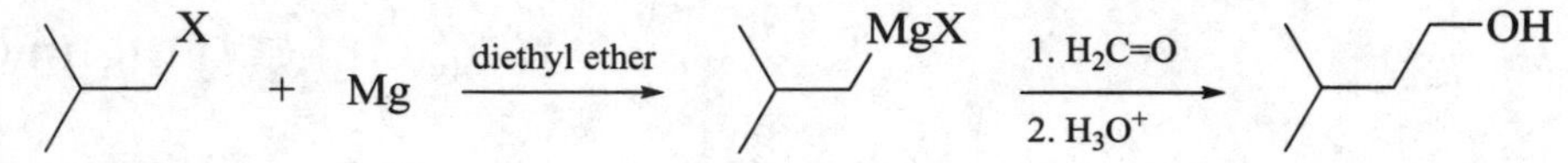

$\rangle-X + Mg \xrightarrow{\text{diethyl ether}} MgX \xrightarrow[\text{2. H}_3\text{O}^+]{\text{1. H}_2\text{C=O}} $ product —OH

(X = Cl, Br, I)

5. (*a*)

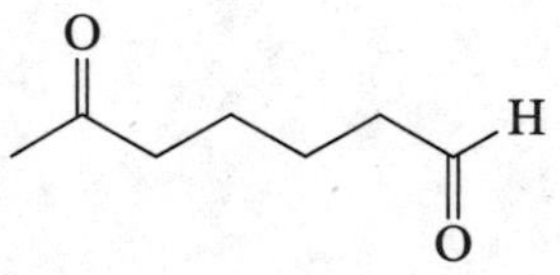

(*b*) HO—chain—OH

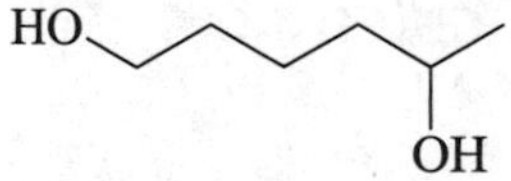

(*c*)

$\begin{matrix} H_3C & & H \\ & C=C & \\ H & & CH_3 \end{matrix}$

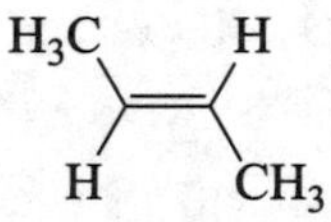

6. (*a*) PCC or PDC in CH_2Cl_2 or Swern oxidation

(*b*) $Na_2Cr_2O_7$, H^+, H_2O, heat

(*c*) 1. $LiAlH_4$; 2. H_2O

(*d*) OsO_4, $(CH_3)_3COOH$, $(CH_3)_3COH$, HO^-

7.

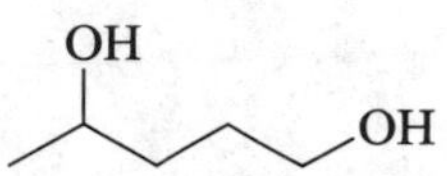

A ($C_5H_{12}O_2$)

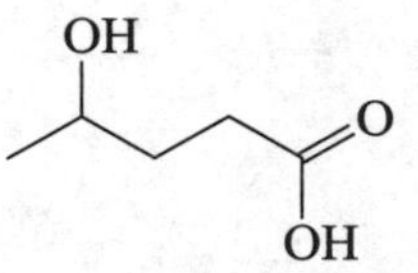

B ($C_5H_8O_3$)

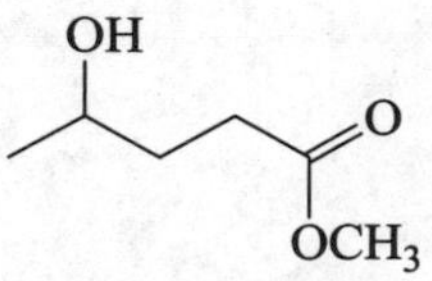

C ($C_6H_{10}O_3$)

8. (*a*)

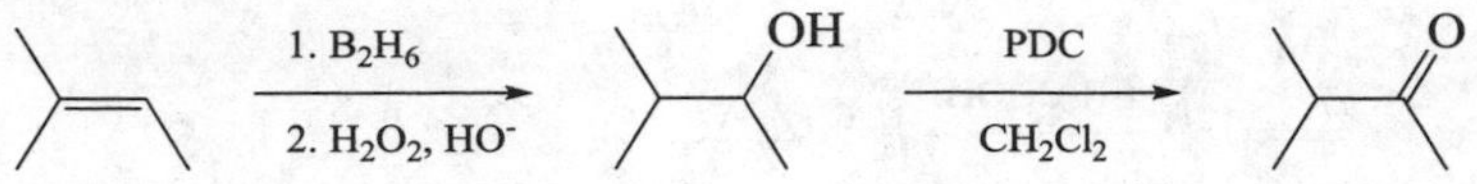

(*b*)

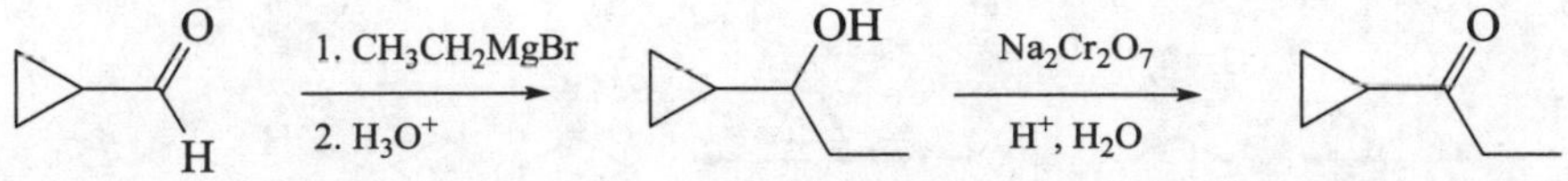

(*c*)

$$C_6H_5CH_3 \xrightarrow[\text{peroxides, heat}]{\text{NBS}} C_6H_5CH_2Br \xrightarrow{\text{Mg}} C_6H_5CH_2MgBr$$

$$\Big\downarrow \begin{array}{l} 1.\ H_2C\!-\!CH_2 \\ 2.\ H_3O^+ \end{array}$$

$$C_6H_5CH_2CH_2CO_2CH_2CH_3 \xleftarrow[\text{H}^+]{\text{CH}_3\text{CH}_2\text{OH}} C_6H_5CH_2CH_2CO_2H \xleftarrow[\substack{\text{H}^+,\ \text{H}_2\text{O} \\ \text{heat}}]{\text{K}_2\text{Cr}_2\text{O}_7} C_6H_5CH_2CH_2CH_2OH$$

CHAPTER 17

1. $CH_3OCH_2CH_2CH_3$ $CH_3OCH(CH_3)_2$ $CH_3CH_2OCH_2CH_3$

Methyl propyl ether   Isopropyl methyl ether   Diethyl ether

2. (*a*)

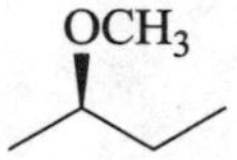

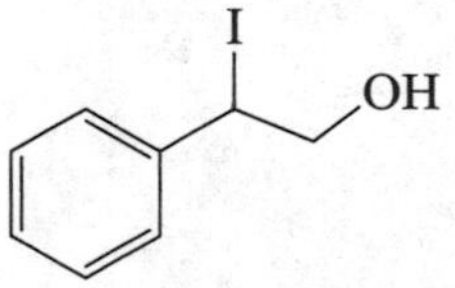

(*d*)

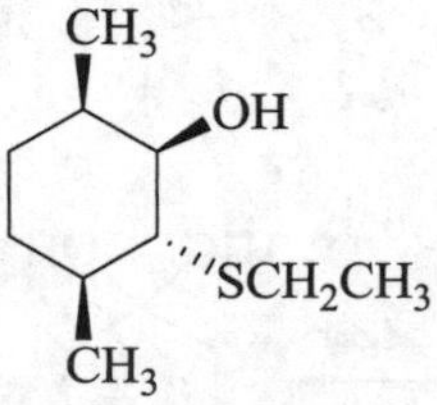

+ enantiomer

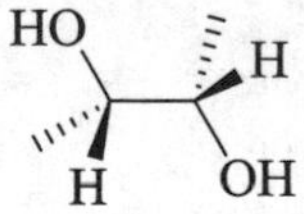

(*b*)

+ enantiomer

(*e*)

(*c*)

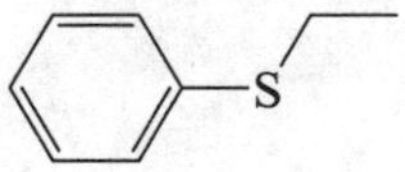

(*f*)

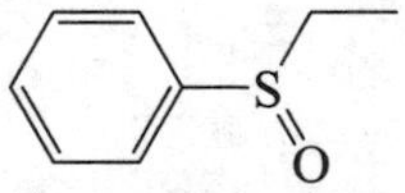

3.

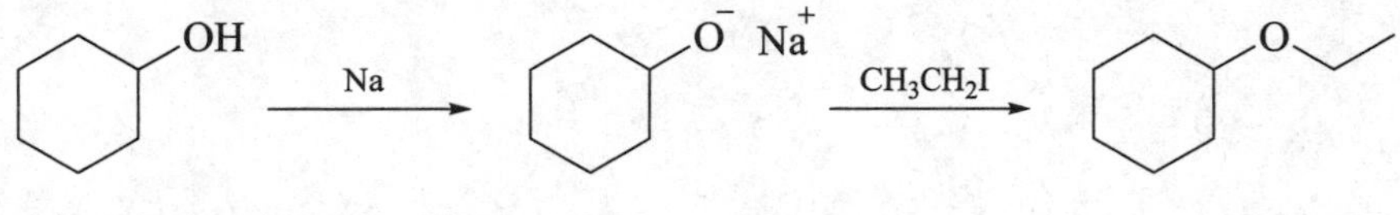

4.

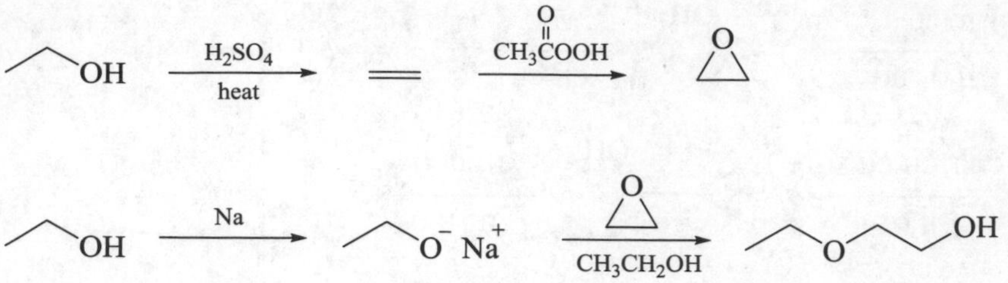

5. (*a*)

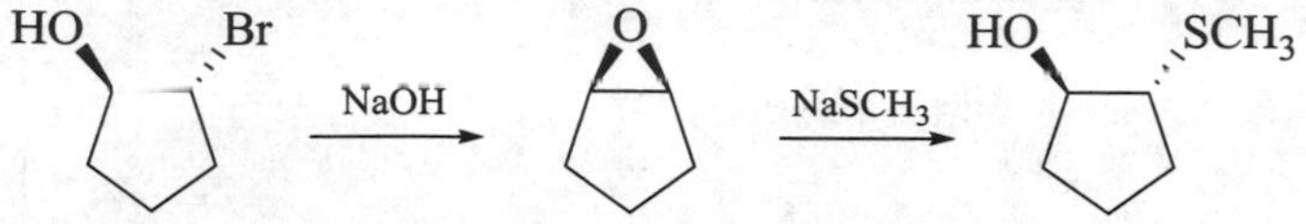

(*b*)

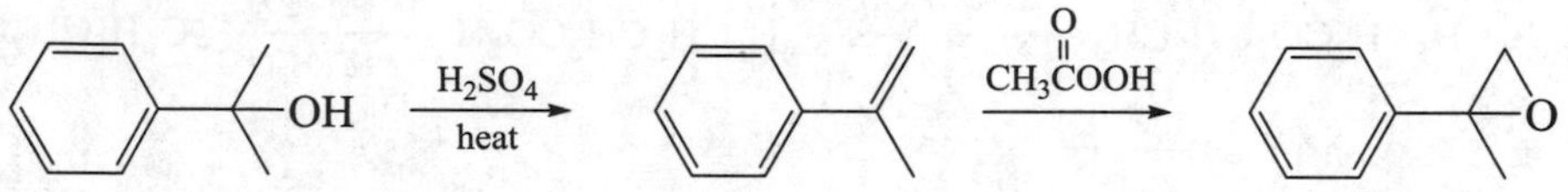

6.

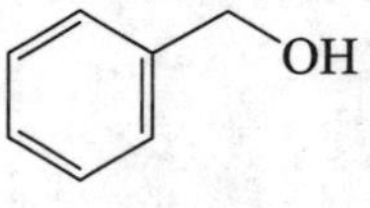

A (C₇H₈O)   B (C₉H₁₂O)

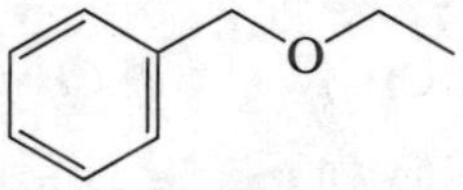

7.

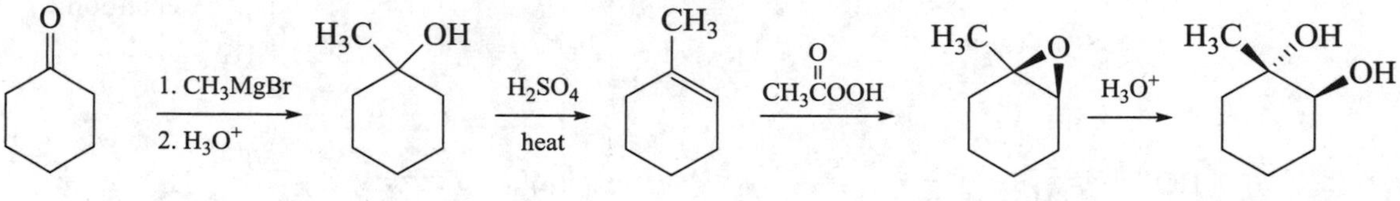

(+ enantiomer)   (+ enantiomer)

8.

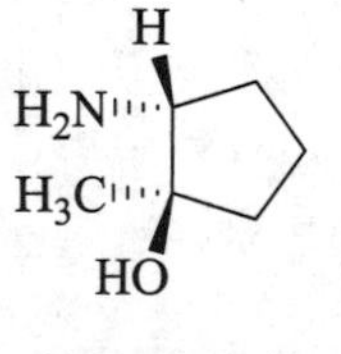

(+ enantiomer)

CHAPTER 18

1. (*a*) 3,4-Dimethylhexanal

(*b*) 2,2,5-Trimethylhexan-3-one

(*c*) *trans*-4-Bromo-2-methylcyclohexanone

(*d*) 4-Methyl-3-penten-2-one

2. (a)

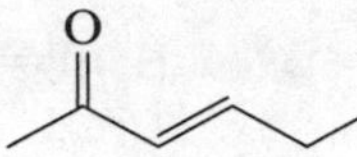

(b)

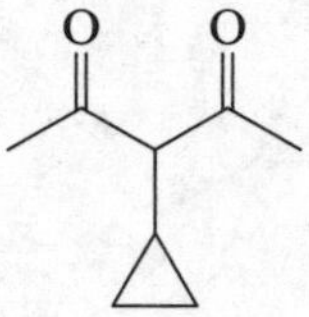

(c)

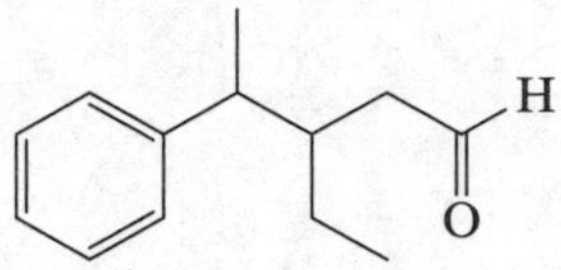

3. (a)

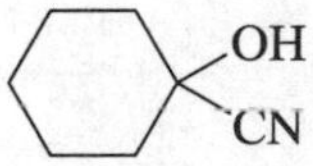

(e)

(b) NH_2OH

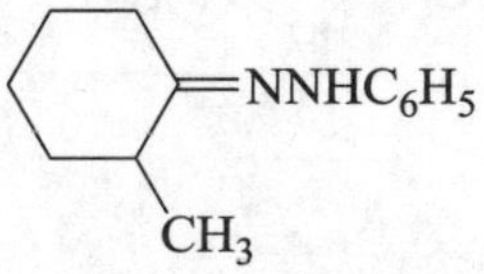

(f) $CH_3CH_2CH_2CH(OCH_2CH_3)_2$

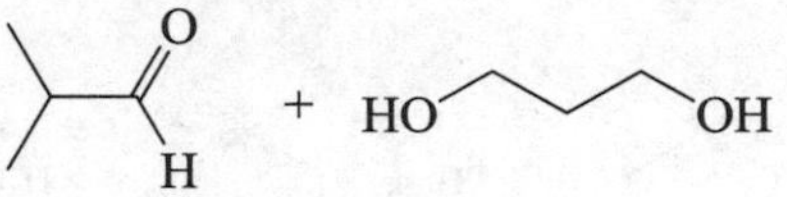

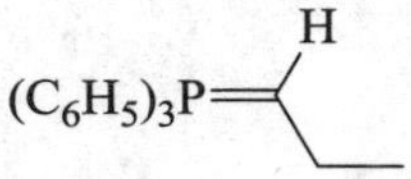

(c)

+

(g)

+ $HN(CH_3)_2$

(d)

$(C_6H_5)_3P$

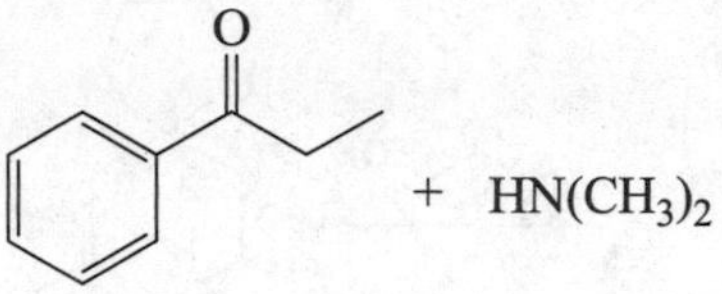

(h)

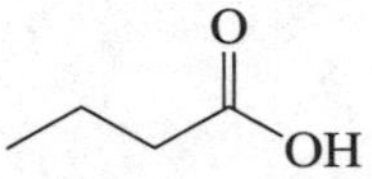

4. (a) $(C_6H_5)_3\overset{+}{P}-CH_2CH(CH_3)_2$ Br^- $(C_6H_5)_3P=CHCH(CH_3)_2$ $(C_6H_5)CH=CHCH(CH_3)_2$

   A   B   C

(b)

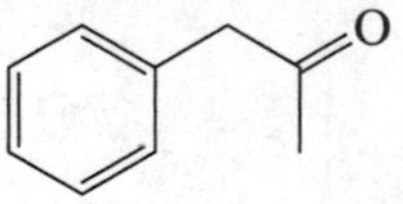

   D

5. (a) 1. CH_3MgI
         2. H_3O^+
         3. H_2SO_4, heat

(c) $HOCH_2CH_2OH$, H^+ (cat), heat

(b) $(C_6H_5)_3P=CH_2$ [from $(C_6H_5)_3P$ + CH_3I $\xrightarrow{}$ $\xrightarrow{C_4H_9Li}$]

6. (a)

+ HO OH

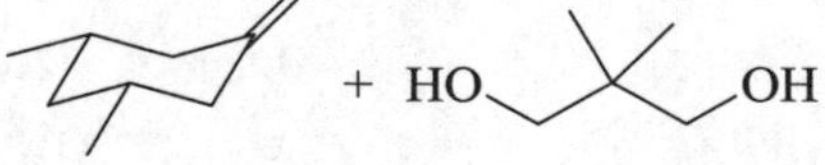

(b)

$\equiv$

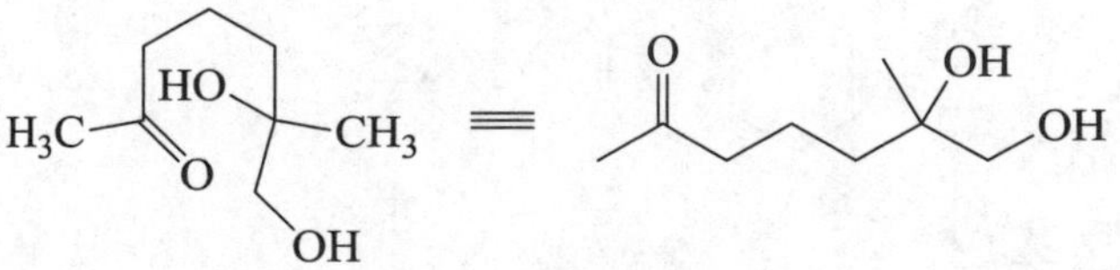

7. (*a*)

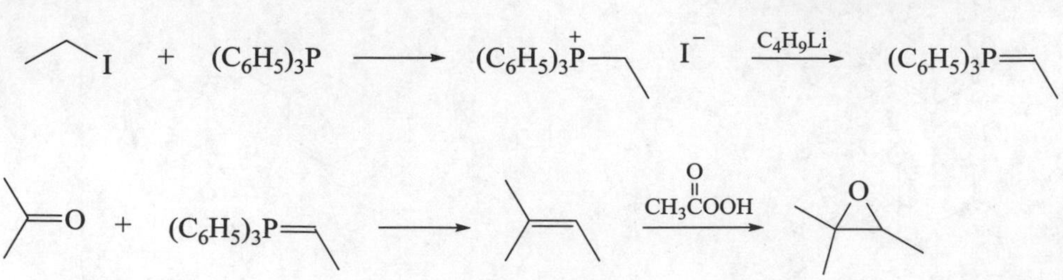
I
$(C_6H_5)_3P$
C_4H_9Li
CH_3COOH

(*b*)

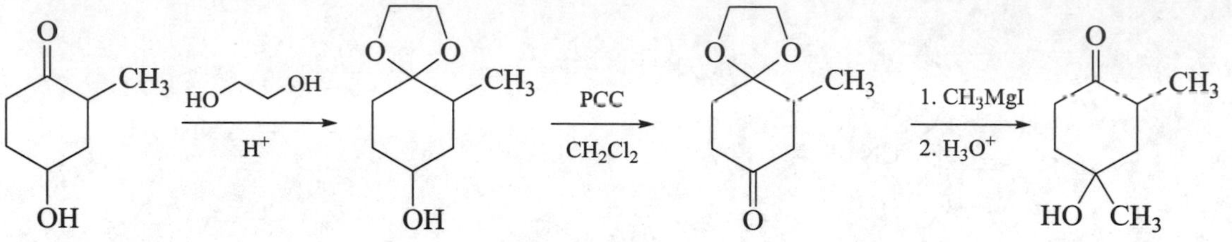
H^+
PCC
CH_2Cl_2
1. CH_3MgI
2. H_3O^+

(*c*)

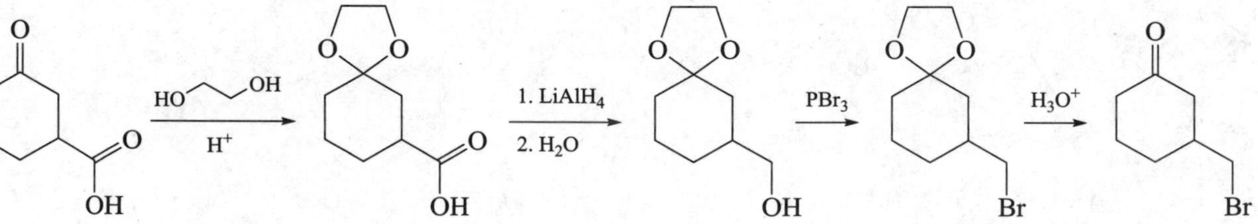
H^+
1. $LiAlH_4$
2. H_2O
PBr_3
H_3O^+

8.

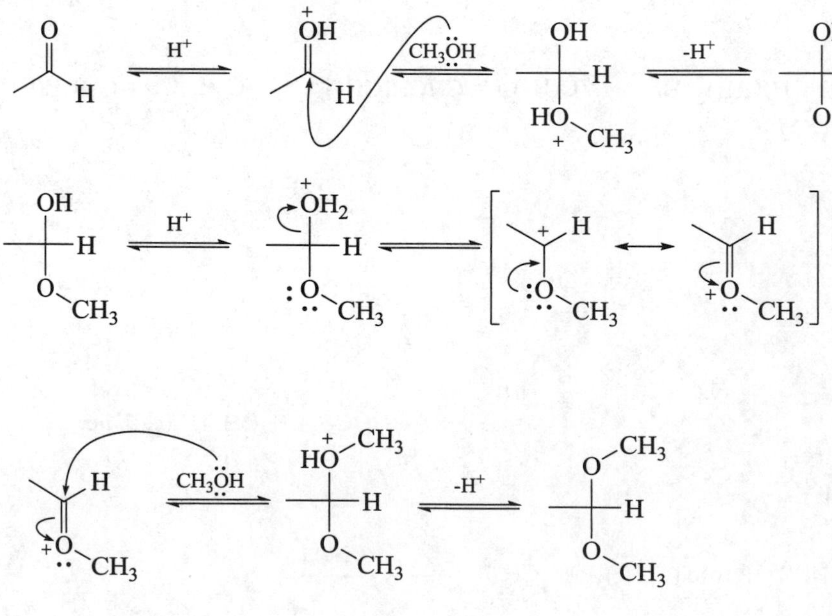
H^+
$CH_3\ddot{O}H$
$-H^+$

9.

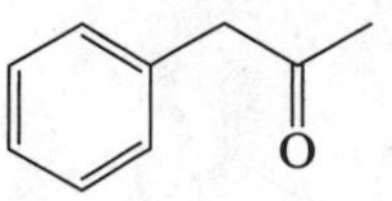

CHAPTER 19

1. (a) 4-Methyl-5-phenylhexanoic acid (c) 3-Bromo-2-ethylbutanoic acid

   (b) Cyclohexanecarboxylic acid

2.

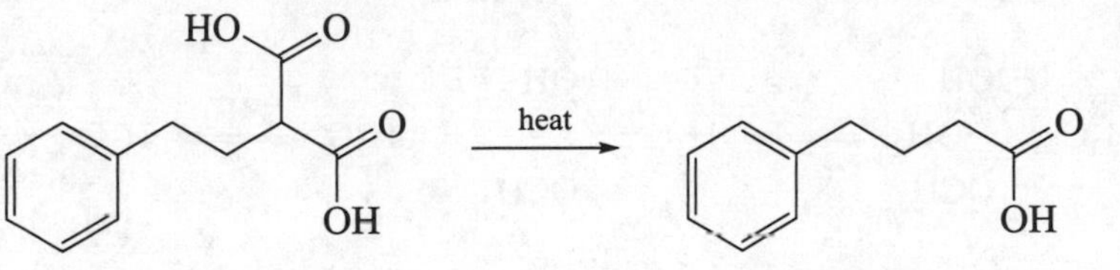

   4-Phenylbutanoic acid

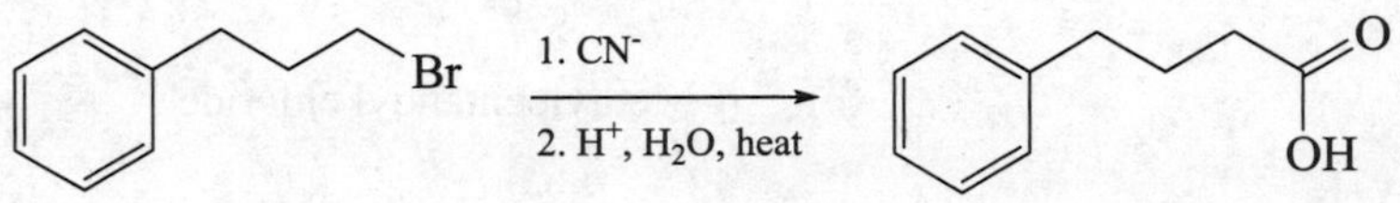

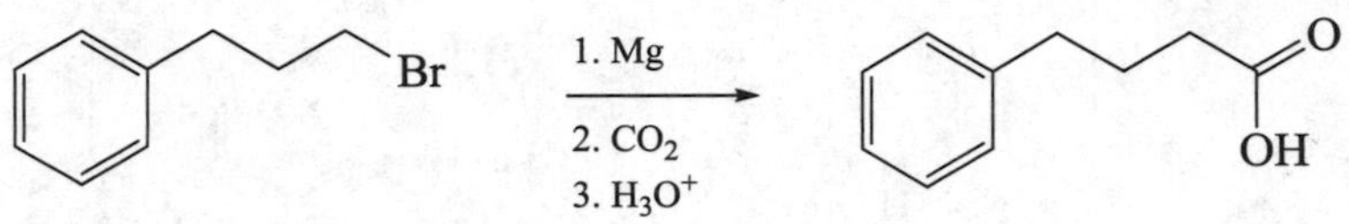

3.

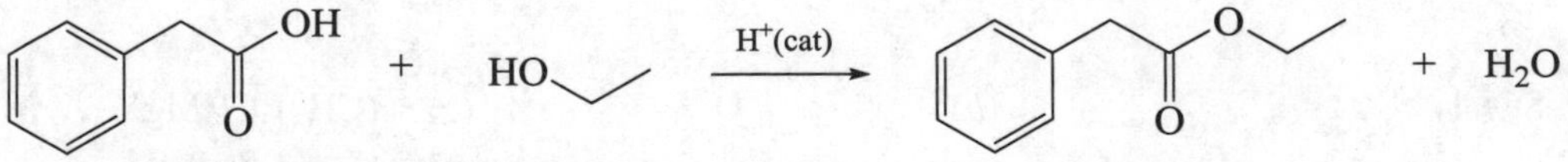

4. (a) (c)

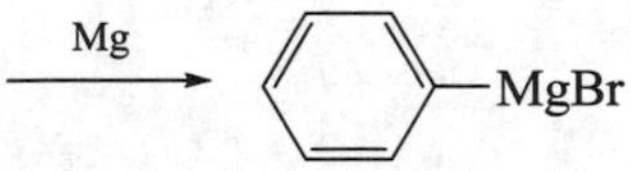

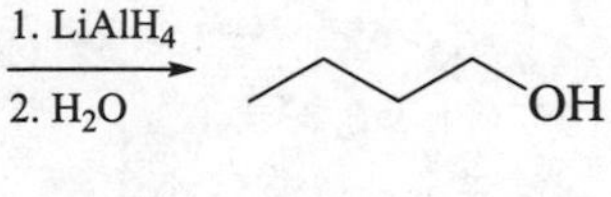

   (b)

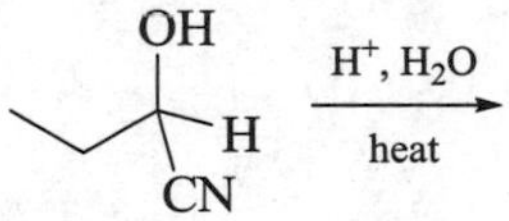

5.

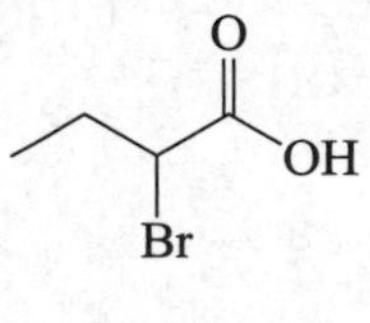

6.

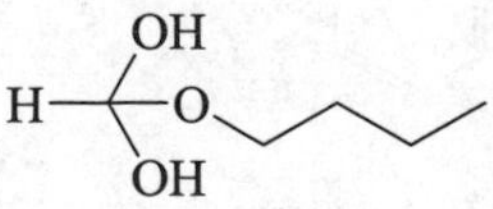

7. 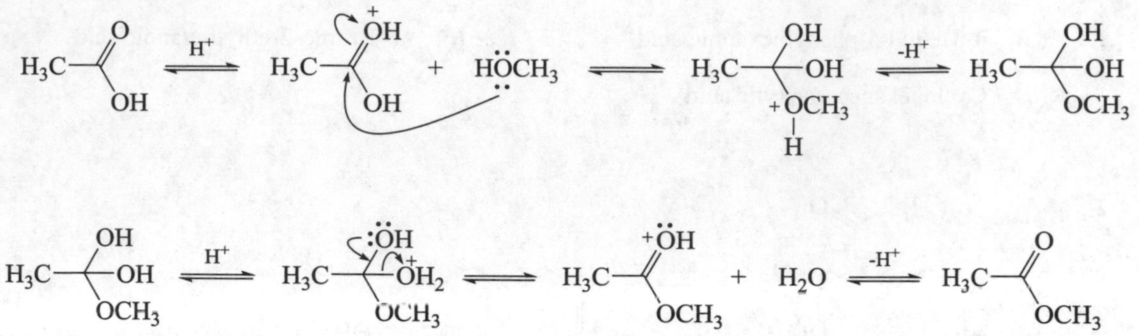

CHAPTER 20

1. (*a*) Propyl butanoate
   (*b*) *N*-Methylbenzamide
   (*c*) 4-Methylpentanoyl chloride

2. (*a*) 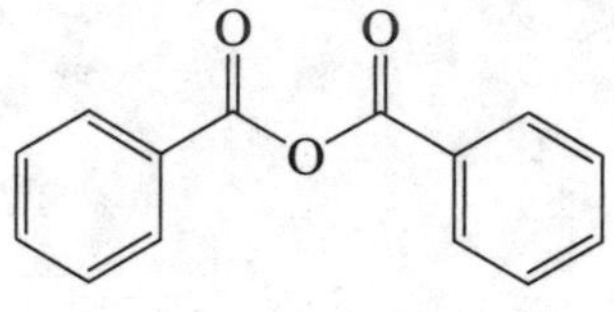
   (*b*) 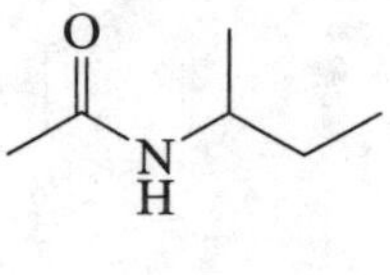
   (*c*) 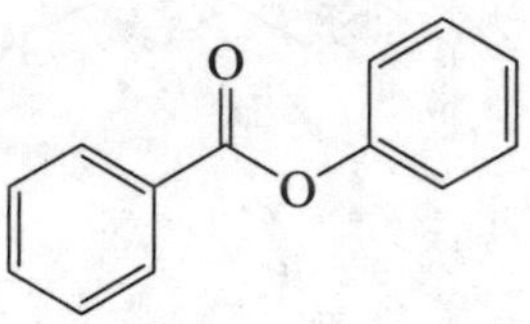

3. (*a*) $SOCl_2$
   (*b*) 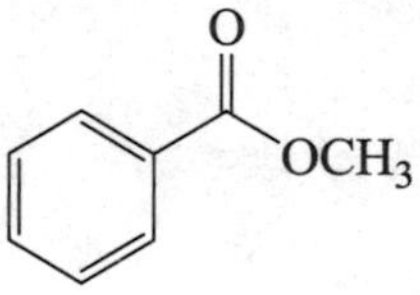
   (*c*) $(CH_3)_2CHMgX$
   (X = Cl, Br, I)

4. (*a*) 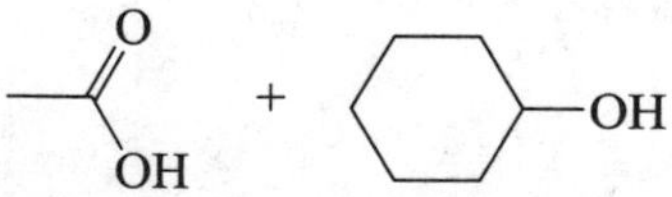
   (*b*) 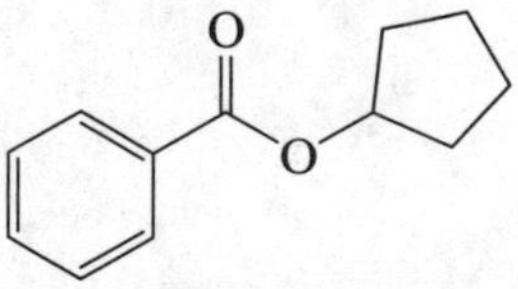
   (*c*) 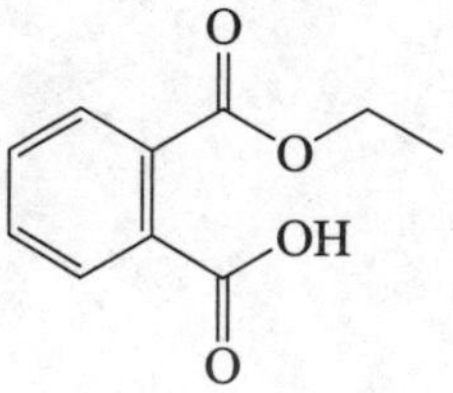
   (*d*) 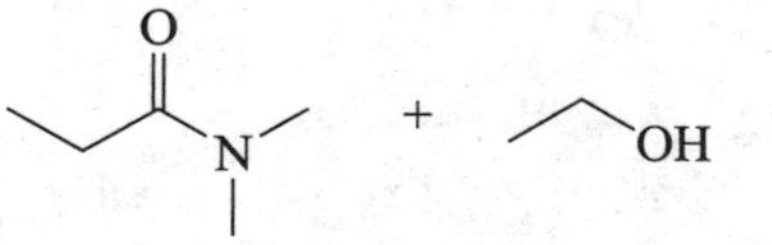
   (*e*) 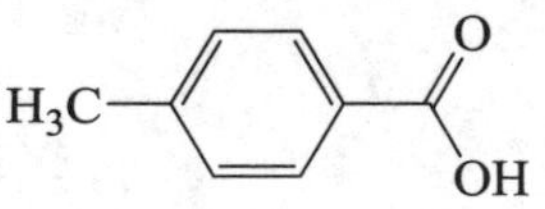 + $CH_3\overset{+}{N}H_3\ HSO_4^-$
   (*f*) 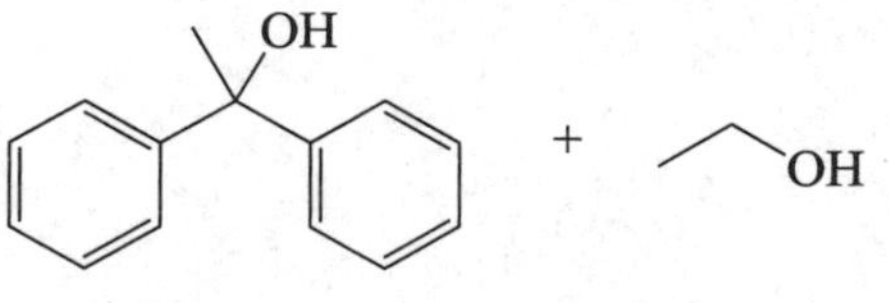

5.

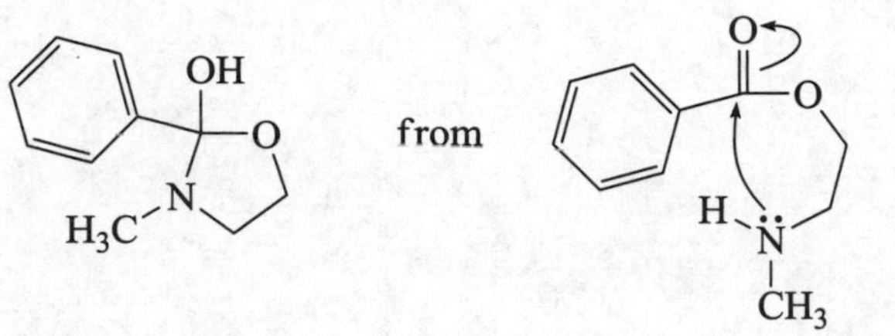
OH
O
N
H_3C
from
O
O
H
N
CH_3

6.

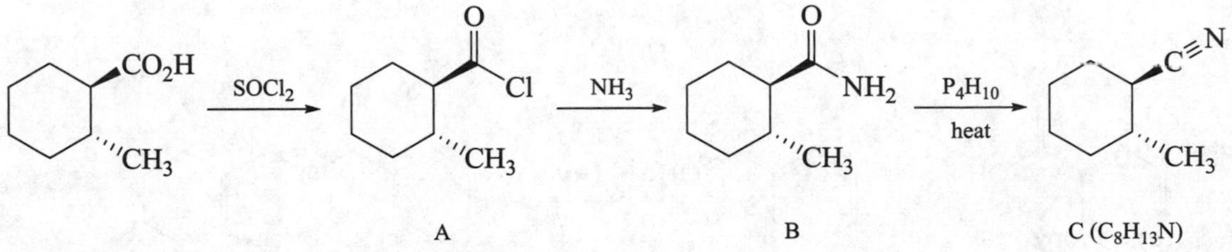
CO_2H
CH_3
$SOCl_2$
O
Cl
CH_3
NH_3
O
NH_2
CH_3
P_4H_{10}
heat
C≡N
CH_3
A
B
C ($C_8H_{13}N$)

7. (*a*)

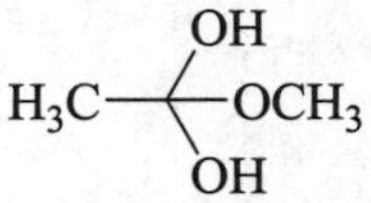
OH
H_3C
OCH_3
OH

(*b*)

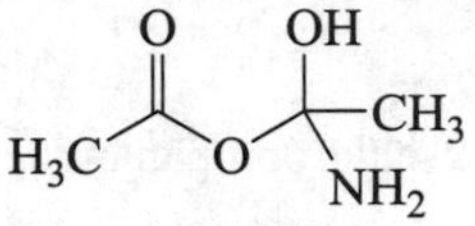
O
OH
CH_3
H_3C
O
NH_2

8.

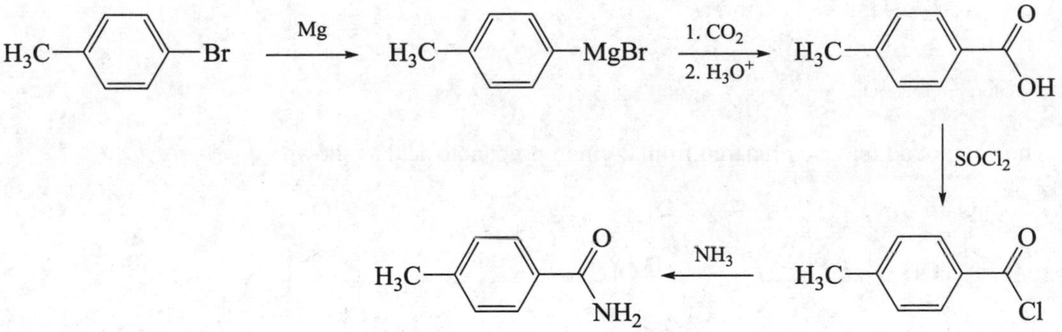
H_3C
Br
Mg
H_3C
MgBr
1. CO_2
2. H_3O^+
H_3C
O
OH
$SOCl_2$
H_3C
O
NH_2
NH_3
H_3C
O
Cl

9.

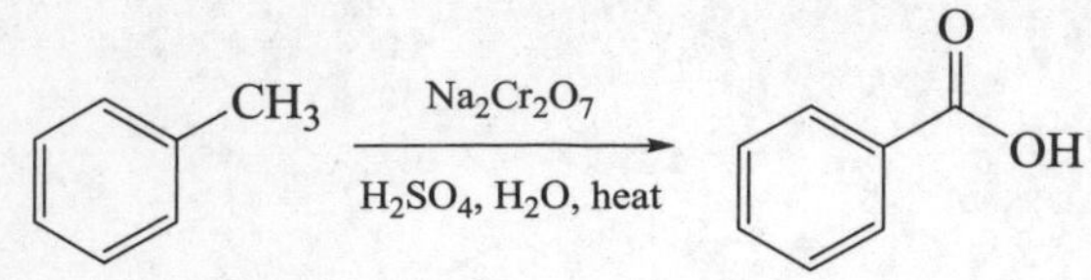

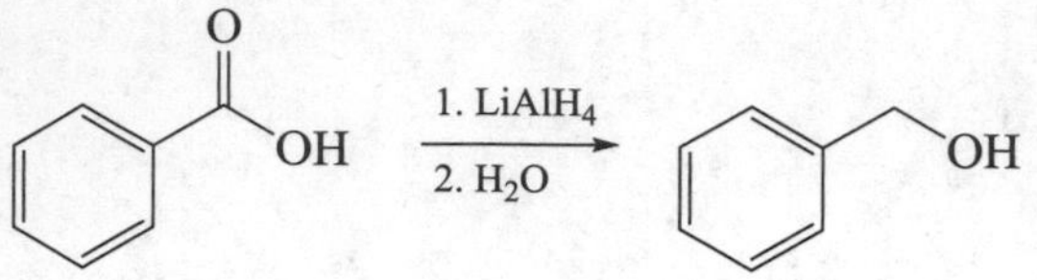

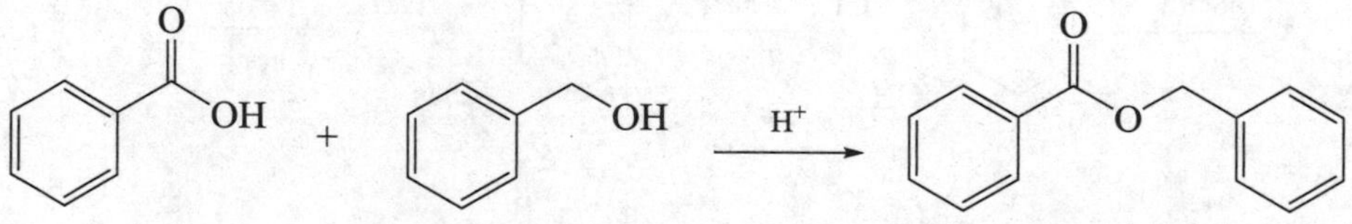

Benzyl benzoate

10. The compound is 2-chloropropanamide.

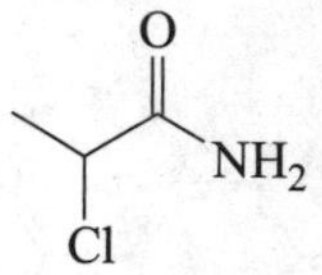

2-Chloropropanamide

The compound may be prepared from 2-chloropropanoic acid as shown.

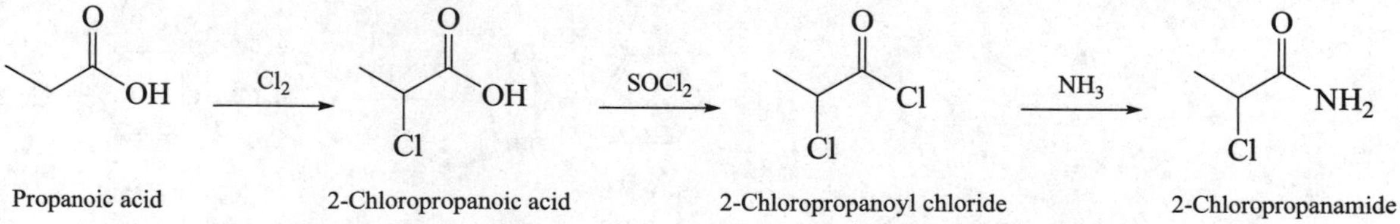

| Propanoic acid | 2-Chloropropanoic acid | 2-Chloropropanoyl chloride | 2-Chloropropanamide |

CHAPTER 21

1. (*a*) 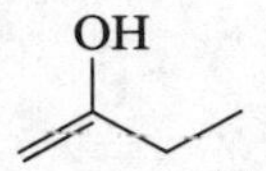 and 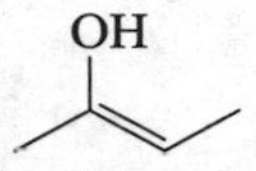   (*c*)

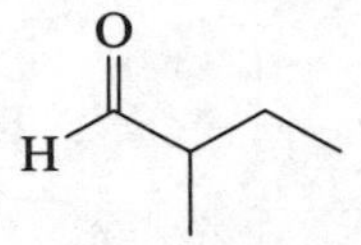

(*b*)

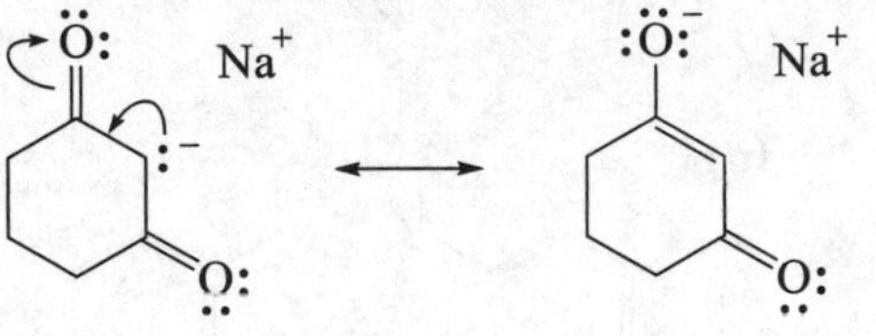

2. (*a*)

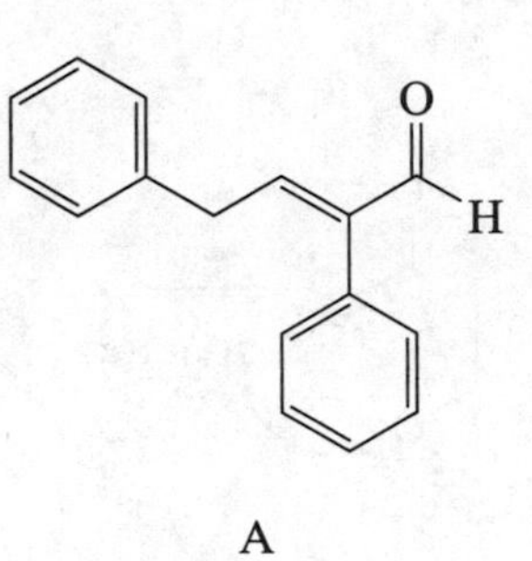

A

(*E*/*Z* mixture)

(*b*)

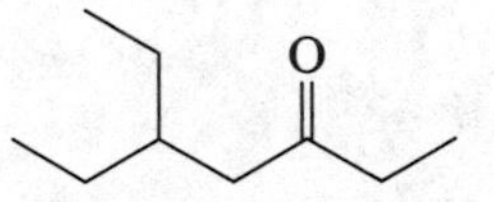

B

3.

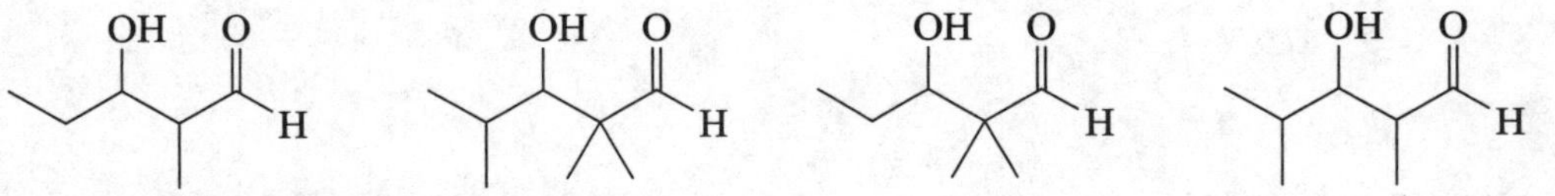

4.

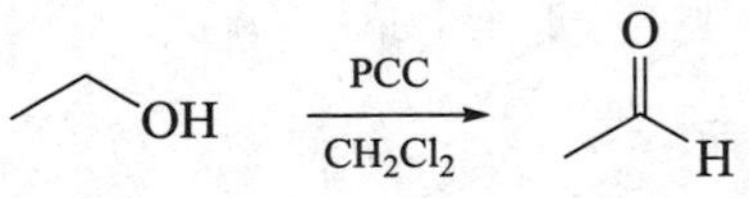

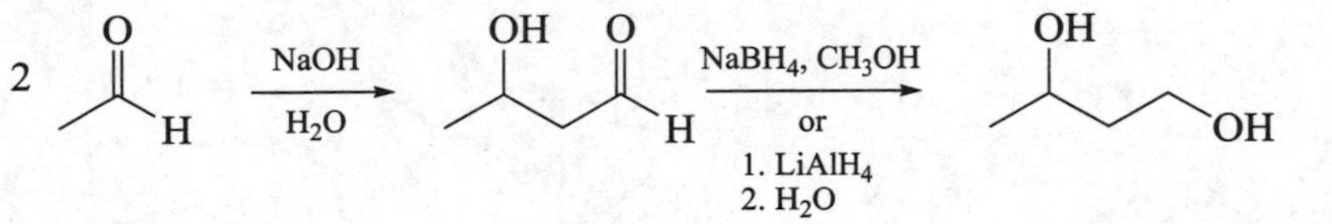

5.

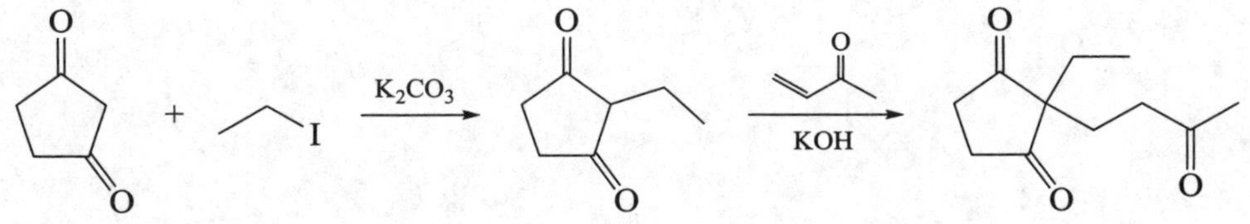

A   B

Note: The order of the two steps may be reversed.

6. (a)

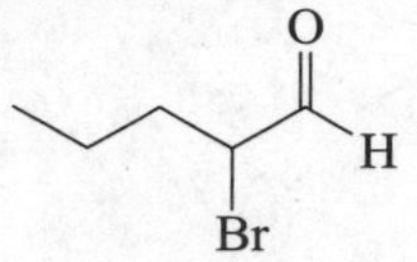

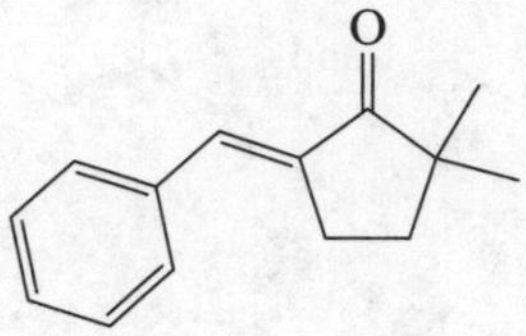

(b)

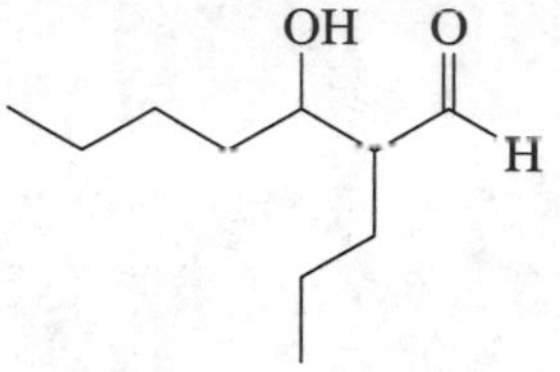

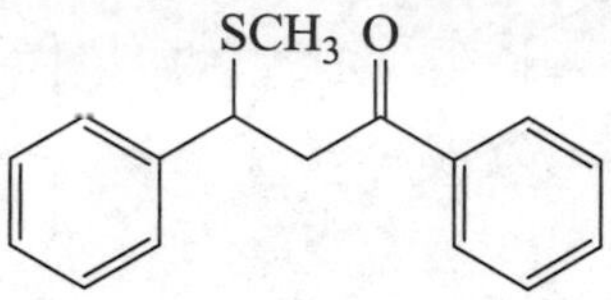

7.

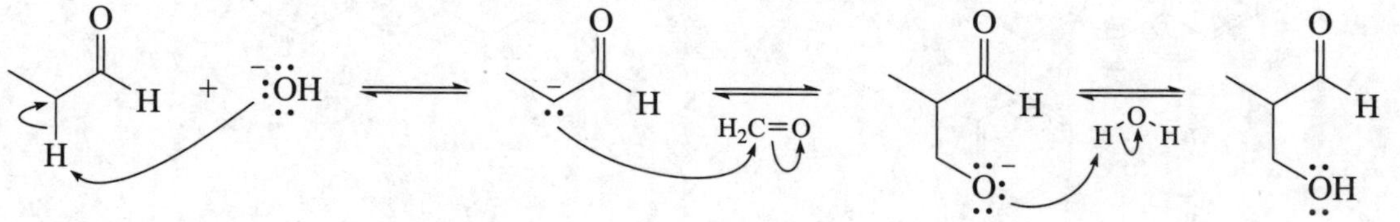

$H_2C=O$

8.

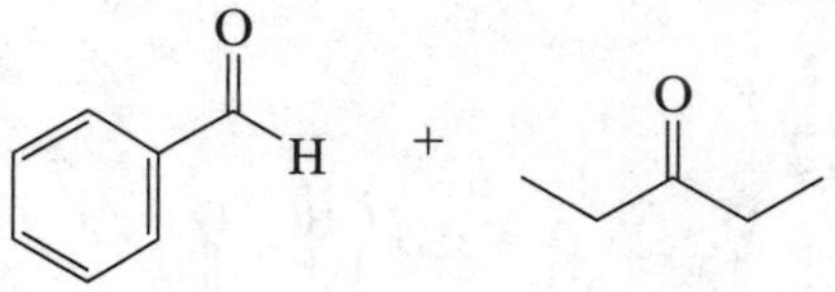

9. (a)

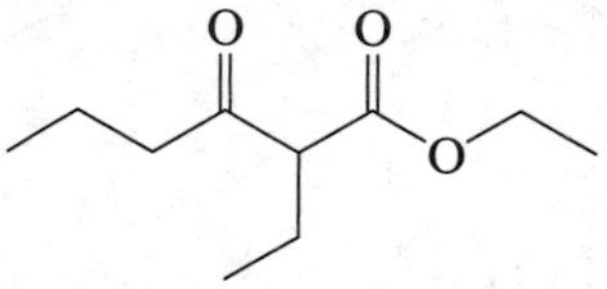

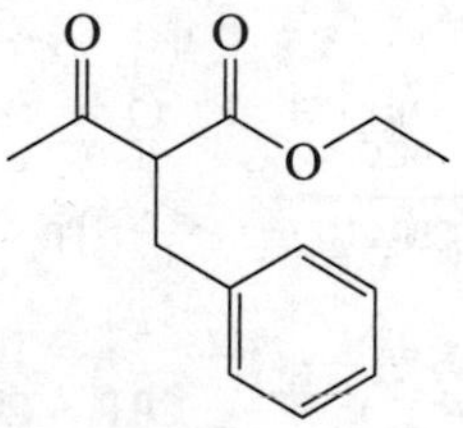

(b)

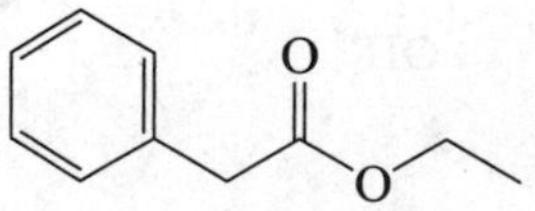

(c)

(c)

(d)

(e)

(f) 1. HO⁻, H_2O
2. H_3O^+
3. heat

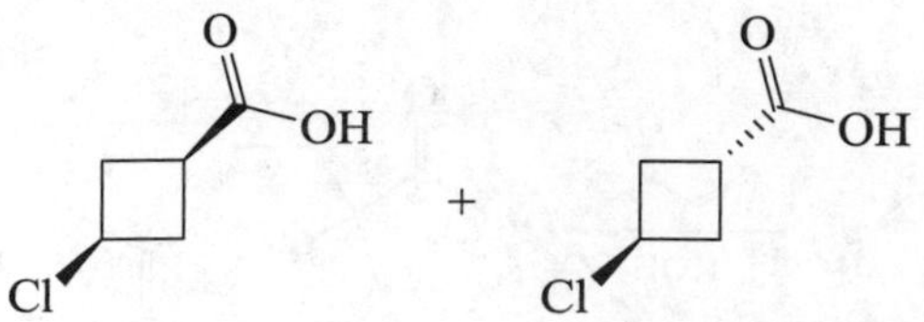

(g)

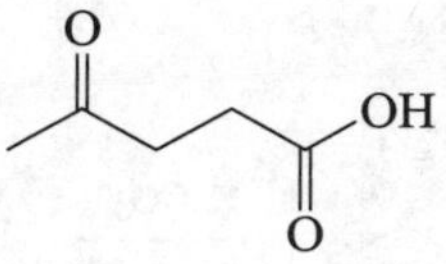

(d)

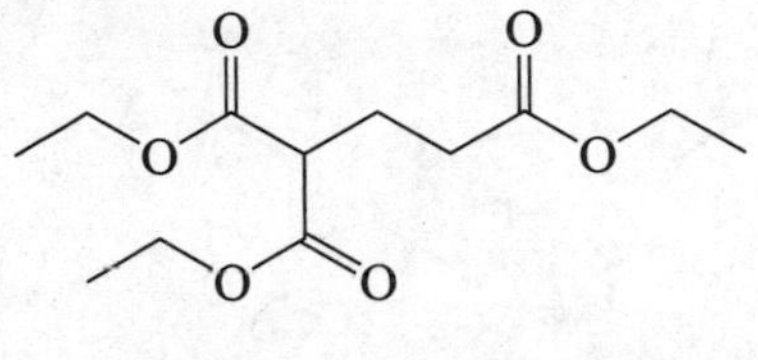

10. (a)

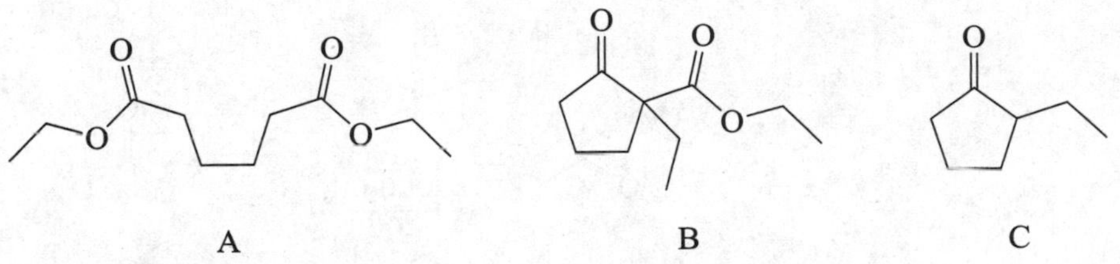

A B C

(b)

D E

11. (a)

1. NaOCH$_2$CH$_3$

2. Br—CH$_2$CO$_2$CH$_2$CH$_3$

1. HO$^-$, H$_2$O
2. H$_3$O$^+$
3. heat

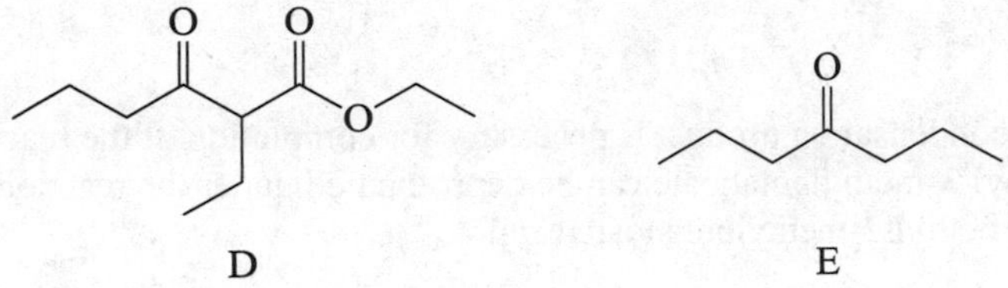

(b)

+

NaOCH$_2$CH$_3$

1. NaOCH$_2$CH$_3$
2.

1. HO$^-$, H$_2$O
2. H$_3$O$^+$
3. heat

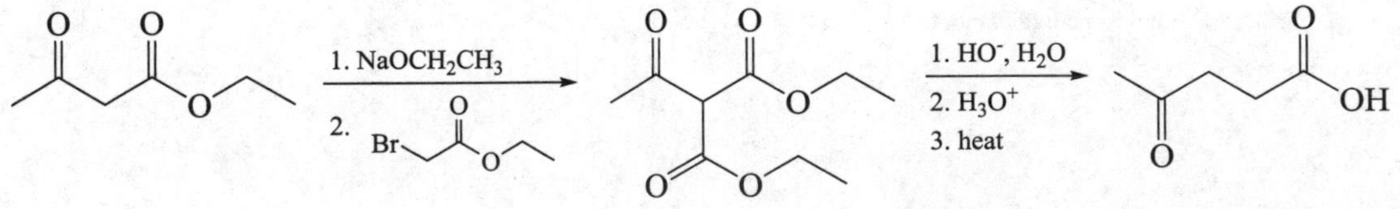

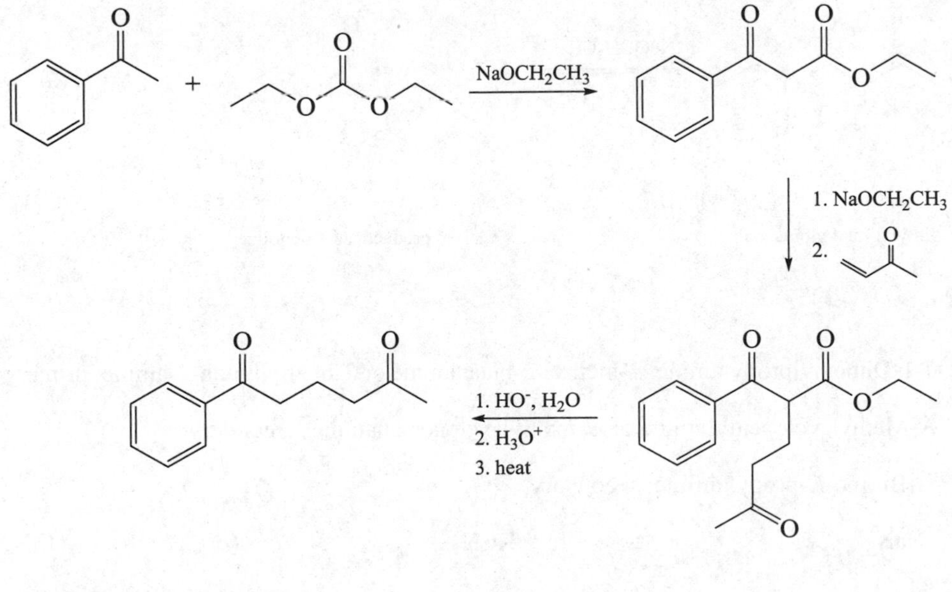

12.

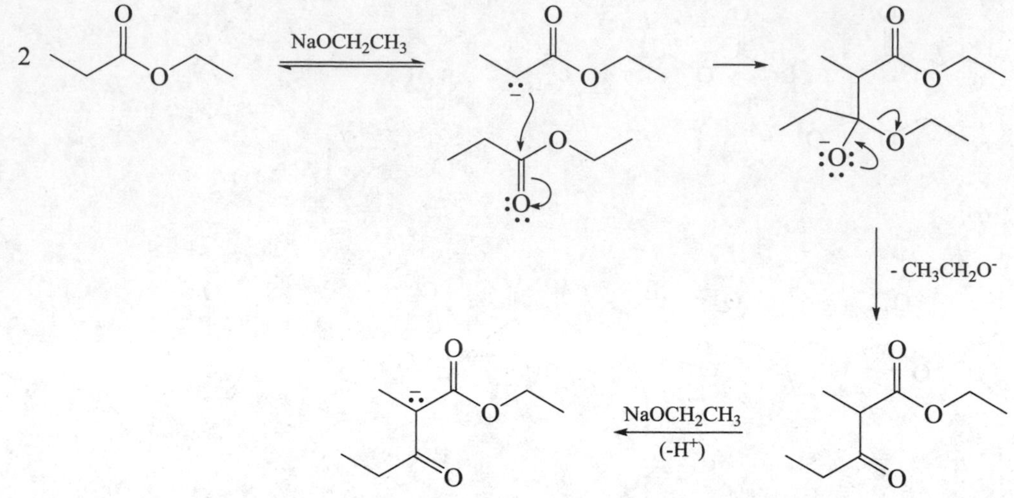

13. α-Deprotonation of the Claisen condensation product is necessary for completion of the reaction. The condensation product of ethyl 3-methylbutanoate can be deprotonated under the reaction conditions; the product from condensation of ethyl 2-methylbutanoate cannot.

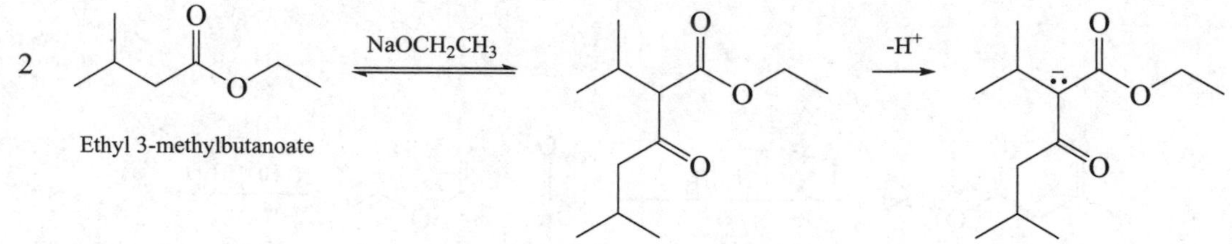

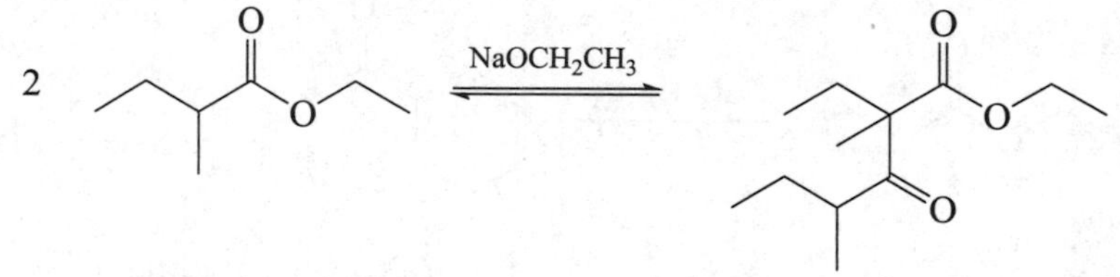

Ethyl 3-methylbutanoate

Ethyl 2-methylbutanoate   Claisen product cannot enolize

CHAPTER 22

1. (a) 1,1-Dimethylpropylamine, 2-methyl-2-butanamine or 2-methylbutan-2-amine; primary

    (b) *N*-Methylcyclopentylamine or *N*-methylcyclopentanamine; secondary

    (c) *m*-Bromo-*N*-propylaniline; secondary

2. (a) NaN_3 (b) KCN   (c)

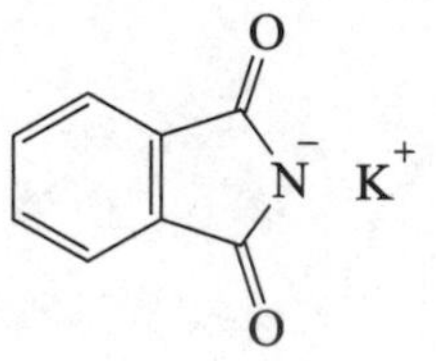

3. (*a*)

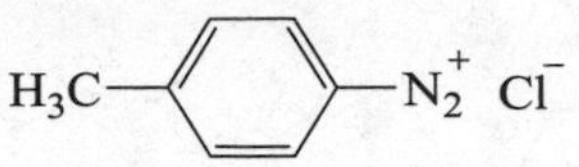

H₃C—⟨benzene ring⟩—N₂⁺ Cl⁻

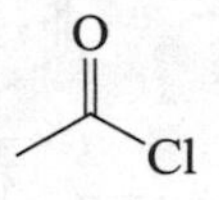

(*e*)

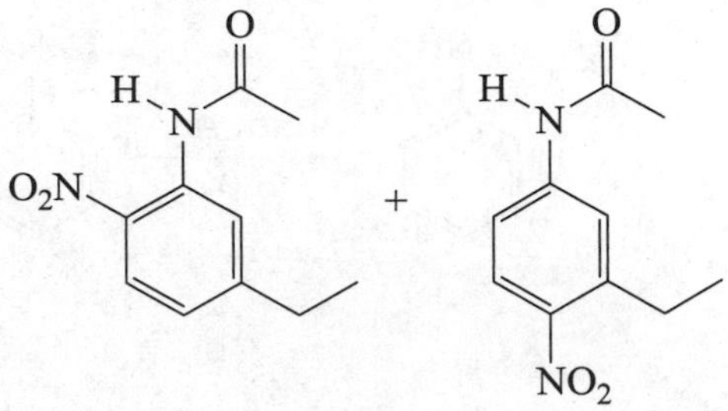

⟨structure: 2-nitro-5-ethyl-acetanilide⟩ + ⟨structure: 4-nitro-3-ethyl-acetanilide⟩

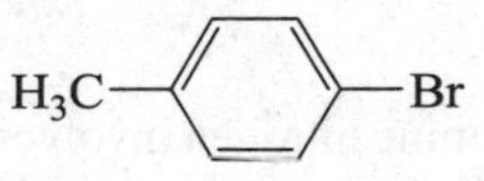

(*b*)

H₃C—⟨benzene ring⟩—Br

(*f*)

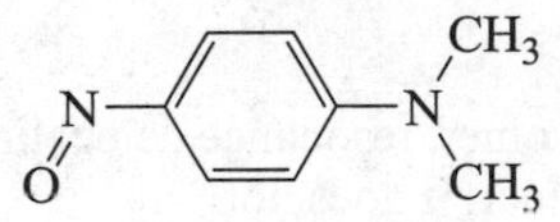

O=N—⟨benzene ring⟩—N(CH₃)(CH₃)

(*c*)   H₃PO₂ or CH₃CH₂OH

(*g*)

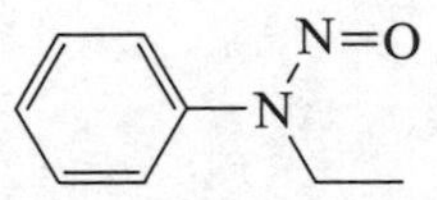

⟨phenyl⟩—N(ethyl)—N=O

(*d*)

⟨acetic anhydride structure⟩ or ⟨acetyl chloride structure⟩

4. (*a*)

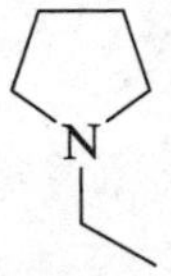

⟨pyrrolidine, N-ethyl⟩   ⟨N-methyl-N-ethyl pyrrolidinium⟩ I⁻   ⟨N-methyl-N-ethyl pyrrolidinium⟩ HO⁻

     A 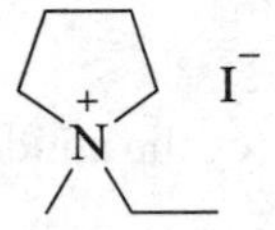   B 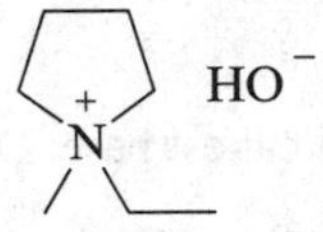   C

(*b*)

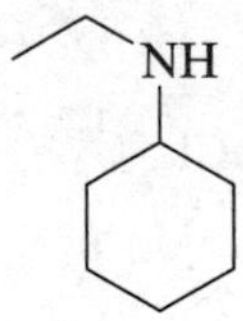

⟨N-ethyl cyclohexylamine⟩   ⟨N-ethyl-N-nitroso cyclohexylamine⟩

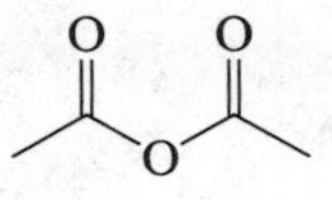

     C 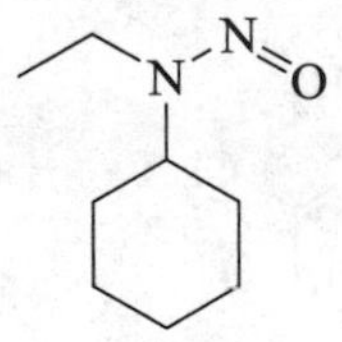   D

5. (*a*)

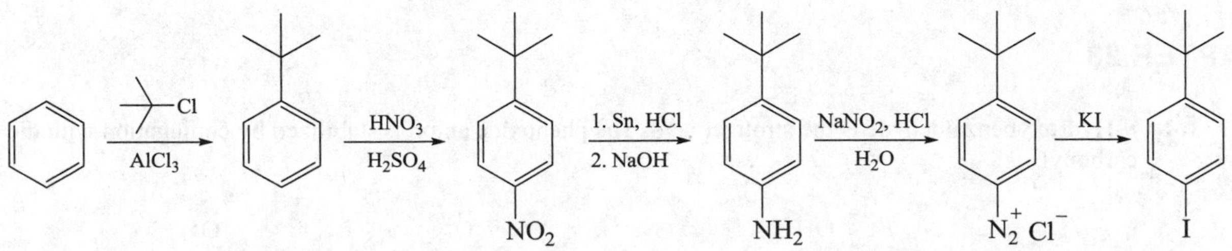

⟨benzene⟩ —[(CH₃)₃CCl / AlCl₃]→ ⟨tert-butylbenzene⟩ —[HNO₃ / H₂SO₄]→ ⟨4-tert-butylnitrobenzene, NO₂⟩ —[1. Sn, HCl / 2. NaOH]→ ⟨4-tert-butylaniline, NH₂⟩ —[NaNO₂, HCl / H₂O]→ ⟨diazonium, N₂⁺ Cl⁻⟩ —[KI]→ ⟨4-tert-butyliodobenzene, I⟩

(*b*)

⟨benzene⟩ —[HNO₃ / H₂SO₄]→ ⟨nitrobenzene, NO₂⟩ —[Cl₂, FeCl₃]→ ⟨3-chloronitrobenzene, Cl / NO₂⟩ —[1. Sn, HCl / 2. NaOH]→ ⟨3-chloroaniline, Cl / NH₂⟩

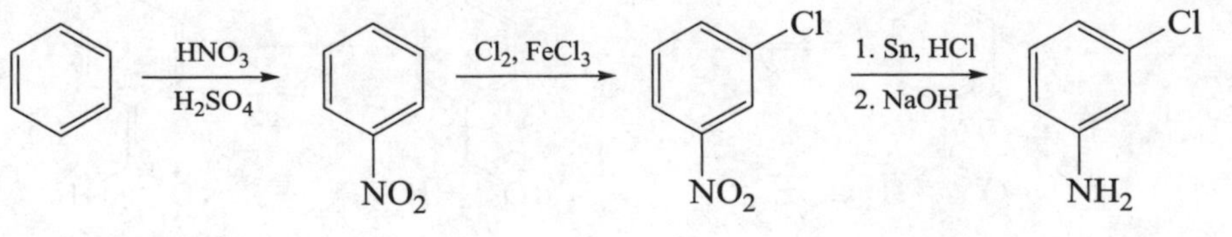

(c)

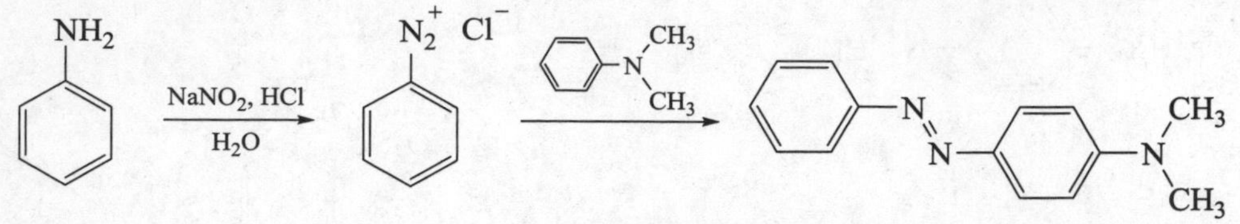

6.   In the para isomer, resonance delocalization of the electron pair of the amine nitrogen involves the nitro group. In the meta isomer, it does not.

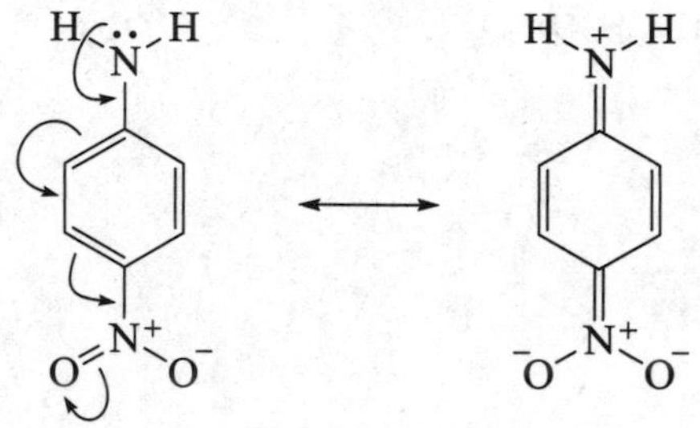

7.   Strongest base: C, an alkylamine; Weakest base: D, a lactam (cyclic amide)

8.

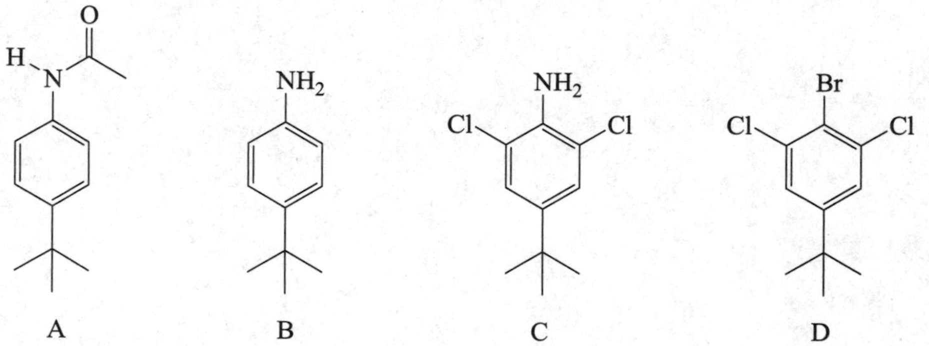

CHAPTER 23

1.   *p*-Hydroxybenzaldehyde is the stronger acid. The phenoxide anion is stabilized by conjugation with the aldehyde carbonyl.

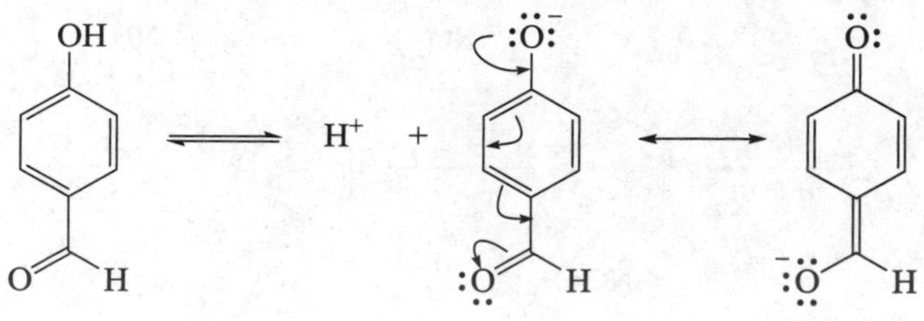

2.

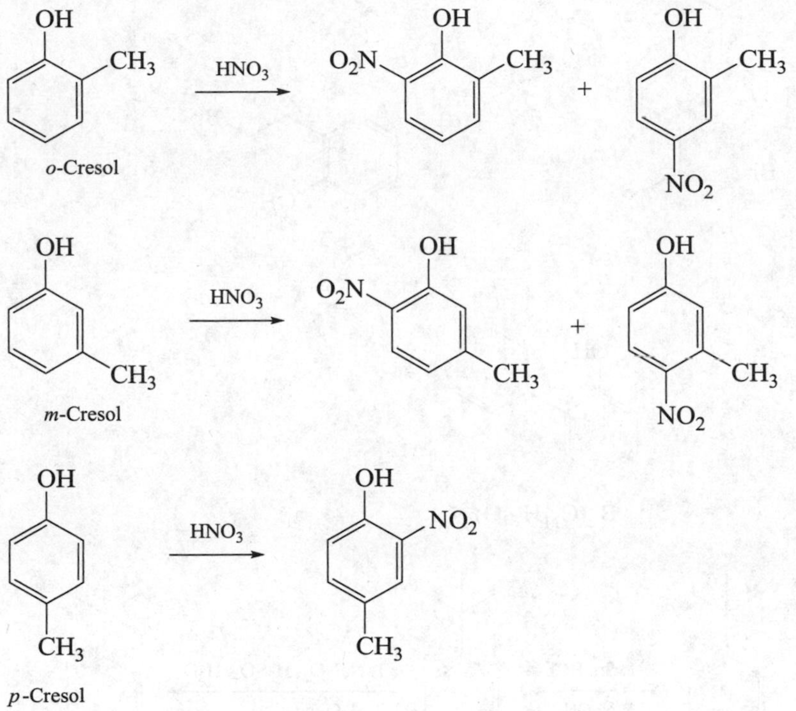

o-Cresol → HNO₃ → (2-nitro-6-methylphenol + 4-nitro-2-methylphenol)

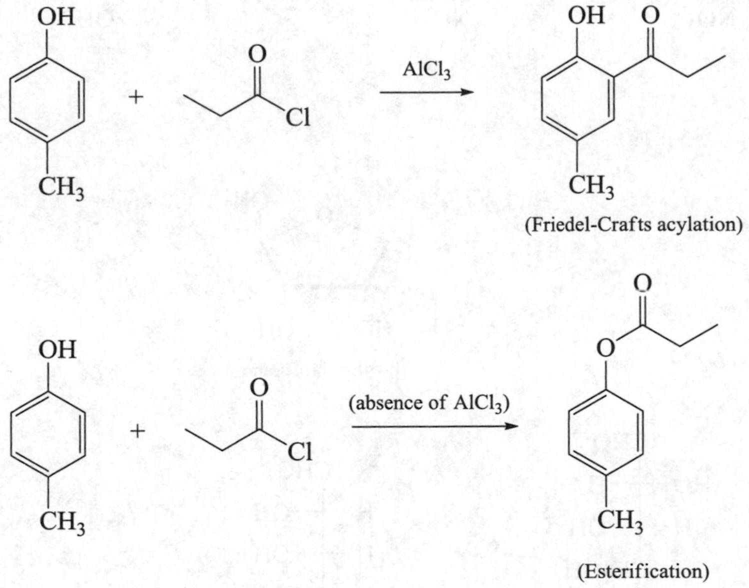

m-Cresol → HNO₃ → products

p-Cresol → HNO₃ → product

3.

(Friedel-Crafts acylation)

(Esterification)

4. (*a*)

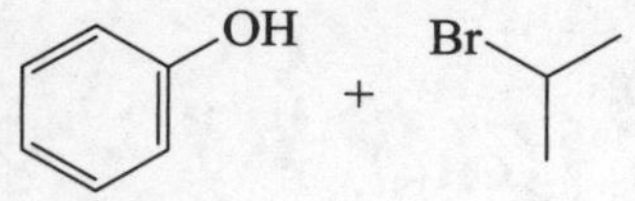

+

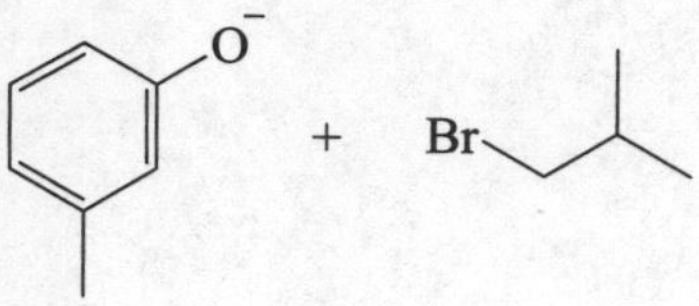

(*c*) CO_2, 125°C, 100 atm

(*b*)

+

(*d*)

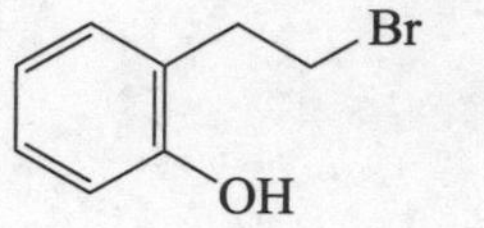

5.

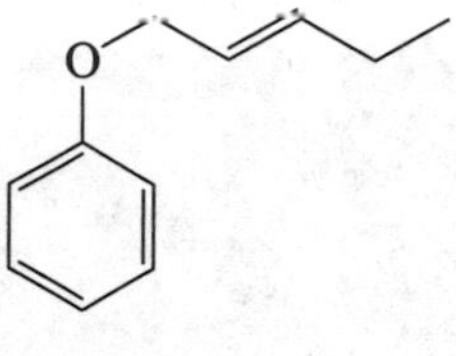

A

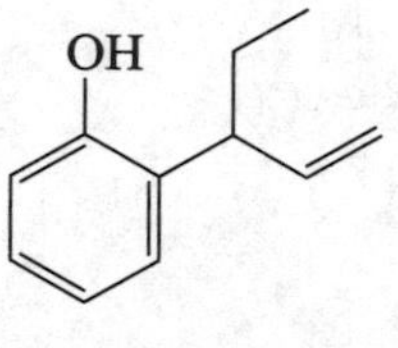

B ($C_{11}H_{14}O$)

6.

$$\xrightarrow[\text{H}_2\text{SO}_4]{\text{HNO}_3}$$

NO_2

$$\xrightarrow[\text{2. NaOH}]{\text{1. Sn, HCl}}$$

NH_2

$$\xrightarrow[\text{2. H}_2\text{O, heat}]{\text{1. NaNO}_2, \text{H}_2\text{SO}_4, \text{H}_2\text{O}}$$

OH

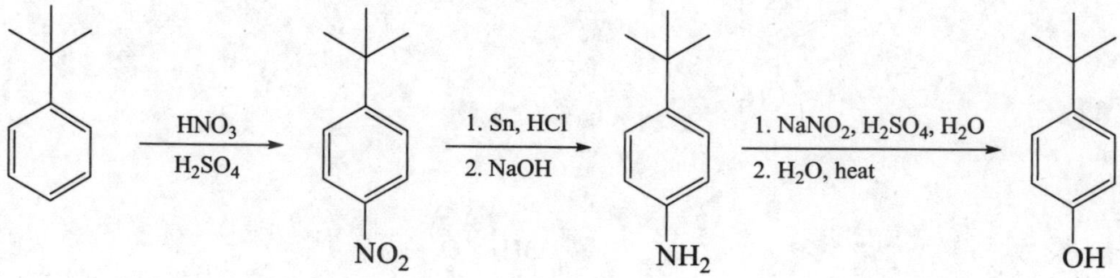

CHAPTER 24

1. (*a*)

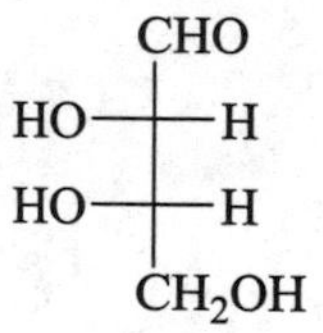

L-Erythrose

(*d*)

β-D-Erythrofuranose

(*b*)

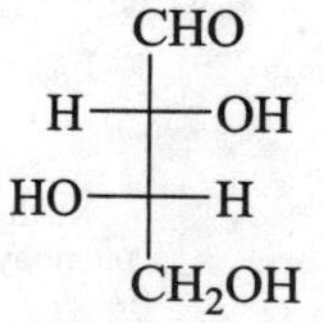

L-Threose

or

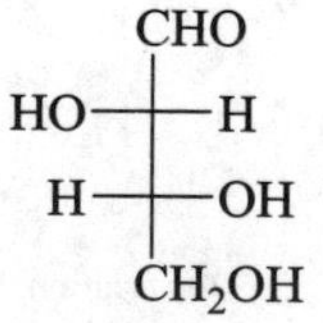

D-Threose

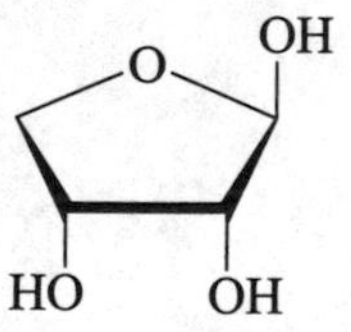

(*e*)

(*c*)

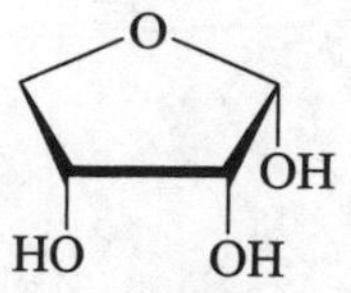

α-D-Erythrofuranose

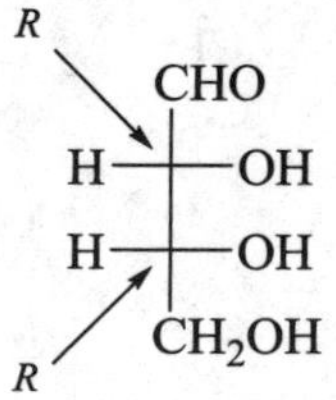

2. (*a*)

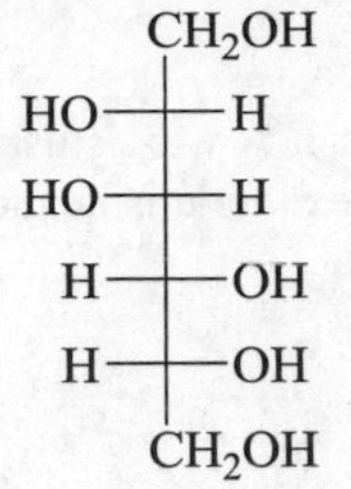
CH₂OH
HO—H
HO—H
H—OH
H—OH
CH₂OH

(*b*) $5\ HCO_2H\ +\ H_2C{=}O$

3. (*a*)

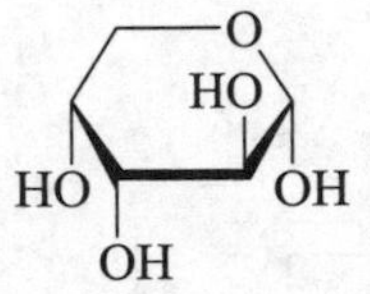
O
HO
HO
OH
OH

(*b*)

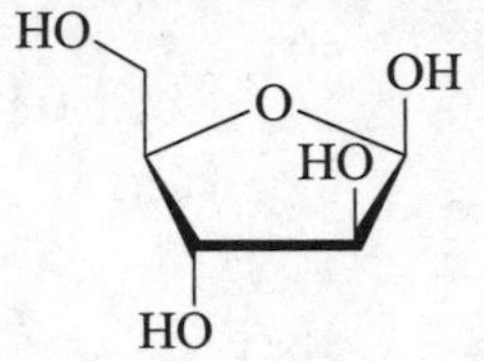
HO
O
OH
HO
HO

(*c*)

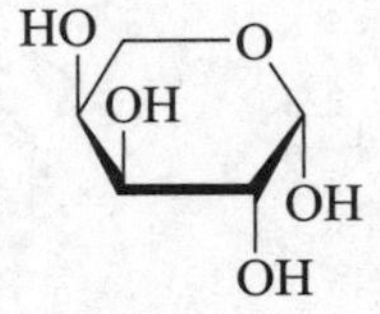
HO
O
OH
OH
OH

4. β-D-Idopyranose (β-pyranose form of D-idose)

5. The products are diastereomers.

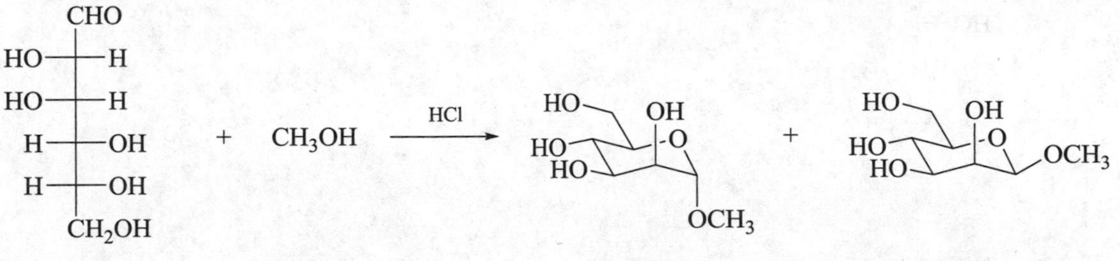
CHO
HO—H
HO—H
H—OH
H—OH
CH₂OH
+
CH₃OH
HCl
HO
OH
O
HO
HO
OCH₃
+
HO
OH
O
HO
HO
OCH₃

D-Mannose | Methanol | Methyl α-D-mannopyranoside | Methyl β-D-mannopyranoside

CHAPTER 25

1.

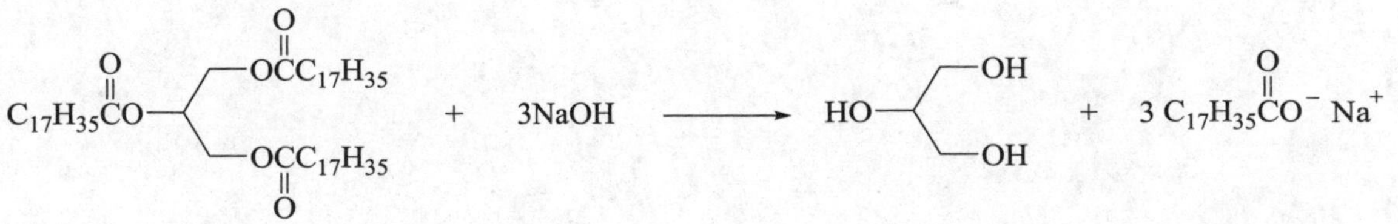
O
C₁₇H₃₅CO
O
OCC₁₇H₃₅
OCC₁₇H₃₅
O
+
3NaOH
HO
OH
OH
+
3 C₁₇H₃₅CO⁻ Na⁺
O

Tristearin

2. Fats are triesters of glycerol. A typical example is tristearin, shown in Problem 1. A wax is usually a mixture of esters in which the alkyl and acyl group each contain 12 or more carbons. An example is hexadecyl hexadecanoate (cetyl palmitate).

$$C_{15}H_{31}\overset{\overset{\displaystyle O}{\displaystyle \|}}{C}OC_{16}H_{33}$$

3. (*a*) Monoterpene;   (*c*) Diterpene;

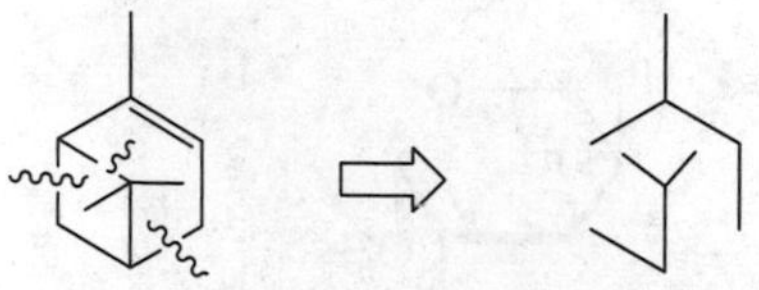

(*b*) Sesquiterpene;

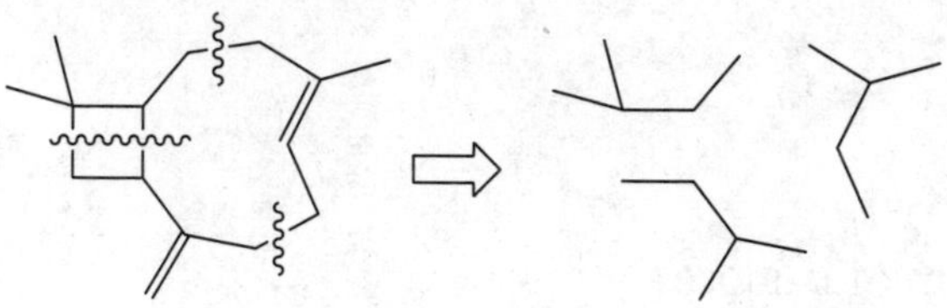

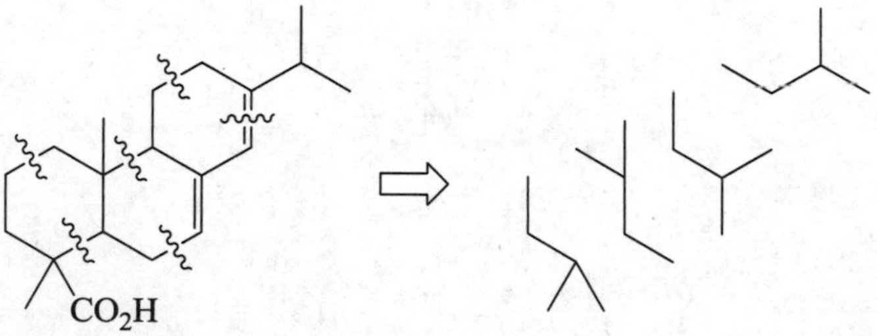

4.

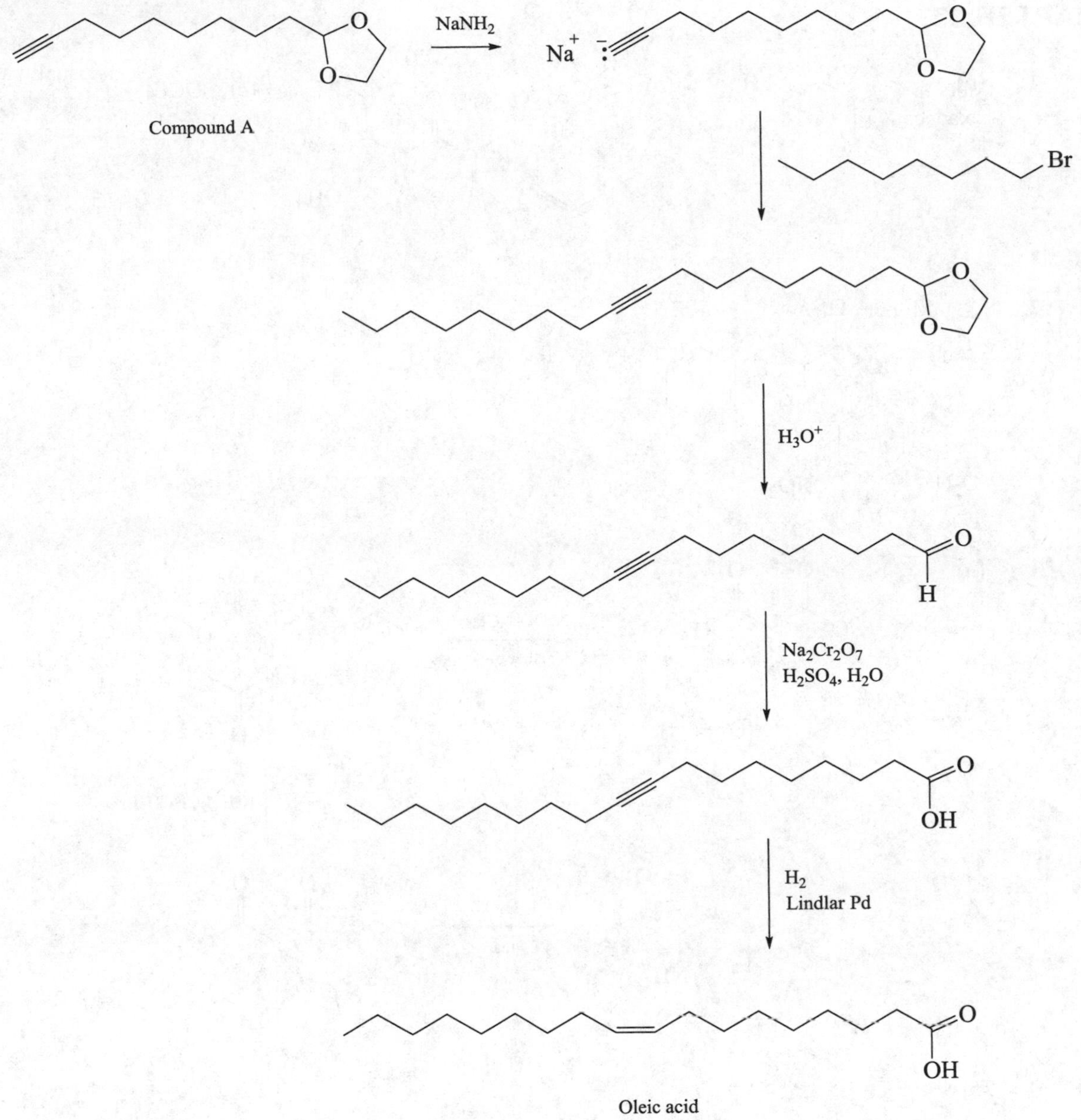
O
O
NaNH2
Na+
O
O
Compound A
Br
O
O
H3O+
O
H
Na2Cr2O7
H2SO4, H2O
O
OH
H2
Lindlar Pd
O
OH
Oleic acid

5.

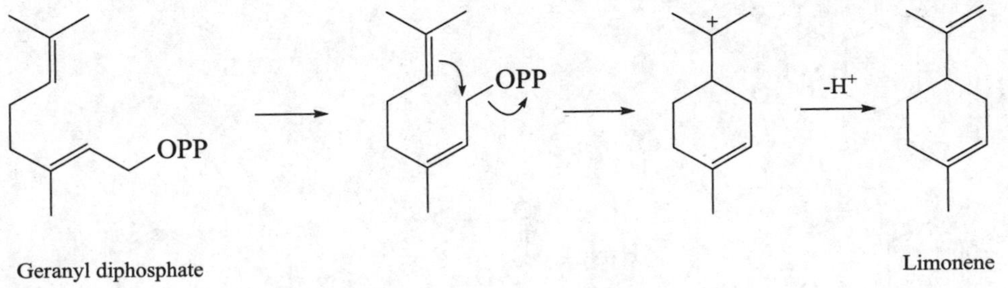
OPP
Geranyl diphosphate
OPP
+
-H+
Limonene

CHAPTER 26

1. (*a*)

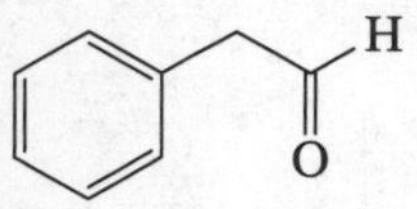

(*b*)

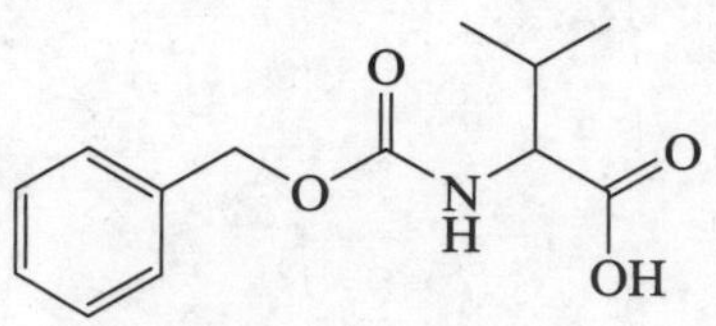

(*c*) DCCI

2.

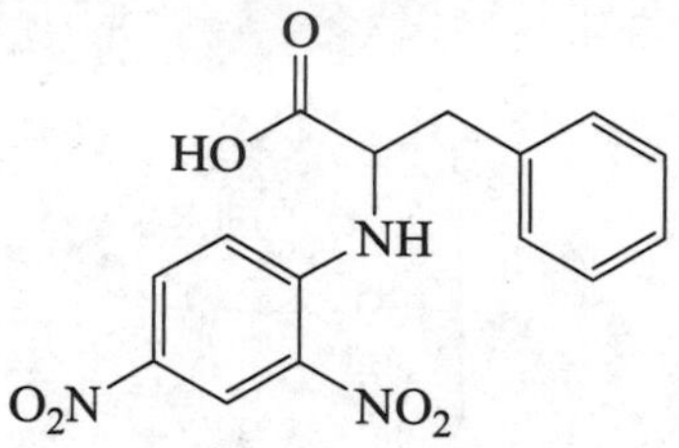

3. (*a*)

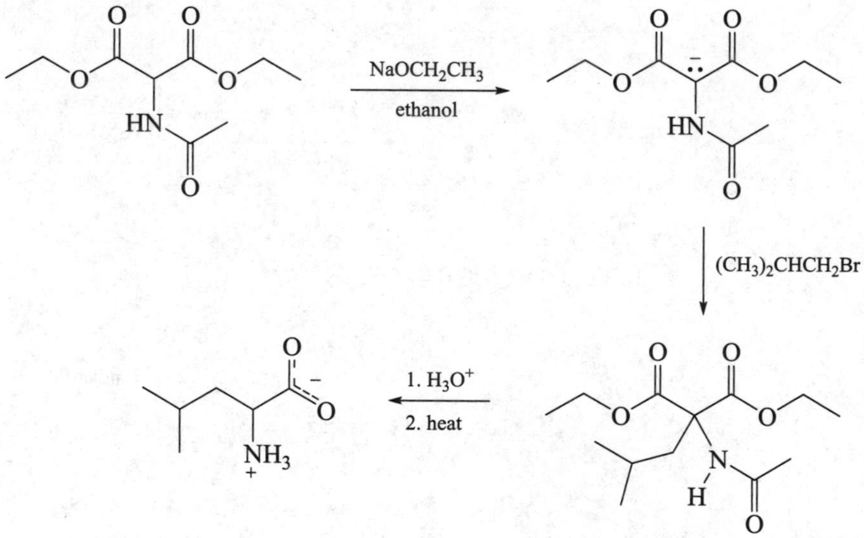

(*b*)

Leu-Val =

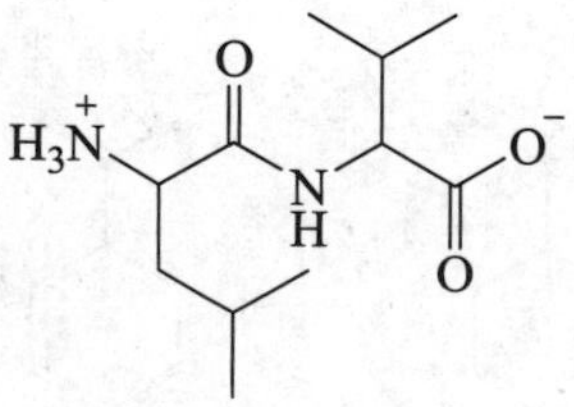

N-Protect leucine:

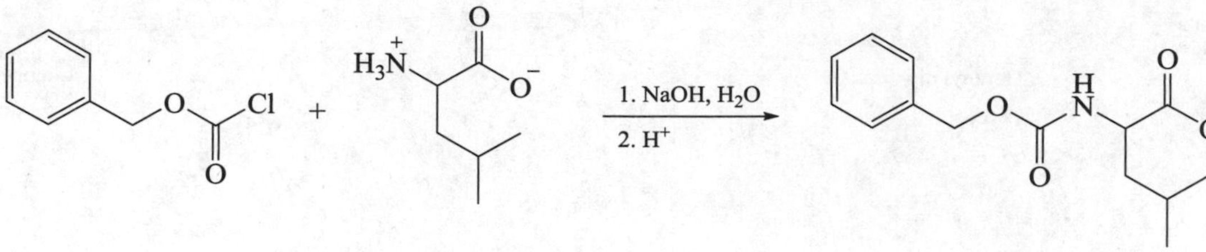

(Z-Leu)

C-Protect valine:

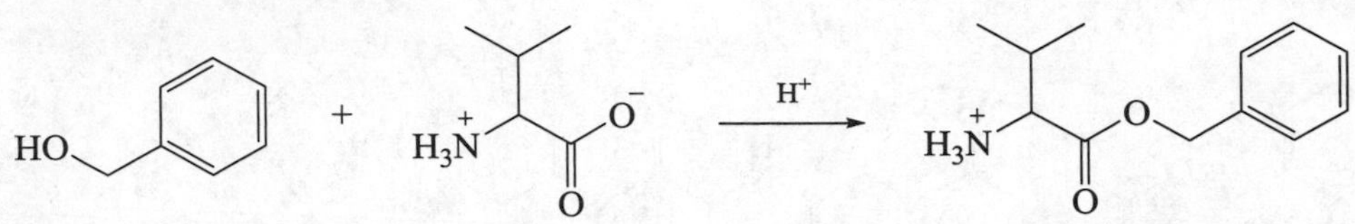

Couple: Z-Leu +

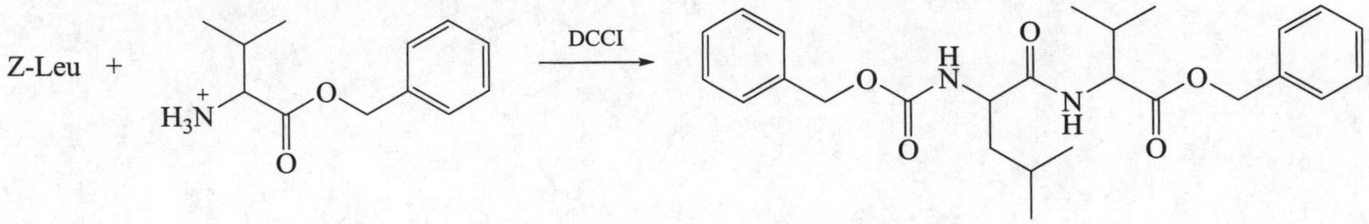

Deprotect:

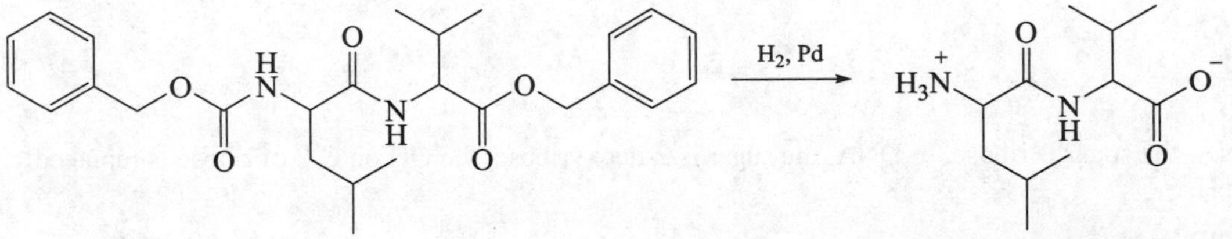

Leu-Val

4. Leu-Val-Gly-Ala-Phe

5. (*a*) Pentapeptide (*c*) Serine (*e*) Ser-Ala-Leu-Phe-Gly

   (*b*) Four (*d*) Glycine

6. (*a*)

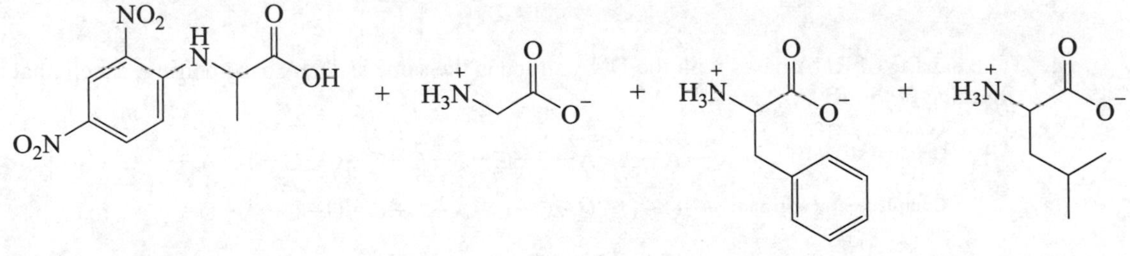

DNP-Ala Gly Phe Leu

   (*b*)

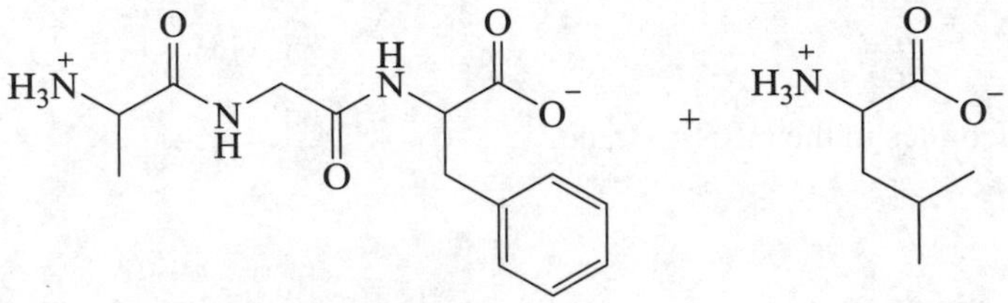

Ala-Gly-Phe Leu

   (*c*) Same as part (*b*); Ala-Gly-Phe + Le

(d)

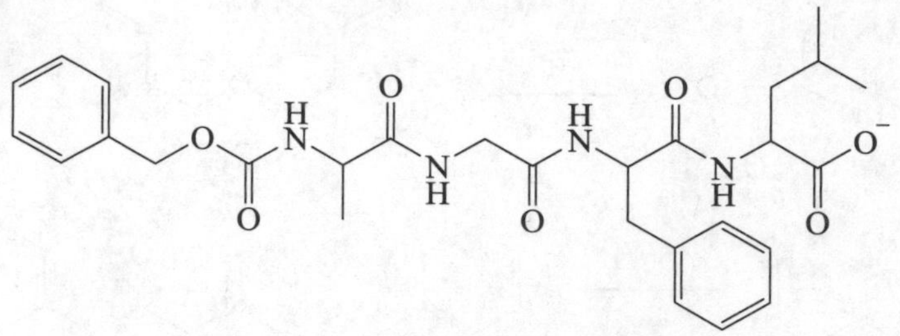

Z-Ala-Gly-Phe-Leu

CHAPTER 27

1. In RNA, the sugar is ribose. In DNA, the sugar is 2-deoxyribose; the OH on C-2 of ribose is replaced by an H.

2. −21.8 kJ

3. Write the symbol for the complementary bases opposite each nucleotide segment in the original strand. Recall that the complementary base pairs in DNA are A···T and G···C.

(a) Original strand: —A–A–A–G–G–T–C–C–C–G–T–A—

Complementary strand: —T–T–T–C–C–A–G–G–G–C–A–T—

(b) Original strand: —T–A–C–T–C–G–C–G–G–A–T–G—

Complementary strand: —A–T–G–A–G–C–G–C–C–T–A–C—

4. The pairing of RNA bases with the DNA strand is the same as DNA base-pairing, except that uracil (U) replaces thymine (T) in RNA.

(a) Original strand: —A–A–A–G–G–T–C–C–C–G–T–A—

Complementary strand: —U–U–U–C–C–A–G–G–G–C–A–U—

(b) Original strand: —T–A–C–T–C–G–C–G–G–A–T–G—

Complementary strand: —A–U–G–A–G–C–G–C–C–U–A–C—

5. The codons are triplets of nucleotides in the mRNA strand.

(a) UUU CCA GGG CAU

(b) AUG AGC GCC UAC

6. Each mRNA codon codes for a specific tRNA and each tRNA delivers a particular amino acid to the growing protein. The codons and the amino acids for which they code are listed in text Table 27.4.

(a)	Codon	Amino Acid		(b)	Codon	Amino Acid
	UUU	Phenylalanine			AUG	Methionine
	CCA	Proline			AGC	Serine
	GGG	Glycine			GCC	Alanine
	CAU	Histidine			UAC	Tyrosine

The amino acid sequences are

(a) Phe-Pro-Gly-His   (b)   Met-Ser-Ala-Tyr

7. An anticodon is a group of three bases in a particular section of tRNA complementary to the bases of each mRNA codon. There is a different tRNA for each amino acid.

(a) Codons: UUU CCA GGG CAU

   Anticodons: AAA GGU CCC GUA

(b) Codons: AUG AGC GCC UAC

   Anticodons: UAC UCG CGG AUG

CHAPTER 28

1. In a chain-growth process, monomers add one by one to the end of the growing polymer chain. Each chain has only one growth point. In a step-growth process, each chain has at least two growth points. The monomer is consumed quickly to form oligomers, and these react with each other to form the polymer.

2. (a) The repeating unit is formed by connecting the end atoms of the monomer. Thus, the repeating unit shown arose from polymerization of ethylene oxide.

Ethylene oxide

   (b) The source-based name of the polymer is poly(ethylene oxide).

3. Reaction of one mole of bisphenol A with one mole of phosgene produces two moles of hydrogen chloride (HCl). The reaction is a condensation process.

4. The monomer used to prepare poly(vinyl acetate) is the alkene vinyl acetate.

Poly(vinyl acetate)   Vinyl acetate

5. Poly(vinyl alcohol) cannot be prepared directly from vinyl alcohol as vinyl alcohol is an enol. The enol form is less stable than the keto form, and the major species present is acetaldehyde.

Vinyl alcohol   Acetaldehyde

Notes

Notes

Notes

Notes

Notes

Notes

Notes

Notes

Notes

Notes

Notes

Notes